INTRODUCED MAMMALS
OF THE WORLD

DEDICATION

I would like to dedicate this book to my closest family, Pat, Melinda and Patrick, Timothy and Joanne and Benjamin John, Gabriella and Nathaniel…

and to my friends Peter, Marion and Ron.

INTRODUCED MAMMALS OF THE WORLD

THEIR HISTORY, DISTRIBUTION AND INFLUENCE

JOHN L. LONG

National Library of Australia Cataloguing-in-Publication entry

Long, John L.
Introduced mammals of the world : their history, distribution & influence.

    Bibliography.
    Includes index.
    ISBN 0 643 06714 0.

    1. Mammals - Handbooks, manuals, etc. I. Title.

599

Published exclusively in Australia and New Zealand,
and non-exclusively in other territories of the world
(excluding Europe, Africa, the Middle East and South America) by:

CSIRO PUBLISHING
150 Oxford Street (PO Box 1139)
Collingwood VIC 3066
Australia

Telephone: +61 3 9662 7666
Freecall: 1800 645 051 (Australia only)
Fax: +61 3 9662 7555
Email: publishing.sales@csiro.au
Web site: www.publish.csiro.au

Published exclusively in Europe, Africa, the Middle East and South America, and non-exclusively in other territories of the world (excluding Australia and New Zealand), by CABI Publishing, a Division of CAB International, with the ISBN 0 85199 736 8.

CABI Publishing
Wallingford
Oxon OX10 8DE
United Kingdom
Telephone: +44 (0) 1491 832 111
Fax: +44 (0) 1491 829 292
Email: publishing@cabi.org
Web site: www.cabi-publishing.org

Front cover
Photo copyright D. Sarson, Lochman Transparencies

Set in 9/11 Minion
Designed by James Kelly
Printed by Ligare

C O N T E N T S

K E Y   T O   M A P S

Introduced successfully

Introduced unsuccessfully

Introduced, outcome uncertain

Native range

Former range

Expanded range

Indicates groups of islands where native or introduced (depending on type of arrow)

A B B R E V I A T I O N S   U S E D

HB   Length of head and body (excluding tail)

SH   Height at shoulder

T   Tail length

TL   Total length (including tail)

WS   Wingspan

WT   Weight

P R E F A C E

■ SCOPE AND ARRANGEMENT
This volume deals with the success or otherwise of mammals introduced, re-introduced or translocated around their ranges or into new environments. It consists largely of a systematic list of introduced mammals containing a brief description of each mammal, a résumé of their distribution, habits and behaviour, and a more detailed section on the history of their introduction(s).

■ CLASSIFICATION USED
The classification for Order, Family and Genus levels followed throughout this volume has been, unless otherwise stated, that outlined in *Mammals of the World* by Walker (1992). At species level, mammals have been placed in alphabetical order under each genus. The reason for this taxonomic procedure is one of author preference rather than one set down by any taxonomic authority. It was necessary to have some order to the systematic list early in the writing and Walker appeared to be the best at the time. The taxonomy of mammals is not yet well set and new works on the subject differ, as do those on either side of the Atlantic.

■ DISTRIBUTIONAL MAPS
Maps are notoriously inaccurate at the scale used in this book and are mainly a guide to the range of each species. They have been drawn using a range of information from numerous texts, thus errors may have been transferred in this manner. Every effort has been made to make them as accurate as possible. There are a host of regional texts in which maps of mammal distributions can be found.

■ MAMMAL INTRODUCTIONS
The detail of introduction(s) for each mammal has been listed by country in alphabetical order, first for region and then for each country. Islands are largely mentioned under the heading of the ocean in which they occur.

■ HABITS AND BEHAVIOUR
This section offers a summary of the species habits, behaviour and status under the headings of habits, gregariousness, movements, habitat, foods, breeding, longevity and status. The intention was to examine the success of introductions with regard to these headings in an effort to show what characteristics contribute to successful introductions.

■ LINE DRAWINGS
The sketches used in this work are original pieces and are subjective in their representation of the species depicted. They are provided for only a small number of species, as there are many readily available sources of drawings and photographs available these days. In this case they serve to punctuate the extensive text and provide relief for the reader. The artwork used in this book was kindly produced by Sophie Moller and Christine Freegard.

ACKNOWLEDGMENTS

Dr David Ransom's encouragement and medical expertise provided me with the time to complete this work. Dr Ron Cameron (deceased) and Mrs J. Cameron helped me with references from *National Geographic Magazine* and introduced me to the Internet. Thanks for information or references to Ron Johnstone (Western Australian Museum), John Darnell, Dr Tony Henson (Albany), Dr Denis King (Perth), and particularly to Dr Peter Mawson (Western Australian Department of Conservation and Land Management), Neil Hamilton (Perth Zoo) and Marion Massam (Agriculture Protection Board, Perth). Thanks also to Dr Mary Bomford and Louise Conibear of Bureau of Rural Sciences, Department of Primary Industry, for help with a number of references and books.

Andrew Nielson (Starfish Technologies, Nedlands, Western Australia) helped with map transfer from LC to PowerMac and with the Internet. Josephine Christmass (Kelmscott) typed a number of hard-to-read drafts onto computer in the initial stages. Howard and Barbara Leach, Sacramento, California, made my wife and I welcome while we were in the United States again.

I am grateful to my daughter Dr Melinda Oxley-Long, who obtained some valuable reference material and books for me while in Canberra and the United Kingdom. My son, Tim Long, also contributed a new edition of *Mammals of Australia* and helped me with Internet searches. Thanks to Dr. Dale A. Wade (formerly US Fish and Wildlife Service) for providing me with reference material and books from the United States; Dr D. A. Scarlato (Zoological Institute, Academy of Sciences, Leningrad), who sent me two volumes of the *Acclimatisation of Game Animals and Birds in the USSR*; Dr Peter Mawson for the *Mammals of Thailand* and Howard Leach for the catalogue of mammals held in the US National Museum. Professor E. W. Jameson, Jr (University of California, Davis), kindly donated a copy of his book *Californian Mammals*.

The following answered my correspondence about introductions: Sir Christopher Lever (UK), Dr M. Dollinger (Switzerland), Jean Roche (Maître-Assistant, Muséum National D'Histoire Naturelle, Paris), Dr Friederike Spitzenberger (Naturhistorisches Museum Wien, Vienna), Dr P. Gruys (Rijksinstituut voor Natuurbeheer, Arnhem, Holland), Anne von Hofsten (National Swedish Environment Protection Board, Solna, Sweden), and Professor Ingemar Ahlén (Institutionen for Viltekologi, Uppsala, Sweden).

Darryl and Trix Blackshaw arranged with John Nortier (deceased) to expertly translate some papers from Dutch to English. Finally, many thanks to my friend and colleague Ron Johnstone who made me welcome at the Western Australian Museum during my time as a research associate there.

John L. Long

PROLOGUE TO THE INTRODUCTION

John Long compiled and wrote this work over a period of 31 years, beginning it in 1969 at the same time that he commenced work on his first major publication, *Introduced Birds of the World*, published in 1981. Life's normal distractions meant that John was not able to return his focus to completing this project until his retirement in 1991. In early 1998 John was diagnosed with cancer and realised the urgency of completing this book. Strength of character and dedicated care from his family and doctors enabled him to continue to enjoy life and also work on the book until the day before he died on 5 January 2000. At that time the main text of this book was almost complete, as was the bibliography. Only the compilation of the indexes and the writing of the introduction remained, along with the final editing.

When John first learnt of his illness he asked me if I would complete the book and see it published in the event that he was unable to achieve that goal. John was my mentor and we shared a passion for the subject of invasive species, and so I agreed to his request. After more than a year of work I stand in awe of John's efforts in writing this book, and its earlier companion. John and I discussed the general issues that he wanted covered in the introduction, but he gave me free reign to write it as I thought best. It has been written in an open style with almost no direct references for a number of reasons. It reads more easily; it is intended to be an introduction to the subject of mammal invasions and not an analysis of this or previous works, and there are sufficient refer-ences provided in the individual species accounts to ensure that any reader will be able to further examine particular issues.

John was a humble man and was always surprised at what other people made of his work on introduced birds of the world. I hope just as much, if not more, is made of this work and that some good will come of it. I also hope that a better approach to managing the introduction of mammals will result. It would be a tragedy if another student of invasive species were able to draw on as much new material and saw the need to compile such a tome again, this century.

Peter R. Mawson
September 2001

Humans have kept and transported mammals for a variety of reasons for thousands of years. Probably the first species to be kept were those that provided a ready source of food when alternative and larger wild game was not available. Typically these were small herbivores such as rabbits (*Oryctolagus cuniculus*) and cavia (*Cavia porcellus*). Later as animal husbandry skills were developed and more permanent settlements were established, larger mammals such as sheep (*Ovis aries*), cattle (*Bos taurus*), goats (*Capra hircus*), pigs (*Sus scrofa*) and camelids (*Camelus dromedarius*, *Lama* spp.) were domesticated. Over time, people began to keep mammals for reasons other than food. Some mammals such as dogs (*Canis familiaris*), and before them, wolves (*Canis lupus*), were kept to assist their human masters in hunting large game. Other species such as cats (*Felis catus*) were kept in a semi-domestic state to help control pest species from invading human living sites and consuming the products of the newly developing agrarian systems. In many parts of the world larger mammals were domesticated and trained to provide heavy transport, for example, donkeys (*Equus asinus*), horses (*E. caballus*), camelids, bovids (*Bos* spp. and water buffalo, *Bubalus bubalis*), reindeer (*Rangifer tarandus*) and elephants (*Elaphus maximus*). In more recent times, as the affluence of human societies has increased, some mammals have been kept purely as companion animals or for show. This habit has developed further into modern zoo collections and the worldwide pet industry that exists today.

As humans began to expand their own range through exploration and colonisation, they brought with them animals as a source of food on sea voyages, and as a basis for a reliable food supply at their final destination. The Romans were probably the most noted practitioners of this activity, introducing rabbits, sheep and pigs to many parts of their empire. Polynesian explorers took pigs and kiore (*Rattus exulans*) with them on their voyages of discovery and colonisation. British and French explorers continued this tradition in more recent times. They often left pigs and rabbits on the oceanic islands they encountered, either as gifts for the local human inhabitants or in the hope that the animals would breed and provide a valuable food source to any unfortunate seamen that might be shipwrecked there at a future time. Commercial whale and seal hunters of the nineteenth century also practised this form of live provisioning.

The transport and introduction of mammals to various parts of the world continues unabated today. In many places it occurs under the banner of developing agriculture, while in other parts it is a consequence of commercial animal keeping and escapes or releases of pets. Regardless of the mechanism by which mammals are being introduced to new ranges, the results are depressingly similar. Native fauna at first seem to co-exist with the new colonisers until the natural resources become depleted or until climatic or environmental conditions become extreme. From then on, the native fauna (mammals, birds, reptiles, amphibians and invertebrates) becomes impoverished and the colonisers begin to proliferate. The net result is a reduction in biodiversity and the creation of regional fauna, and even the beginnings of a global fauna, that is depressingly familiar.

In recent times, attempts have been made to retain some of the species that have been threatened with extinction due to the destruction of their natural habitat or the adverse impacts of introduced species (particularly humans). In some cases, mammal species have been re-introduced to their former range following the control of the key threatening process(es), or the species have been relocated to a site that was not formerly within their natural range because there was no viable alternative. Offshore islands have often been used for this purpose as they are either free of invasive and predatory species or can be made so. As knowledge is gained from each successive re-introduction program, subsequent efforts become more refined and the likelihood of success increases. Such conservation efforts must address several important issues, including small founding populations and a reduced genetic basis, the impact that the (re-)introduced species will have on native flora and fauna in the release site, and the long-term viability of the program, ecologically as well as politically and financially. Not surprisingly, there are supporters and opponents for such approaches to conservation.

THE REASONS FOR INTRODUCING MAMMALS

AESTHETICS

In comparison to birds, there have been a limited number of mammal introductions purely for aesthetic reasons. This is most likely a result of humans usually considering mammals in terms of their potential use rather than their appearance, and the fact that even the smallest mammal species do not lend themselves to captive keeping nearly as readily as caged birds. Possible exceptions to this are the confined keeping of a species of deer (*Cervus* sp.) in parklands on estates of wealthy British and American landowners. Few of these introductions were intended to result in the release of those animals to the wild, but this was often one of the later outcomes. Similarly, a much smaller number of more exotic introductions, such as kangaroos and wallabies (*Macropus* spp.), established wild populations in England, parts of Europe and New Zealand, as a result of the establishment of confined populations kept for viewing.

FOOD, HUNTING AND SPORT

By far the most common motivation for the introduction of mammals into new areas has been to establish new sources of food, new or better hunting opportunities and for recreation (including hunting, but also its modern analogue of photography). This has been particularly evident in those countries where the native mammal fauna was either non-existent (oceanic islands) or impoverished (e.g. New Zealand and New Caledonia). It was also evident in those countries where the European colonists were completely unfamiliar with the endemic mammal fauna and had no concept of how to integrate the use of those native species into their established agricultural systems (e.g. Australia and South America).

Introductions purely for hunting purposes have been less common, and have usually been the preserve of the more wealthy members of societies. Only those who could afford to procure and transport the mammals (typically large species) did so, and they usually protected the animals, at least in the early stages following the introduction. Exceptions to this include some of the most dramatic releases recorded around the world. The rabbits and red foxes that have become so well established in Australia were introduced largely for hunting. The spread of the rabbit and fox across most of the continent has been both rapid and devastating.

Some of these early private introductions led to escapes and the establishment of wild populations. A number of these feral populations subsequently failed or were intentionally eradicated, but the remainder have prospered. In countries with large open rangelands, many of these species (typically ungulates) have established very large populations. Australia now has the largest populations of wild horses, water buffalo, camels and goats in the world. The United States also has significant populations of wild horses and donkeys.

Other species, such as wild boar (*Sus scrofa*), several species of deer (*Cervus* spp.), ibex (*Capra ibex* and *C. pyrenaica*), bighorn sheep (*Ovis canadensis*) and Himalayan tahr (*Hemitragus jemlahicus*), that have more restricted habitat requirements and are highly considered for trophy hunting, are now actively managed as recreational game species. The taking of such species is closely regulated and hunting can only occur under licence, with sex ratios and age classes carefully manipulated to ensure the viability of the hunting stock. The careful management of these introduced populations generates considerable revenue, directly through licence fees and indirectly through the purchase of equipment and services associated with hunting in general.

The preservation of introduced populations of mammals has reached its most sophisticated level in African countries, where species such as ungulates, felids (*Panthera* spp.) and elephant (*Loxodonta africana*) – some of which are now relatively rare – are hunted on a very restricted basis for extremely high fees.

COMMERCIAL ENTERPRISES

Most of the early commercial enterprises involving mammals were focused on domesticated ungulates, since these species formed integral parts of agrarian systems. Although it was seldom the intention of the human keepers of those animals to allow them to escape from domestication, the open range management systems used in many areas during the early phases of colonisation frequently allowed for this. There are several examples in more recent times where domestic stock were abandoned and left to fend for themselves when the human colonists abandoned colonial outposts. This was not uncommon on oceanic island settlements where goats, sheep and cattle were often left untended, but it has occurred at a few continental settlements, such as those in northern Australia where banteng (*Bos javanicus*) and water buffalo (*Bubalus bubalis*) populations subsequently became established.

During the twentieth century a number of intensive commercial industries were established in a range of countries with the aim of producing furs (e.g. mink, *Mustela vison*; sable, *Martes zibellina*; coypu, *Myocastor coypus;* and arctic fox, *Alopex lagopus*). Due to poor cage standards, natural disasters and fluctuations in market prices, escapes and deliberate releases occurred from time to time and introduced populations of these mammals have become established in several parts of the world. Many of those populations have been shown to have had adverse impacts on the local agricultural economies or the native flora and fauna.

There have also been a number of deliberate releases of species into the wild with the intention of establishing commercial quality fur-bearing stocks. For example, more than 30 000 mink were released into the wild in the Russian Federation. In some areas, the introductions involved the release of animals into areas outside their natural range, while in others it involved the mixing of subspecies – at least seven subspecies of red deer (*Cervus elephus*) were introduced or re-introduced in the Russian Federation. In many cases, these efforts failed to establish a viable wild stock of commercial quality fur-bearing animals. In some cases it actually led to a decline in the quality of the furs or trophy antlers produced in whole regions, while in others it resulted in introduced populations becoming established, although not sufficient to support a commercial fur industry. When the wearing of fur-trimmed clothes became less fashionable and synthetic materials were developed in the mid-1950s many of these industries collapsed.

CONTROLLING PESTS

A number of carnivorous mammal species (e.g. mongoose, *Herpestes* spp.; stoats and weasels, *Mustela* spp.; and cats, *Felis catus*) have been intentionally released in some countries in attempts to provide a form of biological control of agricultural crop pests (usually rodents). Most of these introductions occurred in the nineteenth and twentieth centuries, and some of these appear to have been well researched before they were implemented. Most of the introductions were organised by individual farmers or local farmer groups, but a few, such as the introduction of stoats and weasels into New Zealand, had government support.

Many of these introductions failed outright, or failed to achieve the desired control of the target pest species, but more damaging were those that established wild populations and turned to other native fauna as a food source. Ground-nesting birds and native frogs and reptiles appear to have been particularly affected.

ACCIDENTAL INTRODUCTION, ESCAPEES AND PET-KEEPING

There are a number of species of mammal that appear to have been accidentally introduced in many countries. Rodents (*Mus musculus* and *Rattus* spp.), especially those transported in ships' cargoes, have expanded their range via accidental introductions. Most species accidentally transported with cargo and goods are small and relatively cryptic. In comparison, the range of mammal species that have become established in the wild as a result of pet-keeping (including zoos and private wildlife parks) is much more diverse in both the size and range of animals and their taxonomic affinities (e.g. golden hamsters, *Mesocricetus auratus*; musk deer, *Moschus moschiferus;* and red-necked wallabies, *Macropus rufogriseus*).

With the advent of better global, electronic communications and faster freight-delivery services, the potential for the introduction of further mammal species via the pet trade is greatly increased. The only factor likely to reduce this risk is the recent focus on more user-friendly pets, such as reptiles, amphibians and large invertebrates. However, relying on changes in fashion is hardly an ideal way to prevent unwanted introductions.

INTRODUCED SPECIES IN VARIOUS REGIONS OF THE WORLD

Before examining the benefits and harm caused by introduced mammals it is useful to examine the situation in different countries or areas of the world. It appears that the majority of mammal introductions have been successful. However, this needs to be considered with some caution as those introductions that have been successful have been better documented than those that have failed. Certainly many mammals were introduced into North America, the United Kingdom, Australia and New Zealand before the time of 'acclimatisation societies' and this is probably the case for most other parts of the world.

It has been claimed that the formation of the acclimatisation societies, commencing with the first La Société Impériale d'Acclimatation formed in France in 1854, resulted in increased interest in attempts at naturalisation and increased interest in the formation of other societies in other countries. More than likely though, the societies resulted from the increased interest in exotic animal forms. However, there followed such institutions as the Society for the Acclimatisation of Animals, Birds, Fishes, Insects and Vegetables within the United Kingdom in 1860; a number of societies in Australia, commencing with

Numbers of mammal species known to have been introduced (including translocations, re-introductions etc.).

Region	Total introduced	Established	Probably established	Failed or probably failed		Region	Total introduced	Established	Probably established	Failed or probably failed
Africa	64	56	2	6		Atlantic Ocean Islands (contd.)				
Europe						Greenland	2	2	0	0
Albania	1	1	0	0		Sao Tome	1	1	0	0
Austria	11	11	0	0		South Georgia	7	5	0	2
Belgium	2	2	0	0		St. Helena	9	7	0	2
Bulgaria	3	2	0	1		Tristan da Cunha	11	6	0	5
Czechoslovakia	11	11	0	0		Indian Ocean Islands				
Denmark	7	5	0	2		Amsterdam Island	5	4	1	0
Finland	9	6	2	1		Andaman Islands	8	7	0	1
France	21	14	3	4		Assumption Island	5	4	0	1
Germany	30	15	6	9		Chagos Archipelago	4	4	0	0
Greece	5	4	1	0		Cocos-Kealing Islands	1	0	1	0
Hungary	5	4	0	1		Comoros Islands	7	4	2	1
Iceland	4	3	0	1		Crozet Archipelago	5	4	0	1
Ireland	9	7	0	2		Kerguelen Islands	14	9	0	5
Italy	11	6	1	4		Madagascar	14	8	2	4
Luxembourg	1	1	0	0		Maldive Islands	2	2	0	0
Netherlands	10	9	1	0		Marion Island	3	2	0	1
Norway	6	4	1	1		Mauritius Island	10	8	1	1
Poland	11	6	3	2		Nicobar Islands	5	4	0	1
Portugal	4	2	1	1		Reunion Island	3	2	0	1
Romania	5	5	0	0		Seychelles Islands	9	8	0	1
Spain	12	8	2	2		St. Paul	7	5	1	1
Sweden	15	12	2	1		Tromelin Island	3	3	0	0
Switzerland	17	14	2	1		Pacific Ocean Islands				
United Kingdom	60	35	7	18		Antipodes Islands	4	2	0	2
Yugoslavia	8	3	2	3		Auckland Island	6	4	0	2
Arabia/Asia Minor	10	10	0	0		Campbell Island	5	3	0	2
Australasia						Caroline Islands	4	4	0	0
Australia	79	49	6	24		Cocos (Costa Rica)	2	1	0	1
New Zealand	45	31	0	14		Enewetak Atoll	2	2	0	0
Papua New Guinea	28	19	6	3		Federated States of Micronesia	4	4	0	0
Asia						Fiji	9	8	0	1
China	7	6	1	0		Galápagos Islands	11	10	0	1
Indonesian Archipelago	29	20	6	3		Guam	9	9	0	0
Japan	18	12	1	5		Hawaiian Islands	26	16	0	10
Malaysia	5	4	1	0		Jarvis Island	1	1	0	0
Mongolia	1	1	0	0		Johnston Atoll	1	1	0	0
Philippines	6	5	1	0		Juan Fernández Island	11	8	2	1
Taiwan	1	1	0	0		Kodiak Island	1	1	0	0
Thailand	2	2	0	0		Komandorskiye Ostrova	3	3	0	0
Vietnam	1	1	0	0		Kuril Islands	2	1	0	1
Indian sub-continent						Lord Howe Island	5	4	0	1
India	8	5	1	2		Loyalty Islands	1	1	0	0
Nepal	1	1	0	0		Macquarie Island	8	5	0	3
Pakistan	6	5	1	0		Marshall Islands	1	1	0	0
Sri Lanka	4	2	1	1		Marianas Islands	5	3	1	1
Russian Federation and associated independent states	56	42	4	10		Marquesas Islands	2	2	0	0
North America	93	78	5	10		New Caledonia	7	6	0	1
South America	37	25	9	3		Norfolk Island	3	2	0	1
West Indies–Caribbean	37	23	5	9		Phoenix Islands	1	1	0	0
Atlantic Ocean Islands						Ryuku Islands	2	2	0	0
Archipelago Madiera	3	3	0	0		Samoa	1	1	0	0
Ascension Island	5	5	0	0		Solomon Islands	6	6	0	0
Azores Islands	2	2	0	0		Tonga	2	2	0	0
Canary Islands	7	6	1	0		Vanuatu	2	2	0	0
Cape Verde Islands	1	1	0	0		Mediterranean Sea				
Faeroe Islands	3	3	0	0		Balearic Islands	1	1	0	0
Falkland Islands	11	7	0	4		Crete	3	3	0	0
Fernando Poo Island (Bioko)	2	1	1	0		Cyprus	2	1	1	0
Gough Island	1	1	0	0		Malta	4	4	0	0

the Zoological and Acclimatisation Society of Victoria in 1861; the many acclimatisation societies in New Zealand in the 1860s; and also a number of American acclimatisation societies in the 1870s.

The earliest known societies to attempt naturalisation of birds and other animals appear to have been the Zoological Society of London, formed in 1826 and which had as one of its primary objectives the introduction of new and useful animals to the United Kingdom, and the Natural History Society in America, formed in the 1870s. The advent of these societies has, however, resulted in a better, though fragmentary, documentation of events.

Table 1 (see p. xiv) outlines the number of species of mammals that have been introduced to various land-masses and islands. The total numbers of introduced animals included in the table are based on the details documented in this work. The total numbers established rely on the most recent information available and can at best be only an approximation. Some effort has been made to distinguish between those that are well established and those that are not, but the division is arbitrary and based on the latest reports for the species.

The inclusion of translocated and re-introduced species in the table may be irksome for those who wish to consider exotics only, but it is necessary as a large number of the more recent introductions have been of this form. One needs to define 'translocation' in relation to a faunal region, country or part thereof. In some of the cases recorded from Australia, species have become extinct on the entire mainland, with relict populations surviving only on offshore islands. Re-introductions effectively become introductions in such cases. Often it is not reliably known if a species was introduced or arrived unaided by people in some way. This is apparent for several species that were introduced into one part of Europe or South America but subsequently colonised other parts of those land-masses at a later date.

At least 93 species of mammal have been introduced into North America, with a strong emphasis on fur-bearing rodents (including lagomorphs), canids and mustelids. A second group within the total is reflected in introductions of exotic ungulates, such as deer (several species), ovids and caprids, that have formed the basis of recreational hunting industries.

Australia has also been the focus of a large number of introductions, with two distinct phases evident. The early phase consists of attempts to establish exotic species in support of the European colonisation of the continent with mainly domesticated species and those

species native to the United Kingdom. The second phase is much more recent and is almost totally devoted to the re-introduction of native marsupials and endemic rodent species for conservation purposes.

Introductions to Africa have been surprisingly limited, given its lengthy and varied history of colonisation and re-colonisation throughout the continent. A relatively small number of commensal rodent species have been introduced, but most have not expanded their range far beyond the densely populated regions around the coast. The bulk of the remainder are re-introductions of native species, especially ungulates, into areas depleted of large mammalian fauna during the European colonial period from 1800 to the mid-1900s.

The Russian Federation and associated independent states (formerly the USSR) have received a large number of introductions. These introductions are dominated by fur-bearing rodents and lagomorphs, canids and mustelids. The remaining species comprise several exotic deer species and a small number of native gazelles and caprids. The dominance of fur-bearing mammals reflects the efforts of the former USSR government to establish large-scale commercial industries based on wild animal stocks rather than intensive captive operations.

Introductions to the countries making up modern-day Europe are somewhat smaller than might be expected, given the length of time that the various European powers have been in existence, and exploring and colonising other parts of the world. It may be that the lengthy history of Europe and the greater degree of alienation of much of the landscape has made it more difficult for introduced species to successfully become established. Most European countries have had less than 15 introduced species successfully establish, the exceptions being France, Germany and the United Kingdom. The early establishment of acclimatisation societies in these countries may account for the greater number of introductions in these three countries.

With regard to island countries and oceanic islands, there are four notable sites for introductions. The West Indies and the Caribbean, the Hawaiian Islands, New Zealand and Papua New Guinea have similar numbers of successful introductions but very different histories, both in terms of the time since colonisation by humans, and geological history. New Zealand is an ancient landmass with Gondwanan links and, with the exception of a few native bat species, was devoid of mammals until it was colonised somewhere around 800 AD by the Maori people. It did, however, have a very rich and diverse avian fauna that effectively filled every type of ecological niche on the islands. The avifauna was no match for humans and many species were extirminated or pushed to the brink of extinction by the time Europeans first encountered the islands in the seventeenth century. Not surprisingly, the Europeans who colonised the country, and probably the Maori people, were very supportive of attempts to introduce mammal species from other parts of the world.

Papua New Guinea is an old landmass with a very long history of human habitation. However, it was not isolated from other landmasses in the way that New Zealand was. The fauna of the island reflects several waves of colonisation by mammals and birds and it possessed a rich native fauna consisting of both marsupials and eutherians. All of the introduced species recorded in Papua New Guinea are the result of movement of animals (mostly rodents native to the region) with goods traded from the nearby Indonesian Archipelago and Australia, and deliberate introductions of pigs and a number of deer species by European colonists.

The Hawaiian Islands have received a range of introduced species, largely linked with the development of European-style agriculture dating from the 1700s onwards. The introduced mammals include rodents that were crop pests, a small number of carnivorous species that were introduced in attempts to control the rodent pests, and a range of domestic stock species and three deer species introduced as farm animals or for hunting.

The Caribbean islands, while being very small in total area, have a long history of European colonisation, and these islands were also of vital importance during the first phases of colonial activity by the British, the Spanish and the French. They were also significant sources of new agricultural crops and much needed wealth for their parent nations. The mix of introduced species recorded on these islands is quite different to those on other islands. They include the usual commensal rodent species and a mongoose introduced to control the rodents, a range of domestic stock species, but also peccaries (two species) and monkeys (four species).

THE MENACE OF NATURALISED MAMMALS

The introduction of mammals in areas where they do not occur naturally has caused and is still causing problems in many areas of the world. It has not even been necessary for introduced species to be completely successful in establishing for them to cause damage. Even the short-term presence of some species on islands has been sufficient to cause the decline and even extinction of native plants and animals, and also the long-term conversion of the island habitat to a different ecosystem. Exotic species oust native ones by competing for food and habitat, introducing or maintaining diseases and parasites, and damaging agricultural crops and associated infrastructure. These problems have also arisen from the limited release or escape of exotic pet mammals, even from single events involving very low numbers of founding animals.

COMPETITION WITH NATIVE SPECIES

Traditional thinking would suggest that there are many examples of introduced mammals appearing to directly or indirectly affect populations of native ones. However, the qualitative evidence for such claims is usually much harder to find. Even in cases such as the introduction of the rabbit to Australia there is almost no evidence of rabbits having had a direct adverse impact on native herbivores. If anything, there is anecdotal evidence to suggest that rabbits and native fossorial mammals, such as bilbies (*Macrotis lagotis*) and burrowing bettongs (*Bettongia lesueuri*), co-existed with rabbits in the same burrow systems without any conflict. However, this situation probably only lasted while seasonal conditions were good. When a series of major droughts affected large parts of the southern Australian rangelands in the late 1880s the rabbits turned to alternative food sources, consuming the leaves, roots and even the bark of shrubs once the native grasses had gone. This dramatic destruction of the flora probably led to the decline of the native mammals, a decline from which many species were never able to recover when seasonal conditions returned to average. The impact of the rabbit was probably exacerbated in some areas where sheep were being run on open rangelands at stocking rates far above what is now considered appropriate for Australian conditions.

The introduction of predatory exotic species such as mink appears to have reduced the diversity and density of native predators in some parts of Europe and the Russian Federation. Mink have quite a broad diet and have even shown some capacity to alter their food preferences in some parts of their introduced range.

By far the most dramatic competitive impacts of introduced mammals have been in ecosystems where they have competed not with native mammals but with the local mammalian equivalent on oceanic islands. Mammalian herbivores such as goats and rabbits often breed up to extremely high numbers following their introduction, and rapidly convert complex plant ecosystems on islands to simplified grasslands. The loss of habitat diversity leads to a dramatic loss of the endemic flora and fauna, and in some cases ultimately leads to the extinction of the introduced mammal species as well.

EFFECTS OF DISEASE AND PARASITES
There are few documented cases of introduced mammal species introducing diseases and parasites into their new environments. There are a number of reasons why this may be so. It is possible that the successful introduction of most diseases is actually a difficult process to achieve. However, it is just as likely that since most of the really successful introductions occurred prior to the twentieth century, the long sea voyages that were required to achieve most introductions provided an effective quarantine process. There is some evidence to support this idea, with several attempts at introducing camels (*Camelus dromedarius*) failing due to the death of founding stock *en route* or soon after arrival, and the cause(s) being described as 'disease'. The modern situation is now very different with air transport making it possible to ship large animals from one side of the world to the other in only a few days. Modern introductions, be they for agricultural or conservation reasons, now require thorough quarantine screening prior to departure and again on arrival. The value in such systems has been made clear to all by the recent events involving foot-and-mouth disease in the United Kingdom and Western Europe during 2001.

There still remains cause for concern with the pet-keeping industry, which is far more difficult to regulate. It is also less likely to have an immediate impact on agricultural production, but may have an adverse impact on native mammal species in the wild. If this occurs it is likely to go undetected for longer, and governments may be far less willing to devote resources to containing disease outbreaks.

There is one other role that introduced mammals can have in the spread of diseases and parasites. That is one where the host exotic mammal is not significantly affected by the disease or parasite but acts as a reservoir for the disease and provides a means by which it can be spread to domestic exotic animals or native animals when they forage over common ground. One of the best-documented cases of this type of impact is the role that Australian brush-tail possums (*Trichosurus vulpecula*) have had in the maintenance and transmission of bovine tuberculosis to domestic

cattle in New Zealand. Other less well-known cases involve the transmission of canine distemper from domestic and feral dog populations to native pinniped populations in the northern hemisphere and the transmission of toxoplasmosis from feral cats to sheep and native marsupials (e.g. peramelid marsupials).

HYBRIDISATION
The issue of hybridisation has been documented for several species of ungulate (e.g. *Capra ibex* and *Cervus elephas*) and European bison (*Bison bison*). In some cases little consideration appears to have been given to the genetic make-up of the founding stock, or the stock released to bolster failing local populations was different to the endemic gene pool. In other cases hybridisation may be the best course of action to ensure the best possible genetic diversity in new populations established for conservation reasons. In Australia there are several species of native mammal that are now extinct on the mainland and are only represented by populations of island subspecies. Analysis of the genetic diversity of some of these island populations has shown that they are far less diverse genetically than the now extinct mainland subspecies were. In order to both preserve the remaining genetic diversity and possibly improve the chances of successful re-introductions to the mainland it may be worthwhile crossing the subspecies in captivity and then releasing the progeny.

Conversely, concerns about exactly this issue have prompted calls to reverse the introduction of collared lemurs (*Lemur fulvus collaris*) from Berenty Private Reserve for fear that they will hybridise with another subspecies (the red-fronted lemur, *L. f. rufus*) that had previously been established in the reserve. The International Union for the Conservation of Nature (IUCN) has developed guidelines to help with making these decisions, but more research is warranted.

Another issue of importance in mammal introductions relates to the subtle differences in the capacity of wild stock to establish in comparison to domesticated versions. In many species (deer, pigs, goats, donkeys) the differences in capacity to establish are so small that it probably does not matter. In other species there does appear to be some effect of long-term domestication. An example of this is the European polecat (*Mustela putorius*), which is well established in New Zealand and causes considerable damage to the endemic fauna. The domesticated version, the ferret (*M. p. furo*) is widely kept in Australia as a pet and yet has not been successful in establishing in the wild for any length of time. There is currently discussion within Australia as to the possibility of ferret-keepers being given approval to import new blood stock to bolster what are considered declining quality stocks. The focus is on obtaining animals from New Zealand, the nearest source and also a country with an 'A' class quarantine rating. Conservation managers have raised the question of whether the injection of new genes into the Australian domestic ferret gene pool may allow ferrets to become permanently established in the wild, creating a conservation disaster worse than that suffered in New Zealand.

GENETIC CHANGES

While there are many species of mammal that have established successfully in new habitats, a considerable number have not spread far beyond their point of release or have remained in low numbers for decades and even centuries. In most of these cases limited availability of suitable habitat is probably the primary reason for their restricted distribution or low numbers, but in others the reasons are not obvious. The European hare (*Lepus europaeus*) is established in south-east Australia, but is nowhere near as successful there as it is in other introduced habitats, such as New Zealand and the southern parts of South America.

Recent study suggests that the reason for the hare's lack of success in Australia, compared to its native range in Europe and the other introduced populations in the southern hemisphere, is possibly due to a lower pregnancy rate. It is postulated that this lower rate can be attributed to abnormalities of the female reproductive system, presumably limited to the animals in Australia. It has also been suggested that differing levels of reproductive hormones in the hares' food could be contributing to the restricted nature of the species in Australia. If there were to be changes in the genetic make-up of the hare in Australia such that it could overcome these apparent limitations, then the pest potential for this species in Australia would increase significantly.

AGRICULTURAL DAMAGE

There are numerous examples of introduced species causing serious economic damage to agriculture around the world. For the most part those species that cause damage, even minor damage, in their natural range will continue to do so in any introduced range. Even those species that do not cause damage immediately upon establishing may still retain the capacity to do so when agricultural practices and crop types change at some point in the future.

In many instances there is little quantification of the amount of damage sustained for each crop type, partly through the difficulties in devising reliable methods of estimating damage, and partly through a general lack of desire among farmers to bother quantifying what is obvious to them. There are more data available in some countries on the value of the control efforts levelled against the introduced pest species, and a small number of countries where estimates are provided for both the value of the lost crops and the costs of applying controls.

In the United States, for example, the estimated cost incurred as a result of introduced mammal pests is US$25 billion per annum. This is on top of the estimated $36 billion spent on pest plants, $34 billion on microbe pathogens, $23 billion on invertebrate pests, $5 billion on reptiles and amphibians, $2 billion on pest birds and $1 billion on fish.

No comparable figures are available for losses to agriculture in Australia due to introduced mammals, with the exception of the rabbit. Back in the 1950s, when the myxoma virus was first introduced into Australia to help reduce rabbit numbers, it was estimated that the increase in pasture and crop production following the release of the virus in southern Australian regions amounted to A$590 million (*c*. US$320 million) per annum when converted to modern currency rates.

The well-documented mammal control for coypus in the United Kingdom that commenced in the early 1960s and continued until the late 1980s cost in excess of US$7 million at that time. The eradication program was carefully monitored and the cost–benefit analyses consistently indicated that eradication was a worthy goal.

The lack of good quality cost–benefit analyses that can form the basis of any decisions on whether to attempt control, how much control effort to apply and when to stop controls or not bother applying any control efforts is vital to managing mammal pests. Perhaps the lack of such data is caused by the need to integrate the biological sciences with modern accounting techniques, a combination that is rarely catered for in our tertiary institutions, even today.

BENEFITS TO BE HAD FROM NATURALISATION

It would be unfair to present only the negative argument for introductions of mammal species, for there are examples where introductions have had real benefits, not only for humans but also for the mammal species themselves.

RECREATION

One of the benefits of human societies is that for the most part they allow for time to devote to recreational pursuits. Despite our modern developments many cultures still embrace hunting, only now it is practised as a sport and is not a fundamental necessity for life. In many parts of the world the long history of landscape modification has resulted in few areas of natural habitat that can support wild mammal species in any great abundance. Accordingly, those mammal populations that do remain must be managed carefully to ensure that any harvest is conducted in a sustainable manner. This is effectively achieved through regulation and licensing, with both the resource (the mammals) and the licences (the authority to harvest the resource) having commercial value.

Lessons learnt from the past misuse of mammal resources during the last 200 years have now paved the way for better and more sustainable forms of recreational use of mammals. Some conventional hunting still occurs, but it is now often directed at animals that are surplus to the carrying capacity of the area managed for the various game species, and only permitted after careful assessment of the structure of the population.

Even more sophisticated systems have evolved whereby no animals are actually killed, only their photographs are taken or humans savour only the experience of interacting with the animals in their natural environment. In some parts of Africa this form of controlled trophy hunting and eco-tourism now contributes more to local economies than any other industry. There is a genuine incentive for the local people to conserve the native animals and the environment that the animals are dependent on. There are jobs for the local people catering to the tourists, and the resource base is not depleted.

The benefits of trophy hunting and eco-tourism targeting mammalian species also leads to some groups desiring to establish exotic species in new locations purely to allow the development of a particular hunting or tourist industry. Some of these attempts are discussed within the community and the risks weighed against the benefits, but many introductions are still initiated by individuals for personal gain without consideration for the greater community or the environment.

CONTROL OF PESTS

Once exotic species have become widely established few studies have been made of them with the object of deciding what benefits rather that what damage they may do. Few people would now consider introducing European rabbits to Australia or mink (*Mustela vison*) to parts of Europe and the Russian Federation if they were not already there, since studies have shown that large populations of those species can cause considerable loss of agricultural production or mortality amongst native fauna. There have been many attempts at introducing mammal species to control existing pest species (native or exotic). However, there are few if any cases where the introduction has led to successful control of the pest species, and numerous cases where the supposed controlling species has itself become a pest of agriculture or the environment.

PRESERVATION AND CONSERVATION

There are several examples of species that have benefited from being the basis for recreational hunting or sporting ventures or zoological collections. These species have been established in new habitats (some artificial) and then at some later time the original populations have suffered drastic or permanent declines. Without the exotic or captive population, local, regional and even global extinctions of some species would have occurred. This has been most evident among the deer, goat and bovid species (e.g. Pere David's deer, *Elaphurus davidianus*; ibex, *Capra ibex*; bison, *Bison bison*; and Arabian oryx, *Oryx leucoryx*).

There are also examples of remnant populations of some Australian mammals whose distributions contracted so greatly that they were only represented on offshore islands. By good luck rather than good management those islands acted as 'floating zoos', providing the only remaining stock with which to re-establish new populations on the mainland (e.g. western barred bandicoot, *Perameles bougainville*; burrowing bettong, *Bettongia lesueuri*; Shark Bay mouse, *Pseudomys fieldi*; and greater stick-nest rat, *Leporillus conditor*). The lessons learnt about those floating zoos are now being actively applied by intentionally establishing some species on islands where they did not previously occur until secure locations can be identified on the adjacent mainland (e.g. dibbler, *Parantechinus apicalis*; rufous hare-wallaby, *Lagorchestes hirsutus*; and Shark Bay mouse).

SPECULATION ON THE FUTURE

No mammal species has extended its range over such a large area of the world as humans, and neither has any other species had such an impact on changing habitat quality, quantity and diversity for so many other species. Many of the changes we are experiencing now had their beginnings decades or centuries ago when major changes in our industrial practices were first established. Ozone-depleting chemicals released in the air and the combustion of fossil fuels on a massive scale appear certain to alter the climate for decades and possibly centuries to come. These impacts alone and the effect they will have on ultraviolet radiation levels and sea levels may well shape the future of terrestrial life in some parts of the earth.

Continued clearing of natural vegetation, indiscriminate release of toxic pollutants and the ill-conceived introduction of exotic animals will lead to a simplified flora and fauna, represented largely by the weed and pest species we already spend vast sums of money controlling. So how do we learn from the historical message set out in this work? How can it help us predict which species will definitely cause problems in a new environment and which ones could safely be moved around the globe? We need to develop risk assessment techniques.

RISK ASSESSMENT (PAST AND PRESENT)

Science prides itself on developing theories and models that explain observed variation in nature. But can the same process be applied to such complex things as mammal introductions and will decisions made today based on often limited information, be valid in 10 or 50 years time? Researchers warning of the dangers of introducing animals are not a modern phenomenon, with published accounts dating from the late nineteenth century. However, most of those people attesting to the pest potential of introduced animals considered legislation to be the most effective (if not the only) solution to the problem at that time. In many cases no such legislation was forthcoming and introductions continued unabated. Scientists and politicians alike still had to grapple with the problem of how to discriminate between species with pest potential and those with little or no potential to cause harm.

It wasn't until the latter part of the twentieth century that the idea was put forward of examining the biology of species to identify common characteristics that might help define 'pests' from 'non-pests'. Most attempts were limited by the fact that they invariably focused on a small number of species, typically those that were local in a regional sense to the researcher(s). The studies also suffered from the fact that sample cases were often of species with no obvious taxonomic relationships, and that the history of their introductions (how many, how often, to how many places) varied greatly. Documentation supporting many aspects of these important studies was often lacking or difficult to locate, simply because of the basic modes of communication available in those days. Conducting online Internet searches of major libraries' holdings was not an option.

Times have changed. We are now wise enough to attempt this type of analysis. We have the modern communications to support the endeavour and with the benefit of modern statistical techniques we can rigorously test hypotheses and have confidence in the outcomes. Those outcomes can then be presented to decision-makers and law-makers, and rational arguments can be made to regulate, limit or prevent some or all of the potential introductions that could be shown to be detrimental, or likely to be so.

It has been noted with interest the diverse range of detailed studies that have been published based on the large data set provided in John Long's 1981 publication, *Introduced Birds of the World*. More recent studies using more modern analytical techniques will be published soon and give strength to the belief that valid and rigorous risk assessment techniques are possible for birds (Duncan *et al.* 2001). Some of the data provided in this work (relating to introductions of mammals to Australia and New Zealand) were made available for the same kind of analyses and similar predictive capacity seems possible (Duncan *et al.* in prep.). Such techniques would clearly benefit from the testing of data sets relevant to other parts of the world (e.g. North America, South America and Europe), and volunteers are encouraged to step forward – the sooner the better.

POLITICAL WILL

Any lessons learnt from the historical accounts of introductions of mammals to various parts of the world are well worth learning. However, if the lessons are to be translated into positive actions for the better management of our natural biodiversity it will be necessary for scholars to summarise the key elements of the subject and convey those points to those in positions of power among our political representatives. There is no doubt that politicians have many issues to consider, options to weigh and hard decisions to make, but some of the hardest decisions to make regarding our natural resources are those made in ignorance or with limited knowledge.

Decisions relating to the import, keeping, and subsequent escape or release of animals, especially mammals, are likely to attract human sympathy. It follows that any decisions to control or restrict populations of mammals will also attract sympathy from elements of the public. Few politicians wish to be remembered as the people that killed 'Bambi', or 'Black Beauty' or 'Peter Rabbit' (fictional animal characters from children's literature), but sympathy for one group of animals at the expense of many others, let alone agricultural and natural resources, is unjustified.

In most countries there are legislative charters already in place that mandate the protection of the native flora and fauna. Abrogation of those basic charters solely for political gain or fear of voter retribution fails the generation of the day, and diminishes the inheritance of subsequent generations. It may be valid for politicians to argue that absolute control is beyond the financial means of the government of the day. However, it does not excuse governments from developing measures to prevent future adverse mammal introductions, or of them remaining aware of the potential for science and economic circumstances to deliver viable alternatives in the future.

CONSISTENT APPROACH

The introduction and establishment of mammals is a global problem. It is unlikely that there will be any one solution as to how this problem can be managed. Neither will the problem be resolved if only some of the countries develop management strategies and apply them. Successful management is only a possibility if there is an acceptance of the severity of this problem and a consistent approach in dealing with it.

Getting even a few countries in the world to agree on any issue is difficult at times, but history provides some examples of its possibility. Countries form alliances during times of war, for mutually beneficial trade, or for protection against threats of disease and natural disasters. What needs to be accepted is that after habitat loss and fragmentation the threat posed by introduced animals is one of the greatest facing our native flora and fauna.

Again, it will be necessary for the scholars of this subject to summarise the relevant data, and use it to educate the governments and their negotiators, so that when the parties meet they are all conversant with the issues and the task at hand. Education will be as important as scientific research in ensuring that the problem of introduced animals is taken seriously and dealt with accordingly.

MISGUIDED CONCERN

An issue has arisen in recent years that would probably have been unimaginable 30 years ago.

Governments and private individuals with a desire or need have invariably been able to instigate control programs against introduced mammal species. Most control or eradication programs were instigated due to some perceived adverse impact from the introduced species and such actions were either accepted by the public or supported in one way or another without any hindrance. Recently, conservationists, or more specifically animal liberationists, have either undertaken actions to disrupt control efforts in the field or instigated legal proceedings to have court orders issued to stop the control programs. This has occurred in Italy where an unsuccessful legal action lasting three years resulted in wildlife authorities losing the opportunity to eradicate the grey squirrel (*Sciurus carolinensis*) and prevent it from replacing the native red squirrel (*S. vulgaris*). In Australia public protests prevented the removal of rusa deer (*Cervus timorensis*) from Royal National Park on the outskirts of the state capital city of Sydney following major wildfires. In the United States animal rights groups have regularly prevented culling operations of wild horses and donkeys from fragile arid rangelands and conservation estate. Wildlife authorities now have to muster animals and attempt to administer contraceptive drugs to limit populations, but with little success.

There should be concern that this type of conservation ethic will become more widespread and common. What is not clear is why these same animal rights supporters do not seem to be willing to stand up for the rights of the native flora and fauna that is so often adversely affected, if not driven to extinction. It seems to be more a case that if an animal is large, furry and has big eyes that engender sympathy then it must be saved, even if the salvation of those animals comes at the risk of the extinction of other species.

We need to educate our children so that they can develop opinions based on a balanced account of the facts. We must also educate our politicians and lawmakers on these same issues, but, more importantly, we need to educate our judiciary, or at least those aspiring to the judiciary, of complex environmental issues surrounding introduced species.

Family: Tachyglossidae
Spiny anteaters

SHORT-BEAKED ECHIDNA
Spiny anteater, echidna
Tachyglossus aculeatus (Shaw and Nodder)

■ **DESCRIPTION**
HB 300–450 mm; WT 2–7 kg.
Rotund body; dorsal surface and tail covered with long spines with some fur between them; forehead bulbous; eyes small; nose tubular; tongue long and sticky; spines creamy with white tip; fur dark brown to sandy colour; toes five; tail rudimentary. Male has a small non-venomous spur on hind leg and is larger than female.

■ **DISTRIBUTION**
Australia. Throughout the Australian mainland and Tasmania.

■ **HABITS AND BEHAVIOUR**
Habits: nocturnal or diurnal depending on temperature; hibernate in cold areas; live in thick bushes, under logs, under piles of debris, occasionally in rabbit or wombat burrows; curls into ball in defence. **Gregariousness:** solitary; overlapping home ranges; several males pursue one female at mating. **Movements:** no information. **Habitat:** deserts to forests and cold mountain areas. **Foods:** mainly ants and termites, but also earthworms and insects occasionally. **Breeding:** mates July–August; lays single egg 2 weeks after copulation in pouch on female's abdomen; young born hairless, blind, with very short soft spines; pouch life not known, probably up to 3 months; suckling period not known, but probably 3 months; independent adult at *c.* 8–12 months. **Longevity:** no information. **Status:** distribution sparse, but relatively common and locally abundant.

■ **HISTORY OF INTRODUCTIONS**
AUSTRALASIA
Islands off Australia
Forty-three short-beaked echidnas were introduced to Maria Island, off Tasmania in 1971, where the species has become established (Weidenhofer 1977; Summers 1991; Abbott and Burbidge 1995).

Other attempts were made to establish the species on Saddleback Island, Queensland (Abbott and Burbidge 1995) and on Vansittart Island, Tasmania (Whinray 1971; Hope 1973), but the results are not well documented.

Family: Ornithorhynchidae
Platypus

PLATYPUS
Ornithorhynchus anatinus (Shaw and Nodder)

■ **DESCRIPTION**
HB 300–420 mm; T 100–130 mm; WT 0.67–2.7 kg.
Dense underfur dark brown, paler below; long flattened guard hairs; duck-like, sensitive, pliable bill;

cheek pouches; white fur around eyes; sharp, hollow spur on each ankle; hind and forefeet webbed; tail flattened and broad.

■ DISTRIBUTION

Australia: northern Queensland down east coast to South Australia and Tasmania. Now extinct in South Australia.

■ HABITS AND BEHAVIOUR

Habits: aquatic; during day rests in long burrows to 20 m in bank; active dawn and dusk, but may be diurnal or nocturnal at times. **Gregariousness:** solitary. **Movements:** sedentary; males territorial. **Habitat:** near coastal fresh waters from alps to tropical rainforest; riverbanks, lakesides. **Foods:** aquatic invertebrates (insects, molluscs, crustaceans, worms), small fish and amphibians. **Breeding:** August–October; intrauterine gestation 14 days; eggs 2, rarely 3; incubated by female in burrow; hatch in 1–2 weeks; lactation 3–5 months; females breed at 2 years. **Longevity:** 9 years or more. **Status:** common, but vulnerable.

■ HISTORY OF INTRODUCTIONS
AUSTRALASIA
Kangaroo Island

Platypuses were established from introductions in 1928 (three animals), 1941 (10) and 1946 (six) to Flinders Chase, Kangaroo Island, where they were present in Breakneck and Rocky River catchments from 1967–93 (Inns *et al.* 1979; Copley 1995; Strahan 1995).

Western Australia

Introduced to Western Australia in 1984 when a male and a female platypus were released at an unknown location in the Avon River. Two reports of sightings were later made, but neither was subsequently confirmed and they were not seen again (Grant and Fanning 1984; Anon. 1987).

■ DAMAGE

None.

Family: Didelphidae
American opossums

MOUSE-OPOSSUM
South American mouse-opossum
Marmosa robinsoni Bangs
= *M. mitis* Bangs, = *M. m. chapmani*

■ DESCRIPTION
HB 110–185 mm; T 90–280 mm; WT about 95 g.
Upper parts cinnamon or russet; face paler; nose long, pointed; eyes large; dark colouration around eyes gives effect of black mask; tail long, prehensile, densely haired, with white appearance; under parts lighter. Female has 15 mammae and is shorter than male.

■ DISTRIBUTION
South America and West Indies. From Belize, Honduras and Panama to north-western South America (Colombia and Venezuela). On Trinidad, Tobago and Grenada; also on islands of Saboga and San Miguel in Golfo de Panamá.

Mouse-opossum

■ HABITS AND BEHAVIOUR
Habits: mainly nocturnal; agile climber; often shelters under fronds of palms. **Gregariousness:** solitary; density 0.31–2.25/ha. **Movements:** sedentary and occasionally nomadic; home range 0.17–0.36 ha. **Habitat:** llanos, forest and dense scrubland. **Foods:** insects, fruits, small lizards, rodents, birds' eggs. **Breeding:** breeds throughout year, but some areas more seasonally; gestation 13–14 days; oestrous cycle 23 days; litter size 6–13, 15; may breed 2–3 times per year; female builds leaf nest in cavity; carries young for 5 weeks; born blind, naked; attached to teats 21 days; eyes open *c.* 39–40 days; leaves nest 40–50 days; weaned at 60–70 days, then disperse to establish solitary nest site; mature at 6 months. **Longevity:** under 1 year in wild, 3 years in captivity. **Status:** common.

■ HISTORY OF INTRODUCTIONS
WEST INDIES
Grenada and Grenadines
It is thought that the South American mouse-opossum was possibly introduced to Grenada and the Grenadines (Carriacon and Islet Ronde) from Trinidad (de Vos *et al.* 1956; Lever 1985). Other authorities (Hall 1981) make no mention of any introduction, but confirm their presence on the islands.

EUROPE
United Kingdom
A mouse-opossum (*M. m. chapmani* = *M. robinsoni*) was found in the United Kingdom in a shipment of bananas from South America before 1959 (Fitter 1959); individuals often reach USA and other ports in fruit shipments (Walker 1991).

■ DAMAGE
Banana and mango crops are sometimes damaged by these marsupials (Walker 1991) and they are believed to be partially responsible for the disappearance from Dominica of the bridled quail-dove (*Geotrygon mystacea*) (Lever 1985).

AMERICAN OPOSSUM
Virginia opossum, southern opossum, black-eared opossum
Didelphis marsupialis Linnaeus
Some authorities treat D. marsupialis *as synonymous with* D. virginianus. *They are treated as separate species here.*

■ DESCRIPTION

HB 300–584 mm; T 255–535 mm; WT 136–504 g.
Coat varies from grey to black to brown to reddish, or rarely white; muzzle pointed; nose pad naked, pink; facial hair short, white; black patch around eyes; ears naked, or sparsely haired, white with black furred base; legs short, black; toes white; tail prehensile. Female has 13 mammae; fur-lined pouch on abdomen.

■ DISTRIBUTION

South and North America. From northern Argentina, Brazil and Peru north to Mexico and the south-eastern United States. Formerly confined to South America but colonised northwards in prehistoric times.

■ HABITS AND BEHAVIOUR

Habits: nocturnal; mainly arboreal; lives in vacated burrows, hollow trees, brush piles; inactive in cold weather; scavenges at refuse tips and from rubbish

American opossum

bins. **Gregariousness:** singly or in family groups; density 0.25–1.25/ha. **Movements:** sedentary; home range 16–23 ha, but wanders further (up to 11.3 km). **Habitat:** forest, brush areas, woods near streams, cultivated areas with trees, and generally most areas with tree hollows available. **Foods:** insects (grasshoppers, beetles, bugs, moths, flies, ants), earthworms, myriopods (millipedes, centipedes), spiders, molluscs (snails, slugs), amphibians (toads, frogs, salamanders), crustaceans (crayfish), reptiles (snakes, turtles), small mammals (voles, moles, shrews, chipmunks, squirrels, rats, mice, rabbits), birds (ducks, poultry, game birds) and their eggs and nestlings, fruits (grapes, cherries, thornapples, tomatoes, pumpkins, muskmelons), roots, nuts (acorns), seeds, grains (wheat, oats, beans, corn), fungi (mushrooms), green vegetable materials (leaves of grass and clover) and fish. **Breeding:** breeds all year, peaks in January–March and May–July; gestation 8–13 days; litters 1–2 per year; young 4, 8–18, 25 (only those finding mammae survive); female polyoestrous; young in pouch for 2 months; born blind, helpless, completes development in pouch for first 60 days; weaned at 90–100 days; sexually mature in first year, males 8 months, females 6 months. **Longevity:** 2–3 years and possibly up to 7 years (wild). **Status:** common and numerous; extending range; in some areas trapped for fur and flesh.

■ HISTORY OF INTRODUCTIONS

Through introduction and colonisation these opossums have successfully spread northwards in the eastern United States to north-eastern Canada. They have been introduced successfully in western Canada, the United States, and into the Lesser Antilles, but unsuccessfully into the Bahamas.

EUROPE
United Kingdom
Before 1959 an American opossum was once found in a consignment of bananas on a ship in port (Fitter 1959).

NORTH AMERICA
Canada
The opossum is thought to have reached southern Ontario between 1892 and 1906, although there are older records of their presence dating back to 1858. There were subsequently more recent invasions in 1934 and again from 1947 onwards (Peterson and Downing 1956). By the late 1950s they appeared to be in the process of becoming well-established residents of this area. A single opossum was found at Morrisburg on the St. Laurence River in 1952 and others were reported at Landsdowne and Kemptville in 1961 (Banfield 1977).

From introductions into Washington, in the United States, the opossum spread northwards into British Columbia. Two were killed at Crescent Beach in December 1949 (Cowan and Guiguet 1960). Following their entry into the Lower Fraser Valley in 1949, some opossums were recorded north of the Fraser River at Point Grey in 1965 (Carl and Guiguet 1972). The species was later reported to be common north as far as Hope (Banfield 1977).

The opossum is now established on the west coast of North America from British Columbia south to San Diego County, California (Hall 1981).

United States
American opossums were originally confined to South America, but over a period of thousands of years have successfully invaded North America (Morris 1965). In recent times they have spread, probably assisted by introductions, through the south-eastern United States to Ontario, Canada (Hock 1952; Burton and Burton 1969; Hall 1981). They were formerly not found north of the Hudson River valley (Guilday 1958) in historic times. The increase in range is thought to be in part due to the decrease in numbers of predators (Burton and Burton 1969).

Many opossums were introduced and some escaped from captivity a number of times in New York State before 1900, before finally becoming established at a later date. They were established in New England before 1904 (Hamilton 1958). Records of introductions are scarce: some 14 opossums were released north of Tucson, Arizona in 1927, and in 1950 a few escaped at Apache Junction, but there were almost certainly many other releases (Hock 1952; Hall 1981).

On the west coast of the United States the opossum was successfully introduced into California, Oregon and Washington states. They were released in California between 1905 and 1910 (Grinnell 1915; Carl and Guiguet 1972; Deems and Pursley 1978) and at San Jose in 1910 with animals from Tennessee (Jameson and Peeters 1988). In the space of 12 years they were so widespread that it was impossible to exterminate them (Hock 1952). They were well established and spreading in the 1930s (Storer 1934). In Washington state they were released on Comano Island and in the Sedro Valley on the mainland in about 1925. Here, they became firmly established in both areas (de Vos et al. 1956; Cowan and Guiguet 1960) and spread into Oregon and northwards into Canada (Hock 1952; Cowan and Guiguet 1960) from the latter release, although there were apparently early introductions in Oregon about 1914 (Jewett and Dobyns 1929). By 1952 they occurred widely in both Oregon and Washington (Hock 1952). Other introductions may have occurred in Arizona and Colorado (Hock 1952; Hall 1981).

In California they now occur from the eastern edge of the Sacramento and San Joaquin valleys, west to the coast and from San Diego north to Oregon (Jameson and Peeters 1988). They are most abundant in suburban and agricultural habitats (Lidicker 1991).

WEST INDIES
Bahamas
A pair of opossums were released on the island of Grand Bahama in 1923 or 1933, but failed to become permanently established (Sherman 1954).

Lesser Antilles
It is thought that from Trinidad one of the *Didelphis* species (possibly *D. m. insularis*) may have been introduced to Grenada, St. Vincent, St. Lucia, Dominica (de Vos *et al.* 1956; Hinton and Dunn 1967; Eisenberg 1989), and perhaps to Martinique. Others (Hall 1981) make no mention of any introduction to these islands and list them as indigenous.

■ DAMAGE
In the Lesser Antilles it is thought that the American opossum has played a major role in the elimination of certain birds, especially the ground dove (*Geotrygon montana*) (Hinton and Dunn 1967).

The opossum is one of the most heavily parasitised mammals and introductions of them involve public health, poultry and the truck and garden industries. In New York State they were considered to be of minor economic importance to wildlife and small domestic stock in the 1950s (Hamilton 1958). Following their introduction into California, however, it was found that they caused economic losses at times to poultry farms, vegetable crops and cornfields, and attempts were made to control their numbers.

Family: Dasyuridae
Dasyurids

BRUSH-TAILED PHASCOGALE
Black-tailed phascogale
Phascogale tapoatafa (Meyer)

■ DESCRIPTION
HB 148–261 mm; T 160–234 mm; WT 106–311 g.
Upper parts grizzled grey; under parts cream to white; head long and pointed; ears large, naked; feet sharply

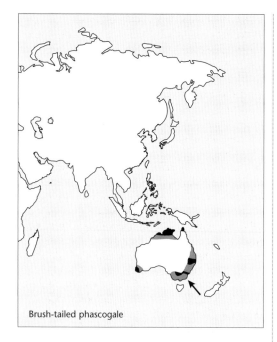

Brush-tailed phascogale

Swamp antechinus

clawed; forefeet with five digits; tail hairy, grey black, bottlebrush shaped, terminal half black. Female has eight mammae and is smaller than male.

■ **DISTRIBUTION**
Australia. Formerly throughout the dry sclerophyll forests and woodland of temperate and tropical Australia.

■ **HABITS AND BEHAVIOUR**
Habits: arboreal and terrestrial; agile; nocturnal; sleeps in nest during day; nests in hollow tree, stump, or bird's nest. **Gregariousness:** males solitary; females with young. **Movements:** disperses in mid summer; female home range 20–70 ha, males twice this. **Habitat:** eucalypt forest and woodland. **Foods:** cockroaches, beetles, centipedes, spiders, bull ants; occasionally small vertebrates; nectar. **Breeding:** mates May–July; gestation 30 days; litter size 3–8; born hairless; lactation 14–20 weeks; sexual maturity 8 months. **Longevity:** females 2–3 years or more, males probably 1 year. **Status:** uncommon to rare, populations localised.

■ **HISTORY OF INTRODUCTIONS**
AUSTRALASIA
Victoria
From 1991–93 three re-introductions of captive-bred phascogales occurred in Gippsland, Victoria (Soderquist 1995).

■ **DAMAGE**
Phascogales cause no damage, but are occasional predators of small vertebrates and even penned poultry (Strahan 1995).

SWAMP ANTECHINUS
Antechinus minimis (**Geoffroy**)

■ **HISTORY OF INTRODUCTIONS**
AUSTRALASIA
Victoria
Ten juvenile swamp antechinus were translocated from Port Campbell to Anglesea for the purpose of monitoring behaviour following their re-introduction. After five days three had died; the remainder appeared established and monitoring was continuing (Aberton *et al.* 1995).

SOUTHERN DIBBLER
Parantechinus apicalis (**Gray**)

■ **DESCRIPTION**
HB 140–145 mm; T 95–105 mm; WT 60–100 g males, 40–75 g females.
Small, rat-sized carnivore, brownish-grey above, freckled with white and greyish-white tinged with yellow. Prominent white eye ring, tapering hairy tail.

■ **DISTRIBUTION**
Restricted to scattered populations in a small area of south coast of Western Australia from Albany east to Fitzgerald River, and two populations on the west coast on Boullanger and Whitlock Islands.

■ **HABITS AND BEHAVIOUR**
Habits: terrestrial and semi-arboreal; agile; crepuscular; sleeps in nests made in burrows constructed by

other animals during day. **Gregariousness:** males solitary; females with young. **Movements:** sedentary; male home range 1.5–6.0 ha, females 0.3–0.6 ha **Habitat:** long unburnt heath shrublands and mallee-heath. **Foods:** insects, spiders; some small reptiles; berries; nectar. **Breeding:** March–April; litter size 1–8; born hairless; dependent on female for 12–16 weeks; young disperse during September–October each year. **Longevity:** females 2–3 years or more, males often only 1 year, occasionally 2–3 years. **Status:** uncommon to rare, populations very localised.

■ HISTORY OF INTRODUCTIONS
AUSTRALASIA
Western Australia
Between October 1998 and January 2001 a total of 94 southern dibblers were translocated from Perth Zoo to Escape Island off the west coast near Jurien Bay, Western Australia. Initial monitoring suggests that they have established a viable population.

WESTERN QUOLL

Chuditch

***Dasyurus geoffroii* Gould**

■ DESCRIPTION
HB 260–400 mm; T 210–350 mm; WT 0.6–2.2 kg.
Upper parts grizzled grey brown, white spotted; under parts pale grey to white; head long and pointed; ears large, rounded; hands and feet sharply clawed; forefeet with five digits; tail hairy lacking spots, grey-black, bottlebrush shaped, terminal half black. Female has eight mammae and is smaller than male.

■ DISTRIBUTION
Australia. Formerly throughout the dry sclerophyll forests and woodland of temperate and sub-tropical Australia.

■ HABITS AND BEHAVIOUR
Habits: arboreal and terrestrial; agile; nocturnal; sleeps in nest during day; nests in hollow tree, stump, or bird's nest. **Gregariousness:** males solitary; females with young. **Movements:** disperse in mid summer; female home range 3–4 km², males up to 15 km². **Habitat:** eucalypt forest and woodland. **Foods:** cockroaches, beetles, centipedes, spiders, bull ants; small vertebrates; nectar. **Breeding:** mates late April–July; gestation 17–18 days; litter size 2–6; born hairless; remain in pouch for *c.* 60 days, lactation 16–20 weeks; sexual maturity 11 months. **Longevity:** females 2–3 years or more, males often only 1 year, occasionally 2–3 years. **Status:** uncommon to rare, populations localised.

■ HISTORY OF INTRODUCTIONS
AUSTRALASIA
Western Australia
In 1987 nine captive-bred animals were re-introduced to Lane-Poole Reserve near Dwellingup, Western Australia. The re-introduction failed primarily due to the lack of protection from introduced red foxes (*Vulpes vulpes*). As part of a formal recovery plan for the species, a series of re-introductions were carried out in south-western Western Australia involving a total of 280 captive-bred animals reared at the Perth Zoo. All re-introductions were to locations where fox numbers were reduced by coordinated baiting programs. From 1992–94, 72 captive-bred western quolls were re-introduced to the Julimar Forest, 81 animals to Lake Magenta Nature Reserve during 1996–97, 46 to Cape Arid National Park during 1997–98, 48 to Mount Lindsay National Park in 1999 and 33 to Kalbarri National Park in 2000. Initial monitoring indicates that the populations at the first four locations have established, with dispersing juveniles recorded from adjacent areas.

MARSUPIAL CAT

***Dasyurus* sp.**

■ HISTORY OF INTRODUCTIONS
PACIFIC OCEAN ISLANDS
New Zealand
Two marsupial cats were introduced by the Canterbury Acclimatisation Society into New Zealand in 1868, but they failed to become permanently established (Thomson 1922; Wodzicki 1950).

Family: Myrmecobiidae
Numbats

NUMBAT

Banded anteater

***Myrmecobius fasciatus* Waterhouse**

■ DESCRIPTION
HB 200–274 mm; T 161–210 mm; WT 300–715 g.
Upper parts red-brown, under parts paler; rump dark, with white transverse bars; head narrow; snout sharp; eye stripe dark; tail hairs long, often erected to give 'bottlebrush' appearance.

■ DISTRIBUTION
Formerly extended from western New South Wales through South Australia and across the southern half

of Western Australia. Now a few isolated populations remain in Western Australia.

■ HABITS AND BEHAVIOUR

Habits: burrows 1–2 m or lives in hollow logs, tree hollows (especially winter); diurnal. **Gregariousness:** solitary. **Movements:** dispersal movements to 15 km or more; home range 25–50 ha, males wander further in breeding season. **Habitat:** dry eucalyptus forest and woodland. **Foods:** termites. **Breeding:** mates December–February; gestation 14 days; litter size 4; weaned 9 months(?); young disperse 11–12

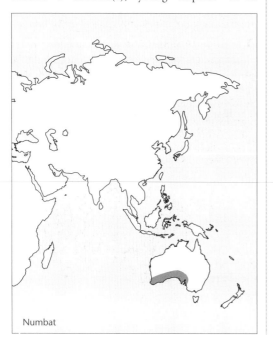

Numbat

months(?); females breed in first year, males in second year. **Longevity:** no information. **Status:** rare and occurs in isolated populations.

■ HISTORY OF INTRODUCTIONS

AUSTRALIA

New South Wales

Forty-three numbats wild-caught at Yookamurra Sanctuary, South Australia, were re-introduced to Scotia Sanctuary, New South Wales, in 1999 (20 animals) and again in 2000 (23 animals), where they appear to be establishing successfully.

South Australia

Numbats were re-introduced to the Yookamurra Sanctuary, South Australia, in 1993 (15 animals), where they survived at least the first 12 months (Friend 1994; Copley 1995) and increased in numbers.

Western Australia

A re-introduction program for numbats has been in operation since 1985, at first translocating them from the wild to areas close to the source location and then to other areas within the species' former range. The first re-introduction to Boyagin Nature Reserve occurred in 1985–96 (35 animals). Establishment, reproduction and population growth were so good that the reserve is fully colonised (Harris 1988; Friend 1989; Shea 1989; Strahan 1995).

From 1985 to 1996 numbats were translocated from the wild at Dryandra to at least seven sites, along with captive-bred animals from the Perth Zoo – besides Boyagin Nature Reserve they have been successfully established at Tutanning Nature Reserve (1987–96, $n = 35$), Batalling Forest (1992–95, $n = 60$), Karakamia Sanctuary (1994–99, $n = 6$), Dragon Rocks Nature Reserve (1995–96, $n = 37$), Dale Conservation Park (1996–98, $n = 62$), and may still persist in small numbers at Karroun Hill Nature Reserve (1986–93, $n = 97$). A more recent translocation was to Stirling Range National Park (1998–2000, $n = 48$). These and earlier translocations are being monitored (Anon. 1994; Friend and Thomas 1995; in press 1985–98).

■ DAMAGE

None.

Family: Peramelidae
Bandicoots and bilbies

WESTERN BARRED BANDICOOT
Marl
Perameles bougainville Quoy and Gaimard

■ DESCRIPTION
HB 171–236 mm; T 60–102 mm; WT 172–286 g.
A small, delicate-looking bandicoot with light brownish grey above and white to slate grey below with three to four alternating paler and darker bands across hindquarters; large pointed and erect ears; white feet; tail white above, except at base.

■ DISTRIBUTION
Australia. Originally found over much of lower southern Australia from the west coast east to the western parts of New South Wales. Now restricted to Bernier and Dorre islands off the central west coast of Western Australia.

Habits: nocturnal, terrestrial; rests during day in grass-lined nest. **Foods:** omnivorous; feeds on invertebrates; roots; seeds; herbs. **Gregariousness:** solitary **Movements:** sedentary; males occupy home ranges of 2.5–14.2 ha, females 1.4–6.2 ha; home ranges decline in size as population density increases; considerable overlap in ranges. **Habitat:** scrub and open steppe. **Breeding:** breeds all year, mainly April–October; gestation 11–13 days; 3–4 litters of 1–3 (usually 2)

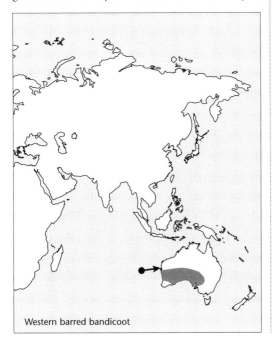

Western barred bandicoot

young per year; up to 5 litters/year; pouch life 55–56 days; young disperse at 3–5 months; females sexually mature 3 months, males 4–5 months. **Longevity:** 2–3 years. **Status:** formerly widespread, now rare.

■ HISTORY OF INTRODUCTIONS
AUSTRALASIA
Western Australia
Wild-caught western barred bandicoots from Dorre Island were translocated to a fenced enclosure on Heirisson Prong (1995–96, n = 14: 3 male, 11 female), Shark Bay, where they bred. The progeny were allowed to disperse from the enclosure and have established a small but growing population.

South Australia
Seven animals were translocated from Bernier Island to a fenced enclosure at Roxby Downs in South Australia in September 2000.

DAMAGE
None.

EASTERN BARRED BANDICOOT
Gunn's bandicoot
Perameles gunnii Gray

■ DESCRIPTION
HB 250–400 mm; T 70–180 mm; WT 500–1450 g.
Grizzled yellow-brown above and slate grey below, with three to four pale bars across hindquarters; tail white above, except at base.

■ DISTRIBUTION
Australia. South-eastern Australia and Tasmania.

■ HABITS AND BEHAVIOUR
Habits: nocturnal, terrestrial; rests during day in grass-lined nest. **Foods:** omnivorous; feeds on soil invertebrates; earthworms, small invertebrates, cockroaches, beetles, field crickets, grasshoppers and caterpillars; also plant material including fallen fruits, bulbs and grasses. **Gregariousness:** density 1.5–8.5/ha. **Movements:** sedentary(?). **Habitat:** grasslands, open woodlands, urban fringes, rubbish tips,

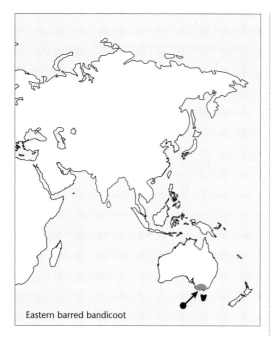

Eastern barred bandicoot

cemeteries, golf courses, farm paddocks, tree plantations, parks, gardens. **Breeding:** breeds all year, mainly July–November; gestation 11–13 days; 3–4 litters of 1–5 young per year; up to 5 litters/year; pouch life 55–56 days; young disperse at 3–5 months; females sexually mature 3 months, males 4–5 months. **Longevity:** 2–3 years. **Status:** formerly widespread, now locally common; declining, range fragmented.

■ HISTORY OF INTRODUCTIONS
AUSTRALASIA
Victoria
Formerly distributed over South Australia, Victoria and Tasmania, eastern barred bandicoots were common and abundant in Victoria until the 1940s, but generally declined with European settlement and were virtually extinct in the wild in Victoria as of early 1994.

Endangered eastern barred bandicoots survive as a single, small, free-ranging population around Hamilton, Victoria, on the mainland, but also in Tasmania. The population is declining rapidly at Hamilton due to loss of habitat, predation, disease, road kills and possibly poisoning.

Several introductions have been initiated in Victoria commencing in the 1980s. The first re-introduction began in Gellibrand Hill Park (645 ha surrounded by an electric fence) near Melbourne as a captive breeding colony in 1989, and releases occurred of a few animals at a time until July 1991. By this time 64 animals had been released. Recently more releases were initiated and will continue until another 30–40 are re-introduced. In all, 88 bandicoots were re-introduced in Gellibrand Hill National Park between April 1989 and April 1993, where they are now established and increasing in numbers (Duffy *et al.* 1995).

Re-introductions at Hamilton Community Parklands (211 ha) commenced shortly after the Gellibrand introductions. Between 1990 and 1991, 47 were released here, which also has a predator-proof fence. Thirty-three were released (nine came from captive stock, and 24 from the wild). The last introduction occurred in February 1991. The population is surviving, but it is thought that the two areas are not big enough to maintain a population over the long term. Further introductions were planned in 1993 (Reading *et al.* 1993).

At Moorawong, near Skipton, some barred bandicoots were kept in pens and some escaped into the surrounding area, but did not survive. A further 24 were released at Gellibrand Hill Park in 1992 and were well established and increasing in 1993. Forty-five were re-introduced at Moorawong over 12 months in 1992 and the results were encouraging. They were still surviving after one year (Backhouse *et al.* 1995).

By 1993 in Victoria over 600 bandicoots were present in three re-introduced populations (Backhouse *et al.* 1995). Beginning with an initial release of 10 animals in 1994, a population of about 50 eastern barred bandicoots is now established at 'Lanark' in the Western Districts of Victoria (O'Neill 1999).

Tasmania
Fifty-five eastern barred bandicoots were introduced to Maria Island in 1971 (Weidenhofer 1977; Rounsevell 1991; Summers 1991).

DAMAGE
None.

LONG-NOSED BANDICOOT
Perameles nasuta Geoffroy

■ HISTORY OF INTRODUCTIONS
AUSTRALASIA
Queensland
Long-nosed bandicoots may have been introduced 'within living memory' to Badu (=Musgrave Is) in Queensland (Garnett and Jakes 1983).

COMMON ECHYMIPERA
Echymipera kalubu (Lesson)

■ HISTORY OF INTRODUCTIONS
AUSTRALASIA
Papua New Guinea
Two specimens of the common echymipera are known from Manus Island, Papua New Guinea, and preliminary investigations of archeological sites suggest that the species has been introduced there (Flannery 1995).

GOLDEN BANDICOOT
Isoodon auratus (Ramsay)

■ DESCRIPTION
HB 190–295 mm; T 84–121 mm; WT 250–670 g.
Grey-brown fur with golden guard hairs, pencilled with black; under parts white; nose naked; forelimbs short; claws strong, curved; tail and upper surface hind feet dark brown; tail short. Female has eight teats.

■ DISTRIBUTION
Australia. Parts of the northern Kimberley and Northern Territory, and also Barrow Island, Western Australia. Formerly more widespread.

■ HABITS AND BEHAVIOUR
Habits: nocturnal; sleeps in grass nest during day. **Gregariousness:** solitary; 10 adults/ha. **Movements:** up to 10 ha/night. **Habitat:** wet vine thickets to sand dunes, sand plain with spinifex and acacia woodland with tussock grassland. **Foods:** termites, centipedes, insect larvae, and plant material, ants, moths, turtle eggs, small reptiles. **Breeding:** all year; litter size 2–3. **Longevity:** no information. **Status:** common, restricted.

■ HISTORY OF INTRODUCTIONS
AUSTRALASIA
Western Australia
Forty bandicoots (*I. a. barrowensis*; 18 male, 22 female) from Barrow Island were released in the Gibson Desert Nature Reserve in 1992. None were found there in 1993, but there were some signs that they were still present (Christensen and Burrows 1995). Foxes and feral cats are considered to have killed all of these animals.

■ DAMAGE
None.

SOUTHERN BROWN BANDICOOT
Quenda, short-nosed bandicoot, brown bandicoot
Isoodon obesulus (Shaw)

■ DESCRIPTION
HB 280–360 mm; T 90–140 mm; WT 400–2000 g.
Dark greyish or yellowish brown above, creamy white below; solid build; ears rounded; head long and tapering; nose naked; forelimbs short; claws strong, curved; tail and upper surface hind feet dark brown; tail short. Female has eight teats.

■ DISTRIBUTION
Australia. South-west and south-east Australia and Tasmania.

■ HABITS AND BEHAVIOUR
Habits: nocturnal; sleeps in nest of grass and other plant material during day; aggressive. **Gregariousness:** solitary. **Movements:** sedentary; home range males 7 ha, females 2 ha. **Habitat:** low scrubby vegetation, forest and woodland with dense understorey. **Foods:** insects, earthworms and other invertebrates; fungi. **Breeding:** mainly winter–spring; mates May–September; young 1–6; 2–3 litters per season; weaned 60–70 days; sexual maturity females 3–4 months. **Longevity:** 3 or more years. **Status:** common in some areas of range.

■ HISTORY OF INTRODUCTIONS
AUSTRALASIA
Western Australia
Southern brown bandicoots (*I. o. fusciventer*) were re-introduced to Tutanning Nature Reserve in 1991–95 (*n* = 106). There have been at least eight attempts to translocate southern brown bandicoots in Western

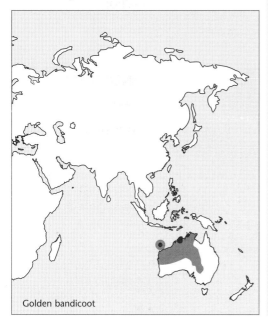
Golden bandicoot

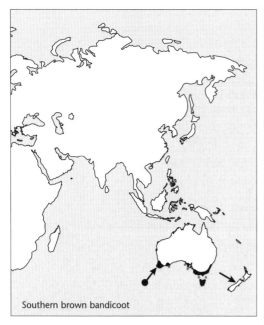

Southern brown bandicoot

Australia between 1991 and 1997: Julimar Forest (1994–95, *n* = 54), Dongolocking Nature Reserve (1994–95, *n* = 37), Leschenault Conservation Park (1995, *n* = 20), Boyagin Nature Reserve (1995–96, *n* = 26), Karakamia Sanctuary (1997, *n* = 34), Hills Forest near Mundaring (1997, *n* = 14), Mount Barker (1997, *n* = 37), Paruna Sanctuary (2000, *n* = 40) and Creery Wetlands near Mandurah (2000, *n* = 18). All of these translocations appear to have been extremely successful and the species has expanded beyond the reserve boundaries in some places.

Tasmania
Apparently southern brown bandicoots were successfully introduced to Maria Island, Tasmania, in 1971 (Storr 1960; Weidenhofer 1977; Rounsevell 1991).

PACIFIC OCEAN ISLANDS
New Zealand
Southern brown bandicoots were introduced to New Zealand, but failed to become established (Thomson 1922). An unknown number of short-nosed bandicoots were introduced by the Auckland Acclimatisation Society in 1873 (Wodzicki 1950).

■ DAMAGE
None.

Family: Thylacomyidae
Bilbies
Now often placed in Family Peramelidae, subfamily Thylacomyinae (see Strahan 1995), but here retained as a full family.

BILBY
Rabbit-eared bandicoot, dalgyte
Macrotis lagotis (Reid)

■ DESCRIPTION
HB 290–550 mm; T 200–290 mm; WT 800–2500 g.
Light delicate build; hair soft and silky; ears long and rabbit-like; tail black on proximal half, then white;

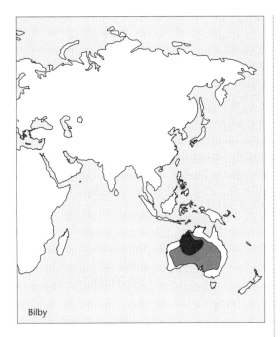

Bilby

prominent dorsal crest; extreme tail tip naked; muzzle long and pointed; hind foot lacks first toe.

■ DISTRIBUTION
Australia. Formerly inhabited most of the arid and semi-arid regions on the Australian mainland. Now confined to Tanami Desert, Northern Territory, west to Broome and south to Warburton, Western Australia.

■ HABITS AND BEHAVIOUR
Habits: nocturnal; terrestrial; sleeps in burrow during day; burrow 3 m long and 1.8 m deep. **Gregariousness:** singly, pairs or small family groups. **Movements:** sedentary, with occasional shifts in range. **Habitat:** arid and semi-arid areas; acacia shrublands, open woodland, shrub steppe, hummock grassland. **Foods:** insects and their larvae (termites and beetles), seeds, bulbs, fruits and fungi. **Breeding:** throughout the year; gestation 14 days; litter size 1–3; pouch life 70–90 days; weaned 13 weeks. **Longevity:** no information. **Status:** rare, declining; range fragmented.

■ HISTORY OF INTRODUCTIONS
AUSTRALASIA
Northern Territory
Having drastically declined since European settlement bilbies were re-introduced in 1983–87 at Simpsons Gap (now West MacDonnell Ranges) west of Alice Springs, in the Northern Territory. Here they were held in pens for a time, then later released. They were also released at Watarrka National Park, south-west of

Alice Springs. Fifteen bilbies were released in 1988 and seven in 1991. These survived for a while, but then began to decline (Southgate 1995).

South Australia
Nine bilbies were released into a fenced enclosure at Roxby Downs in 1999 and numbers have increased to at least 60 by early 2001.

Western Australia
Four bilbies were released in November 1998 into a 10 ha enclosure at Dryandra, north-west of Narrogin in an attempt to re-introduce them into the wild. Three had been either wild-caught or bred from wild stock that came from Kanyana Native Fauna Rehabilitation Centre near Perth, and the other came from a breeding program at the Alice Springs Desert Park. A further 19 animals were being held at Francois Peron National Park, Shark Bay, in enclosures in early 1999, and were being monitored prior to their planned release.

In April–October 2000, 36 animals were released into the wild at Dryandra Forest and in late October–November 2000, 19 animals were released into Francois Peron National Park, Shark Bay. The Dryandra release was affected by fox (*Vulpes vulpes*) and cat predation in the early stages, but the Shark Bay release progressed well. Both populations are surviving despite below average rainfalls in the initial 18 months following release.

■ DAMAGE
None.

Family: Phascolarctidae
Koalas

KOALA
Phascolarctos cinereus (Goldfuss)

■ DESCRIPTION
HB 600–850 mm; T vestigial; WT 4.1–14.9 kg.
Stocky, compact body; head rounded; snout bulbous; ears rounded; fur thick; head and body dark grey; rump dirty white; belly grey white; tail absent. Female has two mammae.

DISTRIBUTION
Australia. From Townsville, Queensland, to Melbourne, Victoria, and inland to the Great Dividing Range.

■ HABITS AND BEHAVIOUR

Habits: arboreal, nocturnal, folivorous, territorial; seldom comes to ground except when moving to another tree. **Gregariousness:** solitary, pairs, or rarely small groups (1 male to few females). **Movements:** sedentary. **Habitat:** dry eucalyptus forest and woodland. **Foods:** leaves of eucalyptus trees (about 20 species). **Breeding:** breeds in June; gestation 25–35 days; oestrus 27–30 days; litter size 1, rarely 2; lactation 6–12 months; pouch life 5–7 months; sexual maturity females end of second year. **Longevity:** 5–20 years (captivity). **Status:** common; range large but fragmented; many management problems because of fragmentation.

Koala

■ HISTORY OF INTRODUCTIONS

AUSTRALASIA

Australian mainland

At European colonisation in 1778 the koala's range was probably from about latitude 20°S in Queensland to latitude 38°S in southern Victoria and south-eastern South Australia, with populations both east and west of the Great Dividing Range.

In the late 1800s fire, disease, and clearing were responsible for the considerable decline in numbers and by about 1925 near extinction of the koala. In 1924 alone two million furs were exported. In 1920 they were introduced to Phillip Island and French Island in Western Port Bay from localities on the mainland. Although the islands supported only limited numbers, some 7000 have been restocked in 50 communities on the mainland (McNally 1960).

Islands off the coast

Introductions of koalas to islands

Island	Date introduced	Notes
Chinaman (Vic)	1930–31, 1957	successful?
French (Vic)	1880–1900, 1957	successful
Goat (SA)	1959–60	established?
Kangaroo (SA)	1923, 1925, 1955–56, 1958	successful
Little Snake (Vic)	before 1971?	successful?
Magnetic (Qld)	?	successful?
Newry (Qld)	?	successful?
Phillip (Vic)	1923, 1945, 1977	very successful
Quail (Vic)	1930–33, 1947	very successful
Rabbit (Qld)	?	successful?
Raymond (Vic)	1953	flourished
Saint Bees (Qld)	before 1968?	?
Snake (Vic)	1945	successful?
Three Hummock (Tas)	before 1973?	failed to become established

References: Abbott & Burbidge 1995; Copley 1995; Hope 1973; Lee & Martin 1987, 1988; Lucas 1968; Martin 1989; Mitchell *et al.* 1988; Norman 1971; Serventy 1987; Wildlife Management Branch 1989.

Koalas were released on Kangaroo Island, South Australia, over successive years – in 1923 (six); 1925 (18); 1955–56 (eight); and in 1958 (12) – where they became well established and thrived (Serventy 1987; Copley 1995).

In 1953, 32 koalas were released on Raymond Island in the Gippsland Lakes from Phillip Island. The population flourished and in 1985 there were in excess of 170 koalas (Mitchell *et al.* 1988).

Introductions to Quail Island occurred in 1930, 1931, 1932, 1933, and in 1947 (Martin 1989). The island was stocked with 165 koalas from French Island between 1930 and 1933. Ten years later the number of koalas had grown to the point where they had killed most of the trees needed for food and many died from starvation. Animals from Quail Island were translocated to the Brisbane Ranges National Park in 1944, Phillip Island in 1945, French Island in 1957 and Phillip Island again in 1977. On all these islands the populations thrived and koalas are now common. The Phillip Island population is currently 300 individuals (Lee and Martin 1987; 1988).

In 1959–60, 13 koalas were introduced to Goat Island, South Australia, where they probably became established (Copley 1995).

They were later introduced to Sanctuary (Renmark) in 1963 (four, not known whether they became established), successfully to Little Toolunka Flat (Wakerie) in 1964 (eight), to Ashbourne (16) and Belair National Park in the Mt. Lofty Ranges in 1965, to Martins Bend Sanctuary (Berri) also in 1965 (four, but failed), to Sleaford on the Lower Eyre Peninsula in 1969 (six), and also in the same year to the Avenue Range (south-east) (six) (Copley 1995).

◼ DAMAGE
On Quail Island, overpopulation of koalas killed the trees and many died of starvation as a result. They caused severe defoliation or at least significant tree defoliation in 1986 on Kangaroo Island (Copley 1995). Again in 1996–97 the same thing happened, resulting in a major program to sterilise males to limit the population size and growth. Culling was considered unacceptable.

Family: Vombatidae
Wombats

SOUTHERN HAIRY-NOSED WOMBAT
Hairy-nosed wombat
Lasiorhinus latifrons (Owen)

◼ DESCRIPTION
HB 770–1000 mm; T 20–60 mm; WT 19–32 kg.
Stout head and body; fur soft, silky, grey; rhinarium white; head broad and flattened; ears narrow and pointed; muzzle hairy; limbs powerful; tail short. Female has two mammae.

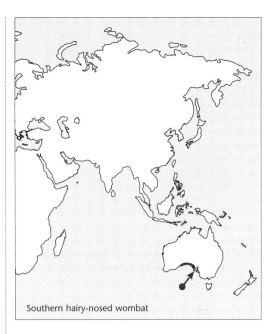
Southern hairy-nosed wombat

◼ DISTRIBUTION
Australia. South-eastern Western Australia and south-western South Australia.

◼ HABITS AND BEHAVIOUR
Habits: burrows, often forming warrens; rests in burrow during day; active evening and night. **Gregariousness:** 5–10 in a burrow; individuals and burrows are frequently grouped. **Movements:** sedentary. **Habitat:** semi-arid shrubland regions with little rainfall. **Foods:** grass and herbs. **Breeding:** births mainly August–December, but may breed only in years of good rainfall; litter size 1; pouch life 6–9 months; lactation *c.* 12 months; sexes mature at 3 years. **Longevity:** 20 years captive. **Status:** common with limited distribution.

◼ HISTORY OF INTRODUCTIONS
AUSTRALASIA
South Australia
In 1971, six wombats were introduced to Wedge Island, South Australia, where they became established and are breeding and increasing in numbers (Robinson 1989; Copley 1995). Six animals were also introduced to Kellidie Bay Conservation Park, but there are no records of what happened to them.

Re-introductions have been made at Pooginook Conservation Park (32 animals) in 1971 and they were still present there in 1993 (St. John and Saunders 1981; Copley 1995), and to Kia-ora Station (18), Glenora Station (21) and Whydown Station (20). They were still present in 1993 at Kia-ora, but it is not known what happened at the other two sites.

One pair was introduced to Kangaroo Island, but failed to become established (Inns *et al.* 1979; Robinson 1989).

■ DAMAGE
Wombats were formerly destroyed as pests, usually because their burrows were a danger to people riding horses. They are now fully protected.

COMMON WOMBAT
Vombatus ursinus (Shaw)

■ HISTORY OF INTRODUCTIONS
AUSTRALASIA
Australia
Twenty-eight wombats from Flinders Island were introduced to Maria Island, Tasmania (Weidenhofer 1977; Summers 1991).

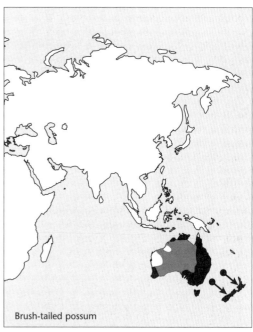

Brush-tailed possum

Family: Phalangeridae
Cuscuses and possums

BRUSH-TAILED POSSUM
Common brush-tailed possum
Trichosurus vulpecula (Kerr)

■ DESCRIPTION
HB 400–700 mm; T 250–405 mm; WT 1.4–6.4 kg.
Head and body silver grey (sometimes black), often with some brown; under parts greyish or yellowish; muzzle blunt, nose pink; ears long and oval; tail bushy, grey at base, tip black or white, prehensile, naked area on underside of terminal half; hind foot with five toes, first toe opposable and clawless; iris brown. Sexes similar in size and weight.

■ DISTRIBUTION
Australia. From northern Queensland through New South Wales to Victoria and South Australia; also south-western and northern Western Australia, northern and southern Northern Territory, Tasmania and Kangaroo Island.

■ HABITS AND BEHAVIOUR
Habits: nocturnal; largely arboreal; males often territorial; occupies den (hollows, logs, buildings), sometimes shared with others (up to 5). **Gregariousness:** solitary, pairs, or family groups; little social interaction except at breeding season; density 0.3–25/ha (NZ). **Movements:** sedentary; home range 0.28–7.45 ha; females 0.6–2.7 ha, but males up to 18.3–24.6 ha; home ranges overlap; dispersal movements of 10–30 km recorded. **Habitat:** forest and open woodland, river valleys with open plains, urban areas with trees, grassland with cover, orchards, swamps, and sand dunes. **Foods:** buds, leaves, bark, shoots, flowers, seeds, fruits, foliage of trees, shrubs, vines, ferns, fungi, invertebrates, grass, herbs, sedges, grain, vegetable crops, horticultural produce, ornamental shrubs, small birds and mice. **Breeding:** breeds all year; gestation 15–24 days; litter size 1 rarely 2; males mature 1–2 years, females slightly later; female

polygamous and polyoestrous; oestrous cycle 26 days; females mature at 9–12 months; some females may breed twice in same year; young attach to teat for 70 days; eyes open 100–110 days; emerge from pouch at 120–140 days; pouch life around 170 days. **Longevity:** 6–14 years in wild. **Status:** common and abundant.

■ HISTORY OF INTRODUCTIONS

Introduced successfully to New Zealand and some offshore islands, and also to some areas in Australia and particularly to some islands off the coast.

AUSTRALIA
South Australia

Re-introduced successfully in South Australia at Willmington in 1985 (five released), Quorn area in 1974 (30+ released), Katarapko Island in the 1970s (several released), Humbug Scrub in the 1980s (several released), and Murray Bridge area in the 1980s (several released), where they are still present (Papenfus 1990). They were re-introduced with little success in the Flinders Ranges National Park, South Australia, in 1961–65 (16 released), Mambray Creek in 1972 (11 released), Ernabella in 1976 (12 released), and Sandilands in the 1970s (several released), and re-introduced successfully to Arkaroola in 1968 when 14 were released, but they were last recorded there in 1987.

Western Australia

They were unsuccessfully translocated in 1993–94 ($n = 31$) in Western Australia when taken from the Perth metropolitan area to Julimar Forest.

There have been several introductions (see table below) of possums to islands off the coast of Australia (Whinray 1971; Hope 1973; Weidenhofer 1977; Rounsevell *et al.* 1991; Summers 1991; Abbott and Burbidge 1995).

Introductions of possums to islands off the coast of Australia

Island	Date introduced	Notes
Dent (Qld)	?	released by lighthouse keepers
East Sister (Tas)	1920s	present?
Maria (Tas)	1971	15 released; present?
Newry (Qld)	?	now occur there
Outer Newry (Qld)	?	now occur there
Prince Seal (Tas)	1920s	present?
Rabbit (Qld)	?	now occur there

PACIFIC OCEAN ISLANDS
New Zealand

The brush-tailed possum was introduced to New Zealand initially to found a fur industry, but in some cases as pets, from Tasmania and mainland eastern Australia (Thomson 1922; Wodzicki 1950; Pracy 1962; Howard 1964; Chisholm *et al.* 1966; Gibb and Flux 1973). Since its introduction, it has become abundant and widely distributed. Brush-tailed possums are deemed a pest because of their depredations in orchards and gardens and the damage caused to the indigenous forests (Tyndale-Biscoe 1955). The first liberations were made between 1837 and 1840, but the first successful release was made by C. Basstian in the forest behind Riverton in 1858 (King 1990).

Between 1837 and 1911 more than 180 possums were imported and subsequently about 470 translocations have been documented within New Zealand. The first successful releases were made in 1858, and most of the releases were made in the period 1890–1900. Initially the importations were made by private individuals and these were later sanctioned by legislation, but the New Zealand government was directly involved in possum acclimatisation from 1895 to 1906. However, the possum had been well established by private enterprise prior to any government action and artificial dispersal by private individuals, and the acclimatisation societies had by far the greatest impact on the overall distribution and consequent spread.

From 1837 to 1875 some 15 known releases occurred, mainly in the Southland and Auckland districts, with individual introductions in Canterbury and on Kawau Island. Between 1875–90 only two importations occurred and only two releases, one by the Auckland Acclimatisation Society and one in the Waikato district. In the next decade, 1890–1900, at least 90 liberations are recorded. From 1900 to 1910 the number of releases decreased, but many were made by the Department of Tourist and Health Resorts at Rotorua. Between 1915 and 1924 importations were mainly restricted to animals as pets and none appeared to have been brought in after 1924. Transfers of possums within New Zealand continued, probably to the 1950s, although the main era was between 1890 and 1930.

The first known introductions in a number of areas of New Zealand and some offshore islands attest to the widespread releases which occurred: first introduction in Auckland 1869, Gisborne 1891, Taranaki 1896, Hawkes Bay 1918, Wellington 1872, Nelson 1890, Marlborough 1927, Canterbury 1865, Westland 1895, Otago 1890, Southland 1837–40, Stewart Island 1890, Auckland Islands 1890, Chatham Islands 1911. There were at least 50 releases in Auckland between 1869 and 1946, 18 in Gisborne 1891–1925, nine in Taranaki 1896–1948, six in Hawkes Bay 1918–36, 60 in Wellington 1872–1932, 40 in Nelson 1890–1929, 11 in Marlborough 1927–33, 23 in Canterbury 1865–1950, 65 in Westland 1895–1952, 29 in Otago 1890–1932, 46 in Southland 1837–1931, and nine on Stewart Island about 1890.

The possum is now widespread and abundant throughout both the main North and South islands of New Zealand, on Stewart Island, Codfish Island, Chatham Islands and on some Cook Strait islands (Wodzicki 1950; Gibb and Flux 1973; Atkinson and Bell 1973).

Possums have been introduced onto at least 19 islands and are still present on at least 13 or 14 of these.

Introductions of possums to New Zealand islands

Island	Date introduced	Notes on status
Allports	<1980	present, eradication failed 1982
Auckland	1890	not now present
Chatham	1890	not now present
Codfish	c. 1890	eradicated by 1987
D'Urville	?	eradicated soon after release
Fortyseven	?	present, eradication attempted 1990
Harakeke (BOI)	c. 1990	present, eradication attempted 1992
Kapiti	1893 and 1932	eradicated 1986–87
Kawau	1869–70	still established
Motutapu	1868	still established, eradication attempted 1990
Peach (Whangaroa)	?	present, eradication attempted 1990
Pig (L. Wakatipu)	c. 1975	present, eradication attempted 1990
Pigeon (L. Wakatipu)	c. 1975	present, eradication attempted 1990
Rangipukea	1920	still present ?
Rangitoto	1868	still established; attempts to eradicate in 1990 unsuccessful
Ruapuke ?	1915	still present
Stewart	1890	still established
Tarakaipa	?	present, eradication attempted 1991
Whanganui	1920	still present ?

■ DAMAGE

The brush-tailed possum was at first a protected animal in New Zealand, but by the 1940s it had become so abundant and widespread that in 1947 this protection was removed. In 1951 a bounty was placed on them and in 1956 the species was declared a 'noxious' animal (Wodzicki 1965).

It seems doubtful that anyone could have foreseen the full extent of the changes that possums were to cause to the vegetation (Howard 1965). The possums damaged poles and trees by biting the bark, breaking the stems and browsing the foliage; stem-breaking results in the stunting of growth and malformation of the tree. In exotic timber forests (willows, pines, eucalypts, poplars, oak and douglas fir) they can cause serious economic damage by ringbarking the leading and upper lateral shoots, which spoils the timber form (Chisholm et al. 1966). They often become so numerous in some areas that they kill many mature trees (Howard 1964). Investigations have shown that high populations in protection forest areas result in the progressive elimination of tree species in order of palatability, and local losses of up to 70 per cent of trees and shrubs has been recorded (Pracy and Kean 1963).

Damage in gardens and orchards can be serious locally. Fruits such as apples, peaches, nectarines, plums, pears, lemons and passion fruit, vegetables such as parsnips, turnips, swedes, carrots, potatoes, beans, peas, silver beet, cabbages and lettuce, and flowers such as roses, polyanthus, carnations, cyclamen, gladiolus and *Godetia* are susceptible to attack from possums. In field crops they also cause damage to maize and rape, and are serious nuisances to homeowners by causing short circuits in power lines. In forest areas they break the leading shoots of pines, and also damage poplar and aspen plantings planted to prevent soil erosion. They enter houses, pollute water supplies and are noisy at night (Wodzicki 1950; Pracy 1963; Pracy and Kean 1963; Howard 1964; Chisholm et al. 1966).

In the East Coast Rabbit District it was estimated that the damage to pole plantings amounted to NZ3500 pounds annually (Chisholm et al. 1966). Damage to pasture is probably of only minor importance.

Bonuses for dead possums were paid from 1951 to 1961, but despite much trapping (in 1963 over one million were taken), this failed to give adequate control because of the high density of animals (density of 30/ha suggested in forest of Westland, but in mixed podocarp-broadleaf forest in Wellington 7–8/ha, which was probably more normal). The bounty system has now been removed (Wodzicki 1965) and 1080 poisoning is the main control. As early as 1960 it was said that attempts to re-forest denuded country was a wiser policy than trying to exterminate these and other introduced animals (Tyndale-Biscoe 1960).

Selective browsing of preferred plant species has intensified the impact of possums on New Zealand forests. The effects are unquestionable, although the consequences for forest dynamics and soil erosion are debatable (King 1990). The reason that the possum is a pest in New Zealand and not in its native habitats in Australia is largely because of its much higher density in New Zealand (King 1990).

Possums have been implicated in the spread of bovine tuberculosis that is causing considerable concern to farmers (Daniel 1984).

ADMIRALTY CUSCUS

Spilocuscus kraemeri **(Schwartz)**

■ HISTORY OF INTRODUCTIONS
AUSTRALASIA
Papua New Guinea
The admiralty cuscus may have been introduced to Manus and (?)Wuvulu islands off Papua New Guinea (Flannery 1995). Archeological evidence suggests that they reached Manus Island recently, perhaps in the last two thousand years.

COMMON SPOTTED CUSCUS

Spotted cuscus, spotted phalanger
Spilocuscus maculatus **(Desmarest)**

=*Phalanger maculatus*

■ DESCRIPTION
HB 348–580 mm; T 310–540 mm; WT 1.5–6 kg.
Large, robust; muzzle pointed; snout short; eye has a reddish rim around it; fur dense and woolly; coat colour and pattern variable, from entirely white or with brown or grey mottling to mottled ginger or ash grey; ears small; dorsal stripe lacking; may have saddle on back; skin often yellow orange; tail prehensile, terminal half naked.

■ DISTRIBUTION
Australasia–Indonesia. Some Moluccan islands, New Guinea Mussau, New Ireland to Cape York, Australia. Includes the islands of Ambon, Banda, Batanta, Biak-Supiori, Buru, Japen, Kai, Kur, Misool, Namfoor, Palau Num, Palau Panjang, Salawati, Seram, Su Mios, Tioor, Wammar, Wokam, Wonoemba and Salyer.

■ HABITS AND BEHAVIOUR
Habits: arboreal; nocturnal; territorial; sleeps most of day in canopy foliage. **Gregariousness:** solitary. **Movements:** sedentary. **Habitat:** primary and secondary forest, tropical forest, lowlands; close to humans. **Foods:** leaves, acorns, fruits, insects, eggs and nestling birds, flowers. **Breeding:** (?) March–September; oestrous cycle 28 days; young 1–3; inter-birth interval 12 months. **Longevity:** 11 years captive. **Status:** wide-

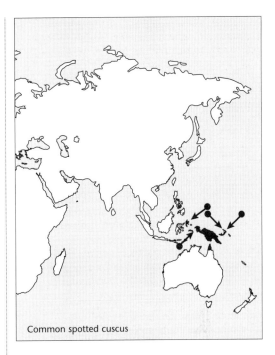

Common spotted cuscus

spread and common to abundant; limited and sparse in Australia.

■ HISTORY OF INTRODUCTIONS
Common spotted cuscuses were probably introduced on some Papua New Guinea islands (?Mussau and New Ireland) and prehistorically to some Maluku islands (?Ambon, Kai Islands and Palau Panjang).

AUSTRALASIA–INDONESIA
Maluku
It seems likely that many of the Moluccan populations of spotted cuscuses were established through introductions by humans (Flannery 1995). They were probably introduced to Ambon, Kai Islands and Palau Panjang.

Mussau
Populations of spotted cuscuses on Mussau were probably introduced by humans over one thousand years ago (Flannery 1995).

New Ireland
Populations of spotted cuscuses were introduced by humans around 1929 when animals from Mussau were released accidentally in New Ireland (Flannery 1995). They are still restricted to the north-west part of the island and have not extended from the Kavieng area, but may be slowly spreading southwards.

Salyer
Spotted cuscuses may also have been introduced to this island (south of Sulawesi), which is some distance from its present distribution (George 1987; Flannery 1995).

St. Mathias Group
Common spotted cuscuses were also introduced to this group, where they have been present for at least two to three thousand years, as bones have been found in the Talepakemelai archeological site on the island of Eloana (Flannery 1995).

■ DAMAGE
None known.

WOODLAND CUSCUS
Phalanger lullulae Thomas

■ HISTORY OF INTRODUCTIONS
AUSTRALASIA
Papua New Guinea
The woodland cuscus has probably been introduced by humans to Alcester Island (Papua New Guinea) some 70 km from Woodlark Island (Flannery 1995).

NORTHERN COMMON CUSCUS
Grey cuscus
Phalanger orientalis (Pallas)

■ DESCRIPTION
HB 377–472 mm; T 278–425 mm; WT 1.6–3.5 kg.
Coat short and slightly coarse; males grey to greyish white; dorsal stripe from head to lower back, distinct, dark; tail white tipped. Female reddish brown. Juveniles reddish or grey.

■ DISTRIBUTION
Australasia and Indonesia. Moluccas, Timor, northern New Guinea, Karkar, Schouten Islands (PNG), Bismark Archipelago and the Solomon Islands.

■ HABITS AND BEHAVIOUR
Habits: little recorded and biology not well known. **Gregariousness:** solitary(?). **Movements:** sedentary. **Habitat:** primary forest, old gardens, close to human habitation, villages. **Foods:** leaves, fruit and bark. **Breeding:** ? April–May, perhaps all year; litter size 1–2. **Longevity:** no information. **Status:** widespread and abundant; commonly kept as pet.

■ HISTORY OF INTRODUCTIONS
Much of the common spotted cuscuses island distribution in Papua New Guinea, Maluku and the Solomon Islands has resulted from prehistoric human introductions.

AUSTRALIA–INDONESIA
Papua New Guinea
The northern common cuscus has been introduced to a number of islands in Papua New Guinea including

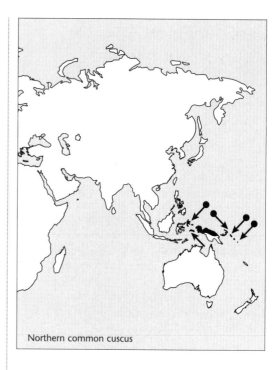

Northern common cuscus

Bougainville, Buka, Mioko, New Ireland and Nissau (Flannery 1995). Much of this species' island distribution is due to prehistoric human introductions, certainly on Mioko and New Ireland. The earliest known introduction was to New Ireland between 10 and 20 thousand years ago (Flannery and White 1991).

Maluku
In this area northern common cuscuses have probably been introduced to ?Ambon, the Kai Islands and to Sanana, and possibly also to Seram and Buru (Flannery 1995). The introduction to the Kai Islands at least seems to have been prehistoric.

Timor
The northern common cuscus was almost certainly introduced to Timor around 6600 years ago (Glover 1986; Flannery 1995).

PACIFIC OCEAN ISLANDS
Solomon Islands
Introduction of the northern common cuscus occurred in the Solomon Islands some time in the past 6600 years (Flannery and Wickler 1990; Flannery 1995). They have been introduced on Choiseul, Guadalcanal, Santa Isabela, Malaita, Mono, New Georgia, Russel Islands, San Cristobal and Velle Lavella (Flannery 1995).

■ DAMAGE
None known.

Family: Petauridae
Possums and gliders
Some authorities now place these species in the family
Pseudocheiridae (see Strahan 1995).

SUGAR GLIDER
Petaurus breviceps Waterhouse

■ DESCRIPTION
HB 110–210 mm; T 120–210 mm; WT 90–160 g.
Small, squirrel-like; head short and rounded; ears
oval; head and body grey; under parts pale; black
stripe from centre of face to centre of back; gliding
membrane along flanks from wrist to ankle; tail grey,
bushy.

■ DISTRIBUTION
Australia–Indonesia: Moluccas, Papua New Guinea
and some nearby islands, Bismarck Archipelago, New
Britain, and northern and eastern Australia.

■ HABITS AND BEHAVIOUR
Habits: nocturnal; arboreal; active; sleeps by day;
carries nest material with tail; nests in hollow branch;
glides *c.* 45 m; territorial. **Gregariousness:** nests in
groups to 20 in holes in trees (up to 7 males, females
and young); density 2.9–10/ha. **Movements:** sedentary (?); home range *c.* 0.5 ha. **Habitat:** dry forest and
woodland. **Foods:** omnivorous; sap, blossom, buds,
nectar, insects and larvae, arachnids, and small vertebrates. **Breeding:** mates June–November; gestation

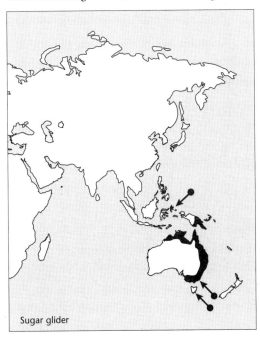

Sugar glider

16–21 days(?); oestrous cycle *c.* 29 days; litter size 1–2,
3; 1–2 litters/year; weaned 14–15 weeks; pouch life *c.*
70 days; deposited in a nest for 30–40 days after
leaving pouch; sexual maturity late in first year or
early second year. **Longevity:** 4–9 years in wild, 14
years captive. **Status:** common over most of range.

■ HISTORY OF INTRODUCTIONS
Sugar gliders have been successfully re-introduced on
the Australian mainland and into Tasmania. They
were also successfully introduced to Maluku,
Indonesia, probably prehistorically.

AUSTRALASIA
Victoria
In southern Victoria sugar gliders have been successfully introduced into re-established habitat and can
inhabit young forest and woodland if nest boxes are
provided (Strahan 1995).

The first release occurred in the forest at Tower Hill
State Game Reserve in 1971 when three adults and
three juveniles were released. There followed in 1979
a program of release of captive-bred sugar gliders that
was initiated with the first release in that year and
subsequent releases in 1980 and 1981 (Suckling and
McFarlane 1983). Twenty-six juveniles (12 females, 14
males) were released in 1979; 12 (six females, six
males) in 1981; all were captive-reared from stock
originating from various areas in south-eastern
Australia.

Tasmania
Introduced to Tasmania in 1835 sugar gliders have
since spread all over the island (Marlow 1962; Smith
1973; Lever 1985).

INDONESIA
Maluku
Sugar gliders were probably prehistorically introduced to Ternate Island (Flannery 1995).

■ DAMAGE
None known.

COMMON RING-TAILED POSSUM
Common ringtail
Pseudocheirus peregrinus (Boddaert)

WESTERN RING-TAILED POSSUM
Western ringtail
Pseudocheirus occidentalis (Thomas)
*P. peregrinus is now usually placed in the separate family
Pseudocheiridae; the western ring-tail possum formerly
included in this species is now given specific status P. occiden-
talis Thomas. Their biology differs only slightly.*

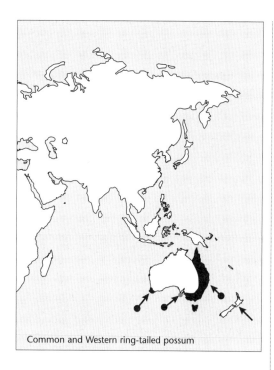

Common and Western ring-tailed possum

■ DESCRIPTION

HB 300–400 mm; T 300–400; WT 700–1100 g.
Head short; muzzle pointed; ears short, rounded, with a white patch behind; head and body rufous grey to black; belly grey-white; tail prehensile, white tipped, naked on lower surface; limbs rufous.

DISTRIBUTION

Australia. Eastern Australia from Cape York, Queensland, to Adelaide, South Australia, and inland to the Great Dividing Range.

■ HABITS AND BEHAVIOUR

Habits: nocturnal, arboreal; rarely seen on ground; utilises hollows in trees, stumps, logs, but can build independent nest or drey of twigs and leaves. **Gregariousnes:** solitary, pairs or family groups; density 0.1–0.37/ha. **Movements:** sedentary. **Habitat:** rainforest, sclerophyll forest and woodland. **Foods:** leaves, flowers, fruits and buds. **Breeding:** breeds April–November; mates May–June; litter size 1–3; 2 litters/year; lactation 6 months or more; pouch life 4 months; females mature at 12 months; polyoestrous; polyovular. **Longevity:** 4–5 years in wild. **Status:** common.

■ HISTORY OF INTRODUCTIONS

Common ring-tailed possums have been introduced successfully to Kangaroo Island and unsuccessfully to New Zealand. Western ring-tailed possums have been successfully re-introduced in some areas in Western Australia.

AUSTRALASIA
Australian mainland
In the late 1980s or early 1990s, 62 hand-reared common ring-tails were released in Ku-ring-gai Chase National Park, New South Wales. More than half were killed by predators such as foxes, cats and dogs, but the fate of the remainder is not known.

Between 1991 and 1996 there have been at least four attempts to translocate western ring-tailed possums. At least two of these, at Leschenault Conservation Park (1991–97, $n = 106$) and Yalgorup National Park (1995–2001, $n = 188$), appear to have been successful. Ring-tail possums have also been re-introduced to Lane-Poole Reserve (1996, $n = $?) and Karakamia Sanctuary (1995--2000, $n = $?) near Chidlow, but the success of these releases is not known at this stage.

Kangaroo Island
Fifteen common ring-tailed possums were introduced to Flinders Chase National Park in 1926 where they became successfully established (Harris 1974; Copley 1995).

Tasmania
Sixty-one common ring-tails were released on Maria Island in 1971 (Weidenhofer 1977; Rounsevell *et al.* 1991).

PACIFIC OCEAN ISLANDS
New Zealand
Two ring-tailed possums were introduced to New Zealand by the Canterbury Acclimatisation Society in 1867 (Wodzicki 1950), but the species failed to become established.

■ DAMAGE
None (?).

Family: Macropodidae
Kangaroos, wallabies and pademelons

WALLABY
Unknown species

■ HISTORY OF INTRODUCTIONS
AUSTRALIA
There was a proposal to stock wallabies on the Kent Island group, Australia, from Toorak in 1909 (Barrett 1918), but it is not known if any were actually released.

LONG-NOSED POTOROO
Long-nosed rat-kangaroo
Potorous tridactylus (Kerr)

■ HISTORY OF INTRODUCTIONS
AUSTRALIA
Long-nosed potoroos were introduced to Maria
Island, Tasmania, when 136 were released there in
1971 (Weidenhofer 1977; Rounsevell *et al.* 1991;
Abbott and Burbidge 1995).

NEW ZEALAND
Potoroos (=*Potorous* (*apicalis*) *trydactylus*) were
introduced to New Zealand by the Auckland
Acclimatisation Society in 1867 (Wodzicki 1950)
without success.

HUON TREE-KANGAROO
**Dendrolagus matschiei Förster and
Rothschild**

■ HISTORY OF INTRODUCTIONS
AUSTRALASIA
Papua New Guinea
Huon tree-kangaroos are likely (Maynes 1989) or
almost certainly (Flannery 1995) to have been intro-
duced to Umboi Island off Huon Peninsula in the
Bismarck Archipelago. Umboi is an oceanic island
with a depauperate terrestrial mammal fauna.

Young wallabies are occasionally kept as pets or
offered for sale in the markets of Papua New Guinea
and it is not unreasonable to postulate that these
records of wallabies on islands off the Sahul Shelf may
be the result of introductions by Melanesian man.
The islands appear to have been always separated by
deep wide water and the animals are unlikely to have
swum there.

TASMANIAN BETTONG
Bettongia gaimardi (Desmarest)
The following Bettongia *spp. are now sometimes placed in a*
separate family, Potoroidae, but under this classification are
included in the Macropodidae.

■ DESCRIPTION
HB 315–332 mm; T 288–345 mm; WT 1.2–2.25 kg.
Upper parts brownish grey pencilled with white;
under parts greyish white; tail furred, white tipped.

■ DISTRIBUTION
Australia. Now confined to Tasmania. Extinct on
mainland, but formerly occurred from south-eastern
Queensland to south-eastern South Australia.

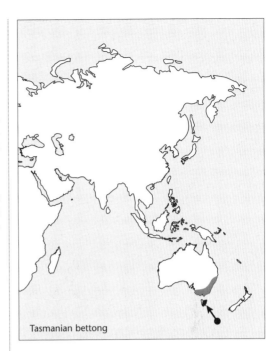

Tasmanian bettong

■ HABITS AND BEHAVIOUR
Habits: nocturnal; terrestrial; territorial; nest of dry
grass under fallen limb or in grass tussock.
Gregariousness: solitary or twos. **Movements:** 1.5 km
to feed; home range 65–135 ha. **Habitat:** open forest
with grassy or heath understorey; grassland. **Foods:**
fruiting bodies of fungi. **Breeding:** all year; gestation
21 days; pouch life 105 days; weaned 40–60 days;
sexual maturity 12 months. **Longevity:** no information. **Status:** extinct on mainland; common in
Tasmania.

■ HISTORY OF INTRODUCTIONS
AUSTRALASIA
Tasmania
The Tasmanian bettong was introduced successfully
when 123 were released onto Maria Island, Tasmania,
in 1971 (Weidenhofer 1977; Rose 1986 in Seebeck *et
al.* 1989; Kennedy 1992).

■ DAMAGE
None known.

BURROWING BETTONG
Boodie
Bettongia lesueuri (Quoy and Gaimard)

■ DESCRIPTION
HB 280–400 mm; T 215–300 mm; WT 1.0–1.5 kg.
Small, thickset; short rounded ears; upper parts
yellow-grey; under parts light grey; tail fat, lightly
haired, sometimes a white tip.

Burrowing bettong

■ DISTRIBUTION
Australia: in Western Australia on Barrow, Middle, Boodie, Bernier and Dorre islands; formerly on Dirk Hartog Island.

■ HABITS AND BEHAVIOUR
Habits: nocturnal; terrestrial; sleeps by day in nest of vegetation in burrow; burrows short, curving. **Gregariousness:** social(?). **Movements:** sedentary(?). **Habitat:** arid to semi-arid woodland, shrubland and grassland. **Foods:** tubers, bulbs, seeds, nuts, green parts of plants, native figs, roots, termites, fungi. **Breeding:** breeds throughout year; oestrous cycle 23 days; gestation 21 days; litter size 1; pouch life 115 days; 3 young in 12 months; sexual maturity 5 months. **Longevity:** no information. **Status:** abundant on islands, extinct on mainland.

■ HISTORY OF INTRODUCTIONS
AUSTRALASIA
South Australia
Burrowing bettongs were introduced to Kangaroo Island, South Australia, in 1924–26, but failed to become established there (Finlayson 1958; Serventy 1987; Short *et al.* 1992). Nine bettongs were released and initially the introduction appeared successful, but the last signs of them were noted in 1948 (Harris 1974; Copley 1995).

In 1995, 20 burrowing bettongs were re-introduced to Yookamurra Sanctuary, where they appear to have established. In 1999, 10 bettongs from Heirisson Prong in Western Australia were re-introduced to

Roxby Downs. In 2000 a further 20 animals obtained from Bernier Island, Western Australia, were added to the colony and by early 2001 the numbers had increased to 70.

Western Australia
Forty-two Burrowing bettongs from Dorre Island were re-introduced to Heirisson Prong, Shark Bay: 12 (four male and eight female) in May 1992, 18 (four male and 14 female) in September 1993 and 12 (four male and eight female) in October 1995 (Short *et al.* 1995). Twenty-two of the 42 animals went into a captive breeding enclosures and 114 animals were subsequently released from the enclosures onto the peninsula (Short and Turner 2000). By October 1999 this population had increased to at least 300 animals (Short and Turner 2000).

Burrowing bettongs were accidentally exterminated on Boodie Island during a campaign to eradicate black rats (*Rattus rattus*) from the island using pindone-impregnated grain baits (Short and Turner 1993). Bettongs from Barrow Island (*n* = 36) were later re-introduced successfully to Boodie Island Nature Reserve in 1993 (Strahan 1995). Animals from Barrow Island, 40 in all, were released in the Gibson Desert Nature Reserve in 1992 and were still there in 1993 (Christensen and Burrows 1995), but are believed to have all been killed by predators.

■ DAMAGE
None known.

BRUSH-TAILED BETTONG
Brush-tailed rat-kangaroo, woylie
***Bettongia penicillata* Gray**

■ DESCRIPTION
HB 300–380 mm; T 290–360 mm; WT 1.1–1.6 kg.
Body plump; head broad, short; upper parts yellowish grey; under parts pale; tail with distal black crest; hind foot longer than head. Male and female similar.

■ DISTRIBUTION
Australia. Now restricted to small areas in Western Australia at Dryandra, Perup and Tutanning. Formerly more widespread in Western Australia, South Australia, Northern Territory and New South Wales.

■ HABITS AND BEHAVIOUR
Habits: spends day in domed nest of grass and shredded bark under a bush; conveys nest material with tail; nocturnal; aggressively territorial. **Gregariousness:** solitary; sexes occupy distinct individual home ranges. **Movements:** sedentary. **Habitat:**

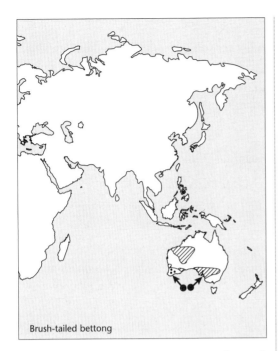

Brush-tailed bettong

open forest, woodland with understorey of tussock grasses. **Foods:** fruiting bodies of underground fungi, bulbs, tubers, seeds, insects, resin and other plant material. **Breeding:** all year; embryonic diapause; litter size 1, rarely 2; pouch life *c*. 90 days; 1 young/14 weeks; young leave nest 17 weeks; female gives birth at age 170–180 days and approx. every 100 days thereafter. **Longevity:** 4–6 years. **Status:** locally common.

■ HISTORY OF INTRODUCTIONS
AUSTRALASIA
South Australia
Following the establishment of a captive breeding program in 1975, a total of 164 brush-tailed bettongs from Western Australia (subspecies *ogilbyi*) were released on five South Australian offshore islands.

Extinct over most of its former range, except for three small areas in the south-west of Western Australia at Tutanning Nature Reserve, Dryandra Forest and Tone/Perup rivers. It has been re-introduced to St. Francis Island and introduced to several other islands off South Australia, where breeding populations have been established on Island A, Venus Bay; Bairds Bay Island; and Wedge Island (Delroy *et al*. 1986 in Seebeck *et al*. 1989).

They were successfully re-introduced to Venus Bay Island A (17 ha) in 1980 (7–8 animals released), Bairds Bay Island (13 ha) in 1982 (10 animals released), St. Peter Island (3500 ha) in 1989 (113 animals released) and Sanctuary Island in 1992

(five) Dryandra, Western Australia, were re-introduced in 1994. They were re-introduced unsuccessfully to St. Francis Island (800 ha) in 1981, 1984, and 1987 (figures vary on number of animals released, either 130 or 172), but failed to become established on all occasions. Here, the island had lost its original population early this century. Six were re-introduced to Bird Club Island (8 ha) in 1979, but failed to become established. They were introduced to Wedge Island (974 ha) in 1983 (11 animals released) and Yookamurra Sanctuary in 1991, both successfully (Delroy *et al*. 1986; Nelson *et al*. 1990; Short *et al*. 1992; Nelson *et al*. 1992; Copley 1995).

Western Australia
Brush-tailed bettongs were introduced to Yendicup Forest, Perup, in 1977, when some 52 were released (Short *et al*. 1992). They have now been established there with the assistance of some fox control for over 22 years.

Fifty-six brush-tailed bettongs were released at Batalling Forest near Collie in 1982–83, but they did not become firmly established until after 1991 when fox-baiting was introduced. Sixty-seven were released at St. Johns Brook, Nannup, in 1983, but all had disappeared in about six months, probably having been eaten by foxes (Short *et al*. 1992).

Between 1992 and 2000 a further 20 re-introductions were made, with nearly all being successful or showing promise of success: Boyagin Nature Reserve (1992, $n = ?$), Julimar Forest (1995, $n = 40$), Hills Forest near Mundaring (1996, $n = 37$), 19 sites in the Northern Jarrah Forest (1995–97, $n = 492$), Lake Magenta Nature Reserve 1997–8 ($n = 35$), Francois Peron National Park (1997–2000, $n = 147$), Poorginup and Chitelup forests (1998, $n = 40$), Easter and Barlee forests (1998, $n = 40$), St. Johns Forest (1998, $n = 40$), Denmark Forest (1998, $n = 38$), Centaur Forest (1998, $n = 39$), Karakamia Sanctuary (1997, $n = ?$), Walpole-Nornalup National Park (1999, $n = 40$), Giants Forest (1999, $n = 40$), Wellington National Park (2000, $n = 30$), Davis Forest (2000, $n = 37$), Kalbarri National Park (2000, $n = 32$), Paruna Sanctuary (2000, $n = 40$), Shannon National Park (2000, $n = ?$), Strickland Forest (2000, $n = ?$), Hadfield Forest (2000, $n = 29$), and two areas of private property along the Harvey River (2000, $n = 19$ at each site).

New South Wales
Woylies were re-introduced to Genaren Hills Sanctuary south-west of Dubbo in 1998 ($n = 12$) with another 12 animals added in 1999. At first they appeared to have established at this location, but later failed.

■ DAMAGE

Brush-tailed bettongs were an agricultural pest during early settlement, and the only known island population (St. Francis Island, South Australia) was exterminated because of the animals' depredations in the gardens. They were also destroyed in huge numbers by mainland farmers (Seebeck *et al.* 1989).

A study of brush-tailed bettongs on a South Australian offshore island found that they were primarily feeding on dicotyledonous plant material, including seeds and fruit. Fungi were not a major food, although endomycorrhizal fungi were dominant in faeces at the end of winter. It was suggested that this may have been ingested as gut content of scarab larvae, which were also a major part of *Bettongia* diet at that time (Green and Nelson 1988 in Seebeck *et al.* 1989).

BANDED HARE-WALLABY

Merrnine

Lagostrophus fasciatus (Pèron and Lesueur)

■ DESCRIPTION

HB 400–450 mm; T 350–400 mm; WT 1.3–3.0 kg.
Slender build; nose naked; ears short; upper parts grizzled grey with three dark bands across lower back and rump; under parts greyish white; tail thinly haired, tapering, grey. Female larger than male.

■ DISTRIBUTION

Australia: formerly from southern Western Australia from Shark Bay to Esperance, except the south-west corner. Now only on Bernier and Dorre islands, Shark Bay.

■ HABITS AND BEHAVIOUR

Habits: shelters beneath scrub in daylight; nocturnal; aggressive. **Gregariousness:** distinctly defined home range; male territories overlap that of several females; solitary. **Movements:** sedentary. **Habitat:** acacia scrub; semi-arid woodland and scrubland. **Foods:** leaves and twigs from shrubs and other plants; grasses. **Breeding:** all year; most births in late summer; embryonic diapause; pouch life 24–26 weeks; lactation 9 months; young 1–2/year; sexual maturity less than 1 year, but breeds in first or second year. **Longevity:** no information. **Status:** range considerably reduced, but still common on islands.

■ HISTORY OF INTRODUCTIONS

AUSTRALASIA

Dirk Hartog Island, Western Australia
A re-introduction program for banded hare-wallabies

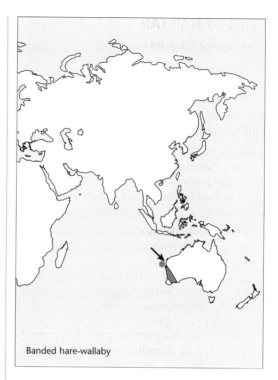

Banded hare-wallaby

commenced in 1974 when 11 were released on Dirk Hartog Island (62 000 ha) and became established there (Short *et al.* 1992) for a short period, but were ultimately unsuccessful probably due to predation by cats (Abbott and Burbidge 1995).

In June 1974, 17 (four adult males, seven adult females plus six pouched young (five male, one female)) were transferred by boat from Dorre Island to two small holding yards on Dirk Hartog Island. By October the population had grown to 25 adults and by December 1976 to 35 adults. In 1977, six were transferred to another enclosure. In 1978 the island was baited in an attempt to eliminate feral cats, although rain probably negated the success of this venture. By June 1978 the group of six had grown by three young (males); one of the originals had died and two females were added to the group. Holes were made in the fence and the animals allowed to disperse into the area beyond. The released group was supplemented by two (females), and later by six females and five males from the captive colony, making a total of 21 adult and independent juveniles.

In June 1979, 13 of the 21 animals were re-trapped, and in 1980 further trapping suggested only 10 remained. The project was then abandoned, the decline being put down to drought and grazing pressure of goats and sheep. No wallabies have been sighted since that period.

BRIDLED NAILTAIL WALLABY

Onychogalea fraenata (Gould)

■ DESCRIPTION
HB 430–700 mm; T 360–540 mm; WT 4–8 kg.
Small sandy wallaby; bridle line from centre of neck down behind forearm on each side of body, white; tail with horny pointed tip.

■ DISTRIBUTION
Australia. Now restricted to a small area near Dingo in central Queensland; formerly from the Murray River in South Australia to Charters Towers in Queensland.

■ HABITS AND BEHAVIOUR
Habits: mainly nocturnal, basks in sun in winter; during day rests under bushes in shallow depression. **Gregariousness:** solitary; females and young; feeding aggregations 4–5. **Movements:** home ranges overlap within and between sexes, 20–90 ha. **Habitat:** semi-arid inland areas; slopes and plains in tall shrubland and grassy woodland; brigalow scrub. **Foods:** forbs, grass and browse. **Breeding:** throughout year, but mainly spring and summer; 2–3 young/year. **Longevity:** no information. **Status:** endangered; about 1500 animals left in protected area.

■ HISTORY OF INTRODUCTIONS
AUSTRALIA
Formerly widespread, bridled nailtail wallabies began to decline in the 1890s. The decline is thought to be associated with the effects of the pastoral industry and perhaps competition with domestic livestock for food or disturbance of the ground cover (Strahan 1995). Whatever the reason, their range collapsed between the 1890s and the 1920s and the species was thought to be extinct until rediscovered in a small area near Dingo in central Queensland in 1973.

Translocations began in 1997 and early 1998 with 16 wallabies into fenced areas, cleared of predators such as cats and foxes, and they were later to be released (Thoday 1999). Six bridled nailtail wallabies were released at Genaren Hills Sanctuary, 100 km south-west of Dubbo, in April 1999.

■ DAMAGE
None known.

NORTHERN NAILTAIL WALLABY

Onychogalea unguifera (Gould)

■ HISTORY OF INTRODUCTIONS
AUSTRALIA
Reported to have been 're-introduced' in the late 1930s to Pulbah Island (64 ha) in Lake Macquarie, New South Wales, northern nailtail wallabies failed to become established there (Harper 1945; Ride 1970; Short *et al.* 1992). There is some doubt about this record and in fact the species referred to may have been *O. fraenata* as *O. unguifera* did not ever occur there.

Bridled nailtail wallaby

Northern nailtail wallaby

RUFOUS HARE-WALLABY
Mala
Lagorchestes hirsutus Gould

■ DESCRIPTION
HB 310–390 mm; T 245–305 mm; WT 780–1960 g.
Fur long and soft; upper parts sandy brown to rufous, under parts paler; head rufous; tail untapered, brownish black above, pale rufous below. Female larger than male.

■ DISTRIBUTION
Australia. Bernier and Dorre islands in Western Australia. Formerly widespread throughout most of the arid and semi-arid regions of central and Western Australia, Northern Territory and northern South Australia.

■ HABITS AND BEHAVIOUR
Habits: shelters in shallow scrape under grass hummock, sometimes developed into a proper burrow during periods of extreme heat; nocturnal and crepuscular. **Gregariousness:** solitary; males occupy exclusive home range overlapping those of several females. **Movements:** sedentary. **Habitat:** spinifex hummock grasslands of sand plain and sand dune deserts. **Foods:** forbs, perennial grasses, seeds, bulbs of sedges and insects. **Breeding:** any time of year, but influenced by rainfall; up to 3 young per year; pouch life 124 days; sexual maturity female 5–23 months, male 14–19 months. **Longevity:** no information. **Status:** rare.

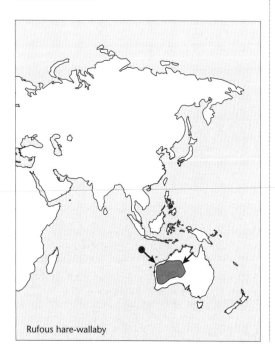

Rufous hare-wallaby

■ HISTORY OF INTRODUCTIONS
AUSTRALASIA
Australia
Rufous hare-wallabies were once common throughout the spinifex hummock grasslands of the Northern Territory, Western Australia and South Australia and probably occurred over at least one-third of the continent. They were reduced to two small populations in the Tanami Desert, Northern Territory, by 1980, but the last remaining mainland population became extinct in the wild in 1991 following a wildfire. A subspecies (recent studies suggest it is the same species) occurs on Bernier and Dorre islands off the Western Australian coast.

Northern Territory
A captive breeding and re-introduction program commenced in the Northern Territory in 1980. Between December 1986 and May 1990, 47 rufous hare-wallabies were transported from Alice Springs and released in enclosures on the Lander River, Northern Territory, where they bred and increased in numbers. In September 1990 some were released into the surrounding bush. A further four groups were released to June 1992 and later 15 individuals were released. These populations were heavily preyed upon by cats (Gibson *et al.* 1995), but none appear to survive in the wild now.

Rufous hare-wallabies were released at two sites, Yinapaka and Lungkartajarra in the Tanami Desert, in 1986 at the former and 1989 at the latter, in enclosures and from there into the wild between 1989 and 1991 (11 at former and 23 at latter site) (Gibson *et al.* 1994). Predation by cats severely reduced the populations at each site and it was found that cat control needed to be continuous to allow their successful establishment.

Of 27 released at Lake Surprise, Northern Territory, none lasted for more than 16 months (Lundie-Jenkins 1989; Short *et al.* 1992). Of 11 released in 1990, all disappeared in about four months, probably having been decimated by cats. All the attempted introductions in the Tanami Desert during the period 1990–92, when 79 were released, have failed because of predation from foxes and cats (Burbidge *et al.* 1999).

Western Australia
Thirty rufous hare-wallabies were introduced to Trimouille Island in the Montebello group off the Pilbara coast in June 1998 and were doing well in 1999 (Burbidge *et al.* 1999). Subsequent monitoring has shown them to occupy the entire island by late 2000.

■ DAMAGE
None.

QUOKKA
Setonix brachyurus (Quoy and Gaimard)

■ **DESCRIPTION**
HB 400–540 mm; T 245–310 mm; WT 2.7–4.5 kg.
Small wallaby; fur long; nose naked; upper parts grey brown; and under parts grey; ears short, round; legs short; tail short, closely haired, tapering. Male larger than female.

■ **DISTRIBUTION**
Australia. In a few localities in south Western Australia and on Rottnest Island and Bald Island off the coast.

■ **HABITS AND BEHAVIOUR**
Habits: nocturnal; terrestrial; sleeps by day in dense vegetation. **Gregariousness:** solitary or small groups 20–150; adult males have linear hierarchy based on age; group territories. **Movements:** sedentary. **Habitat:** forest, dense vegetation around permanent swamps; offshore islands. **Foods:** grass and leaves. **Breeding:** all year; litter size 1; pouch life 30 weeks; weaned about 40 weeks. **Longevity:** no information. **Status:** abundant on Rottnest and Bald Islands, scarce elsewhere.

■ **HISTORY OF INTRODUCTIONS**
Australasia
Western Australia
Quokka numbers declined on the mainland about the 1920s and 1930s, but by the 1960s they still persisted in some coastal habitats of the lower south-west.

Quokka

At least 673 quokkas were re-introduced from Rottnest Island to the Marsupial Research Station of the University of Western Australia (254 ha) at Jandakot from 1972 to 1988. The population fell to nine by 1988 in the absence of restocking, and thus failed to become permanently established. It was suggested that overgrazing by rabbits, disease, re-introduction of inappropriate age/sex classes, and behavioural problems associated with introducing captive-bred stock were the probable causes of failure. However, attempts to modify these causes had little effect on subsequent survival (Short *et al.* 1992), and it was later established that fox predation was the primary cause of the population's demise (M. Massam *pers. comm.* 1988).

Quokkas (three male, two female) obtained from mainland stock were re-introduced to Karakamia Sanctuary near Chidlow in Western Australia in 1997. They had persisted until 2000, but numbers remain low.

■ **DAMAGE**
In the past, quokkas have hindered efforts to restore the flora of Rottnest Island after wildfires by eating off newly planted shrubs and trees. The use of tree guards and better management of the quokka population appears to have overcome this problem.

RED-BELLIED PADEMELON
Tasmanian pademelon
Thylogale billardierii (Desmarest)

■ **DESCRIPTION**
HB 360–370 mm; T 300–483 mm; WT 2.4–12.0 kg.
Small stocky build; fur dense and long; upper parts dark brown to grey brown; under parts buff with a rufous tinge, especially on the lower abdomen; ears rounded; nose naked; tail short, thick, about two-thirds length of body. Males considerably larger than females.

■ **DISTRIBUTION**
Australia. Formerly south-eastern South Australia and Victoria, but extinct on the mainland. Now only in Tasmania and on the larger Bass Strait islands.

■ **HABITS AND BEHAVIOUR**
Habits: mainly nocturnal, uncommonly diurnal; shelters by day in thick undergrowth. **Gregariousness:** solitary; feeding groups to 10 or more; no persistent bond between individuals. **Movements:** up to 2 km to feed; male territorial; home range 170 ha. **Habitat:** dense vegetation in coastal and montane wet sclerophyll forest; rainforest; tea-tree scrubs and open grassy patches in forest. **Foods:** grasses and herbs and

Red-bellied pademelon

some browse from shrubs, leaves. **Breeding:** breeds continuously, but mostly April–June; gestation 30 days; single young; pouch life 200 days; young remains with mother for *c.* 10 months, leaves pouch at *c.* 29 weeks, but will suckle for a further 11 weeks; sexual maturity 14–15 months. **Longevity:** no information. **Status:** common to abundant.

■ HISTORY OF INTRODUCTIONS
Australasia
Australia
The red-bellied pademelon was introduced to Wilsons Promontory, Victoria, in 1911–14, but the number introduced and any success is unknown (Menkhorst and Mansergh 1977). They have also been introduced to Maria Island, Tasmania, where 13 were released in 1971 (Weidenhofer 1977; Summers 1991) and to Three Hummock Island where they have been re-introduced (Hope 1973; Rounsevell *et al.* 1991).

■ DAMAGE
In some areas red-bellied pademelons have caused damage to agricultural crops and it is necessary for their numbers to be reduced (Strahan 1995).

NEW GUINEA PADEMELON
Northern pademelon
Thylogale browni (Ramsay)

■ DESCRIPTION
HB 487–667 mm; T 300–520 mm; WT 3.0–9.1 kg.
Coat dark brown; belly fur grey based; hip stripe lacking.

■ DISTRIBUTION
Papua New Guinea. Northern and eastern New Guinea, Bismarck Archipelago, Buka (extinct), Emirau (extinct), ?Japen, Lihir (extinct), New Britain, New Ireland and Umboi.

■ HABITS AND BEHAVIOUR
Habits: little known of biology. **Gregariousness:** solitary(?). **Movements:** sedentary(?). **Habitat:** disturbed areas, tall cane grass, forest regrowth, abandoned gardens in forest. **Foods:** grass and leaves(?). **Breeding:** ?April–July; young 1. **Longevity:** no information. **Status:** widespread and common.

■ HISTORY OF INTRODUCTIONS
Australasia
Papua New Guinea
New Guinea pademelons were introduced, but are now extinct on Buka, Emirau and Lihir; they were prehistorically introduced to New Britain and Umboi and were introduced to New Ireland (Flannery 1995).

The island distribution of this pademelon is largely or entirely the result of human introductions, but it has since become extinct on some of the smaller islands (Flannery *et al.* 1988; Flannery 1992). They were probably introduced to New Ireland about seven thousand years ago and carried from there to Buka, Lihir and possibly Emirau. They probably did not become established on Buka and Emirau, and became extinct on Lihir about 50 years ago (Flannery *et al.* 1988; Flannery and White 1991).

■ DAMAGE
None known.

DUSKY PADEMELON
Thylogale brunii (Schreber)

■ HISTORY OF INTRODUCTIONS
Asia–Australasia
Indonesia–Papua New Guinea
It is likely that the Melanesians introduced the dusky pademelon to the Bismarck Archipelago (Umboi, Bagabag and New Britain islands) and possibly to the Kai Islands (Maluku) (Maynes 1989). They were probably prehistorically introduced to the Kai Islands (Flannery 1995).

BLACK-FOOTED ROCK-WALLABY
Pearson Island rock wallaby
Petrogale lateralis Gould

■ DESCRIPTION
HB 450–610 mm; T 320–640 mm; WT 2.3–7.1 kg.

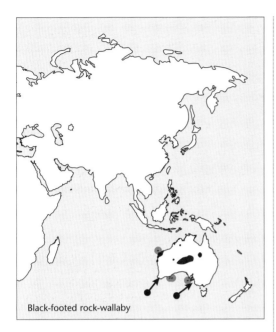

Black-footed rock-wallaby

Thick woolly coat; cheek stripe varies from prominent black to pale; upper parts mostly grey-brown; under parts sandy brown; variable dark mid-dorsal stripe; forearms and hind legs sandy brown; paws black; tail almost buff at base. Degree of ornamentation varies with latitude; southern animals have more noticeable markings. Female smaller than male.

■ DISTRIBUTION
Australia. Western Queensland, south-western Northern Territory, a few islands in Western Australia (Recherche Archipelago) and islands off South Australia. Formerly widespread in Western Australia.

■ HABITS AND BEHAVIOUR
Habits: sleeps during day in shelter amongst rock piles. **Gregariousness:** feeding aggregations common. **Movements:** sedentary. **Habitat:** granite rock piles with mallee or scrub cover in semi-arid and mesic areas. **Foods:** grasses. **Breeding:** litter size 1; embryonic diapause; sexual maturity 1–2 years. **Longevity:** no information. **Status:** generally declining in numbers throughout range.

■ HISTORY OF INTRODUCTIONS
SOUTH AUSTRALIA
Pearson Island
In 1960, five Pearson Island rock-wallabies, *P. l. pearsoni*, were successfully released on central and south Pearson Islands (Robinson 1989; Short *et al.* 1992; Copley 1995). In 1994, six animals were still present there.

Thistle Island
Five rock-wallabies were released in 1974 and 10 in 1975 on this island (Robinson 1989; Short *et al.* 1992; Copley 1995). The species became established and there were 100 of them there in 1994.

Wedge Island
Eleven rock-wallabies were released on Wedge Island in 1975. By May 1993 there were 24 there (Robinson 1989; Short *et al.* 1992; Copley 1995), and the species has become well established.

West Island
In 1973–75, 13 rock-wallabies were released on West Island, but the last recorded there was in 1980 (Paton and Paton 1977; Robinson 1989; Copley 1995), and they are now absent from the island.

WESTERN AUSTRALIA
Seven rock-wallabies, *P. l. lateralis*, were re-introduced to an area known as the 'The Granites' at Querekin (north of Shackleton) in 1990 by the Western Australian Department of Conservation and Land Management. By 1998 the population had grown to 50.

■ DAMAGE
Rock-wallabies, *P. l. lateralis*, have been reported causing damage to cereal crops planted in close proximity to rock outcrops in southern Western Australia.

BRUSH-TAILED ROCK WALLABY
Black-tailed wallaby, western rock wallaby, pale rock wallaby
***Petrogale penicillata* (Gray)**

■ DESCRIPTION
HB 450–600 mm; T 500–700 mm; WT 4.9–10.9 kg.
Back grey brown, but more rufous on rump; belly yellow brown; face dark with white cheek stripe; nose naked; ears short, oval with black patch and whitish margins; black mark under armpit and down side of abdomen; fore and hind feet black; tail long, untapered, rufous at base and remainder black, brush at tip. Female smaller than male.

■ DISTRIBUTION
Australia: Victoria, northern and western New South Wales and south-eastern Queensland.

■ HABITS AND BEHAVIOUR
Habits: mainly nocturnal and crepuscular; agile; rests in den sites under vegetation or rocks or trees. **Gregariousness:** solitary or small groups. **Movements:** sedentary. **Habitat:** mountainous wet and dry forest, cliffs with thick vegetation and with

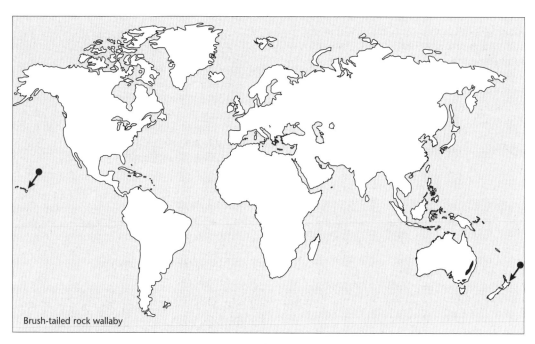

Brush-tailed rock wallaby

adjoining pastures; rocky slopes, rock piles; open woodland. **Foods:** grasses, browses bushes; fallen leaves, fruits, flowers, seeds. **Breeding:** all year; litter size 1; pouch life 30 weeks. **Longevity:** few reach 4+ years in wild, 12–15 years captive. **Status:** uncommon, declining possibly endangered.

■ HISTORY OF INTRODUCTIONS
Introduced successfully to New Zealand, and to Oahu in the Hawaiian Islands.

PACIFIC OCEAN ISLANDS
New Zealand
Introduced in 1870 or 1873, the brush-tailed rock wallaby is now established in small numbers on Kawau, Rangitoto and Motutapu islands, Hauraki Gulf (Wodzicki 1965; Gibb and Flux 1973).

The Auckland Acclimatisation Society received a rock wallaby (*Petrogale xanthopus*) in 1873 and J. Reid liberated small brown rock wallabies on Motutapu Island in Hauraki Gulf (Thomson 1922). A second introduction appears to have been made on Kawau by Sir George Grey, possibly about 1870.

The brush-tailed rock wallabies on Motutapu became established and from there spread to Rangitoto Island (near Kawau), where they were numerous in 1912. Between 1948 and 1950 several hundred were trapped on Rangitoto (Wodzicki and Flux 1967). In the 1960s the species was probably most numerous on Mototapu, where in 1965 some 515 were shot by the New Zealand Forest Service. On Kawau they are more scarce, but occur wherever there are cliffs.

Seven individual rock wallabies from Rangitoto were released illegally on Great Barrier Island in 1981, but all were recaptured or killed (King 1990). They are still present on all three islands in Hauraki Gulf, restricted in distribution on Kawau and Motutapu, but widespread on Rangitoto Island (King 1990).

HAWAIIAN ISLANDS
A Mr R. H. Trent purchased two brush-tailed wallabies (one male, one female and joey) from a consignment of them passing through Honolulu by ship from Australia in 1916 (Tinker 1938). These were temporarily housed in a tent on the island of Oahu, but were harassed by dogs and escaped. Although a reward was offered for their return, little was heard of them until in 1921 it was reported (in press) that there were about 50 on the island of Oahu (Kramer 1971).

In 1937 it was reported (in press) that a number now roamed the uplands of Kalihi on Oahu and in 1939 that there were as many as 100 present on the cliffs above the Material Testing Laboratories of the State of Hawaii (Kramer 1971; Lazell 1980; Lauret 1982). In 1966 it was found that a small population of probably fewer than 100 wallabies was still confined to the Kalihi Valley (Kramer 1971). There may have been a second colony as some were sighted in Moanalua Valley in November 1980, Aiea Loop Trail in 1976 and Waimano Valley in 1979 (Lauret 1982). However, an extensive survey in 1981 found them (at least 11) present only at Kalihi.

Currently they are restricted to Ewa Kalihi in the lower Ko'olau Range on rocky slopes between 90 and 425 m elevation and largely confined to a 7.0-ha valley (Gilmore 1977; Lazell *et al.* 1984). In 1981 the population was estimated to be about 250 animals, but this was reduced as a result of drought conditions in 1983 and 1984 to about 100 animals by 1987 (Lazell *et al.* 1984; Lazell 1987). A survey found only one colony of wallabies after 65 years. The adjacent suitable habitat is limited and it is safe to predict that they will only ever be a small colony. Up to 11 separate individuals have been seen in one day (Lauret 1982).

AUSTRALIA
New South Wales
Brush-tailed rock wallabies disappeared from the Wombeyan Caves between 1929 and 1946. Four were re-introduced in February 1980, and a further six in January 1981, from the Jenolan Caves area by the New South Wales Department of Tourism. In 1986 there were still nine animals near the original release site, but by March 1990 only three remained (Short *et al.* 1992).

■ DAMAGE
In Hawaii, brush-tailed rock wallabies feed mainly on introduced grasses and other plants and have caused no damage (Kramer 1971), probably because of their restricted range and small numbers. In New Zealand, they are considered to be a pest by competing for pastures with stock. On Rangitoto their presence is said to be serious because of their influence in altering and inhibiting regeneration of established plant species (King 1990).

In Australia, brush-tailed wallabies were formerly shot for the skin trade and as a supposed agricultural pest (Strahan 1995).

ROTHSCHILD'S ROCK-WALLABY
Petrogale rothschildi **Thomas**

■ DESCRIPTION
HB 426–592 mm; T 412–704 mm; WT 2.6–6.6 kg.
Upper parts greyish brown; under parts dull brown; upper surface of head and ears dark brown; chest light grey; throat light grey; shoulders and neck greyish. Dampier Archipelago animals are markedly smaller.

■ DISTRIBUTION
Australia. Western Australia in the Hamersley Range area of the north-west, and rocky offshore islands of the Dampier Archipelago (Rosemary, Dolphin, Enderby).

■ HABITS AND BEHAVIOUR
Habits: mainly nocturnal, but also active late afternoon and early morning. **Gregariousness:** no information.

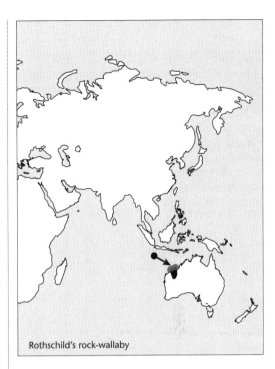

Rothschild's rock-wallaby

Movements: sedentary. **Habitat:** grass steppe, shrub vegetation with rock piles and outcrops, rocky hills and gorges. **Foods:** grasses, herbs, fruits and browse such as leaves. **Breeding:** throughout year. **Longevity:** no information. **Status:** relatively common, but declining; recent studies implicate fox in decline.

■ HISTORY OF INTRODUCTIONS
AUSTRALASIA
Western Australia
Stock from Enderby Island has been successfully re-introduced to West Lewis Island (Strahan 1995). Fifteen (eight males, seven females) were released in 1982 and now the species is widespread on West Lewis Island (Abbott and Burbidge 1995).

■ DAMAGE
None known.

AGILE WALLABY
Sandy wallaby
Macropus agilis **(Gould)**

■ DESCRIPTION
HB 593–850 mm; T 587–840 mm; WT 9–27 kg.
Upper parts sandy brown; under parts whitish; median brown stripe between eyes and ears and a faint cheek stripe; light stripe on thigh; ear and tail edges black.

■ **DISTRIBUTION**

Australia and Papua New Guinea. Northern coastal Australia from Western Australia to Queensland; southern and eastern lowlands of Papua New Guinea.

■ **HABITS AND BEHAVIOUR**

Habits: most active dawn and dusk. **Gregariousness:** groups 3–10; large aggregations at feeding areas. **Movements:** sedentary; moves to higher ground during floods, and sometimes to areas after fire. **Habitat:** open grassland adjacent to low scrub or woodland, along rivers and streams, coastal sand dunes, black soil plains. **Foods:** sedges, grasses, roots, leaves, figs and other fruits. **Breeding:** all year; gestation 30 days; 1 young; pouch life 7–8 months; weaned 10–12 months; sexual maturity females 12 months, males 14 months. **Longevity:** no information. **Status:** common and abundant.

■ **HISTORY OF INTRODUCTIONS**

AUSTRALASIA

Australia

Agile wallabies were introduced to Long Island, Queensland, but are now absent from that island (Abbott and Burbidge 1995).

Papua New Guinea

Agile wallabies may have been introduced by humans into parts of its present island distribution in Melanesia (where it occurs on Fergusson, Goodenough, Kiriwina, ?southern New Ireland and ?Normanby, although archeological evidence is lacking at this time (Flannery 1995). They were probably prehistorically introduced to New Ireland.

On Baniara Island in Milne Bay Province, Papua New Guinea, they are known to have been introduced in recent times, and their presence on Kiriwina, Goodenough and Fergusson islands of the D'Entrecasteaux group has been postulated (by Bass 1956, 1959) as due to the actions of man. The islands may have been joined to other nearby islands and to the mainland 14–17 000 years ago, so their presence may be a natural one, although the actions of man cannot be ruled out (Maynes 1989).

■ **DAMAGE**

Considered a pest in Queensland, the Northern Territory, and the West Kimberley region of Western Australia by pastoralists (pastures) and by growers of some irrigated crops (e.g. rice and corn); many were poisoned with bran baits and poisoned waters in the 1950–70 era in Western Australia (Gooding and Long 1958). A bounty system still operates in Queensland (Strahan 1995).

BLACK-STRIPED WALLABY

Scrub wallaby

Macropus dorsalis (Gray)

=*Wallabia dorsalis*

■ **DESCRIPTION**

HB males 1420–1590 mm, females 1120–1210 mm; T males 740–830 mm, females 540–615 mm; WT males 18–20 kg, females 6.0–7.5 kg.

Head and body brown or greyish, sides paler and greyer; shoulders rufous; distinct dark stripe from back of neck and down centre of back; nose naked; cheeks with white patch behind eye; hip stripe curved, white; belly grey white; tail shortish, sparsely haired, scaly, grey with black tip. Female much smaller than male.

■ **DISTRIBUTION**

Australia: from southern Queensland (Rockhampton) to New South Wales (Tamworth).

■ **HABITS AND BEHAVIOUR**

Habits: rests under cover during day; feeds dusk to dawn. **Movements:** sedentary. **Gregariousness:** groups up to 20 or more of both sexes; old males may be solitary. **Habitat:** rainforest margins, woodland, *Lantana* thickets and brigalow. **Foods:** grasses and herbs. **Breeding:** all year; gestation 33–35 days; embryonic diapause; litter size 1; pouch life 210 days; sexual maturity females 14 months, males 20 months. **Longevity:** 10–15 years. **Status:** common and abundant.

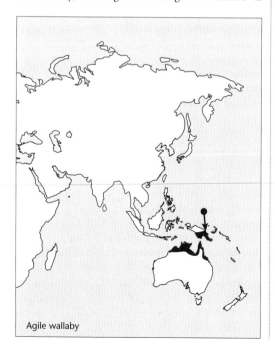

Agile wallaby

Black-striped wallaby

■ HISTORY OF INTRODUCTIONS

Introduced successfully to New Zealand, but now extinct there.

PACIFIC OCEAN ISLANDS
New Zealand

The black-striped wallaby is presumed to have been liberated by Sir George Grey on Kawau Island in the Hauraki Gulf about 1870 (Thomson 1922; Wodzicki 1950; Barnett 1985).

It appears that this species remained established on Kawau for over 80 years; the last authentic record was in 1954 and it now appears to be extremely rare or extinct there (Wodzicki and Flux 1967; Gibb and Flux 1973; King 1990).

The claims for the presence of black-striped wallabies on Kawau have not been substantiated and appear to have been based primarily on early misidentification of specimens of *Macropus parma* (Maynes 1977). Published measurements (Wodzicki and Flux 1967) of a reputed *M. dorsalis* in 1954, clearly do not correspond to the values for *M. dorsalis* in Australia, but do with *M. rufogriseus*. The evidence so far suggests that the other species on Kawau is this latter species.

■ DAMAGE
None known.

TAMMAR WALLABY
Dama wallaby, scrub wallaby
Macropus eugenii (Desmarest)

■ DESCRIPTION
HB 520–680 mm; T 330–450 mm; WT 2.7–10.0 g.
Upper parts silver grey or grey brown; shoulders reddish; nose naked; belly grey white; dorsal stripe faint, dark; tail short, grey with black tip; no distinct face stripe.

■ DISTRIBUTION
Australia: south-western and southern South Australia from tip of Eyre Peninsula, St. Peter's Island, Nuyt's Archipelago and Kangaroo Island, South Australia. Wallaby Island, Houtman's Abrolhos and south-west corner of Western Australia (Geraldton to Hopetoun) and Garden Island. Formerly on Flinders Island, St. Francis, St. Peter, and Thistle. Formerly more widespread across the mainland.

■ HABITS AND BEHAVIOUR
Habits: largely nocturnal, occasionally active late afternoon. **Gregariousness:** groups of up to 5 feeding. **Movements:** sedentary; may move up to 1 km to feed. **Habitat:** dense thickets in sclerophyll forest, shrub woodland, mallee and coastal scrub, grassland, offshore islands. **Breeding:** births December–March; gestation 28 days; embryonic diapause; litter size 1; post-partum oestrus and mates 24 hours after young born; pouch life about 250–252 days; lactation 8–9 months; females sexually mature at 9–12 months, males at 2 years.

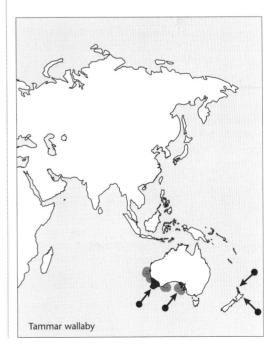

Tammar wallaby

Longevity: 11–14 years at least in wild. **Status:** common some areas only; rare on mainland.

■ **HISTORY OF INTRODUCTIONS**

Tammar wallabies have been introduced successfully to Kawau Island, and to Rotorua on the North Island of New Zealand.

AUSTRALASIA

South Australia

Populations of tammar wallabies from Kangaroo Island have been successfully established on three islands in South Australia (Robinson 1989; Hall 1991; Pool *et al.* 1991; Copley 1995). Five animals (three males, two females) were successfully released on Boston Island near Port Lincoln in 1971 as a tourist attraction. By the early 1980s the estimated population here exceeded 400 animals, but since then numbers have been reduced to about 100.

Tammar wallabies were liberated on Greenly Island in about 1905 by the South Australian government to act as an emergency food supply for possible castaways. The total present population is estimated at 50 individuals.

Tammar wallabies have also been held on Granite Island, near Victor Harbour, but were successfully removed in 1992 by the administering District Council. They were introduced in 1968 when about 12 were released. Skulls have been found on two other islands – Reevesby and North Gambier – and these may represent unsuccessful releases there (Strahan 1983).

Western Australia

Eighty-five tammar wallabies from Garden Island were re-introduced to the University of Western Australia Marsupial Research Field Station at Jandakot, from 1971 to 1981. The exact fate of most of them is unknown, but most are believed to have been taken by foxes and the introduction was a failure (Short *et al.* 1992). They were also introduced to North Island (Abrolhos Islands) probably in about the 1950s from Wallabi Island, but died out and were re-introduced successfully in 1987 (Storr 1960; Abbott and Burbidge 1995). There were several translocation attempts between 1971 and 1988, but only one of these has been successful – that from Perup to the Batalling Forest.

In 1994–95, 39 tammar wallabies were re-introduced to Batalling Forest, where they have established a small population. In 1998–99, 38 wallabies were re-introduced to Warrup Forest where they appear to be established. In 1998, 13 animals (seven male and six female) were re-introduced to Karakamia Sanctuary, where they have bred successfully. In 1998–2000, 35

tammar wallabies were re-introduced to Julimar Forest. In 1999–2000, 46 wallabies were re-introduced to Bennelaking Forest and in 2000, 20 animals were re-introduced to a rehabilitated bauxite mine (Alcoa Australia's Huntley mine site) near Dwellingup.

PACIFIC OCEAN ISLANDS

New Zealand

Tammar wallabies were introduced to Kawau Island by Sir George Grey, who owned the island, in about 1870, and also on the mainland at Rotorua in about 1912 (Wodzicki 1950; Gibb and Flux 1973; Wodzicki and Flux 1967; Lever 1985; King 1990; Poole *et al.* 1991). The origin of mainland animals is obscure. They were possibly liberated by H. R. Benn, but it is uncertain whether they came from Kawau or Australia, at the south end of Lake Okareka about 1912. However, this was denied by a resident who knew Benn. Whatever the manner of their introduction, they were well established by 1930. Their range increased between 1946 and 1966 and by the mid-1960s they were the most numerous species on Kawau. Some may have been transferred from Kawau to Rotorua in 1939, where by 1946 they were distributed over an area of 10–20 km to the north-west of Lake Tarawera and Lake Okatiana.

Tammar wallabies are still established and thriving at Rotorua and on Kawau, and in 1984 occupied an area of some 16.2 km^2 (Barnett 1985; King 1990).

■ **DAMAGE**

Blamed for damage to newly planted pines on Kawau, as many as 3000 tammar wallabies were shot in one year by local landholders (Wodzicki and Flux 1967). In high numbers in forest areas, they are possibly capable of changing the pattern of forest succession or at least altering the local abundance of different species (King 1990). Serious attempts were made to control them on Kawau in the 1960s, but this ceased altogether when farming was abandoned in 1973.

On small islands off the South Australian coast, where introduced, tammar wallabies are reported to be causing damage to the vegetation and vegetated areas, thus causing major changes (Copley 1995). In the early 1920s they were considered a pest in the Denmark and Walpole areas of Western Australia, where agricultural enterprises were being established.

WESTERN GREY KANGAROO
Macropus fuliginosus (Desmarest)

■ **DESCRIPTION**

HB 946–2225 mm; T 425–1000 mm; WT 3.0–53.5 kg.

Large kangaroo; upper parts dark brownish; under

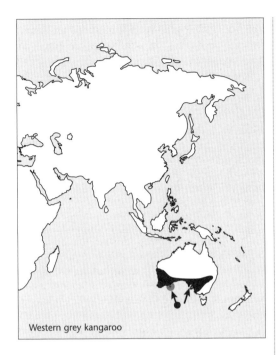

Western grey kangaroo

parts greyish white; tail with dark upper surface and tip. Muzzle finely haired. Males have a strong odour.

DISTRIBUTION
Australia: Southern Australia from about Shark Bay to South Australia, western New South Wales, western Victoria and southern Queensland.

HABITS AND BEHAVIOUR
Habits: rests during day near or under low bushes or low trees; mainly nocturnal. **Gregariousness:** pairs; groups to 20, rarely 100. **Movements:** sedentary. **Habitat:** forest, woodland, heaths, shrubland, grassland. **Foods:** grass and browse from shrubs. **Breeding:** all year; gestation 30 days; no embryonic diapause; oestrous cycle 35 days; pouch life 42 weeks; sexual maturity females 18 months, males 2 years. **Longevity:** up to 16 years in wild. **Status:** common.

HISTORY OF INTRODUCTIONS
AUSTRALASIA
Woody Island, Western Australia
Some western greys were introduced to Woody Island (240 ha) in the Recherche Archipelago before 1948 and became established there (Goodsell *et al.* 1976; Short *et al.* 1992). They still occur on the island. A single animal was released on Boullanger Island, near Jurien in 1985, but was absent in 1991 (Abbott and Burbidge 1995). In 1998 five females were released on Heirisson Island, in the Swan River near Perth, as a tourist attraction. A further female with a male pouch young was added in 2000.

Granite Island, South Australia
In 1971 western greys were successfully introduced to Granite Island, but all were removed in 1984 (Robinson 1989; Copley 1995). They may also have been introduced on Taylor Island (Abbott and Burbidge 1995), but details appear to be lacking.

DAMAGE
Western grey kangaroos are accused of damaging pastures and fences in farming areas and where such damage can be shown, state conservation authorities issue licences to reduce numbers. This generally only happens where over-population occurs. There is commercial harvest in open season areas under management plans in Western Australia, South Australia, New South Wales and Queensland.

EASTERN GREY KANGAROO
Eastern grey, great grey, grey forester
Macropus giganteus Shaw

DESCRIPTION
HB 958–2302 mm; T 430–1090 mm; WT 3.5–66 kg.
Large kangaroo; fur short and woolly, grey to grey brown; belly whitish; tail grey brown, black towards tip.

DISTRIBUTION
Australia. Eastern Australia and Tasmania, from north-eastern Queensland to south-eastern South Australia, and eastern Tasmania.

Eastern grey kangaroo

■ HABITS AND BEHAVIOUR

Habits: crepuscular and nocturnal, occasionally diurnal; rests during day in shade or shelter of trees and shrubs. **Gregariousness:** in groups of 5–20. **Movements:** sedentary. **Habitat:** semi-arable mallee scrub, woodland and forest with grassy areas. **Foods:** grasses and forbs. **Breeding:** all year; gestation 36 days; oestrous cycle 46 days; embryonic diapause; pouch life 11 months; lactation about 18 months; litter size 1; sexual maturity 18 months. **Longevity:** probably to 16 years. **Status:** common and abundant.

■ HISTORY OF INTRODUCTIONS

AUSTRALASIA

Australia

Eastern grey kangaroos were introduced to Heron Island, Capricorn Group, Queensland, but are no longer present there (Kikakawa and Boles 1976). They have also been introduced on Brampton (Woodall 1988), Long (Abbott and Burbidge 1995), Middle Percy (Roche 1989), and South Molle (Abbott and Burbidge 1995) islands in Queensland. Those on Long Island are no longer present, and those released on South Molle appear to have been all the same sex. Those introduced on Middle Percy may not have been eastern greys.

Eastern greys have also been successfully introduced (45 in 1969–70) to Maria Island, Tasmania (Weidenhofer 1977; Rounsevell *et al.* 1991).

PACIFIC OCEAN ISLANDS

New Zealand

Three kangaroos (*Macropus* sp.) were released on Dunrobin Station by C. Basstian in 1863 and three in the same year on Bluff Hills by the Southland Acclimatisation Society (Thomson 1922; Wodzicki 1950). Grey kangaroos did not become established in New Zealand, probably because of the small numbers released.

■ DAMAGE

Following the development of agriculture and the pastoral industries in Australia, eastern grey kangaroos increased their numbers so much so that they were formerly regarded as pests. They are now fully protected, but licences to remove some may be granted where excess animals are causing damage to crops, pastures and fences. Commercial harvesting is allowed under approved management plans.

PARMA WALLABY

White-throated wallaby

Macropus parma Waterhouse

=*Wallabia parma*

■ DESCRIPTION

HB 447–530 mm; T 400–550 mm; WT 2.5–5.9 kg.

Head and body grey-brown; shoulders and back grey-brown; indistinct dark stripe down back; hip stripe absent; throat, chest and belly white; nose naked; upper lip white; cheek stripe poorly defined, white; tail white tipped. Similar in appearance to *M. dorsalis*, but smaller, lacks hip stripe and fainter dorsal stripe. Female smaller than male, with less robust chest and arms.

■ DISTRIBUTION

Australia. Eastern Australia in eastern New South Wales in the Illawarra district and near Coffs Harbour.

■ GENERAL HABITS AND BEHAVIOUR

Habits: mainly nocturnal and at times crepuscular; rests by day in dense vegetation. **Gregariousness:** singly or groups of 1–3 feeding. **Movements:** sedentary. **Habitat:** rainforest and scrubs; sclerophyll forest with thick undergrowth and patches of grass. **Foods:** grasses and herbs. **Breeding:** all year; mainly mates January–May, births mainly February–June; gestation 34–35 days; embryonic diapause; oestrous cycle averages 41.8 days; litter size 1; pouch life 28–36 weeks; lactation 40–60 weeks; female mature 12–36 months,

Parma wallaby

males 20–24 months. **Longevity:** 9.5 years (wild?). **Status:** rare, range reduced, but not in danger of extinction.

■ HISTORY OF INTRODUCTIONS
Introduced successfully to Kawau Island, New Zealand.

AUSTRALASIA
Australia
It was thought that parma wallabies were extinct in Australia in 1957, but it is now known that they exist in an area of coastal New South Wales, north of the Hunter River (Maynes 1977).

A number of individuals were returned to Australia in an attempt to re-establish the species (Lever 1985) before it was known that they were still present here. A conservationist and businessman, P. Pigott set up a captive breeding colony on his property at Mt. Wilson, 90 km west of Sydney in the early 1970s. He imported 30 animals from Kawau Island, New Zealand. Forty-five were released at Robertson in 1988 in an enclosure from which they could wander. Three weeks later most had been killed by foxes (Short *et al.* 1992).

Twenty-four parma wallabies from Kawau Island, New Zealand, were released onto Pulbah Island in Lake Macquarie, New South Wales (area 0.64 km²), in 1972, along with 12 animals from Taronga Park Zoo (originally from Kawau). They were maintained in a yard for a short period, then released. Ten weeks later they had all disappeared for unknown reasons, perhaps predation by dogs (Short *et al.* 1992).

Parma wallabies were re-introduced to Greenly Island, South Australia (128 ha), in 1905 (Mitchell and Behrndt 1947; Robinson 1980; Short *et al.* 1992), where they are still surviving. Twelve parma wallabies were introduced to Granite Island, South Australia (about 25 ha), in 1968 (Short *et al.* 1992), and some are surviving there.

PACIFIC OCEAN ISLANDS
New Zealand
The parma wallaby was probably introduced to the island of Kawau in Hauraki Gulf in about 1870 (Gibb and Flux 1973), but was unknown on the island until discovered in 1965 (Wodzicki and Flux 1967). They are now common on the northern half of the island, where they have been protected since 1968 (Gibb and Flux 1973).

From 1967–75, 736 parma wallabies were captured alive to supply zoos and establish breeding colonies in various parts of the world. They still occur on Kawau Island (Barnett 1985; King 1990).

■ DAMAGE
On Kawau, parma wallabies have significantly curtailed any regeneration of indigenous forests on the island and assisted in the elimination of many plants still found present on nearby wallaby-free islands (King 1990).

Parma wallabies are thought to be more damaging than other introduced wallabies on Kawau Island in eliminating indigenous species of plants (King 1990). They were a pest of young pine plantations during the 1960s and efforts were made to control or eliminate them up until about 1965 (Wodzicki and Flux 1967; Strahan 1995).

WHIPTAIL WALLABY
Macropus parryi **Bennett**

■ HISTORY OF INTRODUCTIONS
AUSTRALIA
Whiptail wallabies were introduced to Heron Island, Queensland, but are now absent from that island (Kikawa 1976; Abbott and Burbidge 1995).

COMMON WALLAROO
Euro, hill kangaroo, eastern wallaroo
Macropus robustus **Gould**

■ DESCRIPTION
HB 1100–1990 mm; T 530–900 mm; WT 6.25–46.5 kg.
Large kangaroo; fur thick; head and body sooty grey; belly grey white; nose naked, black; ear oval-shaped; tail dark grey at base and black tipped; hands and feet black. Other subspecies rufous. Eastern subspecies with grey fur, western and central subspecies with reddish fur. Female half weight of male.

■ DISTRIBUTION
Australia: throughout in rocky ranges in a variety of habitats except the south coast and some north coastal areas.

■ HABITS AND BEHAVIOUR
Habits: mainly nocturnal; during day lies up in rocky outcrops or caves. **Gregariousness:** solitary or small groups; density 0.04–13/km². **Movements:** sedentary. **Habitat:** rocky regions, sclerophyll forest, woodland, desert grassland and stony deserts; arid tussock grassland; rocky slopes with caves and rock-shelves. **Foods:** grasses. **Breeding:** throughout the year; gestation 34 days; embryonic diapause; litter size 1; pouch life 244–261 days; lactation 16 months; sexual maturity 18–24 months. **Longevity:** up to 12 years in captivity. **Status:** common and abundant in most areas.

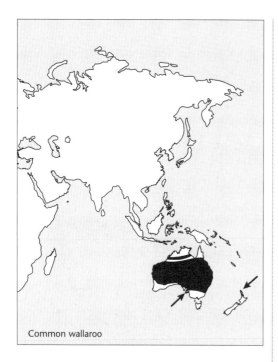

Common wallaroo

■ HISTORY OF INTRODUCTIONS

Common wallaroos were introduced to Kangaroo Island and to New Zealand, but failed to become established.

AUSTRALASIA
Australia
In 1937, two common wallaroos were released on Kangaroo Island, South Australia at Flinders Chase, but there are no further records of them (Copley 1995). They were also introduced at some time to Hook Island, Queensland, but are not present there now (Abbott and Burbidge 1995).

Twenty-nine captive-raised common wallaroos (euros) were successfully translocated from Carrang Station to Useless Loop in Western Australia in 1998.

PACIFIC OCEAN ISLANDS
New Zealand
Roan wallaroos ('*Osphranter erubescens*') were released on Kawau by Sir George Grey between 1860 and 1870, but they failed to become established (Thomson 1922), probably because too few were released.

■ DAMAGE

Although common wallaroos are a protected species, they are sometimes accused of eating pastures available to sheep and of damaging fencing. State authorities issue permits to reduce numbers in some areas. They are hunted commercially in accordance with approved management plans.

RED-NECKED WALLABY
Red wallaby, scrub wallaby, Bennett's wallaby, brush wallaby
Macropus rufogriseus (Desmarest)

■ DESCRIPTION

HB 659–923 mm; T 620–880 mm; WT 11–26.8 kg.
Head and body fawn grey to reddish; shoulders and neck reddish-brown; upper lip with white stripe; under parts grey or white; snout naked; muzzle, paws and largest toe black; ears long; cheek stripe indistinct, white; hip stripe lacking; hind feet black tipped; tail grey-brown with black tip. Females smaller, paler, and weigh less than males.

■ DISTRIBUTION

Australia: from eastern Queensland (Bundaberg) through eastern New South Wales, southern Victoria to south-eastern South Australia (Mt. Gambier); throughout Tasmania and on King Island and Flinders Island in Bass Strait.

■ GENERAL HABITS AND BEHAVIOUR

Habits: nocturnal; uses rest areas in day; moves to feed along well-defined pads; digs with forepaws; regurgitates and re-ingests food. **Gregariousness:** essentially solitary except at mating, but in areas at high density up to 30 may graze in same locality. **Movements:** sedentary. **Habitat:** forest edges, woodland and coastal scrub with grassland. **Foods:** grasses, herbs, leaves from trees, clover, roots, weeds (England heather, bracken, pine and birch scrub, and bilberries). **Breeding:** breeds all year (mainland); in Tasmania January–July; (breeds August–September England); oestrous cycle 33 days; gestation 30–31 days; litter size 1; eyes open 135–150 days; body furred at 165–175 days; pouch life 274–280 days; lactation 12–17 months; sexually mature 14–22 months. **Longevity:** 9 years(?). **Status:** common to abundant in most of range.

■ HISTORY OF INTRODUCTIONS

Red-necked wallabies have been introduced successfully into Tasmania, New Zealand and England, and unsuccessfully to Germany, Czechoslovakia, Hungary and the Ukraine.

AUSTRALASIA
Australia
One hundred and twenty-seven red-necked wallabies were successfully introduced to Maria Island, Tasmania in 1969–70 (Weidenhofer 1977; Summers 1991; Abbott and Burbidge 1995).

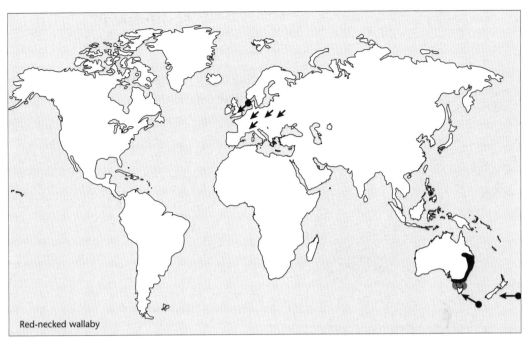

Red-necked wallaby

PACIFIC OCEAN ISLANDS
New Zealand
A number of red-necked wallabies were imported to New Zealand between 1867 and 1870 (Thomson 1922). One male and two females were liberated at Hunters Hills near Waimate in the South Island about 1874 and here they increased to several thousands by 1916 (Studholme 1954). By 1947 they had increased to such an extent that control of their numbers was necessary (Wodzicki and Flux 1967).

A second liberation took place in 1948 (Warburton 1986, gives the date of introduction as 1945) at the head of Quartz Creek, on Mount Burke Station, between Lakes Hawea and Wanaka. In 1914 a population of about 50 was seen on Mount Maude and some attempts were being made to exterminate them (Wodzicki and Flux 1967).

Red-necked wallabies extended their range in South Canterbury and in the 1940s and 1950s occupied an area of some 404 700 ha (Wodzicki 1965). This increased to an estimated half a million on 809 400 ha of range by 1960, but by 1973 they had been reduced by control measures to scattered pockets; a few at Lake Hawea (Gibb and Flux 1973).

The spread in South Canterbury was perhaps regulated by the application of regular control measures as 5000–6000 were destroyed annually during the 1950–60 period. Further reduction in the population was achieved by 1080 poisoning in the Waimate and Rotorua districts (Wodzicki 1965). There is now one thriving population at Hunters Hill near Waimate, and four smaller populations in Kakahu Forest, Pioneer Park, and Peel Forest, and at Quartz Creek (King 1990).

Illegal introductions have occasionally been found in recent years well outside their existing range (Fraser *et al.* 1996), but any success in becoming established is not known.

EUROPE
Czechoslovakia
Red-necked wallabies were released some time before World War 1 in a game park at Podiebrad near Praha (Prague), but disappeared soon after (Niethammer 1963).

Germany
There appear to have been at least four attempts to establish this species of wallaby in Germany in the latter part of the last century and early part of this century. These attempts were all eventually unsuccessful. Their failure to establish in Germany is said to be due to no vacant niche being available and to the fact that they are not regarded as an addition to the game animals.

In 1887 Phillip von Böselager released two males and three females in a 500-ha forest near Heimerzheim in the Bonn district, West Germany. Until 1893 these were carefully preserved and increased in number to some 35–40. At this time, the gamekeepers looking after them died and poachers had decimated the population by 1895. Count Witzleben released some on his estate at Altdöbern in the Calau district of the Frankfurt-Oder region in 1889. Here they also initially increased, but were eventually destroyed because they were said to 'frighten other game animals'. Others dispersed into the surrounding forests and disappeared.

Offspring of red-necked wallabies from a colony established in the Channel Islands (on Herm) were released on a property of G. Blücher von Wahlstatt's at Krieblowitz (Bluchersuh) in Schlesian early in this century. In 1910 it was reported that there were 60–70 of them there and they continued to increase in numbers until World War 1. Following the war they began to decrease in numbers and the few remaining were re-captured and placed in enclosures. After 1920 no further free-living animals were observed. Those on Herm were said to have been eaten as food by English soldiers occupying the island during World War 2 (Niethammer 1963).

A trial release was made on Kühkopf, an island in the Rhine near Oppenheim, in 1910. Cornelius von Heyl released six wallabies in the spring of that year, but they all died of cold in the following autumn (Niethammer 1963).

A release of wallabies (species?) was made near Hamburg in 1940, but they were reportedly exterminated soon after (Boettger 1943).

Hungary
Before World War 1, red-necked wallabies were released in Szenc, but it does not appear to be known what happened to them (Niethammer 1963).

Ukraine
Kangaroos, possibly red-necked wallabies, were living in the wild on an estate in the southern Ukraine owned by Friedrich von Falz-Fein of Askania Nova. By 1945, however, there was no trace of them (Niethammer 1963).

United Kingdom
Although there have been several early escapes and liberations of wallabies and kangaroos in Britain, few have survived. Two separate feral populations of *M. rufogriseus* were reported (Taylor-Page 1970; Yalden and Hosey 1971); however, the currently established population escaped from a private collection near Leek, Staffordshire, in 1939 (Baker 1990).

In 1850 several wallabies (*M. rufogriseus*?) escaped into woods near Norfolk; in 1912 some kangaroos (species unknown) were released on the Isle of Bute by the 4th Marquis of Bute; in the 1920s Mr M. Harman introduced a number of wallabies (species unknown) on Lundy Island in the Bristol Channel (Fitter 1959; Lever 1977). These all failed to produce established populations. Seven escaped in East Grinstead about 1949, but these failed to become permanently established (Fitter 1959).

The ancestors of the present population of red-necked wallabies in the Peak district of Derbyshire and Staffordshire are derived from an escape of five animals from an enclosure in Staffordshire in 1939–40 (Fitter 1959; Lever 1977). Odd animals were noted in 1940, 1944, 1951, 1954, 1955 and 1956, and in 1960 there were somewhere between 40 and 50 of them. Further animals were noted in 1963 and 1970, when it was thought that the total population had dwindled to about 12. The population has declined still further and it is thought that between 1971 and 1975 there were only about four or five present. In 1981–82 it was estimated that 15 were present including two or three pouch young (Lever 1985). Over the last 15 years there have been fewer than 15 present, although about 22 were counted in 1985 (Yalden 1988).

Those red-necked wallabies in the Weald, north central Sussex, are thought to have been present for about 35 years, and to be the progeny from a release by Sir Edmund Loder in 1908 (Lever 1977). One was noted in 1915 and they were reported present in the 1940s; some were reported in 1969 and in 1970; one was found dead and one live animal was seen. They may now be extinct in this area (Baker 1990; Corbet and Harris 1991).

A small colony existed since at least 1975 around Loch Lomond (Corbet and Harris 1991). Escapees from zoos and parks are occasionally reported in other areas. A small colony existed on the Channel Islands (Herm) from about the 1890s to 1910.

In about 1975, two pairs were released on Inchconnachan in Loch Lomond, Scotland, by the Countess of Aran (Mitchell 1983). The first of several animals arrived on the mainland in 1979 and in 1982 one was seen as far away as Balloch Park at the southern end of the Loch. These now appear to have disappeared.

■ DAMAGE
In the 1940s in New Zealand, red-necked wallabies were recognised as a pest of pastures. However, there do not appear to be any quantitative measurements of damage that is probably only local in nature. Between 1947 and 1956 70 000–100 000 were destroyed as pests. Many more were poisoned between 1960 and 1969. From 1969 to 1984 about 2500–3000 were destroyed each year by hunting, but the population remains stable (Warburton 1986; King 1990).

Although red-necked wallabies are reported to cause a considerable amount of damage to agricultural crops and by browsing shrubs and plantations of exotic trees, there appear to be few figures to substantiate other than local damage to remnant patches of indigenous forest and in pine plantations (King 1990). Present control by shooting and poisoning continues to remove about 20 per cent of the population per year without much effect on overall numbers.

In Australia, red-necked wallabies become pests of crops and pastures at times and are killed under licence or during special open seasons if it can be demonstrated that they causing such damage.

SWAMP WALLABY

Black-tailed wallaby, black wallaby

***Wallabia bicolor* (Desmarest)**

=*Macropus bicolor, W. ualabatus*

■ DESCRIPTION

HB 640–850 mm; T 640–862 mm; WT 10.3–20.5 kg.
Fur coarse, upper parts brown to black, but in southern parts brown or greyish-black; head and body dark rufous grey; belly light yellow to rufous orange; snout naked; ears short; face grey; cheek stripe indistinct, light yellow to brown; tail, basal half dark grey, rest black; fore and hind feet dark brown. Female smaller than male.

■ DISTRIBUTION

Australia: eastern Australia from Cape York, Queensland to south-eastern Victoria and inland to the Dividing Range.

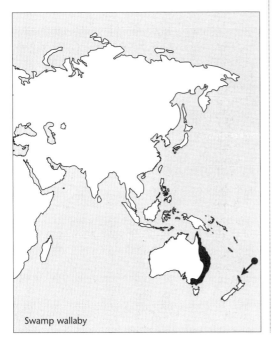
Swamp wallaby

■ HABITS AND BEHAVIOUR

Habits: terrestrial; largely nocturnal, sometimes diurnal. **Gregariousness:** solitary; occasionally groups at feeding areas. **Movements:** sedentary. **Habitat:** rain and wet sclerophyll forest and woodland with dense cover; dense moist thickets; dense grass and ferns, brigalow scrub. **Foods:** shrubs, ferns and grasses; leaves, bark from trees and bushes; pine tree seedlings; bracken and fungi. **Breeding:** breeds throughout the year; gestation 33–38 days; embryonic diapause; females mate again 8 days before first young is born; litter size 1; pouch life 8–9 months but young at foot may still be suckling; lactation 15–16 months; young mature at 15–18 months. **Longevity:** no information. **Status:** common.

■ HISTORY OF INTRODUCTIONS

Swamp wallabies have been introduced successfully to New Zealand.

PACIFIC OCEAN ISLANDS
New Zealand
It is presumed that Sir George Grey liberated swamp wallabies on Kawau Island in Hauraki Gulf about 1870 (Thomsons 1922; Gibb and Flux 1973). They now occur over most of Kawau in damp scrubby areas, but are not abundant. Of 59 wallabies collected in 1966 only four were this species (Wodzicki 1965; Wodzicki and Flux 1967).

By 1973 small numbers could be found over most of the island (Gibb and Flux 1973), and they still occur on Kawau especially at the northern end, but are generally rare (Barnett 1985; King 1990).

■ DAMAGE

In Australia swamp wallabies will graze agricultural crops, especially cereal grain, and pine tree seedlings and can cause some damage (Strahan 1995). On Kawau, in New Zealand, they are a minority species and are thought unlikely to cause too much damage to trees or seedlings (King 1990).

Family: Erinaceidae
Hedgehogs

HEDGEHOG
Northern hedgehog, common hedgehog, European hedgehog

Erinaceus europaeus Linnaeus

Some authorities divide this species into concolor *in eastern Europe and western Asia,* amurensis *in eastern Asia and* europaeus *in western Europe.*

■ DESCRIPTION

HB 87–310 mm; T 10–50 mm; WT 120–1400 g.

Mainly brown (becoming paler with age; albinos uncommon), short, round bodied with no visible neck; upper parts covered with sharply pointed spines (20–22 mm) which are brown to black with a white base and tip, are hard, grooved and erectile; under parts coarse furred, yellowish white to brownish; snout pig-like; eyes black; ears small and hidden by fur; legs short, hidden by fur; toes five-clawed; tail short. Female similar to male, but generally has shorter snout. Immatures have darker noses, spines and footpads than adults.

■ DISTRIBUTION

Eurasia. From Spain, Portugal, Britain, southern Scandinavia and Mediterranean islands (all the larger ones except Balearic), east across mainland Europe to central Asia, and south to Palestine, Iraq and eastern Iran. In eastern Asia in south-central Manchuria and Korea (North and South), and central-eastern China (south to Amur and Yangtze Kiang).

■ HABITS AND BEHAVIOUR

Habits: nocturnal and crepuscular; hibernates in cold parts range (October–April in Europe, June–August in New Zealand); makes nest of grass or other material in hedgerow, burrows in grass tussock, thicket or under a rock; males partially territorial; rolls up in defence. **Gregariousness:** solitary except for mating and for nesting mothers with young; crowd together sometimes at a food source; density 1/ha in open country. **Movements:** sedentary; home range 3–50 ha; travels 0.5–1.5 km/night in open habitat. **Habitat:** occasionally dense woodland and forest; mainly dry open country, hedgerows, woodland edges, cultivation, gardens, roadsides, parks and marshes; coastal farmland. **Foods:** mainly insects and their larvae (beetles, woodlice, moth larvae, ants, bees, cicadas, fly larvae, earwigs), spiders, snails, worms, grubs,

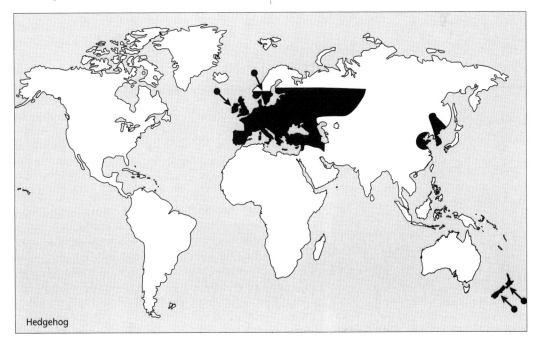

Hedgehog

molluscs, centipedes, millipedes, rats, frogs, lizards, snakes, occasionally birds' eggs, chicks and carrion; acorns and berries, **Breeding:** breeds following hibernation in May–October; gestation 30–35 days; litter size 1–9, usually 4–6; 1–2 litters per year; no postpartum oestrus; young blind at birth, lack spines; eyes open at 14 days; weaned at 4–6 weeks; remain with mother 6–7 weeks; breed at 11 months. **Longevity:** 6–7 years (captivity and wild) and possibly up to 10 years. **Status:** common, often abundant.

■ HISTORY OF INTRODUCTIONS

Hedgehogs have been successfully introduced to New Zealand, Sweden and various British islands (Shetland, Unst, Yell, Foula, Fetlar, Muckle Roe, Bressay, Whalsay, East and West Burra, Ventray, Orkney, Canna, and possibly Ireland). They have probably been unsuccessfully released in Japan and Germany.

ASIA
Japan

Humans were probably responsible for the introduction of the hedgehog from eastern Asia to mainland Japan (Grzimek 1972). However, there appears no other evidence of any introductions and no hedgehogs in that country (Corbet 1978, 1980).

EUROPE

With the exception of some small islands, most of Europe has hedgehogs; they are often moved from place to place by people for different reasons (Niethammer 1963). Their presence on many islands may be due to deliberate introduction since they have been kept in captivity for food or as pets, at least since Roman times (Corbet 1966).

Germany

In Germany in the 1950s it was believed by some that hedgehogs resident in many parks and gardens of some towns such as Berlin had been derived from animals which had escaped or been released (Herter 1952). Some hedgehogs may have been released in the upper Harz Mountains in the early part of this century (Löns 1907).

Hedgehogs were often released in gardens and cultivated areas to control pests, and often these animals were re-transported from areas some distance from their origin. During World War 2 Hermann Goering had a few hundred released at Darss for the control of snakes, although there were hedgehogs present there already. The addition of probably both western and eastern hedgehogs has resulted in the presence of the subspecies *roumanicus* in the area (Niethammer 1963).

North Sea and Baltic Sea islands

In 1937, three hedgehogs from Schleswig-Holstein were released on the island of Sylt for the control of burrowing mice. Whether they were successful does not appear to have been recorded. Other small islands in the North Sea and Baltic Sea to receive introductions of hedgehogs in about 1830 include Borkum, Juist and Spiekeroog (East Friesian Islands in North Sea) and in 1922 Greifswalder Oie (Baltic Sea). On this latter island they had disappeared by 1925, but further attempts were made in 1927–29. On this occasion they remained there until 1934, but they also died out. Small introductions were also made on the island of Ruden (in Baltic Sea, west of Greifswalder Oie) in 1936, with two subspecies, *E. europaeus* and *E. roumanicus* (= *concolor*), being released (Niethammer 1963).

United Kingdom and Ireland

Hedgehogs are known on the Isle of Wight, Isle of Man, Anglesey, Shetland, Orkney, Skye, Bute, Mull, Coll, Canna, Jersey and Guernsey, and are probably introduced on some of these (Southern 1964; Campbell 1955, Fitter 1959). They were probably introduced by humans on Shetland mainland, Unst, Yell, Foula, Fetlar, Muckle Roe, Bressay, Whalsay, E. Burra, W. Burra, Ventray Islands, Orkney mainland, North Ronaldsay and Canna. Introduction was attempted on St. Mary's (1958) (Isles of Scilly) and Sark (Corbet and Harris 1991).

The hedgehog population on Mull is probably an introduced one. On Skye it may be that the present population is the progeny of two pairs liberated at Dunach in or after 1800. Their appearance at Inverbroom, Wester Ross, in 1900 is thought to be due to transport in baled hay, and they may have reached Burra in the Shetlands in ship's ballast or cargo. Some hedgehogs were released on Orkney in 1870 where they survived but did not spread much, and some on Mainland, Shetland, from 1860 onwards. Continuing introductions between 1939 and 1959 appear to have established them on Yell, Foula, Whalsay, Bressay, Burra, Vementry and Fetlar, where an attempt in the 1920s failed. Because there is no common name for the hedgehog on the Isle of Wight the animals there are believed to have been a recent introduction, probably in the early part of the nineteenth century (Fitter 1959; Niethammer 1963).

The range of the hedgehog may have been extended in the Highlands of Scotland during the latter part of the eighteenth century. One hundred years later they had spread north to Ross-shire but there is no direct evidence that introductions aided the extension (Fitter 1959). The presence of hedgehogs in Ireland may also be due to introduction by humans (Lever 1985). An army of people called hedgehog carers is rehabilitating or re-introducing hedgehogs in the

United Kingdom (Mead 1999). Recently 33 mostly captive-raised animals were released at three sites in Devon, Suffolk and Jersey, where they survived well despite some losses from badgers and cars.

PACIFIC OCEAN ISLANDS
New Zealand

The importation of hedgehogs to New Zealand appears to have been prompted at first by pure sentiment. Later it was justified on the grounds of their reputation for eating slugs and snails, which themselves had been earlier inadvertently introduced and become garden pests.

The first hedgehogs to reach New Zealand did so in 1869 when the Canterbury Acclimatisation Society imported a pair of which the subsequent fate is unknown. In 1871 an agent for the Canterbury Acclimatisation Society dispatched 24, but only one survived the voyage. A shipment of 100 hedgehogs made by the Acclimatisation Society in 1885 was similarly unsuccessful, although three arrived and were liberated in a Dunedin garden (the female died and two males were released). Others were probably released shortly after this date as one was found near Port Chalmers in 1890 (Thomson 1922; Wodzicki 1950; Brockie 1975).

The early liberations in Canterbury in the 1870s and 1880s failed to become established and the first hedgehogs in the wild appear to be those noted at Sawyers Valley near Dunedin in 1890.

In the period from 1890 to 1899 hedgehogs were reported at Gore, Omarama and Hakataramea. These may have come from stock established at nearby Dunedin or they may have been a separate introduction. In 1892 Mr P. Cunningham received 12 hedgehogs from England which escaped soon after their arrival at Merivale, Christchurch. The 12 animals introduced at Christchurch in 1892 became established and probably served as a nucleus for introductions to Waiau and Whakapuaka, where they appeared between 1898 and 1900 (Brockie 1975).

From 1900 to 1909 it appears that people continued to transplant hedgehogs to new areas. They were released on the Chatham Islands and on the North Island (Wodzicki 1950). Their appearance at Wairoa, Napier, Hastings, Carterton and Te Wharau between 1907 and 1912 suggested a systematic campaign of liberation in Hawke's Bay and the Wairarapa. A separate introduction was probably made in Taranaki, as they were first noted at Hawera in 1908 and at Stratford in 1909–10 (Brockie 1975).

There appear to have been dramatic increases in the range and numbers of hedgehogs in both the main islands of New Zealand in the period from 1910 to 1940. Further animals were released when 12 were liberated at New Plymouth by W. W. Smith (Wodzicki 1950), and many other releases were probably made from 1910 to 1919 in widely scattered areas, as the species turned up in places to which it could not have spread from existing populations. By 1916 hedgehogs were established in all the districts between Dunedin and Christchurch and were throughout Otago, Southland, South Canterbury and Banks Peninsula. They were increasing in the Nelson and Blenheim areas and were well established on the North Island, spreading at Hawkes Bay and recorded in a number of widely scattered areas (Brockie 1975).

Some hedgehogs were released at Alexandra in 1927 in the hope of reducing earwig populations which were attacking fruit trees, and in 1927 or 1928 some were released at Clyde and also probably some at Cromwell. Some were liberated at Tinopai (Northland) in 1936. Between 1932 and 1939 the North Canterbury Acclimatisation Society paid bonuses on 4752 hedgehog snouts collected in that district and in 1939 the hedgehog was classed as vermin. The North Island Vermin Control Board paid bonus on 823 snouts in the 1934–40 season, 661 of these from the Wellington District. The bounty scheme continued in New Zealand until it was abandoned in 1952 (Brockie 1975).

Between the 1930s and 1971 several liberations of hedgehogs occurred on Kapiti Island and in 1940 several releases were made at Waiouru by soldiers from a nearby army camp. By this latter date hedgehogs had colonised most of the lowland areas to the foot of the mountainous regions of New Zealand. Since 1948 they have increased their range somewhat to include the central Volcanic Plateau of the North Island, areas of inland Nelson, and northern Westland.

Today hedgehog numbers are most abundant in the intensively farmed lowland districts, towns and suburbs, and become less numerous with increasing altitude. They are abundant throughout the North Island and South Island, except in the high mountainous country. Their numbers are reported to have stabilised and they have now colonised all the suitable areas. Hedgehogs are also present on the islands of Waiheke, the Chathams, D'urville, Quail, Rabbit, Motungarara (off Kapiti) and the Stewart islands (Wodzicki and Wright 1984; Barnett 1985; King 1990).

◼ DAMAGE

In New Zealand hedgehogs have been accused of being a nuisance to ground-nesting birds, but there appears to be little evidence (Brockie 1959). The main economic significance is probably their capacity to carry and spread both human and stock infections (*Leptospira* sp.), but there are few proven cases except the transmission of hedgehog ringworm to humans (Wodzicki and Wright 1984; King 1990). They also carry off chickens' eggs, causing a minor nuisance to farmers.

Hedgehogs in the United Kingdom are traditionally persecuted for their predation on the eggs of game birds, although damage is relatively insignificant compared to foxes and crows (Southern 1964; Corbet and Harris 1991).

ALGERIAN HEDGEHOG
Atelerix algirus (Lereboullet)
Atelerix *is often considered a subgenus of* Erinaceus.

■ DESCRIPTION
HB 200–250 mm; T 20–40 mm; WT 900–1600 g.
General appearance speckled black and white; spines banded black and white; ears large; head spines divided into two parts by median parting; muzzle, cheeks, ears and paws brown; forehead, underside and legs white, sometimes brown on underside.

■ DISTRIBUTION
Africa. Confined to north-western Africa from Morocco to Libya, Canary Islands (Fuerteventura) and Balearic Islands (Majorca, Minorca and Ibiza).

Algerian hedgehog

Occurs in a few localities on the Mediterranean coast of France (Lecques, Saint Cyr and Bormes) and Spain (Elche, Tortosa and Barcelona).

■ HABITS AND BEHAVIOUR
Habits: active dusk and dawn; may aestivate in hot weather; hibernates in Europe. **Gregariousness:** solitary? **Movements:** no information. **Habitat:** found near villages. **Foods:** earthworms, insects, snails, frogs, lizards, snakes, eggs, young of ground-nesting birds. **Breeding:** gestation 35–48 days; litter size 3–7; 2 litters/year; eyes open at 8–18 days; weaned at 40 days; independent at 1.5–2 months; sexual maturity 1 year. **Longevity:** probably? 8–10 years. **Status:** no information.

■ HISTORY OF INTRODUCTIONS
EUROPE
The discontinuous distribution of the Algerian hedgehog in France and Spain (a few localities on the coast) suggests that it has been introduced and the same probably applies to populations on Malta, and the Balearic and Canary Islands (Fuerteventura and Tenerife). They have been kept in captivity since Roman times and are particularly liable to introduction by human agency (Corbet 1966, 1978, 1980; Lever 1985; Burton 1991; Wilson and Reeder 1993).

The species is now established in the Balearic Islands and on Malta, but it may be extinct in France (Burton 1991).

WEST INDIES
Puerto Rico
E. algirus caniculus (=*E. krugi* (Peters)) has been found on Puerto Rico (de Vos *et al.* 1956; Corbet 1980), but the species is not established there now and is limited to this one historical record (Wilson and Reeder 1993).

■ DAMAGE
No information.

Family: Tenrecidae
Tenrecs, otter-shrews

TENREC
Tail-less tenrec, common tenrec
Tenrec ecaudatus (Schreber)

■ DESCRIPTION
HB 265–400 mm; T 10–16 mm; WT 634–2.4 kg.
Mostly greyish brown or reddish brown, some animals darker on back and rump; pelage consists of

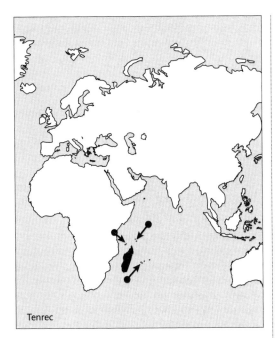

Tenrec

both hair and spines; mane on upper back erectile; head cylindrical; snout pointed; body stout and elongated; forelimbs shorter than hindlimbs. Female has 12 pairs mammae.

■ DISTRIBUTION
Madagascar and the Comoro Islands.

■ HABITS AND BEHAVIOUR
Habits: terrestrial, crepuscular and nocturnal; omnivore; nests in hollow logs, under rocks or in burrows (1–2 m); hibernates in burrow in winter (torpor or seasonal hypothermia) for several months (February–September). **Gregariousness:** family groups to 30, but adults usually forage and hibernate alone. **Movements:** sedentary. **Habitat:** rainforest, forest clearings, brushland, semi-arid scrub, high plateaus and arid areas, but generally in sandy soil. **Foods:** insects (including grasshoppers), worms, snails, arthropods, roots, fallen fruits, and also lizards and eggs of birds. **Breeding:** breeds all year, peak in December; mates October–November (spring); young born November–January (summer wet season); litter size 12–16, 32; gestation 56–64 days; 1 litter/year; stay with female 3–6 weeks. **Longevity:** 6 years 4 months (captivity). **Status:** common and relatively abundant.

■ HISTORY OF INTRODUCTIONS
INDIAN OCEAN ISLANDS
Tenrecs have been introduced successfully to Mauritius, Réunion and the Seychelles. There is some possibility that they were also introduced to the Comoro Islands.

Comoro Islands
Now occurring on the island of Mayotte where they may have been introduced (Grzimek 1972; Walker 1991), although some authorities suggest that tenrecs are native to the Comoros (Wilson and Reeder 1993).

Mauritius
Successfully established from Madagascar, tenrecs were introduced to control insects and also a shrew mouse and a small hare on Mauritius (Encycl. Brit. 1970; Grzimek 1972; Lever 1985).

Réunion (France)
The tenrec was introduced to Réunion from Madagascar probably before 1882 (Encycl. Brit. 1970; Grzimek 1972; Racey and Nicholl 1984).

Seychelles
A single tenrec was captured on the island of Mahé in 1892, at which time it was said to be abundant there (Walker 1967). They are thought to have been introduced about 10 years before this date in 1882. They were introduced as a source of food to a number of Indian Ocean islands, reaching the Seychelles via Réunion about 1882, and now occupy a range of habitats from semi-arid scrubs to the rainforest (Racey and Nicoll 1984; Nicoll 1985).

■ DAMAGE
With the help of the introduced frog (*Rana mascariensis*), tenrecs are reported to have been responsible for the extinction of three endemic frogs – *Nesomantis thomasetti*, *Sooglossus sechellensis* and *S. gardinieri* on the Seychelles (Lever 1985).

Family: Soricidae
Shrews

SPECIES UNKNOWN
■ HISTORY OF INTRODUCTION
THAILAND
A few tree shrews have escaped from captivity and become established in gardens in and around Bangkok, Thailand (Lekagul and McNeely 1988).

DWARF SHREW
Savi
Sorex etruscus (Savi)
=*Crocidura etrusca, Suncus etruscus*

■ HISTORY OF INTRODUCTION
THAILAND
The dwarf shrew occurs in Thailand, but may have been introduced there by humans (Lekagul and McNeely 1988).

SHREW
Crocidura caerulea Kerr

■ HISTORY OF INTRODUCTION
INDONESIA
This shrew may have been introduced to the island of Buru (Flannery 1995). It was noted there in 1929 and two specimens were collected in 1913 (Dammerman 1929; Flannery 1995).

SHREW
Crocidura maxi Sody

■ HISTORY OF INTRODUCTION
MALUKU
This shrew was probably prehistorically introduced to Ambon, the Aru Islands and Kai Islands in Maluku (Flannery 1995). It occurs in Java and on some Lesser Sunda Islands, and Amboina (Maluku).

SUNDA SHREW
Crocidura monticola Peters

■ HISTORY OF INTRODUCTION
INDONESIA
The Sunda shrew of Borneo, Java and peninsular Malaysia may have been prehistorically introduced to Obi and Ambon (Flannery 1995).

HOUSE SHREW
Large musk shrew, musk shrew, money shrew, brown musk shrew

Suncus murinus (Linnaeus)
=*Sorex murinus, Suncus caeruleus, Crocidura murina*

■ DESCRIPTION
HB 50–150 mm; T 46–100 mm; WT males 30–105.6 g, females 20–67.7 g.
Fur generally black or dark brown to pale grey in colour, under parts lighter; snout elongated; whiskers long; ears prominent; tail thick at base and thin at tip, with few scattered hairs; sweat glands on throat and

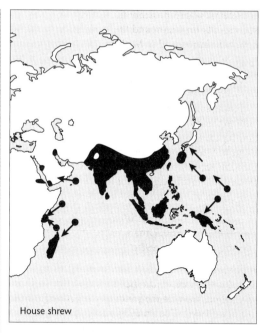
House shrew

behind ears give musky odour; ears round and human-like. Female has three pairs mammae, is smaller and weighs less than male.

■ DISTRIBUTION
Africa–Asia. From Africa, Egypt, Ethiopia, Madagascar, and Asia Minor east to China, Japan, Philippines, and south to Indonesia and Malaysia.

■ HABITS AND BEHAVIOUR
Habits: mainly nocturnal, occasionally diurnal, nests in burrows, musky odour is offensive and lasting; sometimes travels by caravanning when young. **Gregariousness:** solitary and intolerant of each other. **Movements:** sedentary. **Habitat:** human habitation including houses, warehouses, drains and gardens; also open grassy areas, swamps, pond margins, crops (ricefields), grassland and desert areas. **Foods:** mainly insects; cockroaches, crickets and other insects, land molluscs and other animals, humans' food scraps, refuse and stored products such as nuts, grains, vegetables, bulbils, rhizomes and seeds. **Breeding:** capable of breeding throughout year (Malaysia, Thailand, Guam), but elsewhere may be seasonal (India); gestation 29.6–30.3 days; litter size 1–6, 8; young born naked, blind; lactation 17–20 days; sexual maturity 36 days. **Longevity:** 1.5–2.5 years (captivity). **Status:** common.

■ HISTORY OF INTRODUCTIONS
Introduced successfully in East Africa, Arabia, Malagasy, Guam, New Guinea and Japan. In Australia has failed to become permanently established. Occurs on most of the Lesser Sunda Islands and Moluccas,

Indonesia, where possibly introduced. Introduced successfully on Guam and the Ryukyu islands. May have been introduced to Madagascar.

ASIA
The house shrew has probably been spread by humans to many islands and isolated seaports, including some in the Palaearctic region, the Persian Gulf to the Red Sea (e.g. Bahrain, Bastra, Aden, Suez) and Japan, and some Japanese islands (Burton and Burton 1969; Corbet 1978, 1980). It was almost certainly introduced into south-east Asia by humans (Harrison 1950; Lekagul and McNeely 1988).

Arabia
House shrews have been introduced widely in this area to Iraq, Bahrain, Oman, Yemen and Saudi Arabia (Wilson and Reeder 1993)

Indonesia
The house shrew has been spread by humans throughout most of the south-east Asian islands (de Vos *et al.* 1956). They may have been introduced to some of the Moluccas in Indonesia and occur on all of the Lesser Sunda Islands eastwards to Timor (Lever 1985), where they may also have been taken by humans. The species is also known from scattered localities in Sabah and Kalimantan (Payne *et al.* 1985), and may have been introduced here also.

Widely introduced prehistorically and later on the islands of ?Ambon, Aru Islands, Batjan, Buru, Halmahera, Kai Islands, Mangole, Seram and Ternate (Flannery 1995).

Japan
House shrews have reached Japan (Burton and Burton 1969; Corbet 1980). They are established in the Ryukyu Islands, on the island of Fukue (Goto I.) and in two localities on Kyushu (Nagasaki and Kagoshima).

Pakistan
House shrews are present in the Lyallpur region, where they inhabit houses and their surrounds. They are common in the cities, but only occur occasionally in the villages. They were probably introduced to this region (Taber *et al.* 1967).

Thailand
In Thailand a feral population of house shrews occurs in the forest away from human habitation. However, the species is thinly spread throughout the country, mostly in urban areas, and is rare away from these habitats (Lekagul and McNeely 1988).

INDIAN OCEAN ISLANDS
Comoro Islands, Mauritius, Réunion
House shrews are reported to have been introduced

to these islands (Wilson and Reeder 1993), but there appear to be no other records.

Madagascar
The house shrew may have been introduced via trading vessels to Madagascar (Burton 1962; Burton and Burton 1969; Grzimek 1972), where they are now well established.

Maldive Islands
The house shrew has been introduced to the Maldive Islands (Wilson and Reeder 1993).

PACIFIC OCEAN ISLANDS
The house shrew is widespread in the Pacific area as a result of being carried about by the indigenous peoples, much as the house mouse (Carter *et al.* 1945).

Palau Islands (Belau)
House shrews have probably been introduced to the Palau Islands (Flannery 1995).

Guam
The house shrew was recorded on Guam in 1953 (Peterson 1956; Burton and Burton 1969), but may have been present there before World War 2 (Barbehenn 1962, in Flannery 1995), probably arriving by ship in crates or bales from the Philippines where they are relatively common. It has extended its range rapidly on Guam and by 1955 was common from Agat to the Andersea Air Force Base along the western and northern sides of the island. Inland they were common in a number of villages.

AUSTRALASIA
Australia
The house shrew has reached Australia occasionally on ships, but has failed to become permanently established there (Burton and Burton 1969).

New Guinea
House shrews were introduced to New Guinea (Anderson and Jones 1967; Burton and Burton 1969) where they may have arrived accidentally from East Africa or Madagascar (Walker 1967).

AFRICA
Arabia
House shrews are believed to have been introduced to Iraq, Bahrain, Oman, Yemen and Saudi Arabia (Wilson and Reeder 1993).

Egypt
House shrews may have been temporarily introduced to Suez, Egypt, as a new subspecies, *S. m. sacer,* was described from there before 1957 (Corbet 1980).

Pemba and Zanzibar (Tanzania)
House shrews have been introduced and established

on Pemba and Zanzibar islands off East Africa (de Vos *et al.* 1956; Burton 1962; Lever 1985).

South Africa
House shrews have been established on Dyer Island since 1912 or earlier (Lever 1985).

■ **DAMAGE**
The house shrew causes damage to stored products (e.g. fruits) and is a nuisance around human habitation because of its odour (Walker 1967). It was thought that it would cause substantial damage to stored products on Guam, but would also keep down arthropod pests, such as cockroaches and other household pests (Peterson 1956), but it is not clear that this has happened. The species is reported to be sometimes beneficial in Thailand, keeping down insect pests and being intolerant of rats, helping to keep these animals under control (Lekagul and McNeely 1988). They are also believed to kill chickens occasionally.

In parts of Pakistan they sometimes damage lawns by digging out and eating the bulbils of grass (Advani and Roma 1981).

It is probable that gastroenteric sickness in humans in Indonesia and West Malaya, reported by villagers to come from contact with the house shrew, is salmonellosis (Kitchener *et al.* 1996).

GREATER WHITE-TOOTHED SHREW
House shrew
***Crocidura russula* (Hermann)**

LESSER WHITE-TOOTHED SHREW
Garden shrew
***Crocidura suaveolens* (Pallas)**

■ **DESCRIPTION**
Greater: HB 60–90 mm; T 30–60 mm; WT 5.9–11.3 g.
Upper parts greyish or reddish brown; under parts dull yellowish grey; ears prominent; tail with long white hairs; sexes similar; female has six inguinal mammae. Only separable from the lesser white-tailed shrew by dental characteristics.

Lesser: HB 49–78 mm; T 27–50 mm; WT 3–13.3 g.
Fur greyish or reddish brown above and slightly paler ventrally; ears short-haired and prominent; tail short-haired interspersed with fine long white hairs.

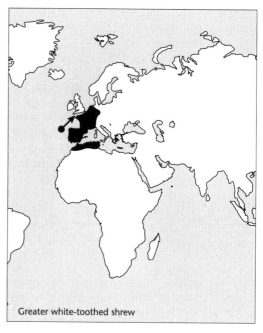
Greater white-toothed shrew

■ **DISTRIBUTION**
Greater: Central and southern Europe, and North Africa. Occurs on some Channel Islands only.

Lesser: South-west Europe from Spain and France to North Africa, Korea, China and Taiwan.

■ **HABITS AND BEHAVIOUR**
Greater: Habits: solitary except at breeding season; home range 75–395 m^2 with some overlap; diurnal and nocturnal; burrow; nest of dried grass.

Lesser white-toothed shrew

Gregariousness: density 77 per 100 ha recorded. **Movements:** sedentary. **Habitat:** woodland, hedgerows, grassland, cultivated areas, urban areas. **Foods:** invertebrates; mainly woodlice, centipedes, moth larvae, gastropods and spiders, but also earthworms, beetles, aphids and millipedes. **Breeding:** breeds February–October; gestation 28–33 days; litter size 2–11, average 4; sexual maturity at 58–71 days; female polyoestrous with post-partum oestrus; probably several litters/year; eyes open 8–9 days; weaned 20–28 days. **Longevity:** 2–2.5 years and up to 4 years captive, few more than 1.5 years in the wild. **Status:** common.

Lesser: Habits: solitary; overlapping home ranges 27–80 m; active day and night; burrow; nest of grass and twigs under logs and rocks. **Gregariousness:** density 1 per 30 m^2 recorded on Scilly. **Movements:** sedentary. **Habitat:** From beaches to floors of pine plantations, heathland, woodland, sand dunes and marquis scrub. **Foods:** crustaceans, amphipods, millipedes, larval flies, beetles, spiders and mites. **Breeding:** breeds March–September; gestation 24–32 days; litter size 1–6; 3–4 litters/year; sexually mature 45–50 days; young born hairless; eyes open 10–13 days; fully haired 16 days; weaned at 22 days. **Longevity:** 12–18 months in wild, 4 years in captivity. **Status:** common.

▪ HISTORY OF INTRODUCTIONS

Greater: Likely to have been introduced to the Channel Islands.

Lesser: Occurs in Channel Islands on Jersey and Sark. In Scilly Islands found on all but some of the smaller islands. Probably introduced on all.

EUROPE

Channel Islands (United Kingdom)
Greater: The greater white-tailed shrew is most likely to have been introduced to the Channel Islands. Its irregular distribution in the islands is difficult to explain and points to human introduction (Corbet and Harris 1991), possibly by Neolithic peoples (Lever 1985).

Isles of Scilly
Lesser: Likely to have been introduced to some British islands by man. Iron Age or earlier traders from France or northern Spain probably introduced them to Isles of Scilly when they came to the Cornish coast in search of tin (Corbet and Harris 1991).

Thought to have been introduced by man to the Channel Islands and to Scilly Islands. *C. russula* on Channel Islands and *C. suaveolens* on Scilly (Lever 1977). {**Note:** No mention in Corbet (1978) of any introduction.}

▪ DAMAGE

The white-tailed shrews are beneficial to humans, as they prey on potential pest invertebrates (Corbet and Harris 1991).

MASKED SHREW
Long-tailed shrew, cinereus shrew, common shrew
Sorex cinereus Kerr

▪ DESCRIPTION
TL 70–125 mm; T 30–50 mm; WT 2.5–7.9 g.
Upper parts pale brown to brown; under parts pale grey or whitish (in winter glossier and greyer); fur is short and velvety; snout flexible; eyes minute; tail long, bi-coloured (brown above, pale below, tip black) and covered with short hairs; feet delicate. Female has six inguinal mammae.

▪ DISTRIBUTION
North America and Asia. Northern North America from the southern Appalachian Mountains and Rocky Mountains (New Mexico and North Carolina), north across Canada and Alaska. Occurs on Prince Edward Island and Nunivak Island.

▪ HABITS AND BEHAVIOUR
Habits: diurnal and nocturnal; territorial; uses runways through vegetation; builds grassy nests; live in burrows made by self or others; density 0.4–5.26/ha. **Gregariousness:** mainly solitary; pairs at mating. **Movements:** sedentary; home range about

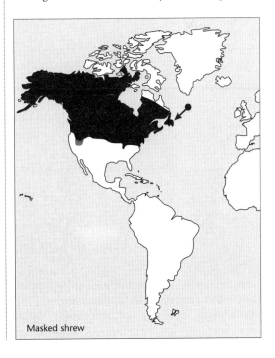

Masked shrew

0.05 ha. **Habitat:** dense forest, woodland, arctic tundra and from seashores to alpine meadows. **Foods:** insects and their larvae, millipedes, centipedes, earthworms, snails, isopods, salamanders, young mice, molluscs, sowbugs, vegetable matter and occasionally nesting birds. **Breeding:** breeds April–October (spring–summer); gestation 14–28 days; litter size 2–10; several litters (to 5) per year, but usually only 1–2; weaned at 1 month, but remain with female another month; females mature 4–5 months. **Longevity:** 12–18 months. **Status:** common.

■ HISTORY OF INTRODUCTIONS
NORTH AMERICA
Successfully introduced and established in Newfoundland.

Newfoundland, Canada
The introduction of an insectivorous mammal to Newfoundland was first suggested in 1942 by an entomologist, R. E. Balch, who was engaged in examining forest insect problems on the island. This suggestion was later reinforced by other workers on the island and further stimulated by severe outbreaks of the larch sawfly (*Pristiphora erichsonii*) in 1954 and 1955. The prospect of achieving control of such outbreaks by the introduction of such a mammal was examined by officials of the federal and provincial governments. The results were attempts to introduce the masked shrew in 1956 and 1957. This species was selected because of its wide habitat range, almost entirely insectivorous diet, and its existence in similar environments to Newfoundland on the nearby mainland (Warren 1970).

Preliminary attempts were made to collect and transport masked shrews from both New Brunswick and Manitoba in 1956 and 1957. Several hundred were captured in Manitoba in 1957, but they all died before they could be released. However, in 1958, 62 shrews (12 females and 50 males) were captured in the Green River watershed, New Brunswick, and liberated in western Newfoundland near St. Georges (Buckner 1966).

At intervals during the first winter supplementary foods such as hamburger meat and salt codfish were supplied (Buckner 1966; Warren 1970), and at least half the original release, 6 females and 5 males, were able to survive (MacLeod 1960). These bred in the spring of 1959 and a trap census showed 130 shrews (11 original and 119 progeny). Estimates were made that the population had increased to 0.9/ha from June to September of that year. By 1960 this figure had risen to 2.5/ha and by 1961 to 3.3/ha (Warren 1970).

A sudden decline in sawfly population levels in 1961 prompted the transfer of masked shrews to central Newfoundland. Nine males and 11 females trapped at St. Georges were taken to Hall's Bay in July and 14 males and 11 females to Exploits Dam in September of that year and released (Buckner 1966; Warren 1970).

Initially the masked shrew occupied an area of some 14.5 ha, but after five years they were found within a radius of 48 km from the release points and the three introduced populations merged in 1966. By 1970 they were distributed over 80 per cent of the island (111 361 km^2) and had dispersed at a rate of between 11 and 19 km/year (Warren 1970).

■ DAMAGE
No serious problems seem to have been caused by the masked shrew introduction into Newfoundland. The only problem of economic significance appears to be that the shrew will feed on snowshoe hares (*Lepus americanus*), themselves introduced, and so earn the wrath of local 'rabbiters' who trap and sell the skins. However, all the evidence available at present suggests that the masked shrew has assisted in the control of the larch sawfly, which has not errupted seriously since the shrew's introduction (Warren 1970).

Family: Talpidae
Moles, shrews and desmans

COMMON MOLE
European common mole, northern mole, mole
Talpa europaea Linnaeus

■ DESCRIPTION
HB 90–165 mm; T 25–40 mm; WT 72–128 g.
Coat varies from velvet black to whitish, under parts sometimes have yellowish markings; fur short and dense; body cylindrical; eyes minute and hidden by fur; muzzle long, pointed, fleshy pink; lacks external ear; limbs short; feet naked and flesh coloured; forefeet turned outwards, with strong claws; hind limbs smaller; feet with five digits; tail stumpy and narrow at base. Female generally smaller than male.

■ DISTRIBUTION
Eurasia. Great Britain, Southern Sweden and Finland, south to northern and eastern Spain, northern Italy, Yugoslavia and east to Siberia and Mongolia. Absent from Ireland, Isle of Man, Outer and Inner Hebrides,

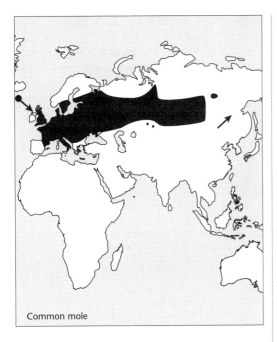

Common mole

Orkney and Shetlands. Present on Isle of Wight, Anglesey, Skye, Jersey, Alderney, but not Guernsey.

Note: Range shown for this species includes that of T. altaica, *which is often regarded as a separate species.*

■ HABITS AND BEHAVIOUR

Habits: spends almost whole life in burrow (184–627 m^2 in extent) with nest; alternating periods of activity and rest during 24 hours. **Gregariousness:** solitary except at mating time. **Movements:** sedentary, but moves to avoid flooding; home range about 1314–1945 m^2. **Habitat:** open ground in woodlands, meadows and cultivation including pastures, fields, gardens and parks. **Foods:** beetles, caterpillars, flies and other insect larvae, wireworms, earthworms, millipedes, centipedes, slugs, molluscs, and occasionally vegetable matter, small mammals, snakes and lizards, and also small birds. **Breeding:** breeds February–June; gestation about 28 days; females polyandrous; promiscuous; litter size 2–4, 7; 1 or 2 litters/year; young born naked; eyes open 22 days; weaned at 3 weeks; leave nest 33–34 days, but remain in vicinity for about 10 weeks; sexually mature in spring following birth. **Longevity:** 1–3, and occasionally to 6 years (in wild). **Status:** abundant and common.

■ HISTORY OF INTRODUCTIONS

Eurasia

Introduced successfully to the Isle of Mull (Inner Hebrides), Britain, and introductions possibly extended the range of the species in Scotland. Probably introduced successfully in two areas in Russia, but failed in others. Introductions in Germany, on Ulva (off Mull) and in West Siberia all failed.

United Kingdom

In the past 200 years or so the common mole has extended its range north and westwards in Scotland. Although there is little evidence, it is thought that the extension may have been assisted by introductions by humans (Fitter 1959).

Common moles are absent from many islands off the coast of Britain except Bute, where they were found in 1777, and on Mull (off west coast of Scotland), where they were accidentally introduced in about 1808. It is thought that moles were taken to Mull in boatloads of soil from Morvern. At the time of their introduction soil was often used as ballast on boats (Fitter 1959; Southern 1964). They also occurred on Ulva (off Mull), but have now disappeared from this island.

Germany

Attempted introductions of common moles in the mid-1800s were apparently unsuccessful in Germany. Efforts to introduce them on Griefswalder Oie (island in Baltic Sea) also failed. In 1862 and later, attempts were made on a small scale to re-introduce them to some areas to destroy cockchafer larvae after floods had wiped out the native animals. However, the success of such efforts was difficult to estimate as the areas concerned were probably quickly colonised from the surrounding countryside (Glaser 1868; Niethammer 1963).

Russian Federation and adjacent independent republics

Common moles were introduced into the Ukraine in 1952 and/or 1953. In 1953, 99 moles were released in the Volnovakhskogo region of the Donetskoi oblast. Here, they became established locally and appear to have persisted, but not to have spread much (Yanushevich 1966; Kirisa 1973). In around 1957 in the Russian Federation some 4760 moles (*Talpa* sp.) and desmans (*Desmana* sp.) were resettled (Lavrov 1957; Niethammer 1963), but there is little information about them.

Attempts to re-introduce common moles in West Siberia appear to have been unsuccessful (Yanushevich 1966), but introductions at Novosibirsk were successful. In 1940 and 1941, 263 moles were released in the Ordnskogo and Suzunskogod regions of the Novosibirsk oblast (Kirisa 1973). Releases have been made in an already inhabited range of the mole in the Barabinski steppes, where 62 were released in 1942 (Niethammer 1963).

■ DAMAGE

It is generally considered that common moles cause both harm and good to agricultural interests (Lancum 1951). They are generally regarded as pests by farmers, horticulturists and green-keepers, but usually do little real damage (Corbet and Harris 1991).

In Europe they are reported to sometimes cause considerable damage to plants by chewing the roots, pulling up seedlings, ruining lawns and causing soil erosion (Ball 1960). They are also said to become a nuisance to farmers by burrowing in pasture lands and are often destroyed for this reason (Southern 1964).

RUSSIAN DESMAN
Desman, muskrat
Desmana moschata (Linnaeus)

■ DESCRIPTION

HB 110–220 mm; T 126–215 mm; WT 50–80 g.

Generally reddish brown, shading to greyish with a silver sheen on under parts; coat composed of short dense underfur with coarse guard hairs; snout long, flexible, grooved above and below; tail flattened, large at base, and encircled by rings of scales with a few hairs between each ring; hind feet webbed; forefeet partially webbed.

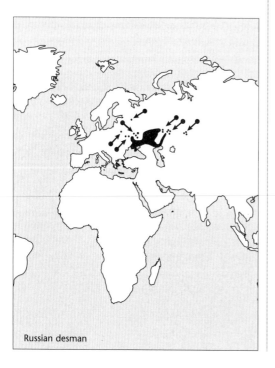

Russian desman

■ DISTRIBUTION

Eurasia. Russian Federation in the river basins of the Ural, Kama, Volga and Don, and small tributaries flowing into the Sea of Azov.

■ HABITS AND BEHAVIOUR

Habits: aquatic, nocturnal, but occasionally diurnal in spring; aquatic; burrows in banks of waterways and lakes to 6 m. **Gregariousness:** small groups of up to 8. **Movements:** somewhat nomadic. **Habitat:** freshwater lakes, ponds, streams, slow-flowing rivers and swampy pools. **Foods:** insects and their larvae, crustaceans, molluscs, leeches, worms, amphibians, frogs, and spawn of fish and frogs. **Breeding:** breeds January–May?; gestation 40–50 days; litter size 1–5. **Longevity:** no information. **Status:** numbers and range reduced; reported as vulnerable.

■ HISTORY OF INTRODUCTIONS

Russian desmans have been introduced successfully into the Tachan and Tartas rivers (Ob basin) and the Dnepr River, and also several other areas in Russia.

EURASIA

Russian Federation and adjacent independent republics

Desmans have been transferred, re-introduced and introduced a number of times in the Russian Federation and in surrounding independent republics, in some instances with success (Yanushevich 1966; Burton and Burton 1969). At least 10 000 desmans have been released but the efforts have generally yielded poor results. The animal is now on the endangered list in the Russian Federation, but successful acclimatisations in Siberia on the Ob River appear to be a successful step towards protection of the species (Sofonov 1981). They have also been successful on the Dnepr River (Corbet 1978, 1980).

At least 9788 desmans have been released in 36 oblasts and republics within and adjacent to the Russian Federation, mainly to increase the range and survival of the species, in areas already inhabited by them, but also in some areas outside their natural range. Some introductions were also made to increase hunting for skins that were once widely used to make coats. Around 1900 some 20 000 skins were processed annually but by 1923–24 this had decreased to some 10 000–12 000 (Grzimek 1972; Kirisa 1973).

From 1929 to 1940 desmans were released in 18 oblasts and republics (about 3411 animals) in the European part of Russia. From 1948 to 1970 about 6377 desmans were released. The first introductions were made in 1929 in the Smolensk oblast, in the Dnepr basin (Ukraine) and in the northern Don basin (Kirisa 1973).

In the Ukraine the desman was possibly established from introductions between 1929 and 1940 when 366 were released mainly in the basins of the Dnepr and northern Don rivers. Introductions occurred between these dates in the Dnepropetrovsk, Zaporojsk, Kiev, Donestsk, Poltavsk and Kharkov areas (Kirisa 1973; Yanushevich 1966). Some were established on the Seim River after their release in the Glushkovo region of the Kursk in 1961 (Serdyu 1978).

From 1955 to 1959, 216 desmans were liberated in Belorussia (Samusenko 1962) and by 1962 the total released had risen to at least 580 animals (Kirisa 1973). Most appear to have been released in the Minsk and Mogilev areas in central and central eastern Belarus, and they appear to have become established in at least two localities.

Areas in which desmans have been released in European Russia include Novgorod (38 in 1940), Litov (eight in 1948, 62 in 1957), Smolensk (675 between 1929 and 1964), Bryahnsk (159 before 1940), Kursk (609 between 1956 and 1961), Voronejsk (241 between 1954 and 1957), Volgograd (26 in 1940), Tambov (30 in 1966), Moskov (67 from 1933 to 1937), Kalujsk (206 in 1959–60), Ryahzan (536 in 1935–38, 139 in 1963–64), Vladimir (619 in 1959–70), Yahroslav (367 between 1959 and 1970), Gorkov (493 from 1940 to 1965), Chuvash (108 in 1959–60), Mordov (193 in 1937–38), Ulyahnov (26 in 1964), Penzen (147 from 1957 to 1964), Saratov (565 from 1951 to 1968), Kuibishev (609 between 1937 and 1970), Tatar (645 in 1931–36), Marii (170 in 1963), Kirov (185 from 1959 to 1965), Orenburg (431 in 1934 and from 1957 to 1966), Baskir (583 in 1939–40) and in Chelyahbinsk (235 in 1953 and from 1961 to 1964) (Kirisa 1973). Introduction in Lithuania in 1957 failed (Yanushevich 1966).

Western Siberian introductions occurred in the Tomsk (238 in 1958, 102 in 1964) and Novosibirsk (114 in 1968) areas (Kirisa 1973).

Introductions in the Volga–Kamar region outside the desmans' normal range in 1958 appear to have been successful and those in West Siberia in 1958 appear to have become fairly widespread (Pavlinin and Shvarts 1961; Yanushevich 1966). Those established in the Dnepr and upper Ob river basins were successful and the species has now been restored in at least some areas (Walker 1991).

■ DAMAGE

The desman is a valuable fur-bearing animal in the Russian Federation, but introductions of the muskrat (*Ondatra zibethicus*) are thought to have caused its decline in numbers in many areas (Skoptsov 1967; Berdov 1987).

In Moldavia it is indicated that the introduced muskrat is a successful competitor of the desman for food and burrows and carries a number of diseases, while the desman has only been implicated as a secondary carrier of leptospirosis (Borodin 1965). Introduced muskrats are said to have displaced the desman from its age-old habitats in the Tambov region (Skoptsov 1964). Following the release of muskrats at Lake Christie in the Tambov region in 1951, they increased rapidly, ate out much of their preferred foods (i.e. water vegetation), then destroyed much of the shellfish and large slow-moving insects that are food for the native desmans. Thus desman populations have been considerably reduced in the Tambov region and probably for the same reason in the Ryazan, Kirov, Vladimir and Markov regions (Scoptsov 1964, 1967).

In the Vyatka river flood plain water reservoirs of the Kirov region, introduced muskrats appeared in about 1960–65 and desmans began to decrease from para-typhoid infection almost immediately. Desmans were found to develop a more severe form of the disease than muskrats and several attempts to restore their numbers have failed because of outbreaks of the disease (Berdov 1987).

Family: not known

BATS
Species not known

■ HISTORY OF INTRODUCTIONS
PACIFIC OCEAN ISLANDS
Hawaiian Islands
Evidently bats of at least two species were introduced to the Hawaiian Islands, but their identity is not known (Kramer 1971). Several trials were made to introduce Japanese bats in 1897, but these were unsuccessful in becoming established (Koebele 1897). Also at this time 600 bats from California were released on Oahu without success.

There was further pressure to introduce bats to the Hawaiian Islands for insect control in 1904 (Blackman 1904; Lowrie 1904), but past failures were pointed out (Perkins 1904) and the matter was dropped. However, further pressure was applied in 1914 and again in 1919 for the same reason, but the stand against their introduction was reiterated and none were introduced (Kramer 1971).

A fruit bat (family Pteropodidae) was reported (in press) to have been found asleep in the rigging of a ship from the Philippines in 1946, but this was captured and destroyed.

Family: Rhinolophidae
Horseshoe bats

GREATER HORSESHOE BAT
Rhinolophus ferrumequinum (Schreber)

■ DESCRIPTION
HB 56–125 mm; T 33–43 mm; WS 330–400 mm; WT 13–34 g.
Fur thick, woolly, pale buff with darker tips, becoming darker and reddish with age; eyes small; nostrils horseshoe-shaped; ears large and triangular; under surface whitish. Female slightly larger than male.

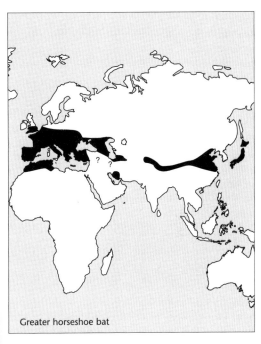
Greater horseshoe bat

■ DISTRIBUTION
Eurasia. Found in most of Europe (except Scandinavia) to south-central Asia and Japan, and south to Morocco. Extinct in the Netherlands and Poland.

■ HABITS AND BEHAVIOUR
Habits: lives in caves, crevices and mines in winter, and roofs, barns, church towers and caves in summer; hibernates late September–May depending on weather; crepuscular. **Gregariousness:** immatures highly social and form colonies with adult males; females and young in large nursery colonies; in winter congregate in large clusters in hibernacula, females more solitary. **Movements:** migrate up to 180 km to winter roosts; return to breeding site annually. **Habitat:** south-facing slopes with mixed deciduous woodland and pastures with caves, mines, cellars, or tunnels as hibernacula. **Foods:** insects; beetles, bugs, moths, tipulid flies, dung beetles. **Breeding:** 1 young born June–July; gestation 70 days; young born blind, with sparse covering of hair; eyes open 9–10 days; flies from roost regularly at 3 weeks; catches insects at 5 weeks; sexually mature at 2–4 years. **Longevity:** females 24 years, males 30 years in wild. **Status:** declined in many areas of Europe including 98 per

cent this century in Britain; uncommon and endangered.

HISTORY OF INTRODUCTIONS
EUROPE
United Kingdom

A pair of greater horseshoe bats was released in county Monaghan in 1930. In 1933 G. Seccombe Hett released nine in passages under Mappin Terraces in Regent's Park. This location is on the edge of the species' northern range and they may have become established (Fitter 1959).

DAMAGE

None known.

MONKEYS – SPECIES NOT KNOWN
Monkeys?

Monkey species not known, uncertain or introduction details not known or not confirmed.

■ HISTORY OF INTRODUCTIONS
COCOS ISLAND (COSTA RICA)
Monkeys (species not known) were introduced unsuccessfully to Cocos Island, Costa Rica.

GALÁPAGOS ISLANDS (CHILE)
Three monkeys were released on the island of Floreana, Galápagos Islands, in the 1930s, but soon disappeared (Lever 1985).

INDONESIA
Macaques (species unknown) are said to have been introduced to Batjan Island and to Timor (Lever 1985).

PAPUA NEW GUINEA
Monkeys are believed to have been introduced by the Japanese during World War 2 to New Britain as a food supply. They have become established on Cape Gloucester in the Limestone plateau country, near Aipati, Aimaya and Siac villages. The monkey is called 'nanukrawa' (Herrington 1977) and may possibly be *Macaca fuscata*. (There appears to be no mention of its presence by Flannery 1995 or other authorities.)

Family: Lorisidae
Loris

SLOW LORIS
Nycticebus coucang (Boddaert)

■ HISTORY OF INTRODUCTION
PHILIPPINES
The slow loris is found in the Tawitawi island group (closer to Borneo than to Philippines) and was possibly introduced to Mindanao, Philippines (Groves 1971).

Family: Lemuridae
Lemurs

CROWNED LEMUR
Lemur coronatus Gray
Until recently considered a subspecies of L. mongoz, *but now regarded as distinct.*

■ DESCRIPTION
HB 400–450 mm; T 400–450 mm; WT 1.8–2.2 kg.
Upper parts brownish to reddish grey; under parts grey-white to pale red; tail red-brown; eye-ring black; cheeks reddish; male has triangular crown of black fur between ears. Female is grey with light brown crown.

■ DISTRIBUTION
Madagascar. Northern and north-eastern Madagascar.

■ HABITS AND BEHAVIOUR
Habits: mainly diurnal, but not uncommonly nocturnal; mainly arboreal. **Gregariousness:** solitary; groups to 10 with several adults of both sexes; density 50–200/km^2. **Movements:** sedentary, but moves locally. **Habitat:** dry forest, dry wooded areas in savannah. **Foods:** fruits and some leaves. **Breeding:** births mid-September to October, but possibly other times also; gestation 125 days; young 1–2; sexual maturity *c.* 20 months. **Longevity:** no information. **Status:** numbers declining; range reduced by agriculture, burning and logging; also hunted.

■ HISTORY OF INTRODUCTIONS
INDIAN OCEAN ISLANDS
Madagascar
A French teacher, B. Le Normand, released a number of crowned lemurs onto the small, uninhabited island of Nosy Hara (Wilson *et al.* 1985; J. Wilson, in litt. in Harcourt and Thornback 1990) where they have possibly become established.

■ DAMAGE
None known.

BROWN LEMUR
Lemur fulvus Geoffroy

■ DESCRIPTION
HB 340 mm; T 440 mm; WT 1.9–4 kg.
Considerable variation within populations; males have 2 distinct colour phases, most are dark grey or grey brown dorsally; *fulvus*: upper parts and tail greyish brown; cheeks and beard white; muzzle and forehead black; under parts creamy tan; females lighter than males; *albifrons*: face black; forehead, crown, ears, cheeks and throat white or cream; tail dark; under parts pale; some lack white colour on head and are black or grey instead; female upper parts grey brown; some have head dark grey while others pale grey; *collaris*: neck, face, ears and top of head black (grey in female); cheeks pale orange, bushy in male; upper parts dark brown or grey brown with darker stripe down spine; under parts paler; *mayottensis*: variable, but similar to nominate and may have been derived from it; *rufus*: variable; upper parts grey; under parts grey brown; head cap bushy, rusty orange; muzzle black; above eyes a pale grey patch; ears grey; cheeks bushy; female reddish brown; under parts pale golden brown or grey; crown grey with light grey or whitish above eye and on cheeks; ears reddish brown.

■ DISTRIBUTION
Madagascar and Comoro islands. Widespread over Madagascar and present on Mayotte Island in the Comoros.

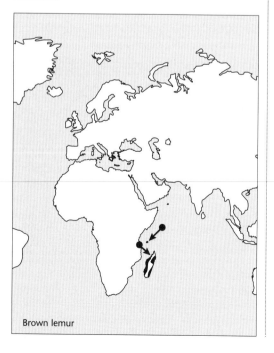

Brown lemur

■ HABITS AND BEHAVIOUR
Habits: mainly diurnal, sometimes nocturnal; mainly arboreal, rarely on ground. **Gregariousness:** groups to 30 (males, sub-adults, females and juveniles); density 40–300/km². **Movements:** daily 125–150 m; monthly 450–1150 m; overlapping group home ranges 7–100/ha or more. **Habitat:** rainforest, forest, secondary growth. **Foods:** fruits, leaves, flowers, sap, buds and bark. **Breeding:** mate April–May, births September–October; gestation 120–135 days; 1 young/year; young independent of mother's back at 11–12 weeks; adult size at 2 years; breeds at 18 months to 2.5 years. **Longevity:** 21 years. **Status:** rare to vulnerable; some subspecies widely trapped for pet trade; probably declining; commonly kept pet in Madagascar.

■ HISTORY OF INTRODUCTIONS
INDIAN OCEAN ISLANDS
Madagascar
The white-fronted lemur, *L. f. albifrons*, was introduced to Nosy Mangabe (now a special reserve) in the 1930s and appears to be thriving there (Constable *et al.* 1985; Harcourt and Thornback 1990).

A few collared lemurs, *L. f. collaris*, have been introduced to Berenty Private Reserve (Jolly *et al.* 1982) and it is suggested that these animals be removed (and taken to Duke Primate Centre) to prevent hybridisation with the introduced *L. f. rufus* there (St. Catherine's Workshop 1986).

Red-fronted lemurs, *L. f. rufus*, were introduced to Berenty Private Reserve in 1974, where the original eight or nine animals imported from Morondava had increased to 62 by 1985 (Jolly *et al.* 1982).

Comoros
The Mayotte lemur, *L. f. mayottensis*, is thought to have been derived from the brown lemur, *L. f. fulvus*, and was probably introduced to the Comoros by man, possibly as long ago as several hundred years (Tattersall 1977; Wolfheim 1983). Its colouration is variable, but similar to the nominate race and there is conjecture that it may not be distinct (St. Catherine's Workshop 1986). It now occurs wherever there is forest on Mayotte, but is rare at altitudes of more than 300 m (Tattersall 1982; Harcourt and Thornback 1990).

■ DAMAGE
None known.

MONGOOSE LEMUR
Lemur mongoz Linnaeus

■ DESCRIPTION

HB 400–450 mm; T 400–450 mm; WT 1.8–2.2 kg.
Coat grey; face pale; cheeks bushy, reddish brown; beard reddish brown; under parts white to pale brown. Female grey brown; cheeks bushy, white; beard white; face dark.

■ DISTRIBUTION

Madagascar and Comoro islands. Western and north-western Madagascar and on Anjouan (Ndzouani) and Moheli (Moili) in the Comoros.

■ HABITS AND BEHAVIOUR

Habits: nocturnal, crepuscular and diurnal (may change seasonally); sleeps in trees in dense foliage, tangled vines or top of tall tree. **Gregariousness:** family groups (adult pair and offspring); occasionally groups 6–8. **Movements:** to food sites; 460–750 m /night; overlapping home ranges *c.* 1.15–100 ha. **Habitat:** dry deciduous forest, humid forest, secondary growth. **Foods:** flowers, nectar, fruits, some leaves and leaf petioles. **Breeding:** births mid-October; gestation 114–128 days; 1 young/year; weaned 5 months; sexual maturity 14–16 months. **Longevity:** captive to 26 years. **Status:** declining; endangered due to habitat destruction.

■ HISTORY OF INTRODUCTIONS

INDIAN OCEAN ISLANDS
Comoro Islands
Mongoose lemurs are found on Anjouan (Ndzouani) and Moheli (Moili) with a few feral individuals on Grande Comoro (Ngazidja Island), which have escaped or been set free there (Tattersall 1977; Thorpe 1989).

The mongoose lemurs on the Comoros were probably (Tattersall and Sussman 1975; Tattersall 1976) or almost certainly taken there from Madagascar, most likely by Mahajanga (Majunga) fishermen some time in the last several hundred years (Petter *et al.* 1977; Harcourt and Thornback 1990) and possibly in the fifteenth or sixteenth century. They are now found on Mohéli and central and eastern Anjouan and Grand Comoro. They are abundant on Mohéli (Tattersall and Sussman 1975; Tattersall 1976).

■ DAMAGE
None known.

RUFFED LEMUR
Varecia variegata (Kerr)

■ DESCRIPTION

HB 510–600 mm; T 560–650 mm; WT 2.4–5.0 kg.
Variable; face black except for short white hair on muzzle below eyes; forehead black; crown black; ears, cheeks, throat tuft white; otherwise white except for ventrum, tail, and lateral aspects of thighs and shoulders, proximal forelimbs and extremities all black; other races have mostly red body with white patch on neck; still others resemble nominate race (above) or are all black with white markings. Female heavier than male.

■ DISTRIBUTION

Madagascar. Eastern and north-eastern Madagascar, but details of distribution are poorly known.

■ HABITS AND BEHAVIOUR

Habits: diurnal and mainly crepuscular, sometimes some nocturnal activity; arboreal, rarely on ground. **Gregariousness:** pairs, territorial groups 2–5; density 20–30/km^2. **Movements:** more than 1 km/day; home range *c.* 197 ha. **Habitat:** humid rainforest. **Foods:** mostly fruit, also nectar, seed, leaves and bark(?). **Breeding:** births October–November; gestation 90–102 days; litter size to 6; young left in nest for 3 weeks; fully mobile at 7 weeks; sexual maturity female 20 months. **Longevity:** 19 years captive. **Status:** endangered due to habitat destruction; declining; hunted for food.

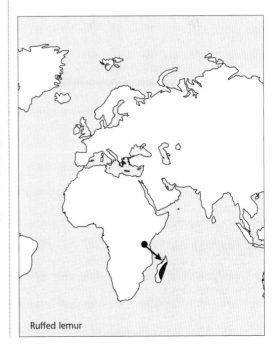

Ruffed lemur

■ HISTORY OF INTRODUCTIONS
INDIAN OCEAN ISLANDS
Madagascar
The black and white ruffed lemur (*V. v. variegata*) is found on the island of Nosy Mangabé where it was introduced in the 1930s (Constable *et al.* 1985). Now as many as 100–150 individuals are on the 520 ha island. In 1984 it was estimated there were between 56 and 84 on the island (Pollack 1984) and in 1983 as many as 175/km^2 (Iwano 1989). They were still present there in 1998 (WCMC 1998).

In October 1997, five black and white ruffed lemurs were returned to Madagascar from the Duke Primate Centre in the United States (Kauffman 1999; Eliot 1999). These have been released in Betampona Reserve where they are doing well with help from supplementary feeding. One was killed by predators and another died from a fall, but another four have since been sent to Madagascar from the Primate Centre (Eliot 1999).

■ DAMAGE
None known.

Family: Daubentoniidae
Aye-ayes

AYE-AYE
Daubentonia madagascariensis (Gmelin)

■ DESCRIPTION
HB 360–440 mm; T 500–600 mm; WT 2–3 kg.
Coat thick, with long guard hairs of dark brown or black over short white hair; head rounded; face flattish; face and throat yellowish white; ears large, naked; nose and spots over eyes, white; hands and feet with opposable thumbs, black; digits elongated; tail bushy, long. Female has two mammae.

■ DISTRIBUTION
Madagascar. Originally found in the coastal areas of north-western and north-eastern Madagascar and on the island of Nosy Bé.

■ HABITS AND BEHAVIOUR
Habits: arboreal, nocturnal, and elusive; nests in hollow or fork of tree during the day. **Gregariousness:** mostly solitary or in pairs, rarely 2–3; home range 4.8–5 ha. **Movements:** sedentary. **Habitat:** rain and deciduous forest, secondary growth, mangroves, bamboo thickets, open bush with low trees; occasionally cultivated areas (e.g. coconut plantations). **Foods:**

Aye-aye

fruit (coconuts, lychees, mangoes and other cultivated fruits), plant galls, plant shoots, bark, small invertebrates, insects and their larvae, grubs, coconut pulp, bamboo pith, birds' eggs. **Breeding:** single young born October–November; gestation not known; inter-birth interval probably 2–3 years; young weaned *c.* 12 months. **Longevity:** 5–23 years (captivity). **Status:** rare or extinct in parts of natural range due to habitat destruction; endangered; declining.

■ HISTORY OF INTRODUCTIONS
Introduced successfully on the island of Nosy Mangabé off the coast of Madagascar and now occurs only on this island.

INDIAN OCEAN ISLANDS
Nosy Mangabé (off north-eastern Madagascar)
In 1966 it was thought that there were less than a dozen aye-ayes left in Madagascar. Nine were subsequently transferred to a special reserve near Maroansetra in the Bay of Antogil on the island of Nosy Mangabé (520 ha) in 1967 (Burton and Burton 1969; Petter and Peyriéras 1970; Grzimek 1972). Previous to the introduction, mango and coconut trees had been planted for them and the reserve became a Special Reserve in 1966. They were still present there in 1975 and 1979 (Wolfheim 1983). Since this time a small population has remained present there (Constable *et al.* 1985; Walker 1992; Mittermeier *et al.* 1992; WCMC 1998).

■ DAMAGE
Ayes-ayes are killed when they raid crops and it has been suggested that local peoples should be compen-

sated for the damage done to their crops; ayes-ayes are also killed because of local superstition (Albgnac 1987; WCMC 1998).

Family: Callithricidae
Marmosets and tamarins

COTTON-TOP TAMARINS
Saguinus oedipus (Linnaeus)

■ HISTORY OF INTRODUCTIONS
Colombia
Cotton-top tamarins have probably been re-introduced to some semi-natural environments in Colombia for the development of conservation education (Savage 1988).

GOLDEN LION MARMOSET
Golden lion tamarin
Leontopithecus rosalia (Linnaeus)

■ DESCRIPTION
HB 190–340 mm; T 260–400 mm; WT 360–710 g.
Fur long and silky, particularly mane on the shoulders; mainly gold colour or glossy golden yellow; tail long and furred.

■ DISTRIBUTION
South America. South-eastern Brazil in the state of Rio de Janeiro from Bahia to São Paulo.

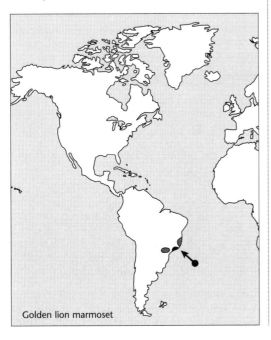
Golden lion marmoset

■ HABITS AND BEHAVIOUR
Habits: arboreal and diurnal; sleeps in holes in trees at night. **Gregariousness:** territorial groups 2–9 (usually 1 adult, and younger animals) or temporary aggregations to 16. **Movements:** sedentary; territories 40 ha. **Habitat:** tropical forest, coastal forest, swamp forest; occasionally secondary forest and cultivated areas. **Foods:** insects, fruits and small invertebrates; lizards, small birds, birds' eggs. **Breeding:** breeds September to March; gestation 125–134 days; young 1–3; born furred, eyes open; weaned at 90 days; sexual maturity males 24 months, females 18 months. **Longevity:** 8–28 years captive. **Status:** reduced in numbers and range and now endangered; formerly a common household pet.

■ HISTORY OF INTRODUCTIONS
Golden lion marmosets have been re-introduced successfully in southeast Brazil.

South America
Brazil
Populations of golden lion marmosets began to decline in the 1960s as agricultural and industrial development in Brazil decimated the Atlantic coastal rainforest that is their habitat (McKinsey 1998). By the 1970s they were almost extinct in the wild due to trapping as pets and the considerable deforestation. In an attempt to reverse the trend the Brazilian government set aside a 5300-ha reserve in rainforest north of Rio de Janeiro for 100 wild survivors. Both the Tijuca National Park and United States National Zoo became involved in breeding them in captivity (Radetsky *et al.* 1993; McKinsey 1998).

In the 1970s a number of animals were collected and taken to Tijuca National Park in Rio de Janeiro for participation in a captive breeding program (Wolfheim 1983; Dietz 1985). In 1984 the first 10 animals were re-introduced into the wild from captive-bred stocks, but most of these failed to survive as they were not able to fend properly for themselves. In 1985 a further seven were released, of which some were trained to survive in the wild, but all except two died. More recently 134 were released, of which 43 survived and have been breeding there. Other introductions have occurred in woods on ranches in the vicinity of Silva Jardim near Rio de Janeiro.

Of the hundreds of zoo-born marmosets re-introduced only 10 to 20 percent have survived and then only with considerable human help. However, their offspring do quite well – the strategy being that these will be self-supporting whereas the parents cannot survive independently (McKinsey 1998).

From about 370 in 1983 the population in 1986 numbered about 600 animals (Kleinman *et al.* 1986;

Burton and Pearson 1987). More than 400 now roam the coastal forest of Brazil (Radetsky *et al.* 1993). In 1996 only seven captive-born animals were re-introduced, but they successfully produced 38 offspring, which are now living on their own. The present population of re-introduced marmosets is probably 220–240 animals (McKinsey 1998).

■ DAMAGE
None known.

TUFTED-EAR MARMOSET
Common marmoset
***Callithrix jacchus* (Linnaeus)**

■ DESCRIPTION
HB 215 mm; T 295 mm; WT 165–360 g.
Fur grizzled yellowish grey with light black bands on body; tail ringed black and grey; ears tufted.

■ DISTRIBUTION
South America. Eastern Brazil from northern Ceará to southern São Paulo.

■ HABITS AND BEHAVIOUR
Habits: diurnal; gather together in hollows at night. **Gregariousness:** large troops. **Movements:** sedentary? **Habitat:** forest and plantations. **Foods:** insects and fruits, leaves, flowers. **Breeding:** 1 young, occasionally 2–3. **Longevity:** 16 years. **Status:** common, but some fragmentation of range; commonly kept pet.

Tufted-ear marmoset

■ HISTORY OF INTRODUCTIONS
Introduced successfully(?) in Guanabara and Rio de Janeiro, Brazil.

SOUTH AMERICA
Brazil
Introduced into Tijuca, Guanabara about 1900, tufted-ear marmosets have since spread into Rio Janeiro (Avila-Pires 1969; Wolfheim 1983).

■ DAMAGE
Tufted-ear marmosets inhabit plantations of cacao, bananas and coconuts and formerly were abundant in orchards and gardens (Wolfheim 1983), but no control is carried out. They are sometimes used in biomedical research in the United States.

Family: Cebidae
New World monkeys

SQUIRREL MONKEY
***Saimiri sciureus* (Linnaeus)**
Two species of Saimiri are currently recognised, S. oerstedii Reinhardt and S. sciureus (Linnaeus), and the exact identification of the species introduced is lacking. Most of the imports to the United States came from the range of S. sciureus.

■ DESCRIPTION
HB 260–360 mm; T 350–430 mm; WT females 500–750 g, males 700–1100 g.
Body greenish yellow; throat, face and ears white; muzzle black; under parts white or light yellow; tail tip black, long, non-prehensile.

■ DISTRIBUTION
Central and South America. Costa Rica and Panama (*S. s. oerstedii*) and Amazonia, Brazil, Colombia, Ecuador, Peru, Bolivia, Venezuela and Guyanas (*S. s. sciureus*).

■ HABITS AND BEHAVIOUR
Habits: active, diurnal, arboreal. **Gregariousness:** large bands, 2–200+; density 7.5–528/km². **Movements:** daily 0.6–1.1 km; overlapping home ranges 17.5–300 ha. **Habitat:** gallery forest, forest edges, palm forest and other forest, cultivated areas, usually along streams or other waters. **Foods:** insects, fruits, leaf material, nuts, flowers, buds, seeds, arachnids, young birds and eggs, and small vertebrates. **Breeding:** mates in dry season; gestation 152–172 days; oestrous cycle 7–13 days; 1 young; clings to

Squirrel monkey

mother's back for first few weeks; independent at 1 year; females mature 3 years, males 5 years. **Longevity:** 15 to 21 years captive. **Status:** declining in a number of areas due to deforestation, locally abundant in others.

■ HISTORY OF INTRODUCTIONS
Squirrel monkeys have been introduced unsuccessfully to the Hawaiian Islands, but successfully to Florida, USA, and possibly to Rio Amazonas, Colombia. They have more than likely been introduced to Costa Rica by pre-Columbian man and to Panama by Indians.

NORTH AMERICA
United States
Between 1968 and 1972 some 173 049 were imported into the United States (Banks 1972). In 1969 alone some 47 096 of these animals were imported into the United States (Jones and Paradiso 1971). It is therefore not surprising that they were introduced as a tourist attraction to Silver Run River, Silver Springs, Florida, with Rhesus macaques (*Maccaca mulatta*) and established there (Lever 1985). There appears no further mention of these monkeys in this area.

PACIFIC OCEAN ISLANDS
Hawaiian Islands
A single squirrel monkey, *Saimiri* sp., was released by the coastguard on Green Island off Kure Atoll in late 1961 and remained there in a semi-wild state until January 1967 when it disappeared (Woodward 1972).

SOUTH AMERICA
Colombia
In an attempt to produce them for export overseas, 5690 individuals were released on an island in the Rio Amazonas near Leticia, Colombia, between 1967 and 1970, but by 1972 the population had decreased to 850–966 and was still declining (Mittermeier *et al.* 1977).

Costa Rica
It has been suggested that an isolated population of squirrel monkeys in south-western Costa Rica may have been introduced there by pre-Columbian man (Hall and Kelson 1959; Hershkovitz 1969).

Panama
Squirrel monkeys now occur in the south-west and on islands south of David, but were previously more widespread; they may have been introduced there by the Indians bringing pets to the area (Bennett 1968; Hershkovitz 1969).

■ DAMAGE
Squirrel monkeys feed on crops in Costa Rica (Wolfheim 1983) and Panama (Baldwin and Baldwin 1976). In Peru and Surinam they are occasionally pests of cacao and citrus plantations (Grimwood 1969; Husson 1957) and will also feed on cultivated bananas (Durham 1972).

BLACK HOWLER
Alouatta pigra **Lawrence**

■ HISTORY OF INTRODUCTION
BELIZE
The black howler monkey has been re-introduced into the Cockscomb Basin, Belize (Howich *et al.* 1994).

RED HOWLER
Alouatta seniculus **(Linnaeus)**

■ HISTORY OF INTRODUCTION
VENEZUELA
Red howler monkeys are kept as a free-ranging species in Venezuela (Agoramoorthy and Rudran 1993).

BROWN PALE-FRONTED CAPUCHIN
White-fronted capuchin
Cebus albifrons **(Humboldt)**

■ DESCRIPTION
HB 300–380 mm; T 380–500 mm; WT 2.3 kg.
Slender, long limbed, with partially prehensile tail;

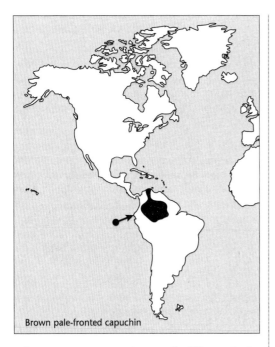

Brown pale-fronted capuchin

colour varies over range, but usually different shades of brown; dorsum brown with white circling the face, and white on forearms; cap darker brown.

■ DISTRIBUTION
South America. Southern Colombia and southern Venezuela, south through Amazon Basin to northern Bolivia.

■ HABITS AND BEHAVIOUR
Habits: diurnal, active; mainly arboreal; territorial. **Gregariousness:** groups 7–30; 1 adult male per troop; density 3.8–45/km². **Movements:** sedentary; home range 60–200 ha. **Habitat:** deciduous and evergreen forest. **Foods:** fruit, seeds, shoots, young birds and eggs, insects and invertebrates. **Breeding:** similar to *C. apella*; 1 young; weaned after several months. **Longevity:** no information. **Status:** declining; little information; hunted for food.

■ HISTORY OF INTRODUCTIONS
SOUTH AMERICA
Trinidad and Ecuador
Brown pale-fronted capuchins have been imported to the island of Trinidad (Roots 1976) and to Ecuador. Almost certainly introduced to Ecuador where there were remnant populations on the eastern central coast including some wildlife sanctuaries (Wolfheim 1983).

■ DAMAGE
Brown pale-fronted capuchins are infamous crop pests in northern Colombia (Green 1976). They often forage in cornfields and are killed as pests (Hernandez-

Comacho and Cooper 1976). In Peru they feed on sugar cane, limes and banana, and in Brazil are also persecuted as pests of crops (Wolfheim 1983).

BROWN CAPUCHIN
Tufted capuchin, mono, black-capped capuchin
Cebus apella (Linnaeus)

■ DESCRIPTION
HB 363–480 mm; T 394–490 mm; WT 2.0–4.5 kg.
Cap on head of short, dark, erect hairs and in male forms ridges on either side of crown; body light brown; males heavier than females by 1 kg.

■ DISTRIBUTION
South America. Southern Colombia, Venezuela, Guyanas, throughout Paraguay and Brazil to northern Argentina.

■ HABITS AND BEHAVIOUR
Habits: no information. **Gregariousness:** groups 1, 5–40; male dominance hierarchy in troops; density 28–111/km². **Movements:** home range 0.25–40 ha. **Habitat:** semi-deciduous to tropical rainforest. **Foods:** fruits, nuts, seeds, insects. **Breeding:** breeds September–December; gestation 160–180 days; polygamous; oestrous cycle 16–20 days; 1 young; sexual maturity females 4 years, males 7 years. **Longevity:** 44 years 7 months captive. **Status:** common and abundant, but declining due to clearing of forest for agriculture.

Brown capuchin

■ HISTORY OF INTRODUCTIONS
SOUTH AMERICA
Margarita Island, Venezuela (Isla de Margarita)
It has been suggested (Eisenberg 1989) that the presence of brown capuchins on this island indicates that they were originally introduced by Amerindians.

■ DAMAGE
Brown capuchins feed on crops, especially immature corn in Colombia and Guyana (Wolfheim 1983). They are destructive to cacao, citrus, palm and corn cultivation in Surinam (Husson 1957), are accused of raiding crops and gardens in Peru (Grimwood 1969), and are shot to protect crops in south-east Bolivia (Wolfheim 1983).

Family: Cercopithecidae
Old World monkeys

Green monkey

TALAPOIN
Miopithecus talapoin (Schreber)

■ HISTORY OF INTRODUCTION
FERNANDO POO AND CANARY ISLANDS
Talpoins from the forests of western Angola, Cameroon and Gabon may have been introduced to Fernando Poo and the Canary Islands (Haltenorth and Diller 1994).

GREEN MONKEY
Vervet monkey, greenish monkey, green gueron, grivet, savanna monkey
Cercopithecus aethiops (Linnaeus)
=*C. sabeus* (Linnaeus)

■ DESCRIPTION
HB 400–830 mm; T 500–700 mm; WT 2.5–9.0 kg.
Slender with long tail; upper parts bright gold green, but varying from silver grey to reddish green; face black; forearms and forelegs grey; underparts, cheeks, sides of neck white to yellowish white; eyelids pale pink; scrotum pale blue and penis red; tail greyish green on basal two-thirds and yellowish distally.

■ DISTRIBUTION
Africa. Sierra Leone, Liberia, Senegal and Somalia south to South Africa.

■ HABITS AND BEHAVIOUR
Habits: diurnal; arboreal and terrestrial; territorial.
Gregariousness: troops of 6–50 (old male, several females, young) and up to 140; density 0.87–153.7/km^2. **Movements:** home range 9.4–518 ha. **Habitat:** forest, woodland savannah, forest edges, thickets, riparian woodland, acacia groves. **Foods:** fruits, berries, grass seeds, leaves, flowers, bark, sap, bulbs, roots, shoots, seed pods, grain, young birds, birds' eggs, insects, spiders, reptiles (lizards), herbs, human food scraps. **Breeding:** breeds all year (August–September St. Kitts); in Kenya mates April–June; gestation 165–203 days; 1 or rarely 2 young; clings to mother for 3 months; weaned 3–6 months; inter-birth interval 1 year; females breed at 3–4 years, males at 4–5 years. **Longevity:** 24 years (captive). **Status:** common and abundant; hunted for meat in many areas.

Note: Behaviour on St. Kitts reported to differ slightly from that of African animals.

■ HISTORY OF INTRODUCTIONS
Green monkeys have been introduced successfully on Barbados, St. Kitts and Nevis in the West Indies, and São Tiago in the Cape Verde Islands.

ATLANTIC OCEAN ISLANDS
Cape Verde Islands
Green monkeys (*C. a. sabeus*) have been introduced by humans to the island of São Tiago (Bannerman and Bannerman 1968), probably from mainland Africa (Osman Hill 1966). They are the only mammal apart from introduced rats in the Cape Verdes. They were once more common, but still inhabited the most inaccessible heights on the island in the 1960s. They were formerly also on the island of Brava, and noted

there in 1987 (Alexander 1898), when they were abundant in the larger valleys causing much damage to the sugar cane.

WEST INDIES
Barbados, St. Kitts and Nevis (Lesser Antilles)
The green monkey is reported to occur on Barbados, St. Kitts (St. Christopher), and Nevis (Sade and Hildrech 1965; Hall 1981; Walker 1992), in the Lesser Antilles, where they are thought to be an accidental introduction associated with the slave trade between Senegal and the West Indies in the 1600s.

C. a. sabeus was first reported on St. Kitts by Father Labat (*Nouveau Voyage aux Isles de L'Amerique*, Paris, 1722) who visited the island in 1700, although there is a doubtful record that they were present on Barbados as early as 1682. According to Labat, they escaped on St. Kitts from the houses of the French when the land was laid fallow under English control. In 1719 (Smith 1745), they were numerous on Mount Misery and were reported again in 1866 to be abundant on the island. Numbers on Barbados initially increased substantially then crashed in the eighteenth century because of loss of forest habitat and bounty hunting, but increased again in the 1950s after some areas had become reforested (Walker 1992).

In 1965 they were present in all parts of St. Kitts where there was some forest cover and were most abundant in the forest of the central mountain ranges, especially in the ravines, with a total population in the vicinity of 1500 monkeys (Sade and Hildrech 1965).

■ DAMAGE
In Africa the green monkey is a frequent agricultural pest and raids orchards, native's crops and villages (MacKenzie 1953; Ansell 1960; Osman Hill 1966; Wolfheim 1983) and is a frequent pest around lodges and campsites (Estes 1993). They damage orchards and market gardens in South Africa, where control of their numbers is carried out (Hey 1964, 1967). In Africa they are known to attack humans in situations where there is overpopulation due to tourists feeding them (Brennan *et al.* 1985). They cause extensive damage to cacao plantations in Sierra Leone, feed in maize patches in Ethiopia and are a notorious crop raider in Senegal, Sierra Leone, Ghana, Nigeria, Uganda, Kenya, Zimbabwe, Malawi, Zambia and Cameroon (Wolfheim 1983). They damage cereal crops, fruits, vegetables and sugar cane ($20 000 cane in one area) in Zimbabwe (Jarvis and La Grange 1984). Their crop raiding has led to extermination programs in several countries (e.g. Sierra Leone and Uganda). They will steal from houses and gardens and from people at picnic spots. In their favour, they are

reported to sometimes eat large numbers of injurious insects (Osman Hill 1966).

On São Tiago, in the Cape Verde Islands, green monkeys are reported to be a pest of fruit plantations and to have formerly raided sugar cane crops (Bannerman and Bannerman 1968). On St. Kitts, West Indies, they may have been responsible for the extermination of the Puerto Rican bullfinch, *Loxigilla potoricensis grandis* (Sade and Hildrech 1965).

MONA MONKEY
Cercopithecus mona (Schreber)

■ DESCRIPTION
HB 400–500 mm; T 540–800 mm; SH 320–350 mm; WT 2.5–7.5 kg.
Upper parts speckled reddish and black, darkest towards the rump; hands and arms black on lateral surface; legs black, speckled with red spots on lateral surface; under parts and medial surface of limbs, greyish white; tail patch to hips white; tail speckled reddish and black, tipped black.

■ DISTRIBUTION
West Africa. Senegal through coastal west and central Africa to western Uganda.

■ HABITS AND BEHAVIOUR
Habits: agile, territorial. **Gregariousness:** groups or family parties of 8–20 and up to 38. **Movements:** no information. **Habitat:** rainforest, islands of forest in

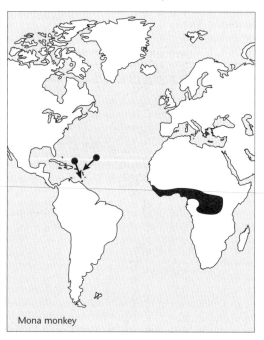

Mona monkey

savanna, mangrove swamps, secondary and lowland forest, plantations, gardens, farmland. **Foods:** leaves, shoots, fruit, insects, tree snails. **Breeding:** births mainly December–February; gestation 6 months; lactation 1 year. **Longevity:** no information. **Status:** common and abundant; hunted for meat; often kept as pets.

■ HISTORY OF INTRODUCTIONS
Mona monkeys have been introduced successfully(?) to St. Kitts and Grenada in the West Indies.

WEST INDIES
St. Kitts and Grenada
It has been concluded (Sade and Hildrech 1965) that records (Hollister 1912) of the Mona monkey on St. Kitts were probably due to error and that the species occurs only on Grand Etang, Grenada. Other authors suggest that it is established on St. Kitts (Osman Hill 1966; Hall 1981) and on Grenada (Hall 1981).

■ DAMAGE
Mona monkeys raid crops and are pests of cacao and maize (Wolfheim 1983). They are frequently shot as pests in Cameroon, Zaire and Sierra Leone.

SILVERED LEAF MONKEY
Brow-ridged langur
Trachypithecus auratus (Geoffroy)
=*Presbytis cristata* Raffles

■ HISTORY OF INTRODUCTION
LOMBOK, INDONESIA
Silvered leaf monkeys occur from Peninsula Burma to

Thailand, Malaysia, Sunda and Java, Bali and Lombok and possibly Borneo. Apparently this species (as *Semnopithecus maurus*) was thought to have been introduced to Lombok by the Balinese Rajahs and is now abundant in the hills from Ampean to Rinjani (Roots 1976; Everett 1896 in Kitchener *et al.* 1990)

STUMP-TAILED MACAQUE
Stumptail macaque, bear macaque
Macaca arctoides (Geoffroy)

■ DESCRIPTION
HB 485–700 mm; T 35–100 mm; WT 8–12 kg.
Upper parts vary from blackish to brownish to reddish, but duller with age; hair shaggy, brown; forehead bald; whiskers under chin form a beard; face pink-tinged with black markings; tail short and naked. Male larger than female.

■ DISTRIBUTION
Asia. Eastern India, Bangladesh, Burma, Thailand, southern China to northern Malay peninsula and Indochina.

■ HABITS AND BEHAVIOUR
Habits: mainly terrestrial; aggressive; diurnal; partly arboreal. **Gregariousness:** group size 5–30 and up to 50. **Movements:** sometimes seasonal migration from one mountain range to another. **Habitat:** monsoon and dry primary or secondary forest. **Foods:** leaves, fruits, roots, seeds, buds, potatoes, insects and small mammals. **Breeding:** gestation 177.5 days; oestrous

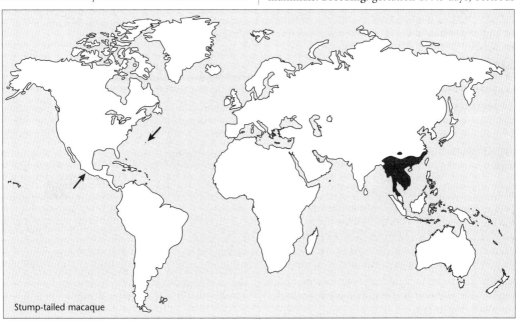
Stump-tailed macaque

cycle about 30.7 days; 1 young every second year; born naked, whitish. **Longevity:** no information. **Status:** uncommon, declining in numbers, rare in most areas.

■ HISTORY OF INTRODUCTIONS
Introduced into Bermuda and Mexico experimentally or for behavioural research purposes.

CENTRAL AMERICA
Mexico
Twenty stump-tailed macaques were released on the island of Totogochillo in Lake Catemaco, Veracruz, in August 1974 and 12 more were released on 12 November the same year. The colony was doing well after 83 days (Estrada and Estrada 1976).

WEST INDIES–CARIBBEAN
Bermuda
Released on Hall's Island, Harrington Sound, Bermuda for behavioural studies and other research by Rutgers University, United States. In May 1983 the free-ranging population was 10 stump-tailed monkeys (Lever 1985).

■ DAMAGE
Invaders of gardens and cultivated fields (Whitehead 1985), stump-tailed macaques feed on crops, including rice crops in Thailand, and in Assam cause damage to potatoes (Wolfheim 1983). At times they play havoc with crops and also even invade isolated huts (Utun Yin 1967).

TAIWAN MACAQUE
Formosan macaque, Formosan rhesus, rock macaque
Macaca cyclopis (Swinhoe)

■ DESCRIPTION
HB 560 mm; T 420 mm; WT ? g.
Appearance similar to other *Macaca* species.

■ DISTRIBUTION
Throughout the island of Taiwan.

■ HABITS AND BEHAVIOUR
Habits: little information on population density. **Gregariousness:** no information. **Movements:** sedentary? **Habitat:** forests, mountains and rocks or inland grassy hills; formerly also sea coasts and beaches. **Foods:** as for other macaques. **Breeding:** as for other macaques. **Longevity:** no information. **Status:** forced to inhabit high elevations by human pressure; declining.

Taiwan macaque

■ HISTORY OF INTRODUCTIONS
Successfully introduced to Oshima Island, Japan.

PACIFIC OCEAN ISLANDS
Oshima Island, Japan
Originally imported from Taiwan after 1942–43 to the Oshima Zoological Gardens and the islets off Shikine on Oshima Island, south of Tokyo, some 36 Taiwan macaques were noted on the island in around 1949–50 (Kuroda 1955). There is some doubt as to whether these macaques occur in the wild at Kiyozumi Prefecture, Chiba, Hondo (de Vos *et al.* 1956), where they were reported to be feral around 1949 (Imaizumi 1949).

■ DAMAGE
No information.

CRAB-EATING MACAQUE
Cynomologus monkey, long-tailed macaque, kra
Macaca fascicularis (Raffles)
=M. irus

■ DESCRIPTION
HB 310–648 mm; T 320–670 mm; WT 1.5–8.3 kg.
Body small, graceful; pelage brown; under parts and also cheeks paler; naked skin on face, hands, feet pinkish brown; tail as long as body or longer, not prehensile.

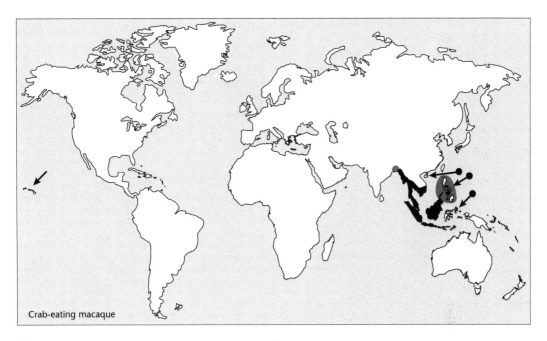

Crab-eating macaque

■ DISTRIBUTION
South-east Asia. From Burma and the Philippines south to western Indonesia (Sumatra and Timor). Now on Sumba, Lombok, Sumbawa, Kanbing, Adonara, Flores, Burma, Thailand, Indochina, Philippines (Luzon, Mindanao, Basilan and Mindoro islands; also Negros Island), south to Sumatra, Java, Borneo and Timor; also on the Nicobar Islands.

■ HABITS AND BEHAVIOUR
Habits: active during day; arboreal and terrestrial; swims well. **Gregariousness:** in troops 6–100 (few males, many females, young); density 5.8–90/km²; linear dominance hierarchy between adults of same sex. **Movements:** home range 25–100 ha. **Habitat:** wide variety of habitats usually near water; forest, secondary forest, mangrove swamps, urban areas, plantations, parks, gardens, woodland, agricultural areas. **Foods:** crustaceans, fruit, insects, amphibians, crabs, shellfish, and other littorial animals, termites, cockroaches, cicadas, moths, bees. **Breeding:** throughout the year; gestation 160–170 days; peak in births in spring; oestrous cycle 24–52 days; sexual maturity females 2.5–4 and males 2–3 years. **Longevity:** 27 years captive. **Status:** locally abundant, but declining some parts of range; frequently kept pet in Indonesia.

Note: Widely used in studies that led to the development of a vaccine for poliomyelitis (Marshall 1967; Walker 1967).

■ HISTORY OF INTRODUCTIONS
Introduced successfully to Sulawesi, the Lesser Sunda Islands, Hong Kong (China), Palau group and Mauritius; unsuccessfully to the Hawaiian Islands.

ASIA
Hong Kong (China)
A group of crab-eating macaques living in the Kowloon area were probably released or escaped from captivity during or shortly after World War 2 (Marshall 1967).

INDIAN OCEAN ISLANDS
Mauritius
Taken to Mauritius by Dutch or Portuguese sailors early in the sixteenth century, crab-eating macaques now number 25 000–30 000 (Sussman and Tatersall 1986). Some were noted there by Cornelius Matelief de Jong as early as 1606. In 1979 they numbered between 12 000 and 15 000.

INDONESIA
Lesser Sunda Islands
Crab-eating macaques were introduced to the Lesser Sunda Islands from the more westerly islands of the Indonesian Archipelago (de Vos *et al.* 1956).

There is controversy as to whether they occur east of Wallace's Line as a result of human introduction (Darlington 1957; Medway 1970) or occurred there naturally (see Fooden 1975). An examination of their morphology (Aimi *et al.* 1982) concluded that on the basis of similarity and ability of the species to swim well, that the distribution on Lombok was a natural one. However, genetically (Kawamoto and Suryobroto 1985) the Lombok animal and those on Sumbawa and Timor are similar. This pattern supports the hypothesis of human introduction (Kitchener *et al.* 1990).

Crab-eating macaques are present on Sumatra, Java, Kalimantan (Borneo), Bali, Lombok, Sumbawa, Sumba, Bangka, Belitung, Littung and Riau archipelago, on Simalur and Nias islands (west of Sumatra), and as far east as Timor.

Sulawesi

The crab-eating macaque has been successfully introduced to Sulawesi (de Vos *et al.* 1956; Roots 1976).

NORTH AMERICA
United States
On several occasions free-ranging colonies of crab-eating macaques have been established in North and Central America for tourism and medical and behavioural research. One such colony has been established as a tourist attraction in Monkey Jungle near Miami, Florida (Lever 1985). Some 1188 monkeys of this species were imported into the United States in 1969 (Jones and Paradiso 1971), and between 1968 and 1972, 8058 were imported (Banks 1972).

PACIFIC OCEAN ISLANDS
Hawaiian Islands
Crab-eating macaques have been introduced to some islands in the Pacific area (Carter *et al.* 1945). Numerous monkeys have escaped, mainly on Oahu, and at least one roamed Hilo, Hawaii, for a few years (Kramer 1971).

Palau Islands
Crab-eating macaques are believed to have been introduced to Angaur Island by German phosphate miners between 1900 and 1914 (Poirier and Smith 1974 in Lever 1985). The population numbered 480–600 in 1973, all supposedly progeny from the introduction of only a single pair of crab-eating macaques.

■ **DAMAGE**
Crab-eating macaques are a pest in some areas because of their raids on fields and gardens (Medway 1978). They feed in cultivated fields and are known to eat rice, cassava leaves, rubber fruit, toro plants and many other crops, and as such are considered a serious pest of agriculture. They will also take food from garbage cans, rubbish dumps and botanical gardens (Wolfheim 1983). In Malaysia, Sumatra, Java and Thailand they are often killed because of their considerable depredations on crops (Wolfheim 1983; Lekagul and McNeely 1988).

Crab-eating macaques are said to pose a threat to a number of endangered endemic birds on Mauritius, and have contributed to the extinction of the blue pigeon, which occurred in 1826.

On Hong Kong they occasionally enter houses and in the past this has led to a number being shot (Marshall 1967).

JAPANESE MACAQUE
Macaca fuscata (Blyth)

■ **HISTORY OF INTRODUCTION**
UNITED STATES
This macaque may have been established at La Moca, in Texas (Lever 1985), but there appear to be few details.

RHESUS MACAQUE
Rhesus monkey, rhesus
Macaca mulatta (Zimmermann)

■ **DESCRIPTION**
HB 470–585 mm; T 205–280 mm; WT 3–12 kg.
Generally brownish or greyish green; hindquarters reddish brown; belly white; forepaws greyish; tail grey green; face, ears, hands pale copper yellow.

■ **DISTRIBUTION**
Southern Asia. From eastern Afghanistan and northern India to south-east China and south to Cambodia, Vietnam and central India.

■ **HABITS AND BEHAVIOUR**
Habits: diurnal, arboreal and terrestrial; dominance hierarchies in both sexes. **Gregariousness:** solitary or groups 8–180; males tend to live in groups with other males or alone; 2–4 times as many females in large groups; density 5–57/km^2, but often considerably larger populations to 753/km^2 in towns. **Movements:** home range 0.05–16 km^2; solitary males nomadic. **Habitat:** semi-deserts to forest and urban areas; agricultural areas. **Foods:** mainly vegetation; fruits, berries, grain, leaves, buds, seeds, grass, fronds, flowers, bark; also insects, other small invertebrates, occasionally eggs and small vertebrates. **Breeding:** October–February; gestation 135–194 days; oestrous cycle about 28 days; 1 young; weaned at *c.* 1 year; inter-birth interval *c.* 14 months; sexual maturity female 3 years, male 4 years. **Longevity:** at least 20–30 years captive. **Status:** common; declining many rural areas.

■ **HISTORY OF INTRODUCTIONS**
Introduced successfully to Puerto Rico, West Indies and possibly near Peking in China. Free-ranging colonies probably exist in India, the United States and Brazil. Unsuccessfully introduced into Germany and Cuba.

ASIA
India
There are apparently some feral troops of rhesus in some towns and forest areas in India (Ciani 1986).

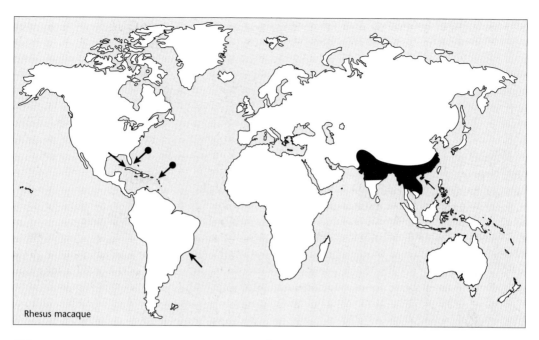

Rhesus macaque

China

An isolated population of rhesus monkeys exists near Peking (Beijing), which is probably introduced (Ellerman and Morrison-Scott 1966; Corbet 1978). However, it is now thought to be one of a relic group representing a former occurrence of the species across northern and eastern China (Walker 1992)

EUROPE
Germany

In Germany there have been at least two escapes that have established for a short period and one successful introduction. An animal escaped in a fruit-growing area near Berlin in 1912 where it remained for a few months before being shot. A second animal lived on the rooftops and in parks of Bonn for a few weeks before being captured (Niethammer 1963).

NORTH AMERICA
United States

A number of rhesus colonies appear to have been established in the United States: at Silver Springs some were released in 1933 on the Silver Run River and there were 78 there in 1968. In 1983, there were 65 rhesus monkeys and squirrel monkeys (*Saimiri* spp.) on the south bank and 200 on the north bank. In 1956–57 a colony of rhesus monkeys was said to have existed on Hilton Head Island offshore from Blufftown, but none remained in 1974.

In the 1970s a free-ranging population was established at Loggerhead Key, Florida, where they were bred for cancer research (Lever 1985). The feral population remaining in central Florida, was probably established as a tourist attraction, and has existed there since the 1930s (Wolfe and Peters 1987). In 1969 alone 27 462 were imported into the United States (Jones and Paradiso 1971), and between 1968 and 1972 some 127 004 were imported into the United States (Banks 1972).

A free-ranging colony of rhesus monkeys was established in Marion County, Silver Springs, Florida, in 1938 (Wolfe and Peters 1987). Although there are no records of the date and circumstances of the introduction, it seems most likely that they were released in 1938, a date supported by a newspaper article at the time. They were more than likely placed on an island in the Silver River by a Colonel Tooey to make a wildlife exhibit for his 'Jungle Cruises' along the river. The monkeys then escaped from the island to other locations along the river. The descendants of those that stayed near the island today inhabit the Ocala National Forest. Others established themselves further up the river near the headwaters and are known as the Silver Springs monkeys. In 1968 a census found 78 living along the north and south banks of the Silver River (Maples *et al.* 1976). In 1981 two large troops were found, each in excess of 50 animals. By 1986 the total in three troops numbered 185 animals (Wolfe and Peters 1987).

SOUTH AMERICA
Brazil

In the 1940s about 300 rhesus monkeys were released for research into yellow fever on Ilha do Pinheiro in Guanabara Bay, Rio de Janeiro; in 1947 around 100 remained at large. It appears that these are still established there (Hausfater 1974; Roonwal and Mohnot 1977; Lever 1985).

WEST INDIES–CARIBBEAN
Puerto Rico
Four hundred rhesus monkeys (Grzimek 1972) were released on Cayo Santiago Island in 1938, but the project was later abandoned because they fought with gibbons introduced at the same time and also attacked human visitors (Carpenter 1942; Altmann 1962; Wilson and Elicker 1976). More recent information suggests that Cayo Santiago may still have a feral population of rhesus, and that they may recently have been introduced to several additional islands (Heatwole *et al.* 1981). They were there in 1965 (Sade and Hildrech 1965) and occurred on at least four cays in the 1970s (Philibosian and Yntema 1977).

A free-ranging population exists at the Carribean Primate Research Centre on Puerto Rico, where they are maintained for behavioural studies.

Cuba
During World War 2 many rhesus monkeys were released for pathological research on Morrillo del Diablo Key, near the Isle of Pines, Cuba. The project was abandoned in the 1950s and some of the monkeys swam to the Isle of Pines, but failed to become established there (Lever 1985).

■ DAMAGE
Rhesus monkeys can become a serious nuisance in gardens and orchards where they steal fruit and other food and damage field crops. They raid fields and gardens and are regarded as pests in the south-east parts of their range (Walker 1992), while in some parts of India they are held to be sacred (Morris 1965).

Rhesus monkeys are known to damage such crops as sugar cane, wheat, grain, pulse, millet, maize, raisins, rice, mulberries, pomegranates and vegetables. However, the extent of the damage is not known and is probably small on a national scale and not as great as that caused by insects, rodents and plant disease (Wolfheim 1983).

PIG-TAILED MACAQUE
Pigtailed monkey, brok
Macaca nemestrina (Linnaeus)

■ DESCRIPTION
HB 470–600 mm; T 125–230 mm; WT 3.5–13.6 kg.
Large, thick set; coat colour variable, but usually uniform mid-brown; crown and forehead with dark

Pig-tailed macaque

brown; under parts paler; light brown hairs on sides of face form conspicuous fringe around head and ears; tail short, carried half erect and arched over body and is one-third to one-half body length. Female weighs less than male.

■ DISTRIBUTION
Southern Asia. Northeastern India (Assam), Burma, Thailand and south to Malaya, Sumatra and Borneo. Also on Bangka and the Mentawi Islands (Siberut), Sipora and the Pagai Islands.

■ HABITS AND BEHAVIOUR
Habits: arboreal, terrestrial, vocal, elusive, shy. **Gregariousness:** troops or parties of 3–50, but occasionally males solitary; density 1–126/km². **Movements:** moves from one feeding area to another; males more often solitary; several parties may travel together or feed together; overlapping home ranges 60–828 ha. **Habitat:** deciduous and evergreen forest, woodland, coastal swamps, plantations and gardens. **Foods:** omnivorous; fruits, grain, leaf shoots, bark and pith, flowers, buds, nuts, seeds, insects. **Breeding:** any time of year; gestation 162–186 days; oestrous cycle 32–42 days; young 1; born brown-haired which turns black in 1 month; weaned 12 weeks; sexual maturity 50 months. **Longevity:** 26 years (captivity). **Status:** fairly common; locally abundant, but declining.

■ HISTORY OF INTRODUCTIONS
Probably introduced successfully to the Andaman Islands, Singapore and Penang Islands.

ASIA

Singapore and Penang, Malaysia

Pig-tailed macaques are not native to either Singapore or Penang Island, or other small islands off the coast of Malaya, although they are often introduced there (Harrison 1966; Medway 1969, 1978).

INDIAN OCEAN ISLANDS

Andaman Islands

Pig-tailed macaques, *M. n. leonina,* have been introduced from India and established in the Andamans (de Vos *et al.* 1956; Burton and Burton 1969; Encycl. Brit. 1976; Lever 1985).

■ DAMAGE

In Borneo, Thailand and Malaya, pig-tailed macaques often raid cultivated crops including grain (padi) and fruit crops (Medway 1978; Payne *et al.* 1985). In Sumatra they are reported to damage corn, papaya and oil palms (Wolfheim 1983).

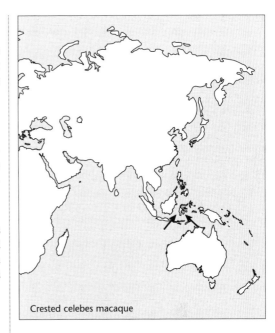

Crested celebes macaque

CRESTED CELEBES MACAQUE

Crested macaque, Celebes crested macaque, black ape, moor macaque, Sulawesi black ape, Celebes black ape

Macaca nigra (Desmarest)

=*Cyanopithecus niger*

■ DESCRIPTION

HB 475–665 mm; T 20–65 mm; WT about 6 kg.

Upper parts dark brown to black; crown with conspicuous conical mass of long erectile hair; face elongated; eyebrows protruding; face, hands and feet hairless; under parts black to nearly grey; buttock pods bright flesh coloured.

■ DISTRIBUTION

Western Indonesia. Northern peninsula of Sulawesi, also on Batjan and some of the smaller adjacent islands; Manadotua and Talise (Talisei) north to Sulawesi; Lembeh east of Sulawesi; on Muna and Butung off south-east shore and other small islands.

■ HABITS AND BEHAVIOUR

Habits: arboreal and terrestrial. **Gregariousness:** 5–25, groups to several hundred. **Movements:** sedentary? **Habitat:** forest and adjacent grasslands, mangrove swamps, bamboo forest. **Foods:** probably roots, buds, leaves, fruits, seeds, insects, worms, eggs, chicks. **Breeding:** oestrous cycle 33.5 days; little information. **Longevity:** no information. **Status:** common, little information.

Note: Natural history and biology in wild appear to be unknown, but probably is similar to other macaques.

■ HISTORY OF INTRODUCTIONS

Possibly introduced to Ambon and Maluku, where it may also be native.

INDONESIA

Ambon

The crested Celebes macaque has possibly been introduced to Ambon (de Vos *et al.* 1956; Lever 1985), but there appears to be little information.

Maluku

M. nigra is presumably a recent introduction to the island of Batjan (Flannery 1995). The date of arrival is unknown and the species has not spread to other islands. They were common on Batjan in 1991.

■ DAMAGE

On Sulawesi the crested Celebes macaque is held to be sacred and causes little trouble, though occasionally they raid plantations (Morris 1965). They will feed in orchards, gardens and cornfields, but the amount of damage appears small (Wolfheim 1983).

BARBARY APE

Barbary macaque

Macaca sylvanus (Linnaeus)

■ DESCRIPTION

HB 550–750 mm; T absent; SH 500 mm; WT 5–17 kg.

Robust body; head rounded; muzzle short; coat greyish brown above with darker face; under parts pale; tail absent. Female is smaller than male and has shorter hair on crown.

Barbary ape

■ DISTRIBUTION
North Africa. Gibraltar, Morocco and northern Algeria. Formerly much of North Africa and Europe.

■ HABITS AND BEHAVIOUR
Habits: diurnal, territorial and arboreal. **Gregariousness:** troops up to 12–25 (males, females, young) and up to 35; multi-male groups; density 12–70/km². **Movements:** sedentary. **Habitat:** wooded mountains and rocky areas, mixed oak forest and fir forest, marquis, garrigue, cedar and palm forests; matorral. **Foods:** insects, scorpions, plants, leaves, fruits, pine cones, leaf tips, seeds, roots, invertebrates, flowers, herbs, grasses, berries, spiders. **Breeding:** breeds throughout the year, but usually only every second year; young born May–September; gestation 180–210 days; 1 rarely 2 young; stays with female to 6 months; weaned 3 months; mature at 3–4 years. **Longevity:** 15–21 years captive. **Status:** declining due to habitat destruction and hunting.

■ HISTORY OF INTRODUCTIONS
During the Pleistocene Barbary apes were widespread throughout Europe and North Africa. By 1920 they only occurred in five regions of Morocco and Algeria, and now probably only in about three. They are now probably restricted to a few isolated areas of montane forest and most of the wild population occurs in the Middle Atlas, Morocco.

Barbary apes have been introduced successfully on Gibraltar in the Mediterranean, possibly to Spain, and to Germany, but in the latter were exterminated for unknown reasons. Planned re-introductions in Algeria–Morocco may have already occurred.

EUROPE
Germany
A soldier (Count Schlieffen) returning from Africa imported Barbary apes and released them on his estate at Windhausen near Kessel in 1763. Here they bred and maintained their numbers (with considerable help) for a period of about 20 years. The population was destroyed for reasons that are now uncertain. Some stories indicate that a child was abducted and another child attacked by the apes, but other accounts suggest that the colony contracted rabies from a dog. Whatever the reason, a monument that still stands today was erected in their honour (Niethammer 1963; Grzimek 1972).

Gibraltar
Introduced to Gibraltar (Sanderson 1955; Corbet 1978), Barbary apes have been there since early times. Possibly the Phoenicians, Carthoginians or Romans took them their originally. Records indicate that they were there in AD 711. They appear to have been there in 1779–83 and may have been there as early as 1704 (Walker 1992). They were first mentioned officially by the British in 1856 when there were about 130 apes there. In 1858 all except three died from an epidemic, but soon after, additional ones were introduced from North Africa. In 1910, 200 apes were there, but since 1913 their numbers have been regulated to about 30–40. In 1943 the numbers had fallen to seven so between 1942 and 1945 a male and six females were imported from North Africa and released (Zeuner 1963; Burton and Burton 1969). Since then a number have been exported to keep the population within bounds.

On the island, where they exist in a state of semi-domestication, their numbers have been as low as seven (in 1924), and as high as 30 animals (in 1955). Their numbers have been supplemented, at least occasionally, by further introductions (seven in 1931, seven in 1943–45) from North Africa. Attempts are made to keep a maximum of 30 monkeys by periodical removal of some to zoos (MacRoberts and MacRoberts 1971).

From 1948 onwards their numbers have fluctuated between 24 and 40. Seventy-six have been exported and a further 10 culled (Lever 1985). Superstition has it that if the apes leave the rock the British will lose the island.

Spain
Fossil remains indicate that *M. sylvanus* occurred in much of Europe during the late Pleistocene and some animals may have survived in southern Spain as late as the 1890s (Walker 1992).

Recently some Barbary apes escaped from captivity in Spain and began to live and reproduce in the wild (Deag 1977; Taub 1977, 1984).

NORTH AFRICA
Algeria–Morocco
Plans have been formulated to re-introduce surplus stocks of Barbary apes to areas where they occurred previously but are now extinct (Oates 1996).

■ DAMAGE
Barbary apes are hunted and shot as pests where they raid crops, gardens and garbage dumps (Deag 1977; Whitfield 1985; Burton and Pearson 1987), although in most areas damage appears to be limited. In Morocco they are shot because of the damage they cause to cedar trees by stripping the bark and eating the cambium layer (Deag 1977). On Gibraltar they are often regarded as a pest when they raid garbage cans and gardens (MacRoberts and MacRoberts 1971).

ZANZIBAR RED COLOBUS
Kirk's colobus
Colobus kirkii (Gray)
=*Procolobus badius kirkii,* and often treated as a subspecies of *C. badius*

■ DESCRIPTION
HB 460–700 mm; T 420–800 mm; WT 5–13 kg.
Face brown, framed in a fringe of long white hair across the head and at back of the cheeks; whiskers white; back and shoulders black two-thirds, remainder of back and upper tail red; forearms and feet darker; remainder pale brown; chest white; underside of hind legs whitish.

■ DISTRIBUTION
Zanzibar Island, Tanzania.

■ HABITS AND BEHAVIOUR
Habits: diurnal; acrobatic, mostly arboreal. **Gregariousness:** lives in groups 5–100 (made up of many family groups) which constantly split and re-unite; density 100/km². **Movements:** sedentary?; overlapping home ranges; territories 25–150 ha. **Habitat:** forest, woodland, secondary forest, mangrove swamps, private gardens with fruit trees. **Foods:** flowers, fruits, leaves, shoots, buds and charcoal. **Breeding:** gestation 4.5–5.5 months; 1 young; weaned 9–12 months, sometimes up to 3 years; inter-birth interval to 3.5 years for some; sexual maturity 2–4 years. **Longevity:** 2? years as captive. **Status:** endangered due to cutting of forest.

Zanzibar red colobus

■ HISTORY OF INTRODUCTIONS
Translocated successfully to the island of Zanzibar and introduced to Pemba Island.

AFRICA
Zanzibar and Pemba Islands
Populations of red colobus have been translocated to Masingini and Kichwele Forest on Zanzibar, and to Ngezi on Pemba Island. The species was rare on Zanzibar in 1978 and the total population only 150–200 (Wolfheim 1983). The total population may now be only 1400–2000, mainly in the Jozani Forest Reserve. The animal is threatened by timber felling, agriculture and hunting (Burton and Pearson 1987).

■ DAMAGE
The Zanzibar red colobus is not generally an agricultural pest, but is shot on Zanzibar because of crop damage probably committed by green monkeys (Wolfheim 1983). However, it continues to be reported to damage village crops (Struhsaker 1998).

Family: Hylobatidae
Gibbons

WHITE-HANDED GIBBON
Lar gibbon
Hylobates lar (Linnaeus)
Specific identification not known, but H. lar *appears to be the animal most commonly traded.*

■ DESCRIPTION
HB 454–473 mm; WT 4.2–5.4 kg.
Mainly black or buff to cream colour; hands and feet, brow band and sides of face white or at least paler than body or forearms; long slender limbs; arms longer than legs.

■ DISTRIBUTION
Asia. From Hainan, Indochina, south-west Thailand, Tenasserim, Cambodia, Malaya and Sumatra.

■ HABITS AND BEHAVIOUR
Habits: arboreal, rarely on ground; crepuscular and diurnal. **Gregariousness:** family parties (males, females, young). **Movements:** home range *c.* 40 ha. **Habitat:** forest. **Foods:** fruits, leaves, new shoots, flowers, insects. **Breeding:** gestation 210–215 days; 1 young; sparsely furred; weaned 4–7 months; inter-birth interval 2–4 years; sexual maturity 6–8 years. **Longevity:** 21–32 years captive. **Status:** no information.

■ HISTORY OF INTRODUCTIONS
There may be free-ranging populations of gibbons in Thailand and elsewhere, but those on the Hawaiian Islands, Bermuda and Puerto Rico have been abandoned or discontinued.

ASIA
Thailand
Short-term releases of gibbons (*Hylobates* sp.) have been made on Ko Klet Kaeo in the Gulf of Siam for reasons of conservation (Berkson and Ross 1969; Wilson and Elicker 1976).

PACIFIC OCEAN ISLANDS
Hawaiian Islands
Short-term releases of gibbons (*Hylobates* sp.) have been made on Laulanui, Hawaii (Wilson and Elicker 1976).

WEST INDIES–CARIBBEAN
Bermuda
Gibbons (*Hylobates* sp.) have been released onto Hall's Island in an attempt to establish a permanent colony (Baldwin and Teleki 1974) for scientific and behavioural studies by the International Psychiatric Research Foundation and the Rockman Research Institute, United States (Lever 1985). The colony appeared to have been present in the locality at least between 1975 and 1977.

Puerto Rico
Gibbons (*Hylobates* sp.) were released on Cayo Santiago Island in 1938, but the project was abandoned when they fought with introduced rhesus monkeys (*Macaca mulatta*) and attacked human visitors (Wilson and Elicker 1976; Carpenter 1972).

■ DAMAGE
None known.

Family: Pongidae
Apes

CHIMPANZEE
Chimp
Pan troglodytes (Blumenbach)
=P. satyrus

■ DESCRIPTION
HB 635–940 mm; SH 700–920 mm; WT males 56–80 kg, females 30–68 kg.
Mainly black (sometimes brown or ginger) with white patch near rump; hairs on head directed backwards or parted; face bare, generally black; nose, hands, ears and feet flesh coloured; brow ridges prominent; tail absent. Female smaller than male.

■ DISTRIBUTION
Africa. Tropical Africa (14°N to 10°S) from Guinea and Sierra Leone to Zaire, Uganda and Tanzania.

■ HABITS AND BEHAVIOUR
Habits: arboreal and terrestrial; diurnal; constructs nests of vegetation in trees for sleeping; territorial. **Gregariousness:** groups or troops of 2–30, and up to 80; dominance hierarchy; density 0.05–26/km². **Movements:** forages over 1.5–15 km; nomadic within home range; home range 5–40 km² and up to 560 km² in marginal habitat. **Habitat:** tropical rainforest,

Chimpanzee

wooded savannah. **Foods:** fruits, nuts, shoots, pith and gum, leaves, roots, vegetables, birds' eggs, insects, small mammals, blossom, seeds, stems, bark, honey. **Breeding:** all year; gestation 202–261 days; promiscuous; inter-birth interval 4–5 years; 1 young rarely 2; stays with female for 2–3 years; sexual maturity females 5.5 years, males 8–9 years. **Longevity:** captive 53 years; may be 60 years in wild. **Status:** declined in numbers through habitat loss and hunting; range fragmented.

■ HISTORY OF INTRODUCTIONS

AFRICA

Several attempts have been made to establish free-ranging colonies of chimpanzees in Senegal, Uganda, the Netherlands and the United States. Only those in Lake Victoria appear to have much chance of long-term success.

Senegal

Beginning in 1973 some re-introductions may have occurred in Senegal with captive-born or captive chimpanzees.

Tanzania

Ten chimpanzees were released on uninhabited Rubondo Island in Lake Victoria. All 10 came from zoos in Europe. It was hoped that the area would become a tourist attraction (Grzimek 1966).

Uganda

Some chimpanzees have been released on Ngamba Island (24 km from Entebbe) in Lake Victoria in about 1998 (Southwell 1999).

EUROPE

Netherlands

A semi-free-ranging colony of chimpanzees was established at Arnhem, in the Netherlands (Van Hoof 1973).

NORTH AMERICA

United States

One male and three female chimpanzees were released on Ossabaw Island, Georgia, in June 1972 in order to establish a free-ranging colony for the requirements of medical research (Wilson and Elicker 1976). The animals are becoming rare and difficult to obtain and so it was decided to establish the colony for future needs. A female died in January 1973, but the remainder appeared to become established with the addition of food and water. In September 1973, four more females were added to the colony and all seven chimps remained on the island to 1975. Ossabaw is one of a series of coastal islands off Georgia and 13 km from the mainland.

Some were also placed on islands in Lion County, Safari, Florida, in 1967, and were reported to be breeding successfully.

■ DAMAGE

Chimpanzees feed on cultivated food crops in Ivory Coast, Cameroon, Equatorial Guinea and Zaire (Burton and Pearson 1987). In Tanzania they eat the stalks of sugar cane and maize, pith of banana stems and nuts of oil palm (Wolfheim 1983). In other areas, such as Uganda, Congo and Gabon, crop damage is used as an excuse to kill them.

Family: Bradypodidae
Sloths

The maned sloth, *Bradypus torquatus* Illiger, an endangered species from south-eastern Brazil is reported to have been successfully translocated in that country (Pinder 1986; WCMC 1998).

BROWN THROATED SLOTH

Grey or brown three-toed sloth, three-toed sloth
***Bradypus variegatus* Schinz**
=*B. griseus* Gray

■ HISTORY OF INTRODUCTIONS
CENTRAL AMERICA
Panama
Introduced to Barro Colorado Island, Panama Canal Zone (de Vos *et al.* 1956), but there is no mention of any introduction by most authorities (Hall 1981; Eisenberg 1989; Walker 1992; Wilson and Reeder 1993).

■ DAMAGE
None known.

Family: Dasypodidae
Armadillos

SIX-BANDED ARMADILLO

White-bristled hairy armadillo, peludo, yellow armadillo
***Euphractus sexcinctus* (Linnaeus)**

■ DESCRIPTION
HB 400–500 mm; T 119–250 mm; WT 3.2–6.5 kg.
Coat generally greyish to reddish brown with sparse hairy cover; ears small; head pointed and flattened and with shield of large plates; fore and hindquarter shields separated by 6–8 moveable bands; tail long, armoured, has two or three distinct bands at base; forefeet with five toes.

■ DISTRIBUTION
South America. Southern Surinam and adjacent eastern Brazil to Bolivia, Paraguay, Uruguay and northern Argentina.

■ HABITS AND BEHAVIOUR
Habits: mainly nocturnal, occasionally diurnal;

Brown throated sloth

Six-banded armadillo

burrows 1–2 m for shelter and rearing young; locates food mainly by smell. **Gregariousness:** occasionally gather into groups at feeding sources, otherwise solitary. **Movements:** sedentary. **Habitat:** dry savannah and drier parts of wet savannah; tropical rainforest; near streams. **Foods:** plant material and insects (ants, termites), invertebrates, carrion, fruits, tubers, palm nuts. **Breeding:** January–October; gestation 65–74 days; litter size 1–3; 2 litters/year; eyes open 22–25 days; sexual maturity at 9 months. **Longevity:** 15.5 years (captive). **Status:** common.

■ HISTORY OF INTRODUCTIONS
SOUTH AMERICA
Chile
Translocated to central Chile (de Vos *et al.* 1956), six-banded armadillos may have become established as an escaped pet (Lever 1985), but there doesn't appear to be any more recent information (not mentioned in Eisenberg 1992).

DAMAGE
Six-banded armadillos are sometimes abundant around plantations where they may damage sprouting corn and other crops (Walker 1967). They are often trapped because of damage to crops and their burrows cause problems for horse riders (Walker 1992).

PICHI
Pichy, small armadillo
Zaedyus pichiy (Desmarest)
=Euphractus pichiy

■ DESCRIPTION
HB 250–400 mm; T 100–150 mm; WT 1.25–2.35 kg.
Head shield and body carapace dark brown with yellowish or whitish lateral edges; ears very small; posterior edge of dorsal plates with blackish hairs interspersed with long yellowish brown and whitish bristles; underparts have coarse yellowish white hairs; tail shield yellowish.

■ DISTRIBUTION
South America. Central and southern Argentina and Chile south of the Aconcagua to the Straits of Magellan.

■ HABITS AND BEHAVIOUR
Habits: hibernates in winter in some localities; makes shallow burrows. **Gregariousness:** solitary. **Movements:** sedentary. **Habitat:** warm sandy soil in pampas; coastal sand dunes. **Foods:** insects and invertebrates (including worms, ants), plant material, other animal food including carrion. **Breeding:** breeds January–February or throughout year;

Pichi

gestation about 60 days; litter size 1–3; young weaned 6 weeks; reach sexual maturity 9–12 months. **Longevity:** 9 years (captive). **Status:** common and abundant; often kept in captivity.

■ HISTORY OF INTRODUCTIONS
SOUTH AMERICA
Pichis have been introduced successfully in Chile.

Chile
Pichis were translocated in 1847 to central Chile (de Vos *et al.* 1956) where they still occur in a wild state (Anderson and Jones 1967; Walker 1967). They are reported to be especially common in the Province of Nuble (Lever 1985).

■ DAMAGE
In some areas pichis are house pets, but they are also hunted for their flesh. They appear to cause no damage to agriculture.

NINE-BANDED ARMADILLO
Peba, Texas armadillo
Dasypus novemcinctus Linnaeus

■ DESCRIPTION
HB 370–440 mm; T 218–395 mm; SH 155–230 mm; WT 4–8 kg.
Body broad, depressed and mottled brownish and yellowish white; upper parts almost lack hair, but under parts have sparse yellowish hairs; snout

tapered; eyes small; ears large, pointed and bases touching; fore and hindquarters with armoured shields (18–20 rows of ossified scales) and 9 hinged bands between; forefoot has four toes; hind foot has five toes, all clawed.

■ DISTRIBUTION
Southern North America and South America. From the southern United States (formerly Rio Grande, now central USA) and Mexico south to Argentina.

■ HABITS AND BEHAVIOUR
Habits: mainly nocturnal, occasionally diurnal; burrows (0.5–3.5 m). **Gregariousness:** small groups, several sharing common burrow (sexes separate). **Movements:** home range 3–15 ha. **Habitat:** dense shady cover and limestone formations, from sea level to 3000 m. **Foods:** insects (ants, roaches, grasshoppers, beetles and larvae, moths, butterflies, flies, bugs, termites, grubs, caterpillars and others), fruits, berries, seeds, mushrooms, arachnids (spiders and scorpions), myriapods, snails, slugs, earthworms,

amphibians, reptiles. **Breeding:** breeds July–August, young born March–April; delayed implantation 120 days; gestation 240–260 days; litter size 4–5; opens eyes at birth; weaned at 3 months; mature at 3–4 years. **Longevity:** 16 years (captive). **Status:** common, increasing range northwards.

■ HISTORY OF INTRODUCTIONS
NORTH AMERICA
Nine-banded armadillos have extended their range northwards and have been introduced successfully into Florida, United States.

United States
Within the last 150 years nine-banded armadillos (*D. n. texanum*) have extended their range northwards from northern Mexico, and then eastwards across the southern United States. This gradual extension of range from the tropics has been favoured by environmental changes, clearing of country for agriculture, and the elimination and diminution of natural predators.

Originally the northern limits of range for the nine-banded armadillo was northern Mexico, and up until 1870 they occurred only in the Rio Grande, Texas. Since this time they have invaded most of Texas and Louisiana (*c.* (1914). or 1926?), parts of New Mexico (*c.* 1905), Oklahoma (*c.* 1944), Arkansas (*c.* 1906), Mississippi (*c.* 1943), and have reached Kansas (*c.* 1943), Missouri (*c.* 1947), Alabama (mid 1930s), Colorado (*c.* 1966), Georgia, South Carolina and Florida by the 1970s (Sherman 1937; Hardberger

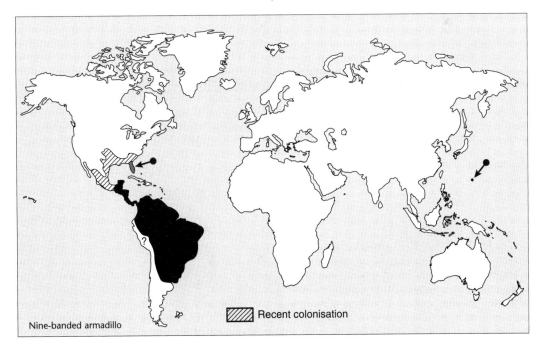

Nine-banded armadillo

////// Recent colonisation

1950; Fitch *et al.* 1952; Neill 1952; NGM 1960; Walker 1965; Anderson and Jones 1967; Layne 1976; Lever 1985; Meaney *et al.* 1987).

In 1905 nine-banded armadillos had reached the Colorado River in western Texas (Burton and Burton 1969). By 1950 they ranged throughout, with the exception of the peninsula and possibly the western panhandle (Neill 1952), and were frequently found in western Louisiana, but were rare from there west to the Mississippi (Hardberger 1950). More recently they have spread northwards with the development of irrigated agriculture in Sinaloa, Mexico (Armstrong and Jones 1971).

Nine-banded armadillos invaded Florida during the natural extension of their range and were as far into north-west Florida as Tallahassee by the mid-1970s. However, earlier introductions from about 1920 onwards probably resulted in their establishment over most of peninsular Florida (Bailey 1924; Sherman 1937 and 1943; Neill 1952; Layne 1976; Newman 1949).

In 1922 one armadillo was caught near Miami (Dade County), Florida, and in 1924 another killed in same area (Bailey 1924). In 1934 one was killed at Flagler Beach, Flagler County and one at Titusville, Brevard County in 1936. By 1936 they were reported at Indian River City, Brevard County and near Crescent City, Putnam County. By 1941 they were known at Pinemount, Suwannee County, and in 1943 one was found in South Jacksonville, Duval County (Neill 1952). At this time (Sherman 1943) it was concluded that they were well established in four adjoining counties (Brevard, Volusia, Flagler, Putnam) east of the St. John River. More were recorded in Flagler and Putnam counties in 1946 and 1950 (Neill 1952). In 1949 they were found established in Brevard, Volusia, Flagler, Putnam, St. John, Indian River, St. Lucie, Martin, Okeechobee, Osceola, Polk, Orange, Seminole, and Lake Cos, with scattered records from many adjoining counties (Hamilton, Alachua, Marion, Sumter, Pasco, Manatee, Hardee, De Soto, Lee, Hendry, Palm Beach and Broward) (Newman 1949).

In 1951 nine-banded armadillos were very abundant in some counties (Volusia and Brevard) and had further expanded their range to reach some offshore islands (e.g. Merritt Island) and Cape Canaveral; some were noted Duval, St. John, Flagler, Putnam, Broward, and Dade counties; also Polk, Lake, Sumter, and Marion counties; with scattered specimens in Pinellas, Hillsborough, Pasco, Hernando, Citrus, Levy, Dixie, and Taylor counties on the west coast and Wakulla County in the Panhandle. It was concluded that they were now over much of Florida, except the swampy south-west portion of the peninsula and possible exception of the western Panhandle (Neill 1952). In 1937 they were established in four counties (Brevard, Volusia, Flagler, Putnam) east of the St. John River (Sherman 1937).

Early releases of armadillos in Florida include two during World War 2 at Hileah and one killed by a dog near Miami in 1922 (Bailey 1924). Some escaped from the Cocoa Zoo, Brevard County, in 1924 (Sherman 1937). The release of the original stock may have taken place in Smyna, Volusia County, in the early 1930s (Neill 1952).

PACIFIC OCEAN ISLANDS
Guam
Nine-banded armadillos were introduced successfully to Guam (Conry 1988), probably some time after 1959.

■ **DAMAGE**
Nine-banded armadillos are responsible for damage in orange orchards, vegetable gardens, nurseries and probably to ground-nesting birds and are thus regarded with disapproval by agriculturists and horti-culturists (Neill 1952). Although they destroy insects and snakes and cultivate and fertilise the soil, they do cause damage by undermining buildings, they start erosion, break dikes and levees, break under fencing and allow stock out, and damage cultivated crops such as canteloupes, watermelons, peanuts and toma-toes (Fitch *et al.* 1952). They are a vector of Chagas' disease and are infected with the bacterium that causes leprosy in humans. They uproot seedlings, eat game birds' eggs and cripple livestock with their burrows (Chamberlain 1980). However, they are generally only considered a minor agricultural pest in the southern parts of the United States (Walker 1992).

Family: Leporidae
Rabbits and hares

EASTERN COTTONTAIL
Sylvilagus floridanus (J. A. Allen)

■ DESCRIPTION

TL 350–463 mm; T 39–65 mm; WT 0.8–1.8 kg.
Coat brownish or greyish; fur long and dense; rump
and flanks washed grey sprinkled with black; between
the ears and shoulders a rufous patch; legs and throat
pinkish buff with dark brown anterior border; tail
brown above, white underneath. Subadults have a
paler and buffer coat. South American forms have a
yellow-brown nuchal patch.

■ DISTRIBUTION

North, Central and South America. From south-
eastern Canada, southern Saskatchewan, Ontario and
Quebec, eastern and central United States, south to
north-western Mexico, Costa Rica and north-western
South America (Colombia and Venezuela). Range
disjunct through Central America.

■ HABITS AND BEHAVIOUR

Habits: crepuscular and nocturnal; terrestrial; aggressive; nest of grass and fur in shallow depression under
cover; terrestrial. **Gregariousness:** male dominance
hierarchy; density up to 8–10/ha. **Movements:** sedentary; home range 0.4–8.9 ha. **Habitat:** meadows,
weedy roadsides, field borders, fence rows, shrubby
areas, borders of forest and woods, farmlands and
orchards. **Foods:** grass, herbs, bark and twigs.
Breeding: breeds mainly February–September; gestation 26–32 days; female polyoestrous; litter size 2–7;
litters 3 or more (5–7) per year; young born naked,
blind; mature at 6–8 weeks. **Longevity:** generally less
than 1 year, probably 6–15 months in wild, and up to
9 years in captivity. **Status:** very common.

■ HISTORY OF INTRODUCTIONS

Eastern cottontails have been widely translocated in
North America (Canada and United States), and
successfully introduced into Europe.

EUROPE
France, Italy and Germany
Eastern cottontails were introduced into Europe from
about 1953 onwards and are now well established in
southern France and northern Italy and are still

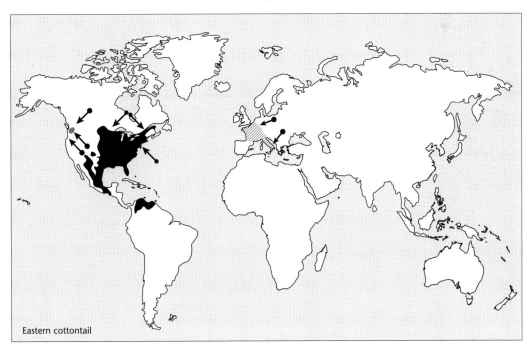

Eastern cottontail

spreading (Burton 1991; Cheylan 1991). Many authorities think the practice of introducing this species as game in Europe should be stopped (Sasse 1983 in Chapman and Flux 1990).

New England cottontails (*S. transitionalis*) have also been introduced into Germany, where they are spreading (Burton 1991).

NORTH AMERICA

Eastern cottontails have been the subject of widespread introduction programs (Chapman and Flux 1990), so much so that the gene pool has been altered and made the species an efficient coloniser. Many island populations may be of recent origin.

Canada

Eastern cottontails (*C. f. similis*) formerly inhabited southern Ontario, but were absent at the time of European settlement. They colonised the area from south-western Ontario in about 1867 or 1870 and were first noted at Niagara Falls in about 1871. One was captured at London, Ontario, in 1883 and four in the Toronto region between 1885 and 1890. They apparently reached Kingston about 1925 and Ottawa in 1931. They also reached Montreal Island about the same time, but have not spread far inland from the Ottawa and the St. Lawrence River valleys. Cottontails appeared in southern Manitoba at Treesbank in 1914 and from there they spread northwards to Worden by 1931 and finally to the Dauphine area by 1940. In the late 1970s some were noted at Estevan, Saskatchewan. It appears as though the spread into Ottawa and Manitoba may be closely linked to the agricultural practices in the region over the last 100 years (Banfield 1977).

Eastern cottontails (*C. f. mearnsi*) were successfully introduced into several counties of Washington, United States, between 1926 and 1933. They spread northwards into British Columbia, becoming established in the Huntington area, the first animal being recorded in 1952, but they may have been there from about 1950 on (Racey 1953; Carl and Guiguet 1972). From here they spread through the Fraser Valley, Cloverdale, Cultus Lake, Longley Prairie and to Tsawwassen. Some were introduced to Vancouver Island in 1964–65 by D. Vandermeer and by the 1970s they were firmly established in the Sooke–Metchosin area, with other records from northern and western areas (Cowan and Guiguet 1960; Carl and Guiguet 1972). They have since become common around Vancouver and are found north to Sayward (Obee 1983).

United States

Shortly after 1900, eastern cottontails were introduced into the western United States for stocking purposes,

despite the protests of some (Chaddock 1938; Hickie 1939; Wilson 1981) who felt it would be likely to introduce diseases not then found among the cottontails already in some areas. It appears they have had no effect on populations of *S. transitionalis* (Wilson 1981).

Introductions were made to Nantucket Island, to Marthas Vineyard, New York, Washington (Dalquest 1948; de Vos *et al.* 1956; Trethawey and Verts 1971), Oregon (Graf 1955), Pennsylvania (McDowell 1955), Ohio (Hickie 1939), Maryland and West Virginia (Chapman and Morgan 1973). Some were also introduced and established on Fishers Island, off New York, in 1924 (Smith and Cheatum 1944).

Massive introductions of middle western subspecies were made into Maryland and West Virginia (Chapman and Morgan 1973; Hall 1981). Between 1922 and 1950 over 206 000 cottontails were introduced by the Maryland Fisheries and Wildlife Administration into Maryland and this figure did not include other releases made by hunting clubs and individuals (Chapman and Morgan 1973). The majority of the animals came from Texas and Kansas and at least six subspecies of *S. floridanus*, four of *S. audubonii*, and also *S. aquaticus aquaticus* could have been introduced. The results of the introductions has been an increase in serum protein patterns of *S. floridanus* from Maryland attributed to the release of *S. floridanus* subspecies or species (Morgan *et al.* 1981).

At least two introductions have been made in western Oregon. In 1941 some sportsmen purchased 19 females and six males from Illinois and released them near Oakville, Linn County, Oregon. Twelve years later a population was still present there and they had spread over an area of 453–518 km^2. In 1937, C. C. Steinel purchased six pairs of cottontails from Ohio, kept them on his farm north of Corvallis, Benton County, Oregon, for two years, then released the colony (*c.* 100). The present population occupies about the same area in Benton County as it did then and there appears to have been no noticeable change in their behaviour (Graf 1955; Chapman and Trethewey 1972).

Between 1915 and 1951 the Pennsylvania Game Commission imported 1 427 317 cottontails, mainly from Missouri and Kansas, which were released during the winters following the hunting seasons and preceding the breeding season. Some were liberated each year apart from 1943 to 1945 until the practice was abandoned in 1951. A program for trapping and transfer of cottontails, also inaugurated in 1937, resulted in 510 759 being trapped and transferred to other areas. During 1915–51 thousands of cottontails were also purchased, imported and released by sportmen's clubs. The effect of such massive introductions on the indigenous population of cottontails was thought to have been little as many were killed in the following hunting season (McDowell 1955). However, despite the injection of about 700 000 cottontails between 1916 and 1936, the population was still declining in the late 1930s (Gerstell 1937). Many indigenous cottontails were translocated between 1933 and 1940, and it was reported that these animals had a better survival rate than imported cottontails (Cramer 1940; Lagenbach and Beule 1942).

From 1928 to 1937 some 46 973 wild (*Sylvilagus* sp). cottontails, mainly from Montana, Kansas and Oklahoma, were introduced to the state of New York by game clubs and private individuals (Bump 1941). These were released in many counties, but with unknown results.

Another cottontail, *S. palustris* (Bachman), was possibly introduced to Hog Island by humans or on natural rafting (Hall 1981).

WEST INDIES
Curaçao and Aruba (Netherlands Antilles)
The black-naped rabbit (*S. f. nigronuchalis*) has supposedly been introduced to Curaçao and Aruba (de Vos *et al.* 1956).

■ DAMAGE
Cottontails occasionally feed on cultivated crops (Storer 1947).

EUROPEAN RABBIT
Rabbit, wild rabbit
Oryctolagus cuniculus (Linnaeus)

■ DESCRIPTION
HB 310–550 mm; T 45–80 mm; WT 0.96–2.5 kg.
Coat colour varies from light sandy to black, but is mainly buff or brown sprinkled with black; nape reddish; ears without large black tips; under parts whitish; front legs short; tail black above, white below.

■ DISTRIBUTION
Europe. Formerly restricted to the Iberian Peninsula; now throughout western Europe, north to southern Sweden and some small Norwegian islands, east to eastern Poland and Hungary, Crete and the Black Sea coast, and south and west to North Africa, Madeira, Azores and Canary Islands.

■ HABITS AND BEHAVIOUR
Habits: mainly crepuscular and nocturnal, but diurnal at times; burrows; 'chin' prominent objects; faeces often deposited in 'latrines' or 'buck heaps'. **Gregariousness:** lives in small groups and sometimes very large groups; strict social hierarchy; numbers 4 to 560/km of transect. **Movements:** sedentary; home range 0.3–3.0 ha; dispersal up to 4 km. **Habitat:** grazing lands from woodland to oceanic islands; farmlands, coastal dunelands, forest, semi-desert. **Foods:** grass, leaves, roots, bulbs and clover. **Breeding:** breeds throughout the year, peak in spring–early summer; gestation 28–30 days; litter size 3–7; non-

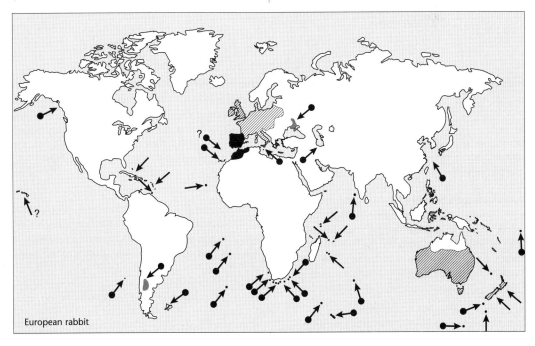
European rabbit

pregnant does 7-day oestrous cycle, pregnant on immediate post-partum oestrous; doe digs short burrow for breeding, up to 1 m; nest grass, lined with fur; can re-absorb embryos in times of food shortages; young born naked and blind in fur-lined underground nest; eyes open at 7–10 days; suckle doe once per night; leave nest at 21–25 days; sexually mature at 3–4 months. **Longevity:** 10–12 years in captivity; up to 7 years in wild. **Status:** common.

■ HISTORY OF INTRODUCTIONS
WORLD DISTRIBUTION
Rabbits have been introduced on more than 800 islands worldwide (see Flux and Fullager 1983 and 1992).

AFRICA
North Africa
The oldest anthropogenic transportation of a mammal could be the introduction of the rabbit to North Africa. Palaeolithic material, probably rabbit, has been found from Algeria and Morocco (Romer 1928; Gobert and Gaufrey 1932), and the abundance of the species in the Neolithic deposits suggests an early introduction from Iberia (Cheylan 1991). They were probably introduced to the island of Zembra, in the Gulf de Tunis off Tunisia by the Phoenicians before AD 200 (Launay 1980). Rabbits are, or were, also present in recent times on the Chafarinas islands (off Algerian coast, but belonging to Spain and including Rey and Congresso), Habibas (off Algerian coast), Conigliera (off Tunisia), Jeziret Jalita (Galite, off Tunisia), Kerkenna (Isles Kerkennah, off Tunisia), and Alborán (between Morocco and Spain) [see table showing introductions of rabbits on Mediterranean Sea islands].

Feral rabbits are found in Lower Egypt and were probably introduced to Morocco where they now occur. Rabbits are believed to have been introduced into Uganda in 1881 by Emil Pasha, and an established colony was found there in 1925 (Thompson 1956).

South Africa
Rabbits were not released on mainland South Africa because of an express prohibition by the council of the Dutch East India Company which feared that they would damage gardens and crops (Bigalke and Pepler 1991), but they were introduced on 13 offshore islands (Cooper and Brooke 1982; Smithers 1983). They have become extinct on at least six of these; Malgas (since 1977), Marcus (since 1960), Meeuw (by 1977), St. Croix (since 1915), Seal, and a small island in the Keurbooms River estuary (before 1865).

Successful introductions which are still surviving occur on Possession (after 1850), Schaapen (since before 1781), Jutten (after 1850), Vondeling (after 1850), Dassen (after 1662), Robben (1656–1658), and on Bird Island, Algoa Bay (after 1852) (McGill 1972; Skead 1980; Cooper and Brooke 1982). They have survived on two larger islands for 300 years and probably less successfully on smaller islands over 20 ha. All the islands were previously used for guano mining and only since 1850 have had soil and vegetation.

On Robben Island rabbits were first introduced in 1654 and again between 1656 and 1658, the first introduction having failed. In 1659 there were 50 present and by 1661 they were described as well established and abundant. Additional domestic stock have been introduced from time to time on the island. Other early reports of their presence were in 1680s and in 1881 (Cooper and Brooke 1982). The original introduction was made to provide food for vessels calling at Table Bay (Bigalke 1937).

Rabbits were released on Dassen Island, probably between 1662 and 1667, with stock from Robben Island. They were present there in 1699, 1705, 1773 and in 1830, but there is little mention of them again until the 1930s (Cooper and Brooke 1982).

Little is known of the rabbits on Bird Island. They were introduced after 1755, and probably after 1852. Some were present there in 1970–71 and some 40 were shot there in 1980 (Cooper and Brooke 1982). Rabbits are still present on Jutten (46 ha), Vondeling (21 ha), but have died out on Marcus (11 ha) and Meeu (7 ha) (Bigalke and Pepler 1991).

ATLANTIC OCEAN ISLANDS
Ascension Island
Introduced from South Africa before 1834, rabbits were still present there in small numbers in the 1980s (Cronk 1980; Flux and Fullagar 1992).

Azores
Rabbits may have been indigenous to the Azores as there appears no reference to any introductions. All the islands in the group have rabbits except Covo (Flux and Fullagar 1992). They have been on Sao Miguel and Terceira at least since 1912 (Miller 1912) and inhabit the southern half of Sao Miguel (Toschi 1965). They were abundant on Terceira in the 1960s (Bannerman and Bannerman 1966).

Beagle Channel Islands (South America)
Some of the rabbits introduced to islands in the Beagle Channel in 1880 came from the Falklands. Here they have been reported on the islands of Tierra del Fuego, Hermite, Isla Grande, Lennox and Rabbit (Flux and Fullagar 1992). They are thought to have been released on a number of islands from the Falkland Islands by the United Kingdom missionary

Thomas Bridges and his sons (Bridges 1949), but may have been in the area as early as 1842 (Ross 1847) [see table under Falkland Islands].

They appear to have been released on a number of islands in Beagle Bay Channel in 1880 by Thomas Bridges (Bridges 1949), and probably to Tierra del Fuego also in 1880 from the Falklands (de Vos *et al.* 1956; Jaksic and Yanez 1983). Others (Thompson 1956; Kirkpatrick 1959) suggest the date was as late as 1910, for sport (Thompson 1956). Some were noted on Rabbit Island in 1902 (Dabbene 1902), but they were fairly rare and not widespread in 1910. The infestation of rabbits that spread all over the northern half of Tierra del Fuego probably originated from two pairs released in 1936 at Punta Santa Maria, near Provenir (Arentsen 1953, 1954).

By 1939–40 they were well established in the Beagle Channel islands and among the Cape Horn islands, notably Lennox Island (Osgood 1943; Olrog 1950). On the Chilean side of Tierra del Fuego they built up in numbers from four to about 30 000 000 in 17 years (1936–53) (Jaksic and Yanez 1983).

On Tierra del Fuego hunting, trapping and gassing has had little effect. In 1951 foxes (*Dusicyon griseus*) were introduced (Goodall 1979; Pine *et al.* 1979; Jaksic and Yanez 1983), and in 1954 myxomatosis (Jaksic and Yanez 1983). Rabbit populations have now been reduced to a low level and are not abundant on the Chilean side of Tierra del Fuego.

Canary Islands
Rabbits have been introduced to these islands, probably as early as the thirteenth or fourteenth century, when many European navigators visited the islands, or earlier by the Romans. They are present on Fuerteventura, Gran Canaria, Hierro, Lanzarote and Tenerife.

Rabbits were abundant on Lanzarote by 1835 (Barker *et al.* 1835), were present there in the 1870s (Mosely 1879), 1970s and 1990s (Aristio 1977; Nogales *et al.* 1990). They were present on Gran Canaria in the 1980s (Santana *et al.* 1986), Fuerteventura in the 1940s (Mohr 1942), and 1950s (Hooker 1958), and present on Hierro in the 1980s (Nogales *et al.* 1988).

Cape Verde Islands
Apparently rabbits were introduced after 1450 and thereafter became abundant on some islands, but have now disappeared (Dost de Naurois 1966; Naurois 1969), and there were none in the 1990s (Flux and Fullagar 1992).

Falkland Islands (United Kingdom)
Rabbits appear to have been introduced to the Falkland Islands by the French prior to 1765 (Strange 1972; Jaksic and Yáñez 1983), although others may have been released later in 1880 (Holdgate and Wace 1961; Niethammer 1963). They could well have been released by Bougainville in 1764 and by John Byron at Port Egmont on Saunders Island in 1765. Rabbits are, or were, present on the islands of Bense, East Falkland, Flat Tyssen, Keppel, New, Rabbit, Saunders and West Falkland (Flux and Fullagar 1992). Those on Keppel Island at one time are reported to have been exterminated by cats that subsequently died out themselves due to disease (Lever 1985).

Introduction of rabbits in the Falkland Islands and on islands in the Beagle Channel

Group and island	Date introduced	Notes
Beagle Channel Islands		
Hermite	1832	from Falklands (9)
Isla Grande	1910 or 1913? and 1926 or 1933–36	increased dramatically despite release of fox and myxomatosis (5)
Lennox (Cape Horn Is)	?	present (8)
Rabbit	before 1902?	present (2; 5)
Tierra del Fuego	1832 or 1880?	in 1950 foxes introduced to control them (3)
Falkland Islands		
Bense	?	present (6)
East Falkland	before 1841	many shot 1832, numerous 1915 (9; 7)
Flat Tyssen	?	present (6)
Keppel	?	one time present, but exterminated by cats (6)
New	early 1800s	released by whalers (1), still present (4)
Pebble	c. 1905	recently stocked by sealers (4)
Rabbit	?	present on most of the 13 'Rabbit Islands' at some time or other (4)
Saunders	1765	landed Port Egmont by John Byron (6)
West Falklands	?	at Port Stephens (6)

References: 1 Cawkell & Hamilton (1961), 2 Dabbene (1902), 3 Duran & Cattan (1985), 4 Flux & Fullagar (1992), 5 Jaksic & Yanez (1983), 6 Lever (1985), 7 Murphy (1936), 8 Osgood (1943), 9 Ross (1847).

Madeira group (Portugal)

Rabbits appear to have been an early introduction to the Madeira group, although at least one early writer (Lataste 1892) suspected that they were indigenous. Prestrello, the coloniser of Madeira, took domestic rabbits of Portuguese stock to Porto Santo in 1418, where they became feral (Hesse 1937; Silverstein and Silverstein 1974). They were present on Madeira and Chāo between 1522 and 1591 when Gaspar Fructuosa wrote that 200 were shot at Machico. The initial colony may have been founded from only one domestic female with young (Niethammer 1963). Rabbits are also reported to have been introduced to Madeira in 1419 (Ognev 1966). Today the rabbits present in the group are wild, but smaller than domestic varieties. They occur on Bugio, Chāo, Deserta Grande, Madeira and Porto Santo. Many residents are reputed to have left the Porto Santo group because of the depredations of the rabbits.

Rabbits inhabited the central valley on Deserta Grande in the Desertas (south-east of Madeira) from the 1950s to the 1980s (Lockley 1952; Bannerman and Bannerman 1965; Cook and Yalden 1980). Some large rabbits were noted on Chāo in the Desertas in 1939 (Lockley 1952). They were present on Bugio in the 1980s (Flux and Fullagar 1992).

South Georgia (UK-dependent territory)

Rabbits were introduced to South Georgia in 1872 from Tristan da Cunha, but failed to become established at this time or at later dates (Holdgate and Wace 1961; Holdgate 1967). Some may also have been introduced about 1906 from Buenos Aires, but these did not thrive (Hodges 1906; Walton 1982). However, a small population survived on Jason Island, off Cumberland Bay, in 1930 (Harrison Matthews 1931), but there have been no rabbits on this island since 1953 (Flux and Fullagar 1992).

St. Helena

Rabbits were introduced to this island (Niethammer 1963; Encyc. Brit. 1973), probably by early Portuguese settlers (Flux and Fullagar 1992). They were still present there in the 1980s (Cronk 1989).

Tristan da Cunha (United Kingdom)

Rabbits were presumably introduced to the island by early seafarers as a source of food. Rabbits occurred on the island in 1829 (Morrell 1832), but appeared to have become nearly extinct by 1873 (Moseley 1879). They were apparently prolific on the island in 1908 (du Baty 1948), but thereafter records of their presence are vague (Wace and Dickson 1965; Wace and Holdgate 1976).

AUSTRALASIA

Australia

Domestic rabbits were introduced with the First Fleet in 1788 and repeatedly thereafter. In early years rabbits were established near Sydney and other towns, but did not spread from these locations (Ratcliffe 1959; Birch 1965). They were reported breeding around houses in Sydney in 1825 and in the 1850s and 1860s were present in all states of Australia following the efforts of acclimatisation societies (Douglas 1986; Stodart and Parer 1988). The Hentys imported rabbits from England to the Swan River in 1829 and from Tasmania to Portland Bay in 1834 (Bassett 1962). Sixteen rabbits arrived by ship in South Australia in 1840 (Wood Jones 1925). There were probably many others in the following years, although little publicity was given to them (Stodart and Parer 1988).

Wild rabbits (24) were imported from England and liberated on a property 'Barwon Park', near Geelong in southern Victoria in 1859 (Ratcliffe 1959; Rolls 1969; Williams and Moore 1989). It is now often accepted that these few rabbits were the most important progenitors, although there is now an accumulation of evidence suggesting a wide range of releases. The 'Barwon' animals appear to be the founders to the mainland population. The characteristics that have failed to persist have been those generally associated with domestic rabbits (Edmonds *et al.* 1981; Stodart and Parer 1988; Long 1988).

From Geelong, rabbits spread northwards and westwards, and by 1880 had crossed the Murray River between Victoria and New South Wales. They were noted on the Queensland border in 1886. Westwards they were noted at Fowlers Bay in South Australia in 1891–92 and at Eucla on the Western Australian border in 1894 (Ratcliffe 1959). In South Australia they arrived at Oodnadatta about 1896, and crossed the border into the Northern Territory in late 1894. At the same time, invasion of the Northern Territory began across the border from Queensland (Strong 1983).

By 1875 they were well established in the western districts of Victoria and in South Australia at the southern end of Flinders Ranges, as well as the earlier colony around Sydney. By 1879 the South Australian and Victorian infestations had amalgamated to cover a large area from Spencer Gulf to north-eastern Victoria (Myers 1986).

Rabbits had reached the Queensland border by 1866 and the Berkeley Tableland and the Gulf of Carpentaria in northern Australia by 1910. Eastwards they reached the border ranges in 1905, Augathella in 1910 and the Winton district in 1922. In South Australia they reached Lake Eyre about 1886 and moved up to the Finke and other rivers to invade the Musgrave, MacDonnell and other ranges and salt lake systems in central Australia. The Western Australian border was reached and crossed in 1894 and they appeared on the west coast near Geraldton in 1906 and Port Hedland in 1912 (Myers 1986).

It was common practice for those engaged in whaling to leave rabbits as a food supply in case of shipwreck or for meat on future visits on small islands within their whaling regions. Whaling commenced in Australia in 1791, bay whaling from Tasmania in 1806, and the whalers were patrolling at least the southern and western coasts shortly after this. American whalers are known from the south coast in the 1830s.

Rabbits are known from islands off the west coast as early as 1827 on Carnac Island and from many others in the 1870s to the 1890s.

The rabbit evolved in Australia in association with none of the species of parasites that it had in its homelands. It is conceivable, therefore, that the reduced parasite fauna could be significant and that population productivity might be higher in Australia at least in some regions, than if a more complete parasite fauna had been introduced with them. There are few natural predators of rabbits in Australia and the resultant lower rate of mortality is generally believed to be the reason for the catastrophic increase in numbers (Dunsmore 1981).

Rabbits are now spread over Australia south of the Tropic of Capricorn although they extend further to the north in coastal Queensland (Wilson *et al.* 1992).

Rabbits have been released on many small offshore islands off the coast of Australia:

Introductions of rabbits to islands off the Australian Coast

Island	Years released	Notes
Actaeon, Tas	?	still occur low numbers (8)
Amadeus (N Aust?)	before 1904	present (41)
Ballee, WA	?	present? (9)
Bennison, Vic	before 1918?	a few black rabbits recorded (6); present? (19; 45)
Betsey, Tas	1825	introduced 1825 (24), present 1827 (51)
Big Dog, Vic	?	present (44; 16)
Big Green, Tas	1862?	8000 destroyed in 10 years to ?1872, now absent (44; 29; 1)
Big Snake, Vic	?	few present (44)
Big, Five Islands group, NSW	before 1843	present 1843 (16), 1849, and 1976; absent 1988 (17), still persist (55)
Bowen, Jervis Bay, NSW	c. 1954	present in 1976 (35); domestics introduced 1954, control 1979–81; none seen since (38)
Breaksea (Middle), Tas	1800s?	released by whalers in 1800s; still present (64), present 1980? (1)
Breaksea, WA	?	present? (3; 41)
Broughton, NSW	1906	used for testing virus control (56), still present (55); descendants of rabbits used by Danyse Rabbit Innoculation Station still numerous in 1976 (35; 16)
Bruny, Tas	?	present? (16, 1)
Cabbage Tree, NSW	c. 1905	still there in late 1970s (15), now being poisoned (55)
Carnac, WA	before 1827, 1934	poisoned 1970s; now absent (66), re-introduced 1934 (4)
Churchill, Vic	before 1946–47	present (11), some control 1974–77 (12)
Citadel, Glennie group, Vic	1913	released by lighthouse keepers in 1913 (4), present 1961–62, gone by 1979 (46)
Clarke, Furneaux Group, Tas	c. 1923	present? (63; 29)
Culeenup, WA	?	present? (9)
Deal, Kent Group	1832	Stokes (1846) released 12 rabbits here as source of food (29, 4)
Doughboy, Vic	before 1908?	present 1908–09 (6), now extinct (45)
Drum, Vic	?	present 1989 (1)
East Kangaroo, Tas	?	present? (29, 16)
Eclipse, WA	?	present? (16)

Introductions of rabbits to islands off the Australian Coast (*continued*)

Island	Years released	Notes
Elizabeth, Vic	?	present? (10; 11)
Erith-Dover, Tas	c. 1865	present 1865, 1872, 1890, absent 1970s (29; 35; 1)
Flinders, Tas	?	now absent (29; 21)
French, Vic	?1862	present (62), present? (44; 11), still present? (Belcher and Hastings in 13)
Gabo, Vic	before 1908?	6 domestic rabbits noted there (6)
Garden, WA	?	present (32)
George Rocks, Tas	c. 1936	there 1979 (43; 23)
Goose Island, Recherche	before 1889	present 1889–1950s (54; 4)
Great Dog, Tas	?	present?
Archipelago, WA	?	(44)
Green Islet, North, WA	?	poisoned, now absent (66; 1)
Green Islet, South, WA	?	poisoned, now absent (41; 1)
Green, SA	?	present? (1)
Green, Tas	?	present (16)
Griffith, Vic	before 1965?	present 1965–1980 (7)
Hareby, SA	?	present? (1)
Hibbs Pyramid, Tas	?	present (16)
Houtman Abrohlos group		see Leo, Morley, North, Rat, Wooded
Hunter, Tas	before 1890	present 1890, now absent (1)
Huon, Tas	?	present (16)
Jeegarnyeejip, WA	?	present? (9)
Jennala, WA	?	present? (9)
Kangaroo, SA	?	present 1920s (61), introduced but failed (29; 30)
Kent group		see Deal
Lady Julia Percy, Vic	1868	flourished; in one month in 1949, 10 000 pairs trapped; decimated by myxo and only few there in late 1970s (49; 12); recently exterminated (4)
Lake Bathurst, NSW	?	site of first release rabbit fleas (65)
Leo, WA	c. 1940 or 1971–72	2 pairs released 1971–72; poisoned 1976; now absent (66); may have been released after World War 2 (58)
Little Yunderup, WA	?	present? (9)
Little, SA	?	present? (1)
Macquarie, Tas	?	present? (31)
Maria, Tas	?	present?, now absent (52; 1)
Meeyip, WA	?	present? (9)
Michaelmas, WA	?	present? (2; 3; 4; 41)
Middle (Abrolhos), WA	?	present?, now absent (1)
Middle Doubtful, WA ?	?	recorded, now extinct (4; 1)
Mistaken, WA	1830	introduced by G. Cheyne; poisoned 1977–80 (66; 4; 41); poisoned 1977, 1978, 1980, re-invaded (2, 3)
Montagu, NSW	before 1967?	present 1967 (28; 14), 1970s (16), may now be extinct (55)
Morley, WA	1970s	poisoned 1973 or 1976, now absent (66; 41)
Mud, Port Phillip Bay, Vic	?	present for many years, nearly exterminated by 1980s, affecting seabird colonies (18, 40)

Introductions of rabbits to islands off the Australian Coast (*continued*)

Island	Years released	Notes
North (Abrolhos), WA	1936	introduced 1936 (16) as food for fishermen, but did not persist (58)
North Bickers, SA	?	? record, skull only (1)
North, SA	?	present? (1)
Orpheus, Qld	1987?	eradication under way 1991 (1)
Partney, SA	?	present? (1)
Pelsaert, WA	1880	early shipwreck, now absent (22; 1)
Penguin, WA	?	present? (1)
Phillip, Vic	1862?	present? (44; 11; 53; 4)
Picnic, Tas	?	present (16)
Quail, Vic	?	present? (44; 11; 4)
Rabbit Rock, Vic	before 1912	present 1912, now absent (25)
Rabbit, Qld	1930s?	chinchilla rabbits farmed in 1930s, now absent (1)
Rabbit, SA	?	present? (1)
Rabbit, Wilsons Promontory, Vic	*c.* 1836	abundant 1832 (51), present 1909 (6), 1912, but later died out or were recently (1968?) exterminated (25; 4).
Rat (Abrolhos), WA	*c.* (1884) and 1940	from shipwreck 1884, not present 1913, re-introduced 1940 (22)
Recherche Archipelago		see Goose, Middle Doubtful
Rodd, NSW	before 1859	present 1888 and in 1960s (47), now absent (1)
Rotamah, Vic	?	present? (1)
Saint Helen's, Tas	*c.* (1920) and 1977	by 1925 caused severe damage to vegetation; exterminated by introduced cats soon after, but re-introduced again in 1977; myxo and rabbit flea introduced and no rabbits seen since April 1978 (39)
Saint Margaret, Vic	before 1920s	present? 1920s (11), still present (16)
Sir Joseph Banks group		see Spilsby, Stickney
Sisters, off Flinders Island	before 1910	there in 1910 (6)
Sloping, Tas	?	present? (33)
Snake, Vic	?	present? (44)
South Mount Dutton, SA	?	present? (1)
South Solitary, NSW	after 1870	introduced for lighthouse keeper's food (55), eradicated by myxo in 1975 (60; 34)
Southport, Tas	?	present (7)
Spilsby, SA	?	present 1980s (50), poisoned? (1)
Stack, Tas	*c.* 1830?	died out of starvation (24)
Steril, Tas	?	present (7)
Stickney, SA	?	present 1980s (50), present? (1)
Sunday, Vic	1876?	present for 100 years (10), uncertain if now present (44; 1)
Swan, Vic	?	present? (1)
Tasmania		present 1825, 1874, 1876 and still present (24; 5; 20; 37), present (59)
Taylor, SA	*c.* 1832	present by 1832 (51)
Three Hummock, Tas	before 1908	one in 1908, now absent (1)
Tollgate, NSW	?	eradicated with poison 1987 (1)
Tullaberga, NSW	1912	present 1912, but since died out (26)
Tumby, SA	?	present? (1)
Venus Bay I., SA	?	present? (1)

Introductions of rabbits to islands off the Australian Coast (*continued*)

Island	Years released	Notes
Wardang, SA	?	present 1938, 1960s (51), 1980s (50), present? (1)
West, SA	1840–44	increased substantially, eradicated by NPWS 1972–73 (48; 16).
Wooded, WA	after 1971–72	introduced from Leo I.; poisoned 1973 or 1976, now absent (66, 41); probably introduced by lobster fishermen after World War 2 (58)
Woody, Qld	1866	present 1866 (16), present? (1)
Worallgarook, WA	?	present? (9)
Wright, SA	?	formerly occupied (48; 16)
Yangie Bay, SA	?	present? (1)
Yunderup, WA	?	present? (9)

References: 1 Abbott & Burbidge 1995; 2 Abbott 1978; 3 Abbott 1980; 4 Armstrong 1982; 5 Barber 1954; 6 Barrett 1918; 7 Bowker 1980; 8 Brothers 1983; 9 Browne-Cooper *et al.* 1990; 10 Edmonds *et al.* 1976; 11 Edmonds *et al.* 1978; 12 Edmonds *et al.* 1981; 13 Flux & Fullagar 1992; 14 Fullagar 1973; 15 Fullagar 1976; 16 Fullager 1978; 17 Gibson 1976; 18 Gillham & Thomson 1961; 19 Gillham 1961; 20 Green 1965; 21 Green 1969; 22 Green 1972; 23 Green 1979; 24 Guiler 1968; 25 Harris & Deerson 1980; 26 Harris *et al.* 1980; 27 Helms 1902; 28 Hindwood 1969; 29 Hope 1973; 30 Inns *et al* 1979; 31 Jones 1977; 32 Jones *et al.* 1966; 33 Kirkpatrick 1973; 34 Lane 1975; 35 Lane 1976; 36 Le Souef 1891; 37 Liederman 1955; 38 Martin & Sobey 1983; 39 McManus 1979; 40 Menkhorst 1988; 41 Morris 1989; 42 Murray 1904; 43 Napier 1979; 44 Norman 1971; 45 Norman 1977; 46 Norman & Brown 1980; 47 Paszkowski 1969; 48 Paton & Paton 1977; 49 Pescott 1976; 50 Robinson 1989; 51 Rolls 1969; 52 Rounsevell 1989; 53 Seebeck 1981; 54 Serventy 1953; 55 Smith & Dodkin 1989; 56 Stead 1935; 57 Stokes 1846; 58 Storr *et al.* 1986; 59 Strahan 1983; 60 Van Gessel & Dorward 1975; 61 Waite & Wood Jones 1927; 62 Wheelwright 1862; 63 Whinray 1971; 64 White 1980; 65 Williams 1971; 66 Young 1981

EURASIA

Originally confined to the Iberian Peninsula, the rabbit probably commenced to spread naturally into other parts of Europe (i.e. south-west France) at the end or soon after the last glaciation. In France fossil evidence suggests that rabbits were present at the end of the last glaciation and in the early post-glacial period. On Gibraltar rabbits were undoubtedly an Upper Pleistocene species.

As well as natural migration, the spread of the rabbit into other parts of western Europe was probably assisted, perhaps initiated, by the Romans, who kept them for food. However, rabbits were kept and reared in captivity long before the Romans. In the first and second century BC they were introduced to Italy and Greece as a domestic animal and kept in leporaria. These walled enclosures were unlikely to have been successful in keeping them enclosed and probably assisted their spread in many areas.

Many rabbits were apparently introduced into central and northern Europe in the Middle Ages. Many were released in Germany and Holland at this time for hunting. However, the rabbit does not seem to have reached far into eastern Europe until the second half of the nineteenth century, when numerous introductions were made.

Formerly it appeared that the rabbit had been introduced to Great Britain by the Normans (1066–1154), but more recent evidence suggests that their arrival was much later, probably between 1154 and 1200. Henry II (1154–89) and Richard I (1189–99) may have been responsible for bringing some rabbits back when they returned from the Crusades. Certainly the Romans brought rabbits to Britain, but there is no evidence until 1176 that any survived in the wild.

Rabbits on islands in Europe

Group or island	Date introduced	Notes
Achill (Co. Mayo Ireland)	?	present (88)
Adriatic Sea islands		see Boban, Brioni, Cres, Iz, Kornat, Lavdara, Levinaka, Losing, Mokan, Pag, Rab, Tremiti
Aegean Sea islands (Greece)		see Chios, Delos, Makria, Pachia
Ailsey (Ayrshire, Scotland)	before 1790	present (53)
Aisla Craig (Firth Clyde, Scotland)	before 1612	present (92; 53; 91)
Alborán (Spain-Morocco, Spain)	?	domestics present 1980s (12)
Algerian islands		see Chafarinas, Habibas
Ameland (Netherlands)	before 1807	origin unknown (37), recorded 1807 (38),
Amrum (German)	c. 1231	probably introduced by Danish King (45), present 1940s, 1960s and 1980s (46; 47; 37)
Anglesey (n. Wales)	before 1790	present 1940s (53; 93), 1950s (94; 95), 1960s (96), 1980s (97)
Aran (Firth Clyde, Scotland)	before 1772 or 1790s	(92; 53), present 1960–76 (44)
Aran Is		see Inisheer, Inishmann, Inishmore
Aranmore (Co. Donegal, Ireland)	?	present (88)
Arø (Isl n.w. of) (Denmark)	?	present (40)
Bagaud (France)	?	present (9)
Balearic Islands (Spain)		see Cabrera, Conejera, Dragonera, Espalmador, Formentera, Ibiza, Mallorca, Menorca, Pitiusas group, Redonda
Baltrum (Germany)	before 1700	present but later exterminated; re-introduced 1963, but now extinct (37)
Bardsey (Wales)	before 1912	present (3; 44)
Barry (Glamorgan, Wales)	c. 17th century	plentiful 17th century (53)
Bass Rock (Firth Forth, Scotland)	before 1584	present 1584 (127; 91), present before 1851 (98)
Bere (Co. Cork, Ireland)	?	present (88)
Bird (Co. Cork, Ireland)	?	present (88)
Blaskets		see Gt. Blasket, Inishnabre, North Blasket
Boban (Adriatic S., Croatia)	?	present 1970s (12)
Bondeholm (S Fyn Is, Denmark)	?	present (48)
Borkum (Germany)	1865 and before 1898	(Kock 1985; 39), released or escaped (37), still present
Bornholm (Baltic S., Denmark)	1975–80	illegally introduced 1975–80 (40)
Brioni (N. Adriatic, Croatia)	?	present since 1960 (12)
Britain	before 1235	(91), still present (99)
Brownsea (Dorset, England)	before 1963	recorded 1963 (44)
Bruray		see Outer Skerries
Burnt (n. of Bute, Scotland)	?	present 1950 (64), 1960s (44)
Cabrera (Balearic Is, Spain)	antiquity	still present 1980s (1)
Caher (Co. Mayo, Ireland)	before 1912	present (3; 90; 88)
Caldey (s. Wales)	?	present (53; 91)
Calf of Man (Irish S.)	before 1790s	present 1960s, 1970s (53; 44; 100), and 1990s (125)
Cape Clear (Co. Cork, Ireland)	?	present 1970s (88; 44)
Capraia (Arch Toscano, Italy)	1967	still present 1980s (12)
Capri (Tyrrhenian S., Italy)	?	formerly present, now died out (25; 12)
Cavallo (s.e. of Corsica, France)	?	present (9)
Cavay (Rabbit) (off Mull, Scotland)	before 1549	present (127)
Cerbicales (Islas)		see Piana di Cerbicale

Rabbits on islands in Europe (*continued*)

Group or island	Date introduced	Notes
Cerboli (Liverno, Italy)	1980s?	recently introduced (12)
Chafarinas (Isles)		see Rey
Channel Is		see Alderney, Brecqhou, Burhou, Guernsey, Herm, Jersey, Jethou, Lihou, Sark
Chios (Aegean S., Greece)	1881?	possibly refers to hares (12)
Clare (Co. Mayo, Ireland)	1675 or before 1911	(3), still present (88)
Clear (s.w. Ireland)	?	present 1980s (128)
Colunbrete Grande (Islas Columbrete, Spain)	1855	removed 1987–88 (7)
Comino (Malta)	?	common 1980s (12)
Conejera (Balearic Is, Spain)	antiquity	(24), still present (1)
Congresso (Islas Chafarinas, Spain)	before 1950	there 1950s (6), 1970s (24), and probably 1980s (12)
Conigli (Isole Pelagie)	?	name indicates early presence; none in 1950s (12)
Conigliera (Tunisia)	before 1920	
Copeland (N. Ireland)	?	present 1980s (12)
Copeland group (N. Ireland)		see Copeland, Lighthouse, Mew
Coquet (Northumberland, England)	?	present (124)
Corsica (Corse) (France)	2nd century BC (5)	present 1990s (12)
Cres (Adriatic S., Croatia)	before 1771	present 1771 (11)
Crete (Kriti) (Greece)	before 1912 or later?	present 1912; not mentioned 1950s (12)
Cross		see Lighthouse
Crowlin Is		see Eilean Mor
Cruit (Co. Donegal, Ireland)	?	present (88)
Cyclades (Greece)	before 1917	present 1917–20s (34; 14)
Davaar (Kintyre, Scotland)	before 1960	present 1960–74 (44)
Delos (Aegean S., Greece)	1970s	origin unknown (12)
Dia (Dhia)(off Crete)	?	present (25; 36)
Doag (Co. Donegal, Ireland)	?	present (88)
Dragonera (Balearic Is, Spain)	?	still present (1)
Drakes (St. Michaels, England)	before AD 1135	present AD 1135 (122; 51)
Dursey (Co. Cork, Ireland)	?	present (88)
Duvillaun (Co. Mayo, Ireland)	?	present (88), in 1971 (44)
Eddy (Co. Galway, Ireland)	?	present (88)
Egadi Islands (Isole Egadi, Italy)		see Favignana, Levanzo, Marettimo
Eilean Ban (White) Skye, Scotland)	?	common 1968 (118)
Eilean Mor (Flannan Is, Scotland)	before 1947	present 1940s (123), 1960s (53)
Eilean Mor (Skye, Scotland)	?	present 1980–81 (120)
Elasa (e. of Crete)	?	present (28)
Elba (Arch. Toscano, Italy)	ancient	still present (12)
Embiez (Golfe du Lion, France)	?	present (9)
Endelave (Denmark)	before 1950	present 1950s (49), and 1980s (40)
Engelsmanplaat (Netherlands)	?	1 record (37)
Eolie Islands (also Lipari Is)		see Salina, Lipari, Filicuda, Stromboli, Vulcano
Eskildsó (S Fyn Is, Denmark)	before 1926	present since 1926 (50; 48)

Rabbits on islands in Europe (*continued*)

Group or island	Date introduced	Notes
Espalmador (Balearic Is, Spain)	?	still present (1)
Eyeries (Co. Cork, Ireland)	?	present (88)
Eysturoy		see Faroes
Fair Isle (between Orkney and Shetland, Scotland)	before 1912	present (3; 116; 44)
Fanø	before 1952	(49), still present (37)
Fara (Orkney Is)	?	present early 1980s (51)
Farne Is		see Brownsman, East Wideopens, Holy (Lindisfarne), Inner Farne, Staple, West Wideopens
Faroes (N. Atlantic)	?	present (62), may still be present (12)
Favignana (Isole Egadi, Italy)	before 1960	present 1960s, 1970s (23) and 1980s (8)
Fedje (Norway)	1875	3–4 pairs introduced from Shetland 1875; still present (52)
Ferkingstad (Norway)	?	present (52)
Fetlar (Shetland Is)	before 1912	present (3), 1970s, 1980s (56)
Fidra (Lothian, Scotland)	?	exterminated by lighthouse keepers 1960–61 (115)
Filfla (Maltese Is)	ancient? and later	died out; later domestics introduced (20)
Filicudi (Eolie Is, Italy)	?	present? (12)
Finish (Co. Galway, Ireland)	?	present (88)
Flannan Is		see Eilean Mor
Flat Holm (Bristol Ch., England)	1492	(91), still present 1952 (44)
Flotta (Orkney Is)	?	present 1967 (44)
Föhr (Germany)	before 1940s	origin unknown (37), present 1940s, 1970s, and 1980s (46; 12)
Formentera (Balearic Is, Spain)	?	still present (1)
Foula (Shetland Is)	1870s	pets released 1870s (53), present 1960–76 (44), 1980s (54)
Foulness (Essex, England)	1183–1220?	(63)
Frioul (Golfe du Lion, France)	?	attempts failed or have disappeared (9; 12)
Furze (Co. Cork, Ireland)	?	present (88)
Gairsay (Orkney Is)	?	recorded since 1960 (55; 44)
Garnish (Garinish) (Co. Cork, Ireland)	?	present (88)
Garvellachs		see Garbh Eileach
Giglio (Arch. Toscano, Italy)	1935–37	present 1980s (12)
Giraglia (off n. Corsica)	?	present 1980s (12)
Gola (Co. Donegal, Ireland)	?	present (88)
Gorgona (Arch. Toscano, Italy)	*c.* 1975–76	present 1980s (12)
Gotland (Sweden)	before 1907?	re-introduced 1907, pest by 1940s (49), present 1950s, 1960s, and 1980s (57; 58; 59)
Gozo (Maltese Is)	?	still present 1980s (12)
Graemsay (Orkney Is)	?	present 1973 (44)
Grand Rouveau (France)	?	present 1980s (8)
Grassholm (Pembrokeshire, England)	?	present but now extinct (108; 12)
Great Blasket (Co. Kerry, Ireland)	before 1756	(117), there 1950s (114), still present (88)
Great Cumbrae (Firth Clyde, Scotland)	before 1612	there before 1612 (92; 3)
Great Mew Stone (Devon, England)	before 1555	present 1555–1850 (51)
Great Saltee (Saltee Co, Ireland)	Middle Ages	present 1960s (12)

Rabbits on islands in Europe (*continued*)

Group or island	Date introduced	Notes
Great Skellig (Co. Kerry, Ireland)	?	present (88), in 1964 (44)
Green Holm (Orkney Is)	?	present 1969 (44)
Greenish (Co. Limerick, Ireland)	?	present (88)
Griend (Netherlands)	?	origin unknown, now extinct (37)
Gruinard (Scotland)	before World War 2	present (51)
Habibas (n.w. Oran, Algeria)	?	present 1980s, early 1990s (16)
Halmø (S Fyn Is)	?	formerly present, extinct in historical times (48)
Hamburger Hallig (Germany)	?	present, origin unknown (37)
Hare (Co. Galway, Ireland)	?	present (88)
Hascosay (Shetland Is)	1900	introduced 1900 (41), present 1969–74 (44)
Havergate (R. Ore, England)	?	there 1940s, but drowned 1949 flood (119)
Hayling (Hampshire, England)	?	present (53)
Hebrides (Inner and Outer)		see table under UK
Heisker Is		see Monarch Is
Helgoland (Germany)	1597	exterminated 1866, later re-introduced; 150 shot 1964 (60); present 1970s (61), 1980s (37)
Hiddensee (Germany)	?	present 1980s (12)
High (Co. Galway, Ireland)	?	present 1980 (12)
Hilbre (Dee Est., England)	before 1540	present 1540, last seen 1939 (63), present 1970s (91)
Hildasay (Shetland Is)	before 1654	warrens present 1654 (62)
Hirsholmene (Denmark)	?	present 1980s (40)
Holm of Melby (Shetlands)	?	died out 1930 (63)
Holy		see Lindisfarne
Holy (Angelsy, Wales)	?	recorded 1960–76 (44)
Holy (off Arran, Scotland)	?	recorded 1968 (44)
Horse (Co. Mayo, Ireland)	?	present (88), in 1965–70 (44)
Housay		see Out Skerries
Hoy (Orkney Is)	?	present 1950s (64),1960–76 (44), increasing 1970s (65)
Hyéres (Isles d')		see Porquerolles, Port-Cross, Levant
I Vow (Loch Lomond, Scotland)	?	present 1967 (44)
Ibiza (Eivissa) (Balearic Is, Spain)	?	still present (1)
Inchfad (L. Lomond, Scotland)	?	recorded 1969 (44)
Inchfad (Scotland)	?	recorded 1969 (44)
Inishbiggle (Co. Mayo, Ireland)	?	present (88)
Inishbofin (Co. Mayo, Ireland)	?	present (3; 88)
Inisheer (Arran Is, Ireland)	before 1960	recorded since 1960 (55)
Inishirrer (Co. Donegal, Ireland)	?	present (88)
Inishkea North (Co. Mayo, Ireland)	?	present (88), in 1971 (44)
Inishleane (Co. Donegal, Ireland)	?	present (88)
Inishmann (Arran Is, Ireland)	?	present (88)
Inishmore (Arran Is, Ireland)	c. 1888	abundant 1888 (3); present (88); recorded since 1960 (55)
Inishtearaght (Kerry, Ireland)	?	present (88), 1966 (44)
Inishtooskert (Kerry, Ireland)	?	present (126), 1966 (44)
Inishturk (Kerry, Ireland)	before 1912	present 1912 (3)
Inishvickillane (Kerry, Ireland)	before 1920	numerous 1920 (114). present (88), and in 1966 (44)

Rabbits on islands in Europe (*continued*)

Group or island	Date introduced	Notes
Inistrahull (Donegal, Ireland)	?	present (3)
Inner Hebrides see Hebrides		
Inshnabro (Blaskets, Ireland)	before 1920	numerous 1920 (114); present (3; 88); recorded 1966 (44)
Inshnabro (Ireland)		present (3; 88)
Inveruglas (Loch Lomond, Scotland)	?	recorded 1962 (44)
Ireland	c. 1282	there 1282 (3), still present (88; 44)
Ischia (Tyrrhenian S., Italy)	ancient	present 1980s (12)
Iz (Adriatic S., Croatia)	?	present 1980s (12)
Jarre (Golfe du Lion, France)	?	present 1980s (9)
Jeziret Jalita (Galite, Tunisia)	before 1920	abundant 1920 (14)
Jordsand (Denmark)	c. 1900	there 1944 (66); now extinct (37)
Juist (Germany)	?	formerly present (67; 68), now extinct (37)
Keeraghs (Keerachs) (Wexford, Ireland)	before 1912	present (90; 3; 117)
Kerkenna (Isles Kerkennah, Tunisia)	before 1920	abundant 1920 (14)
Khios		see Chios
Kid (Mayo, Ireland)	?	present (88)
Kjorholmene (Norway)	?	present, but died out (52)
Koknata		see Kornat
Kornat (Adriatic S., Croatia)	?	present 1980s (12)
Lady Isle (Ayrshire, Scotland)	?	present 1960–74 (44)
Lambay (Dublin, Ireland)	before 1772	24 000 killed 1907 (3), there 1950s (129); still present (88)
Lambholm (Orkney Is)	before 1529	abundant 1529 (69)
Lampedusa (Isole Pelagie, Italy)	recent?, after 1960s	not there 1960 (32; 15)
Langa (Shetland Is)	before 1960	recorded since 1960 (55), 1969–74 (44)
Langeoog (Germany)		originally present (32), now extinct (37)
Langli (Denmark)	?	released or escaped, still there (37)
Lavdara (Adriatic S., Croatia)	?	present (12)
Lavezzi (off e. Corsica)	?	extinct c. 1976 (9; 12)
Lerins (Isle de) (France)		see Sainte Marguerite
Levant (Isle de Hyères, France)	?	present (9)
Levanzo (Isole Egadi, Italy)	?	present and common (23; 8)
Levinaka (Adriatic S., Croatia)	recent?	present from 1980 (12)
Lighthouse (N. Ireland)	?	present (86)
Linga Holm (Orkney Is)	?	recorded 1969–74 (44), present 1980s (70)
Linosa (Isole Pelagie, Italy)	1977	now scarce (12)
Lipari (Eolie Is, Italy)	before 1905	still present (14; 12)
Lipari Islands		see Salina, Lipari, Filicuda, Stromboli, Vulcano
Littke Cumbrae (Firth Clyde, Scotland)	by 1453	present (92), 1845 (51), recorded 1960–74 (44)
Little Roe (Shetland Is)	before 1914	present 1914 (71)
Little Ross (Kirkcudbright, Scotland)	?	formerly abundant, now extinct (12)
Lolland (Baltic S., Denmark)	1975–80	illegally introduced 1975–80 (40), control began 1980 but stopped (72)
Looe (Cornwall, England)	before 1530	formerly present; depleted 19th century (52; 12)
Losing (Adriatic S., Croatia)	before 1771	present 1771 (11)
Losinj		see Losing
Lundy (Bristol Channel)	before 1183	farmed 1183–1219 (91); present 1274–1287 (53); still present (113)
Maîre (Golfe du Lion, France)	?	present (8; 9)
Majorca		see Mallorca

Rabbits on islands in Europe (*continued*)

Group or island	Date introduced	Notes
Makrá		see Makria
Makria (Aegean S., Greece)	*c.* 1914	from Anaphi (27)
Mallorca (Balearic Is, Spain)	?	still present (1)
Malta (Maltese Is)	?	formerly common; present 1950s (13), 1960s (20), 1980s (12)
Man, Isle of (Irish S.)	before 1658	present 1658 (3), 1930s (53); still present 1980s (100)
Marettimo (Isole Egadi, Italy)	before 1970s	present 1970s and 1980s (12)
Marmara (Marmara S., Tunisia)	Frankish Empire	present (26); probably still present (18)
May (Scotland)		1329 (112); present 1549, 1816–87, 1912, 1938, 1959 (112), 1962–68, 1985 (111)
Meda Grossa (Islas Medas, Spain)	?	present (22; 8)
Medas (Islas)		see Meda Grossa
Memmert (Germany)	about 1920	domestics released about 1920 (68; 37)
Menorca (Minorca) (Balearic Is, Spain)	early Neolithic (1400–1300 BC)	from Spain (29), still present (1)
Mew (Ireland)	?	died out 1955–56 (111)
Mezzomare (off Corsica)	?	extinct by 1977 (9; 12)
Middle Bolaerne (Norway)	?	formerly present, origin unknown (52)
Middleholm (Wales)	?	present (109; 108), but extinct 1965
Minorca		see Menorca
Mokan (Adriatic S., Croatia)	?	present 1970s, 1980s (12)
Molen (Norway)	about 1900	introduced about 1900, but now died out (52)
Montecristo (Arch. Toscano, Italy)	ancient and later	present 1960s (33), 1980s (31)
Mureenish (Ireland)	?	present (88)
Nesoy Norway)	?	probably domestics released, data lacking (52)
Nisida (Tyrrhenian S., Italy)	2nd century BC?	present (3; 5)
Norderney (Germany)	?	probably introduced or escaped from captivity (67; 37)
Nordstrand (Germany)	1935?	probably intentionally released, but may have reached there after high dam built about 1935 (45; 37)
North Beveland (Netherlands)	1865?	not reported before 1865, present since 1946 (38; 73)
North Blasket (Ireland)	before 1756	(90), still present 1960–76 (44)
North Ronaldsay (Orkney Is)		recorded 1960–76 (44)
Noss (Shetlands)	1896?–1909	(63), still present (74; 75)
Omey (Ireland)	?	present (88)
Orkney (Orkney Is)	16th century or earlier	(76; 51), numerous 1800 (69); present 1950s (64), 1960s (77), 1970s (78)
Orkney Islands		see table under UK
Out Skerries (Bruray and Housay) (Shetland Is)	?	present 1967–74 (44)
Owey (Ireland)	?	present (88)
Oxna (Shetland Is)	before 1654	warrens present 1654 (62), later exterminated (3; 63), records since 1960 (55), present 1967–74 (44)
Pachia (Aegean S., Greece)	*c.* 1914	present in 1960s and 1970s (27; 12)
Pag (Adriatic S., Croatia)	?	present 1970s, 1980s (12)
Palmaiola (Arch. Toscano, Italy)	recent?	present (12)
Panarea (Eolie Is, Italy)	?	common there (12)
Pantellaria (Sicilian Ch., Italy)	?	present and abundant (8; 12)
Papa (Shetland Is)	before 1960	records since 1960 (55), 1969–74 (44)
Papa Little (Shetland Is)	before 1654?	perhaps there 1654 (62)
Papa Westray (Orkney Is)	?	present 1971 (44)

Rabbits on islands in Europe (*continued*)

Group or island	Date introduced	Notes
Pelagie Islands (Isole Pelagie)		see Lampedusa, Linosa, Conigli
Pellworm (Germany)	before 1870	formerly present (45); origin unknown, possibly extinct by 1870 (37)
Piana di Cerbicale (off se. Corsica)	?	present (9)
Pianosa (Arch. Toscano, Italy)	?	present 1980s (12)
Pitiusas group (Balearic Is, Spain)	before 1970	present 1970s and 1980s (24; 17)
Plane (Calseraigne, France)	?	present (9)
Porquerolles (Îles d'Hyéres, France)	?	present (8; 9)
Port-Cross (Îles d'Hyéres, France)	?	present (8; 9)
Prespansko (Macedonia)	?	present (12)
Puffin (Ireland)	?	present (88)
Puffin (Wales)	*c.* 1784	(12)
Rab (Adriatic S., Croatia)	?	present 1980s (12)
Ramsey (Wales)	13th century	(51); there 1970s and 1980s (110)
Ramsholmen (Norway)	?	present sometimes; colonise from Middle Bolaerne from time to time (52)
Rathlin (Ireland)	1911	there (90); there to 1980s (89; 106)
Rathlin O'Beirne (Ireland)	?	present (88)
Redonda (Balearic Is, Spain)	*c.* 1974	1 found May 1974 (12)
Rey (Chafarina Is, Spain)	?	present (24)
Ringarogy (Ireland)	?	there 1960s (12)
Riou (Arch. de) (Golfe du Lion, France)	?	three islands, two of which have rabbits (12)
Roan (Scotland)	?	been there? (12)
Roaninish (Ireland)	?	there until 1950s (12)
Romo (Germany)	?	origin unknown, now extinct (37)
Rottnumerplaat (Netherlands)	1977	released or escaped since 1977, still present (37)
Rottumeroog (Netherlands)	before 1840 and 1912	exterminated 1869; re-introduced after 1912; now extinct (37)
Rousay (Orkney Is)	?	recorded 1960–76 (44), present 1950s (79)
Rügen (Germany)	before 1940	present 1940s (46), 1980s (80)
S. Pietro		see St. Peter's
Sainte Marguerite (Islas de Lerins, France)	?	present (8; 9)
Salina (Eolie Is, Italy)	?	common (12)
Sanday (Orkney Is)	before 1529	abundant 1529 (69), 1684 (53); present 1950s (79), 1960–76 (44)
Sanguinaires Isles (off Corsica, France)	?	present (12)
Sardegna		see Sardinia
Sardinia (Italy)	before 1912	present 1912 (25), 1960s (33), late 1980s (35)
Scariff (Ireland)	?	present (88)
Scelligs (Ireland)	Middle Ages	(46); there 1960–76 (44)
Schiermonnikoog (Netherlands)	before 1851	present 1930 (43), still present (37)
Scott Head (England)	?	there before 1955 (121)
Sfax (island off town of, Tunisia)	?	present (12)
Shapinsay (Orkney Is)	?	recorded 1960–76 (44)
Shetland Islands		see table under UK
Sicilia		see Sicily
Sicily (Italy)	?	present 1912 (25), 1960s (33), late 1980s (35)

Rabbits on islands in Europe (*continued*)

Group or island	Date introduced	Notes
Skerries (Wales)	c. 1773 and later	(91); there 1960s; re-introduced later (12)
Skokholm (Wales)	12th century	(51); there 1930–1950s (102; 109) and 1970s, 1980s (95; 103)
Skomer (Wales)	c. 1300	there until 1950s (51) and 1970s (104)
Soster (Norway)	1972–77	domestics released about 1972–77 (52)
South Fyn Is		see Bondelholm, Eskildso, Halmø, Vogterholm
South Havra (Shetland Is)	?	present, but eliminated by cats (63; 62)
South Ronaldsay (Orkney Is)	before 1950	present 1950s (79; 81), recorded 1960–76 (44)
Spiekeroog (Germany)	before 1955	disappeared by 1963 (32), now extinct (37)
St. Peter's (off sw. Sardinia)	before 1736	present early times (10)
St. Serf's (Scotland)	c. 1930	there 1930s, until 1957 (105)
St. Tudwal's (Wales)	c. 1536	there 1536–39, 1950s, 1970s and still present 1991 (12)
Steep Holme (Bristol Channel)	?	since 1960 (55), 1968 (44)
Stromboli (Eolie Is, Italy)	?	present and common (12)
Stronsay (Orkney Is)	before 1950	present 1950s (64), 1960–76 (44), 1980s (82)
Sully (Wales)	?	present (101)
Sunk (England)	before 1750	exterminated about 1750 (3)
Swona (Orkney Is)	?	present 1980s (83)
Sylt (Germany)		colonised from Amrum (45), 600–900 shot annually 1940s (46), present 1980s (37)
Terschelling (Netherlands)	before 1400	(38), present 1950s and 1960s (84)
Texel (Netherlands)	before 1400	origin unknown (37); now present (38)
Theodore (Crete)	?	present (12)
Tormore (Ireland)	?	present (88)
Tory (Ireland)	?	present (88), recorded 1960–76 (44)
Toscano (Arch. Toscano)		see Capraia, Elba, Giglio, Gorgona, Montecristo, Palmaiola, Pianosa
Tremiti (Adriatic S., Italy)	?	present for many years (12)
Trischen (Germany)	before 1907	abundant 1907 (68), now extinct (37)
Trondra (Shetland Is)	?	released several times, but failed (63)
Tunisian islands		see Kerkenna, Zembra
Unst (Shetland Is)	before 1912	present about 1912 (3), 1960s (85)
Usedom (Germany)	?	present 1980s (80)
Ustica (n. of Sicily, Italy)	?	abundant, present 1980s (12)
Valencia (Ireland)	?	present (88)
Verte (Golfe du Lion, France)	?	present (8; 9)
Vivara (Campania, Italy)	ancient?	present (30); old introduction (12)
Vlieland (Netherlands)	before 1400 and 1946	re-introduced 1946 (38), now present (37)
Vogterholm (S Fyn Is, Denmark)	?	present (48)
Vulcano (Eolie Is, Italy)	before 1792	present 1792 (10), still present (12)
Wadden Sea islands (Netherlands and Germany)		see Ameland, Borkum, Engelsmanplaat, Griend, Jordsand, Langeoog, Langli, Rono, Rottumeroog, Rottumerplaat, Schiermonnikoog, Spiekeroog, Terschelling, Texel, Vlieland
Wallasea (England)	?	present in early times (63)
Walney (England)	?	recorded 1963 (44) and later (51)
West Burra (Shetland Is)	before 1654	present 1654 (62) or 1684 (53)
West Wideopens (Farne Is)	?	present (89)
Westray (Orkney Is)	before 1950s	present 1950s (79), recorded 1960–76 (44)
Whalsay (Shetland Is)	?	present by 19th century (63). recorded 1960–76 (44), present 1980s (62; 86)

Rabbits on islands in Europe (*continued*)

Group or island	Date introduced	Notes
Whithorn (Scotland)	?	present 1974 (44)
Wight (England)	1225	(91); present 1900–10 (53); there 1950s (64); present 1960–76 (44)
Worms Head	?	present 1977 (12)
Wyre (Orkney Is)	?	recorded 1964 (44)
Yell (Shetland Is)	?	recorded 1960–77 (44)
Zacevo (e. of Krk, Croatia)	?	present (12)
Zacevo (w. of Cres, Croatia)	?	present (12)
Zembra (Gulf de Tunis, Tunisia)	before AD 200 and later	by Phoenicians; domestics released (21), present 1960s (87), 1980s (4; 19)

References: 1 Alcover & Gosalbez 1988; 2 Anon. 1975; 3 Barrett-Hamilton 1912; 4 Ben Saad & Bayle 1984; 5 Bodson 1978; 6 Brosset 1957; 7 Castilla & Bauweus 1991; 8 Cheylan 1984; 9 Cheylan 1988; 10 Donndorf 1792; 11 Dulic 1987; 12 Flux & Fullagar 1992; 13 Gibb 1951; 14 Joleaud 1920; 15 Kohlmeyer 1959; 16 Kowalski & Rzebik-Kowalski 1991; 17 Kuhbier *et al.* 1984; 18 Kumerloeve 1982; 19 Lamine-Cheniti 1988; 20 Lanfranco 1969; 21 Launay 1980; 22 Mas-Coma & Feliu 1977; 23 Massa 1973; 24 Mayol 1974, 1978; 25 Miller 1912; 26 Möbes 1946; 27 Niethammer 1963; 28 Pieper 1976; 29 Reumer & Sandars 1982; 30 Rinaldi & Milone 1981; 31 Spagnesi *et al.* 1986; 32 Toschi 1960; 33 Toschi 1965; 34 Trouessart 1917; 35 Westbury 1989; 36 Zimmerman 1953; 37 Laar 1981; 38 Rijk 1988; 39 Rijk 1985; 40 Strassgaarden & Asferg 1980; 41 Laar 1977; 42 Laar 1974; 43 Rijk 1981; 44 Arnold 1978; 45 Mohr 1929; 46 Lincke 1943; 47 Warnecke 1961; 48 Ursin 1948; 49 Thamdrup 1965; 50 Friis 1926; 51 Booth & Perrott 1981; 52 Myrberget 1987; 53 Sheail 1971; 54 Furness & Hislop 1981; 55 Corbet 1971; 56 Robinson 1986; 57 Anon. 1953; 58 Tjernberg 1981; 59 Englund 1965; 60 Bobak 1970; 61 Heidermann & Vauk 1970; 62 Berry & Johnston 1980; 63 Fitter 1959; 64 Thompson & Worden 1956; 65 Ballard & Goodier 1975; 66 Jepson 1975; 67 Krumbiegel 1955; 68 Tellkamp 1979; 69 Ritchie 1920; 70 Briggs 1981; 71 Hopkins & Rothschild 1953; 72 Lund 1981, 1982; 73 Wijngaarden *et al.* 1971; 74 Stephen 1974; 75 Butler 1982; 76 Martin 1716; 77 Balfour 1968; 78 Jones 1980; 79 St. Aldwyn 1955; 80 Briedermann 1981; 81 Venables 1956; 82 Weir 1981; 83 Hall & Moore 1986; 84 Wijngaarden & Morzer Bruijns 1961; 85 Williamson & Boyd 1963; 86 McKee 1985, 1988; 87 Bernard 1965; 88 Crichton 1979; 89 Armstrong 1982; 90 Moffat 1938; 91 Lever 1977; 92 Millais 1904–06; 93 Allen *et al.* 1947; 94 Hodgin 1984; 95 Lloyd 1970; 96 Lloyd 1965; 97 Bassett 1986; 98 Colquoun 1851; 99 Trout *et al.* 1968; 100 Sumption & Flowerdew 1986; 101 Fern 1981; 102 Lockley 1955; 103 Gynn 1983; 104 Knight 1974; 105 Allinson *et al.* 1974; 106 Greaves 1985; 107 Berry 1979; 108 Gilham 1955; 1953; 109 Lockley 1947; 110 Doncaster 1981; 111 Zonfrillo 1985; 112 Eggeling 1957; 1960; 113 Mead-Briggs 1967; 114 O'Sullivan 1953; 115 Anon. 1961; 116 Kikawa 1959; 117 O'Rourke 1970; 118 Lister-Kaye 1972; 119 Brownlow 1953; 120 Berry 1983; 121 Nicholson 1957; 122 Hurrell 1979; 123 Darling 1947; 124 Day 1980; 125 Walker 1991; 126 Crichton 1989; 127 Ritchie 1920; 128 Newby 1987; 129 Lockie 1956.

Austria

Rabbits were present in Burgenland in the Middle Ages (Wettstein 1955).

Balearic Islands

Strabo (Greek historian and geographer 63 BC to 7 BC?) in 30 BC reported that Emperor Augustus had sent legionnaires to help the inhabitants of these islands destroy the rabbits, which were ravaging their crops. The residents apparently implored the Emperor to either send military aid or resettle the locals elsewhere. A single pair of rabbits introduced some time earlier is said to have multiplied to such an extent that it was impossible to grow any crops.

Rabbits are present or have been present on Cabrera, Conejera, Dragonera, Espalmador, Formentera, Ibiza, Mallorca, Menorca, Pitiusas group and Redonda. They appear to be present still on all except perhaps Redonda and the Pitiusas group [see table above]. Rabbits have been in the Balearics, causing problems there at least since 50 BC (Barrett-Hamilton 1912), and are still present there even though genets were

introduced to control them (Anon. 1975; Alcover and Gosalbez 1988).

Corsica (including other nearby islands in the Mediterranean Sea)

In the second century BC rabbits occurred on Corsica (Bodson 1978), but how they got there is not clear. They were still present on the island in the 1990s (Flux and Fullagar 1992).

Rabbits were also present on Giraglia (off n. Corsica), Lavezzi (off se. Corsica), Mezzomare, Piana di Cerbical (Isles Cerbicales off se. Corsica) and Cavallo (off se. Corsica). They are probably still present on most of these islands, except Lavezzi and Mezzomare where they are now extinct (Cheylan 1988; Flux and Fullagar 1992).

Crete (Kriti; Greece)

On Crete rabbits have been present since at least 1912, but in the 1950s do not appear to be mentioned (Flux and Fullagar 1992). They have, however, been introduced to two small islands off Crete – Dhia and Theodore. These were probably a domestic variety and released at an early date (Miller 1912; Zimmermann 1953). These animals differed so markedly from those established on the European mainland that a separate subspecies *knossius* was described (Niethammer 1963). On Mikronisi some from domestic stock were released in the late 1940s and have maintained their domestic characteristics (Niethammer 1963). East of Crete rabbits have been introduced to the island of Elasa (Pieper 1976).

Croatia
Rabbits have been introduced to the Croatian Adriatic Sea islands of Boban, Brioni (Islands), Cre, Iz, Kornat, Lavdara, Levinaka, Losing, Mokan, Pag, and Rab. They were present on Cres and Losing as early as 771 (Dulic 1987) and were still present on most in the 1980s.

Denmark
Rabbits were released in several places in the early 1900s, but did not increase in numbers substantially. About 1920 they crossed the Danish border from an isolated population released in Germany in about 1900 (de Vos *et al.* 1956).

On islands off the coast rabbits are still present on Bornholm, Fanø, Hirsholmene, Lollard, Arø, Endelave, Langli and the South Fyn Islands of Bordelholme, Eskildso, Halmø and Vogterholm. They are now extinct on Halmø and Jordsand. The introductions to Bornholm and Lolland were recent, 1975–80, illegal releases (Ursin 1948; Strandgaard and Asferg 1980; Laar 1981; Lund 1981, 1982) (see table of introductions of rabbits).

France
A number of islands in the Mediterranean Sea off the coast of France have or have had rabbits introduced including: Embiez, Frioul, Grande Rouveau, Jarre, Levant, Maire, Plane, Pourquerolles, Port-Cros, Sainte Marguerite, Lerins, Sanguinaires, Verte and Bagaud (Cheylan 1984; 1988). Only on Frioul did they fail to become established.

Germany
The first documentary evidence of the arrival of the rabbit appears to be in 1149. These were probably captive animals. In 1407 some rabbits were introduced to an island in Lake Schwerin, Mecklenburg, so that the whole island would become a rabbit garden. They were also noted in 1423 at Buxheim in south Germany, together representing the first records of them in the wild, even though they were known nearly 300 years earlier.

The rabbit appears to have been rare in Germany up until the thirteenth century, and even as late as the sixteenth century, as many notable publications of these times omit mention of them. At the beginning of the fourteenth century a rabbit was said to cost as much as a piglet, which serves to indicate their rarity. Rabbit breeding was certainly in vogue during the fifteenth century. An early introduction of rabbits was made to Amrum in 1231.

Towards the end of the sixteenth century the rabbit was established in the wild in Rhineland, Schlesien (Silesia), and at the beginning of the seventeenth century they were only absent from north-east Saxony. Some were introduced to Hessen in the sixteenth century and some were released in South Thuringia in the second half of the nineteenth century, but later disappeared, except for a remnant population near Reurieth. In 1597 rabbits were recorded in Helgoland and near the beginning of the seventeenth century were released in the sand dunes of Warnemünd where they were very successful. Rabbits were being hunted on Juist (East Friesian Islands) in 1699 (Nachtsheim 1949; Utoth 1956; Niethammer 1963).

Rabbits were uncommon in Saxony until they were introduced more frequently in the second half of the nineteenth century (Zimmermann 1933). A similar situation prevailed in Lipperland, where probably the first release of rabbits took place in 1844 (Goethe 1955). However, in Hanover there is evidence of rabbit enclosures as early as 1700 (Niethammer 1963).

On islands off the coast early introductions of rabbits occurred at Amrum in 1231, Helgoland in 1597 and Baltrum in 1700. They are still present on Hildensee, Rügen, Usedom, Föhr, Sylt, Hamburger Hallig, Nordeney and Nordstrand, but have become extinct on Baltrum, Pellworm, Juist and the Wadden Sea islands of Langeoog, Romo and Spierkeroog. On Memmert, between Juist and Borkum (East Friesians) domestic rabbits were released in 1930 and some of their progeny have retained some signs of domestication at least until the 1960s, but they have since become extinct there. They have presumably been present on Borkum since 1898 (Mohr 1929; Niethammer 1963; Briedermann 1981; Laar 1981; Rijk 1988).

Greece
The descendants of domestic rabbits from the island of Anaphi in the Aegean Sea were released on the islands of Makria and Pachia in about 1914 (Wettstein 1941; Niethammer 1963). The majority of the Aegean islands (Chios or Khios, Delos, Makria, Pachia) have been colonised with rabbits, and the greater number of introductions have been with domestic varieties which have survived with little water. Apart from an early doubtful record of rabbits on Chios in 1881, they appear to have been more recent introductions.

Hungary
The first mention of rabbits being frequently found in Hungary appears to be in 1779 at Zorndorf and Nicolsdorf (Szunyoghy 1959).

Italy
Possibly in ancient times and certainly before 1792 rabbits began appearing on islands off Italy in the Mediterranean Sea including: Capraia, Capri, Cerboli, Elba, Favignana, Filicudi, Giglio, Gorgona, Ischia, Lampedusa, Levanzo, Linosa, Lipari, Marettimo, Montecristo, Nisida, Palmaiola, Panarea, Pantellaria, Pianosa, Salina, Sardinia, Sicily, Stromboli, Tremiti, Ustica, Vivara and Vulcano (Miller 1912; Joleaud 1920; Toschi 1960, 1965; Massa 1973; Westbury 1989; Flux and Fullagar 1992).

Malta
Rabbits have been introduced, date unknown, but on Filfla was most likely ancient (Lanfranco 1969), and have been present from the 1950s on. They were still present on the islands of Gozo and Comino in the 1980s (Flux and Fullagar 1992).

Netherlands
Many rabbits were introduced to Holland in the Middle Ages (Carsdale 1953). Ulisse Aldrovandi (Italian naturalist 1522–1605) mentions that wild rabbits were abundant in the Dutch province of Zealand around 1400.

Early introductions of rabbits, probably before 1400, occurred on some islands e.g. Terschelling, Texel and Vlieland, and in the 1800s on Schiermonnikoog, Rottumeroog and Ameland. They still appear to be present on Ameland, Schiermonnikoog, Terschelling, Texel, Vlieland, and possibly on Rottnumerplaat, but are now extinct on Griend and Rottumeroog, and probably were never on Engelsmanplaat (Laar 1981; Rijk 1981, 1988) (see table of rabbits introduced on islands in Europe).

Norway
There appear to be no records of rabbits introduced to the mainland, but they have been released on a number of islands off the coast. Rabbits have been introduced to the islands of Fedje, Feringstad, Kjorholmene, Molen, Nesoy, Soster, Middle Bolaerne and Ramsholmen. They appear to have now died out on Kjorholmene, Molen, possibly on Nesoy and Ramsholmen, but are still present on Fedje, Ferkingstad and Middle Bolaerne. The earliest releases appear to have been in 1875 on Fedje and there have more recent ones such as on Soster in 1972–77 (Myrberget 1984, 1987)

Poland
The rabbit reached Poland in the second half of the nineteenth century, chiefly as a result of numerous introductions, but there are very few of these left now (Suminski 1963; Nowak 1968).

Romania
Between 1905 and 1907, rabbits from France were released at Jassy where they flourished in areas of woodland-steppe (Niethammer 1963).

Russian Federation and adjacent independent republics
Rabbits were an early introduction into Russia (Naumoff 1950). Domestic rabbits were introduced to the Caucasus and Caspian Sea areas at the end of the nineteenth century. In 1931–32 'Viennese blue' and 'chinchilla' rabbits were released on Bulla and on Zhiloi islands, and in 1958 to Zimbil'nyi Island where they became established (Yanushevich 1966). In 1956, 35 rabbits were set free on Glinyanyi Island in the Caspian Sea and by September 1958 the population numbered 3000. It was found necessary here to limit their numbers because of starvation (Aliev 1960).

Rabbits have also been introduced in the Ukraine (Thompson 1956). Domestic rabbits were released in the Ukraine (Nikolaevsk and Krimsk oblasts), in Uzbekistan (Samarkandsk and Tashkentsk oblasts), in Kazakhstan (Alma-Ata and Balkhashsk regions), in Irkutsk (Sludyahnsk region), and also in the Moskovsk oblast (Kirisa 1974). In the Ukraine they were released at Odessa at the end of the nineteenth century and from the introduction point spread to the north to Baltra and southwards to Nikolajev and Chenson areas (Niethammer 1963). The spread was slow until about 80 years after introduction (Shulyatyev 1987). Between 1949 and 1972, rabbits were released in 13 oblasts in the Ukraine. Fourteen were initially released in 1949 and from 1961 on some were released every year until 1972. The total released was in excess of 2218 (Kirisa 1974).

In 1979 experimental introductions were made in Uzbekistan and later in Lithuania (Shulyatyev 1987).

Efforts with the release of rabbits in the Russian Federation and adjacent independent republics have been somewhat successful and introductions are being continued (Sofonov 1981).

Spain
Some islands off Spain do not appear to have been part of the original range of the rabbit and so they must have been introduced to them. These islands include Meda Grossa in the Islas Medas (Mas-Coma and Feliu 1977; Cheylan 1984).

Sweden
Rabbits were introduced in Scania in southern Sweden about 1904 and here reached close to their

northernmost distribution (Andersson *et al.* 1979). They were introduced to Gotland in 1907 and by 1940 were considered to be in pest numbers (Themdrup 1965), and have remained established on the island (Englund 1975; Tjernberg 1981).

Switzerland
The rabbit was introduced to some areas locally in the nineteenth century, including St. Peters Peninsula, and to the canton of Valais. In the cantons of Basel-Stadt and Basel-Land rabbits have immigrated from the surrounding countries.

The Swiss populations do not spread much as they are culled by game wardens on game preserves, and in other areas the habitat is unsuitable for any expansion (M. Dollinger *pers. comm.* 1982).

United Kingdom and Ireland
Wild rabbits are known from the Scilly Islands in 1176. It seems that they were rare on the mainland at the end of the twelfth century (Veal 1957; Fitter 1959), abundant by the sixteenth century (Thompson 1956) and fairly widespread by the eighteenth and nineteenth centuries (Fitter 1959; Sheail 1971). They were becoming a nuisance to farmers at least by 1845 in England (Thompson 1981).

The rabbit reached Scotland soon after about 1200 and later Wales, probably towards the end of the thirteenth century. They reached Ireland about the same time that they arrived in England or early in the thirteenth century, and were later introduced to virtually every island off the coast from Shetland to the Isle of Wight (Thompson 1956; Fitter 1959; Lever 1977).

In the second half of the nineteenth century the population in Britain increased phenomenally (Sheail 1971), and by 1930 it was estimated that there were up to 30 million of them. By 1950 this had risen to between 60 and 100 million (Sheail 1971). In 1953 the disease myxomatosis spread from France to England and deci-

mated the rabbit population, which has remained at a low level since this time (Sheail 1971; Lever 1977).

On islands off the coast of Britain rabbits were present on Scilly in 1176, and on Lundy between 1183 and 1219, in 1274 and were still there in 1321.

Rabbits on the Scilly Islands

Island	Date introduced	Notes
Annet	?	present 1966 (1)
Gugh	?	recorded in 1964 (1)
St. Agnes	?	there in 1966 (1)
St. Martin's	?	there in 1964 (1)
Samson		there 1960s (3)
Scilly	1176	still present (4; 2)
Tresco	1470s	(4), present 1960s (1)

References: 1 Arnold 1978; 2 Flux & Fullagar 1992; 3 Lockley 1964; 4 Sheail 1971.

Rabbits on the Channel Islands

Island	Date introduced	Notes
Alderney	before 1960?	present (3; 6)
Brecqou (off Sarke)	?	present 1980s (1)
Burhou (off Alderney)	before 1950s?	present 1950s (5)
Guernsey	before 1960?	recorded 1960–76 (7), 1980s (6)
Herm	before 1960s?	numerous 1960s (4; 8)
Jersey	before 1960s?	present (6), in 1960–76 (7)
Jethou	?	present (4)
Lihou	?	present (2)
Sark	before 1960s?	there since 1960 (3; 6)

References: 1 Armstrong 1982; 2 Borwick 1986; 3 Corbet 1971; 4 Cranbrook & Crowcroft 1961; 5 Lockley 1953; 6 Waller 1982; 7 Arnold 1978; 8 Mead-Briggs 1967.

Rabbits on the Orkney Islands

Island	Date introduced	Notes
Burray	before 1684	present 1970s (5)
Caa	1530?	present (5)
Copinsay	?	present 1971 (1)
Eday	before 1955	present 1950s (8) and 1965 (1)
Egilsay	?	present 1971 (1)
Eynhallow	before 1950	present 1950 (2), eliminated by myxo 1955 (6; 4)
Gairsay	before 1960	present since 1960 (3), and in 1965 (1)
Graemsay	?	present 1973 (1)
Green Holm	?	present 1969 (1)
Lambholm	before 1529	abundant by 1529 (7)
Old Man of Hoy	?	present 1973 (1)

References: 1 Arnold 1978; 2 Arthur 1950; 3 Corbet 1971; 4 Dunnet 1975; 5 Lever 1977; 6 Lockie 1965; 7 Ritchie 1920; 8 Thompson & Worden 1956.

Rabbits on the Farne Islands

Island	Date introduced	Notes
Brownsman	?	present 1960s, 1970s (3), still present (1)
East Wideopens	?	present (1)
Holy (Lindisfarne)	1537	there till 1940s (1), still there (5; 6; 8; 4)
Inner Farne	?	formerly present (3), now absent (2)
Staple	?	present 1967 (7), and 1980s (1)
West Wideopens	?	present (1)

References: 1 Armstrong 1982; 2 Corbet & Southern 1977; 3 Cranham 1972; 4 Garson & Haig 1986; 5 Garson 1984; 6 MacDonald 1984; 7 Arnold 1978; 8 Perry 1946.

They were recorded on Wight in 1225. Some were on Little Cumbrae in 1453. They were introduced to North and South Uist and Raasay between 1840 and 1890, and to Hascosay in the Shetlands as late as 1900.

Rabbits on the Shetland Islands

Island	Date introduced	Notes
Bressay	?	present 1969–74 (1)
Burra	before 1654 or 1684	early presence (7), 1970s (1)
Cheynies	?	present 1969 (1)
Hascosay	1900	(6), present 1969–74 (1)
Hildasay	before 1654	warrens by 1654 (2)
Holm of Melby	before 1930	4 introduced but all died about 1930 (4)
Langa	?	recorded since 1960 (3), present 1969–73 (1)
Little Roe	before 1914	present 1914 (5)

References: 1 Arnold 1978; 2 Berry and Johnston 1980; 3 Corbet 1971; 4 Fitter 1959; 5 Hopkins & Rothschild 1953; 6 Lever 1977; 7 Sheail 1971.

Rabbits on the Hebrides Islands of Scotland

Island	Date introduced	Notes
Baleshare	?	present 1974 (2)
Barra	before 1912	present (3), 1960s and 1970s (4; 33; 32)
Benbecula	before 1960	recorded 1960–76 (2)
Berneray	before 1960s	plentiful (13)
Bute	before 1912	present (3; 2)
Canna	before 1904	present (23), 1930s (9), 1970s (30) and 1980s (5; 16)
Cara	?	present 1960–74 (2)
Ceann Ear (Monarch Is)	1914–18	present 1950s and 1970s (16)
Ceann Iar (Monarch Is)	?	present 1970s (18)
Coll	before 1904	present 1904(23; 15), 1950s (7)
Colonsay	before 1764	(22), present 1980s and 1990s (8; 5; 16)
Davarr		
Eigg	before 1955	numerous 1955, declined 1975 (14), still present 1980–81 (5)
Eriksay	before 1960	recorded since 1960 (10), present 1974 (2)
Garbh Eileach (off Jura)	?	present 1970s (22), 1980s (5)
Gigha	1763	(19), present 1890 (22), 1980–81 (8; 5)
Grimsay	?	present 1974 (2)
Handa	before 1912	there 1912 (3), still present 1960s, 1980s (2; 8)
Inch Kenneth	c. 1549	(12; 28)
Inchmarnock (off Bute)	before 1960	present 1960–74 (2)
Iona	before 1912	present 1912 (3), there 1950s (17), recorded 1980–81 (5)
Islay	by 1790	(28), common 1910 (26), still abundant (16)
Jura	by 1790s	(28), recorded 1960–71 (2) and 1980–81 (5)
Kerrera	?	present 1980–81 (5)

Rabbits on the Hebrides Islands of Scotland (*continued*)

Island	Date introduced	Notes
Lewis and Harris	by 1790 and later, or 1865	(28; 4), present 1940s, 1950s (21; 12) and 1980s (11)
Lismore	?	present (2)
Luing	before 1955	present 1950s (31)
Lunga (Treshnish Is)	1867	(19), there 1940s (12), still present (5)
Mingulay	?	present (33), in 1970s (4)
Monarch Is	early in century, see Ceann Ear, Ceann Iar, Shivinish	there 1939–45, 1953–54 (27)
Muck	?	there since 1960 (10)
Mull	1549	(25), common 1912 (3), there 1895 (29), still present 1950s, 1980s (17; 5)
North Uist	before 1790s	(28), still present (24)
Oronsay	?	present 1970s (22)
Pabay	?	there 1960s (20) and 1980s (5)
Pabbay	?	possibly present (4)
Pladda	?	present ancient times (2)
Raasay	1840–1890	(19), present 1967, 1980–81 (2; 5)
Rabbit	?	present (3), there 1956 (16)
Sanda	by 1684	(19), there 1960–74 (2)
Sanday	?	there 1930s (9)
Scalpay	?	there 1960–81 (10; 5)
Scarba	?	there by 1955 (31)
Seaforth	1865	(19)
Shivinish (Monarch Is)	1914–18	there 1970s (18)
Shuna	?	there 1960–81 (10; 2; 5)
Skye	by 1790s	(28), still present (5)
Soay	?	present 1972 (2), 1980–81 (5)
South Rona	?	present 1980–81 (5)
South Uist	by 1790s	(28), 1980s (1)
Tiree	?	none present now (16)
Ulva	?	present 1960s (2), 1980s (5)
Vallay	1905	(6), numerous 1916–27 (28)
Vatersay	?	present (3), 1970s (2; 5)

References: 1 Armstrong 1982; 2 Arnold 1978; 3 Barrett-Hamilton 1912; 4 Berry 1979; 5 Berry 1983; 6 Beveridge 1932; 7 Boag 1987; 8 Booth & Perrott 1981; 9 Carrick 1939; 10 Corbet 1971; 11 Cunningham 1987; 12 Darling 1947; 13 Diamond *et al.* 1965; 14 Evans & Flower 1967; 15 Fitter 1959; 16 Flux & Fullagar 1992; 17 Gillham 1957; 18 Hepburn *et al.* 1977; 19 Lever 1977; 20 Lister-Kaye 1972; 21 Lockie & Stephen 1959; 22 Mercer 1974; 23 Millais 1904–06; 24 Newman 1988; 25 Ritchie 1920; 26 Russell 1910; 27 Shanks *et al.* 1955; 28 Sheail 1971; 29 Simpson 1895; 30 Swan & Ramsay 1978; 31 Thompson & Worden 1956; 32 Thompson 1974; 33 Williamson & Boyd 1963.

On Lunga (Treshnish Isles) they were released there by lobster fishermen in 1867, and on Lewis there were several unsuccessful attempts before 1865, but eventually they became established on an island in Loch Seaforth. On several of the Shetland Islands they have now died out: on South Havra they were killed off by cats; on Trondra they were unsuccessfully introduced several times; on Holm of Melby they were released for shooting purposes and rapidly increased, but died out suddenly in 1930; on Oxna they failed because of a flood. On Hilbre Island, Wales, they were present in 1540, nearly hunted out by 1913 and disappeared about 1939. Rabbits were released on Clare Island, County Mayo in 1906 and in 1911 on Rathlin Island, County Antrim. Rabbits are now present on Inner Farne (now exterminated), West Wideopens, East Wideopens, Staple and Brownsman (Fitter 1959; Sheail 1971; Armstrong 1982).

INDIAN OCEAN ISLANDS
Amirantes group

Rabbits are present on the islands of Desnouefs (Racey and Nicholl 1984; Stoddart and Fosberg 1984) and Poivre (Stoddart and Fosberg 1984).

Amsterdam (France)
Rabbits appear to be established on this island(?) (Watson 1975).

Assumption (Aldabras)
Dupont (1907) suggested introducing rabbits and hares to this island, but no releases are documented.

Crozet Archipelago (Îles Crozet, France)
The exact date of introduction of rabbits to the Île aux Cochins (Hog Island) is not known. Some early writers (Vallaux 1928; Jeannel 1941) suggest about 1820, but more recent studies (Derenne and Mougin 1976) of a number of early manuscripts (including Cecille 1840) indicate that none were found there in 1837–39, and that the introduction was more likely to have been by sealers in about 1840–50. Nearly 50 expeditions, mostly whalers, visited the Crozet Archipelago between 1840 and 1850 (Roberts 1958). These dates for the introduction of rabbits are supported somewhat by the fact that they were intro- duced in numerous localities by sealers and whalers in the sub-antarctic in the second half of the nine- teenth century. On a number of early visits (e.g. 1874, 1876 and 1887) by navigators to Crozet, rabbits were abundant on the Île de Cochins (Kidder 1876; Brine 1877; Richard-Foy 1887).

Rabbits were reported to be abundant in 1873 (Mosely 1892) on Île de la Possession, where they were also noted in 1938 (Jeannel 1941), but had disappeared by 1959 (Holdgate and Wace 1961). Some writers have disputed their introduction (Dorst and Milon 1964), while others record an introduction in the nineteenth century, but say they did not survive (Clark and Dingwall 1985). Rabbits were probably introduced to the island of Est in 1841 and still persisted there in the 1970s (Despin *et al.* 1972).

Rabbit distribution in the 1970s and 1980s on Cochin is discontinuous on the island, but they are particu- larly abundant on the east coast. The population reaches a maximum of 2000 individuals during the summer, mortality is low and is caused chiefly in winter by the predation of introduced cats (Derenne and Mougin 1976; Voisin 1984).

Desroches
Rabbits may have been present in 1905 (Gardner and Cooper 1907; Stoddart and Poore 1970), but there are no further references to them.

Kerguelen (France)
Liberated in 1874 by a British Transit of Venus expe- dition, domestic rabbits became abundant on the island of Grande Terre by 1873 (Kidder 1876; Reppe 1957; Lesel 1967; Watson 1975). They have continued to be remarkably successful and their range extends over all of Grand Terre, except south of Péninsule Rallier du Baty, and they also occur on many other islands and islets (Lesel 1967, 1968; Lesel and Derenne 1977) including Chat, Cimitière, Foch, Mayes, Morbihan Bay Islets and Ouest.

A few rabbits were placed on Chat just before 1955 to save them from extinction when myxomatosis was introduced (Lesel 1967). They are present on Cimitiere, but their origin is obscure, and they occur on the island of Foch and the islets in Morbihan Bay (Lesel 1967). They apparently occurred on Mayes as the modified vegetation attests (Zotier (1990). They are reported to have been introduced on Ouest Island in 1874 (Elliott 1972).

Myxomatosis, which was introduced in 1955, has had no great effect on the population, although there are permanent pockets where it occasionally flares up, but affected areas soon recover. The average density of rabbits on the Courbet peninsula reaches 23/ha (Lesel and Derenne 1977). A report suggests that a comparison of their skulls from 1900 to 1972 shows a gradual regression to the wild-type rabbit (Mougin 1975).

Maldive Islands
Rabbits were introduced and became established in the Maldives (Niethammer 1963), perhaps intro- duced by the Portuguese or Dutch in the sixteenth or seventeenth century.

Madagascar
Rabbits were introduced to l'île Europa (atoll in the Mozambique Channel) about 1860, but no longer occur on the island (Malzy 1966). However, they may occur there now (Flux and Fullagar 1992). Their present status is uncertain.

Mauritius
Rabbits were introduced to Round Island (north-east of Mauritius) in about 1810 and were abundant in 1845. The first introductions are thought to have been *Oryctolagus* but this species is reported to have died out and was later replaced by black-naped hares (*Lepus nigricollis*), which are now established there (North and Bullock 1986). The rabbits present on the island were apparently poisoned and eradication was complete by mid-1987 (Merton 1988 in Flux and Fullagar 1992).

Seychelles
Rabbits are present or have been found on the islands of Cargados Carojos, Chauvre-Souris, Mahé, Marnelle, Praslin and Récife, undoubtedly introduced or escaped from captivity at some time. Rabbits, probably *O. cuniculus,* occur on Chauvre-Souris (off

Praslin), Marnelle and Récife (Racey and Nicholl 1984), and also on Cargados Carojos (Stoddart and Fosberg 1984). On Mahé and Praslin domestic rabbits frequently escape from captivity, but do not appear to become permanently established.

St. Paul (France)
Early seafarers left rabbits on Île St. Paul, probably in 1874 or in about 1880, as a source of food (Hesse 1937; Holdgate 1967; Segonzac 1972). They increased rapidly in numbers causing considerable damage to the vegetation, but latterly decreased, possibly through disease, and in 1957 the vegetation was reported to be recovering (Reppe 1957; Holdgate 1970). However, they were probably abundant again in the 1960s (Gill 1967). Rabbits have colonised the whole island, but are densest on the slopes of Grand Morne, inside the crater (Segonzac 1972)

Tromelin (off Madagascar)
Rabbits were found to be present on the island in 1968 (Staub 1970), probably introduced after the meteorological station was built in 1954 (Flux and Fullagar 1992). None were reported present there in 1953 or in 1962 (Paulian 1955; Morris 1964).

NORTH AMERICA
Alaska
A single male and three female domestic rabbits were released on Middleton Island, Alaska, in 1954 (O'Farrell 1965). At first they lived under the houses and were fed by the residents, but became established in the wild and in 1955 the population was estimated to be about 50 rabbits. They continued to increase in numbers and in the summer of 1956 there were 200 of them. By February 1961 the population was between 3600 and 3700 and by the summer of 1962 some 5000 rabbits were present on the island.

Rabbits were introduced to the Aleutian Islands as a source of food for foxes. Opinions differ as to whether they existed there in the 1980s or before (Flux and Fullagar 1992). They were liberated on Adak by sailors and were present there in the 1960s. They are present on Hay but their origin is unknown. Domestic rabbits were introduced on Popof from Hay shortly after 1955 and were reported to be there in the 1980s (Flux and Fullagar 1992). Rabbits were introduced to Annaniuliak (off Unimak) before 1952 (Jones and Byrd 1979) and were still there in 1981.

Canada
European rabbits have been released a number of times over the years to various islands in British Columbia, Canada. Liberations of rabbits were made on Bare Island, Chatham Island, Strongtide Island, Piers Island, Vancouver Island, South Pender Island, Graham Island, Triangle Island, and the Queen Charlotte Islands, all of which had populations prior to 1910. However, most of these populations had died out by this date. In 1972 only Sidney Island, Triangle Island and Vancouver Island retained populations of any size. On Vancouver Island they were spread over a small area from Sooke to Goldstream (Carl and Guiguet 1972).

United States
There have been many attempts to establish rabbits on the mainland and on many small offshore islands from about 1895 onwards. Fortunately for the farmers these attempts were unsuccessful on the mainland, but some on islands were successful.

The earliest records appear to be those from the San Juan group of islands in Puget Sound, Washington. Rabbits were on Skipjack Island in 1895 and may have been there in the early days of the Hudson's Bay Company (Couch 1929). The original stock introduced was domestic rabbits and may have been introduced to San Juan in the 1880s (Stevens and Weisbrod 1981). They are now widespread on the southern part of the island (Hall 1977). According to Couch (1929), they were probably introduced to Smith Island by the lighthouse keeper in about 1900. Whether they were accidentally or deliberately spread to other islands in the archipelago is not reliably known. By 1924 the vegetation had suffered so much that the population on Smith Island was reduced to a few by poisoning and gassing their burrows. Those rabbits on Long Island were said to have been obtained from Smith Island and probably Colville Island and also part of Whidby Island. In 1929 they were present on the islands of San Juan, Wasp, Jones, Spieden, Flattop, Johns, Skipjack and Mateo, in the San Juan group (Couch 1929; Hall 1977). They have remained on the islands to this day and it was estimated that the population was half a million rabbits at any one time, although efforts to destroy them have continued (Stevens and Weisbrod 1981).

Many rabbits were released in the United States between 1952 and 1958, but none successfully (Kirkpatrick 1955; Presnall 1958). At this time there was much opposition to the introductions (Anon. 1954; Ohio Dept. Nat. Res. 1954; Wildl. Mgmt. Inst. 1954; Davids 1955). Stock from the San Juan group were introduced into Ohio, Michigan, Illinois, Wisconsin (Lemke and Oshesky 1955), Indiana, Pennsylvania (Latham 1954) and New Jersey (McNamara 1955) in this period.

Indiana sportsmen began importing and liberating stock from the San Juan group in about 1949 and nearly 6000 rabbits were released in 50 counties in the next 10 years before further introductions were prohibited (Kirkpatrick 1958, 1960). It was concluded at the end of this time that they had failed because of the lack of suitable habitat.

Feral domestic rabbits became established on the Santa Barbara Islands, off California, during World War 2 (Presnall 1958; Kirkpatrick 1959; Von Bloecker 1967). They have also become feral on South Farallon Island off central California where they were introduced as a source of food by lighthouse keepers about a century ago (Lidicker 1991). On these islands they have caused considerable damage to the vegetation and much control was being carried out in the mid-1950s (Presnall 1958). However, there was still a small population on Santa Barbara between 1975 and 1979 (Murray *et al.* 1983).

PACIFIC OCEAN ISLANDS

New Zealand

The rabbit became established in New Zealand in the 1860s following many importations, beginning in 1838 and continuing up until 1858 (Thomson 1922; Wodzicki 1950; Howard 1958; Gibb and Flux 1973). The exact date is not known (King 1990). Before 1858 there was little or no spread, but between 1864 and 1867 there was rapid spread of wild stock rabbits (Gibb and Flux 1973). In 1873 some 33 000 rabbit skins were exported, in 1877 nearly a million and in 1882 over 9 million. This latter figure increased to nearly 17 million skins and carcases in 1947.

The introduction of rabbits to New Zealand was made for sporting purposes by government agencies, acclimatisation societies, farmers and prospectors (Howard 1958; Lamb 1964). By the 1970s they were firmly established in Otago, Southland, and gradually spread over both islands with the help from people who captured and released them in other areas. Clark (1949) records that by 1869 they were so well established and widespread that a million acres had been damaged in the Marlborough area of the South Island. Between 1920 and 1940, rabbits occupied all of the suitable habitat for them in New Zealand (Wodzicki and Wright 1984).

Rabbits have been present on at least 57 offshore islands and are still present on 27 of these (King 1990). The earliest date to those placed on Motuara by Captain James Cook, who liberated two pairs in 1777.

Some islands off New Zealand with rabbits

Island	Date introduced	Notes
Browns (Hauraki Gulf)	c. 1975	eradicated 1985–91
Korapuki	c. 1900	eradicated 1986–88
Mangere	<1890	eradicated by cats in 1890s
Mokopuna (Leper)	1946	eradicated 1947–54
Moutohora	1968	eradicated 1985–87
Motunau (Cook Straits)	c. 1850	eradicated 1958–63
Native (Stewart)	c. 1932	eradicated 1949–50
Otata	?	eradicated 1945
Stanley	?	eradicated 1991–92
Stewart (part)	1932	eradicated 1948–50
Takangaroa	<1930	eradicated c. 1950
Tiri Matangi	<1894	eradicated 1900–20
Enderby	1865	attempted eradication 1993–
Quail	c. 1855	attepted eradication 1989–
Rose	1851	attempted eradication 1993–
Taieri	?	attempted eradication 1992–
Motuara	1777	2 pairs liberated by Cook
Auckland	<1866	present on Rose and Enderby
Campbell	1883	not now present?

References: Taylor 1967; Atkinson & Bell 1973; Bell 1975; Veitch 1995; King 1990; Holdgate & Wace 1961; Taylor 1968; Challies 1975

Rabbits are still common and well established on both the North and South islands. They have died out or been exterminated on several offshore islands, but still survive on 24 of them. The distribution of rabbits appears to have changed little since the late 1940s (King 1990). Large numbers appear to persist only where the climate resembles that of the western Mediterranean.

Auckland Island (New Zealand)

The first rabbits landed on Enderby Island in 1840 apparently died out, and 12 more from Victoria were landed in 1865 (Bull 1960). These were successful and have remained mostly in the coastal grasslands ever since (Bull 1960; Taylor 1971; Bruemmer 1983). An eradication plan was being considered in the early 1990s (Peat 1991).

Rabbits were initially placed on Friday Island, but because the island was too small they became short of food and so were transferred to Rose Island about 1850 (Bull 1960; Taylor 1968, 1971) or between 1840 and 1860 (Taylor 1968).

On Rose and Enderby islands they were plentiful and had caused severe damage to the vegetation before 1866 (Holdgate and Wace 1961). In the late 1960s their numbers were declining on both islands (Taylor 1968), but some were still present on both in the 1970s (Challies 1975), and mid-1980s (King 1990).

Campbell Island (New Zealand)
Rabbits may have been introduced in 1883 (Holgate and Wace 1961), but there is doubt about this date and it could well have been earlier (Flux and Fullagar 1992). Rabbits are now absent from this island (King 1990).

Fiji
Cook left two rabbits from the Cape, South Africa, on Tonga Taboo (Tongatabu), Fiji (Kippis 1904). Whether these escaped and became established or whether there were later introductions is not known. The introduced rabbits appear to have died out by the 1870s or before (Brenchley 1873).

Galápagos Islands (Ecuador)
In 1965 a small colony existed on an island in the Galápagos, but they are no longer present there (Duffy 1981).

Juan Fernández Islands
Rabbits were present on the island of Morro Vinillo in the 1930s (Lever 1985). They were introduced to Más á Tierra (Robinson Crusoe) in 1935 (Perry 1984; Colwell 1989) and were still present there in the 1990s (Bourne et al. 1992). They were present on the island of Santa Clara in the 1970s (Pine et al. 1979), and in the early 1990s (Bourne et al. 1992). Rabbits were also apparently on Más Afuera (Alejandro Selkirk) in the 1970s (Pine et al. 1979). The removal of rabbits from the Galápagos Islands was recommended in the 1980s.

Hawaiian Islands (United States)
Rabbits have been introduced to a number of small islands in the Hawaiian chain. The earliest introduction appears to have been that on Ford Island (previously called Rabbit Island), in Pearl Harbour, before 1825 (Anon. 1925; Tomich 1986). Other islands where introduced rabbits occurred include Hawaii, Kauai, Laysan, Lehua, Lisianski, Manana, Maui, Mokuola, Molokini, Oahu, and Southeast.

Rabbit tracks were found on the island of Hawaii in 1951 and 1968 (Tomich 1986), and although rabbits escape regularly they do not ever become permanently established (Kramer 1971; Tomich 1986). A small colony of rabbits existed on Kauai, near Hanalei, for several years before finally dying out (Kramer 1971). Another colony more recently in central Kauai also suffered the same fate (Tomich 1986).

Domestic white rabbits (Belgium and English varieties) were released on Laysan Island in 1903–04 by M. Schlemmer, manager of a guano firm (Dill and Bryan 1912; Bryan 1915). Some authorities (Tomich 1969) say 1902–03, but others (Ely and Clapp 1973) agree with 1903–04. The rabbits concerned were originally imported for the purpose of starting a canning business on the island. By 1911, however, they were swarming over the island in thousands (Dill and Bryan 1912) and had already eliminated several species of plants. Efforts at rabbit destruction were made in 1912–13 when many thousands (5024) were killed (Ely and Clapp 1973). In 1914 they were reported to have decreased in numbers, but by 1915 were abundant again (Munter 1915). By 1922–23 only a few hundred remained (Wetmore 1925) and a second expedition was mounted to remove the last of them as the flora on the island had now virtually been destroyed (Kramer 1971; Ely and Clapp 1973). Following this extermination effort no rabbits have been reported from the island.

On the island of Lehua, off Niihau, rabbits were probably introduced before 1915 (Watson 1961). It is suggested that some domestic rabbits were introduced in about 1930, but rabbits are already reported there in 1930 (Caum 1936). They were present in the 1950s (Fisher 1951) and in 1966 the island boasted the largest population in the Hawaiian Islands (Kramer 1971; Tomich 1986).

The exact date that rabbits were introduced to Lisianski Island is not known (Clapp and Wirtz 1975), but certainly before 1914. They are said to have been taken from Laysan to Lisianski Island by M. Schlemmer and released about 1903–04 (Watson 1961; Kramer 1971), or as early as 1902–03 (Tomich 1969), or as late as 1904 and 1909 (Clapp and Wirtz 1975), certainly after 1904. Rabbits were reported there in 1910 and by 1913 were certainly abundant on the island. By 1914 they had destroyed the vegetation and consequently themselves (Elchner 1915). Seven rabbits were removed in 1915 (Munter 1915) and there were none there in 1923 (Watson 1961).

Domestic rabbits were present on Manana Island (Rabbit Island), off Oahu before 1915 (Watson 1961) and were possibly introduced about 1900 (Kramer 1971). At first it was thought that they were cottontails (*Sylvilagus* sp.), but they were positively identified at a later date as *Oryctolagus cuniculus*. The population has remained at moderate levels (Tomich *et al.* 1968), probably between 30 and 60 individuals (Dixon 1973). They were present there in the 1960s (Kramer 1971; Hirai 1979), 1970s (Brown 1974), appeared common in the early 1980s, but none have been seen since 1984 (Swenson 1986).

Six domestic rabbits were released on Maui in 1989 in Haleakala National Park where they increased in numbers rapidly (Cole *et al.* 1991). Attempts to eliminate them have so far not succeeded (Loope *et al.* 1991).

It is not known when domestic rabbits were released on Molikini Island (between Kahoolawe and Maui), but it is thought to be after 1930 (Watson 1961; Tomich 1986). Some were there in 1954 and in 1961 (Kramer 1971), but they now appear to have died out. An early attempt was made to introduce rabbits on Mokuola in Hilo Bay, but this apparently failed (Hall 1873; Tomich 1986). Rabbits were also released on Molokai before 1915 (Watson 1961), but there are no further records of them. Although rabbits are occasionally observed on Oahu, no established colonies have yet been identified (Tomich 1986).

A domestic variety of rabbit was recorded on Southeast Island, Pearl and Hermes Reef, in 1916 and these are believed to have been brought there from Laysan or Lisianski (Amerson *et al.* 1974). A number were noted there in 1923 when 90 or more were killed. In 1924, two or three were killed, 20 in 1927, and in 1928, three were shot, but none have been seen on the island again.

Japan

Domestic rabbits were introduced to Japan in the sixteenth century (Yamada 1991). Some may have been released in some districts near Nagasaki, Kyushu, in the 1840s or earlier (Temminck 1845).

Rabbits have been introduced to the islands of Izu, Jinaito, Kyushu, Mae-jima, Matsushima, Motokojima, Nanatsujima-Oshima, Ohkunojima, Okinosima, Oshima-Oshima, Osima-kojima, and Ushibuku-Oshima.

Two hundred and thirty domestic rabbits were released on Mae-jima, an islet off Okinawa, Japan, in July 1958 (Asahi 1962), 50 more in August and 40 in December. The population began to decline so 50 of the domestic variety were introduced in 1959, but these were reduced to 40 by a cyclone. In 1960, 100 domestic rabbits were introduced, and in 1961 a further 50. By 1990 the population numbered 368 rabbits and seemed stable; the rabbits were introduced for the benefit of sightseers (Yamada 1991).

Some rabbits have been introduced into some small uninhabited islands of the Izu Islands (Seven Islands of Izu, off Tokyo) (Imaizumi 1970). About four rabbits were introduced into Jinaito (Niijima, Tokyo) in 1934 and the 1950s for sightseeing tourists (Yamada 1991). There was a stable population of about 20 in 1970 that was causing some erosion and damage to plants.

In the 1940s rabbits were introduced to Matsushima (off Hyogo) where the population is stable at about 100 individuals; also in 1940, 15 rabbits were introduced to Motokojima (Okayama) where the population has stabilised at 50; two pairs of rabbits that were introduced in 1984 on Nanatsujima-Oshima (Ishikawa) reached a stable population of 200–300 by 1990; on Ohkunojima (Hiroshima) five pairs of rabbits introduced in 1967 reached a population of about 400 in 1990 and are still increasing; two or three rabbits were introduced to Okinosima (Shimane) in 1977–78 and about 100 were present in 1990, but numbers are declining; about 20 rabbits introduced to Oshima-Oshima (Matsumae, Hokkaido) reached a stable population in 1945 of about 300 in 1980, and on Osima-kojima (Matsumae, Hokkaido) 10 rabbits introduced in 1980 reached about 50 in 1990 and are still increasing in numbers despite predators such as cats and rats being present; a pair of rabbits introduced to Usshibuku-Oshima (Kumamoto) in 1982 resulted in a population of 200–400 in 1990, which now seems to be decreasing. On most if not all of these islands, the rabbits are causing soil erosion and/or damage to the plants (Yamada 1991).

Kuril'skiye Ostrova (Kurile Islands, Russian Federation)

Rabbits were released on these islands in 1946 (Kirisa 1974), but their origin and success do not appear to be documented.

Macquarie Island (Australia)

Domestic 'French' rabbits were taken by sealers from Dunedin, New Zealand, and liberated in North East Bay in 1879 (Cumpston 1968) or 1880 (Mawson 1943; Holdgate 1967) to provide fresh meat on subsequent visits (Watson 1975). Here they multiplied rapidly and were abundant at the northern end of the island in 1882 (Scott 1882 in Taylor 1979) and in 1888 (Chamberlain 1888 in Taylor 1979).

Large numbers of rabbits were noted in 1884 and continued to be recorded ('swarming' and 'numerous') on the island until 1906, but thereafter were rare in 1909–10, scarce in 1918 and 1923 (Cumpston 1968), and remained not very abundant until 1930 (Mawson 1943). However, they subsequently invaded all except the northern portion of the island, causing much damage to the vegetation (Taylor 1955; Costin and Moore 1960). They were still well established, widespread and abundant in the mid-1960s (Wodzicki 1965; Johnston 1966) and the late 1970s (Taylor 1979).

In 1968 the rabbit flea, and in December 1978 myxomatosis, were successfully introduced to the island (Copson *et al.* 1981; Brothers *et al.* 1982). In 1956 the population was estimated at 500 000 rabbits, and in 1965–66 some 150 000 rabbits; and in 1974 the population was 50 000 and in 1977–78 was again 150 000 (Copson *et al.* 1981). The population is now reported to have been reduced substantially to allow the flora to recover (Brother *et al.* 1982), although this may not have happened (Scott 1989).

Norfolk Islands (Australia)
Rabbits were liberated on Philip Island, off Norfolk, probably by the early settlers to provide food and sport for the residents of Norfolk (Coyne 1982). They were introduced in 1788 (Boback 1970) or in 1790 (Coyne 1982). A number were shot in 1838 (Taylor 1966) and the island was said to abound with rabbits (Murray 1857). They were present on the island in 1865 (Laing 1915), still present in 1943, and probably two to four dozen were there in 1961 (Watson 1961). Most of the island is now devoid of vegetation and conservation is underway (Coyne 1982). In 1981 myxomatosis was introduced, and in 1983 poisoning followed by trapping, shooting and gassing took place. The last rabbit is thought to have been shot in 1988 (Hermes *et al.* 1988).

One rabbit was released on Norfolk Island, but a hunt was under way to eliminate it (Anon. 1974). Rabbits were once present on the island of Nepean (Fullagar 1978).

Phoenix Islands (Kiribati)
Domestic rabbits were liberated in the 1860s on Phoenix Island by an American guano company (Watson 1961), where they were plentiful in 1889, numerous in 1924 and 1937, but in poor condition in the early 1950s (Maude and Maude 1952). A survey party caught some in 1957 (Watson 1961). The population of 100–1000 is kept in check by crews from copra boats (King 1973). Rabbits are also present on the island of Birnie (Fosberg 1983), but little information about them appears to be documented.

SOUTH AMERICA
In South America there have been two areas of introduction of the rabbit – in central Chile from where they invaded central-west Argentina – and in Tierra del Fuego (Jaksic and Yanez 1983).

Argentina and Chile
Rabbits were introduced to Chile in the mid-eighteenth century (Housse 1953). Rabbits were noted near Provenir and at Punta Arenas on the mainland in 1939–40 (Osgood 1943).

In 1950 rabbits were introduced near Ushuaia by Argentinean Navy personnel and a private rabbit farmer (Goodall 1979). In a few years they had become pests and had devastated the pasture from Bahia Felipe southwest to Bahia Inútil and Cameron, and north-east to the Chile–Argentina border (Anon. 1950). It was estimated in about 1950 that this area supported some 30 million rabbits (Arentsen 1953). By 1960 they were locally abundant in Malleco Province (Greer 1965) and by 1968 were considered a serious pest of agriculture throughout central Chile (Pefour *et al.* 1968; Pine *et al.* 1979; Jaksic *et al.* 1979).

However they obtained their start, the rabbits through natural dispersal and intentional releases penetrated north-west to about latitude 30°S in Chile, which is more than 6°N of where they were first observed in Argentina (Howard and Amaya 1975). They appeared on the west central border of Argentina in Minas in the north-western portion of Neuquén Province between 1945 and 1950. Circumstantial evidence suggests that they emigrated from Chile, but it is always possible that someone released them there. They later spread to Mendoza Province and are spreading northwards towards the Rio Grande and eastwards down the Rio Colorado at about 8 km per year (Howard and Amaya 1975).

It is thought (Jaksic and Yanez 1983) that a reduced predation pressure from native species in central Chile compared to that of rabbit's native range has been responsible for the rabbit's success in this area. There appear to be a few predators in Chile compared with Spain, and what predators there are appear to prefer to prey on the native rodents (Jaksic *et al.* 1979). However, more open microhabitats than in the rabbit's native range may also have aided its survival and spread in Chile and Argentina and probably in Australia (Simonetti 1989).

In Chile rabbits occur mostly around agricultural and grazing lands (Miller 1973). They were introduced to Isla Chanaral at the turn of the century to provide food for stranded fishermen (Arraya 1983). In 1985 they were present on this island off Chile, but are uncommon there (Modinger and Duffy 1987).

WEST INDIES–CARIBBEAN
Rabbits have been introduced to Barbados, one of the Grenadines, Guadaloupe, Dominican Republic,

Cuba, Jamaica and St. Croix (de Vos *et al.* 1956; Flux and Fullagar 1992).

Barbados
Rabbits were introduced to Barbados at some time but did not survive (de Vos *et al.* 1956; Flux and Fullagar 1992).

Cuba
Rabbits were introduced to Cuba about 1880 from the Canary Islands (Varona 1974).

Dominican Republic
Rabbits were introduced to Catalinita between 1931 and 1961, probably in the 1950s (Oliver 1985; Flux and Fullagar 1992).

Grenadines
Rabbits have been introduced to the island of Balliceaux in the Grenadines (de Vos *et al.* 1956) probably about 1880 (Varona 1974). They were reported present in 1903 (Allen 1903) and are still present on the island.

Guadeloupe
Although rabbits were an early introduction to Guadaloupe before 1654 (Allen 1911; de Vos *et al.* 1956), they apparently have not survived and there are no further details (Flux and Fullagar 1992).

Jamaica
Rabbits were possibly introduced to Jamaica before 1851 but had become extinct before 1905 (Flux and Fullagar 1992).

Virgin Islands
Rabbits are kept on free-range farms on St. Croix but are never seen beyond them (Flux and Fullagar 1992).

■ DAMAGE
Damage leading to the devestation of pastures, ruin of sheep farmers and erosion of land by rabbits has occurred in Europe, Australia, New Zealand and Chile.

In England, pre-myxomatosis, the rabbit was a major pest with annual damage costs estimated to be 50 million pounds (Corbet and Harris 1991). Damage to such crops as winter wheat was estimated at 6.5 per cent or 1.6 cwt grain per acre in the United Kingdom (Church *et al.* 1953, 1956). Rabbits extentsively grazed and damaged cereals and grasslands, causing a marked reduction in total yield of herbage. They cause damage in forests and orchards by distorting the growth of trees and by damaging leading shoots, preventing natural revegetation by eating seedlings and killing trees by ringbarking, and causing damage to market garden crops (Thompson 1951, 1956).

On Ramsey Island, off Pembrokeshire, South Wales, it has been reported (Cowdy 1973) that the uncommon chough (*Pyrrhocorax pyrrhocorax*) relies on ants for food. The dispersion and density of the ants is determined partly by the close grazing of the rabbits, which are the only mammalian grazers present (Doncaster 1981). Although there are 20 000 rabbits on the island, their removal may be disastrous for the choughs unless they are replaced by another grazer.

Since their introduction in Chile, rabbits have effected a pronounced change in the spatial distribution of native herbs (Jaksic and Fuentes 1980), probably enhancing the success of introduced weeds, and also causing considerable mortality of the native chaparral (Jaksic and Soriguer 1981).

There is little doubt that the cost of rabbits to Australia has been enormous, but there are few well-researched data. It is estimated that there were 750 million rabbits in Australia before myxomatosis in 1952–53. In 1952 in Tasmania 16 000 were destroyed on a single 1600-acre property. From 1952 to 1960 the sheep wool clip on that property went from 11 770 to 30 654 lbs and the number of rabbits destroyed fell from 16 000 in 1952 to 300 in 1960. Sheep numbers doubled. After myxomatosis it became obvious the way in which rabbits affected the vegetation. Establishment of the rabbit has been a major disaster both ecomomically and ecologically in Australia's biological history (Douglas 1981).

Rabbits are considered the greatest pest of the pastoral industry in Australia. They have a marked effect on the environment, particularly semi-arid and sub-alpine areas. They may alter the floristic competition of pastures by selective grazing (Myers and Poole 1963; Leigh *et al.* 1987, 1989), damaging grain crops (Hone *et al.* 1981), stripping vegetative cover and causing soil erosian and allowing the invasion of weeds (Croft 1986). They damage shrub and tree species in the semi-arid zone (Wood 1984; Fiedal 1985) by ringbarking and consuming seedlings (Cooke 1982). Some native species of mammals, such as bilbies (*Macrotis lagotis*), are reported to disappear after invasion by rabbits (Wilson *et al.* 1992).

Large changes in species composition occurred when rabbits grazed ryegrass–clover pastures. Ryegrass reduced and poor quality palatable grasses and weeds increased, with effects proportional to rabbit density, but even at low densities of 10–12 rabbits/acre. Pasture yield can be reduced by as much as 25 per cent at densities of 10–20 rabbits/acre (Myers and Poole 1963).

The effects of rabbits on islands has been disastrous for vegetation and wildlife. On Laysan and Lisianski islands, Hawaii, rabbits have decimated the flora, but not so much on the other islands. The reasons for this are not known and little studied (Kramer 1971). There has been severe modification of Macquarie Island vegetation and this has indirectly affected the native fauna. Here, the introduction of myxomatosis and the reduction in numbers may allow flora to recover (Costin and Moore 1960; Brothers *et al.* 1982). On Macquarie Island, rabbits virtually eliminated the tussock grassland from many of the steep coastal slopes (Taylor 1955; Costin and Moore 1960).

In Germany rabbits are largely considered a pest (Webb 1960).

On Kerguelen rabbits are said to have altered the vegetation substantially by destroying much of the lowland vegetation (Watson 1975). The plant composition has changed and Kerguelen cabbage (*Pringlea antscorbutica*) has disappeared from areas inhabited by rabbits (Lesel and Derenne 1977).

In the Channel Islands off California rabbits have also threatened the survival of the endemic plants (Presnall 1958).

Following rabbit removal by disease(?) on St. Paul, recolonisation of the barred ground by vegetation was nearly complete by 1957 (Reppe 1957). Prior to this the vegetation had been devestated and there was much soil erosion (Holdgate and Wace 1961).

SNOWSHOE HARE
Varying hare, snowshoe rabbit
Lepus americanus Erxleben

◼ DESCRIPTION
HB 360–520 mm; T 24–55 mm; WT 0.9–2.27 kg.
Summer coat is dusky grey to reddish brown, intermixed with black on back; legs and throat light brown, ears small, brownish with black tips and creamy white edges; under parts white; hind feet large; feet sometimes white, soles hairy. Winter coat is white with black ear tips but some subspecies retain the summer coat in winter.

◼ DISTRIBUTION
North America. Alaska and Canada (northern limit is tree line of Yukon Territory, Northwest Territories and Labrador, and Newfoundland), south to the Allegheny Mountains of North Carolina and Tennessee, and the Rocky Mountains in northern New Mexico and California, United States.

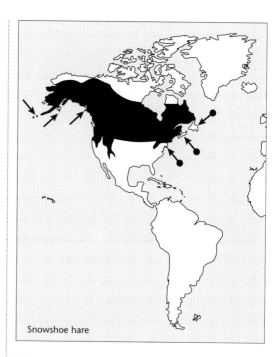

Snowshoe hare

◼ HABITS AND BEHAVIOUR
Habits: crepuscular and nocturnal; non-burrowing, builds form in a thicket to lie up in during day; numbers cyclical 9–11 years; coprophagous. **Gregariousness:** in groups?; dominance hierarchy; young gather at nursing place after sunset to suckle; male territory overlaps that of several females. **Movements:** up to 8 km from feeding grounds to lair known; home range 6–13 ha. **Habitat:** timbered country, clearings, lake shores, semi-open forest, mixed hardwood forest, cut-over forest, swamps. **Foods:** grass, forbs, herbs, tender leaves of shrubs and trees; bark, twigs of birch, spruce, willow, alder, tamarack and pines; sedges, dandelions. **Breeding:** breeds March–September; females polyoestrous; gestation 30–40 days; young (leverets) 1, 2–4, 10; litters 2–4/year; born furred, alert and eyes open; leave nest about 3 weeks; young mature by spring following birth. **Longevity:** 4–5 years in wild, but up to 8 years in captivity. **Status:** locally common.

◼ HISTORY OF INTRODUCTIONS
NORTH AMERICA
Snowshoe hares have been translocated in the United States, and introduced to Anticosti Island and Kodiak Island.

Alaska (United States)
Snowshoe hares have been successfully introduced in Alaska (Burris 1965). Some 18 hares from Washington were taken to Behm Canal in 1923 and 20 were taken from Washington to Admiralty Island

(Alexander Archipelago) and Barlow Islands in 1924. Twenty were taken from Washington to Otstoia Island (Peril Strait) in 1924, and 24 from Anchorage to Village Island (Zimovia Strait) the same year. In 1934, 558 were taken from Anchorage to Kodiak Island (Gulf of Alaska) and Afognak Island (off Kodiak). In 1952, 12 snowshoe hares from Olga Bay, Kodiak, were taken to Woody Island (off Kodiak) and six were taken from Olga Bay to Long Island (Kodiak). In 1955, 15 rabbits from Kodiak Island were taken to Popof Island (Burris 1965).

Aleutian Islands (United States)
Hares, presumably *Lepus* species, were introduced to Hog Island in Unalaska Bay (Unalaska Island) before World War 2, but none were found there in 1964 or 1967 (Peterson 1967).

Canada
The snowshoe hare (race *L. a. struthopus*) was successfully introduced into Newfoundland in 1864 (Keith 1974; Banfield 1977). The introduction to Anticosti Island, Quebec (in the Gulf of St. Laurence), took place in 1902–03, and they were plentiful there by 1936 (Newsom 1937).

United States
From 1927 to 1937, some 58 396 wild snowshoe hares from eastern Maine, Wisconsin and Minnesota were introduced by the New York authorities mainly to the counties of Clinton, Essex, Warren, Franklin, Herkimer, St. Laurence, Sullivan, Delaware, Cataraugus, Albany and Renselaer. Some of these became established and bred, but many disappeared (Bump 1941). Some were released in Ohio before 1954 for restoration purposes (Chapman 1954). Further attempts were made in New York in 1952–53 (Dell 1952–53) and stocking was carried out in New Jersey in about 1953 (Francine 1953).

Snowshoe hares have also been introduced to Nantucket Island and Martha's Vineyard, Massachussets (de Vos *et al.* 1956).

◼ DAMAGE
Although generally not an economic problem, snowshoe hares are pests of forestry and occasionally orchards. They have a 9–11-year cycle of abundance and during peaks can cause significant damage to young coniferous tree plantings by barking and girdling of stems, particularly during the winter (Sullivan and Sullivan 1986).

In California snowshoe hares can be a major obstacle to reafforestation of conifers after forest fires and cause much damage to young pine in montane plantations (Jameson and Peeters 1988). Damage by snowshoe hares in the United States has also been noted in western Oregon, the Pacific northwest, British Columbia and Washington to Douglas fir, Ponderosa pine and other conifers (Cowan and Guiguet 1960; Black 1965; Radwan and Campbell 1968; Black *et al.* 1969).

ALPINE HARE
Arctic hare
Lepus arcticus Ross
Some authorities include this species as a subspecies of the mountain hare, Lepus timidus.

◼ DESCRIPTION
HB 580–790 mm; T 40–73 mm; WT 2.73–5.45 kg.
Pure white in winter, with black tipped ears; in summer bluish grey to white with cinnamon grey wash or cinnamon with pinkish mottling; tail white; feet with yellowish wash; strong curved claws on all feet. Females average slightly heavier than males.

◼ DISTRIBUTION
North America. Arctic Canada and the northern islands; Greenland.

◼ HABITS AND BEHAVIOUR
Habits: no information; probably similar to other *Lepus* spp. **Gregariousness:** occasionally up to 100 or more feeding together. **Movements:** sedentary. **Habitat:** tundra beyond tree line. **Foods:** herbs, twigs,

Alpine hare

roots, tundra plants and willows, seaweed. **Breeding:** summer; gestation 50 days; litter size 2–8; 1 litter per year; young probably mature in second year. **Longevity:** probably up to 5 years. **Status:** uncommon(?).

■ HISTORY OF INTRODUCTIONS
NORTH AMERICA
Canada

Since 1969 the alpine hare (*L. arcticus*) has been released in three areas of the Newfoundland mainland, on 11 islands off Newfoundland and on one island off Nova Scotia. A population became established on Brunnette Island (off south coast), when four hares from the southern Long Range Mountains were released on the island in 1969. From this successful introduction a number (377) were trapped and transferred to the mainland and other islands, mainly in an attempt to test interactions with various mammalian predators and snowshoe hares (Mercer *et al.* 1981).

On mainland Newfoundland alpine hares were released at six sites from 1973 to 1978 on the Avalon Peninsula (68 hares), at three sites from 1973 to 1976 on Burin Peninsula (150) and at one site in 1975 at Pools Cove (21). The hares either failed to survive or produced very low level populations locally in any of these areas.

Following their success on Brunnette Island, nine alpine hares were released on Jude Island between 1974 and 1979, four on Emberley's Island in 1974, 30 on Fogo Island (off north coast) in 1976, 12 on Scatari Island in 1975, 21 on Grey Island in 1975, 17 on Long Island in 1976, seven on Bell Island in 1977, six on Oderin Island in 1979, six on Marticot Island in 1979, six on Isle Valen in 1979, and six on Kelly's Island in 1979. Hares are now known to be established and breeding on Jude Island and to be established on Scatari, Grey and Emberley's islands, but to have failed on the islands of Fogo, Long, and Bell (Mercer *et al.* 1981).

■ DAMAGE
None reported.

BLACK-TAILED JACK RABBIT

Lepus californicus Gray

■ DESCRIPTION
HB 465–630 mm; T 50–112 mm; WT 1500–2500 g.
Upper parts sandy to grey to blackish; ears long; in the south-west of range there is a terminal black patch on the outside of each ear; tail long, with black mid-dorsal stripe extending onto back; legs long.

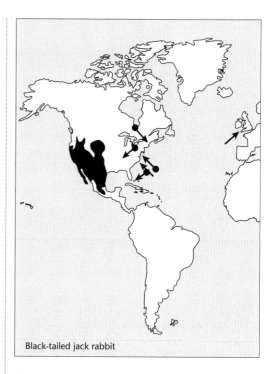

Black-tailed jack rabbit

■ DISTRIBUTION
North America. Central and south-western United States from Washington and South Dakota to Baja California and northern Mexico; and from the Pacific Coast almost to the Mississippi River.

■ HABITS AND BEHAVIOUR
Habits: nocturnal but frequently diurnal; solitary except at feeding sources such as crops; rest under bushes in nests during day; in deserts builds a short burrow. **Gregariousness:** density 0.1–2.7/ha and locally to 34.6/ha. **Movements:** sedentary; home range 16–20 ha. **Habitat:** grassland, meadows, steppe, short-grass rangeland; sagebrush–creosote bush; mesquite–snakeweed and juniper–big sagebrush; agricultural land and orchards, deserts, irrigated pasture, crops. **Foods:** grass, sedges, shrubs and crops. **Breeding:** breeds any time of year; gestation 38–47 days; average litter 2.2, but varies 2–7; may have up to 4 litters/year; may place each young in a separate nest; born furred, eyes open, in shallow depression on ground. **Longevity:** no information. **Status:** locally common.

■ HISTORY OF INTRODUCTIONS
ATLANTIC OCEAN ISLANDS
Falkland Islands (United Kingdom)

Rabbits and hares (species not specified) have been introduced to these islands (Cawkell and Hamilton 1961).

EUROPE

United Kingdom

A species of jack rabbit was released at Woburn, probably between 1920 and 1940, but failed to become established (Fitter 1959).

NORTH AMERICA

Black-tailed jack rabbits have been successfully introduced to at least seven states in the United States (south Florida, Massachusetts, Virginia, New York, Maryland, New Jersey and Kentucky).

United States

Black-tailed jack rabbits have been introduced to Nantucket Island, Martha's Vineyard, in Massachusetts, and to New York (de Vos *et al.* 1956). They are also established in the states of Florida, Virginia (Clapp *et al.* 1976), Kentucky and New Jersey (Knopf 1991).

In Kentucky they have been successfully introduced to Mercer, Pendleton and Hancock counties. In Virginia they were established off the coast on Little Cobb Island, Northampton County, at least in 1973. These were presumedly the descendants of six adults and two young introduced there from Kansas about 15 years before by H. L. Bowen for sport hunting. Some jack rabbits have also been reported from Rogue Island and Hogg Island, just off Cobb Island (Clapp *et al.* 1976).

■ DAMAGE

Jack rabbits (*Lepus* spp.) have been a problem in the desert and plains regions of the United States for over 100 years. Black-tailed jack rabbits (*L. californicus*) are closely associated with semi-desert shrublands and cause most damage on undeveloped land near or in these areas. Most problems are caused in agricultural crops and on rehabilitated rangeland (Evans *et al.* 1970).

In the past in a number of states (Utah, Arizona, Oregon, Washington and California) black-tailed jack rabbits have caused damage to alfalfa, various truck and field crops, such as barley and wheat grass, and they have damaged the bark of young trees and grapevines, Ponderosa pine seedlings and competed with sheep for grass and forbs (Vorhies and Taylor 1933; Storer 1958; Johnson 1964; Currie and Goodwin 1966; Black *et al.* 1969).

Today sport hunting, pest control and better farm management apparently prevents most serious outbreaks (Chapman and Flux 1990).

EUROPEAN HARE

Brown hare, common hare

Lepus europaeus Pallas

Often included with L. capensis Linnaeus as a single species, but most recent works separate them (see Corbet and Hill 1986; Chapman and Flux 1990) on the basis of size, proportion and colouration.

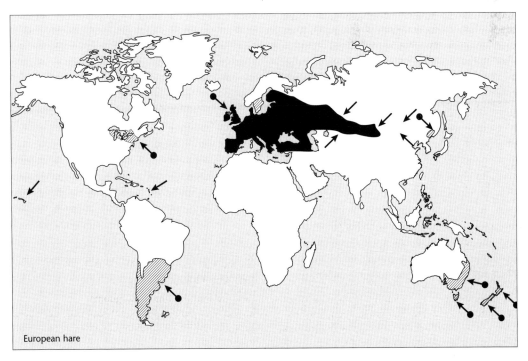

European hare

■ DESCRIPTION

HB 500–760 mm; T 70–120 mm; WT 2–5.0 kg (rarely up to 7 kg).

Upper parts of coat tawny brown; shoulders and flanks reddish; sides of face and outer surface of legs yellowish; ears with black tips; face has long whiskers; fur soft, under parts and under tail white, except for breast and loins; hind legs long; tail black above, sides and below white; upper sides of feet tawny. Colour variation is common and can be grey, sandy, or white. Female usually weighs less than males. Leverets have thick coat of hair and are the same colour as adults when born.

■ DISTRIBUTION

Eurasia. From the British Isles, southern Sweden and Finland, most Mediterranean islands (Balearic, Corse, Crete, Sardegna, Sicily) to western Asia, and southeast to Asia Minor and Persia.

■ HABITS AND BEHAVIOUR

Habits: crepuscular and nocturnal; lives in 'forms' in grass and does not burrow. **Gregariousness:** solitary except at breeding; density to 1/ha or 0.1–9.3/km; dominance hierarchy at food sources. **Movements:** home range 5–330 ha; up to 15 km while feeding; may travel 1.8 km to feeding areas. **Habitat:** cultivated fields, meadows, open areas, moorland, pastures, grassland, alpine grassland, open woodland, steppe, subdeserts, coastal sand dunes, forest clearings, salt marshes. **Foods:** grass and low herbaceous plants, fodder crops, bark and twigs of trees and shrubs, root crops, grain crops, lucerne. **Breeding:** breeds all year, mainly spring (peak in May–July in Britain, August–February Australia); gestation 30–42 days; young 2, 4–7; litters 1–4/year; mating is promiscuous; born well furred, eyes open; leverets grow at 18.8 g/day; eat vegetation at 1 week; weaned 4 weeks; adult mature at 5 months; most young breed in season born. **Longevity:** 7–13 years (wild). **Status:** common to locally common in most of range.

■ HISTORY OF INTRODUCTIONS

European hares have been introduced successfully in Siberia, Finland, Sweden, eastern Canada, northeastern United States, Chile, Argentina, Uruguay, south-eastern Australia, New Zealand, the Hebrides, Orkneys, Isle of Man, Ireland; small introduced populations on several North Sea islands, also on Barbados, Bahamas, Réunion and the Falkland Islands. Many of the introductions have been for sport and this may be the origin of some of the insular populations.

AUSTRALASIA
Australia

Hares were introduced successfully to mainland Australia in the 1860s from Britain for hunting and coursing, following success in Tasmania (Mahood 1983) and later in other areas (Wilson *et al.* 1992).

They were first introduced to Westernport Bay, Victoria, in 1862, and set free for a variety of reasons. They were considered good sport and good eating. Dozens of clubs for 'coursing' were set up in Victoria after 1873 and flourished until the end of World War 1.

Hares transported in the 1930s became established in pockets at Townsville, Ayr and Mackay, Queensland. Hares were first shipped to Western Australia in 1874 and in again in 1896–97 but died *en route*. They were also released on Rottnest Island in about 1900 but failed (Jenkins 1977; Jarman 1986).

A recent study (Stott 1998) suggests that the reason for the hare's lack of success in Australia as compared to Europe is possibly due to a lower pregnancy rate. It is postulated that this lower rate can be attributed to abnormalities of the female reproductive system, presumably limited to the animals in Australia.

EUROPE

Hares have been widely introduced in Europe, in many areas merely back into their original range, mainly as an animal for sporting purposes. Such introductions have occurred in Sweden, Finland, England, Ireland, Scotland, Isle of Man, various islands off the Scottish coast (including Orkneys and Shetland), Switzerland, Italy, France, Poland and the USSR.

Hares are an important game animal in Europe, over five million being shot each year. Massive importations are made each year from Hungary and eastern Europe and taken to western Europe for release.

To the east, hares have expanded their range naturally and by liberation into Siberia and the south Pacific coast of Russia. They now occur over most of Europe except Ireland, the Mediterranean region and Scandinavia, which all have introduced populations (Chapman and Flux 1990).

France

In France it is reported that 90 000 brown hares are imported annually (Niethammer 1963) and Yugoslavia alone in 1961 provided some 10 000 hares. Restocking in France occurs partly by captive breeding and partly by translocation between regions, but chiefly by importation from Hungary and Poland. Between 1970 and 1975, 170 000 annually were imported and released to every region of France (Stuttard 1981).

Germany

Although the European hare is distributed in suitable habitat throughout Germany, there have been many introductions to increase hunting opportunities by the infusion of new blood into the population. Animals from Hungary, Russia, Yugoslavia, Poland and Denmark have supplied large numbers in the past for release for these purposes. So much so that in parts of Germany the hare now present is said to be a hybrid of multiple races. However, this is now difficult to verify as between 1935 and 1939 it is estimated that 2 500 000 annually were imported for release. Since 1936–37 at least 8000 were imported into Germany (Niethammer 1963).

Hares were introduced on Wangerooge (East Frisian Islands) *c.* 1900, and those on Juist and Langeoog were also released towards the end of the nineteenth century (Krumbiegel 1955). On Pellworm (North Frisian Islands) they were taken from Schleswig-Holstein and released *c.* 1870 (Mohr 1931). An attempt to introduce them on the Hallig Hooge failed, however, and repeated attempts on Amrum have remained without permanent success (Niethammer 1963). They have also been successfully introduced on the island of Nordund in the Baltic Sea.

On Griefswalder Oie (Baltic Sea) European hares were released (two females and one male) at the beginning of 1905 and supplemented by further releases in 1907 and 1908 (Muller-Using 1938). They bred well on the island but commenced to decline about the beginning of World War 2 and the last of them were destroyed in the postwar period (Niethammer 1963).

Italy
In Italy there is a small population of European hares, mainly imported from Hungary and Slovakia, supplemented at a large cost (Niethammer 1963). Large numbers are imported from central and eastern Europe and released for hunting (Stuttard 1981). For instance, in the province of Rome about 2000–3000 captive-bred hares are released each year for restocking the local population.

Poland
Hares introduced into the occupied native range of the species did not increase the population (Jezierski 1968).

Russian Federation and adjacent independent republics
Approximately 31 767 hares have been released in Russia for acclimatisation purposes, many of them into areas that already had hares established. They were released in 48 regions between 1928 and 1972.

Number of European hares released in Russia and adjacent republics 1928–72

Western Siberia and Altai		
Novosibirsk	1936–61	835
Omsk	1953–60	113
Tomsk	1953–54	54
Altaisk	1939–69	597
Kemerovsk	1951–52	128
Central and Eastern Siberia		
Krasnoyahrsk	1938–51	315
Irkutsk	1938–56	247
Buryaht ASSR	1956	92
Chitinsk	1938 and 1965	368
Kazakhstan SSR	1929 and 1958–63	399
Far East		
Khabarovsk	1963–64	339
Primorsk	1965	158
European USSR and Urals		
Sverdlovsk	1965–69	192
Chelyahbinsk	1971–	147
Ulyahnovsk	1963–70	343
Chuvashsk ASSR	1961	35
Ubmurtsk ASSR	1966	17
Tatarsk ASSR	1965–66	17
Kuibshevsk	1970–72	450
Saratovsk	1970–72	93
Volgogradsk	1960	47
Astrakhansk	1957–72	337
Moskovsk	1951–70	4536
Kalininsk	1954–69	2266
Novgorodsk	1965	32
Leningradsk	1956–71	541
Pskovsk	1962–67	41
Smolensk	1966–72	552
Kalujsk	1960–72	954
Tulsk	1959–64	355
Ryahzansk	1964–68	327
Gorkovsk	1964	10
Vladimirsk	1956–72	1443
Ivanovsk	1962–69	32
Kostromsk	1969	137
Yahroslavsk	1954–65	837
Kursk	1960–72	974
Voronejsk	1970	100
Tambovsk	1968	171
Krasnodarsk	1960–71	997
Rostovsk	1969–72	2171
Stavropolsk	1972	201
Ukrainsk SSR	1929–69	8374
Total		30 019

Following reclamation of the north Kazakhstan steppes for agriculture and several years of little snowfall in the 1930s and 1940s, the hare rapidly colonised this area and penetrated eastwards as far as Omsk. Subsequently the colonisation slowed down, but in the 1950s and 1960s they had reached the Tatarsk district of Novosibirsk oblast. Some had been deliberatetly released in the Omsk region (Gruzdev 1969).

Attempts in Buryatiya to aclimatise hares were unsuccessful (Izmailov 1969). European hares were also introduced in the Far East (Lindermann 1956).

Hares only naturally occurred in the western part of the Russia, but have been introduced into Siberia. In 1936–39 almost 200 were released near Novosibirsk, over 150 in the Altai, over 300 in Krasnojarsk Territory, 120 near Irkutsk and 200 near Tschita. By 1963 they existed in many parts of Siberia. They have also been re-introduced on Barsa-Kel'mes Island (Aral Sea), and still occured there in 1963 (Tschapskij 1957; Niethammer 1963).

Although repeated attempts to introduce them in Siberia have been made, they have not produced any economically important populations to date (Sofonov 1981).

Sweden
European hares were released at the beginning of the twentieth century and established and increased in numbers. In 1953–54 the population was almost as high as that for the native mountain hare (Niethammer 1963). Hares were introduced to southern Sweden in 1886 and spread over all of the country except the north. The population peaked in 1949–50 and has declined sharply since then (Frylestan 1976).

Switzerland
In Switzerland the European hare is frequently released for hunting, particularly in the cantons of Neuchâtel and Vaud (Niethammer 1963). Of the 8232 hares freed between 1932 and 1959 in Neuchâtel three-quarters were recaptured, and at Vaud about one-third were recaptured.

United Kingdom and Ireland
There are no records of European hares in Britain until the Roman period and the species may well have been introduced there by humans (Corbet and Harris 1991).

Hares have been frequently released in the British Isles in the past, especially in the 1880s in Scotland and Ireland for sport and other reasons associated with hunting (Fitter 1959). They have been released at one time or another on most small islands off the coast of Scotland including: Coll (about 1787), Lewis (before 1797), Mull (1814–15), Islay (before 1816), and Shetland (early 1800s). On Mull they did not survive and died out later in the same century. In the Shetlands there were at least three introductions, but all were killed by farmers because of the damage they caused to their crops (Venables and Venables 1955; Fitter 1959).

Ten or 12 attempts were made to introduce them into Ireland from the 1820s onwards with varying success: 65 from Norfolk were released in Strabane County, Tyrone, in 1876 where they were still surviving in 1910. Hares are also now common in Donegal and Derry where they were imported from England; at Derry where they may be the result of an escape of hares in about 1910 (Fitter 1959). Extensively introduced in Ireland for coursing, populations are established in Donegal, Fermanagh, Londonderry and Tyrone, but records are lacking and they are not common or widespread and are no longer found on many of the original release sites (D'Arcy 1988; Corbet and Harris 1991).

Hares were introduced to the Orkneys and on the Isle of Man (Southern 1964) where they are possibly still established. They are recorded from a number of Scottish islands.

NORTH AMERICA
Canada
Seven females and two males were imported from Germany to Bow Park Farm, near Brantford in 1912 (Reynolds 1955), where they were kept confined. These animals or their progeny escaped from an island by crossing the frozen river in 1912 (Banfield 1977) or in about 1915 (Allin 1950). One was reported in the wild from Aylmer in Elgin County in 1919 (Anderson 1923). They are said to have spread rapidly and in 1921–22 were found in an east–west direction from Sarnia (Saunders 1932) to the Niagara Peninsula in the east and Guelph, Wellington County (Dymond 1922; Howitt 1925).

In 1923 the southern boundary of their range was at Simcoe in Norfolk County (Anderson 1923), and they were reported to occupy an area of some 4500 square miles adjoining Lake Erie (Silver 1924), the shores of which they reached at Port Rowan (Snyder and Logier 1931) in 1924. In 1925 European hares were reported from Woodbridge and Maple (Baillie 1929), and by 1928 they occupied all the south-west of Ontario, with reports of their presence from Highgate in Kent County (Dymond 1928), Walkerton in Bruce County, Flesherton and Meaford in Grey County (Dymond 1928 and 1930), Tottenham, Collingwood and Penetanguishene in Simcoe County (Dymond 1928; Baillie 1928; Saunders 1932). By the end of 1928

Markham, York County and Uxbridge, Ontario formed the eastern boundary (Dymond 1928). In 1930 they were reported further east in Darlington Township, Durham County (Allin 1940), and they were frequently seen in the vicinity of Toronto in 1931 (Thompson 1931) In 1936 some were seen and one shot at Hollowell Township in Prince Edward County (Snyder *et al.* 1941).

More hares were released in the Port Arthur–Fort William area of Thunder Bay, Ontario in 1942, 1943 and 1945 (Allin 1950). These animals were obtained from Meadford, Ontario, and thus originated from the original stock that had now spread over most of southern Ontario. Survivors of these introductions were observed until 1949, but none were seen thereafter (Peterson 1957).

In 1948 a hare was found near Hortington and in 1952 they were reported from Pittsburg Township, Frontenac County (Reynolds 1955). They were first seen at Gananoque in the Kemptville district in 1944–45 (Youngman 1962) and by 1959 were reported to occupy almost the whole of southern Ontario (Reynolds and Stinson 1959). Between 1959 and 1961 they extended their range north in Frontenac County and in 1961 one was collected 19.3 km south of Ottawa in Carleton County (Youngman 1962). During the winter of 1961–62 they extended their range rapidly to the east and more slowly to the north (Dean and de Vos 1965) and there were several reports of hares in the Ottawa region (Youngman 1962).

The European hare now occupies most of the area from the northern shores of Lake Erie, north to Lake Huron (as far north as Perry Sound) and east to Lake Ontario and Ottawa; a spread of some 483 km east and approximately 241 km north (Dean and de Vos 1965; Banfield 1977).

The race introduced in Canada is *L. e. hybridus* (Banfield 1977).

United States
European hares were first liberated at Millbrook, Dutchess County, New York State, in 1893 when a wealthy resident imported several shipments of up to 500 hares from Hungary which were released at intervals of up to four to five years until 1910–11 for hunting purposes (Silver 1924; Dell 1957). According to Bump (1941), several thousand hares from Europe were released at Millbrook from 1893 to 1910. Other imports and introductions occurred on a neighbouring estate at this time, and another some time earlier at Jobtown, New Jersey (*c.* 1888), with hares from England. This latter release was reported to be unsuc-

cessful. Other releases of hares in the 1890s were reported from Bethlehem, Pennsylvania, and at White Plains, New York (Silver 1924). In the 1920s, or shortly before, some were released in Connecticut and Massachussetts, and later in about 1940 in Maryland (Presnall 1958).

By 1903 European hares were said to be scattered over many localities of the northern half of New Jersey and sparingly located in parts of Camden and Burlington counties, and were regularly hunted in Bucks County, Pennsylvania (Rhoads 1903). From 1912 to 1917 hares became a problem by ringbarking orchard trees in Dutchess County, New York. During these years a bonus was paid and some 12 000 hares were harvested (Silver 1924).

Until 1923 the spread of the brown hare was said to have been slow but steady. At this time they occupied an area from southern Vermont to central New Jersey and eastwards some 32–48 km into Connecticut and Massachussetts, and to a limited extent across the Hudson and Delaware rivers into eastern Pennsylvania with odd animals being found further afield (Silver 1924). In 1915–16 the damage caused to orchard trees was estimated as US$100 000. The density of hares in some areas was as high as 10–40 per square mile. In the 1930s, however, a rapid and drastic reduction in range and abundance ensued (Dell 1957).

A single hare was taken in 1929 at Pensselar County, New York (Schoonmaker 1929), and they were reported to still occur scattered over most of Connecticut and numerous parts of Fairfield and Litchfield counties (Goodwin 1935) in 1935. Several were killed in south-west Vermont in 1938 (Osgood 1938), and by the 1940s their population was was said to be static or decreasing in numbers and range (Bump 1941). They were still present in 1947 (Streever 1947) and in the early 1950s were reported to be increasing in numbers again (Dell 1957). Hunting probably prevented much increase and spread and by 1957 the population consisted only of three isolated colonies, occupying less than half that inhabited during their peak abundance. A few still survived in Kent County, Maryland (Presnall 1958), and in the Lower Hudson Valley, New York (Smith 1955), but they had disappeared from Pennsylvania (Whitebread 1952), and from Massachusetts and New Jersey not much later (Dean and de Vos 1965).

The three surviving populations in 1963 were: (a) in the north-eastern half of Dutches County and in two south-eastern towns of Columbia County; (b) on the western border of Herkimer County and east along the county borders of Schohavie and Montgomery, with a small population on the north-eastern edge of Otsego County and north-western edge of Schenectady County; and (c) a small area in western central Washington County. In Connecticut there were three isolated colonies (Litchfield County, New Haven County, and Hartford County) (Dean and de Vos 1965).

European hares have also been introduced to Nantucket Island, Martha's Vineyard and Smith Island (de Vos *et al.* 1956). They were present on East Anacapa (Von Bloeker 1967), but their present status is unknown (Lidicker 1991).

Some have occassionally crossed the border from Canada, the earliest records in 1933–34 in Tuscola County, Michigan (Burt 1954).

PACIFIC OCEAN ISLANDS
Hawaiian Islands
Hares of Russian origin were introduced to two small islets off Oahu (de Vos *et al.* 1956).

New Zealand
European hares were introduced to the South Island in 1851 and 1867 for sport and food, and there were several introductions to both the North and South islands in the 1870s (Wodzicki 1950; Wodzicki 1965; Gibb and Flux 1973). The Canterbury Acclimatisation Committee imported them between 1867 and 1873 from both England and Australia and released them in grounds owned by the society (Lamb 1964). By 1876, 80 were caught at Christchurch for release elsewhere.

Hares were first introduced to Canterbury in 1851 and Nelson Acclimatisation Society liberated more hares in 1863 and 1872. By the 1970s they were widespread throughout the North and South islands, except Northland and Fiordland and the offshore islands (Gibb and Flux 1973). European hares are now spread throughout both main islands except for parts of South Westland, most of Fiordland and north of Auckland for 80 km (Barnett 1985; King 1990).

SOUTH AMERICA
Argentina
In 1888, E. and W. Tietjen imported 36 hares from Germany and released them at 'Estancia La Hansa' near Cañada de Gomez, Santa Fé Province. in Argentina (de Vos *et al.* 1956; Niethammer 1963; Amaya 1981; Grigera and Rapoport 1983). In 1897 E. Dalpech imported hares from France and released them on 'Estancia de Sulpicio Gómez' in Tandil, Buenos Aires Province, and in 1930, nine hares were liberated in the province of Santa Cruz (Grigera and Rapoport 1983). They were spreading rapidly in Buenos Aires Province some 10 years after their introduction (Alsina and Brandani 1981).

By 1907 they had increased in numbers to 'plague proportions' (Niethammer 1963), and had become so numerous that they were declared a pest by the government (Grigera and Rapoport 1983). From the original introduction, and probably others, they spread from the subtropical north to the cold, dry Patagonia in the south (Amaya 1981). It is known that there were further introductions of hares around the beginning of this century near Rio Gallego in Patagonia (Amaya 1981) and some may have colonised some areas of central southern Patagonia from those liberated at Ultima in southern Chile (Grigera and Rapoport 1983).

The European hare is now distributed throughout most of Argentina (Howard and Amaya 1975; Alsina and Brandani 1981), except for a few localities in the central and north-western region, and in the high Cordillera (Grigera and Rapoport 1983). They occupy an area of at least 4 million km^2 in South America (Amaya 1981) and at least 5–10 million are harvested annually (Amaya 1981; Grigera and Rapoport 1983). They have spread at the rate of about 18.6 km/year and the population density in Patagonia where they occur over extensive areas is 2.6–5.1 hares/km.

Brazil
European hares were probably introduced to Rio Grande do Sul, Brazil, in the 1950s (Cardinell 1958; Elton 1958). However, they were noted again in Brazil in 1965 and since 1970 have become abundant in Rio Grande do Sul and Santa Catarina, but not in São Paulo or Rio de Janeiro (Grigera and Rapoport 1983).

Bolivia
Hares were first seen in Bolivia in 1958 in the southern part of the departemento de Tarija, and by 1960 they were found throughout this area having originally colonised from Argentina (Grigera and Rapoport 1983).

Chile
In Chile, European hares were introduced from Germany in 1886 in the zone of Ultima Esperanza in southern Chile (Grigera and Rapoport 1983). They occur mainly around agricultural and grazing lands (Miller 1973). The northern limit is now the Copiapó River and they occur southwards throughout the country, except on the islands off the coast (Grigera and Rapoport 1983).

Paraguay
In Paraguay European hares are occasionally seen in Fortín General Diaz, but their origin is unknown (Grigera and Rapoport 1983).

Uruguay
The European hare is probably now throughout Uruguay, most likely having colonised that country from Argentina. They were abundant in Uruguay as early as the 1920s (Grigera and Rapoport 1983).

WEST INDIES
Barbados, Grenadines, Guadeloupe
European hares are reported to have been introduced to Barbados, one of the Grenadines and to Guadeloupe (de Vos *et al.* 1956).

◼ DAMAGE
Within its native range, brown hares are pests. They cause minor agricultural damage in Britain to grass and cereal crops and to horticultural crops (Corbet 1966; Corbet and Harris 1991). Their taste for agricultural crops is well known and they cause considerable damage by barking trees. Hares are capable of much damage especially in severe weather and in peak numbers (Southern 1964). In northeastern Scotland they damage cereal crops and have been found to damage the outer rows and areas where there were clear spaces around crops (Hewson 1977). In Lithuania European hares are a pest in gardens and orchards, but the amount of wood and bark eaten, and therefore the damage, is controlled by snow depth (Likyavichene 1962). In Denmark they bite the bark off fruit trees and in cold snowy winters appreciable damage is caused by this practice (Westerskov 1952).

In 1915–16 the damage to orchards in Dutchess County, New York, was estimated as $100 000 (Dell 1957).

As an introduced species in South America, hares feed in gardens, orchards and fruit plantations and in some cases cause considerable losses, especially in Argentina (Grigera and Rapoport 1983). In Rio Grande do Sul, southern Brazil, they cause damage to young deciduous fruit trees including apple, plum, and pears, by barking the trunks (Cardinell 1958). Large numbers of hares in Buenos Aires province, Argentina, interfere with forestation programs (Vigiani 1960).

In New Zealand hares cause damage to grasslands, orchards and gardens (Wodzicki 1950). They have never been a significant pest in New Zealand (Wodzicki and Wright 1984), and in recent years have been exported. Hares normally live at low densities and do not affect pastures. Damage to fruit trees, pine plantations and horticulture in New Zealand is significant, but less conspicuous than that of the rabbit (King 1990).

In Australia, hares occasionally cause damage by gnawing bark in orchards or forestry plantations, or by eating cereal, fodder or vegetable crops (Frith 1970, 1973; Mahood 1983; Wilson *et al.* 1992). They are now generally tolerated on farmland, although they can cause damage by gnawing bark in orchards in forestry plantations and by eating some fodder and root crops. Hares are not regarded by some authorities as a substantial pasture pest and probably the least damaging exotic species in Australia (Jarman 1986), although others consider it a major pest causing damage in orchards, plantations and vineyards (Strahan 1995).

BLACK-NAPED HARE
Indian hare
Lepus nigricollis Cuvier

◼ DESCRIPTION
HB 475 mm; T 62 mm; WT 1.8–3.6 kg.
Rufous brown with black on back and face; rufous on breast and legs; chin and under parts white; southern form has dark brown or black patch on back of neck and a black upper surface to tail; a desert form is paler, yellow or sandy.

◼ DISTRIBUTION
Southern Asia. Pakistan, India and Sri Lanka, from the Himalayas to extreme south of India, and from Pakistan to Assam and Bangledesh.

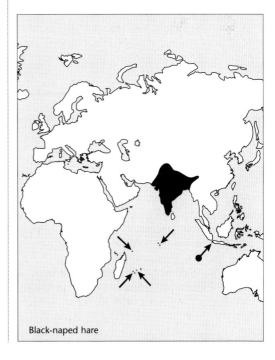
Black-naped hare

■ HABITS AND BEHAVIOUR
Habits: nocturnal occasionally, but feeds mainly in late afternoon and early morning; stays in forms during day. **Gregariousness:** density 0.25–5.8/ha. **Movements:** sedentary; home range 0.7–10 ha. **Habitat:** open desert to thick jungle; scrubland, cultivated plains. **Foods:** grass, sedges, herbs and forbs; sweet potato, lettuce, leaves, stems. **Breeding:** litter size 1–4; breeds all year, but peak is in monsoon season. **Longevity:** no information. **Status:** locally common.

■ HISTORY OF INTRODUCTIONS
Black-naped hares may have been introduced to Java (Chapman and Flux 1990). Introduced successfully to Mauritius and Gunners Quoin, Anelaga, Réunion and Cousin Island (Seychelles).

INDONESIA
Java
In Java black-naped hares are restricted to the extreme west of the island and may have been introduced.

Formerly they were considered to have been introduced to Indonesia by humans (Shortridge 1934; Chasen 1940) from Sri Lanka or India, but there is still doubt. Most early authorities suggest that they were introduced into Java from Sri Lanka or India and became established there (Jerdon 1874; Carter *et al.* 1945; Corbet and Hill 1980). This species has been introduced around Jakarta and now also occurs near Bogor and Bandoeng (de Vos *et al.* 1956). But in view of the long fossil history it must be considered a native species and may have evolved in Indonesia (McNeely 1981).

INDIAN OCEAN ISLANDS
Mauritius
Black-naped hares were introduced to Mauritius from Java by European colonists in the late nineteenth century (Brouard 1963). They were introduced to Round Island, 20 km north of Mauritius, following the failure of the rabbit (*Oryctolagus cuniculus*) some time after 1810 (Jerdon 1874; North and Bullock 1986).

Efforts were made to exterminate them in 1976, but these failed. The population rose from 650 to 1500 in 1975 to between 2450 and 2900 in 1982 (Bullock and North 1984), despite the fact that 883 were shot in 1976 (North and Bullock 1986).

OTHER INDIAN OCEAN ISLANDS
Also reported to have been introduced to Gunners Quoin, Anelaga, Réunion and Cousin (Seychelles).

Seychelles
In the early 1920s or 1930s black-naped hares were taken by coconut plantation workers to the small island of Cousin for food (Kirk and Racey 1992). There is now a high population on the island of 120–170 individuals (Kirk 1981), and they could have an impact on the vegetation and thus the conservation of some endemic species of birds.

■ DAMAGE
Increasing reports of damage to forestry, agriculture and ground nut crops come from India and Pakistan (Jain and Prakash 1976; Brooks *et al.* 1987), but it is not clear whether this is due to the increase in numbers or to greater awareness of pests. More irrigated crops are now grown in desert areas where hares may contibute to any damage.

Since their introduction to Mauritius they have become an agricultural pest (Owadally 1980). On Cousin Island their browsing may prevent the regeneration of *Casuarina equisetifolia,* which is an important foraging tree for several endemic landbirds and nesting tree for some seabirds (Kirk and Racey 1992).

MOUNTAIN HARE
Varying hare, blue hare, white hare, Irish hare
Lepus timidus Linnaeus
Now separated from the Arctic hare, L. arcticus *Ross.*

■ DESCRIPTION
HB 457–650 mm; T 43–80 mm; WT 2.3–3.6 kg (and up to 6 kg).
In northern areas coat is white all year, but in south remains brown all year, elsewhere more or less pure white in winter and greyish brown with blue-grey underfur in summer; ears black tipped; tail short, white; male slightly smaller than female; under parts white in summer and winter. Distinguished from European hare by smaller size and absence of black tip to tail.

■ DISTRIBUTION
Northern Eurasia. Scandinavia to eastern Siberia, south to southern USSR, south-eastern Kazakhstan, the Altai and Sikhote-Alin mountains. Isolated populations in Alps, Ireland, Scotland, and on Sakhalin Island and Hokkaido, Japan.

■ HABITS AND BEHAVIOUR
Habits: crepuscular and nocturnal; simple burrows or cavities among rocks; cyclic population fluctuations of 11 years, or 3–5 years in some areas. **Gregariousness:** group together in breeding season

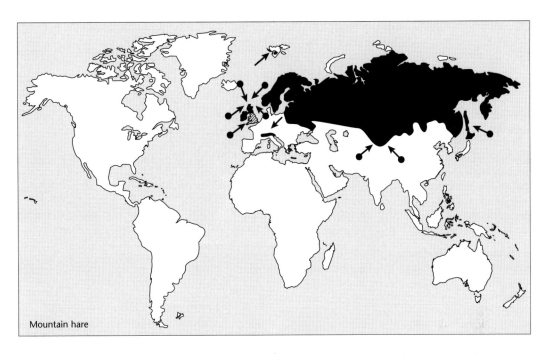

Mountain hare

and to feed; more social than other hares; groups size 20–100; density 0.14–300/km^2, and on islands to 400/km^2. **Movements:** home range 10–305 ha; in mountain areas moves to lower elevations in winter. **Habitat:** lightly wooded areas, open moorland, rocky slopes, tundra, open forest, open steppe, reed belts around lakes. **Foods:** heather, sedges, rushes, grass, lichens, conifer cones, browses bark and twigs on shrubs and trees (birch, juniper, poplar, willow), clovers. **Breeding:** breeds February–July; gestation 47–55 days; litter size 1–5; up to 3 litters per year; young suckle from 10–20 days and up to 6 weeks; mature in second year. **Longevity:** 1–9 years in wild. **Status:** widespread and abundant.

■ HISTORY OF INTRODUCTIONS

Introduced successfully to the Faroes and United Kingdom (southern Scotland, northern England and Wales) and also to the Shetland Islands, Orkney Islands, Inner and Outer Hebrides and the Isle of Man. Introduced unsuccessfully to Svalbard (Spitzbergen).

ATLANTIC OCEAN ISLANDS
Faeroe Islands (Faroes) (Denmark)
Introduced from Norway in the 1820s (Southern 1964) and/or in 1854–55 (Couturier 1955; de Vos et al. 1956), mountain hares became successfully established. Their descendants have given rise to a new subspecies, *L. t. seclusus* (de Vos et al. 1956; de Vos and Petrides 1967), which has lost its winter white coat and retains its brown colour all the year round because of mild winters (Bouliere 1954; Niethammer 1963).

Svalbard (formerly Spitzbergen, Norway)
Mountain hares were introduced unsuccessfully to Spitzbergen in the 1930s, where they apparently died out about 1954 (Lono 1960; Corbet 1978, 1980).

EUROPE
United Kingdom and Ireland
Mountain hares were introduced in southern Scotland, northern England, and in Wales (Burton 1976). Hares have been translocated from the Scottish Highlands to the southern uplands and to the inland areas of the Highlands (Darling 1927 in de Vos et al. 1956). Mountain hares were introduced mostly during the nineteenth century to Shetland, Orkney (Hoy), Outer Hebrides, Skye, Raasay, Scalpay, Eigg (now extinct), Mull (including some from Ireland), Islay (extinct) and Jura.

In historic times, mountain hares only occurred in the eastern Highlands of Scotland and perhaps on Islay and Orkney. In the first half of the nineteenth century they spread westwards and reached Wester Ross about 1830 and Inveray, Argyll, about 1840 in the western Highlands. Southern Scotland was colonised by introductions into Ayrshire and elsewhere in about the mid-nineteenth century, and the Pennine area of southern Yorkshire and Derbyshire by introductions about 1880. Irish hares were also introduced into south-west Scotland about 1923 and Scottish hares to Ireland (Londonderry) in the nineteenth century. In Wales a few remain from introductions near Bangor about 1885. Introductions to the Isle of Man occurred in about 1910 (Corbet and Harris 1991).

The present distribution of mountain hares from Snowdonia, Wales, to the Shetlands is due to the efforts of nineteenth century sportsmen who released many. Most of these introductions were made to increase sport and hunting, except perhaps in Jura where they were introduced as food for eagles. There were so many early introductions that it is now difficult to follow the fortunes of most. Between 1840 and 1914 they were released on more than 12 Scottish Islands including Mainland and Vaila in Shetland; Hoy and Gairsay in Orkney; Lewis, Harris, Barra and North Uist in Outer Hebrides; Skye, Islay, Raasay, Eigg, Mull, Jura and presumedly Scalpay in Inner Hebrides; and on Arran. In 1959 they were still found on all of these except Eigg, Gairsay, Islay and Arran.

In Scotland mountain hares were first introduced to Peebleshire in 1834 and in 1846–47, Cairntable in 1861, Portlands in 1866 and the whole of the hill country of the Scottish Lowlands from Wigtownshire to East Lothian by 1880. They are now widespread in this region. In 1937 some 60 from Inverness-shire were released on Wester Ross in an attempt to halt the decline of populations on the western mainland.

In England mountain hares were successfully released in Northumberland about 1910, although some that were released earlier in Lakeland (in 1892) had failed. Some were also released on Rusland Moors about 1909. The most successful introduction of mountain hares occurred in the southern Pennines. Releases in the 1860s failed, but later ones were successful – near Penistone in 1870, 20 hares near Saddleworth in 1876 and 50 or more at Perthshire and Greenfield in 1880 – and the species became established on the moors of Cheshire, Derbyshire and Yorkshire.

Introductions of mountain hares further south in England in 1894 and 1906 in Staffordshire, in Surrey in 1871 and 1906, and in Bedfordshire in 1935, all failed. There were at least six introductions in north Wales and they were common at Eidda, Caernarvonshire before 1830. Although a subspecies occurred in Ireland, there were many attempts from about 1838 on at least to establish Scottish and English animals, but all these attempts failed. Alternatively, Irish animals were introduced many times in England for coursing between 1881 and 1902, but none are known to have survived anywhere. Irish animals were also introduced to Scotland on the island of Islay in 1818, but died out about 90 years later; Mull in 1863 (disappeared six years later), but did not succeed permanently (Fitter 1959). From 1866 to 1870, 966 mountain hares were introduced to Ireland by a coursing club from Newcastle-on-Tyne, but all were unsuccessful in becoming established (Niethammer 1963), except on two small islands, Rathlin and Clare, where they were successful.

Germany

Towards the end of nineteenth century there were many introductions of mountain hares in central Europe by sportsmen and landowners in efforts to improve local stocks for hunting. Records of these releases date back to at least 1732 and 1734, when some were sent to Berlin, but the peak period for this type of introduction occurred in the 1730s. Most of them appear to have failed to become established permanently and their effects in areas already inhabited by mountain hares appear to have been negligible.

Documented introductions of mountain hares include: in 1893, 54 were released in a hunting preserve at Elsdorf in Rhineland where they rapidly disappeared; in 1895 hunters from Siegen released Russian mountain hares for the improvement of local stock, but there are none now in this area at all; in 1926 A. Lindemann released nine at Dreiborn in Schleiden, but these also disappeared (Anon. 1894, 1895; Niethammer 1963). Most of these introductions were reported to be unsuccessful because of the small number of animals released and the unsuitable habitat.

Carpathians

Originally mountain hares were present in the Carpathians, but not in historical times. A single attempted introduction in 1905, for the Dukedom of Hohenloheschen, where two males and four females were released in the Landoker hunting area, failed to become established because too few hares were used (Niethammer 1963).

Switzerland

The mountain hare (*L. t. timidus*) has been introduced, mainly as captive-bred specimens for hunting, in the cantons of Ticino and probably Valais, but present numbers are not known (M. Dollinger *pers. comm.* 1982). Both this subspecies and the endemic *L. t. varronis* are hunted and in the seven years from 1974 to 1980, between 1400 and 2900 animals were taken each year.

Russian Federation

There was little need to release mountain hares in Russia until the 1940s, when numbers of furs began to decline substantially, particularly in southern areas. The first introduction appears to have been in 1940 when 25 hares were released on Shantar Island (Sea of Okhotsk). From 1946 to 1972 some 10 027 mountain hares were released in Russia, 5419 of them in the Moskovsk oblast. The majority were released in the European part of Russia and were re-introductions and transfers. In central Siberia, mountain hares have been released at Irkutsk (1963–65, 317), Kemerovsk (1965, 23), Novosibirsk (1959, 19) and in Yahkusk (1957, 1960–68, 472), and also on Sakhalin Island (1971–72, 37) (Kirisa 1974).

■ DAMAGE

In Ireland, Scotland and Japan mountain hares cause damage to forestry plantations (Udagawa 1970; Chapman and Flux 1990) by damaging young trees. They also occasionally damage crops (cereals and turnips) in Ireland. In exceptional circumstances they compete with grouse for heather as food (Corbet and Harris 1991).

WHITE-TAILED JACK RABBIT

Lepus townsendii Bachman

■ DESCRIPTION

HB 545–655 mm; T 66–112 mm; WT males 2.15–3.5 kg, females 3.3–3.6 kg.

Upper parts greyish brown in summer, white in winter in northern parts of range, but elsewhere moult to a slightly paler coat; tail entirely white or with a dusky or buffy mid-dorsal stripe; ears black tipped.

■ DISTRIBUTION

North America. Central and western United States and southern Canada. Formerly extended eastwards to Wisconsin, Iowa and Missouri, then declined and became extinct from Kansas and southern Nebraska about 1950. Now from mid Saskatchewan and Alberta to the north of Arizona and from inland Washington and California to Lake Superior.

■ HABITS AND BEHAVIOUR

Habits: nocturnal; spends day in forms; burrows in winter for shelter from snow. **Gregariousness:** density 12.9–295/km^2; may congregate 30–150 in favourite feeding areas. **Movements:** may have to move some distance in winter to avoid hard snow cover; home range 10–89 ha or more. **Habitat:** grassland, steppe, open prairie and plains, montane pastures, sagebrush; cultivated grainfields, sagebrush plains. **Foods:** forbs, grass, wheat, alfalfa, twigs and buds. **Breeding:** breeds February–July; gestation 42 days; litter size 1–9 young; 2–4 litters per year; young mature spring following birth. **Longevity:** not known, probably up to 5 years. **Status:** widespread and common.

■ HISTORY OF INTRODUCTIONS

NORTH AMERICA

United States

White-tailed jack rabbits have been successfully introduced into Michigan (de Vos *et al.* 1956) and Wisconsin (Jackson 1961 in Hall 1981).

The introduction to Wisconsin occurred as a result of humans and agriculture, and the species is now well established in appropriate habitat throughout the state, except in the counties bordering Lake Superior and the upper peninsula of Michigan (Hall 1981).

■ DAMAGE

White-tailed jack rabbits are regarded as pests of agriculture, especially to crops of alfalfa, corn, soya beans and winter wheat (Chapman and Flux 1990). In California they have in the past damaged apple trees by eating the bark from the trunks, but the species has a restricted range and generally only causes minor damage in tree plantings (Evans *et al.* 1970).

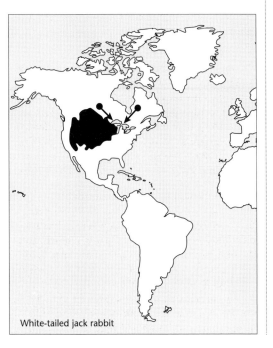

White-tailed jack rabbit

Family: Sciuridae
Squirrels

SIBERIAN CHIPMUNK
Asiatic chipmunk, chipmunk, Siberian ground squirrel
Tamias sibiricus (Laxmann)
=Eutamias sibiricus

■ DESCRIPTION
HB 130–160 mm; T 80–124 mm; WT 50–120 g.
Top of head grey, buff, chestnut to rust brown with mottling; five longitudinal black–chestnut stripes on back alternating with yellowish white stripes; ear tips without tufts; eye ring white; cheek pouches whitish buff; rump reddish and flanks yellowish rust; belly white; tail chestnut grey, white tipped.

■ DISTRIBUTION
Asia: the entire Siberian taiga zone west to the White Sea, south to the Altai Mountains, in western China south through Hopei to Shensi and north-western Szechuan: also on the islands of Sakhalin and Hokkaido, and the southern Kurile Islands. Range has extended westwards during the last century.

■ HABITS AND BEHAVIOUR
Habits: diurnal; mainly terrestrial; burrows (1.5–3 m) between roots of trees or under rocks; nest of dry leaves, ferns, moss or other plant material; hibernates in winter (October–April); stores food for winter. **Gregariousness:** solitary(?). **Movements:** sedentary; occasionally long migrations caused by food shortages; territory 700–3975 m². **Habitat:** forest near steppe; dwarf forest along tundra, deciduous undergrowth, thickets, plantations, areas near field crops. **Foods:** seeds grass, sedges and weeds; seeds, trees and shrubs; pine nuts, grain, flowers, herbs, small fruits and berries, mushrooms, bulbs, amphibians, reptiles, and young birds; some invertebrates (beetles, snails, slugs, ants). **Breeding:** breeds April–May; mates in April, young born May–June; gestation 30–40 days; 1 litter/year; litter size 3–5, 8; young born naked, blind; weaned 50–57 days; sexual maturity 1 year(?). **Longevity:** 6–7 years. **Status:** common.

■ HISTORY OF INTRODUCTIONS
EURASIA
The range of the Siberian chipmunk extended west-

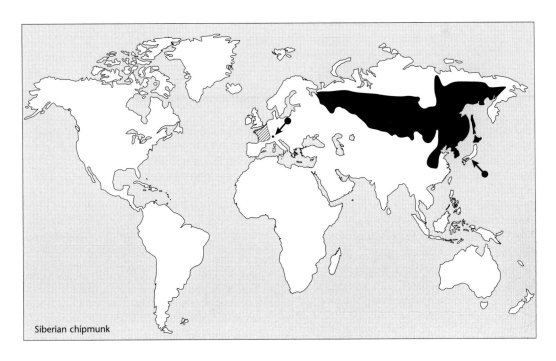

Siberian chipmunk

wards during the nineteenth century. By 1850 they were seen in the Urals and Kasan in the Tartar region. Later they crossed the Volga and in 1935 were seen at Temnikov and Sarov. More recently they were noted in Finland (Grzimek 1975).

Siberian chipmunks have been successfully introduced to France, Germany, Austria, the Netherlands, Switzerland and Japan (Lever 1985; Burton 1991).

Japan
The East Siberian subspecies (*T. s. sibiricus*) has been established in the Tiba Prefecture, Japan (Kaburaki 1940), probably in the 1930s (Kaburaki 1940; Lever 1985).

Switzerland, Germany, Austria, Netherlands, France
As a result of escapes, Siberian chipmunks have become established in parts of France, western Germany, the Netherlands and Austria.

Released by pet fanciers in about 1970 in a park in the city of Geneva, Switzerland, Siberian chipmunks have now established a small, but stable population (M. Dollinger *pers. comm.* 1982).

■ **DAMAGE**
In eastern Siberia chipmunks rob wheatfields and cornfields (Grzimek 1975). In the Amur-Zea Plateau area they are considered to be one of the greatest rodent pests because of damage to the forest, including such valuable trees as the Mongolian oak and hazelnut, and to agricultural crops grown in forest areas (Shigirevskaya 1964).

Siberian chipmunks feed on the seeds of cedar in the Sayon Mountains and destroy half the average forest nut production. Here, they are the most important pest of cedar, but do, however, contribute to the distribution of the tree itself (Shtil'mark 1963).

In autumn, and especially in years when the cedar nuts or other food plants fail, Siberian chipmunks can cause great damage to grain crops. They dig out the seeds at the edges of fields, bite off young stalks of growing crops and take the grain from mature crops. The greatest damage probably occurs when crops are ripe and the chipmunks begin to store food for the winter. Crops affected include wheat, rye, barley, oats, corn, flax, millet and sunflower (Ognev 1966).

Siberian chipmunks also damage gardens and orchards. Crops affected include cucumbers, eggplants, squash, watermelons, melons, gourds, marrow, peas, poppy, beets, carrots, potatoes and in orchards plums, cherries, apricots, pears and apples (Ognev 1966).

EASTERN CHIPMUNK
Hacker
Tamias striatus (Linnaeus)
=Eutamias striatus

■ **DESCRIPTION**
HB 135–190 mm; T 75–115 mm; WT 70–142 g.
Generally greyish with fawn or whitish under parts; muzzle and top of head tawny or greyish brown; cheeks buff and crossed with tawny stripe; has cheek pouches; nape and shoulders grey; back marked longitudinally with five black stripes, alternating with four brown or grey; white stripes above and below eyes; tail russet fringed with black and white tipped hairs, undertail tawny; feet yellow-brown.

■ **DISTRIBUTION**
North America: eastern North America from the states of the USA bordering the Gulf of Mexico, north to eastern Canada from Breton Island, Nova Scotia to southern Manitoba and north to Moosonee, Ontario, and Sept Isles, Quebec, and on the north shore of the Gulf of St. Lawrence.

■ **HABITS AND BEHAVIOUR**
Habits: diurnal; mainly terrestrial, but also arboreal; burrows (3.6 m long and 0.9 m underground); stores food in burrows; torpid at times in very cold weather in winter. **Gregariousness:** solitary except at mating; males and females have own burrow. **Movements:** sedentary; sometimes overlapping home ranges 0.26–0.37 ha. **Habitat:** bush and tall grass on rocky ground or with fallen logs; deciduous forest, gardens, rock piles, hedgerows, thin forest near rural settlements. **Foods:** nuts, bulbs, grain, seeds, small fruits, berries, corn and green vegetation, mushrooms, birds eggs, also small animals (slugs, worms, frogs, mice and salamanders). **Breeding:** February–July; gestation 31–35 days; litter size 1, 3–5, 8; 2 litters/year; female has two oestrous cycles annually; young born blind, naked; lactation 5 weeks; disperse at 6 weeks; sexual maturity 1 year. **Longevity:** 2–5 years in wild, 5–12 years in captivity. **Status:** common.

■ **HISTORY OF INTRODUCTIONS**
Eastern chipmunks may have been introduced successfully into Newfoundland(?), but unsuccessfully to the United States, New Zealand and Britain.

EUROPE
United Kingdom
Eastern chipmunks were released at Woburn, Britain, around 1921 or earlier, but failed to establish themselves (Fitter 1959).

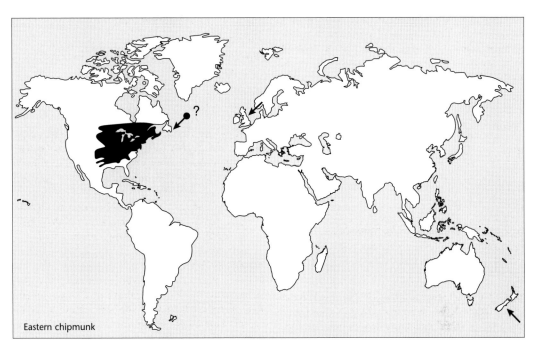

Eastern chipmunk

NORTH AMERICA
Canada
Eastern chipmunks (*T. s. lysteri*) were introduced to Newfoundland (Banfield 1977). [Note: No introduction appears to be mentioned by other authorities and no other details from Banfield (Hall 1981; McDonald 1985).]

United States
For experimental reasons (study of population techniques) eastern chipmunks were introduced to Whatley Island and Pymatuning Reservoir near Linesville, Pennsylvania (Mares *et al.* 1981).

■ DAMAGE
Eastern chipmunks are known to carry rodent-borne diseases transmissible to humans (Ball 1960).

TOWNSEND'S CHIPMUNK
Townsend chipmunk
Eutamias townsendii (Bachman)

■ DESCRIPTION
HB male 139–145 mm, female 145–150 mm; T 96–119 mm, 120–130 mm; WT male and female 71–122.1 g.
Generally brown with dorsal stripes; forehead dark brown to dark grey; cheeks buff and crossed by two brownish stripes; shoulders, flanks and rump dark ochraceous to olive-brown; under parts whitish or greyish; tail bushy, rufous brown bordered with submarginal black band and grey hair tips; underside of tail bright rufous red bordered with black. Female larger than male.

■ DISTRIBUTION
North America: the Pacific coast from California, United States to southern British Columbia, Canada, between the Cascade Mountains and the ocean.

■ HABITS AND BEHAVIOUR
Habits: diurnal; mainly arboreal; burrows; hibernates

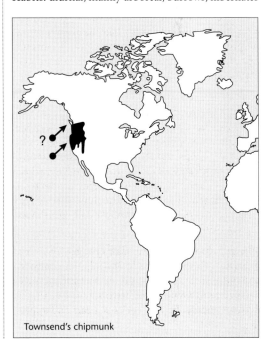

Townsend's chipmunk

in severe winter weather; cannibalistic; caches food in summer nests. **Gregariousness:** no information. **Movements:** sedentary; home range 0.6–0.68 ha. **Habitat:** coastal forests in edge clearings, lake shores, stream sides and old forest roads. **Foods:** nuts, seeds, grasses, fungi, bulbs, roots, berries, fruits insects, and bird eggs. **Breeding:** probably similar to other chipmunks, but not well studied; mate April, young born in May; gestation about 30 days; litter size 2, 3–7; 1 litter/year; mature in spring following birth. **Longevity:** 5–7 years in wild. **Status:** fairly common.

■ HISTORY OF INTRODUCTIONS
NORTH AMERICA
Townsend's chipmunks have been introduced successfully(?) on Vancouver and Sidney islands, Canada.

Canada
Townsend's chipmunks have been introduced to Esquimalt on Vancouver Island (Banfield 1977), where they are now doubtfully established. Chipmunks from the mainland kept in captivity at Beacon Hill Park, Victoria, apparently escaped from time to time. Individuals were seen as early as 1908, and some were apparently present in the wild up until about 1958. Since this time no reports of their presence there have been confirmed (Carl and Guiguet 1972).

In 1965 J. Todd released 36 Townsend's chipmunks from Oregon on Sidney Island where the species was still well established in the early 1970s (Carl and Guiguet 1972).

■ DAMAGE
Townsend's chipmunk is known to carry rodent-borne diseases that are transmissible to humans (Ball 1960).

BOBAK MARMOT
Bobac, steppe marmot, mountain bobak, alpine marmot, baibaka
Marmota bobak (Muller)
=M. baibacina, *which is often given specific rank. See Corbet 1986.*

■ DESCRIPTION
HB 300–600 mm; T 100–250 mm; WT 3–7.5 kg.
Upper parts brownish yellow or light brown (resembles *M. marmota*) or yellowish. Resembles *M. marmota* in size and colour but is heavier. Has shorter tail and legs than *M. marmota*.

■ DISTRIBUTION
Asia: steppes of southern Russia and Kazakhstan,

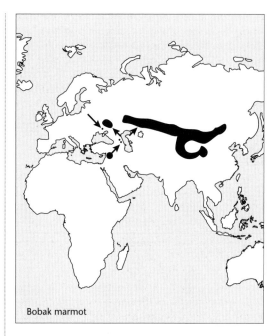

Bobak marmot

mountains of central Asia from Altai south to the Himalayas, and east to central China (Szechuan) and Manchuria. Formerly occurred west to Poland, Hungary and Romania, but their range has now been considerably reduced.

■ HABITS AND BEHAVIOUR
Habits: habits and behaviour similar to the alpine marmot, *M. marmota*; mainly diurnal, terrestrial; live in colonies; burrows 3–7 m and 10–70 m in length; hibernates in cold winters. **Gregariousness:** social groups; density 50–150/km^2; burrow density 80–142/km^2 and up to 400/ha. **Movements:** sedentary (?). **Habitat:** virgin grasslands, herbaceous steppe, alpine meadows, pastures, forest edges, cultivation. **Foods:** grass, forbs, fruits, grains, occasionally insects. **Breeding:** gestation 40–42 days; young born April–May; litter size 4–6; lactation 2 weeks; mature at 2 years. **Longevity:** probably 13–15 years(?). **Status:** range reduced due to agriculture; declining in numbers.

■ HISTORY OF INTRODUCTIONS
EURASIA
Bobak marmots have been introduced successfully to the Caucasus from the Altai Mountains (Corbet 1978, 1980).

Russian Federation and adjacent independent republics
The bobak marmot has been translocated and introduced a number of times within Russia and the adjacent independent republics (Naumoff, 1950; de Vos *et al.* 1956).

The mountain marmot, *M. b. baibacina*, was released on the Gunibsk Plateau near Gunibe, Dagestansk, in 1934 when 113 were introduced (Imshenetskii 1961; Kirisa 1973). They became well established in this area and by 1942 the population had reached some 1000–1200 marmots, but they were still restricted in range to the plateau area (Imshenetskii 1961). The mountain marmot is still established in this area (Kirisa 1973). In 1934 this subspecies of marmot was also introduced unsuccessfully to Bashkirsk where 31 were released (Yanushevich 1966; Kirisa 1973). In 1937, 234 were released in the Onpudaiskom and Alekmonarskom areas of the Caucasus from the Altaisk, and these became well established (Yanushevich 1966; Kirisa 1973; Corbet 1978).

Bobaks were also released in the Ukraine in the Velikoburluskom region (Kharkovskoi) and in the Melovskom and Belovodskom regions (Voroshilovgradskoi), where 100 were released in 1936, 18 in 1950 and 15 in 1961. In these areas they were re-introduced to increase the numbers of furs for market and this appears to have been successful. The number of furs increased substantially in the region soon after (Kirisa 1973), but there is no proof that the increase was due to the re-introductions. However, on the whole the introductions of bobak marmots in the Russian Federation and adjacent independent republics does not appear to have yielded appreciable results as far as hunting and fur resources are concerned (Sofonov 1981).

■ DAMAGE
Although the pelts of the bobak marmot are useful commercially in the Russian Federation and adjacent republics, the species occasionally causes agricultural damage. In Kazakstan they are reported to consume cultivated crops such as wheat, barley, oats, corn and millet. Such damage does not usually exceed 5 per cent (usually less) of the crops, and depends on the population density and distribution of marmot in the area (Tkachenko 1961). They may also cause soil erosion by burrowing in sensitive sloping areas.

BLACK-CAPPED MARMOT
Marmota camtschatica (Pallas)

■ HISTORY OF INTRODUCTION
EURASIA
Russian Federation
Black-capped marmots appear to have been released in the Yahkutsk region of the Russian Federation in 1954 (26 animals) and 1963 (61 animals), without success (Kirisa 1973).

ALPINE MARMOT
Marmot
Marmota marmota (Linnaeus)

■ DESCRIPTION
HB 400–600 mm; T 130–200 mm; SH 180 mm; WT 2.2–8.2 kg.
Heavily built; head broad and rounded; ears small; nose with white bridge; fur long, dense, grizzled reddish brown or reddish yellow with patches of ash grey on head; under parts paler; shoulders, rump and outer half of short tail black; legs short.

■ DISTRIBUTION
Europe: in the Alps, Carpathians and Tatra Mountains. Formerly more widespread.

■ HABITS AND BEHAVIOUR
Habits: diurnal; colonial; alert; hibernates in autumn (October to April); nests in burrows among rocks; burrows 1–4 m, sometimes immense. **Gregariousness:** solitary or in colonies 3–50 or more; family groups; density 1.2/ha. **Movements:** mainly sedentary, but often migrates in fall to lower altitudes; territories 2.5 ha. **Habitat:** alpine pastures, valleys and mountains, valley forests, grassy slopes, boulder regions of mountains. **Foods:** fresh vegetation; grass, sedges, herbaceous plants, leaves, blossom, roots. **Breeding:** breeds April–July; 1 litter/year (perhaps 1 per 2 years); gestation 33–42 days; litter size 2–4, 7; weaned *c.* 40 days; sexual maturity 2–3 years. **Longevity:** up to 20 years captive. **Status:** fairly common.

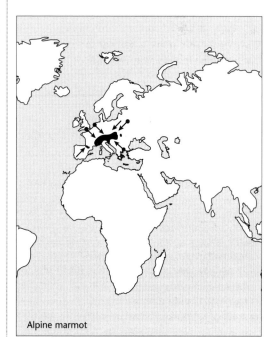

Alpine marmot

■ HISTORY OF INTRODUCTIONS
EUROPE
Alpine marmots (*M. m. monax*) have been restocked and resettled by humans since 1880 in areas where they became extinct as a result of human persecution. It is believed that one-tenth of their present range in the Alps, as well as 10 per cent of the total population (50 000 to 100 000), can be traced to re-introductions (Niethammer 1963; Grzimek 1975).

Restocked, resettled, and re-introduced in a number of areas in the alps region of Europe, they have been re-introduced successfully in the Pyrenees and the Russian Carpathians. Releases have occurred in Germany (1887), Austria, Switzerland (1883), northern Yugoslavia (1953), the Carpathians, the Pyrenees, the northern Apennines, Italy and the Russian Federation (Lever 1985).

Austria
Those alpine marmots occurring in lower Austria, Salzburg, Tyrol and Vorarlberg are descended from resettled animals released at the end of the nineteenth century (Grzimek 1975; Niethammer 1963). In 1955 there were at least 3369 marmots present there.

Czechoslovakia
Resettled in the lower Tatra alpine, marmots were released about 90 years ago (1873?) in the Northern Tatra (Carpathians) and several other places successfully, with stock from the Alps and the High Tatra (Niethammer 1963; Grzimek 1975; Burton 1976).

France, Spain and Andorra
Alpine marmots were resettled in the Pyrenees in some areas in the late 1940s or early 1950s (Couturier 1955; Grzimek 1975; Corbet 1978, 1980; Burton 1991). Six were released from Pragneres in 1948 in the mountains of the Department of Var and later a further seven. After being absent from about the Stone Age in the Pyrenees the alpine marmot colony was 25 in 1954 (Niethammer 1963). In the 1970s they were introduced into Massif Central in France (Burton 1991).

Germany
The original colonies of alpine marmots in Germany at Berchtesgaden and in the western Allgau and elsewhere have been restocked since 1880 (Grzimek 1975). A colony at Hohenaschau in Chiemgau south-east of Munich comes from eight marmots introduced from Berchtesgaden in 1887, and numbered some 100 animals in the 1960s (Niethammer 1963). All those at Steiermark and Karnten are descended from resettled animals introduced at the end of the nineteenth century (Grzimek 1975). In 1954 and 1957 some were resettled in the Feld Mountains in the Black Forest, where the population increased substantially. In 1961 they were resettled in the Swabian Alps near Balingen (Niethammer 1963), where they appear to have become successfully established.

In 1940, seven marmots were released in the territory of the forestry of Ruhpolding-Ost and in two to four years these had increased to 37. This colony numbered 15 in 1954, but became extinct about 1959. Two pairs were also unsuccessfully released near Reinstadt in Thuringen about the turn of the century. They were still present in 1906, but thereafter disappeared. Some were also introduced to Bayerischen in 1912 and these survived until at least 1935.

Italy
Alpine marmots have been translocated to the northern Apennines and Italy (Lever 1985).

Switzerland
Resettlement of alpine marmots has occurred in some cantons in Switzerland including Freiburg and Nuenburg (Grzimek 1975). At Freiburg two were released in 1883 where in the 1960s there were hundreds (Niethammer 1963).

Russian Federation and Kazakhstan
In some areas of the Russian Federation and probably in Kazakhstan, alpine marmots are reported to have been resettled (Grzimek 1975).

Until 1957 some 13 700 marmots were resettled in the Russian Federation and perhaps a number of areas in adjacent republics (Lavrov 1957). In the late 1940s at least, attempts were made to translocate Altai marmots (*M. m. baibacina*) and bobaks (*M. bobac*) in Russia (Naumoff 1950). The results of these attempts are not well documented. In Baskiv releases of marmots have been unsuccessful (Tschapskij 1957 in Niethammer 1963)

Yugoslavia
Resettled in some areas of Yugoslavia (Grzimek 1975), alpine marmots (three females, two with young) were released in 1953 on Triglav-Massiv, but were not observed again (Niethammer 1963).

■ DAMAGE
Alpine marmots are not known to cause any damage (Grzimek 1975).

MENZBIER'S MARMOT
Menzbira marmot
Marmota menzbieri (Kashkarov)

■ DESCRIPTION
No information.

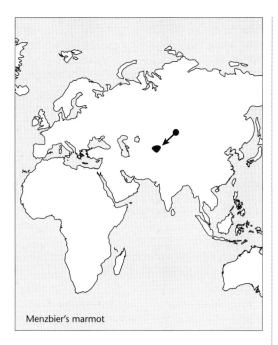

Menzbier's marmot

■ DISTRIBUTION
Asia: Central Asia confined to the western Tien-Shan Mountains of Uzbek and in south-east Kazakhstan.

■ HABITS AND BEHAVIOUR
Habits: diurnal; terrestrial; alert; hibernates in autumn and winter; nests in burrows; burrows 1–4 m. **Gregariousness:** solitary or in family groups. **Movements:** mainly sedentary, but with juvenile dispersal post-breeding. **Habitat:** alpine pastures, valleys and mountains, edges of valley forests, grassy slopes. **Foods:** fresh vegetation; grass, sedges, herbaceous plants, leaves, blossom, roots, fruits. **Breeding:** breeds April–June; 1 litter/year (perhaps 1 per 2 years); gestation 30–32 days; litter size 1–9, avg. 5–6; weaned c. 40 days; sexual maturity 2 years. **Longevity:** no data. **Status:** rare and declining due to agricultural development.

■ HISTORY OF INTRODUCTIONS
Asia
Introductions of Menzbier's marmots in Uzbekistan and Kazakhstan appear to have been unsuccessful.

Uzbekistan and Kazakhstan
In 1956, 61 Menzbier's marmots were released in Uzbekistan, followed by another 94 in 1957, but they failed to become established (Yanushevich 1966; Kirisa 1973). Some were also introduced in Kazakhstan in 1944, when seven were released (Kirisa 1973), but it is not clear whether they were successfully established.

■ DAMAGE
No information.

WOODCHUCK
Ground hog
Marmota monax **(Linnaeus)**

■ DESCRIPTION
HB 300–505 mm; T 100–250 mm; SH 100–190 mm; WT 2.0–7.5 kg.
Body heavy, grey to reddish or brownish; head reddish, but white around nose; ears short, rounded; body heavy; guard hairs with buffs tips giving grizzled appearance; tail bushy, dark brown or black; feet black or dark brown; posterior pad on foot oval. Female has eight mammae; is smaller than male.

■ DISTRIBUTION
North America. Extreme south-western Alaska, throughout most of southern Canada, south to the south-eastern United States.

■ HABITS AND BEHAVIOUR
Habits: diurnal, active mainly early morning and evening; burrows (up to 15.24 m of tunnels, as deep as 4.88 m); hibernates alone in winter; terrestrial; territorial. **Gregariousness:** solitary except at spring mating. **Movements:** mainly sedentary. **Habitat:** alpine pasture, grassland, under wood piles, open woodland, meadows, fields, roadsides. **Food:** grass, bark, twigs, leaves, blossom, roots, tubers, seeds, fruits and green vegetation including cultivated crops (clover); also insects, rodents, birds. **Breeding:** breeds June–July; gestation 28–32 days; litter size 1–4, 9; 1 litter/year; young born helpless; weaned at 42–44

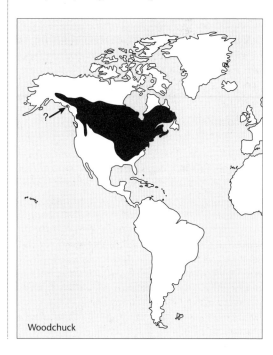

Woodchuck

days; remains with female until disperse in summer; sexual maturity at 1 year, but many don't breed until 2 years. **Longevity:** 4–6 years in wild, 10–15 years in captivity. **Status:** common.

■ HISTORY OF INTRODUCTIONS
NORTH AMERICA
Alaska (United States)
Woodchucks were introduced to Shuyak in 1930, and 13 (*M. m. ochracea*) from Juneau were released on Prince of Wales Island in 1930–31 (Burris 1965), but were probably unsuccessful in becoming established (no mention of them by Hall 1981).

■ DAMAGE
In the United States woodchucks have caused damage to cereal crops, vegetable crops, pastures and fruit trees, and their burrows sometimes become a nuisance to farmers and gardeners (Silver 1928; Farrand 1991).

CALIFORNIA GROUND SQUIRREL
Spermophilus beecheyi (**Richardson**)

■ HISTORY OF INTRODUCTIONS
PACIFIC OCEAN ISLANDS
New Zealand
The California ground squirrel may have been released in Dunedin, New Zealand, in 1906, but failed to become established (Lever 1985).

Hawaiian Islands
California ground squirrels also arrived in Hawaii in the 1980s in freight from California, but they were destroyed (Lever 1985).

SANDY SOUSLIK
Large toothed souslik, yellow spermophile, yellow suslik
Spermophilus fulvus (**Lichtenstein**)
=*Citellus fulvus*

■ DESCRIPTION
Upper parts yellowish, finely speckled due to black-tipped hairs, but lacks any light mottling.

■ DISTRIBUTION
Central Asia: in Kazakhstan from the Caspian Sea and the Volga River east to Lake Balkash, and south through Turkestan to northern Iran and northern Afghanistan. Also in western Sinkiang.

Sandy souslik

■ HABITS AND BEHAVIOUR
Habits, Gregariousness and Movements: probably as for other sousliks. **Habitat:** grassland, steppe, and forest. **Foods, Breeding and Longevity:** probably as for other sousliks. **Status:** no information, probably common.

■ HISTORY OF INTRODUCTIONS
ASIA
Sandy sousliks have been introduced and translocated successfully to a number of areas in Russia and Kazakhstan.

Kazakhstan
A number of introductions were made in Kazakhstan between 1929 and 1952, where sandy sousliks are now established with a restricted distribution (Kydyrbaev 1964; Yanushevich 1966). In the eastern part of Kazakhstan some 2724 sousliks were released and the species has become established in the Karagandinskoi, Semipalatinskoi and East Kazakhstanskoi oblasts (Kirisa 1973).

Russian Federation
Sandy sousliks have been translocated within the Russian Federation (Naumoff 1950; de Vos *et al.* 1956) and their range has been extended by re-introductions and introductions (Sludskii and Afanas'ev 1964).

Between 1929 and 1931 sandy sousliks were intro-duced to Barsa-Kelmes Island in the Aral Sea. Some 3754 sousliks were released in the Irpizskom and Aralskom regions (Kirisa 1973). However, according to some (Shaposhnikov 1960), the value (presumedly furs?) of the results of the introduction to the island is questionable.

■ DAMAGE
No information.

RUSSET SOUSLIK
Altai squirrel, russet squirrel
Spermophilus major (**Pallas**)
=*Citellus major*

■ DESCRIPTION
Dorsal pelage greyish, with slight pale mottling.

■ DISTRIBUTION
Asia: Central Asia from the Volga and Kama rivers to Novosibirsk and the Upper Yenesei, and through most of northern Kazakhstan east to Lake Balkash and the Mongolian Altai.

■ HABITS AND BEHAVIOUR
No information, probably as for other sousliks.

■ HISTORY OF INTRODUCTIONS
EURASIA
Russet sousliks have been introduced successfully into the Caucasus Mountains in the Russian Federation.

Russian Federation
The russet souslik was introduced in the northern Caucasus in 1937 and 1953 and has become well established and widespread (Yanushevich 1966).

■ DAMAGE
No information.

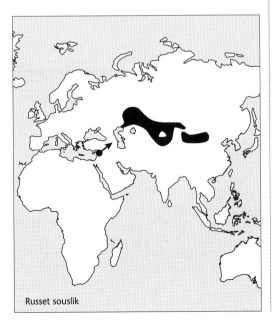
Russet souslik

ARCTIC SOUSLIK
Long-tailed souslik, arctic ground squirrel
Spermophilus parryii (**Richardson**)
=(?) *S. undulatus* (Pallas)

■ DESCRIPTION
TL 332–495 mm; T 77–153 mm; WT 680–910 g.
Head tawny or cinnamon, remainder upper parts reddish brown, cinnamon or fuscous and more or less flecked with whitish spots; eye ring buff; under parts, flanks and legs ochraceous tawny to cinnamon buff in summer and buff or greyish white in winter; tail ochraceous tawny or cinnamon buff mixed with fuscous black; below russet or tawny.

■ DISTRIBUTION
North America. Northern Canada and Alaska, Unimak Island, Kodiak Island, Sumagin Islands, Koniuji Island, Simeonof Island and Marble Island. In the Russian Federation in Tien Shan to the River Lena.

■ HABITS AND BEHAVIOUR
Habits: burrow 0.9 to 20 m; males defend territories; hibernates for 7 months of year. **Gregariousness:** forms colonies (female kin clusters); home range 4 ha. **Movements:** sedentary? **Habitat:** tundra beyond tree line; forest clearings; usually near water. **Foods:** leaves, seeds, stems, flowers, roots of grass, forbs and woody shrubs; also fruit, carrion, eggs and nesting birds. **Breeding:** gestation 25 days; litter size 4–8; 1 litter/ year; males polygamous; young disperse in season of birth; mature spring following birth (11 months). **Longevity:** 8–10 years in wild. **Status:** common.

■ HISTORY OF INTRODUCTIONS
NORTH AMERICA
Arctic sousliks (*S. p. alblusus*) have been successfully introduced to Unalaska, Unimak and Kavalga islands of the Aleutian group (Hall 1981).

Alaska (Aleutian and Pribilof Islands)
Arctic sousliks (*S. p. ablusus*) were introduced to Unalaska Island, Aleutians, from Nushagak, on Bristol Bay, by S. Applegate about 1895–1900 (Osgood 1904). It was the practice of this agent to release squirrels and other animals wherever they did not occur and the subspecies introduced here could be a mixed one (Rausch 1953). They are now widely distributed.

Sousliks (*S. undulatus kodiacensis* or *parryii*) were introduced to Kodiak Island, Alaska (Rausch 1953). At present they appear to be numerous only near the town of Kodiak and other suitable habitat is not utilised (Clark 1958). A number of squirrels were released in the 1920s and perhaps this species was among those introduced.

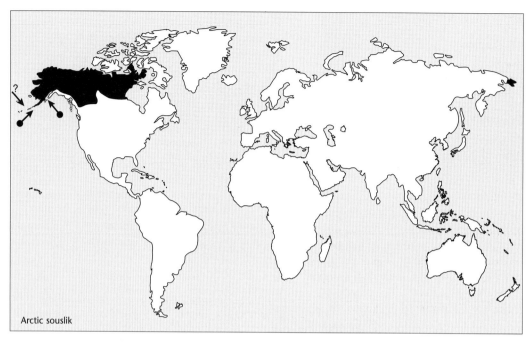

Arctic souslik

Arctic sousliks may also have been introduced to the Pribilof Islands in the Bering Sea (de Vos *et al.* 1956).

■ DAMAGE

In the United States ground squirrels or sousliks of this genera are serious pests of cereal crops and in orchards, and also disseminators of diseases such as bubonic plague. In 1918 damage was estimated to cost US$30 million (Grinnell and Dixon 1918).

Host to many species of fleas, ticks, helminths and coccidea, this souslik is a carrier of tularemia, brucellosis, erysipelus, tick-borne typhus, toxoplasmosis, and bubonic plague in parts of its range in the Russian Federation. In the eastern parts of its range it damages plantings and pastures and only in the extreme north is of value for game and food for fur-bearing predators (Ol'kova 1962).

PRAIRIE DOG
Black-tailed prairie dog
***Cynomys ludovicianus* (Ord)**
Previously in family Echimyidae

■ DESCRIPTION
TL 355–417 mm; T 72–115 mm; WT 0.5–2.2 kg.
Upper parts pinkish cinnamon finely lined with black and buff; upper lip, sides of nose and eye ring buff or whitish; tail black above for proximal two-thirds, black or blackish brown distally; below (tail) vinaceous cinnamon, distal third blackish or dark brown; under parts whitish or buff white.

■ DISTRIBUTION
North America. United States from Texas to the Canadian boundary.

■ HABITS AND BEHAVIOUR
Habits: diurnal; stores food; builds burrows and mounds; multiple exit burrows 3–5 m deep and 16 m in length; inactive in very cold weather; colonial; in towns to 2.3 to several hundred per ha.

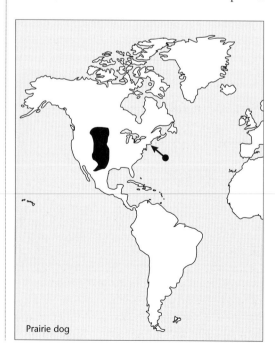

Prairie dog

Gregariousness: social; large towns; colonies 1 male + 3–4 females + 2–25 young. **Movements:** sedentary. **Habitat:** dry grassland, upland prairies, short-grass plains, river flats. **Foods:** weeds, forbs, grasses, roots, leaves, stems and insects. **Breeding:** late winter, April–May; gestation 27–35 days; female monoestrous; oestrous cycle 2–3 weeks; litter size 2–8; young altricial; eyes open 33–37 days; above ground at 6 weeks; weaned 6–7 weeks; males disperse at 12–14 months; sexual maturity at 1–2 years. **Longevity:** 3 years wild, 6–10 years in captivity. **Status:** common, range considerably reduced; some concern may be endangered.

■ HISTORY OF INTRODUCTIONS
NORTH AMERICA
United States
Black-tailed prairie dogs were introduced to Nantucket and Martha's Vineyard islands, Massachusetts, where they cause damage (de Voss 1956). [Not mentioned by Hall 1981.]

Some also existed in Sac County, Iowa; Monroe, Louisiana; and Seneca, South Carolina (Lever 1985). A small colony of five was trapped and destroyed near O'Neill Park, Orange County, California in 1965.

■ DAMAGE
Formerly prairie dogs destroyed crops of wheat, corn, alfalfa and hay and dug up sorghum, beans, potatoes and canteloupes and girdled newly planted fruit trees. They also competed with stock for food and their mounds were hazardous for horse riders. During the first half of the twentieth century there was a constant war against them with poisons, traps and guns resulting in a much reduced range (Banfield 1974) and the species may now be endangered.

BARBARY GROUND SQUIRREL
North African ground squirrel
Atlantoxerus getulus **(Linnaeus)**

■ DESCRIPTION
HB 160–220 mm; T 140–230 mm; WT 300–350 g.
Rigid, sparsely haired coat; upper parts greyish brown with three whitish longitudinal stripes; tail with light and dark rings.

■ DISTRIBUTION
Africa: Morocco and Algeria in the middle Atlas and Antiatlas mountains to the edge of the Sahara.

■ HABITS AND BEHAVIOUR
Habits: diurnal; burrow; does not climb; possibly hibernates at higher altitudes(?). **Gregariousness:**

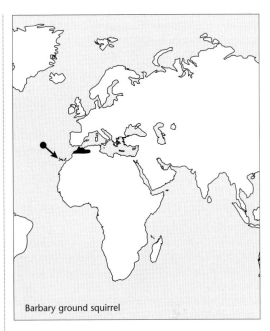

Barbary ground squirrel

sociable; establishes colonies. **Movements:** sedentary(?). **Habitat:** mountains to elevations of 4000 m. **Foods:** vegetation; nuts, conifer seeds, insects, birds' eggs. **Breeding:** no information (unknown(?)). **Longevity:** nine years in captivity. **Status:** common.

■ HISTORY OF INTRODUCTION
PACIFIC OCEAN ISLANDS
Canary Islands
Two or three pairs (later followed by at least two more) native Barbary ground squirrels of Morocco and Algeria were released on Fuerteventura in the Canary Islands between 1966 and 1970. By 1978 they had spread over much of the arid montane area of the island (Lever 1985).

■ DAMAGE
It is thought that they could well become a pest of crops on the Canary Islands (Lever 1985).

ABERT'S SQUIRREL
Tassel-eared squirrel
Sciurus aberti **Woodhouse**

■ DESCRIPTION
TL 463–584 mm; T 195–255 mm; WT 680–907 g.
Back dark grey; sides black; dorsal stripe indistinct, varying from rufous to chocolate brown; tail as for back above, but overlaid with wash of white, white below, and with grey basal band; ear tufts black; belly white (three forms recognised with either white, black or grey under parts).

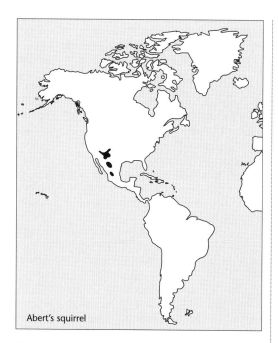

Abert's squirrel

DISTRIBUTION
North America: mountains of south-east Utah, Colorado, New Mexico and Arizona.

HABITS AND BEHAVIOUR
Habits: diurnal; arboreal; quiet; similar to red squirrel. **Gregariousness:** no information. **Movements:** sedentary. **Habitat:** pine and juniper woodland and forests. **Foods:** seeds, pine nuts, acorns, mushrooms, roots, bark, birds' eggs and young. **Breeding:** spring (April–September); litters 3–4; 2 litters/year in south; young born in nest of twigs and needles. **Longevity:** no information. **Status:** common.

HISTORY OF INTRODUCTIONS
NORTH AMERICA
United States
Abert's squirrel (*S. a. aberti*) was introduced in 1940 and 1941 to the Santa Catalina Mountains, Arizona (Lange 1960). They have also possibly been introduced into some places in New Mexico (Findley *et al.* 1975), but no details could be found.

DAMAGE
Not known to cause any damage.

RED-BELLIED SQUIRREL
Mexican grey squirrel, Guatemalan grey squirrel
Sciurus aureogaster Cuvier

DESCRIPTION
TL 418–573 mm; T 206–315 mm; WT no information.
Upper parts light to dark grey with some white

broken up by either nape and rump patches, as well as shoulder and costal patches, or combinations which vary in colour and size; under parts from white to orange to chestnut; tail variegated greyish white or grey-buff to orange or chestnut in those having orange under parts.

DISTRIBUTION
Central America: central Mexico to Guatemala.

HABITS AND BEHAVIOUR
Habits: largely arboreal. **Gregariousness:** no information. **Movements:** sedentary. **Habitat:** forest. **Foods:** mainly frugivorous; mangos, figs, plums, also buds, twigs, seeds, coconuts, nuts. **Breeding:** February–November mainly, but intermittent throughout year; litter size 1–2 (Florida). **Longevity:** no information. **Status:** no information.

HISTORY OF INTRODUCTION
NORTH AMERICA
Introduced successfully into Florida, United States.

United States
In 1938, two pairs of red-bellied squirrels were imported from eastern Mexico by a resident of Elliott Key, Florida and released at two points on the island (Brown and McGuire 1975). They have thrived here for the last 35 years and were common in 1969.

DAMAGE
Reports indicate that red-bellied squirrels occasionally do damage to cornfields in ear and could become a nuisance to fruit areas in Florida (Brown and McGuire 1975).

Red-bellied squirrel

GRAY SQUIRREL

The name has been spelled with an 'a' rather than an 'e' here, as it is in the United States where the species was named.

Grey squirrel, eastern gray squirrel, black squirrel
***Sciurus carolinensis* Gmelin**

■ DESCRIPTION

HB 230–300 mm; T 140–255 mm; weight 340–800 g.
Generally grey or blackish; under parts white or greyish, occasionally red phase or mixtures of all colours; face, feet and flanks dark brownish; ears medium length, without conspicuous tuft; feet greyish; tail long, bushy, black banded, grey brown, often with some white; hind feet with five toes, forefeet with four toes.

■ DISTRIBUTION

North America. From south-eastern Canada south through the eastern United States to Texas and the Gulf States.

■ HABITS AND BEHAVIOUR

Habits: diurnal; mainly arboreal; nest a bulky mass of leaves in trees, holes or dens; does not hibernate; caches nuts in holes in ground. **Gregariousness:** territorial; several together, with strict social hierarchy; density 7.4/ha in England. **Movements:** sedentary, though irruptive or dispersive at times when density high and food scarce; home range 0.77–22.2 ha, females less than males, males overlap that of several females. **Habitat:** mixed woodland and forest, and suburban areas such as open parklands, town parks.

Foods: mast, nuts, berries, flowers, seeds, fruits, fungi, mushrooms, leaves, buds, eggs, small birds, carrion, acorns, bulbs, shoots, catkins, grain and insects.
Breeding: breeds January and June–July, young born March and July; gestation 40–44 days; males promiscuous; female with two oestrous periods (possibly polyoestrous?) and with anoestrus in August–December; litter size 3–5, 7 (in Great Britain breeds December–January and May–June); 1–2 litters/year; young born blind, naked, eyes open 28–30 days; weaned 8–10 weeks; sexual maturity 10–12 months; females mature in 30–36 weeks. **Longevity:** 3–13 years (wild), and up to 15–23 years (captive). **Status:** common and abundant.

■ HISTORY OF INTRODUCTIONS

Gray squirrels have been introduced successfully in Canada, United States, Great Britain, South Africa and Australia.

AFRICA
South Africa

Gray squirrels were introduced near Capetown on the Cape Peninsula by C. Rhodes for aesthetic reasons about 1900 or soon after (Haagner 1920; Bigalke 1937; Davis 1950; Smithers 1983) and possibly as early as the 1890s (Lever 1985). A few pairs were released on Groote Schuur estate on Table Mountain and these had established and spread to overrun the peninsula by 1920.

By the 1950s they were spread over a radius of some 64 km extending from Capetown to Paarl and Klein

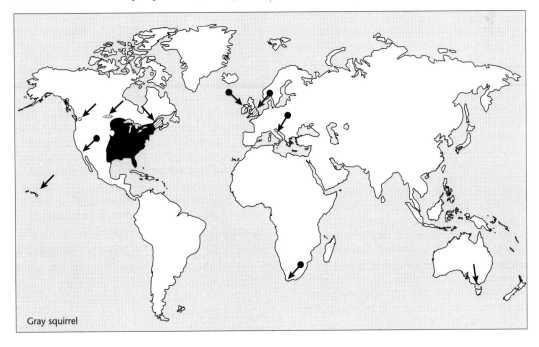

Gray squirrel

Drakenstein in the north-east, and to Elgin in the south-east. From Groote Schuur they spread into the suburbs of Capetown and had reached Tokai in 1908, Wynberg in 1910–11, Plumstead in 1914, Retreat in 1919 and the Eerste River in 1932. By 1930 they had reached the foothills of the Franschhoek and Hottentots Holland mountains in the Somerset West district. In 1933 they reached Oak Valley near Elgin, in 1939 the Steenbras Catchment area, in 1943 Elginvillage, and in 1948 the Lebanon Plantation north east of Elgin. La Motte in the Paarl district was reached by 1945, and by 1949 the gray squirrel had colonised an area of about 259 km^2 (Davis 1950).

The gray squirrel has apparently spread little since the 1950s and is still limited to the Cape Peninsula and areas of the Western Cape (Hey 1974). Continued spread is thought to be unlikely because of the terrain and limits of planted exotic trees.

More recently their range in other areas appears to have extended with alteration by cultivation of the formerly unfavourable habitat (Lever 1985). There have been more recent deliberate introductions in the Ceres Division (100 km from Capetown) and 190 km from Cape Town at Swellendam in 1957 and 1968 respectively. They now occur in patches of suitable habitat within an area of 7000 km^2, but have not invaded the native ecosystems and are confined to urban, agricultural or afforested environments (Macdonald *et al.* 1988; Macdonald and Frame 1988; Bigalk and Pepler 1991).

AUSTRALASIA
Australia
The gray squirrel was introduced to Australia early in the twentieth century by Sir Frederick Sargood, who released them on his estate 'Ripponlea' between Elsternwick and Balaclava, Melbourne. This colony at first thrived and spread and some were reported from Balaclava, Caulfield, East St. Kilda, Camberwell and Malvern (Barrett 1934). They remained common at 'Ripponlea' during the 1930s and became extinct during the 1940s (Watts and Aslin 1981).

A single pair from the 'Ripponlea' colony was released by J. Beaumont at Ballarat in November, 1937. These thrived during World War 2 and persisted until the late 1960s. The last specimen known was noted in 1968, but the reasons for their decline, both here and at Melbourne, are unknown (Watts and Aslin 1981).

EUROPE
Introduced from the United States to about 30 sites in England and Wales between 1876 and 1929, from Canada to three sites in Scotland between 1892 and 1920 and from England to one site in Ireland in 1911 (Corbet and Harris 1991).

Italy
Gray squirrels were introduced to Piedmont from North America in 1948 when two pairs from Washington DC were released at Candiolo, near Turin. This population has thrived and in 2000 occupied an area of about 800–900 km^2, in the provinces of Turin and Cuneo. In 1996 the total population in the Piedmont area was estimated at 2500–6400 individuals, based on drey counts and mark and recapture censuses in sample areas (Genovesi 2000). A second introduction occurred in 1966 at Liguria (north-west Italy) where five individuals from Norfolk, Virgina, were released into a park at Villa Groppallo, Genoa Nervi (S. Bertolino *pers. comm.* 2000). This population appears restricted to a small coastal area at present. Gray squirrels are increasing in numbers and range so quickly that it is suggested that expansion to the rest of continental Europe is likely. The expansion is said to be causing local extinctions of the red squirrel (*S. vulgaris*) and there could be a continental crisis if they continue to devastate hazelnut stocks.

United Kingdom and Ireland
Gray squirrels were introduced to Britain probably as an ornament or curiosity (Shorten 1963) from North America in the latter half of the nineteenth century and the first three decades of the twentieth century at various places in England, Scotland, Wales and Ireland (Middleton 1930). Between 1876 and 1929 they were released on 26 occasions at 20 different sites in the British Isles (Laidler 1980; Lever 1985). The first recorded occurrence was in Llandisilio Hall, Denbighshire, North Wales, in 1828. Some were also seen at Llanfair Caereinion, Llan Eurvyl and Cum Llwyndog in Montgomeryshire (Middleton 1930). The earliest recorded introduction was in 1876 when T. U. Brockehurst liberated a pair at Henbury Park near Macclesfield in Cheshire. A pair was said to have been shot near Highfields, Nottingham, in 1884, and in 1889 five, imported from the United States, were released by G. S. Page in Bushy Park, Middlesex. These latter animals failed to become established, but a release of about 10 squirrels at Woburn Abbey by the Duke of Bedford in 1890 were successful (Middleton 1930).

In 1892 a pair of gray squirrels was released at Fidnnart on Loch Long, Scotland, on the border of Dunbarton and Argyll. These became well established and had spread northwards to Arrachar and Tarbert by 1903, reached Luss in 1904, Innerbeg in 1906, Garelockhead in 1907 and Rosneath in 1915 and Helensburgh, Alexandria and Culdross by 1912. In 25 years they had colonised an area of over 777 km^2 (Boyd-Watt 1923; Fitter 1959; Lever 1977).

In 1902, 100 gray squirrels were released at Kingston Hill, Surrey, and at about this same time others were released at Rougemdont Gardens, Exeter. From 1889 to 1930 it is estimated that at least 33 introductions took place, involving at least 345 or more squirrels. At least 100 from America and 150 from Woburn were released, but accurate records were not kept. Releases occurred at: Rosett, Wrexham, Denbighshire, North Wales, in 1903; Zoological Gardens, Regents Park, in 1905–07; Scampton Hill, Yorkshire, in 1906; Kew Botanical Gardens and Cliveden, Buckinghamshire, in 1908; Farnham Royal, Buckinghamshire, in 1908–09; Dunham, Frimley, Sandling and Chiddingstone, Kent, in 1910; Bramhall, Cheshire, and in Birmingham in 1911–12; Castle Forbes, Longford County (Ireland), and at Corstophine, Edinburgh, in 1913; Yorkshire and Devon in 1914–15; Pittencrieff Park, Dunfermline, Fife, and Bournemouth, Hampshire, in 1919; Yorkshire in 1921; Ballymahon, Longford County, in 1928; Bestwood and Hartsholme in Nottinghamshire and Needwood Forest, Staffordshire, in 1929 (Anon. 1920; Ritchie 1923; Theobald 1926; Middleton 1930; Fitter 1959; Lever 1977).

Apart from those released in 1889, most of the introductions of gray squirrels were successful and they appeared to increase in all these areas. Other small releases in the same period at Ayrshire, at North Queensferry in Fife, and in Berkshire, Northamptonshire, Oxfordshire, Staffordshire, Devon and Warwickshire also helped and by 1930 they were firmly established in the south-east of England as far north as Northamptonshire and Warwickshire. At this time they were reported to occupy some 25 898 to 134 574 km^2 in the south-east of England, the Midlands and Yorkshire (Middleton 1930; Thompson and Peace 1962).

Through the 1930s, 1940s and 1950s the gray squirrel continued to expand its range. Between 1945 and 1955 there was further spread in Montgomeryshire, Cardiganshire, Denbighshire, Glamorgan, Carmarthenshire (Wales) and in England in Shropshire, Staffordshire, Yorkshire and Devon, and in Scotland in Merioneth, Perthshire, Argyll and Mid Lothian. They advanced into Wales on a broad front in both the northern and southern parts, into the south-west areas of Dorset and Somerset and spread outwards from a number of scattered locations in Devon. By 1955 they had reached Cornwall and between 1945 and 1955 had spread considerably further in Cheshire, Shropshire, Staffordshire,

Derbyshire and Nottinghamshire. Some had been reported from Lancashire and Lincolnshire and the spread eastwards in Essex was obvious. They were now found over an area in Scotland twice the size it had been in 1945 and had crossed the River Fife to Perthshire (Shorten 1957).

The area of spread in the British Isles was now reported to be about 101 002 km^2 and the gray squirrel was present in most English and Welsh counties, eight counties in Scotland and 10 in Ireland (Shorten 1946, 1954, 1957 and 1963). Until 1959 the spread continued mainly westwards into Devon, Cornwall and West Wales. The period of greatest advance appeared to be between 1937 and 1945, thereafter continuing at a slower but constant rate. They were still absent from Norfolk, east Suffolk, Anglesey, Westmorland, Cumberland, Northumberland, Isle of Wight and Isle of Ely (Lloyd 1962). The rate of spread of the gray squirrel until this time was estimated as about 8 km/year in England (de Vos and Petrides 1967).

Gray squirrels are now present in most parts of England and Wales south of Cumbria and Northumberland; in Scotland they continue to spread slowly, but are less widely distributed than the native red squirrel (Lloyd 1983). They are now throughout most of England, Wales and lower Scotland (Baker 1990).

Gray squirrels were introduced in Ireland in 1913 at Castle Forbes, County Longford, by the Earl of Granard. By 1956 they had spread to Armagh, Down, Fermanagh and Tyrone and are continuing to extend their range in the central and eastern counties (Lever 1985; D'Arcy 1988).

North America
Canada
Translocations of gray squirrels have occurred at Saskatoon, Saskatchewan, southern Quebec, New Brunswick and Nova Scotia in Canada (Lever 1985; Haltenorth and Diller 1994).

Gray squirrels (*S. c. pennsylvanicus*) were introduced into Stanley Park, Vancouver, British Columbia, when three to four pairs from Ontario were released shortly before 1914. They are still confined to the peninsula of Stanley Park and have maintained a saturation population of 25–60 animals in the 405-ha park since 1920. They are effectively prevented from spreading by the sea on three sides and by the city on the fourth (Robinson and Cowan 1954; Cowan and Guiguet 1960; Carl and Guiguet 1972; Banfield 1977).

Introduced to Vancouver Island, British Columbia (Guiguet 1975).

United States

Successful introductions (mainly translocations) have occurred in Oregon, California, Washington, Montana, North Dakota and Wisconsin in the United States.

The gray squirrel spread over much of north-eastern North Dakota from 1916–26 to the late 1950s, but also occurs in several areas through deliberate introductions. Some were released in Jamestown in 1904; 30 were released at Bismarck in 1914, 12 at Minot in 1915, some in the Killdeer Mountains at about this time and 12 from Wisconsin in 1951 and 1952, and at Valley City some time before 1912. All were apparently successful introductions (Hibbard 1956).

Gray squirrels were also introduced in western Vilas County, Wisconsin, in 1934 (Waggoner 1946) and some from Wisconsin were translocated to West Bay Game Preserve in Allen Parish about 1949 (Washburn 1949). In 1966 and 1968 releases were made in City Park, Great Falls in Cascade County, Montana, where they were still established in the late 1960s. A single animal was obtained in Miles City, Custer County, in 1958 where the species also appears to have been introduced (Hoffman *et al.* 1969).

Gray squirrels have been introduced successfully in Seattle, Washington (de Vos *et al.* 1956; Farrand 1991), and are found in a number of cities in California including Chico, Sacramento and Stockton, where they have also been introduced (Jameson and Peeters 1988). There were probably multiple introductions in California following the gold-rush period of the 1850s (Byrne 1979). Imports certainly began about 100 years ago, but do not appear to have been documented until 1938. At least three eastern United States subspecies were probably involved, as eastern gray squirrels now occur in the San Francisco Bay area, especially on the San Francisco Peninsula, extending as far south as Santa Cruz County (Lidicker 1991). They are also present in some urban parks of the Central Valley and along the Calavaros River. They are mainly restricted to heavily wooded areas and urban parks where they co-exist with and displace the larger fox squirrel.

PACIFIC OCEAN ISLANDS
Hawaiian Islands

A single gray squirrel was caught at Schofield Barracks on Oahu in 1943 and others were reported (in press) to be present at the time. No further animals were confirmed present and the species failed to become permanently established in the Hawaiian Islands (Kramer 1971).

■ DAMAGE

There is little evidence that gray squirrels cause any agricultural damage in North America where they are an important game animal, and many millions are killed annually. However, in Britain they gnaw the bark at the butts of trees, girdle the trunks and strip bark and cambium from branches, and by biting out the leading shoots affect the growth of both hard- and softwood trees (Thompson and Peace 1962). In Jasper Ridge Biological Preserve in California, it is thought that they have now increased and become more widespread probably to the detriment of the fox squirrel (*Sciurus griseus*), whose range has contracted (Macdonald and Frame 1988).

The most serious damage by gray squirrels is done to forestry hardwoods between May and August when the squirrels' food supplies are lowest (Seymour 1961; Thompson and Peace 1962). They strip the bark from the stems of such trees as beech, sycamore and larch, particularly in young plantations, and in severe cases kill the trees (Shorten 1957; Thompson and Peace 1962; Taylor 1963; Southern 1964). The damage caused is generally local, but at times can be both spectacular and serious (Thompson and Peace 1962); however, compared to forestry damage by deer it is said to be less (Thompson 1962).

Besides the damage to forestry, the gray squirrel occasionally digs up newly sown barley crops, attacks nut crops and fruit trees, and damages stacked grain or stooked corn (Thompson and Peace 1962; Southern 1964). However, the damage to agriculture and horticulture is said to be minor (Taylor 1963).

As the gray squirrel spread in Great Britain there was a subsequent decline in numbers of the native red squirrel (Middleton 1930; Shorten 1957; Lloyd 1962). By 1955 in England and Wales approximately 80 per cent of the forests lying within the 1945 range of the gray squirrel had no red squirrels in 1955 (Shorten 1957).

Because of the damage caused by the gray squirrel and its probable effects on the red squirrel population, much effort has gone into attempts at exterminating them, particularly in forestry plantations. From 1953 to 1958 bounty payments on 1 520 304 squirrels destroyed at a cost to the government of 107 500 pounds sterling had little effect on their numbers or checking their spread (Thompson and Peace 1962).

In South Africa it is reported that gray squirrels had become so abundant by 1937 they became pests of fruit crops (Bigalke 1937). Between 1918 and 1922 bonuses were paid on 11 188 squirrels. Because of their limited range and numbers the amount of damage caused in fruit orchards, vegetable crops, introduced oak and pine plantations is minor, but they are also blamed for preying on the eggs and young of some native birds, especially in urban gardens (Bigalke and Pepler 1991). However, the extent of their predation on birds is not known (Hey 1974).

In Italy gray squirrels damage poplars, hornbeams and ceral crops as well as out-competing the native red squirrel.

FOX SQUIRREL

Brown fox squirrel, eastern fox squirrel
Sciurus niger Linnaeus

■ DESCRIPTION

TL 454–700 mm; T 200–330 mm; WT 590–1363.2 g.
Coat generally orange to brown, but three well-marked colour phases of red, black and grey, and combinations; ears, cheeks, feet and under parts pale fulvous to cinnamon; bushy tail, banded black and buff, under tail rufous with black subterminal bands and cinnamon tips; soles of feet naked and black; facial vibrissae and claws black; toes, nose, sometimes ears, and tail tip marked with white.

■ DISTRIBUTION

North America. In eastern North America from the Gulf of Mexico north to the shores of the Great Lakes, New England (and extreme central southern

Canada)?. Range has extended northwards by natural invasion and introductions.

■ HABITS AND BEHAVIOUR

Habits: diurnal and arboreal; lives in leafy nests or holes in trees. **Gregariousness:** solitary or pairs. **Movements:** sedentary; home range of females *c.* 16.2 ha, males wander more widely and juveniles to 64.4 km. **Habitat:** open forest, forest edges, open groves, cypress swamps, farm woodlots, urban streets, and residential areas. **Foods:** nuts, acorns, seeds, fruits, berries, leaves, buds, bark, sap, flowers, catkins, corn, roots, insects, birds' eggs and young, fungi. **Breeding:** breeds January–February and May–June; female polyoestrous; gestation 44–45 days; 2 litters/year of 1, 2–4, 6 young; born blind, naked, helpless; weaned 5–8 weeks; leave nest 6–10 weeks; leave parents 2.5–3 months; females mature at 1 year, males longer. **Longevity:** 6–10 years (wild). **Status:** uncommon and scarce.

■ HISTORY OF INTRODUCTIONS

NORTH AMERICA
Fox squirrels have been introduced to New York, Washington, California, north-eastern Colorado and North Dakota in the United States and to Pelee Island, Ontario, in Canada.

Canada

In 1890 fox squirrels from southern Ohio were released on Pelee Island, Ontario, by C. Mills of Sandusky, Ohio (Wrigley *et al.* 1973; Banfield 1977). These thrived at first, but were decimated by hunting about 1925. However, they are now firmly established and common in the drier, wooded areas of the island (Banfield 1977). In 1972 a single animal was found at St. Claude, Manitoba, and this fox squirrel is thought to have reached this locality from the United States, as by 1948 they had reached Pleasant Lake, Benson County in Minnesota, which is only 72.4 km from St. Claude (Wrigley *et al.* 1973).

United States

In the United States the range of the fox squirrel has been extended northwards with agricultural development into north-eastern Colorado and North Dakota through both natural invasion and by several introductions. They have also been introduced in California, New York and Washington.

The fox squirrel spread northwards from South Dakota and Minnesota into North Dakota in the early 1930s. The spread was aided by general introductions in several different areas (Hibbard 1956). Five fox squirrels from Indiana were released by C. Worst in about 1941 in the Yellowstone River Valley, McKenzie County. These became well established and spread

Fox squirrel

upriver to Sydney, Montana, and along the Missouri River. Six squirrels from the Missouri River in 1953 and 12 in 1954 were released in the Kildeer Mountains in Dunn County, by the North Dakota Fish and Game Department, where they became well established.

Two earlier introductions that may have augmented local stocks, occurred in about 1935 and 1938. Some 50 squirrels from Minnesota were released some time after 1935 along the Missouri River in McKean County by R. Anderson. Two pairs were released at Jamestown in Stutsman County by C. Livesay in 1938. Since the releases in North Dakota the fox squirrel has spread up the Missouri River Valley and by 1969 occurred on the Yellowstone River as far north as Reed Point in Stillwater County (Hoffman *et al.* 1969).

The range of the subspecies *S. n. rufiventer* has extended into north-eastern Colorado through natural invasion and introductions (Hall 1981).

Fox squirrels were introduced to the San Fernando Valley, southern California, before 1904, to the Fresno area about 1900 and were known from the San Fransisco Peninsula in 1921 (Byrne 1979). In some parks in California they became established locally some time before 1934 (Storer 1934). Stock from eastern America (subspecies *rufiventer*) were released in several urban areas such as in Ventura County where they became established in the agricultural areas of Ventura and Oxnard (Wolf 1971). Although they increased sufficiently in the San Fernando Valley to become a pest of agriculture (Storer 1958), they have generally remained in the vicinity of the release area only occasionally spreading into the surrounding countryside (Wolf 1971). They are now present and likely to be the species seen about lawns in a number of Californian cities (Jameson and Peeters 1988). They occur over most of the coastal areas north to Medocino County and parts of the Central Valley where there are orchards or riparian habitats, and in urban parks (Lidicker 1991).

Wild-trapped fox squirrels were introduced to the Cornell University at Ithaca, New York, where they survived for several years prior to 1939. In 1939 the state authorities released 44 wild fox squirrels on Howlands Island Refuge, where they were surviving in 1941 (Bump 1941).

■ DAMAGE
In California the fox squirrel was reported to have become sufficiently abundant in the San Fernando Valley to have become an agricultural pest warranting some control (Storer 1958; Ball 1960). They were reported to damage such crops as walnuts, oranges,

avocados, strawberries and tomatoes (Storer 1958; Wolf 1971) and to gnaw the lead covering (insulation) from communication lines (telephone cables), causing considerable damage (Storer 1958; Ball 1960). In Ventura County they are often found near walnut and orange orchards, but probably eat less than US$20 worth of fruit each per year and because of the low populations the damage is not serious (Wolf 1971; Byrne 1979). Since they have become more widespread, the native western gray squirrel's (*S. griseus*) range has contracted particularly on the coast range on the eastern side of San Francisco Bay (Byrne 1979; Macdonald and Frame 1988).

Although they can be destructive in almond orchards, in cities they may gnaw their way into the attics of houses and live there quite happily (Jameson and Peeters 1988).

RED SQUIRREL
European red squirrel, common squirrel
Sciurus vulgaris Linnaeus

■ DESCRIPTION
HB 180–270 mm; T 140–200 mm; WT 200–480 g.
Winter coat dark greyish brown to reddish; in summer rufous with dark mid-dorsal stripe and in winter brownish-grey to ash grey; under parts white; upper parts of tail brownish red; large pointed ears, tufted, in winter black or brown; hindlimbs with five toes; forelimbs with four toes and sharp claws. Sexes alike. Juveniles darker than adults.

■ DISTRIBUTION
Eurasia. From the British Isles across mainland Europe, and Asia in the northern forested areas south to the southern Urals, the Altai, central Mongolia, Manchuria, Sakhalin, Japan and Korea.

■ HABITS AND BEHAVIOUR
Habits: diurnal, arboreal; nests (dreys) in forks of branches and hollows, 300 mm diameter of twigs, needles, lined soft material; does not hibernate; caches food in summer and autumn. **Gregariousness:** solitary, but has communal nests in winter and spring; dominance hierarchies among and between sexes; density 0.5–0.8/ha. **Movements:** sedentary; overlapping home range 2.2–12.5 ha. **Habitat:** boreal and coniferous forest and woodlands; urban areas, parks. **Foods:** beech mast, pine and other seeds, nuts, acorns, fruits, berries, buds, shoots, flowers, bark, fungi, insects and eggs and small birds. **Breeding:** breeds December–April to August–September, young born March–April and July–August, in stick nest or

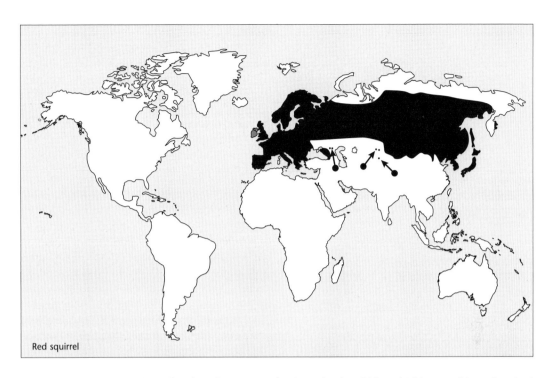

Red squirrel

hollow; gestation 36–42 days; female polyoestrous; litter size 1, 3–4, 6; 1–2 litters/year; young born blind, naked; eyes open 28–30 days; leave nest at weaning 7–10 weeks; sexually mature at 10–12 weeks. **Longevity:** 6 months to 7 years (wild), 12 years (captive). **Status:** common and abundant.

■ HISTORY OF INTRODUCTIONS
EURASIA
Red squirrels have been introduced successfully in the Caucasus, in Crimea and in several parts of Kazakhstan and Kirgizia south of their original native range (Corbet 1978, 1980).

United Kingdom
At one time red squirrels were ubiquitous in woodland throughout Britain and Ireland. They became extinct in Ireland and southern Scotland by the early eighteenth century and rare in the Scottish Highlands in the late eighteenth and early nineteenth centuries. Re-introductions occurred at 10 sites in Scotland, between 1772 and 1782 mainly from England, and at about 10 sites in Ireland from England between 1815 and 1856. They became abundant between 1890 and 1910 throughout the British Isles – but thereafter declined and became scarce in many areas in the 1920s. They declined further as the gray squirrel (*S. carolinensis*) advanced across the country, but the reasons for the decline are still unclear (Corbet and Harris 1991). Several red squirrels were introduced on an experimental basis to Regent's Park in 1984.

The endemic British and Irish race of the red squirrel is characterised by bleaching of the ears and tail. However, introductions of *S. v. vulgaris* from Scandinavia to Perthshire in 1793 (Shorten 1954) and *S. v. fuscoater* from western Europe to the Lothians in 1860 (probably) and Epping Forest about 1910 (Harvie-Brown 1880–81) complicate the picture and now some squirrels exhibit bleaching of the fur and some do not (Corbet and Harris 1991). There is no reason to doubt that it is a native of Britain, but its origin in Ireland is more doubtful and likely the result of introduction. It may have become extinct in Ireland in the eighteenth century and was re-introduced with a series of releases in the early nineteenth century (D'Arcy 1988). It is uncertain if red squirrels were in Ireland before being introduced in 1815–80 (Barrington 1880).

Red squirrels were almost extinct in Scotland during the late eighteenth and early nineteenth centuries, but were then successfully re-introduced (Harvie-Brown 1880–81; Lloyd 1983). In about 1772 some were taken to Dalkeith near Edinburgh and either escaped or were released and within 30 years were found in most of the woodlands of mid and East Lothian and were spreading into Peebleshire. By 1821 they had crossed the Forth and in another 30 years reached the Highland foothills. In about 1827 some of the above animals were released at Minto, Roxburghshire and Selkirkshire, where they became established and spread through the eastern lowlands and adjacent parts of Northumberland. The western lowlands were stocked from a number of sources – partly from natural spread from Cumberland and partly from introductions at Southwick, Wigtownshire, about 1830. A pair were released at Kirkbean, Kircudbrightshire in 1867, and also partly from other escapes and releases, introduced probably in Ayrshire between 1866 and 1872 and Loch Fyne in Argyll, Inverness-shire in 1844, Glen Urquhart and many others.

Between 1815 and 1880 red squirrels were re-established by introduction in 10 places on at least 14 separate occasions (local stock and European animals).

Red squirrels were established in Britain by releases from various sites around the turn of the twentieth century and subsequently spread to colonise most of England and Wales, and parts of Scotland and Ireland (Reynolds 1985). In the early twentieth century red squirrels became scarce and were re-introduced to a number of places in England. They were re-introduced to Epping Forest about 1910, unsuccessfully introduced to Whipsnade in 1931–36 and introduced to Hebden Bridge, Yorkshire, before 1947 (some may have been continental forms). Present populations in England could be a mixture of more than one subspecies in some places (Fitter 1959).

Red squirrels continue to decline as the introduced gray squirrel advances. The remaining populations are located mainly in Cumbria and Northumberland, East Anglia, parts of Wales with smaller colonies, on the Isle of Wight and elsewhere and more widespread in Scotland (Lloyd 1983).

Germany

In Germany red squirrels have become acclimatised on the island of Usedom (Baltic Sea) since 1890 when a few pairs were released, but they may not be truly wild there (Herold 1921). On Greifswalder Oie two were released in 1934, but they have not been permitted to remain permanently established on the island (Herter 1936).

Russian Federation and adjacent independent republics

Deliberate introductions have been responsible for red squirrels becoming a widespread fur animal in the Russian Federation and in adjacent independent republics. Kazakhstan, the Caucasus and the Crimea now have established populations of Siberian red squirrels. They were established in the forests of Zarenzeit in 1911–12 by the efforts of the then government that had released 30 squirrels from Tedeutka, and there were further introductions following the revolution.

Between 1927 and 1940 at least 500 Tedeutka squirrels were released and these succeeded at different localities. Until 1939 there were many translocations from these established populations to other areas, but thereafter the demand for their furs declined (Niethammer 1963).

In 1937, 120 red squirrels (S. v. altaicus) from the Altai were taken to the Caucasus and in five years had increased to a population of 5000. About the same time there were introductions at Novosibirsk and Krasnoyarsk. In 1944 they had been released and established in at least eight different areas in the Russian Federation and adjacent independent republics and in 1958 were still to be found in an equal number of areas. In 1959 the Leningrad fur auction offered for sale some 1 002 000 squirrel pelts, which serves to indicate the importance of the fur squirrel industry in Russia (Niethammer 1963) at this time.

Both Teleut and Altai red squirrels have been released in at least 25 regions between 1911 and 1970. From 1927 to 1941 some 2334 red squirrels and from 1946 to 1970 some 9005 (total of 11 339) are reported to have been released (Kirisa 1973). Introductions of S.v. exalbidus into central Russia from several parts of Siberia have been particularly successful and resulted in thousands being harvested for furs (Schmidt 1954 in de Vos et al. 1956). So many introductions and re-introductions have occurred that the previous range of the species has now been well extended (Sludskii and Afanas'ev 1964).

The race exalbidus has been established in several parts of both Kazakhstan and Kyrgyzstan (Kirgizia), south of its original range. Between 1911 and 1965 some 3011 red squirrels were released in Kazakhstan (Kirisa 1973). Here, 517 were liberated in 1952–53 and in 1960 in the mountain forests of Tien Shan where they established rapidly and began to increase (Afanas'ev 1962). They have also been released in forests of the Tien Shan Mountains, where at least 864 animals were released in the period 1957–60 (Kirisa 1973). Following early successful releases in this area, many were trapped and transferred to other areas in the mountains where their acclimatisation was progressing satisfactorily (Tyurin and Busalaeva 1963; Tyurin 1964), although with a generally restricted distribution (Yanushevich 1966).

Red squirrels (S. v. exalbidus) have been successfully acclimatised in the northern Caucasus, probably from introductions as early as 1937 (Yanushevich 1937; Gorrhkov 1963; Khrustalev 1963). Here they are reported to have a restricted distribution but in 1961 there were 350 squirrels/1000 ha of forest in the area

(Khrustalev 1963). Successful introductions and re-introductions have also occurred locally in Tatarstan where between 1950 and 1953 some 485 animals were released (Yanushevich 1966; Kirisa 1973).

Introductions occurred in the Crimea (Krym) before World War 2, in Lithuania, and re-introductions in the Ukraine and West Siberia have all been somewhat successful while those in Belorussia have failed (Yanushevich 1966).

Introductions were also made in Gruzinsk in 1951 (158 squirrels), Karbadino-Balkarsk in 1954 (73), North Osetinsk in 1952 (120), Checheno-Ingushsk in 1953 and 1959 (169), to Krmskoblast, Ukraine, from 1940 to 1950 (629), and in Stavropolsk in 1937 (120), in Litovsk in 1953 and 1956 (210), Latvia in 1952 (82), Belorussk in 1951–54 (316), Ukraine 1927–50, Bryahnsk in 1947 (248), Orlovsk in 1949 (194), Tambovsk in 1948 (137), Penzensk in 1946 and 1948 (185), Kirovsk in 1949 (184), Chelyahbinsk in 1951 (73), Sverdlovsk in 1949–51 (237), Tumensk in 1958 (135), Omsk in 1957–60 (481), Altaisk from 1939 to 1954 (1002), Krasnoyahsk from 1939 to 1954 (1418) and in Novosibirsk from 1935 to 1940 (618) (Kirisa 1973).

Successful acclimatisation of *S. v. altaicus* in the Teberda State Reservation was achieved and now between 2000 and 6000 squirrels are present at a density of about 800–1000/ha (Bobyr 1978). However, apart from the introductions in the Caucasus, releases of red squirrels have on the whole not yielded appreciable results (Sofonov 1981).

■ DAMAGE

In the British Isles red squirrels may cause serious damage in conifer plantations by stripping the bark from the upper main stems of Scots pine, less often from European larch, Norway spruce and lodgepole pine (Shorter 1957).

Three hypotheses have been advanced to explain the decline of red squirrels with the advance of the intro-duced gray squirrel in the United Kingdom: (1) competition with introduced gray squirrel, (2) environmental change, or (3) disease affecting one species and not other. A number of factors have been examined and have failed to provide a satisfactory explanation, but suggested disease and island effects are important factors in the red squirrel decline and in the gray squirrel invasion (Reynolds 1985). Interactions between red and gray species only partially explain the decline of the red species; in many cases red squirrels became extinct before the area was colonised by grays; in other cases they co-existed for many years and gray presence *per se* did not enhance the probability of red squirrel extinction (Reynolds 1985).

The introduction of red squirrels into parts of London is currently underway to ascertain the factors that may have caused their decline (Baker 1986). Recent computer modelling, however, suggests that competition alone could have been responsible (Okubo *et al.* 1989).

DOUGLAS' SQUIRREL

Douglas's squirrel, Douglas ground squirrel, chickaree
Tamiasciurus douglasii (Bachman)

■ HISTORY OF INTRODUCTIONS
NORTH AMERICA
United States
These squirrels were possibly introduced in California and in the eastern United States, but no records could be found.

RED SQUIRREL

Chickaree, American red squirrel
Tamiasciurus hudsonicus (Erxleben)

■ DESCRIPTION
HB 165–230 mm; T 90–160 mm; WT 140–312.4 g.
In summer upper parts glossy olive-brown to grey or rusty red, flecked with black; under parts white, greyish white or yellowish; eye-ring prominent, white; flank stripe black; backs of ears and limbs cinnamon; tail rufous red (but varies), upper parts with a black submarginal band and tawny or black tip, under tail grey, rufous, or yellowish. In winter fur is longer, silkier, and with grey-buff tipped undercoat and brighter general colour; under parts grey; stripe from head to tail and flanks reddish brown and obscure; ears with prominent red or black tufts; sole hair silvery.

■ DISTRIBUTION
North America. From Alaska and Canada, except the far northern regions, south through the western United States (Rocky Mountains) to southern California, and south in the eastern side of the conti-nent to the Appalachian Mountains.

■ HABITS AND BEHAVIOUR
Habits: arboreal and terrestrial; diurnal, most active after sunrise and before sunset; bold, aggressive and noisy; hibernates in severe cold weather; lives in tree cavities, nests, under rock piles and in burrows.

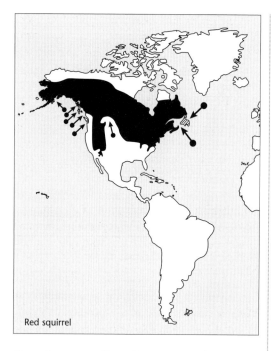

Red squirrel

Gregariousness: solitary. **Movements:** sedentary and territorial; home range 1.1–2.4 ha. **Habitat:** coniferous forests, plantations, swamps; often near human habitation. **Foods:** conifer cones, nuts, seeds, acorns, berries, fruits, buds, flowers, bark, sap, mushrooms, catkins, birds' eggs and nestlings, mice, voles, carrion and insects. **Breeding:** breeds February–August; female polyoestrous; gestation 35–40 days; 2 litters/year, in April–May and August–September of 1–6, 8 young; in northern parts of range only 1 litter/year; young born naked, blind, eyes open 27 days; weaned *c.* 5 weeks; sexually mature at 1 year. **Longevity:** probably less than 3 to 7 years in wild, 9 to 12 years in captivity. **Status:** common.

■ HISTORY OF INTRODUCTIONS
NORTH AMERICA
Red squirrels have been introduced and established on a number of islands off the coasts of Canada and Alaska including Newfoundland, Queen Charlotte Islands, Sidney, Kodiak, Afognak islands and the Alexander Archipelago.

Canada and Alaska
Six red squirrels were transferred from Vancouver Island, Canada, to the Queen Charlotte Islands in 1950 by the Canadian Game Commission (Cowan and Guiguet 1965; Carl and Guiguet 1972). The subspecies *T. h. lanuginosus* is now well established and thriving on Graham Island and Moresby Island in the Queen Charlotte group (Cowan and Guiguet 1965; Banfield 1977). Nine red squirrels were also

taken from Vancouver Island to Sidney Island in 1964, where they also have become well established (Carl and Guiguet 1972).

In Alaska the red squirrel was introduced to the Kodiak Island group (off southern Alaska) in the 1920s (Clark 1958), Baranof Island (Alexander Archipelago) in 1930–31 (93 released), Chichagof Island (Alexander Archipelago) in 1930 (52), Afognak Island (off Kodiak) in 1948 (6) and in 1952 (47), and on Kodiak Island in 1952 (24) (Burris 1965). They appear to have been successfully established on all these islands.

The subspecies *T. h. ungavensis* was introduced to Newfoundland in 1963 and to Camel Island in 1964 (Payne 1973, 1976). Here they became well established and as of 1976 were extending their range (Hall 1981). In July 1964, four male and two female red squirrels were introduced from Labrador to Camel Island in Notre Dame Bay, Newfoundland (Payne 1976). By the autumn of 1967 these had established and bred and the population numbered 68. Between 1969 and 1971 the population fluctuated between 115 and 202 squirrels.

An unknown number of red squirrels from Labrador were released by Main Brook and Roddickton residents in about 1963 (Payne 1976). These became established and some were noted 16 km south of Roddickton in 1967. In 1968–69 some were noted 19 km east of Roddickton at Conche. It is expected that these populations will increase and the species will spread across Newfoundland.

■ DAMAGE
None known.

INDIAN PALM SQUIRREL
Northern palm squirrel, five-striped squirrel
Funambulus pennanti Wroughton

■ DESCRIPTION
HB 115–178 mm; T 110–120 mm; WT 60–180 g.
Body generally greyish or reddish brown; back has three to five longitudinal light stripes contrasting with brown or grey between; belly creamy white; ears small, triangular; tail bushy.

■ DISTRIBUTION
Asia. Northern and central India, Pakistan and Nepal.

■ HABITS AND BEHAVIOUR
Habits: diurnal, arboreal and terrestrial; courtship chasing; nest of sticks and fibres. **Gregariousness:** as many as 10 together; family groups; communal nests.

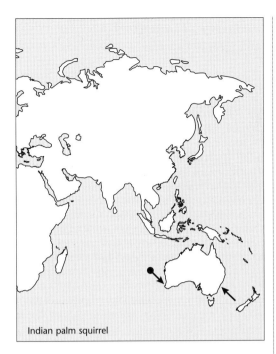

Indian palm squirrel

Movements: sedentary; home range 0.15–0.26 ha. **Habitat:** open palm growths, forest and scrub, gardens and parks, schools. **Foods:** seeds, fruits, nuts, plant stems, bark, buds, leaves, flowers, bread, nectar, birds' eggs, insects and grubs. **Breeding:** breeds in March–September (India) [August–May in Perth]; gestation 40–45 days; litters 1–5; 2 litters/year; lactation 2 months; mature 6–9 months. **Longevity:** no information. **Status:** fairly common.

■ HISTORY OF INTRODUCTIONS
AUSTRALASIA
Australia
Indian palm squirrels have been introduced and established in the cities of Perth, Western Australia, and Sydney, New South Wales, although it is now extinct in the latter city.

In Sydney palm squirrels were confined to an area 5 km in diameter, centred on Taronga Park Zoo. They declined greatly and are now extinct. They are said to have been eradicated (Wilson *et al.* 1992). The last to be seen was at Taronga Park Zoo in 1976 (Watts and Aslin 1981).

Introduced to the Perth Zoological Gardens in 1898, the palm squirrel remained confined to these gardens for many years. However, in the last 25 years they have invaded or been taken to a number of surrounding suburbs. Up until 1960 some had been found up to 4 km from the zoo. In the 1970s and 1980s they were found further afield and in some country towns, but did not remain established anywhere. Their distribu-

tion in the suburbs appears to be limited by the presence of exotic trees, a high mortality rate and a limited food supply. They were released in Perth to 'add colour' to the Zoological Gardens.

■ DAMAGE
In India palm squirrels damage twigs used in lac production. In Western Australia they are accused of damaging fruits in backyards near the zoo, especially citrus and stone fruits, and have been known to ruin electrical wiring in the roofs of houses.

BELLY-BANDED SQUIRREL
Mountain red-bellied squirrel, grey-bellied squirrel
Callosciurus flavimanus (Geoffroy)
This form is variously allocated to grey-bellied or golden-backed squirrel, C. caniceps, belly-banded squirrel, C. erythraeus, and that accepted here, C. flavimanus.

■ DESCRIPTION
HB 180–230 mm; T 150–220 mm; WT 165–315 g.
Body and tail grizzled olive brown except for a pair of broad longitudinal bands of red on under parts from base of forelimbs to hindlimbs on each side and separated by central band the same colour as under parts; tail, some races have black tip; northern animals are paler.

■ DISTRIBUTION
Asia: the eastern Himalayas and Burma east to southern China and south to Malaya; also Taiwan.

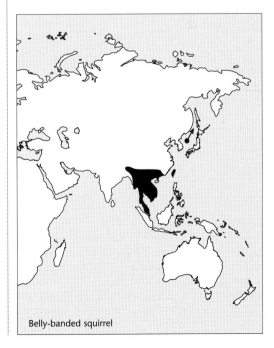

Belly-banded squirrel

■ HABITS AND BEHAVIOUR
Habits: diurnal; arboreal; nests in hollow trees, or spherical tree-nests of leaves, twigs and lined with fibre. **Gregariousness:** solitary or family parties of 2–4. **Movements:** sedentary. **Habitat:** forest, secondary growth, cultivated areas, plantations and gardens. **Foods:** fruit, seeds, nuts, buds, flowers, birds, insects. **Breeding:** breeds year round; litter size 1–5; young born hairless, blind. **Longevity:** 8–9.5 years (captive). **Status:** fairly common.

■ HISTORY OF INTRODUCTIONS
PACIFIC OCEAN ISLANDS
Oshima, Japan
Belly-banded squirrels (*C. c. thaiwanensis* = *C. flavimanus*) were introduced to the island of Oshima, south of Tokyo, Japan, some time after 1930 or 1940 from Taiwan to the Zoological Gardens on the island (Corbet 1978, 1980). Some escaped from the gardens and in 1950 an estimated population of 20 000 squirrels inhabited many parts of the island (Minamino 1950). In the mid-1950s they were found in nearly all the forested areas of the island (Kuroda 1955) and have remained well established there (Udagawa 1970; Corbet 1980).

Izu-shima, Japan
Belly-banded squirrels (Formosan red-bellied tree squirrel, *C. erythaeus* = *C. flavimanus*) were introduced to the island of Izu-shima in Seven Isle of Izu from Taiwan. The species may have been released on two small islands in 1935.

■ DAMAGE
The introduction of the belly-banded squirrel has affected camelia oil production on the island of Oshima as the squirrels destroy the flowers and seeds of the camelia plants (Kuroda 1955). These squirrels can also cause damage to nut crops on the island (Udagawa 1970). On Taiwan the belly-banded squirrel is commonly recognised as an animal that de-barks trees.

AMERICAN FLYING SQUIRRELS
Glaucomys sp. (?)

■ HISTORY OF INTRODUCTION
EUROPE
United Kingdom
American flying squirrels were probably released at Woburn, England, in the 1920s and/or 1930s, but failed to become established (Fitter 1959).

Family: Heteromyidae
Kangaroo rats

KANGAROO RAT
Ord's kangaroo rat
Dipodomys ordii Woodhouse

■ DESCRIPTION
HB 100–275 mm; T 100–163 mm; WT 40–96.5 g.
Upper parts tawny with a few black hairs along mid-dorsal line; under parts white; white spot over each eye; white lines across hips; ventral stripe tapers to point near tip of tail; two grey lines above and below tail and grey terminal tuft; cheek pouches furred; eyes large.

■ DISTRIBUTION
North America. From southern Alberta, Saskatchewan, Canada, and western United States south to northern and central Mexico.

■ HABITS AND BEHAVIOUR
Habits: hops; builds burrows with grass-lined nest at bottom; stores seeds in side chambers; nocturnal(?); territorial. **Gregariousness:** solitary. **Movements:** sedentary; home range 0.43 ha. **Habitat:** sandy soils in open areas with sparse bush or grass; sage brush desert. **Foods:** grass seeds, grasses, forbs, mesquite; also fruits, leaves, stems, buds and insects (grasshoppers, moths). **Breeding:** late winter to early summer,

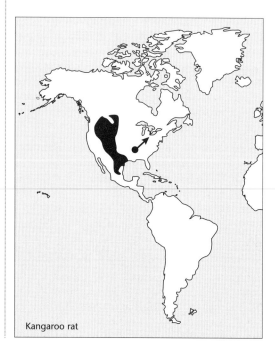
Kangaroo rat

breeding stimulated by rain; gestation 29–30 days; seasonally polyoestrous; oestrous cycle 5–6 days; litter size 1–6; maybe 2 litters per year; young live in nest 4–5 weeks; sexually mature at about 2 months. **Longevity:** at least 2 years in wild, 9 years 10 months as captive. **Status:** common?

■ HISTORY OF INTRODUCTION
NORTH AMERICA
United States
The kangaroo rat was introduced in Ohio on shores of Lake Erie some time before 1956, where they became established in sand dunes near Fairport (de Vos *et al.* 1956; Lever 1985).

■ DAMAGE
None known.

Family: Muridae
Rats and mice

BANANA MOUSE
Sumichrasti's vesper mouse
Nyctomys sumichrasti (Saussure)

■ HISTORY OF INTRODUCTION
UNITED KINGDOM
Before 1959 banana mice were not infrequently found in consignments of bananas in London and Welsh ports (Fitter 1959).

DEER MOUSE
White-footed mouse
Peromyscus maniculatus (Wagner)

■ DESCRIPTION
HB 120–220 mm; T 80–180 mm; WT 11.9–34 g.
Variable species, upper parts pale grey to greyish buff to reddish brown; under parts white; tail short haired, penicillate, bicoloured, dark above and light below.

■ DISTRIBUTION
North America. Labrador to Yukon, Canada south to the southern United States and Mexico.

■ HABITS AND BEHAVIOUR
Habits: builds nest of dry vegetation in hollow logs, under rocks; agile; caches food for winter; enters daily torpor. **Gregariousness:** in winter huddles in groups for warmth. **Movements:** sedentary; overlapping home ranges 1.2 ha or less. **Habitat:** alpine forest, meadows, grassland, scrub, cultivated fields, human

Deer mouse

habitation. **Foods:** plant and animal matter; seeds, nuts, acorns, fruits, mushrooms, flowers, berries, insects (caterpillars) and their eggs and larvae, spiders. **Breeding:** all year, mainly spring April–December; gestation 22–35 days; litter size 1–11; 2–4 litters/year; weaned 1 month; sexual maturity males 40–45 days, females 32–35 days; breed at 7 weeks. **Longevity:** 12–32 months (wild) to 8 years (captive). **Status:** extremely common and widespread.

■ HISTORY OF INTRODUCTIONS
NORTH AMERICA
Canada
A number of deer mice were transferred from Vancouver Island to the Chatham Islands, Discovery Island and Trail Island in 1951 as part of a study of animal populations and have become established on some of them (Carl and Guiguet 1972).

United States
Introduced experimentally between 1962 and 1970 to some small islands in Penobscot Bay, Maine, to study colonisation and extinction of small rodents. They were released onto the 0.8-ha island in order to study their subsequent interactions and persisted there for at least the next six summers (Crowell 1973; Crowell and Pimm 1976).

■ DAMAGE
Deer mice have the reputation of nibbling anything but glass and metal. They enter mountain cabins in winter and make nests of such materials as pillows, mattresses, toilet paper and tampons (Jameson and Peeters 1988).

COTTON RAT
Sigmodon hispidus Say and Ord

■ HISTORY OF INTRODUCTION
UNITED KINGDOM
Before 1959 cotton rats were found in British ports on ships from South America (Fitter 1959).

GOLDEN HAMSTER
Syrian golden hamster
Mesocricetus auratus (Waterhouse)
=*Cricetus auratus*

■ DESCRIPTION
HB 150–180 mm; T 12–13 mm; WT 85–130 g.
Light reddish brown; under parts white or creamy (one form with ashy stripe across breast); domestic variants include long-haired, rex-coated, white, cream and multi-coloured; cheek pouches enormous, with dark stripe; collar mark and pouch patches variable, white; tail short; female with 14 to 22 mammae and slightly larger than males.

■ DISTRIBUTION
Asia Minor (except extreme west) south to Syria and doubtfully to Israel, east through north-western Iran to Caucasus and Kurdistan.

■ HABITS AND BEHAVIOUR
Habits: mainly nocturnal, also crepuscular; hibernates; territorial; burrows. **Gregariousness:** solitary or in families. **Movements:** sedentary(?). **Habitat:** brushy slopes and steppes, sand dunes. **Foods:** omnivorous; fruits, leaves, roots, seeds, and small animals occasionally. **Breeding:** litters 4, 6–7, 15; 2, 3–7, 8 litters/year; breeds all year (captive); gestation 15–19 days; oestrous cycle 4 days; young born naked, helpless; weaned 20–25 days; sexual maturity 6–10 weeks; breeds at 2–3 months. **Longevity:** 1.5–3 years. **Status:** common in captivity as popular pet.

■ HISTORY OF INTRODUCTIONS
Golden hamsters have been unsuccessfully introduced into England, the United States and Australia.

AUSTRALASIA
Australia
More than 40 hamsters were released in dense bush at Cash's Crossing, near Brisbane, Queensland, in May 1981 by a pet owner. Attempts were made to retrieve these animals and the species did not become established. They were illegally imported into the country and more than 700 hamsters and gerbils smuggled in from South-east Asia for sale as pets were seized by investigators.

EUROPE
Germany
Golden hamsters became locally established in the wild in parts of Germany before 1956 (de Vos *et al.* 1956), but do not appear to have become permanently established. In 1950 at Querum two became established and in 1957 at Bielefell seven or eight became established for about seven or eight weeks (Niethammer 1963).

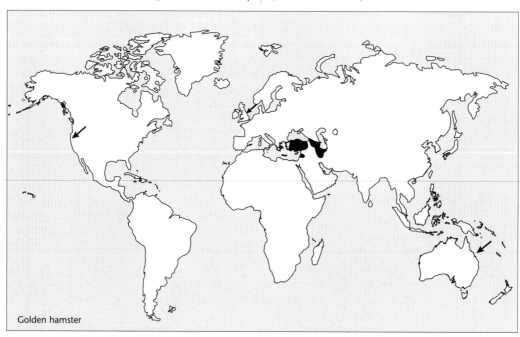

Golden hamster

United Kingdom
Commonly kept in captivity, hundreds of hamsters, if not thousands, escape each year from cages and pens (Baker 1990; Corbet and Harris 1991). Populations of free-living hamsters have been recorded many times (Rowe 1960, 1968; Baker 1968); all were trapped, but it is likely there have been continual deliberate releases, stimulated by media attention (Baker 1986).

In 1957, six golden hamsters escaped in an unheated basement pet shop in Bath, Avon, England, and one year later 52 were caught there (Rowe 1960; Corbet and Harris 1991). They have been found living in the wild since 1957, and a further six cases were reported after that and, like the original, stemmed from escapees from pet shops. The largest outbreak was when some 230 hamsters were accounted for (MAFF 1973; Corbet and Harris 1991). They were reported at Finchley, Middlesex (four escaped in 1960 and 25 caught in 1962); Boothe, Lancashire (17 caught in florist shop 1962); Manchester, Lancashire(?) and Bury St. Edmunds, Suffolk (in colony under shop). In all 230 were captured alive or poisoned.

In 1981 a number became established in Burnt Oak in Barnet on the outskirts of London. By August, 150–180 had been trapped around houses, sheds, and gardens on a council housing estate and the total population estimated as several hundred. They survived the winter in 1981–82 (Lever 1985; Corbet and Harris 1991).

NORTH AMERICA
United States
It was found by experiment that the domestic strain of golden hamster sold by pet stores would become a serious threat to range lands in California if released in favourable habitat and food (Howard 1959).

■ DAMAGE
No information.

LARGE BAMBOO RAT
Rhizomys sumatrensis (Raffles)

■ HISTORY OF INTRODUCTIONS
Singapore
Bamboo rats from Indochina, Malaya and Sumatra occur on Singapore as escapees from captivity (Medway 1978).

MONGOLIAN GERBIL
Meriones unguiculatus (Milne-Edwards)

■ DESCRIPTION
HB 110–120 mm; T 100–120 mm; WT 70–110 g.
Generally brownish or yellowish brown, black tipped hairs on sides and back; under parts whitish or tan; tail well haired, with black terminal brush; eyes large, black, circled by whitish buff ring which extends back to ears; soles haired; feet pale with black claws. Several colour varieties bred in captivity including black, white and pale fawn.

■ DISTRIBUTION
Eastern Asia. Most of Mongolia and China from Sinkiang through Inner Mongolia to Manchuria and adjacent parts of Russia.

■ HABITS AND BEHAVIOUR
Habits: nocturnal and diurnal; lives in colonies in burrow systems; docile; popular pet. **Gregariousness:** pairs or small colonies. **Movements:** sedentary. **Habitat:** steppe, savanna, sandy grasslands, desert regions. **Foods:** leaves, roots, bulbs, tubers, insects. **Breeding:** throughout year, mainly April to September; gestation 24–26 days; oestrous cycle 4–6 days; young 4–6; weaned 21 days; sexual maturity 65–85 days. **Longevity:** 3–4 years. **Status:** common(?).

■ HISTORY OF INTRODUCTIONS
EUROPE
Mongolian gerbils were first bred in captivity in Japan and 11 pairs from there formed the foundation stock

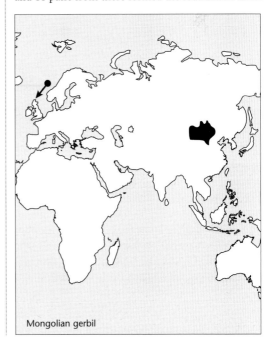

Mongolian gerbil

imported into the United States in 1954 (Corbet and Harris 1991). They reached Britain in the 1960s and have become a common pet.

United Kingdom
Mongolian gerbils are commonly kept in captivity and hundreds, if not thousands, probably escape each year. A number of populations have been discovered in the United Kingdom, but all have eventually disappeared without any control being carried out (Baker 1990).

Escapees can readily establish themselves under floors of houses and out-buildings and less protected environments. There are several records of colonies established in Yorkshire; two isolated areas on Thorne Moor and at Swinfleet Moor early in 1971; at Bradford in 1975 three were found in a burrow under tree roots in woodland near a housing estate; living under sheds at a school in Arnthorpe between 1972 and 1973, and more escaped from the school science laboratory in 1975 and this colony was still in existence in 1977. At Fishbourne on the Isle of Wight, Hampshire, a colony in burrows around a woodyard, under sheds and houses built up to 100 animals by 1976, all the decendants of a few gerbils used in a children's television program and left behind in 1973 (Lever 1977; Corbet 1978; Howes 1983, 1984; Lever 1985; Corbet and Harris 1991)

■ **DAMAGE**
No information.

BANK VOLE
Wood vole, red vole, red-backed vole
***Clethrionomys glareolus* (Schreber)**

■ **DESCRIPTION**
HB 80–123 mm; T 30–65 mm; WT 10–40 g.
Upper parts reddish brown; nose blunt; eyes and ears small; sides and flanks greyish; belly pale silver greyish to greyish yellow or cream buff. Juveniles grey-brown.

■ **DISTRIBUTION**
Eurasia. From the British Isles and south-west Ireland, France and Scandinavia to Lake Baikal; south to northern Spain, northern Italy, the Balkans (absent most of Greece), northern Kazakhstan and the Altai Mountains; isolates in northern Asia Minor and the Tien Shan.

■ **HABITS AND BEHAVIOUR**
Habits: diurnal and nocturnal, less diurnal in winter; makes runways above ground; active burrowers; nest of grass, leaves, moss and feathers in tunnel or tree

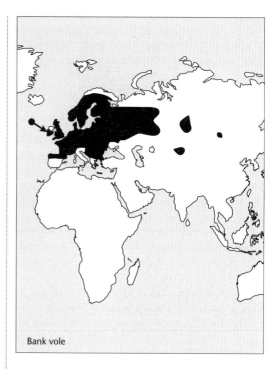

Bank vole

trunk; caches food. **Gregariousness:** density 5–130/ha, but up to 475/ha. **Movements:** sedentary; disperse in spring and summer. **Habitat:** mixed deciduous woodland, grassland, conifer stands, hedgerows, plantations, dry sunny banks. **Foods:** largely herbivorous; fruits, seeds, fungi, moss, roots, flowers, grass, buds, leaves, bulbs, insects, worms, snails and small carrion. **Breeding:** breeds March–October; gestation 18–20 days; litter size 3–7; several litters/year; young born blind, naked, eyes open *c*. 12 days; sexually mature by end breeding season or by next. **Longevity:** 2.2–18 months in wild, 40 months in captivity. **Status:** common.

■ **HISTORY OF INTRODUCTIONS**
EUROPE
Ireland
About 1950, bank voles were introduced into southwest Ireland (Corbet 1978, 1980; Lever 1985), probably as stowaways on shipping docked at Limerick port (D'Arcy 1988). They were first discovered in 1964 in County Kerry, but were subsequently found to be quite widespread in Limerick, Kerry, Cork and Clare (D'Arcy 1988).

By 1971 bank voles were present in the counties of Limerick, Cork, Kerry, Clare and Tipperary, and by 1982 occurred throughout County Limerick, a large part of Cork and Kerry, and a small area of southeastern Clare and western Tipperary (Smal and Fairley 1984)

United Kingdom

The presence of populations of bank voles on the islands of Raasay (Inner Hebrides, Scotland), Skomer (Pembrokeshire, Wales), Ramsay and Jersey (Channel Islands) is also probably the result of early introductions (Corbet 1961; Lever 1985; Corbet and Harris 1991).

■ DAMAGE

In Europe the bank vole is known as an occasional pest of forestry, eating seeds and seedlings and also de-barking small trees such as larch and elder (Corbet and Harris 1991). The main damage occurs in winter and probably when other foods are scarce. In Norway they will enter houses and cause problems, but they are mainly a pest in larch plantations where they gnaw shoots and bark (Burton 1962).

RED-BACKED MOUSE OR VOLE

Northern red-backed vole or mouse

Clethrionomys rutilus (Pallas)

The Gapper's or southern red-backed vole (C. gapperi) is often treated as a separate species; here it is treated as a subspecies of rutilus.

■ DESCRIPTION

HB 80–110 mm; T 23–44 mm; WT 14.2–42.6 g.

Sides of body yellow; upper parts bright reddish to rufous; ears brown; dorsum red, or dark brown to blackish in some animals; tail densely furred; under tail yellow; feet whitish or buff.

■ DISTRIBUTION

North America–Eurasia. In Eurasia from northern Norway and Sweden to north-eastern Siberia and northern Japan, and south to northern Kazakhstan, the Altai, Manchuria and Korea, and Sakhalin; in North America from Alaska and northern Canada south to New Mexico and North Carolina, United States.

■ HABITS AND BEHAVIOUR

Habits: nocturnal and diurnal; burrows and nests under rocks and logs; stores food; hyperactive; terrestrial and arboreal. **Gregariousness:** females more territorial than males. **Movements:** sedentary; male ranges overlap. **Habitat:** forest and taiga zones; tundra, woodland, forest, buildings, around decaying stumps. **Foods:** nuts, fruits, berries, buds, seeds, bark, plant material (buds, sprouts), lichens, fungi, and insects, spiders and snails. **Breeding:** April to October (Europe); gestation 17–20 days; litter size 1–8, 11; 3–4 litters/year; young born naked, helpless; weaned 3–4 weeks; mature at 3–4 months. **Longevity:** usually 1 year but some to 2–3 years in wild. **Status:** common.

■ HISTORY OF INTRODUCTIONS

Red-backed voles have been introduced successfully to Ostrova Beringa (Bering Island), Komandorskiye Ostrova (Commander group), and to Maine in the United States.

ASIA

Komandorskiye Ostrova

Introduced from Kamchatka with firewood in 1870 to

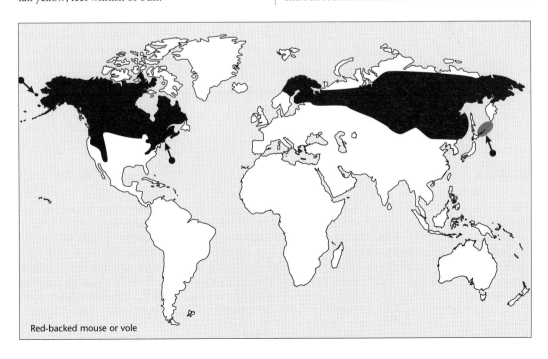

Red-backed mouse or vole

Ostrova Beringa (Bering Island), in 10 years red-backed voles became well established and spread all over the islands from the beaches to the interior mountains (Palmer 1899 in de Vos *et al.* 1956; Barabash-Nikiforov 1938).

NORTH AMERICA
United States
Red-backed voles (*C. rutilus gapperi*) were introduced to some small islands in Penobscot Bay, Maine, between 1962 and 1970 to study their colonisation and extinction. Some were introduced on a 0.8-ha island and were present there for the next six summers (Crowell and Pimm 1976; Crowell 1973).

■ **DAMAGE**
No information.

COMMON VOLE

Water vole, Orkney and Guernsey voles
Microtus arvalis (Pallas)

■ **DESCRIPTION**
HB 80–134 mm; T 20–50 mm; WT 15–50 g.
Brown or black; dorsal pelage dark or light; under parts grey or creamy buff; tail with short tuft of hair. Resembles *M. agrestis*.

■ **DISTRIBUTION**
Eurasia. Northern Spain and Denmark, east through Russia and Siberia to Upper Yenesei, south to the Caucasus, Altai and Lake Balkhash, isolate on Orkney

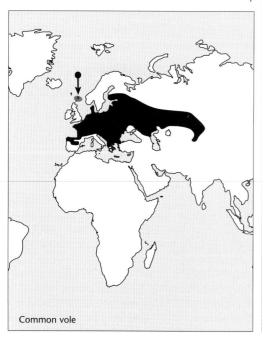

Common vole

Islands, Guernsey, Yeu (France) and perhaps Spitzbergen.

■ **HABITS AND BEHAVIOUR**
Habits: diurnal and nocturnal; extensive runs and burrows. **Gregariousness:** pairs except in winter when much overlap between ranges; lives in colonies; density 29–273/ha. **Movements:** home range 3–3700 m². **Habitat:** lakes, streams, rivers, marshes, moorland, ditches, pastures, meadows, grassland, gardens. **Food:** leaves, stems, roots of water plants, grass, dandelion; bark of willow shoots; freshwater molluscs. **Breeding:** breeds February–September; gestation *c.* 20 days; monogamous; litter size 1–6; 1–2 or more litters per year; lactation *c.* 20 days; young mature in *c.* 3 weeks. **Longevity:** up to 2 years (wild). **Status:** common.

■ **HISTORY OF INTRODUCTION**
EUROPE
United Kingdom
There is some evidence to suggest that the common vole was a prehistoric introduction to the island of Guernsey (Corbet 1966) and it may have been present since the end of the Pleistocene when the island was connected to continental Europe (Corbet and Harris 1991).

It may have been accidently released there by Neolithic peoples or early Bronze Age humans (Lever 1977, 1985).

Common voles were probably introduced to the Orkneys by Neolithic settlers between about 3700 BC, the earliest known human settlement, and 3400 BC, the earliest strata containing the species (Lever 1977, 1985; Corbet and Harris 1991). They now occur on six of the Orkney islands: Mainland, Westray, Sanday, South Ronaldsay, Stronsay and Rousay.

■ **DAMAGE**
None.

CALIFORNIA VOLE

California meadow vole
Microtus californicus (Peale)

■ **HISTORY OF INTRODUCTIONS**
USA
The California vole occurs from Baja California north to Oregon, and there is some evidence to suggest that it (race *M. c. sancidiegi*) was introduced to San Clemente Island, California (Van Bloeker 1967 in Hall 1981).

■ **DAMAGE**
No information.

MUSKRAT
Musquash
Ondatra zibethicus (**Linnaeus**)

■ DESCRIPTION
HB 229–400 mm; T 180–295 mm; WT 541–1816 g.
Coat dark brown or red brown to silvery brown or black, composed of a waterproof underfur overlaid with large guard hairs; under parts olive-grey or tawny shading to silver grey on throat and hips; chin black; lips straw-coloured; hands, feet and tail brown to black; feet broad and flat; hind feet webbed; tail flattened and rudder-like, scattered with hairs and dark grey scales; dorsal feet covered with short grey fur.

■ DISTRIBUTION
North America. Alaska and Canada, except the extreme northern parts, south to South Carolina, Texas, Arizona and northern Baja California.

■ HABITS AND BEHAVIOUR
Habits: mainly nocturnal, occasionally diurnal and crepuscular; aquatic; builds stick and mud houses or lodges usually in water; territorial; can submerge for up to 17 minutes; home territory about 60 m. **Gregariousness:** family units or pairs; density 4.8–14.6 ha. **Movements:** autumn and spring dispersal. **Habitat:** wetlands; salt- and freshwater marshes, lakes, streams, rivers, canals, reservoirs, ponds and sloughs. **Foods:** aquatic vegetation including bulbs and grasses, and also small animals including fish, frogs, mussels, small turtles, salamanders, catfish, snails, tadpoles and crayfish. **Breeding:** throughout year (mainly November–April in south and March–September in north); gestation 21–35 days; female seasonally polyoestrous; oestrous cycle 2–22 days; 2–6 litters per year; litter size 1–6, 12 (varies with latitude); young born naked, blind; eyes open at 14–16 days; weaned 21–28 days; mature at 6–12 months. **Longevity:** 3–4 years in wild, up to 10 years in captivity. **Status:** common and numerous.

■ HISTORY OF INTRODUCTIONS
Muskrats have been successfully introduced into the Palearctic, including Great Britain, northern and central Europe, Ukraine, Russia, parts of China and Mongolia, and Honshu Island, Japan. Also successful introductions have occurred in North America and in South America in southern Argentina and Chile.

ASIA
China and Korea
Since 1945 some muskrats have wandered into northern China (Manchuria), where they appear to be spreading (Niethammer 1963). From 1960 to 1980, three males and one female were caught in northern Xinjang, China, which may have been deliberately introduced or they may possibly have spread from Russia (Ma Yong *et al.* 1981).

Muskrats are established in the Amur region of Hehlung Kiang in northern Manchuria having crossed the border from eastern Mongolia by 1954. Twenty-one were collected in the Hu Ma district

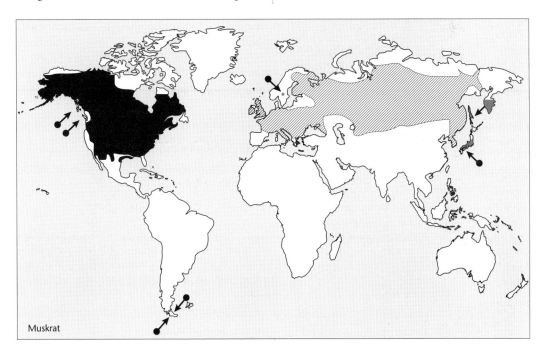

Muskrat

(where they may have arrived as early as the 1940s) in 1955 and more were there in 1956. They also occur on the Kunges River in Sinkiang and at Changwa in Liaoning, probably from Mongolia and in parts of Korea (Lever 1985).

Japan

Introduced from North America as a fur animal before 1940–45, muskrat escaped and became established in the wild in Japan (Kaburaki 1940; Kuroda 1955). After 1945–46, 300 were caught in Tokyo, and some in Chiba, Kamagawa and Saitama Prefectures (Imaizumi 1949; Kishida 1950; Fujiwara 1951). In the mid-1950s they were established, but confined to Tokyo and its environs (Kuroda 1955).

Mongolia

Muskrats wander into northern Mongolia, where they are spreading (Niethammer 1963). In the 1940s and 1950s they spread into this country via the Ili, Ussuri, Irtysch, Selenga and Amur rivers and settled in the Selenga area and in the Onon River System of eastern Mongolia.

In 1967 some muskrats were released at Lakes Charusnur and Has in western Mongolia. In the early 1970s they were spreading in from Kazakhstan via China into Bulgan-gol in south-west Mongolia (Lever 1985).

Eurasia

Muskrats have been introduced successfully into much of central and northern Europe, most of the Russian Federation and some adjacent independent republics, adjacent parts of Mongolia, China and to central Honshu, Japan (Corbet 1978). In the short period of 50 years muskrats have spread over the entire European continent and today number many millions despite many eradication campaigns (Shigesada and Kavasaki 1997).

Albania

Following colonisation of the more northern parts of Europe, muskrats spread into Albania in the 1960s or 1970s.

Austria

In 1914 muskrats penetrated the border between Austria and Czechoslovakia (Mohr 1933). They are now found everywhere, particularly in the Donauauen and Neusiedersee and are not scarce in the Steiermark along the larger waterways (Inn, Salzach, Enns and Mur), but are sparse in alpine regions.

Belgium

In the low countries muskrats occupy nearly all of northern Belgium as a result of introduction for fur farming before 1930 (de Vos *et al.* 1956), by which time they already occupied a considerable area (Niethammer 1963). Colonisation was not completed in some parts of eastern and western Flanders. They are controlled in virtually the whole of Belgium and have disappeared from the eastern part of Limburg province as a result of the measures (Doude van Troostwijk 1976).

Escapees from fur farms established themselves in swamplands between the Nèthe and Demer rivers near Aaschot. From here they spread northwards reaching the Dutch border in 1941. By 1950 they had spread westwards to parts of East Flanders, and in Brabant as far south as the River Dyle near Louvain. After 1965 they spread further west in northern and eastern Hainaut and in the provinces of Antwerp and Brabant. The extreme south of Belgium was infested after 1959 when they spread westwards from Germany. Virtually the whole country was occupied by 1970 (Lever 1985).

Bulgaria

Introduced to Lake Sreburna probably in the 1950s muskrats were well established there in large numbers by the 1960s (Mountfort 1962). They did not, however, spread south in northern Bulgaria until the 1960s and 1970s.

Czechoslovakia

Three female and two male muskrats from Alaska were introduced at Dobrisch, 40 km south-west of Praha (Prague) by Prince Colloredo-Mannsfield in 1905 (de Vos *et al.* 1956; Niethammer 1963). They were said to have been introduced because his wife wished to take home a momento of a visit they made to North America (Kokes 1976). Other later introductions were probably made for sport and economic reasons (Bigalke 1937).

The spread of muskrats from the original release site and the build-up in numbers was apparently rapid, for by 1914 the population in Bohemia alone was estimated at two million animals. Also, in 1914 the colonisation began to penetrate countries bordering Germany and Austria. At this time the whole of Bohemia had been colonised (Mohr 1933) and the radius of expansion was increasing at a rate of from 4–30 km annually (Becker 1972).

By 1933 they occupied most of Czechoslovakia (Mohr 1933) and in the early 1960s around 200 000 pelts were taken annually (Ganzak 1964).

Denmark

Muskrats were still absent from Denmark in 1975 (Doude van Troostwijk 1976).

Finland

Unsuccessful releases of muskrats were made in 1919–20 at Lake Ruuhijärvi near Kajaani. However, they were deliberately introduced to three sites in the lakes region of Finland in 1922 and 1923, from Czechoslovakian stock, and others later from America (Mohr 1933; de Vos *et al.* 1956; Niethammer 1963). Since 1922, over 200 translocations of muskrats have been made in Finland (Elton 1958). During the 1920s and 1930s at least 2300 were released at 293 sites in Finland from Hankoniemi in the south to Inari in the north. After 1937 releases were mainly in the north of the country. By 1942 they were only absent in central Finland. Between 1952 and 1957 they extended their range in northern Finland and across the border into Sweden and by 1960 all the suitable water systems had been colonised (Lever 1985). They probably reached the Åland Islands by themselves.

Muskrats spread at a rate of 29–40 km yearly and were present in most of the country in the 1950s and 1960s, except the extreme north (de Vos *et al.* 1956; Niethammer 1963). At this time some 150 000 to 200 000 pelts were being harvested annually (Hoffman 1952; Westerskov 1952; Schmidt 1954). The number of pelts had built up from about 25 000 muskrats in 1933–37, in 24 years to 603 000 (Niethammer 1963).

The benefits gained from fur far outweigh any damage caused by muskrats in Finland (Mohr 1933) and this still appears to be the case (Doude van Troostwijk 1976).

France

Bred as a fur animal in the late 1920s in several places in France, muskrats were first recorded in the wild in Grenoble by Piraud ('Procès verbaux de la Société dahphinoise d'études biologiques' for March 23, 1930) and again in 1933 by Regnier (*Bulletin de la Société des Amis des Sciences Naturelles de Rouen* for April 1933), in the departments of the Eure and Lower Seine (Bourdelle 1939). A further record indicates that they may have been introduced in Eure by Prince Colloredo-Mansfield in 1924 (Dorst and Giban 1954).

Investigations in 1933 revealed that free-living muskrats were present in at least 12 centres. Five of these were grouped in the basins of the Seine and Somme (departments Eure, Somme, Seine, Oise and Lower Seine), four in eastern France (departments of Ardennes, Meuse, Meurthe et Moselle, and Territory of Belfort) and three others in central south-eastern France (departments Allier, Loire and Isère) (Bourdelle 1939).

Since World War 2 muskrats in Normandy have rapidly extended their range westwards and reached Brittany shortly after 1958 (Laurent 1963). They were recorded on the Vilaine River, south of Rennes and in the marshes near Redon as early as 1957, and appeared on the Oust River in 1958 and around Jugon. In 1960 they were found at Bignon 25 km south-west of Les Forges and 50 km west of Bignon (Laurent 1963).

After 1938 and until 1953 only five of these centres of establishment continued to spread, the remaining seven having remained in much the same areas with little expansion (Dorst and Giban 1954). Muskrats spread rapidly from centres in the Lower Seine, Territory of Belfort, Eure, Somme, and the Ardennes. In the north-west they found the country very suitable in Normandy. They reached Mantes about 1942 and were found in the region of Chartres in 1943 and existed throughout the Eure-et Loir at this time. Towards the south-west they had invaded the basins of the Touques, Dives and Orne (*c.* 1942–44), and also the basins of the Loire. In the east of France they remained mainly confined to the original Territory of Belfort, but had spread steadily north into the department of Vosges and the plain of Alsace (Dorst and Giban 1954).

Muskrat success in France is said to be due to the fact that France is rich in waterways, there are few native predators, and that they are well adapted, robust and hardy animals (Dorst and Giban 1954). Although their furs may be of some value, muskrats also cause much damage to embankments, dikes and other water works, and it has been found necessary to control their numbers (Dorst and Giban 1954). By 1966 efforts were being made to control them in some 18 departments in France (Doude van Troostwijk 1976).

Germany

Muskrats reached Bavaria in southern Germany (formerly German Federal Republic) in 1914 or 1915 at Regen (Baker 1972). In this country in 1922–23 they were reported to be spreading at the rate of 48–72 km annually and by 1933 occupied eastern Germany and in 1928 some 30 000 were killed there for their fur (Mohr 1933).

They reached Prussia in 1924 and, with the exception of a few areas in Schleswig-Holstein and in the north of Oldenburg, occupied all of western Germany. Most of the south was occupied with the exception of areas along the eastern bank of the Rhine and they were very common in eastern Germany by 1975 (Mohr 1933; Doude van Troostwijk 1976).

In 1917 muskrats reached Sachen (Saxony) near Crotendorf and Grumbach in the Erzgebirge and in 1927 reached Würtemburg. Although a bonus was paid for their capture in Bavaria and Thuringen it had little effect in preventing their spread. From 1938 they were in part forbidden and the fight against them was stepped up. However, by this time they occupied some 200 000 km² of Germany. Despite control measures, with few exceptions colonisation was completed by the early 1970s.

Greece
Once the colonisation of central Europe was completed by muskrats they spread into northern Greece by the 1960s or 1970s.

Hungary
By 1933 muskrats occupied a considerable portion of Hungary where they first occurred in 1924 (Mohr 1933). Others suggest the first animals arrived between 1915 and 1924 from Czechoslovakia and completed colonisation of the country by the 1960s (Niethammer 1963).

Italy
Following colonisation of central Europe, muskrats have spread into northern Italy (Lever 1985).

Luxembourg
Muskrats first occurred in the Grand Duchy in the beginning of the 1960s and were found locally in large concentrations (Doude van Troostwijk 1976). One was found as early as 1956 and another in 1957, and later in the same year some 50 near Remich (Niethammer 1963). Their first appearance was probably about 1955 and they more than likely came from France in the south.

After 1960 muskrats began extending their range to the north and west of the country. By 1970 they occurred in all suitable habitats and were approaching the Belgium, Dutch and German borders (Lever 1985).

Netherlands
In the 1940s muskrats reached the southern boundaries of the Netherlands and about 1968 immigrated beyond the Dutch–German Border (Doude van Troostwijk 1978). The first muskrat caught was in 1941, but by 1956 they were slowly penetrating along the southern boundary (de Vos et al. 1956).

In 1975 muskrats occurred in eastern Bath (salt marshes), Zeeuws Vlaanderen, North Brabant to the Maas and along the Dutch–German border from Brabant north to Gelderland and Gronigen (Doude van Troostwijk 1976). Between 1974 and 1980 their range increased rapidly and they were found in 11 provinces. The annual catch had risen from 53 690 in 1974 to 114 814 muskrats in 1980 (Litjens 1981).

Poland
Most of Poland was occupied by muskrats by 1933 (Mohr 1933), the first animals probably arriving in the 1920s. However, before being invaded from Czechoslovakia some had escaped from fur farms and were already established there (de Vos et al. 1956). Most of Poland was colonised by muskrats by the 1960s (Suminski 1963).

Romania
Muskrats occupied most of Romania by 1933 (Mohr 1933), but some areas were not colonised until 1938–39 and in the east (Donau delta), where they came from Russia in about 1955. In 1958, 30 000 furs were harvested from the population in Romania (Niethammer 1963).

Russian Federation and adjacent independent republics
The muskrat was one of the first animals acclimatised in Russia (Skopsov 1967). They became successfully established in 1928 on Karagin Island. In 1932 some were transferred to the mainland in several areas. In 1948, 140 were released at Ivenski Razlivi, on the Isna, and to Lake Linovo in the Tambov region; in 1951, 185 were released in Lake Kochkino and 81 in Lake Christee, where they increased rapidly.

Some (99) were also released on Great Solovetskiy Island in 1928 (Lever 1985). They were probably first released on Solovski Island in the White Sea and Karagin in the Bering Sea and later on the mainland. From 1928 to 1932 some 1646 were released. Later 330 000 were caught and translocated and now muskrats inhabit the entire country including much of western Europe and Asia (Japan, China, Mongolia) and are ranked second or third in value of fur in Russia. However, in recent years the numbers of furs harvested has been declining (Sofonov 1981).

Sixteen muskrats were introduced into Russia in 1927 and between that date and until 1932 some 5232 were imported for release (Farman 1969). From 1927 to 1953 inclusive 117 000 muskrats were released in some 500 localities throughout the Russian Federation and adjacent independent republics (Hall 1963). At least 1650 were probably released from 1928–32 in northern European Russia (Lavrov and Pokrovsky 1967) and up to 1965 some 250 000 in western and eastern Siberia. The original releases bred so well that about 20 million were captured and released again in many new areas (Farman 1969). However, this figure is exceptionally high and needs further substantiation. More recently it has been postulated that there were over 3000 releases in the Russian Federation and adjacent independent republics amounting to about a third of a million animals (Chashchukhin 1987).

Between 1927 and 1945, 79 198 muskrats were released in northern European and Siberian taiga zones, even as far as Kamchatka where a shipment arrived from Ontario in 1928 (Eyerdam 1932). Further introductions occurred in Kamchatka in 1959 and hunting began in 1968, when the harvest was about 25 000 pelts annually (Savenkov 1987).

Within the Russian Federation and adjacent independent republics up until 1970 some 299 687 muskrats have been released, with by far the greater number in the Russian Federation (212 945), but also many in the Ukraine (20 705), Belorussia (2893), Uzbekistan (9401), Kazakhstan (47 600), Gruzinsk and Azerbaidjan (nil), Litovsk (286), Moldavia (1739), Latvia (nil), Kirghizia (2505), Tadjikistan and Armyahn (nil), Turkmensk (747) and Astonsk (866).

In the central forest zone in the European part of Russia some 11 325 muskrats were released between 1943 and 1967. The largest numbers in Kalujsk where 2281 were liberated. In the north European parts of Russia, 18 585 were released between 1929 and 1969 at such areas as Murmansk, Karelsk, Arkhangelsk, Vologodsk, Komi, and Kirovsk. Here their success was spectacular and many skins were being harvested in the 1940s and 1950s. In the southern European parts of Russia some 5041 were released before 1955 and from 1956 to 1970 some 26 293 more. In the Urals and adjacent areas a total of 12 347 muskrats were released from 1930 to 1970, the largest numbers probably in Chelyahbinsk where 4881 were liberated. In the Kazakhsk region they were released (24 152) in some 15 oblasts before 1955. Since 1956 another 23 448 have been released in 14 oblasts. Here, spectacular success has been achieved. Many were also released in a number of Republics in central Asia e.g. Uzbekistan, 9041 from 1944–68; Kirgizsk, 2505 from 1944–62; Turkmensk, 747 from 1955–57; others released in Tadjikistan (Kirisa 1973).

Animals for introduction into Russia came from Canada, Finland and England, mostly *O. zibethicus zibethicus*, but some also of the subspecies *macrodon* (Lavrov and Pokrovsky 1967). Trapping of muskrats for fur began in 1935 and probably about 70 million had been harvested up until 1969 and it was expected that the 1970 harvest would be about eight and a half million in Russia (Farman 1969). In 1941, 150 000 pelts were harvested and in 1954 about 649 000 (Schmidt 1954). In 1957, 12 per cent of the world's muskrat furs came from the Russian Federation and adjacent independent republics (Niethammer 1963).

The release of muskrats began in Murmansk, Leningrad, the Novgorod and Pskov oblasts, and Karelian region in 1931 (Al'tshul 1963), and by 1960 at least 9874 had been imported for introduction. In Murmansk and Karelia they spread almost over the entire areas, although only in small numbers; in Leningrad in 1956 they occupied 309 300 ha, but in Novgorod were not well established.

Some muskrats were introduced in the Kostroma oblast in 1946 and to Udmurtia in 1951, and although they became established their numbers failed to reach a level of commercial significance (Fateev 1960). Also in 1946 they were introduced in the Perm' oblast from Kurgan, and became widely distributed along the Sylva River (Chashchin 1961). From here in 1950, 100 were released along other adjacent rivers and between 1951 and 1959 many intra-regional translocations were carried out. In the early 1960s they were well established in small numbers, but increasing. Muskrats (1145 animals) were also released in the Volga Delta in 1954, where it was hoped that they would become established in the delta area and not along the course of the river, where they could become a pest of agriculture and a menace to the waterways themselves (Zamakhaev 1963). They became established in the Volga-Kama region from introductions about 1924 (Yanushevich 1966).

From 1953 to 1960, 2410 muskrats were released in Belarus (Belorussia) (Samusenko 1962), but the results of these introductions still appear to be uncertain (Yanushevich 1966).

Muskrats were first released in the Ukraine in 1944 when 120 were liberated in the flood plains of the Dneiper River. Subsequently, up until 1959 some 1200 were released in the Ukraine. These releases became established mainly in the lower reaches of the Dneiper, Dnestr and particularly the Danube rivers (Samosh and Razumovskii 1962). The species is now well established and widespread in some areas of the Ukraine (Yanushevich 1966). Some 300 were released in Moldavia in 1947 and by 1956 at least 1040 muskrats were established, mainly on the Prut River (Samosh and Razumovskii 1962). They have remained well established in this area since this time (Yanushevich 1966).

The muskrat became well established and widespread in the Arkhangel'sk oblast from introductions beginning in 1928 (Yanushevich 1966). Eighty-two animals from here in 1954, and 204 from Kazakhstan in 1956 were taken to, and released in Lithuania where they became well established (Mitkus 1962).

Other areas in European Russia where the muskrat has been introduced successfully include: Ivanov region (introductions 1946, 1956–57), the Caucasus (1944 and since 1947), Astrakhan (in 1954), Komi, Tambov oblast, and the Urals. Introductions in Bashkiria in 1946–47 and in 1957 apparently failed (Skopsov 1954; Pavlinin and Shvarts 1961; Yurkin 1961; Yanushevich 1966.)

Muskrats were introduced to Uzbekistan in 1944 and hunting began there in 1946. In 1953 further animals (345) were released in the Tashkent oblast, where they were thinly established in the early 1960s (Ostapenko 1963). Some were released in the Kara-Kalpak area (Aral Sea) in 1944 and these became well established and widespread (Yanushevich 1966). In 1958, 244 muskrats were liberated in Khorezmskaya oblast (Uzbekistan) from a fur farm, and another 1628 in 1954–58. As a result of those releases almost all the lakes in the oblast became populated with small numbers; in 1957, 75 from Tashkent oblast were released and to 1959, 14 250 were harvested (Reimov 1960).

Between 1935 and 1960, 34 605 were released in Alma-Ata and northern Kazakhstan areas and the species is now found in commercial abundance over most of Kazakhstan (Strautman 1963). Also in southern Russia, 715 were liberated in the Tuva area in 1958–62, but the possibilities for acclimatisation in this area are said to be limited (Shurygin and Nikiforov 1964), nevertheless they became established (Volchenko 1960) at least initially. Some have also been successfully introduced in the Buryatiya area (Izmailov 1969). Muskrats have also been introduced, between 1944 and 1954 in Kirghizia (Yanushevich 1966) where they are now widespread.

Large numbers of muskrats were released in western Siberia (93,235) and in central Siberia (5704) from 1929 to 1970 (Kirisa 1973). In western Siberia 200 were released in Turukhansk between 1929 and 1934 (Petrov 1962). These became successfully established and hunting began in 1938. In 1947 they were released near Lake Malkoe in the Noril'sk Lake region, where they became established near the Norilka and Rybnaya rivers and around Lake Glubokoe (Gerke and Krechmar 1963). Subsequently the muskrat became widespread in western Siberia (Yanushevich 1966).

In eastern Siberia and the Far East, 31 132 muskrats were released from 1932 to 1970 (Kirisa 1973). In eastern Siberia they became well established and widespread from introductions from 1930–31 onwards in Yakutia (Yanushevich 1966). The first muskrats liberated in the Far East were those in 1939

when 351 were introduced, and up until 1956, 1600 were liberated in the Amur Foreland (Sapaev 1965). These spread rapidly and within 12 years were appearing in areas 1000 km away. Muskrats now occupy 270 000 km^2 of territory, and although the population has been decreasing since 1961 because of limited food supplies and unfavourable hydrology (Sapaev 1966), they are still well established in the Amur region (Yanushevich 1966). They were established at Khabarovsk from introductions in 1959–61 (Yanushevich 1966).

Muskrats were released in Primorskii Krai in 1947 and up until 1960, 200 animals of local origin were released in some 16 regions (Abramov 1963). They now occupy all of the suitable habitat in the area, particularly the Ussuri-Daubikhe Valley and Lake Khanka.

On Sakhalin Island in 1952, 77 were imported from Primorskii and released, and altogether 1399 muskrats were released between 1952 and 1959 (Ben'kovskii 1963; Kirisa 1973). By 1956 they had become an important fur animal on the island and 600 were harvested. This had risen to a harvest of some 18 000 muskrats by 1962.

Muskrats were released on lakes and shores of Askizskii Raion (before 1968), but not very successfully (Kokhanovskii 1968).

Sweden
Introduced illegally into Sweden, muskrats were released along the River Torne some time before 1944 (Liljestrom 1954). From 1920 to 1955 some 2300 muskrats were released in Sweden resulting in the establishment of the species in a large part of the country (Mareström 1964). They were also wandering into the north of Sweden from Finland as early as 1955 (Niethammer 1963). The spread into Sweden was at the rate of 10 km per year (Lever 1985).

Switzerland
Muskrats entered Switzerland from the Alsace region of France, the first animals being noted in about 1935. By 1950 about 800 had been killed and small numbers were caught in the cantons of Bern and Basel city (de Vos *et al.* 1956; Doude van Troostwijk 1976).

By the 1970s muskrats were found in the northern parts of Switzerland, in the cantons of Aargau, Basel and Jura (Rahm 1976; Rahm and Stocker 1978). More recently they have expanded their range to include much of the Upper Rhine, Lake Constance and the canton of Neuchâtel (M. Dollinger *pers. comm.* 1982). By far the greater numbers appear to occur in the canton of Jura where until 1979 the annual take was about 100 animals a year.

United Kingdom and Ireland
Imported for fur-farming in the 1920s, there were numerous escapes of muskrats from the farms near Shrewsbury, with smaller populations in Surrey and East Sussex in England and near Stirling in Scotland, which all became well established in the wild.

The first introduction to the wild in Scotland was probably in 1927 when nine (five females and four males) escaped from a fur farm and three years later 900 descendants had been trapped (Storer 1937; Burton and Burton 1969). From escapees in 1929–30, two main colonies in Shropshire and mid-Scotland, and three smaller ones in Surrey, Sussex (in 1929) and Ireland (in 1927) became established (Storer 1937; Southern 1964). About 1927(?) six pairs were placed in a pen at Feddal, near Braco, Perthshire, and two years later an established colony was found on a marsh 3.2 km away. Soon after, they had colonised 88 km of the rivers Forth, Teith and Earn and Allan Water.

The earliest release in Ireland was in 1918 at Oban, Argyllshire, where some were apparently deliberately released, but failed to become established for long. In 1927 some were imported from Ontario to Annaberg near Monagh, County Tipperary, for breeding, but these escaped and became established in the wild. Some five years later 389 km^2 of country were infested with muskrats.

In 1929 there were about 85 muskrat farms in England, Scotland and Ireland from which animals had escaped. Around 1930 further import and keeping was prohibited and by 1939 an eradication program had accounted for 4299 muskrats (de Vos et al. 1956). Government control campaigns began in 1932–33 (Storer 1937; Fitter 1959). Extermination campaigns in Ireland were initiated in late 1933, and during these 500 muskrats were harvested, and the last individual was killed in 1935 (Fitter 1959; Corbet and Harris 1991). From 1932 to the end of 1934, 2672 were trapped in seven areas in England and smaller numbers in Scotland. By 1935 in Britain they were almost exterminated and four years later none were found (Fitter 1959). The total killed up until 1937 has been variously estimated at 4388–4500 (Gosling and Baker 1989; Corbet and Harris 1991). The total wild population had been eliminated and muskrats have not occurred in Ireland or Britain since.

Yugoslavia
Muskrats reached parts of Yugoslavia by 1933 (Mohr 1933), but settled in the northern and eastern parts only from 1932 onwards. The yearly harvest for fur in 1959–60 was 13 400 animals.

NORTH AMERICA
Alaska
Introductions of muskrats on the Alaskan mainland appear largely to have been failures (Burris 1965). However, introductions to the Kodiak group, the Aleutians, the Pribolof Islands and to Prince of Wales Island have been successful.

In 1913 an unknown number from Nushagak were released on St. George Island and St. Paul Island in the Pribilof Islands. In 1925, 70 muskrats from Copper River were released on the Kodiak archipelago and in 1929, 21 from Long Island were released near the Afognak Lakes and Buskin River. In 1929–30 some from the Chilka River and 18 from Haines were released on Prince of Wales Island (Burris 1965).

Canada
Before 1837 muskrats were introduced to Anticosti Island (Newsom 1937) where they are now well established (Banfield 1977). They were liberated on Vancouver Island in 1922 when the race *O. z. osoyoosensis* was transferred from the mainland to Cowichan Lake (Carl and Guiguet 1972). In 1924–25 the Game Commission transferred more from the lower mainland to a number of different areas including Shaw Creek, Ucluelet, Jordan River, Port Alice, Hopkins Lake and Comox (Lloyd 1925). The species is now well established on Vancouver Island and on Pender Island, where it was introduced in the 1920s by A. F. Richardson (Carl and Guiguet 1972). Also in 1924 or 1925 some were transferred from the mainland to the Queen Charlotte Islands, where they became established and abundant. They were first taken to New Massett, Graham Island by A. D. Hallett who released about 15, but a number of subsequent introductions were made privately in other areas (Prichard 1934). By 1934 they had spread to areas 48 km from the release point.

United States
In the United States the transfer of muskrats began in a number of areas in about 1900 (Storer 1937; de Vos et al. 1956).

In California they have been translocated in many areas and have become widely distributed (Storer 1933, 1947) by escaping from fur farms and with the increase in irrigated areas (Storer 1958). A single animal was shot west of the Sierra Nevada Range in 1920 (Dickey 1923). Their present occurrence in the counties of Lake and Klamath has been the result of extensive movements (Hansen 1965). Some were liberated at Goose Lake in northern California in 1930 (Twining and Hensley 1943) and many were farmed in Klamath County between 1925 and 1930, where undoubtedly there were many escapes (Hansen 1965). They are now common along watercourses, both artificial and natural, in the Sacramento–San Joaquin Valley (Jameson and Peeters 1988).

Muskrats formerly inhabited a restricted area in California, but through introduction and range expansion of both native and introduced populations they are now widespread (Jameson and Peeters 1988; Lidicker 1991). Colorado River populations have expanded into the Imperial Valley along with extensive irrigation developments (Grinnell 1914). Numerous introductions were made commencing in the early 1920s elsewhere in California, especially the Central Valley and in various coastal areas. Many introductions went unrecorded, but some stock came from Larsen County (Dixon 1929). As a result they now occur throughout the state in aquatic habitats and in isolated patches in southern California and coastal regions. By 1960, 100 000 were being trapped annually for pelts (Seymour 1960). However, in many places they are considered a pest because of their burrowing in levee banks and irrigation channels (Grinnell 1914; Lidicker 1991).

Some muskrats liberated near Lakeview, Oregon, in 1932 are thought to have been responsible for their establishment in that area (Twining and Hensley 1943) and also, some escaped at Jack Lake in Lake County in the 1930s and 1940s (Hansen 1965).

In New York State muskrats were occasionally introduced before 1941 in the central part, where they became established and bred with the native stock (Bump 1941).

Muskrats were possibly introduced in north-west Louisiana (Lowery 1974), and some were introduced to Fish Springs National Wildlife Refuge, Utah, in about 1925, and where in 1978 the population was estimated as 12 000–18 000 animals (McCake and Wolfe 1981).

SOUTH AMERICA
Argentina and Chile
Introduced to Lago Fagnano Isla Grande, Tierra del Fuego, in Argentina between 1940 and 1956, muskrats had spread by 1971 into Chile in the west and south to Isla Navarino (Chile), south of the Beagle Channel (Lever 1985).

■ DAMAGE
In California, where they are abundant, muskrats have caused damage to irrigation structures, railroad fills, earthen dams and to fish culture and other ponds. They can be serious pests in irrigation areas as they burrow in canal banks, levees and ditches. They cause considerable economic losses in irrigation

systems, and damage to rice in the Sacramento Valley is estimated at US$50 000 annually (Storer 1937; Storer 1958; Ball 1960; Marsh 1965; Lidicker 1991). Muskrats not only cause direct and indirect damage to rice in California but in other rice-growing areas as well, where the damage amounts to a considerable figure. They also occasionally cut and eat the young rice plants (Marsh 1965).

In recent years muskrats have become serious pests causing extensive damage to some specific crops, as well as to earthen holding structures in Arkansas. In 1967 the damage to rice crops, food fish and bait reservoirs was estimated at US$900 000 (Miller 1974). In New York and Maryland they infrequently cause pond leakage and impair the physical appearance of ponds. The extent and seriousness of the damage is correlated with the age and length of time the muskrat population has been present (Erickson 1966). On Graham Island, Canada, they are reported to eat salmon fry and damage dikes in the Tlell River area (Pritchard 1934).

The introduction of muskrats in Eurasian waters has disturbed the existing biocoenotic relationships in many localities. Removal of reeds and other plant species, a reduction of aquatic macrophyte thickets and resulting decrease in muskrat abundance have been observed in several regions of Russia and adjacent independent republics and some in western Europe (Chashchukhin 1975). However, opinion on whether they are pests or not appears to be divided, ranging from muskrats appearing to do no harm (Dorst 1965), to destroying water vegetation (Nasimovich 1966), destroying crops (Golubeva 1961), to competing with the indigenous desman (Scopsov 1964, 1967; Borodin 1965).

In Russia muskrats have been found to destroy large masses of aquatic vegetation in large areas of its range (Nasimovich 1966). In Tambov oblast they have displaced the Russian desman (*Desmana moschata*) from its age-old habitats (Skoptsov 1964). It has been suggested that muskrats have caused the decline in numbers of desmans and are responsible for the disappearance of shellfish, insects, fish, water rats and some breeding birds in at least one lake (Skoptsov 1967). The loss of yield in 1958 in agricultural crops due to muskrats in western Siberia was estimated at 50 million roubles and protection of crops was said to be economically desirable (Golubeva 1961). In the Chamzinsk Raion, Mordavia, evidence from observations in 1965 indicates that muskrats successfully compete with desmans for food and burrows and cause diseases such as haemorrhagic fever, tularaemia and leptospirosis (Borodin 1965). They apparently do not cause serious loss to aquatic vegetation in Kazakhstan (Strautman 1963).

Initial surveys in Russia indicated that muskrats could co-exist with the native desman, but more recent studies tend to show that this is not possible. Numbers of desmans decrease as muskrat numbers increase. At Lake Christee, Tambov region, where muskrats were released in 1951, they increased rapidly and their effects on the fauna and flora were considerable. They completely destroyed the shellfish and large slow-moving insects, and a number of fish and the water rats disappeared, as did some of the breeding birds. Here they apparently ate out their own food supply (i.e. water vegetation) then turned to eating shellfish, frogs and occasionally fish (Skoptsov 1967). Their effects in Ryazon, Kirov, Vladinin and Markov regions have apparently been similiar.

The success of the muskrat in Russia is due to common features between it and North America in natural conditions. However it has brought about complicated inter-relations with other animals. In 1960–61 muskrats made up 64 per cent of the diet of golden jackals (*Canis aureus*) compared with only 3.8 per cent in 1947–48 (Lavrov and Pokrovsky 1967).

In 1945, 649 000 muskrat pelts were harvested in Russia (Dorst 1965), and although the harvest of skins is important economically, the damage caused by them outweighs the profit from the sale of skins (Becker 1972). In Russia and the adjacent independent republics muskrats are controlled with poison and vegetable baits (Kucheruk *et al.* 1959; Kuzyakin and Panteleev 1961).

In central and western Europe muskrats are considered a pest because they undermine steep banks and ditches with burrows, feed on aquatic vegetation and animals, and in some areas carry leptospira which causes Weil's disease in man (Dorst 1965; Becker 1972). In Finland they do not appear to do any harm (Dorst 1965), but in France considerable damage is caused by their burrowing in banks, dikes and ponds (Bourdelle 1939).

In the Netherlands muskrats have caused damage by burrowing into retaining banks along rivers and causing them to collapse, damaging drainage ditches, and eating and trampling cultivated crops such as sugarbeet and corn (Doude van Troostvijk 1978; Litjens 1980). In 1979–80 the damage by them was investigated in the mainly agricultural province of North Brabanta. It was found that they caused the stagnation of water by burrowing and blocking drainage systems, caused loss of cultivated ground, made it necessary for re-cultivation of some areas,

undermined fences, caused bogging of machinery, and undermined and blocked culverts and roads. Numbers in the area are controlled by the severe winters and considerable extermination campaigns that keep the pests at acceptable population levels (Litjens 1981).

Damage reports in Switzerland by muskrats appear rather contradictory as considerable damage is reported in the canton of Jura, but little elsewhere (M. Dollinger *pers. comm.* 1988).

In Japan muskrats are considered harmful as they dig holes in banks and stops water flows (Kuroda 1955).

YELLOW-NECKED MOUSE
Apodemus flavicollis (Melchior)

■ HISTORY OF INTRODUCTIONS
The yellow-necked mouse of Europe and Asia could possibly have been introduced accidently by humans to Britain (Lever 1977), but there appears to be little support from other authorities (Corbet 1978, 1980; Corbet and Harris 1991).

FIELD MOUSE OR WOODMOUSE
St. Kilda field mouse, Turkestan rat, long-tailed field mouse
Apodemus sylvaticus (Linnaeus)

■ HISTORY OF INTRODUCTION
EUROPE
Hirta
The field mouse (*A. sylvaticus hirtensis* Barrett-Hamilton) is held by some to have possibly been introduced accidentally by Norsemen from Scandinavia over 1000 years ago (Lever 1977, 1985).

■ DAMAGE
In Scotland field mice nip off the shoots of young beans and scrape the soil away to get at the bean. They feed on stems, leaves and fruits of many plants and serious damage is recorded in barley, wheat, oats, rye, crocus, carnation and strawberries. The damage is so serious at times that farmers have ploughed in the crop (Sneddon 1953). They are considered a pest of nut and fruit crops in southern Kirghizia, Russia (Yanushevich 1966).

Genus Rattus
Rats

The genus *Rattus* appears to have originated in Southeast Asia: the black rat (*R. rattus*) and kiore (*R. exulans*) in Indochina, the Norway rat (*R. norvegicus*) in southern China, and the house mouse (*Mus musculus*) from near the Middle East. Black rats may have been present in Europe as early as the Pleistocene (Grzimek 1975) or, as thought by some, both the black and Norway rats may have arrived in post-glacial times as human commensals (Kurten 1968). They certainly reached the Western Hemisphere during the sixteenth century explorations.

The house mouse and black rat expansions to Europe took place around the first or second millennium BC. On Mediterranean islands the black rat and mouse appeared during the Roman Empire, except in Sardinia where these species were found 2000–3000 BC. The house mouse is known from Pleistocene fossils in Europe and is evidently a natural inhabitant of parts of that continent, as well as much of Asia. The earliest known association of the house mouse with an urban community is at a Neolithic site in Turkey about 8000 years ago (Kurten 1968; Brothwell 1981).

Rats and their diseases spread around the world carried by ships: reaching Suez in 1897, Madagascar in 1898, Japan, east Africa and Portugal in 1899, Manville, Sydney, Glasgow and San Francisco in 1900, Honolulu in 1908, Java in 1911, Sri Lanka in 1914, Paris in 1918, and Marseilles in 1920.

The spread of house mice around the world has been assisted by the construction of buildings that provide shelter, and the development of agriculture that provides food. Their adaptation to commensalism explains the worldwide distribution of *R. rattus*, *R. norvegicus* and *M. musculus* and the local genetic evolution of their populations. The reason why such species possess these adaptations is unknown. It has been suggested that this capacity is probably linked to their behavioural physiology and plasticity.

Both the house mouse and the two rat species are often more abundant in urban sewers and the dwellings of humans than in more natural habitats. There is even a patron saint of rats – St. Servatius's Day, 13th May – which was invoked to ward off rats and bubonic plague carried by their fleas.

Introduction of house mouse and rats to islands

(Islands excluded are those offshore from Australia and New Zealand; for these see the tables under the individual entries).

Note: ? indicates presence at unknown date

Island	House mouse	Black rat	Norway rat	Unspecified
Admiralty Is (Pacific O)	?			?
Aisla Craig (UK)				1800s
Alderney (Channel Is)		?		
Alejandros Selkirk (see Más Afuera)				
Aleutian Is (Bering Sea)	1800s			before 1939
Alor (Indonesia)	?			?
Amboina Is (see Ambon)				
Ambon (Moluccas)	?			?
Amchitka (Aleutians)				World War 2
Amirante Is (Seychelles)		?		
Amsterdam (Indian O)	early 19th century	?	before 1967?	
Anambar Is (Indonesia)	?			?
Andaman Is (Bay Bengal)	9th century BC			
Anegada (Virgin Is)	?	?		
Annobon (Gulf Guinea)				?
Antipodes Is (Pacific O)	before 1907			
Arends	?			?
Arno Atoll (Marshall Is)		?		
Aru Is (Indonesia)	?			?
Aruba (L. Antilles)	?	?		
Ascension (Atlantic O)	?	before 1701		

Introduction of house mouse and rats to islands (*continued*)

Island	House mouse	Black rat	Norway rat	Unspecified
Assumption (Aldabra, Indian O)				before 1906
Astove (Seychelles)				1895?
Azores (Atlantic O)				after 1460?
Bali (Indonesia)	?			?
Baltra (Galápagos)	World War 2	after 1934 or World War 2		
Bangka (Indonesia)	?			?
Banjak Is (Indonesia)	?			?
Barbados (L. Antilles)	1626	before 1654	1536–1626?	
Bardsey (Wales)	? there 1978			
Baru	?			?
Batu Is (Indonesia)	?			?
Bawean Is (Indonesia)	?			?
Belitung (Indonesia)	?			?
Benbecula (UK)		?		
Bering (Commander)	1870		before 1938?	
Berlenga (Portugal)	?	?	there 1939	
Bermuda (Atlantic O)	1612	1613	mid 18th century	
Billiton (see Belitung)				
Bird (Seychelles)	?	1967		
Bonaire (L. Antilles)	?	?		
Bonin Is (Pacific O)	?			?
Borneo (Indonesia)	?			?
Bougainville (Solomons)		World War 2		
British Isles	*c.* 1000 BC	4th–5th century AD		1728–29
Browse (Indian O)	?			
Buck (Virgin Is)		?		
Burnaby (Canada)		?		
Cagayan (Philippines)	?			?
Canary Is (Atlantic O)	?			before 1950s
Capella (Virgin Is)		?		
Caroline Is (Pacific O)	?	*c.* 1912		
Cas (USVI)		?		
Celebes (Indonesia)	?			?
Ceram (Indonesia)	?			?
Cerro Azul (Galápagos)	?			
Chagos Archipelago		by 1813	?	
Channel Is (UK)		?		
Christmas (Indian O)	*c.* 1888?	*c.* 1888?		
Christmas (Pacific O)	?			?
Clyde (UK)			1800s	
Cochin (see Île aux Cochins)				
Cocos-Keeling Gr. (Indian O)	?	shipwreck 1878		
Commander Is (Rus. Fed.)	1870		before 1938?	
Con Son (Vietnam?)		*c.* 1970		
Congo (Virgin Is)		?		
Cook Is (Pacific O)	?			?

Introduction of house mouse and rats to islands (*continued*)

Island	House mouse	Black rat	Norway rat	Unspecified
Copper (Commander)			before 1938?	
Corsica (Mediterranean)	1000–2000 BP	4000–5000 BP	?	
Crozet (Indian O)	*c.* 1772?	19th century?		
Cuba (West Indies)	?			
Curaçao (L. Antilles)	?	?	?	
D'Entrecasteaux (PNG)	?			?
Des Roches (Amirantes)		?		
Deserta Grande (Madeira)	? there 1980s			
Diego Garcia (Indian O)		1813	1890s	
Dog (Virgin Is)		?		
Eaio (French Polynesia)				20th century
East Falkland (Atlantic O)	1764			
Egmont Atoll (Chagos Arch)				1840s
Ellice Is (Pacific O)	?			?
Enderby (Auckland Is)	1840s			
Enewetak Atoll (Marshall Is)	?	1944?		
Enggaño (Indonesia)	?			?
Europa (Mozambique Channel)				?
Faeroe Is (Føroyar) (Atl. O)	250–1000 BP		1768	
Fair Isle (Britain)	?			
Falkland Is (Atlantic O)	*c.* 1764?	with humans?	1764–65?	
Fijian Is (Pacific O)	1840	19th century	early 19th century	
Floreana (Galápagos)	?	after 1934		
Flores (Indonesia)	?			?
Formosa (see Taiwan)				
Foula (Shetlands)	?			
Fuerteventura (Canary Is)	? there 1950s			?
Galápagos Is (Pacific O)	17th or early 19th century	1684–early 1700s or 1830s	World War 2	
Gilbert Is (Pacific O)	?	?		
Gough (Atlantic O)	*c.* 1800			
Graham (Canada)		?		
Great Saltee (Ireland)			there 1990s	
Greenland (Atlantic O)			1780	
Guadalcanal (Solomons)	1965?	World War 2		
Guadeloupe (Mexico)	before 1654 or 1800–30	before 1654		
Guadelupe (Lesser Antilles)				there 1654
Guam (Pacific O)	before 1946?	?	?	
Gunner's Quoin (Mauritius)		?	?	
Hainan (South China Sea)			?	
Hatuta's (French Polynesia)				20th century
Hawaii (Hawaiian)	?	after 1890	1825–35? or later	
Hawaiian Is (Pacific O)	*c.* 1778?	1838–42 or 1870s	1825–35?	
Hebrides (UK)	?	1880s		
Hierro (Canary Is)	? there 1990s			there 1990s
Hog (see Île aux Cochins)				

Introduction of house mouse and rats to islands (*continued*)

Island	House mouse	Black rat	Norway rat	Unspecified
Hong Kong	?		?	
Howland (Pacific O)		?		
Iceland (Atlantic O)			*c.* 1722?	
Île aux Cochins (Crozet)	*c.* 1772?			
Isabela (Galápagos)		1891		
Japan	?	1789		1899
Jarvis (Pacific O)	1858–79?			?
Java (Indonesia)	?			1911
Johnston Atoll (Pacific O)	before 1966?	1960s		
Juan Fernández (Pacific O)	?	before 1945?	?	
Kahoolawe (Hawaiian)	?			
Kai Is (Indonesia)	?			?
Kangean Is (Indonesia)	?			?
Karimata (Indonesia)	?			?
Kauai (Hawaiian)	?	after 1890	1825–35? or later	
Kaula (Hawaiian)	?	*c.* 1938	?	
Kerguelen (Indian O)	before 1874?	*c.* 1956	before 1967?	
Kermadecs (Pacific O)			before 1975?	
Kodiak (Alaska)	*c.* 1920		1920s	
Komandorskiye Ostrova (see Commander Is)				
Koror (Palau)			?	
Kunghit (Canada)		?		
Kuril'skiye Ostrova (Rus. Fed.)	?			?
Kurile Is (see Kuril'skiye)				
Lanai (Hawaiian)	?	after 1890	1825–35? or later	
Leeward Is (West Indies)		1658?		
Lehua (Hawaiian)		?	?1930s	
Lile Europa (Mocam. Ch.)				?
Little Cumbrae (UK)			1800s	
Little St. James (Virgin Is)		?		
Lombok (Indonesia)	?			?
Lord Howe (Pacific O)	19th century	1918		
Los Estados (S. America)			?	
Louisiade Arch. (PNG)	?			?
Luchu Is (see Ryukyu)				
Lulu (Canada)			?	
Lundy (Bristol Channel)		?	before 1959	
Macquarie (Pacific O)	1820	1820	before 1967?	
Madagascar (Indian O)				1898
Madeira (Atlantic O)	*c.* 1420			after 1420
Madura Is (Indonesia)	?			?
Magdalen I. Arch. (Canada)	?		?	
Majuro Atoll (Marshall Is)		?		
Mallorca (Balearic Is)	1000–2000 BP	2000–3000 BP		
Malta (Mediterranean Sea)	7th century or 218 BC	7th century or 218 BC		
Manana (Hawaiian)	?			

Introduction of house mouse and rats to islands (*continued*)

Island	House mouse	Black rat	Norway rat	Unspecified
Manihiki Is (Pacific O)	?			?
Marcus (Japan)		?		
Marianas (Pacific O)	?	?	?	
Marion (Indian O)	by sealers?			
Marquesas (Pacific O)	?	?		
Marshall Is (Pacific O)	?	?		
Martinique (West Indies)	before 1654	before 1654		
Más Afuera (J. Fernández)	?		?	
Más á Tierra (J. Fernández)	?		?	
Matasiri (Indonesia)	?			?
Maui (Hawaiian)	?	after 1890	1825–35? or later	
Mauritius (Indian O)		c. 1500	before 1953?	
Mendanau (Indonesia)	?			?
Mentawi Is (Indonesia)	?			?
Midway (Hawaiian)	?	1943		
Minorca (Menorca?) (Balearic)	2000–3000 BP	2000–3000 BP		
Molokai (Hawaiian)	?	after 1890	1825–35? or later	
Moluccas (Indonesia)	?			?
Mona (Puerto Rico)	?	?		
Monita (Puerto Rico)	?	?	?	
Moresby (Canada)		?		
New Britain (Bismarck Arch.)	?			?
New Caledonia (Pacific O)	after 1774	after 1774	before 1961?	
New Guinea (see PNG)				
New Hebrides (Pacific O)	?			?
New Ireland (Bismarck Arch.)	?			?
New Zealand (Pacific O)	early 19th century	before 1860	1642 or 1769–76, 1835 or later	
Nias (Indonesia)	?			?
Nicobar Is (Bay Bengal)	?	?		
Niihau (Hawaiian)	?	after 1890		
Niue (Cook Is)	after 1900	1900–1920s	before 1969?	
Norfolk (Pacific O)	?	before 1971		
North Natuna Is (Indonesia)	?			?
Nouvelle Amsterdam (see Amsterdam)				
Nouvelle Caledonie (see New)				
Oahu (Hawaiian)	?	1870–90	1825–35? or later	
Orkney Is (UK)		1808		
Palau Is (Pacific O)	?		?	
Palawan (Philippines)	?			?
Papua New Guinea	sailing ships?	sailing ships?	with humans?	
Philippines	?	?		
Phoenix Is (Pacific O)	?			?
Pinzon (Galápagos)		1890s		
Pitcain (Pacific O)	?			?
Ponape			?	
Possession (Crozet Arch)	? there 1990s			there 1990s

Introduction of house mouse and rats to islands (*continued*)

Island	House mouse	Black rat	Norway rat	Unspecified
Prince Edward Is (S. Africa)	? sealers			
Providence (Seychelles)		?		
Puerto Rico (West Indies)	?	after 1658?	1658	
Puffin (Menai Str.)			1816	
Queen Charlotte Is (Canada)		?		
Remire				*c.* 1882
Reunion (Indian O)				?
Rhio-Lingga Arch. (Indon)	?			?
Robinson Crusoe (see Más á Tierra)				
Round (Mauritius)		19th century		
Ryukyu Is (Japan)	?			?
Sable (Canada)				before 1880
Saipan (Palau Is)		?		
Sakhalin (Rus. Fed.)	?		?	
Saleyer (Salgar)	?			?
Samoa (Pacific O)	?			?
San Cristobal (Galápagos)	?	after 1934		
Sand (Hawaiian Is)	before 1963			
Sandpit (Canada)		?		
Sanghir Is	?			?
Santa Cruz (Galápagos)	?	1930–34		
Santiago (Galápagos)	?	1835		
Sardinia (Sardegna) (Italy)	3000–4000 BP	4000–5000 BP		
Sark (Channel Is)		?		
Savu (Indonesia)	?			?
Seychelles (Indian O)	?	after 1770		
Shetland Is (UK)		1904		
Sierra Negra (Galápagos)	?			
Simalu	?			?
Singapore		?		
Skokholm (Irish Sea)	*c.* 1903			
Society Is (Pacific O)	?			?
Solombo	?			?
Solomon Is (Pacific O)	before 1965?			?
South Georgia (Atlantic O)	? there 1960s	1800?	late 18th century	there 1960s
South Natuna Is (Indonesia)	?			?
South Ronaldsay (Orkney)		?		
South Seymour (see Baltra)				
Sri Lanka (Indian O)		1914?		
St. Christopher (St. Kitts)	before 1654	before 1654		
St. Croix (S. Africa??)	?	?		
St. Helena (Atlantic O)	? there 1980s	after 1420		there 1980s
St. John (USVI)	?	?		
St. Kilda (UK)	1000 BP?			
St. Paul (Indian O)	early 19th century	? before 1874	before 1967?	
St. Thomas	?	?		
Ste Barbe	?			?

Introduction of house mouse and rats to islands (*continued*)

Island	House mouse	Black rat	Norway rat	Unspecified
Stevens (Virgin Is)		?		
Sumatra (Indonesia)	?			?
Sumba (Indonesia)	?			?
Sumbawa (Indonesia)	?			?
Taiwan	?			?
Talaud (Indonesia)	?			?
Tambelan Is (Indonesia)	?			?
Tanimbar (Indonesia)	?			?
Tasmania	?	?	?	
Terceira (Azores)				after 1460?
Tierra del Fuego (S. Am.)	?		?	
Timor (Indonesia)	?			?
Tobriand Is	?			?
Tonga Is (Pacific O)	?	before 1987?		
Trinidad (West Indies)	after 1500	1658		after 1500?
Trinidade (Brazil)		before 1967		
Tristan da Cunha (Atlantic O)	*c.* 1816	1882	before 1967?	there 1990s
Tromelin (Indian O)	after 1776			after 1772
Tuomotus (Pacific O)		?		
Unalaska (Aleutians)	early 1800s			1800s
Vancouver (Canada)	?	?		?
Vanua Levu (Fiji)	*c.* 1840			
Vieques (Virgin Is)	?			
Virgin Is (West Indies)	?	?		
Viti Levu (Fiji)	*c.* 1840			
Wake (Pacific O)	?			?
West Falkland (Atlantic O)	1764			
West Indies	1654	1658?		
Wetar (Indonesia)	?			?
Zembra (Tunisia)	? there 1980s			there 1980s

RICEFIELD RAT

Rattus argentiventer (Robinson and Kloss)

■ DESCRIPTION

HB 176–230 mm; T 172–201 mm; WT c. 212 g.
Upper parts grizzled ochraceous tawny and brownish black; under parts creamy white, often traces of a median grey stripe; forefeet brown, hind feet white, broadly brown medially.

■ DISTRIBUTION

Asia. Southern Thailand, Cambodia, southern Vietnam, Malay peninsula, Sumatra, Java, Borneo, Kangean Island, Bali, and islands of Lombok, Sumbawa, Komodo, Rintja, Flores, Sumba, Timor, Nusa Tenggara, and Sulawesi, Mindoro and Mindanao islands in the Philippines, and perhaps Papua New Guinea.

■ HABITS AND BEHAVIOUR

Habits: little known; human commensal; congregates in burrows during day. **Gregariousness:** no information. **Movements:** home range *c.* 273 m. **Habitat:** lowlands, crops, rice fields, tall grass, palm plantations, human habitation, burrows and crevices in slopes of dikes and rivulets. **Food:** omnivorous; grain, flowers of oil palms, fruits, nuts, leaves, shoots, roots, rice plants, plant parts, insects (grasshoppers, termites, ants), land snails, slugs, occasionally lizards. **Breeding:** all year; litter size 5–7. **Longevity:** mean 6.2 months (wild). **Status:** locally common.

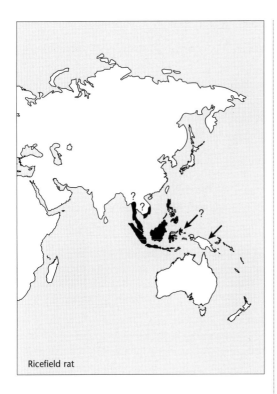

Ricefield rat

■ HISTORY OF INTRODUCTIONS

It has been suggested that the ricefield rat is clearly an element that is native to Indochina and was inadvertently introduced to the Sunda Shelf, the Philippines, Sulawesi, Nusa Tenggara, and to New Guinea, possibly with the spread of rice culture (Musser 1973; Musser and Newcomb 1983; Musser and Holden 1991; Wilson and Reeder 1993).

ASIA
Indonesia–Philippines
Ricefield rats were probably introduced to the Sunda Shelf, Sulawesi, Nusa Tenggara and the Philippines with rice culture (Wilson and Reeder 1993).

AUSTRALASIA
Papua New Guinea
Only six specimens of ricefield rat have been collected, probably in Tanah Merch Bay area, near Hollandia. This bay has been used by navigators and trading vessels for centuries and they were most likely introduced by these means (Flannery 1995).

■ DAMAGE

In Malaya the ricefield rat is considered a major pest in rice fields and palm plantations (Kitchener *et al.* 1990). In Thailand it is a rare species and causes little damage in rice fields, but occasionally eats the seedlings of trees (Lekagul and McNeely 1988).

MAORI RAT OR KIORE
Polynesian rat, little rat, Pacific rat
Rattus exulans (Peale)

■ DESCRIPTION
HB 80–140 mm; T 108–147 mm; WT 30–180 g.
Sleek appearance; upper parts brown or grey brown; under parts whitish with grey underfur; muzzle pointed; ears large; tail dark, generally less than 110 mm and may not extend to snout; tail has fine scales. Distinguished from black and Norway rats by smaller size. Female has two pairs pectoral mammae.

■ DISTRIBUTION
Original range not well defined, but probably southern and South-east Asia. From eastern Bangladesh, Andaman Islands, Burma, Thailand, south to Sumatra, Java, Timor and east through Indonesia (Sulawesi and Sunda Islands) to Papua New Guinea, New Britain and the Philippines; eastwards across the Pacific as far as Easter Island and Kure Atoll (Hawaiian group) and Stewart Island and New Zealand.

■ HABITS AND BEHAVIOUR
Habits: mainly nocturnal; chiefly terrestrial but also arboreal; nests in trees. **Gregariousness**: family ties loose and brief; males and females associate for mating; adult females avoid one another; density 6–188/ha. **Movements**: sedentary; home range 237–1845 m^2; locally 200–280 m common; irruptive (3–5 years). **Habitat**: bush, scrub, houses, clearings,

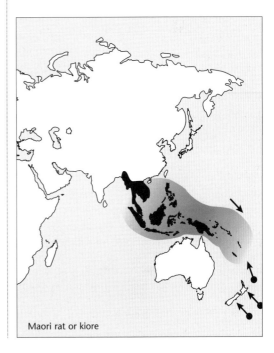

Maori rat or kiore

under logs, rocks, grassland, forest, gardens, paddy fields, and commensal in Asia. **Foods:** omnivorous; snails, crustaceans, insects (cicadas, wetas, ants, beetles), coconuts, fruits, insect larvae, centipedes, spiders, earthworms, weevils, lizards, birds, flowers, stems, leaves and roots. **Breeding:** throughout the year, mainly spring to late summer; gestation 19–30 days; litter size 1, 3–9; 1–6 or more litters/year; post-partum oestrous; female polyoestrous; young born eyes shut, eyes open 2 weeks; young weaned 4 weeks; reach maturity 8–12 months. **Longevity:** 3.2–9 months in wild. **Status:** common and widespread.

■ HISTORY OF INTRODUCTIONS
Maori rats have been introduced, probably prehistorically, by early Polynesian colonists to most of the islands in the South Pacific Ocean.

AUSTRALASIA
Australia
Maori rats are recorded from two offshore islands: Adele Island, north of Derby, and on the Murray Islands, east of Torres Straits (Watts and Aslin 1981), where they were possibly introduced.

PACIFIC OCEAN ISLANDS
Carried as stowaways on sea voyages from one island to the other by the early Polynesian colonists, maori rats have been widespread in the Pacific region since pre-European times (Wiens 1962; Watson 1969). They arrived in Melanesia around 3500 years ago, spread with humans throughout Polynesia, and appear to have arrived in Micronesia during the European period (Flannery 1995). By AD 800 they were well distributed on all the major Pacific islands (King 1984).

Pacific Ocean islands inhabited by *R. exulans* and probably introduced by Polynesians between 1500 BC and AD 600

Island	Notes and references
Adele (Australia)	recorded (23)
Aldermen group (NZ)	present on Middle Chain (6)
American Samoa	introduced and established on Manu`a I (3)
Arapawa (NZ)	present (6)
Bat (PNG)	introduced and established (3)
Batjan (Maluku)	introduced and established (3)
Bay of Is (NZ)	present on Moturua (6)
Belau	introduced and established in Palau Islands (3)
Biak-Supiori (Irian Jaya)	introduced and established (3)
Bougainville (PNG)	present (14); introduced and established (3)
Buka (PNG)	introduced and established (3)
Buru (Maluku)	introduced and established (3)
Caroline Is	present (16), on most atolls in eastern Carolines (11)
Cavalli Is (NZ)	present on Motukawanui and Haraweka (6)
Centre (NZ)	present (6)
Chatham (NZ)	formerly present, now extinct (1, 6)
Chickens (NZ)	present on all islands (6)
Choiseul (Solomons)	introduced and established (3)
Codfish (NZ)	present (1, 6)
Conflict group (PNG)	introduced and established (3)
Cuvier (NZ)	present (6)
D'Urville (NZ)	present (6)
Ducie Atoll (east of Pitcairn)	present (13)
Duke of York (PNG)	introduced and established (3)
East (NZ)	present (6)
Easter	noted by Cook 1744
Efate (Vanuatu)	introduced and established (3)
Esperitu (Vanuatu)	introduced and established (3)
Federated States of Micronesia	introduced and established on Pohnpei (3)
Fergusson (PNG)	introduced and established (3)

Pacific Ocean islands inhabited by *R. exulans* and probably introduced by Polynesians between 1500 BC and AD 600 (*continued*)

Island	Notes and references
Fiji	present; introduced and established on Vanua Levu (3)
Garove (PNG)	introduced and established (3)
Gebe (Maluku)	introduced and established (3)
Gilbert Is	present (14)
Goodenough (PNG)	introduced and established (3)
Great Barrier (NZ)	present (6)
Guadalcanal (Solomons)	introduced and established (3, 14)
Guam	present (19); introduced and established (3)
Halmahera (Maluku)	introduced and established (3)
Hawaiian Is	probably there 1000–1500 years ago; still present (18, 9, 17)
Hen (NZ)	present (6)
Hiu (Vanuatu)	introduced and established (3)
Howland (Line group)	there 1986 (7)
Inner Chetwode (NZ)	present here and on Te Kiore (6)
Irian Jaya	widespread (15); introduced and established on islands of Biak-Supiori, Japen and Owi (3)
Japen (Irian Jaya)	introduced and established (3)
Jarvis	introduced by Polynesians, abundant 1935 (12)
Kai Is (Maluku)	introduced and established (3)
Kapiti (NZ)	present (19, 6)
Kermadecs (NZ)	present (4)
Kiriwina (PNG)	introduced and established (3)
Kure Atoll (Hawaiian group)	present and introduced accidently by Polynesians (5, 17, 6)
Little Barrier (NZ)	present (19, 6, 3)
Long (NZ)	present (6)
Long (PNG)	introduced and established (3)
Macauley (NZ)	present (6)
Malaita (Solomons)	introduced and established (3)
Malakula (Vanuatu)	introduced and established (3, 14)
Maluku	introduced and established, see individual islands (3)
Manam (PNG)	introduced and established (3)
Mangole (Maluku)	introduced and established (3)
Manus (PNG)	introduced and established (3)
Marianas	introduced and established on the islands of Rota, Tinian and Saipan (3)
Marquesas	present, being replaced by *R. rattus* (19)
Marshall Bennett Is (PNG)	introduced and established (3)
Marshall Is	occur there; present on Arno Atoll and Eniwetok Atoll (11, 2)
Mayor (NZ)	present (1, 6)
Mercury group (NZ)	present on all islands except Korapuki where exterminated 1986 (6)
Mioko (PNG)	introduced and established (3)
Misima (PNG)	introduced and established (3)
Mokohinau Is (NZ)	present on all islands (6)
Moratai (Maluku)	introduced and established (3)
Motuara (NZ)	present (6)
Murray Is (Australia)	recorded (23)
Mussau (PNG)	introduced and established (3)
Nendö (Solomons)	introduced and established (3)

Pacific Ocean islands inhabited by *R. exulans* and probably introduced by Polynesians between 1500 BC and AD 600 (*continued*)

Island	Notes and references
New Britain (PNG)	introduced and established (3)
New Caledonia	only rodent 1774; now abundant (22, 19); introduced and established (3)
New Georgia (Solomons)	introduced and established (3)
New Ireland (PNG)	introduced and established (3)
New Zealand	brought with Polynesians (6)
Nissan (PNG)	introduced and established (3)
Niue	only rodent in 1873 and 1900 (20)
Norfolk (Australia)	introduced (3)
Normanby (PNG)	introduced and established (3)
Obi (Maluku)	introduced and established (3)
Ontong Java (Solomons)	introduced and established (3)
Owi (Irian Jaya)	introduced and established (3)
Palau Is (Belau)	introduced and established (3)
Papua New Guinea	widespread (15); introduced and established on many islands, see individual islands (3) and notes following table
Pearl (NZ)	present (6)
Pickersgill (NZ)	present (6)
Pohnpei (FSM)	introduced and established (3)
Putauhinu (NZ)	present (1, 6)
Rangitoto (NZ)	present (1)
Raoul (NZ)	there 1877; still present (4, 6)
Rennell (Solomons)	introduced and established (3)
Rossel (PNG)	introduced and established (3)
Rota (Marianas)	introduced and established (3)
Rurima (NZ)	present but exterminated 1985 (6)
Russell Is (Solomons)	introduced and established (3)
Saipan (Marianas)	introduced and established (3)
San Cristobal (Solomons)	introduced and established (3)
Sanana (Maluku)	introduced and established (3)
Santa Isabel (Solomons)	introduced and established (3)
Shortland (Solomons)	introduced and established (3)
Sideia (PNG)	introduced and established (3)
Sikopo (Solomons)	introduced and established (3)
Slipper group (NZ)	present on Rabbit and Penguin (6)
Solomon Is	introduced and established on a number of islands, see individual islands (3)
Stephenson (NZ)	present (6)
Stewart (NZ)	present (19, 6)
Sudest (PNG)	introduced and established (3)
Three Kings (NZ)	present (1)
Tinian (Marianas)	introduced and established (3)
Tokelau Is	present when discovered 1841 by Europeans (20, 8)
Tolokiwa (PNG)	introduced and established (3)
Tonga	present
Tuamotus	occur on Raroia Atoll where being replaced by *R. rattus* (19)
Uki Ni Masi (Solomons)	introduced and established (3)

Pacific Ocean islands inhabited by *R. exulans* and probably introduced by Polynesians between 1500 BC and AD 600 (*continued*)

Island	Notes and references
Umboi (PNG)	introduced and established (3)
Vanuatu	introduced and established on Efate, Esperitu Santo Hiu and Malakula (3)
Western Samoa	present, occur on all islands (10)
Whakatere-Papanui (NZ)	present (6)
White (NZ)	present (1, 6)
Woodlark (PNG)	introduced and established (3)

References: 1 Atkinson & Bell 1973; 2 Falla *et al.* 1971; 3 Flannery 1995; 4 Gibb & Flux 1973; 5 Kepler 1967; 6 King 1990; 7 Kirkpatrick & Rauzon 1986; 8 Kirkpatrick 1966; 9 Kramer 1971; 10 Marples 1955; 11 Marshall 1975; 12 Rauzon 1985; 13 Rehder & Randall 1975; 14 Rowe 1967; 15 Ryan 1972; 16 Storer 1962; 17 Tamarin & Malecha 1972; 18 Tomich 1969; 19 Watson 1961; 20 Wodzicki 1969; 21 Woodward 1972; 22 Nicholson & Warner 1953; 23 Watts & Aslin 1981.

New Zealand

Maori rats probably arrived in New Zealand between AD 800 and 1350 or as long ago as 12 000 years (King 1984; King 1990). More than likely they came with one or more of the early waves of Polynesian immigrants (Watson 1956) in the fourteenth century (Wodzicki 1965) probably about AD 1350 or earlier (Gibb and Flux 1973). According to Maori tradition the rats came in the canoes of the Great Fleet about 600 to 900 years ago (Watson 1959; Watson 1961; Gibb and Flux 1973). However, it is likely the rats preceded them by some hundreds of years as they appear to have been present on the Chatham Islands before the Maori colonisation (Watson 1959).

Widespread in pre-European times (Watson 1959; King 1984), maori rats remained so until replaced in some parts by European rats after colonisation by whites (Watson 1956). Soon after European colonisation they disappeared from the North Island, but periodically became abundant in the South Island until 1889, but there have been few reports of them since then (Watson 1959). Their disappearance from the North Island coincides with the introduction and spread of Norway rats, *R. norvegicus* (Watson 1961).

Between 1948 and 1956 they were still found in some 11 locations in New Zealand (Watson 1956). They were reported to be in some isolated areas of Fiordland and on Kapiti and Stewart islands in the late 1950s (Watson 1959). In the South Island they persisted in forested parts of the north-west and perhaps the south-west (Watson 1961). In 1965 they were reported to be common, but very local in both the North and South Island and on Stewart Island. In 1961 they were on Little Barrier Island. They were eliminated on the mainland by competition from European rats except in Fiordland, but are abundant on some offshore islands (see table of maori rats on

islands). They formerly occurred on the main island in the Chatham group, but are not present there now (Watson 1961; Wodzicki 1965; Gibb and Flux 1973).

Although widespread in earlier times, maori rats are now largely confined to offshore and outlying islands. On the mainland they are confined to South Westland and Fiordland (King 1990).

Papua New Guinea

Maori rats are found in modified environments (Lidicker and Ziegler 1968; Dwyer 1975, 1978) in New Guinea, but have not reached some remote areas which are suitable for them, and there are no fossil records, which suggests they are a relatively recent arrival (Flannery 1995).

■ DAMAGE

Maori rats are agricultural pests in the Pacific because they damage coconuts, cocoa, sugarcane, and a variety of other crops. They are also a public health risk because they carry leptospirosis, plague, lungworm and other pathogens and parasites (King 1990).

Rats damage coconuts by gnawing either the green fruit on the palms or the ripe ones on the ground (Wodzicki 1972; Mosby and Wodzicki 1973). The damage to the coconuts also contributes to an increase in mosquito population and this is a health hazard as those falling to the ground hold water and provide a place for mosquitoes to breed. The mosquitoes may carry filariasis (Laird 1963). On Kure Atoll these rats are the main cause of nesting failure of red-tailed tropic birds (Fleet 1962).

In the Philippines maori rats cause damage to coconuts (Fieldler *et al.* 1982) and on Nieu cause damage to coconuts, passion fruit, paw paw, cassava and kumaras (Wodzicki 1969). In Hawaii they cause in field damage to sugar cane after lodging (cane 7–10 months) and this damage gradually increases during the remainder of the crop cycle (Fellows and Sugihara 1977).

Rat damage in Hawaii on sugar cane comes from three species of rats and is estimated as 4.5 million dollars annually. Maori rats (*R. exulans*) and black rats (*R. rattus*) primarily eat sugarcane. Maori rats subsist mainly on sugar cane, but black rats also eat other things. Damage starts when the cane is eight to 14 months old and continues until harvest at 22 months (Hood *et al.* 1970).

NORWAY RAT

Brown rat, common rat

Rattus norvegicus (Berkenhout)

◼ DESCRIPTION

HB 165–280 mm; T 122–230 mm; WT 120–580 g and up to 909 g.

Generally large and robust; fur greyish or greyish brown on upper parts and white with greyish under fur on under parts; muzzle blunt; ears small; eyes small; ears, feet and tail flesh coloured, dark above and pale below; tail more than 110 mm, but shorter than body; ears and tail smaller than for black rat. Female has three pairs inguinal and three pairs pectoral mammae.

◼ DISTRIBUTION

Asia. Their original range is assumed to be south-eastern Siberia and northern China, but they are now almost cosmopolitan except for polar regions. Whole of Europe and most islands; Asia Minor, eastwards across southern Siberia to the Pacific; most of China and Japan.

◼ HABITS AND BEHAVIOUR

Habits: nocturnal, diurnal or crepuscular; nest of grass, paper, or similar material in hole or burrow; highly adaptable; aggresssive, mostly terrestrial, territorial, social, swims well. **Gregariousness:** colonial. **Movements:** sedentary; home range as little as 22–46 m, but 0.8–1.8 ha depending on location of food sources; up to 0.5–3.3 km overnight recorded. **Habitat:** ubiquitous, in every habitat except for deserts and polar regions. **Foods:** omnivorous; seeds, fruits, leaves, rhizomes, meat, vegetables, garbage, stored grain, silage, stock feed, root crops, weeds, grasses, insects (beetles), molluscs, crustaceans, annelids, and other invertebrates, birds' eggs, carrion. **Breeding:** throughout the year; gestation 20–26 days; 6 litters per year; 5–10, 22 young; female polyoestrous; young naked and blind at birth; eyes open at 14 days; weaned 28 days; mature at 3–4 months. **Longevity:** 12 months or more for most. **Status:** widespread and abundant.

◼ HISTORY OF INTRODUCTIONS

Introduced widely throughout the world, Norway rats now range from the Antarctic (South Georgia) north to the Arctic (Spitzbergen, Aleutians and Alaska). They may now be found along the coastlines of the entire Palaearctic region having been transported there by humans.

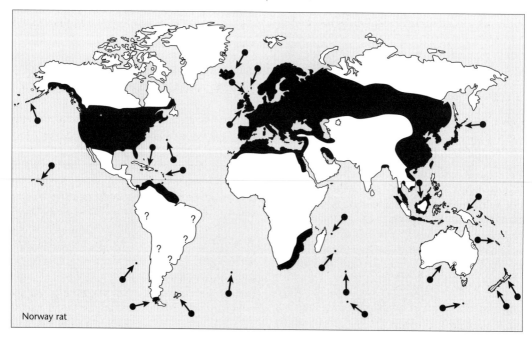

Norway rat

Morocco

Norway rats crossed the Straits of Gibraltar about 1930 and reached Marrakech in the south four years later (Roots 1976; Lever 1985).

AFRICA

Egypt–Sudan

Rats were possibly introduced to the Nile Valley (Khartoum) about 1932(?) and also to Fernando Poo in about 1929(?) (Allen 1954). They arrived in Suez about 1893 and soon after at Giza; in 1918 at Ecnain Upper Egypt, and Sudan about 1909, but for many years were confined to the ports (Shortridge 1934).

South Africa

In southern and eastern Africa, Norway rats are confined to the coastal ports, large coastal towns and their immediate vicinity (Kingdon 1974; de Graaf 1981). They were probably introduced by European ships and are known to have been in the Cape Colony by 1832 (Avery 1985). They are reported to inhabit the shoreline south of Durban (Natal) and near Hout Bay (Cape Town), but there is no evidence of any faunal interactions with native animals (Bigalke and Pepler 1991).

Norway rats were probably widespread by 1930 (Bigalke 1937), and one is recorded collected at Cape Town before 1934 (Shortridge 1934). They are still widespread in the Cape of Good Hope Province (Hey 1974), but occur mainly in coastal ports and larger towns in the vicinity of these ports (Smithers 1983).

ASIA

Borneo (Indonesia)

In Borneo, Norway rats occur in rice fields and towns on the west coast of Sabah; coastal towns in Sarawak, Pontianak and Banjerwasin in Kalimantan (Payne *et al.* 1985)

China

Norway rats occur throughout China except Xinjiang (Sinkiang) and Tibet (Deng and Wang 1984).

Hong Kong

Norway rats have been introduced into Hong Kong, but the black rat appears to have occurred there naturally (Marshall 1967) or was a very early introduction.

India

Only established in some major seaports in India, Norway rats are being displaced by the lesser bandicoot rat, *Bandicota bengalensis* (Fitzwater 1967).

Iran

Norway rats occur in Iran.

Kuwait

Norway rats entered the port some time before 1979 and are now the dominant species (Al-Sanei *et al.* 1984).

Russian Federation

Introduced to Sakhalin Island and in the Far East of Russia, Norway rats are well established and widespread (Yanushevich 1966). They probably arrived in European Russia in about 1727.

Thailand

Norway rats occur in a few towns in Thailand and on Samui Island, and most cities in the south of the country (Lekagul and McNeely 1988).

ATLANTIC OCEAN ISLANDS

Falkland Islands

Probably arriving with the first English and French colonists in 1764–65, Norway rats were certainly present in 1842 around Port Louis. Today they are widespread and also occur on some of the smaller outlying islands (Lever 1985).

South Georgia

Introduced to South Georgia by sealers in the eighteenth and nineteenth centuries (Holdgate and Wace 1961; Bonner and Leader-Williams 1976, 1980), in the late 1950s Norway rats were numerous and widespread in coastal grasslands (Holdgate and Wace 1961). They have thrived on the island (Watson 1975) and are now widespread over most of the vegetated areas and locally abundant in the coastal tussock grassland (Bonner and Leader-Williams 1976).

Tristan de Cunha

Norway rats occur on Tristan (Holdgate 1967).

AUSTRALASIA

Australia

Distribution of Norway rats in Australia is relative to human settlement and the larger seaports; they also occur in eastern Tasmania. They are probably less common in Australia today than they were formerly, but the reasons for this are not known (Watts and Aslin 1981). They are not as widely distributed as the black rat and are mainly found near wharves and heavily developed coastal areas (Wilson *et al.* 1992).

It is possible that Norway rats became established along the coast of Western Australia via early Dutch ships after 1616 (Archer 1984; Hand 1984; Long 1988), but certainly in Australia with the arrival of European colonists from the 1770s.

Papua New Guinea

Recently introduced, the Norway rat is not as widespread as the black rat and is found in the main ports only. It does not appear to be in any rural areas and occurs only in association with humans (Ryan 1972; Herington 1977).

EUROPE

Norway rats invaded Europe in the Middle Ages and this was followed by a major invasion at the end of the eighteenth century (Matheson 1963; Niethammer and Krapp 1978–86). They appear to have reached Europe via Russia or China early in the eighteenth century. They arrived in Copenhagen, Denmark, in 1716, Iceland in 1722, and appear to have spread westwards across the Volga and into European Russia and the Baltic in about 1727, Paris in 1750, were recorded on the Norwegian mainland in 1762, on the Faeroe Islands in 1768, in Brunswick and Greenland around 1780, were in parts of eastern Prussia, Norway and Sweden in 1762–90, Spain and Italy in the mid- to late eighteenth century, and Switzerland in 1809. The exact date of arrival in Great Britain is uncertain, but was most likely between 1714 and 1729 in shipping from Russian ports.

Their rapid spread in Europe is said to have caused the decline of the black rat in many areas (Hinton 1933; Lever 1977; Corbet and Harris 1991). However, skeletal remains from the Middle Ages from West Germany show the species had spread over middle Europe far earlier than previously known, thus population increases in the nineteenth century were not caused by a new immigration, resulting in ousting and replacement of the black rat. Obviously both have different ecological demands and the situation changed in favour of the Norwegian rat (Heirich 1976).

United Kingdom

In Great Britain, Norway rats spread rapidly, reaching Ireland about 1722 and Scotland about 1744 (Fitter 1959; Lever 1977). They were probably everywhere in the British Isles by the latter half of the eighteenth century (Fitter 1959).

Most United Kingdom offshore islands were colonised at some time or other – on Puffin Island in Menai Strait Norway rats came ashore after a Prussian vessel was wrecked off the coast in 1816. They were present on Lundy Island in 1959 (Fitter 1959). Norway rats are now common throughout the United Kingdom (Baker 1990).

INDIAN OCEAN ISLANDS

Amsterdam

Norway rats occur on Amsterdam Island (Holdgate 1967).

Kerguelen

Norway rats occur on Kerguelen Island (Holdgate 1967).

Mauritius

Norway rats are present on the island of Mauritius, where they cause much damage (Williams 1953).

St. Paul

Norway rats occur on the island of St. Paul (Holdgate 1967).

Tromelin Island (east of Madagascar)

Present on Tromelin Island (Staub 1970), Norway rats appear to have been particularly numerous in the southern part of the island in 1953 (Paulian 1955). The island was discovered in 1772 and visited in 1776 and the rats have probably been there since then.

NORTH AMERICA
Canada and Alaska

The date of arrival for Norway rats in Canada is not known, but they are now widely distributed in most settled areas and rural areas of southern Canada (Banfield 1977), throughout the settled areas of British Columbia (Carl and Guiguet 1972) and occur on Lulu Island and Vancouver Island, usually near human habitation (Cowan and Guiguet 1960). They also occur on the Magdalen Islands Archipelago, Quebec, Canada, where they are abundant (Cameroun 1962).

By 1825 they were not west of Ontario, but by 1887 were well established in Vancouver, New Westminster and Victoria. Seven years later they were at Chittiwack on the Fraser River. They crossed into Canada from Manitoba about 1900 and by 1914 or 1919 were as far north as the Assinboine River and had appeared in Saskatchewan (Dorrance 1984). By 1939 most urban areas of both provinces had Norway rats. Colonisation of Canada has continued and they reached Alberta as recently as 1948 (Lever 1985) or 1950 (Dorrance 1984).

Norway rats were introduced in about the 1920s to Kodiak Island and become well established (Clark 1955).

United States

Norway rats reached North America at about the beginning of the American Revolution in 1775 (Lantz 1909, 1910; Silver 1927, 1937; Gottschalk 1967). They gradually spread inland displacing the black rat, and were abundant at several points on the Pacific coast in 1851 including San Francisco, Astoria and Fort Steilacoon (Palmer 1898). The initial inland colonisation followed closely on the heels of the early settlers and they were well entrenched in the larger towns in Colorado and New Mexico by 1890, reaching Wyoming about 1919 and Montana about 1923 (Silver 1937, 1941). They were in every state of the union by 1941 (Silver 1941) and followed closely on the heels of the early settlers except in the high mountains where progress was slower (Silver 1941). They reached south-western Georgia about 1947 and from 1948 to 1950 ousted the black rat in many areas in an invasion that overran 2590 km^2 in six years (Ecke 1954).

In 1956 the most northerly coastal locality reached by Norway rats was Nome, Alaska, where they were restricted to the modified habitats of humans (Schiller 1956). They are now widespread throughout the United States, southern Canada (southern Quebec) and southern Alaska (Howard and Marsh 1976) to Mexico. They were reported to outnumber the human population of the United States in 1910 (NGM 1960) and their distribution is closely associated with humans (Dorrance 1984).

Mexico
Norway rats are present in many coastal areas of Mexico, where they cause much damage (Navarrete 1978).

Aleutian Islands
Possibly introduced in the 1800s Norway rats are widely established now on Unalaska Island, Aleutians (Southern 1964).

PACIFIC OCEAN ISLANDS
Belau
Norway rats occur on Koror in the Palau Islands (Lever 1985; Flannery 1995).

Campbell Island
Probably introduced by whalers and sealers before 1883 (Holdgate and Wace 1961; Holdgate 1967), Norway rats are now the main species on the island (Watson 1961). They still occurred all over the island in the late 1970s (Dilkes and Wilson 1979).

Fatuna Island
Norway rats have been recorded from Fatuna (Wallis and Fatuna Islands) (Tate 1935).

Federated States of Micronesia
Norway rats occur on Pohnpei (Flannery 1995) and on Ponape in the Caroline Islands (Smuts-Kennedy 1975; Lever 1985).

Fijian Islands
Accidentally introduced, Norway rats have caused some damage to ground bird life in Fiji (Turbet 1941). They arrived in the nineteenth century and are now common on all the main islands in association with agricultural, urban, suburban and coastal areas (Pernetta and Watling 1978; Lever 1985).

Hawaiian Islands
Assumed to have arrived with the first sailing ships (Timber 1938) Norway rats may not have become established until between 1825 and 1842 (Kramer 1971; Atkinson 1977). On Hawaii the maori rat (*R. exulans*), black rat (*R. rattus*) and Norway rat (*R. norvegicus*) all live together (Watson 1961). By the 1970s Norway rats inhabited the islands of Hawaii, Maui, Molokai, Lanai, Oahu, and Kaui (Kramer 1971).

Juan Fernández
Norway rats occur on Más á Tierra (Robinson Crusoe Is) and Más Afuera (Alejandros Selkirk) in the Juan Fernández Islands (Lever 1985).

Komandorskiye Ostrova (Commander Islands)
Norway rats were common on Bering and Copper islands in 1938 (Barabash-Nikiforov 1938).

Macquarie Island
Norway rats occur on Macquarie Island (Holdgate 1967).

Marianas
Norway rats occur on Saipan in the Marianas (Smuts-Kennedy 1975; Lever 1985; Flannery 1995).

New Caledonia
Present on New Caledonia (Watson 1961), Norway rats were most abundant in 1953 in Noumea and larger villages, and were not commonly away from human habitation (Nicholson and Warner 1953). They were probably introduced after the discovery of the islands by Europeans in 1774.

New Zealand
Introduced by whalers and sealers early in the nineteenth century or late eighteenth century (King 1990), Norway rats were recorded by Charles Darwin in the Bay of Islands in 1835 (Watson 1959; Wodzicki 1965). Many were probably brought to New Zealand by Europeans in the first half of the nineteenth century (Watson 1961). They were probably well distributed throughout the country by 1800 (Lever 1985).

Norway rats are now widespread and abundant throughout the North and South islands and occur on many offshore islands. Although found throughout they are patchily distributed, mainly in cities, towns, houses, farms (croplands) and along creeks and rivers (Watson 1961; King 1990).

On offshore islands they are present on Stewart, Raoul, Kermadecs, Pearl, Ulva, Bench, Rosa, Chatham, Campbell, Foely, Cook Strait Islands, (see list of islands under house mouse) and many others.

On Raoul they have virtually replaced the maori rat (*R. exulans*). They probably arrived from a shipwreck in 1921, and no maori rats have been seen since 1944 (Watson 1961). In 1868 they were common on the

Chatham Islands, where they are still common; were present on Stewart (in 1874), occurred on Campbell (before 1883), Raoul (in 1921), Chatham (in 1840), Campbell (in 1867), Foely, were eradicated on Breaksea Island in about the 1990s, and probably occur on many others (Watson 1959, 1961; Wodzicki 1965; Gibb and Flux 1973; Lever 1985; King 1990; Allen *et al.* 1994).

Nuie Island
Reported on Nuie Island in 1969, a few Norway rats were later found there (Wodzicki 1969), but may not yet have a permanent foothold.

Samoa
A specimen of Norway rat in the Australian Museum came from Samoa (Flannery 1995). They are recorded from Tutuila in American Samoa (Tate 1935).

Tonga
Norway rats are common on Rangatapu (Rinke 1987).

SOUTH AMERICA
In South America, Norway rats are associated with large urban centres and have not as yet penetrated into many undisturbed habitats (Eisenberg 1989; Redford and Eisenberg 1992) except apparently in Patagonia (Pearson 1983). They are recorded from Argentina, Chile, and south to Tierra del Fuego, but are probably more widespread than this indicates.

Argentina – Chile
Norway rats occur in southern Argentina (Patagonia) (Thomas 1927; Pearson 1983), where they have been found in areas away from human habitation (Pearson 1983). They occur on Tierra del Fuego (Pine *et al.* 1979) and also as far south as the Isla de los Estados off Tierra del Fuego (Lever 1985).

WEST INDIES
In the West Indies, Norway rats occur on Barbados, Bermuda, Puerto Rico and the Virgin Islands, and Trinidad. They probably exist on many other islands in the region, but details were not found at this time.

Barbados
Probably arrived Barbados with or soon after discovery in 1536 or settlement in 1626; rat control started in 1745 and has continued until now (Browne 1982).

Bermuda
Norway rats arrived in Bermuda in the mid-eighteenth century and they are now common in mainly urban areas (Lever 1985).

Puerto Rico – Virgin Islands
Norway rats arrived in Puerto Rico shortly after 1658 and were certainly there in 1877 when the mongoose

was introduced (Pitmental 1955). At this time they were probably the dominant introduced rat species.

Norway rats occur on Monita Island west of Puerto Rico (Dewey and Nellis 1980) and exist on Monita Island in Virgin Islands, but most islands that have rats have black rats (Dewey and Nellis 1989). They occur on all islands in the United States Virgin Islands and Puerto Rico except Mona and Monita Islands and Anegada (Philibosian and Yntema 1977).

Trinidad
In Trinidad Norway rats had replaced the black rat in the Port of Spain wharf area by 1930 (Urich 1931).

■ DAMAGE
For details of the damage caused by *Rattus* species including the Norway rat see the Damage section under Black Rat below. Some specific Norway rat damage is listed here.

Norway rats cause damage to cane fields in Mauritius (Williams 1953) and Hawaii. In Hawaii they cause damage to cane crops, particularly after lodging (at seven months) and damage increases during the remainder of the crop cycle (Fellows and Sugihara 1977).

Surveys in Europe in the 1960s estimated that two million rats, mainly *R. norvegicus,* occurred in Budapest. After seven years control the population was neglible (Bajoni 1980).

Norway rats are reported to be important pest in both urban and rural regions of China (Deng and Wang 1984).

The unintentional introduction of Norway rats in the Aleutian Islands has assisted the decimation of the Canada goose (Franzmann 1988). On Raoul Island in the Kermadecs they have assisted in severely reducing the sea bird population and have almost eliminated the burrowing petrel on the main island in the Campbell Islands (Atkinson and Bell 1973).

BLACK RAT
Ship rat, roof rat, house rat
Rattus rattus Linnaeus
R. rattus *is the name for the Oceanian or European type 2N = 38/40 group and R. tanezumi for the Asian type 2N = 42 form. The two are only distinguishable by biochemical features and to a lesser degree some morphological traits. Both forms can occur without inbreeding (e.g. Fiji), but hybridise in the laboratory (usually sterile offspring) and on some South Pacific Islands (e.g. Chichijima and Eniwetok). Where the Asian type is indigenous the Oceanian form is restricted to ports or ships in harbour. For details of the Asian form see R.* tanezumi *in the systematic list of this volume; for details of their taxonomy see Baverstock* et al. *1983, and Wilson and Reeder 1993.*

■ DESCRIPTION

HB males 165–254 mm; T 140–252 mm; WT 85–350 g.
Sleek, graceful appearance; fur glossy blue-black,
black or brown on upper parts and slate grey, light
grey, buff or whitish on underparts; ears and eyes
large, ears naked; muzzle pointed, snout black; tail
black or brown, almost hairless, with fine scale, at
least the length of snout to end of body and more
than 110 mm. Female similar to male but with four to
six pectoral and six inguinal mammae; feet with five
toes.

■ DISTRIBUTION

Originally inhabited only the Orient (probably Indo-
Malayan region and extending to southern China),
but through introduction and colonisation now
cosmopolitan except for very cold latitudes.
Widespread throughout the southern Palaearctic
including Europe, southern Russia, South-west Asia,
North Africa, southern China and Japan; sporadically
further north, mainly in ports.

■ HABITS AND BEHAVIOUR

Habits: mainly nocturnal, but also diurnal; terrestrial
and arboreal; burrows; highly social, highly adaptable,
territorial. **Gregariousness:** lives in groups, each
dominated by male; density 5–52 rats/ha.
Movements: up to 200 m common; similar to *R.
norvegicus*. **Habitat:** almost anywhere except deserts;
human habitation favoured; likes orchards, gardens,
rivers, streamsides. **Foods:** omnivorous; beetles,
spiders, moths, stick insects, cicadas, fruits, birds'
eggs, stored products. **Breeding:** breeds throughout

year; gestation 21–30 days; polygamous; female polyoestrous; 5–6 litters per year; 1, 5–8, 12 young; nest
spherical of loose, shredded vegetation or other material; oestrous cycle 4–6 days; about 32 days between
litters; young weaned 21–28 days; may reach sexual
maturity at 3–4 months. **Longevity:** males 11 months,
females 17 months in wild; 3.4–4.2 years in captivity.
Status: abundant and widespread.

■ HISTORY OF INTRODUCTIONS

Black rats have been introduced widely throughout
the world, including Europe, Asia, North and South
America, Africa, Australia, and many Pacific, Indian
and Atlantic Ocean islands.

AFRICA

Black rats are known from Iron Age sites in Zambia
(AD 1500–1600), northern Transvaal (about AD 1000)
and Natal (eighth century) (Avery 1985). Whether
they colonised southwards to reach the Cape or were
introduced by European ships is not known, but the
latter seems more likely. However they arrived, the
species is now well established as a commensal in all
but the drier parts of South Africa as well as in many
other parts of Africa (de Graaf 1981).

Black rats have been introduced to Morocco (Allen
1954) and occur in Sudan, Iran and probably the
Congo.

South Africa

Black rats were introduced to South Africa, probably
with European colonists (Bigalke 1937). At Pirie in

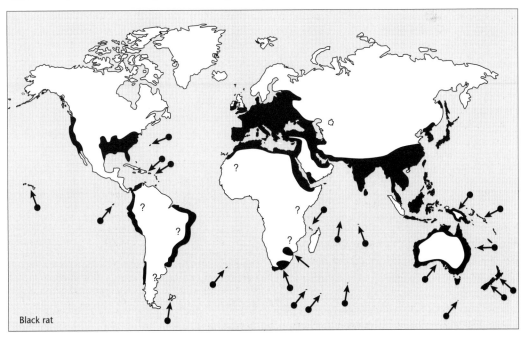

Black rat

Pondo Land in the eastern Cape Province they have been present since the 1930s (Shortridge 1934). They were reported to be widespread in 1937.

In South Africa black rats occur in all the larger coastal settlements and also in some inland areas, mostly as a commensal of man (Hey 1974; Smithers 1983; Lever 1985). They are present in Kruger National Park, Transvaal, and also in Hluhluwe-Umfolozi Game Reserve in Natal (Macdonald and Frame 1988). In South Africa there is no evidence of black rats affecting vegetation or populations of birds or mammals (Bigalke and Pepler 1991).

East Africa – Kenya – Tanzania
Black rats are believed to have penetrated inland areas of East Africa early in this century (Msangi 1975), but only fairly recently, 1976–79, to have reached Seronera village within the Serengeti National Park (Senzota 1982; Macdonald and Frame 1988).

Black rats have also been introduced in Kenya (probably from India), Tanzania and to Zanzibar island (Allen 1954). They are also common in Zimbabwe (Shortridge 1934).

Namibia
In the 1930s in Namibia (SW Africa) black rats were restricted to the larger towns and were only plentiful in the sea ports (Shortridge 1934).

Zambia
Apart from the archeological evidence in Zambia, records of black rats exist from the sixteenth century, but they still do not occur in the western areas (Lever 1985).

Asia
Black rats occur widely throughout Asia and are widespread and most numerous in India (Fitzwater 1967), where they are in most large towns near the coast (Jerdon 1874). They occur widely in Malaysia and Singapore (Dhaliwal 1961), are throughout Thailand and on all the adjacent islands (Lekagul and McNeely 1988). Here in these regions they are mainly present in areas of human activity including towns, villages, houses, rice fields, and oil plantations (Payne et al. 1985).

Black rats also occur throughout Burma and the Philippines and on Marcus island (Japan). They were undoubtedly introduced in the Lyallpur region in West Pakistan (Taber et al. 1967) and have been introduced in Far East of Russia (Yanushevich 1966).

Black rats were probably introduced from mainland Vietnam were recorded on Con Son Island in Con Son town in 1970 when 39 were trapped there (Van Peenen et al. 1970).

Atlantic Ocean islands
Aleutian Islands
Rats (species uncertain) were found on the Aleutians before 1939, but increased rapidly during World War 2 because of increased imports. They are widely established on Unalaska Island, even in some remote areas and have probably been there since the early 1800s; during the war they were such a problem that military forces instituted extensive control campaigns (Peterson 1967).

Together with the introduced fox, the rats are said to have had a greater effect on the flora and fauna of Amchitka than the war, and have nearly caused the extinction of the Atlantic Canada goose (Fradkin 1980).

Annobon Island
Rats (species not specified) were introduced to Annobon, where they are now plentiful (Fry 1961).

Ascension Island
The black rat was present on this island in 1701, but is not now as widespread as the Norway rat (Norman 1975).

Azores
Unspecified rats were on Terceira in the 1960s (Bannerman and Bannerman 1965–68) and they may have been present there since the islands were inhabited in about 1462.

Falkland Islands
Rats (species unspecified) are said to have arrived on the Falklands with humans (Cawkell and Hamilton 1961).

Madeira
Rats (species unspecified) were introduced to Madeira probably after settlement in 1420 (Encycl. Brit. 1970–80).

South Georgia
Both black and Norway rats were reported on South Georgia in the 1960s (Holdgate 1967; Watson 1975). They are thought to have been introduced to Port Olav Harbour at least by 1800 (Lever 1985).

St. Helena
Black rats were introduced to St. Helena, probably after 1420, and still occur there.

Tristan da Cunha
Black rats escaped from the shipwreck, of *Henry B. Paul* in 1882 (Brander 1940), but their establishment was certainly assisted by other later introductions (Hill 1959). Within a few years they were well established and were threatening the local agriculture. They were common on Tristan in 1938, but were not

present on Nightingale or Inaccessible islands. Their numbers decreased after 1961–63 when the residents left the island following a volcanic eruption (Anon. 1963).

AUSTRALASIA
Australia
Black rats occur on the coastal fringe around Australia and are in Tasmania, on Kangaroo Island, and on Lord Howe Island (Watts and Aslin 1981). They are common in urban environments and in some bush habitats to some extent (Wilson *et al.* 1992). (For the presence of rats on islands off Australia see the table under house mouse.)

Black rats were probably introduced to Australia when the early Dutch explorers sailed their ships along the western coast after 1616 (Archer 1984; Hand 1984; Long 1988). This suggestion is reinforced by the finding of rat skeletons in cannons raised from the reefs upon which many of them were wrecked. However, they may not have become permanent fixtures until the colonists set up permanent colonies in the 1770s.

Papua New Guinea
Black rats were introduced in the days of sailing vessels, probably in the last 100–150 years, and have since become widespread and pests of dwellings and agriculture crops in Papua New Guinea (Herington 1977; Flannery 1995). Before the 1970s they were found chiefly in association with humans in coastal and larger inland towns (Ryan 1972) and were reported to be more common than Norway rats. Today they are confined mainly to lowlands around human dwellings and cultivated areas (Flannery 1995).

Introduced on a number of Papua New Guinea islands, black rats are established on the islands of Bougainville (see under island entry), Buka, Manus, New Britain, Nissan, Sideia and Ulna (Flannery 1995).

EUROPE
Originally confined to the Orient, the black rat has invaded the western world. Through introductions and colonisation it now occupies most of Europe except for very cold latitudes. The species was thought to have migrated or been introduced into the western parts of Europe in the twelfth century, possibly with the navies or in the baggage of the returning crusaders or perhaps earlier in trading ships via the main ports of the times (Silver 1937; Fitter 1959; Matheson 1963; Lever 1977). However, more recently bones were discovered in Roman deposits suggesting a third or fourth century introduction pre-dating the Crusades by 800 years (Lever 1985). They may have spread from India to Egypt in the fourth century BC and from there along trade routes into Europe reaching Britain in Roman times (Corbet and Harris 1991). Certainly there is no fossil evidence in deposits before 2000–2200 BP (Armitage et al. 1984), although some island excavations reveal their presence during the fifth millennium BP (Sauges and Alcover 1980). In Egypt black rats appear to be known since 3500 BP (Armitage *et al.* 1984). They were introduced to Menorca (Spain) in the second century BC, at which time the species probably commenced its invasion of Europe (Reumer 1986). Recent archeological evidence shows the presence of rats in Italy and Poland in the Iron Age. New evidence suggests their existence in France in the first century AD and in Switzerland (Lac de Neuchâtel) in the late Bronze Age (eleventh century BC) which seems to be the oldest known proof in Europe (Roguin 1989). The epidemics of plague that struck Athens in 429 BC were due to black rats (Marcuzzi 1990).

Black rats disappeared in many areas of central Europe due to the use of anticoagulants, changes in building structures, shipping, the rural landscape and alterations to older buildings (von Bülow 1981).

Malta
Rats were introduced (both black rat and Norway rat) possibly in the time of the Carthoginians (eighth century BC) or with the Romans (after 218 BC).

United Kingdom
By the fourteenth century they were widely distributed in Britain (Lever 1977) and from the Crusades to the eighteenth century the black rat was the only species established there (Hinton 1933). Their numbers and range declined following the introduction of the brown rat in the eighteenth century and they were considerably less common by 1776 (Hinton 1933; Fitter 1959; Lever 1977). However, they reached the Orkney Islands about 1808, the Hebrides in the 1880s, and the Shetlands in 1904 (Lever 1977). By the end of the nineteenth century they were reported to be nearly extinct in Britain, but became widespread again around the 1930s (Hinton 1933).

The decline in numbers of black rats continued in the 1950s in Britain. In 1951 they were well established in London and in about 40 localities outside it. By 1956 they had decreased in most areas and had disappeared from some (Bentley 1959). By this time also the species had become generally less common in Europe where it was reported to have been ousted by the Norway rat. In Britain at this time they were known to exist for several years on Lundy Island in the Bristol Channel and were also present on Sark and probably Alderney in the Channel Islands (Southern 1964).

However, the decrease continued and the species was present in only 28 localities by 1961 (Bentley 1964). By 1966 black rats were present in only a few localities in areas of the major ports and warehouses and this situation has remained into the late 1970s (Corbett 1966; Lever 1977).

The black rat is now absent except for isolated urban populations in British Isles, Denmark, Scandinavia and Finland (Corbett 1966). Formerly it was widespread but now is confined to a few sea ports (Baker 1990). Today on offshore islands in the British Isles they are confined to docks, warehouses and some major seaports, and are declining (Lever 1985).

INDIAN OCEAN ISLANDS
Aldabra Islands (Seychelles)
Rats (species unspecified) were numerous on the island of Astove in 1895, having been introduced early in the eighteenth century (Bayne *et al.* 1970).

Amirantés
Black rats occur on the island of Desroches in the Amirantés.

Amsterdam
Rats were established on the island of Amsterdam in the 1970s (Watson 1975).

Assumption Island
Rats (species not specified) were abundant before settlement began (before 1906) and were said to be destroying birds' eggs (Stoddart *et al.* 1970).

Chagos Archipelago
Rats (species not specified) occurred on Egmont Atoll in the 1840s and were present on Diego Garcia in 1884 (Bourne 1971). They were probably on Diego and other islands by 1813 (Bourne 1971).

Both black and Norway rats have been collected on Diego. Black rats were not common in the Port of Spain wharf area in the 1890s and were later replaced by Norway rats, but were fairly common elsewhere (Urich 1931). Black rats still occurred there in the 1970s (Hudson 1975).

Crozet
Rats were introduced to Crozet at the time of visits by American and other sealers during the nineteenth century (Watson 1975).

Kerguelen
Introduced by whalers in the nineteenth century, black rats occurred around Port Jeanne d'Arc and perhaps elsewhere in the 1950s and 1960s (Holdgate and Wace 1961).

Madagascar
Rats (species unspecified) occur on L'île Europa (Atoll in the Moçambique Channel) and are still numerous there (Malzy 1966).

Mauritius
Rats appear to have been an early introduction to Mauritius. Black rats were introduced before the Dutch arrived in 1598 and probably with the Arabs in about 1500. Rats are recorded on Mauritius in 1606 and were probably introduced there some time earlier by Portuguese visitors. They were introduced to Round Island in the nineteenth century and are still present there (North and Bullock 1986).

Nicobar Islands
Black rats have been introduced and are present on these islands.

Providence Island (Seychelles)
Black rats occur on the island of Providence.

Rèunion
Rats were probably an early introduction to Rèunion (Encycl. Brit. 1970–80) where they are still present.

Seychelles
Rats were introduced some time after the French settlement was established (on Mahe) in 1770, although the islands were known in the twelfth century (Encycl. Brit. 1970–80). Rats now occur on most of the Seychelles islands.

No rats occurred on Bird Island until 1967, when it was thought they were imported with a consignment of leaves for thatching from Praslin. They increased in numbers rapidly and by 1972 the entire island was infested, but there was no evidence that they were affecting the resident sooty tern colony (Feare 1979).

St. Paul
Rats were numerous on St. Paul in 1874 (Jeannel 1941) and have remained so (Holdgate and Wace 1961), but the identity of those present was by no means clear. Both black and Norway rats were present in the 1960s and black rats were well established there in the 1970s (Holdgate 1967; Watson 1975).

INDONESIA
Irian Jaya
Black rats have been introduced and established on the islands of Bantanta, Biak-Supiori, Japen, Numfoor, Salawati and Waigeo (Flannery 1995).

Maluku
Black rats have been introduced and established on the islands of Ambon, Aru Islands, Bisa, Buru, Halmahera, and the Kai Islands (Flannery 1995).

North America

Canada

The black rat reached Canada on the ships of the early explorers. They were established in Halifax, Nova Scotia, in the early nineteenth century, were exterminated prior to 1861, but occurred on Graham Island, Moresby Island, Kunghit Islands, Queen Charlotte Islands, Burnaby Island and Sandpit Island in the late 1950s (Cowan and Guiguet 1960).

Although less common than the Norway species, black rats were well established on the Queen Charlotte Islands and near Vancouver, and on Vancouver Island in the 1970s (Carl and Guiguet 1972).

At present they appear to be only established on the west coast of British Columbia – Fraser River Delta and southern Vancouver Island and on the Queen Charlotte Islands (Howard and Marsh 1976; Banfield 1977).

United States

Various dates have been advanced for the introduction of the black rat into North America. It is likely that they were introduced from the first vessels reaching these shores (Silver 1937, 1941). Some say they arrived in 1544 (Palmer 1898) and others in 1609 (Smith 1612). They are reported to have reached California in 1851 (Silver 1927). By the mid-nineteenth century they were in San Diego and Humboldt Bay, California (Lever 1985).

Numbers of black rats declined following the introduction of the Norway rat that drove it out of many areas (Silver 1941). At least three subspecies were introduced to North America which have now interbred and it is now impossible to designate any of them (Hall 1981).

Both Norway and black rats occur in major urban centres in the Central Valley in California. The black rat occurs in riparian situations eastward into the Sierra Nevada foothills; Norway rats occur in rice fields and throughout the Sacramento River delta; and both species occur in the San Francisco Bay area, in salt marshes, along dykes and on many islands such as Brooks Island (Lidicker 1973).

Mexico

Black rats are present throughout Mexico, where they cause much damage (Navarrete 1978).

Pacific Ocean islands

Black rats are widespread in Polynesia south to New Zealand (Flannery 1995), but are rare away from grossly disturbed habitats in eastern Melanesia.

Campbell Island

Rats were reported to have been introduced to Campbell Island before 1883 (Holdgate and Wace 1961), but the species was not specified. However, black rats were the main rat species present in the 1960s (Watson 1961; Holdgate 1967).

Caroline Islands

Black rats are present in the Caroline Islands (Storer 1962). They are present on most atolls in the eastern Carolines where they occurred as early as 1912 (Marshall 1975).

Chatham Islands

Black rats are reported present on this island (Atkinson and Bell 1973; Anon. 1980).

Cocos Island (Île del Coco, Costa Rica)

Introduced rats (species unspecified) occur on Cocos Island, Costa Rica.

Eniwetok Atoll

On Eniwetok Atoll black rats probably arrived with American troops in 1944 (Berry and Jackson 1979).

Federated States of Micronesia

Black rats have been introduced and established on the island of Kosrae (Flannery 1995).

Fijian Islands

Accidentally introduced in the nineteenth century, black rats now occur on all the main islands (Turbet 1941). They are locally abundant on all the islands in agricultural, plantation, suburban and coastal areas (Pernetta and Watling 1978). In Fiji black rats are sympatric with *R. tanezumi* (Musser 1993).

Galápagos Islands

Black rats were introduced to Santiago Island about 1835, reached the Isabella Islands about 1891, Santa Cruz Island about 1934 and other islands at later dates. They were already established there when Darwin visited Santiago in 1835. Europeans discovered the islands in 1535 so the rat population could be 150–450 years old. The date of introduction to Pinzon is not known, but they were collected there in the 1890s and arrived on Santa Cruz between 1930 and 1934 (Clark 1980). They arrived on Baltra Island, north of Santa Cruz, with American airmen during World War 2 (Lever 1985).

Black rats are now present on the Isabela islands and the islands of Santiago, Baltra, Pinzon, Santa Cruz, Floreana and San Crisobal, as well as on several smaller islands (Eckhardt 1972; Clark 1980).

Guam

Black rats occur on Guam (Watson 1961).

Hawaiian Islands

It is usually assumed that black rats arrived in Hawaii with the first sailing ships in the eighteenth century (Tinker 1938), but it now seems as though the first stock may have come from Europe or North America on ships at a later date (Atkinson 1977). There are no records of them between 1840 and 1870 and the first specimen collected was in 1899. They probably arrived on Oahu between 1870 and 1890 and spread to other islands in the next 10 to 15 years.

Black rats were present in considerable numbers on Niihau in 1951 (Fisher 1951). The first report from Johnston Atoll was in 1962, but they probably escaped from ships or barges several years earlier and are now uncommon there (Kirkpatrick 1966; Amerson and Shelton 1976).

In 1966 black rats inhabited the islands of Hawaii, Maui, Lanai, Molokai, Oahu, Kauai and Midway. They escaped from ships in 1943 on the latter island and were primarily responsible for the extirpation of the Laysan finch and the extermination of the Laysan rail (Kramer 1971). Unknown species of rats occur on Lehua (Richardson 1963) and they have been present since at least 1938 on barren Kaula Island (Kramer 1971).

Howland Island (mid Pacific)

Black rats are present on Howland Island (Kirkpatrick and Rauzon 1986).

Juan Fernández

Black rats have been introduced to Juan Fernández (Carter et al. 1945).

Kuril'skiye Ostrova (Kurile Islands)

Middle island is inhabited by grey-coloured rats (presumably black rats), which have been introduced (Voronov 1963).

Lord Howe Island

Black rats first appeared in 1918 following the shipwreck of the SS Makambo and are now abundant on the island (Hindwood 1940; Recher and Clark 1974; Flannery 1995).

Macquarie Island

Black rats were probably introduced by sealers (Mawson 1943) in about 1820 (Holdgate and Wace 1961), although others say they did not arrive until early in the twentieth century (Taylor 1979), probably in stores and empty casks for the oil trade. The latter date appears more acceptable because black rats are not mentioned as being present between 1896 and 1900, and the first record appears to be in 1908 (Cumpston 1968).

Numerous and widespread on the island in the early 1840s some black rats were collected there in the late 1950s. They were present in the 1960s, were well established in the 1970s and widespread and abundant in the 1980s (Simpson 1965; Watson 1975; Jones 1977; Taylor 1979; King 1990).

Marshall Islands

Black rats occur in the Marshall Islands (Fall et al. 1971). They are present on Arno Atoll, where they are occasionally found in Ine Village, and occur on two islets on the north side of the island (Marshall 1955). They are also present on Majuro Atoll (Rowe 1967).

Marquesas

Black rats are present in the Marquesas (Watson 1961).

New Caledonia

Widespread throughout New Caledonia, black rats were introduced after the discovery of the islands by Europeans in 1774 (Nicholson and Warner 1953; Watson 1961). They are still established here (Flannery 1995).

New Zealand

The black rat probably arrived accidentally with Captain Cook and other early European visitors in the mid-eighteenth century and the early nineteenth century (Watson 1959; Watson 1961; Wodzicki 1965; Gibb and Flux 1973).

Black rats are now widely distributed in the North and South islands and occur on Stewart and some other offshore islands Barnett 1985). They occur on many southern islands, Foveaux Straits and reached Big South Cape Island in 1955 where they were numerous in 1962 and in plague proportions in 1964 (Atkinson and Bell 1973; Atkinson 1977).

Now black rats are the most widespread rat in New Zealand and will live in plantations of exotic trees (*Pinus radiata*) (Clout 1980).

Black rats occur on many islands off New Zealand including (see table) Macquarie Island, where they are abundant (Brothers et al. 1985) and were probably introduced with sealers or whalers at the end of the nineteenth century (Cumpston 1968; Copson 1986). They are also present on Stewart, Native, Pearl, Big South, Solomon, Pukeweka, Rosa and Chathams and Kawau (Wodzicki and Flux 1967; King 1990). They were successfully eliminated on Maria Island, Hauraki Gulf, by use of warfarin baits (Anon. 1980).

Niue

It is not known when black rats arrived on Niue, but probably between 1900 and the 1920s, and they were widespread there in 1969 (Wodzicki 1969).

Norfolk Island
Black rats have been introduced and established on Norfolk (Flannery 1995).

Solomon Islands
Black rats were introduced to Bougainville Island from US troop and supply ships during World War 2 and probably to Guadalcanal by escaping from US or Japanese ships. They are now well established around Honiara and are also well established on west coast of Malaita. On all of the Solomon Islands rat damage to cocoa, coconuts and garden crops has followed their introduction (Rowe 1967).

Previously unknown, black rats arrived on San Cristobal in 1987 following the shipwreck of a Taiwanese trawler in the early 1980s (Flannery 1995). At present black rats occur on the Florida Islands, Guadalcanal, Tömotu, Neo, Malaita, Russell Islands, San Cristobal, Santa Isabel and Vella Lavella (Flannery 1995).

Tuamotus
Black rats occur on Raroia Atoll in the Tuamotus (Watson 1961).

SOUTH AMERICA
In South America black rats tend to associate with human dwellings, but may extend far into forest regions, but generally occur near coastal settlements (Eisenberg 1989).

Brazil
Black rats occur in many areas in Brazil, and have been recorded (Holdgate 1967) on Trinidade Island, off the coast of Brazil.

Chile
In Chile black rats are relatively common in the central areas including the La Campana National Park region (Jaksic and Yanez 1979; Macdonald *et al.* 1988).

Peru
Arriving in ships of the explorers in about 1544, black rats are today found throughout the country (Lever 1985).

WEST INDIES
Antigua
Rats (species unspecified) have been present on Anigua ever since the arrival of Europeans in about the 1700s. They also occur on Bird Island off the coast of Antigua.

Bermuda
Arriving in Bermuda in 1613 on a captured Spanish grain vessel, within a year black rats were in plague proportions and carried a famine 1615–16. They are still common in rural areas (Lever 1985).

Barbados
Barbados was discovered in 1536, but black rats probably arrived with the first settlement in 1626. Rat control started in 1745, but they were still common and numerous in 1750, and had continued until the present time (Browne 1982; Lever 1985).

Guadaloupe
The first black rat found there was in 1956 on Grande Terre and they now occur throughout the island (Lesel and Derenne 1977).

Leeward Islands
Black rats arrived in the Leeward Islands with the early European colonists, and were certainly present in 1658 (Lever 1985).

Virgin Islands and Puerto Rico
Most of the islands that have rats have black rats. They are present on Cas, Congo, Dog, Buck, Capella, Stevens, and Little St. James. Most of the inaccessible islands have rats and they are reported to be common on Dog Island (Dewey and Nellis 1980).

On Puerto Rico black rats occur on Mona, Monito Islands, Puerto Rico and all islands, St. Thomas and adjacent islands, St. John and adjacent islands, all of the United States Virgin Islands (USVI) and Anegada and St. Croix and adjacent islands (Philibosian and Yntema 1977).

Black rats are believed to have been introduced to Trinidad in 1658 and shortly after to Puerto Rico (Pitmental 1955). They were certainly present in 1877 when the mongoose was introduced. They were the only rats present on Mona Island in 1960 (Pippin 1960). Most of the Virgin Islands are populated with black rats and in 1980, seven of 27 cays had them (Dewey and Nellis 1980).

Remire
Rats (species not specified) were present on this island in 1968 (Stoddart and Poore 1970). The island was discovered in 1770, but was uninhabited until at least 1882 when the rats may have arrived.

■ DAMAGE
One-fifth of the foodstuffs planted every year in the world is eaten or damaged by rodents. Rats damage crops, stored products, structures and materials. Rats eat 10 per cent of their weight each day or 9–18 kg per year, but contaminate much more with their urine and faeces. Pasteurella and murine typhus are transmitted to humans via the rat flea (*Xenophyllus cheops*). Leptospirosis and trichinosis from faeces of infected rats affect humans via foodstuffs and other animals (Howard and Marsh 1976). Most United States cities have as many rats as humans.

In Florida damage by *R. rattus* and other species reached US$95/2.2 ha in one grower's crop of cane (Lefebvre *et al*. 1978).

In the Hawaiian Islands it is suggested that the introduction and subsequent peaks of abundance of black rats coincided with the decline of endemic Hawaiian birds between 1870 and 1930, and that this was a major factor in their decline and extinction. Invasion of Midway Island in 1943 resulted in the decline and subsequent extinction of the Laysan rail and Laysan finches and points to this being the most recent step in the colonisation of the Hawaiian chain and extinction of many birds (Atkinson 1977).

Rats (*Rattus* spp.) cause damage to sugarcane in Hawaii and at population levels of 30 rats/2.2 ha, 29 per cent of the crop is damaged, the damage becoming appreciable as the crops mature (Hood *et al*. 1971).

Rat damage in the Hawaiian Islands has been recognised since the 1800s. Here, three species of rat are involved: the black, Norway and the kiore. They damage cane by chewing out a portion of the stalk internodes and such injury may kill the stalk or severely reduce the sugar production (Hood *et al*. 1971). The main damage is caused at the edges of fields and they generally reside in areas adjacent to the fields (Fellows and Sugihara 1977). Black rats also appear to be a major pest in macadamia orchards on the island of Hawaii (Fellows *et al*. 1978).

In parts of South America rice rats once lived in houses, but have been driven out by introduced black and Norway rats (Burton and Burton 1969). Black rats cause severe damage to coconut crops in Colombia by biting through the shells of coconuts, causing them to fall and not be available for harvest. As much as 24–77 per cent of the crop can be lost in this way (Valencia 1980). In Mexico, Norway rats cause damage to farms, animal husbandry activities and stored materials in marginal inter-urban rural areas near populated areas (Navarrete 1978).

Norway and black rats are two principal reservoirs of typhus in the Carribean (Pippin 1961). Following the introduction of rats and the mongoose on Antigua, the Antiguan racer (*Alsophis antiguae*) has become extinct and now occurs only on Bird Island where it has been introduced.

On islands in the Galápagos where black rats were introduced, rice rats (*Oryzomys* spp.) have become extinct (Burton and Burton 1969; Eckhardt 1972). Wherever black rats came in contact with the native species of rice rats the native species have suffered, and on Santiago, Santa Cruz and San Cristobal the native

species are now extinct (Clark 1980). The nesting success of dark-rumped petrels (*Pterodroma phaeopygia*) doubled during three years of rat control in the Galápagos (Cruz and Cruz 1987). The mockingbird (*Nesomimus trifsciatus*) has been exterminated by black rats on Floreana (Curry 1985) and on Pinzon they have killed every tortoise hatched in the wild in the past 100 years and also eat the eggs and hatchlings of birds and sea turtles (Benchley 1999).

On many other Pacific islands black rats have caused problems. On Majuro Atoll in the Marshall Islands and in the Gilbert Islands rat damage to coconuts been has been severe since 1945 (Rowe 1967). On Okinawa they damage sugar cane and in some years the losses are severe and estimated at 20–60 per cent of production (Udagawa 1970). On Raroia in the Tuamotus and on the Marquesas the black rat is replacing the maori rat, but on Guam, New Caledonia and Hawaii they live together and in the latter two places also with the Norway rat (Watson 1961). The blue-crowned lory (*Vini australis*) may have been exterminated by *R. rattus* and feral cats (Rinke 1987).

Following their arrival on Lord Howe Island in 1918, at least five species of endemic birds and the California quail were probably exterminated by black rats (Recher and Clark 1974; Newsome and Noble 1986). Two birds, the southern boobook owl and barn owl, were taken to Lord Howe Island in the hope that they would eliminate the black rat (Hindwood 1940).

However, black rats on Macquarie Island ate mostly plant material (Copson 1986). In this case the rats were not affecting birds, although this may indicate a failure to recognise eggs in the diet of the rat.

Predation by black rats has reduced the numbers of certain birds (Hill 1959) on Tristan da Cunha, where they have been responsible for the extermination of the endemic flightless rail and have caused a great reduction in the petrel colonies (Hagen 1952). On Amsterdam Island they have also caused damage to bird populations (Aubert de la Rue 1955).

After their introduction to Big South Cape Island (near Stewart Island) in New Zealand in 1955 black rats reached plague proportions by 1964 and vastly reduced bird numbers compared with 1961; lists show at least a bat and eight species of birds have been reduced or eliminated including bush wren (*Xenipes longipes*), snipe (*Coenocorypha auklandica*), robin (*Petroica australia*), fernbird (*Bowdleria punctata*), and brown creeper (*Einschia novaeseelandiae*), and most saddlebacks (Atkinson and Bell 1973; Anon. 1980).

Some authorities (e.g. Norman 1975) suggest the rat's reputation as a predator of birds on islands was based on supposition not always confirmed by food studies conducted on islands where birds and rats co-exist. It was thought the role of rats as predators was overestimated and not supported by direct observation, and that the roles of humans and other predators (dogs, cats, etc.) were generally ignored.

In Indonesia the black rat is the cause of damage to rice crops, particularly in the milk to maturity stages (Marsh 1965). Rats also cause damage to coconuts in Philippines (Fieldler *et al.* 1982) and in Nigeria became a pest of poultry houses and commercial rabbitries (Funmilayo 1982).

HIMALAYAN RAT
Rattus nitidus (Hodgson)

■ DESCRIPTION
HB c. 177 mm; T c. 168 mm; WT 122–200 g.
Short woolly fur, soft, reddish or brownish grey; under parts dull grey; nose long; tail thick, dark; feet white. Intermediate species between *R. rattus* and *R. norvegicus*.

■ DISTRIBUTION
Asia. Northern India (Assam, Bhutan, Sikkim, and Kumaun), Bangladesh, Nepal, to southern China, Hainan Island, Vietnam, Laos, northern Thailand, Burma, Palau, and Luzon in the Philippines.

■ HABITS AND BEHAVIOUR
Habits: climbs well; noisy. **Gregariousness:** no infor-

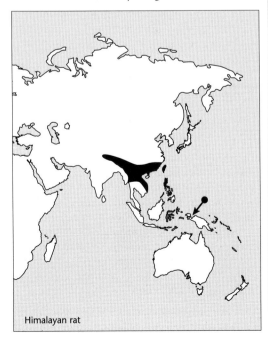

Himalayan rat

mation. **Movements:** sedentary(?). **Habitat:** houses of hill tribes, gardens. **Foods:** no information. **Breeding:** litter size 6 young. **Longevity:** no information. **Status:** common(?).

■ HISTORY OF INTRODUCTIONS
ASIA
Indonesia
In Irian Jaya the Himalayan rat is known only from the Vogelkop area and is found in association with human occupation. Its introduction is unknown, but it appears to be a recent immigrant, perhaps arriving about 400 years ago on Asian trading vessels (Flannery 1995). They may also have been introduced into the Celebes (Lever 1985).

Records from Sulawesi, Luzon Island in the Philippines, Seram Island in the Moluccas, the Vogelkop Peninsula of Irian Jaya, and the Palau Islands are likely to represent early introductions by human agency (Musser and Holden 1991; Wilson and Reeder 1993).

■ DAMAGE
No information, but probably a nuisance around human habitation, as are the black and Norway rats.

LARGE SPINY RAT
Rattus praetor (Thomas)

■ DESCRIPTION
HB 157–245 mm; T 144–181 mm; WT 164–240 g.
Variable species, large, spiny, lacks mottled tail; colour grey to reddish.

■ DISTRIBUTION
Indonesia–Australasia and Pacific Ocean islands. Northern New Guinea, Bismarck Archipelago, New Britain, New Ireland, Admiralty Islands, and to the Solomon Islands (Guadalcanal).

■ HABITS AND BEHAVIOUR
Habits: burrows. **Gregariousness:** no information. **Movements:** sedentary(?). **Habitat:** disturbed areas, offshore islets. **Foods:** no information. **Breeding:** throughout year; litter size 2–7. **Longevity:** no information. **Status:** common.

■ HISTORY OF INTRODUCTIONS
AUSTRALASIA–PACIFIC OCEAN ISLANDS
Papua New Guinea
Large spiny rats have been introduced and established on the islands of Bougainville, New Ireland, and prehistorically introduced on Bat, Manus, New Britain, and also introduced to Buku (Flannery 1995). They were probably introduced prehistorically into

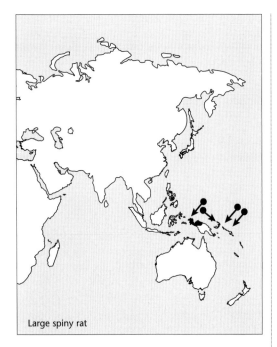

Large spiny rat

Asian house rat

most of their insular distribution, reaching New Ireland by 3500 years ago and the Solomon Islands soon after that. Recent archeological investigations in New Ireland suggest that they were prehistorically introduced, either accidentally or deliberately, into the eastern insular part of their present range during the past 5000 years (Flannery and White 1991; Flannery 1995).

Solomon Islands

Introduced and established on Choiseul, Guadalcanal, Tikopia and Nissan, but are uncommon and most are found in very disturbed habitats or on offshore islets. Only fossils have been found on Tikopia and Nissan, and no live animals have been recorded (Flannery et al. 1988; Flannery 1995).

■ DAMAGE

No information.

ASIAN HOUSE RAT

Rattus tanezumi Temminck

Similar in size, appearance, and ecology and only recently been recognised as a distinct species (Musser and Carleton 1993) from the black rat, Rattus rattus. *See notes under R.* rattus *this volume.*

■ DESCRIPTION

Similar in size and appearance to the black rat (*Rattus rattus*). It belongs to the $2N = 42$ group that is distinguished from the black rat $2N = 38/40$ complex.

■ DISTRIBUTION

Asia. From eastern Afghanistan to Nepal, northern India to southern and central China Korea, and south to the Isthmus of Kra and Hainan; also probably Mergui Archipelago, Andaman Islands, Nicobar islands and south-west peninsular India. Whether native or introduced to Japan and Taiwan is uncertain, and is probably introduced to the Malay Peninsula and some Sunda Shelf islands.

■ HABITS AND BEHAVIOUR

Similar to the black rat.

■ HISTORY OF INTRODUCTIONS

Asian house rats have possibly been introduced to Taiwan and Japan and are most likely introduced to the Malay Peninsula and islands on the Sunda Shelf and to nearby archipelagos just off the shelf, including the Mentawais. They have certainly been introduced to the Cocos-Keeling Islands, the Philippines, Sulawesi, and numerous islands east through the Moluccas and Nusa Tenggara to western New Guinea, Guam, Marianas, and farther east through Micronesia to the islands of Eniwetok and Fiji, and the Federated States of Micronesia.

ASIA

Indonesia

Asian house rats have been prehistorically introduced on Batjan and introduced to the islands of Obi, Sanana and Ternate in Maluku (Moluccas) (Flannery 1995). They have also been introduced to Sulawesi (Musser and Holden 1991) and numerous islands east through

Maluku and Nusa Tenggara (Musser 1970, 1981) to western New Guinea (Sody 1941). They have also been introduced to many islands on the Sunda Shelf (Medway and Yong 1976) and nearby archipelagos just off the shelf, including the Mentawais (Musser and Califia 1982; Musser and Newcomb 1983).

Philippines
Asian house rats have been introduced to the Philippines (Musser 1977)

MICRONESIA–PACIFIC OCEAN ISLANDS
Introduced in Micronesia as far as Eniwetok Atoll and Fiji (Musser 1993), Asian house rats have been present in this region for over 1000 years and pre-date the appearance of the kiore (*Rattus exulans*) there (Flannery 1995). Bones have been recorded from archeological sites on Fais Atoll, Pagan, Rota, Guam, Ngulu Atoll, Chuuk (formerly Truk), Nukuoro Atoll, and Ant Atoll. Their presence in Micronesia appears to have prevented the establishment of the black rat (*R. rattus*) in the region (Johnson 1962).

Federated States of Micronesia
Asian house rats have been introduced and established on the islands of Kosrae, Pohnpei and Uithi (Flannery 1995).

Guam
Asian house rats have been introduced and established on Guam (Flannery 1995).

Marianas
Asian house rats have been introduced and established on Rota, Saipan and Tinian in the Marianas (Flannery 1995).

■ DAMAGE
No information.

GREATER STICK-NEST RAT
Leporillus conditor (Sturt)

■ DESCRIPTION
Upper parts yellowish brown to grey; under parts creamy white; fur fluffy; hind feet with distinctive white markings on upper surface; ears long; eyes large; snout blunt; tail dark brown above and light brown below, usually shorter than body.

■ DISTRIBUTION
Australia. Extinct on the mainland and now exists only on Franklin Island, South Australia.

■ HABITS AND BEHAVIOUR
Habits: communal nests, 1 m high – 1.5 m diameter.
Gregariousness: nests may contain 10–20 animals.

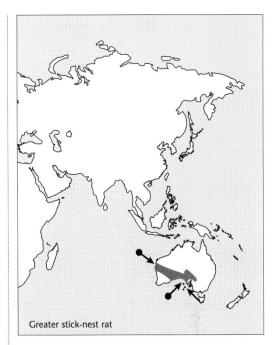

Greater stick-nest rat

Movements: sedentary(?). **Habitat:** Semi-arid to arid perennial shrublands, supporting succulent and semi-succulent plants. **Foods:** herbivorous; leaves, fruits. **Breeding:** throughout year, with a peak in autumn and winter; oestrous cycle 14 days; gestation 30 days; litter size 1–4; lactation *c.* 1 month. **Longevity:** no information. **Status:** rare, limited range.

■ HISTORY OF INTRODUCTIONS
AUSTRALASIA
South Australia
In 1990–92 greater stick-nest rats were released (101 animals) on Reevesby Island in the Sir Joseph Banks group in South Australia. In 1991–92 (two releases), 24 animals were re-introduced to the Yookamurra Sanctuary and in 1993–94 some (101 animals) were released on St. Peter Island in the Nuyts Archipelago Conservation Park, South Australia (Pedler and Copley 1992; Copley 1995; Strahan 1995). All three populations have established successfully. They are now widespread and common on Reevesby Island, increasing in numbers and continuing to expand their range on St. Peter, but have failed to become established at Yookamurra, primarily due to predation from foxes and birds of prey.

In 1998, 101 greater stick-nest rats were introduced to a fenced enclosure at Roxby Downs. By early 2001 the population had grown to 200–300, despite mortalities due to severe heat waves during the two preceding summers.

Western Australia

In 1990, 41 animals from Franklin Island, South Australia were released on Salutation Island in the Small Islands, Shark Bay Marine Park in Western Australia. In 1999–2000, 31 (12 male, 19 female) stick-nest rats taken from Salutation Island were released on Heirisson Prong, Shark Bay. Both populations established but only the Salutation Island population is expanding. Predation by large goannas (*Varanus gouldii*) and cats appears to be a factor on Heirisson Prong.

■ DAMAGE
None.

PLAINS RAT
Pseudomys australis Gray

■ HISTORY OF INTRODUCTIONS
AUSTRALASIA
South Australia
Re-introduced to the Yookamurra Sanctuary in 1991 (167), 1992 (38), 1993 (14) and 1994 (55), but since this time there have been only occasional sightings of plains rats (Copely 1995).

■ DAMAGE
None.

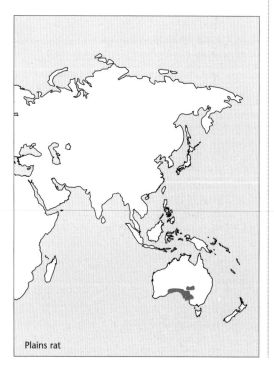

Plains rat

WESTERN PEBBLE-MOUND MOUSE
Pseudomys chapmani Kitchener

■ HISTORY OF INTRODUCTIONS
AUSTRALASIA
Western Australia
Attempts have been made by fauna authorities to translocate pebble-mound mice in Western Australia at four times between 1995 and 1997, with little success.

SHARK BAY MOUSE
Pseudomys fieldi (Waite)

■ HISTORY OF INTRODUCTIONS
AUSTRALASIA
Western Australia
In 1993–98, 149 Shark Bay mice collected from Bernier Island or captive-bred at the Perth Zoo were released on Doole Island, in the Exmouth Gulf. This population has persisted at low numbers. In 1994, 31 mice were translocated from Bernier Island to Heirisson Prong, Shark Bay. This population appears not to have established due to predation by goannas (*Varanus* spp.). In 1997, 26 Shark Bay mice captive bred at the Perth Zoo were introduced to North-west Island in the Montebello Island group, where they have established on one small part of the island.

LAKELAND DOWN MOUSE
Leggadina lakedownensis Watts
Includes the island form known as the Thevenard Island mouse.

■ HISTORY OF INTRODUCTION
AUSTRALASIA
Western Australia
In 1966, 65 (31 male and 34 female) Thevenard Island mice were successfully translocated from this island to Serrurier Island Nature Reserve where they have established successfully. The translocation was made to ensure that the island form was not lost when the house mouse (*Mus musculus*) was accidentally introduced to Thevenard Island.

LARGE BANDICOOT RAT
Indian bandicoot rat, Indian mole-rat, great bandicoot
Bandicota indica (Bechstein)

■ DESCRIPTION
HB 160–360 mm; T 140–260 mm; WT 545–1132 g.

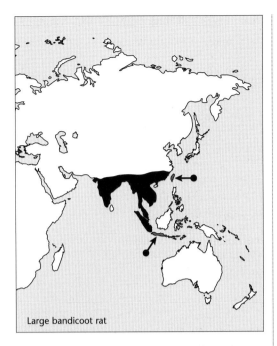

Large bandicoot rat

Rat-like with upper parts light greyish, or brownish to black; under parts whitish; pelage texture varies from dense to coarse fur; muzzle short and broad; tail scantily haired; incisors yellow or orange. Female with 12–18 mammae.

■ DISTRIBUTION

Southern Asia. From India, Sri Lanka and eastern Himalayas to Burma, Indochina and southern China, Hong Kong, Taiwan, and south to Sumatra and Java; also in Perlis and Kedah in western Malaysia.

■ HABITS AND BEHAVIOUR

Habits: mainly nocturnal; nests in burrows with 2–6 entrances and 13 cm diameter; stores food; commensal of man. **Gregariousness:** solitary. **Movements:** sedentary(?). **Foods:** nuts, fruit, grain, tubers and cultivated plants including tapioca, rice, sugar cane. **Habitat:** paddy fields, lowland areas, vegetable plots. **Breeding:** all year; litter size 2–10, 12. **Longevity:** no information. **Status:** common; abundant some areas.

■ HISTORY OF INTRODUCTIONS

Because of their commensal nature and deliciousness to eat, the large bandicoot rat may have been spread by humans in comparatively recent times (Marshall 1977; Wilson and Reeder 1993) to Malaysia, Java and Taiwan, and perhaps other areas.

Asia
Java (Indonesia)
Large bandicoot rats have been introduced and established on the island of Java (Musser and Newcomb 1883; Wilson and Reeder 1993).

Malaysia
Large bandicoot rats were found present in the Kedah and Perlis regions of Malaya in 1946, and may have been spread there by humans in recent times (Lekagul and McNeely 1988; Wilson and Reeder 1993).

Taiwan
Large bandicoot rats (*B. i. nemorivaga*) were introduced by the Dutch and became established in Taiwan (Kaburaki 1940; Walker 1968; Wilson and Reeder 1993).

■ DAMAGE

In southern Nepal large bandicoot rats are a serious problem in some houses, causing considerable damage to stored food stuffs (Chesemore 1970). They are a serious problem in rice and cane fields in China (Deng and Wang 1984). Not only do they spoil grain, but they steal food for their own larders and this makes them a serious pest in agricultural areas (Whitfield 1985).

CAIRO SPINY MOUSE
Acomys cahirinus (Desmarest)

■ DESCRIPTION
HB 90–125 mm; T 80–120 mm; WT 30–86 g.
Fur spiny; upper parts yellowish brown; under parts whitish; ears large; tail long.

■ DISTRIBUTION
Africa. Mauritania across the southern edge of the

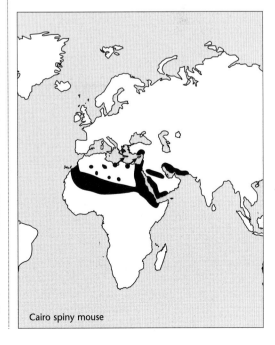

Cairo spiny mouse

Sahara, through much of Egypt and Arabia, north through Palestine to southern Asia Minor and east through southern Iran to Sind; also on Crete and Cyprus; also in Sudan and much of East Africa.

■ HABITS AND BEHAVIOUR

Habits: nocturnal and diurnal. **Gregariousness:** no information. **Movements:** sedentary(?). **Habitat:** desert, savanna, scrub, agricultural land, rocky hillsides, olive groves. **Foods:** seeds, snails, insects. **Breeding:** all year; litter size 1–5; several litters/year; young born naked. **Longevity:** 2–3 and up to 5 years. **Status:** no information.

■ HISTORY OF INTRODUCTIONS

EUROPE

Cyprus and Crete

The Cairo spiny mouse has been introduced successfully to Cyprus and Crete (Burton 1976). They became established as a commensal of humans on Crete and possibly on Cyprus (Lever 1985; Burton 1991).

■ DAMAGE

No information.

HOUSE MOUSE

Common mouse or house mouse

Mus musculus Linnaeus

A revision of the European Mus *has split the taxon previously known as* Mus musculus *into five species or semi-species including* M. domesticus *Rutty and* M. musculus *Linnaeus*

(see Bonhomme et al. *1984). Because of the difficulties in distinguishing between any of them they are here treated as a single species.*

■ DESCRIPTION

Males HB 72–98 mm; T 65–95 mm; females HB 70–102 mm; T 70–91 mm; WT of both 8.5–41.5 g.

Fur brownish grey, but varies from light brown to dark grey; under parts paler, often creamy; muzzle pointed; ears pointed; eyes large; tail semi-naked and as long as head and body.

■ DISTRIBUTION

Originated from an area near Iran and the USSR border in central Asia, but at least from southern Europe and North Africa to China and Japan. In southern Asia from Pakistan to Java, Lombok, Sumbawa and Flores.

■ HABITS AND BEHAVIOUR

Habits: mainly nocturnal, also diurnal; burrows; commensal; agile, climbs but usually forages on ground. **Gregariousness:** either territorial or colonial; dominance hierarchy or groups or shared territories, or individual territories; density 0.55–3.3 per ha. **Movements:** sedentary; irruptive; home range indoors 3.8–6.0 m, outdoors 10–20 m. **Habitat:** found in every habitat; forest, grasslands, semi-desert, and particularly houses, stores, factories, rubbish tips, farm buildings and human habitation generally. **Foods:** omnivorous; especially stored products including cereals, lard, butter, bacon, nuts, meats, chocolate, sweets; both invertebrates and plant mate-

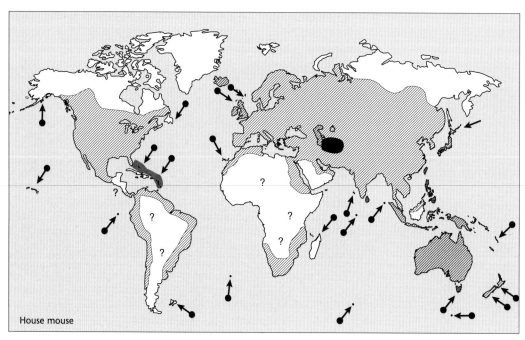

House mouse

rial, seeds, insect larvae, leaves, spores, annelids, arthropod eggs, lizard and bird carrion, grain. **Breeding:** throughout the year; gestation 13–31 days; 6–7, up to 11 litters per year; 1, 6–12 young; polyoestrous, mature 8 weeks of age; oestrous cycle 4–6 days; post-partum fertilisation; newborn naked, eyes closed, become furred in 14 days; eyes open 5–7 days; weaned 21 days; young leave nest 20–23 days; sexual maturity 50–60 days. **Longevity:** 1–3 years in wild and 6 years as captive. **Status:** abundant and widespread.

■ HISTORY OF INTRODUCTIONS
By colonisation and introduction, house mice are now distributed worldwide including the Antarctic.

AFRICA
South Africa, Namibia, North Africa, Egypt and Sudan
The origin and date of arrival of house mice in southern Africa is unknown (Bigalke and Pepler 1991). They are now widespread in the Mediterranean zone of South Africa and elsewhere, but may be only feral in Botswana (Smithers 1971). They occur on the island of Zembra, off Tunisia, in the Mediterranean (Vigne 1988). They are reported as relatively uncommon in central Africa and appear to have been unknown in Uganda and Zambia until the mid-1960s (Lever 1985).

Mice are widely distributed in Africa, particularly the more settled parts of South Africa and Zimbabwe, are plentiful in Namibia in towns and farming districts and occur all over North Africa, Egypt, Morocco and the Sudan (Shortridge 1934; Bigalke 1937; Southern 1964; Smithers 1983).

Mice have been introduced on Marion Island (in Prince Edward Islands, South Africa), where they are common and were probably first introduced by sealers (Smithers 1983).

ASIA
Irian Jaya
House mice are reported to be established on the island of Japen (Flannery 1995).

Hong Kong
House mice have been introduced to Hong Kong, although an indigenous race (*M. musculus bactrianus*) probably occurred there previously (Marshall 1967).

Kuwait
Present (Ali-Sanei *et al.* 1984).

Maluku
Reported to be established throughout Maluku, house mice have been introduced on the island of Buru (Flannery 1995).

Russian Federation
House mice are reported to have been introduced in the Far East, in the Russian Federation (Yanushevich 1966). They have been introduced on the Komandorskiye Ostrova (Commander group) off Kamchatka (see section under individual island).

West Pakistan
Probably introduced to the Lyallpur region in West Pakistan, house mice only inhabit human habitations (Taber *et al.* 1967). They were already widespread and 'indigenous' in early times (Jerdon 1874).

ATLANTIC OCEAN ISLANDS
Aleutian Islands (Bering Sea)
Mice occur in Unalaska village on Unalaska Island and probably have been present there since the early 1800s (Peterson 1967).

Ascension Island
House mice occur on Ascension Island (Lever 1985).

Canary Islands
Introduced house mice occur in the Canary Islands, but no date of arrival was determined. They were probably an early accidental introduction by the many navigators who called at these islands.

Faeroe Islands
Although the introduction of mice is thought to have been about 250 to 1000 years ago, other records suggest a much more recent introduction. The first mice to reach there may have come from Scandinavia with the marauding Vikings (Lever 1985). Certainly they were present on Bordøy about 1838, but were later exterminated by rats in 1915–20; arrived on Hestur after 1939; not on Fugløy and Mykines until some time after 1800. The original establishment may have been on Sandøy and they may have spread from here to other islands although Hestur and Nolsøy may have been colonised from Streymøy (Berry 1981; Lever 1985).

Since their arrival on the Faeroes, house mice have arguably developed into four distinct subspecies (Bouliere 1954; Berry *et al.* 1978).

Falkland Islands (United Kingdom)
Mice arrived with humans on the Falklands, probably with French colonists in 1764 (Cawkell and Hamilton 1961; Lever 1985). They appear to have been recorded in 1774 in the settlement at Port Egmont and in 1842 at Port Louis. They are now widely distributed in areas of human habitation on the East and West Falklands and are on some of the more remote offshore islands including Steeple Jason (Lever 1985).

Gough Island
Introduced before 1887, mice still occur on Gough Island (Holdgate and Wace 1961). They are reported

to have been introduced about 1800, were widespread in the late 1950s (Hill 1959) and are now abundant there (Derenne and Mougin 1976).

Juan Fernández
House mice occur on Juan Fernández (Lever 1985) where they were introduced, probably by early settlers.

Komandorskiye Ostrova (Commander Islands, Russian Federation)
House mice were introduced accidentally in 1870 when they were off-loaded with a cargo of flour from San Francisco (Barabash-Nikiforov 1938; Lever 1985). They became established chiefly on Bering Island (Barabash-Nikiforov 1938).

Madeira
The house mouse was introduced probably some time after 1420 (Encycl. Brit. 1970–80).

South Georgia
During 1975–76 a British Antarctic survey field party based at the south side of Shallop Cove, Queen Maud Bay, reported the presence of house mice not previously recorded from this island (Watson 1975; Bonner and Leader-Williams 1976).

St. Helena
Mice are now present on St. Helena, but details of their introduction do not appear to be documented. They were most likely introduced with early seafarers or colonists.

Tristan da Cunha
The date of introduction for mice on Tristan is unknown (Holdgate and Wace 1961), but was probably by sealing vessels during the late eighteenth or early nineteenth century and subsequent introductions (Hill 1959). It is likely that they were introduced when the first permanent settlements were founded, and the British Garrison was stationed there in 1816. House mice are now abundant there (Derenne and Mougin 1976).

AUSTRALASIA
Australia
Probably arriving with the First Fleet in 1788, if not before on Macassan ships, mice now occur Australia-wide and are also in Tasmania and on numerous islands (see island list following), including Kangaroo Island, Flinders Island and King Island (Watts and Aslin 1981). The oldest specimen in the Australian Museum was lodged in 1841 (Mahoney and Richardson 1988).

Little is known of the colonisation of Australia by house mice. It is presumed that they extended inland from ports and hitch-hiked with humans and produce around the country (Redhead *et al.* 1991). Indeed there were probably multiple introductions from a number of well-used ports of the day. Some may have arrived before the earliest settlers by swimming ashore from shipwrecks along the coast (Wilson *et al.* 1992). They now exist in both a commensal and a feral state.

Rats and mice on islands around Australia

? date not known. * eradicated or extinct.

Island, state	Mouse	Black rat	Norway rat
Abrolhos, WA (see North, WA)			
Alpha, WA		? shipwreck	
Althorpe, SA	?		
Babel, Tas	?		
Badger, Tas	?		
Baird, SA	?		
Barrow, WA	? *	1st record 1990, *1991	
Bathurst, NT		?	
Bedout, WA		19th century; *1981	
Big Green, Tas		?	
Bluebell, WA		? shipwreck	
Boodie, WA		? *1985	
Boomerang, WA		? *1983	
Boullanger, WA	?		

Rats and mice on islands around Australia (*continued*)

Island, state	Mouse	Black rat	Norway rat
Bowen, NSW		<1976	
Boxer, WA		?	
Boydong, Qld		<1990	
Bribie, Qld	?		
Broughton, NSW		<1976, ? species	
Browse, WA	?		
Brush, NSW		<1974	
Burrup, WA	?	?	
Campbell, WA		? shipwreck	
Cape Barren, Tas	?	?	
Capricorn Group (see Heron)			
Carnac, WA	?		
Clonmel, Vic	?		
Culeenup, WA	?	?	
Deal, Tas		?	
Dirk Hartog, WA	?		
Dixon, WA	?		
Dog, Vic	?	?	
Doughboy, Vic		<1977	
East Kangaroo, Tas	?		
East Sister, Tas	?		
Fairfax, Qld		? species	
Faure, WA	?		
Figure of Eight, WA	?		
Fisher, Tas		1971, *1974	
Flinders, SA		?	
Flinders, Tas	?	?	
French, Vic	?	?	
Furneaux Group, Tas		?	
Garden, WA	1960s?	1960s?, *1991	
George Rocks, Tas		<1979	
Gidley, WA		?	
Goose, SA	?		
Great Dog, Tas	recorded 1988	<1988	
Great Glennie, Vic		?	
Green, Qld		?	
Griffiths, Vic		there 1980	there 1980
Groote Eylandt, NT		?	
Hermite, WA	?	*c.* 1966	
Heron, Qld		1926, ? species, *1964–65	
Howick, Qld		?	
Hummock, Vic	?		
Jeegarnyeejip, WA	?		
Kangaroo, SA	?	?	
King, Tas	*c.* 1887?	?	
Lindeman, Qld		?	

Rats and mice on islands around Australia (*continued*)

Island, state	Mouse	Black rat	Norway rat
Lion, NSW	?	? species, *1968	
Little Boydong, Qld		? species	
Little Broughton, Qld		? species	
Little Dog, Tas	recorded 1988	<1988	
Little Goose, Tas	?		
Long, Qld		?	
Long, WA		? species	
Lord Howe, NSW	?	shipwreck 1919	
Macquarie, Tas	?	1880–1908	
Maria, Tas	? there 1980s	?	there 1980s
Meeyip, WA	?		
Melville, NT		?	
Middle (Barrow), WA		?	
Middle Lacepede, WA		? *1986	
Mistaken, WA		there 1980	
Montagu, NSW	?		
Monte Bello, WA		late 1800s; *1996?	
Moreton, Qld	?		
Mount Chappell, Tas	?	?	
Mungary, SA	?		
Mutton Bird, NSW	?		?
Newry, Qld		?	
North (Abrolhos), WA	1970s		
North Bickers, SA	?		
North Double, WA		? *1983	
North Stradbroke, Qld	?	?	
Northwest, WA		? shipwreck	
Pasco, WA		? *1985	
Penguin, WA	c. 1920s		
Phillip, Vic	?	?	
Pigeon, WA		? *	
Prime Seal, Tas	?		
Primrose, WA		? shipwreck	
Quail, Vic	?		
Rat, WA	<1987	c. 1840, *1991	
Recherche Archip. (see Woody), WA			
Reevesby, SA	?		
Rocky, Qld		c. 1937	
Rotamah, Vic	?	?	
Rottnest, WA	?	? species	
Saint Francis, SA		fossil *	
Saint Margaret, Vic	?		
Saint Peter, SA	?		
Sandy (Lacepede), WA		? *1986	
Snake, Vic	?		
Snapper, NSW			1930s

Rats and mice on islands around Australia (*continued*)

Island, state	Mouse	Black rat	Norway rat
South Double, WA		? *1983	
South East (Montebello), WA		? shipwreck or *c.* 1951	
South Molle, Qld		? skull only	
Southport, Tas	recorded 1983		
Sugar Loaf Rock, WA	?		
Sunday, WA		?	
Tasmania	?	?	?
Thevenard, WA	1986		
Three Boys, WA	?		
Trefoil, Tas	?		
Trimouille, WA		? shipwreck or *c.* 1951	
Venus Bay Is, SA	?		
Wardang, SA	?		
West Lacepede, WA		? *1986	
Whitlock, WA	?		
Woody, WA		1950s?	
Wreck, Qld		? species	
Yunderup, WA	?		

References: Abbott & Burbidge 1995; Abbott 1981; Barry & Campbell 1977; Bowker 1980; Brothers & Skira 1987; Brothers 1983; Brothers & Skira 1988; Burbidge & George 1978; Burbidge & Prince 1972; Burbidge 1971; Butler 1975; Campbell 1888; Department Conservation & Environment 1978; Fullagar 1973; Fuller & Burbidge 1987; Garnett & Crowley 1987; Gibson 1976; Green & McGarvie 1971; Green 1969; Green 1979; Hope 1973; Jones 1977; Kikkawa 1976; Lane 1975; Lane 1976; McKenzie *et al.* 1978; Morris 1974; Napier & Singline 1979; Napier 1979; Norman 1970; Norman 1971, 1977; Raines 1985; Recher & Clarke 1974; Seebeck 1981; Serventy 1953, 1977; Skira & Brothers 1988; Storr 1976; Strahan 1983; Swanson 1976; Taylor & Horner 1973; Towney & Skira 1985; Watts & Aslin 1981; Whinray 1971; Young 1981.

Papua New Guinea

House mice probably arrived on European sailing ships and are now found chiefly in association with humans in coastal towns and villages, and some inland towns (Ryan 1972; Flannery 1995). Feral populations occur in grassland around Port Moresby and elsewhere (Menzies and Dennis 1979), and some have been noted at isolated settlements at Telefomin, Sandanin Province where they probably arrived by aircraft (Flannery 1995).

EUROPE

Mouse (*Mus musculus*) fossils are known from the Mindel/Riss interglacial (650 000 to 500 000 BP) in Hungary and Greece (Kurten 1968), and in Israel *Mus* fossils are common throughout the Upper Pleistocene (Tchernov 1984). Re-examination of fossil material has suggested that *Mus musculus* is a much more recent invader, arriving in the Levant not before the Natufian and perhaps even the Aurignacian epoch (22 000–12 000 BP) (Auffrey 1988). The arrival of

Mus musculus on the Mediterranean islands (probably arrived Corsica 1000–2000 BP, Sardinia 3000–4000 BP, Minorca 2000–3000 BP and Mallorca 1000–2000 BP) appears to have been the third millennium BP (Vigne and Alcover 1985), but in Sardinia it is quite possible they invaded in the Neolithic (Sanges and Alcover 1980). No house mice appear to be known in Europe before the Romans (Cheylan 1991), although they may have arrived a little earlier in the United Kingdom.

The house mouse probably originated as a wild species somewhere near the borders of Iran and the Russian Federation and gradually spread from there some thousands of years ago with the practice of agriculture (Davis and Rowe 1963).

Malta

Mice were possibly introduced with Carthoginians in the eighth century or early seventh century BC, or with the Romans about 218 BC.

Skokholm (Irish Sea)

Accidently introduced to Skokholm, Pembrokeshire, about 1903, house mice are now well established there (Fitter 1959). On this island the mice live without the benefits provided by agriculture (Davis and Rowe 1963). With little predation, population size and growth are controlled by the ability of the mice to survive adverse climatic conditions and perhaps making use of adaptations evolved long ago despite their recent history of commensalism (Berry 1968).

United Kingdom and Ireland

House mice have been so extensively transported that it is difficult if not impossible to trace the history of their spread (Corbet 1966). They have been in the British Isles since the Neolithic (Southern 1964), but possibly reached Britain in the pre-Roman Iron Age (tenth century BC–450 BC–AD 54) and may have been introduced to St. Kilda accidentally with the Norsemen from Scandinavia over 1000 years ago (Lever 1977, 1985). There is some thought that they may have arrived about 2000 BC with human colonists (Fitter 1959). They were on the island of St. Kilda at an early date, but died out after the settlement was abandoned in 1930. They also occur on islands such as the Hebrides, Fair Isle and Foula.

House mice are now widespread throughout the United Kingdom (Baker 1990), Ireland (D'Arcy 1988) and occur on Bull Island, Dublin.

INDIAN OCEAN ISLANDS
Amsterdam (Nouvelle Amsterdam)

House mice were possibly introduced early in the nineteenth century when fishermen from Réunion visited the island regularly or in 1949 when a permanent radio-meteorological station was established there. They were present in the 1950s (Reppe 1957) and were abundant there in the 1970s (Watson 1975; Bonner and Leader-Williams 1976; Derenne and Mougin 1976).

Andaman Islands

Now well established in the Andamans, mice were possibly introduced as early as the ninth century BC by Arabs (Encycl. Brit. 1970–80).

Crozet

Mice were introduced some time after the discovery of these islands in 1772. Although the island is unihabited, they still appear to occur there (Bonner and Leader-Williams 1976).

The date of introduction to the Isle aux Cochins (Hog Island) is not known and there is no mention of their presence in the course of the nineteenth century or at the beginning of the twentieth century (Derenne and Mougin 1976). They are not present on Possession or

East islands and appear to have only colonised the Isle aux Cochins in the Crozet Archipelago. Here they have been observed in many areas (north-east coast, Cape Deception, Cape Verdoyant, Bay of Aiguille) and appear to be throughout the periphery of the island to 300–350 m and periodically become abundant in localities close to the sea. It is estimated that there may have been 200 000 individuals on the island at times, but they have not had any effect on the bird life (Derenne and Mougin 1976).

Kerguelen

Introduced before 1874 (Holdgate and Wace 1961), mice were abundant there in 1875 (Kidder 1876). They were introduced about the time sealers and whalers were visiting the islands and now occur all over Grande Terre (Bonner and Leader-Williams 1976; Lesel and Derenne 1977).

Marion Island (Prince Edward Islands, South Africa)

Mice occur on Marion Island (Watson 1975; Bonner and Leader-Williams 1976) and were abundant there in the 1970s (Derenne and Mougin 1976).

Nicobar Islands

Present. Date of introduction and status not known.

Seychelles

Mice are present around the settlement on Bird Island (Feare 1979) and occur on the Seychelles itself (Lever 1985).

St. Paul

Mice were probably introduced to St. Paul early in the nineteenth century. They have flourished and remain numerous on the island (Holdgate and Wace 1961; Bonner and Leader-Williams 1976).

NORTH AMERICA
Canada–Alaska

In Canada the arrival of the mouse coincides with the arrival of the white man. They are now found throughout the north to Alaska and are on Vancouver Island, but not on the Queen Charlotte Islands. Certainly they are throughout southern Canada wherever there is human settlement and rural areas, and north to the Mackenzie Delta and the North West Territories (Cowan and Guiguet 1960). They occur on the Magdalen Islands Archipelago, Quebec, but only near human habitation. It is suggested that two subspecies were introduced – *brevirostris* and *domesticus*.

United States

First introduced from Stephen Harriman Long's expedition of 1819–20 in Iowa (Lever 1985), house mice were present in Manitoba in 1829. By 1855 they were found in many inland localities of the United

States including Kansas, Louisiana, South Dakota, California. They possibly reached Arizona in 1891 (Lever 1985). They have probably been in California for 200 years where they occur throughout and are commensal with humans, and are also found in agricultural areas (Lidicker 1991).

Mexico

House mice are present in Mexico (Navarrete 1978), certainly at Coahuila (Lever 1985), and probably throughout cities and towns.

PACIFIC OCEAN ISLANDS

House mice occur on most islands in the Pacific including the Marquesas (Carter *et al.* 1946). In Micronesia they occur on Hawaii and some islands and a few small atolls (Kirkpatrick 1966; Berry and Jackson 1979).

Antipodes

On the Antipodes mice were probably introduced with stores and were common there in 1907 (Taylor 1979). They were widespread in the 1970s (Gibb and Flux 1973).

Auckland Islands

Introduced to Enderby Island in the 1840s and main Auckland Island in 1820, house mice survived on both islands (Taylor 1968). They occur throughout Auckland Island and are still present on both (Gibb and Flux 1973; Dilks and Wilson 1979).

Belau

House mice have been introduced and established in the Palau Islands (Flannery 1995).

Caroline Islands

Mice were recorded on this island group in 1935 (Tate 1935) and probably still exist there.

Eniwetok Atoll (Marshall Islands)

It is not known how long house mice have been on this island, but some used in atomic tests may have escaped and formed or contributed to the present population (Berry and Jackson 1979).

Federated States of Micronesia

Hose mice have been introduced and established on Pohnpei island (Flannery 1995).

Fiji

Introduced probably by the Wilkes Expedition in 1840 (Lever 1985), house mice are present on Viti Levu and Vanua Levu where they are locally abundant in urban, suburban and agricultural areas (Pernetta and Watling 1978).

Galápagos Islands

House mice were possibly introduced to some islands in the Galápagos by pirates in the late seventeenth century or later in the early nineteenth century when the islands were used as a base for whalers and sealers, and again after 1920 when a few Europeans settled there.

During World War 2 house mice were introduced to the island of Baltra (South Seymour Island) (Davis and Rowe 1963). They were widespread on all the southern islands in the 1960s and 1970s and occurred on Sierra Negra, Cerro Azul (Isabela Islands), Santiago, Baltra, Santa Cruz, Floreana and San Cristobal (Holdgate 1967; Ekhardt 1972).

Guadaloupe Island (Mexico)

Mice are thought to have been introduced by sealers to Guadaloupe between 1800 and 1830 (Huey 1925).

Guam

House mice have been introduced and established in Guam (Flannery 1995). They occurred there in about 1946 (Baker 1946).

Hawaiian Islands

Possibly present before Cook arrived in 1778 (Perkins 1903), mice may have come ashore with him (Kramer 1971), but were more likely to have been introduced by Europeans some time later as they were there in 1816 (Kotzebue 1821), and were well established in towns and villages by 1825 (Pemberton 1925).

Mice were introduced and are abundant on Sand Island and small numbers occur on Johnston Atoll (see individual island entries). Today they occur on Hawaii, Maui, Molokai, Lanai, Oahu, Kaui, Midway, Manana, Kaula Mokuoloe, Kakepa, Kepapa, and probably on Kahoolawe and Niihau (Kramer 1971; Lever 1985).

Jarvis Island (central Pacific)

Possibly introduced with miners in 1858–79 or with colonists in about 1938, mice were certainly present there in 1924 when the Whipporwill Expedition visited the island. They still occur there (Rauzon 1985; Kirkpatrick and Rauzon 1986).

Johnston Atoll

Small numbers of mice occur on Johnston Atoll where they were probably introduced during World War 2 in the cargo of ships or planes (Kirkpatrick 1966; Amerson and Shelton 1976; Berry and Jackson 1979).

Juan Fernández

Mice have been introduced to this island (Lever 1985).

Kodiak Island

Introduced in about 1920, house mice became well established on the island (Davis and Rowe 1963).

Lord Howe Island

Mice have been present on this island since the nineteenth century, but are confined to the area of homes (Recher and Clark 1974).

Loyalty Islands

Mice were recorded on this island group in 1935 (Tate 1935) and probably still occur there.

Macquarie Island

First introduced on Macquarie Island in 1820 (Holdgate and Wace 1961), mice were probably introduced by whalers or sealers at the end of the nineteenth century (Cumpston 1968; Copson 1986). They were noted when they were damaging clothing there in 1890, were recorded again in 1901 (Cumpston 1968) and were well established there in the 1970s (Bonner and Leader-Williams 1976; Taylor 1979) and abundant in the 1980s (Brothers *et al.* 1985).

Marianas

House mice have been introduced and established on Saipan and Tinian and in the northern Marianas (Flannery 1995).

New Caledonia

Widespread throughout, house mice were introduced some time after the island's discovery by Europeans in 1774 (Nicholson and Warner 1953). They are recorded in 1935 (Tate 1935).

New Zealand

Arriving as stowaways with Europeans, house mice were first recorded on Ruapuke Island, Forveaux Straits after a shipwreck in 1824 (King 1990). By the turn of the century they occupied all the most suitable habitats throughout the North and South islands.

Introduced early in the nineteenth century, mice are now widespread and abundant in the North, South and on Stewart and Auckland islands (Wodzicki 1965). In the 1970s they were present on Kawau (Wodzicki and Flux 1967). They were reported in the Bay of Islands in the 1830s and in Dunedin two years after that city was founded. They are now found throughout the main islands and on many offshore islands (Watson 1959; Barnett 1985).

Rats and mice on New Zealand islands

? date not known. * eradicated or extinct.

Island	Mouse	Black rat	Norway rat	Unspecified
'Disappointment'			?	
'Low-lying'			?	
Adele	?			
Allports	c. 1900 *			
Antipodes	?			
Arapawa	?	?		
Arid		?		
Auckland	1820			
Awaiti		?, *1982		
Bare			?	
Bay of Islands	1830s			
Bench			?	
Big South Cape		1955?		
Blumine	?			
Breaksea			1800s, *1987	
Browns	?			
Campbell		<1833	1867	
Chatham	?	?	<1840?	
Coal	?			
Cook Strait Is			<1973?	
D'Urville	?			
David Rocks			<1960, *1964	
Duffers Reef	<1983			

Rats and mice on New Zealand islands (*continued*)

Island	Mouse	Black rat	Norway rat	Unspecified
East and West Atoll			?	
Enderby	?			
Foely			?	
Forsyth	?	?		
Fortyseven		?		
Goat		?		
Great		?		
Great Barrier	? there 1981	there 1981		
Harakeke		?		
Haulashore	?	? *	?	
Hauturu (Whangamata)	?		?	
Hawea			1800s, *	
Kapiti			?	
Kauwahaia		? *		
Kawau	<1967?	?		
Kohangaatara	?			
Leper (Mokopuna)		?, *1961		
Little Rat		?		
Long	?			
Mahurangi			?	
Mana		1800s, *1989?		
Maria		?	<1960, *1964	
Masked	?			
Mayor			?	
Mokoia (Rotorua)	?		?	
Motiti			? *	
Motu		?		
Motu-O-Kura			c. 1930, *	
Motuapo				? species
Motuarohia			?	
Motuhora (Whale)			? *	
Motuhoropapa			<1962, *1990	
Motuihe				? species
Motukahaua		?		
Motukaramarama			?	
Motukiekie			?	
Motumaire				? species
Motumakareta		?		
Motumorirau (Paul)			?	
Motungarara			?	? species
Motuoi				? species
Motuopao				? species
Moturahurahu		?		
Moturako		?		
Moturemu	? *		? *	
Moturoa	?	?	?	

Rats and mice on New Zealand islands (*continued*)

Island	Mouse	Black rat	Norway rat	Unspecified
Moturua			?	
Moturua (Rabbit)			?	
Motutapu	? *	?	? *	
Motuterakihi			? *	
Motuwhakakewa		?		
Motuwheteke		?		
Motuwi (Double)			?	
Motuwinukenuke			?	
Mouse		?		
Native		?		
Ngamotukaraka			?	
Ngawhiti		?		
Noises-David Rocks			<1960 *	
Okahu			?	
Opakau		?		
Otata			c. 1956, *1980–85, *1990	
Oyster		?		
Pearl		?	?	
Phil's Hat		?		
Pickersgill	?			
Pitt	?			
Ponui				? species
Poroporo			?	
Portland	?			
Puangiangi			?	
Pukeweka		?		
Rakino			?	
Rangipukea				? species
Rangitoto	?	?	?	
Raoul			1921?	
Rat		?		
Resolution			?	
Rimariki	?	?		
Rosa		?	?	
Rotoroa			? *	
Rotoroa Stock			? *	
Ruapuke	1824			
Saddle		?		
Shoe			?	
Slipper			?	
Solomon		?		
Somes		c. 1961 *		
Spit				? species
Stewart	early 19th century	?	<1874?	
SW Crater Rim			?	

Rats and mice on New Zealand islands (*continued*)

Island	Mouse	Black rat	Norway rat	Unspecified
Takangaroa			*1987	
Taputeranga			?	
Tarakaipa	?			
Taranaki			? *	
Tauhoramaurea			?	
Tawhitinui		? *1982		
Te Haupa (Saddle)			? *	
Three Kings			?	
Tinui			?	
Titi			? *1975	
Ulva			?	
Unnamed, Bay of Is			?	
Unnamed, Bay of Is		?		
Urupukapuka		?	?	
Waewaetorea				? species
Waiheke	?		?	
Wainui		? *		
Weka				? species
Whakatere-Papanui			?	
Whenuakura	? *		?, *1982–85	
Wood		?		
Wood Stack A	?			

Niue Island
Present on the island in 1969, mice have been reported a number of times since their arrival some time after 1900 (Wodzicki 1969).

Sand Island (Hawaiian chain)
Mice were established and numerous in 1963 (Kirkpatrick 1966) and were abundant there in the 1970s (Amerson and Shelton 1976).

Solomon Islands
First recorded on Guadalcanal in 1965 (Murphy and Pickard 1990), mice have been established in the Solomons for some time (Tate 1935).

Tromelin Island (east of Madagascar)
Present there in 1953 (Paludian 1955), house mice still occurred there (Staub 1970) in the 1970s: the island was discovered in 1772 and 1776 and mice probably arrived soon after.

Vanuatu
Mice have been recorded present on this island (Tate 1935).

SOUTH AMERICA
House mice have been introduced at least into Peru, Chile, Uruguay, Patagonia and other parts of Argentina. They probably also occur in many other areas of South America.

Chile
House mice are relatively common in central Chile and the region of the La Campana Nature Reserve (Macdonald 1988). They have been widely introduced in southern South America and are closely associated with human dwellings and farms and particularly in coastal cities (Eisenberg 1989; Redford and Eisenberg 1992). They are reported to be established as far south as Punta Arenas, Chile (Lever 1985).

Peru
In Peru house mice are said to out compete the native *Phyllotis* rodents (Berry 1981).

Uruguay
Mice were introduced into Uruguay from Spain or Portugal about 1837 (Allen 1954).

WEST INDIES
House mice appear to occur throughout the West Indies, probably as a result of European colonisation in the sixteenth century. They are also present in Cuba.

Bermuda, Barbados, Guadaloupe or Martinique
Mice arrived with European colonists in 1612 in Bermuda (Lever 1985) and probably arrived in Barbados with settlement in 1626 (Browne 1982). They were established in 1654 in the French islands in the West Indies (Lever 1985).

Virgin Islands and Puerto Rico
House mice occur on Mona and Monito islands, Puerto Rico and all Puerto Rican islands, St. Thomas (United States Virgin Islands) and its adjacent islands, St. John (USVI) and its adjacent islands, all the British Virgin Islands and Anegada, and St. Croix (USVI) and its adjacent islands (Philibosian and Yntema 1977). On St. Croix they are reported to live wild among the rocks (Shortridge 1934). They may also occur on Vieques.

◼ DAMAGE
The amount of produce contaminated by house mice is about 10 times greater than the amount eaten by them. They destroy foodstuffs destined for human consumption, destroy housing materials and are implicated in the transmission of a number of diseases including salmonella, rickettsial pox and lymphocytic choriomeningitis. Some of these are spread by the contamination of food with faeces and others by their parasites. Probably none of these diseases is now serious, although the potential for their spread is still there.

In south and south-eastern Australia, house mice periodically erupt to form plagues (Redhead *et al.* 1991; Wilson *et al.* 1992). These plagues can damage standing crops and stored products and are also a social nuisance (Hone *et al.* 1981). In grain-growing regions such plagues can be of major economic importance (Redhead *et al.* 1991). Plagues have cost between A$50 and A$100 million in damage to crops, domestic houses, farm machinery, livestock producers, town businesses and grain stores, and are also a health threat to humans. Plagues can last up to six months and cause high levels of stress to people. However, domestic and industrial losses caused by house mice living in major cities in Australia are not apparent (Redhead *et al.* 1991).

In Papua New Guinea, mice are a serious pest in urban areas (Flannery 1995). In China they are important pests in urban and rural areas (Deng and Wong 1984).

Family: Gliridae
Dormice

EDIBLE DORMOUSE
Fat dormouse, squirrel-tailed dormouse
Glis glis (Linnaeus)

◼ DESCRIPTION
HB 120–225 mm; T 110–200 mm; WT 70–250 g.
Squirrel-like with short, thick, soft fur; upper parts grey to brownish grey; under parts white, greyish white or yellowish, flanks lighter; eyes small, pupil dark and horizontal; ears hairy, large and rounded; ring of black hairs around eye; legs short, with dark stripes on outsides; tail long, flattened dorso-ventrally, bushy, brown with lighter underside, often as long as body.

◼ DISTRIBUTION
Eurasia. From northern Spain, south eastern and eastern France, eastwards to Israel, northern Iran, and also east to the Volga River and also in the Caucasus. Present on the islands of Crete, Corfu, Sicily, Corsica and Sardinia.

◼ HABITS AND BEHAVIOUR
Status: fairly common. **Habits:** mainly nocturnal, but also crepuscular and rarely diurnal; mainly arboreal; hibernates or dormant (period varies with climate); shelters in tree hollows or in burrows; builds nest of plant material and moss in tree; often inhabits human dwellings. **Habitat:** forest, deciduous woodland,

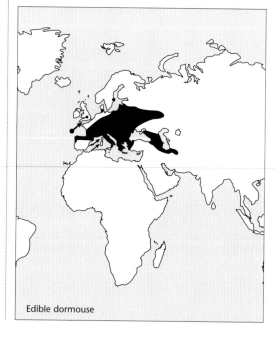
Edible dormouse

sometimes pine plantations, scrub, orchards, vineyards, gardens, dwellings. **Gregariousness:** in colonies (to 8), several families together; density 1–30/ha. **Movements:** migrate in poor acorn years from woods; home range 400–2700 m. **Foods:** beechmast, nuts, acorns, seeds, fruits (including cultivated), berries, buds, leaves, bark, fungi, insects and other small animals, occasionally nestlings and eggs of birds. **Breeding:** breeds June–September; litter size 1, 2–6, 11; 1 litter/year; young blind at birth; eyes open at 21–23 days; leave nest at 30 days; mature in second winter. **Longevity:** 6 years or more in wild, 6–9 years as captive.

■ HISTORY OF INTRODUCTIONS
EUROPE
The edible dormouse has been introduced and established in the United Kingdom.

United Kingdom
In 1902 dormice from Europe (probably Hungary) were released in Tring Park, Hertfordshire, in the Chiltern Hills of England by W. Rothschild (Lloyd 1947; Thompson 1953). These animals multiplied rapidly and caused considerable damage to corn and other crops, and also to thatch and outbuildings, so much so that a campaign to destroy them was mounted shortly after their release (Vesey-Fitzgerald 1936). Such efforts were thought to have exterminated them; however, six were sent to the Zoological Gardens, London, between 1910 and 1924 and they were reported to be numerous in Tring Park between 1925 and 1927. In 1926 Pendley Manor (near Tring) was said to have been overrun with dormice and some 39 were caught there.

From 1927 on, dormouse expansion of range and increase in numbers has been extremely slow. In 1929 one was found at Hastoe, Herfordshire, and by 1913 they had colonised Whipsnade in Bedfordshire (Vevers 1948). In 1933 they were found at Wendover; at Albury in 1935–36 where 75 were caught; between Lion Pit and Holly Findle in 1936 where seven were caught; and in 1938 reports were received from such places as Wiltshire, Berkshire, Northamptonshire, Oxfordshire, Surrey, Hampshire, Gloucestershire, Worcestershire and Shropshire (Middleton 1937; Vesey-Fitzgerald 1938; Anon. 1941; Potts 1942; Carrington 1950; Fitter 1959; Lever 1977). Some dormice were observed at Great Pednor, Buckinghamshire, in 1941 and there were small infestations in a private house and school in Berkhamsted, Hertfordshire, in 1946 (Carrington 1950). Between 1945 and 1951 the rodent control department of the district council caught 215 in 23 houses, and in 1953 a further 83 were caught (Fitter 1959).

The Ministry of Agriculture made an effort to survey the occurrence and abundance of the edible dormouse in 1951. During this survey 24 new colonies were found, including those in Ashley Green, Cholesbury, Hyde Heath, Great Missedon and Pitstone in Buckinghamshire and Ashridge Park, Little Gaddesden, Ringshall and Rossway in Hertfordshire. Between 1902 and 1962 probably 1000 dormice were killed (Lever 1977).

The present range of the edible dormouse in England is probably little different from that in the late 1970s. They inhabit an area of about 260 km^2 in the Chiltern Hills which stretches from Beaconsfield, Aylesbury to Luton, and are locally abundant in some forestry plantations (Lever 1977, 1985; Baker 1990). Their range has been steadily increasing since their introduction.

■ DAMAGE
Domesticated in early times, the edible dormouse was a favourite food of the Romans from second century BC to the middle ages (Zeuner 1963). However, in Europe it can be a serious pest of fruit crops where it occurs in large numbers (Southern 1964). Fairly recent increases in numbers in northern Tuscany (in central Italy) have created problems for the industrial cultivation of *Pinus pinea* (Santini 1978). Between 1969 and 1975 dormice have affected the production of pines by an estimated 1550 tons of pinecones at a cost of 110 billion lire. The dormouse causes damage to the fruitification of the pines.

In France edible dormice are accused of eating all kinds of fruits and are occasionally found in granaries where they cause little damage and are not considered a serious pest. In Germany and Russia there are sometimes depredations to fruit orchards by dormice, but usually only in years with a poor acorn harvest (Thompson 1953).

As an introduced species in England, the edible dormouse has a restricted range and for the most part has caused few agricultural problems. However, they have recently caused considerable damage by barking young conifers, and in large numbers can be a serious pest in orchards (Corbet 1966). Damage occurs after hibernation and takes the form of bark-stripping, mainly on the upper trunks, and may cause local damage to fruit crops (apples and plums) and fruit trees. They can cause damage to property by chewing through electrical cables, roofing felt and ceiling plaster and may also feed on stored food products causing fouling and risk of contamination (Corbet and Harris 1991).

COMMON DORMOUSE
Hazel dormouse
Muscardinus avellanarius (Linnaeus)

■ **DESCRIPTION**
HB 60–90 mm; T 55–80 mm; WT 15–43 g.
Upper coat brownish or orange brown; undersides yellowish white; throat and upper chest white; muzzle large and blunt; eyes prominent; ears rounded; face with long whiskers; tail thick, bushy, as long as body; forelimbs shorter than hindlimbs; forelimbs with three toes and rudimentary thumb; hindlimbs with five toes. Young animals are greyer than adults.

■ **DISTRIBUTION**
Europe. From the Mediterranean to the Baltic (except Iberia and Denmark) and east to 50°E in Russia; isolated populations in England and Wales, southern Sweden, Sicily, Corfu and northern Asia Minor.

■ **HABITS AND BEHAVIOUR**
Status: locally rare to common, but some decline in range; once a common pet. **Habits:** nocturnal; builds globular nests of moss and grass; hibernates at or below ground level from October to April; arboreal. **Habitat:** deciduous woodland with secondary growth, damp woods, marshes, reed beds, copses, hedgerows. **Gregariousness:** solitary or pairs and possibly colonies; density 5–10 adults/ha. **Movements:** sedentary; home range 0.5 ha. **Foods:** nuts, fruits, conifer seeds, shoots and bark of trees; some insects; occasionally eggs and young of birds.

Breeding: June–July and July–September; gestation 22–24 days; litters 3–4, 7; 2 litters/year, young born naked, blind, eyes open after 18 days, independent in about 40 days; sexual maturity about 1 year of age. **Longevity:** 3–4 years in wild, 6 years in captivity.

■ **HISTORY OF INTRODUCTIONS**
EUROPE
The common dormouse has been introduced unsuccessfully into England and Ireland.

United Kingdom and Ireland
The common dormouse is native only to England and Wales where it is widespread but local in suitable habitats (Corbet and Harris 1991). Formerly, it was a more widely kept pet and as such escaped in many areas outside its native range. Some were released in Norfolk in about 1844 and these were established there in three parishes until at least 1879. Despite many introductions, their range in Britain appears to be contracting southwards.

R. M. Barrington released six common dormice at Fassaroe, County of Wicklow, in 1885, but they failed to become established there (Fitter 1959).

■ **DAMAGE**
Common dormice cause no damage and are harmless in Britain and are vulnerable to local extinction (Corbet and Harris 1991). On the Continent they may exceptionally be sufficiently abundant after a good mast crop to cause serious damage to young trees.

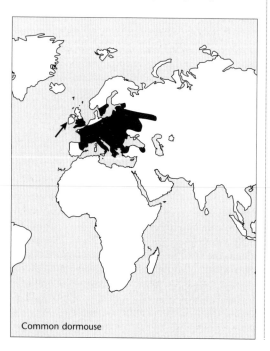

Common dormouse

Family: Castoridae
Beavers

CANADIAN BEAVER
American beaver
Castor canadensis Kuhl

■ **DESCRIPTION**
HB 730–1300 mm; T 210–530 mm; WT 11–35 kg.
Fur reddish brown, brighter on head and shoulders than back; under parts chestnut brown; muzzle blunt; ears small; lips furred; nostrils and ears with valvular flaps; tail broad, flat, thick at base, furred and sparsely haired; forepaws with elongated digits and slender claws; hind toes webbed; five toes on each foot.

■ **DISTRIBUTION**
North America. From Alaska south to northern Mexico.

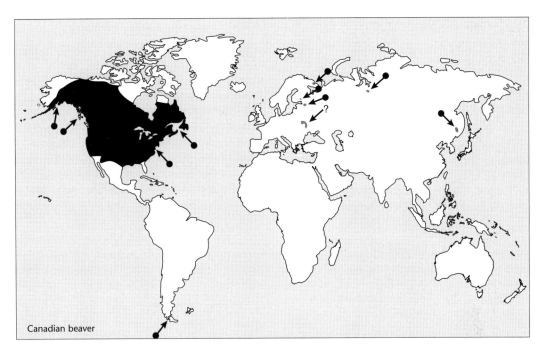

Canadian beaver

■ **HABITS AND BEHAVIOUR**

Habits: mainly nocturnal; burrows; aquatic; builds dams and lodges 1.8–2.7 m high and 3–6.1 m diameter with central chamber, with poles and brush on edges of channels; dams small creeks and waterways with logs, sticks and rocks; can submerge for up to 15 minutes; caches food. **Movements:** sedentary?; home range 1.6–2.4 km of waterway and shoreline; 2-year-old dispersal. **Gregariousness:** colonies up to 12; family units (pair adults plus young previous year); social hierarchy? **Habitat:** slow flowing rivers, streams, marshes and lakes in wooded country. **Foods:** bark and leaves. **Breeding:** mate January–February in mid-winter, young born spring (April–June) in the lodge; gestation 60–128 days; female monoestrous, male monogamous; litter size 1–8; 1 litter per year; remain in lodge 1–2 months when weaned; mature in second winter. **Longevity:** 15–20 years in wild, longer in captivity. **Status:** range reduced, but still common.

■ **HISTORY OF INTRODUCTIONS**

Introduced successfully in North America (United States, Queen Charlotte Islands, Anticosti Island, Kodiak Island group), Finland and Russia.

EURASIA

Finland

Canadian beavers have been introduced and are well established in Finland (Lahti and Helminen 1974).

Seven beavers were imported from the United States in 1937 and released in three regions of southern Finland. In Sääminki in south-east Finland they became well established and by 1945 a number were being transferred to other areas. Ten years later the population had increased to 450–500 and they were expanding their range. The population continued to increase in the 1960s and 1970s and by about 1975 was 6000. Large areas of the eastern parts of Finland now support sizeable numbers of beavers (Lever 1985).

France

In 1975, two pairs of Canadian beavers were released in the Pare Naturael Saint Herbert de Boutissant near St. Fargeau, Loiret; they escaped onto the neighbouring Reservoir du Bourdon (Lever 1985).

Poland

Possibly a population of beavers is established in Poland (Lever 1985), but there are few details.

Russian Federation and adjacent independent republics

Although some beavers were imported into Russia before 1950, the Canadian beaver first appeared in Karelia in 1953–54 and 1956–59, having spread from Finland where it was liberated in 1935–37 (Naumoff 1950; Provorov 1963; Lavrov 1965). To what extent they have now spread or been translocated within Russia is not clear. Up to 500 animals were translocated, mainly to the Far East outside the range of the indigenous beaver (*C. fiber*) (Sofonov 1981). Their establishment in these regions at this time was promising and in north-western Karelia the two species exist together. Twenty were released in Khabarovsk in 1971 and some in 1969 (Kirisa 1972–74).

Seven American beavers were released in the Rovno district of Ukraine in 1933–34, but had died out by 1957 (Lever 1985). In 1969–71, 54 from Leningrad were released on the Obor and Kur rivers in the Khabarovsk Territory. In 1975–76 a further 100 or more were liberated on the Selgon River in Khabarovsk and in the Amur area. More recently in about 1977 some were released in Kamchatka in north-eastern Russia where they are now breeding and slowly spreading (Savenkov 1987). It was also planned to introduce them in Amur-Ussri Territory and on Sakhalin Island (Lever 1985).

NORTH AMERICA

In North America generally, beavers were exterminated by unrestricted trapping in many areas of their natural range. After 1900 some areas were restocked with sometimes different subspecies from the subspecies that occurred there originally and had been extirpated (Hall 1981).

Alaska

Beavers were introduced to the Kodiak Islands, Gulf of Alaska, in the 1920s where they became well established (Clark 1958; Franzmann 1988). In 1925, 24 beavers from the Copper River area were released on Kodiak. In 1927, 10 from Prince of Wales Island, and in 1929, 21 from the Copper River area were also released on Rasberry Island in the Kodiak group (Burris 1965).

Canada

From 1922 on, and almost annually since this date, many beavers were captured and transferred from Bowron Lakes, British Columbia, and released in areas where they did not occur in the province (Carl and Guiguet 1972).

Some beavers were also transferred from the mainland to the Queen Charlotte Islands in 1936, but these failed to become established. To boost low populations wildlife agencies in Ontario, Quebec, and on Prince Edward Island successfully (established viable populations) re-introduced beavers (Deems and Parsley 1978). Beavers (*C. c. leucodontus*) from Vancouver Island were introduced in 1950 and established near Gold Creek and the Tlell River in the Queen Charlotte Islands (Cowan and Guiguet 1960; Carl and Guiguet 1972).

Beavers were also liberated on Anticosti Island, eastern Canada in 1887–98, where they have also become established (Newsom 1937).

United States

The beaver became extinct in many areas of the United States and there have been many re-introductions and translocations. Prior to 1976 at least 20 states had made re-introductions for aesthetic reasons or to boost low populations, which resulted in viable populations (Deems and Parsley 1978). Beavers were once common throughout Louisiana, but settlement brought about a drastic decline to only a few by 1931 (Chalbreck 1958). With increased protection it was found necessary to remove many from areas where agricultural damage became serious.

More than 35 beavers were introduced to New York State between 1901 and 1907 at Litchfield Park, Lake Kora, Whitney Preserve, Moose River, Big Moose Lake, Fulton Chain, Lake Teror, Little Tupper Lake, and Lake Placid by state and private individuals (Bump 1941). Most were of Canadian origin, were wild-trapped animals, and were well established and increasing in the 1940s and 1950s (de Vos *et al.* 1956). Although nearly extinct in Virginia in 1929–30, beavers in that state were saved by introductions before 1941 (Pierle 1941; Johnson 1942). From these re-introductions some 200 colonies in 16 counties were established at that date. The population was extirpated in Tennesseee by about 1911, but re-introductions in 1942–43 re-established them in some areas by 1951 (Goodpaster and Hoffmeister 1952). In Vermont they were extinct for nearly half a century before being re-introduced in October 1921 from the Adirondacks (Kirk 1923). Successful re-introductions in Pennsylvania in 1917 gave rise to a harvest of 6000 beavers over six weeks by the 1940s (Lattinger 1945, 1951).

Translocated to new areas in North Dakota in 1947–53, the beaver became established and spread up to 10 km from the release sites (Hibbert 1958). Between 1951 and 1957, 2200 beavers were trapped live in central Wisconsin and translocated to new sites because they disrupted various human activities by flooding roads and farm fields, raising lake levels or interfering with fisheries (Knudsen and Hale 1965). Translocations have also been carried out in Texas (Lay 1964) and Wyoming (Spriggs 1943; Grasse 1949).

Five beavers from Oregon were liberated on Little River, north of Crannel, Humboldt County, California (Tappe 1942). A second release at Mad River was made in 1946 when three males and one female, of unknown origin, were released at North Fork. They are now widespread along the Mad River system (Yocom *et al.* 1956). They were also liberated in Sequoia and King's Canyon National Parks,

California, but failed to expand their population or range and at present are not a major concern (Macdonald *et al.* 1988).

SOUTH AMERICA
Argentina
Canadian beavers were introduced to Lago Fagnano (Isla Grande, Tierra del Fuego) (Lever 1985), where they are now well established and becoming a significant environmental pest by felling trees and damming waterways (M. Bomford *pers. comm.* 2001).

■ DAMAGE
Beavers cause damage to trees by de-barking them and making the trees more vulnerable to the attacks of insects and to fire damage. They also kill merchantable timber by waterlogging areas with their dams (Chalbreck 1958). They annually cause problems for ranchers, farmers and highway crews in California and south-eastern Idaho and other parts of North America (Leege 1968; Jameson and Peeters 1988). Their dams have even caused the blocking of migratory salmon (Walker 1968).

The successful introduction of beavers onto Kodiak Island may have adversely affected salmon spawning areas (Burris and McDonald 1973). However, it has improved duck nesting habitat and may provide excellent habitat for silver salmon (*Oncorhynchus kirsutch*) in some stream systems.

EUROPEAN BEAVER
Common beaver

Castor fiber Linnaeus
In some recent taxonomic works C. fiber *and* C. canadensis *are held to be only subspecifically distinct.*

■ DESCRIPTION
HB 730–1300 mm; T 210–500 mm; WT 13–35 kg.
Coat blackish brown or yellowish brown; under parts brown to tawny; tail and feet black; body stout; muzzle blunt; ears small; tail scaly, broad, and flat.

■ DISTRIBUTION
Eurasia. Formerly the forested regions of northern Europe south to the Mediterranean and east to Siberia, but now only in scattered colonies in France, Germany, Poland, southern Scandinavia and scattered areas of the central Russia.

■ HABITS AND BEHAVIOUR
Habits: nocturnal and diurnal; nests in lodges and burrows in river banks; coprophagous. **Foods:** aquatic vegetation; shoots, twigs, bark, leaves, buds and roots. **Habitat:** preferably ponds and streams in forested areas but also rivers, lakes, swamps. **Gregariousness:** lives in small colonies; family units to 12. **Movements:**

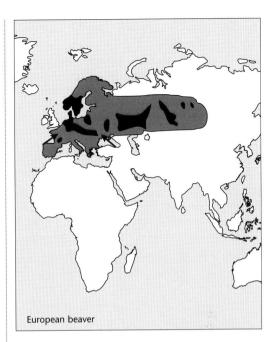

European beaver

sedentary. **Breeding:** in spring, January–February; gestation 60–128 days; 1 litter/year; litter size 1, 2–6, 9; born furred; eyes open; weaned at 6 weeks; mature 1.5–3 years. **Longevity:** 10–17 years in wild, up to 24 in captivity. **Status:** exterminated over much of former range due to hunting for fur; now relatively rare, but recovering.

■ HISTORY OF INTRODUCTIONS
EURASIA
European beaver have been introduced successfully in many areas in Europe and Asia.

EUROPE
Formerly widespread in Europe and Asia, beavers were practically exterminated by habitat loss and overhunting for pelts and meat, but as a result of protection and re-introductions they survived in some areas, such as the Rhône, parts of the Elbe and in Scandinavia and Finland (Lyneborg 1971).

European beavers (*Castor fiber*) died out in Spain in the sixth century, Britain in the twelfth century and Italy in the sixteenth century. By the 1920s they occurred only in parts of Russia, Poland, Germany, and in southern France and south-eastern Norway. However, a number of successful introductions have been made in Sweden, Finland, Switzerland, Austria, Germany, France and in many areas of the Russia (Lever 1985; Burton 1991). In Germany beavers had been exterminated in Rhineland and Saxony by 1840, Bavaria by 1850, Wurtemburg by 1854, Lower Saxony by 1856 and in North Rhine-Westphalia by 1877. Hunting and clearing for agriculture appear to have been the main causes for the decline in numbers and range in these areas (Grzimek 1975).

Following re-introduction programs in at least 14 countries, commencing in the 1920s, beaver populations have been boosted to relative abundance (Taylor 1999). Some half a million are now estimated to be in Europe with rapidly expanding populations in Scandinavia.

Austria
Beaver have been successfully re-introduced into Austria (F. Spitzberger *pers. comm.* 1982).

France
It was proposed to attempt to re-introduce beavers (*C. fiber*) from Rhone to Brittany in the 1960s (Richard 1967) but no further records could be found.

Norway
Beavers (*C. fiber*) occur naturally over much of Norway, but most of them are found in the southern counties. Their range has been slowly increasing due in part to invasion from introductions made in Sweden. However, many have been artificially transplanted and released in other parts of Norway, mainly unsuccessfully because of the small groups of beavers involved (Myrberget 1967), but by about 1970 they numbered several thousand (Lyneborg 1971).

Poland
By 1977 the European beaver was only to be found in north-eastern Poland. However, in 1976–79 some were experimentally released in the Vistula River area from north-eastern Poland and some that had been bred at a beaver farm at Popielno (Zurowski 1979). The re-introduction appears to have been successful (Zurowski 1992).

Russian Federation and adjacent independent republics
European beavers formerly lived over almost the whole of the forest zone of Russia, but today are found only in a few places through protection, re-introduction and re-acclimatisation. From an estimated population of 900 beavers in the 1920s they had increased to about 40 000 by 1964 (Zharkov and Sokolov 1967).

Since 1934 and up until 1966 about 2000 beavers were relocated and since 1948 about 3500 transplanted from new colonies. During the last 30 years 9500 have been successfully moved to new areas and harvesting began in 1963 (Heptner 1967). In western Siberia since 1935 over 1000 beavers have been translocated and some colonies have now existed for over 20 years and a steady colonisation of the river systems has occurred (Zhdanov 1962).

At least 3817 beavers were resettled in the Russian Federation and adjacent independent republics from 1937 to 1957, and about 3323 of these in Russia. By 1960 they inhabited 38 districts in northern Russia, six in Belarus, two in the Ukraine, as well as some in the Baltic republics. Beavers of Belarus origin comprise more than 50 per cent of those distributed and resettled in 24 districts of northern Russia, four in Belarus, two in the Ukraine, as well as some in Lithuania and Estonia. Those from Veronezh oblast origin comprised 44 per cent and were resettled in 31 districts of the Russia, as well as some in Lithuania and Latvia.

From 1927 to 1941 some 316 beavers were released in 12 oblasts of the European parts of Russia and in two oblasts in Western Siberia. From 1946 to 1970 about 12 071 beavers were released in 52 oblasts, krais and autonomous republics in northern Russia, in three oblasts of the Belarus, in eight oblasts in Russia, in Litovsk, Latvia and Estonia (Kirisa 1972–74).

The beaver had disappeared from Lithuania by the end of the nineteenth century, but following World War 2 it was re-introduced to Lake Zhurintas and to the Kertusha, Krempa and Minia rivers (Palionene 1965; Yanushevich 1966). Re-introductions occurred in 1948 and 1959 and by 1962 they were well established and spreading, and the greatest numbers occurred in the south-west of Lithuania.

In the north European parts of Russia beavers were released in Murmansk (67 beavers), Karelskia (6), Arkhangelsk (328), Komi (268), Vologod (431) and Kirov (422) between 1934 and 1970 (Kirisa 1972–74). The 67 released in the Murmansk oblast established a number of small low-density populations on the Chuna River, Nyavka River, in the Olenitsa Basin and on the Ponoi River. South of here in the Novgorod oblast they became established on the Chernaya River, in the basin of the Msta River and in the Udina River system. In the Arkhangelsk oblast the last beaver on the Pechora River was killed in 1817, but from 19 settled in a game reservation a total of 200 were present in an expanding population in 1959 (Yazan 1959). By 1962 this had increased to 1500 and by 1963 some fourteen-fold (Kopytov and Kopytov 1962; Yazan 1963).

In the west European parts of the Russia and areas just west of the Baltic Sea beaver were released in Lenningrad (373), Pskovsk (215), Novgorod (171), Kaliningrad (70), Astonsk (10), Latvia (16), Litovsk SSR (348), Belarus (704), and in the Ukraine (348) between 1930 and 1970 (Kirisa 1972–74). Some were introduced to Estonia in 1957 and by 1960 they had appeared in the Jägala River basin (Ling 1961). Some of those animals released in Latvia were from Norway (Hooper 1945). In the Pskov oblast some were released on the Chernaya River, at the mouth of the Ludavka River, at the mouth of the Volosna River and on the Velikaya River. By 1960 there was a population of at least 500 beavers in the oblast. North of these areas in the Lenningrad oblast more than 10 liberations were made between 1952 and 1970 and the beaver became established in many areas.

The beaver in Mogilevsk oblast, Belarus, were exterminated by 1900. In 1904, four pairs were released and by 1926 there were at least 50 families present. By 1960 they inhabited some 60 rivers and creeks in the Soza River basin and were still spreading (Lyarski 1961). Further releases were made in Belarus between 1948 and 1949 (Samusenko 1962) and these became established, but with restricted ranges (Krapiynyi 1963; Yanushevich 1966).

Introductions of beavers in eastern European Russia and the Urals

Area	Dates	Numbers
Astrakhan	1946–48	19
Bashkir	1963–67	84
Bryahnsk	1947–57	183
Chelyahbinsk	1948, 1961–66	86
Chuvash	1951–67	91
Gorkov	1939–69	648
Ivanov	1954–69	382
Kalinin	1936–70	397
Kalujsk	1952–65	78
Kostrom	1958–70	417
Kursk	1961–62, 1970	66
Lipetsk	1957–58, 1970	74
Mariisk	1947–61	124
Mordovsk	1936–68	208
Moskov	1946–56	183
Orenburg	1958–68	69
Orlovsk	1951	29
Penzensk	1961–67	128
Permsk	1947–69	267
Rhyazan	1937–40, 1963–65	117
Saratovsk	1964–69	109
Smolensk	1950–68	345
Sverdlovsk	1953–70	471
Tambov	1964–68	108
Tatar	1949–70	90
Udmurtia	1947–67	365
Vladimir	1940–67	195
Volgograd	1965–69	97
Voronejsk	1937–47	103
Yahroslav	1955–67	156
Yulyahnov	1965–70	19
Total		5708

Some 5708 beavers were released in this region. Those released in the Vladimir district were successful and had increased to 745 by 1967 (Sysoev 1967). Beavers were exterminated in the Ivanov oblast prior to the revolution but were re-introduced from 1940 with animals initially from Veronezh. By 1960, 130 colonies with an estimated total of 535 head were established (Pankratov 1961). Releases in the Vologda oblast in 1948 and 1960 also became established and were spreading in the early 1960s (Belozertsev 1962). In the Volga Delta region, Astrakhan and in the Volga-Kama region they became established from a number of small introductions (Yanushevich 1966). By 1960 several releases in the Muryginoi River, using some animals from Veronezh Preserve, were well established in 21 small communities and increasing (Rukorvskii 1948; Zamakhaev 1963). By 1960 beavers occurred in many areas of the Ural Mountains as a result of introductions (Koryakov 1962), mainly with animals from Veronezh (Tsetsevinskii 1963). Releases in the Shalinsk raion of the Sverdlovsk oblast were sucessful in the basin of the Sylva River, along the Vogulka River, and along the Dikaya Utka River (Bakeev 1963). In the Permsk oblast by 1959 there were 530 present (Chashchin 1961).

East of the Ural Mountains in Western Siberia and the Altai there were a number of introductions of beavers in Russia (Kirisa 1972–74). These included Tumensk (442 in 1935–68), Tomsk (409, 1941–62), Omsk (435, 1953–70), Kurgan (31, 1961), Novosibirsk (299, 1956–67), Kemerovsk (145, 1960–62) and in the Altai (188, 1952–55, 1964–70). Between 1935 and 1957, 210 were released in the Uvat, Vagai and Tyumen raiones of the Tyumen oblast. By 1956–57 they were well established throughout the basin of the Aitka and Pyshma rivers. Further releases were made on the Ityugas River in 1956 (48) and the Taim-Tashet River (59) and on the Sig and Tizeva rivers (48) in the Omsk oblast. Here they were increasing in 1960 (Borisov 1963). Releases from Veronezh (80) were liberated on the Bol'shoi Kemchug River in 1948–50 and had increased to 600 by the early 1960s (Zharov and Vinichenko 1962; Zharov 1963). Beavers have also been acclimatised in the Republic of Tuva (Volchenko 1964).

Releases of beavers were also made in central Siberia in Tuvin (38 in 1953) and Krasnoyahsk (792 in 1948–68), in eastern Siberia in Irkutsk (274, 1950–63) and in the Far East in Khabarovsk (90, 1964–69).

Between 1975 and 1986, 233 beavers were re-introduced into tributaries of the Vistula River (Zurowski 1987). Of these 14 per cent were farm-raised and the others were captured in low-lying lake land habitats.

The result was the establishment of small beaver colonies; in total 168 were released in lowland regions and established 16 populations; 55 beavers were introduced into mountainous regions and made up four populations. There was very little expansion of mountainous region colonies until at least 1986, but the lowland populations expanded by increments of 20 per cent per annum.

Since 1934 about 14 000 beavers have now been re-introduced and their former range has now nearly been restored (Sofonov 1981).

Sweden

Beavers invaded Sweden about 6000 BC and expanded into all the suitable habitats, but after the seventeenth century they became scarce and the last was killed in 1871 (Curry-Lindahl 1967). A series of re-introductions was initiated in 1922 and continued up until about 1940 when some 80 beavers from Norway had been released in various parts of Sweden. These became well established and dispersed along the rivers and the rapid expansion of their range was said to have been spectacular in some areas. A census in 1961–63 indicated that the population in Sweden was then about 2206.

Switzerland

Beavers probably occured naturally in Switzerland as late as the sixteenth century. The last animal was possibly killed in 1705 in Birs, near Basel.

Beavers were re-introduced in 1956 and by 1962 a small colony was established (Dottrens 1965). A pair were also introduced on the Versaix River (Canton de Geneve) in early 1958 and bred there in 1959 (Blanchet 1960). Some were re-introduced at Neuchâtel in the early 1960s or earlier and were reported near that time as the third successful re-introduction to Switzerland (Anon. 1963).

Between 1956 and 1968 approximately 25 beavers from France and 30 from Norway were imported and released in different rivers (in about 15 areas) in northern and western Switzerland. Today there are several stable populations, totalling about 300 animals in Switzerland (Rahm 1976; M. Dollinger *pers. comm.* 1982).

United Kingdom

Extermination in the twelfth century (Grzimek 1975) led to a number of introductions later in the 1870s (Lever 1985). Beavers were introduced in England in 1870 at Sotterley Park, Suffolk, but these later died out. In 1874 some were released on the Isle of Bute, in Scotland, where they are said to have survived until 1890 (Fitter 1959).

A proposal to re-introduce beavers to England in 1977 found little support (Lever 1985) and there are none at present in Britain. However, a pilot scheme to bring back the beavers to Scotland has recently commenced and is expected to start re-introductions within three years (Taylor 1999).

■ DAMAGE

In Norway the beaver is locally considered a pest of forestry and some extent to agriculture (Myrberget 1967). They cause little damage in Switzerland because populations are small, but locally they damage poplar crops (Dollinger *pers. comm.* 1982).

In the Rhone Valley of France they undermine river banks and occasionally the banks of irrigation ditches and also destroy trees and garden vegetables (Lagaude 1961).

Family: Hystricidae
Porcupines

Since 1969 various species of *Hystrix*, including the Himalayan porcupine, have been found in the wild in Britain having escaped from captivity or been deliberately released (Baker 1986). Over the past 15 years, 10 individuals have been involved in escapes on six occasions.

CRESTED PORCUPINE

Porcupine, North African crested porcupine
Hystrix cristata Linnaeus

■ DESCRIPTION

HB 600–800 mm; T 50–90 mm; WT 13–30 kg.
Crest mainly white; rump dark; long spiny black and white quills; quills on body and tail.

■ DISTRIBUTION

Northern half of Africa (Morocco, Libya and probably Egypt) from Tanzania to upper Egypt and south of the Sahara westward to Senegal and Gambia.

■ HABITS AND BEHAVIOUR

Habits: nocturnal; den or nest in burrows or rock dens to 10 metres; activity reduced in winter but no hibernation. **Gregariousness:** family groups or solitary; at mating 1 male to 1–2 females. **Habitat:** rocky areas, open woodland and dry savanna scrub; wooded river banks, wooded gullies, cultivated hillsides, coastal woodland, open cultivated plains. **Foods:** largely roots; bulbs, tubers, rhizomes, but also seeds, sprouts, twigs and bark of herbaceous plants.

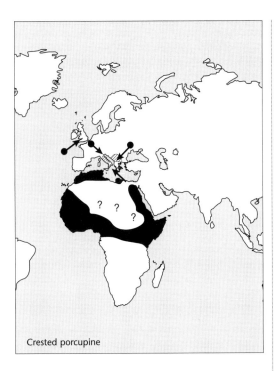

Crested porcupine

Breeding: breeds throughout year; mates in spring; gestation about 90 days; litter size 1–2, 4; 1–3 litters per year; mature in second year. **Longevity:** up to 20 years (captive). **Status:** present status uncertain in Europe.

■ HISTORY OF INTRODUCTIONS

Probably introduced into southern Europe crested porcupines are confined to Italy, Sicily and the Balkans (Albania and southern Yugoslavia).

EUROPE

It seems likely that crested porcupines were introduced to Europe (Corbet 1978, 1980). Their presence in Italy, Sicily, Albania and Yugoslavia is difficult to otherwise explain (Burton 1976).

Albania–Yugoslavia

It appears that the introduction of the crested porcupine to the Balkans was fairly recent (Corbet 1966). They may also have occurred on the Adriatic coast of Yugoslavia, possibly south to Greece, but there is no recent evidence for this (Burton 1991).

Britain

A pair escaped in 1972 from the Botanical Gardens at Alton Towers, Stoke-on-Trent, Staffordshire, England, and were later reported over an area of about 7 km^2 for at least two years (Lever 1977; Corbet and Harris 1991). Attempts were made to live trap them but their numbers appear to be unknown.

More recently a single animal escaped from a collection in County Durham and was recaptured 20 months later close to where it escaped (Baker 1986).

Italy

It seems probable, on taxonomic grounds, that Italian populations of crested porcupines owe their origin to deliberate introductions (Corbet 1966) and were probably introduced to Italy from North Africa by the Romans. Crested porcupines were extending their range north and east into the slopes of the central Apennines in the 1980s (Santini 1980; Burton 1991). They are still found in southern Italy and Sicily, but their present status is uncertain (Burton 1991).

■ DAMAGE

Porcupines can be a serious pest of crops, especially among root crops and in orchards (Corbet 1966). In Italy they damage crops in cultivated areas, e.g. maize, potato and chickpeas (Santini 1980).

In Britain they have caused damage to trees in plantations (Corbet and Harris 1991). A survey in 1973 estimated that four hectares of Norway spruce at Folly Gate valued at 500–1000 pounds sterling had been damaged by this species.

HIMALAYAN PORCUPINE

Hodgson's porcupine, Malayan porcupine

Hystrix brachyura Linnaeus

=*Hystrix hodgsoni* Gray

■ DESCRIPTION

HB 600–930 mm; T 80–170 mm; WT 8–30 kg.

Front half covered with short, dark brown spines while hindquarters have long pointed whitish quills usually with one distinct black ring; crest on upper neck and upper back whitish; tail short with both long pointed quills and rattle quills.

■ DISTRIBUTION

Asia. Central and eastern Himalayas, north-eastern India, southern China and parts of Malayasia; also Sumatra, Borneo, southern Thailand and Singapore (rare or extinct).

■ HABITS AND BEHAVIOUR

Habits: mainly nocturnal; terrestrial; burrows. **Gregariousness:** largely solitary? **Movements:** sedentary. **Habitat:** forest, plantations. **Foods:** roots, tubers, bark and fallen fruit. **Breeding:** 1 young, 2 rare. **Longevity:** 27 years and 3 months captive. **Status:** no information.

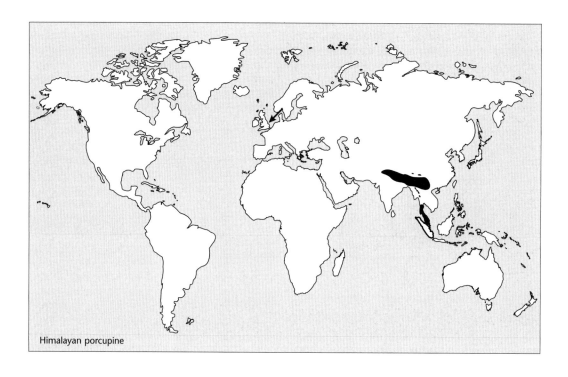

Himalayan porcupine

■ HISTORY OF INTRODUCTIONS
EUROPE

United Kingdom

In 1969 a pair of adults that had been obtained from Calcutta escaped from a Wildlife Park two miles from Okehampton in Devon (MAFF 1975; Gosling 1980; Baker 1990). One was killed in 1971, one was found dead and a third was trapped in 1973.

Himalayan porcupines became established in 3 km² of country from Oaklands in the south through Hook, Abbeyford, North, Springetts and Parsonage Woods, past Risdon, Folly Gate, and Inwardleigh, as far as Hayes Barto south of Jacobstowe where they were largely confined to conifer plantations.

In 1974 the population was thought to be at least 12 or more and appeared to be still there although trapping had been carried out to remove them (Lever 1977, 1985).

A number of records were reported within 16 km² of Okehampton, but by 1980 a total of five adults and one sub-adult had been accounted for (Corbet and Harris 1991), and the species was almost certainly eradicated from Britain (Baker 1990).

■ DAMAGE

Himalayan porcupines often raid tapioca plantations in rural areas in Malaysia (Medway 1978).

Family: Caviidae
Guinea pigs

GUINEA PIG

Cavia porcellus (**Linnaeus**)

The wild form is often referred to as C. aperea *Erxleben and distribution shown here is of this form.* Cavia tschudii *is probably the ancestor of the domestic species.*

■ DESCRIPTION

HB 225–275 mm; WT 400–1200 g.

Coat variable dull fawn or brown, greyish buff or drab with short, smooth hair; tailless; four toes on front feet, three toes on hind feet. Wild form closely resembles domestic guinea pig.

■ DISTRIBUTION

South America. Colombia south through Brazil and into Argentina and Uruguay. Also in eastern and northern Paraguay, north-eastern and east-central Argentine, south to Buenos Aires province.

■ HABITS AND BEHAVIOUR

Habits: nocturnal and diurnal; burrows; colonial; little known of wild animal. **Gregariousness:** small groups of 5–10; density 38.7/ha. **Movements:** sedentary; home range males 1387 m², females 1173 m². **Habitat:** grassland, rocky regions, savannas, edges of forest, and swamps, gardens and waste ground.

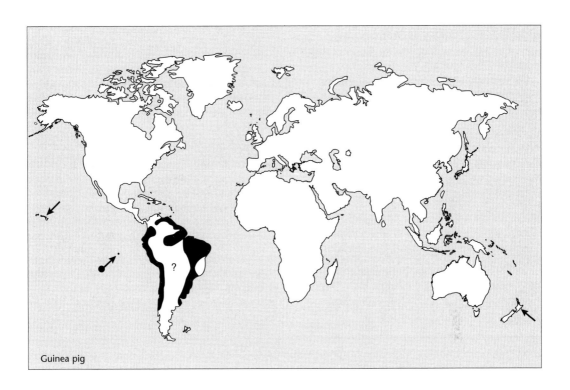

Guinea pig

Foods: herbivore; leaves and stems of plants. **Breeding:** breeds throughout the year; gestation 59–72 days; oestrous cycle 15–17 days; litter size 1–9 in wild (up to 12 in captivity); up to 5 litters/year; lactation 14–28 days; young born furred and active, eyes open; sexual maturity at 2–4 weeks. **Longevity:** 3–4 years in wild, 5–9 as captive. **Status:** little information, widely introduced in captivity as a cage animal and pet.

■ HISTORY OF INTRODUCTIONS
Guinea pigs were introduced unsuccessfully to Hawaii and successfully to the Galápagos Islands. They have been widely introduced in captivity and escape or are liberated regularly, but have not become feral anywhere.

EUROPE
Guinea pigs occur sporadically throughout Europe as an escapee around human habitation, but have not become permanently established (Burton 1991).

PACIFIC OCEAN ISLANDS
Galápagos Islands (Ecuador)
In about 1969 guinea pigs escaped on Santa Cruz in the Galápagos, where by the 1980s the population had increased to between 300 and 500 animals (Lever 1985).

New Zealand
Guinea pigs were unsuccessfully introduced to New Zealand by the Auckland Acclimatisation Society in 1869 (Thompson 1922).

Hawaiian Islands (United States)
Introduced to Laysan Island in 1903–04 by the manager of a mining company, guinea pigs were possibly abundant there on the south end of the island in 1911 (Dill and Bryan 1912; Bryan 1915), although one report mentions only four animals were seen at this time (Ely and Clapp 1973). These, according to one source (Ely and Clapp 1973), were killed and so the population exterminated. Certainly by 1923 they had vanished and it was thought that they were unable to compete with rabbits introduced at the same time (Kramer 1971).

SOUTH AMERICA
Argentina, Chile and Peru
In Peru, Chile and Argentina, the exact origin of the northern colonies of guinea pigs is obscure, but some may have been transported to their present locations by humans (Eisenberg 1989).

■ DAMAGE
None known. Apparently no effects where introduced in the Galápagos, although there may be some potential in this regard.

Family: Dasyproctidae
Agoutis

A Dasyprocta *sp. may have been introduced by early Amerindians to Auba, Curacao (Eisenberg 1989), but apparently no longer occurs there.*

BRAZILIAN AGOUTI

Cutia, picure, orange-rumped Agouti, aguti
Dasyprocta leporina (Linnaeus)
=D. leporina aguti

■ DESCRIPTION
HB 410–620 mm; T 10–30 mm; WT no information.
Body slender; fur long and thick; tail obsolete; hind feet with hoof-like claws.

■ DISTRIBUTION
South America. Venezuela, Guianas, Amazonia and eastern Brazil and the Lesser Antilles.

■ HABITS AND BEHAVIOUR
Habits: diurnal; burrows; jumps well. **Habitat:** forest and savannah. **Foods:** leaves, fruits and roots. **Breeding:** 2–4 young. **Status:** common.

■ HISTORY OF INTRODUCTIONS
WEST INDIES
The taxonomy of West Indian agoutis is at present based on geographic distribution and some species are questionable. It is currently thought that those inhabiting the Lesser Antilles (e.g. St. Croix, St. Kitts, Antigua, Monserrat, Dominica, St. Vincent, Grenada, Martinique, St. Lucia, Guadaloupe and possibly Barbados) were probably introduced from the mainland in pre-Columbian times (Varona 1974). The pattern appears to be *D. leporina agouti* from Brazil to the Virgin Islands; *D. l. albida* on St. Vincent and Granada; *D. l. fulvus* on Martinique and St. Lucia; and *D. l. noblei* on Guadaloupe, St. Kitts, Dominica and Montserrat (Wilson and Reeder 1993).

Grenada
The Brazilian agouti was introduced by humans to Grenada (Eisenberg 1989), probably in pre-Columbian times (Varona 1974).

St. Thomas (Virgin Islands, United States)
D. leporina. aguti was possibly introduced to St. Thomas, Virgin Islands, where it may have become established for a period (Miller 1918; de Vos *et al.* 1956), although it is not known to have bred there (Philibosian and Yntema 1977).

A Brazilian agouti was obtained in the winter of 1916–17 on St. Thomas and the species was reported there as early as 1852, but there is no archeological evidence of them previously. It seems probable that it was introduced from Brazil and the Lesser Antilles (Miller 1918).

■ DAMAGE
None known.

MEXICAN BLACK AGOUTI

Mexican agouti, cerreti
Dasyprocta mexicana Saussure

The Mexican black agouti *D. mexicanus* was introduced in the late nineteenth century to the Sierra de los Organos, Pinar de Rio, and the Sierra Cristal, Oriente, in Cuba (Wilson and Reeder 1993).

RED AGOUTI

Central American agouti, picure, rojizo
Dasyprocta punctata Gray

■ DESCRIPTION
TL 490–620 mm; T 10–35 mm; WT 3.1–4.0 kg.
Upper parts variable; blackish (especially head and nape) mixed with buff or russet, or distinctly reddish, usually brightest on posterior of body; under parts paler, sometimes tawny; hair on rump longer than on upper parts, and is erectile.

■ DISTRIBUTION
Central and South America. Southern Mexico (Chiapas and Yucatan Peninsula), through Panama,

Red agouti

south to Ecuador, southern Bolivia, south-western Brazil and north-western Argentina.

■ **HABITS AND BEHAVIOUR**
Habits: diurnal; cursorial, territorial; caches food in small pits. **Gregariousness:** density 0.1/ha. **Movements:** sedentary; home range 2–4 ha. **Habitat:** forest and wooded areas. **Foods:** seeds, fruits, leaves and roots. **Breeding:** breeds throughout year; gestation 120 days; young 1–2; 2 litters/year; digs burrows and builds nest; young born furred, in burrow or crevice; eyes open; mature at 16 months. **Status:** common? **Longevity:** 10 years?

■ **HISTORY OF INTRODUCTIONS**
CARIBBEAN
Cayman Islands (United Kingdom)
The red agouti (subspecies unknown) was introduced and established in the Cayman Islands at the close of the twentieth century (de Vos *et al.* 1956; Varona 1974; Wilson and Reeder 1993).

Cuba
A subspecies of the red agouti (*D. p. yucatanica* or *D. p. nelsoni ?*) was introduced into western and eastern Cuba from central America in the nineteenth century (Varona 1974 in Hall 1981; Corbet and Hill 1980; Wilson and Reeder 1993).

■ **DAMAGE**
None known.

PACA
Lapa; tepizicuinte
Cuniculus paca (Linnaeus)
=Agouti paca

■ **DESCRIPTION**
TL 600–795 mm; T 10–25 mm; WT 4.3–10.5 kg.
Upper parts red-brown to almost black; two to seven lines of white dots on each side of body; under parts white to buff; legs short; muzzle square; large lips; eyes large; vibrissae stiff; fur-lined cheek pouches; four toes on fore and hindfeet. Male larger than female.

■ **DISTRIBUTION**
Central and South America. Southern Mexico south to southern Brazil, northern Argentina and eastern Paraguay.

■ **HABITS AND BEHAVIOUR**
Habits: nocturnal; makes own burrows or enlarges others; territorial. **Gregariousness:** solitary or females and young together; density 84–93/km². **Movements:** sedentary. **Habitat:** lowland forests, moist areas, arid areas with streams or watercourses. **Foods:** fruits,

Paca

nuts, seeds, and other vegetation. **Breeding:** breeds throughout year; gestation 115–116 days; 1 young, rarely 2; may have 2 litters/year; young born furred, eyes open; weaned at 90 days; may breed in second year. **Status:** fairly common. **Longevity:** no information.

■ **HISTORY OF INTRODUCTIONS**
Introduced and established in Cuba (Lever 1985; Wilson and Reeder 1993).

■ **DAMAGE**
None known.

Family: Chinchillidae
Chinchillas

CHINCHILLA
Long-tailed chinchilla
Chinchilla laniger Molina

■ **DESCRIPTION**
HB 225 –380 mm; T 75–180 mm; WT 500 g to 1 kg.
Fur silver grey, soft and silky, dense; upper parts bluish, pearly or brownish grey with faint dusky or blackish markings; eyes and ears large; ears almost naked; whiskers of moustache long and bristly and whitish or black; cheek pouches vestigial; under parts

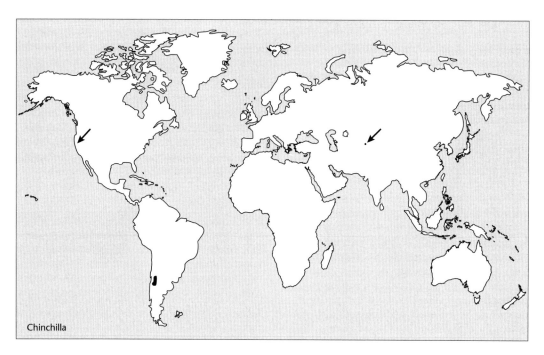

Chinchilla

yellowish white; tail coarse, bushy and squirrel-like, well haired and heavily marked black or brownish; forefeet with four digits, hind feet three. Females larger than males.

■ DISTRIBUTION
South America: Formerly the Andes of Bolivia, Peru, Chile and Argentina. Now probably only survive in northern Chile (and perhaps northern Argentina and Bolivia).

■ HABITS AND BEHAVIOUR
Habits: little known in wild; nocturnal and crepuscular; shelters in holes and crevices in rocks; hunted for fur. **Gregariousness:** colonies up to 100 or more. **Movements:** sedentary. **Habitat:** rocky areas in foothills; rocky barren slopes in mountains at 3000–6000 m. **Foods:** herbivorous; vegetable matter; grass and herbs. **Breeding:** breeds September–April; gestation 105–114 days; monogamous; breeds 1–3 times per year; litter size 1–4, 6; eyes open at birth; lactation 6–8 weeks; mature at 8 months. **Longevity:** 10 years. **Status:** almost extirpated in wild, now rare; numbers and range considerably reduced.

■ HISTORY OF INTRODUCTIONS
Chinchilla have been unsuccessfully introduced into Tajikistan and the United States. Restocking in the Andes in South America may have had some success.

ASIA
Republic of Tajikistan
In 1960, 200 chinchillas were imported and released

in the Republic of Tajikistan, but they failed to become established (Lavrov and Pokrovsky 1967).

Chinchillas were released in the eastern Pamir ranges of Tajikistan from 1964 to 1969, when 211 animals were liberated. Initially these were successful (Kirisa 1973), but later failed to become established (Sofonov 1981).

NORTH AMERICA
United States
There were two separate introductions of chinchillas in the United States in Los Angeles and Humboldt counties of California, with culled animals from fur farms in August and September 1952, but they failed to become established (Voris *et al.* 1955).

Four female and eight male chinchillas were released in the Whittier Hills, Los Angeles County – one was seen nine months later. Thirteen male and one female chinchilla were released in Humboldt County in September 1952: two were seen a few weeks later and one was reported a year later, but none have been seen since.

SOUTH AMERICA
Chinchillas were hunted until the 1930s when their numbers had been decimated, but thereafter they have been protected. Recently, attempts have been made to restock them in the Andes with animals bred in captivity on fur farms, but it is not known how successful these attempts have been (Grzimek 1975).

■ DAMAGE
None known.

Family: Capromyidae
Hutias and coypus

JAMAICAN OR BROWN'S HUTIA
Capromys brownii Fischer
=Geocapromys brownii

Jamaican or Brown's hutias are being bred in captivity for re-introduction on Jamaica (Oliver 1985; Oliver *et al.* 1986), but recent information on any success is not known.

COYPU
Nutria, swamp beaver
Myocastor coypus (**Molina**)
In some classifications the coypu is included in a monotypic family (Myocastoridae).

■ DESCRIPTION
HB 357–635 mm; T 224–425 mm: SH 125–140 mm; WT 1.4–17 kg.
Coat black to reddish brown or yellowish brown, but may appear dark brown or even blackish as coarse guard hairs are dark and dense underfur is yellowish; stout body; muzzle squarish with white tip; chin white; incisors orange or orange-yellow; eyes and ears small; under parts greyish; undersurface of feet black, hairless; hind feet webbed; tail round, scaled and scantily haired. Female has six pairs mammae situated laterally.

■ DISTRIBUTION
South America. Central and southern South America from the Straits of Magellan north to Bolivia, southern Brazil and Peru.

■ HABITS AND BEHAVIOUR
Habits: aquatic; lives during day in short burrows (up to 15 m) where builds nest of grass and reeds; colonial; crepuscular and nocturnal; coprophagous. **Gregariousness:** in colonies or solitary; females in kin groups at high density. **Movements:** mainly sedentary; occasionally moves long distances up to 80 km. **Habitat:** lakes, streams, swamps, tidal waters, marshes, slow flowing rivers with shore vegetation. **Foods:** mainly aquatic plants; reed shoots, sedges, seeds, grass, leaves, stems, roots, bark, rhizomes, tubercles and freshwater mussels. **Breeding:** breeds all year; female polyoestrous; gestation 100–132 days; 2 litters/year; young 2, 3–12; precocious, weaned at 5–8 weeks; mature at 4–6 months, females produce young at 8–9 months. **Longevity:** at least 2–5 years in wild, and up to 15–20 in captivity. **Status:** common.

■ HISTORY OF INTRODUCTIONS
Coypus have been introduced successfully in Great Britain, other parts of Europe, Asia, Japan, East Africa and North America. Now extirpated in Britain.

AFRICA
Kenya
Introduced to a farm near Hanynki (140 km north of Nairobi) in about 1940, coypus had by 1950 become established in a moat on the farm and two years later

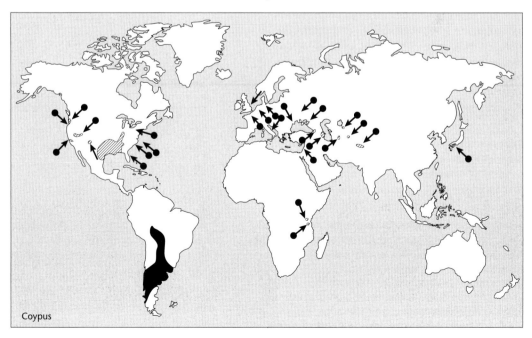

Coypus

in a nearby dam. This population increased and expanded and was noted on the western shore of Lake Ol Bolossat in 1970, but were probably there much earlier. From here they infested the headwater of the Malewa River and on south to Lake Naivasha (Lever 1985; Haltenorth and Diller 1994).

Tanzania, Zambia and other areas

Coypus are feral in the coastal swamps of Tanzania and have become established in Zambia (Lever 1985; Haltenorth and Diller 1994). They were being farmed in Zimbabwe and in the Republic of South Africa, but at this date were not established in the wild (Lever 1985).

ASIA

Japan

Seven or eight coypus from Europe were brought to Tokyo in 1931 for a fur farm, and in the same year eight from France to Oji in Tokyo (Kaburaki 1940; Kuroda 1955). After 1940 they were raised on several fur farms in several places in Japan (Kuroda 1955). In 1942 there were some 1427 animals being kept in captivity (Imaizumi 1949) and in 1944 about 40 000 were being kept in Tokyo and westwards to Shikoku and Kyushu (Oka and Takashima 1947). Before 1955 a few had escaped and were established along the coast of Kojima Bay, south of Okayama, and in 1949 about 500 coypus were known to be present in the area (Kuroda 1955). These were last reported as still established and increasing in numbers (Udagawa 1970).

At least 48 fur farms existed between 1938 and 1955 in the Okayama Prefecture but at the latter date, decreasing demands for pelts caused farming of coypus to be discontinued and large numbers were released. Colonies became established with small numbers in the network of the Yoshii and Asahi rivers. Animals established in the Kojima area increased and the population expanded gradually north to Yoshii and Asahi and then west along the Oda River (Miura 1976).

EUROPE

Coypus are also established in Austria, Norway, Czechoslovakia, Poland, Yugoslavia, Greece and Spain (Lever 1985).

Belgium

Some coypus escaped from a fur farm at Lanaken, after 1963, and some spread here from the Netherlands (Litjens 1980), but presumably they were decimated in the severe winter of 1979.

Denmark

Since World War 2 coypus have been frequently seen in the wild in areas where they are farmed, but lasting colonies do not appear to have resulted (Niethammer 1963). However, coypus now live in the wild in Denmark (de Vos et al. 1956; Corbet 1966).

France

Coypus were first introduced in France in 1882 as a fur animal and they became popular again in 1925–33. In 1939 they occurred in the wild as a result of escapees and abandoned animals in several places in France, principally in the central section (Bourdelle 1939). In the 1950s and 1960s they appeared to be maintaining their numbers (de Vos et al. 1956; Elton 1958; Corbet 1966). Now apparently the coypu is only established in France in areas where the muskrat is absent (Dorst and Giban 1954).

In 1954 coypus became established in the wild from escapees at a number of places at Sologne, the valley of the Somme, the swamps and embankments of the Seine and many canals of the north.

Germany

By 1960 it was estimated that there were some 5000 coypus established in western Germany. Principal areas of infestation were in the Rhenish-Palatinate, especially along the Glan River (Lever 1985) and on the Rhine between Worth and Speyer south of Mannheim. A sequence of mild winters between 1970 and 1975 saw a build up in numbers and some subsequently crossed into the southern Netherlands (Lever 1985).

Coypus living in the wild in Germany became established from animals imported for fur farming in 1926 (de Vos et al. 1956; Niethammer 1963; Corbet 1966). They were found established near Bonn following World War 2, but gradually disappeared, as did a small colony at Niederrhein, discovered in 1955. These latter ones disappeared following a severe winter in 1955–56. Several other small colonies on the Ruhr and other places also were short-lived. These include Schleswig-Holstein, near Ashaffenburg, Euskirchen, Laacher Lake in Eifel, near Göppingen in Wurtemburg, and in the Leine in Neidersachen. The longest lasting colony, which existed for 10 years, was that at Schwalm in Hessen, where about 300 were present in the early 1960s and seemed to be spreading. Some also existed at Altwassern and Lippemündung where in 1961 there were about 70 animals. The main concentrations were present in Sachsen-Anhalt (Niethammer 1963). By the early 1970s, with the exception of a few areas, colonisation of Germany was complete.

Italy

Introductions for fur farming began in 1928 and some coypus later escaped. They are now well established in southern Tuscany where they are spreading. Elsewhere they occur in Umbria and Latium, along the River Tiber and its tributaries, at Lago di Bracciano (30 km west of Rome) and along Volturno, Sele and Tusciano rivers in Campania (Lever 1985).

Netherlands

The first reports of coypus in the wild in the Netherlands were made before 1940 at such places as Bunde, Diepenveen and in East-Gelderland. In 1937 or thereabouts they are reported to have been breeding in the wild at Diepenveen. Following World War 2 the coypu fur industry in this country expanded, but by 1950 because of the poor quality of fur produced and the influx of large numbers of furs from eastern European countries, commercial rearing ceased shortly after this date. After 1945 the number of reports of coypus began to increase rapidly and between 1950 and 1955 they were reported from nearly all the provinces in the Netherlands. Many of these animals were thought to have been due to a mass release by commercial breeders when the industry floundered because of low fur prices in 1949 (de Vos *et al.* 1956; Koenders 1964; Litjens 1980).

During a severe winter in 1955–56 the greater part of the population of coypus in the Netherlands perished. Remnant populations remained in such places as the Roer basin in Mid-Limburg, where the surface water did not freeze, and a colony was later formed at Vriezenveen. However, in a further severe winter in 1962–63 all the coypus in the country perished (Koenders 1964; Van Wijngaarden *et al.* 1971).

Re-population of the Netherlands by coypus began at the end of 1963 from Germany where a small founder population had survived the severe winter in the Roer river basin (Koenders 1964). A series of mild winters up until 1979 assisted the spread, which extended along the Roer into the Maas river basin, southwards to Obbicht and northwards to Sanbeek, and along the Niers River. At least 20 other small colonies were formed in other parts of the Netherlands as far away as Willemstad, Vleuten and Englelbert. In January 1979 a severe winter again reduced the distribution and numbers of the coypu (Litjens 1980). From a total population of several hundreds in 1977 (Broekhuizen 1977) they were reduced to some tens and survived only in the Maas and Roer basins (Litjens 1980).

Russian Federation and adjacent independent republics

Coypus were imported into Transcaucasia, the Kuban delta, the downstream parts of the Kura and Terek rivers and the southern part of the Amu-Darja Basin, where they have multiplied. Efforts to introduce them in central Russia have failed because of the unsuitable climatic conditions. However, they are now successfully established in Armenia, Georgia, and the steppes of Shirwan (Lindermann 1956). A release in Kuban, Sea of Azov region, in 1930 and 1950 apparently failed (Pavlov 1958).

Coypus were apparently unsuccessfully introduced in Kazakhstan (Sladskii and Afanas'ev 1964). In 1932, 123 released in the stream network in the Kazakhian district of Azerbaidzhan where they bred, but numbers were reduced by predation, but a few spread downstream from the release point. In 1940, 50 were released in Armenia and became established and spread throughout the entire valley where released. Some dispersal of coypus was also observed in central Asia, the northern Caucasus and in the Ukraine (Aliev 1965).

The species was established in western Georgia before 1941(?) (Vereschagin 1941) and efforts were made before 1939 (Shaposhnikov 1939).

Coypus were introduced in 1930–32 from Argentina, England and Germany, and became established in a number of areas. In this period some 2656 animals were released, of which 676 came from Argentina and 1980 from England and Germany. Altogether from 1930 to 1942 some 1107 were released, and from 1942 to 1963 some 5163, making a total of 6270 released in the Russian Federation and adjacent independent republics. Introductions include Tajikistan in 1949, where they became established but with a restricted range; Uzbekistan in 1931, failed, but successfully established locally in 1950; North Caucasus, escaped or were released since 1930; Kirghizia, experimentally introduced in 1954 when 62 were released, but all were later re-trapped; Armenia in 1940, (Azerbaidzhan, Krasnodar, Georgia and Daghestan), becoming established but with a restricted distribution; and Transcaucasia, where they became established (Yanushevich 1966).

In 1930, 113 from Argentina were introduced and a further 2500 from England and Germany in 1931–32 but these failed to become established. In 1930–41, 1100 were imported and released in Caucasus and central Asia where they became established. In 1962 more coypus were released (Lavrov and Pokrovsky 1967).

In the Caucasus, releases occured in Azerbaidjan (1931–59, 3468), Gruzinsk (1932–68, 606), Armyahn (1950, 40), Dagestan (1932, 22), and Krasnodarsk (1930–59, 552). All these releases appear to have had some success.

In Kazakhstan and Middle Asia some 14 were released in Kazakhsk in 1930 and in 1952 some 170, but they did not become well established. Other releases included Tadjiksk (1949–58, 899) where they became well established, Turkmensk (1931, 1957, 263), Kirgizsk (six in 1954, and 220 in 1962–63) and Uzbeksk (150 in 1953) (Kirisa 1973).

Sixty coypus were imported into Tajikistan from Kyudamir, Promkhoz, and Azerbaidzhan (before 1953), but only 28 were set free; in 1954 more from Turkmenia may have been released and in 1956 about 1000 were present (Nernyshev 1959). Coypus were unsuccessfully introduced into Kazakhstan (Sludskii and Afanas'ev 1964).

In the Tashkent oblast of Uzbekistan, coypus were introduced in 1931 and became established along the middle reaches of the Amu-Darja River. In 1958, 18 males and 20 females were released at Lake Malyi, Kalgansyr, apparently successfully (Mukhamedkulov 1963).

Although several thousand were introduced (Lavrov and Pokrovsky 1967), efforts to establish coypus have not been entirely successful and they are no longer established in the wild in Russia (Sofonov 1981).

Switzerland
In the early 1980s, two coypus were captured in the canton of Jura, but as yet no population appears to be permanently established there. These two animals are thought to have wandered in from France (M. Dollinger *pers. comm.* 1982).

United Kingdom
Imported to Britain in 1929–30 as a fur animal (Davis 1956; Fitter 1959), and between 1932 and 1937, coypus escaped from a number of farms in Surrey, Sussex, Hampshire, Norfolk and Devon. Those first recorded in the wild came from a farm in Sussex in 1932 and this was followed by others dating back to 1937, noted at Tiverton, Devon, Horsham, Hampshire, Walvesley, Huntingdonshire, Buckinghamshire, Cheshire, Essex, Gloucestershire and Staffordshire (Fitter 1959; Lever 1977). By 1939 there were about 40–50 fur farms, which were abandoned or had closed down by 1940 or during World War 2 (Davis 1956; Gosling 1989). However, by this time many had escaped and the species was well established in the wild (Laurie 1946; Gosling and Baker 1989).

In other parts of the British Isles some coypus escaped from fur farms in Scotland in 1934 at Morayshire and some in Perth, and also Wales, Montgomeryshire, in 1936 (Lever 1977), but these apparently did not become permanently established. By 1939 they had escaped from 37 places in 11 counties in England.

Between 1940 and 1945 there were 23 reports of escaped animals at large in nine counties (Fitter 1959).

Initially it was thought that although the coypus' range was extending slowly they would not build up in numbers nor spread much. Although capable of damage, it was thought that they might prove useful in opening waterways and eating vegetation (Davis 1956). Their range steadily increased in the period 1945–62, but not until the late 1950s did their numbers increase to the point where excessive damage occurred (Norris 1967).

Coypus escaped from fur farms in Norfolk in 1937 and were well established in that area by 1943–44, by which time they had spread along the Wensun and Tas rivers. In 1943 they had spread along the Yare to Keswick and Cringleford and by 1944–45 had reached Bairburgh and Marlingford. By 1945 they had spread east to Cantley, Reedham and Langley marshes and south-west to Northwold on the Missey River, and were found in small numbers at Wroxham. In about 1955 they had reached Glaren in northern Norfolk and were beyond Guist, and in 1956 were colonising Suffolk and were a firm feature of both Norfolk and Suffolk (Laurie 1946; Fitter 1959; Davis and Jensen 1960; Davis 1963; Southern 1964, Norris 1967; Lever 1977).

In 1954 there was some public alarm over the continued spread of coypus (Davis 1963) and the serious damage being caused (Gosling 1989) and so eradication campaigns began in 1962 (Norris 1967). In the initial campaign some 40 461 were killed, but later up to 97 000 were exterminated. Around this time the population in Britain was estimated at 200 000. In 1962 they were abundant in Norfolk and Suffolk and adjoining counties. After 1962 they continued to spread westwards and were increasingly found outside Norfolk and Suffolk in such areas as Lincolnshire, Cambridgeshire, Huntingdon, Petersborough, Bedfordshire, Hertfordshire and Essex (Norris 1967; Newson and Holmes 1968). Control campaigns continued from 1962 until 1965, when there were only a few coypus left in marshy areas of East Norfolk where they were expected to remain. By 1969 only about 5000 coypus remained in Britain (Norris 1967; Newson 1969). From 2000 animals in mid-1970 the coypus population expanded to nearly 19 000 in late 1975 because of mild winters and low trapping intensity. However, it declined to less than 6000 following cold winters in 1978–79 (Gosling *et al.* 1981).

By the mid-1970s they were mainly confined to marshy tracts in the Broads of eastern Norfolk and

Suffolk and did not seem to be spreading, although the situation was carefully watched and control work was still proceeding (Lever 1977).

In 1981 the Ministry of Agriculture began an extermination campaign designed to eradicate them by 1990 at an estimated cost of 1.7 million pounds sterling. Some 121 862 were destroyed in control programs or killed in other ways between 1970 and 1987. Coypus are probably now eradicated from Britain (Baker 1990) as none have been caught since April 1987, although there may be a few still in East Anglia (Lever 1985; Gosling 1989; Gosling and Baker 1989; Corbet and Harris 1991).

Middle East
Israel
Coypus from Chile were introduced to Kafr Masaryk in northern Palestine (Israel) in 1953 and 181 were released onto two ponds and a drainage canal of the Kafr Rupin in 1957. Within a year animals from each introduction had escaped (Lever 1985).

At Kafr Masaryk eight escaped into nearby swamps at Afequ, when by 1960 they were regarded as established throughout the Naaman region (Lever 1985). In the Kafr Rupin area escapees were established in nearby swamps and marshlands (Lever 1985).

Turkey
Coypus have colonised the waters of Kora Su near Kars, Arahk, in eastern Turkey (Lever 1985).

North America
Canada
Coypus, *M. c. bonariensis*, were imported to British Columbia as a fur animal from 1938 on and although a few escaped from time to time the species did not become permanently established (Carl and Guiguet 1972) until recently. By the mid-1970s escaped and established feral populations apparently existed in the Lower Fraser Delta area (Banfield 1977).

Prior to 1960, escapees from fur farms were caught at Burnet Creek in Bunaby and some had been seen near Crescent Beach, also on the Cowichan River, and a single animal at Courtenay (Cowan and Guignet 1960).

In other parts of Canada, escaped and established feral populations of coypus exist in the Whitefish River drainage of the Thunder Bay district, Ontario, and in the Ottawa River drainage of western Quebec and eastern Ontario (Banfield 1977). Two that were caught in Thunder Bay, Ontario, in 1953 were thought to have crossed from the United States where

introductions occurred in Minnesota in 1941 and 1945 (Allin 1955). The two noted previously were presumed escapees from fur farms. Introductions by fur farmers in Nova Scotia did not succeed (Deems and Parsley 1978).

United States
Coypus are found in Texas, Louisiana, and east along the Gulf Coast, in isolated colonies in New Jersey, Maryland, Great Plains and the Pacific north-west (Knopf 1991). In North America in 1945, some 8000 pelts were taken for the fur market. By 1950 this had risen to 40 000 and in 1961 there was an annual take of over 1 million pelts (Evans 1970).

The coypus was first introduced to the United States at Elizabeth Lake, California, in 1899 without success and again in the 1920s and 1930s when farmed for fur (Sanderson 1955; Evans 1970). Ranches were established in Washington, Oregon and Michigan in the early 1930s, New Mexico (mid 1930s), Louisiana and Ohio (1932), Utah (1939) and Maryland (Willner *et al.* 1979), and elsewhere later (Evans 1970). Shortly after World War 2 prices became low due to competition with beaver pelts and other reasons, and many ranchers released them or let them escape (Evans 1970). In this latter period, or soon after, coypus became established in many areas of the United States including Mississippi, Louisiana, Texas, Washington (Sanderson 1955), Michigan (Ashbrook 1948), Oregon (de Vos *et al.* 1956), Missouri, Alabama, Florida, North Carolina, Virginia, California (Howard 1953), Idaho, Ohio, Kentucky (Presnall 1958), New Mexico (Elton 1958), Minnesota (Gunderson 1955), Oklahoma (Ashbrook 1948), Iowa (Petrides and Leedy 1948) and Montana. By 1970 they had been reported in 40 states since 1889 and still persist in at least 20 states (Evans 1970).

As early as 1941, sportspeople and trappers were translocating coypus into marshes from Port Arthur, Texas, and the Mississippi River in Louisiana. Some were sold to hunters who released them as far away as North Carolina. The biggest dispersal occurred when get-rich-quick promoters sold them as 'weed cutters' in the south-east United States. State and federal agencies transplanted them to Alabama, Arkansas, Georgia, Kentucky, Maryland, Missippi and Oklahoma, and inland in Louisiana and Texas (Evans 1970).

Those occurring in Texas are the descendants of 20 coypus imported by E. A. McIllhenny to Avery Island, Iberia Parish, Louisiana, in 1937. Over the next few years some escaped or were released in surrounding marshes where they became established (Lowery 1974). Some escaped from this farm during a hurricane in 1940 (Harris 1956). By the early 1960s this species was the most important fur-bearer in the Louisiana fur industry (Shirley *et al.* 1981). As early as 1946 a few were trapped near Port Arthur in south-eastern Texas. In 1946 A. C. Lively obtained 20 females and four males and released them near Slocum in Anderson County (Swank and Petrides 1954). About the same time C. N. Campbell released three in a small pond near Grapeland in Houston County (Petrides 1950). Four pairs were introduced to Mobile Delta in 1948 to test the ability to control water vegetation (Lueth 1949). Wholesale introductions in 1949–50 occurred in about 22 counties of Texas when 195 coypus were released at the rate of two females per male (Petrides 1950; Swank and Petrides 1954). The coypus has now spread along the entire western gulf coast, inland to Texas. In 1939, 12 escaped by burrowing under a fence and another 150 escaped in 1940 during a storm-caused flood, and these reproduced and spread (Evans 1970).

Coypus were found to be feral in Stanislaus County, California, in 1945 as a result of escapees from fur farms (Howard 1953) south-west of Oakdale. Thirty coypus were purchased from Ramsey, New Jersey, in August 1942 by a fox farm operator and farmed at a location near Oakdale. These were sold to a neighbour who purchased eight more from Louisiana in 1944 and between 1945 and 1948 some of these escaped. By 1951, 100 or more wild coypus had been trapped or killed. Although they are not now known to exist in the area, they were present in the area for a while (Howard 1953; Schitoskey *et al.* 1972; Howard and Marsh 1984). Coypus are possibly established on the Hearst Ranch, San Luis Obispo County, California (Lidicker 1991).

In Oregon coypus were introduced as a fur animal from about 1930 to the 1950s, but some escaped and some were released and they quickly spread through western Oregon (Kulin and Peloquin 1974). Some were released as early as 1937 in Tillanook County (Larrison 1943), and by 1946 they were well established in several localities. In the 1950s and early 1960s a depressed fur market led to the release and/or escape of many animals (Kulin and Peloquin 1974).

Specimens and reports of feral coypus were received by authorities in Minnesota from 1947 to 1949 (Gunderson 1955). They were released along the Rat Root River, near Ray, between 1941 and 1945 as a result of an unsuccessful effort to raise them from 1939 on. Some were liberated at Rainy River, Ontario, across the border from Baudette, Minnesota, in 1948, and some of these may have crossed into the state. In the early 1980s Louisiana led the United States in production of wild furs with two million pelts, of

which coypus account for 65 per cent of the harvest and US$9 million annually (Linscombe *et al.* 1981).

Coypus were first reported to be exotic in Florida in the 1950s when feral animals were captured at Panhandle and Hillsborough River drainages off the west coast of the state (Griffo 1957). Much of Florida's panhandle populations resulted from an eastward expansion of range along the Gulf Coast from Louisiana marshes (Atwood 1950; Lowry 1974). Colonies at Hillsborough River and other locations in peninsular Florida resulted from escapees or releases from abortive fur farming attempts. The species is now exceedingly abundant in both the above areas (Brown 1975).

A fur farm on the Pecos River, Chaves County in New Mexico lost about 500 coypus during a flood in 1937. A flood in the late 1930s was also the cause of early escapes in Oregon on the Nestucca River in Tillamook County. Several small colonies were found in the Mobile Delta area of Alabama after 1948, in western Florida, on the outer banks of North Carolina in Currituck County after 1941, and in Black Bay, Virginia. On the west coast, in addition to the population in Oregon, small groups existed around Lake Washington, Washington, some along Grande Ronde River, Union County, east Oregon, a few on St. Maries River Watershed, Benewah County, Idaho, a small colony in Gallia County, Ohio, several small colonies in southern Michigan and possibly a few near Fort Knox, Kentucky (Presnall 1958).

Coypus were believed to have been introduced into the Maryland marshes in the late 1930s or early 1940s. The first recorded introduction was in 1943 when some escaped in the Blackwater National Wildlife Refuge (BNWR) and later three were killed. None were reported between 1944 and 1950, but in 1951 a private owner released five on Coles Creek and in 1952 another owner released 20 on Gibbs marsh at Meekins Creek. Between 1952 and 1955 only a few were noted. In 1956 an estimated 20 were recorded on Meekins Creek marsh. In 1961, 68 were found dead during a heavy freeze. The population began to increase throughout Dorchester County in 1969, at which time an estimated 2075 were at BNWR. At present the major populations are in Dorchester, Somerset, Talbot and Wicomico counties in eastern Maryland (Willner *et al.* 1979; Morgan *et al.* 1981).

At present coypus are well established over south-eastern United States and in many other areas.

■ DAMAGE

In the United States a study in Florida found that coypus were extremely adaptable to a wide range of aquatic conditions and exhibited the potential of being ubiquitous in aquatic systems as is the black rat under more terrestrial situations in many parts of the world (Brown 1975). In North America generally, damage to trees and canal banks by coypu burrowing has caused considerable economic losses in irrigation systems (Ball 1960). Increasing complaints of damage to vegetation and competition with other species and some damage to rice, cane and other crops has been reported (Presnall 1958). By the 1960s in Oregon, damage was common to severe to grain crops, forage, hay and trees. Burrowing damage in stream banks, field borders and farm ponds was also reported (Kulin and Peloquin 1974). In the Louisiana–Texas coastal marsh region and agricultural lands coypus were a serious problem as they invade rice fields and burrow in canal levees and also ate the rice plants (Marsh 1965).

In Britain coypus attacked a wide variety of crops including sugar beet, fodder beet, kale and other brassicas, cereals and occasionally potatoes, damage trees (larch etc.), and where numerous have damaged the banks of waterways and roads (Davis 1963; Norris 1967; Anon. 1978; Gosling and Baker 1989). They caused damage to sugar beet in early summer and to other crops, such as marigolds, swedes, kale, brussel sprouts, and cereals are sometimes grazed. Damage was mainly to agricultural crops in Britain, especially sugar beet, kale, cereals in green stage and occasionally potatoes and pastures (Southern 1964). Potentially more serious was the damage from burrowing in dykes and river banks causing flooding (Norris 1967). In the Norfolk Broads they changed the vegetation by selectively feeding on a reed (*Phragmites australis*) fringing the open waters (Boorman and Fuller 1981; Baker 1986).

The damage in Britain was tolerated when coypus were cropped for fur or meat, but controlled elsewhere because of the damage to crops, drainage systems and native plants (Gosling 1989).

In the Netherlands coypus cause damage by undermining banks that caused cave-ins and machines to become bogged. They caused substantial losses in sugar beet fields and to riverbank vegetation used to consolidate the banks, and caused changes in species composition in pastures (Litjens 1980). Fortunately they do not like heavy frosts and freeze-ups which to date together with control campaigns have prevented them from spreading throughout the Netherlands.

In Japan coypus have caused damage to rice fields and vegetables (Undagawa 1970), and in the Russian Federations they have been used for controlling coarse vegetation in reservoirs (Oleinikov and Vasil'eva 1963).

Within their native range in South America, coypus are an important fur animal, where nine million pelts were exported from Argentina between 1976 and 1979 (Redford and Eisenberg 1992).

Family: Echimyidae
Spiny rats

RED-NOSED TREE RAT
Spiny rat
Echimys armatus (Geoffroy)
=*Makalata armata*

■ **DESCRIPTION**
HB 170–264 mm; T 182–220 mm; WT 147–317 g.
Upper parts dark yellowish brown, heavily lined with black; back dark brown, furry and spiny posterior third of back is spectacled yellow; spines are pale grey at base and darker distally and with distinct pale yellowish terminal band; sides of body lighter than back; under parts pale yellowish or grey-brown; tail short, furred at base, remainder sparsely haired.

■ **DISTRIBUTION**
South America. Central South America from Ecuador, Peru, Colombia, Venezuela to the Guianas, north eastern Brazil and the island of Trinidad.

Red-nosed tree rat

■ HABITS AND BEHAVIOUR

Habits: nocturnal; arboreal or scansorial; uses nests or hollows in trees or among roots;. **Gregariousness:** small groups. **Movements:** sedentary? **Habitat:** evergreen forest in moist habitats; river banks and flooded areas. **Foods:** grass, sugar cane, bananas, fruit, seeds and nuts. **Breeding:** breeds November; litter size 1–2; young precocial. **Longevity:** 3 years 1 month in captivity. **Status:** no information.

■ HISTORY OF INTRODUCTIONS

WEST INDIES

Red-nosed tree rats have been introduced unsuccessfully to Martinique in the Lesser Antilles.

Martinique

The red-nosed tree rat occurred on Martinique apparently as an introduction by humans (Walker 1968), but any introduction appears to be based on a single specimen that was brought in by a vessel and died without reproducing (Hall 1981).

■ DAMAGE

Red-nosed tree rats are harmful to the cultivation of bananas in Surinam, climbing trees at night and eating the green bananas (Walker 1992).

Family: Canidae
Dogs, wolves and foxes
The following notes are on species not treated fully in the text.

JACKAL
Canis aureus Linnaeus

■ HISTORY OF INTRODUCTIONS
AUSTRALIA
In 1856, 20 common or Asiatic jackals from India were imported by a Colonel Roberts, who was the agent for the purchase of Indian 'remounts'. These were released at locations near Melbourne for the purpose of hunting by foxhounds (Fitzpatrick 1878). It can be surmised that they were never released more than one at a time, and that their survival before the hounds seems unlikely.

FALKLAND ISLAND WOLF
Dusicyon australis (Kerr)

■ HISTORY OF INTRODUCTIONS
FALKLAND ISLANDS
It has been suggested by some authors that the Falkland Island wolf arrived in the Falklands as a domestic species of animal brought in by prehistoric Indians. However, still others have suggested that lowered sea levels would have allowed natural movement to the island and that the wolf's distinguishing characteristics resulted from subsequent isolation rather than domestication. The species has always exhibited remarkable tameness towards people. Large numbers were killed in the 1830s by fur traders and farmers and the animal was rare by 1870 and last seen in 1876 (Walker 1992).

WOLF
Dusicyon cf. thous

■ HISTORY OF INTRODUCTION
WEST INDIES
This species may have been transported to Aruba in the West Indies by early Amerindians (Eisenberg 1989).

CAPE FOX
Vulpes chama (Smith)

■ HISTORY OF INTRODUCTION
SOUTH AFRICA
The Cape fox of South Africa has been released (re-introduced) in Mountain Zebra National Park (Penzhorn 1971), but no further details were obtained.

ISLAND GRAY FOX
Vulpes littoralis (Baird)
=*Urocyon littoralis*

■ HISTORY OF INTRODUCTIONS
UNITED STATES
The Island gray fox or Island fox (*Vulpes littoralis*), a resident of six islands in the Santa Barbara Islands off California, was thought to have been introduced to San Clemente Island by humans in 1875. However, a study in 1975 concluded that the race (*V. l. clemente*) inhabited the island in prehistoric times, although a pair of *V. l. catalinae* may have been introduced there by S. Ramirez in 1875 (Johnson 1975).

CORSAC FOX
Cosac fox, korsac
Vulpes corsac (Linnaeus)

■ DESCRIPTION
HB 500–600 mm; T 250–350 mm; SH c. 305 mm; WT no information.
Coat soft, reddish grey or reddish brown with a silver shade given by tips of guard hairs; back of ears reddish grey; chin and lip white; middle of back and shoulders darker than flanks; belly off-white or yellowish; tail tip dark brown or black (not white). Similar in appearance to the red fox (*Vulpes vulpes*).

■ DISTRIBUTION
Central Asia. From the Lower Volga, Kazakhstan, and Ural Steppes east to Mongolia, northeastern China and Manchuria south to Turkestan, Tibet and Sinkiang (perhaps also extreme northern Afghanistan and northern Iran?).

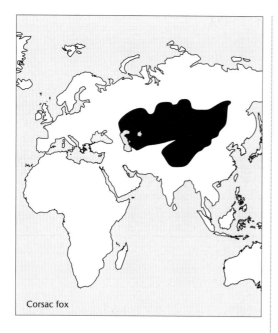

Corsac fox

■ HABITS AND BEHAVIOUR
Habits: nocturnal; lives in burrows, occasionally colonially. **Gregariousness:** lives in small groups and hunts in packs. **Movements:** nomadic(?); no fixed home range; may move south in winter in deep snow areas. **Habitat:** steppe and desert, semi-deserts, foothills, ploughed steppe and inhabited areas. **Foods:** rodents, rabbits, pikas, carrion, birds and eggs, insects, fruits and plant material. **Breeding:** oestrus in January–March; gestation 50–60 days; litter size 2–11, 16; females reach sexual maturity at 3 years. **Longevity:** no information. **Status:** range reduced and fragmented with spread of agriculture.

■ HISTORY OF INTRODUCTIONS
ASIA
Russian Federation
Corsac foxes were introduced in central Asia and released on Barsa-Kelmes Island in the Aral Sea in 1930 (Naumoff 1950 in de Vos *et al.* 1956), where they don't appear to have been very successful (Kirisa 1974).

■ DAMAGE
No information.

SWIFT FOX
Kit fox
Vulpes velox (Say)

■ DESCRIPTION
HB 375–525 mm, T 225–350 mm, WT 1.6–3.0 kg.
Long thick fur; greyish brown above (more reddish in summer); flanks orange brown; legs, underside of tail and belly whitish; ears small; tail black tipped.

■ DISTRIBUTION
North America: Canada through central and western United States to Mexico.

■ HABITS AND BEHAVIOUR
Habits: nocturnal; lives in burrows by day. **Movements:** sedentary(?). **Gregariousness:** solitary or in pairs with young; sometimes 1 male and 2 females. **Habitat:** open prairies, grassland, shrubby deserts. **Foods:** small mammals; rabbits, rodents, birds, insects, lizards, fish, grasses, berries, and carrion. **Breeding:** mates December–January, young born March–April; gestation 50–60 days; female monoestrous; litters 3–6, 8; 1 litter/year; eyes open 10–15 days; weaned 6–7 weeks; disperse August–September; males mature first year, females 10 months, but all do not breed in first year. **Longevity:** wild 8–10 years, to 13 years in captivity. **Status:** range declined, but still common.

■ HISTORY OF INTRODUCTIONS
NORTH AMERICA
Canada
In 1983 the swift fox was re-introduced to the Canadian prairies after an absence of 45 years. It was once common on portions of the prairies and extended from Manitoba across Saskatchewan to the foothills of Alberta. As many as 117 025 were

Swift fox

harvested for pelts between 1853 and 1977. A combination of intensive trapping and poisoning appears responsible for the decline in the northern parts of its range. By 1900 it was rare in the northern United States and Canada and officially designated extirpated in Canada in 1978 (Herrero *et al.* 1986). Some 250 foxes have now been released and they are surviving (Carbyn 1989).

There have been several releases of captive swift foxes in Alberta and Saskatchewan. They were released near Mayberries, Alberta, during 1983–84. A further release was made in Alberta in the autumn of 1985. Founder populations came from northeastern Colorado (seven) and south-west South Dakota (three). Releases were made in Alberta and Saskatchewan with animals trapped in Colorado, Wyoming and South Dakota.

Populations are now established and being monitored by Canadian Wildlife Service personnel (Canadian Wildlife Service 1997). Introductions are continuing and it is hoped to remove the species from the endangered list by 2000.

■ DAMAGE
Questions of appropriateness of introduction have been raised and it has been suggested genetic infection may cause a decline of more southern populations of swift foxes because hybrids may be less desirable in some way (Stromberg and Boyce 1986).

RED FOX
Common fox, European fox, American red fox
***Vulpes vulpes* (Linnaeus)**

■ DESCRIPTION
HB 460–900 mm; T 222–600 mm; SH 300–400 mm; WT females 3.5–4.5 kg, males 4.5–10 kg.
There are three colour phases: red, red-brown and silver (many more in domestication). Red phase: chest, abdomen, insides of ears and tail tip creamy white; face and flanks ochraceous; vibrissae black; cheeks, dorsum, rump and tail rufous; black guard hairs along mid-dorsal line and prominent in tail; back of ears and anterior portions of legs black; muzzle sharp and pointed.

■ DISTRIBUTION
Eurasia, North Africa and North America. From Spain, Ireland, Norway and North Africa, northern Europe and Asia, south island of Novaya Zemlya, south to North Africa (n.w. Africa), Arabia, Iran, northern India, Bangladesh, Pakistan, central India, Egypt (Nile Valley), Tunisia, Algeria and Morocco, northern Indochina, southern China, Japan, Sakhalin, and the Kurile Islands. In North America from Alaska, Baffin Island, Ellesmere Island, and northern Canada south to southern United States.

■ HABITS AND BEHAVIOUR
Habits: mainly nocturnal and crepuscular, but also may be diurnal in quiet areas; lives in burrow or

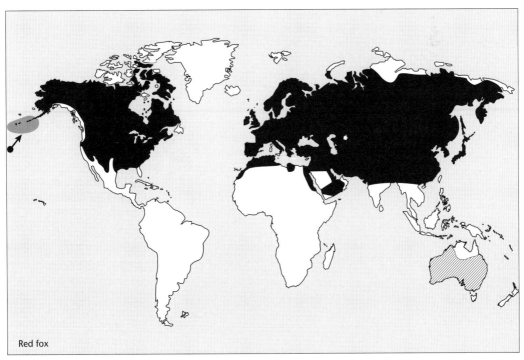

Red fox

natural cavity (dens 3.5–10 m); caches food. **Gregariousness:** solitary except at breeding; family groups which share joint territories; territorial pairs and small groups 4–5 adults; density 0.2–32/km². **Movements:** sedentary, but autumn dispersal of young; sometimes moves to lower altitudes in winter; territories 20–100 ha; home range 3.6–8.1 km². **Habitat:** alpine and arctic tundra to semi-deserts including sea cliffs, forest, woodland, cultivated areas, river valleys and towns. **Foods:** omnivorous; small mammals (rats, mice, voles, rabbits, lambs, hares), domestic livestock, insects, fruits, carrion, birds, offal, invertebrates, grass and vegetable matter. **Breeding:** mate December–February (Europe); young born March–May; largely monogamous; gestation 51–63 days; litter size 4–6, 12; breeds at 10 months; cubs born blind, furred; vixen nurses for 2–3 weeks and fed by dog fox; eyes open 11–14 days; weaned 4–5 weeks and emerges from den. **Longevity:** probably 1–4 years in wild, but some individuals to 8 or even 12 years; in captivity 10 years or more. **Status:** common and numerous.

■ HISTORY OF INTRODUCTIONS
Introduced successfully in Australia, North America, northern Europe and to the Russian Federation.

AUSTRALASIA
Australia
Individual foxes were imported by hunt clubs in the first half of the nineteenth century for recreational hunting, but were unsuccessful (Wilson *et al.* 1992). Newspaper accounts indicate that foxes were introduced as early as 1855 and that it is likely the first successful releases occurred in southern Victoria in 1871 (Rolls 1969). They were hunted in Victoria from about 1865 (Jarman 1986). An early import was made by T. H. Pyke in 1845 to Victoria, but it is not known whether these were actually released. The Melbourne Hunt Club set some free in 1854 (Terry 1963). More arrived or were released in Victoria in 1855 (two), 1864 (three), 1868 (one), and 1869 (two). None of these releases appear to have been successful. More were released in 1871 (two) at Ballarat and in the early 1870s at Point Cook, and these were certainly successful.

In the space of 100 years foxes became distributed all over the mainland, except for the tropical north, at a colonisation rate of 160 km/year (Redhead *et al.* 1991). They reached South Australia by the 1880s, New South Wales by the 1890s, Queensland by the 1900s and Western Australia by 1911 (Jarman 1986; Long 1988). By 1885 they were abundant in Victoria and in the 1890s their numbers in central and northern Victoria produced complaints in the Legislative Assembly (Jarman 1986). Foxes had by 1880 spread to the North Shore of Corio Bay and shortly joined the spread of those from Ballarat. At this time (1880) they had spread over 13 000 km² of Victoria and were established in parts of South Australia. They continued to spread and were seen at Bendigo in 1886 and in Cobar and Armidale by 1900. In 1911 they were reported in southern Queensland. In 1920 one was seen at Longreach 1600 km north of Melbourne and by 1933 they had reached Julia Creek. Foxes reached the Western Australian border by 1911–12, Esperance in 1916 and by 1925 had spread along the coastline to Geraldton and reached the Kimberleys in the 1930s.

The early spread can be closely linked with the spread of the introduced rabbit. The spread was rapid across saltbush plains and mallee and slower in wooded country. Foxes reach the highest densities in the southern agricultural areas where fragmentary habitats and secure food supplies mimic to some extent the situation in their native range (Mitchell *et al.* 1982; Jarman 1986; Redhead *et al.* 1991).

Foxes are now distributed over the southern half of Australia, the present range being reached in the 1930s in the east and probably as late as 1950 in the west (Jarman 1986). They occur over the entire southern half of Queensland, except for the Dividing Range area (Mitchell *et al.* 1982), throughout Western Australia, except for the north Kimberley and many offshore islands (King and Smith 1985; Long 1988); an isolated population exists as far north as Killarney Station in the Victoria River District of the Northern Territory. There are none in Tasmania or on Kangaroo Island (Wilson *et al.* 1992), although a single fox escaped from a ship at Hobart harbour in 1998 having recently sailed from Melbourne, Victoria. An extensive eradication program is believed to have removed this animal.

Foxes are present on Benison Island, Corner Inlet, Victoria, where they are a predator of the seabird colonies (Norman 1977). An attempt was made to introduce them to Tasmania in 1890, but they were destroyed by authorities before any releases took place. A fox was recorded on Garden Island, Western Australia, in July 1996 when it crossed the causeway to the island. It killed at least 25 tammar wallabies (*Macropus eugenii*) in one week before being poisoned (Wykes *et al.* 1999).

New Zealand
A pair of foxes were taken to Christchurch, New Zealand, in 1864, but they are not known to have been released (Lamb 1964). An Act passed in 1867 prohibited further importations (Lever 1985).

EURASIA

Finland

Alaskan silver foxes (*V. v. fulva*) were released in Finland in 1938 with the idea of producing a cross fox hybrid between them and the native red fox (*V. v. vulpes*). Interbreeding occurred and has been reported to have improved the fox fur industry (de Vos *et al.* 1956; Niethammer 1963; de Vos and Petrides 1967).

Russian Federation

Large numbers of North American silver foxes have been released (Schmidt 1954) in European Russia (de Vos *et al.* 1956). Silver foxes were released on Chechen Island, Caspian Sea, in the Caucasus region in 1932 (Yanushevich 1966) and eight silver black foxes were released in the Komi area in 1954 (Yurkin 1961). All of the numerous attempts to introduce silver foxes in Russia have failed (Novikov 1962).

From 1929 to 1934 some 251 Canadian foxes (*V. v. fulva*) were released in 61 areas of Russia. These areas were Kamchatsk, Moskovsk, Dagestansk, Irkutsk, Buryahtsk, Krasnoyahrsk, Komi, Arkhangelsk, Voronejsk, Karelsk and Turkmensk. They were released because of their silvery-black fur, which it was hoped would improve the local stocks of foxes for fur (Kirisa 1974). What effect they have had is not clear, but probably very little and they may not have become established.

Sweden

Foxes have been introduced from Britain to Sweden (de Vos *et al.* 1956; de Vos and Petrides 1967).

United Kingdom and Ireland

Foxes (*V. v. vulpes*) from Scandinavia have been introduced in Scotland (Tetley 1941; Hattingh 1956). They were released on Anglesey a number of times in the nineteenth century, but were destroyed by the inhabitants. Although a native species in the United Kingdom, a number from the European mainland have been introduced. In 1884 Spanish foxes were said to have been introduced into Epping Forest and in 1845 some were released on the Isle of Wight, where foxes have been present ever since. They may also have been introduced on Skye and Mull (Fitter 1959).

Until recently foxes were absent from Isle of Man, but several reliable reports and extensive debate in Manx press in 1988 suggests that a recent illegal introduction may have occurred. Whether foxes will become established is unknown at present (Corbet and Harris 1991).

Foxes were present on Anglesey until some time in the nineteenth century, but died out and the island was free of foxes until 1960. At this time three were

released near Holyhead Island. Three adults and a litter of seven cubs were killed and then no more were noted until 1967. In 1973 over 100 were killed and in 1974 over 340 adult foxes were killed.

Twelve foxes were released on Great Saltee Island in Wexford, Ireland, but all of them died (Fluzx and Fullagar 1992)

NORTH AMERICA

Alaska

There have been at least 46 translocations of red foxes in the Aleutian Islands; most occurring in the early 1900s and many unrecorded ones prior to that (Burris and McKnight 1973). Russian explorers and settlers introduced red foxes onto some Aleutian Islands with the hope of adding to their fur harvests (Elkins and Nelson 1954). Unimak appears to have had red foxes introduced (Flux and Fullagar 1992).

Canada

Foxes (*V. v. fulva*) escaped from fur farms on Vancouver Island, British Columbia, and have become established in the Sayward forest north of the Campbell River. Large numbers were noted there in 1948 and since 1960 the sparse population has expanded south to Courtenay and west to the head of the Alberni Canal (Cowan and Guiguet 1960; Carl and Guiguet 1972).

The islands of Baffin, Cornwallis and Ellesmere appear to have been colonised in fairly recent times. In 1918–19 the fox reached southern Baffin Island (MacPherson 1964) and by the late 1940s occupied the entire island. In 1950 they crossed Fury and Hecla Strait to Melville Peninsula and expanded south to Repulse Bay. In 1962 they crossed Lancaster Sound north of Baffin Island and reached Resolute Bay on Cornwallis Island and Grise Ford on the south coast of Ellesmere Island.

In Canada red foxes have also been introduced to Anticosti Island (de Vos *et al.* 1956). Foxes have been introduced to Sable Island, Nova Scotia, but the species released is not identified (Flux and Fullagar 1992)

United States

In the nineteenth century populations of eastern red foxes (*V. v. fulvus*) were introduced into the lowlands of California. Those in the Sacramento Valley are believed to have descended from foxes that either were released or escaped from fur farms and most closely resemble red foxes from the northern plains states (Deems and Pursely 1978; Jameson and Peeters 1988). They became established in the Sacramento Valley in the vicinity of Marysville Buttes (Grinnell

1933; Seymour 1960). This population is still separated from others and has spread widely (Gray 1977) in the central valley and in recent years has extended to coastal areas of central and northern California (Lidicker 1991).

Red foxes were absent from the eastern deciduous forests of New England at the time of European settlement. Numerous accounts substantiate the introduction of English red foxes into New England in the eastern United States between 1650 and 1750. Thus the evidence tends to suggest that red foxes in the eastern United States are a direct descendant of English foxes or hybrids between the two (de Vos *et al.* 1956; Ables in Fox 1975).

Foxes from Britain were introduced to Maryland in the middle of the eighteenth century and later to Long Island, New Jersey, Virginia, and other eastern states. These may have had some influence on the genotype that exists there today.

The southern-most limit of the native fox range in the time of pre-European settlement is not known. It is conceivable that the introduced British foxes expanded in the eastern and southern areas of North America as the forests were cleared and agriculture took over (Seton 1929 in Lloyd 1981; Lloyd 1981). Foxes have certainly spread northwards to Baffin Island in recent times and in some areas of the United States they have extended their range also in recent times. They are not native to Tennessee, but were either introduced or migrated into western Tennessee in 1845 (Goodpaster and Hoffmeister 1952).

European foxes from England were introduced numerous times to the eastern states from 1650 to 1750 and have possibly crossed with the native species of fox (Presnall 1958). Some were imported from England and released in Virginia in this period and were reported to have become established and crossed with the local race (Gottschalk 1907). Many were also released near New York, Nantucket Island and on Martha's Vineyard also during these times (de Vos *et al.* 1956). Populations of foxes now in eastern central and south-eastern United States may be entirely from introduced animals (Gilmore 1946) or mixed. A population has become established in the Point Reyes Headlands, California, as a result of an unauthorised introduction (Gogan *et al.* 1986). The red fox is also considered to have inter-bred with the gray fox (*Urocyon cinereoargenteus*) (Howard and Marsh 1984).

Prior to 1941 red foxes were introduced to New York State on more than one occasion and in Dutchess County are said to be established and increasing (Bump 1941). They were introduced to San Juan Island, Washington, in 1963 and again in 1965 by the local inhabitants, presumably to control rabbits, and are still present there (Hall 1977).

Between 1964 and 1967 foxes (*V. v. fulva*) were introduced to small islands off the coast of Massachusetts to control populations of herring gulls (*Larus argentatus*). Annual predator introductions on the islands (all between 13 and 40 acres) of Outer Brewster, Middle Brewster, Calf, Kettle, Straitsmouth and Spectacle for two to four years caused major reductions in colony size and occasionally total abandonment of the island site. The introduction of the foxes effectively eliminated production of young gulls, but the foxes generally died out through lack of food on most islands, although some lasted more than one year. The herring gulls were a major problem to airports along the coast (Kadlec 1971) and this appears to have prompted the release of foxes.

Some 180 foxes were translocated in Iowa in around 1970–72 to study the movements and survival of foxes (Andrews *et al.* 1973). Foxes were also introduced to Texas by fox hunters and formed a viable population there (Deems and Pursley 1978).

■ DAMAGE

Predatory habits resulting in depredations on small game birds and mammals, domestic poultry, pigs and lambs and the role of the fox in rabies epizootics bring about control efforts at considerable monetary cost to people. Millions of dollars in the United States were paid out in bounties in the last 30 years but this was ineffectual in reducing populations. Foxes provide recreational hunting for people (Ables in Fox 1975). Changes in habitat by humans have also increased fox numbers in many parts of its natural and introduced range, further exacerbating the economic costs of fox control.

The number of active nests of two species of gull declined on South Manitou Island, Lake Michigan, when subjected to nine years of fox predation (Southern *et al.* 1985).

The Aleutian Canada goose (*Branta canadensis leucopareia*) has been severely reduced in numbers and this has been related to some translocations of red foxes which have had a detrimental effect on ground-nesting birds (Murie and Scheffer 1959; Franzmann 1988).

The impact of fox predation on indigenous fauna in Australia has been the subject of much conjecture but very little scientific evidence. The greatest losses in vertebrate fauna have occurred in the past 120 years and anecdotal and circumstantial evidence implicates the fox to a considerable extent (Hubach 1981;

Jarman 1986). Dietary studies from heavy bush country (Coman 1973; Brunner 1975) indicate that small native mammals comprise the bulk of food intake. However, it is difficult to partition the impact on fauna between predators (cats, dogs, foxes), land clearing for agriculture, sheep and cattle grazing, and changes in vegetation cover because of alteration in the timing and intensity of wildfires (and forest burning off) (Redhead *et al.* 1991). In Western Australia one study implicates foxes as a major factor in the decline for remnant populations of black-flanked rock-wallabies (*Petrogale lateralis*) (Kinnear *et al.* 1988). The spread of the fox in Western Australia appears to coincide with the disappearance of several medium-sized marsupials and a marked reduction in numbers of others. This effect is compounded by fragmentation of areas of native vegetation and forest by farmland (Christensen 1980; Christensen and Burrows 1986; Kinnear *et al.* 1988). The fox is not regarded as a serious agricultural pest, although there is some evidence of significant predation on new-born lambs and goats under certain circumstances (Cohen 1980; Hone *et al.* 1981), but other studies have shown that predation is negligible and that most lamb deaths could have been prevented by better management (Hubach 1981; Long *et al.* 1988).

The introduction of foxes from Britain to Sweden is reported to have resulted in the appearance of 'Samson foxes', which have an inherited deficiency of guard hairs (de Vos and Petrides 1967).

ARCTIC FOX
Blue fox, Siberian polar fox
Alopex lagopus (Linnaeus)

■ DESCRIPTION
HB 430–850 mm; T 225–550 mm; SH 250–300 mm; WT 1.4–6.0 kg rarely to 9 kg.
Summer, coat brown or greyish yellow; winter coat white or cream (also blue-black to pearl grey colour phase which remains uniform all year); head rounded; muzzle short and blunt; ears small and rounded; soles furred; tail thick black brush; outer sides of legs brown; belly and flanks yellowish white.

■ DISTRIBUTION
Holarctic. In northern Europe the arctic regions of northern Scandinavia, south to southern Norway and Sweden, and to northern Russia. In northern Asia to Kamchatka and also in Greenland, Iceland and Spitzbergen as far north as 85°N. In North America in Alaska, and arctic Canada including the northern Yukon, the Mackenzie district, northern Manitoba, Ontario, northern Quebec and coastal Labrador. Also on the islands of Jan Mayen and Bear, and the Kurile Islands(?).

■ HABITS AND BEHAVIOUR
Habits: mainly nocturnal or diurnal; lives in excavated burrow or den or under rock piles; no hibernation, active in temperatures down to −50°C in northern winters; caches food; 4 year population cycle

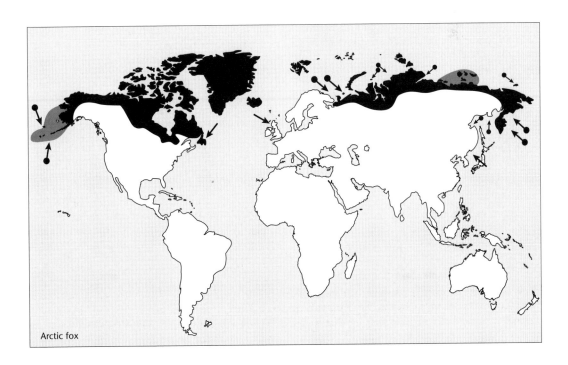

Arctic fox

of abundance; dens may be colonial. **Gregariousness:** solitary or pairs, but congregates in larger groups at food sources. **Movements:** wanders extensively and somewhat nomadic; irruptive migrations or immigrations at commencement of winter when tends to move south; some areas have regular seasonal movements governed by food availability; home range 16–25 km². **Habitat:** arctic and alpine tundra, forest borders, frozen polar seas, offshore ice flows; rocky coasts. **Foods:** small rodents (lemmings, squirrels, hares, voles), fish, molluscs, crabs, sea urchins, eggs and fledglings of ground-nesting birds, crustaceans, insects, berries and carrion, shellfish. **Breeding:** mates January–April, young born May–August; gestation 49–57 days; male monogamous, female monoestrous; litter size 4–7, 11 (up to 25 in times of food abundance); born blind, naked, helpless; female stays in den for few days after birth while male brings food; weaned 2–4 weeks; young disperse late summer after birth; mature 9–10 months. **Longevity:** about 7–10 years in wild to 14–15 years in captivity. **Status:** widespread and generally common.

■ **HISTORY OF INTRODUCTIONS**
Introduced successfully to the Kuril'skiye Ostrova, Aleutian Islands, Russian Federation, and the United States.

EUROPE
Russian Federation
The Arctic fox has been released on the Kola Peninsula, in northern Russia (Lindermann 1956; de Vos *et al.* 1956), on the Kuril'skiye Ostrova between Hokkaido and Kamchatka, on the Komandorskiye Ostrova between Kamchatka and the Aleutians, and on the Pribilof Islands in the Bering Sea.

The race *A. l. groenlandicus* was introduced in Russia (Naumoff 1950), but they were unsuccessful in becoming established (Shaposhnikov 1960). *A. l. lagopus* was released in the Komi region and Arkangel'sk, in 1926 to Ostrov Kolguev in the Barents Sea, and in 1928 (18 released) on Ostrov Mednogo near Ostrov Karaginskiye (Karaginskii Island) (Lavrov 1946; Kirisa 1974).

In 1925 (20 released), 1927 (11) and 1928 (16) *A. l. beringensis* were released on Shantarskiye Ostrova in the Sea of Okhotsk (Khabarosk area) In 1927, 13 were released on Ostrov Zavyalova (off the Magadan coast, Sea of Okhotsk). In 1929, 43 were released on Furugelin Island (Primorskii area) and seven more in 1939. In 1929, 14 were released on Anzer in the Solvetskiye Ostrova in the White Sea (Arkangel'sk region). Also in 1929, 96 were released on the Kil'dii Ostrova (Murmansk region).

UNITED KINGDOM
In the middle of the nineteenth century Arctic foxes may have been introduced in north-western Scotland, as one was trapped at North Point near Wester Ross in 1878 and others were obtained in the general area in 1848 and 1871 (Fitter 1959). More recently 40 animals bred for the fur trade were released by 'animal liberationists' in the United Kingdom (Baker 1986).

NORTH AMERICA
Alaska
Arctic foxes were introduced on Middleton Island, Alaska, in the 1920s and became well established there (de Vos *et al.* 1956; Rausch 1958).

United States
Arctic foxes have also been introduced in Minnesota (de Vos *et al.* 1956), but failed to become established.

PACIFIC OCEAN ISLANDS
Kuril'skiye Ostrova (Kuril I.) (Russian Federation)
Acclimatised in the Kuril'skiye Ostrova from releases in about 1915, arctic foxes occur on Ushisir, Sinsiru, Uriri and Kharukaru (Novikov 1926; Kirisa 1974), and are now relatively abundant there (Lavrov 1962), especially on the middle islands (Voronov 1963).

Aleutian Islands (United States)
On small uninhabited islands off the coast of Alaska breeding stocks of Arctic foxes were often released and allowed to breed naturally, and food was provided so the animals could be later harvested for fur (Burton 1962).

The Arctic fox was in this manner introduced successfully on many of the Aleutian Islands (Murie 1941; Rausch 1958) from Siberia and the Komandorskiye Ostrova (Gottschalk 1967). Originally they occurred only on the extreme western end of the Aleutian chain of islands, but they have been extensively introduced (especially blue-phase animals for fur) to other islands. They are native to Attu Island (Near I. group), but not to the islands to the east of here, where they have been spread for commercial purposes by humans (Fradkin 1980), including Amchitka Island (Rat I. group). None were observed on Unalaska Island (Fox I. group) in the late 1960s (Peterson 1967). Although some authors (Murie 1959) have concluded that the fox is a native on Attu Island and introduced to the remainder of the islands, it is hypothesised by others (Buskirk and Gipson 1981) that it is native to none of the Aleutians as no pre-Russian remains of foxes have yet been found.

Most populations in the Aleutians have been introduced for fur production. A number of introductions were made by the Russian-American Co. from 1819 onwards, but it is not known when they first began. It may have been as early as the 1750s on Attu and Atka (Andreanof I. group). Following the United States' purchase of Alaska in 1867 further introductions were made and confirmed up until World War 1 (see Gray 1937; Burris and McKnight 1973). By 1925 Arctic foxes had been released on about 80 islands throughout the archipelago and by 1936 more than 25 600 pelts had been harvested (Swanson and Hudson 1980; Buskirk and Gipson 1981; Lever 1985; Schmidt 1985).

Because of drastic changes in the fauna since the fox was introduced, attempts to eliminate them began on Amchitka Island in 1949. In the following 30 years attempts extended to Agattu, Nizki, Alaid, and Konga islands (Near I. group) (Springer *et al.* 1978).

Today the Arctic fox inhabits more than 40 of the Aleutian Islands and is also on St. Lawrence, St. Matthew (Bering Sea), St. Paul, St. George (Pribilof I, Bering Sea), Nunivak (Bering Sea) and the Komandorskiye Ostrova (Commander Islands) between Kamchatka and Aleutians (Buskirk and Gipson 1981; Lever 1985).

The poison 1080 has, or seems to have, successfully removed the Arctic fox from Kiska Island (Rat Is. group) in the Aleutians and the population of several native nesting birds have increased (USDA Report 1990).

■ DAMAGE
On Amchitka Island, Aleutians, the introduction of the Arctic fox and rats are said to have had a greater effect on the island's flora and fauna than Word War 2. Fox farming reached its heights between the two World Wars and, along with rats, nearly caused the extinction of the Aleutian Canada goose (Fradkin 1980). On Middleton Island, Alaska, they may have been responsible for the extermination of the original land mammals (Rausch 1958).

Drastic changes have been noted in the avifauna since Arctic fox introductions began. Fox predation is a major mortality source for some birds, particularly those with low predator avoidance capabilities (Buskirk and Gipson 1981) on the Aleutian Islands such as the Aleutian Canada goose (*Branta canadensis leucopareia*). A recent survey indicates that of more than 100 fox-infested islands south of the Alaskan peninsula, all had a complete absence of nocturnal seabirds (Lever 1985). More recently a study found that fox removal improved the nest success of the black brant (*Branta nigricans*) (Anthony *et al.* 1991).

PATAGONIAN FOX
Chico grey fox, Argentine grey fox
Dusicyon griseus (Gray)
=*Pseudalopex griseus*

■ DESCRIPTION
HB 446–670 mm; T 202–427 mm; WT 2.5–5.45 kg.
Small grey body; mid-dorsal stripe and tail tip black; some have chin black; under parts cream; underside of tail pale tawny and black; sometimes nearly all black.

■ DISTRIBUTION
South America. Patagonia and Ecuador south to southern Chile and southern Argentina to Tierra del Fuego.

■ HABITS AND BEHAVIOUR
Habits: mainly crepuscular and nocturnal, but occasionally diurnal; den among rocks, tree bases, low shrubs and burrows of other animals. **Gregariousness:** density 0.95 –4.35/km^2 to 1/43 ha. **Movements:** sedentary(?). **Habitat:** low open grassland and forest edges; lowlands and foothills of coastal ranges, low scrubs. **Foods:** omnivorous; rodents, rabbits, berries, reptiles, birds and eggs, insects, sheep carrion and plant material (grass). **Breeding:** November?; gestation 53–58 days; litters 2–5. **Longevity:** probably only several years in wild, to 13 years as a captive. Related *D. gymnocerus*. **Status:** scarce and rarely seen because of hunting.

Patagonian fox

■ HISTORY OF INTRODUCTIONS
SOUTH AMERICA
Falkland Islands (United Kingdom)
Patagonian foxes were introduced to some southern islands to control rabbits. In the early 1930s J. Hamilton released them on Weddell, Statts and Beaver islands off the west coast of West Falklands. Today they still occur in these islands and also on Tea Island in the Weddell group, as well as River Island near Pebble and on Split near Roy Cove (Lever 1985).

Isla Chanaral (Chile)
The foxes *D. griseus* and *D. culpaeus* were both introduced to this island in 1941 by Ramon Callejas with the idea of building a fur industry. At first they flourished but then died out. It was reported that after annihilating the diving petrel population they died out three years after their introduction. Millie, who visited the island, saw seven foxes in 1943. Arraya (1983) was told that only two foxes were introduced to control introduced European rabbits. Visits to the island in 1982–85 failed to find any sign of foxes. (Johnson 1965; Mödinger and Duffy 1987).

Tierra del Fuego (Chile and Argentina)
Patagonian foxes were possibly introduced first in Estancia Cullen (Goodall 1979; Pine *et al.* 1979), but others (Jaksic and Yanez 1983) say this is probably incorrect. Twenty-four young foxes of both sexes from Magallanes and perhaps also adjacent Argentina were released at Onaisin (65 km ESE Porvenir) in 1951.

Because the rabbit disease myxomatosis was released soon after the fox release, it is not known what effect the foxes may have had on the rabbits. They are now spread all over the Chilean part, but probably their greatest abundance is in Bahia Inutil in areas where they were first released. Patagonian foxes now outnumber the indigenous fox (*Dusicyon culpaeus*) by 10:1. It has also been found that the native fox (*D. culpaeus*) is a better predator of rabbits and it is doubted that introduction of Patagonian foxes (*D. griseus*) would have helped against the introduced rabbit (Jaksic and Yanez 1983).

■ DAMAGE
Foolish introductions of Patagonian foxes to Weddell Island (off Falklands) must have played havoc with the avifauna as it has done with the sheep (Cawkell and Hamilton 1961). They are thought to be having an effect on the goose (*Chloephaga picta*) populations, and some sheep farmers claim losses due to fox predation in the lambing season (Walker 1992).

RACCOON-DOG
Raccoon-like dog
Nyctereutes procyonoides (Gray)

■ DESCRIPTION
HB 500–800 mm; T 100–260 mm; WT 3–10 kg.
Coat generally dirty brownish grey or yellowish-black tipped with a blackish brown shade; dark band on back forming a collar, wider at the shoulders; head small; muzzle short and pointed; face mask black; ears short but large; hairs elongated at sides of head to form 'side whiskers'; neck ash coloured; chest brownish black; belly yellowish-brown; legs short, blackish brown; tail short, bushy, and dark tipped.

■ DISTRIBUTION
Asia. Eastern Siberia from the Amur River, northern Manchuria and Sakhalin, south to Japan, southern China and North Vietnam, west to Shansi and eastern Szechuan.

■ HABITS AND BEHAVIOUR
Habits: mainly nocturnal, rarely diurnal; lives in burrows up to 2 m deep and with two or more entrances, or in natural cavities; dormant in very cold weather in northern regions. **Gregariousness:** solitary, pairs or family groups of 5–6; larger aggregations at food sources; density 1–20/1000 ha. **Movements:** sedentary and partly nomadic?; home range 8–200 ha, but only as little as 8–48 ha in introduced range. **Habitat:** river valleys, near lakes, grassy plains, forest slopes with underbrush, rocky banks,

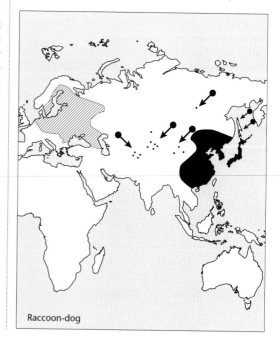

Raccoon-dog

marshes, rocky wooded ravines. **Foods:** omnivorous; small rodents (voles, gerbils, mice) and other small mammals (hedgehogs, shrews, moles), reptiles, amphibians (toads, frogs), molluscs, tortoises, fish, crabs, sea urchins, birds, carrion, insects (dung beetles, cockchafers), nuts, acorns, eggs, fruits (grapes) and berries, grains, refuse, garbage and roots. **Breeding:** mates February–April, young born May–June: gestation 56–64 days; litter size 5–8, 19; 1 litter/year; young born blind, toothless, haired, weigh 60–90 g; eyes open 9–10 days, weaned 1.5–2 months; sexual maturity 9–11 months. **Longevity:** 7–8 years in wild, 10 in captivity. **Status:** common, spreading in Europe.

▆ HISTORY OF INTRODUCTIONS

Successfully introduced and or colonised west to Germany and parts of the Russian Federation. Originally the species was native to eastern Asia, but was introduced to Russia in the 1930s as a fur animal and has subsequently spread to Finland, Sweden, Poland and Rumania (Lyneborg 1971).

EURASIA

From 1927 to 1957 over 9000 raccoon-dogs were released in some 40 regions, territories and autonomous republics in the European and Asian parts of the Russian Federation and adjacent independent republics west of the species' natural range (Yanushevich 1966; Walker 1992). They thrived in many areas particularly around Moscow, Leningrad, Kalinin, Smolensk, and in the Pripet Marshes on the western border. From these and other introductions the species spread westwards into central Europe to Romania, Poland, Hungary, Czechoslovakia, Switzerland, Finland, Austria, Netherlands, Germany, Sweden and Norway (Lindermann 1956; Niethammer 1963; Kirisa 1974; Burton and Burton 1976).

By the early 1970s the raccoon-dogs had spread across Germany through Turkey, Greece, Yugoslavia, Denmark and Norway (Lever 1985). They had now reached Switzerland, the low countries and central Sweden.

In the early 1960s raccoon-dogs first appeared, probably as a result of range extension from western China in the People's Republic of Mongolia near Lake Buir and Mount Chorbalsan (Lever 1985).

Bulgaria
Raccoon-dogs reached Bulgaria in 1967 (Lever 1985).

Czechoslovakia
Raccoon-dogs reached Czechoslovakia in 1959 having travelled north from Poland and south-east from Romania (Robben 1975; Lever 1985).

Finland
The raccoon-dog's first appearance in Finland was in about 1934 when a single animal was recorded. More were reported in other areas between 1935 and 1939, mainly in southern Finland (Siivonen 1943, 1953, 1958; Suomalainen 1950; Robben 1975). In 1941 some had reached south-western Finland (Siivonen 1958) and they appeared in northern Finland in 1945–46 (Notini 1948).

Germany
Raccoon-dogs reached eastern Germany in 1961 and western Germany in 1962 (Robben 1975; Lever 1985).

Hungary
Raccoon-dogs reached Hungary in 1961.

Poland
The raccoon-dog was first encountered in Poland in 1955 at Bialowieza Forest and by 1965 they were frequently found in this and the Olzztyn Voirodeships and were proceeding to colonise westwards (Dehnel 1957; Suminski 1963; Dudzinski *et al.* 1965; Robben 1975).

Romania
Raccoon-dogs appeared in Romania in 1951 (Robben 1975; Lever 1985).

Russian Federation and adjacent independent republics
Introductions of the raccoon-dog in the Russian Federation and adjacent independent republics appear to have spanned the period from about 1926 to 1963. They were introduced in the Kalinin area of central Russia and since 1934 have increased tremendously (Lindermann 1956). Here they are looked upon as a desirable addition to the fauna and by the 1950s were looked upon as one of the principal fur-bearers.

Translocated into western Russia, by 1944 raccoon-dogs had been released in about 36 locations in the west of the Ukraine where they had spread to Karelien(?). From here they wandered westwards to Finland and Norway (Siivonen 1953, 1958), Poland (Dehnel 1957), Baltic republics (Anon. 1959) and Romania. Finland stocks were partly from fur farm escapees (Niethammer 1963) and partly from colonisation.

They are now widespread in many habitats north of 63°N and south of the Caucasus, and have recently spread as far as Sweden, Czechoslovakia (Roben 1975), Hungary in 1962 and France in 1979.

Originally raccoon-dogs occurred only in an area of the Far East. From 1929 to 1955, 8850 (before 1944 some 3288 released and from 1949 to 1955 some

5567) were introduced into 82 oblasts, Krais and republics in the European section, middle Asia and in Siberia. As a result of these introductions, their introduced range is now several times larger than their native range (Sofonov 1981).

Releases of raccoon-dogs in Russia and adjacent republics

Place	Date	Number
Northern and Central Europe–Russia		
Murmansk	1930	30
Pskovsk	1947	80
Komi	1954	100
Latvia	1947	84
Arkhangelsk	1950–53	219
Astonsk	1950	88
Karelsk	1957	?
Belorussia	1936–53	370
Leningradsk	1936–53	82
Smolensk	1936	62
Novgorodsk	1935	50
Bryahnsk	1936	38
Southern Europe–Russia		
Azerbaidjan	1938–39	147
Gruzinsk	1938–39	168
Amyahn	1934, 52	118
Astrakhansk	1936–52	406
Volgogradsk	1947–49	269
Saratovsk	1936–48	276
Kuibshevsk	1954	118
Orenburgsk	1934, 54	147
Ukraine	1935–54	1529
Moldarsk	1949–54	365
Central European USSR		
Kalininsk	1934	50
Yahroslavsk	1957	58
Kostromsk	1936–54	234
Gorkovsk	1936, 40	104
Kirovsk	1950, 54	155
Udmurtsk	1954	59
Mariisk	1948–9	88
Chuvashsk	1948	95
Tatarsk	1934–52	223
Mordovsk	1948	150
Penzensk	1934, 54	109
Ryahzansk	1936	100
Tulsk	1958	61
Orlovsk	1954	100
Voronejsk	1936	100
Northern Caucasus		
Stavropolsk	1934–53	275
Krasnodarsk	1936–53	325
North-Osetinsk	1951, 53	91
Karbardino-Balkarsk	1952	102
Dagestan	1934–62	?
Southern Urals		
Bashkirsk	1935, 52–55	142
Chelyahbinsk	1953	63
Southern RSFSR (Asia)		
Kazakhstan	1936–37	386
Kirgizstan	1934–53	149

In Siberia and the Far East a number of translocations were made to increase the range and introductions in new localities in these regions from 1929 to 1955 including: Altaisk, Gorno-Altaisk, Tomsk, Buryahtsk, Yahkutsk, Tuvinsk, Krasnoyahsk, Irkutsk, Maiminsk, Primorsk and on Sakhalin Island.

In 1936 they were released in Kostroma (Fateev 1960), where they became widespread and by 1953 were becoming established in Udmurtia. Between 1936 and 1953 some 346 were released in Belorussia, where they became established and widespread (Samusenko 1962). Other successful releases commencing in the 1930s include those in Kirghizstan in 1934, 1944 and 1950, Astrakhan between 1936 and 1939, Karelia in 1936, Tatar in 1934–35, the northern Caucasus between 1934 and 1960, the Ukraine in 1936 (Yanushevich 1966), and in Azerbaidjan in 1938–39 when 147 were released (Aliev 1962). Unsuccessful introductions occurred in Bashkivia in 1935, 1952 and 1955 and in Moldavia between 1949 and 1954.

Later introductions include the release of 100 raccoon-dogs in the north-eastern part of Latvia in 1948, where by 1962 the population was estimated to be in the vicinity of 10 000 (Lapin 1963). There was also the release of 101 in the Komi in 1954 (Yurkin 1961), and others in the Arkhangelsk region in 1950–53, where they became established locally, on Sakhalin with animals from the Far East in 1955, and in the Caucasus and West Siberia (Yanushevich 1966). Other introductions which appear to have been successful but about which details are lacking include: Krasnodar Krai (Kotov and Ryabov 1963), the Urals (Pavlinin and Shvarts 1961), Tuva (Volchenko 1964) and Kazakhstan (Sludskii and Afanas'ev 1964). The raccoon-dog was apparently unsuccessful in Buryatiya (Izmailov 1969).

By 1966 the highest numbers of raccoon-dogs were to be found in Novgorod, Kakinin and Moscow regions. Acclimatisation had been less successful in the Asiatic part of the country (Yanushevich 1966). Lindemann (1956) indicated that introductions into Siberia were a dismal mistake as the fur was found to be of no value.

Following introduction to north-west Soviet Union between 1935 and 1953 it has since spread over northern and eastern Europe (Lavrov 1971) and is now established throughout southern and central parts of Finland during the past three to four decades. Numbers reached a peak in the mid-1980s and since then declined slightly (Helle and Kauhala 1993).

Sweden

By 1945–46 some raccoon-dogs were recorded in

Sweden across the border from Finland (Siivonen 1958; Robben 1975).

Switzerland
Some time after 1967 raccoon-dogs reached Switzerland and the Low countries (Lever 1985).

■ DAMAGE
The results of the acclimatisation of raccoon-dogs in the Russian Federation are considered questionable (Shaposhnikov 1960), as their fur is of little value.

Raccoon-dogs cause damage to game animals in European Russia, especially Astrakhan where they destroy the nests of game birds and are carriers of rabies (Yanushevich 1966). In the Ukraine damage is caused to vineyards and to waterfowl. The results of their establishment in Latvia has been the displacement of badgers from their burrows (1940–58, 50 per cent burrows occupied by badgers; 1961–62 only 20 per cent) and the extermination of useful birds and animals which formed 61.9 per cent of the contents of 60 stomachs of raccoon-dogs (Lapin 1963).

Opinion as to whether the raccoon-dog is a pest where introduced appears to vary somewhat. Some consider it is not a serious pest, but may, however, be a carrier of rabies and other diseases (Burton and Burton 1969). Others report that as it becomes established in Europe in a new area it displaces or destroys native species (Sayre 1983). Still others considered it a nuisance west of the Soviet Union, destroying small game animals and fish, and that their fur is not as long and dense as it should be and as such is worthless (Walker 1992).

FERAL DOG
Domestic dog, dingo, wild dog
Canis familiaris Linnaeus

■ DESCRIPTION
There are 400 breeds of dog that could become feral, but probably few do, except for the larger breeds.

HB of typical feral dogs: 360–1450 mm; T 130–510 mm; SH 150–840 mm; WT 9–80 kg and up to 150 kg (TL 1054 – 1397 mm; WT 9.5–23.1 kg in Alabama).

New Guinea dog (*C. f. hallstromi*): short reddish brown hair; head broad; ears pricked; tail feathered; pelage varies as for dingo.

Kuri (*C. f. otahitensis*): short legged fox-like creature with bushy tail.

Dingo (C. f. dingo): HB 1170–1240 mm; T 300–330 mm; SH c. 500 mm; WT 10–20 kg.

Tawny yellow, sometimes white or black, brown or shades of these colours; feet and tail often white.

■ DISTRIBUTION
Widespread as a domestic animal (originally that of *Canis lupus*? see note below).

Dingo: throughout mainland Australia, but disappeared from many settled southern districts in both western and eastern Australia. Formerly widespread throughout southern Asia. *Kuri:* introduced by Polynesian explorers and colonists widely in the western Pacific, including the Society Islands, Tonga, Hawaiian Islands, Marquesa Islands and New Zealand.

■ HABITS AND BEHAVIOUR
FERAL DOGS
Habits: mainly nocturnal and crepuscular; often feral only at night; often lives in abandoned buildings or under cars or stairways or parks; dominance hierarchy. **Gregariousness:** density 150–230/km^2 in large cities in United States; groups 2–5, 7, or solitary. **Movements:** home range 0.1–11.1 ha or 444–1050 ha (Alabama). **Habitat:** urban fringes, agricultural areas, towns. **Foods:** garbage, domestic stock, carrion, small animals and vegetation. **Breeding:** births in heavy cover; oestrous twice/year, lasts 12 days; gestation 63 days; litter size 3–10; lactation 6 weeks; sexual maturity in 10–24 months. **Longevity:** 12 years in wild, a few to 20 years.

DINGO
Habits: Mainly nocturnal in hot temperatures. **Gregariousness:** solitary, pairs occasionally, groups to 6–7. **Movements:** Home range dingo 30–200 km^2 in Queensland. **Habitat:** semi-desert to forest and alpine heaths. **Food:** rabbits, rats, kangaroos, domestic stock. **Breeding:** mates April–June, pups July–September. Pups born late winter/spring. **Status:** Common in north of range, greatly reduced in the south. **Longevity:** 4–8, rarely to 14 years.

Note: Recently there has been a tendency to call the dingo Canis lupus dingo and in the text I have endeavoured to clearly identify to which dog is referred. At present it is thought that the dingo and domestic dog were derived from a wolf, although some authorities claim multiple domestication and others a single ancestral form for the dog. It is estimated that the dog was first domesticated some 10 000 to 12 000 years ago (Scott 1968) or in Mesolithic times (Zeuner 1963).

The relationship, one to the other, of domestic dogs, dingoes and other members of the canidae has been of considerable interest and argument for some time. Recent information suggests that the plains wolf (Canis lupus pallipes) is the likely ancestor of the dingo (Bodenheimer 1958; Marlow 1962; Fiennes and Fiennes 1968). Certainly domestic dogs and dingoes share a common pool of genes and all canids except foxes possess genes in common (Clark et al. 1975).

The New Guinea dog (C. f. hallstromi *Troughton*) also falls into the familiaris group. It is superficially similar to the dingo, although much smaller and was probably introduced to Papua New Guinea about 200 years ago (Flannery 1990). Evidence suggests it is a breed of this species, although many zoologists do not agree (Ryan 1972; Gollan 1984). Most consider the dingo and C. hallstromi *were carried by natives to Australia and Papua* (Fox 1975).

The Kuri or Polynesian dog of New Zealand (C. f. otahitensis) can be included in the C. familiaris group although its bona fides as a subspecies may be in doubt. Here it is treated as a subspecies.

From the Balkans to North Africa to South-east Asia, dogs known as pariahs lead semi-domestic or even feral existences around villages. These dogs have a primitive physical appearance and are probably closely related to the earliest dogs as well as the dingo (Gryzimek 1975; Fox 1978). There are also wild dog populations in New Guinea and Timor that are related to the primitive pariah–dingo group (Troughton 1971). It is possible that these dogs were spread by maritime peoples of south central Asia rather than by migrating aboriginal peoples (Gollan 1984). There are many other populations of feral dogs, notably on islands in Italy, but these are descended from domesticated individuals (Lever 1985).

HISTORY OF INTRODUCTIONS

Feral/domestic dogs on islands

Island	Date introduced/present	Notes
Alejandro Selkirk (see Más Afuera)		
Aleutian	?	?
Amami Oshima	1980s	still present
Anderson (Aust)	1830?	present 1830, now absent
Auckland	1840s, 1860s	present for a time
Augustus (Aust)	?	present?
Bahamas	1990s	present
Bermuda (WI)	1980s	now removed
Bigg (Aust)	?	present?
Borneo	1980s	still present
Breaksea (Aust)	1900	pet dog *c.* 1900, now absent
Bremer (Aust)	?	domestic dogs?
Bribie (Aust)	?	present?
Bruny (Aust)	1829	domestic dogs, present with aborigines
Burrup (Aust)	?	domestic dogs
Canary Is	?	present
Cape Barren (Aust)	1831	present 1831, ? pet 1874
Cayman	1980s	present and causing damage
Cayman Brac	1980s	present and causing damage
Centre (Sir Ed Pellew) (Aust)	?	present?
Chagos Archipelago	1840–1975	possibly still present
Chatham (NZ)		present at one time
Clarke (Aust)	*c.* 1830	present 1830
Cocos-Keeling	?	present?
Croker (Aust)	?	domestic dogs
Cuba	1511–1980s	still present?
Curtis (Aust)	?	present?
Deal (Aust)	?	? pet
Elcho (Aust)	?	domestic dogs
Enderby (NZ)	1840s,1890s	survived for a while, none now
Flinders (Qld, Aust)	?	present?
Flinders (Tas, Aust)	*c.* 1872	wild dogs, 'kangaroo dogs' present 1872
Floreana (Galápagos)	1842-	still present 1986
Fraser (Aust)	?	dingo introduced?
French Frigate Shoals	1859, 1942–43	possibly not feral

Feral/domestic dogs on islands (*continued*)

Island	Date introduced/present	Notes
Fuerteventura (Canary Is)	?	possibly present in 1950s
Galápagos	1842–1986	still present
Gough		introduced?
Great Dog (Aust)	*c.* 1830	wild dogs plentiful 1830
Great Inagua (Bahamas)	1990s	present
Green (off Kure)	1837–43, 1960s	now probably only pets?
Groote Eylandt (Aust)	?	present?
Hawaiian	arrival Polynesians 1960s	possibly still present
Heard	1947–55	there until 1955
Heron (Aust)	?	present?, now absent
Hinchinbrook (Aust)	?	dingo
Hispaniola	1526–1809	exterminated, none present 1950s
Hunter (Aust)	*c.* 1830	present 1830, left by sealers; 10 'kangaroo dogs' in 1851
Inaccessible		lived there for a time
Isabela (Galápagos)	1835–1986	still present 1990s
Johnston Atoll	1963–65, 1966–69	semi-feral, not present now
Juan Fernández	1686–1830	exterminated
Kangaroo (Aust)	?	wild dogs present?
Kerguelen	1902–28	there until 1928?
King (Aust)	*c.* 1887	wild dogs, present in 1887
Lady Elliott (Aust)	?	dog, one ?pet removed 1969
Little Barrier (NZ)	<1896	present for short period
Little Cayman	1980s	present and causing damage
Little Dog (Aust)	*c.* 1831	wild dogs, several in 1831
Lizard (Aust)	*c.* 1880	two in 1880
Lord Howe (Aust)	?	wild dogs
Macquarie	*c.* 1810–1820	there until 1820?, but not after
Marchinbar (Aust)	?	present?
Maria (Aust)	*c.* 1884	dogs present 1884
Marianas		
Más Afuera (Juan Fernández)	1618	exterminated by 1830
Más á Tierra (Juan Fernández)	1618	exterminated by 1830
Melville (Aust)	?	present?
Middle Osborne (Aust)	?	present?
Milingimbi (Aust)	?	domestic dogs?
Mornington (Aust)	?	present?
Mutton Bird (Aust)	?	wild dogs, natural spread
New Caledonia	1770s–1950s	possibly still present
New Zealand	Maori settlers	
Rabama (Yabooma?) (Aust)	?	dingoes & domestic dogs
Robinson Crusoe (see Más á Tierra)		
Rotamah (Aust)	?	wild dogs
Ryukyu	1980s	present on one island(?)
San Cristobal (Galápagos)	1842–	still present 1986
Santa Cruz (Galápagos)	1925–45?	present 1986?
Sims (Aust)	?	present?
South Georgia	?	there for a time
South Solitary (Aust)	?	? now absent
South West, Pellew (Aust)	?	present?

Feral/domestic dogs on islands (*continued*)

Island	Date introduced/present	Notes
St. Croix (West Indies)	1966–	possibly still present
St. Francis (Aust)	1922	present?
St. Helena	early 17th century	no longer present
Sunday (Aust)	?	wild dogs present?
Swan (Aust)	*c.* 1830	30 wild dogs in 1830
Tasmania (Aust)	after settlement	wild dogs present
Three Hummock (Aust)	*c.* 1830	wild dogs present 1830
Tristan de Cunha	1817–1963	all but two destroyed 1963
Vanderlin (Aust)	?	present?
Vansittart (Aust)	*c.* 1840	40 wild dogs in 1840, now absent
Walker (Aust)	*c.* 1830	sealers dogs in 1830
Whitsunday (Aust)	1930s	present 1930s, 1974, now absent
Wigram (Aust)	?	domestic dogs, one dingo introduced

AFRICA

South Africa

Feral dogs are present in the Cape of Good Hope Province in many areas (Hey 1974).

ASIA

Borneo

Feral dogs are occasionally found near human settlement and around abandoned logging camps and other places in Borneo (Payne *et al.* 1985).

AUSTRALASIA

Australia

The oldest confirmed remains of dingos date from 3500 years ago and were found at Madura Cave in Western Australia (Milham and Thompson 1976). It is assumed that the introduction of the dingo occurred after the inundation of Bass Strait, 11 000 years ago, as they are absent from Tasmania, and it is assumed they arrived with Asian seafarers rather than the aborigines (Corbett 1985).

Dingoes are now widespread in Australia except for Tasmania and most offshore islands, but have been introduced to some such as Fraser Island, Queensland (Wilson *et al.* 1992).

In Western Australia there is always a steady drift of domestic dogs into the wild, particularly around the cities and large towns. The extent of cross-breeding with the dingo is a matter of conjecture, but they will mate and produce fertile offspring (Tomlinson 1955).

Present indications are that the introduction of dingo is fairly recent geologically and possibly arrived about 9000 years ago (MacIntosh in Fox 1975).

While the dingo is found throughout most of Queensland, wild domestic dogs fail to survive inland but do so near the coast (Mitchell *et al.* 1982).

Wild dogs are a common nuisance in the urban fringe around Adelaide, South Australia (Burley *et al.* 1983).

Wild dogs occur in Victoria (Stevens 1981), and in eastern New South Wales hybrids comprise a large proportion of the wild dogs (only 25 per cent pure dingo) (Newsome *et al.* 1973). Damage in the form of stock losses (primarily sheep) varies between 1 and 7 per cent (Hone *et al.* 1981).

Feral dogs appear to exist at times in all states, including Tasmania, and the Northern Territory of Australia.

Papua New Guinea

Dogs were probably introduced with early indigenous peoples about 2000 years ago (Flannery 1995). They appear to have been introduced to Papua New Guinea more recently than the dingo was to Australia as the oldest archeological evidence is less than 2000 years. Feral populations are known from certain subalpine and alpine grassland regions such as Star Mountains and Wharton Range (Flannery 1995).

European domestic dogs were brought in after the German colonisation of Papua New Guinea. Some have bred with the 'village dog', there prior to the Europeans, and in turn reached the packs of wild dogs found in many parts of Papua New Guinea (Herrington 1977). Many dogs have deserted the villages and roam the mountain ridges as feral animals (Ryan 1972).

ATLANTIC OCEAN ISLANDS

St. Helena

An abundance of feral dogs was noted on the island in the seventeenth century (Temple 1914), which were assumed to hunt the many feral goats there at the time (Cronk 1986).

EUROPE

Feral dogs are often numerous in many parts of Europe, particularly the south. If breeding in the wild, they have a tendency to revert to a jackal-like form though often variable in colour (Burton 1991).

Russian Federation

Feral or stray dogs are a problem in some reserves in the Russian Federation (Filonov 1980). Some were removed from the Hopersky Reserve, Veronezti oblast, when wolves (*Canis lupus*) moved into the reserve. The dogs had been there since the mid-1960s (Ryabov 1979). In the late 1940s feral dogs were present in western Georgia (Vereshchagin 1950).

Italy

The largest population of feral dogs in Europe may be in Italy where it is believed there are 80 000 or more (Lever 1985).

INDIAN OCEAN ISLANDS

Chagos Archipelago

Feral dogs were reported on Egmont Atoll in the 1840s (Bourne 1971). They were reported as widespread and feral on Diego Garcia in 1972 when an extermination (shooting) campaign was under way (Hutson 1975).

PACIFIC OCEAN ISLANDS

The spread of the dog throughout the south-west Pacific and Moluccas is poorly understood. It is clear it reached the Halmahera area and most of Polynesia by 1000 years ago. Domesticates exist on most islands, but feral populations are more restricted and not known with any certainty (Flannery 1995).

Auckland Islands

Dogs were introduced in the 1840s to Enderby Island (off Auckland) and again in the 1890s, when they were present for a few years. They were introduced to Auckland Island in the 1840s, 1860s, 1880s and in 1900, and survived for a while, but there are none there today (Taylor 1968).

Galápagos Islands

Domestic dogs were first introduced to the Galápagos with the colonisation of Floreana (Charles) Island by General Jose Villamil in 1832, and later in 1842 when he relocated the settlement and the dogs on San Cristobal (Chatam). Since this time feral dogs have occurred on both islands (Melville 1856; Salvin 1876;

Martinez 1915; Slevin 1931, 1959; Thornton 1971; Barnett 1986).

Little appears to be known of the introduction of dogs on Santa Cruz (Indefatigable) Island, except that the first permanent settlement was established in the 1920s and members of the Norwegian Ulve expedition shot several dogs in 1925. Feral dog tracks were noted there in about 1935 at Tortuga Bay and also some dogs were introduced during American occupation during World War 2 (Salvin 1876; Heller 1903; Beebe 1923, 1924; Kastdalen 1982; Barnett 1986).

Several dogs are reported to have been abandoned by General Jose Villamil on Isabela (Albemarle) Island in 1835 while on a hunting trip to the island. The first wild dogs were seen in 1868 long before the first settlement was established in 1897–1903. Predation by dogs on tortoise eggs was noted in 1898 and several wild dogs were observed by passengers on a passing ship in 1906. By 1913 they were said to be preying heavily on wild cattle populations in the highlands (Salvin 1868; Martinez 1915; Slevin 1931, 1959; Barnett 1986).

Although there have been claims of up to 5000 wild dogs (Naveda 1950) on Isabella Island, more recent estimates of 200–500 animals appear more reasonable (Kruuk 1979; More 1981; Barnett 1986). On Santa Cruz Island there have probably never been more than 40–50 (Naveda 1950; Kruuk 1979; Barnett 1986).

Wild dogs still occur on the Isabella Island, Santa Cruz, Floreana and San Cristobal (Eckhardt 1972; Barnett 1986; Benchley 1999).

Hawaiian Islands

Dogs were taken to the Hawaiian Islands by the Polynesians who used them as food, but after the introduction of the European dog they quickly lost their identity by cross-breeding (Kramer 1971). As early as 1840 wild dogs were numerous in the interior of the island of Hawaii (Wilkes 1845) and in 1848 a poisoning program was carried out to rid the island of the ever increasing numbers.

Dogs still range the high mountains of Hawaii, where for 90 years they have been predators of feral livestock. They run free from time to time on all the islands, but their abundance is not well known. It was reported (in press) that during a seven-year period on Oahu in the late 1940s and early 1950s that at least 1000 were captured, and in a program in 1959–60 some 100 were exterminated. They still occurred on many islands throughout the 1950s and 1960s and there were many reports (in press) of them killing or maiming domestic stock and wildlife (Kramer 1971).

There were two dogs on Johnston Atoll in 1963–65 and from 1966 to 1969 the numbers were higher. Although these animals may not be truly feral they did prey on the seabirds (Kirkpatrick 1966; Amerson and Shelton 1976). One was found abandoned on French Frigate Shoals in 1861, presumably left in 1859; introduced in 1942–43 with humans but not feral there (Amerson 1971). One was also found on Green Island, off Kure Atoll in 1843 by the crew of the 'Parker' which was apparently left there following the wreck of the 'Gladstone' in 1837 (Woodward 1972). Dogs have more recently been kept (since 1960s) on the island as pets of the human occupants.

Juan Fernández

Mastiffs were introduced by the Spaniards in 1686 to control goats, which were being utilised by buccaneers as a source of food. The dogs increased to such an extent that they became a danger to humans and were exterminated by 1830, but kept the goats down while they existed (Holgate and Wace 1961).

Kerguelen

Wild dogs have been reported on the islands from 1902 to *c.* 1928 and are reputed to be the descendants of sled dogs abandoned by the Gauss Expedition in 1902 (Mawson 1934; Jeannel 1941), but this has been denied (von Drygalski 1935) and the existence of the animals doubted (Reppe 1957).

Little Barrier Island

The Maoris kept dogs on this island prior to 1896 and these wandered free over the island. However, they did not survive when the Maoris left in 1896 (Watson 1961).

Macquarie Island

As early as 1815 it was reported that dogs were numerous on the island. They were present in 1820 but there are few references to them after this date (Taylor 1979). They were reported wild on the island in 1821 (Bellinghausen 1948), but other reports indicate that they died out before 1820 (Holdgate and Wace 1961).

New Caledonia

Feral dogs were present on the island in the 1950s (Barrau and Devambez 1957). Captain Cook, probably on his second voyage, left a dog and a bitch on Mallicollo (16°25′20″S, 167°57′23″E) (Kippis 1904).

New Zealand

The Polynesian dog or kuri (a small to medium canid) arrived with the Maori settlers and was a domesticated pet and source of food when the Europeans arrived. It seems unlikely that the dog ever became truly independent; at least not in any numbers, although a dog-like creature was seen by James Cook's party at Pickergill Harbour (King 1984, 1990).

In the eighteenth century the species was found on the Tuamotus, Society Islands, Hawaiian Islands, and New Zealand. It was formerly more widespread and occurred on Tonga and the Marquesas (King 1990).

Their relationship to other varieties of domestic dog in the south Pacific is uncertain (Corbet and Hill 1976).

Ryukyu Islands

Feral dogs are present on Amami Oshima in the Ryukyus (Hayashi 1981).

Tristan da Cunha

Dogs were introduced before 1824 to Tristan and are occasionally left on Inaccessible Island by Tristan islanders (Holdgate and Wace 1961). When the residents left the island following a volcanic eruption in 1961 many dogs were left behind. These ran wild and killed nearly all the 740 sheep also abandoned on the island. When the residents returned in 1963 all but two of the dogs were destroyed (Anon. 1963).

NORTH AMERICA
Canada

Feral dogs are present and established in British Columbia (Carl and Guiguet 1972), and probably occur in other parts of Canada.

United States

Surveys in the United States have found feral dogs to be present in almost every state (McKnight 1964; Scott and Coney 1973). Feral dogs were causing problems in Pennsylvania about 1952 (Sand 1952) and Virginia about 1953 (Bowers 1953). Growing packs of homeless dogs were reported in New York State (Petruska 1949) in the late 1940s. One such study found 32 000–54 000 free-ranging dogs (or 450–750/2.59 km^2) in Baltimore (Beck 1970). Feral or free-ranging dogs are also present in California (Howard and Marsh 1984) in the 1970s in Crab Orchard National Wildlife Refuge in Illinois (Nesbitt 1975), in Alabama (Scott and Causey 1973), Colorado, and in Coeur d'Alene River drainage area, northern Idaho (Lowery and McArthur 1978). In these areas they are sometimes predators of deer, but not enough is generally known about their predator–prey relationship (Lowry and McArthur 1978).

It is estimated that up to half of the dogs in Baltimore, Maryland (10 000–100 000) are free-ranging, at least at times (Beck 1973; 1975). They feed largely on garbage and have been implicated in the spread of several diseases, attack people and may assist rat populations by overturning garbage cans so these animals can obtain food.

The city of Los Angeles, California, spends US$1.25 million each year collecting and killing stray dogs and cats. They occur widely in this state menacing both wildlife and sometimes humans (Lidicker 1991). In New York City it is thought that between 40 000 and 60 000 dogs were running loose and in the state of Georgia there may be as many as 500 000 free-running dogs (Caras 1978).

WEST INDIES
Bahamas
Feral dogs are present on Great Inagua, Bahamas.

Bermuda
Feral dogs were present in Castle Harbour Islands National Park until their removal (Lever 1985).

Cayman Islands
On Cayman Brac and Little Cayman, between Jamaica and Cuba, feral dogs have reduced some reptile populations (Lever 1985) and they appear to be particularly involved in the extermination of snakes.

Cuba
In Cuba feral dogs are said to be descended from those introduced by the early Spanish conquistadors in 1511 (Lever 1985).

Hispaniola
In 1535, de Oviedo y Valdes (1851–55) reported that feral dogs were numerous in Hispaniola. The Parmentier brothers (1883) reported many near Santo Domingo in 1526 and observed that packs of them kept cattle numbers down. A report in 1561 (Coll y Toste 1914) contained a statement to the same effect. Evidently large dogs were bred to hunt cattle and often escaped and bred with wild ones to exacerbate the problem of predation.

In 1701 feral dogs were reported to be again numerous in Hispaniola. Descourtilz (1809) in the early nineteenth century said that wild dogs existed in the French colony 15 years before, but had subsequently been exterminated by human hunting and poisoning. They were still abundant in many districts of sparsely settled, pastoral Hispaniola in 1785. There were no feral dogs in Hispaniola in 1952–53 (Street 1962).

St. Croix
So-called 'wild dogs' were present on the island in 1966, 'travelling singly or in packs and taking heavy toll of the deer population' (Seaman 1966).

■ DAMAGE
Free-roaming dogs or cats in the United States are potential ecological, medical and social threats in several ways: (a) harbouring disease transmittable to man; (b) inflicting bites; (c) damaging property and wildlife; (d) causing accidents; and (e) creating nuisance and pollution (Feldmann and Carding 1973). In 1975 there were over 33 million dogs in the United States.

Free-roaming dogs come from (1) pets released for unsupervised exercise; (2) escapes of pets; (3) pets abandoned when families move; (4) pets that run away; (5) births as feral animals (listed in order of magnitude).

Feral dogs cause some losses in the United States, but mainly losses are from unrestrained animals. In the United States feral dogs have hybridised with the grey wolf (*C. lupus*), red wolf (*C. rufus*) and coyote (*C. latrans*) (Howard and Marsh 1984).

In New York city dog bites are a major health problem, dogs litter streets, spread disease to other pets, and help other vermin to spread by overturning garbage cans and ripping open garbage bags. In Georgia they are said to inflict enormous losses on dairy cattle and beef cattle by inciting them to run. In beef breeds this causes loss in weight and abortion of calves and in dairy breeds the loss of milk production. In some areas they may affect the wildlife, particularly where there are no domestic animals to prey on. One of the main concerns is human indifference to the problem (Caras 1978).

There are numerous popular accounts of feral dogs being serious predators of deer, but studies in Alabama found that they are not an efficient predator of these animals and only a nuisance by chasing them (Causey and Cude 1980). They do not seem to prey on white-tailed deer or cattle in Alabama (Scott and Causey 1973). Only a low incidence of leptospirosis and tularaemia was found in feral dogs in Alabama, whereas other studies showed a high incidence (Scott and Causey 1978).

In Idaho they were found to be responsible for killing and maiming many deer, particularly in areas where residential areas are expanding into deer range. Some 39 incidents were witnessed resulting in 12 deaths of deer (Lowry and McArthur 1978).

In Virginia little evidence has been found to indicate that stray dogs are a problem of great magnitude statewide (Perry 1971).

It has been estimated that there is one free-roaming dog for every nine humans in Baltimore, where the average dog home range is 0.26 km^2.

Free roaming dogs are involved in the disruption of efficient garbage collection by overturning bins, which subsequently attracts rodents. Dog bites also represent a health problem, with 6227 bites reported in 1969.

Dog faeces in streets and parks are offensive and are a potential health hazard. Salmonella may be transmitted from dog to humans via flies feeding on faeces. Dog faeces is the second most important breeding ground for flies after garbage. In last five years dogs have been implicated in spread of leptospirosis to people and is transmitted via dog urine.

Miscellaneous problems include noise, hindering traffic and killing trees (Bech in Fox 1975).

Sheep are the animal most commonly attacked in Australia around urban fringes. Most properties with such damage are within 5 km of urban development. A wide variety of breeds are involved and larger breeds are most likely to be involved. German shepherds are the breed mostly involved (26 per cent) in attacks (Burley *et al.* 1983).

The dingo is persecuted in Australia because farmers and graziers report predation on sheep. However, some studies suggest that such claims are overly exaggerated (Macintosh 1975; Whitehouse 1977). Certainly dingoes are capable of causing damage in sheep flocks, but the overall amount is small when compared to the total numbers of sheep.

Dingoes and feral dogs are sometimes a serious threat to livestock in certain circumstances. Calves and sheep of all ages may be harassed, maimed and killed by dogs and many sheep may be killed and not eaten (Green and Catling 1977; Thomson 1984; Fleming and Robinson 1986; Wilson *et al.* 1992).

In the Galápagos Islands feral dogs prey on iguanas, tortoise eggs and other wildlife (Barnett 1986; Benchley 1999). A study of their food habits on Isabela Island showed that they ate marine iguana (*Amblyrhynchus cristatus*), penguin (*Spheniscus mendiculus*), sea lion (*Zalophus californianus*), fur seal (*Arctocephalus australis*), shearwater (*Puffinus thermintieri*), brown pelican (*Pelicanus occidentalis*) and booby (*Sula nebouxii*) (Barnett and Rudd 1983). It is thought that some of these populations are endangered by the continued presence of dogs on the islands. Work is continuing on their extermination.

Introduction of the New Guinea dog may have helped the extinction of the local thylacine (*Thylacinus*) species and two wallaby (*Thylogale*) species in New Guinea (Flannery 1995).

COYOTE
Canis latrans Say

■ DESCRIPTION
HB 700–1000 mm; T 250–400 mm; weight 6.8–23.2 kg (occasionally individuals to 33 kg).
Dog-like carnivore; coat colour varies grey to rufous or brown with black tipped guard hairs; muzzle, outer ears, forelegs and feet reddish brown to yellow; throat and belly white; dorsal stripe and shoulder stripe dark; tail fawn, tip glack. Has a narrower build, larger ears and narrower snout than *C. lupus*.

■ DISTRIBUTION
Alaska, Canada, United States (except SE) to Costa Rica and western Panama. Present range Arkansas, most of Alabama, Louisiana and Mississippi, western parts of Tennessee and Kentucky, north-western Florida, south-western Georgia, and scattered locations in Virginia, Maryland, West Virginia, North and South Carolina and peninsular Florida.

■ HABITS AND BEHAVIOUR
Habits: will mate with domestic dogs; climbs fences easily. **Gregariousness:** loose packs and family packs; larger numbers sometimes at food sources; density 0.2–2.0/km^2. **Movements:** overlapping home range varies 0.38–80 km^2; disperse in winter; in some localities moves to higher ground in summer, 80–160 km. **Habitat:** grasslands, semi-arid lands, desert to alpine regions, broken forest. **Foods:** rabbits, poultry, domestic livestock, small rodents, young deer, and

Coyote

other mammals, carrion, invertebrates, birds and some plant material, insects, fruits, seeds, amphibians, snakes, fish, acorns, crayfish, turtle eggs. **Breeding:** mates January–March; gestation 60–65 days; female monoestrous; litters 1–6, 12; eyes open 14 days; emerges from den 2–3 weeks; starts solid food c. 3 weeks; weaned 5–6 weeks; pups raised by parents and other pack members; mature 1–2 years; young disperse at 1 year. **Longevity:** wild 6–14 years 6 months; captive 21 years 10 months. **Status:** common, but reduced in numbers.

■ HISTORY OF INTRODUCTIONS
EUROPE
Great Britain
A pet coyote is said to have escaped at Leytonstone about 1880, but was later recaptured. Another animal may have been released about 1862, but there appears to be no definite proof (Fitter 1959).

NORTH AND CENTRAL AMERICA
Coyote populations have increased dramatically in numbers in the south-eastern United States since 1972. They have substantially extended their range since the arrival of European colonists and are a recent coloniser in Panama where land clearing has made this possible (Eisenberg 1989).

United States
Introduced in West Virginia, Virginia, North Carolina, South Carolina, Georgia, Mississippi, Alabama, Tennessee, Florida, Maryland, Ohio and on Rhode Island (Deems and Pursley 1978; Hall 1981). Some stock from translocated animals may still exist in Lake Wales region of Polk County, Florida (Lever 1985). Few introductions have resulted in low or variable populations except those in Ohio and Georgia.

Introduced in New York State (de Vos et al. 1956). About eight were introduced in Ontario County about 1928 and in about 1934 some to Saratoga, Columbia, Franklin and Albany counties. The former animals escaped from captivity, but were said to have been killed a few years later, and the latter were said to have gradually disappeared (Bump 1941).

There are at least 20 occasions where coyotes are reported to have been released or escaped from Florida (5), Georgia (5), Alabama (3), Tennessee (3), Mississippi (2), North Carolina (1) and Virginia (1). Seven were for hunting, three escaped and 10 are not determined. Several authors have suggested that the coyotes' presence has been facilitated by releases by humans (Schultz 1955; Galley 1962; Hill et al. 1987). (See table of liberations in the south-east United States.) Several of these releases appear to have gener-

Coyote liberations in the south-eastern United States

Year(s)	Location	No. released	Source	Success
1924	Barbour, Alabama	?	?	yes
1925	Palm Beach, Florida	4	Nebraska	no?
1925	DeSoto, Florida	10	Texas	no?
pre 1930	Hardeman, Tennessee	?	?	possibly?
1930–31	DeSoto, Florida	16	?	no?
1935	Hickman, Tennessee	?	?	possibly?
late 1930s	Turner, Georgia	6	Texas	yes
pre 1950	Gadsden, Florida	11	captives	possibly?
1950	Polk, Florida	?	?	no?
1952	Tazell, Virginia	2	Oklahoma	possibly?
c. 1955	Sequatchie, Tennessee	?	?	possibly
1959	Turner, Georgia	12	Texas	yes
late 1950	Madison, Alabama	?	?	possibly
1967	St. Clair, Alabama	11	Iowa	yes
mid-1970s	Bullock, Georgia	?	Georgia	probably?
1979	Oktibbeha, Mississippi	3	Mississippi	possibly?
1981	Lincoln, Mississippi	?	Arkansas	possibly?
not known	Gaston, North Carolina	2–4	?	possibly?

References: Sherman 1937; Kellogg 1939; Young & Jacksic 1951; Schultz 1955; Cunningham & Dunford 1970; Hill et al. 1987.

ated expanding populations, but some have not. Reductions in wolf numbers have allowed coyotes to invade the south-eastern United States where they were previously absent. Certainly coyotes have benefited greatly by changes made by humans to the environment.

A general range expansion has occurred eastwards, and through the north-eastern United States and Quebec to New Brunswick and Nova Scotia. Introductions appear to have occurred concurrently, beginning in Louisiana in the late 1940s and Arkansas in the 1960s. In New Brunswick the number of coyotes collected increased from five in 1973–74 to 338 in 1979–80 and 368 in 1980–81, and they were first recorded in Nova Scotia in 1977 (Moore and Millar 1984). Their spread in southern states east of the Mississippi River appears to have been expedited by merging local populations established through the escape of captive coyotes or the release of coyotes for chase with hounds (Moore and Miller 1984; Hill *et al.* 1987).

Coyotes now inhabit nearly all of North America from Yukon, Alaska, to Central America, though there is some evidence to suggest that they were there at least intermittently in prehistoric times (Walker 1992).

From the 1930s to the 1960s coyotes established themselves in New England and New York, and have now pushed to Nova Scotia and Prince Edward Island and down the Appalachians to southern Virginia. Others colonised towards Missouri, Arkansas in 1920s, Louisiana in 1950s, Mississippi in 1960s and subsequently Florida and the Carolinas.

■ DAMAGE

The food habits of coyotes are not detrimental to human interests and may be beneficial, but consumption of domestic stock, poultry, deer and wild birds reveals the serious economic importance of the coyote (Sperry 1941). In Wyoming a study showed that only about 11 per cent of the diet could be detrimental to humans and it was concluded that the animal was not a great pest in this area (Murie 1935). Economic status appears dependent on the locality, on sheep range they are a menace, but not all bad, and control rather than extermination is probably justified (Dixon 1920).

Whether coyotes are an economic pest in the United States still appears a controversial issue. Currently widespread control is not carried out in the United States, but individual coyotes are destroyed when they are found causing problems (Walker 1992).

GREY WOLF
Gray wolf
Canis lupus Linnaeus

■ DESCRIPTION
TL 1000–2046 mm; T 300–500 mm; SH 700–1000 mm; WT 18.2–80 kg.
Colour varies from white to black and all intermediate degrees of cream, grey, brown and orange black, but grey tones most common; under parts whitish or greyish; tail thick and bushy.

■ DISTRIBUTION
Formerly entire Palaearctic except for North Africa. Now extinct in western Europe, but still present in some parts of Iberia, Italy, the Balkans and Scandinavia. Extinct in Japan.

North America, including Arctic islands and Greenland south to Mexican Highlands, but absent in south-eastern United States. Now rare in continental United States. In northern Washington, northern Idaho, northern Montana and Great Lakes region. Formerly from central Mexico northwards.

■ HABITS AND BEHAVIOUR
Habits: mostly nocturnal; diurnal activity increases in cold weather; den among rocks and tree roots; strict social hierarchy. **Gregariousness:** packs all year 5–36 individuals (often adults and young of one or more years); density 1/73 km². **Movements:** sedentary in spring and summer, but wanders extensively following migration of reindeer (hunting grounds 1036–12 943 km²); may cover up to 80 km/day. **Habitat:** all habitats including forest, woodland, open plains, forest edges, lake shores. **Food:** carnivorous; beaver, deer, hare, moose, caribou, wapiti, bison, calves, rabbits, shrews, voles, mice, squirrels, muskrats, carrion, garbage, domestic animals, squirrels, grouse, sheep, horse, pigs, birds, fish, occasionally fruits, berries, grass, and insects. **Breeding:** winter; mate February–March–April (Europe), cubs in April–May; gestation 62–63 days; litter size 3–11; 1 litter/year; born blind, hairless; eyes open 11–15 days; emerge from den at 3 weeks; weaned at 5 weeks; mature at 22 months to 2 years. **Longevity:** 10–16 years in wild, 18 years in captivity. **Status:** much reduced in range and numbers.

■ HISTORY OF INTRODUCTIONS
EUROPE
The wolf became extinct in many parts of Europe by the 1960s. They were present only in northern, southern and eastern Europe with small populations still surviving in Portugal, Spain, France, Italy, Poland, the Balkans and in northern Scandinavia, but numbers

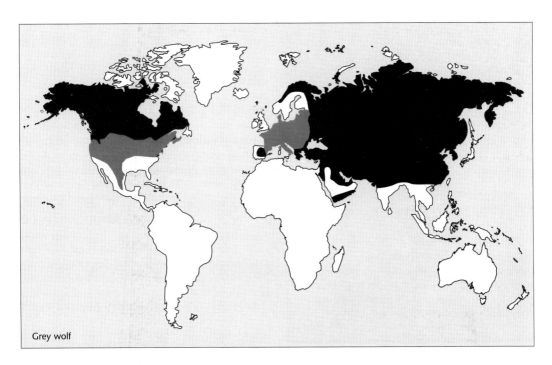

Grey wolf

were slowly decreasing. Occasionally they were reported from Germany, but these were thought to be just wandering animals (Lyneborg 1971).

United Kingdom

Wolves probably became extinct in England during reign of Henry VII (1485–1509) and in Scotland in 1743, and the last in Ireland about the 1760s. There may have been some re-introductions to reinforce declining populations in about 1465 or earlier (Fitter 1959). Recently suggestions have been made to re-introduce the wolf into Britain, but so far none have been released.

NORTH AMERICA

At the same time as some Indian tribes were being 'pacified', war was also being waged against the grey wolf in North America by the new white settlers because of their menace to people and livestock. This, together with a demand for pelts, led to killing of wolves on a massive scale in the late 1860s. The near elimination of the bison slowed the killing, but it was revived in the 1880s and 1890s as livestock owners successfully lobbied for bounties on dead wolves (Cheater 1998). In Montana alone more than 80 000 wolves were destroyed by bounty hunters between 1883 and 1918 and many more after 1915 when the government hired professional hunters and trappers to kill wolves and other predators which were thought to endanger livestock. Grey wolves had been elimi-nated from most of the 48 states by the early 1930s. In 1978 they were listed as endangered.

United States

Grey wolves were introduced in New York State, in southern Franklin County, by local residents when imported animals escaped in about 1930. They are reported to have crossed with dogs and to be increas-ing in numbers (Bump 1941; de Vos *et al.* 1956). In Minnesota 104 wolves were translocated from an area where they were causing damage to livestock to the north and east for 50–317 km, three others west. The translocation was largely unsuccessful at keeping problem wolves out of livestock production areas. Other wolves were translocated to Alaska, Michigan, and Minnesota (Fritts 1984).

Once common in the western United States, grey wolves have been reduced to near extinction in the Rocky Mountains region. The wolves in Yellowstone National Park became extinct 69 years ago. In 1985 it was suggested that a re-introduced population be established in Yellowstone National Park (McNaught 1987). Wolves were finally released into the park in 1995 where they are established and thriving and are at present under study.

In the winters of 1995 and 1996, 31 wolves were released in Yellowstone National Park and 35 Canadian wolves (15 in an initial release and 20 in 1996) were released into Idaho's remote Frank Church – River of No Return Wilderness (Chadwick and Satore 1998; Cheater 1998; Highley 1998; McNamee 1998). As of 1996 there were 40 wolves roaming Yellowstone with at least five packs fully established and two others nearly so. Ten breeding packs have now been established in each of two re-introduction areas and have been together now for three years. Their numbers have grown to 120 wolves in Yellowstone and about 70–75 in Idaho. Ten packs are established in the Montana region where natural colonisation continues. All three ecosystems had at least half a dozen packs by 1977. The established packs have nearly doubled their populations every year since the re-introductions. The wolf program is costing about US$300 000 annually.

Re-introduction of the Mexican grey wolf (*C. l. baileyi*) to a portion of its former range in Arizona was carried out in early 1998 following its extinction in the 1950s (Highley 1998, 1999). Eleven were released into eastern Arizona at Apache National Forest in March 1998, the culmination of a recovery effort which began in 1977 (when five Mexican wolves were captured to start a captive breeding program). Additional wolves were to be liberated annually in the Blue Range Wolf Recovery Area in the Apache and Giln National Forests of eastern Arizona and western New Mexico until reproduction in the wild is adequate to sustain expanding populations. This time is estimated as about nine years for 100 individuals.

The Mexican grey wolf recovery program has been set back by the loss of nine of the original 11 animals released, as last year five were shot, one is unaccounted for, and three were recaptured when they failed to re-adapt to life in the wild. The two remaining males were recaptured in late 1998, were paired with two new females and released in December (Highley 1999).

■ **DAMAGE**
A study in Alaska in 1969 showed that wolf utilisation of ungulate prey, moose, caribou and sheep, did not interfere significantly with human recreational use of the same resource. However, competition between the two predators (human and wolf) could create problems in human utilisation approaches and the net animal increase of ungulates (Rausch 1969).

In Minnesota wolves prey on domestic animals from May through October, but the extent of depredation varies considerably from year to year (Mech *et al.* 1988) and is related to the severity of the previous winter.

Although the wolf re-introduction program has met with much success in the United States it has also stirred up controversy in some states. Idaho, Montana and Wyoming farm bureaus (made up largely of cattle ranchers) have recently challenged the restoration programs resulting in the federal judgement that the

wolves have been illegally re-introduced. The court has ordered their removal, but at present this ruling is open to appeal by the Department of the Interior which oversees the restoration program. To October 1999 at least eight of the Yellowstone wolves were involved in livestock depredations and all but one of these was killed (Highley 1998). At least five cattle and 53 sheep have been taken by wolves in Idaho so far and the ranching community is concerned about how seriously they may affect their livelihoods (Cheater 1998).

In Bulgaria wolves attack and kill grazing sheep and destroy many game animals such as deer and wild boar (Genov 1987).

RED WOLF
Wolf
Canis rufus Audubon and Bachman

■ **DESCRIPTION**
HB 950–1300 mm; T 250–350 mm; SH 660–790 mm; WT 18–41 kg.
Generally greyish with buff or reddish and blackish tinge on upper parts; under parts white or pale buff; tail tipped black.

■ **DISTRIBUTION**
North America. Formerly throughout south-eastern United States from Florida to central Texas. Now extinct in the wild.

Red wolf

■ HABITS AND BEHAVIOUR
Habits: mainly nocturnal, but some activity in daylight in winter; den in hollow trees, stream banks, sand knolls or under rocks to 2.4 m. **Gregariousness:** basic social unit a pair, or groups 2–3 and larger packs. **Movements:** home range 44–78 km². **Habitat:** forest, prairies, wetlands, swamps. **Foods:** mammals (small deer, pigs, coypus, rabbits, raccoons) and carrion. **Breeding:** litters 4–7, 12; gestation 60–63 days. **Longevity:** 4 years wild, 14 years captive. **Status:** endangered; currently considered extinct in wild; decline due to habitat loss, hunting and hybridisation with coyotes.

■ HISTORY OF INTRODUCTIONS
NORTH AMERICA
United States
Formerly widespread in the United States, but by the 1960s the only red wolves left were in south-east Texas and southern Louisiana. By the 1970s few remained because of persecution and habitat destruction. By the 1980s they were believed to be extinct in the wild and the current wild population is descended from re-introduced animals. In 1992 the total population including captive animals was 204 (Smith and Phillips 1987; Wilcove 1987; Mech 1992; WCMC 1998).

Captive-breeding started in 1973 and by 1980 there were over 50 in captivity (Burton and Pearson 1987). Conservation efforts were not very successful at this time as the red wolves were gradually hybridising with the coyote (*C. latrans*). After 1975, 14 were captured for a breeding program. Since 1977 offspring have been produced from this stock and by 1989 there were some 83 living descendants. Experimental re-introduction was then made on Bull Island off South Carolina and Horn Island off Mississippi.

The first re-introduction was made in 1977 when red wolves were re-established on Bull Island in Cape Romain National Wildlife Refuge, South Carolina (Burton and Pearson 1987; Smith and Phillips 1987), and Horn Island off Mississippi. Since then, more attempts have been made in North Carolina (Smith 1987) and South Carolina (Kleiman 1989).

Of the 73 red wolves left in North America, 40–50 are in the Tacoma program which is assisted by the United States Fish and Wildlife Service. Seven red wolves to be re-introduced in the United States are part of a red wolf survival plan. These animals seven will spend six months at Alligator Pines Wildlife Refuge in North Carolina before release. They have been raised in captivity and are two to three years old (Hudson 1986).

Starting in September 1987, pairs have been released as part of a large-scale effort to re-establish a perma-nent population at Alligator River National Wildlife Refuge in north-eastern North Carolina. Results have been inconclusive (Carley 1979; McCarley and Carley 1979; Nowak 1979; Smith and Phillips 1987; Phillips 1988; Rees 1989; WCMC 1998). Twenty-one red wolves have been released, five were killed by vehicles. Since the start, 17 wolves have been recaptured on 24 occasions and the re-introduction has now been going for three seasons (Phillips 1988; Parker and Phillips 1991). The population of 50–100 is still thriving and progressing well in North Carolina (Highley 1999).

From 1987 to 1988 five groups (the majority in pairs) were released at two sites in South Carolina and appear to be surviving without help (Kleiman 1989). South Carolina was chosen because it was thought less likely that the wolves would hybridise with coyotes there.

An eight-year effort to restore red wolves in the Great Smoky Mountains National Park, Tennessee has failed. All but 11 of 37 introduced wolves either died or were captured after straying onto private lands (Highley 1999).

■ DAMAGE
Despite its near extinction, public views of the red wolf are still poor (WCMC 1998). They are still thought of as a pest of livestock and game and a possible threat to humans.

BAT-EARED FOX
African big-eared fox, big eared fox
***Otocyon megalotis* (Desmarest)**

■ DESCRIPTION
HB 460–700 mm; T 230–350 mm; SH 300–400 mm; WT 2.5–5.4 kg.
Long-legged fox; general coat colour yellowish-brown or yellowish; ears large, black tipped; long slender muzzle; head broad; snout pointed; eyes large; under parts white to buff; mask, lower legs, feet, and tail tip black; feet black; legs slim, short.

■ DISTRIBUTION
Africa. Arid regions of eastern and southern Africa from southern Angola and Zimbabwe to South Africa. East Africa from Ethiopia and southern Sudan to Tanzania.

■ HABITS AND BEHAVIOUR
Habits: mainly nocturnal, but often diurnal in winter; den in burrow, rocks, under bushes or enlarges burrows of others; burrow several meters. **Gregariousness:** singly, pairs or groups (adults and young) up to 2, 5–8; density 0.8–0.9/km². **Movements:**

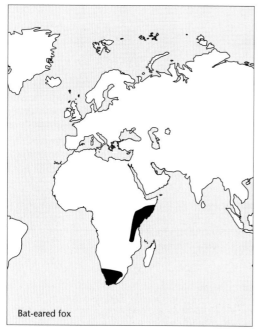

Bat-eared fox

12 km/night while foraging; home range 0.5–3.5 km². **Habitat:** steppe and grassland, savanna and brush country. **Food:** primarily insectivorous; insects, termites, fruits, berries, roots, tubers, small mammals (mice), earthworms, insects, scorpions, snakes, and occasionally carrion, beetles, rodents, eggs of birds, lizards. **Breeding:** all year, but mainly November–April; gestation 60–70 days; young 2–5, 6; eyes open at 9 days; emerge from den 17 days; lactation 15 weeks; mature 5–9 months. **Longevity:** 5–14 years captive. **Status:** fairly common and often seen in parks.

■ HISTORY OF INTRODUCTIONS
Africa
South Africa
Bat-eared foxes have declined in numbers in the settled parts of South Africa, but have extending their range eastwards into Mozambique, Zimbabwe and Botswana. Some have been re-introduced to Mountain Zebra National Park (Penzhorn 1971).

■ DAMAGE
Bat-eared foxes cause no damage (Hey 1964) and rarely attack domestic animals (Walker 1968).

CAPE HUNTING DOG
Wild dog, hunting dog, African hunting dog
Lycaon pictus (Temminck)

■ DESCRIPTION
HB 760–1120 mm; T 300–410 mm; SH 600–750 mm; WT 17–36 kg.

Long-legged; head broad; variably dark brown, black or yellowish coat, mottled with light patches; muzzle short and powerful; ears black, erect, large, rounded; tail bushy, white tipped.

■ DISTRIBUTION
Africa: Formerly all Africa except deserts and forests, now savanah and sub-desert zones south of Sahara extending northwards as far as Tanezrouft, southern Algeria.

■ HABITS AND BEHAVIOUR
Habits: nocturnal and diurnal; hunts in packs. **Gregariousness:** socially complex packs 6–30 or more and occasionally 90–100. **Movements:** nomadic; travels 40–70 km/day on foraging trips. **Habitat:** open savanah, plains, semi-desert, bush, lowland forest, mountains. **Foods:** flesh of medium-sized ungulates (gazelles, impala, waterbuck) and calves of larger antelope and foals of zebras. **Breeding:** breeds all year, peak after rainy season; litter size 2–16; gestation 69–73 days; inter-birth interval 12–40 months; pups born in burrow; blind at birth, eyes open at 2 weeks; weaned 10–12 weeks; learn to hunt 6 months; sexually mature at 1.5 years. **Longevity:** 10–12 years. **Status:** endangered in South Africa through persecution, disease and habitat destruction; numbers reduced and in many areas already extinct.

■ HISTORY OF INTRODUCTIONS
Africa
South Africa
Hunting is the leading cause of the decline of Cape hunting dogs, but disease, including canine distem-

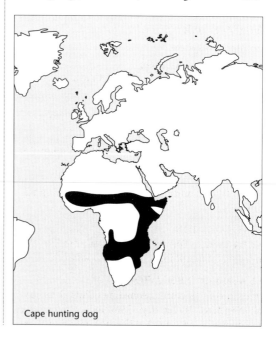

Cape hunting dog

per, rabies and anthrax, have decimated many populations in Africa (WCMC 1998). This decline in numbers has generated the few re-introductions. Twenty Cape hunting dogs were re-introduced to the Hluhluwe Umfolozi Park (HUP) in Natal in the early 1980s. Since this time numbers have fluctuated from a minimum of three in 1988–89 to about 29 in 1986. They first bred in 1984.

In 1983 dogs were noted in Itala Game Reserve about 100 km northwest of HUP and since 1988 there have been regular reports of them in the game/cattle farming area north of HUP. None of these appear to be the resident HUP dogs, but could be escapees. In 1988 a pack escaped from the park and only three returned.

Tanzania and Namibia
Re-introduction programs in an area around Mkomazi, Tanzania and also into Etosha National Park, Namibia, are currently underway (WCMC 1998).

■ DAMAGE
The Cape hunting dog is undeservedly perceived as a voracious killers of game and livestock in Africa and so hunted and poisoned (WCMC 1998).

Family: Ursidae
Bears

BLACK BEAR
American black bear
Ursus americanus Pallas
In some classifications this species is included under U. arctos.

■ DESCRIPTION
HB 1300–1900 mm; T 70–180 mm; SH c. 1000 mm; WT males 113.2–270.5 kg, females 92.3–204.1 kg.
Coat varies from black or cinnamon to bluish, white or yellowish white; body large and stout; muzzle cinnamon brown; chest with white V; claws on forefeet about same length as those of hind feet; tail minute.

■ DISTRIBUTION
North America. Formerly throughout most of the United States except for some western states.

■ HABITS AND BEHAVIOUR
Habits: mainly nocturnal, sometimes diurnal; territorial; hibernates 74–126 days. **Movements:** sedentary; overlapping home range 2–173 km². **Gregariousness:**

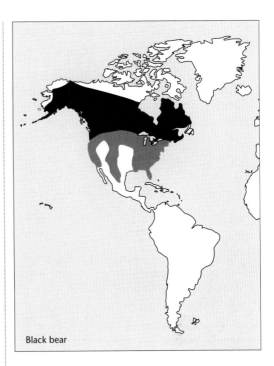

Black bear

solitary, except in mating season; density 1 per 0.67–14.5 km². **Habitat:** forest, swamps, marshes, tundra, mountains, orchards and garbage dumps. **Foods:** mainly vegetarian; berries, fruits, nuts, acorns, fungi, insects, (ant and beetle larvae), mice, ground squirrels, and occasionally ground-nesting birds, twigs, leaves, tubers, roots, eggs, carrion, honey. **Breeding:** mates summer, mainly June–July, young born January–February; gestation 210–220 days including delayed implantation; litter size 1–5; breeds once every 2 years; 1 litter every second year, sometimes 3–4 years; young naked, blind, weaned 6–8 months; disperse spring after second winter; males mature at 5–6 years, females at 3.5–5 years. **Longevity:** 10–30 years wild, 30 years captive. **Status:** range reduced, but still common.

■ HISTORY OF INTRODUCTIONS
NORTH AMERICA
United States
During the 1960s wildlife agencies in Louisiana and Arkansas imported a number of bears from Minnesota to their respective states (Lowery 1974), presumably for release into the wild. Some were re-introduced to Kentucky by wildlife agencies for hunting before 1976 and this resulted in a low population of bears (Deems and Pursley 1978).

Bears were introduced to southern California from Yosemite National Park, California, in 1933 (Burghduff 1935). Sixteen were released into the San

Bernardino Mountains and for over 40 years no information was gathered on them. In 1974 baseline data were collected on the animals established there (Novick and Stewart 1982).

From 1958 to 1968 the Arkansas Game and Fish Commission translocated about 260 black bears from Minnesota and Manitoba to the Ouachita and Ozark mountains of western and north-western Arkansas into areas where bears had been extirpated earlier this century. Investigations in 1987–91 indicate that the population in these areas is 2500 bears, making it one of the most successful re-introductions of large carnivores. Several factors are said to have contributed to the success: a long period of release (11 years), the large numbers of bears released each year (20–40), high-quality habitats and the use of remote release sites.

PACIFIC OCEAN ISLANDS
Hawaiian Islands (United States)
A single bear escaped in the mid-1950s and from time to time was seen in the Koolau Mountains, but was last reported in 1966 at the Aiea Heights Trail (Kramer 1971).

■ **DAMAGE**
Bears are frequently killed because of depredations on domestic animals and crops. Attacks on livestock are probably negligible (Lowery 1974), but bears can cause serious damage to cornfields and honey production. In 1973 bears caused $900 000 damage to bee-keeping in Alberta, Canada (Gilbert and Roy 1977).

BROWN BEAR
Grizzly bear, big brown bear
Ursus arctos Linnaeus

■ **DESCRIPTION**
HB 1700–2800 mm; T 60–210 mm; SH 900–1500 mm; WT 70–780 kg.
Coat shaggy brown or black to pale fawn; prominent hump on shoulders; face dish-shaped; claws long; white tips on guard hairs give grizzled look; distinguished from *U. americanus* by hump and snout rises more abruptly into forehead; longer pelage and claws; tail vestigial.

■ **DISTRIBUTION**
Europe, Asia and North America. In Europe in Pyrenees, Carpathians, Italian Alps, Balkans and Scandinavia; formerly widespread from British Isles to Japan. Now extinct in western Europe except for a few isolated populations in Spain, the Pyreees and Italy. In North America formerly from Alaska and western and central Canada through the United States to northern Mexico; now Alaska, north-western Canada to California.

■ **HABITS AND BEHAVIOUR**
Habits: mainly nocturnal; hibernates in winter in den among roots, in a cave or hollow tree; shy; aggressive; strong home instinct. **Movements:** moves to high mountains in summer. **Gregariousness:** solitary; females and cubs; males establish territories which overlap several females; density 1 per 1.5–2600 km².

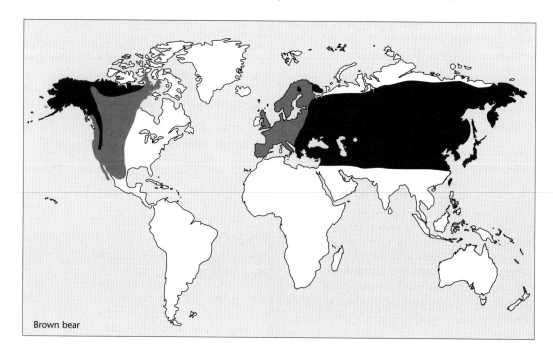

Brown bear

Habitat: often in mountains; forest, woodland, alpine tundra and meadows. **Foods:** omnivorous; fruits, berries, honey, grubs, vegetables, invertebrates, beetles, fish, eggs, small mammals, roots, tubers, bulbs, nuts, fungi and grain, grass, sedges, moss, rodents and other mammals. **Breeding:** ruts May–July; cubs in January–February (winter–early spring); promiscuous; gestation 180–266 days; litter size 1–4; young born every second year; born naked and blind; weaned 5 months; mature 3rd year; mature 3–6 years or older. **Longevity:** 30–40 years in wild, probably 50 years in captivity. **Status:** reduced in numbers and range; probably endangered.

■ HISTORY OF INTRODUCTIONS
EUROPE
Germany
In the seventeenth and eighteenth centuries bears still lived in Prussia, but elsewhere were probably captives or those introduced in hunting parks and zoos. In 1625 about 15 young bears were noted in the Neumark. In 1525 some were released near Grimmitz (Muller-Using 1938). Between the two World Wars some people were in favour of re-introducing bears in Germany, but as far as is known, none were released (Niethammer 1963).

Poland
It is not reliably known whether those bears occurring in Poland after 1700 were captives, escapees or re-introduced animals (Niethammer 1963).

Russian Federation and adjacent independent republics
Attempts were made to acclimatise bears in northern Kazakhstan and near Moscow in the mid-1960s, without much success(?) (Kirisa 1974).

Spain–France
Three Slovenian bears were introduced in the central Pyrenees in 1966 to boost the local population (Weyndling 1998).

NORTH AMERICA
Alaska (United States)
Conflicts between bears and humans are frequently resolved by translocating bears away from the area of conflict. However, it was found that with distances of up to 200 km the bears were likely to return home. In a program to evaluate the effects of translocation, 47 bears (*U. arctos*) were captured and translocated in Alaska in 1979 because of conflict. Later it was concluded that translocation does not appear to be a reliable management procedure as the threshold distance may be in excess of 258 km (Miller and Ballard 1982).

■ DAMAGE
Bears have long been persecuted as a predator of domestic livestock, especially sheep and cattle and they are still hunted as a game animal in Europe and North America (Walker 1992). Farmers in Navarre in the Spanish Pyrenees claim that bears introduced in the Pyrenees have caused them severe sheep and goat losses. However, the introduced bears were released over 200 km away and it is not likely that it was one of these. Farmers in these areas receive compensation for sheep killed but many reject the idea of co-existing with bears and view further planned introductions with suspicion (Weyndling 1998).

Family: Procyonidae
Raccoons, coatis, kinkajou

KINKAJOU
Potos flavus (Schreber)

The kinkajou was thought to have been introduced to Más á Tierra (Robinson Crusoe Island) in the Juan Fernández group (de Vos *et al.* 1956), but this was in mistake for the coati (*Nasua nasua*) which is established there.

RACCOONS
Genus – *Procyon*

The Guadeloupe raccoon (*P. minor*) Miller from Guadeloupe Island, Lesser Antilles and *P. gloveralleni* of Barbados may have been introduced (de Vos *et al.* 1956). *P. gloveralleni* resembles *P. minor* and the Bahaman race *P. maynardi*. They are usually treated (Hall 1981) as separate species.

P. gloveralleni itself is now rare or extinct but there is no evidence of any introduction. *P. maynardi* is probably not a valid species and represents an introduced population of *P. lotor* (Olson and Pregill 1982). The latter species is now established on Grand Bahama Island (Burden 1986).

RACCOON
American raccoon, common raccoon
***Procyon lotor* (Linnaeus)**

■ DESCRIPTION
HB 410–600 mm; T 192–405 mm; SH 230–300 mm; WT 1.8–28 kg (males 4.0–9.2 and females 3.9–7.9 kg is usual).

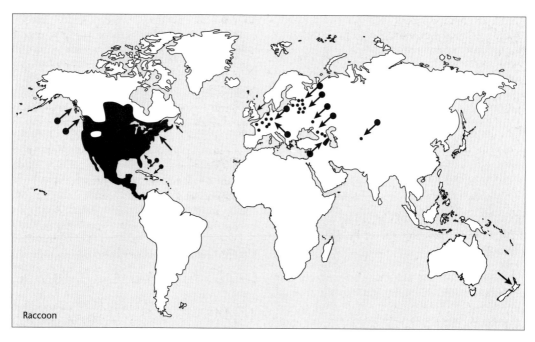

Raccoon

Upper parts grey black to grey brown; face pointed and framed by ruff of grey hairs behind cheeks; black facial mask extends across cheeks, eyes and nose; sides of muzzle, lips and chin white; facial mask bordered by white line from forehead under ears to sides of neck; eyes black with pale grey bars above and below; ears rounded, greyish and white tipped; tail short, bushy, and has four to seven conspicuous black rings and black tip; hind legs blackish near keels; forefeet whitish; hind feet usually whitish; soles naked; feet have five toes. Female with four pairs of mammae.

■ DISTRIBUTION
North, Central and South America. From southern Quebec, central Ontario, the prairie provinces and southern British Columbia, Canada, south to Colombia, South America.

■ HABITS AND BEHAVIOUR
Habits: nocturnal; partially hibernate or dormant for period in cold regions; terrestrial and arboreal; dens in hollow tree, rock crevices or abandoned burrows of other mammals. **Gregariousness:** solitary or family groups; density 1 per 5–100 ha. **Movements:** more or less sedentary; overlapping home ranges 806–1139 ha. **Foods:** omnivorous; grain, corn, acorns, nuts, berries, fruits, frogs, fish, crayfish, shellfish, mussels, molluscs, crabs, snails, turtles, tortoises, snakes, lizards, insects (beetles, grasshoppers, grubs, crickets), walnuts, apples, chickens, small mammals and birds, occasionally small rodents, human food scraps, salamandas, earthworms, eggs, chicken feed. **Habitat:** forest near watercourses, river valleys, coasts, irrigated areas, swamps, orchards, mountain forests. **Breeding:** mates January–March; young born April–June; female monoestrous; gestation 60–73 days; litter size 1, 2–6, 8; 1 litter per year; young born blind; eyes open 3 weeks; kept in den 7–9 weeks; weaned 7 weeks to 4 months; accompany female on short trips 10–11 weeks; mature at 6 months, males at 2 years or older, females often breed as yearlings. **Longevity:** 20 years 7 months (captive), 13–16 years in wild, most less, but often over 5 years. **Status:** numerous; widely kept as pet; extending range; hunted for fur.

■ HISTORY OF INTRODUCTIONS
Introduced successfully to France, Germany, Netherlands, Russia and some adjacent independent republics.

EUROPE
Escape from captivity has resulted in well-established populations in western Germany (Hessen, Eifel), adjacent parts of France and the Netherlands, White Russia, Caucasian region, and Fergana region of Turkestan (Corbet 1978, 1980).

United Kingdom
There are at least 20 records of free-living raccoons in Britain: Brecon, Powys, Wales, in 1977; Strathclyde, Scotland, in 1981; Heywards Heath, Sussex, in 1978. More recently raccoons have been recovered from Norfolk, two from Yorkshire, including a pair of wild-born cubs near Sheffield in 1984, and three from Somerset in 1985. Some found were at liberty for over one year and captured four years later (Corbet and Harris 1991).

France and Netherlands

Raccoons have colonised parts of France and the Netherlands adjacent to Germany (Corbet 1978; Lever 1985), where they have been recorded in the valley of Mosel (Corbet 1966).

Germany

The raccoon appears to have been a fairly recent introduction to Germany (Federal Republic of Germany or West Germany) (Niethammer 1963; Aliev and Sanderson 1966). They escaped from captivity and became feral in the Eifel district (Corbet 1966) and in western Hessen (Aliev and Sanderson 1966), and from here have spread into Luxembourg (Corbet 1966; Burton 1976). In 1966 it was indicated that there were between 4000 and 5000 feral animals in Germany.

Originally imported for fur farms, since 1927 raccoons have frequently escaped and been found in the wild. In 1929, three and in 1930, two escaped at Ahrdorf in Eifel and by 1936 a population of 20–25 animals existed there. Following World War 2 raccoons were found to be well spread over the districts of Schleiden, Ahweiler and Daun (Niethammer 1963). There may have been further introductions following World War 2, when pets of US servicemen escaped or were released (Lever 1985). In Hessen they have been particularly successful from the introduction of two pairs in 1927 near Altenlotheim in the district of Frankenberg on the Eder River. By 1952 the population was spread from Weser and Fulda to Laasphe, Berleburg, Kirchhein near Marburg at the Lahn and towards the Knull Mountains. Since 1958 they have colonised south-eastern Westphalen and parts of the Sauerlands. In 1962 they occupied a region of approximately 5000 km^2. Another population has become established in the woods in Seenahe, east of Berlin, where they escaped in 1945. In 1961 there were several hundred in the area (Niethammer 1963).

In 1970 they reached Schleswig-Holstein. By 1970, 20 000 were in Hessen alone. By 1980 most of eastern and western Germany was populated with raccoons.

Russian Federation and adjacent independent republics

The acclimatisation of raccoons in the Russian Federation began when they were introduced from western European fur farms into the Russian Federation for fur between 1929 and 1936 (Lavrov and Pokrovsky 1967). The first release of 22 animals was made in the walnut forests in northern Fergana and Kirghizia (Arslanbob Preserve), where they became established (Aliev and Sanderson 1966; Kirisa 1973).

Although there are considerable differences in the numbers reported released over the years, it appears that there were initially at least 26 or 27 releases of between 1240 and 3200 raccoons (Aliev and Sanderson 1966; Naumov 1972; Kirisa 1973; Baker 1986). A later report suggested that between 1936 and 1986 there were 28 releases in central Asia, the Caucasus, Belorussian Polesye, and the Far East (Gineyev 1987). These animals originated from some 37 pairs imported from fur farms and 26 from zoological gardens in western Europe. Of these, 504–526 were released in the Caucasus, 98–120 in central Asia, 127–130 in Belorussia and 489–490 in Dal'niy Vostok in the Far East (Aliev and Sanderson 1966; Kirisa 1973). Another 1100–1200 raccoons were trapped between 1949 and 1966 for re-settlement in the Caucasus, central Asia and the Far East (Redford 1962; Kirisa 1973).

Apart from the original 22 released at Arsanbob, Kirghizia, in 1936, a further 33 were released in that region in 1952 (Kirisa 1973). In the following years the raccoon had increased its range in this area by 40–50 km (Aliev and Sanderson 1966). They are now settled in 100 000 ha of the hazelnut and Kara-Alma forests (Gineyev 1987). Introduction in the Pekem basin in Uzbekistan failed.

By about 1956 they inhabited an area of 12 000–15 000 ha in the western part of the Achinsk district in the Dzhalalabad region (Novikov 1962).

In 1941, 10 female and 11 male raccoons were released in the Ismaillinskiy region of Azerbaidjan. These became established in the eastern part of the Zakatalo-Nukhinskaya (Vereshchagin 1947; Aliev 1955; Rukovskiy 1963). By 1956 they occupied an area of 900 km^2 in the eastern part of the valley in the territory of Ismailly, Kutkashen, and to some extent the Vartashen districts (Novikov 1962). Several from this area were transferred to other areas of Azerbaidjan (Aliev 1963) and other adjacent areas, including 18 districts, territories and republics of the Russian Federation (Aliev and Sanderson 1966). In five regions of the Azerbaijan area releases occurred in 1949–50, 1952–53 and 1957, making a total of 202 raccoons introduced since 1941. The quantity of furs harvested rose from 36 in 1954 to 3279 in 1967, but this had fallen to some 1610 by 1970 (Kirisa 1973).

The first release of four raccoons (Pankrat'ev 1959) on Petrov Island, Maritime Territory, was unsuccessful (Novikov 1962; Yanushevich 1966). However, subsequent attempts with 55 animals in 1954 and 73 in 1955 in the Suchan river basin with raccoons from the Caucasus were successful. By 1959 they had spread over an area of 5000 km^2 (Pankrat'ev 1959).

Raccoons were released in Uzbekistan in 1953, when 43 were set free in the Bostandyk raion on the Pskem River (Mukhtarov 1963). Here the population was studied in 1957–59 when they were well established and spreading, and attempts were still being made to increase their range. Between 1954 and 1958 some 127–128 raccoons were released in Belorussia (Samusenko 1962) in two regions. They have also become well established in this area.

Other successful releases of raccoons in the Russian Federation have occurred in Dagestan (23 released in 1950 and 30 in 1965), Kabardino-Balkar (16 in 1953), Stavropol (100 in 1954), and Primorsk RSFSR (four in 1937, and 486 in four regions from 1954–58). In Krasnodar 28 were released in 1951 in the Apsheronskii regions (Kirisa 1973). Fifty-two raccoons were released in the Poles'es in 1954 and 75 in 1958 where they became well established and the population increased ten-fold (Vasil'kov 1966).

The population of raccoons in the Russian Federation was estimated at 40 000–45 000 in 1964, of which the greatest population of 24 000–25 000 was in the Caucasus (Aliev and Sanderson 1966). Here, at least five commercially viable populations are established (Gineyev 1987) at Talysh; Alazan-Avtoranskaya valley and forests along the Kura tributaries; Samur-Divitchinskaya lowland and forests on the Galagerychai and Rubas rivers; Tersko-Sulakski forests; and Krasnodar Territory forests in the west.

The raccoon is now established in Azerbaidjan, Dagestan, Krasnodarskiy Kray, Karbardino-Balkarskaya, Belorussia, Kirghizistan, Uzbekistan, Primorskiy Kray and in Stavropol'skiy Kray. Almost everywhere they have adapted, reproduced and extended their range. Density in the present inhabited range is over 1.6 raccoons per 1 km^2 (Aliev and Sanderson 1966; Kirisa 1973). Attempts to establish them in Kazakhstan appear to have failed (Sludskii and Afanas'ev 1964).

Attempts were made initially to establish in mountain forests of middle Asia, Caucasus and southern parts of Far East (Petrov Island in 1937) when 47 captive-bred animals were released. They became established and bred only in the Caucasus (Azerbaidjan). Since 1947 raccoons have been translocated to many other regions. At least 1200 were moved to the Far East, middle Asia and Belorussia. In the 1980s the only stable populations existed in the Caucasus and Belorussia and elsewhere their numbers are low and potential expansion was limited (Sofonov 1981).

Switzerland

In Switzerland there have been no known releases of raccoons, but the species has wandered in from the Baden-Wurtemberg area (Germany) since about 1972. Numbers in Switzerland are not known, but it is thought that their range there is slowly expanding (M. Dollinger *pers. comm.* 1982).

NORTH AMERICA
Alaska

There appear to have been at least three attempts to acclimatise raccoons in Alaska. In 1935 some were released on Long Island; in 1941 eight from Indiana, United States, were released on Singer Island; and in 1950 some were released at Japonski (Burris 1965). These introductions were apparently successful (de Vos *et al.* 1956; Burris 1965), and were made by fur farmers to establish a fur industry (Deems and Parsley 1978). An introduction to Kodiak Island may have adversely affected ground-nesting birds (Burris and Macdonald 1973; Franzmann 1988)

Canada

Stocks of raccoons (*P. l. vancouverensis*) (Cowan and Guiguet 1960) were transferred by trappers from Vancouver Island to Cox Island in the late 1930s, and the species is now well established on that island (de Vos *et al.* 1956; Carl and Guiguet 1972). This subspecies was also introduced on Graham Island (Cowan and Guiguet 1960) and to Prince Edward Island (Cameron 1950; Banfield 1977) in the Queen Charlotte Islands in about 1949, and the raccoon is now established throughout these islands (Carl and Guiguet 1972). On Prince Edward Island the introduction was made by fur farmers as a fur resource and resulted in the formation of a viable population (Deems and Parsley 1978).

United States (excluding Alaska)

The Western New York Coon Hunters Association released 50 pen-raised raccoons in New York State in 1939, but the result of their efforts is not known (Bump 1941). Some were also translocated in South Carolina in about 1955 (Nelson 1955), the result of which does not appear to be documented. An introduction in Kansas by fur hunters for hunting resulted in the establishment of a viable population (Deems and Parsley 1978).

Between 1964 and 1967 a number of raccoons were released on small islands (0.8–2.0 acres) off the coast of Massachussetts for the control of herring gull populations, which at the time were a major nuisance on coastal airports (Kadlec 1971). Islands on which the raccoons were released included Eagle, Ram, Green, and northern Gooseberry. The annual introduction of both foxes and raccoons to islands off the coast reduced the herring gull colony sizes and occasionally caused the total abandonment of the nesting site by eliminating the production of young gulls through predation.

By the 1950s raccoons were extending their range in the northern prairies (Laycock 1982).

PACIFIC OCEAN ISLANDS
New Zealand
Raccoons were introduced (two animals) to New Zealand as escapees in Rotorua in 1905, but they failed to become established (Wodzicki 1950).

WEST INDIES
Barbados and Guadeloupe
It is thought that raccoons may have been introduced to Barbados (before 1750) and to Guadaloupe (Lever 1985). See note under *Procyon* at beginning of this section.

Bahamas
For many years it was generally accepted that an indigenous species of raccoon inhabited the Bahamas, but in 1957 it was shown on examination of skulls that the animal present was at most an island race of *P. lotor* (McKinley 1959). It is now thought that *P. lotor elucus* was introduced to Grand Bahama (Hall 1981) and they appear to have been a fairly recent introduction (de Vos *et al.* 1956; Petrides 1959).

It is recorded that one pair of *P. l. elucus* from Florida, United States was turned loose in 1932–33 by J. Morris. The raccoons are reported to have established and increased enormously in numbers, so much so that they became a pest of agriculture (to peanut and corn farmers) in 19–20 years (Sherman 1954).

The raccoon *P. l. maynardi* is thought to have been introduced to New Providence prior to 1784 (Petrides 1959), as at that time a naturalist, J. D. Schoepf, indicated that they had been released (McKinley 1959). From possibly one or more pairs of tame animals they have now become established on the island and increased substantially in number.

■ DAMAGE
In the United States the raccoon is said to cause problems for game management organisations in at least five states (McDowell and Pillsbury 1959). In the eastern United States in the 1940s there were large build-ups of raccoon numbers on many game refuges (Dozier *et al.* 1948). These increased numbers caused problems with their predation on waterfowl, muskrats, quail and many other forms of wildlife. Usually the damage was not serious (Jackson 1961).

On Grand Bahama raccoons have become so numerous that they are a great nuisance to farmers trying to grow corn and peanuts (Sharman 1954).

Opinion on whether the raccoon has been detrimental to wildlife in Russia appears somewhat divided. Some authorities indicate that they do not compete with native species because of the development of different food habits and hunting methods since their introduction, and have not introduced any exotic diseases (Redford 1962; Lavrov and Pokrovsky 1967). However, they have been recorded to attack game birds and are accused of being the cause for the reduction of the golden-eye duck in the Poles'es (Vasil'kov 1966), but they appear to do little damage to cultivated plants (Redford 1962).

BROWN-NOSED COATI
Coatimundi, northern coati
Nasua nasua (Linnaeus)

■ DESCRIPTION
TL males 340–890 mm; T 420–680 mm; SH to 305 mm; WT 1.0–7.75 kg.
Upper parts reddish brown to black; ears small and short, white tipped; forehead flat; snout long; nose tip black; face has black and grey markings; muzzle, chin and throat whitish; under parts yellowish brown; tail often banded with yellow and brown; feet blackish.

■ DISTRIBUTION
There are two subspecies, *narica* which ranges from the southern United States to Panama, and *nasua* which occurs over South America, except for Patagonia and parts of Venezuela.

■ HABITS AND BEHAVIOUR
Habits: mainly diurnal; roost in trees; terrestrial and somewhat arboreal; scansorial. **Gregariousness:**

Brown-nosed coati

bands or groups of 2–20 and up to 40 females and young males; males solitary except in the breeding season; density 1.2–42/100 ha. **Movements:** 1.5–2.0 km/day foraging; home range 35–270 ha. **Habitat:** forest, thorn scrub, wooded areas. **Foods:** mainly frugivorous; fruits, berries, insects, millipedes, worms, snails, spiders, lizards, mice, crabs, occasionally poultry and other meat, slugs, beetles, ants, termites, palm nuts and figs. **Breeding:** breeds October–February, births April–June; in dry season; gestation 70–77 days; litter size 2–6, 7; young stay in nest in cave or tree 2–5 weeks then accompany female; young born blind, eyes open 11 days; weaned 4 months; sexual maturity at 2 years. **Longevity:** 9–17 years 8 months (captive). **Status:** common.

■ HISTORY OF INTRODUCTIONS

Introduced successfully to the island of Juan Fernández (Chile) and perhaps successfully translocated within Chile. Unsuccessfully introduced in the United States.

NORTH AMERICA
United States

A subspecies of the coati *Nasua n. nasua* may have been introduced to Oklahoma and Indiana (de Vos *et al.* 1956). Those in Oklahoma and Indiana may have been occasional wanderers or released or escaped individuals as there are no other records of their introduction. However, the species has extended its range northwards in the twentieth century (Walker 1992).

PACIFIC OCEAN ISLANDS
Juan Fernández (Chile)

Formerly it was thought that the kinkajou (*Potos flavus*) had been introduced to Juan Fernández, but it is now known it was the coati (see note under family).

Coatis were introduced on Juan Fernández (de Vos *et al.* 1956) off the coast of Chile for rat control (Hinton and Dunn 1976) and became established on the island.

In 1935 two pregnant females escaped from captivity on Más á Tierra (Robinson Crusoe Island) in the Juan Fernández group and in 1972 the population was estimated to be 4000 coatis (Eisenberg 1989–92). In 1976 the population was reported to be between 2500 and 5000 (Lever 1985).

SOUTH AMERICA
Chile

Coatis were translocated in 1940 in Chile (de Vos *et al.* 1956), but further details are not known.

■ DAMAGE

In the Huachuca Mountains coatis are so numerous as to have caused considerable controversy resulting from alleged depredations in orchards and in chicken houses (Wallmo and Gallizioli 1954). However, they rarely damage crops and infrequently take chickens (Walker 1992).

On Juan Fernández coatis are reported to have devastated the island's avifauna (Lever 1985), but there appears little proof of this except that they are said to cause damage (de Vos *et al.* 1956) by preying on endemic species of birds.

Family: Mustelidae
Weasels, badgers, skunks and otters

OTTERS AND SKUNK

Otters from Tierra del Fuego, *Lutra felina* or *Lutra provocax*, and skunks, perhaps *Conepatus humbolti* a species of hog-nosed skunk from Patagonia, were released on a number of islets off West Falkland by an emigrant Scottish shepherd, Mr John Hamilton, in the early 1930s. Of these only the otter appears to remain. In 1962 one was shot at East Bay and in 1965 droppings were found in a creek on the south-east coast of East Falkland (Strange 1972).

Since 1976 possible sightings have been reported from Weddell Island and from West End of Pebble Island and some more recently (Lever 1985) on Sealion Island off East Falkland.

The spotted skunk or civet, *Spilogale putorius* (Linnaeus), may have been introduced in the Russian Federation.

STOAT OR ERMINE
Short-tailed weasel
Mustela erminea Linnaeus

■ DESCRIPTION

HB 175–367 mm; T 57–140 mm; WT 140–454 g.
In winter in northern areas, white except for black tail tip and sometimes yellowish on rump; soles furred. In summer fur reddish brown with white to whitish yellow under parts and throat; lips, undersides of legs and toes creamy white; terminal third of tail tipped with stiff black hairs; short oval ears thinly furred, legs short, body long. In some areas brown all year. Female has 10 mammae and is usually smaller than male.

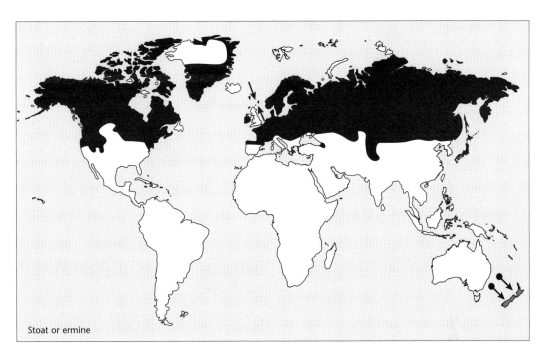

Stoat or ermine

DISTRIBUTION
Eurasia and North America. Britain, Ireland and
northern Europe and northern Asia south to the
Pyrenees, Alps, Caucasus and western Himalayas,
northern Mongolia, Manchuria, Sakhalin Island and
Japan (Hokkaido and northern Honshu). Alaska and
Canada south to the 40th parallel (New York,
Maryland, California and New Mexico). Also north-
eastern Greenland.

HABITS AND BEHAVIOUR
Habits: mainly nocturnal (winter), but frequently
diurnal (summer); terrestrial, occasionally climbs;
lives in den, usually rodent burrows or under logs,
rocks etc.; home range 20–50 ha. **Gregariousness:**
solitary or family parties (females and young); adults
live on separate home ranges for most of year; pair
bonds not established; density 0.3–2.2/10 ha.
Movements: male home range 2–50 ha but can be up
to 100–200 ha; young disperse 6–23 km (males only),
females stay near natal area. **Habitat:** farmland, woodland, moors, marshes, mountains. **Foods:** small
mammals (rats, mice, possums, voles, rabbits), birds
(incl. poultry), eggs, amphibians (frogs), reptiles
(snakes and lizards), insects (beetles), fish, earthworms, berries, carrion and human refuse, crayfish.
Breeding: mates March–July, young born April–May;
female polyoestrous; gestation 20–28 days, with
delayed implantation extends to 280 days; litters 1 per
year; young 3, 9, 14 (18 also recorded Russia); lactation *c.* 5 weeks; post-partum oestrus; males mature at
10–12 months; females 2–3 months, may breed in

their first summer; young are blind, deaf, toothless
and downy, eyes open at 5–6 weeks; female alone rears
young. **Longevity:** 3–8 years in wild. **Status:** numbers
reduced.

HISTORY OF INTRODUCTIONS
AUSTRALASIA
Australia
Apparently stoats were introduced to Australia at an
unknown early date, but failed to become established
(de Vos *et al.* 1956).

EUROPE
Stoats were probably used by central and northern
Europeans before the introduction of the cat (i.e.
from about first to ninth century AD) to control small
rodents (Grzimek 1975).

Denmark
Stoats were introduced to Strynoe Kalv (46 ha),
Denmark, in 1980 to control small mammal pests
(King 1990), but exterminated a population of water
voles (*Avicola terrestris*) on the island (Corbet and
Harris 1991).

Netherlands
In 1931, six to nine stoats were released with
102–104 weasels (*Mustela nivalis*) on Terschelling
Island (210 km²), for the control of water voles
(*A. terrestris*), which were causing damage to planted
pines, oaks and alders (King and Moors 1979), and
possibly also for rat and rabbit control (de Vos *et al.*
1956). They increased prodigiously and by 1934

numbered at least 180 animals, but later decreased in numbers as the water voles disappeared. By 1937 they had established a fluctuating population and the water voles had become extinct (Van Wijngaarden and Bruijus 1961 in King and Moors 1979; Corbet and Harris 1991). In 1953 it was reported that the stoat had increased to a high population level, but that the weasel introduced at the same time had disappeared (de Vos *et al.* 1956).

United Kingdom (Shetland–Orkney Islands)
Stoats were introduced by humans in the seventeenth century or earlier to the mainland Shetland Islands (Venables and Venables 1955; Southern 1964; Lever 1985).

Some stoats may have been released on Shetland prior to 1680, as it is said that Her Majesty's falconer released two because he was refused rabbits for his hawks, but more likely they were imported to control the rabbits. Some stoats were released on Whalsay in the nineteenth century to control rabbits and rats, and on Colsay for rat control, but they failed to become established on both islands. Stoats were also released in the Orkneys probably in the seventeenth century (Fitter 1959).

Pacific Ocean islands
New Zealand
In 1884 a private shipment of both stoats and weasels was imported and during 1885 and 1886 the New Zealand government imported some from London and liberated both species (Marshall 1963). However, it appears that there may have been other introductions and liberations; a number were imported and released between 1884 and 1899 for the control of rabbits (Gibb and Flux 1973). In 1885 alone about 3000 stoats and weasels were sent from Lincolnshire to New Zealand (King and Moors 1979).

Stoats and weasels were first noted established at Tutira in 1902 and by 1904 had reached Poverty Bay some 65 kilometres to the north (Marshall 1963). At least by 1950 they were widespread and common in New Zealand (Wodzicki 1950) and are still widespread in both the North and South islands (Wodzicki 1965; Gibb and Flux 1973), although 1080 poisoning for rabbits and deer poisoning in the 1960s are said to have reduced their numbers (Marshall 1963).

Now stoats are throughout the two main islands and have reached some small islands near the mainland (islands in Fiordland and Marlborough Sounds), but do not occur on any offshore islands (Barnett 1985; King 1990). They occurred on Maud Island where they were introduced about 1980, but were eradicated 1980–83. They subsequently re-invaded the island by

swimming in about 1989 and were eradicated again in 1990–93. They also occurred on Otata Island where they were eradicated in 1955 and on Adele island where they were introduced in about 1977, and where eradication efforts failed in 1980.

■ **DAMAGE**
The stoat is persecuted in Europe because it preys on poultry and game birds, but this is offset by the destruction of rabbits, rats and mice (Southern 1964). A recent study (Erlinge 1981) found that the usual prey was small rodents and lagomorphs, the preferred food was field voles, and that birds were only alternative prey.

In New Zealand predation by the stoat has been shown to be the most important factor in the continuing decline of some hole-nesting (e.g. kaka, *Nestor meridionalis,* and yellow-crowned parakeet, *Cyanoramphus auriceps*) and ground-nesting birds (e.g. all four species of kiwi, *Apteryx* sp. (McLennan *et al.* 1996)). They have also contributed to the decline of others (e.g. both subspecies of New Zealand dotterel, *Charadrius obscurus*) (Dowding and Murphy 1996; Dowding 1999).

STEPPE POLECAT
Light polecat, Asiatic polecat
Mustela eversmanni Lesson
Although M. eversmanni *is almost indistinguishable from* M. putorius, *except that it is invariably lighter in colour, it continues to be separated from that species in most recent taxonomic works.*

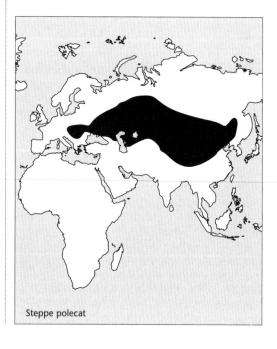

Steppe polecat

■ DESCRIPTION

HB 290–562 mm; T 70–183 mm; WT 0.57–2.05 kg.
Back light yellow-brown in winter and duller and shorter in summer; muzzle and ears whitish; mask around eyes yellow-brown; chest brownish black; dark spots on abdomen; paws brownish black; base of tail straw-yellow, with tip black-brown; hind limbs dark brown; under parts lighter. Difficult to distinguish from *M. putorius*.

■ DISTRIBUTION

Eurasia. The southern Russian Federation east to eastern China, Tibet and Mongolia and south to the Himalayas. Isolated populations in eastern Europe in Hungary, Austria, Czechoslovakia, Poland, Yugoslavia, East Germany and possibly Romania. In recent times has extended range northwards with the clearing of forests.

■ HABITS AND BEHAVIOUR

Habits: Lives in burrows made by other animals; behaviour and ecology similar to *M. putorius*; mainly crepuscular, but also nocturnal; occasionally stores food. **Movements:** occasionally mass movements in times of food shortages; up to 18 km while foraging. **Gregariousness:** solitary? **Habitat:** steppe, open steppe, semi-deserts, meadows and fields, open grassland. **Foods:** squirrels, hamsters, pikas, marmosets, voles and other rodents, hedgehogs and other small mammals, birds and eggs, amphibians, reptiles, fish, insects and frogs. **Breeding:** oestrous in February–March; young born April–May; gestation 36–41 days, litter size 4, 8–11, 18; 1 litter/year; young born blind; eyes open 4 weeks; weaned 6 weeks; disperse at 3 months; sexual maturity 9–10 months. **Longevity:** no information. **Status:** rare in Europe, range fragmented.

■ HISTORY OF INTRODUCTIONS

EURASIA
The steppe polecat (*M. eversmanni*) lives in Asia from northern Urals to Siberia, to Amu River, and southwest through Manchuria to Yangtze Kiang and west to Himalayas, Kashmir, and Altai Valley and Caspian Sea. It did not occur west of Urals in the beginning of the nineteenth century but today is found as far west as Austria and Czechoslovakia (Grzimek 1975).

Russian Federation
The steppe polecat was introduced or translocated in the Russian Federation (Naumoff 1950 in de Vos *et al.* 1956) a number of times mainly in the 1940s. Considerable numbers were released in 1940–41 in the Novosibirsk region where they became well established (Lavrov 1946 in Novikov 1962). Re-introductions in West Siberia failed to become established (Yanushevich 1966).

Steppe polecats (*M. eversmanni*) were released in the Tomsk region in 1940–41. Some 179 were liberated in Chainskii, Parabelskii and Kolpashevskii districts (Kirisa 1973), but the results have not been spectacular (Sofonov 1981).

■ DAMAGE

Steppe polecats are considered to be beneficial to agriculture because of their destruction of rodents (Walker 1992).

EUROPEAN MINK

Mink

Mustela lutreola (**Linnaeus**)

■ DESCRIPTION

HB 280–430 mm, T 120–200 mm, WT 500–1500 g.
Coat reddish brown to dark cinnamon, darker on limbs and tail and lighter on under parts; under fur grey; muzzle with white tip; chin, lower lip, often on chest and upper lip white; ears short and greyish brown. Similar in size and colour to American mink, but distinguished by having white on upper and lower lip (American mink on lower lip only).

■ DISTRIBUTION

Europe. Formerly from north-western France, north-western Spain and from northern and eastern Germany, Poland, eastern Austria, Czechoslovakia, Hungary and eastern Yugoslavia, Bulgaria, northern Romania, central and southern Finland and western

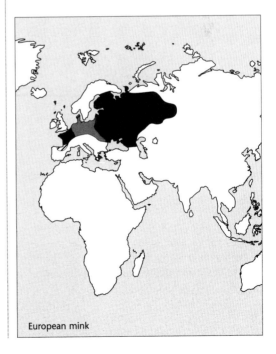

European mink

parts of the Russian Federation, east to the Ural Mountains and Tobal and Ob rivers.

Note: Range now difficult to define because of the establishment of American mink which are similar in appearance. Confirmed recent records only from Russia, Romania, France, Spain and Finland. There maybe isolated populations still in Germany.

■ BEHAVIOUR AND HABITS

Habits: nocturnal and crepuscular; home range in summer 15–20 ha, in winter up to 10 km; lives in burrows or nests in grass and reeds. **Gregariousness:** males solitary. **Movements:** sedentary except for winter wanderings. **Habitat:** forests near streams, wooded marshland, rivers and lakes and marshes. **Foods:** mainly rodents (voles, shrews etc.), frogs, fish, crabs, molluscs, occasionally birds (including domestic poultry), crayfish, water insects, berries, newts, birds' eggs. **Breeding:** mates February–April; young born April–June; gestation 35–42 days; polygamous; litter size 2–7, 10; young born blind; eyes open 4 weeks; weaned 10 weeks; disperse in autumn; sexual maturity following year. **Longevity:** up to 10 years. **Status:** now extinct or greatly reduced in numbers over much of range; decline due to habitat destruction and introduction of American mink.

■ HISTORY OF INTRODUCTIONS
ASIA
Russian Federation
In the last 100 years mink have spread naturally somewhat eastwards in the Russian Federation (Corbet 1966). Introductions have been made in the Far East (de Vos *et al.* 1956; Lindermann 1956), but probably with little success.

Kuril Islands
Middle Island is reported to be inhabited by mink which have been introduced there (Voronov 1963).

In 1983 European mink were released in the Kunaschir Islands and later also on Urup Island by the Biological Institute of the Siberian Department in USSR Academy of Science. The Kuril Islands are outside the natural range of mink, but they have adapted well to the conditions there. The introduction is thought to threaten the local herpetofauna (Schreibet *et al.* 1989).

INDIAN OCEAN ISLANDS
Kerguelen
A single male and two female European mink were released on Kerguelen in 1956 (Lesel 1967) on the Île du Chat, a small island in Golfe Morbihan, but did not reproduce and in 1965 there were no traces of their presence (Lesel and Derenne 1977).

■ DAMAGE
No information.

BLACK-FOOTED FERRET
Mustela nigripes (Audubon and Bachman)

■ DESCRIPTION
HB 380–500 mm; T 110–150 mm; WT 0.75–1.08 kg.
Coat generally yellow buff, paler under parts; forehead, muzzle and throat nearly white; top of head and mid back brown; face mask, feet and terminal quarter of tail black. Female with three pairs of mammae.

■ DISTRIBUTION
North America. Formerly from Alberta, Canada south to the south-western United States.

■ HABITS AND BEHAVIOUR
Habits: nocturnal; lives in burrow dug by prairie dogs. **Gregariousness:** solitary except during breeding season. **Movements:** sedentary; home range 10–120 ha, avg. 36 ha defined by prairie dog town size; density 1 ferret per 50 ha of prairie dog colonies. **Habitat:** short and mid grass prairies. **Foods:** prairie dogs, mice, voles, squirrels, and other small mammals. **Breeding:** mating March–April; gestation 42–45 days; 1–6, avg. 3.3 young per litter. Young emerge from burrow early July, dispersing September–October. Sexual maturity 12 months. **Longevity:** to 12 years. **Status:** extinct in wild.

■ HISTORY OF INTRODUCTIONS
NORTH AMERICA
United States
In the 1800s black-footed ferrets were widely distributed in 10 states (Crane 1990). In Canada they are not recorded since 1937. Ferrets were extinct in the wild by 1981. The reason for decline is loss of prey due to extensive campaigns to eradicate prairie dogs, mainly through poisoning and canine distemper (Crane 1990).

The last wild population grew steadily to 60 in 1982 and 129 in 1984 as a result of protection. However, canine distemper infected the colony and reduced it to 31 in 1985. The remaining animals were captured for a captive breeding program – six died in captivity and global population totalled 25 (Crane 1990).

Since 1991 over 200 have been released back into the wild (WWF-Canada 1997). Captive breeding by the Wyoming Game and Fish Department in co-operation with US Fish and Wildlife Service has now bred large numbers of ferrets. Research has been undertaken into how to achieve re-introduction without effect from keeping in captivity (Godbey and Biggins 1994). In 1991 initial re-introduction took place when ferrets were released into the wild at Shirley Basin, south-eastern Wyoming (US-FWS

1997). Other populations are being established in Montana, South Dakota and Arizona (Clark *et al.* 1987: WWF-Canada 1997).

The US Fish and Wildlife Service is aiming to establish 10 free-ranging populations comprising a total of 1500 animals in the wild by 2010 (US-FWS 1997), but at present they are restricted to the re-introduction sites in Arizona, Montana, South Dakota and Wyoming (WCMC 1998).

■ DAMAGE

Now fully protected by law, but were formerly thought to be farm pests and large numbers were destroyed for this reason. This destruction and loss of natural prey through clearing has caused the species to decline and it is now considered endangered (Whitfield 1985).

WEASEL

Least weasel, snow weasel

Mustela nivalis Linnaeus

Includes M. rixosa *of North America.*

■ DESCRIPTION

HB 160–230 mm; T 30–75 mm; WT 35–185 g.
Colour in summer red-brown above and white under parts and throat; in winter in northern latitudes may be pure white with a few black hairs on tail tip, but in some more southern areas coat may be mottled brown. Body slender; legs short; face blunt, brown spots sometimes on cheeks; feet whitish, soles furred; ears short, oval; tail stubby, without black tip. Female smaller than male, with three to four pairs of mammae.

■ DISTRIBUTION

Eurasia, North Africa and North America. In North America throughout continental Canada (except southern Ontario and Quebec, coastal British Columbia and north-eastern Northwest Territories), Alaska and extreme northern United States. In Eurasia throughout western Europe and east throughout Russia (except the northern Taimyr Peninsula, Sakhalin and the central part of Kyzyl-kum and Kara Kum areas) to Japan. In North Africa in northern Morocco, Algeria and Tunisia, and extreme north-eastern Libya and north-western Egypt.

■ HABITS AND BEHAVIOUR

Habits: nocturnal and diurnal; usurp burrows of prey species; 2 annual moults. **Gregariousness:** sexes live on separate home ranges; immatures tolerant of others, adults antagonistic; family groups break up at 9–12 weeks. **Movements:** home range 1–25 ha (males on largest home ranges). **Habitat:** mixed forest, woodlands, moorlands, mountains, stubble fields, meadows, parkland, river banks, suburban gardens. **Foods:** small mammals (rats, wood mice, mice, bank voles, field voles, moles, rabbits), frogs, small birds, amphibians, eggs,

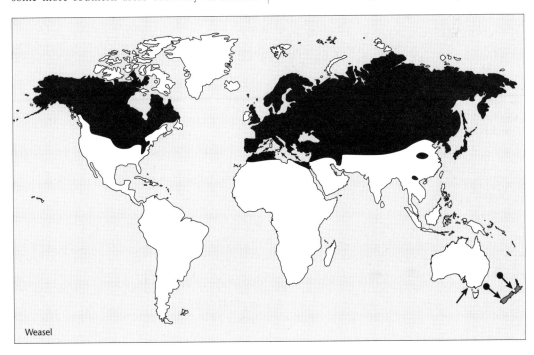

Weasel

occasionally poultry, carrion, reptiles (geckos and skinks), and insects. **Breeding:** breeds mainly March–August; gestation 34–37 days; female poly-oestrous; two or more litters per year; young 3, 4–8, 10; born blind, naked, but develop covering of white hair in 2 weeks; teeth erupt at 2–3 weeks; eyes open at 4 weeks; weaned at 4–5 weeks; female mature at 3–4 months, males 8 months. **Longevity:** 6–8 years in captivity, probably 1–3 years or less in wild. **Status:** numbers reduced.

■ HISTORY OF INTRODUCTIONS
Australia, New Zealand and the Netherlands. Distribution on Atlantic Islands and in Mediterranean probably influenced by introductions by humans (Corbet 1978, 1980).

EUROPE
The weasel was evidently introduced by human agency on certain Mediterranean islands, on the Azores and Sào Tome off west-central Africa (Corbet 1978) and to Malta (Haltenorth and Diller 1994).

Terschelling Island (Netherlands)
In 1931 some 102 to 104 weasels (*M. n. nivalis*), together with six to nine stoats, were introduced on Terschelling Island to control water voles (*Avicola terrestris*) which were causing damage to planted pines, oaks and alders (de Vos *et al.* 1956; King and Moors 1979). By 1934 the weasels and the other animals introduced had disappeared.

PACIFIC OCEAN ISLANDS
New Zealand
The weasel was introduced to New Zealand with the stoat between 1884 and 1899 (Gibb and Flux 1973). The most popular dates for their introduction are 1885–86 (Wodzicki 1950, 1965; Hartman 1964; Hinton and Dunn 1967). In 1885 alone about 3000 stoats and weasels were sent to New Zealand from Lincolnshire to be introduced to control rabbits.

The weasel is not as widespread as the stoat in New Zealand (Marshall 1963; Wodzicki 1965; Gibb and Flux 1973), and is still patchily distributed over most of the two main islands, with the possible exception of the south and west of the South Island. It is not known to have reached any of the outer islands (King 1990).

AUSTRALASIA
Australia
Weasels were introduced to Australia to control rabbits in 1885, but failed to become permanently established (Hinton and Dunn 1967).

■ DAMAGE
In New Zealand control of weasels has been considered unnecessary as they present no threat to the survival of native species (King 1990). Prior to human settlement the endemic fauna evolved in the absence of mammalian predators. However, because of this evolutionary history it is suggested that the endemic fauna is especially susceptible to predation. One study found no significant difference between native and introduced birds in frequency of predation (Moors 1983). In New Zealand, despite trapping and persecution they have survived and increased in numbers.

In Britain weasels are regarded widely as vermin, but probably cause little damage apart from occasional attacks on poultry (Southern 1964). They appear to have no observable effect on density or survival of small rodents in English woodlands (King and Moors 1979; King 1980).

EUROPEAN POLECAT
Polecat, ferret, foul-marten, forest polecat, domestic ferret
Mustela putorius Linnaeus
The domestic ferret is similar to the European polecat, but is slightly smaller and is often elevated to species status. Its origin is uncertain, but cranially it seems closer to M. evers-manni; however, the karyotype is identical with that of M. putorius and is different from M. eversmanni. Domestic ferrets (M. p. furo) appear as a domestic animal about 1000 BC and were certainly used by the Romans to catch rabbits. M. eversmanni and nigripes are treated separately.

■ DESCRIPTION
HB 225–460 mm; T 85–190 mm; WT 400–1850 g.
Generally dark or black-brown in colour (body colour of *M. p. furo* can vary from brown to white); guard hairs blackish, under fur cream-yellow; long narrow body with short legs; sides of neck, throat, chest and forelimbs darker; muzzle greyish white; ears rounded, flattened against head, edged greyish white; dark mask between and around eyes; tail bushy, dark brown or black. Male is hob, female is jill or jen, young are kits or (pole) kittens.

■ DISTRIBUTION
Eurasia. Western Europe from the Mediterranean north to central Scandinavia and Finland, and east to about central Kazakhstan, Russia, Romania, Hungary, Czechoslovakia, Yugoslavia, eastern China and Mongolia, south to the Himalayas. In Britain but absent from Ireland.

■ HABITS AND BEHAVIOUR
Habits: mainly nocturnal but also diurnal; uses old burrows; single moult/year; eyesight poor; smell and

hearing main senses; swims but climbs poorly. **Gregariousness:** males establish dominance relationships which determine access to females; solitary usually; density 1 per 1000 ha. **Movements:** mostly sedentary; home range 100–150 ha. **Habitat:** wooded and rocky areas, broken woodland and glades, mixed forest clearings, re-growth areas, margins of lakes and marshes, often in villages and cities; steppe, semi-arid areas, river valleys (rough grassland and scrubland and fringes of forest in NZ). **Foods:** small mammals (moles, shrews, voles, hedgehogs, rats, mice, rabbits, possums), toads, frogs, fish, eels, lizards, insects, snails, worms, carrion, snakes, birds (occasionally domestic poultry), eggs and carrion. **Breeding:** oestrus (mate) March–May, young born May–June; gestation 40–42 days; young 2, 3–6, 8 and up to 12 recorded (Russia); 1 litter per year (possibly a second?); young at birth have sparse hair covering; eyes open at about 30 days, weaned 6–8 weeks; disperse from territory at 3 months. **Longevity:** 8–14 years captive; probably 4–6 years in wild. **Status:** common most areas; extended range to north, east and south during recent decades through clearing of forests and extension of agricultural areas.

■ HISTORY OF INTRODUCTIONS

Australia, New Zealand, West Indies, Japan, Great Britain. As widely kept pets, escaped domestic ferrets may be encountered almost everywhere and this makes it difficult to detect well-established populations.

ASIA
Japan

European polecats were introduced and established in the Tohoku district in the 1930s (Kaburaki 1940), but evidently they failed to become established, as the species is not mentioned at later dates (Kuroda 1955).

Russian Federation

Introduced or translocated (Naumoff 1950 in de Vos *et al*. 1956). Considerable numbers were released 1940–41 in the Novosibirsk region and apparently became established there (Lavrov 1946 in Novikov 1962).

AUSTRALASIA
Australia

Ferrets were introduced to Australia initially to control rabbits in 1885 (de Vos *et al*. 1956; Hinton and Dunn 1967), but failed to become permanently established in the wild. However, they escape into the wild on many occasions and continue to do so (Myers 1986; Long 1988). Populations are occasionally found in the wild, but so far have not remained permanently

established. An isolated population existed to the south of Launceston, Tasmania, in the early 1990s (Wilson *et al*. 1992). Other small populations have survived for periods in Western Australia (Long 1988).

PACIFIC OCEAN ISLANDS
New Zealand

The ferret (*M. p. furo*) was introduced to New Zealand in 1867–68 (Thomson 1922), and some were also released in 1882 (Marshall 1963; Wodzicki 1965) or 1886 (Wodzicki 1950). The New Zealand Department of Agriculture bred ferrets for release until about 1897 (King and Moors 1979). They are now widely distributed and common in the North and South islands (Wodzicki 1950 and 1965).

Five ferrets were imported by the Canterbury Acclimatisation Society in 1867 and an additional one in 1868; however, there are no records of any liberations until the 1880s (Marshall 1963). The first consignment of ferrets was introduced by the government in 1882, and soon after further shipments were made from both London and Melbourne. These were liberated in large numbers.

Many thousands of ferrets were liberated in the 1880s and the species became widely distributed, its range linked with the distribution of the introduced rabbit. From the 1950s on the numbers of the three introduced mustelids in New Zealand declined as rabbit numbers were controlled (Gibb and Flux 1973). There have been no distribution surveys since 1962 when they were present throughout the two main islands except in Northland, in eastern Bay of Plenty and Poverty Bay, in large areas of Taranaki, western parts of Nelson, and the whole of Westland (King 1990). No ferrets appear to occur on offshore islands, but they were released on Haulashore in Nelson Harbour to control rabbits in the 1960s, and 40 were released on Rangitata (island in river near Christchurch) also for rabbit control (Flux and Fullagar 1992).

WEST INDIES
Jamaica

Ferrets were introduced into Jamaica, in the West Indies, for rat control, but died out before becoming well established (Milne and Milne 1962; Silverstein and Silverstein 1974).

AFRICA
Morocco

It is not known how long the polecat has occurred in Morocco, but most appear to agree that it was an anthropogenic introduction (Cheylan 1991).

EUROPE

Escaped animals are often found almost anywhere in Britain. Wild populations of ferrets (*M. p. furo*) are found on Sardinia and Sicily (Brink 1967) and in many places on continental Europe. The first description of ferrets being bred for bolting rabbits was at least 2000 years ago (Blandford 1987) by Aristotle in the fourth century BC, and 350 years later they were mentioned by Strabo as having been introduced into the Balearic Islands to counter a plague of rabbits (Thomson 1951).

United Kingdom

There have been some successful introductions of ferrets (*M. p. furo*) to Scotland (Corbet and Harris 1991). They have been introduced to the islands of Bute, Man, Mull, Arran and Lewis (Thomson 1951; King and Moors 1979; Blandford 1987), where they once occurred in large numbers (Corbet and Harris 1991). They were introduced to the island of Mull about 1933–34 and soon escaped and are now firmly established and a pest. They were also introduced to the island of Harris in the Outer Hebrides to control rabbits (Fitter 1959). Ferrets often escape and feral populations are known on several offshore islands in Britain (Man, Anglesey, Lewis, Arran and Bute) (Corbet and Southern 1977; Lever 1985; Blandford 1987). Two ferrets introduced to Great Saltee, island in Wexford, Ireland, died in the first winter (Flux and Fullagar 1992).

What is known as the polecat ferret (thought to be a hybrid between the European polecat (*M. putorius*) and the domestic ferret (*M. p. furo*)) exists in parts of the British Isles.

The exact date of introduction of ferrets as domestic animals to Britain is not known, but they have been known there since the late thirteenth century, probably having arrived with either the Romans or Normans. They were most likely brought in for hunting and the control of rabbits (Lever 1977; Corbet and Harris 1991). Escaped domestic ferrets are constantly being reported, particularly in northern England.

The remaining populations of the native Euroopean polecat in Britain are holding their own, and expanding their range in the face of dilution of their gene pool through interbreeding with feral domestic ferrets (Blackman 1990). European polecats were once common but began to decline about 1850. Decreased persecution after World War 2 has allowed their recovery and their range is slowly expanding in contrast to the present decline elsewhere in Europe.

ATLANTIC OCEAN ISLANDS

Canary Islands

Ferrets have probably been introduced successfully to this island (Encycl. Brit. 1976–78).

■ DAMAGE

In Europe polecats are heavily persecuted because of their depredations on game and poultry (Morris 1965; Corbet 1966).

In New Zealand in 1986 there were 127 registered farms. Farming originally commenced with feral stock, but most have now imported superior stock from Europe. The real effects of mustelids introduced into New Zealand is unknown (King 1990), but recent accounts indicate that ferrets contribute to the rapid decline in the numbers of kiwis (*Apteryx* sp.) on both main islands (Chapple 1999). Polecat farming still continues there and recently 10 farms imported ferrets from England to improve fur quality and colour (Wodzicki and Wright 1984).

SIBERIAN WEASEL
Siberian mink, kolinsky
Mustela sibirica Pallas

M. lutreolina *of Java and Sumatra is sometimes included as a subspecies of the Siberian weasel,* M. sibirica.

■ DESCRIPTION

HB 250–390 mm; T 133–210 mm; WT males 650–820 g, females 360–430 g.

In winter, coat light red except muzzle tip which is brown above; lips and chin white; small spot sometimes on neck; flanks darker and duller; body slender and elongated; head slightly elongated; ears broad; tail red, long and furry; vibrissae brownish with red tips. In summer coat darker; under fur greyer and brown of head darker. Female generally smaller than male; has four pairs of mammae.

■ DISTRIBUTION

Asia. Extreme eastern Europe and central Russia (west to 50°E) from about the Urals east to Siberia and the Far East, Japan and south to Mongolia, Korea, southern China, Nepal, and the Himalayas, northern Thailand, northern Burma (Assam), India (Kashmir), Afghanistan and Tibet; all main Islands of Japan and Taiwan.

■ HABITS AND BEHAVIOUR

Habits: mainly nocturnal and crepuscular, occasionally seen in day; lives in burrows or rock crevices, or under logs, buildings; lines nest with fur, feathers and vegetation; inactive during severe cold; swift and agile. **Gregariousness:** pairs. **Movements:** 8 km/night foraging; may move in winter from cold uplands to

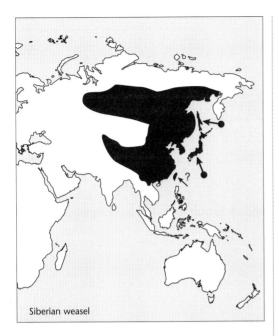

Siberian weasel

The Japanese race, *M. s. itatsi,* of the Siberian weasel was introduced to Hokkaido before 1901 and was increasing rapidly in numbers and range up until at least the mid-1950s. Here, there has been some inbreeding between the Siberian and Japanese races (Lever 1985). Some were translocated from Hokkaido to the small islands of Rishiri and Rebun off the northern tip of Hokkaido in 1933 (Kuroda 1955), and in 1948 to the island of Okujirijima for the extermination of *Rattus norvegicus* and *Apodemus speciosus* (Inukai 1949 in Kuroda 1955).

Between 1958 and 1968 severe rat damage was caused to agricultural products, chiefly sugar cane and pineapple, in many parts of the Ryukyu Islands. A program was begun in 1966 to introduce the Japanese weasel into a number of islands by Dr T. A. Uchida for the World Health Organisation. Results were far from promising where control was left to the weasels alone. On Ishigaki-jima 1600 weasels were introduced. On islands where rodenticides were also used the results were more encouraging and sugar cane damage was reduced from 30 per cent to almost nil (WHO 1968).

Russian Federation

Siberian weasels have been introduced or translocated in the Russian Federation (Naumoff 1950 in de Vos *et al.* 1956). In 1932 or 1933 Japanese weasels (*M. s. itatsi*) were introduced into southern Sakhalin for rodent control and became established in a restricted area (Inukai 1949 in Kuroda 1955; Yanushevich 1966; Kirisa 1973; Sofonov 1981).

Some 30 Siberian weasels were released in 1937 in the Semonovsk district of the Gorkov region of Kirgizistan (Lavrov 1946 in Novikov 1962) and some in the (Djet-Oguzov) Dzhety Oguz district of the Issyk-Kulsk region, Kirgizia, in 1941 (Lavrov 1946; Yanushevich 1966). The results of the first release do not appear to be known, but in the latter area they became successfully established in low numbers and were breeding (Lavrov 1946). Later again they were reported to be established locally (Yanushevich 1966; Kirisa 1973).

Pacific Ocean islands
Amami–Oshima (Ryukyu Islands, Japan)
The Japanese weasel has been introduced and established. Two thousand Japanese weasels were introduced to Amami–Oshima for rat control and snake control from 1954–58, but fortunately they failed to become established (Hayashi 1981).

■ DAMAGE

In Japan, the Korean weasel has interbred with the native race which it is gradually replacing (Kuroda 1955; de Vos *et al.* 1956).

valleys; reports of mass migrations with food shortages. **Habitat:** forest, along streams and river valleys, lake shores, swamps, forest steppe, taiga, occasionally in towns and even cities. **Foods:** small rodents and other mammals (voles, rats, chipmunks, squirrels, pikas, hares), small birds, amphibians, fish, lizards, birds' eggs, slugs, berries, nuts and occasionally domestic poultry. **Breeding:** mates late winter and early spring (February–April), but may breed twice per year in some areas; litters 2–4, 12, gestation 28–30 days; young born April–June; eyes open 1 month; lactation 2 months; young leave female in August. **Longevity:** 8 years 10 months in captivity. **Status:** fairly common; important fur bearer in the Russian Federation.

■ HISTORY OF INTRODUCTIONS
Asia
Introduced from Kirghizistan, to Sakhalin Island in the Russian Federation and in 1965 to Iriomote Island, Ryukyu Islands (Obaba 1967). They were introduced to the Ryukyu Islands to control rats, but have preyed upon the indigenous small fauna (Lever 1985).

Japan
Korean weasels, *M. s. coreana,* were imported from South Korea in about 1930 (Tokuda 1951 in Kuroda 1955) for fur farming, escaped and became established in Japan. They have been recorded in Kobe, Akashi, near Osaka, south-western Hondo (Chugoku) and also in Tokushima (Shikoku). Their range in the mid-1950s appeared to be confined to south-western Hondo (Kansai area) and the eastern parts of Shikoku (Kuroda 1955).

MINK
American mink
Mustela vison Schreber

■ DESCRIPTION
HB 320–430 mm; T 125–220 mm; WT 0.68–2.31 kg.
Body slender with fur dark brown to black (many colour forms under domestication); under parts paler; neck long; face pointed; splashes of white on lower lip, chest and occasionally on abdomen; ears small; legs short. Females generally half to two-thirds size of males and have three pairs of inguinal mammae.

■ DISTRIBUTION
North America. Alaska and northern Canada south to Florida, New Mexico and California.

■ HABITS AND BEHAVIOUR
Habits: mainly nocturnal and crepuscular (females with young tend to be more diurnal); semi-aquatic; dens under trees, banks, logs or usurps burrows; riparian. **Gregariousness:** solitary except at mating time; density 1–8/km². **Movements:** males may travel long distances in dispersal and mating season (1–20 km recorded), females more sedentary; home range females 7.7–20.2 ha and males up to 7.8 km². **Habitat:** near rivers, streams, lakes, wooded marshland, swamps, tidal flats along forest edges and woods; edges of cultivated pasture and fields. **Foods:** muskrats, voles, mice, hares, shrews, rabbits, moles, squirrels, rats, (small mammals), birds, insects (beetles), amphibians, frogs, fish, reptiles, snakes, crayfish, snails, pond mussels, molluscs, worms, eels, earthworms, marine crustaceans. **Breeding:** mates February–April; young born April–July; males promiscuous; females polyoestrous; gestation 39–76 days; varying period of delayed implantation; kits 2–12, 17; 1 litter per year; born blind; lactation 6–8 weeks; young breed following year; males mature 18 months, females 12 months. **Longevity:** 3–5 years in wild, and 10 or more as captive. **Status:** common but numbers reduced (widely farmed for fur).

■ HISTORY OF INTRODUCTIONS
Europe, the Russian Federation, Kuril Islands, Japan, Iceland, France, Germany and Scandinavian countries.

EURASIA
Introduced as a fur animal in the 1920s and 1930s in Germany and elsewhere. Mink have escaped in Denmark, Netherlands and at Brandenburg, but have been unable to establish lasting colonies in central Europe to date, although in Scandinavia and Iceland they have done so and made themselves a thorough nuisance to game and domestic fowls. In 1957 they were known in five districts in Sweden.

North American mink now live in Sweden, Norway, Denmark, Iceland, Great Britain, Ireland, Finland, central Urals, the Altai, the Ussuri region and other parts of the Russian Federation from escapees (Grizmek 1975). Farmed commercially in Germany,

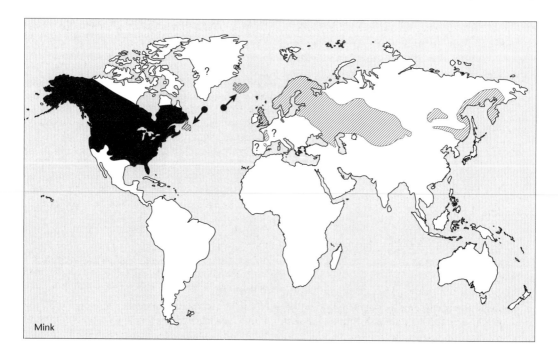

Mink

Switzerland, Belgium, Austria, Italy, Czechoslovakia, Poland, Bulgaria, Hungary, Romania and feral populations are occasionally found established in these countries.

Denmark
Fur farms were set in the mid-1920s in Denmark, and mink became established as escapees in 1930. They have gradually extended their range (Thompson 1952; de Vos *et al.* 1956) until in the 1960s were well established in Denmark (Thompson 1964).

France
Occasionally small remnant populations of mink are found in north-west France (Lever 1985).

Finland
The first ranch minks were probably imported to Finland in the 1920s and the industry increased rapidly after World War 2. By 1964, there were 2000 farms with stock of one and a half million mink. They became established in the wild in 1930 and have gradually extended their range since then (Thompson 1962; Corbet 1966).

Since 1950 escaped ranch stock mink have become established in wide areas of the Finnish coast; and by the 1960s they were abundant in some places (Tenovuo 1963; Thompson 1964).

Between 1951 and 1964, there were 734 reports of the species in the wild. They are now found over a wide area mainly in south-west Finland, are increasing in the Aland Islands and in inland areas occur regularly only in Eastern Lapland and around Jyraskyla (Corbet 1966; Westman 1966).

Iceland
Mink escaped from fur farms in Iceland (de Vos *et al.* 1956), where they were first farmed in the 1930s, and are now established and widespread along the coast and rivers in the south-west of Iceland (Thompson 1962; Corbet 1966). By the 1950s they were distributed throughout south-west Iceland from Myrdalur in the south to Breidafjordur area in the west (Lever 1985), and by mid-1970s almost the whole mainland was colonised.

Kuril Islands (Kuril'skiye Ostrova, Russian Federation)
Mink were introduced to the Kuril'skiye Ostrova, where they are now relatively abundant (Lavrov 1962).

Netherlands
American mink have for some years been caught in the Netherlands during control campaigns against the muskrat. This work dates back to World War 2 and mink have been known for many years. Initially they were probably escapees and it was considered that there was little hope of them becoming permanently established. In recent years, however, the numbers caught have increased, especially in Noord-Brabant and Zeeuws-Flanders. In Noord-Brabant the increase was from one caught in 1971 and 19 in 1974 and 50 in 1979. They were later found in the provinces of Dreuthe, Overijssel, Gelderland, Utrecht, Noord-Holland, Zuid-Holland, Zeeland, Noord-Brabant and Limburg. The largest concentrations appeared to be in the Limburg (13 caught in 1979), Noord-Brabant (50 in 1979), Gelderland (15 in 1979) and Utrecht (eight in 1979) provinces. Observations and known areas appeared to be following the same pattern of distribution as the muskrat (based on catches by muskrat exterminators), and little is known of minks in non-muskrat areas (Litjens 1980).

There is some evidence (in 1979) that they were breeding in the wild in the eastern part of Brabant (Asten and Deurne) and in the Zeeuws-Flanders (at Zuiddorpe) and in fact in June 1979 young minks were observed for the first time in Noord-Brabant (at Helenaveen) (Litjens 1980).

Norway
Mink escaped from fur farms and became feral by about 1930 in Norway (de Vos *et al.* 1956; Wildhagen 1956).

Mink imported from North America for fur farming in the 1920s escaped and became established in the wild. By 1948 they were found in a great many areas in the south of Norway (Wildhagen 1956; Thompson 1962; Pedersen 1964) and 480 were shot as escapees in the wild (Westerskov 1952).

In 1954 there were half a million mink in fur farms in Norway and it was quite common for animals to escape and at this time their distribution appeared to be governed largely by the distribution of fur farms. Between 1948 and 1956 there was little change in distribution in southern Norway but they had extended their range north of Trondheimsfjord (Wildhagen 1956).

In Norway they have now colonised all areas except the Arctic and highest mountains (Thompson 1964).

Russian Federation and adjacent independent republics
Wild mink in Europe are all derived from escapees from fur farms except in the Russian Federation where some 30 000 have been released to breed and provide pelts (Thompson 1967).

Mink were imported to Russia in 1928 and kept on farms until 1933 when some were released (Lavrov and Pokrovsky 1967). From 1928 to 1944, 3140 mink were released in 17 areas mostly at the limits of the range of the indigenous European mink (Kirisa 1973).

In the European Russia, since 1930, there have been more than 30 releases with more than 2000 animals and in the Trans-Ural more than 160 releases with 14 000 animals (Popov 1964) (see Lever 1985 for further notes).

In the Russian Federation from 1933 to 1961, 11 000 mink were released and the species was acclimatised successfully in the mountainous Siberian forests of the Kemerov oblast and Altaisk, Khabarov, Primorsk and Krasnoyarsk, and trapping for fur began two to three years after the first releases (Berger 1962). From 1946–70 some 17 311 were released in 32 oblasts, krai and republics, about 21 of which had not had earlier releases (Kirisa 1973).

European Russian Federation

Attempts were made to acclimatise them in the Kvarel region in 1950–51 (Ekvtimishvili and Gamarashvili 1951). In 1953–57, 725 were released in Belorussia (Samusenko 1962). Successful attempts have also been made in Tuva (Volchenko 1964) and in the Urals (Pavlinin and Shvarts 1961). Acclimatisation attempts in Buryatiya were unsuccessful (Izmailov 1969).

Although successfully established and widespread in the Russian Federation, numbers have declined in recent years due to unknown causes (Sofonov 1981). In north-western Russia mink were released in Murmansk (83 in 1935–36), Karelsk (328 in 1934–70), Leningrad (colonised from Karelii), Pskovsk (colonised from others) and Archangel (44 in 1957). In this area mink were released in Litovsk (113 in 1950–53) and Belorussia (895 from 1953–58).

In the central and wooded regions of the Federation mink were released in Kalinin (60 in 1948), Gorkov (119 in 1957–58), Tatar (570 from 1934–64), Kuibishev and Ulyahnov (probably colonised?) Mariisk (192 in 1948–49), Bashkir (1245 from 1935–67). In the forest-steppe and steppe zone of the Federation mink were released in Vononej (19 in 1933), Penzensk (42 in 1964), Volgograd (100 in 1959) and possibly the Ukraine.

Mink were also released in the Azerbaidjan region (46 in 1938), in Gruzinsk (63 in 1939) and the north of Osetinsk (56 in 1951–53).

Acclimatisation of mink in the Far East began in 1936 and by 1959, 3830 had been released. The most successful release was in the Khabarovsk krai area where the annual numbers trapped reached 4000 by 1958 (Kazarinov 1963). By 1969 some 5730 released in five main areas Amur (539), Khabarovsk (2679),

Primorsk (1004), Magadansk (1328), and in Kamchatsk (180) (Kirisa 1973).

Mink acclimatisation in the Russian Federation commenced in 1933 and by 1948 over 3700 had been released at over 50 locations including Murmansk, Karelo-Finnish, Kalinin, Veronezh, Tatar, Mari and Bashkir, Georgian and Azerbaijan, Sverdlovsk, Omsk, Novosibirsk, Kemerov, Altai, Gorno-Altai, Krasnoyarsk, Irkutsk, Buryat-Mongolia, Chita, Khabarovsk, Amur and Maritime regions (Lavrov 1946 and 1950 and Popov 1949 in Novikov 1962). By 1956 mink had not been successful in all regions but while becoming well established they had failed to reach commercial densities. They were not successful in Tatar and Baskiv, in Altai, eastern Siberia, and the Far East (Novikov 1962).

Acclimatisation began in the Far East in 1936 and was widely developed after 1947 (a total of 3244 released) when favourable results were obtained and the species became widely established (Vaseneva 1964).

Mink were most successfully established in the Far East and Altai, where populations reached 1.06/km^2. In western Siberia, however, the mink density was low (Popov 1964).

Mink were established in Siberia and the Far East for fur. They are now widespread in the Far East from introductions between 1938 and 1961; in Sakhalin from escapes from fur farms and introductions; Karelia from introductions in 1934; Volga-Kama region from introductions in 1934, 1935, 1948, and 1960–62. Following introduction to the Arkhangelsk region in 1947 they are now established but locally restricted. They became established in Belorussia from introductions in 1953–58; West Siberia, Lithuania, from introductions in 1953 (having previously failed in 1950); and Kirghizia from introductions in 1956 and 1962. They were introduced and established in Bashkiva from introductions in 1935; Maritime Territory from introductions in 1936; introduced unsuccessfully in the northern Concausus in 1960 and the Caucasus (Yanushevich 1966).

Mink were imported into Kirghiziya in 1956 from the Choisk Raion of the Gorno-Altaisk Autonomous oblast and some were released (13 males and 33 females) at the middle reaches of the Kara-Ungur River and a second release (15 males and 33 females) in the lower reaches of the Balyk-Sai River. These areas were surveyed in 1958 and 1961 and considerable numbers of mink were found and they have now spread over an area of 30 000 km^2. (Beishebaev 1963).

ASIAN RUSSIA
Urals and Western Siberia
In this area mink were released in Sverdlovsk (653 from 1934–70), Chelyahbinsk (478 in 1960–66), Tumensk (1984 from 1935–68), Omsk (711 in 1948–64), Tomsk (1030 from 1937–58), Novosibirsk (361 in 1950–69), Altai (1089 from 1937–64), Kemerov (638 from 1948–56). In 1935 American mink were released in the upper reaches of the Konda River. Several groups were released between 1937 and 1940 on the taiga rivers of Agan, Jugan, Vasiugan, and Ket in the Ob river basin. Another stage of acclimatisation occurred in the 1950s when during two decades 40 more groups were released (total of 3500 animals) in Tyumen, Omsk, Novosibirsk and Tomsk regions (Sinitsyn 1987).

Kazakhstan
In 1952 some 156 mink were released in the Leninogoskom region of eastern Kazakhstan in four localities.

Central Asia
From 1956 to 1967 releases made in four regions in Kirgizistan. At least 336 American mink were released.

Central Siberia
Mink were released in Krasnoyahrsk (1159 from 1936–59) and Tuvin (99 in 1951).

Eastern Siberia
Released in Irkutsk (367 during 1936–42 and 48 in 1951), Chitinsk (70 in 1939), Buryaht (69 in 1939) and Yahkut (686 in 1961–64).

Introduced to Kamchatka in 1960 and by 1970 2500 pelts were being taken annually (Savenkov 1987).

Sakhalin
From 1956 to 1971, 809 mink were released in seven regions (Kirisa 1973).

EUROPE
Spain
Some mink escaped from fur farms in 1982 (Lever 1985; Smal 1988), but it is not certain whether they are established in the wild.

Sweden
Feral mink which had escaped from fur farms were found in Sweden in 1928 (de Vos *et al.* 1956; Gerell 1967). Fur farms were first set up in the 1920s and by the 1960s mink were breeding in the wild all over Sweden (Thompson 1962; Niethammer 1963; Gerell 1968).

The first Swedish mink farm was started as early as 1925 and the industry increased until by 1939 when there were some 1224 farms with a total of 107 232 mink in captivity. Although production fell considerably during World War 2 due to lower prices and forage rationing, there were still 714 farms and about 98 507 mink in 1945. The industry again rapidly expanded in the post-war period (Gerell 1967).

Mink managed to escape from fur farms in the first few years of their breeding in Sweden, and the likelihood of escapes increased during the war, when materials and manpower decreased. The first records of them in the wild came from the southern parts of Sweden in 1928 and in the southern part of the northern half of Sweden in 1929. From these dates until 1939 there were at least 43 records from different localities in these areas of escaped, trapped or observed mink (Gerell 1967).

In 1928 the first free-living mink was noted near Limedsforsen, Kopparbergs län (administrative district), the second in 1929 near Ullanger, Vasternorrlands län, and there were probably escapes in the vicinity of Anundsjo some 50 km from Ullanger. There were two mass escapes in the 1930s when 20 mink escaped from a farm at Bjurholm, Vasterbotten län in 1932, and during 1933–34 about the same number escaped from a farm at Hjartum near the river Gotaalv, Göteborgs and Böhus län. It is thought that in the 1930s individual mink escaped from farms over the greater part of the country, but only became established in favourable habitats such as the eutrophic lakes on the plains of southern Sweden. In 1941–42 some may have invaded Sweden from Norway (Gerell 1967).

Following World War 2 the wild mink populations continued to increase and invade further mink-free areas. In 1958–59 the annual catch was 14 000 animals and they became established on the archipelago in Göteborgs and Böhus län (Gerell 1967).

Populations of mink continued to increase and expand and by about 1964 only the mountainous regions of the northern-most parts of Sweden were free of them. One of the last areas to be invaded in the south of the country was the island of Oland, where they became established in 1963. Although populations are still increasing the rate of increase has slowed (Gerell 1967).

The wild mink now in Sweden are more than likely the results of crossing at least three subspecies as they were imported into Europe originally as fur animals from North America.

In 1948 in Sweden some 2000 mink were shot as escapees in the wild (Westerskov 1952).

By 1964 the only mink-free areas were in the mountainous regions of northern Sweden. An indication of the increase in mink numbers is given from the figures from the period 1944–49, when the catch of mink for fur was estimated at 6530, and in the period 1959–64 the estimated catch was 84 895 (Gerell 1967).

United Kingdom and Ireland

Mink were first brought to Britain in 1929 and interest in mink farming grew after World War 2 when there were about 700 farms. These mink farms were widely distributed by 1962. The first records of mink found in the wild in Britain were in 1929 with further escapes from fur farms from 1930 onwards. Most escapes from fur farms occurred after 1962 (Thompson 1962, 1968; Clark 1970). Some mink were also released by disenchanted owners and vandals, or escaped due to inadequate pens. In the 1950s the industry expanded and large numbers were imported from the United States and Scandinavia (Lever 1977). By the late 1950s there was evidence that they were living and breeding in the wild in Britain, and by 1962 they were found to be well established in the south-west counties, especially Devon, but also Lancashire, Yorkshire, Sussex, Hampshire and south-west Wales, and in Aberdeenshire and Bamffshire in Scotland (Thompson 1968). By 1962 there were 600–700 farms and by 1968 the number had fallen to 240 (Thompson 1962). Since 1962 mink have increased in numbers and by the 1980s were found in most counties of Britain. Even though they were thinly distributed in some areas it is expected that further expansion will occur.

In England mink have been breeding in the wild at least since 1957 and since 1961 some 1620 were caught in the wild and the species was probably more widespread in southern England than believed at the time (Thompson 1967). Between 1953 and 1967 many were caught in the wild in the United Kingdom, but they continued to survive in most areas and to increase their range in some. The first reports of breeding in the wild after World War 2 were in Lancashire, where there were a large number of fur farms, and then odd reports were received from 1953 to 1958. In the late 1950s and early 1960s they began to spread considerably and were reported in many areas of Britain (Lever 1977). In 1964 they were known to be established and breeding in Hampshire, Wiltshire, Pembrokeshire, Carmarthen, Cardigan and around Banffshire and Aberdeenshire (Southern 1964).

Wild mink were known to be present along the whole length of the Teifi River from Cardiganshire to Tregaron and on the larger tributaries and also on the Western Claddan in Pembrokeshire (Hill 1964). Between 1962 and 1969, 2700 were trapped (Clark 1970) and up until 1970, 5025 were trapped (Thompson 1971). In Lancashire from 1954 to 1961, 70 were caught in the Fylde area and from 1963 to 1964, 51 were caught. In 1965 the population extended north to Lune and south and west up the Ribble to West Riding. In Devon 80 were captured and in Sussex 87 (Clark 1970)

By 1965–67 mink had been caught in 31 counties in England, 24 in Scotland and four in Wales. The principal concentrations were on the rivers Exe, Teign, Axe, and others on Dartmoor, and now Somerset, Avon, in Hampshire and Wiltshire; Stour in Dorset, Lune and Wharfe in Yorkshire/Lancashire; Ouse in Cuckmere in Sussex; Teifi in Carmarthenshire. From 1962 to 1969 some 2700 mink were trapped (Clark 1970).

Today mink are present in significant numbers in England in Devon, Hampshire, Wiltshire, Dorset, Yorkshire, Lancashire and Sussex, and in Wales in Carmathershire, Cardinganshire and Pembrokeshire (Lever 1977). They are now found throughout Britain (Baker 1990) and on the islands of Lewis and Arran (Corbet and Harris 1991).

The first mink fur farm was established in Scotland in 1938 and until 1946 remained the only one. By 1948 there were five and this increased to 100 in the 1940s and 1950s, but from 1962 the number of farms declined and in 1971 only 29 farms remained (Cuthbert 1973). A single mink was caught in Berwickshire in 1938 and the earliest post-war record was one in 1955. There were only 14 records of them in the wild in the next six years (Cuthbert 1973). The first record of them breeding in the wild was in 1962 in Aberdeenshire and Banffshire (Hewson 1971), and from 1964 onwards the Department of Agriculture trapped wild mink until about 1970 (Cuthbert 1973). Early distribution in Scotland was related to the distribution of fur farms, and mink establishment in the wild was the result of lack of security and the inefficient management of casual breeders. There were three mass escapes in 1967; 19 escaped from a Clackmannanshire farm, in 1968, 250 in East Lothian, and in 1966 some on Shetland (all recovered) (Cuthbert 1973). In Scotland mink occur in the river systems mainly the Tweed in Berwickshire, Roxburghshire and Selkirkshire; Forth and Teith in Stirlingshire and Perthshire; Tay and Earn in Perthshire; Dee, Don Deveron and Ugie in Aberdeenshire and Bonffshire; Spey in Morayshire and Bamffshire, Doon in Ayrshire and Urr Water in Kirkurdbrightshire (Thompson 1968) and at Midlothian (Lever 1977).

Mink now occur over most of southern Scotland (Corbet and Harris 1991).

American mink were first introduced into Ireland for commercial fur farming in 1951 and successfully established themselves in the wild from escaped stock (Smal 1988). The first substantiated escape was of 30 animals in 1961 in County Tyrone (Smal 1988). By 1973 mink had been noted in 34 10-km squares (Crichton 1974) and in 94 10-km squares by 1979 (Smal 1988). Extensive trapping failed to reduce the numbers of mink and they are now widespread (D'Arcy 1988). Since 1961 mink have been reported in numerous areas in Ireland and by 1977 were breeding in numbers in Tyrone, and had spread throughout the watershed areas of the rivers Dodder and Liffey, County Dublin (Lever 1977). In the 1960s they were well established in Tyrone and there were records of them from counties of Devon, Armagh, Antrim, Fermanagh, Dublin, Kildare, Wicklow, Laoise, Kerry and Meath (Dean and O'Gorman 1969).

Mink now occur over most of Ireland, except for a small part of Northern Ireland (Smal 1988; Corbet and Harris 1991).

ASIA
Japan
Mink were farmed for fur on Hokkaido in the 1930s and some were occasionally found in the wild (Kaburaki 1940), but they are not known to be permanently established there.

NORTH AMERICA
Alaska
Domestic mink were introduced to the Kodiak Islands in the 1920s but failed to become established there (Clark 1958). In 1951 a second attempt was made when 24 domestic mink were liberated on Montague Island, and in 1952 when 24 were released on Kodiak Island (Burris 1965).

Canada
In British Columbia some mink were introduced by trappers to Lanz Island and the species is now well established here and on adjacent Cox Island (Cowan and Guiguet 1960). The introduction is believed to have been made in the late 1930s (Carl and Guiguet 1972).

Mink were also liberated on Anticosti Island in 1912, where they became established for a while but later died out (Newsom 1937).

Mink were imported from Nova Scotia to Newfoundland in 1934 and 1935 for fur farming and by 1952 there were about 70 farms and by 1954 about 100 farms containing 50 000–60 000 animals producing 25 000 to 35 000 skins annually (Cameron 1959; Banfield 1974; Northcott et al. 1974). The number of farms declined after 1954; the last was closed in 1971. However, the descendants of escapees had by this time become established. Two deliberate introductions were also made in 1948, when 18 mink were released on Chapel Island in Notre Dame Bay and 13 on Swale Island in Newman Sound, Bonavista Bay, and these also contributed to the wild population.

Mink began escaping from fur farms in Newfoundland shortly after the fur industry was founded and the first wild-trapped animals were taken as early as 1938–39 near Springdale. By 1944 mink were being trapped in areas southwards of the interior and around St. John. In the 1940s and early 1950s there were mass escapes in the Corner Brook area and some were trapped in the Bonavista Peninsula in 1961. In Newfoundland the mink now occupy most of the suitable habitat, having spread at a rate of 1.6–9.6 km per year, but have probably not as yet reached their maximum density.

In the 1990s they occurred over most of southern and central Newfoundland and were only absent from north and north eastern parts (Banfield 1974; Forsyth 1985).

SOUTH AMERICA
Chile
Attempted introductions of mink that were made in Chile (Lake Todos los Santos) in 1940 and perhaps at other times but have repeatedly failed (de Vos et al. 1956; Niethammer 1963).

■ DAMAGE
In Iceland mink have turned to eating birds as food in the absence of mammals and have become a serious menace to breeding ducks and waders. Here, they are considered by some to be one of the most serious factors in the decline of waterfowl numbers (Fjeldsa 1975).

In Europe it is uncertain to what extent introduced mink have displaced the indigenous species of mink (Corbet 1966). In Norway mink cause damage to poultry farms and destroy freshwater fish and other wildlife (Pedersen 1964). In Great Britain they have caused losses to poultry locally since 1951 and it is thought that potentially they could become serious pests (Thompson 1967). In some situations mink are potentially harmful with the main types of damage caused to fisheries and ornamental waterfowl (Swan

1981). A limited study in Britain concluded that their damage was overrated and that they co-existed well with the other fauna (Linn and Chanin 1978; Chanin and Linn 1980). Commercially their impact appears to be negligible (Corbet and Harris 1991). Although regarded as a pest by fish farmers, game and poultry keepers, the overall impact of mink is negligible; the nature of impact on native wildlife has been subject of debate for some years (Baker 1986; Corbet and Harris 1991). However, a study in Yorkshire found that mink reduced the population size and fragmented colonies of water voles, *Arvicola terrestris*, to such an extent that it posed a serious threat to the survival of water voles on British rivers (Woodroffe *et al.* 1990).

In Finland opinion on mink as a pest appears to differ again. Where they are abundant mink are a most serious predator of game (Tenovuo 1963).

In Sweden also mink appear to be endangering fish and small game (Thompson 1964). Here, they have become successfully acclimatised because of the similarities in climate, habitat, variety of food habits and possibly the lack of many predators. They appear to have caused no harm, although the otter (*Lutra lutra*) has decreased in numbers since the mink's introduction and there is some evidence that others do not breed where mink populations are dense (Gerell 1967). In Sweden by 1981 about 20 000 per year were trapped for fur (Swan 1981).

In the Russian Federation and adjacent independent republics mink do not appear to be having much effect (Lavrov and Pokrovsky 1967) and are not disturbing the existence of other fur-bearing animals such as ferrets and beavers (Vaseneva 1964). However, they are said to have displaced the European mink in Tatar (Popov 1964).

MARTEN
American marten or sable, pine marten
Martes americana (Turton)

■ DESCRIPTION
HB 318–512 mm; T 135–240 mm, WT males 450–1500 g, female 280–1000 g.
Coat golden brown to blackish brown on feet and tail tip; orange or yellowish on throat or chest; long slender body; small head; head greyish in some subspecies; ears rounded, edged white; claws semi-retractable; eyes dark brown; short pointed muzzle; tail bushy. Female similar, but smaller than male.

■ DISTRIBUTION
North America. Cananda and northern United States from Alaska to Newfoundland, south in mountainous areas to central California and northern New Mexico.

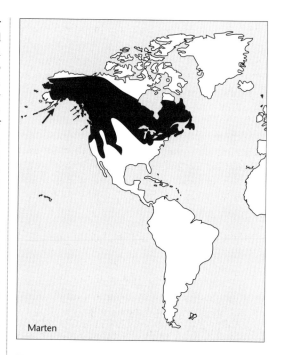

Marten

■ HABITS AND BEHAVIOUR
Habits: climbs; territorial; dens in hollow trees, stumps and cavities in wind falls or snow; nocturnal and often diurnal; partly arboreal; active all year. **Gregariousness:** solitary except at mating season; density 0.5–20/km. **Movements:** home range 2.5–38/km^2. **Habitat:** coniferous and mixed forest, cedar swamps, logging camps and dumps, woodland. **Foods:** rodents and other small mammals (mice, chipmunks, rabbits, shrews), birds, fruit, insects, carrion, amphibians, reptiles, fish and shellfish, nuts and berries. **Breeding:** mates June–August; delayed implantation; welps March–April; gestation 220–275 days; oestrus 2 weeks; litter 1–5, usually 3–4 kits; mature in second year; young haired at birth; eyes open 30–36 days; weaned 6–7 weeks; leave den 3–5 months; sexual maturity 15–24 months. **Longevity:** 5–8 years, up to 15 in wild, 18 years in captivity. **Status:** now uncommon in many areas.

■ HISTORY OF INTRODUCTIONS
NORTH AMERICA
Alaska
Martens were introduced on a number of islands off the coast of Alaska. They were released on Kodiak Island in 1920s, but failed to become established (Clark 1958). Ten from Behm Canal were released on Prince of Wales Island in 1934; seven from Cape Fanshaw on Baranof Island in 1934, six from Baranof Island in 1949, and 15 from south-eastern Alaska in 1952 on Chichagof Island, 20 from Minchumina Lake on Afognak Island in 1952 (Burris 1965; Franzmann

1988). Also an unknown number from the mainland were released on Koyak Island and Patterson Island in about 1940 (Burris 1965). Some 171 martens were also translocated to the south coast in the Yukon Territory between 1984 and 1987 (Slough 1989). Most were trapped 100–300 km east of the release sites and were released immediately upon arrival.

Canada
Ontario and a number of states, including the Cypress Hills Provincial Park, Saskatchewan, have had some success with translocating and re-establishing marten populations (Northcott 1977; Hobson *et al.* 1989).

United States
Martens were also introduced in New Hampshire and Wisconsin (de Vos *et al.* 1956). Re-introductions were attempted in Michigan and Wisconsin (Knap 1975), New Hampshire (Strickland and Douglas 1987).

Formerly martens were indigenous in many wooded areas of mid-west United States, but their range was reduced during the 1800s and 1900s due to excessive trapping, fires and logging. They were extirpated in Wisconsin by 1925, but were later re-introduced (de Vos *et al.* 1956; Knap 1975). Several introductions into new areas were attempted in Wisconsin in the 1940s and 1950s with varied success (Mitchell *et al.* 1971). Between January 1975 and April 1976, 130 animals trapped in Ontario were taken to Wisconsin and subsequently 124 (97 males and 27 females) released in the Nicolet National Forest (560 km^2), but these failed to establish a viable population (Davis 1983).

Attempts to re-introduce martens in upper Michigan were made in 1969–70, but failed because of trapping pressure and emigration. Several re-introductions were also made in New Hampshire for aesthetic reasons (de Vos *et al.* 1956).

■ DAMAGE
No information.

BEECH MARTEN
Stone marten, white-breasted marten
Martes foina (Erxleben)

■ DESCRIPTION
HB 420–560 mm; T 220–320 mm; SH c. 120 mm; WT 1.1–2.3 kg.
Coat yellowish brown to cinnamon brown, under fur grey or greyish white; muzzle pale; neck patch pure white or yellowish and extending to forelimbs on either side; tail and limbs blackish brown; soles lightly haired. In summer coat more brownish. Similar in appearance to pine marten (*M. martes*), but with smaller ears and shorter legs.

Beech marten

■ DISTRIBUTION
Eurasia. Western and central Europe, Caucasus and Asia Minor east to Iran, the Altai, Afghanistan, Syria, Palestine, northwestern Pakistan and India (Himalayas), Tibet, Mongolia and north-western and western China. Also occurs on the islands of Crete, Rhodes and Corfu.

■ HABITS AND BEHAVIOUR
Habits: mainly nocturnal but also crepuscular; shelters and dens in crevices, hollows and stone piles, abandoned burrows of other animals, barns and under floors of summer houses. **Gregariousness:** probably solitary. **Habitat:** mountain ravines and canyons, bush-covered slopes, forest edges and woodland, suburbs, rocky and open areas; often enter towns and buildings; cultivated areas. **Movements:** sedentary; home range to 80 ha. **Foods:** small rodents (rats and mice), small birds, frogs, insects, reptiles, amphibians, birds' eggs, fruits, berries and occasionally domestic poultry. **Breeding:** ruts in July; litters 1, 3–4, 8; gestation + implantation 230–275 days; 1 litter per year; born blind; mature at 1–2 years. **Longevity:** 14 to 18 years in captivity. **Status:** common; not hunted as much as pine marten as fur not as commercially valuable.

Note: Both behaviour and reproduction are similar to the pine marten (Martes martes).

■ HISTORY OF INTRODUCTIONS
Introduced successfully(?) in the Russian Federation.

EUROPE
Russian Federation
Apparently beech martens have been introduced or translocated at times in the Russian Federation (Naumoff 1950 in de Vos *et al.* 1956). In 1936 approximately 60 beech martens were released in the Ryazan region (Novikov 1962) where they may have become established.

■ DAMAGE
Beech martens raid poultry runs and dovecotes (Lyneborg 1971).

PINE MARTEN

Forest marten, European pine marten, marten cat
Martes martes (**Linnaeus**)

■ DESCRIPTION
HB 365–580 mm; T 185–280 mm; SH to 150 mm; weight males 0.67–1.95 kg, females 0.48–1.48 kg.
Summer coat short; fur dark brown with chest orange or yellowish; throat patch orange; muzzle dark brown; cat-like with pointed face; ears large and rounded; lower legs and feet blackish; paws covered with hair; tail long and fluffy. Females have four mammae.

■ DISTRIBUTION
Eurasia. Europe except southern Iberia and Greece, east to western Siberia; Ireland, Britain (Wales, Lake

Pine marten

District and north-west Scotland), Sardinia, Corsica and Sicily; the Caucasus region, Elburz Mountains and north-eastern Asia Minor.

■ HABITS AND BEHAVIOUR
Habits: mainly nocturnal; dens under rocks, fallen trees, tree roots; partly arboreal. **Gregariousness:** solitary, overlapping home ranges. **Movements:** 20–30 km during hunting; young disperse in autumn and may wander extensively in winter; home range 5 km^2. **Habitat:** conifer and mixed forest, woodland, pasture, scrub, coastal areas moorland, clear felled areas in forest. **Foods:** small rodents (voles, lemmings), berries and fruits, mushrooms, carrion, rarely birds, other mammals (squirrels, hares, stoats), eggs, and some invertebrates. **Breeding:** mates July–August; young born March–April; gestation and delayed implantation 260–275 days; litter size 2–5, 8; 1 litter/year; born blind; whitish fur; eyes open 32–38 days; leave den about 6 weeks; weaned 6–7 weeks; sexual maturity 2–3 years. **Longevity:** up to 17–18 years. **Status:** once common, but now locally rare due to hunting for fur.

■ HISTORY OF INTRODUCTIONS
EURASIA
Russian Federation
In 1954, 201 martens were released in the forests along the Tiryakhtyakh River (Yakutia) where the release was considered successful (Mel'chinov 1958). Pine marten were also released in Arkhangel oblast in 1962 (Kirisa 1973), but generally introductions in the Russian Federation have not yielded appreciable results (Sofonov 1981).

United Kingdom and Ireland
Due largely to persecution, pine martens survived only in north-west Scotland, northern Wales, Lake District and parts of Northumberland, northern Yorkshire and Ireland by the turn of the twentieth century. By 1939 they had expanded their range and spread south and east into their former range and they still appear to be expanding (Corbet and Harris 1991). A pair of pine martens was released at Ardverikie Forest, Loch Laggan, Inverness-shire in 1930, where they had become extinct (Fitter 1959).

■ DAMAGE
Pine martens may take chickens if given the opportunity, but are of no economic significance as vermin (Corbet and Harris 1991).

JAPANESE MARTEN
Martes melampus (Wagner)

■ DESCRIPTION
HB 470–545 mm; T 170–223 cm; WT up to 1.5 kg.
Yellowish to dark brown; sandy brown; neck patch white; darker on legs and tail; paler on under parts.

■ DISTRIBUTION
Japan on the islands of Honshu, Kyushu, Shikoku and Tsushima and in South Korea(?)

■ HABITS AND BEHAVIOUR
Note: Little known or recorded, but presumed similar to other martens.

Habits: arboreal. **Habitat:** broad-leaf deciduous forests. **Status:** declining due to overhunting for fur.

■ HISTORY OF INTRODUCTIONS
ASIA
China, Japan and Korea
Japanese marten were introduced on Sado Island (off the west coast of Honshu, Japan) and possibly to South Korea. Perhaps also introduced into part of China (Corbet 1978, 1980). There is disagreement as to whether or not the populations on the Asian mainland are derived from introductions by humans (Schreiber *et al.* 1989).

■ DAMAGE
No information.

FISHER
Pekan
Martes pennanti (Erxleben)

■ DESCRIPTION
HB 325–745 mm; T 253–422 mm; WT 1.3–5.5 kg.
Body with buff-tipped under fur and long brown guard hairs with grey sub-dominal band; mantle grey; nose black and button-like; ears short and rounded; eyes small; rump, long bushy tail, feet and belly dark chocolate brown, almost black in some; chest and abdomen sometimes with irregular white spots; feet large with narrow curved claws. Males larger than females.

■ DISTRIBUTION
North America. From the southern Yukon, Alaska and the southern Northwest Territories, Canada south to New England, the Adirondacks and California, United States.

■ HABITS AND BEHAVIOUR
Habits: nocturnal and diurnal; terrestrial, occasionally arboreal; dens in hollow logs, rocks and brush. **Gregariousness:** solitary, pairs (in breeding season) or family groups (female and young); density 1 per 2.6–7.5 km^2 but can be as low as 1 per 200 km^2. **Movements:** sedentary; 1.5–3.0 km per day while hunting; home range 15–35 km^2. **Habitat:** coniferous forest near watercourses and mixed forest below snow line. **Foods:** small rodents (squirrels, voles, shrews and

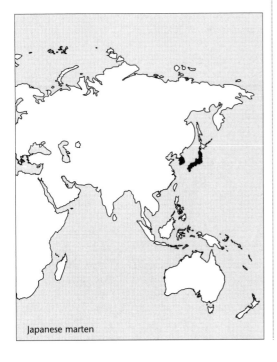

Japanese marten

Fisher

mice), and other small mammals (porcupine, hares, rabbits, small deer), birds (grouse), berries, fruits, fish, frogs and insects, seeds, fern tips. **Breeding:** mates March–May; young born following March–April; gestation including delayed implantation 338–358 days; litter size 1–4, 6; polygamous; kits altricial, helpless; stay with female 3–4 months then disperse; young born blind and partly furred; eyes open 7 weeks; walk at 8–9 weeks; weaned 8–10 weeks; males mature 2 years, females 1 year. **Longevity:** up to 10 years in wild and captivity. **Status:** common; range reduced in southern areas.

■ HISTORY OF INTRODUCTIONS
NORTH AMERICA
Alaska
Efforts were made before 1937 to introduce the fisher into other parts of Alaska where it does not occur, but these were unsuccessful (Newsom 1937).

Canada
Fishers declined in the nineteenth and twentieth centuries because of excessive trapping for fur and habitat destruction. They were almost entirely eliminated in the United States and were greatly reduced in eastern Canada. Re-introductions in the 1950s and 1960s and conservation has restored their range somewhat (Walker 1992).

Formerly a native of Nova Scotia, the fisher became extinct there in the early part of the twentieth century. In July 1947, two males were released on the shores of Hobiatic Lake in north-eastern Queens County, and 10 (four males and six females) were released at the same site in July 1948. The former animals came from ranch stock. Two fishers were caught in 1958, one in Digby County and one in Queens County, and at this time they appeared to be reproducing and spreading (Benson 1959). The species is now well established in Nova Scotia as a viable population (Weckwerth and Wright 1968; Deems and Pursley 1978).

Re-introductions in Ontario by the Natural Resources Ministry to boost low populations have also assisted to restore viable populations (Deems and Pursley 1978). Fishers have also been introduced to Anticosti Island, Canada (de Vos *et al.* 1956).

United States
Fishers have been re-introduced into Idaho, Maryland, Massachusetts, Michigan, Montana, Oregon, Rhode Island, Vermont, West Virginia and Wisconsin mostly by wildlife agencies (Deens and Pursley 1978; Davies 1983). These re-introductions were mainly successful in establishing viable or low

populations for aesthetic reasons and for the control of porcupines.

Fishers (36 animals) from British Columbia were released at three sites in western Montana in 1959–60 in an effort to re-establish the species for fur where it had been extirpated (Weckwerth and Wright 1968). At least one of these translocations (in the Swan area) was successful and some were captured as late as 1968 (Mitchell *et al.* 1971). The three translocations of 1959–60 were made at: Pine Creek Drainage in Lincoln County in 1959 (four males and five females), Holland Lake, Missoula County in 1959–60 (seven males and eight females), and Moose Lake, Granite County in 1960 (four males and eight females).

In West Virginia the fisher was believed to be rare at the turn of the twentieth century and by 1912 very rare or extinct (Pack and Cromer 1981). Releases were made at Canaan Mountain in Tucker County and Cranberry Glades in Pocahontas County, both in the Monongahela National Forest. In 1969 the West Virginian Division of Wildlife Resources re-introduced 23 fishers (15 at Blackwater Falls State Park) from New Hampshire Fish and Game Department, and 15 were released at Canaan and eight at Cranberry Glades. These releases resulted in legislation to prevent further releases of fishers in that state. Since that time 18 have been captured in West Virginia and tracks have been noted in Maryland in 1974–75 and in 1977 (Cottrell 1978). A single female was caught near Mountain Lake Park in western Maryland in 1977. Fisher range largely remains in the mono-forest, but has also expanded into Maryland. They are now established in eastern West Virginia and probably western Maryland where the population is expanding slowly (Pack and Cromer 1981).

■ DAMAGE
No damage reported

SABLE
Marten
Martes zibellina (Linnaeus)

■ DESCRIPTION
HB 350–560 mm; T 110–200 mm; WT 700–1810 g.
Coat light yellowish brown to dark brown, almost black and sometimes tinged with light grey; summer fur darker; head cone-shaped and whitish; cheeks lighter; some black on muzzle, but sides whitish; body slender with short, stout limbs which are darker than body; eyes large and black; ears large and blunt and lighter coloured than body; a small yellowish neck patch; belly duller than remainder of body; in winter soles are haired.

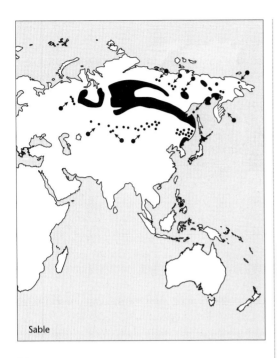

Sable

■ DISTRIBUTION
Asia. From the Ural Mountains, Russian Federation east to Mongolia, northern Mongolia, Manchuria, the Altai, Kazakhstan, north-eastern China, North Korea and Japan (Hokkaido) and Sakhalin Island.

Originally occurred from Scandinavia to eastern Siberia, but now extinct in Scandinavia and Finland. Range in the Russian Federation is complicated by local exterminations and subsequent re-introductions.

■ HABITS AND BEHAVIOUR
Habits: nocturnal and diurnal; mainly terrestrial; dens in logs, under rock piles, among roots. **Gregariousness:** density 1 per 1.5–25 km². **Movements:** home range to 3000 ha; moves sometimes to higher country in summer and occasionally during food shortages. **Habitat:** forest in river valleys, ravines, plateaus, and taiga. **Foods:** small mammals (rabbits, pikas, hares, lemmings, mice, squirrels), small birds, birds' eggs, fish, insects and plant food (fruits, nuts, berries, honey). **Breeding:** births April–May; young 2–4, 6; gestation 250–300 days; delayed implantation; eyes open 30–36 days; emerge from den 38 days; weaned *c.* 7 weeks; sexual maturity 15–16 months. **Longevity:** 10–15 years. **Status:** range reduced, numbers decreasing; farmed for pelts.

■ HISTORY OF INTRODUCTIONS
EURASIA
Russian Federation
From the fifteenth to the eighteenth centuries, sable played a leading role in the fur trade (Sofonov 1981). In 1900 as many as 48 000–53 000 sable furs were sold on the world markets. Such was the demand for the fur of this animal that populations began to decline. By 1914 only 5000 sable furs were marketed (Grzimek 1975). The extermination of such a valuable fur resource in many areas of the Russian Federation prompted their re-introduction, with translocations and introductions of sable taking place after 1925. Most of the releases since this time have been aimed at restoring the animals to their former range or for improving the quality and quantity of furs for market.

The earliest release of sable appears to have been on Karagin Island in the Far East in the early 1900s. After 1927 restoration began in earnest, particularly in this region and in Siberia. From 1901 until 1970 some 19 187 sable were released in 20 oblasts of the Russian Federation mainly in these two regions (Kirisa 1973; Sofonov 1981), resulting in the restoration of the population and fur harvest to the levels that they were 300–350 years ago (Sofonov 1981).

By the end of the 1920s there were no sable left in the Baraba steppe, but in 1940, 40 from Barguzin were released in the upper reaches of the Nyurol'ka River where they quickly became established (Zhdanov 1963). From here in the 1950s some were taken and released in the taiga of the Tomsk oblast, where between 1940 and 1958 some 2000 were liberated. In this area in the early 1960s sable were being hunted and taken at the rate of 3000 per year (Zhdanov 1963; Kirisa 1973).

Other releases in western Siberia include those in Tumen (15 sable in 1933 and 1042 in 1952–59), Novosibirsk (34 in 1953), Kemerov (460 in 1947–55), Altai (538 in 1940–54) and in Kazakhsk (181 in 1952–53 and 377 in 1962–65) (Kirisa 1973). In Kazakhsk their range has now been extended somewhat (Sludskii and Afanas'ev 1964).

Since 1930 some 8338 sable have been released in eastern Siberia in oblasts in Irkutsk, Buryaht, Chitinsk and Yahkutsk. Most of these were released in small groups of 20–30 animals that became well established. There are now no fewer sable in these areas than there were 200–250 years ago (Timofeev 1961; Izmailov 1969; Kirisa 1973).

In central Siberia sable have been released in Krasnoyahrsk and Tuvin regions. From 1949 to 1958, 846 were liberated in Krasnoyar Krai to improve the fur of the local sable population which had been restored to their previous range by bans on trapping and shooting (Numerov 1958; Kirisa 1973). Between 1952 and 1954 some 287 sable were released in Tuvin.

A number of introductions in the Far East have been successful in establishing sable into limited areas (Yanushevich 1966). Kamchatka sable were re-acclimatised in the Penzhina River area in 1951 when 66 males and 52 females were released in the lowland forests of the middle Penzhina (Vershinin 1962). By 1956–57 these had repopulated the whole basin area of the upper Penzhina and some 1170 sable were counted there. Other releases in the Far East, most of which appear to have resulted in successful establishments, include those in Amur (690 in 1951–58), Khabarovsk (1481 between 1927 and 1958), Primorsk (1513, between 1940 and 1962), Magadansk (816, in 1951–58) and on Sakhalin (80, in 1951–59) (Kirisa 1973).

Sable have also been translocated from the Trans-Baikal and probably other areas to the Ural region, where they appear to have been successfully established. Releases have been made in such oblasts as Perm (96 in 1953), Sverdlovsk (226 between 1940 and 1953) and in Chelyahbinsk (14 in 1955). In 1950–53 sable were released in the south-west part of the Verkhoyansky mountain range, where by 1954 they were well established and widely distributed (Gryaznukhin 1958).

The Kamchatka sable (*M. z. kamshadalica*) has formed the basis of most introductions, particularly in western Siberia (de Vos *et al.* 1956).

■ DAMAGE
Although the introductions, re-introductions and translocations of sable in the Russian Federation are considered successful (Shaposhnikov 1960) and the animal has been restored to some of its former range, the results were not always entirely successful. Kamchatka sable introduced into western Siberia inter-bred with the native form to produce hybrids with heavier furs (Lindermann 1956), but not so Siberian sable introduced to the Urals to improve the fur of the local animal (Pavlinin and Shvarts 1961). However, the moving of Baikal sable (dark form with fine fur) into the Yenisei region, Middle Siberia, is said to have been profitable (Nonakhov 1987).

BADGER

Meles meles (Linnaeus)

■ DESCRIPTION
HB 600–900 mm; T 110–190 mm; SH c. 300 mm; WT 6.6–16.7 kg.
Upper parts greyish; under parts black; limbs black; each side of face has black stripe from snout through

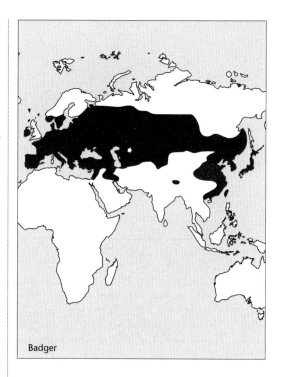

Badger

eyes to ears; white stripe borders black stripe; stocky body; eyes small; ears short; legs short; tail short.

■ DISTRIBUTION
Eurasia. Britain and Ireland across Europe, Asia except north-eastern Siberia, North Africa and Arabian region. Southern boundary Palestine, Iran, Tibet, and southern China; Balearics, Crete, Rhodes, Quelpart (Korea) and all large islands of Japan.

■ HABITS AND BEHAVIOUR
Habits: hibernates in cold weather; communal burrows (sets) 0.2–several ha; nocturnal. **Gregariousness:** social groups 2–23; territories 30–150 ha; density 20 adults/km² to 1 or 2/km². **Movements:** no information. **Habitat:** woodland, scrub, hedgerows, quarries, sea cliffs, moorland, open fields, mines, coal tips, rubbish dumps. **Foods:** omnivorous; small mammals, birds, reptiles, frogs, molluscs, insects, larvae of bees and wasps, carrion, nuts, acorns, tubers, rhizomes, mushrooms, berries, fruits, and seeds. **Breeding:** mates spring; gestation 7 weeks; delayed implantation until December; oestrus 4–6 days; 1–5 young in February; born pink, silky hair; eyes open 5 weeks; emerge April at 8 weeks; weaned 12 weeks; mature 12–15 months; sexual maturity 1.5–2 years. **Longevity:** up to 6–11 years in wild and 19 as captive. **Status:** declined over much of its European range, some recovery of numbers in the United Kingdom in recent years.

■ **HISTORY OF INTRODUCTIONS**
EUROPE
United Kingdom
In the past badgers have occasionally been introduced for sporting purposes, or have escaped from captivity and established themselves in areas where they no longer occur. It is claimed that in some cases the animals came from the European mainland, but badger hunting is not a widespread sport.

Many badgers have been taken to the Isle of Wight where they escaped to become established. However, the population was exterminated in 1899. It is known that when scarce in Essex in 1866 E. N. Buxton released a few pairs in Epping Forest and that those there are now possibly their descendants. Others were introduced in Essex at the same time and some were released in 1894. Some ex-zoo animals were released in south Devon, but this area still had badgers. In 1892 it was indicated that they may have been introduced in Lakeland some time before. Other areas in these times where they were deliberately released included Castleteads, Edenhall and Gowbarro Park, also in North Yorkshire in 1874 and later, where they survived for nearly 25 years.

In Scotland there were numerous releases of badgers at the end of the nineteenth century including Aisla Craig in 1876 (failed after a while); Dalmeny Park, Westlothian (deliberate) in 1889; Cambo, Fife and in Wigtownshire. Some were caught at Jura in 1856 where it was believed that they were introduced. About 1925 some escaped at Tongue on north coast of Sutherland and re-populated an area where they had been extinct for 50 years or more (Fitter 1959). Attempts to introduce badgers to Jyrra and Aisla Craig failed (Corbet and Harris 1991).

■ **DAMAGE**
No damage of any economic importance is reported, but badgers can kill poultry near habitation. Occasionally they cause damage to high value fruits or vegetables (e.g. grapes, sweet corn, strawberries) (Corbet and Harris 1991).

STRIPED SKUNK
Common skunk
***Mephitis mephitis* (Schreber)**

■ **DESCRIPTION**
TL 512–800 mm, T 184–393 mm; WT 0.95–4.5 kg.
Generally black with white stripes; fur long, harsh with soft under fur; head small, shiny black with narrow white frontal stripe between eyes; ears small; body stout with two broad white stripes from neck to

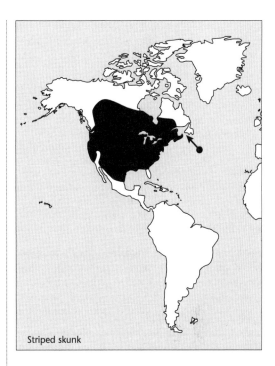

Striped skunk

base of bushy tail; tail black and white; legs short; hind feet plantigrade; soles naked; claws curved. Females have 10–14 mammae and are smaller than males.

■ **DISTRIBUTION**
North America. From southern Canada throughout the United States except Florida and some southern parts to northern Mexico.

■ **HABITS AND BEHAVIOUR**
Habits: terrestrial; mainly nocturnal and crepuscular; lives in natural rock crevices or in underground burrow (1.8–3.7 m) usurped from other species; hibernates December–March in north, but active in mild spells and in south dormant short period in cold weather; sluggish; discharge musk from anal glands 2–3 m. **Gregariousness:** family groups, pairs or solitary; communal winter dens (2–19); density 0.7–18.5 per km². **Movements:** 1.5 km between dens; disperse in summer up to 22 km; home range 110–370 ha. **Habitat:** forest, woods, plains, desert, agricultural lands, open fields, river valleys, marshes, stream sides, suburban areas. **Foods:** omnivorous; rodents (mice), other small mammals, reptiles (snakes, lizards), amphibians (frogs), fish, molluscs, crayfish, insects (beetles, grasshoppers, caterpillars), berries, buds, fruits, corn, nuts, leaves, grain, grass, eggs, nesting birds, mushrooms, carrion. **Breeding:** mates February–April, young born May–June; gestation 42–77 days; males polygamous; females monoestrous; young 2, 4–8, 10; 1 litter/year, rarely 2; open eyes at 3 weeks; weaned 6–10 weeks; disperse in autumn;

mature at 9–12 months (females). **Longevity:** 4–13 years captive, about 4 years or more in wild. **Status:** common.

■ HISTORY OF INTRODUCTIONS
Introduced successfully to Prince Edward Island, and colonised Nova Scotia?, Canada, and possibly to Petrov Island, Ukraine and the Caucasus, Russian Federation and adjacent independent republics.

NORTH AMERICA
Canada
Striped skunks were introduced on Prince Edward Island as a fur-bearer in captivity, but escaped and became established in the wild (Cameron 1959; Banfield 1977; Deems and Pursley 1978).

Some were found on Vancouver Island, probably having been turned loose by pet owners who had become tired of them. They have been found to eat quail eggs and voles (Obee 1983).

On mainland Canada they have been expanding their range (Obee 1983) and have spread into Nova Scotia since 1850 (Banfield 1977).

EURASIA
Russian Federation and adjacent independent republics
Efforts have not been very successful where striped skunks have been introduced or translocated in the Russian Federation and adjacent independent republics (Shaposhnikov 1941; Naumoff 1950 in de Vos *et al.* 1956; Sofonov 1981; Lever 1985).

Skunks have been released in the Russian Federation, Ukraine, Kirghizstan, Dagestan and Azerbaidjan. In the Russian Federation 26 were released in the Voronejskaya oblast near Usmansk in 1933, and three in Partizanskii, Primorskii Krai, in 1936. In 1936 (five) and 1937 (24) were released in the Pechenejskii region, Kharkovsk in the Ukraine. Twenty-nine were released in the Karavanskii region of the Oshsk oblast of Kirgizstan in 1937, and 58 in two places in the Kayahkentskii region of Dagestan in 1939. At least 70 skunks were released in five areas of Kutkashensk in Azerbaidjan in 1939 (Kirisa 1973).

Skunks have also been introduced in the Ukraine, and in the Caucasus but failed to become established in both areas. They have been introduced into the northern Caucasus since 1930 and may be established (Yanushevich 1966).

Three skunks of uncertain identity (*Mephitis* or *Vormela*) were set free on Petrov Island in July 1936 and survived there with supplementary feeding for more than a year but then all died out (Bromlei 1959).

EURASIAN OTTER
Otter
Lutra lutra (Linnaeus)

■ DESCRIPTION
HB 500–950 mm; T 260–550 mm; SH c. 300 mm; WT 3.1–15 kg.
Colour uniform brown; throat paler; sometimes white patch on chin; brownish above, pale below; tail long, flattened and tapering, thick at base; body long; legs short; fore and hind limbs have broad web between toes; head flat, eyes and ears small; muzzle broad; whiskers prominent.

■ DISTRIBUTION
Eurasia. Europe and most of northern Asia (except Siberian tundra), North Africa east of Algeria, Arabia, and southern Iran to Malaya, southern India, Sri Lanka, Taiwan, Sumatra. Formerly more widespread over most of Europe and northern Asia and southern Asia to Java and India.

■ HABITS AND BEHAVIOUR
Habits: agile swimmers; mainly nocturnal, occasionally diurnal; dens in cavities under roots, rocks or rabbit burrows. **Gregariousness:** mainly solitary, except during courtship; groups females and young of year; home range 2–39 km of river length. **Movements:** up to 9.5 km over night; overlapping home ranges; density 0.7–1.0/km^2. **Habitat:** wide variety of aquatic habitats; coasts, rivers, lakes, wetlands, streams, salt or fresh marshes. **Foods:** water voles, fish (eels, perch, butterfish), water birds, aquatic insects, crabs, but mainly fish. **Breeding:** peak births May–August; gestation 60–63 days; females polyoestrous; litter size 1–5; born furred; eyes open 4–5 weeks; suckle for up to 14 weeks; stay with female for 7–12 months; females breed at 3 years; mature second or third year. **Longevity:** up to 12–20 years. **Status:** numerous and widespread in remote areas; declined in Europe.

■ HISTORY OF INTRODUCTIONS
EUROPE
United Kingdom
Occasionally Eurasian otters are translocated for sporting purposes and some apparently from Norway were released in Dumfriesshire to augment local stock for hunting in earlier years (Fitter 1959). Since the mid-1950s European otters have declined in numbers throughout most of England through habitat destruction and persecution (Chanin and Jefferies 1978).

Captive-bred otters were released in selected river areas of East Anglia to restock depleted areas and to re-introduce them to areas in which they are absent.

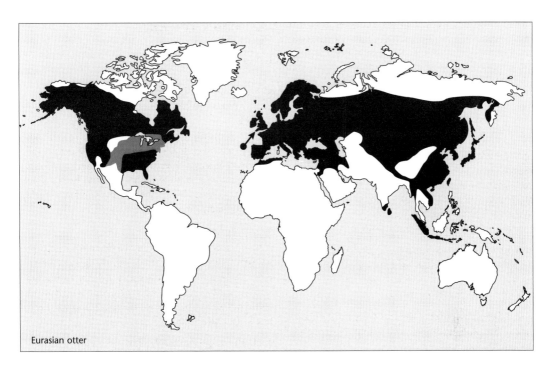

Eurasian otter

Five groups, bred from wild stock, of either one male and one female or one male and two females (total 13 otters) were released from 1983 to 1986 (Jefferies and Wayre 1983; Jefferies *et al.* 1983; Wayre 1985; Corbet and Harris 1991). They were thought to be breeding one year later. In July 1984 three more were released in another river in East Anglia. Animals came from the Otter Trust in Norfolk where they were bred from wild stock (Wayre 1985).

In 1998 England's 100th captive-bred otter was released, the twentieth animal to be released that year by the Otter Trust (Lawson 1998).

■ DAMAGE
Eurasian otters are branded as pests of fishers, but this is doubtful except in fish hatcheries and farms that may be vulnerable if otters are not excluded (Corbet and Harris 1991).

RIVER OTTER
North American otter
***Lutra canadensis* (Schreber)**

■ DESCRIPTION
HB 870–1300 mm; T 300–510 mm; WT 5–13.64 kg.
Coat short, oily, dense under fur, brownish; throat whitish or silvery; head broad and flattened; legs short; toes webbed; soles furred;

■ DISTRIBUTION
North America. From the southern United States north to Alaska and northern Canada. Extirpated from a large portion of their original range in the United States (Hall 1981).

■ HABITS AND BEHAVIOUR
Habits: almost completely aquatic, dens on land; forages ashore. **Gregariousness:** male and female establish separate territories; groups of females and young. **Movements:** no information. **Habitat:** marshes, wooded streams, estuaries, and other inland waters. **Foods:** fish, aquatic invertebrates, amphibians, mammals and birds also eaten. **Breeding:** gestation plus implantation 245–380 days; litter 6; breeds late winter–early spring; kits born spring, blind, helpless; eyes open 3 weeks; swim at 6–9 weeks; disperse at 2 years; mature 2 years. **Longevity:** 13 years in wild and 14–23 in captivity. **Status:** numbers reduced, but moderately common.

■ HISTORY OF INTRODUCTIONS
NORTH AMERICA
Canada and United States
At present 14 states and one Canadian province are planning on, or completing programs of translocation of otters including: Alberta, Arizona, Colorado, Indiana, Iowa, Kansas, Kentucky, Minnesota, Missouri, Nebraska, Oklahoma, Ohio, Pennsylvannia, Tennessee, and West Virginia.

Otters have been introduced to locations in Grand and Lamine River watersheds and Swan Lake National Wildlife Refuge (SLNWR) and Lamine River Wildlife Area (LRWA), Missouri. Nineteen otters were released at SLNWR in March and May 1982; 20 were released at LRWA in April 1983. A year later both colonies were doing well (Erickson and McCullough 1987). An introduction in Colorado by the wildlife department for aesthetic reasons has resulted in a low population (Deems and Pursley 1978). Some success has also been achieved at Pine Creek in north central Pennsylvania (Serfass and Rymen 1985) and in the Obed Scenic River areas of Tennessee (Griess and Anderson 1987).

SEA OTTER
Enhydra lutris (**Linnaeus**)

■ DESCRIPTION
HB 1000–1200 mm; T 250–370 mm; WT 14–45.1 kg
Coat varies from rusty red to dark brown and to almost black, but is paler on throat, chest and head; head broad and flat; neck short and thick; snout and tail short; limbs short, toes webbed to form flippers which are furred on both sides; nose blunt; eyes black; ears short and naked; vibrissae stiff, long and whitish yellow. Females have two inguinal teats, and are usually smaller than males.

■ DISTRIBUTION
Restricted to Kiril Islands and Komandorskiye Islands and to the western Aleutians and central California coast. Formerly northern Japan, Sakhalin, Kamchatka Peninsula, Commander Islands, Bering Island, Aleutian Islands and to the Pacific coast of North America as far south as Baja California. Original population extirpated along the coast of British Columbia after 1900.

■ HABITS AND BEHAVIOUR
Habits: aquatic, diurnal, sleeps on shore at night; may dive to 91 m; males territorial; territories 20–50 ha. **Gregariousness:** congregates in 'pods' of various ages and sex up to several hundred individuals; females form nursery groups in summer. **Movements:** non-migratory; sedentary; home range 8–16 km coastline, males further at times; annual movements 50–100 km. **Habitat:** seas off rocky islets, reefs and rocky coasts, and kelp beds; rarely more than 0.8 km offshore. **Foods:** sea urchins, molluscs, mussels, crabs, limpets, snails, (epibenthic?) slow-moving fish, starfish, octopus, seaweed and abalones, clams, crabs, fish eggs, scallops, chitons, annelids, anemone, barnacles, kelp and algae. **Breeding:** mates in spring and summer in the water; young 1 (2 rare) born ashore in spring at intervals of 1 or more years; polygamous; gestation 6.5–9 months; probably delayed implantation; male can breed at 5–6 years, female at 4 years; pups unable to swim or dive until 2 weeks; female carries pup on chest and back while in water; stays

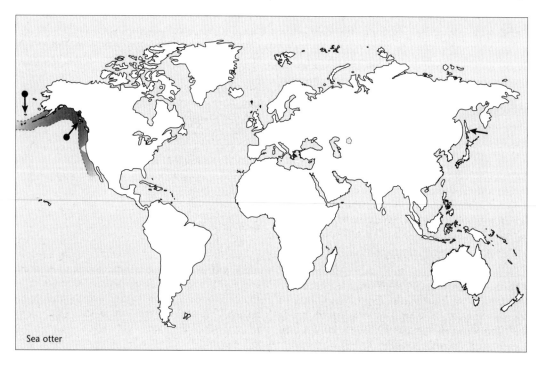

Sea otter

with female for about 1 year. **Longevity:** 20–23 years (captive and wild). **Status:** considerably reduced in numbers and range.

■ HISTORY OF INTRODUCTIONS
EURASIA
Russian Federation
Sea otters have been introduced and translocated in some areas of the Russian Federation. In 1937 acclimatisation began on the Murmansk coast of the Kola Peninsula, where some success appears to have been achieved. Releases have also been made on islands off the south-west coast of Sakhalin in the Tator Strait but appear to have had little success (Naumoff 1950; Novikov 1963; Kirisa 1973).

NORTH AMERICA
Alaska, Canada, United States
One hundred and seventy years of exploitation by fur traders between 1741 and 1911 (when it was protected) eliminated the sea otter from most of its original range. By the latter date only remnant populations in Alaska and on the central coast of California survived (Lensink 1960; Estes *et al.* 1978). The Alaskan population had been reduced to probably several hundred otters in seven areas, whereas in 1740 their numbers had been estimated as about 200 000. Between 1741 and 1867 about 800 000 sea otters had been harvested for the fur trade (Harris 1968; Johnson 1982). The otter population on the Californian coast was thought to have been completely exterminated until its rediscovery in 1938 (Howe 1983). A populaiton of 50 was found at Point Sur, California, in about 1914 (Shigesada and Kawasaki 1997).

Protection extended to the sea otter in 1911 allowed the population to steadily increase. From a possible population of 200–500 in 1911, the population reached an estimated 40 000 by the early 1960s (Johnson 1982). However, introductions commencing in 1955 have assisted the spread and the continued rise in numbers.

Re-introduced populations have established off south-east Alaska (now number 500–1000), Vancouver Island (about 350), and Washington (100). Re-introduced groups off Oregon and in Pribilof Islands have not done as well and all have disappeared (Estes 1980; Jameson *et al.* 1982; Rotterman and Simon-Jackson 1988).

The early experiments (1955–59) at translocation are thought to have failed mainly due to the methods of capture–release. The only success achieved in this period appears to have been with those released on St. Paul in the Pribilofs (Jameson *et al.* 1982).

Through protection and translocation the sea otter population has recovered and in 1982 was estimated to be between 150 000 and 200 000 (Johnson 1982). From an insignificant resource in the twentieth century it now appears that the situation is rapidly changing. The sea otter is again being recognised as a valuable economic fur industry.

The restoration of the sea otter has, however, presented some problems. In recent times it has not occurred near major commercial shell fisheries, or has been encountered only at low population levels. Because of a liking for shellfish its impact on this fishery is causing some concern. In some areas the sea otter is accused of decreasing the availability of the shellfish. The importance of the conflict with humans is difficult to evaluate at this stage but will need careful management in the future.

A further problem has arisen along the coast of California where considerable drilling for oil is planned in the near future. Biologists say that the risk of oil spills places the colony of sea otters at risk. Plans to translocate animals to form further colonies are at present being considered should efforts to prevent the oil drilling be thwarted.

Alaska
In 1951 attempts were made to translocate 35 otters from Amchitka to various other localities but failed due to the mortality in captivity (Griffiths 1953; Stullken and Kirkpatrick 1955).

In 1955 some 16 or 19 sea otters from Amchitka Island were released on Otter Island in the Pribilof Islands, Alaska (Coolidge 1959). This was followed by two further releases in 1956 and 1959. Five or six otters from Amchitka were released on Attu Island, Near Islands, in 1956 and some seven on St. Paul Island, Pribilof group, in 1959. These animals were not seen again and the attempts were deemed unsuccessful, although natural immigration was responsible for some in the area at later dates (Kenyon and Spencer 1960; Lensink 1960; Burris 1965).

Several attempts were made in south-eastern Alaska in the 1960s when 403 otters were released (Schneider 1972). From 1964 to 1973 several attempts to introduce otters were made and which have proven to be fairly successful (Franzmann 1988). During this period approximately 708 sea otters captured in Alaska were translocated to parts of their range where they had been previously extirpated. Some 467 were translocated in Alaska between 1965 and 1969, 89 to British Columbia between 1969 and 1972, 29 were released in Washington 1969–70, and 93 were released in Oregon 1970–71 (Mate 1972; Jameson *et al.* 1982).

Canada

Releases of sea otters in British Columbia were made on Bunsby Island, north-west of Vancouver Island. The three groups were made up of 29 from Amchitka Island released in 1969, 14 from Prince William released in 1970 and 46 also from Prince William released in 1972 (Carl and Guiguet 1972). The success of these animals remained uncertain until in 1977 some 70 sea otters in two colonies were located (Bigg and MacAskie 1978).

United States

In Washington sea otters also appear to have been successfully established and the population in 1981 was 36 (Bigg and MacAskie 1978, Jameson *et al.* 1982). Those released in Oregon, however, appear to have disappeared. In 1973 they were reported to be breeding and the population to number about 23, but by 1981 there was only one of the original 93 left (Jameson *et al.* 1982).

In 1987, 63 otters were taken from the Californian population and released around San Nicolas Island (Brownell and Rathburn 1988).

Family: Viverridae
Civets, genets and
mongooses

VIVERRIDAE

The banded mongoose (*Mungos mungo* (Gmelin)) is alleged to have been introduced to Zanzibar in about 1935 or more recently from Africa (Lever 1985; Haltenorth and Diller 1994).

MALAY CIVET

Oriental civet

Viverra tangalunga **Gray**

■ DESCRIPTION

HB 621–672 mm; T 315–482 mm; WT about 4.7 kg.

Coat dark grey to yellow or brownish grey; head long and fox-like; sides and lower surfaces of neck banded with distinct black stripes with pale or white inter-spaces; bands on remaining parts broken up and forming a spotted effect; mane of long hairs along top of back from shoulders to base of tail is black; tail with 10 black and white or dark brown rings from base and joined for half the length on upper (dorsal) surface by a dark line.

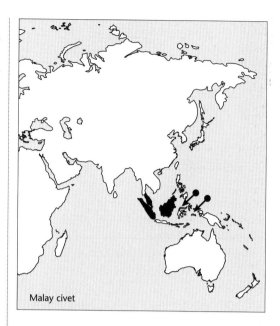

Malay civet

■ DISTRIBUTION

South-east Asia. Malaya, Singapore, Langkawi, Riau Archipelago, Sumatra, Bangka and Borneo.

■ HABITS AND BEHAVIOUR

Habits: terrestrial, nocturnal, predatory. **Gregariousness:** generally solitary. **Movements:** sedentary. **Habitat:** dense cover of forest or brush, and grasslands. **Foods:** invertebrates and small vertebrates; birds, frogs, insects, eggs, fruits, roots and snakes. **Breeding:** litter size 2–3. **Longevity:** 11–15.5 years (captive). **Status:** common.

■ HISTORY OF INTRODUCTIONS

INDONESIA

Malay civets have been introduced to and become established in Sulawesi and perhaps other islands in Indonesia.

Maluku

The Malay civet has been widely introduced as a source of civet, which is used in the production of perfume, to the islands of Batjan, Buru, and probably to Halmahera (Flannery 1995).

The species was recorded on Buru before 1954 (Laurie and Hill 1954). It was found to be present, but rather uncommon on Batjan and possibly Halmahera in 1991 (Flannery 1995).

Sulawesi

The Malay civet was imported into Sulawesi (de Vos *et al.* 1956). They have been introduced to several islands in the East Indies (Laurie and Hill 1954; Groves 1976).

■ DAMAGE

None known.

LARGE INDIAN CIVET
Asiatic civet
Viverra zibetha Linnaeus

■ DESCRIPTION
HB 760–850 mm; Tl 380–495 mm; WT 8–9 kg.
Grey to yellowish grey or fawn; flanks marked with indistinct spots or mottling of black or dark brown; tail marked with five to six dark rings separated by pale rings; basal tail ring joined to dark line along the spine.

■ DISTRIBUTION
Southern Asia. Eastern Himalayas, Burma, and southern China south to Malaya. Also recorded on Singapore Island.

■ HABITS AND BEHAVIOUR
Habits: terrestrial; frequently lives in burrows; nocturnal; climbs with ease. **Gregariousness:** generally solitary. **Movements:** sedentary. **Habitat:** forest, brush, grassland. **Foods:** largely carnivorous; lizards, shrews, insects (cicadas), oil palm seeds, fish, crabs, snakes, frogs, fruits, roots, small mammals. **Breeding:** breeds all year; litter size 1–3, 4; born with eyes closed, open at 10 days; weaned *c.* 1 month. **Longevity:** 15.5–20 years captive. **Status:** declining in some areas.

■ HISTORY OF INTRODUCTION
INDIAN OCEAN ISLANDS
Andaman Islands
Large Indian civets have been introduced and become established in the Andaman Islands where they are now locally abundant (Lever 1985). They are believed to have been introduced to the islands for the sake of the perfume obtained from their musk glands.

■ DAMAGE
Large Indian civets may raid poultry farms or scavenge in garbage dumps (Lekagul and McNeely 1988).

SMALL INDIAN CIVET
Indian civet, rasse, lesser oriental civet, little civet
Viverricula indica (**Demarest**)
=*V. malaccensis* (Gmelin), =*Viverra indica* Desmarest

■ DESCRIPTION
HB 440–645 mm; T 290–430 mm; WT 1.8–4 kg.
Coat is harsh, coarse and buffy, brownish or greyish; muzzle sharp and pointed; thin black stripe runs through line of each eye; forequarters with small spots; flanks with large spots which tend towards longitudinal lines; back has six to eight dark stripes; feet black; tail ringed black and white, six to nine rings of each colour.

■ DISTRIBUTION
Southern Asia. From India east to southern China and Hainan, and south to Sri Lanka, Sumatra, Java, Bali, Lombok, Sumbawa, Bawean, Kangean, Penang and Malay Peninsula.

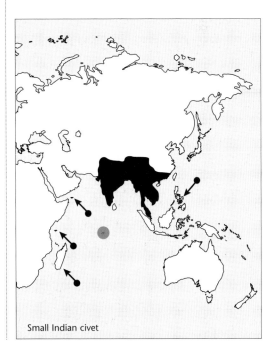

Large Indian civet

Small Indian civet

■ HABITS AND BEHAVIOUR

Habits: nocturnal, occasionally diurnal; terrestrial; may also climb well?; digs burrows under rocks, tree stumps or in drains; shelters in clumps of vegetation or in buildings. **Gregariousness:** solitary, occasionally pairs, or females with young. **Movements:** no information, probably sedentary. **Habitat:** forests, grassland, plantations, vicinity of villages. **Foods:** small vertebrates; crabs, frogs, snakes, small mammals, birds, eggs, tubers, insects, grubs, fruits, roots and carrion. **Breeding:** breeds all year (Sri Lanka) [mates July–August, young born September–December in Madagascar]; 2–5 young; young born in chamber at end of burrow usually under rocks or tree stump or in thick bush. **Longevity:** 8–10.5 years (captive). **Status:** common.

■ HISTORY OF INTRODUCTIONS

Small Indian civets have been introduced to Madagascar, Socotra, Comoro Islands, Philippines, Zanzibar, Pemba and possibly to Tanzania. Their presence on the Indonesian islands of Sumbawa and the Lesser Sundas is probably due to introductions (McDonald 1984; Lever 1985).

Asia
Indonesia
The occurrence of small Indian civets on Sumbawa and to the east of Bali is thought to be due to introductions (Laurie and Hill 1954).

Philippines
Small Indian civets have been introduced to the Philippines, date unknown (Laurie and Hill 1954; Haltenorth and Diller 1994).

Indian Ocean islands
Comoro Islands
Small Indian civets were introduced to the Comoro Islands, date unknown (Laurie and Hill 1954; Haltenorth and Diller 1994).

Madagascar
Small Indian civets were introduced by Indo-Malayan traders who kept them on their vessels to hunt rats. The introduction may also have been a deliberate one as the production of civet from them was widely recognised (Laurie and Hill 1954; Burton 1962; Burton and Burton 1969; Haltenorth and Diller 1994).

Socotra Island
Small Indian civets were introduced to Socotra Island, probably for the production of civet at an unknown date (Laurie and Hill 1954; Burton and Burton 1969; Haltenorth and Diller 1994).

Zanzibar and Pemba
The small Indian civet (*V. i. rasse*) was introduced on both Zanzibar and Pemba Island at unknown dates (Laurie and Hill 1954; de Vos *et al.* 1956; Burton and Burton 1969; Haltenorth and Diller 1994).

■ DAMAGE
No damage is reported, but small Indian civets readily take domestic fowls (Lekagul and McNeely 1988; Haltenorth and Diller 1994).

SMALL SPOTTED GENET
Genet, feline genet, common genet
Genetta genetta (**Linnaeus**)

■ DESCRIPTION
HB 400–600 mm; T 400–510 mm; SH 150–200 mm; WT 1.3–2.4 kg.
Slender cat-like; body long, heavily spotted; short legs; snout pointed; ears rounded; brownish, greyish or sandy yellow with dark spots on body, tending to be in rows; tail long, banded; eye mark on cheeks below eye white; black line up back; female has two pairs of abdominal mammae.

■ DISTRIBUTION
Africa, south-west Europe, Arabia, to Asia Minor. In Europe from France to Iberia, Majorca; Palestine, south-west Arabia. North Africa from Morocco to

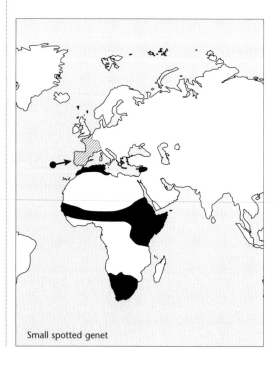

Small spotted genet

Algeria and Tunisia and north-west Libya; also from Senegal, Nigeria to Ethiopia and Somalia and south to Tanzania; in southern Africa in Angola, Zambia, Namibia, Botswana, South Africa, Zimbabwe and southern Mozambique.

■ HABITS AND BEHAVIOUR
Habits: nocturnal; terrestrial; possibly territorial; lair in hollow trees, rock clefts or roots or abandoned burrows. **Gregariousness:** solitary or pairs. **Movements:** sedentary; home range 1.4–50 km². **Habitat:** forest, woodland, scrub, grassland, savanah, semi-desert. **Foods:** birds and their eggs, small mammals, rodents, insects, spiders, scorpions, millipedes, crabs, mussels, reptiles (lizards, snakes), frogs, fish, grass, fruits; occasionally game birds and poultry. **Breeding:** gestation 56–77 days; litter size 1–3, 4; 1–2 litters/year; born blind in hole in tree, ground, or under rocks; eyes open 5–12 days; weaned *c.* 3 months; independent at 9 months; mature at 2–4 years. **Longevity:** 13–21.5 years captive, less in wild. **Status:** common, but declining due to persecution; kept as a rodent killer in ancient Egypt.

■ HISTORY OF INTRODUCTIONS
EUROPE
The occurrence of the small spotted genet in Spain, Portugal, the Balearic Islands and southern and central France may well have been result of introductions (Hagues 1928; Corbet 1978; Lever 1985).

Spain, Portugal and France
On zoogeographical grounds, it is highly probable that the small spotted genet (*G. g. isabelae*) was introduced to Europe possibly in Roman times or earlier (Corbet 1966, 1978). It is also possible that the first introduction was made by the Arabs (Hagues 1928), possibly to Ibiza Island, Spain (Schreiber *et al.* 1989).

Introduced from North Africa, the species has a restricted distribution in south-west Europe (which tends to suggest it was artificially introduced), but is widespread from the Iberian Peninsula west through most of France where its range is still increasing (Burton 1991). It also occurs on the island of Majorca.

■ DAMAGE
Small spotted civets raid poultry yards and dovecotes, but are said to be useful, killing many rats and mice (Lyneborg 1971).

PALM CIVET
Common palm civet, toddy cat, Indian palm civet
Paradoxurus hermaphroditus (Pallas)

■ DESCRIPTION
HB 380–710 mm; T 400–660 mm; WT 2.4–4.0 kg.
Body long; coat varies seasonally and individually from olive-grey to almost cream with dark bases to hairs; head and cheeks white; back marked with three distinct black or dark brown longitudinal stripes in the mid-line; flanks with few dark spots which sometimes form longitudinal stripes; tail long, terminal half black, sometimes white tipped; legs short, black; feet black. Females have three pairs of mammae.

■ DISTRIBUTION
Southern Asia. From Sri Lanka, India, Burma, Thailand, Indochina and southern China south to Malaya, Sumatra, Java, Lesser Sunda Islands and the Philippines. Recorded on the islands of Pinang, Tioman, Langkawi, Singapore, Enggano, Timor, Kangean, Sumbawa, Flores, Komodo, Selayer, Sipora, Pagai, Riau, Bangka, Anamber, Simeulue, Seram, Aru and Kai Island groups.

■ HABITS AND BEHAVIOUR
Habits: largely arboreal, but often on ground; nocturnal; often lives in eaves of houses, or hollow trees and in rocks. **Gregariousness:** adults generally solitary. **Movements:** sedentary? **Habitat:** forest, plantations; often near human habitation. **Foods:** flesh, fruits

Palm civet

(mangos, coffee, pineapples, melons, bananas), small vertebrates, insects, seeds, palm juice, molluscs. **Breeding:** breeds throughout the year; litter size 2–4, 5; eyes closed at birth; sexually mature at 11–12 months. **Longevity:** 14–22 years 5 months (captive). **Status:** common.

■ HISTORY OF INTRODUCTIONS
Introduced successfully to the Maluku and Lesser Sunda Islands.

INDONESIA
Maluku and Lesser Sunda Islands
The palm civet is thought to have been introduced throughout the Maluku and the Lesser Sunda Islands (de Vos *et al.* 1956; Lekagul and McNeely 1988). They are carried about from island to island by people who use them as a rat catchers and this may be the reason for their presence on Sulawesi, Timor and other islands (Walker 1992). Presumably they were introduced to the North Maluku to help control rodents (Flannery 1995).

Palm civets have been introduced to the Aru Islands, Batjan, Halmahera, Kai Islands, Seram, and Sula Islands (Lever 1985; Flannery 1995)

■ DAMAGE
Palm civets are clearly regarded as a pest of bananas in parts of Indonesia (i.e. Desakuta) (Kitchener *et al.* 1990).

MASKED PALM CIVET
Himalayan palm civet
Paguma larvata (Hamilton-Smith)

■ DESCRIPTION
HB 415–762 mm; T 392–640 mm; WT 3.0–5.0 kg.
Generally grey, tinged buff, orange or yellowish-red; face region dark but has a white stripe from tip of head to nose; below each eye a white mark; feet blackish; tail tip darker than body or white.

■ DISTRIBUTION
Southern Asia. The Himalayas, Burma, southern China (to Peking), Taiwan and Hainan, south to the Andaman Islands, Sumatra, Malaya and Borneo.

■ HABITS AND BEHAVIOUR
Habits: not well known, probably similar to palm civet; mainly arboreal, and often in trees; mainly nocturnal, occasionally active during day; terrestrial; sleeps in holes or forks in trees. **Gregariousness:** solitary. **Movements:** none known; sedentary. **Habitat:**

Masked palm-civet

forest and brush; rice fields, secondary growth, dumps. **Foods:** omnivorous; rats, insects, fruits, roots, seeds, fish, figs and other small vertebrates (small mammals, birds, reptiles). **Breeding:** gestation *c.* 56 days; young 1–4; born in hole in tree; 2 litters per year; eyes open at 9 days. **Longevity:** to 15.5 years (captive). **Status:** fairly common.

■ HISTORY OF INTRODUCTIONS
Introduced successfully to Honshu, Japan.

ASIA
Japan
Masked palm civets were imported to Japan in ancient times and again more recently (about 1940) as a cage animal for their fur (Kuroda 1955; Udagawa 1970). They were possibly imported from Taiwan (subspecies *taiwana*) or from southern China (subspecies *larvata*), although the animal now established is the gem-faced civet (*P. larvata taiwana*). Some of the captive civets evidently escaped, according to the sporadic records, in central Hondo (on Amakusa-Shimo-Shima off Kyushu) and Shikoku Island (Udagawa 1954, 1970). They were recorded in Yamanashi (Honshu, west of Tokyo) and Shikoku in 1951 (Udagawa 1951), in Shinano (Honshu) about 1953 (Shimoake 1953) and in the Nagano, Yamanashi and Shizuoka (in Honshu) prefectures before 1955 (Kuroda 1955) and more recently (Furuya 1973).

Until 1970 they were increasing in numbers in eastern and southern Honshu (Udagawa 1970) and the species is still established there (McDonald 1985).

■ DAMAGE

On the island of Honshu palm civets damage oranges and mandarins and other local fruits, and also vegetables (Udagawa 1970). In Thailand they occasionally raid chicken coops and are said to be great rat catchers (Lekagul and McNeely 1988).

SMALL INDIAN MONGOOSE

Indian mongoose

Herpestes auropunctatus (**Hodgson**)

=H. javanicus *(Geoffroy). Included here because it is similar in appearance to the small Indian mongoose, but in some works is considered a separate species.*

■ DESCRIPTION

HB males 250–350 mm; T 200–310 mm; WT 312–1300 g. Coat speckled buff and black, but varies, desert populations palest; fur soft, silky, olive-brown; tail shorter than head and body length. Female has three pairs of mammae.

■ DISTRIBUTION

Asia. From Arabia, Iraq, Iran, Afghanistan to northern India, southern China and Hainan, south to the Malay Peninsula (rare in Malaya) and southern Thailand.

■ HABITS AND BEHAVIOUR

Habits: diurnal; terrestrial, but climbs well; lives in burrows and crevices or under rocks, logs and roots. **Gregariousness:** solitary, pairs or family groups of females with young to 14. **Movements:** sedentary; overlapping home range 0.25–1.0 km^2; individuals may move up to 1.6 km in search of food **Habitat:** open dry bush and savanah; near villages. **Foods:** small mammals (rats, mice), birds (pheasants, chickens, pigeons), amphibians (toads, frogs), reptiles (lizards, snakes, geckos, skinks), crustaceans, fish, crabs, asteroids, freshwater prawns, insects (mole-crickets, crickets, moths, grasshoppers, stick insects, dragonflies, flies, weevils, bugs, cockroaches, and beetles and their larvae, caterpillars, cutworms), arachnids (tarantulas and other spiders), centipedes, isopods, myriapods, and plant remains (seeds including roots and fruits), arthropods. **Breeding:** peak in August–February in Fiji and January–October in Puerto Rico; gestation 42–43 days; litter size 2–4; 2 litters per year (in Peurto Rico); young born blind, hairless; eyes open at 16–17 days; weaned 4–5 weeks; sexually mature in first year. **Longevity:** 6–8 years (captive). **Status:** common.

■ HISTORY OF INTRODUCTIONS

Introduced successfully to the West Indies, Hawaiian Islands, South America, Fiji, Mafia Island, and Africa, and unsuccessfully to Australia.

AUSTRALASIA

Australia

In 1884 the mongoose (thought to be this species?) was introduced from Sri Lanka to northern Queensland to cope with a plague of rats which threatened to ruin the local sugar cane plantations.

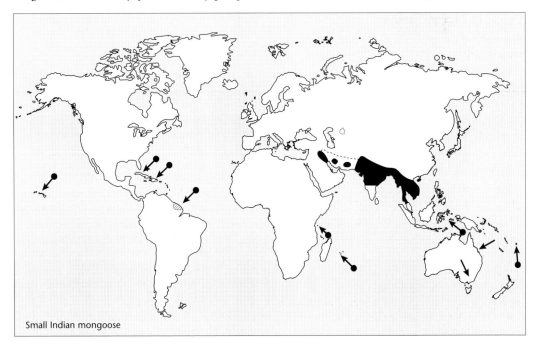

Small Indian mongoose

However, the species failed to become permanently established (Anon. 1946). There is also a report of several experiments to establish mongooses in Australia in the 1880s, with more than 100 animals being liberated near the Murray River, and others in New South Wales, all of which were unsuccessful (Palmer 1898).

CARIBBEAN–WEST INDIES

Between 1872 and 1900 the mongoose was introduced to most cane-growing islands in the Caribbean (Nellis and Everard 1983) including: Jamaica, Trinidad, Leeward Islands group, Puerto Rico, St. Kitts, St. Croix, Virgin Islands, Cuba, Barbados, Hispaniola, St. Lucia, Antigua, St. Vincent, Grenada, Dominica, Viequas, St. John, St. Thomas, Tortola, St. Martin, Nevis, Marie Galante, Desirade (Varona 1974; Hall 1981). On Martinique and St. Lucia they were introduced to control pit vipers, *Bothrops atrax* (Lever 1985).

Mongoose introductions in the West Indies–Caribbean Islands

Island	Date introduced	Notes
Antigua	after 1870	well established
Barbados	1870s, possibly 1877–79	well established
Buck (St. Croix)	1952	?
Cuba	after 1870	well established
Desirade	1870s or 80s	present
Dominica	1880s	unsuccessful
Grenada	1870s	well established
Guadeloupe	c. 1880	still there
Hispaniola	after 1870	well established
Jamaica	1872	widespread
Leeward Is group	c. 1880	present
Marie Galante	1870s or 80s	present
Martinique	?	?
Nevis	1870s or 80s	present
Puerto Rico	c. 1888	?
St. Croix	1884	?
St. John (USVI)	1880s?	still there
St. Kitts	1884	still there
St. Lucia	after 1870	well established
St. Martin	1870s or 80s	present
St. Thomas	?	?
St. Vincent	after 1870	well established
Tortola	1870s or 80s	present
Trinidad	1870s	widespread
Vieques	1870s or 80s	present
Virgin Is	1880s	?

References: Espeut 1882; Urich 1931; Milne & Milne 1962; Seaman & Randall 1962; Sade & Hildrech 1965; Phiibosian & Ytema 1977; Nellis *et al.* 1978, Hall 1981; Lever 1985.

Antigua

Introduced to Antigua some time after 1870 the mongoose is now well established (Hinton and Dunn 1967; Varona 1974; Hall 1981).

Barbados

Mongoose were introduced to Barbados some time in the 1870s, probably between 1877 and 1879, and are now well established (Hinton and Dunn 1967; Varona 1974; Hall 1981; Lever 1985).

Cuba

The small Indian mongoose was introduced in Cuba some time after 1870 and is now well established (Hinton and Dunn 1967; Varona 1974; Hall 1981). Mr W. B. Espeut indicates he exported them from Jamaica to Cuba, Puerto Rico, Grenada, Barbados and Trinidad in the 1870s (Espeut 1882). In Cuba they were well established by 1886 and by 1929 occupied some 2600 km^2 around Havana and today they still are confined to this area (Lever 1985).

Dominica

Mongooses were unsuccessfully introduced to Dominica in the 1880s (Milne and Milne 1962; Hall 1981).

Grenada

Mongooses were introduced to Grenada some time after 1870 and are now well established (Hinton and Dunn 1967; Varona 1974; Hall 1981).

Guadeloupe (Leeward Island group)

The mongoose was introduced to Guadeloupe about 1880 to control rats, but also caused the extermination of several species of animals. The mongoose still occurs on the island (Varona 1974; Hall 1981).

Hispaniola

Introduced to Hispaniola some time after 1870 the mongoose is now well established (Hinton and Dunn 1967; Varona 1974; Hall 1981).

Jamaica

Four male and five female mongooses from Calcutta, India, were released in Jamaica in 1872 by W. Bancroft Espeut, a sugar cane grower, to combat losses in cane fields caused by black and brown rats (Espeut 1882; Bigalke 1937). These animals became established and bred. They increased in numbers and spread rapidly, so much so that after the first 10 years it was estimated that they had saved the planters some £45 000 per year in cane losses from the rats (Morris 1882). In 1883 the Jamaican government prohibited further imports of the mongoose (Silverstein and Silverstein 1974). However, 20 years after the first introduction the rats had become scarce and the mongoose had turned to eating domestic poultry and native wildlife and was generally regarded as a pest (Bigalke 1937). The species is now widespread on the island (Varona 1974; Hall 1981).

Puerto Rico

The mongoose was introduced to Puerto Rico from Jamaica in about 1888 (Philibosian and Yntema 1977) for rat control (Colon 1930), although there may have been some releases in the 1870s (Espeut 1882). They may have been well established by 1877 on the coast of Arecibo, San Juan, Fajardo, Arroyo, Ponce and Mayaguez and inland at Utuado and Adjuntas (Lever 1985).

At first the mongoose appeared successful in controlling the rats and for the space of some 15 years they were described as an effective predator of introduced rats (Pitmentel 1955). Later the limited benefit gained for rat control was offset by the complaints of damage to agriculture.

The mongoose still occurs in Puerto Rico (Varona 1974; Hall 1981) on all the islands except Mora and Monito islands, and Anegada (Philibosian and Yntema 1977). See US Virgin Islands section, opposite.

St. Croix

The mongoose was introduced to the island of St. Croix in 1884 from Jamaica and soon became established (Seaman and Randall 1962). Two pairs were released on Buck Island in 1952 (Nellis *et al.* 1978) and they became established there also. The island became a national park in 1962 and efforts were made to eliminate the mongoose. Following 10 years of trapping and poisoning the control work was discontinued with the mongoose still firmly established on the island. The mongoose is still established on St. Croix and adjacent islands (Philibosian and Yntema 1977).

St. Kitts

The small Indian mongoose was introduced to St. Kitts in 1884 (Sade and Hildrech 1965) and is still present (Varona 1974; Hall 1981).

St. Lucia

Introduced to St. Lucia some time after 1870 to control pit vipers, the mongoose is now well established (Hinton and Dunn 1967; Varona 1974; Hall 1981).

St. Vincent

Introduced to St. Vincent some time after 1870 and are now well established (Hinton and Dunn 1967; Varona 1974; Hall 1981).

Trinidad

The small Indian mongoose was introduced to Trinidad in 1870, probably from Jamaica, and had colonised the entire island by 1930 (Urich 1931; Hinton and Dunn 1967), but were found mainly in the cultivated and disturbed areas rather than the forests (Urich 1931).

By 1912 they had spread widely but had not reached Cedros, Mayaro, Orapuche and La Brea, and were rare at Toco and Blanchisseuse. Between 1902 and 1908 bonuses were paid on 30 895 mongooses. When the bonus was again applied between 1927 and 1930 some 142 324 mongooses were presented for payment (Urich 1931).

Virgin Islands (USVI)

Mongoose were introduced in the US Virgin Isalnds to Vieques, St. Thomas, St. John, Tortola and Guana before 1898 (Lever 1985).

The importation of the small Indian mongoose to the Virgin Islands was for the purpose of rat (*Rattus rattus*) control and not for destroying snakes as is sometimes thought (Seaman and Randall 1962).

In the Virgin Islands, all the main inhabited islands and several of the smaller private islands now have the mongoose (Dewey and Nellis 1980). Control efforts for the protection of wildlife began in St. John in 1983 on an experimental basis to prevent predation on hawksbill turtle (*Eretmochelys imbricata*) nests (Coblentz and Coblentz 1985).

The mongoose is still established on St. Thomas and adjacent islands and also on St. John and adjacent islands (Philibosian and Yntema 1977).

Indian Ocean Islands
Mafia Island (Tanzania)

Mongooses have been introduced on Mafia Island, off Tanzania (de Vos *et al.* 1956; Corbet 1978).

Mauritius

Mongooses were introduced and established to control rats in sugar cane plantations (Lever 1985).

Indonesia
Ambon

Mongoose were introduced from other islands of the Indonesian Archipelago (de Vos *et al.* 1956; Haltenorth and Diller 1980, 1994).

North America
United States

Mongooses have been introduced to the United States, although there appear no details of any releases or escapes. A single animal was trapped in Woodford County, Kentucky, in 1921 (Jackson 1921) and a supposed fisher skull found by Van Bloeker (1937) in material from San Benito County, California, has been identified as that of a mongoose (Van Gelder

1979). Probably three animals were destroyed on Dodge Island, Port of Miami in southern Florida in 1976–77 (Nellis *et al.* 1978).

PACIFIC OCEAN ISLANDS
Fijian Islands
The small Indian mongoose was introduced to Viti Levu in 1883 when imported from India to control rats (Turbet 1941) in sugar cane plantations (Gorman 1974, 1975). Since this time the species has spread to other Fijian islands (Gorman 1974, 1975). In the 1970s they were present on Viti Levu and Vanua Levu where they are common and widespread in all habitats (Pernetta and Watling 1978; Flannery 1995).

Hawaiian Islands
In 1883, J. Tucker returned from Jamaica with 72 live mongooses as a result of a trip paid for by the Hilo Planters' Association. These were divided among planters of Hamakua coast (Tinker 1938) and liberated on the Hilo-Hamakua coast of Hawaii to control rats (Bryan 1938). Two years later another consignment (numbers unknown) were brought back to Hamakua planters by J. Marsden from Jamaica (Anon. 1885). One year later a plantation owner claimed no evidence of rat damage and concluded that the mongoose saved them at least US$50 000 (Walker 1945) and in 1888 reported (in press) that not a single stalk of cane was damaged in the Hamakua coast area (Anon. 1888). Following this, mongooses were shipped to other islands including Maui, Molokai, Oahu and Kauai but an accident at the dock prevented their release on the latter island (Tinker 1938). The species was subsequently released on Maui and Oahu to control rats (Baldwin *et al.* 1952) and later introduced to Molokai.

The mongoose rapidly became adapted to conditions in Hawaii (Tomich 1969) and in 1952 occurred on all four of the islands where releases had occurred (Baldin *et al.* 1952). In 1969 they were locally abundant in the north-east sector of Hawaii (Tomich 1969). During a rabies scare in 1967 some 7000 were killed (Kramer 1971).

In 1984 a program to reduce numbers to reduce predation on endangered seabirds was commenced (Stone and Keith 1986). These studies are continuing into testing trapping and poisoning methods.

SOUTH AMERICA
Surinam, Guyana, French Guiana and Colombia
The mongoose is now well established on Surinam where it was imported from Barbados (Hinton and Dunn 1967). Soon after its introduction to Surinam it was imported to Guyana and French Guiana (Lever 1985).

Colombia, Guyana, and French Guinea
The small Indian mongoose was introduced via the West Indies to South America after 1872 (de Vos *et al.* 1956). In 1951 they were introduced from Jamaica to northern Colombia (Roots 1976) when 80 were released. They were also introduced to British Guiana in 1872 where they spread through the coastal cane country but did not occupy the undisturbed forest areas (de Vos *et al.* 1956; Milne and Milne 1962; Corbet 1978).

■ DAMAGE
On Viti Levu and other islands mongooses have exterminated the ground-dwelling species (Turbet 1941). Where they were introduced in Viti Levu, Fiji, the mongoose is reported to have played a part in the near extinction and diminution in numbers of native frogs (*Platymantis vitianus* and *P. vitiensis*) and several ground-nesting birds including the banded rail (*Rallus philippensis*), sooty rail (*Porzana tabuensis*), white-breasted rail (*Poliolimnas cinereus*) and the purple swamp hen (*Porphyrio porphyrio*) (Gorman 1975, 1979).

In the West Indies and Hawaiian Islands the mongoose not only caused significant damage by raiding poultry and preying on the wildlife, but at best only partially depressed the rat population (Hinton and Dunn 1967). Since their introduction in the Hawaiian Islands, several native animal species have been reduced in numbers or locally extirpated, but the part played by the mongoose in this regard is probably small. However, they are regarded as notorious raiders of poultry pens and small birds and their eggs (Baldwin *et al.* 1952), and were once observed killing a nesting Hawaiian goose (Baker and Russell 1979). According to Tomich (1969) they are of significance in problems of public health, agriculture and game management.

Following their introduction on St. Croix, their effect on the rats appeared severe, but after predator–prey adjustment the rat population recovered, became numerous again and continued to cause damage. The mongooses began to prey on domestic fowl and small stock and their effect on other wildlife became deleterious, resulting in a serious reduction in numbers or even extinction of some species, particularly reptiles. The ground lizard (*Ameiva polops*) was eliminated on St. Croix except for islets off the coast where the mongoose was not present (Seaman and Randall 1962). By 1930 on St. Croix it was reported that the snake, *Alsophis sanctae-crucis*, and ground lizard (*Ameiva polops*) were believed to be extinct (Barbour 1930) but were later found on two small keys off the coast where there were no mongooses (Nellis *et al.*

1978). On Buck Island off St. Croix since the discontinuation of a long-term control program against the mongoose a re-introduced population of the lizard *A. polops* has been eliminated (Nellis *et al.* 1978). In the West Indies their role in the decline and extinction of native fauna is probably not as great as is sometimes attributed (Heatwole *et al.* 1981).

On Jamaica the rat, *Oryzomys antillarum*, was a pest of sugar cane but may now be extinct due to the predation by the mongoose (Burton and Burton 1969). They have also caused reduction in the numbers of the short-tailed hutia (*Capromys brownii*) and the Jamaican petrel (*Aestrelata caribboea*) (Bigalke 1937). Schmidt (1928) found that they had exterminated both ground lizards and snakes on Vieques Island, and Myers (1931) reported that they had had similar effects on other West Indian islands.

On Puerto Rico there is some evidence that the mongoose is not an effective predator on rats and in 1950 they were shown to be an important vector and reservoir of rabies. On Puerto Rico it is reported to be impossible to run poultry in yards because of mongoose depredations (Pitmentel 1955).

Mongooses are claimed to have caused a reduction in the numbers of birds and lizards on Trinidad and the effect on reptiles is borne out by a food habits study (Williams 1918). Walcott (1953) says it is natural for an animal immortalised as a snake eater to turn to lizards in West Indies. Sixty years after the mongoose's introduction rats are still numerous on Trinidad but lizards and snakes of many species had become rare. Some lizards were rare on the Trinidadian mainland but were still abundant on Bocas Islands (nearby) where the mongoose was not present (Urich 1931).

Food habit studies in the Virgin Islands show mongooses are generalist feeders, consuming all classes of terrestrial vertebrates and a wide variety of invertebrates. Native wildlife is affected on most islands. Predation caused near-extinction of the ground-nesting quail dove (*Geotrygon mystacea*) in the Virgin Islands (Nellis and Everard 1983) and extirpated ground lizards (*Ameiva polops*) from Buck Island and St. Croix (Philobosin and Ruibal 1971). Heavy predation was also noted on hawkesbill turtle (*Eretmochelys imbricata*) nests on St. John (Small 1982) and so a trapping study commenced in 1983 to see if turtles could be saved by removing mongooses (Cobletz and Cobletz 1985).

Islands in West Indies that now have mongooses do not have nesting seabirds (Dewey and Nellis 1980). All the main inhabited islands and several of the smaller private islands have mongooses. Introduction of mongooses to Guadaloupe to control rats caused the extermination of several bird species, lizards and snakes, and threatened the existence of native mammals such as the agouti (Encycl. Brit. 1970).

Introduced mongooses have had a negative impact on native biota of Caribbean islands (Coblentz and Coblentz 1985).

GREY MONGOOSE
Indian grey mongoose
Herpestes edwardsi (Geoffroy)

DESCRIPTION
HB 373–450 mm; T 282–375 mm; WT males 1340–1790 g, females 896–1120 g.
Coat pale grey to light brown and finely speckled with black, legs darker in colour than body; plantigrade feet; five toes on all four feet; three to four pairs of mammae.

■ DISTRIBUTION
Asia. From north-eastern Arabia through Iran to Baluchistan, India (Assam), Bahrain and Sri Lanka.

■ HABITS AND BEHAVIOUR
Habits: terrestrial; mainly diurnal; lives in holes under rocks and burrows. **Gregariousness:** solitary or pairs. **Movements:** sedentary. **Habitat:** cultivated areas, villages. **Foods:** small mammals (rats, mice), carrion, fruits, roots, reptiles (lizards, snakes), birds, eggs, insects. **Breeding:** breeds in spring; gestation 60–65 days; litter size 2, 3–4; males mature at 6 months. **Longevity:** 11 years 2 months (captive). **Status:** still numerous.

■ HISTORY OF INTRODUCTIONS
Introduced to Malaya, Mauritius, and the Ryukyu Islands, and Tonaki Islands, Japan.

ASIA
Japan
Grey mongoose were introduced in 1910 to Loochoo (Ryukyu Islands), Japan, for rat and snake (pit viper) control (Kaburaki 1934) and also to Tonaki, Japan, where they prey on poultry.

Ryukyu Islands (formerly Loochoo islands)
Grey mongoose were introduced in 1910 from India to the Ryukyu Islands and Tonaki Island (Hinton and Dunn 1967) for rat and snake control (Kaburaki 1940).

Malaya
The grey mongoose has been recorded from Wellesley Province, Parak, Selangor and Malacca (Medway 1978), where they are thought to have been introduced (Walker 1992).

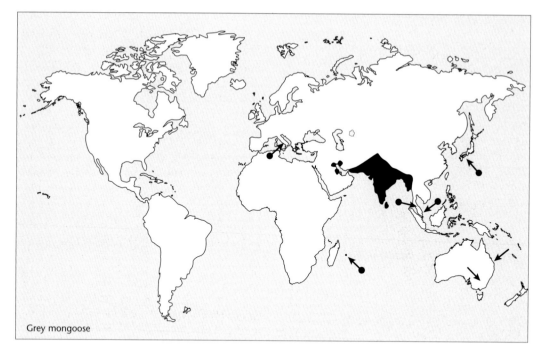

Grey mongoose

AUSTRALASIA
Australia
A mongoose, probably *H. edwardsii* or *H. skithi*, from Sri Lanka was introduced in northern Queensland to control rats in sugarcane fields, but failed to become established (Hinton and Dunn 1967). Palmer (1898) also reports that several experiments with mongooses were carried out in Australia, including the release of more than 100 individuals near the Murray River, and others in New South Wales. All of these introductions failed.

EUROPE
Italy
In the 1960s grey mongooses were released around Monte Circeo, 100 km south of Rome to control the viper (*Vipera aspis*) (Lever 1985; Walker 1992). This introduction has been erroneously attributed to *H. ichneumon* (Egyptian mongoose) by other authors (e.g. Hinton and Dunn 1967; Roots 1976).

INDIAN OCEAN ISLANDS
Mauritius
The grey mongoose was introduced to Mauritius from India (Hinton and Dunn 1967; Walker 1992) in 1899 to control rats, but they then became a pest (Enc. Brit. 1970–80). The mongooses are said to have contributed to the decimation of partridge, quail and black-naped hare (*Lepus nigricollis*), and also to the extinction of Timor deer (*Cervus timorensis*) that was introduced to island in 1639 from Batavia (Hinton and Dunn 1967; Lever 1985).

PACIFC OCEAN ISLANDS
New Zealand
Fourteen mongooses were released on a farm in Southland in 1870, but they failed to become established (Thomson 1922).

■ DAMAGE
None reported.

ICHNEUMON
Egyptian mongoose, large grey mongoose
***Herpestes ichneumon* (Linnaeus)**

■ DESCRIPTION
HB 450–650 mm; T 350–550 mm; SH 190–210 mm; WT 2–8 kg.
Coat a uniform grizzled brown and grey, but greyer on forequarters and flanks, fur soft and fine; body long; head pointed; ears short and rounded; muzzle and legs brownish grey; lower parts of limbs black; tail tapering, black tufted.

■ DISTRIBUTION
Europe and Africa. In Europe confined to southern Spain and Portugal, Italy and Yugoslavia where may have been introduced. Also occurs on the Yugoslavian island of Mijet. Formerly more widespread. Most of Africa from Morocco to Tunisia, Egypt and perhaps eastern Libya through Palestine to southern Asia Minor, most of the savannah zone of Africa south of the Sahara.

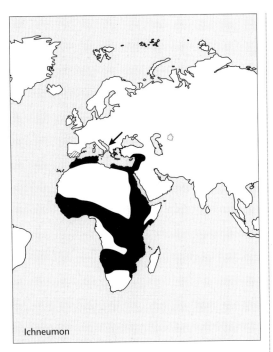

Ichneumon

HABITS AND BEHAVIOUR

Habits: terrestrial, mostly nocturnal, sometimes diurnal; nests in rock cleft or burrow. **Gregariousness:** solitary, pairs or groups 3–7. **Movements:** sedentary. **Habitat:** reed beds, savannah and steppe, dense undergrowth near water, semi-desert. **Foods:** omnivorous; small mammals (rabbits), reptiles (snakes, lizards), fish, crabs, birds (Guinea fowl) and eggs, crayfish, fruit, earthworms, and insects. **Breeding:** seasonally April–May; gestation 60–84 days; young 2–4; 1 litter/year; young born blind; eyes open 2 weeks; emerge from cover 6 weeks; weaned 10 weeks; independent 4 months; mature in second year. **Longevity:** 5–7 years (wild) to 20 years (captive). **Status:** uncommon and range reduced.

■ HISTORY OF INTRODUCTIONS

Probably introduced to Italy, Yugoslavia and Madagascar

EUROPE

The ichneumon was domesticated by the Egyptians, but the animal lost its popularity from about the Hellenistic period onwards when the domestic fowl was introduced. On zoogeographical grounds it seems probable that the European population is the result of human introduction, perhaps in antiquity (Carter 1978).

Iberia

The ichneumon appears to have either invaded Iberia as a post-glacial immigrant or to have been intro-

duced by the Arabs in ancient times (Cheylan 1991; Walker 1992). It may also have been introduced into Iberia by the Romans as early writings mention the introduction of what are assumed to be ferrets from North Africa to control rabbits. It is conceivable that it was introduced there for this reason (Corbet 1966).

Italy

The ichneumon has recently been introduced to Italy to control vipers (Corbet 1966; Hinton and Dunn 1967). *H. auropunctatus* was introduced there in the 1960s and some were released around Monte Circeo, 100 km south of Rome (Lever 1985) (see *H. edwardsi*).

Yugoslavia

Ichneumons were released by the Austrian government on the island of Mljet, off the Dalmatian coast, for control of the sandotter, *Vipera ammodytes* (Kuhn 1935). Their effect on the snakes is not known, but the animal became a nuisance. However, they were apparently extinct or rare there in 1961 (Niethammer 1963). Niethammer records that they were released on the island of Korcula, north of Mljet, in 1910.

INDIAN OCEAN ISLANDS

Madagascar

Ichneumon are said to occur in Madagascar (Lever 1985; Walker 1992) as an introduced species, but does not appear to have been recorded recently (Haltenorth and Diller 1994).

■ DAMAGE

Ichneumons are persecuted because they often raid poultry runs, killing the inhabitants (Corbet 1966; Lyneborg 1971; Smither 1983).

Family: Felidae
Cats

ONCILLA

Felis cf. *tigrina*
=*Leopardus tigrinus* (Schreber)

This species may have been transported to Aruba by early Amerindians (Eisenberg 1989).

SERVAL

Felis serval Schreber
=*Leptailurus serval* (Schreber)

Servals have been re-introduced into a number of South African reserves (Van Aarde and Skinner 1986; Anderson 1992).

f 2002

DOMESTIC CAT
Feral cat, cat

Felis catus Linnaeus

All forms of the domestic cat show specific taxonomic criteria, hence a common ancestor in the wild (F. lybica) at the species level judged on cranological criteria, os penis and karyotypes of all three basic domestic forms (Kratochvil and Kratochvil 1976). They will interbreed with the European wild cat (F. silvestris) and other forms (Zeuner 1963). F. silvestris was found in towns in Palestine 7000 years ago; domestication occurred about 4000 years ago. Introduction to Europe began around 2000 years ago and there is some inter-breeding between it and F. s. lybica. Domestication appears to have had a religious basis (Grzimek 1975) and was certainly the object of a passionate cult in ancient Egypt.

■ DESCRIPTION
TL male 535–885 mm; T 160–330 mm; WT 1.1–5.8 kg.
Six coat colours readily distinguished: striped tabby, blotched tabby, black, grey, ginger and tortoiseshell. Belly, throat and limbs often with white. White is rare in feral cats and most feral cats are short-haired and not fancy breeds (Siamese etc.).

■ DISTRIBUTION
Worldwide. Feral cats are distributed throughout the world wherever humans have colonised, and also occur on most of the world's islands. (Note: the map does not show all of the island introductions of cats around the world; these are contained in the table under the History of Introductions.)

■ HABITS AND BEHAVIOUR
Habits: active mainly at night, but greatest at sunrise and sunset; little hierarchy and not asocial. **Gregariousness:** complex social hierarchy; solitary or adult females may live in groups with adult male and range overlapping with other female groups; group size 1–50; young males leave groups at sexual maturity (1–3 years); density varies 1 cat/km^2 to 2350/km^2 and depends largely on food abundance [density 0.7/km^2 in Victorian mallee; 3.65 cats/km^2 on Macquarie Island; 1 per 20–30 acres in Sacramento Valley; in south-eastern Australia 0.74–2.4 cats/km^2; in Portsmouth docks in England 2.4–220/km^2;

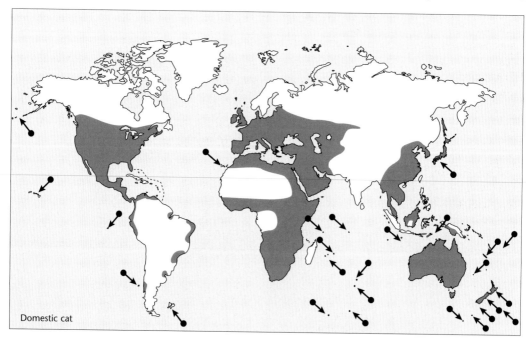

Domestic cat

Presence of cats on islands

Island/country	Date introduced	Notes
20th Day (Aust)	*c.* 1910	?
Alcedo (Galápagos)	?	present
Aldabra	after 1888	present
Aleutian Is		present
Althorpe (Aust)	before 1910	present?
Amirantes		present Eagle I.
Amives (Carolines)	about 1912	present
Amsterdam	before 1930	still present
Anacapa (California)	?	present
Anchor (NZ)	sealers 1792	present?
Anderson (Aust)	before 1830	present 1830
Angel (Aust)	natural spread?	present ?1972
Annobon (Gulf of Guinea)		plentiful there 1960
Arapawa (NZ)	?	still present
Ascension	?	present
Auckland (NZ)	1806–40 by sealers or whalers	present main island and Masked I.
Babel (Aust)	1960s?	present?
Beata (off Dominica)	?	present
Berlenga (off Portugal)	before 1927	present 1927
Bernier (Aust)	before 1906–07	? pet, absent in 1959
Borneo (Indonesia)	?	present
Bribie (Aust)	1970s?	present?
Britain	Middle Ages	present and widespread
Broughton (Aust)	before 1911	present?
Bruny (Aust)	?	present?
Burrup (Aust)	?	present?
Caldey (Wales)	?	monks introduced many cats to control rabbits
Campbell (NZ)		still present
Canary Is	?	present
Cape Barren (Aust)	1970s?	present?
Cape Verde Islands	?	present on some islands
Caroline Is	about 1912	present on Namoluk Atoll, Toinom and Amives
Cerro Azul (Galápagos)	?	present
Chagos Archipelago	1840s	present Egmont Atoll
Channel (California)	1800s	present some islands
Chappell (Aust)	?	present
Chatham (NZ)	before 1840	still present main island
Choiseul (Solomons)	?	present
Christmas (Indian O)	1888	present
Clarke (Aust)	?	present *c.* 1970s
Cocos (Costa Rica)	?	present
Cocos-Keeling	?	present?
Columbrete Grande (Medit. Sea)	*c.* 1855	present ?
Croker (Aust)	?	present?
Crozet (Île aux Cochins)	about 1887	still present
Culeenup (Aust)	?	present?
Curtis (Aust)	?	present?
Cuvier (NZ)	after 1889 by lighthouse keeper	exterminated 1961–66

Presence of cats on islands (*continued*)

Island/country	Date introduced	Notes
D'Urville (NZ)	by settlers	still present
Darwin (Galápagos)	?	present
Dassen (Africa)	late 19th century	present
Deal (Aust)	before 1890	present 1890
Deliverance (Aust)	before 1888	present *c.* 1888, still present 1928
Desertas	before 1920s	present?
Desroches	1905	present there in 1905
Dirk Hartog (Aust)	before 1917	present
Dolphin (Aust)	?	natural spread?
Eagle (Amirantes)		present
East Intercourse (Aust)	1963	causeway to mainland
East Sister (Aust)	?	present 1970s?
Eau (Tonga)	?	present
Egmont Atoll (Chagos Arch.)	1840s	present
El Hiero (Canary)	?	present
Falkland Is	?	present
Farquar Atoll	?	present N and S Is
Faure Is (Aust)	1930s or before	present
Fiji	before 1870	present on many islands
Flinders (Aust)	?	present?
Flinders (Tas, Aust)	before 1872	present 1872
Floreana (Galápagos)	?	present
Forsyth	by settlers	still present
Fraser (Aust)	?	present?
Frégate (Seychelles)	?	present
French (Gippsland, Australia)	?	present 1990s
French Frigate (Hawaiian)	1948	?
Fuerteventura (Canary Is)	?	present 1950s
Gabo (Aust)	1972	present
Galápagos Is	18th century and since	present on Isabela, Santiago, Santa Cruz, Floreana and San Cristobal
Galito (Fiji)	?	present
Garden (Aust)	1960s	present; one in 1991, now absent?
Garden (WA, Aust)	?	present
Gidley (Aust)	?	natural spread
Grand Terre (Kerguelen)	1951	control, but still present
Great Barrier (NZ)	by early settlers	still present 1980s
Great Dog (Aust)	?	present 1970s
Great Mercury NZ)	by settlers	still present
Great Saltee (Ireland)	1950	present
Great Saltee (Ireland)	?	1950 introduced, but died out after about 8 years or so
Griffith (Vic, Aust)	?	present 1980
Guadalcanal (Solomons)	?	present
Guadaloupe (Mexico)	1800–30	present
Guam	?	present
Haiti (Hispaniola)	before 1535	now throughout
Hawaiian Is	with European settlers	present on all larger islands
Hebrides (UK)	?	present
Herekopare (NZ)	1911, 1924–26, 1931	eradicated 1970
Hermite (Aust)	before 1912	? shipwreck

Presence of cats on islands (*continued*)

Island/country	Date introduced	Notes
Heron (Aust)	?	now removed
Hog (see Île aux Cochins)		
Holm of Melby (Shetlands)	before 1930	released before 1930; present?
Howland	1966	still present?
Île du Chat (Kerguelen)	before 1874	died out
Îles Glorieuses		possibly present
Ireland	Bronze Age	present and widespread
Isabela (Galápagos)	?	
Isla del Coco (Cocos Is)		present?
Iwo Jima (Japan)	World War 2	present?
Jarvis	1855 and/or about 1935	eradicated 1983
Jeegarnyeejip (Aust)	?	present?
Johnston Atoll	early 19th century	present?
Juan Fernández		present
Kadavu (Fiji)	?	present
Kahoolawe (Hawaiian)	?	present
Kangaroo (Aust)	?	present
Kapiti (NZ)	before 1905 by settlers	eradicated by 1934
Kawau (NZ)	by settlers	still present
Keppel (Falklands)	?	present 1980s
Kerguelen	early 19th century–1874 and again 1951–52 and 1956	died out 1850–1900, but accidentally re-introduced. Several attempts to eradicate 1969–1974
King (Aust)	before 1887	present 1887
Kinsha (South China Sea)	1982	present?
Legendre (Aust)	?	natural spread
Line Is	?	present?
Little Barrier (NZ)	1867–1880 by settlers	eradicated by 1980
Little Dog (Aust)	?	present
Little Green (Aust)	?	present
Lord Howe (Aust)	soon after discovery	present and thinly dispersed 1970s
Macquarie (Aust)	about 1810	still present
Magnetic (Aust)	?	present?
Malden (Line)	?	present?
Malolo Laila (Fiji)	?	present
Mangere (Chatham Is, NZ)	before 1893 by settlers	released to control rabbits; exterminated rabbits then died out in 1950s
Maria (Tas)	?	present 1980s
Marion	1949	efforts exterminate in 1970 with feline panleucopaenia
Masked (Auckland)	about 1820	present
Matacawalevu (Fiji)	?	present
Matakohe (NZ)	?	eradicated 1991
Mauritius		present
Mayor (NZ)	by settler before 1926	still present
Milingimbi (Aust)	?	present?
Montutapu (NZ)	?	still present
Moreton (Aust)	?	present?
Motuihe (NZ)	19th century	eradicated 1978–79, but re-introduced and eradicated 1981
Mount Chappell (Aust)	?	present?
Mutton Bird (Aust)	?	natural spread
Namoluk (Carolines)	about 1912	present

Presence of cats on islands (*continued*)

Island/country	Date introduced	Notes
Navassa, between Jamaica & Hispaniola	?	present
New Caledonia		present
New Zealand	1769	present on North and South Is
Niihau (Hawaiian)	?	present
Ninth (Aust)	?	present?
North (NZ)	1769 on	present
North Reef (Aust)	?	present?
North Stradbroke	?	present?
North Uist (Hebrides)	?	present 1990s
Northwest (Aust)	before 1925?	present 1925, eradicated 1986
Noss (Shetlands)	1890	present
Nouvelle Amsterdam	early 19th century	numbers reduced 1950s, but still present 1985
Oleron (France)	?	present 1980s
Otaheite (Society)	1774	?
Ovalau (Fiji)	?	present
Papua New Guinea	with Europeans	present
Pelsart (Houtman Abrolhos group)	before 1913	now absent
Phillip (Aust)	?	present?
Pitt (Chatham Is)	before 1868 by settlers	still present
Ponui (Chamberlains) (NZ)	by farmers since 1850s	still present
Possession (Crozet Arch.)	?	formerly present, not in 1990s
Prime Seal (Aust)	?	present?
Putauhina (NZ)	mutton birders	died out?
Queen Charlotte (Canada)	?	present
Rabama (Aust)	?	may not be feral?
Rakitu (Arid)		now eradicated
Rangitoto (NZ)	?	still present
Raoul (NZ)	c.1850 or 1836–1870	eradication 1972 on, still present
Rat (Aust)	c. 1900	present?
Reevesby (Aust)	before 1990?	eradicated 1990
Remire (see Eagle)		
Revilla Gigedo	?	present on San Benedicto
Robben	late 17th century	?
Rocky (Aust)	?	now absent
Rodrigues	before 1803	present?
Rosemary (Aust)	1989	single animal introduced
Rotamah (Aust)	?	causeway to mainland
Rottnest (Aust)	?	still present
Ruapuke (NZ)	by settlers	still present
Sable (Canada)	1880	present?
Sable (Nova Scotia, Canada)	?	introduced to control rats and rabbits
San Benedicto (Revilla Gigedo)	?	present
San Clemente (California)	?	?
San Cristobal (Galápagos)	?	present
Sand (Hawaiian)	?	caught in 1964
Santa Barbara (California)	late 1800s	eliminated 1978
Santa Cruz (Galápagos)	?	present
Santiago (Galápagos)	?	present
Sao Tiago (Cape Verde Is)	? after 1642	present
Serrurier (Long) (Aust)	1987	single animal introduced

Presence of cats on islands (*continued*)

Island/country	Date introduced	Notes
Seychelles	after settlement 1770	present Frégate I. and ? others
Shivinish (Hebrides)	?	there in 1973
Siera Negra (Galápagos)	?	present
Society	?	?
Solomon		present Choiseul I.
South (NZ)	1769 on	present
South Georgia		present at times
South Havra (Shetlands)	?	present
South Molle (Aust)	?	unsuccessful eradication 1980s?
St. Francis (Aust)	before 1922?	present 1922, now absent
St. Helens (Aust)	1920s, *c.* 1945	absent 1930s; re-introduced?; eliminated by disease; absent?
St. Kilda (UK)	1930	died out 1931
St. Paul	early 19th century to 1874	has since died out
St. Vincent (Windward Is)	?	present
Starbuck (Line)	?	present?
Stephens (NZ)	1892 or soon after	eradicated 1925
Stewart		present
Sunday (Aust)	?	present?
Swan (Aust)	?	? pet
Swan (off Honduras)	?	present
Tahiti	1774	left by Cook in 1774?
Tasman (Aust)	?	some control, present
Tasmania	?	present
Taveuni (Fiji)	?	present
Tenerife (Canary)	?	present
Tern (Hawaiian)	1965	tamed
The Brothers (Aust)	?	present?
Thevenard (Aust)	1970s?	single animal
Three Hummock (Aust)	?	present?
Timor (Indonesia)	?	present
Tiritiri Matangi (NZ)	early 1960s	extirpated by 1970s
Toinom (Carolines)	about 1912	present
Tokelau Is	about 1841	present on all
Tonga		present on 'Eua
Tori Shima (Japan)	World War 2	present?
Trimouille (Aust)	?	? shipwreck
Trinidad	?	?
Tristan da Cunha	about 1810	still present; exterminated 1990s?
Troubridge (Aust)	?	single animal removed 1980s?
Vanderlin (Aust)	?	present?
Vanua Levu (Fiji)	?	present
Vatua Vara (Fiji)	?	present
Viti Levu (Fiji)	before 1870	present
Waiheke (NZ)	by settlers	still present
Wardang (Aust)	?	present?
Whale (NZ)	before 1925	died out by 1956
Wolf (Isabela)	?	present
Yadua (Fiji)	?	present
Yaqaga (Fiji)	?	present
Zembra (Tunisia)	?	recently introduced 1980s

free-ranging domestics 6.3/100 ha in Illinois, United States]. **Movements:** home ranges overlap, range 0.7–15.0/ha to 0.7–9.9 km^2 [home range adult males 228 ± 100 ha, females 112 ± 21 ha (Illinois); adult males 30–41/ha on Macquarie Island, females 0.03–990 ha (to 10 km^2)]. **Habitat:** most terrestrial habitats; sand dunes, desert, scrub, forest, tussock grasslands; urban, suburban; edge cover in agricultural areas (Illinois). Sub-antarctic islands to temperate farmland and urban areas. **Foods:** small mammals (rabbits, mice, rats, hares, voles, hedgehogs, possums), birds (quail, poultry, pheasants), invertebrates and insects (grasshoppers, beetles, mantids, caterpillars, moths, grass grubs, grasshoppers, cicadas, scarab beetles, dragonflies), spiders, centipedes, scorpions, plant items (grass), carrion, human refuse, reptiles (snakes, lizards, skinks, geckos), fish, freshwater crayfish. **Breeding:** breeds throughout year, but mainly spring and summer; mainly October–March with peak in November–December (Macquarie Island); gestation about 65 days; oestrus 4–6 days; 2–3 litters/year; litter size 2–10; average 1.6 litters/year (Illinois); eyes open 8–13 days; weaned 8 weeks; independent 6 months; sexual maturity 7–12 months. **Longevity:** 14–25 mostly to 31 years (captive); 3–5, but up to 11 years (free-ranging domestics). **Status:** common and abundant. Hybridise with wildcat (*F. silvestris*) in Europe.

■ HISTORY OF INTRODUCTIONS
AFRICA
South Africa
Feral cats occur in many areas of the southern African sub-region, but usually around human habitation. They have been introduced to Hluhluwe-Umfolozi Game Reserve, South Africa, but are actively controlled as they are considered likely reservoirs of cat flu (Macdonald and Frame 1988). They occur in Kalahari, Botswana, and Cape Province of South Africa, coming as domestics with the early settlers, and then becoming feral (Smithers 1983).

AUSTRALASIA
Australia
Cats have been in Australia since the earliest settlements by Europeans and may possibly have arrived with Dutch shipwrecks in the seventeenth century (Burbidge *et al.* 1988). They are now found throughout the mainland and on many offshore islands (WIlson *et al.* 1992).

A study of feral cats in Hobart, Tasmania, suggested that animals present there have closest genetic affinities to those of southern England and New Zealand rather than to mainland Australia (Dartnell and Todd 1975).

Feral cats occur throughout the Northern Territory, even in remote uninhabited areas (Letts 1964). They occurred on Broughton Island, New South Wales, before 1911 and are still present there, where they may cause some damage to the breeding sea birds, but there is no direct evidence (Lane 1976). They occurred on Heron Island, Capricorn Group, Queensland, at some time, but were removed long ago (Kikkawa and Boles 1976). They are present on Gabo Island, Victoria (Reilly 1977).

Thirty to 35 cats were found on Tasman Island off the south-east coast of Tasmania in 1978, but many have been eliminated by shooting (Brothers 1979). Some cats were introduced about 1925 or soon after to St. Helens Island to control rabbits and were reported to have been exterminated by 1930, but were eliminated themselves by an infectious feline virus disease (McManus 1979).

Ninth or 20th Day Island off Launceston had introduced cats in 1910 (Barrett 1918). They are also present on Great Dog Island, Little Green Island, Little Dog Island and Chappell Island in the Furneaux Group, Tasmania (Brothers and Skira 1987, 1988; Skira and Brothers 1988).

Cats were present on Hermite Island in the Montebello Group where they are believed to have contributed to the extermination of the spectacled hare-wallaby (*Lagorchestes conspicillatus*) and the golden bandicoot (*Isoodon auratus*). They still occur on Dirk Hartog Island in Shark Bay, where they were responsible for the extirmination of the original population of banded hare-wallabies (*Lagostrophus fasciatus*) and a re-introduced population in the 1970s.

Papua New Guinea
Feral cats are generally scarce in New Guinea and are restricted to the immediate vicinity of human settlements and re-growth areas (Flannery 1995). They are present around towns and villages and are established in the wild adjacent to these areas in many parts (Ryan 1972). Probably some feral populations exist and some have been released on islands off the coast with rat problems without success (Herington 1977).

Feral cats occur in areas of cane grass and early re-growth on Mount Erimbari (Dwyer 1983) and are feral at Telefomin. A decrease in numbers of small mammals was recorded following the introduction of domestic cats into a village in Yapsiei area Sandanin Province during 1984–85 (Flannery 1995).

ATLANTIC OCEAN ISLANDS
Annobon Island (Gulf of Guinea)
Feral cats were plentiful on the island in about 1960 (Fry 1961).

Canary Islands

Cats have been introduced and are feral in the Canary Islands (Encycl. Brit. 1976–78), certainly on El Hiero (Nogales *et al.* 1987) and Tenerife (Nogales *et al.* 1990).

Cape Verde Islands

Feral cats occurred on Sao Tiago in 1951 (Bannerman and Bannerman 1968) and probably occur on the other islands in the group. They may have been there since the early days of settlement, which took place in 1462.

Desertas

Feral cats occurred on the Desertas in the 1920s (Bannerman and Bannerman 1965) and probably still do.

Falkland Islands

Cats arrived on these islands with humans, but little appears to be known about their status (Cawkell and Hamilton 1961).

CENTRAL AMERICA

Costa Rica

Cats are feral on Cocos Island, Costa Rica.

EUROPE

The cat was originally spread by the Egyptians as a protector of granaries from mice and rats (Niethammer 1963). It was certainly domesticated by them from the sixteenth century BC onwards (Zeuner 1963) and may have spread to other areas from Egypt. They probably reached Greece about the fifth century BC and other parts of Europe in Roman Imperial times. They were introduced into Switzerland by the Roman legions (Schauenberg 1970). Domestic cats first appeared in Europe in about the fourth century AD (Rome). Hybridisation with the wild cat was not a problem until the Middle Ages when rapid deforestation caused populations of the wild cat to decrease dramatically. Cats colonised England in the tenth century and in the twelfth century were at Kiev (Heptner 1992).

United Kingdom and Ireland

Cats were introduced to Britain some time prior to the Middle Ages, originally to assist with the control of vermin and as a companion of humans, and are now feral throughout the British Isles (Lever 1977). They were probably introduced to Ireland in the Bronze Age (Stelfox 1965).

In about 1900 there may have been 80 000 to 100 000 feral cats in London and in Cardiff in 1944 there were 6600 in that city among some 23 000 owned cats (Fitter 1959). Few urban and rural areas are today without their complement of feral cats, many of which exist independently of humans (Lever 1985).

Feral cats occur on islands off the British Isles where they were deliberately set free on Noss in the Shetlands to control rats in 1890. They were introduced at various times to the island of South Havra and to Holm of Melby in the Shetlands, and were reported to have been released on an island in the Hebrides for the control of rats. Twelve were released on St. Kilda in 1930, but most died out and the remainder was shot in 1931.

In Ireland three dozen cats were released on Great Saltee Island, off Ballyteige Bay, County Wexford in 1950 to control rats (Fitter 1959; Lever 1977).

INDIAN OCEAN ISLANDS

Aldabra

James Spur introduced cats to this island some time after 1888 (Bourne 1971).

Amirantes

Cats are present on Eagle (formerly Remire) Island (Lever 1985).

Amsterdam (Nouvelle Amsterdam)

Before the 1930s feral cats were reported to have caused damage to the native bird populations on the island (Aubert de La Rue 1930). An attempt to reduce the population is said to have caused a rise in the numbers of rats and mice and so was abandoned (Reppe 1957). Feral cats were still present on the island in the mid-1960s (Holdgate 1967), 1970s (Derenne and Mougin 1976) and in 1989 (Furet 1989).

Ascension Island

At times cats are reported to be present on the island (Watson 1975; Lever 1985).

Chagos Archipelago

Cats were reported on Egmont Atoll in the 1840s (Bourne 1971) and were found to be feral around areas of settlement in 1972–73 (Hutson 1975).

Christmas Island

Cats were taken to the island in 1888 when the first settlement began, and a feral cat population was established there by 1904 (Tidemann *et al.* 1994).

Crozet Archipelago

Cats were introduced to the Île aux Cochins (Hog Island) at an unknown date and are first mentioned on the island in 1887 (Richard-Foy 1887). The rarity of the animal at this time suggests it was then a recent introduction (Derenne and Mougin 1976), as about this time (second half of nineteenth century) they were introduced on many other sub-antarctic islands.

Feral cats have now colonised most of the island except for the central high elevations. Their numbers are still fairly low, being about 100 during the winter and about three or four times this total in the summer (Derenne and Mougin 1976).

Dassen Island (Africa)
Cats occur on Dassen Island and are probably feral on Robben Island. They were introduced to Dassen in the second half of the nineteenth century (Green 1950), although they could have been there long before in the seventeenth century as the island was populated by humans at that time (Skead 1980).

Desroches
Feral cats may have been present on this island in 1905 (Gardiner and Cooper 1907; Stoddart and Poore 1970).

Farquar Atoll
Feral cats are present on north and south islands of this atoll (Lever 1985).

Îles Glorieuses
Cats are now present on this island where they have been for some time (Lever 1985).

Kerguelen
Cats were unsuccessfully introduced to control rodents in the early 1800s, but died out by 1850. Cats were re-introduced about 100 years later, and are now abundant there (Watson 1975).

Feral cats are thought to have been introduced by seal hunters at the beginning of the nineteenth century (Derenne 1974) or some time before 1874 (Holdgate and Wace 1961) as they were present on Île du Chat at this date (Kidder 1876). Thereafter they became widespread on Kerguelen, but disappeared between 1850 (Derenne 1974) and the beginning of the twentieth century (Lesel 1971; Lesel and Derenne 1977). However, they were accidentally re-introduced in 1951–52 and 1956 at the time of the installation of permanent bases on the archipelago (Lesel 1971; Derenne and Mougin 1976).

Two or three cats were taken to Grand Terre by a relief vessel at the end of 1951 or early in 1952. These became wild and had disappeared from Pont aux Français by 1954. Several more were released to control the all too plentiful rats and mice in 1956 (Derenne 1976; Lesel and Derenne 1977).

The first attempts to destroy the feral cat population were made in 1958 and by 1960 the population was limited to a few (Lesel 1971). However, these increased in following years (Derenne 1974) and the feral cat became well established again. Several campaigns have been made in the three years before 1977 to eradicate them without total success (Lesel and Derenne 1977). From 1969 to 1972 some 1080 were killed and in 1973–74 some 1712 were killed

(Derenne 1974). At the end of April 1974 it was estimated that 2000–3000 cats were on the island (Derenne 1976). Their diet on the island consists of 70 per cent birds and 35 per cent rabbits and so they exert some pressure on the local fauna. Furthermore, the island population has a high growth rate (55 per cent) and is expected to increase and spread further (Derenne 1976).

Rodrigues
Feral cats were reported to be present on Rodrigues in 1803 according to the Civil Agent, Marragon, at the time.

Marion Island
Feral cats are widespread on Marion Island (Holdgate 1967; Watson 1975; Derenne and Mougin 1976), where they were introduced to control mice and rats around the meteorological station (Derenne 1976). Five cats were introduced in 1949, and by 1975 the cat population was estimated to be killing 450 000 petrels each year.

Over 2000 cats were present by 1975 and in 1977 it was estimated that there were 3409 there. In 1977 feline parvo virus was introduced in an effort to create an epidemic of the disease feline pan leucopaenia. This highly contagious, host-specific disease reduced numbers to 615 (s.e. ± 107) during 1982, but the disease is no longer spreading effectively (Van Aarde 1979; Van Rensburg *et al.* 1987). Thereafter hunting was instigated and eight two-man teams killed 807 cats with 12-bore shotguns. Hunting and trapping continued, and by 1991 only 80 cats were killed. There were no further record of cats from 1991 to March 1993 despite searches. The current belief is that they have been eradicated from the island (Bloomer and Bester 1990, 1992; Bester and Skinner 1991).

Mauritius
Cats are present on the island (Lever 1985).

Seychelles
Cats have been introduced to the Seychelles (Crook 1961). At present they occur on Frégate Island (Lever 1985).

St. Paul
St. Paul had numerous feral cats in 1874. They were living on the native birds (Vélain 1877; Jeannel 1941), but have since died out on the island (Holdgate and Wace 1961; Holdgate 1967).

South Georgia
Cats are present on South Georgia (Lever 1985).

Tristan da Cunha

Brought in by the early settlers about 1810, cats soon ran wild and became widely distributed on the island (Holdgate and Wace 1961). They had already had an effect on the local birds by 1832 (Earle 1832 in King 1990).

During the period when the residents of the island left in 1961–63, due to volcanic eruptions, the cats decreased in numbers considerably, probably due to hounding by feral dogs (Anon. 1963). However, some were still present and feral in 1967 (Holdgate 1967) and in the 1970s (Watson 1975; Derenne and Mougin 1976).

INDONESIA
Borneo

Cats are associated with humans in Borneo and are found near most settlements.

Timor

Feral cats are present on Timor (Carter *et al*. 1945).

NORTH AMERICA
Canada

Cats arrived with the first settlers and escaped or were abandoned and are now widespread as a feral species near the haunts of humans. They are abundant on the Queen Charlotte Islands (Carl and Guiguet 1972).

On Sable Island, off Nova Scotia, in 1880 introduced cats exterminated a rabbit population which had been introduced half a century earlier (Lever 1985).

United States

Introduced by Europeans, domestic cats spread across the United States with settlement and then many became feral. Their numbers are not large except in urban areas because of the presence of other large carnivores.

In Illinois rural areas cat density was 6.3/100 ha, but there was no evidence of truly feral cats which avoided humans and bred in the wild; however, transient cats, particularly sub-adults, were common. The Illinois population of cats was probably 5–6 million and in the United States an estimated 42 million free-ranging domestic cats were present in 1981 (Warner 1985).

Cats occur throughout California, and on San Clemente Island (Howard and Marsh 1984; Jameson and Peeters 1988) and are reported to be having a major effect in the San Francisco Bay region (Pearson 1985).

Feral cats were abundant on Santa Barbara Island, California, in the late 1800s and early 1900s (Howell, 1917). These had been reduced to one animal by 1975

and the last animal was eliminated in 1978 (Murray *et al*. 1983). They also occur on Anacapa Island in the Channel Islands, California (Presnall 1958; Lever 1985).

PACIFIC OCEAN ISLANDS
Mid-Pacific Ocean islands

First introduced to most islands from early in the nineteenth century, feral populations of cats are now established on most of the inhabited islands from Hawaii to Johnston Atoll (Kirkpatrick 1966). However, many unpopulated islands also have populations of feral cats (King 1973; Parry 1980; Kirkpatrick and Rauzon 1986).

Aleutians

Cats are present on at least some islands (Lever 1985).

Auckland Island

Possibly introduced about 1820 (between 1806 and 1840 (King 1990)) by sealers or whalers, feral cats have been present since then (Taylor 1968), although the exact date of introduction is not known (Holdgate and Ware 1961). They were numerous in the early 1950s (Eden 1955) and still occurred on the island through the 1970s (Holdgate 1967; Atkinson and Bell 1973; Gibb and Flux 1973; Jones 1977; Dilkes and Wilson 1979). They occur throughout the island and are still present on main Auckland Island and on Masked Island (King 1990).

Caroline Islands

At present some 50 semi-feral cats are on Namoluk Atoll and about 20 on Toinom and Amives in the eastern Carolines, where they were recorded possibly as early as 1912. There are none on Lukan and Unman (Marshall 1975).

Chatham Islands

Feral cats are present (Anon. 1980).

Cuvier Island

Feral cats have now been exterminated on this island (Anon. 1980). They were introduced by the lighthouse keeper after 1889, but were eradicated by 1964 (King 1990).

Fiji

Cats were introduced in the early nineteenth century and by 1870 had become numerous in the interior of Viti Levu. In the 1970s it was reported that only domestic cats existed in Fiji, where they were uncommon (Pernetta and Watling 1978). However, feral cats were later reported to be established on the two main islands and also on Taveuni, Yaqaga, Ovalau, Kadavu, Matacawalevu, Yadua, Galito, Malolo Laila and possibly Vatua Vara (Lever 1985).

Guadaloupe (Mexico)

It is thought that feral cats were introduced to Guadaloupe by sailors between 1800 and 1830 (Huey 1925). They are still present there (Lever 1985) and also on San Benedicto (Revilla Gigedo group) (Lever 1985).

Galápagos Islands

Cats were probably introduced to the Galápagos Islands by sealers, whalers or buccaneers who visited the islands in the eighteenth century. Numbers were probably supplemented when permanent settlements were established on the islands.

Feral cats are present on the Isabela Islands (Wolf, Darwin, Alcado, Siera Negra and Cerro Azul), on Santiago Island, Santa Cruz Island, Floreana Island and San Cristobel (Eckhardt 1972; Konecny 1987).

Great Barrier

Cats were probably introduced by early settlers, and feral cats are still present on the island today (King 1990).

Guam

Cats have been introduced and are established on this island (Flannery 1995).

Hawaiian Islands

The first European settlers introduced cats to these islands where they quickly became feral (Perkins 1903; Bryan 1915). They are found on all the eight larger islands including Kahoolawe which has a thriving population (Kramer 1971). They are occasionally found at altitudes exceeding 2000 metres in the Hawaiian Islands (Tomich 1969) and many were reported on Niihau in the 1950s (Fisher 1951).

House cats reached French Frigate Shoals with humans in 1948, 1953 and 1960 and a single feral animal was on Tern Island in 1965, but was later tamed (Amerson 1971).

Domestic cats have been introduced on Johnston Atoll where they are allowed to roam, but have not yet become feral. On Sand Island, two were present (feral) until 1964 when they were caught (Amerson and Sheton 1976).

Herekopare Island (near Stewart)

Cats may have been introduced in 1924; other reports and records indicate it could have been in 1926 or in 1931. Efforts were made to exterminate them in the 1940s and from 1940 to 1944 more than 111 were killed. From 1965 to 1970 about 157 were killed. In 1970 a total population of 33 cats, at a density of 1.2 cats/ha, was killed because of the effect on the bird population on the island (Fitzgerald and Veitch 1985). Feral cats have now been exterminated on the island (Anon. 1980).

Howland Island (central Pacific)

The present population of cats was introduced in 1966 (King 1973) on this uninhabited island. They are still present there (Kirkpatrick and Rauzon 1986).

Japan

Cats were left behind on Tori Shima, south of Honshu after World War 2 and on Iwo Jima further south (Lever 1985).

Jarvis Island (central Pacific)

Introduced to this uninhabited island about 1935 (King 1973), cats were eradicated in 1983 because of the damage to bird life (Kirkpatrick and Rauzon 1986).

The earliest record of a cat on the island was in 1885 (MacFarlane 1887), but they may have been there with the first phosphate miners between 1858 and 1879. The Whippoorwill Expedition found none there in 1924 and there were none in 1935. They were introduced again in 1938, brought in by a second wave of occupants when Britain relinquished control of the island to the United States (Rauzon 1985). These settlers left cats when they left the island (King 1973).

Efforts to eradicate the cats since 1957–58 have resulted in several hundred being killed at this time and over 200 more in 1964–65. In 1967–68 only nine could be located; however, in 1973 at least 14 were counted. In 1976, 12 were killed and over 50 more sighted on the island. Further eradication efforts in 1977 resulted in 102 shot, but 50–75 remained. In 1978 another 160 were shot. Eradication attempts were continued in 1982–83 and feral cats may now be absent from the island (Rauzon 1985).

Juan Fernández

Feral cats are present on this island (Lever 1985).

Kinsha Islands (South China Sea)

Cats were introduced by Chinese soldiers about 1982 to control rats on these islands.

Line Islands

Cats were introduced on two uninhabited islands – Starbuck and Malden – but there are few details (Perry 1980).

Little Barrier Island

Introduced between 1867 and 1880 by settlers to the island (King 1990), feral cats were present on Little Barrier before 1895, and 20 were shot there in 1896, and they were reported to be rare in 1901. In 1932 they were common and during the next 11 years 360 were destroyed. From 1945 on about 15 were shot annually, but they remained widely distributed on the island (Watson 1961). They were eliminated from the island about 1980 (Anon. 1980; Wodzicki and Wright 1984) and this has resulted in a dramatic increase in native bird species (King 1990).

Lord Howe Island

Cats were introduced on this island soon after its discovery (Recher and Clark 1974). In 1978–80 feral cats were found thinly dispersed on Mt. Lidgbird and 80 were destroyed 1976–78 (Miller and Mullette 1985).

Macquarie Island

Introduced shortly after the discovery of the island by sealers in 1810 (Jones 1977), cats had become feral and well established by 1820 (Debenham 1945; Bellinghausen 1948; Holdgate and Wace 1961; Taylor 1979; Wharton and Demspster 1981). Some cats were noted in 1888 or thereabouts (Chamberlain 1888 in Taylor 1979) and following the introduction of rabbits the cats increased markedly in numbers and many were reported in 1890–94, when at least two species of birds became extinct (Hamilton 1894; Falla 1937; Mawson 1943). Their numbers apparently remained high until at least 1900, but they were rare again by 1909 (Taylor 1979), and were said to be diminishing in 1913 (Cumpston 1968), but have since remained relatively scarce for many years (Taylor 1979).

Cats were not numerous on the island in 1930 (Mawson 1943), but were common all over in 1955 (Taylor 1955). Their presence was reported in 1967 (Holgate 1967) and their numbers in 1973–75 were estimated to be about 250–500 distributed over 65 per cent of the island (Jones 1977). On Macquarie Island the total population of feral cats was 169–252 adults. However, between December 1976 and February 1981, 246 were collected on the island (Brothers 1985).

Mangere Island (Chatham group)

Cats have been present on Pitt Island since before 1868 (King 1990). On the main island of Chatham they were introduced by settlers before 1840 and are still present (King 1990). Some cats may have been introduced to Mangere in the 1890s to control rabbits (Bell 1975). Feral cats inhabited Mangere last century, but died out in the 1950s following the removal of rabbits (Bell 1975; Anon. 1980).

New Caledonia

Feral cats are present on New Caledonia (Barrau and Devambez 1957) where they are reported to be well established (Flannery 1995). They cause some damage to native bird and animal populaitons (Barrau and Devambez 1957).

New Zealand

Cats were an early introduction to New Zealand, probably by sealers and whalers from 1769 onwards

and later settlers for the control of rabbits, rats and mice (Gibb and Flux 1973). They appear to have been introduced early in the nineteenth century and to have become widespread and common in both the North and South islands (Wodzicki 1950, 1965). Some were released in south Nelson by a landowner to control rabbits in 1866 (Lamb 1964) and they were probably established in the North Island by the 1830s (King 1990).

Feral cats are now common in the wild and well established in the North and South islands and on a number of offshore islands including Stewart, Auckland, Campbell, Chathams, and Raoul (Gibb and Flux 1973). They are also present on Little Barrier, on Stephens Island in Cook Strait after 1894, on some southern islands in Foveaux Straits, and were introduced to Herekopare Island after 1911 and exterminated in 1970; and were also on Cuvier Island where they were eliminated in 1961–66 (Atkinson and Bell 1973; Dilks and Wilson 1979).

Populations of feral cats in New Zealand are thought to be self-maintaining, although strays may continually augment feral numbers (Fitzgerald and Karl 1979). They are still present on 18 islands, but have been eradicated from six others and have died out on three islands (Veithch 1985; King 1990).

Raoul Island (Kermadecs, NZ)

Introduced by settlers between 1836 and 1870, feral cats were first observed there in the 1860s (Smuts-Kennedy 1975) and are still present on the island.

In 1955 the Wildlife Service exterminated 45 cats, in 1972 another 61 were destroyed, in 1973 15 cats and in 1977 a further 61 were destroyed. The policy of destruction of the cats is continuing in the hope of eliminating them.

Society Islands

Captain Cook gave 20 cats to the natives on Otaheite Island, probably on his second voyage (Kippis 1904). What became of them does not appear to have been recorded.

Solomon Islands

Feral cats are present on Choiseul (Lever 1985) and Guadalcanal (Flannery 1995).

Stewart Island

Cats are present on the island and threaten the survival of some birds (Wodzicki and Wright 1982).

Tahiti

Feral cats are present on this island (Lever 1985). James Cook is reported to have left 20 cats at Tahiti in 1774 plus others at Ulietea and Huaheine (Beaglehole 1961 in King 1990).

Tokelau Islands
Cats were introduced some time after 1841 and are now found feral on all the islands (Kirkpatrick 1966).

Tonga
Cats are found on 'Eua where they may have assisted in the extermination of the blue-crowned lory, *Vini australis* (Rinke 1987).

WEST INDIES–CARIBBEAN
West Indies
Cats have been introduced to some Caribbean Islands (de Vos *et al.* 1956), probably for the control of rats and mice.

Hispaniola
Feral cats were numerous on Hispaniola in 1535 (de Oviedo y Valdes 1851–55) and appeared to be fairly numerous in about 1769. They now occur over most of Haiti (Street 1962).

Other islands
Feral cats are also present on Swan Island (off Honduras), St. Vincent (Windward Islands), and Navassa Island (between Jamaica and Hispaniola).

■ DAMAGE
Domestic cats are important predators and due to semi-domestication avoid regulation from variable prey abundance. Conservationists regard them as pests that destroy indigenous wildlife and hunters persecute them as unacceptable competitors of game (Libery 1984). Free-roaming cats are said to be a potential ecological, medical and social threat because they harbour diseases transmissible to humans, damage property and wildlife, create a nuisance and cause pollution, and inflict bites and cause accidents (Feldmann and Cording 1973).

Some food studies in Europe tend to indicate that at least cats are not as damaging to wildlife as some authorities suggest. In Switzerland it has been shown that feral cats relied strongly on the presence of humans, but could not be looked upon as harmful to game animals or birds, nor as controllers of vermin (rats and mice etc.) (von Goldshmidt-Rothschild and Lueps 1976). Stray cats in Westphalia (West Germany) were found to be much less harmful than widely believed in hunting circles (Spittler 1978).

In the United States it has been found that free-ranging cats preyed on a number of native mammals such as shrews, chipmunks, moles, squirrels and mice and rabbits, but that birds were not an important part of diet (Parmelee 1953; Eberhard 1954; Toner 1956).

One study found free-ranging and feral cats obtained 15–90 per cent of their food from natural prey and that the natural prey related to prey abundance, annual production, and availability. Prey choice of feral cats was similar to that of house-based cats, but the former lived almost entirely on natural prey and their absolute intake was four times that of average house-based cats (Libery 1984). Another early study suggested that cats living in residential areas and those frequenting roadsides appeared to be a greater menace to birds than those hunting in fields or woods, though the prey was largely the most abundant and available forms (McMurry and Sperry 1941). Others found non-field cats took few birds and that for field cats, birds were only a secondary item (Eberhard 1954). Cats were unlikely to affect bobwhite quail (*Colinus virginianus*) populations in east central Texas (Parmalee 1953) or affect numbers of game bird species in the Sacramento Valley, California (Hubbs 1951). Others are of the opinion that feral cats are more beneficial than detrimental (Korschgen 1957).

Feral cats on Anacapa Island in the Channel Islands, California, are reported to be detrimental to a nesting colony of brown pelicans (*Pelecanus occidentalis*) (Presnall 1958).

In Australia cats may have eliminated the golden bandicoot (*Isoodon auratus*) from Hermite Island off Western Australia (Burbidge 1971), but there is little convincing evidence to support the claim that cats can be blamed for large scale changes in abundance of native animals on mainland Australia (Wilson *et al.* 1992).

Feral cats are opportunist predators and scavengers and the level of predation depends on relative availability. Predation on native mammals is probably limited to those undeveloped areas of bush and scrub where such species are still plentiful (Coman and Brunner 1972).

Cats are probably not involved in the epidemiology of cysticercosis in domestic livestock in Australia (Coman 1972).

Most studies indicate that feral and free-roaming cats are involved in the taking of small native animals to some degree (McMurry and Sperry 1941; Parmalee 1953; Eberhard 1954; Hubbs 1960; Archer 1972; Coman and Bruner 1972; Coman 1975; Martin *et al.* 1996) and that the effects are more obvious on islands (Smuts-Kennedy 1975; Derenne 1976).

On Macquarie Island feral cats have greatly reduced the numbers of burrow-nesting petrels and together with weka (*Gallirallus australis*) were probably responsible for extinction of the parakeet (*Cyanoramphus novaezelandiae erythratis*) and barred rail (*Rallus phillippensis*) before 1900 (Jones 1977; Taylor 1979; Brothers *et al.* 1985).

In New Zealand the indigenous birds evolved in the absence of carnivorous predators and it is probable that cats are a serious hazard to the continued existence of ground-frequenting species (Bull 1953). Assisted by rats, they have severely reduced the sea bird populations on Raoul Island in the Kermadecs, on Herekopare Island, and on Campbell Island where they assisted rats in the elimination of burrowing petrels on the main island (Atkinson and Bell 1973).

On Stewart Island cats threaten the survival of the kakapo (*Strigops habroptilus*). It is estimated that they eat or kill 25–50 per cent of the population of 100 kakapo, reducing an already endangered species (Wodzicki and Wright 1984). As a result, all the remaining kakapo were removed to cat-free islands during 1980–97 (Merton 1998). A recent survey found cats are definite hosts for several sporozoa parasites which are intermediate parasites of sheep (Wodzicki and Wright 1984), therefore they may be of economic importance to the meat industry.

On Mangere Island cats have almost certainly contributed to the extinction of about 12 species of birds (Bell 1975) and are estimated to kill 65 per cent of chicks of petrels on Little Barrier Island (Imber 1973). Where no mammals lived on Herekopare Island, examination in 1911, the 1940s, 1968 and 1970 indicated vast populations of diving petrels and thousands of brood-billed prions were probably exterminated by cats (Fitzgerald and Veitch 1985). By 1973–74 on Kerguelen cats were said to be seriously affecting the indigenous birds, particularly the blue petrels, and so eradication campaigns were commenced (Lesel 1971; Derenne 1974).

Cats are reported to have exterminated the puffin (*Puffinus pacificus*) from Jarvis Island (mid Pacific Ocean) (Kirkpatrick and Rauzon 1986), and are blamed for the demise of ground-nesting birds on many Pacific islands, including Fiji (Watling 1982) and Tahiti (Thibault and Rives 1975), together with other introduced predators such as the mongoose and rats.

Feral cats were also responsible for the eradication of re-introduced populations of burrowing bettongs (*Bettongia lesueur*) and golden bandicoots (*Isoodon auratus*) in the Gibson Desert Nature Reserve, Western Australia, in 1992 (Christensen and Burrows 1995).

MOUNTAIN LION
Puma, cougar, panther
Felis concolor Linnaeus
=*Puma concolor* (Linnaeus)

◼ DESCRIPTION
HB 966–1959 mm; T 534–814 mm; WT 34–120 kg.
Slender; ears small, rounded; upper parts vary from grey or dark brown to shades of buff, cinnamon or rufous; colour usually more intense along mid-dorsal line from head to tail; shoulders and flanks lighter; under parts dull whitish with buff; forelegs sometimes with faint horizontal stripes; sides of muzzle black; chin and throat white; ears black; nose pad pink bordered with black; tail lighter below, tipped black. Young spotted black on buff. Female smaller than male, has three pairs of mammae.

◼ DISTRIBUTION
North and South America. In mountains from southern Canada and Alaska south to Patagonia.

◼ HABITS AND BEHAVIOUR
Habits: mainly nocturnal and crepuscular; shy and rarely seen; except for tracks, presence undetected; males territorial; dens in rock caves and under fallen trees and in thick vegetation. **Gregariousness:** home range 80–629 km^2; male range/territory overlaps with that of several females; density 0.5–5 individuals/100 km^2; essentially solitary. **Movements:** home range less than 100 km^2 to several 100 km^2; tends to shift down mountains away from heavy snowfall following

Mountain lion

seasonal movement of ungulates. **Habitat:** forest, scrub, variety of habitats from swamps, wooded river valleys to forest and high mountains; arid desert; sometimes occur in intensive agricultural cultivation. **Foods:** skunks, porcupines, capybara, squirrels, rabbits, mice, deer, beaver, hares, muskrats, raccoons, occasionally domestic stock. **Breeding:** all year; gestation 87–96 days; male polygynous; litters 1, 2–4, 6 young (kits) in den mostly in summer; kits weaned 4–5 weeks, but stay with female for up to 2 years, but do not breed until 3 years; lactation 3 months or more; inter-birth interval 1 year; sexually mature, males 3 years, females 2–3 years. **Longevity:** 8–13 years in wild, 18–21 years in captivity. **Status:** extinct or rare over most of North American range; reduced numbers in most areas.

■ HISTORY OF INTRODUCTIONS
The mountain lion is now found in areas colonised by deer that were originally outside its historical range (e.g. Great Basin Desert in the western United States) (Berger and Wehausen 1991). Essentially they were eliminated from most of the eastern United States within 200 years following European colonisation (Wright 1959).

NORTH AMERICA
United States
Mountain lions have been re-introduced into several western American states (Jordan 1991; Hornocker 1992). In the San Andres Mountains, Chihuahua Desert, 13 of 20 mountain lions have been translocated to another area some 483 km away as an experiment. Four became settled, two were killed, two died of hunting injuries and one returned the 483 km to where it came from.

Captive-bred mountain lions released in Florida have not only survived and reproduced (Jordan 1991, 1994), but also appear to be settling down in the release area more readily than translocated wild-caught animals (Beldeu and McCowan 1993). They were released in the everglades in the 1950s and early 1960s (O'Brien *et al.* 1990), but the population now appears to be affected by health problems and is in danger of declining. It consists of 30–50 adult animals (Jordan 1994) which are confined to fragmented habitat in the Everglades National Park and Big Cypress Swamp ecosystem.

Seven wild-caught animals from Texas have been released in the Osceola National Forest in northern Florida on the boundary with Georgia. All were recaptured earlier than planned due to conflicts with humans. The latest attempt involves 10 Texan mountain lions, three captive-bred and seven wild-caught,

released in the same area in February 1993 (Belden and McCown 1993). However, the same problems as above are happening again. Four have been recaptured and others relocated after conflicts with humans (Nowell and Jackson 1996).

■ DAMAGE
In the past, stockmen suffered heavy losses through the predatory attacks of lions on young domestic stock, particularly foals, lambs, kids and even full-grown horses and cattle (Young 1945). More recently however, in North America, mountain lions were only found in the more remote mountainous regions (Forsyth 1985), where they were less likely to come into conflict with domestic stock. In some areas this is now being reversed by decisions to give more protection to mountain lions and their numbers are again increasing.

Many farmers view mountain lions as threats to livestock and poison them and tree them with dogs for shooting. In some places in Chile they may be significant predators of sheep on ranches (Nowell and Jackson 1996).

LYNX
Northern, Spanish, European, or Canadian lynx, pardel
Felis lynx Linnaeus
=*Lynx lynx* and *Lynx canadensis*

Some authorities treat F. lynx and F. canadensis as two separate species. F. canadensis Kerr is remarkably similar in appearance, although generally smaller, and is reported to show marked adaptive differences for prey capture. Here they have been combined as F. lynx, but it is noted in the text to which subspecies is being referred.

■ DESCRIPTION
HB 800–1300 mm; T 40–80 mm; SH 600–750 mm; WT 8–38 kg.
Fur long and silky; coat fawn, rufous or yellowish orange or bluish with variable number of dark spots (three main coat patterns: predominantly spotted, predominantly striped or unpatterned); tail short, latter third tipped black; feet pads furred; dorsal margins ears pointed, tufted, black; under parts pale buffy brown or whitish; prominent ruff around face; black stripes on forehead and around ruff; feet hairy. Female slightly smaller than male.

■ DISTRIBUTION
Eurasia and North America: Scandinavia to eastern Siberia, Iberia, Balkans and Carpathians (but formerly more widespread in Europe), Caucasus, Asia Minor, Kopet Dag and all main ranges of central Asia

from the Altai to Kashmir and east to Manchuria, Kansu, Tsaidam, and south-east Tibet, on Sakhalin Island and perhaps Sardinia. In North America from Alaska to the northern United States.

■ HABITS AND BEHAVIOUR

Habits: nocturnal and diurnal; mainly terrestrial; lives in dens in hollow tree or under rocks; probably territorial; shy; female and kittens hunt together; dens in crevices under bushes, in hollow logs; carnivorous. **Gregariousness:** solitary; home rages overlap; social relationships not well known; density 0.3–37.2/100 km^2; home range 4–300 km^2 (male) female smaller. **Movements:** moves considerable distances; nightly foraging 5–19 km; wanders during food shortages

otherwise sedentary(?). **Habitat:** forest, dense undergrowth, open forest, rocky areas, tundra, wooded areas, scrub woodland, barren rocky areas above tree line, rocky hills in desert regions, unusual in cultivated areas. **Foods:** hares, rabbits, ground birds, chamois, deer, young domestic stock, carrion, fish, rodents, hedgehogs, and carrion. **Breeding:** female monoestrous; mates early spring (February–April, births May–June in Europe), gestation 60–76 days; young born March–April; 1 litter/year; litter size 1, 2–4, 8; born blind, furred, in den; lactation 6 months; independent at 10 months; mature 13–30 months; female raises young alone; young stay with female until ready to mate following year. **Longevity:** 15–26 years 9 months captivity, less in wild. **Status:** formerly more widespread in western Europe and North America; some races in danger of extinction.

■ HISTORY OF INTRODUCTIONS

EURASIA

Efforts were made to re-introduce the lynx in some European countries between 1970 and 1976 (Stehlik 1979). Recent introductions have occurred in several European countries, including parts of Germany, Austria, Switzerland, Italy and Yugoslavia (Breitenmoser and Breitenmoser-Wursten 1990; Stahl and Artois 1991; Walker 1992). Some management problems are reported to have followed re-introductions into Austria (Gossow and Honsig-Eslenburg 1986).

The spotted and striped types predominate in present re-introductions of European lynx populations (orig-

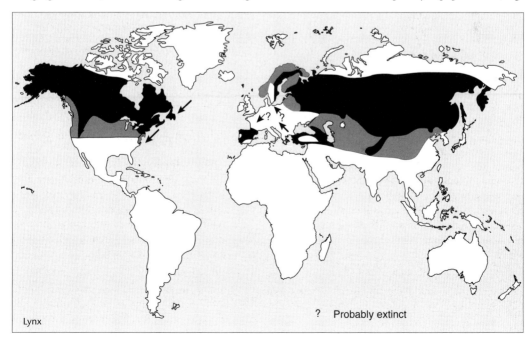

? Probably extinct

Lynx

inating mainly from the Carpathian Mountain regions) (Ragni *et al.* 1992).

In the Russian Federation there has been a major population increase and range expansion (including the colonisation of the entire Kamchatka Peninsula) which took place in the 1930s to 1940s (Hepner and Sludskii 1972) during a period of social unrest when not much hunting took place.

France
Three captive-bred European lynx from Czechoslovakia were re-introduced to the Vosges Mountains, Alsace, from where they have been absent for over 100 years. In the following September another pair from Riber Zoo in England were released in the same region (Lever 1985).

Germany
Isolated attempts have been made to re-establish the European lynx in Germany and they occasionally wander to this country from their range further east.

A single animal released in 1938 remained in the heaths near Rominter for many years, but eventually disappeared. Later more were found there, but it is thought they may have wandered there naturally from their present range further east in Europe. Three animals escaped from an animal park at Hellbabrun near Munchen and evidently became established for a period, as they were still known in the environs of that city in 1950 (Niethammer 1963).

Slovenia (Yugoslavia)
The European lynx has been re-introduced and is rapidly increasing in numbers and considerably expanding its range (Cop 1992).

Sweden
European lynx have been successfully re-introduced into Sweden.

Switzerland
Between 1970 and 1980, 18 *F. l. lynx* from the Russian Federation were imported and released in several parts of Switzerland. They became established and the present population is thought to be expanding in central and western Switzerland. In 1981 it was estimated that the population was 40–70 adults. In central Switzerland their range extends from Altdorf west to about Leysin and in western Switzerland from La Choux-de-Fonds and Neuchatel south-west almost to Geneva. A small population is also established in the region of Zernez in eastern Switzerland (Stehlik 1979; M. Dollinger *pers. comm.* 1982). While the introduction has been considered a success the population has stopped expanding and is threatened by an unbalanced sex ratio (lack of males) (Breitenmoser *et al.* 1994).

NORTH AMERICA
Canada
Two lynx (*F. l. canadensis*) were released on June Island off Newfoundland in March 1975, but had disappeared by April 1979 (Mercer *et al.* 1981).

United States
Restoration of the lynx population is currently under way in New York's Adirondack Park. Eighteen lynx were released in the High Peaks during the winter of 1988–89 with plans to release 30 or more in 1989–90. The lynx came from near Whitehorse, Yukon Territory, Canada (Brocke *et al.* 1990). From 1988 to 1990, 83 Canadian lynx (48 males, 35 females) wild-caught in Yukon were released in the Adirondack Mountains. Twenty-three had died by 1992, 12 hit by cars, five were shot, and six died of miscellaneous causes. Three raided livestock pens and some migrated from the release site. There was no direct evidence of breeding, but there were some unverified sightings of kittens (Brocke and Gustafson 1992; Nowell and Jackson 1996).

◼ DAMAGE
In Switzerland the effects of the re-introduction of lynx are thought to be negligible by hunters, but some foresters think that they may adversely affect roe deer populations (Dollinger *pers. comm.* 1982).

Problems of predation are most severe in western Europe where lynx have been re-introduced. After native wild ungulates re-adapted to the presence of predators, livestock killing increased, but later declined as the lynx dispersed and became less concentrated. Overall losses are generally low and are compensated for by government or environmental groups (Nowell and Jackson 1996). Switzerland pays about US$7000 annual compensation.

BLACK-FOOTED CAT
Felis nigripes Burchell

◼ DESCRIPTION
HB 337–500 mm; T 150–200 mm; SH c. 250 mm; WT 0.8–2.75 kg.
Body short; legs long; ears rounded; coat dark yellowish to sandy; darker on back, paler on belly; flanks, throat, chest and belly with dark brown to black spots in rows; under parts pale; cheeks with two streaks across; forelegs with two transverse bars; up to five bars on haunches; bottoms of feet black.

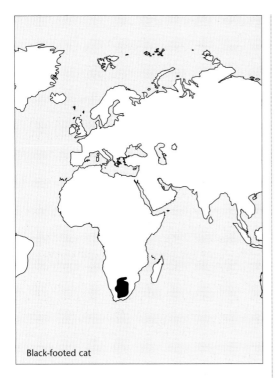

Black-footed cat

■ DISTRIBUTION

Africa. South-west Botswana, western Namibia, and in Orange Free State and Cape Province, South Africa.

■ HABITS AND BEHAVIOUR

Habits: shelters in old termite mounds and abandoned burrows; mainly nocturnal; secretive. **Gregariousness:** usually single; highly unsocial. **Movements:** home range *c.* 12–13 km^2. **Habitat:** arid areas in steppe and savannah with stands of tall grass or scrub. **Foods:** rodents and other small mammals (mice, gerbils), birds and reptiles, spiders and insects (termites, grasshoppers). **Breeding:** November–February; gestation 59–68 days; litter size 1–3; eyes open 6–8 days; leave nest 28–29 days; sexually mature 12–21 months. **Longevity:** up to 13 years. **Status:** uncommon to common.

■ HISTORY OF INTRODUCTION

AFRICA

South Africa

Black-footed cats are reported to have been re-introduced into Mountain Zebra National Park (Penzhorn 1971).

■ DAMAGE

In South Africa black-footed cats frequently raid poultry yards (Hey 1964).

BOBCAT

Felis rufus **Schreber**

=*Lynx rufus* (Schreber)

■ DESCRIPTION

HB 650–1100 mm; T 100–200 mm; WT males 6.4–18.3 kg, females 4.1–15.3 kg.

Coat light grey, buff or brownish above, paler below with dark spotting; ear tufts short, black; tail tip blackish above whitish below, with sub-terminal black bars; prominent streaked ruff on each cheek extends down the side to below lower jaw; under parts white with black spots; limbs tawny with black horizontal bars on them.

■ DISTRIBUTION

North America. Southern Canada south through the United States to southern Mexico. There has been some northward expansion of range in the past century (Banfield 1974).

■ HABITS AND BEHAVIOUR

Habits: dens in cavities under rocks or brush; prefers rough terrain with rock, caves and ledges; territorial; active day and night, crepuscular peak. **Gregariousness:** solitary; density 1/25 km^2. **Movements:** nightly 3–11 km; home range 0.6–326 km^2. **Habitat:** forest, mountain areas, semi-deserts, brushland, rocky hillsides, coastal swamps, agriculture land with woody cover. **Foods:** rabbits, hares, cotton rats, wood rats, kangaroo rats, beaver, peccaries, birds, occasionally deer, reptiles, insects, snails

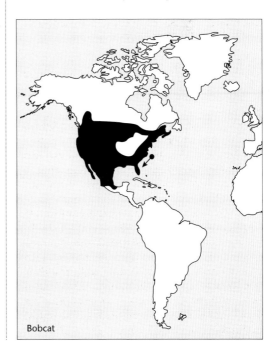

Bobcat

and carrion. **Breeding:** late winter (February–March), births in April–May, but all year in southern latitudes; gestation 50–70 days; male polygynous, females seasonally polyoestrous; litter size 2–8; inter-birth interval 1 year, 2 litters/year in southern range; eyes open 9–10 days; lactation lasts 2 months; females mature at 9–12 months, males in second year. **Longevity:** up to 33 years captive, 12–14 years wild. **Status:** common and abundant, but declined in southern parts of range due to exploitation and agriculture.

■ **HISTORY OF INTRODUCTIONS**
NORTH AMERICA
United States
Efforts are being made to re-establish bobcats in areas where they have been extirpated (e.g. New Jersey) (Burton and Pearson 1987).

Bobcats were last reported on Cumberland Island, Georgia, United States, in 1907 when for unknown reasons they became extinct there. Three bobcats were released on the island in 1972–73, but these had disappeared by 1988. In 1988, 14 wild-trapped adult bobcats (three males and 11 females) were released. Four were released on 13 October, six on 3 November and four on 28 November. One was found dead in January 1989 and one female swam back to the mainland in February 1989. All the others survived and remain on the island. Kittens were found in dens in April 1989.

In Georgia 32 bobcats (15 male, 17 female) wild-caught on the coastal plain were re-introduced in 1988–89 to Cumberland Island. All were radio collared. Evidence of breeding was found in the first year and the population doubled in the following year (Baker 1991; Brocke and Gustafson 1992; Diefenbach 1992). Eighteen bobcats were released in 1989 (12 males and six females); six on 5 October, six on 25 October, and six on 6 December. One male swam off and drowned, all the others have survived to date bringing total on the island to 29 (Warren *et al.* 1990). The releases were made to help control the deer population on the island.

■ **DAMAGE**
Occasionally bobcats prey on domestic animals, especially poultry, turkey and on sheep ranches (Banfield 1974; Walker 1992), but are not generally persecuted as a pest species in North America. They may occasionally raid poultry, but are uncommon, although bounties were paid in earlier years. In central Mexico they are reported to be a major predator of sheep (Govt. US 1983) and are persecuted by ranchers (Gonzales and Leal 1984).

EUROPEAN WILDCAT
Forest cat, Asiatic wildcat, wildcat
***Felis silvestris* Schreber**
=*sylvestris*, =*F. lybica*

It is often suggested that this species should be separated into African wildcat (F. lybica) and European wildcat (F. silvestris). The Asiatic form tends to be smaller.

■ **DESCRIPTION**
HB 470–800 mm; T 235–370 mm; SH 350–400 mm; WT 2.5–8.0 kg (up to 15 kg known).
Fur generally long, soft, thick, yellowish-grey or reddish and with seven to 11 brown or black vertical stripes; tail about half the length of body with three to five distinct rings and terminates in blunt, black tip. Differs from domestic cat by stouter, longer body; limbs longer; squarish head; has high arch to nasal bones, longer carnassial teeth and long gut; four to seven transverse bands on forelegs; four longitudinal stripes extend from nape over forehead; dorsal stripe from shoulder to base of tail. Females have eight mammae.

■ **DISTRIBUTION**
Eurasia–Africa. Central, south and south-east Europe and Africa and east into Asia as far as western China and India.

■ **HABITS AND BEHAVIOUR**
Habits: mainly crepuscular and nocturnal, peaks at dawn and dusk; territorial; makes den (lair) in rocks,

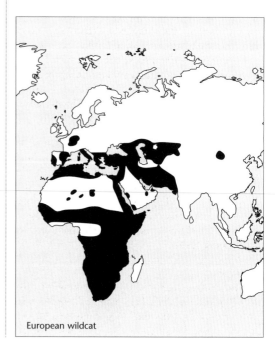

European wildcat

hollow tree or burrow during day; often lives in trees for long periods. **Habitat:** woodland, rocky mountains, forest, scrub, open rocky habitats, scrub desert, wet swampy areas, grasslands; often near cultivated areas and settlement, and environs of towns. **Gregariousness:** solitary or pairs; density 1/3.3 km^2, 1/0.7–10 km^2. **Movements:** young enter roaming stage at 5 months while establishing home range; home range 50–175 ha. **Foods:** small mammals (mice, shrews, hares, rabbits, gerbils, voles), birds, fish, insects (beetles), amphibians (frogs), reptiles (lizards, snakes), grass and carrion. **Breeding:** mates January–March; births April to mid-May–August. Second litter August; hybridise freely with domestic cats; gestation 58–69 days; males sexually active December–June; females polyoestrous; oestrous cycle 5–8 days; 2, 4–8 kittens; one litter/year (truly wild form), 2 litters/year (mixed with domestics); young born blind, furred, pads pink (dark at 3 months); eyes open 10–13 days, blue, but become gold at 5 months; weaned at 30 days, but lactation may last 2.5–3.5 months?; emerge from den 4–5 weeks; hunt at 12 weeks; separate from female 5 months; sexual maturity females at 9–10 months to 1 year, males 12–36 months. **Longevity:** 13–15 years captive. **Status:** common in some areas, uncommon others; widespread hybridisation with domestic cats is leading to increased rarity.

■ HISTORY OF INTRODUCTIONS
EUROPE
Wildcats have been re-introduced into several European countries (Breitenmoser and Breitenmoser-Wursten 1990; Stahl and Artois 1991). They were eradicated from much of Europe between the late 1700s and early 1900s, but recolonised many countries between 1920 and 1940 including, Belgium, Czech Republic, Slovakia, France, Germany, Switzerland and the United Kingdom. This expansion has now ceased or continues at a very low level.

Wildcats occur on Crete, Corsica, Sardinia, and the Balearic Islands, as well as numerous other small Mediterranean islands, and some authorities consider these populations to be discrete subspecies related to the *lybica* group. On the other hand, some (Vigne 1992) consider them to be feral forms of domestic cats introduced centuries before by humans (Nowell and Jackson 1996).

Germany
In Bavaria, Germany, 237 wild-caught and captive-bred European wildcats (130 male, 107 female) were released between 1984 and 1993 in at least three sites in state-owned forest. Those released were monitored by radio tracking and initially suffered a high degree of mortality due to high numbers of road kills and the survival rate was around 30 per cent. However, there was evidence of establishment and reproduction (Buttner and Worel 1990; Nowell and Jackson 1996).

A review of several re-introduction attempts in Europe concluded that this long-term project by the Bavarian Conservancy Association was probably the best; however, opinion was that more effort should be put on conservation of small isolated pockets of population already in existence (Stahl 1993).

Switzerland
The European wildcat may or may not have been introduced (or re-introduced) in the Swiss Alps (Herren 1964; Meylan 1966; Hainard 1969; Schauenberg 1970).

A number of wildcats caught by trapping in Dijon, France, were released in a reserve on the slopes south of Brienzergrat, north of Lake Brienz, Berne, in 1962. Two more were released in the same region in October 1963. It is not known how successful this introduction became, but any success is doubted (Schauenberg 1970).

■ DAMAGE
Wildcats were formerly persecuted, but are still regarded as pests by hunters and gamekeepers (Corbet and Harris 1991). The main complaint against them was damage to poultry, lambs and kids (Burton 1962); however, now that they are less numerous probably little damage is caused by them.

JAGUARUNDI
Jaguarondi
Felis yagouaroundi Geoffroy
=*Herpailurus yaguarondi* Lacepede

■ DESCRIPTION
HB 550–770 mm; T 320–690 mm; SH 305 mm; WT 2–9.0 kg.
Body long, slender, elongated, weasel-like; legs short; ears small; tail long; fur blackish to brownish grey or chestnut; lacks spots; three colour forms – black, dark brownish grey and reddish chestnut (darker forms most common in forest).

■ DISTRIBUTION
North and South America. From Arizona and Rio Grande, Texas, south through Central America, mainly eastern South America to northern Argentina.

Jaguarundi

■ HABITS AND BEHAVIOUR

Habits: mainly diurnal, somewhat nocturnal and crepuscular; terrestrial; territorial; often near water. **Gregariousness:** pairs, female and young, or solitary. **Movements:** sedentary?; 13–20/100 km². **Habitat:** deciduous and evergreen forest and scrubs; lowland forest, secondary growth, savannah woodland, thorn forest, swampy grassland; riparian and old fields. **Foods:** small mammals (rodents, rabbits) and birds, reptiles, frogs, insects, arthropods, and chalcid fish. **Breeding:** gestation 63–78 days; litters 1–4 young; 1 litter/year; mature at 2–3 years. **Longevity:** to 15 years. **Status:** declined in many areas; rare in south-western United States.

■ HISTORY OF INTRODUCTIONS

NORTH AMERICA

United States

Released in Chiefland and Hillsborough River State Park, Florida, about 1942 and currently established in parts of central Florida where they cause damage to poultry (de Vos *et al.* 1956; Lever 1985; Walker 1992).

■ DAMAGE

As predators of domestic poultry jaguarundis are notorious (Bisbal 1986; McCarthy 1992). Where introduced in Florida they cause damage to poultry.

LION

African lion, Asiatic lion
Panthera leo (Linnaeus)

■ DESCRIPTION

HB 1400–2500 mm; T 700–1050 mm; SH 1070–1230 mm; WT males 150–272 kg, females 103–182 kg.
Coat colour varies from buff to yellowish or reddish brown; under parts pale; tail tuft black; mane long and hairy, light yellow to black. Female weighs less than male.

■ DISTRIBUTION

Africa–Asia. Formerly occurred widely in Europe, over much of Asia and all of Africa. Now only in Africa south of the Equator and in the Gir Forest of India.

■ HABITS AND BEHAVIOUR

Habits: mainly nocturnal and crepuscular, occasionally diurnal where persecuted; territorial. **Gregariousness:** density 0.17–30/100 km²; in prides 5–40 (females, young, and 2–3 males); groups 2–4, 37; bachelor groups or solitary, single or multiple males hold tenure over one or more prides. **Movements:** nightly moves 0.5–11.2 km; some animals nomadic following migration of ungulates; overlapping home range 20–400 km², nomadics to 4000 km². **Foods:** large mammals; sable, kob, roans, impala, hartebeast, waterbuck, buffalo, zebra, wildebeest, springbok, and

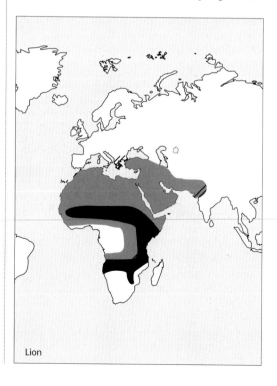

Lion

small rodents. **Habitat:** grassy plains, savanna, open woodland, scrub country. **Breeding:** throughout year, mainly March–July; females polyoestrous; oestrus lasts 4 days; gestation 100–119 days; litter size 1, 2–3, 6; inter-birth interval 11–25 months; eyes open at birth to 2 weeks; weaned 6–7 months; sexual maturity 2 years females, 2.5 years males. **Longevity:** 12–18 years wild, 13–30 years captive. **Status:** range considerably reduced and fragmented; increasingly rare outside protected areas.

■ HISTORY OF INTRODUCTIONS
AFRICA
Botswana
Concern has been expressed over later releases of lions from Kenya into Botswana (Anon. 1991, 1993) as the animals were not of the local race and some were genetic 'cocktails' due to cross-breeding.

Kenya
The release of lions by Joy and George Adamson (Adamson 1960, 1969, 1986) and probably others have been marred by the deaths of humans, probably due to the cats' familiarity with people and the lack of caution. The Adamson releases also allowed 'foreign' genes into resident populations.

South Africa
Lions were re-established in the Hluhluwe-Umfolozi Game Reserve, South Africa, in the 1960s (Macdonald and Frame 1988). A nomadic male took up residence in the Reserve in 1958 and seven years later two adult females and two young cubs were released there. In 1963, three males moved out onto nearby ranches where they killed livestock valued at £500 before they were shot (Anderson 1981). As the population increased so did the dispersion of lions and numbers of livestock killed, so much so that fencing and population control were instigated. The population today is thus descended from a very small gene pool: the two females came from the same pride and it is presumed that they were related, while the original cubs were likely to have been killed by the male (Anderson 1981, 1992). In 1993 male lions from the reserve were found to have low sperm quality (Nowell and Jackson 1996). Translocated from Kruger National Park more than 100 km away the lions are the founders of a population that is now estimated to be 800 animals (Anderson 1981, 1992).

Lions have been re-established in Kruger National Park (Macdonald and Frame 1988) and re-introduced to various reserves (Van Aarde and Skinner 1986; Anderson 1992). In 1992, thirteen lions left the security of enclosures and were released into the Rhinda Resource Reserve, a privately owned site in South Africa (Hunter 1998).

Zimbabwe
Partly successful lion translocations occurred in north-eastern Zimbabwe in about 1977 when six from the Rukomedlin Tsetse Research Station were translocated to the Cheware Controlled Hunting Area (distance 45 km) and two from farming land in the Matetsi area were translocated to Kazuma National Park (distance 27 km). None of the six returned to their original area initially, but two were back in five months to the original area (Vander Meulen 1977).

ASIA
India
Lions were unsuccessfully re-introduced to Sitamata in northern India in the 1960s (Nowell and Jackson 1996).

■ DAMAGE
Generally lions are considered to be serious problems to human settlements and cattle culture and are likely to prey on stock if wild ungulate prey is scarce (Nowell and Jackson 1996).

Two to three years after their introduction into Umfolozi they were escaping from the park into surrounding farmland where they were killing domestic livestock. Attempts were made to prevent such escapes by culling and these so far have been successful in maintaining a small population there.

LEOPARD
Panther
Panthera pardus (Linnaeus)

■ DESCRIPTION
HB 910–1910 mm; T 580–1400 mm; SH 450–780 mm; WT 26–91 kg.
Greyish on pale yellow to rich buff or chestnut with black rosette-shaped spots; melanistic (black panthers) form common, especially in moist forest, dense forests; head, lower limbs, belly spotted with solid black. Male larger than female.

■ DISTRIBUTION
Formerly all over Africa, throughout Near and Middle East and southern Asia south to Sri Lanka, Malay Peninsula, Java, Kangean Island. Now extinct in most of North Africa, the Middle East, Near East and reduced to scattered populations over most of Asia and West Africa; fairly widespread in Central and East Africa.

■ HABITS AND BEHAVIOUR
Habits: normally nocturnal, sometimes diurnal or both; territorial. **Movements:** 25–75 km/night; home range 8–63 km^2. **Gregariousness:** solitary; density

Leopard

1/20–30 km²; **Habitat:** lowland forest, mountains, brush, desert, open grassland. **Foods:** mammals (antelope, deer, eland, monkeys, pigs, hares, rodents), livestock, carrion, birds and fish. **Breeeding:** probably throughout year; inter-birth interval 1–2 years; oestrous cycle 46 days; heat 6–7 days; gestation 90–112 days; litter size 1–3, 6; young born in rock crevice or hole in tree; eyes open 10 days; weaned 3 months; independent 13–24 months; sexual maturity no information, but breed at 30–36 months females, 2–3 years males. **Longevity:** generally 12–15 years, but to 23 years (captive). **Status:** common, range reduced, several subspecies very rare.

■ HISTORY OF INTRODUCTIONS
AFRICA
South Africa
Recently leopards have been re-introduced in some Transvaal and Natal reserves (Estes 1993). One translocated animal in South Africa travelled more than 540 km back to its original home range (Jewell 1982).

Kenya
Seven problem leopards were released in the 8000-km² Tsavo National Park, but none survived to occupy a territory (Panwar and Rodgers 1986).

ASIA
India
Suggestions to re-introduce tigers and leopards in India have been considered, but so far have been rejected on the grounds of area size, difficulty in establishing a viable population, and inadequate prey base (Panwar and Rodgers 1986). However, two leopards were introduced or re-introduced in India's Dudhwa National Park by Arjan Singh (Singh 1981, 1984).

Andaman Islands
Two female leopards were introduced to the Andaman Islands in 1952–53 in an attempt to control the burgeoning population of axis deer (*Cervus axis*) that had been established on the islands since World War 1 (Lever 1985). The leopards died out and the deer remain on the islands.

INDONESIA
Kangean Island
Leopards are found only on Java and the small island of Kangean where they are suspected (Van Helvoort *et al.* 1985) of being introduced. Kangean is situated further from Java than Bali, which has no leopards (Novell and Jackson 1996).

■ DAMAGE
Leopards have a widespread reputations as killers of domestic stock and will take sheep, goats and young camels. Local peoples sometimes go to great lengths to hunt down and kill them when they are reported in the vicinity (Nowell and Jackson 1996).

TIGER
Panthera tigris (Linnaeus)
=Felis tigris

■ DESCRIPTION
HB 1800–2800 mm; T 910 mm; WT 75–258 kg.
Fawn but vary from reddish orange to yellowish brown, rarely whitish; under parts white; black stripes over upper parts and sides, stripe pattern of individuals differs.

■ DISTRIBUTION
Asia. From Siberia, India and to Java and Bali.

■ HABITS AND BEHAVIOUR
Habits: hunts mainly during night; climbs occasionally; lies up in heat of day; readily enters water. **Gregariousness:** usually solitary; females with cubs; sexes associate to breed; occasionally in groups. **Movements:** sedentary? **Habitat:** dense cover, especially reedy swamps; mangrove swamps; tall grass jungle; forest to scrub oak and birch woodlands. **Foods:** small game to large animals; buffalo, deer, pigs, monkeys, birds, reptiles, fish and carrion. **Breeding:** mates year round; gestation 103–113 days; males polygynous; litters 2–3 to 6; inter-birth interval

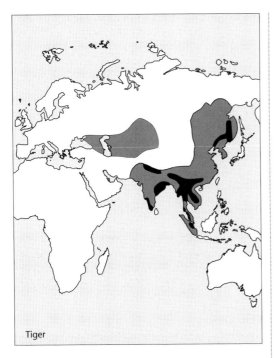

Tiger

20–24 months; young blind at birth; eyes open at 14 days; weaned at 6 weeks; mature at 3 years, reproduce at 3.4–6.8 years. **Longevity:** 15–25 years. **Status:** common, but some subspecies (e.g. Bali, Javan and Caspian) extinct and range considerably reduced in historic times.

■ HISTORY OF INTRODUCTIONS
ASIA
India
There have been several attempts to translocate tigers in India, about which there are few details. A tiger responsible for killing a women at the edge of the Indian Sundarbans mangrove delta was released in the interior of the Sundarbans Tiger Reserve, but was soon killed by a large tiger (Seidensticker *et al.* 1976). Another animal which had taken livestock was released in the Sundarbans, but disappeared immediately, and its fate is unknown (Ghosh 1988).

■ DAMAGE
Although lions and leopards also kill humans, tigers have the greatest reputation as man-eaters, especially in India. Generally man-eating is the result of the tiger's incapacity through injury or age to catch normal prey. However, in India most deaths these days appear to be the result of people illegally entering tiger reserves (Nowell and Jackson 1996).

Large numbers of tigers were formerly killed in Russia and China where they were officially considered pests, and bounties were paid for their destruction.

CHEETAH
Hunting leopard
Acinonyx jubatus (Schreber)

■ DESCRIPTION
TL 1120–1500 mm; T 600–940 mm; SH 700–900 mm; WT 35–72 kg.
Body long and muscular; coat short, yellowish or buff, heavily spotted with small close black spots; ears short, rounded; under belly white; shoulders with erectile crest of hair; tear lines black, extending from inner corner of each eye to the outer corner of mouth; legs long and slender; males usually larger than females.

■ DISTRIBUTION
Southern Asia and Africa. Baluchistan through Iran and Turkestan to north-eastern Arabia, and Africa south of the Sahara (Sudan, East Africa, Senegal, Zimbabwe, Botswana, Transvaal). Formerly throughout Africa, the Near East, Middle East to the Russian Federation and north-west India. Extinct in Arabia by 1950.

■ HABITS AND BEHAVIOUR
Habits: fastest land mammal; mostly diurnal; males territorial, females non-territorial. **Gregariousness:** females solitary, or small groups (females with cubs); pairs or family parties; density 1/5–250 km²; non-territorial males nomadic; male litter mates in coalitions; other groups 14–19 reported. **Movements:** home range 50–130 km² or as large as 800–1500 km² depending on prey; sedentary; daily 3.7–7.1 km. **Habitat:** savannah woodland, open grasslands and scrubs, semi-desert. **Foods:** small to medium mammals (gazelles, impalas, calves of ungulates, kudu, warthogs, hares), birds (guinea fowl, francolins,

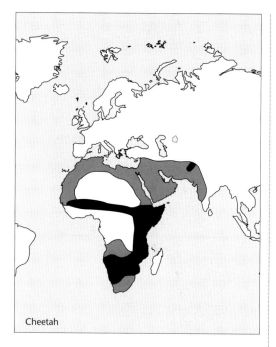

Cheetah

bustards, young ostriches). **Breeding:** all year; gestation 90–98 days; oestrus to 15 days; litter size 1–8 usually 3–5; inter-birth interval 15–19 months; born blind; open eyes 4–11 days; weaned 3–6 months; follow female at 6 weeks and stay with her up to 2 years; sexual maturity 12–36 months. **Longevity:** 10–14 years wild, 15–19 years captive. **Status:** rare, range reduced and much fragmented.

■ HISTORY OF INTRODUCTIONS
AFRICA
South Africa and Namibia

An endangered species in Africa, the cheetah's chances of survival may have been enhanced by the results of relocation experiments in various reserves (Van Aarde and Skinner 1986; Anderson 1992). Between 1975 and 1976, five males and three females were relocated in the Suikabostrand Nature Reserve. Apart from the species survival they were also released in the hope of controlling large numbers of blesbok (*Damaliscus dorcas*) and springbok (*Antidorcas marsupialis*) in the reserve.

The cheetahs established well and within two years there were 24 present in the reserve and they have appeared to control somewhat the large numbers of 'antelope'. In 1979 the cheetah population had reached 29–31. Since this time a further 14 cheetahs have been released into native reserves in Cape Province, Natal, and in eastern Transvaal between 1978 and 1979 (Pettifer 1981).

In Natal cheetahs were exterminated by the 1930s. They were re-established in the Hluhluwe-Umfolozi Game Reserve in Natal from Namibia in the late 1960s (Macdonald and Frame 1988). Inadvertently, a mite causing sarcoptic mange was introduced with them and the mange has been linked with the decline of black-backed jackals in the reserve area over the last two decades (Keep 1970; Macdonald and Frame 1988). Re-introductions occurred in Hluhluwe, Umfolozi and Mkuze Game Reserves in 1965 from Namibia and in 1975 to the eastern shores of Lake St. Lucia (Smithers 1983).

To increase population size 30 cheetahs were released into South Africa's Kruger and Kalahari Gemsbok National Parks and into Namibia's Etosha National Park in the 1970s (du Preez 1970; Anderson 1992). Few were subsequently re-sighted although they were marked for identification. A small number were still present in Etosha National Park in 1994, where they were being studied using radiotelemetry. The effects, both intermediate and long term, of these introductions are unknown.

Between 1981 and 1984, six cheetahs were re-introduced into Pilanesberg National Park, Bophuthatswana. Within a year there were 17 and 10 were subsequently removed. The re-introduction was successful but the impact of their predation became unacceptably high and their numbers have to be controlled (Anderson 1986).

Botswana

In 1992, 17 cheetahs were released from enclosures at the Rhinda Resource Reserve, a privately owned site in South Africa where they became established (Hunter 1998).

■ DAMAGE

Cheetahs are hunted as a predator of domestic animals (Burton and Pearson 1987), in particular preying on young camels, sheep, goats and other livestock, especially in Niger and Namibia (Nowell and Jackson 1996). Cheetahs released in several South African reserves have had to be removed because of their high levels of predation on small populations of antelopes in fenced reserves which were not exposed to large predators (Pettifer 1981; Van Dyk 1991; Anderson 1992). Captive-raised and released cheetahs repeatedly invaded chicken houses in South Africa even with people present on the game farm throwing stones at them (Van Dyk 1991).

Family: Elephantidae
Elephants

ASIAN ELEPHANT
Asiatic or Indian elephant
Elaphus maximus Linnaeus

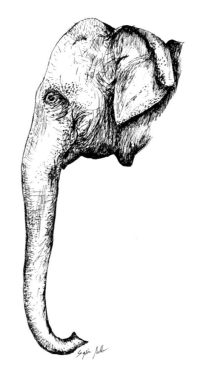

■ DESCRIPTION
HB 5500–6500 mm; T 1200–1500 mm; SH 2500–3000 mm; WT 2700–5400 kg.

Large size with large tusks and elongated nose (trunk); tusk length 3000 mm, weight 18–44 kg; colour slate grey (albinos rare); skin loose and almost naked; trunk smooth with one-lip (single finger); feet flat and rounded; hind foot with four nails, front feet with five nails. Has smaller ears than African elephant. Female smaller than male; hair covering scant; tuft on tip of tail; skin dark grey or brown, often mottled; tusks rudimentary; one pair pectoral mammae.

■ DISTRIBUTION
Southern and South-east Asia. South of the Himalayas, India to Thailand and Cambodia, Malay Peninsula, Sri Lanka, Sumatra and north-east Borneo. Formerly over southern China, now confined to southern border of Yunnan Province. Originally from Syria and Iraq east across southern Asia. Range now fragmented (see map).

■ HABITS AND BEHAVIOUR
Habits: most active at dawn and dusk; territorial. **Gregariousness:** herds (probably large family groups), old bulls solitary: couples, or small herds of females to 30–40 and all male groups to 7; density 0.12–1. 0/km². **Movements:** seasonal movements 30–40 km; home range 10–17 km². **Habitat:** forest and jungle to grassy plains. **Foods:** vegetable material; palm and other tender shoots, vines, grass, foliage, leaves, bark, roots, fruits, cultivated crops of bananas, paddy and sugar cane. **Breeding:** gestation 607–641 days; young 1, occasionally 2; oestrous cycle 22 days; oestrus 4 days; heat period called musth; young weigh 50–150 kg at birth; weaned at 18–24 months; sexual maturity at 8–12 years, but don't give birth until about 15–16 years. **Longevity:** 69–77 years (captive).

Status: considerably reduced in numbers and range, endangered in some areas.

■ HISTORY OF INTRODUCTIONS
ASIA
Elephants may have been native in Java. It is generally believed that they were introduced and became successfully feral in Sabah, Borneo.

INDIAN OCEAN ISLANDS
Andaman Islands
A small introduced population of elephants is present on the Andaman Islands (Walker 1992).

SOUTH-EAST ASIAN ISLANDS
Borneo (Sabah, Malaysia)
Elephants are probably not a native species in Borneo, but are believed to have been taken there by humans and reverted to a wild state (Carter *et al.* 1945). Populations have inhabited Borneo for at least several hundred years, probably as a result of human introduction (Walker 1992). A considerable herd of feral animals roamed the Kinabatangan area of the north-east of Sabah in the 1970s (Encyc. Brit. 1970–80).

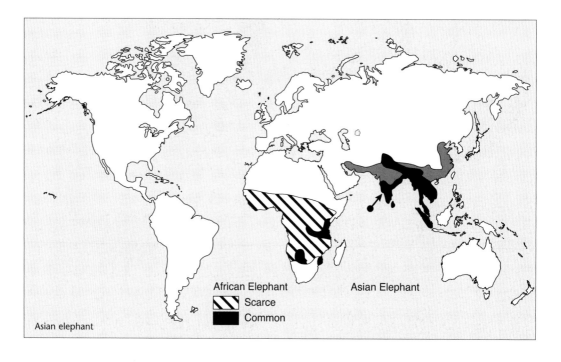

African Elephant

Asian Elephant

Scarce

Common

Asian elephant

Sri Lanka

A program to release elephants back into the wild began in Sri Lanka in about 1995. The Elephant Transit Home project was planning to release three female and two male young elephants back into the jungle in southern Udawalawe National Park in early 1998.

◼ DAMAGE

In southern Nepal elephants cause damage to field crops grown near jungle edges (Chesemore 1970). In Sumatra they have been largely driven from their natural habitat into conflict by the spread of cultivation and human settlement.

Changes in the range of elephants have occurred in China due to human pressure and lack of protection (Yaoting 1981). In 1989 in southern China losses of 250 000 kg of rice were caused as a direct result of elephants. However, elephants have now become scarce and are considered to be vulnerable and deserve some protection. For years around Lampung Province in the south to Aceh Province in the north-west they have pillaged towns and trampled crops but in some instances they can be herded away by helicopter (Anon. 1985).

Elephants often become a pest of agricultural crops in Sri Lanka and India (Walker 1992).

AFRICAN ELEPHANT
Loxodonta africana (Blumenbach)

◼ DESCRIPTION

HB 6000–7500 mm; T 1000–1500 mm; SH 1600–4000 mm; WT 2400–7500 kg.

Larger than Asian elephant; skin greyish brown to slate; sparsely scattered black and bristly hairs; trunk long, with two finger-like extensions at end; tusk length to 3500 mm; tusk weight 61–130 kg; front feet with four nails, hind feet with three; tail, end flattened and with tuft of crooked hairs; large ears. Female smaller and with more slender tusks.

◼ DISTRIBUTION

Africa: Formerly over most of Africa south of the Sahara in forest and savanna zones. Now extinct except for isolated pockets. Extinct north of the Sahara, but believed to survive in north-west Africa until first century AD.

◼ HABITS AND BEHAVIOUR

Habits: diurnal; nocturnal; non-territorial. **Gregariousness:** matriarchal clan society; herds occasionally to 1000; but more often 20–400; batchelor herds to 144; old males solitary; density 0.26–5.00/km². **Movements:** annual migrations of several hundred km; daily movements of 12 km; old bulls tend to be more sedentary; home range 14–3520 km². **Habitat:** forest, open savanna, wet marshes,

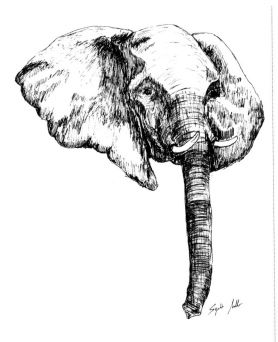

thorn bush, semi-desert scrub. **Foods:** grass, reeds, herbs and shrubs; foliage, leaves, bark, twigs, roots, fruits and occasionally cultivated crops. **Breeding:** breeds at any time of year, but peaks during rains (October–November); males over 25 years enter musth and seek females; female polyoestrous, oestrous cycle 2 months; oestrus 2–6 days; gestation 17–25 months; litter size 1–2 (twins rare); young weaned 6–18 months occasionally to 24; inter-birth interval of 2.5–9 years; sexual maturity 10–11 years and sometimes 20–22 years. **Longevity:** 50–80 years. **Status:** numbers declining; range reduced; catastrophic decline 1970s, 1980s; vulnerable.

■ HISTORY OF INTRODUCTIONS
AFRICA
Elephants have probably been re-introduced in a number of areas in Africa.

Rwanda
In 1975, 26 calves and immature elephants were translocated from southern Rwanda to Akagera National Park where none had been sighted for some decades, and became established there (Mountfort and Mountfort 1979).

South Africa
Between 1981 and 1984, 54 were re-introduced to Pilanesberg National Park, Bophuthatswana. By 1984, 44 were still present and the re-introduction appeared to be successful (Anderson 1986). Fifty-six elephants were released in the Phinda Resource Reserve, a privately owned site in South Africa between 1990 and 1992 (Hunter 1998).

■ DAMAGE
Once occurring all over Africa, elephants had by Roman times been exterminated from most of North Africa. Throughout West Africa populations are fragmented and probably only in Central Africa are there substantial numbers (Burton and Pearson 1987). In some areas, however, they are still abundant enough to be pests of agriculture. In Botswana in the Northern Tuli Game Reserve elephants are coming out of the reserve onto farms to raid crops (du Toit 1998).

In Ghana they are regarded as pests and cause damage to cocoa, oil-palm and food crops such as plantain, cocoyam, and cassava (Burton and Pearson 1987), and in Zimbabwe cause much damage by raiding farm crops (Jarvis and La Grange 1984). In Kenya they are fence breakers and crop raiders in the Laikipia district (Thouless and Sakwa 1995).

Family: Equidae
Horses and asses

Grevy's zebra, *Equus grevyi* Oustalet, was introduced to Tsavo National Park in south-eastern Kenya in 1963 (Haltenorth and Diller 1994), but there appears little recent information on any successful establishment.

MOUNTAIN ZEBRA
Cape mountain zebra
Equus zebra Linnaeus

■ DESCRIPTION
HB 2100–2600 mm; T 400–800 mm; SH 1160–1500 mm; WT 230–386 kg.
White with black stripes, stripes narrower on body than on rump; gridiron pattern on rump; under parts white with mid ventral black stripe on chest and belly. Distinguished from other zebra by the broader black stripes on rump, a dewlap under chin and larger ears.

■ DISTRIBUTION
South Africa and Namibia. Small colony in South Africa, but occur mainly in mountains of south-west Africa including Angola and Namaqualand. Formerly southern Angola to Transvaal and south-west Cape Province.

■ HABITS AND BEHAVIOUR
Habits: mainly diurnal or crepuscular; rank hierarchy; non-territorial. **Gregariousness:** small non-territorial breeding bands 2–13 with overlapping home ranges (1 stallion and 5 mares); foraging bands (to 30); home range 3.1–16.0 km². **Movements:** move to lower levels in winter; movements to 100–120 km between wet and dry season ranges; daily 1–5 km; home range 5–20 km². **Habitat:** slopes and plateaus of mountain areas; barren, rocky uplands and arid plains. **Foods:** grass, leaves, bark, and some browse. **Breeding:** breeds throughout the year, peaks in summer or rainy season; gestation about 362–395 days; 1 young; lactation 2 months; weaned at 10 months(?); young leave natal group 13–37 months; inter-birth interval 1–3 years; sexual maturity 2–3 to 5 years. **Longevity:** 25 years captive. **Status:** range and numbers reduced; a few hundred left; endangered or vulnerable.

■ HISTORY OF INTRODUCTIONS
Introduced unsuccessfully to Kawau Island, New Zealand and possibly in Australia. Re-introduced successfully in South Africa.

AFRICA
South Africa
The Cape mountain zebra (*E. z. zebra*) was rescued from possible extinction in the 1930s by creation of the Mountain Zebra National Park, near Craddock. From 13 head in 1950 there were about 474 in 1985 thanks to the re-introduction of surplus stock to other protected areas (Eloff and Van Rooyen 1987; Estes 1993). Twenty-three were translocated to Karoo National Park in 1978 and 1979 and added to a small colony there, where the population in the 1980s was some 215 head (Smithers 1983).

AUSTRALASIA
Australia
Zebras (sp. unknown) are reported to have been released at Newmarracarra, Western Australia, after 1900, but not to have survived (Allison 1969).

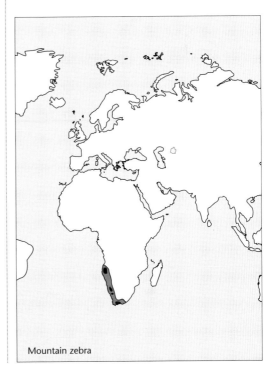

Mountain zebra

Ukraine

A zebra species has been acclimatised in the Askanya-Nova Preserve in the Ukraine (Treus and Lobanov 1963).

■ DAMAGE
No information.

COMMON ZEBRA
Burchell's zebra
***Equus burchelli* (Gray)**

■ DESCRIPTION
HB 1900–2460 mm; T 430–570 mm; SH 1100–1450 mm; WT 175–385 kg.
Base colour of body varies white to yellowish; body stripes are broad, especially on the flanks and extended to midline on belly; some animals have stripes extending down legs; some have brown 'shadow' lines between black on flanks; much geographical variation in pattern and darkness of stripes; (light to dark brown to black); small erect mane on back of neck.

■ DISTRIBUTION
Africa. East and southern Africa from south-eastern Sudan to South Africa and west to Angola.

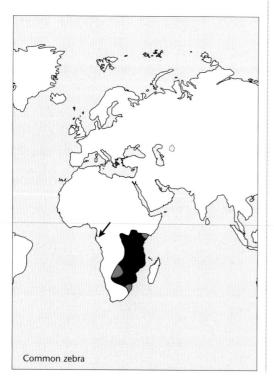
Common zebra

■ HABITS AND BEHAVIOUR
Habits: diurnal and nocturnal, more active in daylight; rank hierarchy. **Gregariousness:** family units (1–6 adult mares and young and 1 stallion) to 15; bachelor stallion groups to 16; density 0. 7–2. 2/km^2 and up to 19. 2/km^2; family units join to form herds. **Movements:** sedentary and migratory; seasonally 100–150 km; daily 13–17 km; overlapping home ranges 19–600 km^2. **Habitat:** steppe and savanna, woodland, open scrub, grassland. **Foods:** grass, leaves, bark, and some browse. **Breeding:** breeds throughout year, peak in rainy season; gestation 360–396 days; female seasonally polyoestrous; oestrus 2–19 days; 1 young, 2 rarely; on feet 15 minutes after birth; weaned 7–11 months; disperse at 1–3 years; females reach puberty 16–22 months; inter-birth interval 1–3 years. **Longevity:** 9 years in wild and 40 years as captive. **Status:** numbers and range considerably reduced.

■ HISTORY OF INTRODUCTIONS
AFRICA
South Africa
The common zebra (*E. burchelli*) is caught throughout south-west Africa and sold to farmers for restocking purposes (Hofmeyr 1975). In the 1970s and 1980s they were re-introduced to private properties and parks on a country-wide basis (Smithers 1983).

They have been re-introduced to the Golden Gate Highlands National Park successfully (Penzhorn 1971; Novellie and Knight 1994), the Nduma Game Reserve, Natal (Smithers 1983), in the Transvaal (Haltenorth and Diller 1994) and to Pilanesberg National Park in Bophuthatswana (Anderson 1986). At Pilanesberg 679 were re-introduced between 1981 and 1984 and there were 800 there in late 1984.

Gabon
Zebras from Kenya have been introduced to the Wonga-Wongue Presidential Reserve (4800 km^2) in north-western Gabon where they met with little initial success in becoming established (Nicoll and Langrand 1986; Blom *et al.* 1990).

Zimbabwe
Unspecified species of zebra were released in McIlwaine and Metapos National parks (five in each) before 1963 by South Rhodesian authorities and they were surviving there in 1963 (Riney 1964).

■ DAMAGE
No information.

DONKEY

Feral burro, wild ass

Equus asinus Linnaeus

Some authorities prefer E. asinus *for the domestic animal and* E. africanus *for the wild animal (see distribution).*

■ DESCRIPTION

HB 800–2500 mm; T 450 mm; SH 1200–1800 mm; WT 250 kg.

Colour predominantly grey, but also black, white and piebald, black to light grey and intermediate shades of grey, brown and dun; muzzle, eye patch, belly, medial aspects of legs usually lighter in colour; sometimes zebra stripes visible on legs and often grey types have a dark dorsal stripe and another across the withers; mane long, thin; tail tufted; hooves long and narrow.

■ DISTRIBUTION

Africa. North-east Africa.

■ HABITS AND BEHAVIOUR

Habits: mainly crepuscular and nocturnal; dung heaps of territorial significance (up to 3 m). **Gregariousness:** unstable groups or harems (15) or herds (60–130) of mixed sex (females, foals and mature jacks) (mainly females and young); older males and very young males in smaller groups (to 11), but some tend to be solitary; males tend to be seasonally territorial; groups maintained throughout year;

density 0.5–4.9/km^2 (United States) to 10/km^2 (Australia). **Movements:** largely sedentary?; home range 1–32 km^2. **Habitat:** desert, semi-desert; hilly and rocky country, grassy flats, dry cactus woodland. **Foods:** grass, sedges, forbs and browse. **Breeding:** breeds mainly in spring and summer; 50 per cent of conceptions occur before wet season in Australia; gestation 330–365 days; seasonally polyoestrous; females sexually mature late in first or second year, but probably do not breed until 3–4 years; 1 young. **Longevity:** males to 20.5 years in wild, females to 15.5 years in wild; 40–47 years in captivity. **Status:** not well known; probably declining in wild and rare. Many herds are crossbred with domestic stock.

■ HISTORY OF INTRODUCTIONS

DOMESTICATION

The ancestor of the domestic ass (*E. asinus*) is derived from the African ass (*E. africanus* (Fitzinger)) and probably the Nubian wild ass (*E. africanus nubianus*) (Zeuner 1963), which is now confined to the Danakil Desert region of Ethiopia and to northern Eritrea and northern Somalia, but formerly occurred in the eastern Sudan and north-west Africa, probably in Upper Egypt or Mesopotamia or Lybia in predynastic times. Wild asses now in Sudan and the eastern Sahara are doubtfully of pure wild stock and are more likely feral animals from domestic stock (Corbet 1978). Possibly 2000 Somali wild asses are left in Danakil, Ethiopia (Woodward 1979). Domestication

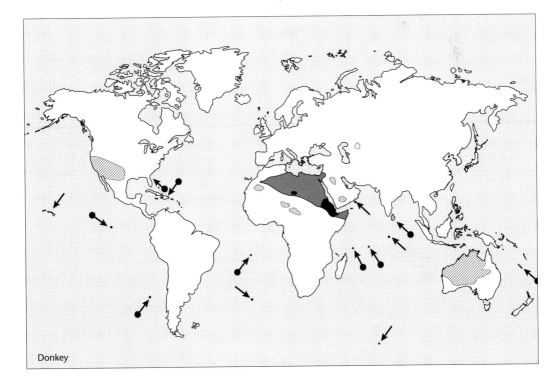

Donkey

occurred about 6000 years ago (Zeuner 1963; Corbet 1978; Walker 1992).

Donkeys have been successfully introduced to Australia, the United States, Galápagos Islands, Tristan da Cunha, Kerguelen, Juan Fernández, West Indies, Chagos Archipelago and Africa.

AFRICA
The so-called wild asses found in Ethiopia today are probably derived from donkeys released by the Ethiopians as they retreated in front of the Italians during the last Abyssinian War. In North Africa the race became extinct in Roman times (Burton and Burton 1969).

Algeria, Sudan, Chad, Namibia
Feral donkeys roam the central Sahara of Algeria. In the eastern Sahara (Chad and Sudan) in mountainous regions free-ranging donkeys are also found (Lever 1985). Within the former range of *E. a. africanus* isolated troops of free-living domestic asses, such as those found on the island of Socotra in the Gulf of Aden, now occur (Haltenorth and Diller 1994). They also inhabit parts of the Namib Desert, Namibia (Krecek *et al.* 1995).

ASIA MINOR
Israel
A small herd of donkeys were recently introduced on a reserve in Israel (Clark 1983). An attempt to establish them in the desert at Hal-Bar initially was encouraging, but the breeding rate of this group became depressed and by 1992 only 18 remained (Duncan 1992).

Saudi Arabia
Feral donkeys roam the desert regions of Saudi Arabia (Lever 1985).

Yemen (Socotra)
Feral animals on Socotra in Indian Ocean may be derived from domestic donkeys introduced by the ancient Egyptians (Lever 1985).

ATLANTIC OCEAN ISLANDS
Ascension Island
A population of 60 donkeys remain on an island where they were introduced by British marines in the nineteenth century (Lever 1985).

AUSTRALASIA
Australia
The first donkeys were imported into Australia in 1866 and used extensively as pack and haulage animals until the early 1900s when they were superceded by motor-driven vehicles. The earliest

reports of large numbers in the wild appear to have been in the 1920s and 1930s. There are probably now 2–5 million feral donkeys in Australia (Strahan 1995).

They occurred in plague numbers in the eastern goldfields of Western Australia in 1959 but also in north-eastern pastoral districts and north-west, and in the Kimberley where have become the greatest menace. Some stations claim more donkeys than cattle (Tomlinson 1959).

Initially used as pack animals and draught animals, the feral descendants are now scattered throughout every pastoral district in the Northern Territory (Letts 1964). They now have only limited use as pet food and a few are occasionally trapped and domesticated (Wilson *et al.* 1992).

Variations in colour patterns in Australia tend to suggest a population made up of a number of breeds including French Poiton ass, Spanish giant ass, Nubian wild ass, Somali wild ass and others (McCool *et al.* 1981). They occasionally occur in small colonies in western regions of Queensland and north of Charters Towers and a few small colonies in the East Coast River system. Herds to 50. Some came from the Northern Territory and some are due to releases of domestic stock (Mitchell *et al.* 1982).

There are an estimated 55000–65000 in the Victoria River District of Northern Territory (Anon. 1987).

INDIAN OCEAN ISLANDS
Chagos Archipelago
It is recorded that donkeys were on Danger Island at the beginning of the twentieth century (Bourne 1971). In 1975 there were 200 feral donkeys on the island of Diego Garcia (Hutson 1975). There are now more than 100 surviving in the uninhabited areas of Diego, with smaller numbers on Boddam Island in Saloman Atoll, on Coin de Mire Island in Pecos Banhos Atoll and elsewhere in the Chagos group (Lever 1985). These animals were abandoned when the human inhabitants were evacuated in the early 1970s.

Desroches
There are now about 70 semi-feral donkeys present on the island, the progeny of animals formerly used in the coconut mills (Stoddart and Poore 1970).

Kerguelen
Donkeys or mules were imported in 1949, but the last one was killed in 1953 (Migot 1956). Although mules were kept on the island for short periods in recent years near the scientific station, they have not become established in the wild (Watson 1975).

Seychelles
One single donkey was present on Bird Island in 1972–73 (Feare 1979).

Sri Lanka
A feral population of donkeys lives on the north-west of the island and is strictly protected by wildlife authorities (Lever 1985).

Tristan da Cunha
Donkeys were introduced to the island before 1867 (Holdgate and Wace 1961) and occurred near the settlement (Munch 1945; Holdgate 1961), but probably were not truly wild. When the residents left the island following a volcanic eruption in 1961 some donkeys were left behind and when they returned in 1963 some 70 were present there (Anon. 1963). Some of these were still present in 1967 (Holdgate 1967), but are apparently absent now.

NORTH AMERICA
United States
Donkeys were introduced to North America by the Spanish in the 1530s, but probably not until the latter half of the nineteenth century did they become widely dispersed, in south-western United States primarily through use as pack animals by prospectors. Following the mining boom and advent of the rail road the need for pack animals declined and many were abandoned or became feral, principally in desert mountains along the Colorado River in Arizona, California and Nevada (McKnight 1958; Weaver 1974; Carothers *et al.* 1976; Seegmiller and Ohmart 1981; McGill 1984).

By the end of the nineteenth century they were established in many isolated regions of the United States. In the early 1920s they were well established in Grand Canyon National Park, Nevada, and from 1924 to 1931 some 1467 were destroyed and the population reduced to 50–75 head. Between 1932 and 1956 an additional 370 were removed and between 1956 and 1958, 771 more were destroyed and 252 removed from the Park. Since 1969 there has been no control because of public sentiment for the donkey (Carothers *et al.* 1976). Domestic animals were brought to the Grand Canyon in about 1878 by early prospectors and miners who abandoned them when they left. About 1926 they were reported to be becoming a problem in the area. By 1951, 1500 had been shot; between 1951 and 1969 some 252 were removed, 1200 destroyed and 400 shot from helicopters (Behan 1978).

In 1958 donkeys were reported to exist in all the western states of the United States except Washington and Montana. The principal concentrations were in

California, Arizona, Nevada and New Mexico, but there were also small numbers in Utah, Idaho, Colorado, Wyoming, Oregon and Texas, and their distribution was restricted to rough terrain (McKnight 1958).

Donkeys were present on public lands in south-eastern California, southern Nevada, southern Utah and western Arizona in the mid-1960s. About 12 000 occurred on Bureau of Land Management and US Department of the Interior lands in special management areas set up after 1965 (Howard and Marsh 1964). Donkeys were still present in Nevada, Arizona, California and New Mexico in 1986 (Kovach 1986). In the Bill Williams River area of western Arizona there is a population of 60–90 donkeys (Seegmiller and Ohmart 1981).

In California, where they descended from donkey stock discarded by miners, they were well established in the 1930s when many were used for pet food. In the early 1970s the population was estimated to be some 3400 (Weaver 1974; Jameson and Peeters 1988). The Bureau of Land Management in California records that there are probably 12 000 of them on public lands in California alone and some 22 000 in California and Arizona together (Steinhart 1981). At present they are in arid lands from Inyo to Imperial County (Jameson and Peeters 1988). The largest concentrations occur in Death Valley where there are about 2100 head (Miller 1982), but attempts were being made to remove them in the 1980s (Rothfuss 1986).

Under state and federal protection donkey populations expanded in the 1970s. Donkeys are present in the Naval Weapons Centre, Mojave Desert, 190 km north-east of Los Angeles and by 1979 the area was overrun with them. In 1980 the population was estimated at between 3500 and 5700 animals. To improve safety the navy initiated an emergency removal program beginning in 1980–81, when some 47 were removed by the Fund for Animals Incorporated (Anon. 1981). A total of 864 were later rounded up and 649 others were shot. In 1982, 2441 were removed. In 1983 removal reduced the population to 1656. Adverse publicity stopped the shooting phase of the program and after negotiation with Congress and conservation groups 4387 were removed alive in two years. Now there are only 200 left and native bighorn sheep are being re-introduced (McGill 1984; Perryman and Muchlinski 1987; Johnson *et al.* 1987).

They are widespread throughout the Panamint Range, California, and small numbers are present in the Grapevine Mountain section of the Amargosa Range. The total population in 1980 was estimated as

1600 donkeys. From 1939 to 1961 some 3600–4100 were removed from this area by trapping and shooting (Blake *et al.* 1981). Recently, to reduce the risk of aircraft colliding with donkeys on the China Lake Weapons Centre airstrip, the navy wanted numbers reduced.

There is a population of 80 feral donkeys in the Chemehuevi Mountains, California, descendants from releases after the mining boom at the end of nineteenth century, but also possibly derived from donkeys used as draft animals by a homesteader in 1925 (Woodward and Ohmart 1976; Woodward 1979).

Records and status of feral donkeys in the United States

State	Area	Past	Present
Arizona	Grand Canyon, south of Colorado River	1960s	
	Black Mountains		
	Lake Mohave and L. Havasu		
	Bill Williams River drainage		1980s
	Kofa, Chocolate, Trigo and Castle Dome Mtns to Colorado River		
	Wickenburg		
	Castle Hot Springs		
	Mohawk Mountains		
	Organ Pipe National Monument		
	Black and Galiuro Mountains		
California	Panamint Mountains, Death Valley		1980s
	Death Valley National Monument	1930s	1980s
	Saline Valley, Inyo County		
	Coso and Argus Mountains		
	San Bernadino County		
	Ravendale, Smoke Creek Desert		
	Naval Weapons Centre, Mojave Desert		1980s
	southern Sierra Nevada		
	Avawatz, Clark, Chemehuevi and Whipple Mountains		
	Merced River Valley		
	Eureka Valley		
Colorado	Fort Garland, San Luis Valley		
Georgia	Ossabaw Island		
Idaho	Owyhee Canyon		
Nevada	Colorado River ranges		
	Grand Canyon National Park	1920s	1980s
	Goldfield		
	Wassuk Range		
	Smoke Creek Desert		
	Virgin Valley, Charles Sheldon National Antelope Range		
New Mexico	Luera and Elk mountains		1980s
	Copper Canyon		
	Bandelier National Monument		
	Mesa Prieta		
	Tularosa Basin		
Texas	Big Bend National Park		
Utah	Colorado and San Juan rivers		
Wyoming	Vermilion Creek/Salt Wells area		
	Bandelier National Monument		

PACIFIC OCEAN ISLANDS

Galápagos Islands

Donkeys occur on islands of Sierra Negra and Cerro Azul (Isabela Islands), Santiago, Floreana and San Cristóbal (Eckhardt 1972). Donkeys were introduced in the early 1830s by tortoise oil seekers and/or sulphur miners, and later became feral. On Alcedo donkeys now number between 500–700 animals (Fowler de Neira and Johnson 1985).

Donkeys were probably introduced when settlements were started by Ecuadorians in 1832–1900. In the early 1960s it was estimated that between 500 and 700 donkeys were on Volcan Alcedo, some 100–300 on Volcan Sierra Negra (Isabela Island), 300 on San Cristóbal, 200–300 on Santa Cruz, 500–700 on Santiago and 2000–5000 on Floreana (Lever 1985; Fowler de Neira and Johnson 1985).

Juan Fernández

Donkeys were introduced soon after 1547 to Más á Tierra (Robinson Crusoe) in probably about 1580 (Holdgate and Wace 1961).

New Caledonia

Donkeys have been introduced to Walpole Island (Lever 1985).

Marquesas

Donkeys have been introduced on Uapon (Lever 1985).

Hawaiian Islands

Donkeys were first introduced to these islands as domestic animals from England in 1825 (Anon. 1925), and some were reported to be running at large by 1851 (Cummins and Meek 1851). Donkeys were in demand in the late 1800s and early 1900s and a few may have become feral, but there are none left today (Kramer 1971).

In a 25-year period from 1921 to 1946, 357 were eliminated from a fenced forest reserve (Bryan 1947) on Hawaii. Some were feral on Eastern Island (Midway group) in 1915 (Elschner 1915), but none are present there now. A population of 50–100 donkeys still ranges over parts of Huehue Ranch and a few are still present on the Hualalai Ranch (Kramer 1971) on the main island of Hawaii. Some were also present in the Kaupulehu–Kiholo area of the North Kona district, and on the McCandless Ranch in South Kono district, also in the 1970s and on Molokai in the upper Halawa Valley and in the Waimanu Valley on Hawaii, and in the upper Hamakua irrigation system on Kohala Mountain and in the Waipio Valley (Lever 1985). Probably few of these remain now.

Some may have been on Laysan Island between 1905 and 1910; mules were introduced to the island by a guano company operating there in about 1891 and a small herd was noted in 1902; a single survivor was removed in 1910 (Ely and Clapp 1973).

WEST INDIES–CARIBBEAN

Bahamas

Apparently donkeys occur on some remote islands in the group as feral animals occasionally (Encycl. Brit. 1978–80).

Hispaniola

Valverde (writing in 1785) recorded that since 1725 Dominicans had hunted donkeys and sold them. They were still abundant in many districts of sparsely settled pastoral Spanish Hispaniola in 1785. There is no evidence in 1952–53 of feral donkeys in the region (Street 1962).

Virgin Islands

Feral donkeys (=burros) have existed on St. John, US Virgin Islands, since the mid-1950s. It is not known whether and what damage they may cause (Turner 1984).

■ DAMAGE

Feral donkeys are variously accused of causing declines in perennial grasses, declines in plant species susceptible to trampling damage, increased soil compaction and reductions in small mammal populations (Turner 1984).

In the United States feral donkeys are reported to compete with stock for food and water in arid environments, assist in denuding areas of vegetation and to accelerate soil erosion (McKnight 1958). The alleged impact that donkeys have on the native fauna and flora has been the subject of much controversy in the United States for some time and there appears to have been a general lack of definitive studies of the problems. Recently a number of studies have assisted in this direction. From 1975 onwards there has been controversy and conflict over the impact of feral donkeys in the United States (Behan 1978).

Investigations in Grand Canyon National Park in 1976 showed that they had a negative effect on the natural ecosystem and that the principal impact was the habitat destruction through grazing and trampling (Carothers et al. 1976). In other areas it was thought that they became established at the expense of bighorn sheep and perhaps mule deer. Weaver's (1974) study indicated that they caused devastating damage to vegetation and soil, causing wildlife numbers to decline (bighorn sheep especially) in desert ecosystems. Management practices were said to be necessary because of the large increases in donkey numbers that can occur in a predator-free environment (Woodward and Ohmart 1976).

In Arizona it has been found that the donkey has a substantial impact on the vegetation of desert systems by decreasing the canopy cover for all species (Hanley and Brady 1977).

Seegmiller and Ohmart (1981) studied the impact of feral donkeys on bighorn sheep. They found that there was a high dietary overlap between the two species and that the desert vegetation was not resistant to utilisation by the donkey and eventually the bighorn will suffer from this depletion. Their view was that the donkeys should be controlled or eliminated before they cause the bighorn's extinction.

In California donkeys compete with bighorn sheep for grass and forbs and drive the sheep from waterholes (Jameson and Peeters 1988; Lidicker 1991).

On the Galápagos feral donkeys eat grass and sedges similar to the giant tortoise (*Geochelone elephantopus*), and competition occurs when the plants are in short supply. One can only speculate as to whether the donkeys deplete the food reserves needed for tortoise survival on Volcan Alcedo. Donkeys may also trample on the tortoise nests (Fowler de Neira and Johnson 1985).

In Australia it is considered that donkeys are a pest in pastoral areas, probably competing with stock for food and water and also are blamed for extensive erosion in hilly country (Strahan 1995).

ONAGER OR ASIATIC WILD ASS

Kulan, kiang, hemione, khur, dziggetai, half-ass
Equus hemionus Pallas

■ DESCRIPTION

HB 2000–2500 mm; T 300–425 mm; SH 1100–1400 mm; WT 200–260 kg (the onager, E. h. onager, has a SH 1200 mm and WT approx. 290 kg).
Coat grey or brown, muzzle, flanks and belly lighter; mane coarse, hair short; ears long; dorsal stripe black with white border and occasionally a transverse shoulder stripe; lacks forelock; tail with tuft of long hair at tip; ears long; front hooves narrow. Female always white on underside and has streaks of white on rump, underside of neck and on back of head.

■ DISTRIBUTION

Asia. Formerly from Syria, Arabia and the Black Sea to Iran, Iraq, southern Turkestan, north-western

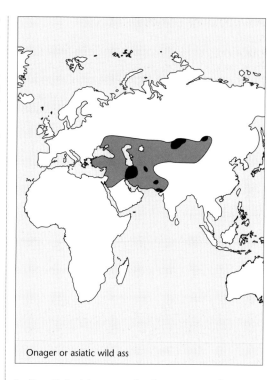

Onager or asiatic wild ass

India, Baluchistan and the Rann of Kutch, Transcaspia, Mongolia, western Manchuria and to the Yellow River of northern China. Their range is much fragmented and they are now found only in the Karakum Desert in Turkestan, Russia and deserts of Iran, India, Tibet and Mongolia.

Note: Used as a domestic animal by ancient Sumarians (c. 2000 BC) but not after the 2nd millennium BC (Zeuner 1963).

■ HABITS AND BEHAVIOUR

Habits: older males territorial; perhaps nocturnal; weak social system and spatial organisation. **Gregariousness:** unstable herds or troops of 30–50, small groups or solitary; troops of 10–12 (usually one male and several females and young); occasionally troops band together to form large herds (200–300) for migration. **Movements:** from high summer grasslands to lower levels in winter, up to 30 km for food. **Habitat:** desert or semi-desert areas in steppe, plains, river margins, mountains and gorges. **Foods:** grazes and browses; grass, sedges, leaves and pods, herbs and shrubs. **Breeding:** mates in spring and summer; gestation 11–12 months; breeds once in 2 years, possibly all year in some areas, and in some areas peak in winter; 1 young; lactation 1–1.5 years; sexual maturity 2–3 years. **Longevity:** 35 years 10 months as captive. **Status:** range and numbers considerably reduced by human settlement and grazing of domestic stock and hunting.

■ HISTORY OF INTRODUCTIONS
ASIA
Russia

Eight Asiatic wild asses imported from Badkhyz Reserve were released in the Barsa Kel'mes Preserve on Barsa Kel'mes Island in the Aral Sea in October 1953. Here they were kept in a semi-feral state and an additional male yearling was released in 1955. Breeding occurred in 1956 and 1958 and there were 34 there in 1963 and the species appeared to be well acclimatised on the island (Rashek 1959; Bannikov 1963). By the 1980s they were still surviving, and this herd was thought to be kulans (*E. h. kulan*) (Burton and Pearson 1987).

Onagers (*E. h. onager*) have also been introduced successfully in Kazakhstan (Sludskii and Afanas'ev 1964).

ASIA MINOR
Israel

Several herds of onagers (*E. h. onager*) from the deserts of Iran, where there were fewer than 800 living in 1988, were introduced into a reserve in Israel in the early 1980s (Clark 1983).

■ DAMAGE
No information.

FERAL HORSE

Feral pony, brumby, mustang, tarpan, wild horse

Equus caballus Linnaeus

E. caballus *is here retained for the feral horse as most of the animals treated here are feral domestic horses. The alternative* E. ferus *Boddaert is the wild horse or tarpan and is the ancestor of the domestic horse. Przewalski's horse* E. ferus przewalskii *or* E. caballus przewalskii *no longer exists in the wild.*

■ DESCRIPTION

Feral horse, E. caballus: *HT 1520 mm; WT 410–540 kg; SH 1020–1650 mm. Wild horse or tarpan,* E. ferus przewalskii: *HB 2200–2800 mm; T 920–1110 mm; SH 1200–1460 mm; WT 200–300 kg. Connemara ponies height 1400 mm.*

Feral horse: numerous colour patterns from dun, bay, pintos, palaminos, sorrel, chestnut, and others.

Tarpan: upper parts yellowish red-brown; flanks paler; under parts whitish; mane and legs dark brown or blackish.

■ DISTRIBUTION

Eurasia. The wild horse *E. ferus* was formerly distributed across Eurasia from eastern Poland and Hungary east to northern Turkestan and Mongolia. The last wild survivors of Przewalski's horse (*E. f. przewalskii*)

existed in Dzungaria, south-east Chinese Turkestan, probably until about 1969.

■ HABITS AND BEHAVIOUR

Habits: dominance hierarchy with separate orders for males and females; diurnal or nocturnal; home range 0. 9–303 km^2 (52–88 km^2 Australia). **Gregariousness:** harem groups with dominant male 2–20, 100; bachelor male groups or bands; occasionally solitary males; bands of juveniles 2–3 years old of both sexes; overlapping home ranges 0.8–78 km^2; density 0.1–11/km^2. **Movements:** seasonal movements governed by habitat, water and altitude; daily movements 1–3 km. **Habitat:** forest, open forest, grassland, steppe, semi-desert. **Foods:** grass and grass-like plants and weeds. **Breeding:** breeds throughout year, peaks April–June (United States); gestation 315–387 days; reach sexual maturity at 2–3 years; most come into season in spring or 9 days after foaling; oestrous cycle about 3 weeks; young 1–2 (twins rare); at birth covered with hair, eyes open, and stands in 1 hour; weaned at 5–12 months. **Longevity:** 20–50 years captive, 7–25 years in wild; Przewalski's horse 34 years captive. **Status:** extinct in wild; reduced in numbers and range as feral animal.

■ HISTORY OF INTRODUCTIONS

Became feral in Europe, Asia, Australia, New Zealand, North America, South America and on many islands.

AFRICA
Gabon

Ponies from Norway have been introduced to the Wonga-Wongue Presidential Reserve in north-western Gabon, where they have met with initial success in becoming established (Nicoll and Langrand 1986; Blom *et al.* 1990).

AUSTRALASIA
Australia

Seven horses came with the First Fleet in 1788, but only two survived more than a few years. With more imports and breeding a small export trade in live horses had begun by the 1820s, particularly to the Indian Army (Wilson *et al.* 1992).

Timor ponies have been breeding in the Northern Territory for 120 years. They were introduced to the Coburg Peninsula at Port Essington in the 1830s, where in the mid-1960s there were at least 300 present. Brumbies have been feral in the Northern Territory since 1870 and occur on practically every station (Letts 1964; McKnight 1969).

The domestic horse population in Australia reached its peak in 1918 when 2.5 million were present. Today the figure is around 1.2 million. Horses were widely used for exploration and development, and escape

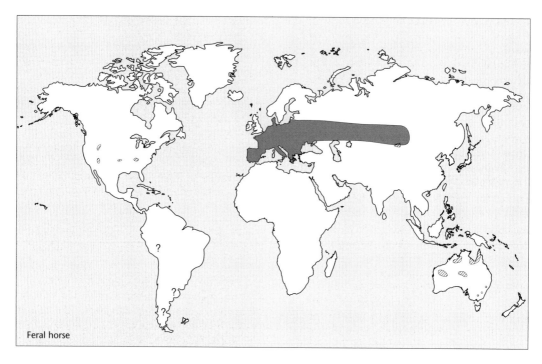

Feral horse

and releases were a common occurrence. Such events are responsible for the current feral populations that occur in many areas of Australia (McNight 1969; Wilson *et al.* 1992).

Major concentrations of feral horses now occur in the north of the Northern Territory, around Alice Springs, east of Birdsville in south-west Queensland, and in the Alps of south-eastern Australia (Wilson *et al.* 1992).

They occur mainly in the north and north-west regions of Queensland with an estimated 100 000 (Mitchell *et al.* 1982).

There were possibly 200 000 horses in the Northern Territory in 1985, with a greater density in central Australia where an estimated 80 000 occurred. Abbatoirs in South Australia and one at Tennant Creek in the Northern Territory killed and processed some 8000 horses in 1985 for human consumption and further 6000 for pet food. The average on-station price in 1986 was A$87. Shooting probably cost A$18 per head (Anon. 1987).

Only a few feral horses are found in Victoria where they have little effect on the wildlife (Wharto and Dempster 1981).

Papua New Guinea

Horses are reported to be feral on Maron and Luf islands of the North Western Island group, as well as on Manus Island (Herrington 1977).

Torres Straits Islands

A few feral horses persisted on Badu and Moa and also a substantial population on Prince of Wales Island in the early 1980s (Draffan et al. 1982).

EURASIA

The horse was domesticated about 2500–5000 years ago (Clutton-Brock 1981 in Berger 1986), and from this time on was widely introduced as a domestic animal throughout the world.

Once widespread across Europe and northern Asia, the horse's range commenced to contract in early times. By the thirteenth century they were extinct in western Europe, but survived in the steppes in Ukraine until the eighteenth century. By the beginning of the nineteenth century they were more or less confined to Mongolia and adjacent China. By the 1970s they were probably extinct in wild. Western populations, known as tarpans, became extinct in 1851 in Ukraine. A tarpan bred from primitive Polish Konik ponies has been re-introduced to the Bialowieza Forest, Poland. The eastern horse, or Przewalski's horse, remains only in captivity.

There were about 40 Przewalski's horses in Tachin Shara Nura Mountains, Mongolia, in the 1940s. Between 1942 and 1945 some were caught. As late as 1983 between 480 and 500 Przewalski's horses survived in zoos. Efforts are now afoot to breed them up and re-introduce them into the wild (IUCN Surv. Serv. Comm. 1982).

The last of the European wild horses, the Mongolian or Przewalski's wild horse (*E. przewalskii* Poljakov), is now an endangered species (Sayr 1980). A plan has been made to restore captive descendants back to the species natural range in Mongolia (FAO 1986). All the captive specimens living today trace their ancestry back to 12 animals brought out of Mongolia at the turn of the century, and to a single mare captured in 1947 (FAO 1986). At present about 260 males and 348 females are known to have been bred in captivity (kept in 74–80 zoos and private collections around the world). The World Wildlife Fund and other bodies were raising funds to return some into the wild.

In the 1940s large herds of Przewalski's horses were still observed in Dzunger Gobi, Mongolia, but by the late 1940s they had almost disappeared. No confirmed reports have been provided since 1968–69, and it is concluded that the animals no longer exist in Mongolia or China (FAO 1986). The reasons for the rapid decline in numbers are thought to have been shooting, severe winters, use of the land by grazing stock and use of natural waters by stock (FAO 1986).

Feral horses, however, exist or have existed in many areas of Europe and Asia. In 1963 wild horses still existed in the deserts of Mongolia and Manchuria, mountains of Norway, Corsican scrublands, Breton moors, swamps and peat bogs of Iceland, Shetland and Falklands (Anon. 1963).

France
A primitive breed known as the Potiok survives in small numbers in a semi-feral state in the Basque country of south-west France (Lever 1985) and the Camargue Delta of France (Duncan 1980).

Germany
Horses at semi-liberty occurred near the village of Dulmen, Westphalia, in north-west Germany and have possibly existed there since 1316 on a 10 000 acre marshland (Anon. 1963). At the beginning of the nineteenth century wild horses were still found in the Westphalia lowlands. Most disappeared after the reclamation of marshlands between 1840 and 1850 and also through growing development. In 1963 there were about 180 in Merfeld Swamp. These are mustered yearly and branded.

Greece
Small semi-feral horses (9–11 hands) live on the island of Skyros in the Sporades in the Aegean Sea (Lever 1985).

Poland
Some efforts were made in the past to preserve the steppe tarpans in a reserve in Poland (Prusskii 1965).

Portugal
A feral herd of about 30 ponies survives in the Peneda-Gerês National Park in north-western Portugal (Lever 1985; do Mar Oom and Santos Reis 1986).

Spain
About 500 semi-feral horses roam the Sueve Mountains of Astivias in north-west Spain (Lever 1985). In the south-west, in Galicia, feral ponies inhabit mountainous regions of Hugo and Ponteredra provinces (Lever 1985). A number of these latter animals are slaughtered annually for meat.

Sweden
On the fringe of the Arctic Circle in the north of Sweden feral horses were present in the early 1960s (Cornelius 1963).

United Kingdom and Ireland
In the past semi-wild ponies have occurred in the New Forest, on Dartmoor and Exmoor, in the Lakes District and on Northumberland and Cumberland fells, in the Welsh mountains, on Shetland, Western Islands and Connemara (Fitter 1959; Southern 1964). Those on Dartmoor in the eleventh century and those in the New Forest were thought to be descended from Spanish horses dating back to escapees from the Armada (Fitter 1959). Most are free-ranging but are not really wild as they are herded twice a year and the population is strictly managed (Baker 1990).

Free-roaming but managed populations occur in New Forest (Tyler 1972; Pollock 1980) and Exmoor preserves (Gates 1979).

Connemara ponies are believed to have descended from animals brought to Ireland by the Celts. They now traditionally run wild in the Connemara countryside and have developed the characteristics of the semi-wild or feral horses seen in other parts of the world (D'Arcy 1988).

Russia
Feral horses (or mustangs) exist on one of the islands of the Manych-Gudilo or Great Manych Lake, near Rostov-on-Don, where they have been present for about 30 years (FAO 1986). These were experimentally introduced in the 1950s and now a herd of up to 80 is established. At present the herd numbers 40 (FAO 1986).

MIDDLE EAST OR ASIA MINOR
Iran
In 1965 a small population of semi-feral 'Caspian' horses were discovered in northern Iran on the Caspian Sea (Lever 1985).

Island populations of horses

Area	Date introduced	Status
Pacific Ocean		
Fatuhiva, Marquesasa		
Eau, Tonga		
Easter Island		
Walpole, New Caledonia		
Más á Tierra (Robinson Crusoe), Juan Fernández group		
Agalégas		
Tristan da Cunha		
Galapágos Islands	between 1832 and 1900	introduced by settlers
Hawaiian Islands	1778–1803	exterminated by 1970s
New Zealand	early 19th century	Kaimanawa Ranges and Aupouri Forest
Macquarie Island	before 1923	two present, not estabished as feral animal
Chatham Islands	1880s	gone by early 1950s
Raivavai, Tupuai & possibly Rurutu in Austral Islands		present
Rapa Islands		present
Atlantic Ocean		
South Georgia	after 1905	existed for some years
Falkland Islands	1764	abundant 1845, none now
Sable Island	1738?; 1898	present in 1997
Indian Ocean		
Kerguelen	since 1950	horses present, not feral
Sri Lanka	?	small feral horses in north-west of island

References: Darwin 1845; Falla 1937; Wodzicki 1950; Holdgate & Wace 1961; Holdgate 1967; Berger 1972; Eckhardt 1972; Atkinson & Bell 1973; Watson 1975; Lever 1985; King 1990.

ATLANTIC OCEAN ISLANDS
Falkland Islands
The explorer Bougainville took horses to the East Falkland in 1764 and in the 1840s there were between 2000 and 3000 feral (Darwin 1845; Lever 1985). They were present in the late 1950s and early 1960s (Holdgate and Wace 1961), but there are none present now (Lever 1985).

South Georgia
Feral horses existed on South Georgia for some years after 1905 (Holdgate 1967).

INDIAN OCEAN ISLANDS
Kerguelen
There have apparently been horses on Kerguelen since 1950 (Holdgate 1967). Certainly some were kept on the islands for short periods in recent times near the scientific station (Watson 1975).

Sri Lanka
A population of small feral horses known as 'Mannor ponies' exists in the north-west of the island (Lever 1985) where they are protected.

PACIFIC OCEAN ISLANDS
Chatham Islands
Horses have been introduced (Atkinson and Bell 1973) and were plentiful there in the 1880s but were gone by the early 1950s (King 1990).

Galapágos Islands
Settlers from Ecuador introduced horses between 1832 and 1900. They were reported on Charles Island (Floreana) in 1876. Wild horses occurred on Sierra Negra and Cerro Azul (Isabella Islands) (Eckhardt 1972). There were wild horses on San Cristóbal in 1868, although they appear to have been absent by 1875. They are now present on both Isabela and San Cristóbal (Lever 1985).

Hawaiian Islands

Horses became feral in the Hawaiian Islands between 1778 and 1803, became well established there, but were exterminated in the 1930s (Berger 1972).

The first horses were brought to Hawaii in 1803 as domestic stock. By 1852 there were 1200 wild horses on Hawaii, 2500 tame and wild on Maui, 200 on Molokai, 6500 on Oahu and 1300 on Kaui and Niihau (Bishop 1852). Numbers in the wild and in captivity are said to have increased until about 1928 when petrol-driven vehicles began to replace them (Kramer 1971).

The last record of feral horses appears to be of a herd driven from Haleakala National Park in 1942 (Yocom 1967), other than a few animals near Wapio and Waimanu (Kramer 1971). The last known feral animals may have been some in Waimea Canyon on the island of Kauai in the 1970s (Lever 1985).

Macquarie Island

Horses were introduced before 1923 (Holdgate and Wace 1961), as two were present there at this date (Falla 1937), but they have not become established as a feral animal (Watson 1975).

New Zealand

Introduced to New Zealand early in the nineteenth century, horses became feral soon after. However, by the mid-1960s they were local and rare on the North Island (Wodzicki 1950, 1961, 1965) and more recently were restricted to a few localities around Lake Taupo in the central North Island (Gibb and Flux 1973). Feral horses now exist only in the Kaimanawa Ranges where a reserve has been established to protect them (Wodzicki and Wright 1982). They have been present in these ranges from about 1876. However, they have now virtually disappeared from the rest of New Zealand except for two remaining refuges on the North Island (Kaimanawa Ranges and Aupouri Forest (Northland)) (King 1990).

West Indies–Caribbean

The first horses that were introduced to the New World were brought by Columbus from Spain on his second voyage when one to three dozen were unloaded in Hispaniola in December 1493 (McKnight 1958). By 1511 they had been introduced to Puerto Rico, Jamaica and Cuba, where they quickly became established. From these areas they were taken to other islands.

Bahamas

Apparently horses occasionally occur on some of the more remote islands as feral animals.

Hispaniola

Following the first introduction in 1493 for the next three decades almost every ship from Spain brought more horses and Hispaniola became a major horse-raising centre (McKnight 1958).

The first wild horses recorded in Hispaniola were in 1526 (Parmentier and Parmentier 1883) when many were found on the coast near Santa Domingo (in the Dominican Republic now). However, it is more than likely that they had been present there for many years.

In 1701 wild horses ranged the Plaine de Léogane, and in 1785 Valverde (1947) stated that the Dominicans had hunted and sold them since 1725. In that year hunters travelled from Port de Paix to Port-à-Piment in the north-west to capture wild horses (Moreau de Saint-Méry 1797–98). By 1785 they were numerous in many sparsely settled districts of pastoral Spanish Hispaniola (Street 1962). More recently, in 1947 some feral horses were sighted at Savane Philippe, Masif de la Selle, and it was reported that in 1952–53 they were occasionally seen in the area of the Massif de la Selle near the Dominican frontier (Street 1962).

South America

From the West Indies horses were taken to Peru after 1511 (McKnight 1958) and probably to many other parts of South America. They have been found in Argentina, Brazil, Chile, Peru, and Venezuela (Nichols 1939; Wyman 1945). In 1838 Don Felix de Azara commented on the problem of Paraguayan horses in immense herds (Berger 1986) since the founding stock were introduced in 1535 (Redford and Eisenberg 1992).

In 1691 a chronicler wrote that oxen, cows and horses roamed the plains in such prodigious numbers that 'in some places the fields are covered with them as far as your eyes will reach'.

Argentina, Chile, Patagonia

Following the early exploration of South America in about 1500, the Spaniards founded Buenos Aires. This colony was forced to move to Asuncion, Paraguay, in about 1537, for economic reasons. During the move a group of horses was liberated at Buenos Aires in 1537, founding a population that extended to the Straits of Magellan by 1580 (a rate of spread of 48 km/year) (Darwin 1845). Between 1545 and 1580 there was a brisk trade in horses across South America and nobody knows how many escaped into the wild (Redford and Eisenberg 1992). By 1699 horses had become so abundant that Patagonian Indians had domesticated them and founded a horse culture similar to that developed by the Plains Indians of North America.

In the late 1970s feral horses were reported from areas south of Buenos Aires (Chapman and Chapman 1980).

Colombia

Small wild horses are established in the páranos of the Sierra Nevada de Santa Marta in Colombia and are believed to have descended from introductions made by Spanish conquistadors (Lever 1985).

CENTRAL AND NORTH AMERICA

Horses became extinct in America about 8000 years ago and all the present stock are derived from releases or escapees from captivity of stock introduced by the Spanish and later colonists.

Alaska

In 1958 a small number of feral horses were reported from Alaska (McKnight 1958), but there appears to be no recent information.

Canada

Horses reached Canada via the southern Indian tribes in about 1750 (Worcester 1945) or earlier. When the white man arrived herds of wild horses were present in the interior of British Columbia and were believed to have originated from stock which was spread north by Indians from the Santa Fe region about 1600 and arrived in the Flathead area about 1700 (Haines 1938). By the end of the eighteenth century, feral horses were spread from the United States border north to the Athabasca River (McKnight 1958).

In 1958 there were hundreds of feral horses in western Canada, mainly in the Rocky Mountain foothills in Alberta and basins of the Thompson and upper Fraser rivers of British Columbia (McKnight 1958). More recently they created range problems in southern and central grasslands of British Columbia and many were destroyed by ranchers, but small populations still exist in the region (Carl and Guiguet 1972).

About 200 feral horses were present in western central Alberta in the late 1970s (Salter and Hudson 1978, 1979, 1980).

Mexico

Horses were brought to Mexico from the West Indies some time after 1511 (McKnight 1958). The Cortes expedition (in 1519) first brought them in, but 16 animals had died within two years of their arrival (Carera 1945). However, others were soon imported and it was not long before they were widespread as a domestic and a feral animal (McKnight 1958).

Unites States

Horses were brought to North America as domestic stock by the Spanish explorers and later settlers, and some were released and some escaped from captivity (McKnight 1958). All of the present stock were derived from those held by ranchers, miners, Indians, explorers, and other travellers. Few show any affinity to the Spanish horses that escaped from Hernando Cortés in 1519 at Vera Cruz, Mexico, or from Hernando de Soto's 1543 travels on the Mississippi River (Wyman 1945; Berger 1986).

Possibly the first horses in the United States may have been those with the Coronado Expedition of 1540, but many more followed. Although early expeditions lost and abandoned horses, it is doubtful that there were enough to form a breeding nucleus (McKnight 1958). The missions established by Oñate were probably the initial source of strayed and stolen horses that became the first feral animals in the American southwest (Cabrera 1945). However, the initial formation of feral bands soon increased and they were augmented by horses frightened off during Indian attacks, worn-out animals turned loose and escapes from prospectors, miners, ranchers and travellers. It was originally presumed that the Indians obtained horses from feral bands (Barnes 1924), but more recent evidence suggests that they acquired them before the spread of wild populations through the United States (McKnight 1958; Berger 1986). The Indian population obtained their horse stocks by trade and stealing in the southern areas near Mexico, and by about 1750 the use of horses had reached the Canadian tribes (Worcester 1945).

In the early eighteenth century there was a gradual northward movement of horses from Mexico to the central Rocky Mountain states. They were being used in Idaho about 1700, the Dakotas about 1750, the northern central valley of California about 1775 and by the mid to late eighteenth century. Feral herds were soon running over most of the western United States and parts of southern Canada. Early travellers reported thousands in single bands in Texas, Colorado and other parts of the south-west (Bursey 1933; Chamberlin 1945; Worcester 1945). Early travellers of the 1830s and 1840s in south Texas reported observing great herds of wild horses (Inglis 1964).

By the eighteenth century feral horses were spread from the Rio Grande in the south to the Athabasca River in the north, and from the Mississippi in the east to the Pacific Ocean in the west (McKnight 1958). Around 1870 there may have been 2–5 million wild horses in the United States (Silverstein and Silverstein 1974), with the greatest concentrations in the southwest (Texas, Oklahoma, Colorado and New Mexico). The most populated range was probably in west-central Texas (Denhardt 1947).

As settlement and fences spread westwards in the United States, feral horses decreased in some areas and increased in others. During the nineteenth century numbers fluctuated considerably and towards the end of the century the principal areas were west of the Rocky Mountains (McKnight 1958).

In the 1890s both numbers and range increased when more were released as a by-product of increased horse ranching (Taylor 1957). However, many were removed between 1899 and 1903 for use during the Boer War in South Africa, and from 1914 to 1918 during World War 1 (Wyman 1945; Taylor 1957). In the 1920s chicken food processors used many for meat meal (Wyman 1945) and in 1924 their use as pet food affected the numbers of feral horses. During the late 1940s and early 1950s over 100 000 were taken from Nevada range lands and much smaller numbers from other western states (McKnight 1958).

By the mid-1950s several thousand feral horses still ranged over the remote parts of 13 western states. It was estimated that total numbers in the United States were between 17 000 and 34 000. The largest herds at this time existed in north-eastern Nevada (Elko County), and in central and western Nevada. Other herds occurred in the Upper Rio Grande Valley, the Mescalero Apache Reservation north-east of Alamogordo, and the Tularosa Basin in the south; in Yakima County, Washington, the Colville and Spokane Indian reservations; in Bend, Oregon; in north-eastern Utah in the Uintah and Ouray Reservations, and in various semi-arid basins in the western parts; in the north-western corner of Colorado and the adjacent parts of Wyoming and Utah; in the Salmon River drainage basin of Idaho; and in south-western Colorado and eastern Arizona (McKnight 1958; Frei *et al.* 1978).

In 1974 some 16 000 feral horses in 11 western states were estimated to be present (Silverstein and Silverstein 1974). They then (1982) occurred in at least 10 states, most within the Great Basin Desert, with estimated 44 930 head, although this figure is hotly disputed (Berger 1986). Feral horses occur on islands along the Atlantic coast (Welsh 1975; Keiper 1976; Rubenstein 1981), in remote desert ranges of western North America (Berger 1977; Miller and Demiston 1979; Salter and Hudson 1982), Wyoming (Boyd 1979), the northern Great Basin Desert, Death Valley, California (Berger 1986), and since 1738 on Sable Island, north Atlantic coast (Welsh 1975). There were populations in Washoe, Mineral, Lyon, Nye and Churchill counties, Nevada, and Harney County in Oregon in 1986 (Bowling and Touchberry 1990).

The origin of horses in the Great Basin is obscure. Free-ranging horses were there in 1841 and by 1911 they were widely distributed there and possibly numbered 70 000 (Berger 1986).

Feral horses are present on a number of islands off the coast of the United States including Assateague, Shackleford, Ocracoke and Sable islands. On Assateague Island off the coast of Maryland and Virginia, in 1976–77 there were 150–200 of them in about 20 herds on the 35-mile-long island (Keiper 1976; Keiper and Keenan 1980). They also inhabited Shackleford and Ocracoke islands along the coast of North Carolina, and Cumberland Island, along the coast of Georgia (Keiper 1976). In 1984 the population on Shackleford Island reached 92 horses (Rubenstein and Hohmann 1989). The population on Cumberland Island, Georgia, has increased from about 144 in 1981 to 154 horses by 1983, and 180 in 1985 (Turner 1988).

Feral horses inhabit the sandy islands of the Rachel Carson Estuarine Sanctuary near Beaufort, North Carolina (Stevens 1988).

Stories vary as to how horses arrived on Sable Island; some were shipped there in 1738 and mainland stallions were introduced to increase the size of animals already there in 1898 (Grosvenor 1965). In 1965 there were 200 head on the island.

Populations of feral horses on islands off the coast of North America

Island	State	Notes
Unalaska	Alaska	feral
Little St. Simons	Georgia	semi-feral
Little Cumberland	Georgia	semi-feral, there 1988
Assateague	Maryland	semi-feral, 150 in 1980
Chincoteague	Virginia ?	semi-feral
Ocracoke	North Carolina	semi-feral, there 1976
Shackleford	North Carolina	semi-feral, there 1990
Cedar	North Carolina	semi-feral, there 1980
Sable	Nova Scotia	200–300 feral, there since 1738

References: Keiper 1976; Keiper & Keenan 1980; Welsh 1975.

Feral horses were present in early 1900s at Beaty's Butte and Jackie's Butte in Oregon, but today only a few remain. In late 1940s there were 2500–3000 present at Beaty's Butte but by 1950 only 50 head remained.

Numbers increased again to 419 by 1980 (Eberhardt *et al.* 1982).

At least three herds of feral horses exist near Salmon, Idaho (Seal and Plotka 1983).

Feral horses increased rapidly in numbers with protection from 1971 (Wolfe 1980). Most are now on Public Lands (in Nevada). Others states with 1000 or more head include California, Colorado, Idaho, Montana, Utah and Wyoming. At present there are about 45 000 horses on Bureau of Land Management and US Department of the Interior lands. A number of special wild horse ranges called Special Management areas were set up in various states after 1965 (Howard and Marsh 1984).

Before 1971 feral horses had no status with respect to ownership or management. Under provisions of the Horse and Burro Act of 1971 that established federal ownership on public lands, they have increased rapidly on western range lands since their protection (Wolfe 1980; Garrott and Taylor 1990; Garrott 1991). The Bureau of Land Management has been attempting to control numbers by periodically gathering animals from wild herds. The 'adopt-a-horse' scheme has resulted in the removal of over 100 000 animals from western range lands (Garrott 1991). Currently other control methods, such as reproductive inhibitors or fertility control, are under investigation.

Feral horses are present in north-eastern California (Jameson and Peeters 1988). Feral herds also occur in southern California, northern and western Nevada, south-eastern Oregon, south-western Wyoming and north-western Colorado (Wolfe *et al.* 1989).

■ DAMAGE

In north-eastern Colorado, elk and wild horses have been increasing in recent years and mule deer numbers have been decreasing. Studies have shown that the foods of elk, wild horses and cattle are more similar to each other than to mule deer. Cattle and wild horses have very similar diets and would compete for forage (Hansen and Clark 1977). Dietary overlap between feral horses and cattle was found to be high in desert biomes (Miller 1983; McInnis 1985) in Oregon and Wyoming. They shared 60 per cent of the same plant communities in all seasons (McInnis 1985) in south-east Oregon.

Feral horses take advantage of water points provided by pastoralists in Australia for their stock and also use natural waterholes in mountainous areas which are inaccessible to cattle and musterers (Berman and Jarman 1987). The diet of feral horses has consider-

Feral horse populations in five states of the United States in the 1980s

Area	States
Twin Peaks Management Area	n.e. California / n.w. Nevada / California
Douglas Mt., Dinosaur National Monument	n.w. Colorado
Massacre Lake Management Area	n.w. Nevada / California
High Rock Management Area	n.w. Nevada / Susanville Distr., California
Tuledad Management Area	n.e. California / n.w. Nevada / California
Humboldt National Forest	Ely, Nevada
Cold Spring Management Area	Vale District, Oregon
Merger Allotment	Vale District, Oregon
Sheepherd's Basin / Barren Valley	Vale District, Oregon
Jackie's Butte	Vale District, Oregon
Three Fingers	Vale District, Oregon
Cedar Mtns	n.w. Utah
West Desert Area	w. Utah
Seven Lakes Planning Unit	Rawlins District, Wyoming
Adobe Town	Rawlins District, Wyoming
Atlantic Rim	Rawlins District, Wyoming
Little Colorado Area	s.w. Wyoming
Northeast Area	Rock Spring District, Wyoming

Reference: Wolfe 1980.

able overlap with that of cattle (Berman 1987) and they probably compete with them for food. When they are present in large numbers they can be a pest, breaking fences and damaging watering points, as well as consuming pastures otherwise available to stock (Strahan 1995). In Queensland they damage fencing and compete with stock. On one station staff shot 3000 horses (Mitchell *et al.* 1982).

In western Alberta, Canada, it was found that potential for competition was highest between horses and cattle (Salter and Hudson 1980)

In Red Desert, Wyoming, it was found that the possibility of direct competition between cattle and horses was strongest for forage in autumn and in severe winters, and for water during summer (Miller 1983). Management of horses has been a subject of controversy for some years in United States.

In Oregon it was found that more than 88 per cent of mean annual diets of horses and cattle consist of grass. Overlap was high in all seasons (61–78 per cent). Horses and cattle showed more than 60 per cent of same plant community each season. The study suggests that there is a high potential for exploitative competition under conditions of limited forage availability (McInnis 1985).

Family: Rhinocerotidae
Rhinoceroses

GREATER INDIAN RHINOCEROS
Asian one-horned or Indian rhinoceros
Rhinoceros unicornis Linnaeus

■ DESCRIPTION
HB 3100–4200 mm; T 700–800 mm; SH 1480–2000 mm; WT 1600–2200 kg.
Bare skin grey brown with pinkish skin folds and large tubercles; body hair occasionally apparent; ear fringes and tail brush always present; single black horn 200–600 mm.

■ DISTRIBUTION
Southern Asia. Foothills of the Himalayas from northern Pakistan east through India and Nepal to Assam and Bengal. May also have occurred in Burma, Thailand and other parts of South-east Asia until the Middle Ages, but the exact extent of its range is not now known.

■ HABITS AND BEHAVIOUR
Habits: solitary; mainly nocturnal and crepuscular. **Gregariousness:** cow–calf pairs; groups rare; density 0.4–2.0/km^2 and up to 4.85/km^2. **Movements:** sedentary; home range 2–8 km^2. **Habitat:** forest, grassland, swamps, reed-beds. **Food:** grasses, reeds, twigs, fruits, leaves and cultivated crops. **Breeding:** throughout year; females polyoestrous; oestrus 21–42 days; gestation 462–491 days; 1 young; lactation 12–18 months; inter-birth interval about 3 years; sexual maturity female 5–7 years. **Longevity:** 47 years captive. **Status:** reduced in numbers and range, mostly surviving in sanctuaries and parks; endangered.

■ HISTORY OF INTRODUCTIONS
ASIA
Re-introduced successfully in India and Nepal and recent projects to re-introduce more widely.

India and Nepal
By the 1900s the population of the greater Indian rhinoceros in India was considerably reduced.

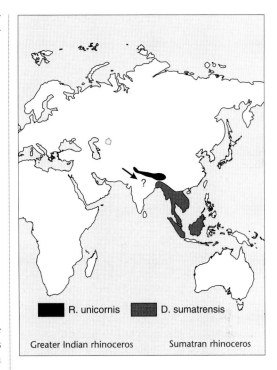

R. unicornis   D. sumatrensis

Greater Indian rhinoceros   Sumatran rhinoceros

Hunting was banned as early as 1910. By the early 1960s, 1000 were surviving, mainly in sanctuaries and reserves (Burton and Pearson 1987). The Chitawan National Park, which has the largest surviving population, is protected by armed guards to prevent poachers from obtaining the horns (Walker 1992).

In 1984, six were captured by drug immobilisation and were transported to a stockade and a few days later five remaining animals were crated and flown and trucked to Dudhwa National Park in central northern India. From here they were unloaded into stockades. One female died but the remaining four animals were released about a month later. In 1985 four animals captured in Sauroha, north of Chitwan National Park, Nepal, were trucked to Dudhwa to join those already established there. They were released in the wild about one week later. This population was surviving well in 1986, but they were confined by an electric fence. Two female rhinoceroses from brought from Assam died (Sale and Singh 1987). At present the total in the area numbers 12 and there are plans to re-introduce more animals (Javed 1993).

At present there are a number of projects underway in both India and Nepal to re-introduce breeding populations to areas of former occurrence (Mishra and Dinerstein 1987; Sale and Singh 1987). Animals from Chitwan National Park have also been re-introduced in the Bardia Wildlife Reserve, in western Nepal (Sale and Singh 1987; Bauer 1988).

■ DAMAGE
In Assam greater Indian rhinoceroses can cause considerable damage to cultivated crops.

SUMATRAN RHINOCEROS
Hairy rhinoceros, Asian two-horned rhinoceros
***Dicerorhinus sumatrensis* (Fischer)**

■ DESCRIPTION
HB 2350–3180 mm; T 600–650 mm; SH 1000–1500 mm; WT 800–2000 kg.
Two horned; armour-plated appearance; leathery skin, dark grey brown; facial skin wrinkled around eye; muzzle rounded; body covered with short coarse blackish hair; two short horns, anterior horn 150 mm in females up to 450 mm in males, posterior horn *c.* 50 mm in females and up to 150 mm in males.

■ DISTRIBUTION
Southern Asia. Assam to Bangladesh, south through Burma, Thailand, Vietnam, through to Malay Peninsula, and to Sumatra and Borneo.

■ HABITS AND BEHAVIOUR
Habits: mostly nocturnal, wallows in pools during day; visits salt licks regularly. **Gregariousness:** density 13–14/km²; males usually solitary. **Movements:** seasonal movements to hills during floods and lower elevations in dry periods; overlapping home ranges 10–30 km². **Habitat:** swamps; near water in secondary forest, often hilly country. **Foods:** leaves, twigs, fruit, bamboo shoots, and cultivated crops. **Breeding:** gestation 7–8 months; 1 young, born haired; separates from female at 16–17 months; inter-birth interval 3–4 years. **Longevity:** 32 years 8 months captive. **Status:** range and numbers reduced by hunting and clearing of land; endangered.

■ HISTORY OF INTRODUCTIONS
ASIA
Re-established in at least one area in India.

India
At the end of the nineteenth century the Sumatran rhinoceros was still a widespread species. However, it now occurs only in scattered populations numbering a few hundred animals and there may be fewer than 1000 left in the wild. The species has been re-established in the state of Utar Pradesh (Sale 1986).

■ DAMAGE
No information.

BLACK RHINOCEROS
Hooked-lipped rhinoceros
***Diceros bicornis* (Linnaeus)**

■ DESCRIPTION
HB 2950–3750 mm; T 600–700 mm; SH 1400–2250 mm; WT 700–1800 kg.
Low powerful build; head long; eyes small and well formed; horns two, prominent, longest 500–1200 mm; rear horn 350–400 mm; lip prehensile, dark, long and pointed; upper lip protruding; muzzle pointed; hide dark grey to dark brown; skin with swollen folds on neck, breast and top of forelegs; tail round with terminal bristled tassel.

■ DISTRIBUTION
Africa south of the Sahara. Formerly Lake Chad and Cameroun to Sudan and Ethiopia, and south to Mozambique and Natal.

■ HABITS AND BEHAVIOUR
Habits: solitary, aggressive, wallows; unpredictable; territorial or non-territorial(?); active day and night. **Gregariousness:** temporary aggregations to 13; females and calves together; adult males solitary; density 0. 5–1/km². **Movements:** sedentary; daily to drink up to 18–25 km; overlapping home ranges 252–4400 ha. **Habitat:** semi-desert, thornbush, grassland, wetlands, bush country, open forest. **Foods:** twigs of woody growth and legumes; herbs, leaves, buds, shoots of trees and bushes. **Breeding:** throughout year; promiscuous; gestation 530–550 days;

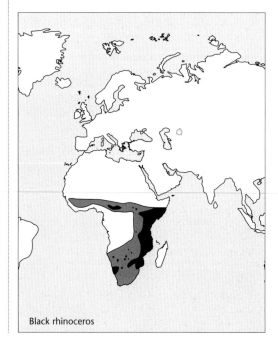

Black rhinoceros

1 young; oestrous cycle 17–60 days; newborn mobile 10 minutes after birth; calf weaned at 2 years; independent at 2.5–3.5 years; inter-birth interval 2–5 years; reproductive maturity: females 4–6 years and males 7–9 years. **Longevity:** 45–50 years captive. **Status:** declining in range and populations mainly in parks; now endangered.

◼ HISTORY OF INTRODUCTIONS
AFRICA
Successfully re-introduced into some national parks in South Africa and Tanzania and Zimbabwe.

Formerly widespread in southern Africa, the black rhinoceros were by the 1970s declining because of poaching. By the 1980s there were fewer than 30000 remaining. Now five populations of fewer than 500 animals occur mainly in parks within their range (Burton and Pearson 1987). Numbers fell in the wild from 65 000 in 1970 to 3800 in 1986 and now only about 2000–3000, mainly due to poaching for horns. Between 1970 and 1994 black rhinoceros suffered a 95 per cent population decline (WCMC 1998). Capture and translocation of rhinoceros began in 1961 and has been successfully achieved in a number of areas in Africa since this time (Borner 1988).

Recently de-horning programs, to make rhinoceros unattractive to poachers, have been initiated in Namibia and Zimbabwe (WCMS 1998).

Malawi
In 1994 two pairs of black rhinoceros from South Africa were translocated to Liwonde National Park in Malawi (Newton 1999).

Rwanda
Black rhinos have been re-introduced to Kagera National Park (Haltenorth and Diller 1994), but there appear few details.

South Africa
Black rhinoceroses were resent in small numbers in Kruger National Park in the first few decades of the nineteenth century but it was accepted by 1945 as having become extinct there. It has been re-introduced from Natal (Penzhorn 1971).

In March 1961 a pair was released in Addo Elephant National Park from Kenya and in February 1962 a further five were added. They were also successfully re-introduced into Kruger National Park (Penzhorn 1971; Novellie and Knight 1994).

In Natal Province a small number have been re-introduced to Ndumu and Itala Game Reserves (Smithers 1983). From 1981–84, 19 black rhinoceroses were successfully translocated to Pilanesberg National Park in Bophuthatswana (Anderson 1986).

Namibia
Black rhinoceroses were successfully translocated to Etosha National Park, Namibia, when 39 were released there in 1970–72. They are now seen not infrequently, whereas before were rarely noted (Hoffmeyr 1975). Between 1970 and 1972, 43 were transferred to Etosha National Park, where they are now well established (Hoffmeyer et al. 1975).

Tanzania
Black rhinos were introduced to Rubondo Island (240 km^2) in Lake Victoria, Tanzania, before 1966, principally to make the area a tourist attraction (Grzimek 1966). Sixteen animals were introduced in 1964 and have done well despite poaching (Borner 1988).

Zimbabwe
In about 1974, 19 were moved from the Zambesi Valley, Zimbabwe, to Gonar-re-Zhou Reserve, in south-east Zimbabwe, and seven young were born there (Encycl. Brit. 1970–80).

In 1975, seven (four males, two females and a juvenile male) were captured south of Lake Kariba and released in Zambesi National Park (Booth et al. 1984), and some were still present there in 1980. They have also been re-introduced in Rhodes-Matopos National Park in Zimbabwe (Haltenorth and Diller 1994).

Fifty-nine black rhinoceroses were released with 12 white rhinoceroses in the Hwange National Park and Matetsi Safari Area, Zimbabwe, in 1984–85 (Booth and Coetzee 1988).

In 1991 the World Wide Fund for Nature initiated a conservation program for black rhinoceroses, translocating animals from areas of high poaching activity to areas of relative safety and de-horning them. There is evidence that de-horned animals are left unharmed by poachers (MCMC 1998). By August 1993, 122 black rhinoceroses had been de-horned.

■ DAMAGE
No information.

WHITE RHINOCEROS
White square-lipped rhinoceros
Ceratotherium simum (Burchell)

■ DESCRIPTION
HB 3350–5000 mm; T 500–1000 mm; SH 1500–2000 mm; WT males 2000–3600 kg, females 1400–1700 kg.
Head massive with wide, square mouth; two horns to 600–1580 mm, front ones three times the length of the rear ones; lip square; body yellowish brown to grey; almost naked except for ear fringes and tail bristles; ears pointed; pronounced shoulder hump.

■ DISTRIBUTION
Africa. Formerly most of Africa; Chad to Central African Republics, north-eastern Zaire, southern Sudan and north-western Uganda; also southern Africa from south-eastern Angola to Cape Province. Now in few areas.

■ HABITS AND BEHAVIOUR
Habits: mainly crepuscular, but also nocturnal; wallows during heat of day; non-aggressive; territorial.

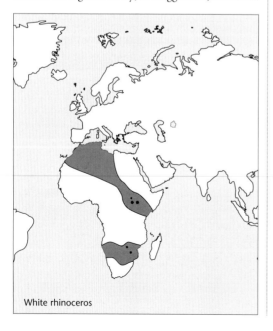

White rhinoceros

Gregariousness: temporary associations to 14; territorial bulls solitary; several females and calves in groups; density 0.03–0.08/ km^2 and locally to 5.0/km^2. **Movements:** territories 80–260 ha; overlapping home ranges 6–20 km^2; daily movements 4–15 km. **Habitat:** open grassland and lightly wooded areas, bushy savanna with thickets. **Foods:** grass and herbage. **Breeding:** throughout year, peak in rainy months; gestation 520–550 days; weaned 2–12 months; inter-birth interval 2–3 years; 1 young; calves alone in thicket; stays with female 2–3 years; sexual maturity 4–7 years. **Longevity:** 36 years in wild, but potentially 40–50 years. **Status:** much reduced in numbers and range by poaching for horns; endangered.

■ HISTORY OF INTRODUCTIONS
AFRICA
In 1882 the white rhinoceros was believed extinct in the southern parts of its range, but a small population still existed in Umfolozi, Zululand. Here they increased and were being translocated to other sites in the 1960s and by 1980 over 3000 were present in South Africa and elsewhere.

In 1900 the northern subspecies of white rhinoceros (*C. s. cottoni*) was widespread, but by 1980 only 1000 remained. By 1985 only a single population of fewer than 20 existed (Burton and Pearson 1987).

Botswana
Southern white rhinoceroses (*C. s. simum*) were re-introduced to Chobe and Motemi National Parks in Botswana from South Africa (Haltenorth and Diller 1994).

Kenya
Southern white rhinoceroses were introduced to Meru and Tsavo National Parks in Kenya, which are areas outside the natural range of this subspecies (Walker 1992; Haltenorth and Diller 1994); however, all animals have been eliminated by poachers.

Mozambique
Small groups of southern white rhinoceroses have been introduced to this country outside the natural range of the species and but have been eliminated by poachers (Walker 1992).

Namibia
Southern white rhinoceroses from South Africa were re-introduced to the Cunene area of Namibia (Haltenorth and Diller 1994).

South Africa
Widely translocated, there is hardly a country in southern Africa that has not re-introduced white

rhinos (Smithers 1983). They have been re-introduced to a number of national parks and to Mkuzi, Itala and Nduma game reserves in Natal.

Massive re-introduction programs in South Africa and other countries (Groves 1972; Owen-Smith 1981; Smither 1983) resulted in the occurrence of 4404 white rhinos, mainly in South Africa, in the 1980s.

Southern white rhinoceroses (*C. s. simum*) had disappeared from Transvaal by 1898, but have been successfully re-introduced from populations in Hluhluwe and Umfolozi game reserves in Natal into Kruger National Park (Penzhorn 1971; Haltenorth and Diller 1994; Novellie and Knight 1994).

Between 1981 and 1984, 248 white rhinoceroses from Umfolozi Game Reserve were translocated to Pilanesberg National Park in Bophuthatswana, where in 1984 there were 230 (Anderson 1986). Thirty white rhinoceroses were released in Phinda Resouce Reserve, a privately owned site in South Africa between 1990 and 1992 (Hunter 1988).

In 1991 the World Wide Fund for Nature initiated a conservation program for white rhinoceroses, translocating de-horned animals to areas safe from poachers. By August 1993, 111 animals had been de-horned (WCMC 1998).

Uganda

A most successful translocation of white rhinoceroses (*C. s. cottoni*) was made in Uganda where the population had declined to 80 animals in 1962. These were taken from West-Madi, west of the Nile, to Kabalega (Murchison Falls) National Park where they were established. Fifteen were captured and moved to Murchison Falls National Park, where their numbers grew to 80, but all were killed in 1980 (Walker 1992; Haltenorth and Diller 1994).

Zimbabwe

Four white rhinoceros were introduced into Rhodes-Matopo National Park in southern Zimbabwe before 1963, and where they were surviving at that time (Riney 1964). Others from Umfolozi Game Reserve, Natal, were taken in 1962 to Zimbabwe and became established there (Dorst 1965). Twelve were released in Hwange National Park and Matetsi Safari Area in 1984–85 in Zimbabwe along with black rhinos (Booth and Coetzee 1988). In 1975, 10 white rhinoceroses (one adult male, one adult female, two juvenile males and six juvenile females) were released in Kazuma Pan National Park (Booth *et al.* 1984) where they established.

■ DAMAGE

No information.

Family: Suidae
Pigs

WILD BOAR

Feral pig, European wild pig, wild pig, swine or hog

Sus scrofa **Linnaeus**

■ DESCRIPTION

HB 1100–1650 mm; T 150–350 mm; SH 550–1000 mm; WT males 30–190 kg (and up to 350 kg); females 15–110 kg (and up to 150 kg).

Body colour varies from grey to brown or black or a mixture of colours; tusks in males to 150 mm; tail has short hairs at end; ears ovate and pointed backwards. Females with 8–14, 16 mammae; generally smaller in size and weigh less than males. Feral animals may be white, black or red or shades and mixtures, but predominately black.

■ DISTRIBUTION

Eurasia. From western Europe and northern coast of Africa eastwards to Japan (Honshu) and south to Sri Lanka, Sumatra, Malaysia (including Singapore, Penang, Langkawi and Pangkor islands) and Indonesia (Java and Sunda islands). Formerly in southern Scandinavia and Great Britain. Also occur on Sardinia and Corsica.

DOMESTICATION AND HYBRIDISATION

There is some evidence to suggest that pigs became domesticated, probably in the Neolithic, variously and independently in Europe, Asia Minor, the Far East and various parts of South-east Asia (Zeuner 1963; Groves 1981; Oliver 1985). Domestication may have taken place in China around 4900 BC, and possibly as early as 10 000 BC in Thailand (Lekagul and McNeely 1977).

The feral pig populations of New Guinea and Ceram and some of the smaller islands in the Moluccas, which are to a large extent genetically continuous with the domestic pig populations in the region, appear to have resulted from hybridisation between introduced stocks of the Celebes warty pig (*S. celebensis* Müller and Schlegel), and the wild or feral pig (*S. scrofa*) (Groves 1981). The Celebes warty pig of Sulawesi and neighbouring small islands (Indonesia) and other pigs are not treated separately in this works, but *S. celebensis*

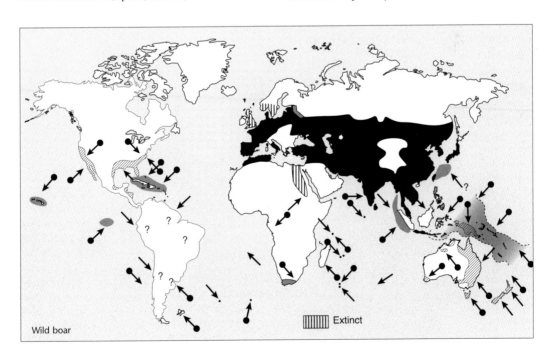

Extinct

Wild boar

appears to be feral at least on Halmahera and the Simaleue Islands. Feral and domestic populations of *celebensis x scrofa* appear to occur on the Moluccas, in New Guinea and on the Solomon Islands and probably on other islands. (See Groves 1981 for the distibution of *S. celebensis* and other pigs, and their hybridisation in the Indonesian region.)

A general view, as expressed by Honacki *et al.* (1982), is that *S. verrucosus* includes *S. celebensis* and that the resulting species, and not *S. barbatus*, is found in the Philllipines. *S. verrucosus* is restricted to Java and the nearby islands of Madura and Bawean, while *S. celebensis* is indigenous to Sulawesi and some nearby small islands.

■ HABITS AND BEHAVIOUR

Habits: largely diurnal (particularly morning and evening), but also nocturnal. **Habitat:** forest, woodland, pine plantations, scrubland, mangroves, grassland, swamps, river beds, streamsides, usually not far from water. **Movements:** relatively sedentary, home range varies 28 ha–50 km^2 (United States, Europe, Australia). **Gregariousness:** old males solitary; females and young or immature in family groups or parties to 30 to 50 (rarely exceed 12 in Australia but can be as large as 400); density 1–80/km^2 (Australia, United States, NZ and Europe). **Foods:** omnivorous but primarily herbivorous; bulbs, roots, tubers, grass, forbs, shoots and leaves of plants, grass, ferns, seeds, fruits, berries, nuts, acorns mast, beech mast, fungi, mushrooms, lizards, snakes, frogs, young rabbits, hares, fawns, mice, voles, fish, crabs, leeches, insects, earthworms, carrion, turtles, snails, slugs, isopods, birds' eggs, cultivated cereal and root crops, tapioca, podi, sugarcane, grains, potatoes, beet, turnips, corn, molluscs, arthropods, coconuts, crustaceans. **Breeding:** throughout the year, peak in autumn and winter and decline in spring and summer; ruts November–February (Europe, United States), and young born March–May; most areas 2 litters/year; litter size 1, 6–10, 12; gestation 110–115 days; females polyoestrous; oestrous cycle 21 days; young weaned 2–4 months; boars mature 10–12 and sows 7–12 months. **Longevity:** to 27 years. **Status:** reduced in numbers and range.

■ HISTORY OF INTRODUCTIONS

Pigs have been successfully introduced and re-introduced in a number of areas in Eurasia. They have been introduced in a number of regions of Africa, Indonesia and Papua New Guinea, Australia, New Zealand, North America, South America, the West Indies, and on many islands in the Pacific, Indian and Atlantic oceans (see below).

Introductions of pigs on islands

Island or group	Date introduced	Notes
Aldabra Is	<1878	died out *c.* 1878
Amami Oshima (Riukius)		present?
Amsterdam	1823	still present
Andaman Is	?	still present
Aorangi (Poor Knights, NZ)	*c.* 1820	eradicated 1936
Aore (Vanuatu)	?	
Aquijan (Marianas)		
Arapawa (NZ)	?	still present 1990
Aru Is (Maluku)		present
Astove		present
Auckland (Auckland Is)	>1840	still present
Auckland Is (NZ)	1807, 1840, 1842–43, etc.	still present 1990
Bahamas (West Indies)		still present
Batanta		present
Batjan (Maluku)		present
Beaver (Falkland Is)	>1765	?
Biak-Supiori (Irian Jaya)		present
Bird (Seychelles)	1970s?	still present?
Bismarch Archipelago		still present
Bleaker (Falkland Is)	>1765	?
Blumine (NZ)	<1957	eradicated 1988–89
Borneo		failed

Introductions of pigs on islands (*continued*)

Island or group	Date introduced	Notes
Bougainville (PNG)		present
Bruni (Australia)	1770s	did not survive?
Buka (PNG)		present
Buru (Maluku)		present
Campbell (NZ)	1865, 1883	died out *c.* 1960s
Cavalli (Motukau anui, NZ)	?	died out
Cerro Azul (Galápagos)		
Chagos Archipelago	1840s?	present?
Channel Is (United States)	mid-1800s or late 1500s	present
Chatham (NZ)	?	still present, declining 1990
Chetwode (NZ)		control in 1970s
Choiseul (Solomons)		present
Clarion (Baja California)	?	present 1979
Cocos (Costa Rica)		still present
Columbrete Grande (Medit. S)	*c.* 1855	still present?
Credner (PNG)	1901	present?
Crozet	<1820	exterminated late 19th century
Cuba (West Indies)	*c.* 1493	still present
Cumberland (United States)	>1900	still present
Curtis (Australia)		
D'Urville (NZ)	?	still present 1990
Duke of York (PNG)		present
East (French Frigate Shoals)	1867	disappeared about 1872
East Falkland (Falkland Is)	1764	still present?
Egmont Atoll (Chagos Arch.)	1840s?	present?
Eiao (Marquesas)	early 19th century	still present
Enderby (Auckland Is, NZ)	1843, 1867	died out after 1894
Enggano (Sumatra)		
Espiritu Santo (Vanuatu)		present
Facing (Australia)		
Falkland Is	1493 and later	still present
Fergusson (PNG)		present
Fijian Is (all main Islands)	pre-European	still present
Flinders (Australia)		
Floreana (Galápagos)	1832	still present
Flores (Sunda Is)		still present
French Frigate Shoals (see East)		
Futuhiva (Marquesas)		
Galápagos Is		
Gebe (Maluku)		present
Goodenough (PNG)		present
Grand Isle (Aldabra Is)	<1878	died out *c.*1878
Great Barrier (NZ)	?	still present
Guadalcanal (Solomons)		present
Guam	1672	still present
Halmahera (Maluku)		present

Introductions of pigs on islands (*continued*)

Island or group	Date introduced	Notes
Hammond (Australia)		
Hatutas (French Polynesia)	early 19th century	still present
Hawaii (Hawaiian Is)		still present
Hawaiian Is	1000 AD, 1778–1803	still present
Hispaniola		still present?
Horn (United States)	?	still present?
Huahine (Society Is)		
Hull (Phoenix Is)		
Île aux Cochons (Crozet)	<1820	exterminated late 19th century
Inaccessible (Tristan)	<1873	exterminated *c.* (1938)
Inner Chetwode (NZ)	*c.* 1900, 1954	eradicated 1926, and 1959–63.
Iromote (Riukius)		present?
Isabela (see Cerro Azul)		
Ishigaki (Riukius)		present?
Jack's (West Falkland)	1932	died out *c.* 1952
Jamaica (West Indies)	<1838	still present
Java (Indonesia)		present
Juan Fernández	1574	present
Kahoolawe (Hawaiian Is)	*c.* 1823?	eliminated by 1931
Kakerome (Riukius)		present?
Kangaroo (Australia)		
Karkar (PNG)		present
Kaui (Hawaiian Is)		still present
Kermadecs (NZ)	1836	died out or eradicated
Kiriwina (PNG)		present
Kure Atoll (Hawaiian Is)	<1966	removed 1966
Kusaie (Carolines		
Lady Julia (Australia)	1884	eliminated?
Lanai (Hawaiian Is)	early 1900s	shot out by about 1930
Laysan (Hawaiian Is)	<1891	eliminated before 1902
Line Is		
Little Andaman (Andamans)	?	still present
Little Barrier (NZ)	?	died out
Lord Howe (Australia)	efforts to eliminate 1980s	
Louisiade Archipelago		still present
Macauley (Kermadecs, NZ)	1836	died out
Madagascar	?	present 1970s
Maitea (see Osnaburg I.)		
Malaita (Solomons)		present
Malden (Line Is)		
Maluku (Indonesia)		present many islands
Mangole (Maluku)		present
Manus (PNG)		present
Marianas	1672–85	present
Marmot (Alaska)	1984–85, *c.*1987	still there
Marquesas	<1770s?	present?
Más á Tierra (Juan Fernández)	*c.* 1580	still present?

Introductions of pigs on islands (*continued*)

Island or group	Date introduced	Notes
Maui (Hawaiian Is)		still present
Mauritius	1512	still present
Mayor (NZ)	?	attempted eradication 1963 failed, still present 1990
Mehetia (Society Is)		
Melville (Australia)	1827	
Middle Andaman (Andamans)	?	still present
Midway Is (see Sand I.)		
Misima (PNG)		present
Molokai (Hawaiian Is)		still present
Moluccas (Indonesia)		still present
Moreton (Australia)		
Motuara (NZ)	?	eradicated *c.* (1950)
Motuoruhi (Goat) (NZ)	?	eradicated about 1970
Namoluk Atoll (Carolines)		present
Nansei Is (see Ryukyus)		present?
Native (NZ)	?	died out 1940s
Nendo (Solomons)		present
New Britain (PNG)		present
New Caledonia	1770s	?
New Georgia (Solomons)		present
New Hebrides		
New Ireland (PNG)		present
New Quaker (Falkland Is)	>1765	?
New Zealand	1773	
Nias (Indonesia)		still present
Nicobar Is	?	still present 1980s
Niihau (Hawaiian Is)		still present
Nissan (PNG)		present
Norfolk (Australia)		
Normanby (PNG)		present
Numfoor (Irian Jaya)		present
Oahu (Hawaiian Is)		still present
Obi (Maluku)		present
Okinawa (Ryukyus)		present?
Osnaburg	<1770s?	
Ossabaw (United States)	*c.* 1559?	still present?
Outer Chetwode (NZ)	*c.* 1948, *c.* 1955	eradicated 1953 and 1964
Palau (Carolines)		
Papua New Guinea	4000–10 000 years ago	still present
Penryhn Atoll (see Tongareva)		
Phillip (Australia)	early 1900s	
Pickersgill (NZ)	?	eradicated *c.* 1950
Pitt (NZ)	?	still present 1990
Ponape (Carolines)		
Poor Knights (NZ)		exterminated 1936
Prince of Wales (Australia)		still present

Introductions of pigs on islands (*continued*)

Island or group	Date introduced	Notes
Puerto Rico		still present
Raivavae (Tabuai Is)		
Rakitu (Arid) (NZ)	?	died out 1960s
Raoul (Kermadecs, NZ)	1836	eradicated in 1960s
Rennell (Solomons)		present
Ruapuke (NZ)	?	still present 1990
Rurutu (Tabuai Is)		
Salawati (Irian Jaya)		present
San Clemente (Channel Is)	1950s	present
San Cristobal (Galápagos)		present?
San Cristobal (Solomons)		present
Sanana (Maluku)		present
Sand (Midway group)	<1915	disappeared soon after
Santa Catalina (Channel Is)	mid-1800s or late 1500s	present
Santa Cruz (Channel Is)	mid-1800 or 1920s	control in 1980s and 1990s
Santa Cruz (Galápagos)		present?
Santa Isabel (Solomons)		present
Santa Rosa (Channel Is)	late 1500s or 1800s	present
Santiago (Galápagos)		still numerous
Saunders (Falkland Is)	1765	?
Seram (Maluku)		present
Seychelles	?	present?
Siberut (Sumatra)		still present
Sierra Negra (Galápagos)		still present
Simeulue (Sumatra)		present
Society Is		
Solomon Is		still present
Speedwell (Falkland Is)	>1765	?
St. Helena		eliminated?
St. Simon (United States)	>1697	?
St. Paul (Indian O)	<1823	still present; may now be extinct?
Stewart (NZ)	?	control 1948–65, but still present?
Sula Is (Indonesia)		present
Sulawesi		failed
Sumba (Sunda Is)		still present
Sumbawa (Sunda Is)		still present
Sunda Is		still present
Suvarov (Cook Is)		
Swain's (Tokelau)	<1939–42	few in 1965
Sydney (Phoenix Is)		
Tabuai (Tabuai Is)		
Tabuai Is.		
Tahati (Society Is)		present?
Tasmania (Australia)	1903?	still present
Taveuni (Fiji)		present
Tikopia	prehistoric	present
Timor (Sunda Is)		present

Introductions of pigs on islands

Island or group	Date introduced	Notes
Tofua		
Tokelau Is	>1841	none now?
Tokunoshima (Ryukyus)		present?
Tonga		
Tonga Taboo (Fiji)	1770s?	present
Tongareva (Cook Is)		
Tongareva Atoll (Penryhn At.)	1853	?
Trinidade (Brazil)		present
Tristan da Cunha	1790–1810	exterminated
Tuputupunahau (NZ)	1950s	eradicated c. (1960s)
Vanu Levu (Fiji)		present
Virgin Is (West Indies)		still present
Viti Levu (Fiji)		present
Walpole (New Caledonia)		
West Indies	1493	still present
West Point (Falkland Is)	>1765	?
Woodlark (PNG)		present

AFRICA

Gabon

Wild boars from Europe have been introduced to the Wonga-Wongue Residential Reserve (4800 km^2) in north-western Gabon. They have become well established and have interbred with the indigenous bush pig (*Potamochoerus porcus*) (Nicoll and Langrand 1986; Blom *et al.* 1990).

South Africa

Domestic pigs were released by the Forestry Department in its plantations in the south-west parts of Cape Province and in the George area in the 1920s in an endeavour to control the pine tree Emperor moth, *Nudaurelia cytherea*, whose larvae defoliate pine trees. The pigs flourished in both areas. There were 200 feral pigs on a property in Cape Province (Piketberg District) in 1973. A herd existed in the mountains near Broekhuizens Poort, 14 km west of Grahamstown in the 1980s. At least up until the 1940s feral domestic pigs were living in the vlei at Kleinmond. Some were also reported from farms in the Transvaal in the 1980s (Siegfried 1962; Smithers 1983).

Subsequently, Austrian and Bavarian wild boars were imported to improve stocks (Siegfried 1962). Some of these populations have become extinct, but others have persisted (Botha 1985) and utilise the pine plantations and raid surrounding farmland, where they are shot when possible (Bigalke and Pepler 1991).

Pigs were introduced into the Kluitjieskraal plantation in South Africa before 1942 to control insect larvae (Thomas and Kolbe 1942).

Zimbabwe

In north-eastern Zimbabwe indigenous peoples kept free-ranging pigs and it was not unusual to find them feral there in the 1980s (Smithers 1980).

ASIA

Indonesia

Introduced to Indonesia (de Vos *et al.* 1956) pigs are present on many of the Sunda Islands and others, e.g. Sumba, Sumbawa and Flores (Oliver 1984, 1985), where they are often poisoned as vermin.

Borneo and Sulawesi

Feral pigs failed in both places, but both islands have an endemic form of pig (Oliver 1985).

Enggano, Simeulue, Siberut and Nias (islands off w Sumatra)

Pigs are present as an introduced variant of *S. scrofa vittatus* and another *S. scrofa mimus*. On Simeulue Island, off north-west Sumatra, the pig species present is believed to be feral *S. celebensis* (Oliver 1984, 1985). A native tradition quotes them as swimming ashore after a shipwreck in the late nineteenth century (Sody 1940).

Those pigs occurring on Siberut island are introduced and feral (Oliver 1985).

In the Baluran National Park it was suspected that the pig (*S. scrofa*) was out-competing and possibly hybridising with the endemic Javan warty pig (*S. verrucosus*). However, this does not appear to be happening as the two appear to be segregated by different patterns of habitat utilisation (Blouch *et al.* 1983; Macdonald and Frame 1988).

The feral pigs on Nias are probably *S. celebensis* and not *S. scrofa niadensis* (Oliver 1985).

Irian Jaya

Pigs have probably been introduced to the islands of Batanta, Biak-Supiori, Numfoor and Salawati off Irian Jaya (Flannery 1995).

Java

It is suspected that the pig is an introduced species on Java (Blouch *et al.* 1983).

Lesser Sunda Islands

S. celebensis is introduced and has hybridised with *S. scrofa* types which have similarly spread, with human assistance, along the Lesser Sunda chain into Melanesia (Oliver 1985). Those on Timor and Flores have variously been accorded sub-specific status, but both are certainly feral (Oliver 1985). Probably most of the populations are mixed and both *S. celebensis* and *S. scrofa* types are present and represent probably two separate introductions.

Maluku

Pigs have probably been introduced on the Aru Islands, Batjan, Buru, Gebe, Mangole, Obi, Sanana, Seram and Halmahera (Flannery 1995).

The Moluccas has both types of pig (*S. celebensis* and *S. scrofa*) and intermediates between.

Pigs are possibly indigenous to the Sula Islands in the North Moluccas, although it is shown that pigs introduced to Tikopia in the prehistoric period were extremely rare or had become extinct before European contact (Kirch and Yen 1982; Flannery 1995).

ATLANTIC OCEAN ISLANDS

Falklands Islands

Pigs were introduced to East Falkland in 1764 by the French explorer de Bougainville. One year later they were released on Saunders Island. Later whalers and sealers landed them on Beaver, Bleaker, New, Quaker, Speedwell and West Point islands to provide a source of food. Some were released in 1932 on Jack's Island, West Falkland, although they have not been reported since 1952 (Oliver 1985; Lever 1985).

St. Helena

Pigs were once feral in the Great Wood, but are now only reared in pens (Cronk 1986) on the island. They are considered to have eaten the ebony seedlings and probably did not assist regeneration on the island (Cronk 1986).

AUSTRALASIA

Australia

Feral and deliberately released colonies of wild pigs have existed in Australia since the days of the early settlements. Temporary colonies existed in a number of localities in the early years because of the escape of unrestrained stock. However, it is difficult to trace any of the present colonies back to these times. Certainly pigs brought out for food arrived with the First Fleet in 1788 and shortly after they landed some pigs were permitted to roam around the settlement at Sydney Cove (Pullar 1950, 1953). By 1795 these animals had become a nuisance and could be shot if found trespassing on anyone's property (Robertson 1932).

Captain Cook released two pigs on Bruni Island, off Tasmania, in the 1770s, but it appears doubtful that these would have survived. Some were later deliberately released in Tasmania about 1903 (Pullar 1953), probably for hunting purposes. Pigs were taken to Melville Island in 1827 and later transferred to Raffel's Bay, Coburg Peninsula, on the Northern Territory mainland where they were later abandoned (Pullar 1953). Some were turned loose on Lady Julia Island, Victoria, when farming failed in 1884, but were rounded up some time later by fishermen and taken to market and none occur there now (Pescott 1976). Some were introduced on Phillip Island by early settlers for food and sport (Coyne 1982), off Norfolk Island and by 1912 had caused severe damage to the vegetation (Watson 1961).

In more recent years pig introductions have been made by hunters rather than spread by natural dispersal (Tisdell 1982; Auld and Tisdell 1986; Wilson *et al.* 1992).

Certainly colonies of wild pigs existed in Queensland, the Northern Territory, New South Wales, Flinders Island, Kangaroo Island and in the Darling Ranges of Western Australia prior to 1870. Some of the earliest records can be traced to the Dawson River, Queensland, about 1885, and near Broome, Western Australia, in 1894–96. Stokes (in 1837–43) saw pigs on an island in Bass Strait and Jukes (in 1847) liberated a boar and sow on an island near the Queensland coast, but shot them a year later (Pullar 1953).

In the early 1950s feral pigs were throughout the greater part of Queensland, except the low rainfall and closely settled districts. They also occurred on Prince of Wales, Hammond, Curtis, Facing and Moreton islands, and in New South Wales along the

Darling River and its tributaries, and in isolated colonies on the Lachlan–Murrumbidgee and Murray rivers. In Victoria small colonies of pigs existed from time to time. In South Australia pigs occurred on Flinders Island and Kangaroo Island and in the Northern Territory they were recorded on the Coburg Peninsula, eastern Arnhemland, the King and Daly rivers and at Maranboy. In Western Australia a number of isolated colonies were in the Kimberleys and in swamps and along rivers in the south-west of the state (Pullar 1953).

In the 1960s they were reported present in large colonies on the sub-coastal plain from Darwin to Arnhemland and extended eastwards along the Daly and Katherine River tributaries, with smaller colonies on the Roper River (Letts 1964). Generally, feral pigs reported in Victoria in 1959 were believed to have been caught in New South Wales and deliberately released in Victoria. In 1978, however, an investigation showed that they were in 23 districts and probably in 11 of these feral pigs had become established since 1970 (Stevens 1981).

In the south-west of Western Australia they were distributed from near Jarrahdale south to Boddington, Boyup Brook, Kirup and Harvey, but control work was reducing their numbers (Masters 1979; Anon. 1981). They also occurred in farming regions from Northampton south to Geraldton and along the Hill River near Jurien Bay (Oliver *et al.* 1992).

In the Australian region most feral pigs are in the higher rainfall country extending from New South Wales north through Queensland and the Northern Territory to the Kimberleys. Smaller numbers occur in the south-west of Western Australia.

Between 1.7 and 2.3 million feral pigs were estimated to be in Queensland, 66 per cent inhabiting the northern areas (north of the twentieth parallel), but there has been a considerable decrease since the 1970s due to dry weather. Some were reported liberated at the Daintree River in northern Queensland during the gold-rush days early in century (Mitchell *et al.* 1982).

A few pigs are present on Badu and Moa, Torres Strait, and a large population is on Prince of Wales Island (Draffan *et al.* 1982).

Feral pigs are now widely distributed and abundant in Queensland, the Northern Territory, New South Wales and the Australian Capital Territory; isolated populations occur in Western Australia, South Australia and Victoria; also on Flinders Island. Tasmania is free of them (Wilson *et al.* 1992).

Currently in Australia annual losses are estimated at $100 million (Choquenot *et al.* 1996) and the feral pig game meat industry is worth in excess of $20 million annually, mainly in Queensland and New South Wales.

Papua New Guinea

The antiquity of pigs in New Guinea is the subject of much debate. It is probable that they were introduced from Asia (de Vos *et al.* 1956) between 4000 and 10 000 years ago, as those now occurring there are an intermediate type between the wild boar *S. scrofa vittatus* and the Celebes wild boar *S. celebensis* (Ryan 1972; Groves 1981; Oliver 1984). Pigs were present as an introduced animal on the Bismarck Archipelago, Louisiade Archipelago and on the Solomon Islands in the 1960s (Anderson and Jones 1967). They have probably been introduced to the islands of Bougainville, Buka, Duke of York, Fergusson, Goodenough, Karkar, Kiriwina, Manus, Misima, New Britain, New Ireland, Nissan, Normanby and Woodlark (Flannery 1995).

Since white settlement, European pigs have been introduced and now it is difficult to find any without some European origin. In 1901 the government of New Guinea released one boar and two sows from Sydney, and two local sows, on Credner Island for swine breeding (Ryan 1972). Domestic pigs are kept at varying densities throughout New Guinea. Feral pigs are now abundant in many parts and are most common where people are scarce: in swamps, forest and alpine grassland. Certainly they are abundant in grassland on Mount Albert Edward (Flannery 1995).

EUROPE

Populations on Corsica and Sardinia, and those formerly in Egypt and northern Sudan, are, or were, of old feral origin. Feral populations are also widespread in Norway and Sweden (Wilson and Reeder 1993).

Russian Federation and adjacent independent republics

Wild pigs have been translocated in some areas of Russia (de Vos *et al.* 1956; Bannikov 1963), and have been established as a game animal. A number of introductions and re-acclimatisation attempts have been made in the Russian Federation, but these have mainly been unsuccessful. However, introductions between 1960 and 1964 in the Ukraine have possibly been successful (Yanushevich 1966). Introductions appear to have increased the range of pigs, at least in European Russia (Kirisa 1974).

In 1961 some pigs were introduced into the Barsa-Kelmes Preserve, where they initially became a nuisance to the native aquatic animals (Bannikov 1963), but they later failed to become permanently established as the release was made in an unsuitable locality (Sludskii and Afanasev 1964). Some were introduced in the Crimea where the herd of 35 expanded to 2100 over a period of 10 years (Kormilitsin and Dulitskii 1972 in Bratton 1975).

Some 4992 pigs were released in the Russian Federation for acclimatisation purposes between 1937 and 1972, in 21 regions in Russia, six regions in the Ukraine and in eight other republics.

Releases in Russia include: Atrakhansk (1969–72, 30 animals), Vladimirsk (1954–55 and 1968–71, 173), Volgogradsk (1969, 26), Gorkousk (1963 and 1969, 51), Kolininsk (1937–71, 1361), Kalujsk (1964–71, 347), Kursk (1971, four), Moskovsk (1947–70, 472), Novgorodsk (1971, 22), Orenburgsk (1971, 31), Penzensk (1970–71, 46), Rostovsk (1970–72, 99), Ryahzansk (1948–50, 47), Saratovsk (1970–72, 63), Osetinsk (1966–68, 160), Smolensk (1966–71, 251), Stavropolsk (1969–72, 66), Tambursk (1969, 19), Tatarsk (1972, 30), Ulyahnovsk (1969, 18), and Yahroslavsk (1961–70, 505), a total of 3821.

In the Ukraine some 505 were released from 1957 to 1972.

In other republics releases occurred in Armyahsk (1969–72, 404 animals), Belorussia (1954 and 1961, 19), Gruzinsk (1960–71, 164), Kazakhstan (1961 and 1972, 34), Kirgizstan (1971, 15), Latvia (1956–58, 17), Litovsk (1956 and 59, four) and Astonsk (1966, eight).

United Kingdom

Wild pigs have been extinct as a native species in Great Britain since the late seventeenth century, but they have been the subject of many deliberate re-introductions, particularly on private estates during the nineteenth century (1820 to 1840s). More recent inadvertent liberations have occurred in 1976 near Nairn and in 1977 in Kent, with most of those animals shot or killed (Corbet and Harris 1991).

Wild pigs survived in the Scottish Highlands until the middle of the nineteenth century and were still present in 1617 in the south of England, but there-after disappeared. A few years later they were being released in the New Forest from France and Germany. Two escaped animals lived in woods near Colchester in the 1820s and some were released in Hampshire and Surrey in the seventeenth century, but because they became a nuisance to the local farmers they were eventually destroyed. In the 1830s the Earl of Fife released some, but these failed to become established (Fitter 1959).

INDIAN OCEAN ISLANDS

Agalégas

Feral pigs occurred on L'ile du Nord in the Agalégas at one time or another (Lever 1985).

Aldabra Islands

Pigs occurred on Grande Isle many years ago, but died out before 1878; they were possibly introduced and exterminated more than once (Diamond 1981).

Amsterdam

Wild pigs were abundant on this island in 1823 (Goodrich 1843) having probably been introduced there by sealers for food (Oliver 1985). They were later exterminated (Holdgate 1967).

Andaman Islands

An aberrant and dwarf wild pig, *Sus scrofa adamanensis,* has long been thought to be an endemic subspecies on the Andamans. It is now contended that it is a feral species and that the sub-specific status is invalid.

Remains of pigs have been found in the earliest midden deposits of the original negrito tribes by whom the pigs were probably introduced (Schreiber *et al.* 1984; Oliver 1985). There are two types of pigs on the Andamans (Abdulali 1962): a large-snouted animal from Little Andaman and a short-snouted form on Middle Andaman. Now both are endangered and to complicate things further, introduced domestic pigs have now run wild on the islands.

Astove

Pigs were introduced at an early date and apparently still occur there (Bayne 1970).

Chagos Archipelago

In the 1840s Egmont Atoll was overrun with an estimated 600 pigs (Bourne 1971).

Crozet

Wild pigs were established on Île aux Cochons some time before 1820, as by that year they had become numerous there (Goodrich 1843). The island was so overrun in 1840 that you could hardly land for them (Ross 1847). In all probability they were introduced by sealers as food (Oliver 1985).

Because of the damage caused to bird life they were exterminated in the mid- or late nineteenth century (Holdgate and Wace 1961; Holdgate 1967), but the small numbers and limited amount of time they were on the island appears to mitigate against much damage to fauna or flora (Derene and Mougin 1976).

Madagascar

Wild pigs occurred on Madagascar in the 1970s (Brygoo 1972).

Mauritius
Pigs were introduced to this island before the arrival of the Dutch in 1598. The Portuguese navigator Pedro Mascarenhas released 'hogs' on Mauritius in 1512, although the islands were not colonised until 1638 (Hatchisuka 1953).

Nicobar Islands
Feral pigs are present (Oliver 1985), and it seems doubtful that they are indigenous to these islands (Encycl. Brit. 1976).

Seychelles
At one time or another, pigs have been established in the Seychelles (Lever 1985). About 400 feral pigs were present on Bird Island in 1972–73 (Feare 1979).

St. Paul
Wild pigs were abundant on this island in 1823 (Goodrich 1843), where they were probably introduced by sealers as food (Oliver 1985).

Tristan da Cunha
Domesticated pigs were introduced to Tristan da Cunha before 1810 (Holdgate and Wace 1961), and may have arrived with the Portuguese in about 1790. They had certainly run wild by 1824 (Earle 1832) and some were reported to be present there in 1829 (Morrell 1832).

The wild pigs on the island have now been exterminated (Holdgate and Wace 1961), although some occur near the settlement (Munch 1945; Holdgate 1955) on the island, but are presumably confined by the inhabitants.

Inaccessible Island, off Tristan, also once had feral pigs (Moseley 1892), which were abundant in 1873 (Moseley 1892; Lockhart 1930), but only one animal was present there in 1938 (Hagen 1952) and there appear to have been none since this time (Holdgate 1967).

New World
Pigs were carried by the early European navigators and colonists to the New World. Thus Christopher Columbus introduced them to the West Indies in 1493 and Spanish settlers brought them to Florida in 1539. By the end of the sixteenth century, Spanish colonial settlement was well established in Mexico, parts of Central America, West Indies, Peru, and Chile and there were Portuguese settlements in Brazil. There was a widespread practice of a free-range system of keeping domestic pigs that led to the early establishment of feral populations. They now occur in 11 states of the United States and in most countries of Central and South America (Oliver 1984).

North America
Alaska
In 1984 private individuals obtained a permit from the Alaska Department of Natural Resources to introduce wild boar onto Marmot Island (near Kodiak I.). They survived the winter of 1984–85 and appeared to have established themselves (Franzmann 1988), although further introductions were made in 1987 (Lloyd *et al.* 1987).

United States
The first wild pigs in North America were those that escaped in the southern United States from the Spanish colonists during the early sixteenth century (Towne and Wentworth 1950; Belden and Frankenberger 1989). As early as 1526 these colonists started a large settlement called San Migual near Georgetown, South Carolina (MacDonald 1975), and although domestic pigs are not mentioned specifically, this was probably the time of their first introduction to the United States (Wood and Brenneman 1977).

Hernando De Soto landed at Charlotte Harbour, Boca Grande, Florida, in May 1539 (Lewis 1907) with, among other stock, some 13 sows which were more than likely the descendants of those left by Columbus in the West Indies. Some of these animals are thought to have escaped from De Soto as he travelled in the southern United States. Another early introduction occurred in Florida in 1565, when Admiral Pedro Menendez landed and brought with him some 400 domestic pigs (Towne and Wentworth 1950).

One hundred years after Menendez landed with his pigs, some eight towns, 72 missions and two royal haciendas possessed descendants of these, from which many escaped and became well established (Towne and Wentworth 1950). The Spanish missionaries and others continued to introduce pigs in the seventeenth and eighteenth centuries (Hanson and Karstad 1959). With the spread of agriculture and the practice of free-range stocking, which continued until the mid-twentieth century, feral pigs became fairly common in the forests of at least the south-east United States (Wood and Lynn 1977; Wood and Barrett 1979).

The Indians are thought to have assisted the pig's naturalisation in the sixteenth and seventeenth centuries by acquiring animals which they then allowed to roam free (Hanson and Karstad 1959). When the French attempted settlement in Florida in 1560 the Indians supplied them with pork from feral herds (Hanson and Karstad 1959). In 1697 Dickinson (1790) who probably visited St. Simon Island, Georgia, reported that the Indians were hunting both deer and pigs.

In 1989 feral pigs were still present in 60 counties in Florida, but the highest concentrations were in Levy, Dixie, Lafayette and Taylor counties, where slash pine flatwoods interspersed with coastal salt marsh dominate (Belden and Frankenberger 1989).

Feral pigs have probably been present in Texas at least since 1689 (Benke 1973). Many have escaped from domestication to become established (Ramsey 1968). In 1900 they were run on Blackjack Peninsula (now the Aransas Wildlife Refuge), where their descendants were noted in the wild in 1919 (Halloran and Howard 1956). In 1930–33, 11 wild pigs from the southern Appalachians were introduced and the progeny of these and others were still there in 1955, although from 1936 to 1938 some 2500, and from 1938 to 1939 another 891, were removed (Ables and Ramsey 1973). Some wild pigs were apparently introduced from Europe in the 1930s and many others have since been released in new areas by ranchers, as well as in those already containing feral pigs (Ramsey 1968).

Wild pigs existed, mainly in the eastern part of Texas, on the Rio Grande Plain and on the Edwards Plateau (Ramsey 1968; Ables and Ramsey 1973), where in 1967 it was estimated that there were about 10 000 of them (Ramsey 1968).

Domestic pigs arrived in California with the Spanish in 1769 (Hutchinson 1946; Barrett 1977; Van Vuren 1984), and no doubt offspring from these escaped and became feral. Some pigs were probably released by Russians at Fort Ross, Sonomo County, perhaps as early as 1812 (Hutchison 1946). After 1850 pigs were frequently released by ranchers to forage in woodland (Shaw 1950) and most populations descended from these (Barrett 1978). Settlers hunted feral pigs in the foothills of Red Bluff in the 1880s (Leslie 1966). With the spread of agriculture they were commonly released to forage in woodlands (Shaw 1940) to fatten on acorns (Barrett 1980), and most of those that are now wild are the descendants of free-ranging animals. Both European wild boar and feral crosses are wild in California, some being introduced as early as 1889 and 1912 (Gottschalk 1967), and more at later dates (de Vos *et al.* 1956). European wild pigs were introduced to Carmel Valley, Monterey County, in 1925 (Barrett 1977). From there they spread south through the Santa Lucia Mountains and stock from this source were released elsewhere in California. At Dye Creek in Tehama County, they have been present as a feral animal since about 1900 and the population had expanded to 700 in 1966, 1000 in 1970 and had reached 1500 by 1971 (Patten 1974; Barrett 1978). Between 1968 and 1970 the owner bred feral sows with European wild boars and released about 20 of

their progeny (Barrett 1980). Feral pigs are now common over extensive areas particularly along the Sierra Nevada foothills and coast ranges where they currently have a major negative impact on crops and rangelands and native biota (Lidicker 1991).

European wild pigs from North Carolina were introduced in the Carmel Valley, Monterey County, California, in 1923 (Bruce 1941). These dispersed through the Santa Lucia Mountains to San Luis Obispo County and interbred with the feral pigs already present (Barrett 1977). In 1924 some were released in Los Padres National Forest, California, and it was estimated in 1940 that there were about 100 there (Shaw 1941). Some from Monterey County were translocated to several areas of California, including Tehama County (Barrett 1980). Between 1965 and 1975 feral pigs expanded their range throughout the oak woodland zone of California, assisted by unregulated translocations, and European wild pig traits are to be found throughout (Barrett 1977).

Pigs were also introduced to the Channel Islands off California, where they became established and have persisted (Storer 1934). They have inhabited Santa Cruz since at least the 1920s and possibly since the mid-1800s. They were introduced to Santa Catalina Island in the mid-1930s and are now widespread there. They are reported to have come from Santa Rosa Island where they have been feral since the late 1800s or late 1500s (Baber and Coblentz 1987). The origin of the Santa Rosa population is uncertain. In 1964 several thousand were reported on Santa Cruz, Santa Rosa and Santa Catalina islands (Van Vuren 1984). The Department of Fish and Game introduced them to San Clemente Island in the 1950s, together with white-tailed deer (Howard and Marsh 1984). In the late 1960s it was estimated that 12 000 of them existed on the mainland and some 7000 on the islands of Santa Catalina, Santa Cruz and Santa Rosa (Nelson and Hooper 1953). Efforts were being made to remove them from Santa Cruz Island in the late 1980s and early 1990s (Sterna and Barrett 1991). They occur in Pinnacles National Park where they have been controlled since the 1970s (Macdonald and Frame 1988). Feral pigs are harvested annually by hunting, but remain abundant and have contributed significantly to alteration of native insular communities (Baber and Coblentz 1986, 1987).

In 1964 it was estimated that there were between 5000 and 10 000 wild pigs in California (McKnight 1964). More recently, estimates have increased to 18 000 in 1970, to 27 500 in 1976, and they are now harvested in 27 of the 58 counties in that state (Barrett 1977). In

San Benito County alone it has been estimated that some 7000 wild pigs range over 54 per cent of the county and the Californian estimate has risen to some 30 000 harvested (Barrett and Pine 1980).

Wild pigs were introduced to New Hampshire in 1889 (de Vos *et al.* 1956). A small population was established in north-west Sullivan County in the late 1950s, the descendants of escapees from Corbin's Park during a hurricane in 1938, where they numbered about 20–40 head (Presnall 1958).

Feral pigs were reported from southern South Carolina early in the twentieth century (Salley 1911). Some have inhabited the hardwood swamp forests of the United States Department of Energy's Savannah River Plant, located near Atiken, for more than 35 years. They are largely descended from free-ranging domestic animals kept by farmers in the area before the site closed to the public in 1952 (Jenkins and Provost 1964). Since this time they have been relatively free from outside introductions of domestic pigs and protected, except when harvested as game during public hunts (Smith *et al.* 1980). Wild pigs are now well established in South Carolina (Sweeney *et al.* 1979) and are abundant in the lower coastal plain area in numerous localities (Wool and Roark 1980). They are still present there (Stribling and Brisbin 1984).

Feral pigs are present in the Great Smoky Mountains National Park (in Swain and Haywood counties), North Carolina, and in Tennessee (in Sevier and Blount counties) (Smith *et al.* 1980). These are the descendants of European wild pigs released at Hooper Bald, North Carolina, in the southern Appalachians (Stegeman 1938; Jones 1959), although during their spread into the park they interbred with feral pigs and are thus no longer pure (Rary *et al.* 1968; Henry 1970; Smith *et al.* 1980). The original European stock from the Hartz Mountains, Germany, was taken in 1912 to the Hoopera Bald hunting preserve and kept in an enclosure where they were allowed to breed. In the early 1920s many escaped (approximately 100) into the surrounding forest to become established and eventually expand their range into Tennessee (Stegeman 1938; Shaw 1941; Conley 1977). In the 1940s or 1950s they entered the Great Smoky Mountains National Park from the increasing numbers in North Carolina (Fox and Pelton 1977), where they were well entrenched in 1936, when the Cherokee National Park was established (Bratton 1975). From this latter park, some 200–300 were removed between 1965 and 1971 (Fox and Pelton 1977), but the population had expanded to occupy about three-quarters of the area (Howe and Bratton

1976). Although the expansion of the population in the Great Smoky Mountains National Park was slowed by control programs, wild pigs occupied the western half by 1971 and expansion continued in 1972–73 (Bratton 1975). Wild pigs are now found in the Southern Appalachians of eastern Tennessee over an area of 2040 km^2 (Conley 1977).

'Rooshians' have been established in Tennessee since 1912 (Vinson 1946). Twenty-six pen-reared European wild pigs were released near Crossville, Tennessee, in 1962 to establish a hunting population, but they were found to be too tame and by 1965 had all disappeared (Lewis 1966).

In about 1900, 15–20 European wild pigs from Germany were released in Litchfield Park, New York, where they maintained their numbers for a period of 20 years (Bump 1941). A second introduction occurred in New York State in 1924 when a Mr Moore shipped some from North Carolina and released them on the San Francisquito Ranch, Carmel, where they became established and spread into the adjacent national forest. Two dozen were translocated in 1932 from the Moore Ranch to the Carmelo Creek watershed for hunting purposes and these had increased to 100 in the national forest in about 1940 (Shaw 1941).

On Ossabaw Island (Chatham County), Georgia, the wild pig population is said to have been there for several hundred years (about 400 years) (Hanson and Karstad 1959; Wilson and Flicker 1976). There have probably been no other introductions during this time (Brisbin *et al.* 1977) and at present several thousand inhabit the island (Tipton 1977; Stribling and Brisbin 1984). Pigs are also feral on Horn Island (off the coast of Mississippi), Gulf Islands National Seashore (Bratton 1977).

Feral pigs now occupy significant portions of 11 states. Stable populations occur in the south-eastern coastal plain and in Hawaii, but are rapidly expanding in Texas and California (Wood and Lynn 1977; Wood and Barret 1979). A population is expanding to the north-west in North Carolina and Tennessee at about 2.5 km/year (Singer 1981). In the southern United States they are now found in approximately 20 wildlife refuges in some 10 states (Thompson 1977). It was estimated in 1959 that there were probably 32 000 pigs in 14 national forests in the south-east of the United States and in 1964 some 33 000 (Lucas 1977). They are reported to be present in 66 of the 67 counties in Florida (Frankenberger and Belden 1976), where regular re-stocking with several hundred trapped for this purpose occurred annually between 1960 and 1976.

Surveys in the 1970s in the south-eastern states found them in 11 states (Alabama, Arkansas, Florida, Georgia, Lousiana, Mississippi, North Carolina, South Carolina, Tennessee, Texas and Virginia) where they occupied an area of 109 626 km^2 (Wood and Lynn 1977).

Illegal stocking of pigs by hunters occurs in Tennessee, North Carolina and California (Wood and Barrett 1979), and has certainly aided the continuing expansion of their range in California (Barrett 1977).

European wild boar were reported to have been released in central Tennessee in 1971 (Conley 1977), West Virginia in 1975 and western Tennessee in 1979, where they have prospered (Singer 1981).

Feral pigs now occur in 13 areas in the National Parks Service system. All have stable populations except Great Smoky Mountain National Park where they are rapidly expanding. The impact of animals is related to pig density and sensitivity of ecosystem which overall is a minor problem, but is severe in three parks. There is potential for further invasion of several Appalachian Mountains areas (Singer 1981).

Mexico
Feral pigs have been present in Mexico since shortly after the introduction by the Spanish as a domestic animal in the fifteenth century.

PACIFIC OCEAN ISLANDS
Pigs were introduced to Melanesia, probably arriving some 3500 years ago, and thence to Polynesia. Human settlers and their pigs reached Fiji by 1300 BC and spread into most of Polynesia by 1000 BC. By 1000 AD they had been introduced throughout much of Oceania, including the Hawaiian Islands (Oliver 1984, 1985; Flannery 1995).

European expansion into the Indian and Pacific Oceans also contributed greatly to the spread of domestic and feral pigs. Colonial explorers, traders and navigators carried domestic pigs and other animals for food and trade and in the absence of settlements often marooned stock on islands for the benefit of shipwrecked crews or later voyagers. Thus the Portuguese navigator Pero Mascarenhos released 'hogs, goats and fowls' on Mauritius in 1512, although it was not colonised until 1638 (Oliver 1984).

Pig have been introduced to the following Pacific Islands (see list at beginning of History of Introductions):

Juan Fernández, Galápagos, Guam, Aquìjan (Marianas), Palau, Ponape and Kusaie (Caroline Islands), Solomon Islands, Santa Cruz; New Hebrides and Walpole Island, off New Caledonia; Tofua and Tonga; Hull and Sydney (Phoenix Islands), Malden (Line Islands), Suvarov and Tongareva (Cook Islands), Raivavae, Tubuai, and possibly Rurutu (Tabuai Islands), Mehetia, Tahati and Huahine (Society Islands), Eiao and Fatuhiva (Marquesas) (Oliver 1984).

Captain James Cook mentioned the presence of hogs on (Maitea) Osnaburg Island and Huaheim Island in the Society Islands and on Amsterdam Island, on Tonga Taboo in the Friendly Islands and on the Marquesas (Kippis 1904).

Pigs were introduced in French Polynesia on Eiao and Hatutas islands and assisted with the modification of the islands during the twentieth century (Thibault 1989).

Auckland Islands
Pigs were liberated at Port Ross in 1807 by Captain Bristow of the ship *Sarah* (Ross 1847; Norman and Musgrave 1866). This introduction was supplemented by others in 1840 by the British Antarctic Expedition (Ross 1847); in 1842–43 by Maori settlers (McLaren 1948), and on several occasions since 1850 (Waite 1909; Thomson 1922). Soon after their introduction they ran wild and began preying on the colonies of ground nesting birds (Holdgate and Wace 1961) and decimating the vegetation (Challies 1975).

Pigs were introduced to Enderby Island about 1843 and in 1867, where they survived for a while, but soon died out (Taylor 1968, 1971). About 1894 they were reported in large numbers on both Auckland and Enderby islands (Challies 1975).

Wild pigs were well established on the northern end of Auckland in 1840 (McCormick 1884) and during the next 50 years spread throughout the island. After 1894 their numbers declined in some areas, but they have been consistently reported from this island since then. In 1941–45 they were found scattered throughout, and in 1972–73 were in all areas of the island visited.

In 1840 they were numerous at Port Ross (McCormick 1884). Ten years later large numbers were reported on both Auckland and Enderby islands. On Enderby as many as 100 pigs could be seen feeding along the shoreline at any one time and on Auckland they were even more numerous (Enderby 1875). Within the next 15 years they declined on the north-eastern part of Auckland Island and probably died out locally. Since this time densities in the northern part of the island have remained relatively low. In 1865 evidence of them was found at all east coast inlets from Chambres Inlet south to Smiths Harbour, but there was little sign of any in the inlets north of Chambres or on the tussock grasslands above Port Ross (Norman and Musgrove 1866). They reached Carnley Harbour during the next decade (Musgrave 1866; Challies 1975). In 1886 they were noted at Waterfall Inlet near the south-east end of the island (Challies 1975). Survivors of the shipwreck *General Grant* saw no signs of pigs at Erebus Cove in 1866–67, but found a few at the north-west end of the island (Raynal 1874; Sanguilly 1899). However, in 1876 they were regularly seen around Port Ross and the adjacent high country (Newton 1876 in Challies 1975) and have since been consistently reported there at low density (Challies 1975, 1976).

Campbell Island

Pigs were introduced by Captain Norman in 1865 (Sorensen 1951; Oliver 1985) and were released in 1883, but none survive on the island today (Holdgate and Wace 1961; Holdgate 1967; King 1990).

Caroline Islands

Pigs are present on Namoluk Atoll in the eastern Carolines, but all appear to be enclosed and none are feral (Marshall 1975).

Fiji

Pigs were introduced by pre-European voyagers (Turbet 1941) before 1300 BC and now occur on all the main islands where they are common in some areas (Pernetta and Watling 1978). There is a popular belief that they were introduced by Captain Cook (Tubet 1941), who certainly left pigs on Tonga Taboo, Fiji, on his third voyage (Kippis 1904).

Introduced pigs now occur on Taveuni, Vanu Levu and Viti Levu (Flannery 1995) and may occur or have occurred in the past on Yaqaga, Ovalu, Kadavu, Yadua and perhaps Matacawalevu (Lever 1985).

Galápagos Islands

Probably introduced in 1832 when settlement was established on Floreana (Charles), wild pigs are now present on the islands of Sierra Negra and Cerro Azul (Isabela I.), Santiago, Santa Cruz, Floreana and San Cristóbal, but are probably most numerous on Santiago (Eckhardt 1972).

Early introductions of pigs probably followed as the other islands were settled in the 1860s and 1890s. Darwin reported them on Floreana in 1835, and when Salvin (1877) visited these islands in 1875 they were present on Floreana, Chatham and Santiago (Lever 1985). An expedition in 1905 reported them present on Santa Cruz and Santiago and many were left when a farmer abandoned his stock in 1927.

Guam

Early Spanish colonisers introduced domestic pigs to the Marianas between 1672 and 1685 (Intoh 1986). The stock probably came from the Philippines and was introduced for food. Feral pigs were established by 1772 and were abundantly distributed throughout the island by the 1900s. Today they are widely distributed throughout the island. Since 1980 they have increased dramatically in the northern secondary limestone forests. They cause damage to agricultural crops, such as watermelon and taro, and also cause damage in forest areas by wallowing, trampling and rooting (Conry 1988).

Hawaiian Islands

Wild pigs were one of the first animals to be introduced to the Hawaiian Islands (Walker 1967), initially by the early Polynesians and later by Captain Cook and others (Bryan 1937; Griffin 1977) between 1778 and 1803 (Berger 1972). The first introductions by the Polynesians, most likely for food, occurred about 1000 AD and probably came from Tahiti (Smith and Diong 1977). The Asian pig varieties introduced by the Polynesians were later absorbed and replaced by those introduced by the Europeans (Tomich 1969).

Wild pigs were present on Kahoolawe in 1841 (Wilkes 1845), where they may have been present since 1823 at least, but there were none there in 1931 (Kramer 1971). Semi-wild animals occurred on Laysan in 1891 (Walker 1909), but the last was killed and eaten before 1910. Pigs were also present on Sand Island (Midway group) in 1915, but have not been present there for many years (Kramer 1971). Pigs were introduced to Laysan Island by a guano company operating there about 1890 and were present on the island for a few years; they were reported roaming there in 1891, but not in 1902 (Ely and Clapp 1973).

Pigs were left on East Island (French Frigate Shoals) in 1867 by the crew of the USS *Lackawanna* and five years later two were seen there, but they were not seen again (Amerson 1971).

In the early 1970s pigs occurred on Hawaii, Maui, Molokai, Oahu, Kaui, Niihau, and were present on Lanai from the early 1900s until the last was shot in about 1930 (Kamer 1971). A single animal was present on Kure Atoll in 1966 when introduced, but was removed later in the same year (Woodward 1972). Numbers were increasing on Niihau in 1951 (Fisher 1951).

Removal of feral pigs from national parks is continuing (Stone and Keith 1986) in the Hawaiian Islands through hunting, fencing and systematic searching.

Efforts to control them continued through the 1980s (Taylor and Stone 1986).

Japan
Little appears to have been documented on feral pigs in Japan. Some occurred in the Rokko Mountain area of Japan in the 1980s (Hirotoni and Nakatani 1987).

Juan Fernández
Wild pigs were introduced soon after 1574 on the island of Más á Tierra (Holdgate and Wace 1961). Juan Fernández stocked the islands with pigs between 1563 and 1574 and lived there until about 1580 when he left and abandoned the pigs and other animals (MacKenna 1883). Descendants of these pigs were abundant when Jakob le Mâitre and William Schouten visited the islands in 1616 for water and provisions (Encycl. Brit. 1970–80).

Kerguelen
Some domestic pigs were kept on Kerguelen for short periods in recent years near the scientific station (Watson 1975), but they are not known to have become feral there.

Kermadecs
Introduced to both Macauley and Raoul islands in 1836, pigs died out on Macauley and were eradicated on Raoul in the 1960s (King 1990).

Lord Howe Island
Introduced in the nineteenth century, feral pigs are still present on the island (Recher and Clark 1974; Flannery 1995). They were released there as a source of meat around 1800, with separate groups being placed on Big Slope in about 1900 to open up human access to commercially valuable stands of Howea palm.

Between 1978 and 1980 they ranged throughout the southern mountains, except for the summit of mounts Gower and Ligbird, and Little Slope. They were thought to perhaps be affecting woodhen (*Gallirallus sylvetris*) populations (Miller and Mullette 1985). Efforts to remove them were made from 1972 to 1981 when 180 were shot. There was, however, evidence that at least one pig remained in 1984.

New Caledonia
Pigs have been introduced and now occur on New Caledonia (Flannery 1995). Captain James Cook left a boar and a sow on New Caledonia on one of his voyages (first encountered in 1774) (Kippis 1904).

New Zealand
Captain James Cook gave pigs to the Maoris on all three voyages and noted some pigs had survived at least to the second voyage (Kippis 1904). Some were liberated (one boar and two sows) by Captain Cook in 1773 and there were many more later introductions (Wodzicki 1950). The first to arrive in New Zealand were two pigs which were given to natives at Doubtless Bay, Northland, in 1769 by de Surville, when the *Saint Jean Baptiste* called there (King 1990). They were plentiful in the Nelson province by 1840.

Many of the feral pigs in New Zealand descended from animals introduced by European visitors and settlers during the eighteenth and nineteenth centuries (Thomson 1922). Many feral colonies also came from pigs kept by the Maoris, who transported them about as trade, gifts and food (Challies 1976).

In the 1960s and 1970s feral pigs were common on both the North and South islands (Wodzicki 1965; Gibb and Flux 1973) and still occur in many areas of both islands (King 1990) and on many offshore islands.

Pigs were present on the Chathams and eliminated on Poor Knights Island in 1936 (Atkinson and Bell 1973). They were liberated on Raoul and Macauley islands in 1836 and on Enderby and Campbell islands in 1867 (King 1990). In the early 1970s they were widespread on the Chathams, Great Barrier, Auckland (Gibb and Flux 1973), Mayor, D'Urville and many other islands (Challies 1976). A control program was being carried out to remove them from Chetwode Island (Anon. 1980) in late 1970s (see table of introductions of pigs on islands for others).

Although pigs have been eradicated on some islands and have died out on others, their range has expanded in New Zealand since it was last mapped in the 1970s and 1980s. The main recent expansion has occurred in Otago, Southland, Canterbury and the West Coast of the South Island, partly from dispersal, but mainly from illegal liberations (Fraser *et al.* 1996).

Ryukyu Islands (Nansei Shoto)
The pigs on the Ryukyu Islands are thought by some (Imaizumi 1973) to be a separate species (*S. riukiuensis*), but most consider them to be feral pigs *Sus scrofa* (Oliver 1985). Most agree that they are probably endemic, and the balance of the evidence, including fossil material, suggests that they are (Groves 1981). However, introduced domestics have now run wild on the islands which complicates the issue further (Oliver 1984). Pigs are found on Amami Oshima, Kakerome, Tokunoshima, Okinawa, Ishigaki and Iromote (Oliver 1985).

Solomon Islands
Pigs have been introduced to and now occur on the islands of Choiseul, Guadalcanal, Malaita, Nendö, New Georgia, Rennell, San Cristobal and Santa Isabel (Flannery 1995)

Tokelau Islands

Pigs were introduced to the island in 1841 when Captain Hudson left three there (Wilkes 1845; Hales 1846). They ran at large on some islands such as Swain's where 500 existed in 1939–42, but in 1965 there were only a few (Kirkpatrick 1966). Most of the animals now belong to the natives and are largely semi-domestic.

Tongareva Atoll (Penryhn Atoll)

It is suggested that pigs were introduced in 1853 from a shipwrecked vessel the *Chatham* (Clapp 1977).

Vanuatu

Pigs have been introduced and now occur on the islands of Aore and Espiritu Santo (Flannery 1995).

SOUTH AMERICA

Pigs have been introduced into various parts of South America. On the mainland they have become feral in Argentina, Brazil, Peru, Colombia, Chile and Ecuador (Petrides 1975; Lever 1985). Domestic pig stock was introduced to South America by the Spanish in the fifteenth century. As happened in North America the practice of allowing domestic pigs free-range probably contributed to the initial feral populations occurring in South America.

Argentina

Pigs are now reported to be feral in many parts of Argentina and Chile. In the latter place they may have invaded from Argentina (Miller 1973). They have been reported from La Pampa (Lever 1985).

Brazil

Feral pigs are reported established in parts of the Matto Grosso on the Bolivian border, and from the states of Santa Caterina and Paraná (Lever 1985). Pigs have been introduced on Trinidade Island, off the coast of Brazil (Holdgate 1967).

Colombia

It is reported that pigs introduced 200 to 400 years ago are hunted for meat and for the tusks of the boars in the Llanos Orientales of Colombia (Lever 1985).

Costa Rica

Feral pigs have been successfully introduced on Cocos Island, Costa Rica.

WEST INDIES

Christopher Columbus introduced pigs to the West Indies in 1493 (Towne and Wentworth 1950; Beldon and Frankenberger 1977).

Bahamas

Feral pigs apparently still occur on some of the remote islands (Encyl. Brit. 1970–80).

Barbuda

Feral pigs occur on Barbuda in the Leeward Islands (Lever 1985).

Bermuda

Pigs were probably released on Bermuda as food for shipwrecked voyagers some time after 1515, when the island was discovered by Juan de Bermudez. They were abundant there in 1609, but appear to have disappeared between 1620 and 1630, possibly exterminated by the early colonists (Lever 1985).

Cuba

Columbus on his second voyage in 1493 carried some eight pigs that were to be released in the West Indies to become a food source for future voyagers (Towne and Wentworth 1950). These were apparently left or released in Cuba. The early Spanish settlers also introduced pigs and some of these also became feral (Oliver 1985). They apparently live sympatrically with collared and white-lipped peccaries (*Tayassu tajacu* and *T. pecari*) (Lever 1985), which have also been introduced.

Hispaniola

Pig husbandry during the Spanish period allowed for the easy escape of animals, some of which became established (Street 1962). The pigs were hunted by the Spanish stockmen for their meat and the earliest record indicates they were present in 1535, at which time they were said to be numerous (de Oviedo y Valdes 1851–55 in Street 1962).

In the seventeenth century pigs were apparently feral near Morne à Mantegue, near Limonade, which name commemorates a former abundance of them. During this century pig-hunting was apparently popular in many areas, particularly in western Hispaniola. In 1760 mention is made of buccaneers hunting them with the aid of dogs (Jefferys 1760 in Street 1962), especially in the Fort Liberté area. In 1724 (Labat 1724) there were complaints about the price of pork and its scarcity and in 1725 (Moreau de Saint-Méry 1797–1798 in Street 1962) it was reported that hunters travelled from Port de Paix to Port-à-Piment in the north-west to kill wild pigs. According to the same report they were numerous in 1780 and in the closing years of the eighteenth century in numerous areas.

Wild pigs now inhabit the Peninsula of Baradères, the Tapion de Papye region of Northwest Peninsula and on the coast near Grand-Grosier and Anse-à Pitre and also in sparsely settled districts in the high mountains e.g. pine forests of Massif de la Salle. They recently disappeared from Île de la Tortue and in the Massif de la Hotte north of Pic de Macaya (Street 1962).

Jamaica

Feral pigs have descended from stock originally introduced by the Spanish at about the end of the fifteenth century. Many domestic pigs were abandoned about 1838 with the abolition of slavery on the island. Feral pigs still survive on Jamaica (Oliver 1985).

Mona Island

Feral pigs occur on Mona Island (Lever 1985).

Puerto Rico

Introduced to Puerto Rico, pigs are recognised as a game animal on the island (Wood and Brennerman 1980). Some were reported present in the late 1970s (Wood and Barrett 1979).

Virgin Islands

Feral pigs occur on a number of the Virgin islands (Lever 1985). Some were noted present in the late 1970s (Wood and Barrett 1979).

■ DAMAGE

Feral pigs have a propensity towards negative effects on ecosystems in which they are introduced (Bratton 1975; Challies 1975; Spatz and Mueller-Dombois 1975; Wood and Barrett 1979; Diong 1982; Coblentz and Baber 1984; Singer et al. 1984; Baber and Coblentz 1986). They have a wide ecological plasticity and an ability to exist near humans (Timofeeva 1978).

In Europe pigs cause agricultural damage (Brierderman 1967; Snethlage 1967; Mackin 1970) to such crops as potatoes, corn, oats, turnips, beet and corn (Hvass 1961; Corbet 1966). In Poland in 1959 it was estimated that the population was about 40 000 pigs and it was planned to reduce their numbers because of the damage caused to crops (Haber 1961) such as oats, potatoes, rye, wheat and barley. In Europe they are also known to raid vineyards with drastic results (Corbet 1966), and in some forests in Russia they have caused considerable damage by destroying the seedlings of oaks, nuts and apricots (Dinesman 1959). Damage in Europe may depend somewhat on the natural availability of beech and acorn mast and when this is not available pigs invade cultivated crops (Mackin 1970).

In Malaysia feral pigs have caused damage to sweet potatoes and other crops (Medway 1978) including sugar cane, tapioca, padi, coconuts and rubber. A herd of 25 pigs completely uprooted a 15-acre tapioca plantation in Perak (Diong 1973). In Pakistan they have caused damage to a variety of crops of wheat, sugar cane and potatoes.

On islands pigs have been accused of causing serious damage to fauna and flora, but conversely on many Pacific Ocean islands and in South-east Asia they are an important food source. Pigs are reported to have seriously degraded natural forests and other vegetation on Ossabaw Island, Georgia, United States (Graves and Graves 1977), in the Hawaiian Islands (Tomich 1969; Jacobi 1976; Lamoureux and Stemmerman 1976), and on Santiago and other islands in the Galápagos Islands (Eckhardt 1972). On the Auckland Islands they virtually eliminated large-leaved sub-antarctic endemic species of *Pleurophyllum*, *Stilbocarpa* and *Anisotome*, and changed the plant composition of the high country. There is little evidence that pigs have modified the vegetation since the 1940s and control is not needed, as if left undisturbed the balance will be maintained between them and the modified environment (Challies 1975). In the Hawaiian Islands feral pigs are reported to have damaged the ecosystem by enhancing the growth of exotic plants (Tomich 1969; Spatz and Mueller-Dombois 1972; Jacobi 1976; Lamoureux and Stemmerman 1976).

Wild pigs are also reported to have been responsible for changes in the vegetation of Phillip Island (off Norfolk) before 1912 (Watson 1961), and on Lord Howe Island where they are causing extensive damage to herbaceous ground cover and may be responsible for the disappearance of a number of forest floor and soil invertebrates (Recher and Clark 1974).

In New Zealand pigs are now generally considered to be of minor importance in protection forests and are only locally a problem in production forests and on farm lands (Challies 1976).

Feral pigs have caused soil erosion on Santiago Island in the Galápagos (Eckhardt 1972).

A number of reports other than those already mentioned refer to the destruction of native animals by feral pigs. They are reported to have damaged the penguin rookeries early in the nineteenth century on Crozet Island (Holdgate and Wace 1961). On Mona Island, Puerto Rico, they destroyed 100 per cent of iguana nests in excessively dry years (normally only 25 per cent) on the coastal plain (Wiewandt 1977). On Auckland Island they have undoubtedly had an effect on the distribution and number of birds, but their role is difficult to separate from that of man and domestic predators, such as the cat and dog. The islands are now untenable to some species and others are adversely affected, especially sea and oceanic species that nest there (Bell 1963).

In Australia feral pigs are considered serious pests in some areas. They compete with stock, trample and graze cereal crops and improved pasture, degrade land and damage fences, roads, and water points (Hone 1980; Hone *et al.* 1980; Stevens 1981; Wilson *et al.* 1992). They can be a significant predator of lambs in some areas (Plant *et al.* 1978; Pavlov *et al.* 1981; Pavlov and Hone 1982). The total agricultural damage has been estimated to cost more than $70 million annually (Tisdell 1982; Pavlov 1983; Auld and Tisdell 1986).

In New South Wales pigs are a serious pest and are reported to damage sorghum, wheat, sunflower, barley, oats, potatoes, sugar cane, corn and Japanese millet (Hone *et al.* 1980). In Victoria they damage grain crops by feeding and trampling, and are known to have destroyed whole potato crops; more recently they have caused serious damage in vineyards by eating grapes from vines and drying racks (Stevens 1981). They damage flora and fauna by destroying habitats but there is little quantification of such damage (Auld and Tisdell 1986). In forest areas they adversely affect mountain ash (*Eucalyptus regnans*) re-growth by rooting up young trees (Stevens 1981).

Pigs are the basis of a significant commercial harvesting industry in Australia – with exports to lucrative European markets for consumption of game meat (O'Brien 1987).

In Papua New Guinea pig density appears to have a marked effect on the forest understorey and probably affects the abundance of other animals (Dwyer 1978). Hunters are sometimes killed or maimed by wild boars (Flannery 1995).

Although they are recognised as a valuable game animal in many parts of North America, wild pigs are also reported to cause considerable damage to the environment. Historically their effects have been controversial and in recent years there have been many studies in the United States in an effort to resolve the problems, particularly for wildlife managers. Certainly, feral pigs have no place in many of the areas that are presently inhabited.

In Texas pigs have caused damage to agricultural interests by damaging fencing and stock, and also the local wildlife. In the latter respect they have been accused of competing with the wild turkey (*Meleagris gallopavo*) for food and of also destroying their nests (Ables and Ramsay 1973). In South Carolina and other states they are considered a pest of longleaf pine (*Pinus palustris*) regeneration and agricultural crops, and are seen to compete with wildlife for the autumn mast crop, and to act as a reservoir of disease, particularly swine brucellosis and pseudorabies (Wood and Brenneman 1980). They are considered a valuable game animal in California, but

are also pests because of the damage to rangeland, crops and watering facilities (Barrett and Pine 1980). Here they have been reported to damage artichokes and grain fields (Pine and Gerdes 1973).

Pigs have affected the vegetation in the Great Smoky Mountains National Park (Bratton 1977), where their rooting activities have damaged the herbaceous understorey of several types of forest (Bratton 1976). Their success in this park has been attributed to similar plant associations to their native range and the lack of natural predators (Bratton 1975). They cause a reduction in the understorey cover and cause soil erosion from which the forest cannot recover while they are present (Bratton 1975). Rooting by pigs was found to accelerate the leaching of some minerals (Ca, P, Zn, Cu and Mg) from the leaf litter, and soil nitrate levels were also higher in the soil, soil water and stream water in rooted areas (Singer *et al.* 1984).

In national wildlife refuges, wild pigs damage crops grown for waterfowl, pine plantations, seedlings of trees for regeneration, and consume pine and hardwood seed available for wildlife (Thompson 1977). However, the exact role played in this damage is still not well known or documented. In the southern Appalachians it has been reported that competition between pigs and wildlife for most is dependent on the quantity and quality of the crop (Henry and Conley 1972).

Wild pigs are reported to be predators of ground-nesting birds (Thompson 1977), but their effects are thought to be minor. They certainly prey on the nests of wild turkeys (Matschke 1965; Henry 1969), but the degree of impact is not clear (Wood and Barrett 1979).

Relatively little longleaf pine is now being regenerated for commercial sale and it is this species which pigs are notorious for damaging, however, timber regeneration damage alone may not be a good measure of the total effect that pigs may be having on the forest flora (Wood and Lynn 1977). Damage to natural plant communities has been greatest in areas where public hunting is not allowed or is restricted because the area is a national park (Wood and Barrett 1979).

The potential for wild pigs to harbour diseases transmissible to man or domestic animals is considerable. Cholera, swine brucellosis, trichinosis, foot and mouth disease, African swine fever and pseudorabies can all be transmitted by wild pigs (Wood and Barrett 1979). From 1973 to 1977 there were 24 cases of trichinosis in humans in the United States who had contracted the disease from wild pigs (Centre for Disease Control 1974–78). Brucellosis presently occurs in feral pigs in South Carolina (Wood *et al.* 1976), Florida (Becker *et al.* 1978) and Hawaii (Griffin 1978) and is transmissible to humans (Siegmond 1973). Pseudorabies has been discovered in pigs in Florida, and pigs in one area of California provided a reservoir for bovine tuberculosis that infected the cattle on a ranch at San Simeon in 1965 (Wood and Barrett 1979).

In Australia feral pigs are thought to be an important link in the chain of transmission of certain infectious diseases between indigenous animals, domestic stock and humans (Keast *et al.* 1963). They are an important source of infection of leptospirosis (Keast *et al.* 1963). Sparganosis (Appleton and Norton 1976), and also *Brucella suis* (Norton and Thomas 1976) have been found in feral pigs. Many have abscesses caused by mycobacteria (Cornes *et al.* 1981). They are a reservoir for Murray Valley encephalitis and Ross River virus (Strahan 1995).

Feral pigs may be capable of playing a major role in Australia in the spread of an exotic disease such as foot and mouth (Garner and O'Brien 1989). They are also known to carry and spread anthrax, brucellosis, hydatids, leptospirosis, measles, sparganosis and tuberculosis (Stevens 1981).

In 1956–57, over 40 000 pigs were destroyed in a bounty scheme in New Zealand because it was thought that feral pigs had a great potential as a vector of various exotic diseases (Martin 1973).

AFRICAN BUSH PIG
Red river hog, bosvark, bushpig
Potamochoerus porcus (Linnaeus)

■ DESCRIPTION
HB 1000–1500 mm; T 300–450 mm; SH 550–965 mm; WT 45–130 kg.
Short rotund body; back rounded; coat short haired, reddish or reddish brown (black in old males); face patches white; muzzle narrow; ears long, pointed, white and tufted; pair of warts on face below eyes; white or black line along back; snout elongated; upper tusk 75–76 mm, lower tusks 165–190 mm; female has three pairs of mammae.

■ DISTRIBUTION
Africa. Africa south of the Sahara except West Africa; southern Ethiopia and southern Somalia south to south-west Cape. Presence on Madagascar and Mayotte (Comoros) as a result of introduction. Also on Zanzibar and Mafia islands(?).

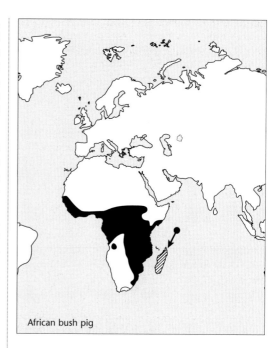
African bush pig

■ HABITS AND BEHAVIOUR
Habits: mainly nocturnal, but also diurnal; sounders use communal latrines. **Gregariousness:** families, harems, herds or sounders 4–24 (1 male, several females and young) and sometimes larger groups to 40; old males may be solitary; density 1.29/km². **Movements:** sedentary; feeding to resting areas up to 4 km; home range 20–1000 ha. **Habitat:** reed beds, thick cover with nearby water, well-wooded ravines, broken country with patches of forest and thickets; swamps, marshes. **Foods:** omnivorous; grass, roots, bulbs, tubers, corms, leaves, fallen fruits, berries, seeds, insects, snakes, frogs, birds' eggs, young birds, carrion, fungi, earthworms and insect larvae. **Breeding:** breeds all year, but mainly September–March, peaks in summer rainy season; gestation 120–175 days; litters 2–6, 10; probably 2 litters/year; young born in nest of grass or other vegetation; sexual maturity females at 3 years, males 2 years. **Longevity:** 12–20 years (captive). **Status:** common.

■ HISTORY OF INTRODUCTIONS
INDIAN OCEAN ISLANDS
Madagascar and (?) Comoros
The bush pig may be a recent introduction to Madagascar (George 1962). Its presence on the Comoros and Madagascar is probably through introductions (Oliver 1985). In the literature it is often suggested that it is native to these islands and that perhaps it is even a separate race. It may have colonised these islands naturally since bush pigs sometimes venture into extensive papyrus beds that

may detach and float out to sea (Kingdon 1979). It has probably been introduced to Mayotte in the Comoros (Ansell 1971 in Schreiber *et al.* 1984).

■ DAMAGE

The bush pig is referred to as one of Africa's most serious pests; stockade type fences are required to protect crops such as manioc and maize (Estes 1993). They are a menace to cultivated crops (Burton and Burton 1969) because they feed by extensively rooting up an area (Morris 1965).

Bush pigs damage many kinds of cultivated crops and have been known to wipe out entire peanut crops. They cause damage to cultivated crops such as maize, sweet potatoes and potatoes in Transvaal, and to pineapples, bananas, mango, young fir trees and also to sugarcane plantations in Natal (Naude 1962; Burton 1962). In Zimbabwe damage is caused by bush pigs to maize, potatoes, ground nuts, pineapples and irrigated fruit tree roots (Jarvis and La Grange 1984), and they can do considerable damage in a short period of time (Walker 1992).

Warthog

WARTHOG

Cape, Somalia or desert warthog
Phacochoerus aethiopicus (Pallas)

■ DESCRIPTION

HB 900–1500 mm; T 250–500 mm; SH 630–850 mm; WT 48–150 kg.
Long legs; head large; muzzle broad; grey to greyish brown, sparsely covered with bristles which form yellowish or brown mane; mane extends to mid back, then continues on rump; long ridge-like folds on cheeks have white hairs; males have prominent wart-like protuberances on either side of head. Females smaller than male and with shorter tusks.

■ DISTRIBUTION

Africa. Africa south of the Sahara from Ghana south to South Africa.

■ HABITS AND BEHAVIOUR

Habits: usually diurnal (boars somewhat less diurnal); makes burrows in the ground or usurps those dug by others; **Gregariousness:** clans, sounders or family groups (pair and 2–4 young); bands 4–16; groups occasionally to 40; old adult males often solitary; density 0. 2–30/km². **Movements:** daily up to 7 km; overlapping home ranges 64–420 ha. **Habitat:** savannah, open treeless plains, light forest. **Foods:** grass, roots, tubers, carrion, berries, fruits, bulbs, bark of trees. **Breeding:** gestation 170–175 days; females seasonally polyoestrous; oestrus lasts 72 hours at intervals of 6 weeks; litter size 1–8; born hairless, remain in burrow 6–7 weeks; young weaned 3–6 months; sexual maturity 17–20 months. **Longevity:** 18 years 9 months (captive and wild). **Status:** declining, but still fairly common.

■ HISTORY OF INTRODUCTIONS

AFRICA
South Africa

Warthogs were eliminated from South Africa, but occur elsewhere in Africa. They were unsuccessfully re-introduced to the Golden Gate Highlands National Park after 1963 from Natal (Penzhorn 1971).

Zimbabwe

Six warthogs were introduced to Metopos National Park before 1963 by Southern Rhodesian authorities, but it is not known if they are surviving there. Fourteen were also introduced to McIlwaine National Park, but it is not known if they became established (Riney 1964).

■ DAMAGE

No information.

BABIRUSA

Babyrousa babyrussa (Linnaeus)

■ DESCRIPTION

HB 870–1070 mm; T 270–320 mm; SH 650–800 mm; WT to 100 kg.
Skin rough brownish grey or smooth and sparsely haired with whitish grey to yellowish hairs; upper

Babirusa

tusks grow upwards through muzzle and curve backwards; males also have prominent lower tusks; underside and legs sometimes lighter; females have two pairs of mammae.

■ DISTRIBUTION
Sulawesi. Northern and eastern Sulawesi, Lifamatola?, Mangole, Toliabu nearby Togiean Islands (Malengi Islands), Sula Islands, and Buru and Lembeh in the Moluccas.

■ HABITS AND BEHAVIOUR
Habits: diurnal; elusive; active. **Gregariousness:** small parties, groups or solitary. **Movements:** sedentary. **Habitat:** moist forest, canebrakes, shores of rivers and lakes. **Foods:** foliage and fallen fruits, roots, berries, tubers, leaves. **Breeding:** gestation about 125–160 days; 2 litters/year; litter size 1–2. **Longevity:** 24 years (captive). **Status:** declining; uncommon and probably endangered.

■ HISTORY OF INTRODUCTIONS
INDONESIA
Sulawesi
On account of the impoverished nature of the mammalian fauna on the island of Buru (south Molaccas) and the Sula Islands (Taliabu and Sulabesi), it is believed by some authorities that the barbirusa was introduced to these islands by native peoples in prehistoric times (Dammerman 1929; Groves 1980; Schreiber *et al.* 1984; Flannery 1995).

■ DAMAGE
No information.

Family: Tayassuidae
Peccaries

COLLARED PECCARY
Peccary, muskhog, javelina
Tayassu tajacu (Linnaeus)
=Dicotyles tajacu (Linnaeus) =Pecari angulatus

■ DESCRIPTION
HB 644–1050 mm; T 10–106 mm; SH 450 mm; WT 16–23.5 kg.
Coarse black or dark brown fur mixed with white giving a greyish appearance; cheeks have yellowish tinge; collar a semicircle of whitish to yellowish hairs from jaws over shoulders to throat; mane of long stiff hairs along mid-dorsal line from crown to rump giving the back a black stripe; tail vestigial; feet blackish; four front and three hind digits.

■ DISTRIBUTION
Central and South America: south-western United States (Arizona, New Mexico, Texas), and northern Mexico to Patagonia and northern Argentina.

■ HABITS AND BEHAVIOUR
Habits: active day and night; hide in burrows of armadillo or in fallen tree trunks; territorial.

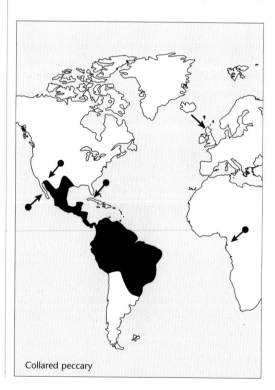

Collared peccary

Movements: sedentary?; home range 38–126 ha. **Gregariousness:** groups of 5–15 (mixed sexes). **Habitat:** rainforest, arid woodland, dry scrub, semi-desert. **Foods:** roots, bulbs, tubers, fruits, fruit and stems of cacti, berries, green grass, shoots, herbs, small invertebrates, insect larvae, grubs and worms. **Breeding:** may breed year round; gestation 142–158 days; young 2, but varies 1–4; young weaned at 6–8 weeks. **Longevity:** up to 20 years. **Status:** common.

■ HISTORY OF INTRODUCTIONS

AFRICA
Peccaries (*Tayassu* sp.) from the United States have been introduced to the Wonga-Wongue Presidential Reserve in north-western Gabon where they have met with some initial success in becoming established (Nicoll and Langrand 1986; Blom *et al.* 1990).

AMERICA–CARIBBEAN
Cuba
The collared peccary (*T. t. yucatanensis*) and white-lipped peccary (*T. pecari* (*T. p. ringens*)) were introduced by humans in 1930 to western and eastern Cuba (Varona 1974 in Hall 1981; Corbet and Hill 1980). They were introduced for hunting they are now sympatric with feral pigs (*Sus scrofa*) which were introduced by Spanish settlers (Oliver 1985).

The peccaries are now established in the areas of Pinar del Río and Sierra Cristal, Oriente, Cuba (Lever 1985).

Mexico
The collared peccary (*T. tajacu = P. angulatus*) may have been introduced to Cozumel Island, Mexico (de Vos *et al.* 1956). However, more recent authorities make no mention of any introduction.

United States
The Arizona Game and Fish Department translocated collared peccaries into Arizona by moving over 400 animals. However, they dispersed in all directions on release and have not as yet become established anywhere, but the experiments are to be continued (Day 1985).

EUROPE
Scotland
In 1972 a pair of collared peccaries were released by the Countess of Arran on Inchconnachan in Loch Lomond, Scotland. One disappeared and the other was removed in 1984 (Lever 1985).

■ DAMAGE
No information.

WHITE-LIPPED PECCARY
Tayassu pecari (Link)

■ DESCRIPTION
HB 885–1380 mm; T 10–65 mm; SH c. 530 mm;
WT 20. 5–49.5 kg;
Colour dark brown to black; cheeks to lower jaw white; dorsal hairs erectile; some also have white on tip of muzzle and under eye as well as in pelvic region; snout long, mobile; ears white haired on inside; legs black and tan.

■ DISTRIBUTION
Central and South America: discontinuous distribution from southern Mexico to Argentina.

■ HABITS AND BEHAVIOUR
Habits: musky smell; noisy; active in cooler hours of day; agile. **Gregariousness:** large groups 50–100 (90–138 of both sexes and ages); density 1.06/km². **Movements:** some seasonal movements; home range 60–200 km². **Habitat:** forest. **Foods:** palm nuts and seeds; plant material, figs, bulbs, roots, and small animals. **Breeding:** breeds July–September; gestation 156–158 days; young 2; born reddish, adult colouration in second year; females mature at 18 months. **Longevity:** probably similar to *T. tajacu*, up to 20 years. **Status:** common?

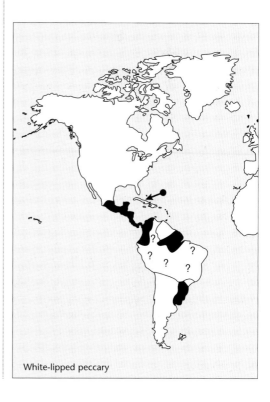

White-lipped peccary

■ HISTORY OF INTRODUCTIONS
CARIBBEAN
Cuba
White-lipped peccaries have been introduced for hunting in Pinar del Río and the Sierra Cristal, Oriente, Cuba (Lever 1985) in 1930, where they have become established.

■ DAMAGE
No information

Family: Hippopotamidae
Hippopotamuses

HIPPOPOTAMUS
Hippopotamus amphibius Linnaeus

■ DESCRIPTION
HB 2900–5050 mm; T 460–560 mm; SH 1300–1650 mm; WT 655 to 4500 kg.
Large, smooth, naked skin; inflated looking body; colour slaty copper brown with shades to dark brown above and purplish below; body covered with sparse covering of hair; upper and lower canines enlarged into tusks up to 300–600 mm long and weighing up to 3 kg; skin glandular; eyes protruding; ears well back on head; tail paddle-like, muscular and short.

■ DISTRIBUTION
Tropical Africa. River systems almost throughout Africa south of the Sahara; extinct in Egypt and the northern Sudan, Sierra Leone to Nigeria. Also occurs on Mafia and Bijangos islands.

■ HABITS AND BEHAVIOUR
Habits: rests during day, grazes by night on land; amphibious; territorial. **Movements:** sedentary; foraging trips to 3–33 km; will graze 3.2 km from water. **Gregariousness:** solitary; aggregations to 2–150; usual group size 10–15; groups may be bachelor or females and young; density 7.7–19.2/km^2. **Habitat:** forest by fresh water; adjacent reed beds and grassland, swamps. **Foods:** grass. **Breeding:** mainly mates in dry season, calves in rainy months; females polyoestrous; gestation 225–257 days; litter size 1–2 (twins rare); inter-birth interval 2 years; born mostly on land, active shortly after birth; female stays with calf 10–44 days then joins herd; calves sometimes créched; weaned at 8–12 months; sexual maturity females 3–4 years, conceive at 7–15 years, males at 4–5 years. **Longevity:** 40–45 years wild, 54 years 4 months captive. **Status:** declining, range reduced and largely confined to protected areas, but still in many major rivers and swamps.

■ HISTORY OF INTRODUCTIONS
AFRICA
In historic times hippopotamuses were exterminated over the entire northern parts of their range; the last was found in Egypt in 1816.

South Africa
Hippopotomuses captured in the Kruger National Park were unsuccessfully translocated to Aldo Etosha National Park (Penzhorn 1971; Novellie and Knight 1994). The captures and translocations continued in the 1980s (Henwood and Keep 1989).

Between 1981 and 1984, eight were re-introduced successfully to Pilanesberg National Park, Bophuthatswana (Anderson 1986).

■ DAMAGE
Damage is caused to rice crops in Gambia, but the extent of destruction is exaggerated in order that hippos may be shot for food. Because they are less palatable than other animals (i.e. baboons, monkeys and weaver birds), they are overlooked although separately they may do as much damage (Clarke 1953).

Hippos cause extensive damage, especially in areas where crops are near river banks, to such crops as corn and sugar cane by eating and trampling the crops (Hvass 1961; Jarvis and La Grange 1984; Walker 1992).

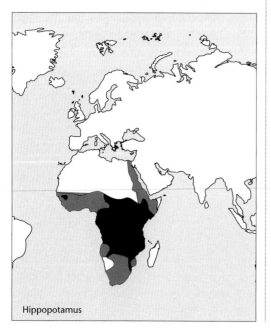
Hippopotamus

Family: Camelidae
Camels

NOTE ON SOUTH AMERICAN CAMELIDS

There are four forms or species of South American Camelidae, the llama (*Lama peruana* or *Lama glama*), alpaca (*Lama pacos*), guanaco (*Lama guanicoe*) and vicuña (*Lama vicugna*). They will breed with one another. The llama is the largest standing *c.* 1. 2 m at the shoulder, weighs 66–151 kg and has coarse wool which is black, white or brown or shades between. The alpaca is smaller than the llama, has a woolly face, short ears and a rounded rump; with spongy, crimped or straight locks; is *c.* 1m in height at the shoulder and averages 62–64 kg. The llama and alpaca exist only under domestication; they were probably domesticated in the Andean 'puna' by 4000 BC, but the nature of the ancestral forms from which they were domesticated remains a matter of debate. The most recent evidence suggests that the guanaco is the ancestor of the llama and that the vicuña is the ancestor of the alpaca, and that their domestication occurred between 6000–7000 years BP. When the Spanish arrived in 1532 there were probably 10 million alpacas and almost as many llamas – together with a similar number of the wild species. The guanaco exists only in the wild and has a shaggy coat of dark chestnut brown and looks somewhat like a llama. The vicuña is smaller and has thick, fine wool and averages *c.* 38. 5 kg; the toes end in broad elastic pads.

The Incas rounded up the wild vicuña and guanaco for shearing about every four years. With the advent of Spanish rule this ceased and within a century the populations of alpaca and llama, as well as the wild species, had been decimated in the Andes. Competition with domestic stock may have been the main reason for the disappearance from the greater part of their range (Summer 1988; Wheeler 1988; Torres 1992).

VICUÑA
Alpaca
Lama vicugna (**Molina**)
=*Vicugna vicugna* and *L. vicuna*

■ DESCRIPTION
HB 1250–1900 mm; T 150–250 mm; SH 700–1100 mm. WT 35–65 kg.
Head small, ears prominent; body slender; slightly smaller than guanaco, paler and lacks dark face; coat thickest and longest on sides; wool is thick and fine; upper parts of coat light brown or cinnamon and with yellowish, whitish or yellowish red bib on lower neck and chest; chest with long, off-white fur; under parts whitish; inner thigh surfaces white; toes end in broad elastic pads.

■ DISTRIBUTION
South America: Southern Peru and Bolivia to northern Chile and north-western Argentina. Formerly a more extensive range.

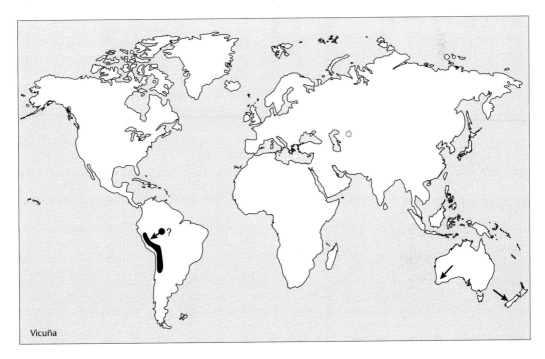

Vicuña

■ **HABITS AND BEHAVIOUR**

Habits: males territorial at *c.* 3–4 years old; defecate on traditional dung piles. **Gregariousness:** groups to 20; territorial males maintain family groups (male, adult females and young less than 1 year); troups 15–25; multi-female groups and one male/and bachelor male groups; density 14–87/km². **Movements:** sedentary, territorial; territories 7–30 ha. **Habitat:** arid montane grassland and plains at high altitude (above 3500 m). **Foods:** forbs and grasses. **Breeding:** mates April–June, births February–March; gestation 10–11 months (330–352 days); young 1/year; young born in spring; lactation 6–10 weeks; females mature 12–14 weeks, but most breed at *c.* 2 years; polygynous. **Longevity:** 10 to 24 years 9 months (captive). **Status:** range and numbers considerably reduced through hunting for meat and wool.

■ **HISTORY OF INTRODUCTIONS**

AUSTRALASIA

Western Australia

Possibly introduced unsuccessfully into Western Australia (Allison 1969; Long 1988).

PACIFIC OCEAN ISLANDS

New Zealand

Alpacas were introduced to New Zealand in 1869 and a herd was kept on private property (W. B. Rhodes) in the 1870s. The original five animals increased to 13 in four years. They were present there until 1874 when sold to J. Matson who kept them for many years, but there are no further records of them (Lamb 1964).

SOUTH AMERICA

Peru

Some family groups of vicuñas have been successfully transferred to other parts of the Peruvian Andes with some success (Andrews 1982).

■ **DAMAGE**

No information.

GUANACO

Llama

Lama guanicoe (Muller)

=*L. huanaco*

■ **DESCRIPTION**

HB 1200–2800 mm; T 150–250 mm; SH 900–1300 mm. WT 48–96 kg.

Head grey; ears pointed and deeply cleft; lips mobile; ear edges and around lips white; upper parts of short, woolly coat are reddish brown or tawny brown; under parts white; face blackish; lower neck with white collar; limbs and neck slender; inner sides of legs white; hooves modified. Female has four mammae.

Distinguished from the vicuña by larger size and have callosities on inner sides of fore limbs and lack whitish or yellowish bib of vicuña.

■ **DISTRIBUTION**

South America. Equador south to Tierra del Fuego and east across Patagonia. Domestic forms in Peru and Bolivia. Formerly inhabited northern Peru, Chile and Argentina and perhaps southern Colombia south to southern tip of Chile. Now found in southern Peru along the Andean zone of Chile and Argentina and thence to Tierra del Fuego and Navarino Island; also occur in southern Paraguay and southern Argentina south of 40°.

■ **HABITS AND BEHAVIOUR**

Habits: family groups have territories defended by male; communal defecation piles. **Gregariousness:** herds or family bands (1 male, several females (4–18) and young); male troops or herds to 200 (unstable in size), solitary males. **Movements:** in some areas seasonally migratory (spends winter in more sheltered areas). **Habitat:** semi-arid and arid grasslands at high elevations, pampus; mixed forest, shrublands, grasslands. **Foods:** generalist herbivore; grazes and browses grasses, lichens, forbs, seeds, fruits, rushes and sedges. **Breeding:** mates August–September; male polygamous; gestation 10–11 months; inter-birth interval 2 years; young 1;

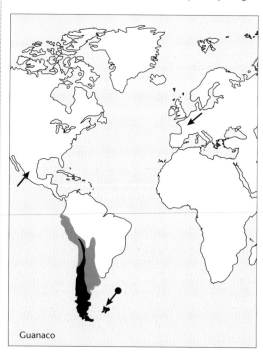

Guanaco

lactation 15 months; female mature in second year. **Longevity:** about 20 years. **Status:** rare in wild; reduced to a few scattered populations through hunting and habitat destruction.

■ HISTORY OF INTRODUCTIONS
AUSTRALIA
Guanacoes were unsuccessfully introduced into Western Australia (Long 1988). Their introduction was recommended soon after European colonisation in 1839, as it was thought that such an animal would be useful in arid areas as a source of wool and meat. The Acclimatisation Committee of Western Australia obtained a single animal in 1898–99, but its fate is unknown. In the late 1980s the species was imported into Australia as a domestic animal and is now widely farmed here.

ATLANTIC OCEAN ISLANDS
Falkland Islands
First introduced in 1862, guanacoes were certainly well established there in 1871 when some were shot by the then Duke of Edinburgh near Mare Harbour (Lever 1985). Guanacoes are still present in the vicinity of Mare Harbour.

In the 1930s J. Hamilton placed a small number on Statts Island (off Weddell Island), where they increased to a herd of several hundred. To prevent overgrazing and starvation, in 1956–59 nearly 400 of them were shot. In 1970 there were about 60 there and their numbers have fluctuated since then, but there were 150–200 in 1983 (Lever 1985).

EUROPE
Germany
Guanacoes were introduced in 1860 to Vogesen (?) and the Pyrenees. Some were introduced to Germany in 1911, but did not become established (Niethammer 1963).

NORTH AMERICA
Mexico
Guanacoes may have been introduced into Mexico in the sixteenth century (de Vos *et al.* 1956), but do not now appear there except as a farmed animal.

■ DAMAGE
In South America guanacoes are hunted for their flesh and because sheep farmers regard them as competitors of sheep in areas where there is little grass. Sheep have been largely replacing the guanaco and llama as domestic animals, leaving them to roam. The herds that once roamed the countryside, have now been considerably depleted by hunting and the species is endangered.

DROMEDARY
Arabian camel, one-humped camel
Camelus dromedarius Linnaeus
=*Camelus ferus* Przewalski

BACTRIAN CAMEL
Two-humped camel
Camelus bactrianus Linnaeus

The two camels are here retained as separate species although they are treated together. The two hybridise and the progeny will breed. The dromedary has no wild counterpart and may have been derived in part from the Bactrian. It is indicated that the earliest name applied to the Bactrian is Camelus bactrianus *L. There is no reason other than established useage for not including the two as subspecies i.e.* C. b. dromedarius *and* C. b. bactrianus *(Corbet 1978; Gauthier-Pilters and Dagg 1981; Wilson 1984).*

■ DESCRIPTION
HB 2200–3450 mm; T 350–550 mm; SH 1700–2300 mm. WT 300–1000 kg.
Long eyelashes; long legs; long neck; feet are large pads with two anterior toe nails. Bactrian has longer, darker hair, shorter legs and a more massive body. Colour varies from deep brown to dusty grey; *dromedarius* has a single hump, *bactrianus* has two; *dromedarius* has a short fine coat, *bactrianus* has long hairs, thickest on head, neck, humps, forelegs and tail tip.

■ DISTRIBUTION
The dromedary is not known in the wild in historic times, although some suggest that they may have survived into the early Christian era 2000–1800 years BP in the wild. It was probably domesticated in central or southern Arabia about 4000 years BP, and may originally have inhabited an area from North Africa to India. In historic times it occurred as a domestic animal in North Africa south to about 13°N. In the eastern parts of its range it occurs in northern Sudan, northern Kenya, eastern Ethiopia, and Somalia. In Asia it occurs on the Arabian Peninsula, in Syria, Lebanon, Israel, Jordan, Turkey, Iran, Iraq, Afghanistan, Pakistan, north-west India, China (Sinkiang) and in the south-western parts of the Russian Federation (mainly Turkestan).

Bactrian camels are present in north-eastern Afghanistan, the Steppes of southern Russia, and in Siberia east of Lake Baikal, Mongolia, and northern China. Wild animals may still persist in the Trans-Altai Gobi Desert at Lop Nor and in southern Mongolia. However, they interbreed with feral animals and it is not known how pure the stock may now be. In the late 1970s it was estimated that there

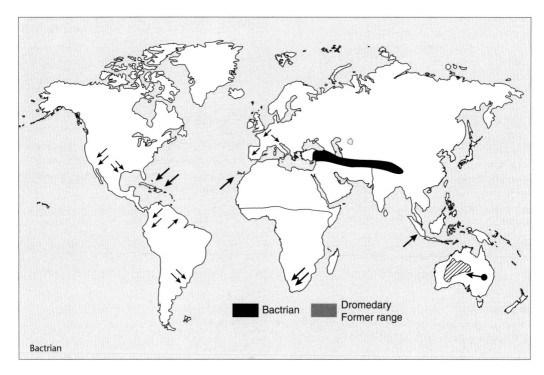

Bactrian ■ ▨ Dromedary Former range

Bactrian

were about 900 animals. (Zeuner 1963; Burton and Burton 1969; Gauthier-Pilters and Dagg 1981; Wilson 1984; Whitfield 1985).

■ **HABITS AND BEHAVIOUR**
Habits: diurnal, mobile forager (50–70 km/day); can withstand extreme heat. **Gregariousness:** groups of 6–45 (bachelor males; female with new born; several females with young and 1 male; also small herds of mixed sexes), surplus and older males often solitary; density 5/100 km^2 (Australia 1/7.5 km^2). **Movements:** generally sedentary, but wanders over large areas. **Habitat:** desert and semi-desert regions; generally dromedary inhabits the dry arid climates and flat terrain. The bactrian inhabits the more mountainous, rocky regions. **Foods:** desert vegetation; grasses, herbs and browse from shrubs and trees, and often very spiny plants; fruits, leaves, stems. Distinct preference for succulent and semi-succulent plants high in moisture. **Breeding:** most mate November–May (May–October in Australia); young born in January–March; gestation 370–440 days; females may be pregnant only every 18–24 months, but in some areas feral reported to breed annually(?); female seasonally polyoestrous; oestrous cycle 13–14 days; litter size 1 rarely two; lactation 1 year or more; mature at 3–4 years. **Longevity:** dromedary at least 40 years; bactrian 50 years. **Status:** *dromedarius* common in captivity; *bactrianus* now less common and may be endangered.

■ **HISTORY OF INTRODUCTIONS**
AFRICA
Southern Africa
British, Portuguese and German colonists introduced camels into south-western Africa (Kalahari) with some success towards the end of the nineteenth century (Wilson 1984; Lever 1985). The earliest introductions may have been about 1898 to Namibia. Those introduced by the Germans came from the Canary Islands, while some of those brought in by the British were imported from Somalia.

Some camels are still in use in Botswana for police work (Wilson 1984), but none appear to be feral in Africa. Arabian camels have also been introduced into fenced areas on private farms in Orange Free State in recent years (Lever 1985).

ASIA
Java
An early attempt to introduce camels to Java failed when they died from a liver disease (Saint-Hillaire 1861; Lever 1985).

China
It has recently been discovered that some 120 bactrian camels live in the desert of Xinjiang Province and in 1997 a sanctuary was established for them in the area. They may occur there naturally or may have been introduced.

EURASIA

Camels appear to have been first used by humans in about 5500 BC. Domestic forms (dromedary) evolved during the period about 2500–1700 BC. The bactrian was domesticated separately some time before 2500 BC, probably in northern Iraq and south-western Turkestan (Zeuner 1963; Gauthier-Pilters and Dagg 1981). As the original range of the wild animal is not known, it is impossible to say to into which areas the animals may have been introduced. Certainly the dromedary was taken to other areas in western Europe where archeological evidence to date suggests that they were not native. Certainly in historic times camels were introduced to parts of southern Europe (Hvass 1961).

Egypt

Camels disappeared or became extinct in Egypt in pre-historic times, but were re-introduced again at later dates.

France

A number of camels were used in France in the mid-nineteenth century (Wilson 1984).

Germany

Attempts were made to introduce camels to Germany (Wilson 1984), but these were unsuccessful.

Italy

In 1622 some camels from Tunis or India (Cochi 1858) were taken to Tuscany, where a herd existed on the grassy plains near Pisa (Burton and Burton 1969). These are reported to have been imported by Ferdinand II de Medici and placed at San Rossore, near Pisa (Lever 1985). Further imports were made between then and the mid-eighteenth century, when there were 196 of them and they survived in Tuscany until World War 2 (Wilson 1984).

At the end of the nineteenth century an effort was made to establish bactrian camels in Sicily, but this was unsuccessful (Lever 1985).

Portugal and Spain

The Muslims are reported to have introduced the first camels to southern Spain (Lever 1985). A breeding herd derived from Canary Island stock was established on the royal estate at Aranjuez from 1786 until the start of the Spanish Civil War.

Some 110 dromedaries were introduced to Spain in 1829–31 (Graells 1854; Grzimek 1972) and some 80 were released in Coto Donana in the Guadalquivier delta (Niethammer 1963; Grzimek 1972) by the Moors (Burton and Burton 1969). Here their descendants lived freely in swampy country until 1950 when the last five animals were stolen (Grzimek 1972).

Thirty camels were introduced to Barcelona in 1831 by the Captain-General of Catalonia, but these did not survive for long (Lever 1985).

An unsuccessful effort was made to establish bactrian camels in Portugal at the end of the nineteenth century (Lever 1985).

Russian Federation, Poland, Macedonia and Bulgaria

Towards the end of the nineteenth century unsuccessful attempts were made to establish bactrian camels in the Russian Federation, Poland, Macedonia and Bulgaria (Lever 1985).

ATLANTIC OCEAN ISLANDS

Canary Islands

Camels were also taken to the Canary Islands (Gauthier-Pilters and Dagg 1981) from Morocco in 1402–06 by Jean de Béthencourt (Lever 1985). They were certainly there in 1840 when some were imported to Australia. They continue to be used there as a beast of burden.

NORTH AMERICA

Canada

Ninety bactrian camels were imported to the United States in 1860, and some of these animals were taken to Canada (Hutchinson 1950). Twenty-two were imported into British Columbia in 1862 for use during the gold-rush times. These proved unsatisfactory for this type of work and several were turned loose. Some were apparently released at Lac La Hache and some at Westwold (Hutchinson 1950; Carl and Guiguet 1972). In the latter area they survived for several years in the wild, and the last animal in the Okanagan Valley died in about 1905 (Carl and Guiguet 1972).

United States

Camels were taken to Virginia as early as 1701, when they were imported as curiosities (Dareste 1857). Further imports were made in the period 1856–58, when about 120 camels were shipped to Texas during the Civil War for use as pack animals by the army (Lesley 1929; Emmett 1932; Greenly 1952; Fowler 1954).

The idea of introducing camels to the United States was supported by some army officers as early as the 1840s. In 1851 efforts to import 50 camels was thwarted by the army administration (Carson 1980). In 1855 a camel bill was passed by the House of Representatives but rejected by the Senate. At last in 1855 sufficient money (US$30 000) was secured and officers were sent to Europe to procure stock. In February 1856 the ship *Supply* headed for Texas with nine dromedaries from Egypt, 20 burden camels and four others of mixed breed, which were landed in April 1856 (Carson 1980). The same ship landed a further 44 camels at the mouth of the Mississippi River in 1857. A third load was delivered to Indianola by private persons. These latter animals were turned loose in coastal country, but were killed off by cattlemen (Carson 1980). Some camels that were used by the government for survey work were either sold or released in Arizona when the work was completed in 1857 (Gauthier-Pilters and Dagg 1981).

The imported camels were tested on various projects and at the beginning of the Civil War were kept at Val Verde, 60 miles south-west of San Antonio near the Guadelupe River. Some were taken to California in 1857 but apparently were never used and during the war were sold to an entrepreneur for circuses and zoos.

In 1861 the Confederate forces seized the Val Verde herd. Some of them were used as work animals, but some were also turned loose. During the reconstruction years following the Civil War some of the Val Verde camels ended up in zoos and circuses, but the government turned some loose in Arizona, where they thrived and bred in the wild (Anderson 1934; Carson 1980).

Twenty to 30 camels were caught near Tuscon, Arizona, in 1877 and used as pack animals (Legge 1936). Some were noted in Texas in 1891 at Kingsville (Gauthier-Pilters and Dagg 1981). In 1870 a salt miner rounded up 25 in the Carson River district of Nevada (Carson 1980). A herd was seen near Oatman's Flat on the Gila River, Texas in 1875 (Carson 1980). In 1885 Douglas MacArthur (later General) saw one at Fort Selden in New Mexico (Carson 1980). In 1890–91, nine were noted on the edge of Death Valley, California (Carson 1980).

The last of the feral camels in the United States may have been in about 1902 (Carson 1980), or perhaps two noted in Nevada in 1907 (Legge 1936). Some reported roaming the Arizona and Nevada desert areas until at least 1905 (Gauthier-Pilters and Dagg 1981), possibly as late as 1915 (Burton and Burton 1969), with some unconfirmed reports received up until 1929 (Legge 1936).

WEST INDIES
Cuba and Jamaica
Attempts were made to introduce camels into both Cuba and Jamaica, but all attempts were apparently unsuccessful (de Vos *et al.* 1956; Gottschalk 1967). Camels were imported into Barbados as early as 1675 and to Jamaica before 1774, where they survived for

about 50 years before dying out (Legge 1936). In 1841 some 70 were used in carrying copper ore near Santiago, Cuba, and were subsequently used on sugar cane plantations (Dareste 1857). All the attempts known were to introduce camels under domestication and as far as is known none ever became feral in these areas.

SOUTH AMERICA
Peru, Brazil, Bolivia and Venezuela
Camels were taken to South America (from the Canary Islands to Peru) by the Spanish in the sixteenth century, but only for domestication and no releases in the wild are known. Eventually the domesticated ones died out (de Vos *et al.* 1956; Gottschalk 1967; Burton and Burton 1969; Wilson 1984). The first imports to Brazil were in 1793 and later another attempt was made in 1859 when 14 from Algiers were landed, but these later died of disease (Legge 1936). Other attempts were made in Venezuela with animals from the Canary Islands, but these were said to have died of snake bite (Dareste 1857). Bolivia imported 30 head in 1845 and by 1864 these had increased to about 100, but their subsequent fate is not clear (Legge 1936).

AUSTRALASIA
Australia
Introductions of camels were made to Australia in the mid-nineteenth century to assist with exploration work and development of agriculture in arid regions (Letts 1964). Their value was largely as a beast of burden and this included carting supplies, wood carting, and the pulling of ploughs. Many were used in the surveying, construction and maintenance of the overland telegraph line (1870–72), the east–west transcontinental railway (1900–17), and the rabbit-proof fence systems in Western Australia (1901–08). Between 1850 and 1920 camels were the main mode of transport for supplies to settlements and stations (McKnight 1969).

The explorer Horrocks used a camel in 1846, which had originally been landed in Australia in about 1840 (McKnight 1969), on an expedition in South Australia. Other explorers, such as Burke and Wills (1860) and also McKinley (1861), used camels on other expeditions. However, they were not numerous until Elder imported 124 in 1866, of which 121 were landed in South Australia. Five years later Warburton used camels on his expedition. Warburton (1873–74) left two camels behind on his expedition (Burton-Cleland 1911).

The first camels to be imported arrived at Port Adelaide in 1840 but only one of four from Canary

Islands was landed, the others died on the voyage. Other attempts to import camels include two imported from Teneriffe by J. Thomson in December 1840 (fate unknown). The third shipment arrived in May 1841 – three camels – one died on the voyage and the other two females landed. These were sold to the government in Sydney together with a male replacement that arrived in 1841–42. These three grazed the Sydney domain until 1845 but their ultimate fate is unknown (McCarthy 1980). Economic exploitation began in 1866 with 121 camels imported by Elder. By the late 1880s between 10 000 and 20 000 camels were present in Australia. Types initially introduced were said to be Mekrana, Scinda and Candahar camels.

One-humped camels were common in domestic service in Australia by 1895. In contrast only five two-humped or bactrian camels were imported during the same period (Rolls 1969).

The first feral camels in Australia were probably two animals abandoned from the Burke and Wills Expedition shortly after the first imports (Gauthier-Pilters and Dagg 1981). With the advent of motor transport in the 1920s and 1930s, many camels were turned loose because it became uneconomical to use them (Letts 1964; McKnight 1969; Long 1972, 1988; Gauthier-Pilters and Dagg 1981; Siebert and Newman 1989; Wilson *et al.* 1992). Thus 40–50 camels were abandoned by Afghan camel drivers some 30 miles from Marble Bar towards the end of 1927 (Barker 1964). Twenty-five camels were also freed by a cameleer on open country near Barramine Station (on edge of Great Sandy Desert, 100 miles north-east of Marble Bar) in 1929 as there was no work for them and no sale for them could be found (Barker 1964).

During World War 2 the feral camel population in Australia may have been as high as 30 000–100 000. A more conservative estimate in about 1981 indicated between 15 000–20 000 feral camels (McKnight 1969; Gauthier-Pilters and Dagg 1981), but others indicate about 100 000 (Burbidge 1989).

They occur in the far-western districts of Queensland, where it is estimated there are about 1000 (Mitchell *et al.* 1982). In the 1980s it was estimated that there were 31 570 camels in the Northern Territory (Anon. 1987) with the highest density in the Simpson Desert where 10 700 were estimated to occur in 1986 (Anon. 1987). They are now widely distributed over arid and semi-arid areas of central and western Australia (Strahan 1995).

■ **DAMAGE**

The camel's ability to survive in a feral state in arid environments frequently brings it under attention as a potential source of protein in areas which are limited for other domestic animals, especially cattle. Despite having obvious attributes its propagation for meat production in Australia has had many problems and the realisation of market potential has become uneconomic at times (Newman 1975). At present there appears to be a small live animal market for camels in the Middle East and revived interest in a meat trade.

Camels may have an impact on arid zone vegetation because they are selective browsers. However, because of their mobility, cushion-like foot pad and ability to range further from water, it is suggested that in the short term camels could have a less deleterious effect on ecosystems in terms of pasture degeneration and soil erosion than cattle. However, given their preference for shrub and tree material, and the longer regeneration time for these plants compared with grasses, camels could have a more serious long-term effect than cattle (Newman 1975). Research carried out in Central Australia (Heucke *et al.* 1989) suggests that at a density of 0.05–0.3/km^2 the impact on vegetation is negligible, but at a density of 2.0/km^2 several shrub and tree species can be severely damaged. The possibility exists that the browsing and grazing of camels has reduced shelter for small desert mammals (Newman 1983) and that damage to vegetation may be high relative to available pastures during times of drought (Doerges and Heucke 1989).

There is extensive evidence implicating camels in human plague epidemics. In Libya it has been found that camels are subject to natural infections and assume a significant role in the dissemination of plague (Christie *et al.* 1980). Camels may also be affected by rabies, Rift Valley fever, Surra, *Brucella mellitensis*, haemorrhagic septicaemia, bubonic plague, foot and mouth disease, and screw worms (Anon. 1987).

In Australia camels damage fencing and watering points, but the overall damage is small because of the low density and remoteness of camel populations (Long 1988; Wilson *et al.* 1992).

Family: Cervidae
Deer

MUSK DEER

Siberian musk deer
Moschus moschiferus Linnaeus
=*M. sibiricus* Pallas

This species is sometimes accepted as being in a separate family – Moschidae.

■ DESCRIPTION
TL 800–1000 mm; T 40–60 mm; SH 500–700 mm; WT 7–17 kg.
Coat brownish, lightly marked with reddish or yellowish spots (hair is dense, stiff and pithy); chin, inner borders of ears, and insides of thighs are whitish; occasionally a white spot is present on each side of throat; upper (canine) teeth developed as tusks (*c.* 75 mm); musk gland on abdomen. Both sexes lack antlers; females lack musk gland and canine tusks, and have two mammae.

■ DISTRIBUTION
Central, north-eastern and eastern Asia. From Siberia, Manchuria, Sakhalin and Korea south to the Altai, northern and western Mongolia, China and the Himalayas.

■ HABITS AND BEHAVIOUR
Habits: mainly nocturnal but sometimes crepuscular; territorial, shy, timid. **Gregariousness:** solitary, 2–3 together, or small groups (females plus young); density 73–78/100 ha. **Movements:** sedentary; occasional migrations 12–35 km in Siberia; home range 200–300 ha. **Habitat:** dense cover in mountains (e.g. forest and brush land at elevations 2600–3600 m), often steep slopes with rocky outcrops and in summer grassy valleys. **Foods:** grass, moss, tender shoots, twigs, buds, lichens, bark, leaves and pine needles. **Breeding:** ruts November–January, calves April–July; gestation about 160 days; young 1 or rarely 2; mature at 15–17 months. **Longevity:** 20 years captive. **Status:** probably still fairly common; little information most areas; declining in China.

■ HISTORY OF INTRODUCTIONS
Musk deer have been unsuccessfully introduced in Russia and the United Kingdom. Attempts have been made to farm musk deer in China since 1958 and in Russia since 1976. Musk from the males is used extensively in the manufacture of perfume, soap and medicinal preparations (WCMC 1998).

EURASIA
Russian Federation
The musk deer has been introduced and re-introduced to some areas of Russia (Bannikov 1963; Yanushevich 1966) including the northern Urals, where it has failed to become established (Yanushevich 1966). In 1954, 11 musk deer were released in the Sverdlovsk area (Kirisa 1974), but the introduction appears to have been unsuccessful.

United Kingdom
Many musk deer were released in the woods near Woburn before 1959 and although they bred there, did not persist for any length of time (Fitter 1959).

■ DAMAGE
No information.

CHINESE WATER DEER

Water-deer
Hydropotes inermis Swinhoe

■ DESCRIPTION
HB 750–1000 mm; T 50–80 mm; SH 425–550; WT males 9–16 kg (to 30 kg); females 2–3 kg.
Summer coat pale reddish brown, winter coat dark brown; chin and upper throat whitish; face reddish-brown; head buff; ears large, hairy, erect, buff; white around nose and insides of ears; eyes rounded and button-like; back and sides uniform yellowish brown, finely stippled black; under parts white, tail short, hairy stump; upper canines tusk-like in males. Female has four mammae. Both sexes lack antlers.

Musk deer

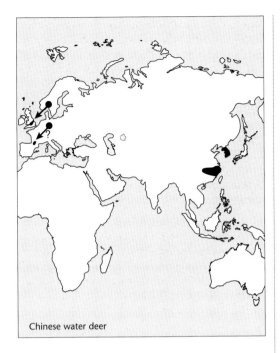

Chinese water deer

◼ DISTRIBUTION

Eastern Asia. Formerly in eastern China from the lower Yangtze Kiang Basin, west to Hupeh and north to Korea. Probably now extinct or considerably reduced in numbers except in the northern, central and southern parts of range.

◼ HABITS AND BEHAVIOUR

Habits: nocturnal and diurnal; territorial. **Gregariousness:** singly or pairs, rarely in herds; occasionally several does and young together. **Movements:** sedentary; territories about 4 ha. **Habitat:** grassy river valleys, marshy river beds, swamps, reed beds, open grassland, cultivated country occasionally, woodlands, marshes, open parkland with woodland, fields. **Foods:** reeds, grass, vegetables, root crops, sedges, brambles. **Breeding:** ruts November–January, fawns May–July (UK); in native habitat breeds autumn and early winter; gestation 170–210 days; young 2–5, 7; born in summer; fawns spotted for several months; young left in sheltered places; weaned at 2 months; mature at 7–8 months for females 5–7 months for males. **Longevity:** 6 years in wild, 10–12 years in captivity. **Status:** range reduced declining in numbers.

◼ HISTORY OF INTRODUCTIONS

Chinese water deer have been successfully introduced in the United Kingdom and France and probably unsuccessfully to Australia.

AUSTRALIA

Chinese water deer may have been introduced at some time into Australia (Bentley 1978), possibly on the Yorke Peninsula in South Australia, but there are no records of any releases and no records of any established in the wild.

EUROPE

Chinese water deer are established as feral animals in the south-east of England and in France (Corbet 1966; Burton 1977). In Britain they originated from escapees from deer parks or private collections (Baker 1990) from about the 1920s, and now occur in several scattered populations in damp habitats in East Anglia and central and southern England.

France

Between 1960 and 1965 some Chinese water deer were released for sporting purposes near Solignac-le-Uigen, south of Limoges. Here, they became well established and a few years later one was killed 80 km to the south-west (Corbet 1978; Lever 1985).

United Kingdom

In England Chinese water deer may have escaped or were released from parks since about 1850 (Matthews 1952 in de Vos *et al.* 1956), however, there appear no records before 1900. Some were introduced at Woburn Park, Bedfordshire, by the Duke of Bedford in about 1900 and some time later some escaped (Fitter 1959; Southern 1964; Whitehead 1964). During World War 1 some escaped from Woburn into the surrounding woods, and their numbers were augmented by a few escapees from Whipsnade (Whitehead 1964; Corbet and Harris 1991).

In 1929–31 the Duke of Bedford gave 32 water deer to Whipsnade and within four years there were 200 of them (Fitter 1959; Whitehead 1964). They were introduced into parks including Cobham, Surrey (1934); Basingstoke and Stockbridge, Hampshire (1944); Rippon, Yorkshire (1950); Bishop's Castle, Shropshire (1950); and Ludlow, Shropshire (1956) (Corbet and Harris 1991).

In 1944 the Duke of Bedford sent water deer to two places in Hampshire, from which some escaped soon after and they were reported on the Hampshire-Berkshire border in about 1948–49 (Fitter 1959). Between 1945 and 1950 they were reported in the wild from Buckinghamshire, Northamptonshire and Oxfordshire (Fitter 1959; Lever 1985). A few also escaped from a park at Ludlow, Shropshire, in 1956 and in 1963 it was estimated that there were 20 feral outside the park (Whitehead 1964). Some were introduced to Studley Royal Park, Ripon, Yorkshire in 1950 and from which there were escapees and in 1954 three or four were known outside the park (Whitehead 1964; Fitter 1959). They were also reported in the wild in Shropshire in 1956, Norfolk in 1968,

Cambridgeshire in 1971, and Bedfordshire in 1976 (Corbet and Harris 1991).

By the 1960s Chinese water deer were reported to be fairly widespread in Bedfordshire, Hartfordshire, Buckinghamshire and some Woburn stock introduced into northern Hampshire and at Walcot Park, Shropshire, were feral in woods in these areas (Southern 1964). They were present in Bedfordshire, Buckinghamshire, Hampshire, the Norfolk Broads and Shropshire in the 1970s, but others were noted in surrounding counties from time to time.

Chinese water deer are now established in the wilds of Bedfordshire, Cambridgeshire, and in the Broads of Norfolk (Corbet and Harris 1991), and there has been little expansion in range or numbers in recent years (Lever 1985).

■ DAMAGE
In China water deer are killed for meat and skins and also as a pest of crops. In England they are not numerous enough to cause any damage (Corbet and Harris 1991).

COMMON BARKING DEER
Red muntjac, barking deer, kakar, Indian muntjac
Muntiacus muntjak (Zimmermann)

■ DESCRIPTION
HB 900–1100 mm; T 170–190 mm; SH 500–560 mm; WT 20–28 kg.
Small with short antlers; pelage deep brown to yellowish or grey brown with creamy white markings; body hairs short, soft, except for ears which are sparsely haired; dorsal surface of tail as brown as rest of body; antlers 73–130 mm.

■ DISTRIBUTION
Asia. Sri Lanka, India, north-east Pakistan, to southern China, Indochina and Hainan, and south to Lombok, Bali, Java, Kangean, southern Sumatra, Nias, Borneo, Bawal, Matisiri, Bangka, Belitung, Bintan, Riau, and Linnga.

■ HABITS AND BEHAVIOUR
Habits: active day and night, most active sunrise and sunset; shy. **Gregariousness:** solitary or pairs; density 1.5–2.5/km^2. **Movements:** home range 4–5 km^2. **Habitat:** forest. **Foods:** leaves of shrubs and herbs; fallen fruit, buds, flowers. **Breeding:** all year; gestation *c.* 7 months; oestrus 48 hours. **Longevity:** 10–17 years 7 months captive. **Status:** common.

■ HISTORY OF INTRODUCTIONS
Common barking deer have been introduced success-

fully to Lombok and unsuccessfully to the United Kingdom.

EUROPE
United Kingdom
Introduced, but failed to remain established (see also notes under *M. reevesi*).

INDONESIA
Lombok
It has been suggested that common barking deer were introduced into Lombok by the Balinese Rajahs (Everett 1896). Some were present there in 1950, 1954 and in 1986, but none were noted in 1987 (Kitchener *et al.* 1990).

■ DAMAGE
No information.

CHINESE MUNTJAC
Barking deer, Reeves' muntjac
Muntiacus reevesi (Ogilby)

■ DESCRIPTION
HB 800–1070 mm; T 110–200 mm; SH 400–450 mm; WT 9–22 kg.
Small brown deer with simple antlers (*c.* 150 mm sheds May–June); fur red brown above and white below, in winter duller, greyer brown; upper canine teeth protrude below lip in males; forehead in male has black stripes forming 'v' pattern, in female is black kite-shaped pattern; forelegs almost black on front;

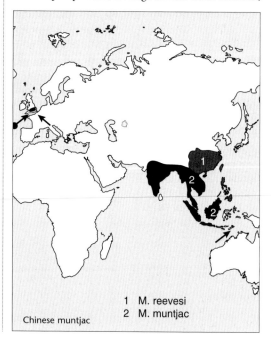

1 M. reevesi
2 M. muntjac

Chinese muntjac

tail chestnut dorsally, white below. Fawns have female face pattern and are spotted for eight weeks.

■ DISTRIBUTION
Asia. Shensi, Kansu, south-east China and Taiwan.

■ HABITS AND BEHAVIOUR
Habits: mainly crepuscular and diurnal, sometimes nocturnal; territorial. **Gregariousness:** largely solitary or pairs; occasionally aggregations at feeding sources; seldom form herds; density 1/6.8 ha. **Movements:** home range *c.* 14 ha; some natal dispersal. **Habitat:** woodland, coppice, scrub, young plantations, undisturbed gardens, parkland, hillsides. **Foods:** ivy, ferns, fungi, buds, leaves and shoots of trees and shrubs, nuts, fallen fruits, grass; also cultivated plants. **Breeding:** throughout year; gestation 209–220 days; female polyoestrous; oestrous cycle 14–21 days; oestrus 2 days; 1 fawn, rarely 2; sexual maturity 11–12 months. **Longevity:** 16–19 years captive. **Status:** common.

■ HISTORY OF INTRODUCTIONS
EUROPE
Chinese muntjacs have been successfully introduced in England and possibly to France.

France
Chinese muntjacs were introduced to France in 1890, but failed to become established in the wild, although feral populations are often reported (Lyneborg 1971). They are still maintained in some large private estates and in some zoological parks (Lever 1985).

United Kingdom
Chinese muntjacs are now established over most of southern England, East Anglia and as far north as Cheshire, Derbyshire, Nottinghamshire and Lincolnshire in the Midlands. Wandering individuals are not infrequently reported in urban areas, even central London (Corbet and Harris 1991). They now frequent woods and copses through most of southern England and in south Wales (Baker 1990). All those established in England are the Chinese muntjac and not the common barking deer.

The first introduction of Chinese muntjac was in Woburn Park and neighbouring woods, Bedfordshire, early in the twentieth century to replace Indian muntjac introduced a few years earlier. The latter were then shot out, although a few may have escaped elimination. Chinese muntjac spread from here; there are reports from 1922 on from the immediate area and from adjacent counties. Some additional escapes/releases occurred of small numbers from Whipsnade Park (liberated 1929–31), Bedfordshire, Broxbourne in Hertfordshire (1930s) and Northamptonshire (1937). In the first 60 years their range extended to a radius of 72 km from Woburn; it

is now approximately 300 km to the south-west, 200 km to the north and north-east and 120 km to the south-east (Fitter 1959; Willet 1970; Corbet and Harris 1991).

■ DAMAGE
In England Chinese muntjacs cause very little damage, if any, to conifers or arable crops (Corbet and Harris 1991).

AXIS DEER
Axis, spotted deer, chital
Cervus axis Erxleben
=*Axis axis*

■ DESCRIPTION
HB males 1190–1850 mm, T 250–450 mm; SH 800–1010 mm; WT 33.6–113 kg; females 1140–1470 mm, 190–300 mm, 670–870 mm, 25–64.5 kg.
Coat reddish brown, fawn or tan, and spotted with white; white spots, in rows, retained throughout the year; under parts, under tail and inner hind legs white; dorsal stripe chocolate; in males a black or dark brown diamond shaped spot in mid-foreneck; black line from antlers to eyebrow and bridge of nose; black band from rear of mouth forward over muzzle; ear edges with rim of dark hairs; muzzle dark brown, face buff, chin and throat white; antlers reddish-brown, three-tined, lyre-shaped, curved backwards and outwards 760–1000 mm, six to 10 points, sheds August–September (October–February United States). Fully antlered at five to six years of age. Female lighter on neck and fore-shoulders and lighter facial markings; white throat patch extends along lower jaw to below nostrils.

■ DISTRIBUTION
Asia. The Himalayas, Nepal, Bhutan, east Pakistan and down the east coast of India to the Malabar coast and Sri Lanka.

■ HABITS AND BEHAVIOUR
Habits: nocturnal and diurnal. **Gregariousness:** groups or herds, 2–38 but varies, 2–300; male bachelor groups, in breeding season 2–25; family groups; congregates at food sources; density 3–5/km^2 (India) and up to 23/km^2. **Movements:** sedentary; home range about 180–890 ha. **Habitat:** semi-open savanna (Texas), coastal thickets and grasslands, deciduous forest, woodlands, forest with grassy clearings, grasslands and edge of forest, open country, secondary forest, open forest, dry scrub to moist, deciduous forests, semi-deserts shrub to rainforests. **Foods:** grass, fruits, and browse, leaves, shoots and flowers of trees and shrubs, seed pods, forbs, sedges, mushrooms, woody pieces, mast. **Breeding:** breeds

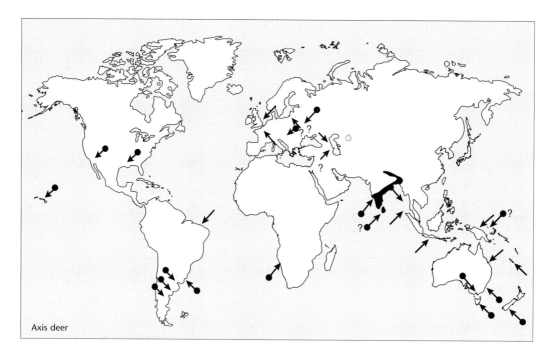

Axis deer

throughout the year with distinct peaks; major breeding season May–August, June peak (United States, Hawaii, Texas); ruts in May, fawns February (Uttar Pradesh), all year in Hawaii (peak November–April); varies October–April (India) or March–June (Australia April–May and September–November); gestation 210–240 days; 1 young, rarely 2; female diestrous; female sexually mature at 14–17 months. **Longevity:** 12–20 years (captive). **Status:** range greatly reduced, but still fairly common; abundant some wildlife sanctuaries and locally.

■ HISTORY OF INTRODUCTIONS

Axis deer have been successfully introduced to New Zealand, Hawaiian Islands, Australia, New Guinea, Russia, the United States (California and Texas), Argentina, Brazil, Uruguay, Yugoslavia and the Andaman Islands. In New Zealand and New Guinea they have died out.

AFRICA
South Africa

Axis deer occur on private ranches in fenced areas only in the western Cape, South Africa (Lever 1985).

ASIA
Java

Introduced unsuccessfully to Java (Bentley 1978).

West Pakistan

Axis deer have been introduced in the Changa Manga plantation forest in the Lyallpur region and there are now about 500 present there (Taber *et al.* 1967).

AUSTRALASIA
Australia

There were many liberations of axis deer in Australia during the nineteenth century. Most of them were successful in establishing populations in the wild (Strahan 1995). The Australian animals were derived from the nominate race *C. a. axis* from peninsula India.

Axis deer were reported to have been released in New South Wales in 1803 near Sydney by Dr J. Harris (Terry 1963; Wilson *et al.* 1993) and were firmly established prior to dispersal in 1812–13 (Whitehead 1972). Today they are also established in Sydney National Park, Glenn Innes. Some were imported from Batavia in 1867 by the Acclimatisation Society of Queensland, some were released on the Darling Downs in 1872, and some were also released at Charters Towers some time after 1886 (Roff 1960; Bentley 1978) or 1866 (Bentley 1957). In 1862, three were liberated by the Victorian Acclimatisation Society at Sugar Loaf Hill near King Lake. An unknown number were also released at the source of the Yarra. In 1863, four were released on Wilson's Promontory (Waterloo Bay) and in 1870, five at Yering. There was a major effort to establish them at Longerenong between 1864 and 1866 or earlier, where they became established and spread to the Grampians. They appeared well established there in 1872. In 1953 about 600 were established at Charters Towers (Whitehead 1972).

Probably axis deer were imported into Tasmania in 1829 and later three escaped into the bush, but there is no further mention of them (Bentley 1978).

In the 1950s two herds were reported to be established in Australia, that at Charters Towers and another in the Grampians in Victoria (Bentley 1957). The herd at Charters Towers is still established (Cowling 1975; Wilson *et al.* 1992), but there are none in Victoria today, although they are occasionally reported (Bentley 1978).

The only herd now established is that released at Maryvale Creek, some 130 km north-west of Charters Towers, Queensland, in 1886. Here they are well established and have made some extension in range (Strahan 1995).

Papua New Guinea
Axis deer were introduced to New Guinea by German settlers in about 1900 (Lindgren 1975; Bentley 1978). They were established in the Madang area before World War 1 and in the mid-1960s still existed in small numbers on the outskirts of Madang (Bentley and Downes 1968; Downes 1968). About 150 head were there in 1953 (Bentley 1978). In the 1980s they were reported to be restricted to the Madang area of Papua New Guinea, where they were scarce to moderately abundant (Ziegler 1982; Flannery 1995).

Axis deer were also introduced to the Hermit Islands before World War 1, where they were established (Herington 1977), but there do not appear to be any recent records.

EUROPE
France
Axis deer were introduced in France in 1890, but were unsuccessful in becoming established (Dorst and Giban 1954; Lever 1985).

Russian Federation and adjacent independent republics
In the Russian Federation axis deer were introduced to the Caucasus in 1939 and again in 1954 and between 1953 and 1958 (Yanushevich 1966). These introductions became established locally.

In 1939, 39 axis were released in Teberda ravine in the Caucasus. Within three years the population had increased to 150 head, but by the end of World War 2 the population had halved due to shooting by German troops. By 1956 a small herd, which had been translocated from the Altyagach region in Azerbaijan two years earlier, had increased to about 100. However, this herd was later reduced by foot and mouth disease (Lever 1985).

In 1953, 16 axis hinds and four stags were transferred from a stud in Ussuri Territory to Yerevan Zoo in Armenia. These were released in 1954 in the Kohoosnov State Forest in the Vedi District, where they disappeared, probably due to predation. A second introduction of 14 hinds and five stags in 1958 proved successful and by 1961 there were 58 deer there (Lever 1985).

Axis deer kept in captivity in Askania Nova, Ukraine, before World War 2 escaped and became established in Burkuty. From here they were released in hunting reserves in Kiev, Cherkassy, Vinnitsa and Kherson regions in 1956–57 and subsequently in the Dnnepropetrovsk and Volga regions. Most of these were allowed free-range.

Two dozen axis deer were released in the forests in Lithuania in 1954 and adapted well, and by 1961 had increased to 67. In 1959, 1960 and 1961, 97 were released in woodlands in central Kodry, Moldavia.

United Kingdom
At one time (1944–45) axis deer were reported to be feral in Buckinghamshire, England, but there is no evidence that they have been present outside a deer park and there is only one record of an escape by a single animal (Fitter 1959, Whitehead 1964). In 1888 one was shot in West Sussex (Fitter 1959).

Yugoslavia
A feral population of axis deer is reported to exist in the peninsula of Istria, in northern Yugoslavia, derived from animals which escaped from captivity in 1911 and have increased in numbers substantially (Niethammer 1963; Corbet 1966, 1978, 1980; Wilson and Reeder 1993).

Axis deer were introduced on the island of Brioni (Brionski Otaki) (Lever 1985) on the Adriatic but were not very successful (Whitehead 1972).

INDIAN OCEAN ISLANDS
Andaman Islands
After World War 1 axis deer were introduced to the Andaman Islands for meat (Whitehead 1972; Lever 1985; Wilson and Reeder 1993). They increased rapidly in numbers and by the end of the World War 2 had become a pest, destroying seedlings of trees planted for forest regeneration. Two female leopards, *Panthera pardus,* were introduced in 1952–53 to help control them, but they had little effect as the deer are still locally abundant in the Andamans.

Nicobar Islands
Introduced from India in 1846, axis deer seem to have disappeared from these islands for no apparent reason (de Vos *et al.* 1956).

Sri Lanka

Although axis deer are said to occur in Sri Lanka as part of their natural range, it is thought by some that they may have been introduced there by the Portuguese in the sixteenth century (Lever 1985).

NORTH AMERICA
United States

Axis deer were introduced in Texas in 1932 on private lands, such as the King Ranch (Ramsey 1968; Whitehead 1972), where they became established (Presnall 1958; Jackson 1964; Dasmann 1968; Ramsey 1968; Schreiner 1968). In 1972 there were 11 000 in 45 counties, mostly on the Edwards plateau, lesser numbers on the south Texas plains, gulf prairies and marshes, and cross timbers and prairie regions (Ables 1974). Earlier they had been reported captive in 11 counties with over 50 animals each, and 22 counties with less than 50 animals each (Ramsey 1970). They escaped some time later and became established outside the boundaries of the private lands and free-living herds roamed in Bexar, Kendall and Comal counties.

There have been some more recent introductions into Texas (Gottschalk 1967); at least six animals were introduced on the H. B. Zachry Ranch, Laredo, Texas, in 1959–60 (Sanders 1963). In the 1960s and early 1970s there were at least two populations of more than 1000 in semi-captivity and an unknown number had become established as free-ranging wild animals on the Edwards Plateau area (Ables and Ramsey 1973). In 1979 free-ranging animals occurred in 20 counties with an estimated 7877 head (Lever 1985).

Axis deer have also been established in Florida (de Vos *et al.* 1956; Presnall 1958). They escaped from a yard in Volusia County some time in the 1930s and since then have spread into Flagler, St. Johns and Duval counties (Presnall 1958). Since 1951 they have been protected (Whitehead 1972). There are still herds of undetermined status in Volusia and Marion counties, Florida that apparently have not expanded since the introductions in the 1930s (Feldhamer and Armstrong 1993).

Axis deer were introduced to Point Reyes Peninsula, Marin County, California: two males and two females in 1948, and four other animals in 1947, by M. Ottinger on his ranch 'Inverness Ridge'. The stock came from the San Francisco Zoo (Jameson and Peeters 1988). Bucks were shot by local ranchers in 1956–67, but there was no population control exercised. Many were shot by ranchers in the late 1960s as they were perceived to be competing for range for cattle (Wehausen and Elliott 1981, 1982). Reductions in numbers became apparent between 1971 and 1981,

and at this time there were only 250 axis left (Gogan *et al.* 1986). Now about 364 head exist in the area, but they are controlled by culling to prevent excessive numbers. Efforts are now made to keep the herd at about 350 animals (Feldhamer and Armstrong 1993).

PACIFIC OCEAN ISLANDS
Hawaiian Islands

Axis deer from India have been introduced and established in the Hawaiian Islands (Gottschalk 1967; Tomich 1969). In 1867, seven axis (three males and four females) were shipped to the reigning monarch of Hawaii as a gift from his envoy in Japan. Some of these came from the Upper Ganges River, India, but some died on the voyage and were replaced at Hong Kong with animals of unknown origin. Further animals may have arrived in the Hawaiian Islands from Japan in the same year (Tinker 1941), but it is not known whether there really was a second shipment (Kramer 1971) as details are hazy. In January 1868, eight were shipped from Honolulu to Molokai where they were released into the forests above Kaunakakai (Nicholls 1962; Kramer 1971). By 1898 the herd had increased to 6000–7000 and in 1900 hunters were allowed to remove 3500 to reduce their numbers as they were reported to be competing with cattle which were grazing there at the time. Numbers then dwindled and by 1950 only about 2200 remained on Molokai.

Sometime following the original introduction in the Hawaiian Islands, some were released on Oahu (possibly Tinker's animals from Japan?) where they increased to form two herds. One of these herds became established on Diamond Head, but was wiped out in the early twentieth century. The other herd established at Moanalua Valley and had 300 removed in 1940. They numbered about 12 in 1950, and about 20 were known there in the 1960s.

In the period 1920–23 a few (12 head) from Molokai were transferred to Lanai by G. Munro, where they were released onto Palawai plateau. Some effort was made to exterminate them in the 1930s and they were few in numbers in 1935, but built up during the 1939–45 period to around 400 animals. In 1953 public hunting commenced on Lanai following an outcry over the damage they caused to pineapple plants grown on the island (Nicholls 1962; Kramer 1971). In the 1960s there were 1200 on Lanai (Graf and Nicholls 1966), 2900–3000 on Molokai, and 25 on Oahu (Ables 1974).

In 1961 the population on Molokai was reported to be 600 animals, and that on Lanai to be 1700, and there has been hunting of axis every year since 1959–60 (Nicholls 1962). This has had little effect on numbers, as shortly after it was reported that there were 5000 to 6000 on Molokai and Lanai (Graf and Nicholls 1966).

A pair was released at Kanoehe Marine Corps Air Station, Oahu, in 1954, where they survived for several months before disappearing. Two males and three females were released in the 'Red Hill' area of Maui in 1959, and an additional male and three females were released on the Kaonoula Ranch in 1960. The population in this area was estimated at 85–90 deer in 1968 (Kramer 1968). Although they fluctuated in numbers, two huntable populations existed on the islands of Lanai and Molokai in the 1970s (Ables 1972).

New Caledonia
Probably two axis deer were introduced to New Caledonia in 1862. They arrived on the frigate *Isis* from India and if they were released they did not become permanently established there (Barrau and Devambez 1957).

New Zealand
Seven axis deer were introduced from Melbourne and released in the area between Oamarn and Palmerston (South Island) in 1867 (Thomson 1922). Several other introductions, including those to the island of Kapiti in 1893, Otago area in 1902(?), Tongariro National Park (near Lake Taupo) in 1906, Dusky Sound in 1907 or 1909, but none were seen after 1920–30 (Thomson 1922; Donne 1924; Lever 1985).

SOUTH AMERICA
More than likely axis deer only exist in semi-confinement in Argentina and Brazil (Whitehead 1972).

Argentina
Axis have also been introduced from India to Argentina where they have become established (de Vos *et al.* 1956; Petrides 1975). They have become established in Buenos Aires, Santa Fé, Neuquen and Rio Negro provinces as a result of translocations from India in 1906, 1910 and 1932 or thereabouts (Petrides 1975).

Brazil
Introduced to Brazil (de Vos *et al.* 1956), axis deer appear to have been unsuccessful in becoming established.

Uruguay
In 1930 Sn. Aaron de Anchorena introduced axis deer to his estate near the mouth of the San Juan River, Colonia, 200 km from Montevideo. Axis deer are now established in the wild along the Negro River north of here in the Department of Soriano (Lever 1985).

◼ DAMAGE
Throughout its native range, axis deer are frequently seen near villages and do considerable damage to standing crops (Whitehead 1972).

It has been reported in the United States that axis deer have some game ranch potential in Texas provided that ranges are not overgrazed. When overgrazing occurs the deer compete with livestock and native wildlife, particularly the native white-tailed deer (*Cervus virginianus*) (Ables 1974, 1977). Food studies and comparison with white-tailed deer preferences for food suggest that there is direct competition between the two species. Additionally axis deer will browse when grass is scarce and white-tailed deer can not (Armstrong and Hamel 1981).

FALLOW DEER
Dama
Cervus dama Linnaeus
=*Dama dama*

Opinion as to whether Dama *should be considered a distinct genus appears to be equally divided. Here,* Cervus *is retained for reasons mentioned earlier.*

◼ DESCRIPTION
HB 1300–2350 mm; T 140–240 mm; SH 720–1050 mm; WT 30–200 kg.

In summer reddish brown with numerous white spots and white line along flanks; in winter uniform grey-brown with less spots (species variable and many colour varieties ranging from black, sooty, blue, silver and whitish); under parts white; buttocks with white area margined black; long tail brown or black above, white below and with black tip and long terminal hairs; male with antlers 630–940 mm, palmate, large brow tine, pattern variable but usually six tines in sixth year, sheds in about April–May (October in Australia). Female lacks antlers.

◼ DISTRIBUTION
Western Europe–Asia Minor. Originally in Mediterranean region of southern Europe and Asia Minor, but now difficult to define because of introductions and re-introductions, and in a wild state throughout most parts of western Europe, western Ukraine, Baltic countries and Great Britain. Now extinct in Africa and Asia, except for a few survivors in western Iran.

◼ HABITS AND BEHAVIOUR
Habits: mainly diurnal and crepuscular; territorial.
Gregariousness: aggregations of 70–100 at feeding

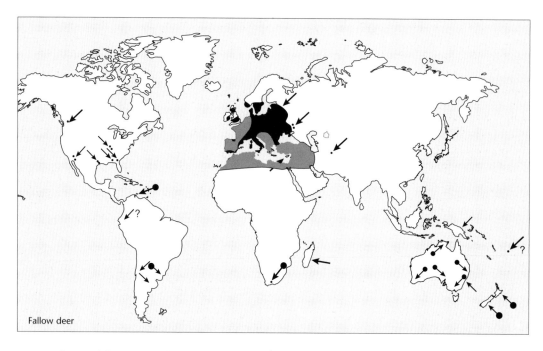

Fallow deer

sources; does and fawns in small herds; bucks form separate groups; sexes together at rut e.g. males gather harems; at rut males tend to be solitary; herds break up in September–October; density 1 per 4–43 ha (20.3/km^2 California). **Movements:** home range 50–110 ha. **Habitat:** open woodlands with under-growth, adjacent grasslands, parklands, plains, light hilly country with dense grassy covers and sparse woods or brushy areas. **Foods:** grass, forbs, sedges, browses leaves and bark of shrubs and trees, young shoots, beech mast, chestnuts, acorns, roots, vegetables, flowers and cultivated crops, dried leaves, bark of trees and bushes, mosses and lichens crops, roots, vegetables and flowers. **Breeding:** annually; ruts September–December (Russia, United States); (Australia ruts March–April), calves May–July (Australia November–December); gestation 225–270 days; male polygamous, female seasonally polyoestrous; oestrous cycle 22–26 days; 1 calf, occasionally twins; males mature 7–14 months, female sexually mature at 16 months. **Longevity:** 16–20 years, males rarely 8–10 years; 20–30 years in captivity. **Status:** reduced in numbers and range as a truly wild animal, but still common as a feral park and wild animal in Europe.

■ HISTORY OF INTRODUCTIONS

The present worldwide distribution of fallow deer is due almost entirely to humans. Most were introduced in the nineteenth and twentieth century. In Europe they have been introduced over a longer period from the eleventh to twentieth centuries. During the last interglacial period they were widespread in Europe from England to Russia. During the Wurm glacial period which followed (lasting 60 000 years) their range diminished and they only existed in few places at the end (about 10 000 years ago). Unlike other deer, they did not re-colonise Europe after the Ice Age and it is now thought most likely that their present distribution is largely human-made.

In many places the free-ranging populations of fallow deer owe their origins to park deer, usually by accidental escape rather than deliberate release. This correlation between wild herds and the proximity of enclosures or former parks is well known in Britain, but also is reported from other countries including Argentina, the Netherlands, United States and Germany (Chapman and Chapman 1980).

AFRICA

Algeria, Tunisia, Libya and Egypt

Introductions of fallow deer have been made to North Africa, but it is doubtful if they have continued to survive (Ellerman and Scott 1951). Some may still be present, but most liberations have gone unrecorded (de Vos et al. 1956). There are reports of them from Algeria, Tunisia and Libya late in the twentieth century (Millais 1906). The origin of these animals is unknown, but they may have been introduced by the Phoenicians or Romans to Egypt in the sixteenth century (Joleaud 1935; Chapman and Chapman 1980).

South Africa

In South Africa some fallow deer were introduced to Groot Schuur, Table Mountain, Capetown, presumably by Cecil Rhodes in 1887 (Siegfried 1962). In 1937 there were 400 head present on his estate. Some were also kept in the park of Newlands House, Capetown, in 1869 (Lever 1985). Both these areas still have managed herds. Another herd occurs at Bredasdorp that is also behind fences. However, after a century fallow deer have not succeeded in invading the surrounding country (Bigalke and Pepler 1991).

In 1914, three were introduced to the Vereeniging Estate and by 1937 there were 50 present on this 3500 acre well-wooded area. Mr D. Baxter obtained some from Groot Schuur and liberated them on his estate at Somerset West and by 1937 there were 25 there. C. Newberry liberated fallow deer on his estate in 1910 between Clocolan and Gumtree and there were about 40 in this area at this time (Bigalke 1937). There were still about 50 head on the Vereeniging Estate in the 1960s (Niethammer 1963). In 1970 they were present in 32 of the 113 provinces from Capetown to Mafeking in the north-east. At Groote Schuur in 1974 about 350 were present on the lower slopes of Devil's Peak (Chapman and Chapman 1980). In Orange Free State they are present on a number of farms in at least the Harrismith and Heilbron districts since 1914. Earliest releases occurred at Buckland Downs, Harrismith in the early 1900s, and elsewhere more recently (Chapman and Chapman 1980). They have also been released in the Franklin Game Reserve in Bloemfontein (Van Fe 1962) and were introduced to the Transvaal soon after 1900, but all appear to occur only on farms. They became temporarily established in Kruger National Park in the 1940s, but died out in the early 1950s (Macdonald and Frame 1988), and also became established in the Hluhluwe-Umfolozi Game Reserve, but also died out in the 1950s. They were established on Robben Island when three from Groote Schuur were taken there in 1963 and by 1977 the population had increased to about 40 (Chapman and Chapman 1980) Fallow deer are now widely farmed and many of them are still fenced and not really wild.

AUSTRALASIA
Australia
Scattered populations of fallow deer are present in Australia from south-east South Australia to Stanthorpe in Queensland, and in Tasmania, where they are well established with a total number in the vicinity of 8000–10 000 head (Caughley 1988; Wilson *et al.* 1992; Strahan 1995). They were introduced to Australia as a source of food and for recreation (Searle and Parker 1982).

The earliest introductions of fallow deer in Australia appear to be those in 1829 in Tasmania (Wapstra

1975) and about 1844 on the mainland at Albury in New South Wales (Bentley 1978). From these dates until about 1924 there were several more introductions in different parts of the country. Further releases certainly occurred in Tasmania about 1846, and on the mainland in 1850, and probably shortly after the latter date (Bentley 1978). By 1850 several herds were thriving on and about the properties of some influential settlers (Strahan 1995). Other introductions occurred in Victoria on Phillip Island in 1860; in Queensland in 1865, 1870, 1872 and 1890 (Roff 1960; Bentley 1978); in South Australia in about the 1880s, 1917 and 1936 (Bentley 1978; Castle 1989); in Western Australia in 1899 and 1903–06; and in New South Wales in 1924 (Bentley 1978).

From the early introductions in Tasmania they become established, and by 1863 there were thought to be some 600–800 fallow deer in Tasmania. They have continued to spread only slowly and now occupy some 400 000 ha of the Midlands and the population in the late 1970s was probably 7000–8000 head (Cowling 1975; Wapstra 1975; Bentley 1978).

Six fallow deer from Tasmania were imported to Queensland in 1865 and later released. The Queensland Acclimatisation Society liberated some at Westbrook in 1870, on the Darling Downs also in 1870, and six near Warwick in 1872. In 1890 some were released at Stanthorpe. At present they appear to be still established at Warwick (Cowling 1975; Bentley 1978) and may still be present west of Stanthorpe (Bentley 1978).

Six releases of fallow deer occurred in Queensland: the Westbrook area (failed); Maryvale Station near Cunningham's Gap, where they spread well and were plentiful between 1930 and 1980 and were still occasionally reported thereafter; Canning Downs via Warwick (destroyed after damage to crops); Main Range near Toowoomba, where they were reported until the 1950s; Pikedale Station where they were successful and now extend over an area of 65 km (details of release unknown); and McPherson Range south of Beaudesert (details of release lacking) and reported recently (1980?) at Rathoowney. Fallow deer are now common in the Pikedale area and elsewhere are uncommon (Searle 1980).

Fallow deer were released before the 1880s at Pewsey Vale and descendants of these are reported to have roamed the Adelaide Hills, South Australia, until they were shot out in the 1940s and early 1950s. In the 1880s they were released near Penola where they survived until 1914, but thereafter disappeared. Some were also released at Two Wells on the Gambier River in 1917, but were largely confined to a property

(Bentley 1978). It is thought that fallow deer still exist in parts of south-east South Australia (Cowling 1975). The majority, about 800 animals in 1989, are in three main regions and are known as the Clare, Buckland Park and upper south-east herds (Castle 1989).

For many years fallow deer survived near Delaware, Burgowanah, Jindera and Albury in New South Wales as a result of an early introduction by Sir Rupert Clarke, and around Lake George between Queanbeyan and Goulburn from an introduction in 1886. Two hundred head were reported there in 1910 (Bentley 1978), but it appears there may be few at this time. Fallow deer however are reported to have survived in New South Wales from six animals released at Glenn Innes in the New England ranges in 1924 by C. Campbell (Cowling 1975; Bentley 1978). In 1978 fallow deer were present in an area from Wallangarra in the south to Cunningham's Gap in the North.

Those fallow deer on Phillip Island, Victoria, survived for 60 years, but the last was noted in 1920 (Bentley 1978). Many deer were released by the Victoria Acclimatisation Society in scattered localities throughout that state. They are still reported to be found in small numbers at Narbethong-Healesville, Kinglake, Yarra Glen, in the Grampians, and possibly in the Blackwood and Brisbane ranges and isolated localities in western Victoria (Bentley 1978). Three were released in the Narbethong-Healesville area from Tasmania in the 1850s and by 1919 they were numerous there (Bentley 1978).

Introductions in Western Australia have been largely unsuccessful. A pair was released at Cape Leeuwin in 1899 where at first they appeared to thrive, but began to decline in numbers between 1924 and 1930 (Le Souef 1912; Colebatch 1929; Bentley 1978; Long 1988). Some were also introduced to the Porongorup range in 1903–06, when several species of deer were released (Bentley 1978; Long 1988) and a few may have been released at Gingin (Allison 1969). A population recently (1995–2000) became established in the Gidgegannup area in the Avon Valley just east of Perth, probably as a result of animals escaping from deer farms. A local farmer is believed to have deliberately released a stag and three hinds near Pinjarra, 65 km south of Perth in 1997–98 to establish a population for hunting. The stag was shot by local agriculture department staff but the hinds were not seen again.

In the Northern Territory fallow deer were released at Port Essington about 1912, but failed to become established (Bentley 1978).

Now fallow deer are established in Australia in scattered populations from south-eastern South Australia to Stanthorpe in Queensland, and a large population is present in Tasmania (Wilson et al. 1992).

Papua New Guinea
Fallow deer were probably introduced to the Madang area with other deer in the 1920s, but they are not now known to be established anywhere (Bentley and Downes 1968; Ziegler 1982; Flannery 1995). It is possible that these were misidentified as fallow deer and may have been *C. timorensis*, that are widely established in Papua New Guinea.

EUROPE
Since the Middle Ages fallow deer have occurred in almost every country in Europe. Originally they inhabited only the Mediterranean countries, but were transported by the Phoenicians who carried them to their newly founded colonies, because of a religious cult, and the Romans likewise spread them as the holy deer of the goddess Diana (Niethammer 1963; Lyneborg 1971).

Many introductions have occurred in Europe since the Middle Ages, mainly from the Mediterranean provinces, particularly Asia Minor (Hesse 1937; de Vos and Petrides 1967). They now occur in nearly every country of western Europe, but not in Norway or Finland, and also in southern Sweden (Corbet 1966; Chapman and Chapman 1980).

Some fallow deer occur on the island of Rhodes in the Aegean Sea, but it is not known if they were introduced or are indigenous. They are believed to have come from Asia Minor (Whitehead 1972). Fallow deer became extinct in Greece in the nineteenth century and on Sardinia in the 1950s (Burton and Pearson 1987).

Austria
Fallow deer have been introduced successfully into Austria (F. Spitzenberger *pers. comm.* 1982) and wild fallow deer now occur in at least three localities.

Between the World Wars feral herds existed in Leithagebrige in northern Burgenland and in Dunkelsteiner Wald in central lower Austria (Amon 1931). However, by the 1940s these had all been exterminated (Kerschagl 1964). From introductions with German stock in 1932 they have been present near the River Salzach, north of Salzburg. Further releases were made here in 1943 from Czechoslovakia. Fallow deer also occur near Horn, 65 km north-west of Vienna, where there were 60 in 1972 (Dobschova 1972). Similar numbers occur in central Austria near Mautern in Steiermark (Chapman and Chapman 1980; Lever 1985).

Belgium

Fallow deer were introduced to the Ardennes near Rochefort in south-east Belgium about 1850. The population living in the 1500-ha forest of Ciergnon is maintained at about 100 head or so in recent years (Chapman and Chapman 1980; Lever 1985).

Bulgaria

The first introductions of fallow deer in the wild occurred in 1908–11 by Czar Ferdinand I, but they probably existed here also from the Middle Ages or earlier (Niethammer 1963). Some were introduced in 1904 when a buck and two hinds were imported from Germany. They were enclosed in the Kricim Hunting Reserve to the east of Plovdiv, where a herd of 80 still exist. Fallow deer from here were released in the wild in a number of areas. They now occur at Voden (near Yambol) at the Ropotamo Reserve near the Black Sea, in central Bulgaria on the south side of the Balkan Mountains and further south, and in western Bulgaria in the Vitoscha Nature Park south of Sofia (Chapman and Chapman 1980; Lever 1985).

Cyprus

Attempts were being made in 1980–82 to re-establish fallow deer on Cyprus, where they have been extinct since medieval times (Lever 1985).

Czechoslovakia

Fallow deer have been introduced in the Pavlov Hills, South Moravia, where they have become established (Grulih 1979). They are now widely distributed in the wild and are kept in about 24 enclosed reserves. Although kept in enclosures since early times, the first releases occurred in the last century. Bohemia has a number of milk herds south-west of Plzen, south of Prague in the Dobris area, and at Melnik, near Varnsdorf, near Jan Nisou and Dvur Kralove and in eastern Bohemia. They now occur in six areas in Moravia and a number in eastern Slovakia (Chapman and Chapman 1980; Lever 1985).

Denmark

In Denmark fallow deer are mentioned in the literature as early as 1231, having been introduced by Danish kings for hunting. They now occur in many deer parks as well as in the wild (de Vos *et al.* 1956; Chapman and Chapman 1980), and are present on the Jutland Peninsula and on four of the major islands – Langeland, Holland, Funen and Zealand – and four other very small islands (Chapman and Chapman 1980). During the period 1945–70 they disappeared from the island of Samsø, from Kastrup on Zealand, and from Rosenwold and Sostrup in Jutland. The largest populations are now in the forest between Hillerød and Esrum.

Some fallow deer were introduced to the island of Livø in 1876; new stock was introduced in 1957 and 20 were counted a few years later. About 1970 the herd was much reduced. A further buck and two does were introduced in 1975, but by April 1978 only three remained (Chapman and Chapman 1980; Lever 1985).

Finland

Some uncertainty exists as to whether fallow deer occurred in Finland in the wild. Probably there were one or two enclosed herds, but the herd at Hyvinkää, near Helsinki, may be free-ranging from deer obtained from central Europe in 1938. They numbered about 50 in 1973 (Chapman and Chapman 1980; Lever 1985).

France

Fallow deer occur in a number of areas in France today. Some were introduced in the Rouvray Forest, near Orival, Somme, in about the 1860s. They multiplied and spread towards Rouen (Pennetier 1905). They now occur east of Longueval, 30 km north-east of Amiens. A small number occur in the Samoussy Forest near Laon, north-east of Paris in the Department of Ainse, between St. Quentin and Reins, having escaped from a nearby enclosure, but their present status is not well known (Chapman and Chapman 1980; Lever 1985).

Germany

Recorded in Germany since the Middle Ages (Lindermann 1956; Neithammer 1963), fallow deer have been introduced in selected forest areas where they have a fairly limited distribution (Webb 1960). The first historically established attempted releases occurred as early as 1577 when 30 head were liberated near Salaburg, but there were many others in the early seventeenth and eighteenth centuries (Niethammer 1963). In 1713 they were introduced to eastern Germany in the Osterburg area and west of Stendal in the Altmark region. The largest continuous distribution is in the area bounded by Demmin, Neubrandenburg, Prenzlau, Eberswalde, Oranienburg, Neuruppin, Malchow Lube and Gustrow. Releases were attempted in this area in about 1755. More recently, introductions have been made in the south of Germany (Siefke 1977).

Although fallow deer have been widely distributed in western Germany since the sixteenth century, they were found mainly in parks and not until much later escaped and established in the wild. Subsequently they were released deliberately in various places. In Westphalia, they escaped from parks or hunting enclosures (Ueckermann and Hansen 1968) and were first reported as feral in 1883 in the Kottenforst district. Populations at Hoxter and adjacent areas owe their origin to deer that escaped from a park near Bad Dridburg. North-west of Hoxter, near Detwold, some escaped in 1938 and became established. Fallow deer occuring at Dulmen are the descendants of deer liberated from a park in 1890. They were also introduced in Schleswig-Holstien from Denmark and are now well established in many districts, particularly in the eastern parts (Ueckermann and Hansen 1968; Sartorious 1970). Some (10 head) were released near Schleswig in 1939 and are still established (Heidemann 1973) between Satrup and Dannewark. Some were also present near Salzau where they have been for at least a century (Chapman and Chapman 1980).

The majority of fallow deer in western Germany are now in Schleswig-Holstein (since about 1883), Lower Saxony and Westphalia (Ueckermann and Hansen 1968). In eastern Germany they are mainly in the central and northern regions. They are established in Rugen and Darss-Zingst Peninsula to the west of Rostock. There are also a few on islands of Usedom and between Greifswold, Anklam and Uechkermunde.

Greece
There are now no fallow deer wild in Greece. Those on the island of Rhodes in the Aegean Sea may have been taken there from Asia Minor to stamp out snakes. Fallow deer were probably introduced by the Knights of the Order of St. John of Jerusalem in the fourteenth century. These were exterminated during the Turkish rule in 1522–1912, but were re-introduced by the Italians between 1912 and 1945 (Dicks 1974). It was estimated in 1973 that there were 300–400 head present (Chapman and Chapman 1980).

Hungary
Fallow deer occur in only a few areas of Hungary (Chapman and Chapman 1980). East of Budapest near Kecskemét and Szolnok there are some 600 fallow on an unfenced hunting preserve around Pusztavacs, where they were introduced in the late nineteenth century.

In other areas they occur near Gyula between the River Körös and the Romanian border, around Taktakenez and Taktaharkany, between Budapest and Komáron west of the Danube (Lever 1985).

Italy
Fallow deer occur in several areas of Italy within state reserves and forests. In central Italy near Trieste they were introduced from Yugoslavia in 1971. Numbers are small and poaching appears to prevent much increase in population (Calligaris et al. 1976). Most of the areas with fallow deer occur between Rome and Bologna (Lever 1985). Some occur on Sardinia, but only within enclosures (Chapman and Chapman 1980).

Netherlands
Early in the seventeenth century Prince Maurice imported 100 fallow deer from England and released them on the dunes near the Hague. Twenty years later they were killed because the level of damage to the vegetation was unacceptable. In 1647 Willem II introduced fallow deer to Hof te Dieren near Rheden, but they were shot out by end of eighteenth century. In about 1880 Willem III put some in the royal forest at Hetloo and in 1912 these were released into the forest. Some still exist in this area and from here became established a few kilometres north-west near Elspeet.

Fallow deer were released in Deelerwoud, north of Arnhem in about 1915. Before World War 2 the population reached 600 head, but only about 200 were there in 1980. Some escaped in 1945 when the fence was damaged and this led to the establishment of a herd at Hoge Veluwe, however the last animal was shot here in 1954. Some occurred near Leersum where they escaped from nearby parks in 1963 and 1968. In Zeeland some were living wild after escaping from enclosures in 1968. Also in the late 1960s some escaped and became established around Haamstede. Fallow deer were living in the forest at Kennemerduinen after 1958 when park escapees appeared there; about 20 head were present in 1974 (Chapman and Chapman 1980; Lever 1985).

Norway
It is uncertain when the first fallow deer were introduced to Norway. Probably it was as early as the seventeenth century. In the first few years of the twentieth century several liberations were made: three from Denmark were released on Rauø Island in Oslo Fjord and 50 were there in 1911, however, they subsequently died out. Fallow deer from Hankö were released at Hurdalen in 1903, but these did not last long. Those on Skorpø Island in 1904 swam to the larger island of Stronen, but died out after 1911–12. Today the only free-ranging population occurs on Hankö, originating from seven deer introduced from Denmark in 1901 and 1902. By 1936 this herd had increased to 300 head, but was much reduced during World War 2. There were about 100 head in 1968 and 40 in 1972.

Attempts to establish fallow deer elsewhere in the south and west of Norway failed probably because of the deep snow (Chapman and Chapman 1980; Lever 1985).

Poland

Introduced to Poland probably in the seventeenth century, fallow deer are now established in many areas, particularly in the western half of the country. More recently, introductions have been made in some central and eastern districts (Haber, Pasawski and Zaborowski 1977; Oko and Wodek 1978) and they now occur in at least 30 of the 49 voivodeships (Chapman and Chapman 1980).

Fallow deer were introduced from the Mediterranean region in about 1890 to Bialowies Forest in eastern Poland (Lindemann 1956), but they were unsuitable and disappeared about 1930 because they were unable to compete with red (*C. elephus*) and roe (*Capreolus capreolus*) deer.

Portugal

In Portugal fallow deer are not well established in the wild, although escapees are often seen in the vicinity of parks. However, a number of enclosed herds occur there (Chapman and Chapman 1980; Lever 1985).

Romania

Fallow deer were acclimatised in the middle of the nineteenth century in Romania (Almeshan 1959). Wild populations were established in at least 16 areas, with a total population of about 3800 head in 1975 (Chapman and Chapman 1980).

Russian Federation and adjacent independent republics

Fallow deer were released in western Russia from 1932 until 1968. A total of 124 animals were released in Belorussia in 1932 (eight), Gruzinsk in 1949 (six), Kirgizstan in 1962 (eight), Latvia from 1955 to 1958 (15), Moldavia in 1961 (20), and in four areas in the Ukraine from 1948 to 1963 (67) (Kirisa 1974). They have been successful in some areas (de Vos *et al.* 1956), especially Moldavia and the Ukraine where they are established locally (Yanushevich 1966). Some were introduced to the Caucasus in 1888, but were all killed in 1919–20 (Yanushevich 1966). Introductions occurred in the Bialowies Forest in western Russia in about 1890 and were partially successful, but none have been seen there since 1930 (Lindemann 1956).

A number of early introductions occurred in European Russia, but they were extinct there by 1920–30 (Lindemann 1956; Heptner *et al.* 1966). Fallow deer from Germany were released in 1932 in the Voloshin-Ivenez-Molodetscho area near Minsk (Heptner *et al.* 1966), but by 1974 there were none left. In Russian Moldavia 40 fallow deer were present near Kishinev and these may have been the result of a liberation in 1961–62 (Treus and Lobanov 1966). In 1888, 54 fallow were liberated in the mountains of Borhomi

game reserve in the Georgian Republic (Caucasus). More were released in the following two years and by 1918 several hundred were in an enclosure, but these were exterminated in the next two years (Vereschagin 1967). Later attempts to acclimatise fallow deer in the Caucasus also failed (Aliev 1970).

Spain

Fallow deer occur on the Coto Doñana, where at least six fairly recent introductions have been added and these may have been responsible for restoring the species distribution in the region (Niethammer 1963). They are present in several provinces, but there are no large herds, although numbers appear to be increasing (Hingston 1975). Introductions in Coto Doñana occurred in 1917 (Sartorious 1970) by the Duke of Tarifa and these were being hunted by 1924 (Fernandez 1975). Now probably the largest population of fallow deer occurs in the Coto Doñana (Chapman and Chapman 1980; Lever 1985).

Sweden

Introduced in the sixteenth century, the exact date of the first fallow deer introductions are uncertain. Some were sent from England in 1579, but this was not the first. Now they are occasionally found in the wild in Skåne, Halland, Småland, Västergötland, Dalsland, Östergotland, Södermanland, Närke, Västmanland, and Uppland (Siivonen 1972).

Switzerland

Occasionally individual fallow deer escape from deer parks, or immigrate from surrounding countries and are observed or shot, but the species has not become permanently established anywhere in Switzerland (M. Dollinger *pers. comm.* 1982).

United Kingdom and Ireland

Fallow deer were almost certainly re-introduced to Britain by the Normans in the eleventh century when they were released in forests as highly prized quarry for hunting. All the free-living deer in the British Isles descended from medieval introductions to a forest or were escapees from deer parks, especially in the twentieth century during World War 2 (Corbet and Harris 1991).

The status of fallow deer as an introduced or indigenous animal still carries some doubts. There is circumstantial evidence that their origin can be ascribed to the Romans or to the Bronze or Iron Age Phoenicians (Millais 1906; Fitter 1959; Southern 1964; Whitehead 1964; Chapman 1977; Lever 1977; Baker 1990). It is generally agreed that they were well established in Britain in Roman times (Ingersoll 1906; Whitehead 1964; Christie and Andrews 1966; Lever 1977). They were well established in many parts of the country by the time of the Norman Conquest (Whitehead 1964) and were well established and recorded as such in the Doomsday Book (Southern 1964). By the early to middle seventeenth century there were more than 700 parks containing fallow deer (Whitehead 1964; Lever 1977). During the Civil Wars many escaped, or were released, to form feral herds in the surrounding countryside (Whitehead 1964; Lever 1977). During the eighteenth century there appears to have been renewed interest in preserving fallow deer and it was estimated that by 1892 some 72 000 were in English deer parks (Whitehead 1964).

Early introductions include an unsubstantiated one by monks in about AD 900 to the Isle of Islay. They were certainly on the Scottish mainland in 1283 and were well established in Scotland by 1564 (Fitter 1959; Whitehead 1964). They probably arrived in Ireland with the Normans in 1244, were certainly present in 1599 in small numbers, and well established by the end of the seventeenth century (Fitter 1959; Whitehead 1964). Fallow deer have been present in Wales from about 1250 on (Whitehead 1964).

Numerous importations of fallow deer were made to England from 1603–25 onwards, including some to Epping Forest by James I in 1611. In more recent times there have been many escapes and deliberate releases of fallow deer in the British Isles. Escapes have occurred in Shropshire in about 1889, Dumfrieshire in 1898, Suffolk in 1914, Hertford in 1916, Chilterns in about 1928, Norfolk in 1952 and in the Chilterns in 1957. Deliberate releases occurred in Caithness in about 1900, Yorkshire–Lancashire in 1906 and Sutherland in 1940. Early introductions to islands off the coast occurred on Mull in 1868 and on Lambay (County Dublin) in 1889. Those on Lundy (Bristol Channel) and on Harmetray (Outer Hebrides) failed to become established (Fitter 1959). Towards the end of the nineteenth century descendants of escaped animals from parks established themselves in most of the west coast counties of Ireland (Whitehead 1964).

Further escapes and releases occurred during the two World Wars (Whitehead 1964; Willett 1970; Lever 1977). Between 1920 and 1922 many were deliberately released during the troubled times in Ireland and by 1930 they were found in most counties (Whitehead 1964; Lever 1977). During World War 2 some escaped (in 1940) into the Castor Highlands Nature Reserve (Collier 1965).

Surveys in the 1970s indicated that fallow deer were the most widely distributed and common species in England and Wales, and that most of the herds owed their origin to escapees from parks (Chapman and Chapman 1969, 1980; Willett 1970; Lever 1977). However, some herds had been present for extremely long periods (Chapman and Chapman 1969), and were probably the descendants of the original herds (Southern 1964).

In 1977 it was reported that fallow deer were present in 37 counties in England, 10 in Scotland, 18 in Ireland, and seven in Wales (Lever 1977). Of the various introductions to islands off the west coast of Scotland, only three now survive – on Mull, Islay and Scarba all in Argyllshire (Chapman and Chapman 1980). In North Wales in the county of Gwynedd a mass escape of 90 deer from Nannan Park near Dolgellau in 1963 resulted in the establishment of several feral populations (Vaughan and Carne 1971). Fallow deer now occur throughout England with the greatest concentrations in the south-east, and also in Wales and Scotland (Baker 1990).

The Normans introduced fallow deer to Ireland in the thirteenth century, originally to parks, but since then many have escaped and formed feral herds; probably the best known are those in Dublin's Phoenix Park (D'Arcy 1988). Today they are established in six counties in Northern Ireland (Tyrone, Armagh, Down and Antrim) and are widespread in County Waterford, Tipperary, Cork, Clare, Offaly, Leix, Wicklow since 1244, Wexford, Sligo, Monaghan and occasionally elsewhere in the Republic of Ireland (Lever 1985).

Yugoslavia

Fallow deer are generally few in number and are mostly found in northern Yugoslavia, occuring in Slovenia, Croatia and Serbia. In the latter area, south-west of Belgrade, about 70 fallow deer exist from introductions in 1958. In 1965 some were also released into a reserve at Boranja, south of Loznica, but these appear to be enclosed animals. Most of the herds established in Yugoslavia came from stock on Brioni Island off Pula (between gulfs of Kvarner and Venice) (Chapman and Chapman 1980; Lever 1985).

INDIAN OCEAN ISLANDS
Madagascar

Fallow and sika (*C. nippon*) deer were successfully introduced into the forests of Madagascar (Whitehead 1972). In 1932 fallow deer from Czechoslovakia were released on the plateau the Massif de L'Ankaratra near Mandjakatompo (60 km south of Antananarivo). In 1962 a number occurred in the forests to the east of this area (Vincent 1962), but they appear to have been exterminated some time prior to 1974 (Whitehead 1972; Chapman and Chapman 1980; Lever 1985).

MIDDLE EAST

Iran

Fallow deer were re-introduced to Iran from Germany in 1973 (Smithers 1983), but there appear to be few details.

NORTH AMERICA

Canada

In Canada fallow deer have been introduced to James Island in the Straits of Georgia, British Columbia, where they have been flourishing since their release in 1895 (Banfield 1977). Some were trapped and transferred to Saltspring and Pender islands and also to the Alberni district on Vancouver Island. The first transfer took place in 1931 and additional ones to Saltspring and near Alberni in 1934 and 1935 (Carl and Guiguet 1972). In the late 1950s they were abundant on James Island (Cowan and Guiguet 1960) and individuals were occasionally seen on the Saanich Peninsula. In about 1970 they were still present on James, Saltspring and Sidney islands. There have been no further records of animals on the mainland, and no records from Pender Island since the 1940s (Carl and Guiguet 1972). There may have been a second introduction to James Island in 1907 from England. In 1977 they were believed to number several hundred (Chapman and Chapman 1980).

United States

Fallow deer have been released in several areas of the United States since the 1870s and have become established in some (de Vos *et al.* 1956; Presnall 1958; Dasmann 1968). They appear to be still established in localities in Alabama, California, Colorado, Kentucky, Massachusetts, Maryland, Nebraska, New Mexico, Oklahoma, Texas, Georgia, and are established on estates in at least five states: Texas, Michigan, Nebraska, Kentucky and Alabama.

In California 12 fallow deer were released on Point Reyes Peninsula, Marin County, in 1930 and 20 in Mendocino County in 1937; in 1957, or thereabouts, there were 60 head present in the former and 80 present in the latter area. Fallow deer are still increasing on Point Reyes Peninsula and the density is now 20.3/km^2 (Gogan *et al.* 1986). Some were also reported to have been released from the Hearst Ranch in San Luis Obispo County in 1940, and these were maintaining their numbers in the late 1950s (Presnall 1958). A herd still persists on Point Reyes Peninsula (Point Reyes National Seashore), Marin County, originating from introductions in the 1940s by Dr M. Ottinger, who obtained stock from the San Francisco Zoo and released them on his ranch on Inverness Ridge. In 1967 ranchers began shooting these animals because of the reported competition for range with

their cattle. In 1982 there were still 523 present, although their numbers had been supplemented by further releases in 1942 (15); 1947 (11); and 1954 (two) (Wehausen and Elliott 1982). Fifty-one were introduced (white) from the Hearst Ranch on Ridgewood Ranch, Mendocino County, in 1949 and these became well established. In 1970 there were at least 130 head there. Fallow deer were kept on the Hearst Ranch from the 1930s and in 1953, 85 escaped but by the mid-1960s only one remained. There were also fallow deer in Santa Clara County, originally from two pairs released there in 1934, but there were only about 20 head in 1976 and these were being culled annually. Some were also on the Pomponio Ranch, but their present status is not known.

Fallow deer may also be in Tehama County where a captive herd is maintained, and they may still exist in San Mateo County (Jameson and Peeters 1988). In 1976 some escaped when fences were vandalised and some may still survive along the Sacramento River.

White fallow deer were released in 1967 and 1970 at Lone Pine Ranch in Trinity County, California, and by 1971 there were still 20–30 there. Thirteen were liberated in 1960 on Butos Ranch near Call Mountain, San Benito County, but only two were left in 1970. Some were released in Siskiyou County in the 1950s at Yreka, but these have since disappeared (Jurek 1977; Chapman and Chapman 1980).

A few fallow deer were released in Colorado in the late 1920s or early 1930s east of La Jura, Conejos County. Another small group was liberated in Chenokee Park, Larimer County in about 1935 and a third release in 1944 near Bulford in Rio Blanco County. In 1958 there were 50–75 head wild in Colorado (Presnall 1958), but there are no recent records.

Fallow deer were released in Kentucky about 1900 and in the late 1950s a herd of nearly 200 existed in the Kentucky National Wildlife Refuge in Trigg County. This herd, which is reported to be the oldest in the United States (Presnall 1958), was released in 1918, and in 1975 had increased to 700–800 head (Terpening and Hawkins 1975). The population peaked at 800–1000, but is currently 200–300 head; they tend to stay in a limited area and do not disperse from it (Feldhammer and Armstrong 1993). Others were released at Fort Knox and Fort Campbell Military Reservation in the 1970s, but did not survive for long, as did some released at Grayson Lake Wildlife Refuge (Chapman and Chapman 1980).

In Maryland a few fallow deer remain from introductions between 1920 and 1930 in Worcester County and between 1935 and 1945 in Talbot County. Few survived to the 1950s and now there are none (Presnall 1958).

An introduction to Boone County, Nebraska, occurred in the late 1930s when 60 fallow deer were taken to a ranch near Petersburg. These increased in numbers and extended their range. The population in the mid-1950s numbered about 350–400 head, originating from two introductions to Ray Hall Ranch in Beaver Creek Valley: 20 in 1939 and 53 in 1940. Some were noted in 1946 and again in 1954. Some were also found in the adjoining Greely and Howard counties (Packard 1955). In 1973 there were about 50–60 there, but these were originally culled to prevent conflict with agricultural interests and competition with native deer (Chapman and Chapman 1980).

Several private herds occurred in New Mexico, some of which spread into adjacent areas, but there is little information about them. Some in the Sacramento Mountains in eastern central New Mexico originated from animals that escaped from a park. These were still there in 1973 (Chapman and Chapman 1983). A small herd of less than 50 maintained itself in Muskogee County, Oklahoma, from 1956 to 1958 (Presnall 1958), and more are present now (Chapman and Chapman 1980). In Tennessee no fallow deer are known to be established, but occasionally an animal is shot (Anon. 1968).

In Texas fallow deer have been introduced on the King Ranch on open range (Gottschalk 1967). Animals from Milwaukee, Alabama and San Diego Zoo were liberated on Blackjack Peninsula, now part of the Aransas Wildlife Refuge, between 1930 and 1936, but by 1938 only 22 out of 500–600 imported actually survived. These were either captured or disappeared by about 1940 (Halloran and Howard 1956; Ramsay 1968; Schreiner 1968). By 1970 they were present in six counties with over 50 animals and in 27 counties with less than 50 animals, all fenced. The majority of fallow deer in Texas are still on ranches behind deer-proof fences, but some are still free-ranging. Most information in recent reports does not indicate which are free-ranging and which are enclosed (Young 1972; Armstrong 1975; Chapman and Chapman 1980).

Fallow deer have also been introduced on Nantucket Island and on Martha's Vineyard, Massachusetts (de Vos *et al.* 1956). The introduction occurred in 1932 on Martha's Vineyard and in the 1950s there were 150 head (Presnall 1958). Further releases occurred in 1968 and 1969 (Keith 1967; Godin 1977); however, they are now extinct on Martha's Vineyard (Chapman and Chapman 1980). They have not been present on Nantucket for the past 28 years (Chapman and Chapman 1980).

In 1925, 12 were released in Black Warrior National Park in Wilson and Lawrence counties, Alabama, but did not survive (Allen 1965). In 1930s some escaped in Wilcox County from an enclosed herd and during next 30 years the population reached 200–300 head and spread up to 32 km (Allen 1965). Since then they have not increased much. Thirty escaped in Mobile County in 1946 (white fallow): their fate is uncertain, but up to 1965 it was thought that some may still be there. Other escapes in Chapman and Butler counties failed to become established (Allen 1965). A small release in Wilcox County, Alabama, became established in the 1950s and was maintaining its numbers at this time (Presnall 1958).

Early in the 1920s and the 1950s an unknown number of fallow deer were introduced to Little St. Simon's Island off Brunswick, Georgia. In 1974 there were about 500–600 there and there are currently about 500 (Feldhamer and Armstrong 1993). Some were released on the mainland near Tocoa in the Lower Mountains, but these were exterminated within a year (Chapman and Chapman 1980).

In the United States fallow deer are now free-ranging in nine states. Small populations, about which little is known, exist in Alabama (Wilcox and Dallas counties), Nebraska (Boone and Wheeler counties), the Sacramento Mountains of New Mexico, and Maryland (Talbot County). Larger populations occur in Trigg County, Kentucky, on land between the Lakes, Little St. Simon's Island, Georgia, and Point Reyes in California. In Texas about one half of the 14 000 fallow in the state are free-ranging (Feldhammer and Armstrong 1993).

PACIFIC OCEAN ISLANDS
Fiji
Fallow deer were imported in 1880 by Captain Padd to the island of Wakaya, Fiji, where by 1929 there were about 1000 of them (Spencer 1929). For many years there was some confusion about the identity of the deer on the island. Initially they were reported as red deer introduced early in century (Derrick 1951). Later they were believed to be Timor deer (*C. timorensis*) (Whitehead 1972), and more recently it was confirmed that they were fallow deer based on skulls, antlers, and photos (Chapman and Chapman 1980). It is possible that the other species were also present at some time, but none have been seen in recent years. By 1977 they had decreased to about 400 head (Chapman and Chapman 1980).

New Zealand
The first import of fallow deer to New Zealand occurred in 1864 from Surrey, England, for sport and the animals were released near Nelson. There were

subsequently 25 known introductions involving at least 60 fallow deer (Christie and Andrews 1966). Fallow deer were introduced to Kawau in the 1860s and to the mainland in 1867 (Thomson 1922). Some (Wodzicki 1950; Gibb and Flux 1973) suggest the date of the first introductions to the mainland occurred in 1864, while others (Clark 1976) say they were first liberated at Nelson in 1861. Some were certainly released in Otago in 1867–70, on Kawau and Motutapu islands in 1870 and 10 other areas between 1887 and 1900 (Gibb and Flux 1973).

All the introductions of fallow deer were partially successful except for those on Kapiti Island. In the mid-1960s there were herds present in about 14 localities, mostly around or near the liberation points, at Wairoa River, Great Barrier Island, Coramandel Peninsula, Kawau Island, Kaipara, Manakau, Rangitoto and Motutapu Islands, Matamata, Wanganui, Mt. Arthur (Nelson), Aniseed Valley (Nelson), Paparoa Range (Westland), Opihi River in South Canterbury and the Blue Mountains in Otago (Christie and Andrews 1966).

Peak densities appear to have been reached between 1923 and 1940 when they were declared vermin by various acclimatisation societies (Christie and Andrews 1966; Clark 1976). Numbers remained static, but in 1947 fallow deer spread explosively and many were killed for economic reasons (Wodzicki 1961).

However, fallow deer generally have not spread much in New Zealand, although they are common in some areas (Wodzicki 1965; Gibb and Flux 1973). Only in the Wangan area in the 1970s did they spread much (Gibb and Flux 1973). They still occur in both the North and South islands, with the largest herds probably in the Wanganui, Westland and Wakatipu areas (Barnett 1985), in small forested areas surrounded by private farmland with little scope for any expansion in range or numbers (Fraser *et al.* 1996).

SOUTH AMERICA
Fallow deer have been introduced and established in Argentina, Chile and Peru (Whitehead 1972; Petrides 1975), mainly on private estates.

Argentina
Fallow deer were introduced to Argentina from Europe in the 1920s (de Vos *et al.* 1956). They were imported from Spain and released into the Buenos Aires, Santa Fé and Rio Negro provinces and other importations on other occasions to other parts of the country (Petrides 1975).

The first introductions appear to have been to Park Pereyra Iraolo early in the nineteenth century, and

most of the present populations of fallow deer owe their origin to those imported by E. Tornquist in 1905 from Spain and Poland. About 1930 these were released into the surrounding country where they prospered. The area of range is now bounded by Buenos Aires, Mar del Plata and Bahia Blanca and possibly north-west into Santa Fé Province. Other herds exist on the border of Neuquen and Rio Negro, where they were introduced by Sr. Aaron de Anchorena. East of San Carlos de Bariloche, Neuquen province, they have been established for over 40 years at Estancia La Primavera, Lago Trafal, near Confluencia. There were 180 there in 1977. A few small populations exist north of here at San Martin de Los Andes and Junin de Los Andes (Chapman and Chapman 1980).

Chile
Fallow deer were first introduced about 1887 from Europe to the national park near Lota, south of Concepcion, in central Chile, where they became well established. They caused some damage and were transported to a reserve near Coronel. From here some escaped into the Nahuelbuta Mountains. The remainder were taken to Santa Maria, a small island in Golfo de Arauco, where they thrived for 15 years. A plan to round them up exterminated many when they fell over the cliffs. A few survivors were taken to Collipulli between Concepcion and Osorno (Schneider 1936). None are now present. Most fallow deer in Chile live on farms or other privately owned land. Free-ranging populations occur on two islands – Isla Altue-Huapi in Lake Ranco near Rio Bueno, Valdivia province, and an island in Lake Ruponco, Osorno province (Chapman and Chapman 1980; Lever 1985).

The introduction of fallow deer is thought to have been good for the country (Miller 1973). In the early 1960s they were reported to be present between Temuco and Puerto Montt, Valparaiso (Niethammer 1963) and in the mid-1970s as widespread over the country (Petrides 1975). Other introductions appear to have occurred in Chile in the 1920s from Germany (de Vos *et al.* 1956).

Peru
Fallow deer were released in 1948 on Huacraruco Ranch at Cajamarca province, north of Lima (Petrides 1975), and apparently earlier in the 1920s on the Hacienda Casa Grande in the northern part of Peru (Niethammer 1963). They have not multiplied greatly at Casa Grande, possibly due to predation by pumas (*Felis concolor*) (Niethammer 1963) and persistent poaching at Cajamarca, so that only 20–25 remained there in the 1970s (Petrides 1975). These

were later released in the Chicamo river basin, where they became established (Chapman and Chapman 1980).

Uruguay

During the 1930s some fallow deer were kept in an enclosure in the Department of Colonia. The enclosure no longer exists, but fallow deer have been reported in the vicinity (Chapman and Chapman 1980).

WEST INDIES
Lesser Antilles

In the Leeward Islands, Lesser Antilles, two islands have free-ranging populations of fallow deer. They were introduced on Barbuda in the eighteenth century, and probably before 1725 by the Codrington family (Aspinall 1954). It was presumed for many years that there were no fallow deer there, but recent photos and specimens indicate they are still present.

Guana or Iguana, off Antigua, has fallow deer believed introduced by the same family. It was initially thought that they were white-tailed deer until in 1978 an experienced hunter from Barbuda identified them as fallow (Chapman and Chapman 1980).

■ DAMAGE

In Europe fallow deer occasionally eat turnips, beet and other crops and may cause considerable damage. Damage to forests and farm crops in Britain is sometimes severe (Southern 1964). In England the most common damage is grazing of early spring grass and corn, and in the latter they can cause considerable havoc. Fallow deer eat the fruit from trees and retard plant growth by feeding on the leaves and shoots (Marshall 1970).

It is probably only in areas of high density that fallow deer can become pests of forestry and agriculture. In woodland they damage young plantings or prevent regeneration of coppice. In agricultural areas they feed in farmland and may constitute a problem by competing for feed with stock (Corbet and Harris 1991). They are now farmed extensively for venison and velvet.

In New Zealand fallow deer are able to build up into large numbers and cause severe damage to the vegetation, but do not occupy high altitude forest and alpine grassland where the erosion risk is severe (Christie and Andrews 1966). They do, however, cause some damage to watershed protection forests and pasture lands (Daniel 1962).

Where fallow deer have been introduced on Little St. Simon's Island, Georgia, United States, the native white-tailed deer has disappeared (Feldhamer and Armstrong 1993).

SWAMP DEER
Barasingha
Cervus duvauceli Cuvier

■ DESCRIPTION

TL about 1800 mm; T 120–200 mm; SH 1000–1350 mm; WT 160–181 kg.

Coat reddish brown or yellowish red, with a row of white spots on either side of a dark dorsal line; antlers

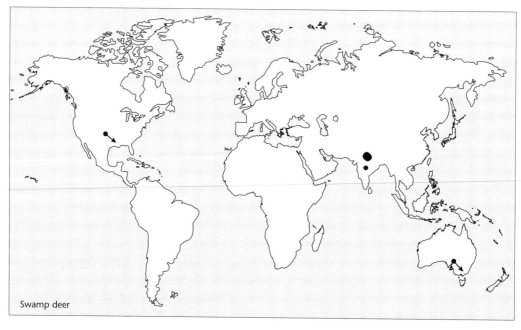

Swamp deer

800–1000 mm, lack bez and trez tines, each tine of terminal fork divided in two, 10–15 points, sheds February–May; neck of male maned; underside of tail white or pale yellow. Females generally lighter in colour.

■ **DISTRIBUTION**
Asia. Central and northern-central India, and south-western Nepal. Extinct in Pakistan.

■ **HABITS AND BEHAVIOUR**
Habits: nocturnal and diurnal; male dominance hierarchy. **Gregariousness:** groups or herds 13–19 (females with young, or males); in breeding season males join female groups; density 0.2/km². **Movements:** sedentary(?). **Habitat:** swamps and grassy plains, marshes, grassy areas close to water, woodland, forest. **Foods:** grass and herbs. **Breeding:** ruts September–April, fawns May–June; gestation 240–250 days; 1 young, or rarely 2; female sexual maturity at more than 2 years. **Longevity:** 21–23 years (captivity). **Status:** formerly common, but now rare in some areas; numbers reduced by cultivation and hunting; endangered.

■ **HISTORY OF INTRODUCTIONS**
AUSTRALIA
Swamp deer were released in Australia in the nineteenth century and were reported to be established (Bentley 1957), but there are no recent records of them. They were imported by the Victorian Acclimatisation Society in 1867–68 and probably released between 1871 and 1885. The exact location is not known, except that it was in the mountainous part of Gippsland where they apparently survived for at least a few years (de Vos *et al.* 1956; Bentley 1957, 1978). They were also released at Port Essington, Northern Territory, in about 1912, but are not known to have survived (Bentley 1978).

■ NORTH AMERICA
United States
Swamp deer exist only in small numbers on ranches in Texas. They were probably introduced in the 1930s or 1940s to some estates (Schreiner 1968; Whitehead 1972; Ables and Ramsey 1973).

■ **DAMAGE**
No information.

RED DEER OR WAPITI
Elk, Roosevelt elk, Rocky Mountain elk, maral
***Cervus elaphus* Linnaeus**
=*C. canadensis* (Erxleben)

Red deer C. elaphus and wapiti C. canadensis are treated here as a single species with canadensis *a subspecies. Because they have been traditionally treated separately the specific names have been retained in the text on introductions for each of them (for relationship see Lowe and Gardiner 1989 and Walker 1992).*

■ **DESCRIPTION**
HB 1650–2650 mm; T 80–270 mm; SH 750–1500 mm; WT 75–509 kg.
Coat reddish brown, yellowish brown or greyish brown, darker on the face, belly, neck and legs (in winter coat is greyer and browner on back, flanks, face, neck, belly and legs); neck with ventral mane (during rut); caudal disc or rump patch yellowish brown; occasionally a dark dorsal stripe; tail short and glandular; tuft of stiff hair inside each heal over tarsal gland; antlers usually six to eight tines (more than 10 exceptional), sheds spring (March–April); rump patch buff or greyish; inner sides of thighs creamy yellow. Female lacks antlers, generally darker on flanks than male. Newborn fawns have white spots on dark brown coats.

Note: C. elaphus is generally smaller and darker than C. canadensis.

■ **DISTRIBUTION**
Eurasia and North America. From southern Scandinavia to the Mediterranean and from Britain, Ireland, Corsica and Sardinia, and north-west Africa (Tunisia and Algeria) east to Manchuria, Korea and western China, and south to the Himalayas and Yunnan. Also southern Canada, most of conterminous United States. Now extinct in much of Russia and present range in Europe fragmented due to local extinction. The subspecies (*C. e. merriami*) known from south-western United States and northern Mexico is now considered extinct.

■ **HABITS AND BEHAVIOUR**
Habits: diurnal, crepuscular and partly nocturnal; crepuscular activity greatest. **Gregariousness:** lives in discrete groups 9–40, 50; bands hinds and calves to 25; males in bands or solitary; large herds after rut to 100 or more; hinds singly at calving and form harems in breeding season; larger groups at food sources or when migrating; density 5–15/km². **Movements:** migratory; altitudinal migration in North America; home range 25–60 ha to 200 ha. **Habitat:** open areas, grasslands, alpine pastures, marshy meadows, river flats, open prairie, parklands with trees, moors, open

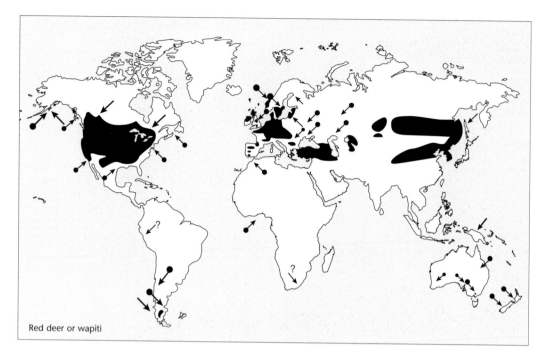

Red deer or wapiti

forest, forest margins, occasionally coniferous forest, forest, and rainforest. **Foods:** browses and grazes; grass, herbs, forbs, shrubs, sedges, terminal buds, leaves, twigs, ferns, berries, heather, moss, mushrooms, lichens, clover, agricultural crops. **Breeding:** ruts September–November, calves May–July (Europe and North America) (Australia and New Zealand ruts March–April and calves November–December); gestation 210–262 days; oestrous cycle about 18 days; male polygamous in breeding season; female seasonally polyoestrous; calves 1–2 rarely 3, generally 1; accompany female at 7–10 days; sexually mature at 16 months, females mate at 3 years and stags at 4–5 years age. **Longevity:** 15–26 years (captive and wild). **Status:** range considerably reduced, but re-established in many areas.

■ HISTORY OF INTRODUCTIONS

Red deer (*C. e. elaphus*) have been introduced to Europe (Austria(?), Finland (unsuccessful), Norway, Sweden, Switzerland, United Kingdom and Ireland); Morocco; North America (Alaska, United States); South America (Argentina, Peru, Chile); Fernando Poo; South Africa (unsuccessful?); Australia and New Zealand. Wapiti (*C. e. canadensis*) have been introduced in Russia (Urals and Volga regions), New Zealand and Mexico (unsuccessful).

AFRICA

Fernando Poo
Spanish red deer (*C. e. hispanicus*) were introduced to Fernando Poo in the Gulf of Guinea in 1954 (Lever

1985; Haltenorth and Diller 1994), where they are still presumably established.

Morocco
Spanish red deer (*C. e. hispanicus*) were introduced into Morocco between Ceuta and Tangier in 1952 (Lehmann 1969; Haltenorth and Diller 1994). Five stags and 10 hinds from Spain were released in the vicinity of Ksar-es-Seghir on the north coast by Sn. Garcia Valino, Governor. A number were killed by poachers, but by 1969 two small herds were established (Lever 1985).

South Africa
Red deer were introduced in 1895 by C. Newberry to an estate between Clocolan and Gum Tree, Orange Free State. In the mid-1930s there were at least 50 present (Bigalke 1937). In 1975 some escaped from a zoo or private collection near Vereeniging, Transvaal, and in 1981 three strays were established there (Lever 1985).

AUSTRALASIA

Australia
Many red deer were imported into Australia between 1860 and 1888. In Victoria, until the early years of the twentieth century, some roamed the plains west of Geelong, but these died out with the advance of settlement (Bentley 1978; Lever 1985).

In western Victoria the largest populations in the 1970s and early 1980s lived in the Grampians and probably originated from escapees as early as 1859.

Some were liberated in Linton Forest about 1914 and some red deer may still survive there. They were established at Gembrook for a time, probably as a result of escapees or releases in the 1890s. These, however, had all disappeared by about 1946. In eastern Victoria some are established on the border with New South Wales and these may have originated from escapees and releases about 1918 or before. A small herd may still exist in the shire of Granville south of Ballarat (Bentley 1978).

Two males and four females were imported from England and released at Cressbrook Station in September 1873 (at Scrub Creek (Roff 1960)) near Toogoolawah in the Brisbane River Valley, Queensland (Bentley 1978; Searle 1981). More were imported (one male and two females) in 1874 and apparently released in the same area (Roff 1960; Bentley 1978). By 1878 these were reported to be increasing in numbers and spreading (Bentley 1978), and in the 1880s large numbers were said to be at Black Jack, Scrub and Waterfall creeks at the head of the Brisbane River (Bentley 1978).

Releases of red deer were also made in the Conondale Range area in Queensland in the late twentieth century, and also at Cunningham Gap at Maryvale in southern Queensland in 1903 (Searle 1981). Seven were released at Warwick in 1903 and some in the Stanley Range in 1923 (Bentley 1978). From these areas red deer spread south to the head of the Condamine River and north to Mt. Castle/Liverpool Range area (Searle 1981). In 1978 they ranged from just north of Toowoomba north to Wide Bay Highway west of Gympie. In the west they reached Gobongo, Nanango and Cooyar, and eastwards reached Northbrook, Conondale and Kandanga (Searle and Parker 1982). They are now throughout the watersheds of the Brisbane, Stanley and Mary rivers (Searle 1981).

The population of red deer in the Brisbane Valley area of the Queensland east coast was reasonably well established in the late 1950s (Roff 1960) and now is estimated to be 8000–10 000 red deer (Mitchell *et al.* 1982). They were also released on Hinchinbrook Island in 1915 by the Queensland government (Bentley 1978), and in later years an additional pair was released (Searle 1981). Two male and two female red deer were released on the island about 1900 to provide food for shipwrecked sailors (Bentley 1978). More were released in 1906 and in 1915, but all had disappeared by 1918.

Red deer were presented to the State of Western Australia by Queen Victoria and were kept in the Zoological Gardens and later liberated at Cape Leeuwin and other areas. Red deer were introduced at

three localities near Albany, close to Cape Leeuwin, and in an unoccupied tract of land between Pinjarra and Rockingham. One pair was released at Cape Leeuwin in 1899 and two years later those and others were said to be thriving. However, they began to decline for reasons unknown at Cape Leeuwin between 1924 and 1930. Four red deer were liberated on an estate near Pinjarra in 1903 and by 1912 a herd of 30 were there. Red deer were released on the property of D. Paterson, 'Creaton', Pinjarra, about 1915 and the herd in this area numbered about 150 in 1920. They became so well established and multiplied to such an extent that it was necessary to 'destroy many as they entered cultivated land' (Colebatch 1929). At one time they ranged widely between the Dandalup and Murray rivers. A diminishing herd appears to have remained in the area until the last was shot in about 1960 (Le Souef 1912; Kingsmill 1920; Colebatch 1929; Bentley 1978).

Introductions of red deer may also have occurred at the Porongorup Ranges, where several varieties of deer were released by the Western Australian Acclimatisation Society in 1903–06. Herds of red deer reported at Menzies, north of Kalgoorlie, and in timber south of Coolgardie may or may not have been true (Allison 1967). Red deer have certainly been found as escapees spasmodically around Pinjarra, Harvey, Waroona, Byford, and many other areas in Western Australia up until the present time. Red deer were reported from jarrah forest south of Pinjarra in 1988, and 25 animals escaped from a deer farm in an area between Rockingham and Mandurah in 1999

In South Australia red deer were released at Yallum Park, near Penola, about 1880 and some were reported from this area in 1914, but the herd disappeared soon after (Bentley 1978).

Red deer are now moderately common in the headwaters of the Brisbane River in Queensland and in the Grampian Mountains in Victoria. A small population liberated near Aston, New South Wales, about 1914 exists on the New South Wales–Victorian border near the headwaters of the Snowy River and appears to be spreading southwards. An unconfirmed population lives in the Otway Ranges of Victoria; a remnant herd may be established south-west of Ballarat, Victoria; herds in South Australia and Western Australia appear to be extinct (Wilson *et al.* 1992; Strahan 1995).

Papua New Guinea
Red deer were probably introduced to Papua New Guinea at the same time as other deer species in the late nineteenth century (Bentley and Downes 1968), but are not now established anywhere in that country.

EUROPE

In continental Europe there have been many translocations and re-introductions of red deer (*C. e. elaphus*) within their natural range (Lever 1985).

Austria

Red deer (*C. e. canadensis*) have been introduced in Austria (de Vos *et al.* 1956), but appear to have had little success and have probably hybridised with the nominate race (*C. e. elaphus*). A few hundred were introduced to Austria by Franz Joseph I (Lorenz 1953).

Finland

The Province of Åland in Finland was reserved in 1537 as a royal hunting park to which red deer were introduced. They survived in this area until 1778, when the last was recorded (Salo 1976).

Norway

Red deer (*C. e. elephus*) from Germany have been introduced into Norway (de Vos *et al.* 1956) and are reported to have replaced the indigenous stock.

Russian Federation and adjacent independent republics

Almost 4000 red deer (*C. e. elaphus*) of seven subspecies have been introduced, re-introduced or translocated in Russia and the adjacent independent republics since the late 1880s. The results of these efforts do not appear to have been spectacular, but a number of small populations have been established locally. Introductions of the race *C. e. canadensis* in the time of the Czars is said to have reduced the value of some herds, as hybridisation with the native subspecies (maral *C. e. maral*) provided animals with less desirable antlers (Romanov 1932 in Lindemann 1956; de Vos *et al.* 1956).

Some 2949 red deer (*C. e. elaphus*) have been released, mainly in the European part of the Russian Federation and adjacent independent republics, in areas where red deer already occur or have occurred in the recent past. Between 1918 and 1972 releases occurred in 23 regions of the Russian Federation, eight in European Russia, five in European republics and five in Asian republics. The subspecies involved in these introductions included: the Middle European (*C. e. hippelaphus* Erxleben), the Carpathian (*C. e. montanus* Botez), the Caucasian (*C. e. maral* Ogilby), and the Crimean subspecies (*C. elaphus*).

Some 793 maral (*C. e. maral*) were released in five regions in the Russian Federation and also in Kazakhstan, Latvia, Ukraine and in the Astonsk region between 1937 and 1970. The Izubr (*C. e. xanthopygos*) was released on Sakhalin Island (133 in

1965–70) and at Odessa in the Ukraine (five in 1967). Bukhara red deer (*C. e. bactrianus*) were released in Tajikstan (12 in 1960–61) (Kirisa 1974).

In the Ukraine re-acclimatisation has been fairly successful in some areas from introductions in 1918 and again in 1952, and these were still established locally in the 1960s (Yanushevich 1966). First introduced to the Caucasus in 1888, red deer were established there for many years. However, further introductions were necessary to re-establish them in this area in 1960 (Yanushevich 1966). Introductions occurred in Moldavia between 1954 and 1960 and appear to have resulted in the successful establishment of red deer locally (Averin and Uspenskii 1962; Yanushevich 1966). Local success has also been achieved with re-acclimatisation of red deer in Tadzhikistan in 1960–61 and on the Black Sea Preserve in 1957 (Yanushevich 1966).

Maral have been successfully introduced in the Urals (Kaznevsky 1956) without much economic effect in the area (Pavlinin and Shvarts 1961) and appear to be still established in the region (Corbet 1978). The range of the Caspian red deer has been extended in Kazakhstan by introductions and translocations (Sludskii and Afanas'ev 1964).

Introductions of red deer in Russia 1940–1972

Area	Number	Date
RSFSR		
Armyahsk	20	1971
Belgorodsk	32	1971
Bryansk	58	1965,1972
Gorkovsk	45	1963
Kabardino-Balkar	44	1957
Kaliningradsk	33	1962
Kalujsk	66	1962–63, 1967
Krasnodavsk	90	1958–61
Kuibshevsk	91	1964–65
Moskovsk	191	1960–70
Orlovsk	21	1971
Osetinsk	30	1964–65, 1971
Penzensk	18	1957
Rostovsk	150	1968–72
Ryahzansk	114	1965–71
Saratovsk	135	1962–70
Smolensk	189	1967–71
Tulsk	30	1965
Veronejsk	86	1966–72
Vladiminsk	101	1962–63, 1969–70
Volgogrodsk	60	1961, 1967

Introductions of red deer in Russia 1940–1972
(*continued*)

Area	Number	Date
USSR		
Boroshilovgradsk	16	1962
Jitonirsk	16	1956
Kharkovsk	50	1952–56
Khersonsk	30	1918
Kievsk	106	1957–69
Lvovsk	44	1961, 1970
Odessa	50	1963, 1967
Volnsk	20	1968
Republics		
Astansk	12	1965
Kazakhstan	30	1962–63
Kirgizstan	14	1962, 1965
Latvia	25	1956, 1964, 1969–72
Litovsk	37	1956
Belorussia		
Bitebsk		
Gradensk		
Minsk	857	1956, 1962–72
Mogilevsk		
Neizvestnask		
Maral releases		
Kalininsk	359	1937–70
Kalujsk	74	1967–69
Moskovsk	127	1955, 1960–67
Sverdlovsk	7	1959, 1969
Yahroslav	114	1955, 1967
Baskirsk	66	1941
Kazahkstan	21	1940–62
Latvia	2	1957
Ukraine	14	1967–68
Astansk	9	1957

Reference: Kirsta 1974.

Sweden

From 1957 to 1959 red deer (*C. e. canadensis*) were introduced to parts of central Sweden (Lavsund 1975). By 1968 they were present in 10 areas of southern and central Sweden, and the population was estimated to be 800–1000 head.

Switzerland

Red deer have been immigrating into Switzerland from surrounding areas for at least 75 years. Some were introduced into several parts of the country in the 1920s and 1930s. Those introduced into the Aletsch Forest, canton of Valais, in 1934 failed to become established. Some were also released at Val Ferret, canton of Valais, from Austria in 1926, and some in the canton of Schwyz from zoo stock in 1934.

Red deer (*C. e. hippelaphus*) now inhabit about 50 per cent of Swiss territory and the expanding population was estimated to number about 20 500 head in 1980. Since 1977 about 4000 head are taken annually by hunters (M. Dollinger *pers. comm.* 1982).

United Kingdom and Ireland

In Great Britain indigenous stocks of red deer (*C. e. elaphus*) are probably now confined to a few areas. However, through introduction, re-introduction and escape from parks they still flourish in many areas. It is estimated that in the early 1960s herds of wild deer derived either wholly or partly from indigenous stock occurred in Devon, Somerset, Cumberland, Westmorland and North Lancashire and that others of recent origin were present in parts of some 15 counties in England (Whitehead 1964). In Scotland they were present in 15 counties and on 14 islands off the coast and although probably indigenous to Ireland many had been introduced from both England and Scotland (Fitter 1959; Southern 1964; Whitehead 1964).

Red deer were introduced to County Wicklow, Ireland in about 1246. By the end of the nineteenth century they appear to have been fairly rare but were re-introduced many times (Fitter 1959; Whitehead 1964).

On the island of Mull red deer were almost extinct at the end of the eighteenth century when more were imported from Scotland, England and Ireland. The islands of Rum, North Uist, Jura, Raasay, Harris and Padbay have all been replenished with imported stock. Lundy Island was stocked in 1927 with 10 red deer from Derbyshire and there were nine still there in 1955 (Fitter 1959). Islay in the Hebrides also had an escaped population of red deer (Fitter 1959). Twenty-eight red deer were introduced to Ramsey Island in Wales, in 1979 (Pratt 1980). In the 1960s they were found throughout Scotland, through the Midlands and southern counties and West Country (Willet 1970). In Ireland the largest concentrations were in Donegal, Kerry and Wicklow (Mulloy 1970). Probably the only indigenous red deer in Ireland are those in County Kerry. Elsewhere they have been affected by introductions and hybridisation with sika deer (*C. nippon*) (D'Arcy 1988).

There have been several attempts to introduce red deer (*C. e. canadensis*) from North America into Great Britain to 'improve the quality of heads'. Some were imported into England between 1781–94 and probably the first attempt to establish them in the wild was made at Dunkeld (Perthshire) at the beginning of the nineteenth century. Sir Arthur Grant introduced some in Aberdeenshire during the 1880s and some of these were reported to have crossed with the native red deer. These animals were later taken to Inverness-shire where they apparently failed to breed well. In the early part of the nineteenth century 30 red deer (*C. e. canadensis*) were released in the Meoble Forest (Inverness-shire) in a park (Whitehead 1964). At present no herds of wild North American red deer are known to exist in the British Isles. Wapiti were introduced to Ireland at the turn of the century at Caledon and the present herd of red deer there still has some wapiti characteristics (Mulloy 1970).

NORTH AMERICA

Red deer (*C. e. canadensis*) are widely distributed and fairly common animals in North America (see map). There have, however, been a number of translocations, re-introductions and introductions in this area including some with the European red deer, *C. e. elaphus*.

Alaska

Since the 1920s there have been about nine translocations/introductions of red deer into Alaska (Franzmann 1988).

Some were taken from Washington and released on Afognak Island, Kodiak group, Alaska, in 1929, and by 1960 there were about 800 on the island (Batchelor 1955; Troyer 1960; Jones 1966). Eight were released between 1926 and 1937 on Kruzof Island (Alaska); three in 1926–37 and 24 in 1963–64 on Revilla Island; eight in 1929 on Afognak Island; and eight in 1962 on Gravina Island (Burris 1965). In 1928, eight deer (five males and three females) were translocated from Olympic Peninsula in Washington to Kalsin Bay on Kodiak Island. These were kept in semi-domestication, but because of conflict with a local grazing enterprise, were transferred to Afognak Island in 1929, where they were thriving in 1936 (Murie 1941). In 1941 about 64 were counted and by 1958 the population had risen to a count of 599 with possibly up to 800 head present (Troyer 1960). From Afognak Island some were moved to Raspberry where they established well, and have been hunted since the 1950s (Franzmann 1988). Elk (*Alces alces*) were noted on Kodiak in 1946 and more recent reports suggest some may still be established in the north-east of the island.

Canada

Red deer (*C. e. canadensis*) have been re-introduced in a number of areas in Canada some with considerable success. They were re-introduced to Wood Buffalo Park and in the southern Rockies (Banfield 1977). In 1948–49 they were re-introduced in the Lake Claire district in northern Alberta and two years later at Braeburn Lake in southern Yukon Territory, and have also been re-introduced in some areas of Ontario (Whitehead 1972).

In 1918 red deer (*C. e. canadensis*) were released in Roscommon, Alpena, Otsego and Cheboygan counties and there was a subsequent release of 16 to Roscommon in 1932. The Otsego release became established and by 1941 some 300–500 were present, but the other releases did not succeed and their numbers dwindled; by 1941 only occasional report of them was received (Ruhl 1941).

In 1932, 25 red deer (*C. e. canadensis*), probably from Jasper, Alberta, were transferred to Adams Lake in British Columbia. Five from Stanley Park Zoo were released at McNab Creek on Howe Sound in 1933. Some 35 were transferred from near Penicton to the Princeton area in 1931–32. There was probably an early introduction to the Yalakom River area. In 1971, 30 red deer were transferred by the fisheries and wildlife branch from Banff National Park to the Grand Forks region, where they appeared to become established (Cowan and Guiguet 1960; Carl and Guiguet 1972; Banfield 1977). Native stock in the southern Rockies has been augmented by introductions from Yellowstone Park to Banff National Park in 1917–20 (Banfield 1977).

Red deer (*C. e. canadensis*) have also been introduced with some success to islands off the coast of Canada. Eight were placed on Graham Island in the Queen

Introductions of deer into Alaska

Date	Place	Origin	Success
1926, 1928	Kruzof I.	Washington State	unsuccessful
1929	Afognak I.	Washington State	harvestable
1937	Revillagigedo I.	Washington State	unsuccessful
1962	Gravina I.	Afognak & Raspberry Is	unsuccessful
1963, 1964	Revillegigedo I.	Afognak I.	unsuccessful
1986, 1987	Etolin I.	Oregon State	persisting

Charlotte Islands (Cowan and Guiguet 1960; Carl and Guiguet 1972), where they became successfully established, but were exterminated by 1947. Red deer were taken to Anticosti Island in 1903 and 1911, but only three remained by 1937 (Newsom 1937).

European red deer (*C. e. elaphus*) were introduced to Graham Island in the Queen Charlotte Islands, but were exterminated between 1942 and 1946 (Banfield 1977). The release occurred in 1914 by the Game Commission which imported a male and three females from New Zealand. They were kept on a game farm at Chilliwack until 1918 when they were released near Masset. These were reported to have formed a large herd that declined during World War 2 (Carl and Guiguet 1972). There have been no records of any since this time.

Mexico
In 1941, 18 red deer (*C. e. canadensis*) from Wichita National Wildlife Refuge, Oklahoma, United States, were liberated in northern Coahuila, Mexico, but they failed to become established as all had died or been killed by 1943. In 1952 and 1955 about 30, on each occasion from Yellowstone National Park, were liberated at Coahuila, but the results of this attempt are not known (Whitehead 1972). Some may have been released again in more recent years (Petrides 1975), but any success is not known.

United States
As a result of introductions in the twentieth century, red deer (*C. e. canadensis*) now inhabit many of their former haunts. By 1955 it was estimated that they were present in over 20 states (Whitehead 1972). Translocations have occurred in various areas of the United States including New York and Michigan (de Vos *et al.* 1956).

Red deer (*C. e. canadensis*) were introduced to the Big Horn Mountains, northern Wyoming, in 1910 and there have been open seasons in that area since 1925 (Bagley 1938). Some were introduced in Arizona in 1913 and there have been open seasons in that state since 1935 (Kartchner 1940). Other introductions were made in Wisconsin in 1913 and 1917, where up until at least the 1940s they were said to be increasing in numbers (Reese 1944), and some have been transferred from Montana and Wyoming into Colorado (Rogers 1947). In Pennsylvania red deer were exterminated early in the century and were re-introduced again before 1936 (Gerstell 1936).

In 1895, 12 and between 1896–1902 a further 60 red deer (*C. e. canadensis*) were released in Litchfield Park, New York, but these had disappeared by 1910. In 1893–94, 66 red deer from Wyoming were liberated in the Nehasane Preserve by W. S. Webb, but were decimated in a fire in 1903. From 1901 to 1903, 178 red deer were released in various areas of New York State including Hamilton County, Little Tupper Lake, Raquette Lake, Bay Pond Preserve, Suranac Inn and Big Moose Lake. These were released by state authorities, W. C. Whitney, W. A. Rockefeller and others. Most herds gradually disappeared for one reason or another. The Adirondacks Guides Association also released 17 deer in Essex County in 1906 and local resort owners released four at Lake George and five at Tongue also in 1906. State authorities released some in the Adirondacks and in 1917 E. H. Harriman released 60–75 deer on an estate in Orange County where they became established and were increasing in numbers in the late 1930s. Six red deer were liberated on the De Bar Mountain Refuge in 1932 and there were 14 there in 1937 (Bump 1941). New York still had some herds of introduced red deer in the 1950s (de Vos *et al.* 1956).

More than 60 000 red deer (*C. e. canadensis*) have been translocated into Montana since the first releases in 1910. At least 4140 of these were released between 1941 and 1970 (Rognud and Janson 1971). Two of 12 introductions of red deer in Colorado have failed, probably due to parasites (Gogan and Barrett 1987).

Red deer were successfully re-introduced to Michigan at Pigeon River in 1918 (Stephenson 1942). Here, they increased in numbers rapidly and by 1950 were causing some deterioration to the habitat and to agriculture (Moran 1973). In 1964–65 a herd-reduction program was commenced and the herds were depleted by 1970. Development in the area for oil and gas did not affect the few remaining animals, but seismic activity did have some effect (Knight 1981). North Maniton Island, Michigan, was stocked in 1925 and 12 years later overgrazing was causing the deer to die of starvation (Bartlett 1944).

By 1860–1900 the native race of red deer was nearly extirpated from California. Recovery has subsequently occurred and many translocations have been made in this state. Rocky Mountain elk (*C. e. nelsoni*) were introduced and now persist at three locations: Hearst Ranch, San Luis Obipo County and adjacent Monterey County; Shasta Lake area, Shasta County; and Tejon Ranch, Kern County, and there is a privately managed herd on Santa Barbara Island (Lidicker 1991). Some were also introduced at Owens Valley, Inyo County, California, and to Cache Creek, Lake Colusa County, and to a state park at Tupman, near Bakerfield, and now about 500 of them are present (Nelson and Hooper 1953). Some are well established and reproducing in Kern County, California (Thomas 1975).

Two re-introductions made to Point Reyes and Grizzly Island, California, with *C. e. nannodes* began about 1977 (Gogan and Barrett 1987). Ten adult elk from Merced County were kept in a holding yard and released in 1978. Three adults from Inyo County were released there in 1981, but these subsequently disappeared. Seven adult elk were relocated on Grizzly Island from Kern County also in 1977. Both these introductions were successful. In California several (*C. e. elaphus*) escaped from the Hearst Ranch, San Luis Obispo County (Presnall 1958), but their fate is not known.

Between 1969 and 1972, 335 (*C. e. canadensis*) were translocated from the Wichita Mountains National Wildlife Refuge to eastern Oklahoma. Here, release of 157 deer was made in north-eastern Oklahoma, and five releases (184 deer) were made in south-eastern Oklahoma. After a year they were still present in most areas (Stout *et al.* 1973).

European red deer (*C. e. elaphus*) were introduced at Bay Pond Preserve, New York, in 1905–08 by W. M. Rockefeller, but strays were shot and the remainder disappeared (Bump 1941). Some (*C. e. elaphus*) are present in Texas in captivity (Ables and Ramsey 1974), at least since 1958 (Schreiner 1968). Red deer (*C. e. elaphus*) have also been established in Kentucky Woodlands National Wildlife Refuge in Twigg County and in Bernheim Forest in Bullitt County (Presnall 1958). Sixty-seven were released in 1934 and the first open season was held in 1945, but only 29 were shot from an estimated population of 500 (Vinson 1947). In 1955 there were still about 120 head in Kentucky (Whitehead 1972).

SOUTH AMERICA
Argentina
Sr. Pedro Luno imported some European red deer from Hamburg in 1906 and stocked his park near Santa Rosa, La Pampa province. In 1922 his estate was sold and fences broken and the deer escaped into the surrounding countryside. A few still existed in this area to the mid-1970s at least. In 1922 also, 18 red deer were brought by Sr. Robert Hohmann and taken to his estancia at Collunco, near San Martin de los Andes, in Neuquen province. These became established, spread across the border into Chile, and south to Bariloche. They were throughout Lanin National Park and in the northern part of adjoining Nahuel Huapi National Park, where about 13 000–15 000 existed in 1965 (Cresswell 1972).

European red deer may have been imported from Germany, Austria and Hungary on several occasions between 1902 and 1971. They have certainly become established on a number of ranches in Neuquen, La Pampa and Chubut provinces (Petrides 1975).

Chile
Red deer (*C. e. elaphus*) were introduced to Chile from Germany in 1916 (de Vos *et al.* 1956) and are now found over much of the country (Petrides 1975). They also invaded Chile from Argentina (Creswell 1972; Miller 1973) and have become established in several localities between Temuco and Puerto Montt. In 1975 a herd of 200 was on an island in Lake Rupanco, Orsono province, and another herd to the north of an island in Lake Rauco, Valdivia (Lever 1985).

Peru
Thirty red deer (*C. e. elaphus*) were imported from Argentina in 1948, and were at first kept confined in an enclosure on a ranch in San Juan in northern Peru, but have since escaped and in the early 1970s numbered some 200–300 (Petrides 1975).

PACIFIC OCEAN ISLANDS
New Zealand
European red deer (*C. e. elaphus*) were introduced to Nelson on the South Island, in 1851 (Wodzicki 1950, 1961) and are now widespread and abundant in the North and South islands and on Stewart Island (Wodzicki 1965; Gibb and Flux 1973). Further introductions were made in Nelson in 1854 and 1861, Wairarapa in 1862, Otago in 1871 and many more subsequently (Gibb and Flux 1973). Some were imported to Canterbury in 1862–64 and nine (three stags and six hinds) were liberated on the Rukaia Range in about 1897, where they became established and increased in numbers (Lamb 1964). By 1907 these had spread throughout the Rukaia Valley (about 150 head). More were imported (eight) and released in the Poulter River district in 1908. In 1909 red deer crossed the Clarence River from Nelson into the Waiau district of Canterbury (Lamb 1964). It is estimated that between 1851 and up to 1923 well over 100 liberations were made (Riney 1956; Wodzicki 1961) involving about 1000 deer (King 1990).

Eight red deer were imported and landed at Port Chalmers in the 'City of Dunedin' in January 1871 and a second shipment of nine in February 1871. Some were released in the Morven Hills and some at Bushey Park, Palmerston. By 1890 they had found their way into the catchments of Timaru Creek and the Ahuriri River and along the Grand View Range. Stags from Melbourne, Australia, were introduced about 1895 and further introductions occurred in 1899 and 1900. In 1913, four were introduced by the Waitaki Society to Maitland Stream and Ben Avon. By 1916 they had spread from Lake Hawea to Lake

Pukaki. In 1921 crop damage was reported and in 1925 landholders were granted licences to shoot them. By 1927 the deer were causing concern to farmers by reducing sheep carrying capacity. In 1930 protection was removed (Lamb 1964).

By 1908 red deer were well established in the Hunter Valley and by 1915 in the headwaters of the Hunter and Makarora. In 1926 they were established along the Burke River, a tributary of the Haast River, and arrived at the Landborough River in 1912–14 in South Westland. They found their way into Waiatoto about 1930. In 1925 they were established in the upper reaches of Moeraki, Paringa and later reached Mahitahi. In 1937 they were found at Makawhio and reached Karangarua and Copland. In 1968 they reached the northern extremity in the Cook River system (Banwell 1968). By the mid-1920s red deer were established discontinuously in most mountain areas of the Bay of Plenty in the North Island to Southland in the South Island, and by 1947 there were few deer-free areas (Wodzicki 1961). Control of deer was taken from the acclimatisation societies by the Department of Internal Affairs in 1931, and thereafter many thousands were removed by government and commercial hunters (Banwell 1970).

Wapiti (*C. e. canadensis* = *C. c. nelsoni*) were first introduced in 1870 (Wodzicki 1950). They were introduced in 1905 in George Sound, Fiordland, and spread from there to Lake TeAnau, South Island (Gibb and Flux 1973). They were well established and spreading in Fiordland in the late 1950s (Miers, 1961). They were released at Dusky Sound in 1909 and several sites to the east and south several years earlier. Peak population was reached in 1945–60, but they subsequently declined following depletion of the vegetation, but high numbers still exist locally (Tustin 1974). Wapiti are now locally common in South Island (Wodzicki 1965; Barnett 1985).

Red deer are the most widespread species of deer in New Zealand. Since 1990 they have spread into several new areas in Northland, Auckland and the western King Country, where they did not occur previously (Fraser *et al.* 1996).

■ DAMAGE
In Germany red deer cause damage to forest trees and feed on crops planted adjacent to forests. On the forest trees they damage terminal buds and small branches, and peel the bark from such species as spruce and beech; in agricultural crops the damage is sometimes extensive (Webb 1960). In West Germany supplementary feeding with silage significantly reduced the damage of bark stripping and browsing, particularly in the winter in periods of food shortage (Ueckermann *et al.* 1977).

In Germany one study found that the factors affecting red deer damage are the kinds and quality of the forage available, size and age structure of the herd and the possibility of deer being able to maintain normal daily rhythm. The study concluded that healthy deer would have neither the inclination nor the time to peel and browse trees (Hennig 1960).

Translocation of red deer from Germany into Norway has apparently resulted in the virtual extermination of the native subspecies there because the German strain is said to be less hardy (de Vos and Petrides 1967). During the nineteenth century there were many attempts to introduce wapiti or maral into the range of European red deer with the object of obtaining more impressive trophies, but this crossing has now been abandoned (Krzywinski *et al.* 1987) although in practice it appears to produce significantly heavier animals.

Studies of red deer in Hungary indicate that as many as 89.2 per cent of poplar trees may be damaged. High and moderate damage did not affect the commercial value of the timber produced, but severe damage certainly did. Reducing populations, fencing, and winter feeding were suggested to overcome the problems (Bencze *et al.* 1977). Damage to agricultural crops and forestry is reported to be considerable and in some regions in Switzerland they are thought to compete for food and habitat with the alpine chamois (*Rupicapra rupicapra*) (Kuster 1966).

In Britain the most common damage by red deer is grazing of early spring grass and corn, but they can affect root crops (e.g. sugarbeet, potatoes, cabbage and brussel sprouts). In corn crops they can cause considerable havoc (Marshall 1970), but are considered to have a major impact on commercial forestry and conservation of some native plants and woodland (Corbet and Harris 1991). In Scotland red deer readily feed in cornfields, meadows, pastures and arable crops (Mitchell *et al.* 1977). However, they do not appear to be a serious agricultural pest.

Sometimes red deer cause damage in forest nurseries in Russia, but supplementary feeding prevents deer damage, except when red deer feed exclusively on trees (Dinesman 1959).

Monocultures of coniferous trees are very poor deer habitat in the Netherlands. Here, planting of mixed forests and artificial feeding (with hay) have prevented much bark stripping if it is combined with good forest management (Van der Veen 1973).

Conflict in North America occurs with ranching interests when red deer descend onto private land in winter or spring and use forage needed for livestock. Competition with sheep occurs when both deer and sheep graze summer weedy patches or in winter where both graze and browse the same areas (Rognrud and Janson 1971). In some areas of Montana, competition with both cattle and sheep is indicated on areas used by deer in winter (Stevens 1966).

In South America red deer have probably driven out the native huemul (*Hippocamelus bisulcus*) which is now found only beyond red deer range (Creswell 1972).

In New Zealand red deer cause economic concern as they damage forests protecting watersheds and pasture lands (Daniel 1962). They have caused irreversible changes to the vegetation in some national parks and protected natural areas and may have increased the potential for soil erosion; they have modified the understorey of some forest areas to such an extent that their own numbers are now declining (King 1990). There is still a substantial commercial trade in skins, meat and live animals for export in New Zealand and many animals are now confined to farms. The trend on farms of crossing wapiti with red deer is a worry as many of the escapes or liberations may be crosses (King 1990).

It is thought by some that as red deer eat the same foods as the takahe (*Porphyrio mantelli*), it may explain the disappearance of that species in some areas of New Zealand (Mills and Mark 1977).

In Queensland, Australia-introduced red deer are known to cause damage to re-forested pine plantations and crops such as maize, oats, sorghum and lucerne (Roff 1960). Most damage is caused to crops and pastures during severe winters and so numbers of deer are destroyed occasionally (Searle 1981).

BLACK-TAILED DEER

Mule deer, coast deer, Columbian blacktail

Cervus hemionus (**Rafinesque**)

=*Odocoileus hemionus*

■ DESCRIPTION
TL 1160–1880 mm; T 106–230 mm; SH 675–900 mm; WT 31.5–150, and up to 215 kg.
Coat reddish brown or yellowish brown in summer and dark brownish or rufous grey and speckled whitish in winter; forehead dark brown; face white; muzzle black; sides of nose with brown patch; ears black-rimmed; chest dark brown or blackish; inside ears, thighs, belly, throat and rump patch white; tail black tipped. In winter coat longer and greyer or darker brown. Antlers multi-forked, dichotomous sheds January–March.

■ DISTRIBUTION
North America. From northern Mexico through western and central North America to the southern Yukon (Alaska) and MacKenzie district, North West Territories and northern British Columbia, Canada. Also on Vancouver Island and Queen Charlotte Islands.

■ HABITS AND BEHAVIOUR
Habits: nocturnal and crepuscular. **Gregariousness:** herds; small groups males and females in summer, mixed sexes and ages together in winter; feeding groups to 50; density 5–50/km^2. **Movements:** sedentary; some subspecies altitudinally migratory (down in winter) in heavy snowfall areas; home range similar to *C. virginianus*. **Habitat:** open coniferous forest, aspen parklands, woodlands and bush, river valleys, chaparral, shruby grassland, steep broken terrain. **Foods:** grass and herbaceous plants, twigs, forbs, herbs, foliage from shrubs and trees; buds, leaves, acorns, mushrooms, nuts, lichens. **Breeding:** ruts autumn (September–January); fawns spring (April–July); male polygamous, female seasonally polyoestrous; gestation 180–212 days; fawns 1–3, generally 2; weaned at 4 months; sexual maturity females 1.5 years, males 2.5 years. **Longevity:** up to 10–20 years in wild, 20–25 years captive. **Status:** very common.

■ HISTORY OF INTRODUCTIONS
Black-tailed deer have been introduced successfully to New Zealand, Alaska, Queen Charlotte Islands and Kodiak Islands. Translocated/re-introduced successfully in the United States. Introduced unsuccessfully in the United Kingdom.

EUROPE
United Kingdom
Some black-tailed deer were liberated about 1910 in woods near Woburn Park and for a few years they flourished there, but ultimately dwindled and disappeared before World War 2 (de Vos *et al.* 1956; Whitehead 1964).

Other reports indicate that in 1900 a stag and three hinds were observed in a park in South Berwickshire, and in 1947 another stag was shot by a farmer in Northumberland (Fitter 1959). However, the species has not become permanently established in the British Isles (Lever 1985).

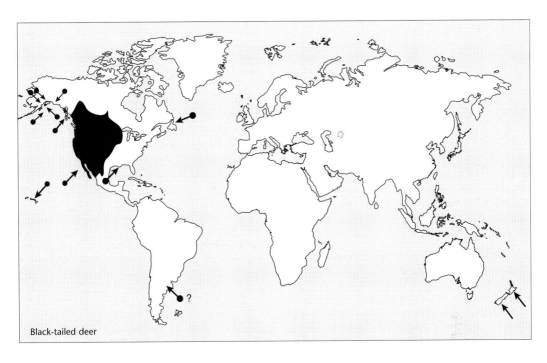

Black-tailed deer

Known tranlocations of black-tailed deer in Alaska

Date	Place	Origin	Results
1916–23	Hinchinbrook and Hawkins Is	Sitka	harvestable
1923	Homer	Sitka	unsuccessful
1924, 1930	Long I., Kodiak Arch., and Prince of Wales I.	Sitka	harvestable
1934	Kodiak I.	Petersburg	harvestable
1934	Yakutat Bay islands	Petersburg	failed
1952, 1956	Taiya Valley, Skagway	se Alaska	unsuccessful
1951–54	Sullivan I., Lynn Canal	se Alaska	harvestable
1977	Admiralty I.	Kupreanof I.	few persist
1979	north Kupreanof I.	?	few persist

NORTH AMERICA

Alaska

Black-tailed deer have been introduced successfully into Alaska between 1916 and 1956 (de Vos *et al.* 1956; Burris 1965; Burris and McKnight 1973). In 1916 (eight head) and 1917–23 (16 head) were released on Hinchinbrook and Hawkins islands in Prince William Sound; in 1923 (seven) at Homer; in 1924 (16) on Long Island in the Kodiak group; 1930 (two) and in 1934 (nine) to Kodiak Island; in 1934 (12) to islands in Yakutat Bay; in 1951–56 (13) to Taiya Valley near Skagway; and in 1951–56 (seven) to Sullivan Island in Lynn Canal, and to the Kenai Peninsula (Troyer 1960; Burris 1965; Burris and McKnight 1973; Rue 1978; Franzmann 1988).

The introduction on Long Island thrived and they became numerous there, but many died in 1935 due to overgrazing (Murie 1940; Troyer 1960). Black-tailed deer were also successfully introduced from south-east Alaska to Yakutat, Prince William Sound, and to Kodiak Island, where well established populations are now present (Merriam 1965). The Taiya Valley introduction has now failed although it lasted until at least 1973. Hunting began in all these areas in 1935 and has continued until 1984 (Franzmann 1988). A more recent introduction occurred in March 1979 when seven does and six bucks were released on north Kupreanof Island, where they are surviving (Franzmann 1988).

Canada

Black-tailed deer, *C. h. sitkensis,* were introduced to Graham Island and Moresby Island (Hall 1981), and were successfully translocated to Sidney Island, British Columbia (Lever 1985).

United States
There have been several translocations, re-introductions and introductions in the United States. Between 1941 and 1956 more than 1300 were trapped and translocated in Montana, many of them in the period between 1947 and 1950 (Egan 1971). They were introduced to Blackjack Peninsula, Texas, from 1931 to 1935 (from Salt Lake City), when about 50 were released, but all had died by 1939 (Halloran and Howard 1956). Some were also translocated in Pecos County, Texas (Etheredge 1949). Some that were released in Palo Duro Canyon, Texas, did not multiply significantly and local landowners asked the Texas Game and Fish Commision to introduce other suitable ungulates (Dvorak 1980).

In 1894, two black-tailed deer were released in the Nehasane Preserve, New York, by W. S. Webb, but these were killed in a fire in 1903. Between 1895 and 1900 a few were released at Litchfield Park, New York, by E. H. Litchfield, but these disappeared (Bump 1941).

Black-tailed deer from Angel Island, San Francisco Bay, California, were relocated without much success on the mainland (Cooker Recreation Area) because of overcrowding on the island, but did not adapt well (O'Bryan and McCullough 1985). The results of these relocations were poor and expensive, and it was found not to be a satisfactory means of reducing numbers. Some were, however, successfully translocated to Santa Catalina Island in the Channel Islands, California.

Mexico
There have been some translocations of black-tailed deer in Mexico (Morrison *et al.* 1992). In December 1985 the New Mexico Department of Game and Fish, in cooperation with the Autonomus University of Nuevo Leon, re-introduced 15 mule deer captured in the Hondo Valley of New Mexico, into the Sierra Madre Oriental mountain range near Iturbida, Nuevo Leon, Mexico (Morrison *et al.* 1987). Initially they were released into an 18 ha enclosure for university research studies.

PACIFIC OCEAN ISLANDS
New Zealand
There appear to have been releases of black-tailed deer at Mercer and Piako on the North Island in 1877 when 11 deer were released, but these were all later shot.

Liberated at Runuanga, Hawke's Bay, North Island, in about 1905, black-tailed deer were increasing there in 1915, but appear to have died out some time later (Wodzicki 1950; Christie and Andrews 1965). Some may also have been released (nine head) in 1905 near

Wakatipu, in the South Island, but these also failed to become established (Donne 1924).

Hawaiian Islands
Five pair of black-tailed deer, *C. h. columbianus*, from Oregon were introduced on Kauai in the Hawaiian Islands for sport and hunting in 1961. Some 40 head were released between 1960 and 1966 on Kauai where they became established in brushy forest on the west side of the island (Walker 1967). Originally 10 (five males and five females) were released on Polihale Ridge on the western slopes of Kauai after being kept in a pen for nine days. Other introductions of 30 animals followed – 10 (two males, eight females) in 1962, five (all female) in 1965 and 15 (13 female, two male) in 1966 – all released in the same area (Kramer 1971).

By 1965 it was known that there were at least 100–150 animals present on Kauai (Swedberg 1965; Walker 1967), this had risen to 120–160 by 1966–67 (Swedberg 1966, 1967); in 1968 there were 150–200 (Kramer 1971), and in 1981 about 350 (Lever 1985).

SOUTH AMERICA
Argentina
Black-tailed deer are reported to have been introduced to Argentina (Petrides 1975), but there appear few details of any introductions. They have certainly been stocked on several ranches in Neuquen Province, but do not appear to have become established in the wild.

■ **DAMAGE**
In California black-tailed deer, *C. h. columbianus*, suppress the growth of timber reproduction in *Pseudotsuga menziesii* and *Sequoia sempervirens* and unprotected trees required about 20 years to reach a height where they can no longer be damaged (Browning and Lauppe 1964).

Damage by *Cervus* species to apple trees has been an increasing concern in last 20 to 30 years. Browsing causes young trees to become stunted and misshapen and useless for future production (Eadie 1961). It is likely that planting dwarf or semi-dwarf trees will become the trend and on these smaller trees browsing by deer retards the growth more severely especially on M9 dwarfing rootstock (Cummins and Norton 1974). The advent of high-density dwarf or semi-dwarf planting has now changed the situation to include orchards as high value crops (Caslick and Decker 1977, 1979).

In Montana after 1949, agricultural depredations were frequently reported to alfalfa haystacks and growing alfalfa, when browse became inadequate and caused food shortages for the deer following restrictions of range by snowfall (Wilkins 1957).

SIKA DEER

Sika, Japanese deer, spotted deer
Cervus nippon Temminck

■ DESCRIPTION

HB 950–1400; T 75–200; SH 640–1100 mm; WT 25–131 kg.
In summer, coat buff-brown or reddish-brown, darkest along back and covered with yellowish white spots; in winter uniform blackish brown or grey-brown, female lighter; neck and back with black dorsal stripe; tail long, white below with a faint black line; caudal patch white, edged above and on sides with black; antlers (400–775 mm) similar to red deer, but invariably lack bez or bay tine and generally have four points each (eight tines); young are spotted.

■ DISTRIBUTION

Eastern Asia. South-eastern Siberia (Ussuri region) to eastern China, Japan (main islands), Taiwan and south through Manchuria and Korea. Formerly from Hopei south to North Vietnam. Now probably extinct in China apart from Manchuria.

■ HABITS AND BEHAVIOUR

Habits: mainly nocturnal and crepuscular; territorial; behaviour and life cycle similar to red deer; ability to cope with harsh conditions, especially severe winters. **Gregariousness:** solitary, small groups at rut, and small single sex herds; aggregations of 40–50 known; family groups 2–8; at rut males collect harems; density 2.85–40/km^2. **Movements:** juvenile stags wander widely; seasonal altitudinal movements reported; territories 2.7–7.7 ha. **Habitat:** broadleaf and mixed forest, mixed woodland, estuarine reed beds, heaths, fields, cultivation, plantations, taiga, clearings, marshes and grassland. **Foods:** grass, herbs, and grass-like plants, forbs, browse from trees and shrubs (bark, shoots, twigs, buds, seeds and leaves), also algae, acorns. **Breeding:** seasonal; ruts September–November (Europe and United States; April–May in New Zealand), calves May–June (Europe); gestation 210–246 days; males polygamous; oestrous cycle 21 days; 1 young, occasionally 2; sexually mature (females) at 6 months, but most do not breed until 16–18 months or more. **Longevity:** 20 (in wild)–25 years (captive). **Status:** formerly more widespread; greatly reduced in range and numbers and wild in few spots; some subspecies endangered.

■ HISTORY OF INTRODUCTIONS

Sika deer have been introduced into central and western Europe from Japan and northern China and have become feral in several countries. Feral herds due to successful introductions now exist in Morocco, New Zealand, Ireland, Great Britain, Denmark, France, Austria, Poland, Czechoslovakia, Germany, Russian Federation, South Africa, Madagascar, the

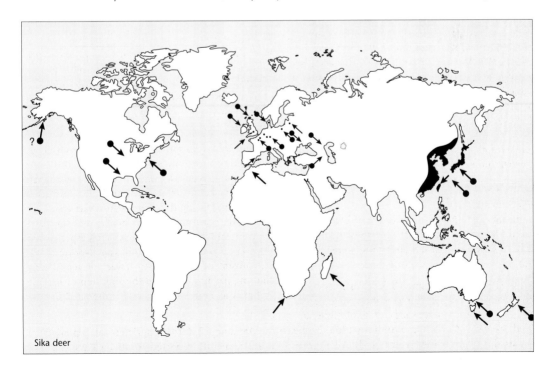

Sika deer

United States, and the island of Oshima, near Tokyo, Japan; also Solo Islands (Philippines, still extant?), and Kerama Island (Ryukyu Islands).

AFRICA
South Africa
Sika deer were introduced by Cecil Rhodes to his estate at Groot Schuur in 1897 and some were still present there in 1937, when the population numbered 20 (Bigalke 1937). Sika deer were also introduced in other areas, but are now not present elsewhere in South Africa.

Morocco
Some sika deer were released at Maroc in 1952–53 in the forests of Mamora, but were later exterminated (Dorst and Giban 1954; Niethammer 1963). There have been several reports of sika releases in Morocco (Lever 1985), but none appear to be established there today.

AUSTRALASIA
Australia
Sika deer were introduced in the 1850s and in 1890 in Tasmania (de Vos et al. 1956). It is not unlikely that they are established somewhere in Australia (Bentley 1957), but none have been found.

From 1863 to 1890 the Acclimatisation Society of Victoria imported and bred them. Between 1887 and 1900 they were released at Gembrook where they were said to be increasing, but there have been no reliable reports for many years (Bentley 1978).

Papua New Guinea
Sika deer were probably introduced into Papua New Guinea with other deer, but they are not known to be established (Bentley and Downes 1968) and there have been no further records of them.

EUROPE
In Europe feral populations of sika deer derived from escapees from parks occur in several parts of Britain, Ireland, France, Austria, Denmark, and Germany. Other populations resulting from the deliberate release occur in Poland, Czechoslovakia and Russia. The first introduction was probably that to Regents Park, Britain in 1860.

Austria
Sika deer were introduced successfully into Austria (Spitzenberger 1982). The race *C. n. hortulorum* has been introduced into the Brdy Forests where about 150–200 head existed in the early 1960s (Ganzak 1964).

Czechoslovakia
Sika deer were kept in many parks and game reserves prior to World War 2. Many have escaped since, and in 1977 they were reported to be established in 11 localities in small numbers in Czechoslovakia (Lever 1985).

Denmark
In Denmark sika deer reached there around 1900 (probably from Germany) and were later released from captivity. About 500 head were reported to be living in the wild in the mid-1950s (de Vos et al. 1956). They are reported to have caused so much damage that their reduction in numbers became necessary and in 1947–48 about 252 were shot, but in 1955 there were still 700 there (Westerskov 1952; Niethammer 1963). They were apparently still there in the 1970s (Burton 1977). Free-ranging herds probably still exist at Frijsenborg, Katholm and Zealand and Jutland, the largest populations are most likely in Jutland and Zealand (Lever 1985).

France
Sika deer are reported in France in the wild (de Vos et al. 1956; Dorst 1965) and apparently still exist there (Burton 1977) as escapees from park herds. They were recently reported from at least two localities in northern and southern France (Lever 1985).

One male and three females of the Japanese race were introduced to France in 1890, a gift from the Emperor of Japan to M. Sodi Carnot, the then president. They were released in the presidential hunting reserve at Marly and by 1895 were well established and breeding. In 1898 following the destruction of the forests of Marly, some seven males and 11 females were transferred to Rambouillet where they prospered so well that a number were transferred to other domains from 1913 onwards. By 1924 there were at least 76 head there in a 900 ha fenced area. Animals produced at Rambouillet and in the Chambord Reserve (177 sika) were released in 1952–53 in forests in Pas-de-Calais, Lower-Loir, Upper-Vienne, Upper-Savoire, Ardennes, Marne, Aub, Meur-et-Moselle, Upper-Rhin, Côté de Nord, Isère, Alps-Maritimes, Ariege and a herd at Maroc in the forests of Mamora. In all these areas they were successfully established. In 1936–37 some were introduced at Estérel where they survived the war and its accompanying depredations on them, and in 1954 two males and seven to eight females were breeding regularly there (Dorst and Giban 1954).

Germany
Since 1893 sika deer have been held in the forest of Arnsberger; in 1930 they were allowed their freedom and had increased to some 3000 head in 1958, although there were probably many there in 1937 (Niethammer 1963). Later introductions include some in 1910 in Cadinen where in 1941 there were 150 head and in 1963 several were still present. Some

were released at Gerolzhofen, Bavaria where a few still existed in 1957; two small colonies exist in Schleswig-Holstein probably introduced about 1929 where they were seen occasionally for 20 years (Niethammer 1963).

Several populations are still established in the wild in Germany (Grzimek 1972; Lever 1985).

Russian Federation and adjacent independent republics

Sika deer have been successfully acclimatised in a series of preserves in Russia and adjacent independent republics, where more than 240 000 were distributed (Flerov 1960).

Sika deer introduced in the 1930s are now common in Mordovsky, Oksky, Ilmensky and other reserves in the Russia (Filonov 1980). They have been re-acclimatised and acclimatised successfully, except in the European part where they have failed (Yanushevich 1966). They were introduced in Lithuania in 1954, Moldavia 1954–60, Armenia in 1953–54, and the Ukraine in 1956–57, where they became established locally. They were also introduced in the northern Caucasus in 1930.

In 1953, 16 females and four males, and in 1958, 15 females and five males were released in the Khosrovskii Forest in the Vedinskii Raione, Armenia from Ussuriisk (Airumyan 1962). These became established but did not increase much in numbers and there were only 58 there in 1960. Some 20 of the race *C. n. hortulorum* were released in the Black Sea Preserve in 1957 and by 1964 the population had built up to 150 animals (Berestennikov and Ardamatskaya 1964).

Other introductions include: to the Karelian Isthmus, Leningrad district (Timofeeva and Fedotova 1973), and in the Khoper Reserve, where they increased in the absence of predators to a herd of 2000 by 1972 (density 117/1000 ha) (Ryabov 1975). In the latter area there have been some morphological changes in the animal (Petrashov 1977). In 1950 the race *C. n. dybowskii* were introduced and established in Azerbaidjan (Grzimek 1972). The race *C. n. hortulorum*, was introduced to the Cherkassy oblast in about 1962 from Primor'ye and became successfully established and increased from 25 head to 473 by 1977 (Evtushevskii 1977).

In Russia and the adjacent republics from 1933 to 1972, some 2393 sika deer were released for acclimatisation purposes. In Russia releases have occurred in Vladimirsk (1955, 1960 and 1970, 48) Voronejsk (1938, 27), Kalininsk (1933–66, 143), Kalujsk (1955–72, 124), Kuibshevsk (1938, 26), Kursk (1960, 12), Leningradsk (1958–59, 15), Mordovsk ASSR (1940–44, 137), Moskovsk (1940–70, 293), Orenburgsk (1938, 16), Penzensk (1970, 1972, 41), Ryahzansk (1939, 1967, 36), Rostovsk (1971, 65), Saratovsk (1971, 12), Srerdlovsk (1969, 12), Stavropolsk (1938, 54), Chelyahbinsk (1939–40, 42), Yahroslavsk (1950–66, 171), Azerbaidjan (1952 and 1960, 18) Armyahnsk (1954, 1959, 1970, 288), Kazakhstan (1960, 10), Oshsk (1959, 1962, 23), Latvia (1954, four), Litovsk (1954, 24) and Moldavia (1960–61, 97). In the Ukraine 655 were released in 13 oblasts from 1941 on. Other releases include: released at Mordovskii (Temnikovskii region, Mordovsk) in 1940–44; 26 released at Okskii in the Spasskii region, Ryahzansk oblast in 1939; they were also released at Ilmenskii in the Chebarkulskii region, and the Chelyahbinsk oblast; in the Armyahnsk area sika deer were released in the Khosrovskii Forest, Vedinskii region (Kirisa 1974).

Switzerland

Sika deer were introduced in the canton of Appenzell in 1915, when six to eight animals were released. These increased to a population of up to 120 head, but completely disappeared in the early 1960s (Kuster 1966). Shortly after World War 2 animals from the Baden-Wurttenberg area of Germany immigrated into Switzerland and became established north of the Rhine in the cantons of Schaffhausen and adjacent Zurich. The population there is stable or possibly increasing slowly (M. Dollinger *pers. comm.* 1982).

United Kingdom and Ireland

Sika deer were introduced about 100 years ago to Britain from the Far East and now several populations are thriving and increasing their range rapidly (Ratcliffe 1987). Large populations of sika deer now occur in Sutherland, Argyll, Peebles, Ross-shire and Inverness, with smaller populations in Hampshire, Lancashire and Cumbria (Corbet and Harris 1991). They were introduced in several locations in Britain, mainly between 1860 and 1920. The 1860 introduction to Powerscourt, Ireland, was the basis for many subsequent introductions to Britain. Only the Peebles population was known to come directly from Japan. Sika deer are currently increasing their range at most Scottish locations. Three subspecies occur in Britain: *C. n. nippon*, *C. n. mantchuricus* and *C. n. taisuanus* (Fitter 1959), which interbreed, and other subspecies have been kept in captivity.

The first sika deer imported into Britain were probably those presented to the Zoological Society in London in 1860 (Whitehead 1964) and some introduced to County Wicklow in Ireland in the same year

(Mulloy 1970). In about 1907 sika deer were introduced in Yorkshire on an estate, but later escaped and became wild.

By 1939 sika deer had become established and were numerous in Challock, Kent, and in the New Forest, Hampshire, and a few were seen in Wareham, Dorset (Taylor 1939). In the late 1940s they did not appear to be numerous in the New Forest or at Wareham, but were occasionally noted at Puddletown in Dorset (Taylor 1948). The Hampshire animals are thought to have been derived from escapees in about 1904 (Lever 1977). They continued to increase in numbers and range, especially during the two World Wars and by the 1960s feral populations were known in Essex, Devon, Dorset, Hampshire, Kent, Lancashire, Somerset, Surrey, western Yorkshire, Westmoorland and Buckinghamshire (Southern 1964). Those in Yorkshire and Lancashire are thought to have descended from some released in about 1904–06 (Lever 1977). In the 1970s they were present in the New Forest, Hampshire and at Wareham to Poole Harbour in Devon, a few in east Sussex and into Kent and on the Devon and Somerset borders (Willett 1970). There are a few in Oxfordshire and Buckinghamshire and a small herd in Kent dating from about 1939. The Sussex animals may date from early in the century and Yorkshire and Lancashire animals may be descendants of those released in about 1904–06. A number escaped during World War 2 at Herts and Essex, but both herds disappeared some time ago (Fitter 1959). In the 1950s they were numerous in the Midlands and in the south-east counties from Kent to Dorset (Matthews 1952).

Sika deer were also introduced on Lundy Island by M. Harman in 1927 or 1929, when seven were released (Whitehead 1964; Lever 1977). These became well established and numbered about 33 in 1949, about 80 in 1956, but only 25 head in 1977.

In Scotland sika deer were released a number of times and at various places. Introductions occurred at Tulliallan, Fifeshire, about 1870, Achanalt Forest, Ross and Cromardy in 1889, Mull of Kintyre, Argyllshire about 1893, Inverness-shire in about 1900 and Sutherland also about 1900. Some escaped in 1912 in Peebleshire and possibly still exist there, Caithness in 1920 and 1930, and in Angus about 1940–45 (Fitter 1959; Whitehead 1964). In the late 1940s they occurred as far north as Loch Shin, and were in small numbers in Ross and Cromardy and two areas of Inverness-shire, occasionally in Mull of Kintyre in greater numbers and occasionally reached the borders of Fife and Clackmanannonshire (Darling 1947;

Taylor 1949). By the 1960s sika deer were established in Argyll, Caithness, Fife, Inverness-shire, Peebleshire and Ross and Sutherland (Southern 1964). An introduction in 1893 still survived at Carradale, Kintyre, in the 1940s (Darling 1947).

In 1959 sika deer were present in 12 English, nine Scottish and five Irish counties. In Scotland there have been at least eight deliberate introductions for sporting reasons but in England only one or two, the rest originating from escapees from parks. Sika deer were first introduced into Fife in 1870s and by 1914 there was a herd of 160–200, but by 1958 only about 30 deer were left. There were further successful early introductions at Loch Rosque, Ross-shire in 1887 and Mull of Kintyre in 1893 and others (Whitehead 1972).

Sika deer were present in Irish deer parks for over 100 years and have subsequently escaped and are now feral in a few areas. They are now established in Dublin, Fermanagh, Kerry, Tyrone and Wicklow (Fitter 1959; Whitehead 1964). They are also reported present near Inniskillen (de Vos *et al.* 1956). The first herd was probably established in Ireland in County Wicklow in the mid-nineteenth century. They have now spread to Leinster, Munster and the border counties. In places they are common and widespread, as in Wicklow, and continue to hybridise with red deer (D'Arcy 1988).

Sika deer have become feral in a few areas in Ireland where the species has been kept in parks for over 100 years. In the 1960s they were established in Dublin, Fermanagh, Kerry, Tyrone and Wicklow (Fitter 1959; Southern 1964; Whitehead 1964). Some of those introduced to Powerscourt in 1860 were later sent to counties Kerry, Fermanagh, Tyrone and Limerick. Those at Limerick died out, but in Kerry they spread and by 1970 occupied a large area of country. Those at Wicklow remained in the Powerscourt area until the mid-1930s, but since then have spread to Ballinglen and Glen of Imaal some 48 km away. Also by 1970 the Tyrone population had reached 250 head and those at Fermanagh 150 head. The total population in Ireland in 1970 was about 2500 head (Mulloy 1970).

Today fragmented populations of sika deer exist in a number of areas of Britain. They are numerous over much of eastern Scotland where the population is expanding (Lever 1977, 1985; Ratcliffe 1987; Baker 1990).

INDIAN OCEAN ISLANDS
Madagascar
Sika deer are reported to have been introduced and established on Madagascar (Grzimek 1972;

Whitehead 1972), but the only deer present on the island are fallow (*C. dama*) and Timor deer (*C. timorensis*).

NORTH AMERICA
Alaska
Sika deer were introduced to the Kodiak Islands in the 1920s where they became established(?) (Clark 1958).

United States
Hog deer were thought to have been introduced to James Island, Maryland, in 1916 or 1922 by the then owner of the island (Presnall 1958; Flyger 1960), however, it is now known that the species involved was sika deer. From here they spread to nearby Taylor's Island in 1935, became well established, and in the late 1950s the two populations and those on the adjacent mainland were estimated at 150 animals.

On James Island they reached a density of two per hectare in 1955 (Christian *et al.* 1960). A 60 per cent die off in 1958 occurred due to physiological derangement resulting from high population pressure. The original release was probably made by C. Henry in 1916 and the animals increased and spread to nearby Taylor's Island and also to mainland Dorchester County (Flyger 1960). It was estimated that there were 270 on James Island in 1957–58 (Flyger and Warren 1958).

W. M. Rockefeller released eight male and 12 female sika deer in Bay Pond Reserve, New York State, in 1904–10 from Germany (Bump 1941). Here they bred and maintained their numbers for several years, but apparently later disappeared or were shot out. More recently they have been introduced in other parts of the United States mainland (Dasman 1968) such as Virginia, Maryland (Presnall 1958) and Texas (Ramsey 1968), where they appear to be established.

They were introduced in Texas as early as 1932 (Ramsey 1968), and again in 1959–60, when 16 sika deer were introduced on the H. B. Zachry Ranch, Laredo, Texas (Sanders 1963). Recent figures suggest that there are now 12 000 sika in Texas, of which about 5600 are free-ranging (Feldhamer and Armstrong 1993).

C. Law of Berlin, Maryland, purchased five sika deer in 1920, which he kept for several years before selling them to another person who released them on Assateague Island (Flyger 1960). Others suggest (Anon. 1935) that this introduction was in about 1930 and that within five years there was an estimated 100–125 there. Still other versions report that two or three pairs were given to a scout troop in 1923 by Captain Will Powell and were released on the north end of Assateague Island, Worcester County, Maryland; or they escaped from a nearby island in 1930 (Presnall 1958). Whatever, they moved south and are now found on Chicoteague National Wildlife Refuge, Virginia, and continue to expand in both states (Flyger 1960; Mullan *et al.* 1988).

In 1962 sika deer occupied the western third of Dorchester County, Maryland, and in Virginia remained confined to Assateague Island (Flyger and Davis 1964). Although found only in the south-west part of the county until 1964, they have since crossed the Nanticoke River into Wicomico County (Feldhammer and Chapman 1978) and the annual harvest in the area averages 1460 head. Since their introduction to James Island 60 years ago they have dispersed into Dorchester County at the rate of 0.8 km/year. In many areas where they are introduced sika are apparently capable of out-competing sympatric species of deer, probably due to more diverse and adaptable feeding habits. Now larger bag limits have been set to diminish or reverse the expansion trend in the next few years (Feldhamer and Chapman 1978). Until 1970 distribution and density were restricted, but since then there has been a steady upswing in numbers and a consequent diminution of native white-tailed deer (*C. virginianus*) (Feldhamer and Armstrong 1993).

There have been several other introductions of sika deer including those in Texas, Nebraska and Michigan (Whitehead 1972). In Texas about 50 range the Rickenbaker Ranch.

PACIFIC OCEAN ISLANDS
Japan
Formosan sika deer (*C. n. taiouanus*) were released on Oshima Island (Ō-shima), the northern-most island of the Seven Islands of Izu (Izu-shotō), south of Tokyo, some time after 1942–43 (Kuroda 1955). These animals were imported from Taiwan for the Oshima Zoological Gardens, but escaped and became established in the wild. Some 50 sika were observed on the undeveloped eastern slopes of the island in 1950. Apparently they have not bred with the native race on the island (Whitehead 1972).

New Zealand
The first sika deer in New Zealand were three imported by the Otago Acclimatisation Society in 1885 and liberated on the Otekaiki estate, near Oamaru in the South Island, where they were said to be increasing in numbers in 1890 (Donne 1924). This introduction was unsuccessful (Kiddie 1962; Wodzicki 1965) as the animals were thought to have all been shot (Donne 1924).

In 1904 the Duke of Bedford presented the New Zealand government with six sika, probably three males and three females, and these were released at Taharua Station in 1905 (Donne 1924) on Merillees Clearing in the Kaimanawa Range, 32 km east of Lake Taupo (Kiddie 1962) on the central North Island. This introduction was successful and they spread slowly southwards along the Kaimanawa State Forest and were noted in the Otupua Valley in the 1930s. The spread continued down the Mohaka River and along the Makahu Stream and reached the Kaweka Range in the early 1940s. At this time they were also numerous in the northern part of the Kaimanawa Forest and were common over a range extending about 2590 km^2 around the liberation point. The total area of their range was estimated as 11 136 km^2 (Kiddie 1962), and they were still spreading in the 1960s (Wodzicki 1961).

The sika deer are now common and restricted on the North Island (Wodzicki 1965), but their range is still extending (Gibb and Flux 1973). They are still found on central North Island south of Taupo (Barnett 1985), but have not yet expanded fully into the habitat available to them (Fraser *et al.* 1996).

Ryukyu Islands

Sika deer (*C. n. keramae*) were introduced from Japan prior to 1757, as they were well established at this time in the Ryukyu Islands. They are currently established in small numbers on Yakabi-shima, Kuba-shima, and Keruma-shima (Lever 1985; Wilson and Reeder 1993).

Philippines

Sika deer have perhaps been introduced in the distant past to Jolo Island, south of the Philippines, as an isolated subspecies *C. n. soloensis* is present (Groves and Grubb 1987; Wilson and Reeder 1993).

■ DAMAGE

In Russia, where sika deer have been introduced in the Khoper Reserve, they increased to a density of 117 animals per 1000 ha. The overbalance in population has led to an adverse effect on the woody vegetation and degradation particularly to the young growth of plants in the area (Ryabov 1975).

Feral herds occasionally damage trees and crops in Britain (Southern 1964), where large populations cause considerable damage to crops, particularly in spring to spring grass or winter-sown corn (Willett 1970). They occasionally raid cultivated fields of soya beans and oats (Whitehead 1972). The browsing of the leading shoots of young trees and the removal of bark is important to commercial forestry. Also, the constant hybridisation of sika deer with red deer

threatens the genetic integrity of the native red deer (Corbet and Harris 1991).

In France, although sika deer compete with the indigenous red deer, they cause no damage to agriculture or forests (Dorst and Giban 1954) and they are less destructive than red deer (Dorst 1965). However, in Switzerland sika deer are reported to cause considerable damage to monoculture forests (M. Dollinger *pers. comm.* 1982), and where introduced in New Zealand they have caused damage to watershed protection forests and pasture lands (Daniel 1962).

Food studies conducted in Kerr Wildlife Management areas in the United States have shown that sika deer are in direct competition with the native white-tailed deer for food. Sika deer have the ability to shift their preference to grass on severely grazed areas which white-tailed deer cannot do (Armstrong and Harmel 1981).

In Japan the sika deer (*C. n. yesoensis*) causes crop damage (Onoyama *et al.* 1990).

HOG DEER

Para

Cervus porcinus (Zimmermann)

=*Hyelaphus* or *Axis*

■ DESCRIPTION

HB 1050–1500 mm; T 120–210 mm; SH 530–750 mm; WT 25–50 kg.

Coat reddish or yellowish brown, spotted on back and neck in summer; squat appearance with stout, small body and short legs; dark dorsal stripe; under parts white; antlers short (300–600 mm), to 6 points, sheds March–May.

■ DISTRIBUTION

Asia. From Pakistan, northern India, Burma to Thailand, Indochina (Vietnam) and southern China (Yunnan).

■ HABITS AND BEHAVIOUR

Habits: nocturnal and crepuscular, sometimes diurnal. **Gregariousness:** solitary or groups 2–7 (all sexes and ages) and up to 18; not generally in herds. **Movements:** sedentary. **Habitat:** open grassy plains, swamps, paddy fields, swampy plains, marshes. **Foods:** grazes and browses; fallen fruits, grass, flowers of trees. **Breeding:** ruts August–December (Australia peak rut in February–March), calves March–May (Australia calves August–October); gestation 220–240 days; 1 young, occasionally 2; hide for 10–14 days after birth; maturity 8–12 months. **Longevity:** 10–15,

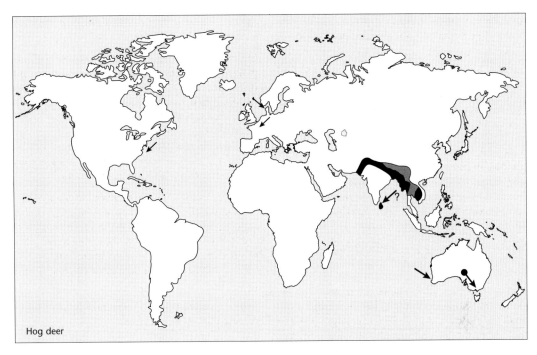

Hog deer

rarely to 20 years. **Status:** range and numbers reduced, but locally common many areas and rare others.

■ HISTORY OF INTRODUCTIONS

Hog deer have been successfully introduced to Australia, Sri Lanka (may be extinct now?) and the Philippines. They have been unsuccessful in Europe.

AFRICA
South Africa

Hog deer are kept on several large fenced ranches in Orange Free State, South Africa (Lever 1985), but none are free-ranging.

AUSTRALIA
Australia

Hog deer were introduced by the Acclimatisation Society of Victoria from India to Victoria in about 1866. They now thrive on the south Gippsland coast (Barret 1955; Bentley 1957; Cowling 1975) from Wilson's Promontory and the Tarwin River to Orbost (Wilson *et al.* 1992) and occur on several islands of the Noramunga Wildlife Reserve at Corner Inlet (Strahan 1995).

In 1866, 12 were released at Cape Liptrap by the Acclimatisation Society and in 1870 some were released at Gembrook, but apparently these did not survive. They were liberated on Snake Island where they increased and there may have been 3000 there in 1947. In 1976 some were caught on Sunday Island (seven male, 15 female) and in 1977 a further 21 (18 from Snake Island and three from Sunday Island) hog

deer were transferred to the mainland. The population was about 500 on Wilsons Promontory in the 1950s, but low in 1977 (Bentley 1978).

A pair was released at Cape Leeuwin on the south coast of Western Australia in 1899, but did not succeed in becoming permanently established (Le Souef 1912; Bentley 1967; Allison 1969; Long 1988). Some may also have been released in the Porongorups, where several deer species were released in 1903–06 (Bentley 1967) and perhaps in other areas (Le Souef 1912).

EUROPE
Denmark

Hog deer were introduced from India to Samso Island in 1880, but no longer exist there (de Vos *et al.* 1956; Niethammer 1963).

France

Introduced for hunting in France, hog deer were unsuccessful in becoming established (Dorst and Giban 1954).

INDIAN OCEAN ISLANDS
Sri Lanka

Hog deer were introduced to the western parts of Sri Lanka during the Portuguese occupation in the sixteenth century (Whitehead 1972) or during the Dutch occupation in the eighteenth century. They persisted and increased there until about 1920. With the increasing human population and subsequent hunting pressure they have now been practically exterminated (de Vos *et al.* 1956; Niethammer 1963). They were scarce there in the 1970s (Whitehead 1972).

NORTH AMERICA
United States
Formerly it was thought that hog deer were introduced to Maryland in 1922, but it is now known that sika deer, *Cervus nippon*, was the species released in that state.

■ DAMAGE
On account of damage to crops, hog deer are slaughtered ruthlessly by the local people in Burma (Whitehead 1972).

An attempt was made to remove an introduced population from Wilsons Promontory, Victoria, Australia, in 1953, but only 34 out of about 500 were shot and the venture proved unsuccessful (Whitehead 1972).

RUSA DEER
Javan rusa deer, Timor deer, Sunda sambar
Cervus timorensis Blainville

■ DESCRIPTION
HB 1420–1850 mm; SH 800–1100 mm; T 100–300 mm; WT 41–125 kg.
Coat long and shaggy, greyish brown with reddish tinge; belly buff on yellowish white; light grey patch under neck and from throat to chin; inner legs buff, tail long and shaggy with thin terminal brush, greyish

brown. Generally male is maned, which in female is less developed or absent. Antlers 115–690 mm, cast in January–February (Australia).

Note: Positively distinguished from C. unicolor *only by form of antlers in mature males (Payne et al. 1985). There has been some difficulty in distinguishing between* unicolor *and* timorensis *in the text because of the overlap in common names in general use.*

■ DISTRIBUTION
Indonesia. Timor, Java, Sulawesi, Butung, Muna, Peleng, Sula, Banggai(?), Saleyer(?) (or Salajar), Prinsen Eiland, Nusa Barung (off south-east Java), Nusa Barung, Karimon Djawa, Kamudjan (Java Sea), Pulu Genteng(?), Sepandjang (Kangean Archipelago or Sepandjang?), Bali, Lombok, Sumbawa, Rintja, Komodo, Flores, Adonara, Solor, Sumba, Roti (and Pulau Ndana off Roti), Senan, Pulu Kambing, Alor, Pantar and Pulu Rusa (west of Pantar); Moluccas (Ternate, Halmahera, Mareh, Moti and Batjan and Parapottan in Batjan group; also Buru, Seram, Banda, Saporua). Aru Isles (perhaps only imported stock).

■ HABITS AND BEHAVIOUR
Habits: mainly nocturnal, occasionally diurnal. **Gregariousness:** in herds of 6–100 or more; outside rut stags and hinds in separate herds. **Movements:** sedentary. **Habitat:** light forest, woodland, open plains, grassland subject to flooding, swamps, parkland, coconut plantations. **Foods:** grass, shrubs and herbs, leaves, sugarcane shoots. **Breeding:** breeds all year (mainly September–March in Papua New Guinea) (mates June–October and calves March–May

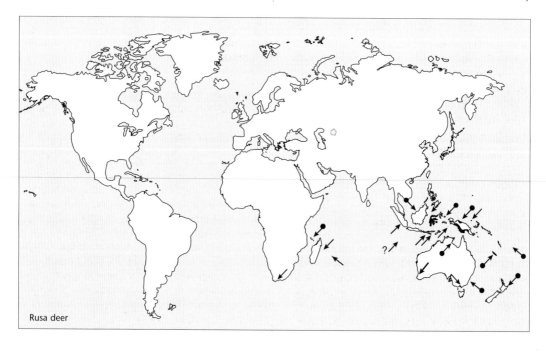

Rusa deer

in Australia); gestation 249–284 days; 1 young; lactation 3 months; sexual maturity 1–2 years. **Longevity:** 8–10 years(?). **Status:** common.

■ HISTORY OF INTRODUCTIONS

Rusa deer have probably been introduced in antiquity to the Lesser Sunda Islands, Moluccas, Sulawesi and Timor. Introduced to Kalimantan (extinct?), Papua New Guinea, New Britain, Aru Islands, Mauritius, Comoro Islands, Madagascar (extinct?), Australia, New Zealand and New Caledonia (Reeder and Wilson 1993). They also occur in south-eastern Borneo, Obi Island, Ambon Island, Amboin, Hermit Islands, and Fiji (Wakaya).

Populations were established in Papua New Guinea in about 1900, and in various parts of Australia in the early nineteenth century. There are also introduced populations on Mauritius, Comoro Islands, New Zealand, New Britain, New Caledonia and Horsburgh Island in the South Pacific.

AUSTRALASIA
Australia
Introduced to various states in Australia in the latter half of the nineteenth century, rusa deer are established near Sydney, New South Wales, and on Friday Island, Prince of Wales Island, Possession Island and Groote Eylandt (Wilson *et al.* 1992; Strahan 1995).

In 1890 rusa deer were released in the ranges at Gembrook by the Victorian Acclimatisation Society and 10 years later they were reported to be established in various parts of Victoria. They were released in Gippsland, but were said to be rare in the 1950s (Bentley 1957). Until 1948 animals were occasionally shot in west Gippsland and rumour still persists that they are still established in Victoria. However, the last confirmed sighting appears to have been in about 1940. The early liberations in southern Australia are said to have failed because of cold winters (de Vos *et al.* 1956; Bentley 1978; Strahan 1995).

In 1952 Mr F. Gray shipped a male and three females from Friday Island to North East Island off Groote Eylandt in the Gulf of Carpentaria, Northern Territory, where they became established (Whitehead 1972; Cowling 1975; Bentley 1978; Wilson *et al.* 1992; Strahan 1995). The herds were decimated by troops on the island during World War 2 (Roff 1960).

In New South Wales rusa deer were released in 1907 in Royal National Park, not far from Sydney. These were obtained from a shipment en route to New Zealand from New Caledonia, where they had been introduced some time earlier. Rusa deer are still to be found in Royal National Park (Bentley 1957; Cowling

1975; Strahan 1995). Between 1809 and 1812, 400 rusa deer are said to have escaped from Surgeon John Harris from a property near Bathurst, New South Wales. Some were also wild near Cumberland in 1827. Rusa deer were also among several species turned out in Royal National Park near Sydney in 1885 and shortly after were still there (Bentley 1978). The population was badly affected by a major wildfire that burnt through the park in the late 1990s, and it is recovering slowly. The New South Wales National Parks and Wildlife Service considered culling the population after the wildfire and while numbers were reduced, there was considerable public opposition to the proposal so no action was taken.

Rusa deer were also established for a time on the tip of Cape York by stragglers swimming from Possession Island (3 km) (Wilson *et al.* 1992).

A pair of rusa deer was released at Cape Leeuwin, Western Australia, in 1899, and in 1901 these were said to be thriving, and others were also released among several varieties in the Porongorups in 1903–06 (Bentley 1967). They did not become established in Western Australia.

In 1912 (or possibly 1910?), eight rusa deer, from the Moluccas (*C. t. moluccensis*), were released on Friday Island, off Cape York Peninsula, by H. N. Hocking. This population increased and some crossed the 1.2 km to Prince of Wales Island, and although few remained about 1967, the main herd is now located there (Roff 1960; Cowling 1975). In 1914 rusa deer from Friday Island were released on Possession Island, 26 km south-east of Thursday Island, where they became established and increased in numbers. Their progeny are still there today (Cowling 1975; Bentley 1978; Strahan 1995).

Papua New Guinea
Rusa deer were introduced to the Port Moresby area by W. Gorse in about 1900, where a small population has persisted since then (Downes 1968; Lindgren 1975; Herington 1977; Bentley 1978).

In the Trans Fly River area where the rusa crossed the border from Irian Jaya (where they were introduced from 1913 to the 1920s), they now inhabit over 260 km^2 of country. The main concentrations are on the plains between the Bensbach and Morehead rivers, where it was estimated that there were 12 000 on the Bula Plains in 1968 and about 10 787 in the same area in 1973 (Downes 1968; Herington 1977; Lindgren 1975). There are possibly two races introduced, *C. t. timorensis* and *C. t. moluccensis* (Whitehead 1972). Their range now covers most of the south coastal plains, from the Gulf of Papua to the Onin Peninsula.

They also occur in the northern Vogelkop and in the Tamran and Arfak mountains and have more recently been introduced into the Sentani region of Irian Jaya. There are now probably over 100 000 head present in Papua New Guinea today. In Irian Jaya, and to a lesser extent in south-west Papua New Guinea, the meat of this deer is used extensively as food (Petocz and Raspado 1984; Flannery 1995).

Rusa deer are now present on Boigu and Saibai islands in Torres Strait, to which they swam the three kilometres from Papua New Guinea to become established there (Draffan *et al.* 1982; Wilson *et al.* 1992).

New Britain
In about 1910 German settlers introduced rusa deer, probably the race *C. t. moluccensis*, near Rabaul, New Britain (Whitehead 1972; Lindgren 1975; Bentley 1978). A small population still inhabits the Gazelle Peninsula (Downes 1968; Herington 1977; Flannery 1995).

INDIAN OCEAN ISLANDS
Cocos–Keeling Islands
Rusa deer may have been introduced to Horsburg Island (Van Bemmel 1949), but there appear to be few details. One report suggests that they may still be present there (Lever 1985).

Comoro Islands
Rusa deer were introduced by the Dutch in 1870 to Anjouan (Lever 1985; Haltenorth and Diller 1994).

Madagascar
Released experimentally in 1928 (or 1930?) at Perinet, east of Antananarivo, rusa deer were spreading there in 1955. Since this time they have declined in numbers because of hunting and may now be extinct (Lever 1985; Haltenorth and Diller 1994).

Mauritius
Imported to Mauritius by the Dutch as source of food in 1639, rusa deer became established in the wild (Van Bemmel 1949; Haltenorth and Diller 1994). The population is now said to be about 3000 head (Lever 1985).

It has been reported that sambar deer (*C. unicolor*) were introduced to Mauritius by the Dutch colonists between 1598 and 1710 (Encycl. Brit. 1970–80), but references to their presence almost certainly refer to the somewhat similar rusa deer (Owadally and Bitzler 1972, in Lever 1985).

Indonesia
Rusa deer have been introduced successfully from 1680 on in a number of areas of Indonesia including Sulawesi, Kalimantan (Borneo), the Obi group in the

Moluccas, Ambon and Irian Jaya. Introductions to Sumatra have failed and introductions to some other islands may have been made, but the species is at present thought to be native to them. Rusa on Timor are introduced (Lever 1985).

Aru Islands
Rusa deer were imported from Seram to Wasior and Wammer in 1855 by Governor Cleerens (Van Bemmel 1949). The original six animals had increased to 80 by 1867 and in the 1940s they numbered in their thousands (Rosenberg 1867; Van Bemmel 1949). The present rusa (*C. t. moluccensis*) population were derived from the pair released at Wasior.

Hermit Islands (west of Bismarck Archipelago)
In 1909 H. R. Wahlen, a planter on Maron Island, liberated rusa deer (*C. t. timoriensis*) obtained from an Australian zoo. These later crossed to Arkeb Island, where they increased substantially in numbers and were transferred or driven to Luf Island. The bulk of the herd is now on Luf Island and in 1954 there were about 200 head, but the species is probably rare (Downes 1968; Whitehead 1972; Herington 1977; Bentley 1978).

Rusa deer may also have been introduced to Ninigo Island, off Wewak, as they formerly occurred on this island in the north-west of the Hermit group (Bell 1975; Lever 1985). They do not occur there now (Bell 1975).

Irian Jaya
Rusa deer (*C. t. moluccensis*) were imported from Seram and released on the western part of the Onin Peninsula in 1913 by R. van Oldebarnevelt (Van Bemmel 1949). Some were imported by Lulofs from Halmahera in 1920 and these were released on the west coast of Geelvink Bay near Manokwari, Momi, Muturi River, and on Runberpon Island, and also near Djayapura (Hollandia) in north-east Irian Jaya (Westermann 1947; Van Bemmel 1949; Lindgren 1975). They were apparently released in the Merauke area in south-east Irian Jaya in the 1920s (Downes 1968), possibly in 1928 (Flannery 1995), where they are now well established.

Kalimantan (Borneo)
Rusa deer were introduced to Mataram in southern Kalimantan from Java in 1680 (Van Bemmel 1949). These had increased to enormous herds in about 1840 on the grassy plains near Pulu Lampej and in the Tanah Laut near Bandjarmasin (Muller and Schlegel 1839). The large herds have since declined, but rusa were still present in the area in about 1949 (Van Bemmel 1949). Their status in the 1980s was unknown (Payne *et al.* 1985).

Sulawesi

There appears to be some doubt as to the origin of the Rusa deer on Sulawesi. A number of authorities have reported them to be introduced (Sarasins 1905; Raven 1935; Dammerman 1939), but others (Mohr 1920; Beaufort 1926) report that they are indigenous. More recently it is indicated that *C. t. macassaricus* is probably native to the islands, but that *C. t. rusa* appears to have been imported and released several times (Van Bemmel 1949). An early report (Graafland 1898) mentions that the deer from Minahassa in northern Sulawesi were imported from Java.

In the late 1940s rusa deer were numerous in south-western Sulawesi, but scarce in the south-east (Van Bemmel 1949).

Maluku (Moluccas)

Rusa deer have been introduced to Batjan, Bandar Islands, Ambon, Buru, Halmahera, Mangole, Sanana, Saparua, Seram, Taliabu and Ternate in Maluku. Archeological evidence is absent from sites dating to around 3000 years ago on Halmahera and rusa were probably carried to the northern Moluccas after this time (Flannery 1995).

Rusa deer (*C. t. moluccensis*) were introduced to the Obi Islands when imported by Diepenheim in about 1930 and released on the island of Belang-Belang (Van Bemmel 1949).

Valentijn (1726) mentions an import of rusa deer from Java and Makassar, Sulawesi to Ambon, probably in the seventeenth century (Bentley 1978). The race introduced to this island appears to be *C. t. russa* (de Vos *et al.* 1956), although at least *C. t. macassaricus* may also have been introduced.

Sumatra

Rusa deer have occasionally been found in Sumatra, e.g. in 1915 and 1926, but have failed to become permanently established in the wild (Van Bemmel 1949; Lever 1985).

Sumba

Records available suggest that rusa deer may have been introduced to this island (Dammerman 1926, 1928). However, later material collected and examined from the island suggested that the species was indigenous (Van Bemmel 1949).

PACIFIC OCEAN ISLANDS

Fiji

Rusa deer were thought to have been introduced to Wakaya Island, probably in about 1920 (Lindgren 1975). The exact identification of those released was uncertain (Whitehead 1972), but it has finally been resolved that those present there now are fallow deer (*Cervus dama*), which are still present on the island (Pernetta and Watling 1978).

New Caledonia

Rusa deer (race probably *C. t. russa*) were introduced to New Caledonia in 1870 where they are still well established (Carter *et al.* 1945; Whitehead 1972, Lindgren 1975). Within 13 years of their introduction complaints were being received of their damage to crops (Whitehead 1972).

Although a pair of deer from the Philippines (probably sambar) was imported between 1861 and 1870, the introduction which led to the establishment of the rusa (*C. t. russa*) was probably made in 1870. At this time 12 arrived on the steamer 'Guichan' from Java. These were said to have been released by an agronomist by the name of Boutan, who thought they would make good hunting. It was also said that he was unable to feed so many and that the government park would not accept them. Whatever the reason, they were extremely successful on the island and in 70 years the original 12 became an estimated 200 000 or more.

By 1882 there were complaints about the amount of damage they were causing to cultivated plants and native flora. Between 1929 and 1932 a deer meat butcher sold nearly one tonne of meat per week to butchers in Noumea at a price cheaper than beef. The species had become so common and the damage so great that it was suggested that tigers be introduced to control them. Export of deer skins from New Caledonia reached a peak just before World War 2, when the total soared to some 8000 in one year. During the war the populations were reduced somewhat by the hunting activities of American soldiers stationed there.

In the 1950s they were well established in the plains and foothills and valleys of the central chain of the island and especially abundant on the western slopes. Elsewhere they were present on Île Lepredour in the Bay of St. Vincent on the east coast (Barrau and Devambez 1957).

New Zealand

Rusa deer (14 in number) were introduced to the North Island, New Zealand (Galeata, south-east of Rotorua), from New Caledonia in 1907 (Wodzicki 1965). They were thought to be sambar deer (*C. unicolor*) and were not correctly identified for many years. They still inhabit manuka and fern scrublands in the Rotorua district (Barnett 1985), where they were thriving in the 1920s.

In the 1970s a herd was established locally in the Galatea-Urewera district (Gibb and Flux 1973). They have not spread much since being introduced (Wodzicki 1961) and in 1960 occurred in a narrow belt of country not far from their liberation point (Wodzicki 1961). At present they are not increasing in numbers or spreading further (Fraser *et al.* 1996).

■ DAMAGE

Rusa deer cause damage to crops, cultivated plants and native flora in New Caledonia, but little specific information was found.

SAMBAR DEER

Rusa deer, sambhur, Indian sambar
***Cervus unicolor* (Kerr)**

■ DESCRIPTION

TL 1540–2700 mm; T 210–350 mm; SH 1000–1600 mm; WT 109–318 kg.
Uniform light brown to dark brown, under parts paler; male has short mane on neck and antlers which are replaced periodically; ears large, rounded and approximately half length of head; antlers with up to three tines, terminal forward facing fork and two long brow tines 700–1270 mm; sheds March–July; inner legs and chin buff; tail and rump fringed with orange coloured hairs; tail short; yellowish tinge under chin, inside of limbs, buttocks, and underneath tail.

Note: The Philippine sambar, Cervus unicolor mariannus, *from the Marianas resembles* Cervus unicolor, *but is considerably smaller, HB 1620–2460; T 80–120 mm; SH 550–700 mm; WT 40–60 kg. It is uniformly dark brown, paler below, and has a more compact build, short antlers (200–400 mm), and a shorter tail. In most recent works it is regarded as a separate species* Cervus mariannus, *however it has been retained here as it is considered the taxonomy of the unicolor-timorensis group is in need of review.*

■ DISTRIBUTION

Asia. India and Burma and Sri Lanka to southern China and Taiwan, Hainan and south to Sumatra, Java, Borneo, Siberut, Sipora, Pagi, Nias, Sulawesi (Muna and Buron) and Malaya, Timor, Flores and Lombok; Marianas (Guam, Rota, Saipan?) and Bonin(?), Moluccas, and Philippines (Mindano, Mindanao, Basilan, Luzon). Formerly on Pinang Island and Singapore.

■ HABITS AND BEHAVIOUR

Habits: shy, largely nocturnal and crepuscular. **Gregariousness:** usually solitary, but also pairs or small groups 3–4, occasionally 6–8; males solitary outside rut; density 0. 8/km². **Movements:** sedentary. **Habitat:** diptercarp forest, secondary forest, swamp forest, praires and marshes, wooded areas, mountain slopes, plantations and gardens. **Foods:** grass, herbs, bark, leaves, buds, berries, fallen fruits, and browse of shrubs and trees. **Breeding:** ruts October–January (in Australia September–October, and March–April); ruts variable (December India) (October–November Sri Lanka) (July–November Borneo); young born May–June; gestation 240–270 days; young 1; females sexually mature in second year. **Longevity:** 26 years 5 months (captive). **Status:** relatively common; locally abundant some areas.

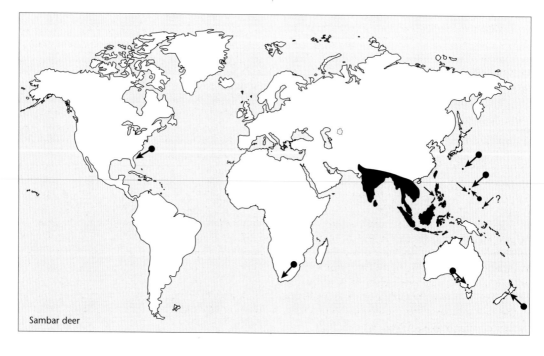

Sambar deer

◼ HISTORY OF INTRODUCTIONS

Successfully introduced to New Zealand, Guam, United States, Africa, and Australia. Failed on Bonin(?).

AFRICA
South Africa
Cecil Rhodes introduced sambar deer to Groot Schuur, Republic of South Africa, in 1897 and some were also released in 1937. There were about 50 of them there in the 1930s (Bigalke 1937). They currently exist on several enclosed estates in the Western Cape (Lever 1985).

AUSTRALIA
Introduced in the 1850s or 1860s, sambar are the most successful deer released in the wild in Australia (de Vos *et al.* 1956; Wilson *et al.* 1992; Strahan 1995). They were first imported in 1857, with the first release by the Victorian Acclimatisation Society in 1863 at Mt. Sugarloaf (five released), north-east of Melbourne. Probably many more sambar were released in the period 1863–70, including probably seven near Burrumbeet in western Victoria in 1872; some released in 1866 in the Grampians at Mt. Zero, and at Snake Island (off Wilson's Promontary); three released on the property of W. Lyall, near Tooradin in 1868, and more (four) in 1873, and hunting began there 10 years later. These spread and populated areas of the foothills of the Great Dividing Range and the population reached a peak in the 1920s.

In the 1950s sambar deer were reported established in Gippsland, eastern Victoria, and in two other localities, the Grampians and at Mt. Cole. They were plentiful on French Island off the south coast where they may have been since 1859. There were 4000–5000 on the island in the 1970s (Bentley 1978).

Stock from India, Sri Lanka and Sumatra were released in Victoria and are now widely distributed throughout Gippsland, and have penetrated deep into New South Wales and the Australian Capital Territory (Mourik 1987; Strahan 1995). They are now continuously distributed throughout Gippsland and the Victorian Alpine district and extend into the southern part of Kosciusko National Park; also two isolated populations occur in the Grampians and Otway ranges (Wilson *et al.* 1992). In the eastern highlands in Victoria they occur within an area bounded by Whittlesea, Tallarook, Euroa, Benalla, Myrtleford, east of Lake Hume and into New South Wales. The southern boundary extends from a line south of Healesville in the west to Cape Howe in the east and includes French and Snake islands (Harrison 1989).

Sambar deer were liberated at Port Essington on the Coburg Peninsula in the Northern Territory in 1912 (Bentley 1957; Cowling 1975; Strahan 1995), and now appear to be scattered over the peninsula, but restricted to monsoon forest (Bentley 1978; Wilson *et al.* 1992). There may also have been releases in Western Australia and in New South Wales (Bentley 1957).

Australian populations of sambar deer appear to be mainly *C. u. unicolor* from Sri Lanka (Strahan 1995).

NORTH AMERICA
United States
Several introductions of sambar deer have been made in Texas and Florida, mainly on ranches (de Vos *et al.* 1956; Dasmann 1968). Some were introduced in 1932, some in 1939, and in the early 1940s (Schreiner 1968; Ramsey 1968; Whitehead 1972). They now exist in small numbers in Texas, probably the largest single population of 50 in the prairies and marshes on the south-east coast (Ables and Ramsey 1973).

In Florida sambar deer are established on St. Vincents Island near Appalchiola and are restricted to this privately owned island (Presnall 1958; Dasmann 1968). Sambar may also exist in California, where they are reported to have escaped from the Hearst Ranch in San Luis Obispo County, but are now only questionably established (Presnall 1958; Lidicker 1991).

PACIFIC OCEAN ISLANDS
Caroline Islands
Recent indications are that an established group of sambar is present on Ponape Island in the Carolines (Groves and Grubb 1987), where they were probably introduced.

Fiji Islands
The introduced deer on Wakaya Island were believed to be sambar deer (Turbet 1941), however, they have now been positively identified as fallow deer *Cervus dama* (see under rusa deer, *C. timorensis*).

Guam
The early Spanish colonists may have carried sambar deer (*C. u. marianus*) from the Philippines to Guam (Baker 1946 in de Vos *et al.* 1956; Burton and Pearson 1987; Flannery 1995), although the evidence for introduction is not conclusive (Whitehead 1972). They appear to have arrived during the 1770s, became well established, and are now distributed throughout the island. They are hunted, although their numbers are generally low over most of the island (Conry 1988).

Marianas (Rota and Saipan)
Introduced to Rota by the Spanish from the Philippines (Baker 1946 in de Vos *et al.* 1956), although such evidence may not be conclusive (Whitehead 1972), sambar deer no longer occur there (Whitehead 1972; Lever 1985). It is likely that they were also introduced to the island of Saipan (Burton and Pearson 1987), but they no longer occur there.

New Zealand
A pair of sambar deer introduced in 1875–76 from Sri Lanka to a property at the mouth of the Rangitikei River, Manawatu, flourished and bred to around 100 head. In 1915 some of these (one stag and six hinds) were transferred to the Hot Lakes district near Rotorua, where they also became established (Thompson 1922; Wodzicki 1950, 1961, 1965). They were restricted in range, but common there in the North Island in the 1950s and 1960s.

Sambar deer are now restricted to the coastal Manawatu and near Rotorua in the North Island and have not increased and dispersed little over the years (Gibb and Flux 1973; Barnett 1985; Fraser *et al.* 1996).

Ogasawara-shotō (Bonin Island, Japan)
Introduced in 1853 sambar deer became established on the island for many years, but have now been extinct there for some time (Kaburaki 1934).

■ DAMAGE
On Guam, at their present density, sambar deer may pose a threat to forest resources in some areas and heavy browsing is evident on grass and shrub species (Conry 1988).

WHITE-TAILED DEER
Virginian deer, whitetail deer
Cervus virginianus Zimmerman
=Odocoileus virginianus

In some North American taxonomic works the genus Odocoileus *is replaced by* Dama *which in most recent European works is changed to* Cervus. *For these and other reasons I have preferentially followed the use of* Cervus.

■ DESCRIPTION
HB 850–2100 mm; T 150–330 mm; SH 550–1143 mm; WT 41.0–141.4 kg (some South American forms weigh as little as 18 kg and larger form males to 215 kg).
Upper parts of coat reddish brown to grey (greyish or greyish brown in winter); belly, throat, eye ring, inside of ears and legs, chin, and underside of tail white; face with dusky wash and a black bar on lower jaw; under parts white; tail brown above and with wide white fringe; antlers short, upright, branching and curved forwards with tines which are not forked, sheds in January–February.

■ DISTRIBUTION
North and South America. Throughout southern Canada from Cape Le Breton Island to south-eastern British Columbia, and north to northern Alberta, and south to Peru and northern Brazil in South America. Have disappeared from much of Mexico and Central America due to habitat destruction.

■ HABITS AND BEHAVIOUR
Habits: nocturnal and crepuscular; browses and grazes. **Gregariousness:** generally males solitary, groups 2–4 in summer; density 2.2/km² to 25–50/km². **Movements:** partly altitudinally migratory (high levels summer, low in winter); home range 16.2–356 ha. **Habitat:** forest edges and clearings, swamp edges, stream banks, prairie with thickets, cedar swamps, open brushy areas. **Foods:** grass, leaves, herbs, needles, fruits, forbs, buds, twigs of shrubs and trees, and mushrooms. **Breeding:** ruts October–December (North America), February–March (South America); calves May–June; gestation 195–210 days; male polygamous; female seasonally polyoestrous; calves 1–2, occasionally 3 or 4; fawns not able to travel for several days after birth; weaned at 4 months; males mature at 1 year, females at 7–12 months. **Longevity:** 10–16.5 years in wild, 20 years as captive. **Status:** very common.

■ HISTORY OF INTRODUCTIONS
In Europe white-tailed deer have been successfully introduced in Czechoslovakia and Finland, but have failed in the United Kingdom and Germany. In North America they have been translocated/re-introduced successfully in some areas. They have also been successfully introduced in the West Indies, although they have failed on many islands, and to New Zealand.

EUROPE
Czechoslovakia
White-tailed deer were introduced and became established in the Brdy Forest, Czechoslovakia where there were 150 animals in the 1960s (Ganzak 1964). Some were imported about 1890 by Prince Colleredo-Mansfield and kept on his estate for many years. Some time later they were allowed to run free and the population is now about 200 head in the Brdy Forest and at Hrebeny in central Bohemia (Lever 1985).

Finland
Introduced in south-western Finland (de Vos *et al.* 1956; Lyneborg 1971) in the last three decades (Neithammer 1963), white-tailed deer are rapidly increasing in numbers and are spreading mainly towards the west and south (Koivisto 1966).

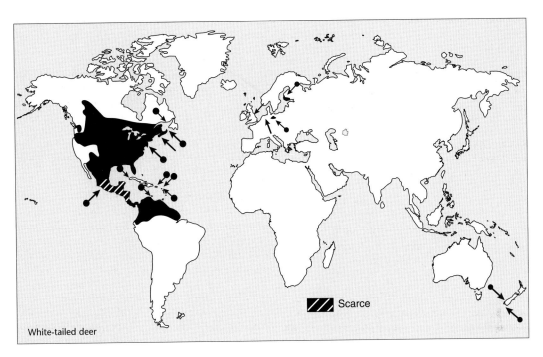

Scarce

White-tailed deer

In 1934, five (one male and four females) were released in south-west Finland at Laukko in Vesilahti (in Tavastland). Until 1939 they had only increased by two calves, and one cow had become lost. Around 1949 there were 100 animals. The herd continued to develop and by 1961 about 1000 head occupied the area in southern Finland. Some animals have turned up as far away as Lapland and the adjoining parts of the Russian Federation. They do not do well in deep snow and dogs are probably their chief enemy (Niethammer 1963). The population in the late 1960s was 2500 (Lever 1985).

Germany
White-tailed deer were possibly introduced into Germany (Lever 1985), but there are no details of any introductions and no deer in the wild today.

United Kingdom
A number of white-tailed deer were released at Arran about 1832, where they thrived for a time, but later decreased in numbers and died out after 1872. Some were also released at Woburn and thrived for several years, but ultimately they also failed to become permanently established (Fitter 1959).

NORTH AMERICA
White-tailed deer have been introduced successfully to Anticosti Island and Prince Edward Island, Canada, and have been translocated/re-introduced to a number of places in the United States including New York, and Texas, and also to Mexico.

Canada
White-tailed deer were introduced to Anticosti Island, Quebec, in 1894 or 1896 and are now common there (Newsome 1937; Banfield 1974). Some were introduced on Isle Royale, Lake Superior, Michigan, in 1906 and persisted there until about 1936 when unfavourable conditions are presumed to have led to their disappearance (Karns and Jordan 1969).

White-tailed deer were introduced and re-introduced into the Liscombe Sanctuary, Nova Scotia, in 1864, 1894, and in 1910 (Benson and Dodds 1977). They reappeared by natural immigration in New Brunswick about 1918 and were introduced to Prince Edward Island in 1949, where a small herd is now established at the western end of the island.

United States
Widespread translocations of white-tailed deer have occurred in the United States (Whitehead 1972). A trapping and translocation program was initiated in 1945 to re-introduce them in habitats east of the Divide. By 1951 a total of 426 had been released in nine counties and these supplemented a natural range expansion that was taking place at the time (Allen 1971).

In 1886, 15–20 deer were released in Tuxedo Park, New York, by the Tuxedo Park Club. They increased rapidly and about 50 were turned loose in 1905. In 1896, 45 were liberated in State Park, Ulster County, where they increased substantially in numbers. In 1917, 50 were released in the Adirondacks by the state authorities, but the results are unknown (Bump 1941).

From 1925 to 1937 white-tailed deer were re-introduced successfully to Blackjack Peninsula, Texas (now Arkansas Wildlife Refuge) (Halloran 1943; Halloran and Howard 1956). Some were also introduced (60 head) on the H. B. Zachry Ranch, Laredo, Texas, in 1959–60 (Sanders 1963). Some trapping and translocating occurred in Arkansas about 1952 (Hunter 1952).

White-tailed deer have long been exterminated in western Tennessee, but were re-introduced in 1932–33 and by 1951 a herd estimated at 125 animals was present (Goodpaster and Hoffmeister 1952). In New Jersey white-tailed deer were down to 200 head by the end of the 1800s and many private persons introduced hundreds from Virginia, Maine, Michigan, Wisconsin and possibly elsewhere. The deer were released or escaped in various parts of the state and many have since been translocated to other areas of the state. This situation is similar to a number of other eastern states (Rue 1978).

Some 167 white-tailed deer were released at three sites in Indiana to study the effects of their dispersal (Hamilton 1962) and 28 were translocated in southern Illinois in the 1960s for telemetry experiments (Hawkins and Montgomery 1969).

PACIFIC OCEAN ISLANDS
New Zealand
The use of white-tailed deer for sport prompted several introductions to New Zealand from North America. At least four releases occurred: in 1901 in the Takaka Valley, Nelson, where they failed to become established; in 1905, 22 were purchased from North America and nine liberated at Port Pegasus, Stewart Island, nine on the western side of Lake Wakatipu and one at Takaka. Those on Stewart Island became established and in 1917 were reported to be increasing in numbers. By 1947 they were throughout the coastal bush covered parts of the island and by 1961 were everywhere. The herd at Wakatipu remained confined to the Rees Valley and have only recently become established in adjacent Dart Valley (Thomson 1922; Wodzicki 1950, 1961; Christie and Andrews 1965).

White-tailed deer are now thriving on Stewart Island and a few areas near Lake Wakatipu, South Island (Wodzicki 1950, 1965; Gibb and Flux 1973; Barnett 1985). However, the present environment in New Zealand does not appear to be entirely suitable for them (Christie and Andrews 1965) and they have not expanded their range in recent times (Fraser et al. 1996).

WEST INDIES
From 1790 onwards white-tailed deer were successfully introduced widely in the West Indies (Varona 1974), including Cuba, Curaçao and other islands (de Vos et al. 1956), St. Croix (Seaman and Randall 1962) and Puerto Rico (Philibosian and Yntema 1977). They are also said to have been imported to the Dominican Republic (Hispaniola), Jamaica, Dominica, Grenada, Leeward Islands, Virgin Islands and some other islands, but there are few or none there today (Lever 1985).

Antigua and Barbuda (Leeward Islands)
White-tailed deer are reported to have been introduced as game animals to Antigua and Barbuda in the seventeenth century, but only fallow deer (C. dama) now appear to occur on Barbuda and Guana (near Antigua) (Lever 1985).

Cuba
White-tailed deer were introduced to Cuba and several other Caribbean islands about 1850, but now are not as plentiful as they were formerly due mainly to increased clearing of the country (de Vos et al. 1956; Niethammer 1963; Whitehead 1972).

Curaçao (off Venezuela)
White-tailed deer were introduced to Curaçao from the Guajira Peninsula, Colombia (de Vos et al. 1956; Lever 1985).

Puerto Rico
White-tailed deer were introduced in the eighteenth century to the island of Culebra and a number of nearby small islands where they are still established (Philibosian and Yntema 1977).

Virgin Islands (St. Croix and St. Thomas) (USVI)
Introduced from the United States to St. Croix prior to 1800, probably before or about 1790, and certainly by 1840, white-tailed deer were numerous there until recently when the illegally commercially exploited population was estimated as 3000 head. This had been reduced to 200–300 head in the 1960s (Smith 1840; Beatty 1944; Seaman and Randall 1962; Hinton and Dunn 1967; Webb and Nellis 1981).

A memorandum of the former government of St. Croix (Danish West Indies) indicates that they were introduced before 1790, when five were brought in by the captain of a schooner trading between the West Indies and America (Seaman 1966). They inhabited the mountainous parts of the island in 1840 (Smith 1840). In 1922 the population was estimated at 3000 head. In the mid-1920s this was reduced by spotlight shooting and between 1938 and 1941 efforts were made to eradicate them because they harboured cattle fever tick. A census in 1942 indicated that there were only about 600 left and in 1966 the few surviving individuals were reported to be in danger of extinction (Seaman 1966).

■ DAMAGE
Orchard damage by deer is one of the leading wildlife depredation problems in United States. Farmers, orchardists, nursery workers and foresters in many areas have experienced heavy crop losses because of the imbalance between deer and their natural food supply and many methods are used to control the damage (Harris *et al.* 1983). White-tailed deer cause losses that necessitate control measures in gardens and in crops such as vegetables, soy beans, corn, wheat and young trees in orchards (Carpenter 1967). In orchards they browse the trees, eat the fruit and rub their antlers on the trees.

Damage from browsing is the commonest form. Young trees are most vulnerable and the popularity of growing dwarf or semi-dwarf fruit trees has increased the potential for damage and devaluation in Wisconsin (Katsma and Rusch 1980). In western Colorado they cause damage to fruit trees, particularly young trees, by browsing and antler rubbing (Harder 1970; Anthony and Fisher 1977). In North Carolina they cause damage to soybeans (a major crop) – the plants are browsed and most damage occurs in the first weeks after sprouting and thereafter declines. Most damage to crops occur in fields near woodland and forest and can result in a reduced yield of 80 per cent per plant (de Calesta and Schwendeman 1978).

In Pennsylvania white-tailed deer cause widespread losses to crops, where alfalfa (*Medicago sativa*) is a major crop. Studies show the damage to be equal to one-fifth of the crop over a two-year period and losses of 17–22 per cent (Palmer *et al.* 1982). White-tailed deer are also a major concern on airfields in Pennsylvania, especially between dusk and dawn. Their feeding in pastures around landing strips provides potential for accidents. There were 23 collisions between aircraft and deer at 13 Pennsylvania airports in a 12-year period (Bashore and Bellis 1982).

In the south-eastern United States white-tailed deer can damage soybeans by grazing, but their grazing does not affect yield greatly unless 67–100 per cent of the leaves are being eaten from young plants (Garrison and Lewis 1987). Here and in New York State particularly, farmers are willing to sustain crop damage of several hundred dollars in exchange for the presence of deer (Brown *et al.* 1978) as the damage to crops can be offset by the income from hunting leases (Garrison and Lewis 1987).

In the United States, population control of white-tailed deer through hunting is said to offer the best and cheapest method of control (Harder 1968).

PÈRE DAVID'S DEER
Milu
Elaphurus davidianus (Milne-Edwards)

■ DESCRIPTION
HB 1830–2160 mm; T 220–500 mm; SH 1050–1370 mm; WT 150–214 kg.
Coat tawny red mixed with grey; flanks and throat with dark patch; mane on neck and throat; dorsal stripe dark; tail tufted and longer than in other deer; antlers with no forward pointing tines; 700–875 mm; sometimes sheds twice/year (usually December–January); hooves large and spreading.

■ DISTRIBUTION
Originally north-eastern and east central China, but known only in captivity and there are no records of wild animals. In historic times the only known colony were descendants of a herd formerly kept at the Imperial Hunting Park, Beijing, China. Present animals occur only in zoos and parks.

■ HABITS AND BEHAVIOUR
Habits: no information; probably similar to red deer.
Movements: no information; probably sedentary.

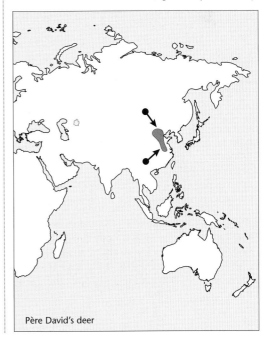

Père David's deer

Gregariousness: formerly hundreds in herds; males solitary before and after breeding; harems with dominant male during rut. **Habitat:** originally swampy and marshy habitats. **Foods:** grass and water plants. **Breeding:** ruts June–August, fawns April–May; gestation 250–290 days; females seasonally polyoestrous; oestrous cycle 20 days; 1–2 young; sexually mature at 2 years. **Longevity:** 20–23 years (captive). **Status:** extinct in wild; small numbers in captivity; endangered.

■ HISTORY OF INTRODUCTIONS

Père David's deer have been re-introduced to part of their former range, near Beijing and Shanghai. Escapees have been unable to establish themselves permanently(?) in Britain.

Asia
China

Formerly Père David's deer were thought to inhabit the swampy plains of northern China until clearing for agriculture wiped them out, but they were kept in hunting parks by some emperors. In 1865, 100 were in an Imperial Hunting Park, but these escaped in 1894 when a flood damaged the walls of the park, and most were killed. The survivors were killed during the Boxer Rebellion in 1900 and by 1911 only two remained. However 10 years later these were dead (Burton and Burton 1970). In captivity a population of about 600 remained at Woburn Park in the United Kingdom and also over 100 at Wadhurst in Sussex. Around 50 captive populations existed in the world (WCMC 1998).

There probably have not been any truly wild in China for 200–300 years (Louden and Fletcher 1983), as they may have survived to the Ming Dynasty (1368–1644) (WCMC 1998). Only captives survived in the Imperial Hunting Park until the end of the nineteenth century. However, some Père David's deer were sent to Europe by Abbe Armand David for breeding purposes (Anon. 1963; Burton and Pearson 1987).

Those at Woburn, England, had built up to a herd of 250 by 1963 (Whitehead 1964). In 1956 after World War 2, some were returned to the Beijing (Peking) Zoo in China (Anon. 1963; Burton and Burton 1970).

Recently some semi-wild animals were re-established in Beijing (Louden and Fletcher 1983) and this colony has had several more added to it in the last few years. In the mid-1980s the re-introduction of the deer was successfully undertaken at two locations: at Nanhaizi Park near Beijing, five male and 15 females were released in 1985 in a 100 ha area of original habitat with the current population being about 100 individuals; 39 were re-introduced to the Da Feng Reserve,

Jiangsu Province, north of Shanghai in 1986 (Thouless *et al.* 1988: WCMC 1998). This group had grown to 68 individuals by the end of 1989 and they are now fully protected (Sitwell 1986; Ohtaishi and Gao 1990). By 1992 there were 122 in the Da Feng Reserve (WCMC 1998).

Europe
United Kingdom

Several Père David's deer have escaped from zoos and parks in Britain: e.g. Aston Abbotts, Buckinghamshire in 1963–64; Loch Lomond in 1952–53; a group of 12 near Swindon, Wiltshire in 1981, and several of this group may still be at large (Baker 1986; Corbet and Harris 1991).

ELK OR MOOSE

European or American elk or moose
***Alces alces* (Linnaeus)**
=*A. americana* (Clinton)

■ DESCRIPTION

HB 2000–3100 mm; T 40–120 mm; SH 1500–2250 mm; WT 202–1295.4 kg.

Upper parts blackish brown to reddish brown or greyish brown, greyer in the winter; head long; ears large; upper lip pendulous; throat with tassel of skin and hair 150–750 mm in length; antlers palmate, extending laterally from head, two to five tines, and up to 1950 mm in length, sheds November on; tail small; rump lower than shoulders; under parts brown; lower legs grey. Female lacks antlers. Dewlap well developed in American form (moose) and short in European (elk).

■ DISTRIBUTION

Eurasia and North America. Northern Eurasia from Scandinavia and eastern Poland east to Manchuria, Mongolia, Siberia and the Pacific coast, and south to the Ural Mountains, the Altai and Lower Amur. Formerly west to northern Germany and France. In northern North America from Alaska to Quebec and Nova Scotia, Canada and south to northern New England, Michigan, Minnesota and the Rocky Mountains.

■ HABITS AND BEHAVIOUR

Habits: diurnal and crepuscular; principally a browser; swims well. **Gregariousness:** mainly solitary; density 0.1–1.1/km^2, but sometimes reaches 100/km^2 or more locally. **Movements:** in North America altitudinally migratory (down in winter) and in northern Europe migrates regularly south in winter (179–300 km); individual seasonal home ranges 2.2–16.9 km^2. **Habitat:** well-watered forest, marshes, lake shores,

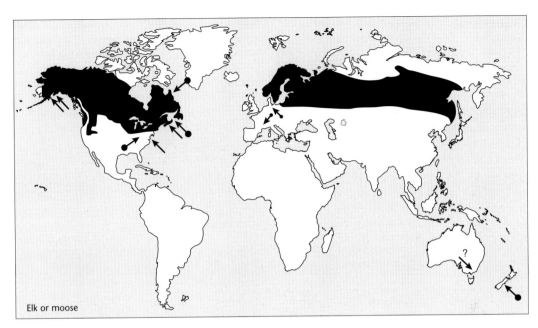

Elk or moose

wooded areas, and in summer arctic tundra. **Foods:** leaves, shoots and bark from trees and shrubs; also grass, forbs, heather, and marsh and water plants. **Breeding:** mate September–October, calve in spring (May–June); gestation 226–264 days; male polygamous; female seasonally polyoestrous; oestrous cycle 20–22 days; calves 1–3, usually 1; weaned 5–6 months; sexually mature 2–3 years, males breed at 5–6 years. **Longevity:** 20–27 years. **Status:** still fairly common, but range considerably reduced.

■ HISTORY OF INTRODUCTIONS

Introduced successfully to New Zealand and unsuccessfully to Australia. They appear to have had little success where re-introduced in Europe, including the Russian Federation and Latvia. Re-introduced successfully in parts of North America.

NORTH AMERICA

Alaska

Twenty-three moose from Kenai Peninsula, Anchorage, Susitna and Matanuska were released in the Copper River Delta in 1949–58, and 21 from Chickaloon, Susitna and Matanuska at Berner's Bay in 1958–60. Also six from Anchorage were released on Kalgin Island, Cook Inlet, in 1958–66, and 14 from Chickaloon Flats were released at Chickamin River in 1963–64 (Burris 1965). Except for Chickamin, where a few still persist, these translocations from south-central Alaska have resulted in harvestable populations (Franzmann 1988).

In 1966 moose were also introduced to Kodiak Island from south-central Alaska but they were unsuccessful in becoming established (Franzmann 1988).

Canada

By the beginning of the nineteenth century moose range in North America had been much reduced in size. However, since 1920 they have increased both their range and numbers, assisted by introductions and natural immigration (Banfield 1977). Their range had reached the international boundary with the United States by at least the late 1950s (Cowan and Guiguet 1960). They expanded naturally into northern Quebec and Ontario and even the tundra regions of the North West Territories (Banfield 1977).

A pair of moose from New Brunswick were introduced to Grander Bay, New Foundland, in 1878 with doubtful results and a second release of four (two males and two females) from Nova Scotia at Howley in 1904 appeared much more successful (Pimlott and Carberry 1958; Banfield 1977). In the mid-1950s the population here was estimated at 30 000–40 000 head (Pimlott and Carberry 1958), and a large population still occurs there (Walker 1992).

Several introductions of moose were made to Anticosti Island between 1895 and 1913 (Newsom 1937; Pimlott and Carberry 1958; Banfield 1977; Hall 1981). They were reported to have been released at Juniper River (20 animals) in this period, 12 in 1897 and six in 1898 (Pimlott and Carberry 1958). In 1953, seven females and five males trapped in Newfoundland were released at St. Lewis River, in southern Labrador (Pimlott and Carberry 1958; Banfield 1977). They were exterminated on Cape Breton Island, Nova Scotia, early in the last century (by 1924), but were successfully re-established in

1947–48 when 18 were liberated in Cape Breton National Park. In 1928–29, two adults and five calves were introduced and 35 from Elk Island National Park, Alberta, were released there 20 years later. In the 1970s these herds were increasing in numbers (Whitehead 1972).

United States

Approximately 10 moose were released in the Nehasane Preserve, New York, in 1894–95, but these were thought to have been exterminated in a fire in 1903 (Bump 1941). Further introductions were made in this state in 1902–03 when six males and six females from Canada were released in the Adirondacks by state authorities, but the project failed (Bump 1941; Pimlott and Carberry 1958). Also in 1903 a Mr W. C. Whitney released a few moose at Suranac Inn, where they were reported for several years (Bump 1941).

In 1922, six calves from the Kenai Peninsula, Alaska, were taken to Oregon, then shipped to San Francisco and then to Portland and released on Tahkenitch Lake in Douglas County (Shay 1976). Five survived the trip to be released, but they were eventually unsuccessful and all disappeared by 1931 (Franzmann 1988).

The moose on Isle Royale, Michigan, had increased to large numbers by 1929, so between 1934 and 1937 about 71 head (38 female and 33 male) were transferred to the upper Peninsula where there were only a few native moose left (Ruhl 1941; Pimlott and Carberry 1958). This re-establishment appeared successful (Ruhl 1941), but after a few years the population was once again back to its former low level (Pimlott and Carberry 1958).

Moose were also translocated to Wyoming in 1934, 1948 (eight) and 1950 (eight). The 1934 release failed, but the later ones appeared to be successful (Pimlott and Carberry 1958).

PACIFIC OCEAN ISLANDS

New Zealand

Moose (*Alces machlis* (=*A. americana*)) were introduced to New Zealand in 1900 and 1910 (Thomson 1922) and are now local and rare in the South Island (Wodzicki 1965). The first release was unsuccessful, the second was successful, but did not extend their range much beyond the release site (Donne 1924; Pimlott and Carberry 1958; Wodzicki 1961). In 1900, two bulls and two cows were released near Hokitika, but failed to become established; in 1910, four bulls and six cows from Saskatchewan were released in Dusky Sound, Fiordland, but the population did not spread much. Probably there were less than 25 head in the area in 1972 and they were considered unlikely to survive in New Zealand much longer. A few still remained in the 1970s (one shot April 1971) (Gibb and Flux 1973), but in contrast to other deer in New Zealand they seemed to be dying out (Wodzicki 1961). Competition from introduced red deer was probably a factor in their demise (Tustin 1974).

EUROPE

Austria, Czechoslovakia, Germany and Poland

North American moose from Alberta were sent to Berlin Zoo in 1936 and placed in a game reservation (Pimlott and Carberry 1958), but there are no details about them.

Occasionally elk reach central Europe as wandering animals from their range further east. However, a series of re-introductions were made in West and central Europe from about 1530 on with little success. Probably the only really successful introduction was in the National Park of Kampinos, 26 km from Warsaw in 1951, when four from a reserve in east Poland were placed in a 140 ha park. In 1957 there were 28 animals, and half were released in a 80 000 ha forest. In 1960 there were 47 there and by 1965 about 80 animals were present (Niethammer 1963; Pielowski 1969).

After the fences of Kampinos National Park, Warsaw, were opened elk have appeared in eastern Germany, Czechoslovakia, western Germany and in Austria up until 1967. Movements of up to 1000 km were known among a dozen animals, but unsuitable habitat made it impossible for them to become established or to extend their range (Briedermann 1968).

Russian Federation and adjacent independent republics

Some 105 elk were released in eight regions of Russia between 1954 and 1970 for acclimatisation purposes – Kalininsk (1969, 15), Kirovsk (1967, one), Kursk (1967–70, two), Moskovsk (1954–1968, 59), Smolensk (1966, 16), Tumensk (1968, one), Yahroslavsk (1966, 1968, 10), and in Latvia (1956, one) (Kirisa 1974). They appear to have had little success.

■ DAMAGE

Where they are native in the European part of the Russian Federation, elk become pests in some areas (Yanushevich 1966). They most often damage aspen, pine, oak, spruce, elm, linden and poplar. Damage to pine and oak appears to be of the greatest importance, with the damage becoming obvious where there are less than 25 hectares of young trees per elk (Dinesman 1959). In the Bryansk oblast, where the elk population is 1.6/1000 ha, the heaviest damage is inflicted on pine forests, which constitutes 40 per cent of the area in enclosed plantations. Damage is concentrated at the periphery and affects 5–15-year-old trees, and aspen, willow and mountain ash are eaten and damaged, but spruce, birch and oak rarely so (Fedosov 1959).

As a result of over-population in the Oka Reserve young trees and shrubs are repeatedly damaged and in some places completely destroyed. Elk damaged 70–99 per cent of pine and 50–91 per cent of oak shoots and ate 100 per cent of mountain ash, aspen and willow. The degree of damage in different areas depends on the height of the snow cover (Borodin 1959).

In the central regions of European Russia, elk are reported to reduce the amount of marketable pine trunks by 22–60 per cent and 9 per cent of the trees were found to be dying. Where there was one elk/18 ha, 35–69 per cent of all the trees were damaged and 32–60 per cent of the pines. At one elk/10 ha, 85–91 per cent of pines were damaged (Kozlovskii 1959). In the Moscow oblast they cause damage to forests especially to aspens, mountain ash and European bird cherry (Anon. 1959) and on the Volga Plains near the Knibyshev Sea, elk cause damage to buckwheat and oats (Shmit 1959).

In Russia at a density of 9 elk/100 ha of forest they caused heavy damage to spruce trees by gnawing stems and biting shoots of young trees (Il'yushenko and Smirnov 1979). Studies in the Tatar region suggest that elk populations should not exceed 5 per 1000 ha to keep damage to a minimum (Aspisov 1959; Nazarova 1959). In the Tatar region, pine and broad leaf forest plantations up to the age of 20 years are damaged and in a number of forestries, damage of 87 per cent of pines has been noted.

High populations of moose in central Newfoundland cause browsing damage to white birch (*Betula papyrifera*) and balsam fir (*Abies balsamea*) (Bergerud *et al.* 1968). This browsing damage to these trees is caused by a density of 7–12 moose/2.56 km^2. Heavy moose damage halts the growth of birch and kills regeneration, and the damage to balsam fir is extremely severe, resulting in the suppression of terminal growth in many trees and reduced stocking by uprooting seedlings. Where there are 6 moose/2.56 km^2 damage is not serious (Bergerud *et al.* 1968).

A study in the United States found that the foods of moose, wild horses and cattle were more similar to each other than either one to mule deer in northeastern Colorado (Hansen and Clark 1977).

REINDEER OR CARIBOU

Arctic caribou, woodland caribou, barren ground caribou

Rangifer tarandus (Linnaeus)

=*R. arcticus* (Richardson)

■ DESCRIPTION

HB 1200–2540 mm; T 100–220 mm; SH 680–1400 mm; WT 52.3–184 kg (large animals, probably males, to 318 kg. Caribou of North America are usually slightly larger).
Coat generally brown (varies dark brown to blackish to some very pale); face, chest and dorsal surface of tail darker; neck and main creamy white, which extends above darker bar in band across lower shoulders and flanks; belly, rump and under tail white; legs brown, with white socks above hooves; hooves cleft; antlers 530–1300 mm, palmate in form, brown, sheds December. Female generally smaller than male; antlers 230–500 mm, sheds February.

Note: Arctic subspecies predominantly white with blue-grey saddle; forest subspecies brown-black with whitish under parts.

■ DISTRIBUTION

Eurasia, Greenland and North America. In Eurasia confined to the mountains in southern Norway and in Karelia, Russia, but formerly more widespread; in Greenland confined to parts of west coast and in the far north; in North America range from Alaska, most of northern Canada and Arctic Islands, south to Newfoundland, the Great Lakes Region and northern British Columbia. Occur on islands of Graham (Queen Charlotte group), Unimak, Spitzbergen, Novaya Zemblya and Sakhalin. Their range has much altered due to local extinctions and replacement by domestic reindeer that occur from Scandinavia to eastern Siberia. Most of those found in Europe and Asia are now domesticated or at least semi-domesticated.

■ HABITS AND BEHAVIOUR

Habits: mainly diurnal; dominance hierarchy. **Gregariousness:** bands or harems (1 bull, 20 or more cows) or loose herds to 1000; mature males solitary in summer; on migration large herds 50 000–100 000; some subspecies form large herds but others do not; density originally 0.4–0.6/km^2, on migration up to 19 000/km^2. **Movements:** nomadic, and northern population seasonally migratory from winter to summer feeding grounds (movements up to 300–900 km known); traditionally move south in winter; daily 19–55 km/day; home range of non-migratory animals 100–200/km^2. **Habitat:** tundra, subarctic taiga, alpine meadows, open moors, and boreal coniferous forest. **Foods:** aquatic plants, grass, sedges, forbs, flowering plants, twigs, buds, leaves and shoots of trees and shrubs, lichen and fruits; will chew dead

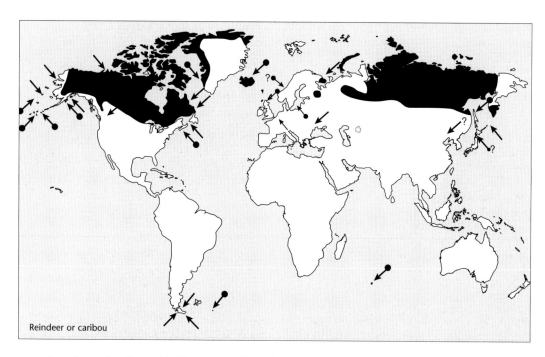

Reindeer or caribou

animals and occasionally eat birds' eggs. **Breeding:** ruts September–November, calves April–July; gestation 215–246 days; oestrus 10–24 days; male polygamous, female seasonally polyoestrous; calves 1–2; calf precocious, runs within hours of birth; females mature 16–41 months, males mate at about 3 years. **Longevity:** 2–15 years in wild, about 20 years captive. **Status:** much reduced in range and numbers, but common under domestication.

■ HISTORY OF INTRODUCTIONS

Domestic reindeer have been introduced into Iceland, Orkney Islands, Scotland, South Georgia, Kerguelen Islands, Alaska, Canada, Greenland (Banfield 1961; Smit and Von Wijngaarden 1981), China (now feral), Sakhalin Island, United States, Pribilof Islands and St. Matthew Island (Wilson and Reeder 1993).

ASIA
China
Wild reindeer occur in China in the north-east near the border with Russia. However, domestic reindeer have been released in the wild in certain areas. In the extreme northern Heilongjiang and Nei Mongol A. R. the Evenki(?) nomads raise reindeer under semi-wild conditions. Some say that wild reindeer are now extinct in China and that those that appear in the wild are actually released domestics (Ohtaishi and Gao 1990).

Japan
In 1924, 10 domestic deer were introduced to Shinshiru Island in the central island group by Fisheries Bureau of Tokyo. They increased and by 1940, 224 had been recorded, but there are no recent records (Whitehead 1972).

ATLANTIC OCEAN ISLANDS
Aleutian Islands
Reindeer were introduced to Atka and Unmak islands in the Aleutians for the use of the native peoples (Murie 1941; Klein 1968). Calves captured in 1958–59 in the interior of Alaska were introduced to Adak Island, in the central Aleutians, where they were held in semi-domestication for two months and then released. After release the original 23 expanded to a herd of 83 animals in five reproductive seasons and the herd is now wild in disposition (Jones 1966). They were also once abundant on Nunivak Island, but were extirpated around the end of the nineteenth century and replaced in 1920 with semi-domestics; there are about 10 000 there now (Spencer and Lensink 1970).

Greenland
Domesticated reindeer from Norway have been introduced into Greenland. Since 1952 they have become established in the wild in parts of western Greenland (Lever 1985).

Iceland
Between 1771 and 1787, domestic reindeer were introduced on four occasions from Norway to Iceland and released (de Vos et al. 1956; Lever 1985). In 1777, 23 were released in south-west Iceland, where by the

middle of the nineteenth century they had increased to 100 head on the Reykjanes Peninsula. These became extinct at some point between 1920 and 1930.

In 1784, 35 from Finmark were landed in northern Iceland. Here, they increased and by the beginning of the nineteenth century locals were complaining about the effect they were having on the lichens. This herd began to expand in the 1820s and spread eastwards to Thingeyjar Sysla. The population reached its peak in 1850, but thereafter declined and finally disappeared about 1936. In 1787, 35 from Finmark were landed in north-eastern Finland at Vopnafjordur in Nordur Mula. From these few descendants the reindeer population is now established in Iceland today. They reached peak numbers in the mid-nineteenth century, but thereafter declined, and by 1939 only between 100–300 remained at the north-eastern corner of the Vatnajokull icecap. In 1984 a census revealed that about 1200 were present, although it is thought that the total may exceed 3000 (Lever 1985).

Pribilof Islands (St. Paul and St. George, Bering Sea)
The United States government landed four males and 21 females on St. Paul and three males and 12 females on nearby St. George Island (66 km south) in 1911 to provide the indigenous population with fresh meat (Klein 1968; Whitehead 1972). By 1922 St. George Island had 222 head, but the numbers declined to a stable herd of 30–40. On St. Paul the original introduction increased steadily until the early 1930s when the population suddenly erupted. By 1938 there were over 2000, but 12 years later only eight head remained. The decline is thought to have been due to the drastic effects that a large population had on the availability of lichen, which was the principal item in their diet.

South Georgia
Reindeer were first introduced to South Georgia in 1911–12 by the Larsen brothers for food and sport. They became established and by the 1930s hundreds ranged over a small part of the island (Harrison-Matthews 1931; Klein 1968). At least 10 animals from Norway were released on Barff Peninsula in 1911 by whalers. This herd was confined to the peninsula for many years, but spread to an area north of Royal Bay in about 1961–65 and formed a second herd.

From the 1930s to the 1950s whalers shot up to 100 reindeer per year from the Barff herd. Following closure of the whaling station in 1964–65 a few were shot up until 1972. A third herd arose from an introduction of seven to the Busen area in 1925. From 1973 to 1976 about 500 were shot for research studies, but in 1976 about 2500–2600 reindeer were present on the island in three herds (Leader-Williams 1980).

Some reindeer were shot by Argentinians during their occupation of the island in 1982 (Lever 1985).

St. Matthew Island (Bering Sea)
In August 1944 a group of 29 reindeer (24 males and five females) from Nunivak Island, Alaska, were taken to St. Matthew in the Bering Sea, for the purposes of studying herd development, and released by the United States Coast Guard (Klein 1959, 1968; Skuncke 1968). Initially the population responded to the high quality and quantity of forage and increased rapidly, with high birth rates and low mortality, and reached a density of $46.9/2.5 \text{ km}^2$ (Klein 1968). By 1957 there were 1350 head (Skuncke 1968) and this had increased to 6000 by 1963 (Klein 1968; Skuncke 1968), although between these two dates some 105 were taken by hunting (Klein 1968). The population then crashed and by August 1964 there were only 45 left (Skuncke 1968), falling further to 42 (all males) by 1966 (Klein 1968).

The crash in numbers on St. Matthew was thought to have been brought about by overgrazing of lichen, with the animals dying of starvation (Klein 1968; Skuncke 1968).

EUROPE
Finland
Most of the truly wild reindeer had gone from Finland by 1900 and now only large herds of domestic and semi-domestic animals exist there (Whitehead 1972). Ten wild reindeer were translocated in 1979–80 from eastern Finland to central Finland, where they were kept in enclosures (Nieminen and Helminen 1987). Yearlings were released in 1981–83 and the enclosed animals in 1984. Now a herd of 60–65 inhabits two protected areas, Salamajarvi National Park and Salamanpera Natural Park.

Germany
Many attempts were made to establish reindeer in Germany in the sixteenth century between 1520 and 1582, including releases in Hessen East Prussia and some since this time (Niethammer 1963). Today reindeer occur only in zoos and wildlife parks.

Norway
Between 1771 and 1787 some 103 reindeer were introduced to Finmark, Norway, from Iceland (Neithammer 1963). These quickly became wild and reproduced. However, they had disappeared by 1817. In the 1970s in Norway, both wild and domestic herds of reindeer occurred in many parts of the country and appeared to be increasing. The present population is probably a mixture of domestic and wild reindeer in many of the areas in which they occur (Whitehead 1972).

Russian Federation and adjacent independent republics

At the present time there are probably 900 000 wild reindeer in the Russian Federation, but there are tens of thousands of feral ones that have escaped from domestic herds (Klein and Kuzyakin 1982).

Reindeer have been released for acclimatisation purposes in five regions of mainly Asiatic Russia. Some 50 were released on the Soloveshkie Islands in the Arkhangelsk region in 1962; 47 in the Gorkovsk region in 1965; and eight in the Khersonsk area of the Ukraine in 1958 (Bannikov 1963; Kirisa 1974).

Many domestic animals now occur on Sakhalin Island, where it is now not possible to distinguish between these and the wild reindeer. Moderate numbers are feral in Chukotka in far eastern Siberia where intensive herding has resulted in the almost complete disappearance of the wild herds.

Sweden

In Sweden the last truly wild reindeer was killed about 1865, but in the northern part many domestic herds and semi-feral animals are herded by the Lapps (Whitehead 1972).

Some were released on Gotska Sandon, north of Grotland, in the Baltic Sea in the seventeenth century where they were successfully established until about 1825 when there were few left (Niethammer 1963).

United Kingdom

Reindeer may have survived in Scotland to about the Roman period and some may have been introduced in the Middle Ages to north Britain, as there was considerable trade with Scandinavia at the time (Fitter 1959). However, there is also good evidence to suggest that the wild animal did not survive into the post-glacial period (Lever 1985).

During the eighteenth and nineteenth centuries several attempts were made to re-introduce reindeer to Scotland without success. Fourteen were brought to Dunkeld by the Duke of Atholl and released in Atholl Forest, Perthshire, at different times (Fitter 1959; Whitehead 1964). These animals survived for a period of about two years. In 1820 the Earl of Fife introduced some to Mar Forest, Aberdeenshire, but they all died (Fitter 1959; Whitehead 1964). The largest attempt appears to have been in about 1820 when a Mr Bullock released 200 over a period of time in the Pentland Highlands, but these gradually disappeared (Harper 1945). Other attempts were made with three imported from Archangel to the Orkneys in 1816, and by Sir Henry Liddell who introduced five on his Northumberland estate in 1786 (Fitter 1959; Whitehead 1964). These attempts were unsuccessful.

In 1952 a small domestic herd from Lapland was introduced to the Cairngorms in Inverness-shire by Mikel Utsi (Perry 1963; Southern 1964; Whitehead 1964). These animals were breeding at least by 1957 (Fitter 1959) and following the initial introduction more were released, until by 1963 a herd of 25–30 occupied some 2000 ha of a reserve in the Cairngorms (Perry 1963). By 1972 there were 100 head (Whitehead 1972). Although they are free-ranging they are not really feral and the population is managed (Baker 1990). In 1988 the herd numbered 80, which are free-ranging on the northern slopes where they remain above the tree line throughout the year (Corbet and Harris 1991).

INDIAN OCEAN ISLANDS

Kerguelen

Two pairs of reindeer were imported to the island of Grand Terre in 1955, and one male and five females in 1956 from Sweden, and these appeared to become well established (Reppe 1957). However, another report indicates that two herds were released in 1956, one on Grand Terre of two males and five females, and the other on Île Haute of one male and two females (Lesel and Derenne 1977).

The herd on Île Haute was apparently successful and by 1965 numbered some 34 head (Lesel 1967). By 1968 the number had reached 90, but dwindled to 60 a year later. However, in 1971 there were 93, in 1972 some 99, and in 1973 from 70 to 80 (Lesel and Derenne 1977).

Those reindeer on Grand Terre were also successfully established but there is no accurate estimate of their numbers (Lesel and Derenne 1977), although there may be as many as 2000 (Lever 1985).

NORTH AMERICA

Woodland caribou (*R. t. caribou*) formerly occupied the boreal forests of the entire North American continent. They declined drastically in numbers in the late 1800s and early 1900s and this trend has continued into the 1980s. Originally there were about two to three million caribou, but by 1949 only 668 000 were left and in the 1970s about half a million (Anon. 1973).

Domesticated strains of reindeer (*R. t. tarandus*) from Eurasia have more than once been introduced into North America (Alaska, Canada and Greenland) after 1890, but by 1975 most or all of them had died out in the United States.

Alaska

Some 171 domestic reindeer from Siberia were introduced to the tundra on the Seward Peninsula, Alaska, in 1891 while the wild animal (caribou) was absent. They were imported to supply a stable food source and promote industrial education for local Eskimos. Further imports up until 1902 increased the total number of reindeer introduced to 1280 head. By this time there were 5000 reindeer in Alaska from imports and natural increase. The industry proved of little benefit in early years, but by 1920 it had been expanded into much of north-western Alaska by white people. At this time there were probably 180 000 reindeer. Between 1920 and 1929 reindeer meat was exported to the United States, but this industry then folded through lack of demand for the meat. The herds of reindeer, however, kept increasing, and there were 600 000 or more there in 1934; range deterioration was widespread and many were being lost through disease. In 1937 restricted ownership of reindeer was turned over to the indigenous Alaskans. Numbers declined and many herds were released to wander and become feral or join wild caribou herds. From flourishing and increasing herds of about one million reindeer in the 1930s, they declined through mismanagement to about 120 000 by the mid-1940s. By the 1950s this had fallen to about 25 000 (Gottschalk 1967; Burton and Burton 1969; Whitehead 1972; Siverstein and Silverstein 1974).

Caribou were introduced to Kodiak Island in the 1920s where they became established (Clark 1958), but their recent status is uncertain. Two successful introductions of caribou have occurred in Alaska: to Adak Island and on the Kenai Peninsula. Caribou calves were airlifted from Nelchina Basin to Adak Island and reared for two months prior to release (Jones 1966). In 1958, 31 were captured and 10 released; in 1959, 45 were captured and 14 released. They became successfully established and harvesting, which began in 1964, has continued until the present time (Franzmann 1988).

Re-introduced to Kenai from the Nelchina Basin in 1965 (32 animals) and 1966 (29), caribou harvesting began in 1974 and continues. In 1984, 28 were translocated (16 died out of original 44) to Kenai from Nelchina by the Alaska Department of Fish and Game and the United States Fish and Wildlife Service to establish populations in areas not permanently occupied (Bailey and Bangs 1985). These translocations – probably successful – are still being monitored (Franzmann 1988)

In 1965, 15 Caribou from Christchurch(?) were released at Chickaloon River (Kenai Peninsula), Alaska (Burris 1965).

Canada

In 1924, 145 reindeer (Norwegian stock from Newfoundland) were introduced to Anticosti Island, Quebec (Bergerud and Mercer 1989). By 1941 there were only nine survivors and the reason for the decline is not known, but competition with introduced white-tailed deer may have been the reason for the failure.

The Canadian government introduced reindeer into its North West Territories in 1929 to build up an industry for the Eskimos there. A herd of 3400 was driven from Kotzebue Sound, Alaska, to the MacKenzie River, the trip taking some five years and only 2370 surviving (Gottschalk 1967; Burton and Burton 1969; Whitehead 1972). By 1952 there were 7000 (Whitehead 1972).

Populations of caribou have been restored to portions of Quebec and a remnant population along the British Columbia–Idaho border was supplemented about 1986 (see United States). Eighty-two caribou were released at Laurentide Park, Quebec, between 1966 and 1972, and in 1983 the herd numbered 80–90 animals (Vandal 1984; Vandal and Barrette 1985).

Caribou introductions in Newfoundland 1961–82

Blow-me-Down Mts	1964	13	increasing
Bonavista	1964, 67, 68	33	decreasing
Brunette Island	1962	17	stable
Burin Foot	1964	13	stable
Burin Knee	1964, 65, 67	44	stable
Butterpot	1969	4	extinct
Cape Roger	1965	9	extinct
Cape Shore	1976, 77	27	increasing
Change Island	1964	5	extinct
Englee	1982	15	?
Fogo Island	1964, 65, 67	26	increasing
Gregory Plateau	1965, 67	25	increasing
Grey Island	1964	8	increasing
Horse Island	1964	6	extinct
Jude Island	1964, 65	6	stable
Merasheen Island	1961, 63, 65	35	increasing
Port au Port	1964, 65	20	decreasing
Random Island	1964, 65, 67	20	increasing
Sound Island	1961, 63, 64	13	stable
St. Anthony	1976, 77, 82	32	increasing
Twin Lakes	1964, 67	9	?
Weir's Pond	1964	4	extinct
Total		384	

Ten caribou from Newfoundland were released in the Liscome Game Sanctuary, Nova Scotia, in 1939 (Tufts 1939), but failed to become established (Benson and Dodds 1977). It was suspected that parasites were responsible for the failure. Fifty-one more caribou were released in the Cape Breton Highlands National Park, Nova Scotia, in 1968 and 1969 (Dauphine 1975). By 1973 this herd had disappeared and the failure was thought to be due to meningeal worm infections.

From 1961 to 1982, 384 caribou were released at 22 sites in Newfoundland. By 1982, 17 of the 22 had resulted in viable herds, and these herds numbered about 1500 animals (Bergerud and Mercer 1989).

In 1970, 12 caribou were released on Great Gloche Island in Lake Huron, Ontario. Sixteen months later they were all dead of a neurological disease (Anderson 1971, 1972; Bergerud and Mercer 1989).

In 1982, eight caribou were introduced to Michipicoten Island, Lake Superior, from the Slate Islands, Ontario. After six seasons these had increased to about 26 animals (Bergerud 1985; Bergerud and Mercer 1989). Eight caribou were also released on Montreal Island, Lake Superior, in 1984. These were still present in 1988, when 14 were seen. Six were released on Bowman Island, Lake Superior, in 1985, but after moving to other islands these eventually disappeared, except for one animal still alive in 1986 (Bergerud and Mercer 1989).

United States
In the early 1800s caribou resided in Maine, northern Vermont, New Hampshire, the lake states north of 45–46°N, and throughout Atlantic Canada. They disappeared from these areas between 1830 and 1930. Small herds still ranged between southern British Columbia, Washington, and Idaho (Bergerud and Mercer 1989). A small herd of about 25 still occurs here and has recently been augmented with animals from British Columbia (Bergurud and Mercer 1989).

Until 1905 caribou lived on Mt. Katahdin, Maine, but there are no reports after 1907. Some were re-introduced in 1963 (23 caribou) from Newfoundland (Rogers 1964), 17 on Mt. Katahdin and six nearby (Dunn 1965). Fourteen were seen in 1964, but all of them had disappeared by 1966. In December 1986, 25 were transferred from Newfoundland to Maine (Bergurud and Mercer 1989). There may have been further re-introductions in 1989.

Attempts to acclimatise reindeer in Michigan have ended in complete failure (Whitehead 1972). In 1922, 60 reindeer from Norway were imported and released in Michigan, but within five years they were consid-ered unsuccessful as the herd did not increase – most young died or were born dead (Ruhl 1941). Later another attempt failed when 14 were released in a 2640 ha enclosure in Wisconsin, but all died of disease (Trainer 1973).

By 1937 probably only eight caribou remained in Minnesota (Cox 1939), but 10 (two bulls, eight calves) were re-introduced from Saskatchewan in 1938 to Red Lake (Cox 1941). One was released in the wild to join three native females and the remainder enclosed, but these were later released with their descendents in 1942. All, however, had disappeared by 1946 and the range of the caribou continues to retreat northwards (Gogan *et al.* 1990).

Overall there have been 33 introductions of caribou into eastern North America from 1924–1985, but only 20 have resulted in sustaining populations and 13 have failed. Generally introductions where white-tailed deer were common have failed, probably because of meningeal worms.

Pacific Ocean islands
Komandorskiye Ostrova (Commander Islands)
In 1882, 15 were introduced to Bering Island (Niethammer 1963) where they continued to increase until 1896 and were still well established there in the 1930s (Barabash-Nokiforov 1938). In 1928 there were 28 there, and in 1963 there were 150 head present on the island (Niethammer 1963).

Kurilskiye Ostrova (Kurile Islands)
An unknown number of reindeer occurred on the Kurilskiye Ostrova in 1965 (Kirisa 1974). They appear to have been introduced to these islands more than once. Releases may have been made in 1928 (Kirisa 1974), in the 1930s (de Vos *et al.* 1956), and as early as 1924 (Whitehead 1972). Those introduced in the 1930s are reported to have died out in the 1950s (de Vos *et al.* 1956).

South America
Argentina
In 1947 and 1948 reindeer were released on Tierra del Fuego, on Isla de los Estados and on Georgia del Sud island (de Vos *et al.* 1956; Petrides 1975) in southern Argentina, but are reported to have been extermi-nated by local hunters (Petrides 1975). In 1971 some were again reported to have been released on Tierra del Fuego, but on Chilean territory, but it is not known what happened to them (Lever 1985).

■ DAMAGE
Whilst caribou were in a state of decline there was no conflict, but since they have increased and returned to many of their former areas, overlapping ranges of caribou and reindeer have caused problems. Such

problems occur because reindeer are loose herded and are inclined to join caribou, thus competing for range and forage, and also because of the importance of the transmission of disease from reindeer to caribou. The two populations of the same species, one domestic and one wild, competing for high quality range has created a unique grazing situation. It is a complex problem for land managers as both animals are important resources in the area (Adams and Robus 1981).

Reindeer introduced on South Georgia have had a serious impact on the vegetation of the island (Leader-Williams 1978). Extensive stands of tussock grassland have been overgrazed probably because the species is non-migratory on the island and does not depend on lichens for winter forage as it has to in the Northern Hemisphere (Leader-Williams 1988).

ROE DEER

European roe deer
Capreolus capreolus (**Linnaeus**)

■ DESCRIPTION

HB males 950–1510 mm; females 950–1350 mm;
T 20–40 mm; SH 600–1000 mm; WT 11–50 kg.
In summer, coat short and red-brown; ears blackish; under parts white; lacks white tail patch; muzzle dark, but has white spot on either side; chin white. In winter, coat long, thick, dark brownish speckled, greyish fawn; conspicuous white tail patch; throat white; chest and legs tawny; ears whitish inside, rump patch white in winter; antlers short (200–300 mm), three tines, knobbed at base, sheds October–December, on male only, but occasionally females grow small antlers. Black band from nostril to angle of mouth. Young dappled with white spots on flanks and sides, but gone by six weeks.

■ DISTRIBUTION

Eurasia. British Isles and southern Scandinavia south to the Mediterranean, and east across central Asia to south-eastern Russia (south-eastern Siberia), Manchuria, Korea and central China; south through western China to Szechuan and eastern Tibet; also in Caucasus, Asia Minor, northern Iraq and northern Iran; and from Altai to the Tien Shan and mountains of Turkestan.

■ HABITS AND BEHAVIOUR

Habits: mainly crepuscular, also nocturnal; shy. **Gregariousness:** usually solitary, pairs, or family groups (female and young 2–10); forms larger herds in winter (8–60); density 8–25/km^2 and up to 40/km^2. **Movements:** sedentary and migratory (300 km to winter sites); territories 7–25 ha for males and 3–180 ha for females; home range 7.5–151 ha. **Habitat:** low mountain slopes, open woodland, wooded valleys, woods near grassland, open moorland, agricultural land near cover, forests. **Foods:** grass, leaves of trees and shrubs, acorns, beech mast, corn, clover, buds, pine shoots, heather, moss, fungi and cultivated crops. **Breeding:** ruts July–August, fawns March–July; gestation 4–5 months; with delayed implantation

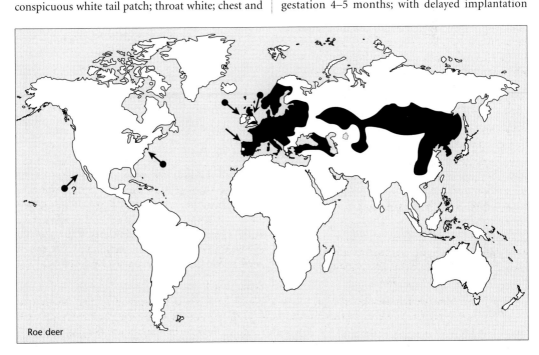

Roe deer

9 months; young 1–2, rarely 3; fawns lie up, but follow mother at 6–8 weeks; sexually mature at 1 year 4 months. **Longevity:** 10–12 years (in wild) to 15 years (captive). **Status:** common.

◼ HISTORY OF INTRODUCTIONS
EUROPE
Since 1938 there have been many introductions of roe deer in central Europe, often to improve the bloodline of local stock, and this has resulted in the Siberian race *pygargus* now being more widespread. In Slovakia hybrids between *pygargus* × *capreolus* are often found (Niethammer 1963).

Some roe deer were released on Fehmarn island about the time of World War 1 (Niethammer 1963). Attempted introductions occurred on Juist in the East Freisian Islands before 1932, but there are few there today.

Czechslovakia
The Siberian race of roe deer, *C. c. pygargus*, was released in Czechoslovakia prior to World War 1 (de Vos *et al.* 1956), where they crossed with the native race in some localities (Turcek 1951). This gave rise to an animal with perceptibly abnormal, tall and thick antlers.

Germany
Attempts to establish roe deer from 1894 to 1938 were made in Germany with 32 introductions of well over 183 animals in about 21 different areas (Beninae 1941). Probably only four of these were successful, the majority failing for various reasons such as disease, badly chosen release sites, domestic stock used, for no apparent reason, and some stock simply wandered off.

Spain
Two pairs of roe deer were released at Siera de Cazorla, Spain in 1952 (Yebes 1959), but the results of the introduction do not appear to have been documented.

Sweden
In about 1870, 300 German roe deer were released in Sweden (Niethammer 1963). The species has recently extended its range northwards in Scandinavia by several hundred kilometres (Danilkin 1986), although this may not have been due in any way to the early introduction.

Switzerland
Roe deer have been successfully introduced into Switzerland (Anon. 1963).

Russian Federation and adjacent independent republics
In Russia and adjacent independent republics roe deer have been translocated in a number of areas (Naumoff 1950; de Vos *et al.* 1956) and a number of re-establishments were being carried out in the 1950s and 1960s (Bannikov 1963). Introductions in the European part of the Russia have failed, but re-acclimatisations and acclimatisations in other areas, and introductions in 1960–64 into new regions have met with some success (Yanushevich 1966; Fadeyev 1969).

The Siberian race of the roe deer was introduced to Baskivia in 1941 and became established locally (Yanushevich 1966). They were introduced successfully in the Serpuklovskii Raione in the Moskovskoi oblast in 1950 and 1954, where their numbers were increasing in 1960 with some supplementary feeding (Zablotskaya 1961). Some were released in Lithuania in 1957 and they became locally established there (Yanushevich 1966), and in the early 1950s in the Lena-Vilyuisk Valley, and on the east bank of the Lena River in Yakutiya (Popov 1963). Here they are spreading slowly and in 10 years have advanced some 10 kilometres, but their numbers are still few.

About 2172 roe deer were released between 1929 and 1972 for acclimatisation purposes in the Russian Federation and adjacent independent republics in 26 regions of Russia and in five republics: Belgorodsk (1960, 20), Vladiminsk (1950 and 1962, 15), Volgogradsk (1968, 33), Ivanovsk (1963, 16), Irkutsk (1963, 20), Kalininsk (1931–1970, 429), Kaliningradsk (1962 and 1970, 17), Kalujsk (1967–70, 35), Kemerovsk (1960, 9), Krasnodarsk (1958 and 1962, 23), Kuibshevsk (1971, 8), Kursk (1967–70, 9), Leningradsk (1963, 7), Mordovsk (1940, 10), Moskovsk (1940–69, 612), Nougorodsk (1965, 12), Penzensk (1956–57, 29), Rostovsk (1970–72, 35), Ryahzansk (1949 and 1966, 28), Saratovsk (1960 and 1970–71, 72), Sverdlovsk (1965–69, 40), Stavropolsk (1956, 15), Tulsk (1938, 12), Tuvinsk (1962, 9), Udmurtsk (1957, 26), Yahroslavsk (1948–54 and 1967–69, 62), three areas in the Ukraine (1929–69, 528), Latvia (1955, 13), Kirgizsk (1967, 4), Litovsk (1956, 18), and Astonsk (1967, 4) (Kirisa 1974).

United Kingdom and Ireland
Roe deer have been re-introduced in many parts of England (Corbet 1966), where they have been escaping from captivity since about 1850 (Mathews 1952). Some were introduced to Thetford Chase in 1880 from the continent and they have now colonised the Chase and are spreading into the surrounding woodland (Willett 1970).

It was widely held that the roe deer disappeared in the wild in England in the latter part of the eighteenth century and were re-introduced later in widely separate parts of the country (Burton and Burton 1969).

However, in the last 100 years they have increased from 100 to 2350 animals in Northumberland and Durham in northern England, and the reason may be that they did not become extinct as formerly believed (Cowen *et al.* 1965).

Siberian roe deer were locally established, but not numerous in Yardley and Hazelborough in Northamptonshire, and at Ampthill in Bedfordshire in 1939 (Taylor 1939). At this time there were also feral herds in Northumberland and Cumberland, the Lake District, the eastern Midlands and in the south of England, particularly Dorset, Hampshire, Surrey and Sussex. They were reported to be numerous in East Anglia as a result of introductions about 1899. Generally in southern England they are probably more numerous in recent times due to reafforestation in the last 25 years, than they ever were in the past (Burton and Burton 1969).

Roe deer were formerly common in north of England, Wales and in Scotland, but by 1809 few were left in England or Wales although some existed until 1892. Some were introduced to Dorset in about 1800. However, attempts during the last century to re-establish the species in the wild succeeded and they subsequently spread until they were resident in at least 13 English counties and spreading (Fitter 1959; Whitehead 1972). The most successful introductions were at Petworth in Sussex (about 1800), Milton Abbas in Dorset (about 1800) and Thetford in Norfolk (about 1884) (Whitehead 1972).

These days roe deer in England are all considered to be derived from introductions. Some of unknown origin were introduced to Milton Abbas, Dorset, in 1800, and subsequently colonised much of south-eastern England; six pairs of deer from German stock were introduced to East Anglia around 1884; roe deer in the Lakes District are thought to be of Austrian origin (Corbet and Harris 1991). Some were released in Epping Forest in 1883, where they thrived until 1901 and then declined and disappeared about 1923 (Fitter 1959). Siberian roe deer were released at Woburn, Bedfordshire in 1887, where they became established, but were eventually exterminated by 1939–45. In 1874 they were released at Carnarvonshire, Wales, where they thrived and colonised Snowdonia and a large part of central Wales.

Although continuously native in Scotland, roe deer have not descended from indigenous stock on the west coast islands of Bute, Islay and Seil. They were unsuccessfully introduced on Raasay and were released on Bute, Islay and Seil where they survive today, and on Mull, Arran and Tura where they failed

to become established (Fitter 1959). In Scotland they were released in Ayrshire at the beginning of the nineteenth century and spread rapidly. Some were also liberated in Dumfriesshire about 1860 and became established there. By 1900 they occupied the greater part of the Scottish Lowlands (Whitehead 1972).

Despite some early historical illusions to the contrary, most agree roe deer were introduced to Ireland in the nineteenth century. There is no early archaeological evidence for their presence and a few introduced in the latter part of the nineteenth century did not prove successful (Millais 1906). Some appear to have been introduced at Sligo in 1870 and some were later introduced in counties Sligo and Mayo (Moffat 1938), but these only survived until shot out early in the nineteenth century (Fitter 1959; Whitehead 1972).

NORTH AMERICA
United States
In 1902–03, W. A. and W. Rockerfellow released 18 German and Siberian roe deer from Germany into Bay Pond Preserve, New York State (Bump 1941). Those of the race *capreolus* disappeared, the Siberian stags *pygargus* were shot, and the remainder of this group also disappeared. However, in 1957 a small herd was found established on an estate near Millbrook, Dutchess County, New York, which were believed to have been introduced about 1900 and have been there ever since (Manville 1957). At this time the herd numbered six animals.

There may have been earlier introductions of roe deer in the United States, as around 1900 there were reports of releases by sports people on certain unnamed ranches (Manville 1957). Some snow deer (*C. c. pygargus*) have been present on Santa Rosa Island, California (von Bloeker 1967), but their current status is not known (Lidicker 1991).

■ DAMAGE
In Europe roe deer become a problem to agriculture where their habitat adjoins these areas. In Denmark they attain their maximum numbers in woods surrounded by arable land and where agricultural crops become a prominent part of their diet (Andersen 1961). Costly counter-measures are necessary to keep such damage by the deer within acceptable limits.

Roe deer also cause damage to young trees in forest areas (Westerskov 1952). Damage occurs by the deer browsing and fraying young trees and can be important in areas where regeneration is in progress or on new plantings (Corbet and Harris 1991), and also in forest nurseries at times.

Increasing numbers of roe deer in Scotland have caused a number of problems for foresters (Peterle 1958; Gibson and MacArthur 1965). Damage to forest plantings is heaviest in areas that are replanted after partial failure, and in plantations of sapling-sized trees at the time the bucks are rubbing off the velvet (Peterle 1958). They often cause more damage to young forestry plantations than to agricultural crops in England, but certainly far less than other deer species (Willett 1970). Here, the main damage in young forests occurs from the fraying and browsing (Fooks 1958).

A study of the roe deer populations in Denmark, England and Scotland suggested that to achieve the control necessary for good forest management it would be necessary to kill a quarter or more of the roe deer each year (Loudon 1978).

In Czechoslovakia where the Siberian roe deer (*C. c. pygargus*) has hybridised with native roe deer, many females of the latter covered by larger bucks died giving birth, because the size of the young made parturition difficult (Dorst 1965).

Family: Giraffidae
Giraffes

GIRAFFE
Giraffa camelopardalis Linnaeus

■ DESCRIPTION
HB 3000–5300 mm; T 760–1100 mm; SH 2000–3700 mm; WT 550–1932 kg.
Coat boldly spotted and blotched (pattern variable) irregularly with chestnut or dark brown on pale buff; neck long; head tapered; lips hairy, mobile; tongue extensile; legs long; back sloping; tail long and tufted; horns small, solid bone, skin covered, 130–220 mm.

■ DISTRIBUTION
Africa. Sudan and Somalia to South Africa and west to Nigeria. Range now reduced and becoming fragmented.

■ HABITS AND BEHAVIOUR
Habits: mainly crepuscular and diurnal; shy, timid; non-territorial. Gregariousness: loose open herds or troops to 70, but mostly 6–15; occasionally in large herds; adult males often solitary or in bachelor herds; density 0.1–3.4/km². Movements: seasonal movements 20–30 km; home range 23–653 km². Habitat:

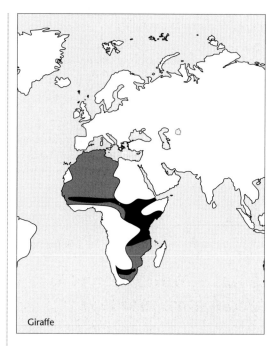

Giraffe

dry tree savanna, semi-desert and open woodlands. Food: leaves, buds of Acacia, thorny scrub and other trees and shrubs; mimosa, grass, fruits including wild apricots, and grain crops. Breeding: all year, peaks in rainy season, mates July–September; gestation 420–468 days; litter size 1, rarely twins; inter-birth interval 16–23 months; weaned at 6–13 months; calves lie out 1 week; remain with female further 2–5 months; sexual maturity females 2.5–3.5 years and males 3.5–4.5 years. Longevity: 15–26 years in wild and 28–36 years 2 months in captivity. Status: range and numbers considerably reduced, but still common some areas

■ HISTORY OF INTRODUCTIONS
Introduced successfully in a number of parks and reserves in South Africa and Zimbabwe, and probably to Rubondo Island, Lake Victoria, in Tanzania.

AFRICA
South Africa
Giraffe have been introduced to some South African game reserves (Haltenorth and Diller 1994). They have been introduced to the Hluhluwe-Umfolozi Game Reserve, South Africa (Macdonald and Frame 1988), and the Transvaal Department of Nature Conservation released 42 between 1952 and 1961 somewhere in South Africa.

Seventy-seven giraffe were re-introduced to Pilanesberg National Park, Bophuthatswana, in 1981–84 and there were 85 there in late 1984 (Anderson 1986). Between 1990 and 1992 some were also released on Phinda Resource Reserve, a privately owned site in South Africa (Hunter 1998).

Tanzania
Some giraffe were released on Rubondo Island, in Lake Victoria, Tanzania, before 1966 in the hope that they would become a tourist attraction (Grzimek 1966).

Zimbabwe
Seven were released in McIlwaine National Park and eight in Matapos National Park, Zimbabwe, by authorities before 1963 and they were surviving in both areas (Riney 1964).

■ DAMAGE
No information.

Family: Antilocapridae
Pronghorn antelope

PRONGHORN ANTELOPE
Pronghorn, American antelope
Antilocapra americana (Ord)

■ DESCRIPTION
HB 1000–1500 mm; T 60–170 mm; SH 810–1040 mm; WT males 45.4–70.6 kg, females 32.2–47.7 kg.
Upper parts of body rich red-brown; hair is coarse and brittle; muzzle, neck and patch on throat black;

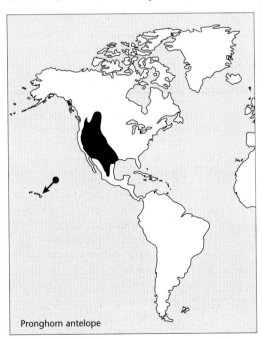

Pronghorn antelope

eyes large; mane hair erectile; lower jaw, sides of head, flanks, belly and two throat bars, white; horns (380 mm or more) branched and outer sheath shed annually (October after rut). Female usually smaller and weighs less than male; horned; muzzle only is black; four mammae.

■ DISTRIBUTION
North America. From south-western Canada (southern Alberta and southern Saskatchewan) south through the western United States to northern Mexico. Formerly more widespread, at least in Canada.

■ HABITS AND BEHAVIOUR
Habits: nocturnal and diurnal; agile; dominant males territorial at rut. **Habitat:** scrub and desert grassland, plains, steppes, foothills, sagebrush grasslands. **Gregariousness:** small bands in summer, large troops in winter up to 100; form harems (male and 7–15 females) in autumn (mating season). **Movements:** sedentary, but seasonally migratory between summer and winter range. **Foods:** forbs, grass, clover, weeds and browse from shrubs and trees; cacti, sagebrush. **Breeding:** ruts August–October, calves in March–June; male polygamous; female seasonally polyoestrous; gestation 230–250 days; young 1–3, generally 2; young precocious; kids hidden for first three weeks; young sexually mature at 16 months, males at 1.5 years, but rarely mate until 3 years. **Longevity:** 4.5 years to a maximum of 14 years in wild; 12 years known in captivity. **Status:** rare, numbers and range reduced; but more recently increased under conservation.

■ HISTORY OF INTRODUCTIONS
Pronghorn antelope have been introduced successfully to the United States, Mexico and the Hawaiian Islands.

NORTH AMERICA
The continental North American population of pronghorn antelope had been reduced from a possible 30–40 million to about 30 000 head by the 1920s (Mitchell 1983). This fact probably prompted the translocations and re-introductions to several states of the mainland United States, Mexico and the Hawaiian Islands, where they are now established successfully.

Mexico
Pronghorn translocations have been made in Mexico (Barker 1948), but the results do not appear to be well documented. Some were re-introduced in remote regions of San Luis Potosi, central Mexico and have spread as far as Durango, Golfo de California (Lever 1985).

United States

Beginning in 1910 there were a number of translocations of pronghorns for conservation reasons in the United States. Re-introductions were attempted between 1910 and 1937 including 12 animals that were released in 1938 and 22 in 1940, but the species did not become well established; there were 71 head present in the Wichita Mountains Wildlife Refuge, Oklahoma, in 1943 (Buechner 1950).

Some time before 1941 two antelope were liberated in Nehasane Park, New York, by W. S. Webb, but they disappeared (Bump 1941).

By 1941 pronghorns had become restricted in range to central and south-eastern Montana by the destruction of their habitat. Trapping and translocations began in this state at this time and 3554 were released in 33 areas where none had existed previously, and in nine other areas to supplement existing herds (Compton *et al.* 1971). Pronghorns are now widespread in the eastern two-thirds of Montana as a result of these translocations (Compton *et al.* 1947).

Early in the twentieth century pronghorns became extinct at Big Pine and Bodie, California. The population that now exists there is the result of an introduction in 1949 by the Department of Fish and Game (Jameson and Peeters 1988).

Other states to have translocation programs include New Mexico before 1937, Colorado in the mid-1940s, South Dakota in 1950 and Washington before 1956 (Russel 1937; Elliott 1948; Anon. 1951; de Vos *et al.* 1956). They appear to have been successful in New Mexico and Washington.

PACIFIC OCEAN ISLANDS
Hawaiian Islands

Introduced from Montana in 1959, a small herd of pronghorns is now established on the island of Lanai (Walker 1967). Of 56 trapped in Montana only 44 survived the trip to the islands and a further two died en route to Lanai, thus some 38 were released. However, many were lost in the ocean soon after release and the survivors were reduced to about 18.

The survivors adapted to the grassland plateau and the herd in 1965–68 numbered some 150 head (Burris 1965; Kramer 1971); by 1983 there were 130 on Lanai. Their numbers on Lanai are probably only limited by the limited habitat available.

◾ DAMAGE

None known. Pronghorn do not appear to cause any agricultural damage in Hawaii or to compete with native species of wildlife, but their numbers and range are small.

Family: Bovidae
Antelope, cattle, sheep and goats

ARTIODACTYLA (UNGULATES)

In southern Africa from 1954 to 1964 thousands of introductions or re-introductions of between 20 to 25 species were made, to the effect that in 1964, scattered throughout were hundreds of nucleus populations of one or more species of large mammal in some stage of becoming adjusted to its environment (Riney 1964). Most of these introductions involved the re-establishment of species in areas where they had disappeared in the preceding 50–100 years. In the Transvaal alone the Department of Nature Conservation had distributed 2398 animals of nine different species. Additionally, many ranches and farms that formerly held wild animals had been restocked with one or more wild species, an activity that has increased since 1950. Until 1963, the Southern Rhodesian (now Zimbabwean) authorities had released 164 animals of 14 mammal species in McIlwaine National Park and 108 animals of 11 mammal species in Matapos National Park.

BOVIDAE

A small domestic form of the gaur or Indian bison, *Bos gaurus*, known as the gayal, *Bos frontalis*, has been introduced to the Jaldapara Wildlife Sanctuary from Manipur by the Forests Department. They are semi-tame and frequent the area around the Baradabri tourist lodge (Spillett 1966).

Gayal are also found in a feral state in Thailand, eastern India and Burma (Grzimek 1975). They sometimes occur in a feral state with some 50 000 in Arunachal, Pradesh, India (Burton and Pearson 1987).

Duikers (*Cephalophus* sp.) were released in McIlwaine National Park (15) and Matapos National Park (six) in Zimbabwe (Southern Rhodesia) by authorities but any success is not known (Riney 1964).

Two klipspringers (*Oreotragus oreotragus*) were introduced to McIlwaine National Park, Zimbabwe (Southern Rhodesia), by authorities before 1963, but only one was left in 1963 (Riney 1964).

There have been at least three attempts to translocate Dall's sheep (*O. dalli*) from the Kenai Peninsula to Kodiak Island, Alaska. These introductions were made in 1964, 1965 and 1967, but had only limited success and few animals persisted to the mid-1970s (Franzmann 1988).

Addax (*Addax nasomaculatus*) and Dorcas gazelle (*Gazella dorcas*) were being prepared for re-introduction into Tunisia (Bertram 1988).

BUSHBUCK
Tragelaphus scriptus (**Pallas**)

■ DESCRIPTION
HB 1050–1500 mm; SH 650–1100 mm; T 200–350 mm; WT 24–80 kg.
Variable species; back rounded; hind quarters powerful; back and sides light tawny to reddish in female to dark brown and almost black in male; under parts slightly darker; white patches on throat and lower neck; line down middle of back and stripes or rows of dots vertically on sides; some have pronounced manes full length of back; tail bushy; ears large; horn keels twisted into spirals, length 350–635 mm. Female smaller than male, and generally lacks horns.

■ DISTRIBUTION
Africa; south of the Sahara from Mauritania to at least George, Cape Province, in South Africa.

■ HABITS AND BEHAVIOUR
Habits: found near water; crepuscular or nocturnal; not always territorial. **Gregariousness:** usually solitary; females and young or courting pairs; density 4–26/km². **Movements:** sedentary, but may undertake seasonal movements in search of food; territories 15–35 ha; home range 0.2–1.0 km². **Habitat:** edges of swamps and rivers and other areas near water in forest and scrub; savanna woodland, plantations, secondary bush, forest clearings, abandoned cultivation,

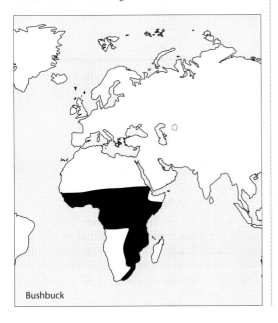

Bushbuck

gardens. **Food:** browses leaves, twigs of trees and shrubs; grass, buds, shoots, fruits. **Breeding:** throughout year, but mainly dry season; gestation 6–7 months; inter-birth interval 8 months; young 1; concealed for first few weeks of life; weaned at 6 months; sexual maturity 11–12 months. **Longevity:** 12 years or more. **Status:** range reduced by hunting and habitat destruction; locally common to common.

■ HISTORY OF INTRODUCTION
AFRICA
South Africa
Bushbuck were re-introduced to Bontebok National Park after 1961 from Humansdorp (Penzhorn 1971), but they failed to become established because the habitat was unsuitable (Novellie and Knight 1994).

■ DAMAGE
Bushbuck will eat vegetables and garden flowers and the bark of citrus trees, but the extent of any damage is not known.

GREATER KUDU
Kudu
Tragelaphus strepsiceros (**Pallas**)

■ DESCRIPTION
HB 1800–2450 mm; T 300–550 mm; SH 1000–1500 mm; WT males 190–315 kg, females 120–315 kg.
Coat ranges reddish to pale slaty blue-grey with six to 10 vertical white markings on the sides of body; body narrow; face with white markings on nose, cheeks, and around eyes; ears large; legs long, dark garters on upper parts; tail long and tufted, black tipped, white under; throat with fringe of long hair; horns greyish brown with pale tips, about 1016–1222 mm, wide open spirals of two and a half turns. Female similar in colour to male, occasionally horned, lacks neck mane.

■ DISTRIBUTION
Africa. East and South Africa south of Zambezi River, west to Angola and east and north to Ethiopia (Chad to Ethiopia to South Africa).

■ HABITS AND BEHAVIOUR
Habits: nocturnal and diurnal; old males territorial in breeding season. **Gregariousness:** after mating males live in unstable bachelor groups 2–10; adult males solitary or; female groups 5–12 (adults and young); females solitary at birth; family groups 20–40; density 1.9–3.2/km². **Movements:** extensive seasonal movements; home range 3.6–11.2 km². **Habitat:** savanna, mixed scrub woodland, acacia, thickets, rocky

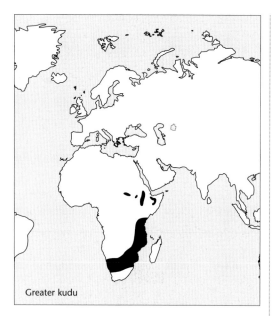

Greater kudu

country with heavy brush in lowland hills and mountains, cultivation where thick cover nearby. **Foods:** feeds mainly by browsing (acacia and other plants), flowers, pods, twigs, fallen fruits, shoots, tubers, and leaves, but also grazes grass at times. **Breeding:** annual; young usually born January–May; gestation 210–270 days; 1 young; hide calf for 2 weeks; young weaned at 6 months; females sexually mature at 17–24 months, males 5 years; males disperse second to third year. **Longevity:** 20 years 9 months (captive), probably 7–8 years in wild? **Status:** still common South Africa, but elsewhere habitat loss leaving isolated populations.

■ **HISTORY OF INTRODUCTIONS**
AFRICA
South Africa
Greater kudu were re-introduced successfully to Addo Elephant National Park from Grahamstown and Jansenville conservation parks (Penzhorn 1971). They have also been re-introduced to Loskop Dam Reserve in Transvaal (Haltenorth and Diller 1994). Between 1981 and 1984, 160 were released in Pilanesberg National Park, Bophuthatswana; by late 1984 they had increased to about 800 and the introduction was considered successful and annual removals were necessary to prevent overpopulation (Anderson 1986).

Zimbabwe
Southern Rhodesian authorities released 25 greater kudu in McIlwaine National Park before 1963 and they were still surviving (seven) there in 1963 (Riney 1964).

NORTH AMERICA
United States
New Mexican authorities imported greater kudu in 1962 or in about 1965–67 and kept them in the Albuquerque Zoo. The progeny were to be released later if they could be successfully bred (Gordon 1967; Haltenorth and Diller 1994).

An unsuccessful attempt was made to introduce two kudu (1 female and 1 male) in August 1967. The male died immediately and the female later in 1967 (Wood *et al.* 1970).

■ **DAMAGE**
Greater kudu allegedly damage crops (Walker 1992), making night raids on cultivated crops, clearing two-metre fences easily to invade fields (Whitfield 1985). Where they thrive in settled areas, greater kudu will also raid gardens (Estes 1993).

NYALA
Tragelaphus angasi Gray

■ **DESCRIPTION**
HB 1350–2000 mm; T 400–550 mm; SH 800–1210; WT 85–140 kg.
Large slender animal; ears large; coat shaggy; adult males grey; facial chevron white; fringe of long pendant hairs that form line around lower neck, lower shoulders, sides of belly, lower thighs and back of thighs; tail bushy; horns 835 mm. Females and males under one year are red-brown.

Nyala

■ DISTRIBUTION
Africa. Malawi, Mozambique, Zimbabwe to Natal, South Africa.

■ HABITS AND BEHAVIOUR
Habits: mainly crepuscular; overlapping home ranges 0.65–0.83 km^2; non-territorial. **Gregariousness:** herds or troops to 30 (1 bull, cows and young); young male parties; female groups 6; older males solitary; in dry season several troops together to 50. **Movements:** sedentary. **Habitat:** lowland forest, savanna thickets, plains and mountains near water and dense cover. **Foods:** mainly browse; leaves, shoots, bark, buds, fruits, tender new grass growth. **Breeding:** throughout year; female polyoestrous; oestrous cycle 10–34 days; oestrus 2–3 days; post-partum oestrus 2–7 days following birth; inter-birth interval 231–297 days; gestation about 220–255 days; 1 young; nurse for 7 months; sexual maturity females 20–36 months (some authorities say 12 months), male puberty 12 months. **Longevity:** to 16 years. **Status:** range and numbers reduced, but little information.

■ HISTORY OF INTRODUCTIONS
AFRICA
South Africa
Nyala are thought to be a native of South Africa, but had become locally extinct there. They were successfully introduced and re-established in Kruger National Park (Macdonald and Frame 1988), and have been introduced in Loskop Dam Nature Reserve in Transvaal and on farms in Natal, and in the Adelaide regions of Cape Province (Mozambique localities before Civil war) (Haltenorth and Diller 1994). Nyala became locally extinct in the Hluhluwe-Umfolozi Game Reserve but have been re-introduced there (Macdonald and Frame 1988).

■ DAMAGE
No information.

ELAND
Common eland
Taurotragus oryx (Pallas)
=*Tragelaphus oryx*

■ DESCRIPTION
HB 1800–3450 mm; T 500–900 mm; SH 1000–1800; WT 300–1000 kg.
Greyish fawn to tan, and becoming blue-grey in males with age; whitish or cream vertical stripes (10–16) on upper parts; head short, broad; neck thick; dewlap between throat and front of breast; hump on withers;

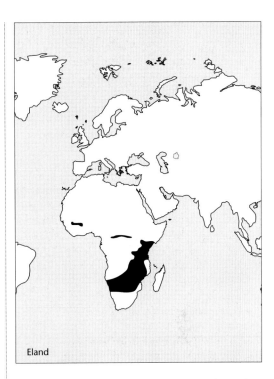

Eland

short mane on nape and longer hairs on throat; legs with white markings and dark garters; rounded hooves; tail with terminal tuft; spiral horns in both sexes, 430–700 mm.

■ DISTRIBUTION
Africa. South of the Sahara, southern Sudan and Ethiopia to South Africa, and west to Namibia and Angola, except desert and deep forest.

■ HABITS AND BEHAVIOUR
Habits: lies up in shelters in heat of day; largely crepuscular, but also nocturnal and diurnal at times. **Gregariousness:** unisex herds 3–5; mixed herds 10–12; herds up to 25–30 (cows, sub-adults, and 1 bull); larger herds (temporary aggregations) to 400–500 females; old bulls solitary; bachelor troops; density 1/9 ha to 1.2/km^2. **Movements:** during drought wanders widely; overlapping home ranges 11.7 km^2–422 km^2. **Habitat:** montane forest, semi-desert, savanna, open plains, woodland, grassland, undulating country with bush and scattered trees, and mountains. **Foods:** grazes and browses; leaves, shoots, tubers, bulbs, melons, onions, thick-leaved plants, fruits, seed, seed pods and herbs. **Breeding:** all year; calves born July–August; gestation 254–277 days; 1 calf; calf hides for 2 weeks before joining herd; lactation 4–6 months; sexual maturity about 2.5–3 years females, 4–5 years males. **Longevity:** 15–25 years. **Status:** drastically reduced in numbers and range.

■ HISTORY OF INTRODUCTIONS

AFRICA

South Africa

Widely re-introduced on farms in South Africa (East 1993), eland were successfully re-introduced to Addo Elephant National Park in the early 1960s and also to Bontebok National Park after 1961, to Golden Gate Highlands National Park after 1963 and Mountain Zebra National Park (Penzhorn 1971; Novellie and Knight 1994). All of these animals came from Kalahari Gemsbok National Park. Many attempts were made to re-establish eland in the Kruger National Park in the late 1960s and early 1970s (Macdonald and Frame 1988) and to Bontebok National Park (Novellie and Knight 1994), but they failed to become established (Macdonald and Frame 1988). Repeated attempts were made to re-establish them in Hluhluwe-Umfolozi Game Reserve in the 1960s and 1970s, but all failed mainly because of predation from a newly established lion population (Bourquin *et al.* 1071). Some 366 eland were released in Pilanesburg National Park between 1981 and 1984 and by late 1984 there were 450 (Anderson 1986).

Namibia

Eighty-five eland were moved in 1972 to Waterberg Plateau Park in Namibia (Hofmeyer 1975).

Zimbabwe

Southern Rhodesian authorities released 13 eland in McIlwaine National Park and 18 in Matopos National Park before 1963, and where they were surviving in 1963 (Riney 1964). Eland have also been widely re-introduced on farms in Zimbabwe (East 1993).

NORTH AMERICA

United States

Many eland were introduced on ranches in Texas in 1951 (Schreiner 1968), but the species does not appear to have been released there or to have become feral.

AUSTRALASIA

Western Australia

Introduced to Western Australia many years ago, eland did not survive the many native poison plants (*Gastrolobium* sp.) (Colebatch 1929). At the insistence of the Governor Sir Gerald Strickland, a pair was presented to the Acclimatisation Committee by the Duke of Bedford, and these were kept in the zoological gardens and the young were to be used for acclimatisation purposes (le Souef 1912; Long 1988).

■ DAMAGE

Eland appear to cause no damage to agriculture in Africa.

NILGAI

Indian nilgai or nilghae, blue bull, nylghaie
Boselaphus tragocamelus (Pallas)

■ DESCRIPTION

HB 1800–2100 mm; T 456–535 mm; SH 1200–1500 mm; WT c. 200–300 kg.

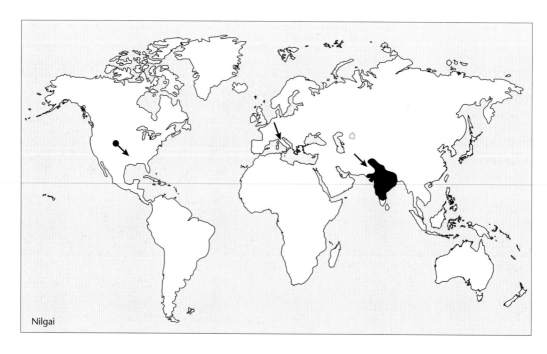

Nilgai

Short wiry coat, reddish brown to slate grey; throat with soft, black tuft of hair; mane on neck; ears with white stripes inside; head long and pointed; front legs slightly longer than back legs; under parts white; tail, lower surface white, tip black; fetlock with white rings; horns short, have curved spikes 200–230 mm long and triangular base. Female tawny and lacks horns.

■ DISTRIBUTION
Asia. Peninsular India from the Himalayas to Mysore.

■ HABITS AND BEHAVIOUR
Habits: diurnal and crepuscular; grazes and browses; males territorial in breeding season(?). **Gregariousness:** male solitary (sometimes), females in herds with calves. parties 15–20; non-territorial male groups to 18; old males frequently solitary; density 0.07/km^2–1/15–20 ha. **Movements:** home range 4.3 km^2. **Habitat:** open forest, low jungle, open grassy plains, and occasionally scrub. **Foods:** browse shrubs and graze grass; fruits, sugar cane. **Breeding:** breed throughout year, mainly March–April (in United States peak August–September); gestation *c.* 243–247 days; 1–3 calves; female mates again immediately after calving; females reach sexual maturity at third year, males about 5 years. **Longevity:** *c.* 10 years (wild) to 21 years 8 months (captive). **Status:** widespread, stable and increasing.

■ HISTORY OF INTRODUCTIONS
Successfully introduced into the United States (Texas) and to ranches in South Africa (Orange Free State). Re-introduced into at least one area in Pakistan. Unsuccessfully introduced into Europe (Italy).

Africa
South Africa
Introduced for sport on several large ranches in the Orange Free State (Lever 1985), nilgai are not known to be feral anywhere.

Asia
Pakistan
Nilgai were re-introduced to Lal Suhanra National Park, Pakistan (East 1993).

Europe
Italy
Some nilgai were introduced near Rome, but they disappeared during World War 2 (Niethammer 1963).

North America
United States
Nilgai were released between 1924 and 1949 on the King Ranch Inc. in south Texas and from here were translocated to nine counties in Texas and one in Mexico (Sheffield *et al.* 1971). Two females and one male were released in 1924, but records of subsequent releases are vague. Until 1941 small groups obtained from the zoological gardens, primarily San Diego, were released on the Norias division of the King Ranch. In 1949 a final release of eight females and four males was made on Norias. Since 1960 acquisitions from the King Ranch have been involved in 16 small translocations of from one to eleven animals each.

In the 1950s they were well established and numbered about 500 head (Presnall 1958), but have now expanded naturally throughout a 2600 km^2 area of Kennedy and Willacy counties and in 1970 it was estimated that there were 2149 animals (Sheffield *et al.* 1971). However, estimates in 1970 varied from two counties with more than 50 animals and in three counties with under 50 animals (Ramsey 1970) all behind fences. In the mid to late 1970s their range was still expanding and the population was then estimated to be 2400–2786 in 15 counties (Ables and Ramsey 1974; Lever 1985). Now an estimated 8000–9000 in 2600 km^2 range in Kennedy and Willacy counties (Sheffield *et al.* 1983), but the population appears to be declining (Lever 1985).

■ DAMAGE
Nilgai cause considerable damage to sugar cane crops in their native range (Walker 1968; Whitfield 1985). They cause extensive damage to agricultural crops such as gram, wheat seedlings and moong, especially in the Haryana area of India (Chauhan and Singh 1990).

WATER BUFFALO
Indian buffalo, carabao, swamp buffalo
***Bubalus bubalis* (Linnaeus)**

■ DESCRIPTION
HB 2400–3000 mm; T 600–1000 mm; SH 1500–1900 mm; WT 250–1200 kg.
Coat generally ash-grey to black and short haired; tuft on forehead; long narrow face; ears small; horns (to 1925 mm) wide spreading, triangular shaped (flat fronted), wrinkled and curved backwards. In captivity can be grey, black or white in colour, otherwise similar to domestic animal.

■ DISTRIBUTION
Asia. Original range probably some parts of southern Asia (and may have been confined to India and Sri Lanka). Reported to be wild in Nepal, Bengal and Assam, but they are possibly feral animals. Now

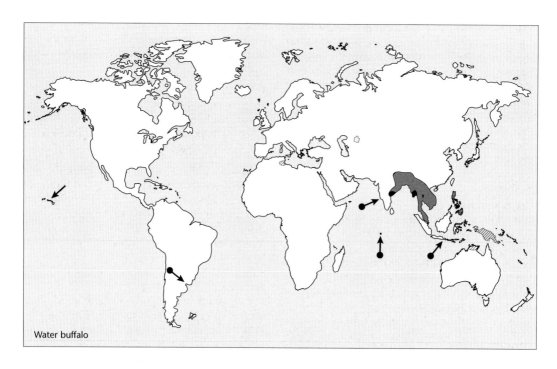

Water buffalo

widely distributed as a domestic species in southern Europe and from Egypt to the Philippines, Indochina and Borneo. Wild water buffalo are probably now confined to the Brahmaputra Valley in Assam, the Nepal Terai near the Saptakosi River, the state of Orissa in India, and in the mountainous areas of Amphur Ban Rai, Uthai Thani Province, where there are a few small bands left and there may still be considerable numbers in northern Cambodia (Lekagul and McNeely 1988).

■ DOMESTICATION
Probably domesticated as early as 2500–3000 BC.

Domestication began about 5000 years ago in the Indus Valley. They were in use in China 4000 years ago and spread through the Middle East around AD 600.

Probably first domesticated as a work animal in Mesopotamia during Akkadian dynasty about 2500–2100 BC and in the Harappan culture of Mohenjo-Dano about 2500 BC. Some identified in Chang remains around 1400 BC in China. World population about 150 million, of which China has 30 million, India has about 55 million and Pakistan has about 67 million. Numbers declining with advance of mechanisation in under-developed areas.

■ HABITS AND BEHAVIOUR
Habits: nocturnal and diurnal; dominance hierarchy. **Gregariousness:** segregated herds in dry season; groups and herds ('clans') of various sizes (10–20 common,

30–500 known); bulls solitary; density 13–34 per km^2. **Movements:** move locally in relation to water and food only; home range 6 km^2. **Habitat:** swampy areas, dense tall grass, reed beds, moist areas and mud holes, water courses, grass jungle. **Foods:** water plants, grass and other vegetation near rivers and lakes, grass, leaves, bark of shrubs and trees. **Breeding:** throughout the year, but mainly wet season or early in dry; gestation 300–334 days; young 1 per year or 2 years, rarely twins; usually mate October–April in wet season, calve May–September; young cows breed at 2–3 years age; born with eyes open, fully haired; weaned at 6 months; sexual maturity 2–4 years. **Longevity:** 18(?)–30 years (captive). **Status:** probably eliminated as a wild species, but numerous under domestication and as a feral animal.

■ HISTORY OF INTRODUCTIONS
Successfully introduced as a domestic animal to southern Europe, north Africa, southern Asia and many Pacific Islands. Feral populations in Australia, New Guinea, Tunisia, north-eastern Argentina and other places.

■ EUROPE
Water buffalo are now present in many parts of Europe including Greece, Bulgaria, Hungary, Romania, Yugoslavia, Italy and Spain as a domestic animal. They were in Italy or Sicily in AD 723, but were probably introduced from Hungary in the sixth century AD, but there are now few there.

Napoleon made an unsuccessful attempt to introduce them to Western Europe on his return from Egypt, when he released some at Landes, near the Pyrenees and Bay of Biscay (Garland 1922; Zeuner 1963). Other early experiments at introduction failed in Spain and Germany. They were introduced to Greece and Hungary in the fifth century AD. There were plenty in Bulgaria and Macedonia in AD 1200 and from here they spread to Hungary (Zeuner 1963).

AFRICA
Egypt and Congo
Water buffalo were probably introduced to Egypt as a domestic animal from Iraq or Syria in the ninth century AD. They were used by Egyptians in the fifth century and reached there in the Middle Ages, and were present in the Jordan Valley in AD 723. They were introduced to the lower Congo from Italy as a domestic animal (Zeuner 1963).

Eritrea
A small number of water buffalo were taken to Eritrea from Egypt in 1933.

Tunisia
Introduced and feral in Tunisia (Carter *et al.* 1945; Lever 1985; Walker 1992), water buffalo have probably been established since Roman times or before, as there is a free-ranging herd at Lake Ischkeul, near Bizerta, northern Tunisia (Haltenorth and Diller 1980, 1994). In 1957 all but three were shot, but since then the herd has built up again and they are kept in a special reserve at Ischkeul (Haltenorth and Diller 1994).

ASIA
Water buffalo are now present in Turkey, Iraq, Syria, trans-Caucasia, Azerbaijan, Malaysia, Indonesia, Indochina, southern China, India and Pakistan as a domestic animal; also the Philippines and Japan, and spread through Indonesia at an early date.

In the nineteenth century colonies were established in Brazil, Bulgaria, Indonesia, Malaysia and the Philippines. Probably introduced to Burma, Cambodia, Laos, Thailand and Vietnam in second millenium BC, if not earlier (Epstein 1969 in Cockrill 1976).

In the last 20 years some types have been introduced from Pakistan to improve breeds in China.

India
Large numbers of semi-wild water buffalo were present in the Kaziranga Wildlife Sanctuary, Assam, in 1966 (Spillett 1966).

Java
Water buffalo have been introduced to Baluran National Park in Java (Anon. 1971; Macdonald and Frame 1988).

Sri Lanka
Those water buffalo now found in this country were introduced by humans, though they may have once occurred there naturally and certainly occur in the wild form (Lekagul and McNeely 1988).

Vietnam
Domestic water buffalo have been on Con Son Island since the eighteenth century. At the present time some wild animals exist there and these are thought to be descendants of some that were freed during the Japanese occupation (Van Peenen *et al.* 1970).

AUSTRALASIA
Australia
Buffalo were first introduced to Australia from Timor in 1825 (Ford and Tulloch 1982) or 1826 (Campbell 1834; Tulloch 1969, 1986). In 1826, 16 were landed at Fort Dundas, a military settlement on Melville Island, which was later abandoned in about 1829 (Tulloch 1969). Some of these animals or their progeny were subsequently taken to the mainland, at Port Essington, Coburg Peninsula, in 1828 or 1833 after the settlement closed (Ford and Tulloch 1982; Wilson *et al.* 1992). When they first became feral or were released is not known, but wild buffalo were encountered by Leichhardt's exploration party in 1844 (Wilson *et al.* 1992; Strahan 1995).

The original introductions were reinforced by further introductions to the Victoria Settlement in the 1840s from Timor, Kisar, and other Indonesian islands (Letts 1964; Tulloch 1969; Ford and Tulloch 1982). In 1886 a few milking types from India were taken to Darwin. These animals, two cows and one bull, were to be used to establish a buffalo butter industry that was unsuccessful. In 1838, 18 were obtained from Kisar (Indonesia) and in 1843, 49 were imported and taken to Port Essington. These latter animals were freed when the settlement was abandoned in 1849. Introductions may have continued up until 1866 (Letts *et al.* 1979).

By the mid-1870s and 1880s at least 22 000 feral buffalo ranged over the Coburg Peninsula alone (Considine 1985). Shooting for hides commenced in 1886 and between this time and 1911 over 100 000 buffalo hides were exported. It was estimated that there were 60 000 head on the mainland and 6000 on Melville Island in the 1880s (McKnight 1971). Shooting continued after World War 1 and apart from the war years an average of 7000 bull hides were exported annually from 1911 to 1956 (Ford and Tulloch 1982). Up until 1940, 280 000 hides were taken and a further 100 000 between 1946 and 1965 (Tulloch 1969). The record season for hides occurred in 1937–38 when 16 549 were exported.

In the mid-1950s the market for hides collapsed due to inexpert curing and competition from Asian countries. Pilot projects in domestication and meat production then began in 1958–59. Within 10 years six small abattoirs were established and the value of production was in the order of A$1 million per annum (Ford and Tulloch 1982). During nine decades of exploitation for hides and meat, a total of 700 000 buffalo were killed (Considine 1985).

Since the exploitation for skins began in the 1880s several abattoirs (between 1959 and 1965 there were eight abattoirs) have opened to supply game meat to Europe and Taiwan and some herds have been re-domesticated as interest in buffalo farming grows (Considine 1985; Tulloch 1986; Strahan 1995). In 1975, and again in 1984, the government of the Northern Territory released land for the sole purpose of farming domesticated buffalo (Strahan 1995).

In 1985–86 it was estimated there were 350 000 head in the Northern Territory (Bayliss and Yoemans 1989), but since then total numbers have been affected by control programs to eliminate bovine tuberculosis from Australia (Stoneham and Johnson 1987). There are now 250 000 buffalo spread over 100 000 km^2 margin of coastal flood plain and adjacent woodland between the Daly River west of Darwin and Arnhemland to the east (Considine 1985). Lone bulls have reached Townsville, Queensland in the east, Tenant Creek in the south and Broome, Western Australia in the west.

The buffalo now occupies a considerable part of the Van Dieman Gulf drainage basin, but is mainly confined to the sub-coastal plains, the Marrakai and Koolpinyah Land Systems, but individuals are distributed more widely. The largest numbers now occur on the Adelaide River East Plains, the plains of the Mary River, Whim Creek plain, Camor plain, and on the plains of the South Alligator and East Alligator rivers, Cooper Creek and the Murgenella plain, particularly in association with permanent waters (Ford and Tulloch 1982). Small herds and lone bulls sometimes wander more widely and have been recorded in Queensland (Mitchell *et al.* 1982) and Western Australia (Long 1972). There are probably now between 100 000 and 200 000 buffalo in the Northern Territory (Tulloch 1975).

In 1996 or 1997 a Mr Norwood released more than 20 domesticated buffalo into vacant crown land north-east of Esperance, Western Australia, when market prices made buffalo farming unprofitable. The buffalo survived in country comprising mallee-heath interspersed with brackish lake systems. Following instructions from the state agriculture agency the owner hunted down and shot all of the buffalo, removing the last animal in 1999.

Papua New Guinea

Water buffalo were introduced as a domestic animal from the Philippines between 1900 and 1903 during the German administration (Carter *et al.* 1945; Lever 1985).

They were introduced to Papua New Guinea before World War 1 at a time when cattle were suffering badly from tick fever. Buffalo are resistant to the tick and could be used as draught animals in coconut plantations. They were first recorded in literature there in 1891. The few that are left now are feral animals (Ryan 1972; Walker 1992). After their introduction by the German administration in 1900–03, there were further imports in 1906 from Indonesia and in 1913 again from Philippines (Lever 1985). Some buffalo are still feral on the Gazelle Peninsula in New Britain and in New Ireland.

INDIAN OCEAN ISLANDS

Andaman Islands

Water buffalo were introduced to the Andamans some time before 1956. They became well established and were slowly extending their range due to the absence of predators and the presence of a considerable food supply (de Vos *et al.* 1956; Roots 1976).

Guam

Buffalo were introduced by Spanish missionaries some time in the 1600s and were most likely introduced using domestic stock from the Philippines as a beast of burden. A large free-ranging herd exists on the Naval magazine in central Guam, where they are protected. Numbers have declined since 1982, probably due to illegal hunting. Some localised habitat damage has been recorded (Conry 1988).

INDONESIA

Java

Some 200 feral water buffalo are established in forest and savannah around Baluran in eastern Java (Lever 1985). They are present as a feral population in Baluran National Park in north-east Java (McDonald and Frame 1988).

PACIFIC OCEAN ISLANDS
Hawaiian Islands
The first pair of water buffalo were brought to the Hawaiian Islands in the early 1800s (Oberline 1940) to assist the Chinese with rice growing. They were plentiful in the islands by 1892 and had reached a peak population about 1900, but thereafter declined in numbers.

After the failure of taro growing in 1921 on Molokai and the decline in rice growing in the 1930s many were turned loose and the species began to increase their numbers in the wild (Tinker 1941; Kramer 1971). However, the last was shot on Molokai in 1936 and those in the Waipio Valley, Hawaii, were removed a few years later. There are now no feral animals in the Hawaiian Islands (Kramer 1971).

SOUTH AMERICA
Water buffalo are said to have been introduced to Guayana, Cayenne and Brazil as domestic animals from Italy.

Argentina
Water buffalo have been established in Corrientes Province for sport since 1900 (Petrides 1975).

Brazil
Water buffalo were introduced on Marejo Island, at the mouth of the Amazon River, where some have become feral (de Vos *et al.* 1956). More recently they have been advertised as available for hunting (Petrides 1975) and are possibly still established there.

Peru
Water buffalo were introduced to the Amazonas area in northern Peru, where they are presumably still confined (Petrides 1975).

Venezuela
Water buffalo were imported on several occasions in the last 35–40 years as domestic stock from Trinidad (Petrides 1975). Between 1935 and 1936 some escaped in the Rio Limon Valley, but are believed to have been exterminated by hunters.

■ DAMAGE
Buffalo damage flood plain environments by trampling and grazing flood-plain plants (Considine 1985). They cause extensive damage to freshwater swamps and farm trails between tidal rivers and floodplains, eat out native grasses (Calder 1981; Considine 1985) and change the structure of monsoon forest (Braithwaite *et al.* 1984). Buffalo also trample nesting grounds of the rare pig-nosed turtle (*Carettochelys insculpta*) on the sandy shores of billabongs and rivers (Georges and Kennett 1990).

In Australia water buffalo have caused environmental changes and acted as reservoirs of bovine tuberculosis (Letts *et al.* 1979). A national campaign to eradicate tuberculosis and brucellosis was established in 1970 and experiments to reduce herd numbers in some areas have been carried out (Ridpath and Waithman 1988).

Assisted by Balinese cattle, the buffalo was probably responsible for the introduction of buffalo fly and cattle tick into Australia (Ford and Tulloch 1982).

Buffalo in Australia have caused the decline of pastures by overgrazing, trampling and soil erosion and salinity problems, and are not compatible with areas set aside for conservation of indigenous fauna and flora (Stocker 1971).

Overgrazing and trampling have damaged many areas of the Northern Territory and seriously affected potential productivity for timber, grazing and wildlife (Stocker 1972). The buffalos dependence on water confines them to areas close to remaining water during the arid phase of the dry season and this results in intense grazing of these areas (Ridpath and Williams 1982).

The sub-coastal plains and wooded lowlands used by buffalo appear to be undergoing rapid change. The floristic composition around water sources has probably changed with the expansion of the buffalo population. Impact reports to date have indicated substantial changes to floristics, vegetation structure and micro-topography of the floodplain and adjacent communities. However, it is suggested that the elimination of the buffalo might benefit some animal species, but be to the detriment of others (Board of Inquiry into Feral Animals in the NT 1979; Ridpath and Williams 1982; Friend and Taylor 1984).

AFRICAN BUFFALO
Cape buffalo
Syncerus caffer (Sparrman)

■ DESCRIPTION
HB 1700–3400 mm; T 500–1100 mm; SH 1000–1700 mm; WT 250–900 kg.
Coat reddish brown to brownish black; skin sometimes naked in patches; bulky ox-like form; head large; horns large, spread outwards, downwards then upward, bases may meet across forehead; ears drooping and fringed with soft hairs. Both sexes horned males 1170–1500 mm; females usually with a tinge of red.

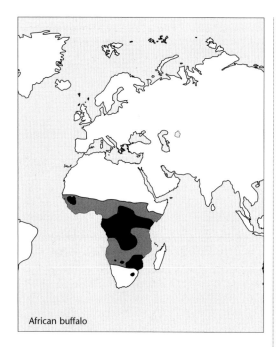

African buffalo

■ DISTRIBUTION

Africa. Formerly all Africa south of the Sahara Desert, but now reduced and fragmented, particularly in western parts, and there are now large areas where they do not occur. Exterminated in South Africa and Fernando Poo.

■ HABITS AND BEHAVIOUR

Habits: mainly nocturnal, occasionally diurnal; non-territorial. **Gregariousness:** large mixed herds of 10–100s or even 1000s, containing smaller groups of bachelors to 50, females and young; occasionally solitary old males; density 0.17–3.77/km^2–18/km^2. **Movements:** daily movements between food and water to 27 km; home range 5–1075 km^2. **Habitat:** forest, savannah, clearings, swamps, floodplains, secondary growth; prefers habitats near water with dense cover nearby. **Foods:** browses on leaves and twigs of shrubs, grazes mainly grass. **Breeding:** breeds most of year, but largely seasonal; gestation 340–346 days; oestrous cycle 23 days; oestrus 5–6 days; 1 young; inter-birth interval *c.* 15 months; sexual maturity 2–5 years, males may be longer 8–9 years?; cows calve every alternate year; young suckle for about 6–15 months. **Longevity:** 16–18 years (wild) to 26–29 years 6 months (captive). **Status:** range restricted and numbers considerably reduced in wild.

■ HISTORY OF INTRODUCTIONS

Africa

South Africa

African buffalo have been re-introduced in parts of Cape Province (Smithers 1983). Unsuccessful re-introductions have occurred in Bontebok National Park after 1961 and in Golden Gate Highlands National Park after 1963 with animals sourced from Addo National Park on both occasions (Penzhorn 1971; Novellile and Knight 1994). Between 1981 and 1984, 19 buffalo were released in Pilanesburg National Park, Bophuthatswana, where they became established and numbered 31 in late 1984 (Anderson 1986).

Zimbabwe

Six were released in McIlwaine National Park and 10 in Matapos National Park, Zimbabwe (Southern Rhodesia) before 1963 and some were still surviving there in 1963 (Riney 1964).

Australasia

Western Australia

African buffalo are reported to have been released at Newmarracarra, Western Australia, some time after 1900 but were unsuccessful in becoming established (Allison 1969; Long 1988).

■ DAMAGE

African buffalo are relocated regularly in Zimbabwe for disease control (Jarvis and LaGrange 1984).

BANTENG

Balinese cattle, Bali cattle, Bali banteng, tembadan

Bos javanicus d'Alton

=*B. banteng* Wagner, =*B. sondaicus*
Domestic form known as Bali cattle.

■ DESCRIPTION

HB 1800–2250 mm; T 650–700 mm; SH 1200–1900 mm; WT 400–900 kg.
Similiar to domestic cattle, but slimmer and smaller; mature males black to brownish grey, cows and young pale brown; crown between crescent-shaped horns (600–750 mm) in male is hairless; forehead slightly convex; stockings and rump patch white.

■ DISTRIBUTION

Southern Asia. From Burma, Thailand and southern parts of Indochina, the Malay Peninsula, Java, Bali and also Kalimantan. Probably domesticated in prehistoric times in Bali and Sumbawa and subsequently taken to Sumatra, Java, Borneo, Sulawesi, Lombok and Timor.

■ HABITS AND BEHAVIOUR

Habits: nocturnal in some areas but in others active any time; wary and shy. **Gregariousness:** 2–40 in herds

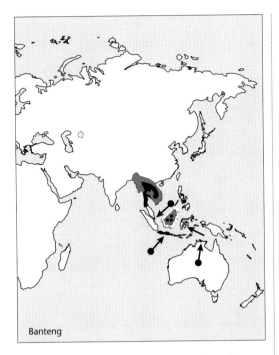

Banteng

or groups (1 male, cows and young); males solitary or in bachelor herds to 40. **Movements:** moves to higher ground in monsoons. **Habitat:** dense forest and clearings, logged areas, thickets with open areas, swamps. **Foods:** grass, sedges and herbs, bamboo shoots and other woody herbage. **Breeding:** seasonal, mates in wet season, calves June–September; gestation 285–300 days; sexual maturity, cows 2 years, males 5 years; 1–2 calves each year; young weaned 10 months; oestrus 6–8 weeks after parturition; lactation 9 months; mates freely with cattle. **Longevity:** 15–25 years (captive). **Status:** declined in range and numbers, but still locally common; possibly endangered.

■ HISTORY OF INTRODUCTIONS
AUSTRALASIA
Australia
Domestic strains of banteng cattle were imported from Timor and possibly Bali, with buffaloes introduced to the Coburg Peninsula between 1829 and 1849. They were abandoned when the settlement closed and have remained on the Peninsula, mainly in the north-east part. The population has not dispersed beyond the introduction site and numbered some 1500 in 1961 (Letts, 1964) and still varies between 1500 and 3500 (Strahan 1995). They have in the past reached as far south as Murgarella to the north-east of Kakadu National Park (Wilson *et al.* 1992).

At present the largest extant wild herd in the world occurs in Gurig National Park in north-western Arnhemland, Northern Territory (Choquenot 1998).

The herd is considered to be of some conservation value, but there is increasing evidence that the Banteng are having a significant impact on the vegetation in the park.

INDONESIA
Bali, Sangihe, and Enggano islands
Banteng currently are established in the wild in Bali Barat Game Reserve in western Bali; in Sangihe Islands off north-eastern Sulawesi; and on Enggano Island off Sumatra, where a special hunting reserve is being formed (Lever, 1985).

■ DAMAGE
Banteng are accused of causing adverse effects on coastal grasslands by overgrazing and are now fenced from leaving the peninsula in Australia. They are also a source of food for the indigenous inhabitants (Strahan 1995).

FERAL CATTLE
Wild domestic cattle, wild cattle
Bos taurus Linnaeus
=*original ancestor* Bos primigenius; *zebu,* Bos indicus, *now widely introduced (see Walker 1968).*

■ DESCRIPTION
SH 900–1100 mm; WT 450–1000 kg (400 kg in Spain).
Colour varies from whitish to black with shades of red, brown and buff; size and conformation vary and feral cattle can only be distinguished by their location.

■ DISTRIBUTION
Worldwide under domestication. Probably first domesticated in south-west Asia. Aurochs domesticated about 8000 years ago.

■ HABITS AND BEHAVIOUR
Habits: browse and graze; diurnal. **Gregariousness:** stable groups (cows and young), bachelor groups of males, and old solitary bulls; density 6/100 ha (NZ). **Movements:** sedentary. **Habitat:** thick scrub, open grassland. **Foods:** leaves of shrubs and trees, ferns, grass, sedges, herbs. **Breeding:** throughout year; gestation 277–290 days (9.5 months); oestrous cycle 3 weeks; young 1 rarely 2; calves born with eyes open; stand and suckle soon after birth; weaned before next calf is born; males mature 10 months, females 6–10 months. **Longevity:** more than 20 years (as domestic).

■ HISTORY OF INTRODUCTIONS

Introductions of cattle on islands

Island	Date introduced	Status
Amsterdam (France)	1871	removed about 1988
Antipodes (NZ)	before 1887	later died out
Auckland (NZ)	1849	killed by sealers
Bahamas	after 1656	on some remote islands
Campbell (part) (NZ)	1902	eradicated 1984
Cerro Azul (Galápagos)	?	still there
Chatham (NZ)	?	being removed 1976?
Cumberland (United States)	?	removed
Enderby (Aucklands)	1895–96	still there?
Enderby (NZ)	1894	eradicated 1991–93
Falkland (UK)	1764	few left by 1880s
Felicite (Seychelles)	?	Feral herd?
Floreana (Galápagos)	?	still there
Galápagos (Ecuador)	?	still occur some islands
Graham I. (Canada)	prior World War 1	still there?
Hawaiian (United States)	1778–1803	still exist some islands
Inaccessible (Tristan)	?	none there?
Inagua (Bahamas)	?	still there?
Juan Fernández (Chile)	?	possibly no longer present
Kapiti (NZ)	c. 1837	eradicated 1916–17
Kauai (Hawaiian Is)	?	?
Kerguelen (NZ)	1950	died out after 1975
Lanai (Hawaiian Is)	1928	few left by 1970
Macquarie (Australia)	?	unsuccessful
Molokai (Hawaiian Is)	?	eradicated about 1934
New Zealand	c. (1819) on	still present few areas
Nicobar (Indonesia)	?	occasionally feral
Oahu (Hawaiian Is)	1793	doubtful if any left?
Orkney (see Swona)		
Pitt (NZ)	?	?
Rose (Aucklands)	1895	still there?
Salt (Puerto Rico)	?	died out
San Cristobal (Galápagos)	?	still there
Santa Ana (Solomons)	?	reported there
Santa Cruz (Galápagos)	?	still there
Seychelles (see Felicite)		
Sierra Negra (Galápagos)	?	still there
Solomon Islands	?	present on one island
St. Paul	early seafarers	died out?
Stewart (part) (NZ)	?	eradicated 1940s
Swona (Orkney Is)	1974? 1980s	feral herd present
Tristan de Cunha	c. 1817	none now?

ATLANTIC OCEAN ISLANDS

Falkland Islands

Cattle were introduced in 1764 and soon ran wild on the islands (Holdgate and Wace 1961). L. A. de Bougainville founded a French colony in the East Falklands at this date at the head of Berkeley Sound and cattle (seven females, two males) were among a number of animals imported at this time. In 1842 a number of British colonists were attracted to the Falklands and began to exploit the wild cattle which were descended from those introduced in 1764. By the 1880s sheep farming had replaced most of the cattle (Encycl. Brit. 1970–78). (See Lever 1985 for more details.)

Orkney Islands

A feral herd of 33 cattle exists on the island of Swona. The people left the island in 1974 and no intensive husbandry has been practised since then and no hay crops have been grown since 1977. In 1977 some animals were removed for sale and the entire herd may well be removed in the near future.

St. Helena

Cattle were introduced to St. Helena in the sixteenth century and some feral cattle inhabited the interior in 1600 (Hakluyt 1600).

AUSTRALASIA

Australia

Feral cattle occur from time to time in many places throughout the pastoral rangelands of Australia (Long 1988; Wilson *et al.* 1992).

Cattle exist in many areas of Queensland and it is difficult to say which are feral and which are not. In northern areas probably eight per cent are regarded as feral (Mitchell *et al.* 1982).

Papua New Guinea

Some feral cattle were established near the coast north of Madang in eastern Papua New Guinea (Lever 1985).

EUROPE

France

There is a small herd of feral cattle in Haut Ariègo region of the French Pyrenees. A second population formerly occurred on the lower slopes of Mt. Rhune in the Atlantic Pyrenees, but were destroyed in 1924 in the interests of tourism (Lever 1985). East of Andorra in the eastern French Pyrenees, 300–400 small cattle are established near the Massane River on the northern slopes of the Alberes Mountains in the Pyrénées Orientales. Across the border in Spain larger numbers occur on the southern slope of Albères.

Spain

Small herds exist in forests around Trucios and Goizueta in northern Navarra; in 1975 they numbered 14. Other small groups occur at Urnieta in Guipuzcoa and on the Marquis de los Arcos' estate at Olague in Vavarra. How long they have roamed the Pyrenees is unknown.

About 100 wild cattle are established near Entrino in the south of Orense Province of Galicia in north-west Spain. A small population of cattle occur in the Doñana National Park in Andalusia (Lever 1985) where there are now about 145 head (Lazo 1992, 1994).

United Kingdom

Feral cattle are present on Swona in the Orkney Islands (Hall and Moore 1986).

INDIAN OCEAN ISLANDS

Kerguelen

In 1950 some were landed and thereafter they were kept for short periods near the scientific station, but have since died out (Watson 1975; Lever 1985).

Nicobar Island

Occasionally cattle are abandoned and become feral in the Nicobars (Encycl. Brit. 1970–80).

Nouvelle Amsterdam (France)

Cattle were introduced to Amsterdam when a farm was established on the island in 1871 (Holdgate and Wace 1961), but they ran wild some eight months later when it was abandoned. Although their numbers are regulated somewhat by winter starvation (Reppe 1957), there were some 1000–2000 head present in about 1960 (Holdgate and Wace 1961), and they were still present in 1967 (Holdgate 1967), 1975 (Watson 1975), and in 1990 (Daycard 1990).

In 1988 a count estimated that there were about 2000 head on the island. The island was then divided by a fence and shortly after, about 1059 cattle were removed (Berteaux 1993).

St. Paul

Early seafarers left cattle on St. Paul as a source of food (Hesse 1937), but they are not mentioned at later dates (Holdgate and Wace 1961).

Tristan da Cunha

Cattle were introduced to Tristan in about 1817, some six years after the island was colonised, or in 1820, certainly before 1824 (Holdgate and Wace 1961). Some were present there in 1829 (Morrell 1832) and in the 1940s and 1950s they were grazed there, but only partly confined to enclosures (Munch 1945; Holdgate 1958).

Cattle were apparently established there as a semi-wild species (Holdgate 1967) and a few may still remain on the southern lowlands, although other reports indicate they have gone (Watson 1975). Apparently cattle were also introduced to Inaccessible Island (Lever 1985).

PACIFIC OCEAN ISLANDS

Antipodes

Feral cattle were present in 1887, but died out at a later date (Atkinson and Bell 1973).

Auckland Islands

In 1849 cattle were introduced to Auckland Island, but were killed off by sealers a short time later (Holdgate and Wace 1961; Holdgate 1967; Taylor 1968). Others were introduced to Rose Island and Enderby Island in 1895 and by 1916 had apparently increased in numbers substantially. They had over-grazed the range and were suffering from starvation (Holdgate and Wace 1961). In 1960 a small number survived on these islands.

The herd on the Enderby–Auckland group is derived from 12 shorthorn cattle landed as food for castaways and an abortive attempt at farming in 1895 and 1896. There were 10 in 1903, 20 in 1925, 29 in 1954, 50–60 in 1963, 48 in 1966, 40 in 1971 and 39 in 1973 (Taylor 1976). Cattle now exist on Enderby Island only (Challies 1975).

Campbell Island

Campbell Island became a pastoral lease in 1894 and when abandoned in 1931 some 30 cattle and many sheep were left to run wild. The population has remained static for 40 years and fewer than 30 cattle have remained in a small area in the south of the island (Bell and Taylor 1970). In 1971 they were still present, but by 1976 only a few remained grazing on the slopes of Menhir and the lower eastern slopes of Mt. Paris (Dilks and Wilson 1979).

Eight Ayrshire and shorthorn cross cattle were taken to the island in 1902 and the population remained small and stable over the years. There were 27 in 1927, 12 in 1941–45, 18 in 1961, 16 in 1966, 18 in 1967, 22 in 1969, 20 in 1971, 22 in 1975 and 12 in 1976. Their range was restricted to the corner of the island (Taylor 1976).

Small populations of feral cattle existed on Campbell and Enderby islands (Wodzicki and Wright 1982), but were exterminated on Campbell Island in 1984.

Chatham Islands

Feral cattle are present in the Chathams (Atkinson and Bell 1973; Gibb and Flux 1973). In the mid-1970s they were present on the southern tablelands, and on

Pitt Island, where efforts were being made to remove them (Taylor 1976).

Galápagos Islands

Feral cattle now occur on Sierra Negra and Cerro Azul (Isabela Islands), Santa Cruz, Floreana and San Cristobal (Eckhardt 1972).

Hawaiian Islands

Cattle were apparently given their freedom between 1778 and 1803, and some still exist there on some islands (Berger 1972). First brought to the islands by Captain Vancouver on his second voyage in 1793, cattle were released on the slopes of Hualalai where they soon increased to a sizeable herd. By 1830 they were numerous in the hills and valleys of Mauna Kea. They were recognised as a menace to forest growth as early as 1815 (Hall 1904), and in 1851 it was estimated that there were 12 000 of them wild on Hawaii (Henke 1929). Between 1859 and 1869 many were hunted for hides and between 1850 and 1875 the tallow industry flourished.

There appears to be little mention of wild cattle in the 1890s, but at this time 100 were killed in one week in Kauai (Judd 1939). In 1904 it was reported that there may have been at least 10 000 wild cattle on Mauna Kea alone (Hall 1904).

Wild cattle were heavily hunted for the next several decades (Kramer 1971), and 738 were recorded to have been killed on fenced forest reserves between 1921 and 1946 (Bryan 1947). The last feral animals on Mauna Kea were eradicated in 1931 (Tinker 1941), although several small herds (200 animals) existed on the south-eastern slopes. On most of the other islands there were few left by the 1930s. The last animal was reported to have been shot on Molokai in 1934. There were none on Lanai until after cattle ranching commenced in 1928 and in 1952 a hunting season was held to get rid of the remaining 50 animals, but only 12 were destroyed (Kramer 1971).

The last major herds of feral cattle, except for a few strays, appear to be in the Honaunau Forest Reserve. An aerial census in 1956 indicated that there were 1000 there, but by 1958 some 900 had been shot by hunters and ranch employees. There were still about 100 left in the area in 1968 (Kramer 1971).

Cattle were introduced to Laysan Island in the early 1900s and a few dairy cows were noted there in 1902 and 1905, but these either died or were removed by 1910 (Ely and Clapp 1973).

Japan

For about 100 years a feral herd of cattle have existed on Kushinoshima in the Tokara Islands, south-west of Kyushu. In 1979 the population numbered about 24 head (Kimura and Ihobe 1985; Lever 1985).

Juan Fernández

Cattle have been introduced to the island (Encycl. Brit. 1970–80), but little appears to be documented about them.

Macquarie Island

Cattle were introduced, but did not become established (Watson 1975).

New Zealand

Cattle have been feral since the early nineteenth century and are now locally common on the North and South islands (Wodzicki 1965), where small herds are established in remote areas (Gibb and Flux 1973). Feral cattle have existed in New Zealand since the early days of settlement when domestic stock was allowed to roam. They were in the bush at the Bay of Islands in 1819, in Marlborough Sound in 1839, and on Kapiti Island by 1840 (Thomson 1922). They were most numerous on the main islands from the 1860s to the 1880s (Thomson 1922; Wodzicki 1950). In the 1960s they were reported from the forests in remote country of the North Island and in the north-east and north-western parts of the South Island (Wodzicki 1961). Thirty years ago they were still common in remote areas and are now found in widely scattered areas including Northland, Coramandel, the Volcanic Plateau, Wanganui River region, valleys of main North Island ranges from East Cape to the Ruahines, the South Island back country from Farewell Spit to Fiordland and near Ruggedy Range on Stewart Island. They also occur (or occurred) on Chatham, Pitt, Campbell and Enderby islands (Wodzicki 1950; King 1990). With the advancement of settlement and farming generally they have gradually been eliminated; the last on Stewart Island was shot out in the 1940s, and many of the other herds on the North and South islands have now disappeared; the last at Farewell Spit in 1975 (Taylor 1976).

In the last 30 years feral cattle have declined in numbers. They still occur on Chatham and Pitt islands, but are frequently reinforced with captive escapees. A small herd on Enderby (Aucklands) is the only one to remain completely free of domestic stock for the last 100 years (King 1990). There are now 15 distinct populations still present.

Solomon Islands

Wild populations of cattle are reported from Santa Ana in the south-east Solomons (Flannery 1995).

NORTH AMERICA

Canada

Domestic cattle were abandoned on the west coast of Vancouver Island, British Columbia, in the Tofino area, where they became established as a feral population and were hunted until about 30 years ago when they disappeared. Feral cattle have also been reported established in the vicinity of Estevan Point and at Cape Scott, but there is no definite information. They were said to be numerous on the north and east coasts of Graham Island following the early settlements prior to World War 1. Those on the north coast disappeared about in the 1920s, but a small population still exists on the east coast. Some 18 head were reported at Cape Ball in 1963 and 40 head at Oeanda River in British Columbia in 1962 (Carl and Guiguet 1972).

United States

Formerly vast herds of feral cattle occurred in the west of the United States, also in the pampas of Argentina (Jackson and Langguth 1987). They did not persist in California and other areas in the United States, probably because of the presence of large predators (Mooney *et al.* 1986). Some were established in Jasper Ridge Biological Preserve in California but have now been removed because it was thought that they had some impact on the indigenous plant species (Macdonald *et al.* 1988).

SOUTH AMERICA

Ecuador

Some feral cattle are still found in areas of the Páramos on Mt. Antisama in Ecuador originating from bulls brought over by the Spanish for fighting centuries ago (Andrews 1982).

Colombia

Feral cattle exist high in the Sierra Nevada de Santa Marta in Colombia. They are probably descended from cattle introduced by the Spanish Conquistadors. Most live in Parque Nacional Natural Sierra Nevada Santa Marta and possibly also in Llanos Orientales (Lever 1985).

Argentina, Uruguay, Paraguay, Chile

During the abandonment of Buenos Aires by the Spanish in favour of Asuncion, Paraguay, in 1537–38, some cattle were left behind at Buenos Aires (Redford and Eisenberg 1992). By 1699 cattle by the millions were roaming free on the pampayan grasslands and by the end of the seventeenth century the trade in cattle hides between Argentina, Paraguay and Europe was already well established.

A chronicler wrote in 1691 that oxen, cows and horses roamed the plains in such prodigous numbers that in 'in some places the fields are covered with them as far as your eyes will reach'. They were so numerous that one had only to ride out into the countryside and round them up. They belonged to no individual.

By the mid-1700s Buenos Aires and Montevideo were exporting 800 000 cattle hides per year and there were similar trends in Chile, where large herds were noted in central areas during the seventeenth century.

Some feral cattle were established recently as far south as Isla los Estados, Argentina, where about 12 were reported in Bahia Crossley in 1971 (Lever 1985).

WEST INDIES

Bahamas

Feral cattle occur on some remote islands, particularly Inagua, probably as a result of human settlement after 1656.

Puerto Rico

Feral cattle were present on Salt Island, but have since died out because the vegetation has disappeared through the presence of goats and other grazing animals (Heatwope 1981).

Hispaniola

In 1535 feral cattle were numerous in Hispaniola (de Oviedo y Valdes 1851–55). In 1526 they were reported as numerous near Santo Domingo (Parmentier and Marmentier 1883). The first cattle were brought there by the Spanish and in 1851–55 they were so numerous that some were exported and many killed and left to rot. The Spanish settlers opened large ranches with vague boundaries and minimal supervision of stock and most cattle were wild or semi-wild. Under these circumstances feral stock readily bred. Spanish stockmen hunted the cattle for meat and hides. The term 'buccaneer' (French *boucanier*), originally designated a hunter of wild cattle, was applied to the scourge of the Spanish Main. Las Casas (1875–76) recorded feral cattle were especially abundant at Aquin and Leogane. Spanish lancers staged there near the close of seventeenth century held a great hunt in 1614 to kill the cattle whose chase attracted the buccaneers (Moreau de Saint-Mery 1797–98). Rapid development for agriculture of some areas after 1697, especially the Plaine du Nord and Plaine de Leógane, somewhat reduced the number of feral cattle. But as early as 1703 Leógane was supplied with beef imported from the Spanish part of the island.

In 1701 feral cattle were still widespread. In 1725 hunters went from Port de Paix to Port-à-Piment in the north-west to hunt cattle, but they appeared to be scarce in the 1780s. At the close of eighteenth century some feral cattle are mentioned at Anse-à-Pitre, at Saltrou, in the Massif de La Selle in the south-east corner of what is now Haiti, and in the country of the Mornes de Plymouth north of Les Cayes. Descourtilz (in 1809) said that feral cattle were a hazard to the crops. They were still abundant in many districts of sparsely settled pastoral Spanish Hispaniola in 1785.

Feral cattle are still found in arid, limestone areas of sparse soil with vegetation of xerophytic brush such as grows on the Peninsula of Baraderes, in Tapion de Papaye region of the Northwest Peninsula and on the coast near Grand-Gosier and Anse-à-Pitre. As recently as 1920, feral cattle were recorded on the Plateau of Segin or south flank of Massif de La Selle, but with the advance of agriculture they vanished (Street 1962).

■ DAMAGE

Overgrazing by semi-wild cattle on Tristan has caused erosion on the southern areas of lowland plain, at Cave Point and Stony Beach, and the areas of unstable sand near the main landing place in the north-west may once have been covered by tussock grass which is no longer present on the island (Holdgate and Wace 1961).

On Amsterdam Island feral cattle have eliminated the trees from much of the northern part and have checked the regeneration by biting off the young saplings (Aubert de la Rue 1955).

Overgrazing is evident in the Auckland Islands where it led to the starvation of the feral cattle population in the late 1950s and early 1960s (Holdgate and Wace 1961). Cattle can modify the native vegetation by browsing, grazing and trampling (Wodzicki 1950) and on Enderby Island have prevented the regeneration of tussock grassland (*Poa* sp.) and a variety of endemic herbs (King 1990).

BISON

North American bison or buffalo, European bison, American bison, wisent

Bison bison (Linnaeus)

It has been accepted here that the North American bison (B. bison bonasus) is but subspecifically distinct from its European counterpart (B. bison bison). This may not please all taxonomists, however, it is made clear in the text to which subspecies is being referred.

■ DESCRIPTION

HB males 2100–3800 mm; T 430–800 mm; SH 1670–1820 mm; WT 460–1000 kg; females 1950–2250 mm; 450–525 mm; 1520 mm; 360–500 kg. European subspecies may weigh up to 1364 kg.

Generally dark brown; head, shoulders and forelegs covered by shaggy, dark chocolate-brown mane; hind quarters coppery brown; massive head, and beard, black; rhinarium naked and black; tongue slate-blue; horns black, short and curving out, forwards then inwards; tail with terminal tassel of long hair; massive hump on shoulders; hooves rounded. American subspecies has longer coat on neck, shoulders and forelimbs, the horns are smaller and less curved, and the hind quarters are smaller.

■ DISTRIBUTION

North America and Europe. *B. b. bison* formerly from northern Mexico to the Great Lakes and from Washington to the Rocky Mountains, United States, north to southern-central Canada. Now occur as a wild herd only in north of Wood Buffalo Park and various semi-wild populations in a number of national parks and reserves. *B. b. bonasus* formerly occurred throughout the deciduous woodlands of Europe from southern Sweden south. Now extinct in the wild apart from re-introduced populations.

■ HABITS AND BEHAVIOUR

Habits: mainly diurnal and crepuscular, but nocturnal on moonlit nights. **Movements:** migratory; daily movements of about 3 km, but several hundred km south in winter; home range 30 km^2–100 km^2. **Gregariousness:** cohesive herds females and juveniles 4–20 or 30; mature males solitary except at rut; formerly formed large herds of thousands, perhaps millions; sexes separate except at breeding time; density 12/1000 ha to 3 or 4/1000 ha. **Habitat:** open plains, arid plains, meadows, river valleys, aspen parkland, coniferous forest. **Foods:** grasses, forbs and sedges, acorns, heather, and the leaves and shoots on trees and shrubs. **Breeding:** mates summer–autumn (July–September), calves spring–summer (April–June); gestation 260–300 days; oestrous cycle 21 days; oestrus 9–28 hours; females seasonally polyoestrous, males polygamous; 1 young, rarely 2, every 1–2 years; calf can run in 3–4 hours; stays close to female for 3 weeks; weaned 12 months; males mature 2–3 years, females 2–4 years. **Longevity:** 40 years (captive). **Status:** almost extinct in Europe; reduced from about 60 million to a few herds of 1000 in North America.

■ HISTORY OF INTRODUCTIONS

Re-introduced into some small areas in both North America and Europe.

EUROPE

The European bison (*B. b. bonasus*) originally inhabited a vast area from the Caucasus to France and Belgium (Dorst 1965). The population in the Caucasus became extinct in 1925. The population found in the Bialowieza Forest (on the Russia–Poland border) was exterminated during World War 1 (Lasoki 1963). In 1892 it was estimated that only 375 animals survived in the wild (Dorst 1965). They had been extirpated completely by 1921 (Kleiman 1989).

In 1929 bison were re-introduced from European zoos to a game reserve in the Bialowieza Forest (Zabinski 1949; Lasocki 1963), and in 1931 Poland had 30 of the surviving animals (a further 35 were in Germany), which were reported to number only 96 in total (Dorst 1965). By 1944 there were 44 bison in Poland, all from the original nucleus of three (Zabinski 1949). In Germany a herd of 28 existed at Springe (Mohr 1949).

There are now, however, nearly 4000 wild animals as a result of carefully controlled captive breeding programs among zoos in Russia, Poland and Germany, although the Russian population has an admixture of *B. b. bison* genes.

Poland
Although a number of breeding centres have been established in Europe, Bialowieza appears to be the only area in which pure stock have been released into the wild. This occurred in 1952 when two bulls, and in following years a herd of 27, were set free (Lasocki 1963; Krasinski 1963). At the end of 1962 some 40–50 bison were established in the wild while at least another 33 remained under protection. More recent estimates place the herd size at around 120 head (Dorst 1965). There are now two herds: 271 and 315 bison in the Polish and Belorussian parts of the forest (Kozlo 1993). In fact the 4000 bison now in Poland are all descended from 12 animals held at the Bialowieza bison centre in the 1920s (Sykes 1998).

Russian Federation, Ukraine, Kazakhstan
Re-acclimatisation of the bison began in Russia in 1929 from Germany, Sweden and Poland (Yanushevich 1966) and possibly as early as 1926 (Heptner 1967). Further introductions were made in 1946–51 when 17 were obtained from Poland and placed in stud farms, but were later established in some parks (Heptner 1967). However, re-introductions in the western Caucasus were made with one bull and four females of American origin (i.e. *B. b. bison*) (Jenkins 1963). By 1963 these had established a flourishing herd of more than 200 head and occupied an area from the upper reaches of the river Terek to the Black Sea coast, and from the river Kuban to the Inguri. From here many were transported to other areas, such as Moscow and the forests of Kharkov, in the hope of establishing herds there. Other attempts made in the Caucasus in 1940–56, the northern Caucasus since 1930 and in the Ukraine in 1913 failed

to become established (Yanushevich 1966). So at this stage it would appear that those re-established in Russia are *B. b. bison* or a mixture of *bison* and *bonasus*.

More recently some have been introduced successfully in Kazakhstan (Sludskii and Afanas'ev 1964) and it is proposed to re-establish them in Belarus, the Ukraine and in part of Russia (Pereva 1987).

NORTH AMERICA
The North American bison (*B. b. bison*) originally occurred over the Great Plains in vast numbers from Lake Eerie to Louisiana and Texas, and north to western Canada. It was estimated that there may have been 50–75 million bison when Europeans began to colonise North America. The populations were decimated by hunting, clearing and slaughter, so that there were few left by 1900. In fact, an estimate in 1889 records that there were only 540 live animals left. Near-extinction in the late 1800s left remnant herds at only two locations: one at Yellowstone National Park, Wyoming, and the other at Wood Buffalo National Park, Alberta. Other unconfirmed populations were subsequently initiated in Alaska, southern Utah, Grand Teton National Park (Wyoming), Santa Catalina Island (California) and in the North West Territories in Canada (Van Vuren and Bray 1986). Both re-introductions and subsequent translocations have led to healthy bison populations which today number in the thousands (Kleiman 1989). Those occurring on Santa Catalina Island are fenced and managed (Lidicker 1991).

Alaska
In 1928, 23 bison from Moise, Montana, were released in the Delta River area south of Fairbanks in central Alaska (Murie 1940; Burris 1965). In about 1940 the herd numbered some 200 and has been established in the area since its introduction (de Vos et al. 1956; Burris 1965). There has been a harvestable population there ever since (Burris and McKnight 1973; Franzmann 1988).

In 1950, 17 from this area were released at Slana, and in 1962, 35 at the Chitina River, where they appear to have become established (Burris 1965). These releases have also resulted in harvestable populations of bison as has one at Farewell in 1965–68 (Burris and McKnight 1973; Franzmann 1988). Those at Chitina River may not have been from the translocation effort.

Anticosti Island
Some bison were liberated on this island in 1896, but they failed to become established (Newsome 1937).

Canada
In 1974 a pure-blood remnant herd of *B. b. athabascae* persisted in the MacKenzie District in the Nyarling River–Big Buffalo Lake area of the Wood Buffalo Park. After 1957 stock from here were transported 160 km west north-west and established in an area near Fort Providence (Banfield 1974). Now several herds of *B. b. bison* have been successfully established in wildlife preserves in the United States and Canada (Hall 1981).

Mexico
The bison has been re-introduced into Mexico (de Vos *et al.* 1956), where it is now established in the Sonora and Chihuahua regions (Petrides 1975).

United States
In 1907 a herd of 15 captive-bred bison from the Bronx Zoo were re-introduced to a reserve in Oklahoma and additional captive animals were released periodically through 1917 into empty ranges in South Dakota, Nebraska and Montana (Kleinman 1989; Chan 1993).

The Henry Mountains, Utah, population became established in 1941–42 when 15 cows and eight bulls were obtained from Yellowstone and released about 50 km north-east of the Henry Mountains. Three bulls vanished soon after release, so the present herd descended from about 20 animals. In 1963 about 80–82 were there; in 1977 some 205 and by 1983 about 343, and the herd appeared to be expanding exponentially. They occur over about 300 km^2 of combined summer and winter range (Van Vuren and Bray 1986).

Eighteen bison were released in 1963 into MacKenzie Bison Sanctuary, where they had increased to 645 in 1979 (Calef 1984 in Van Vuren and Bray 1986)

In 1902 bison from ranched herds were introduced to Yellowstone National Park (Meagher 1989). These bred with the remnant wild population. In the early 1980s the population expanded and some animals began to move beyond the park boundaries. The expansion was interrupted because of conflicts with human interests.

■ DAMAGE
No information.

KOB
Buffon's kob, kob antelope
Kobus kob (Erxleben)

■ DESCRIPTION
HB 1250–1800; T 180–400 mm; SH 700–1050 mm; WT 50–120 kg.
Stocky appearance; coat glossy, golden to reddish brown; throat patch white; eye ring and inside of ears white; legs black fronted; horns S-shaped and strongly ridged, 400–690 mm. Female paler than male.

■ DISTRIBUTION
Africa. Senegal to south-eastern Sudan, south to southern Uganda.

■ HABITS AND BEHAVIOUR
Habits: mainly crepuscular, but also nocturnal and diurnal; males territorial at breeding time. **Gregariousness:** small herds or troops 5–40 females and young; form leks when population high; bachelor males 30–40, rarely 100. **Movements:** sedentary and migratory(?). **Habitat:** savannah, floodplain, borders marshland, open grassy plains, woodland edges. **Food:** grazes; grass and herbage, and occasionally foliage of trees and bushes. **Breeding:** all year; gestation 261–271 days; 1 young; inter-birth interval 21–64 days; weaned 6–7 months; sexual maturity females 13–14 months, males 3 years. **Longevity:** Not recorded but probably similar to *K. lechwe*. **Status:** common, but range declining.

Kob

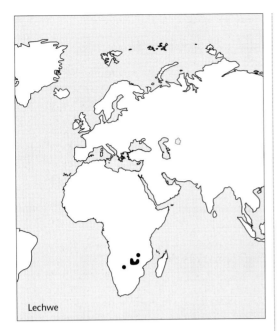

Lechwe

HISTORY OF INTRODUCTION
AFRICA
Senegal
Buffon's kob, *K. k. kob*, has been introduced in Basse Casamance National Park in south-west Senegal and now occurs there in small numbers (East 1990).

DAMAGE
No information.

LECHWE
Waterbuck, black lechwe
***Kobus leche* Gray**
The common waterbuck (Kobus ellipsiprymnusi) was re-introduced to Pilanesberg National Park, Bophuthatswana, in 1981–84 when 96 animals were released. There were about 150 there in late 1984 (Anderson 1986).

DESCRIPTION
HB 1300–1800 mm; T 300–450 mm; SH 850–1100 mm; WT 60–130 kg
Coat greasy; colour varies, brownish or reddish yellow, to greyish brown; neck mane shaggy; throat, belly and backs of legs white; forelegs with conspicuous black markings; tail tipped black; male horned, lyrate shaped, thin, strongly ridged, 450–920 mm; white patch around eyes.

DISTRIBUTION
Africa. Zaire, eastern Angola, the Caprivi Strip of Namibia, Zambia and northern Botswana.

HABITS AND BEHAVIOUR
Habits: crepuscular, nocturnal and diurnal; partly territorial; spends most of time wading in water and only come out to calve or rest. **Gregariousness:** solitary; herds 10–50 females; groups or leks 50–100 males; congregate in large loose herds males and females; sometimes 1 male with cows and young; density 1000/km^2. **Movements:** moves back and forth with rising and falling of annual floods. **Habitat:** reed beds, swampy tracts and shrubby growth near water; wet grassland, marshes, swamps and around lakes, floodplains and adjacent ground. **Foods:** grass, water plants and sedges. **Breeding:** mainly rainy season, peak calving July–September; gestation 210–240 days; 1 young; young lies up in tall grass; weaned 3–4 months; males mature 2–4 years, females 1.5 years. **Longevity:** 15 or more years. **Status:** numbers and range reduced, endangered in Zaire.

HISTORY OF INTRODUCTIONS
AFRICA
South Africa
The Transvaal Department Nature Conservation introduced six lechwe somewhere between 1952 and 1961. Seven were introduced to McIlwaine National Park before 1963 and were surviving and increasing there in 1963 (Riney 1964).

Lechwe have been introduced in the Adelaide District of Cape Province and also in Orange Free State (Haltenorth and Diller 1994).

Zambia
Lechwe have been found in an area around Bangweola Lake, northern Zambia, but became extinct and were re-introduced in the mid-1970s (Burton and Pearson 1987)

The black lechwe (subspecies *smithemani*) occurred in the Bangweolo Lake region of northern Zambia and also earlier extended north-east to River Chambeshi, but was re-introduced there in 1975 (Haltenorth and Diller 1994).

DAMAGE
No information.

SOUTHERN REEDBUCK
Common reedbuck
***Redunca arundinum* (Boddaert)**

DESCRIPTION
HB 1200–1600 mm; T 180–300 mm; SH 650–1050 mm; WT 50–95 kg.
Body colour ranges from brown to yellowish or

Southern reedbuck

greyish brown; nose and forehead darker; eye stripe, lips, chin, under parts tail whitish; back darker than upper parts; throat patch crescent-shaped, grey-white; chest greyish white; distinct dark brown band down front of each foreleg; males horned, 250–460 mm.

■ DISTRIBUTION

Africa. Southern Africa savannah zone from Gabon and adjacent parts of the Congo to Zaire, Tanzania and South Africa (Cape of Good Hope).

■ HABITS AND BEHAVIOUR

Habits: largely nocturnal but also active during day; territorial. **Gregariousness:** singly, pairs or family groups; aggregations to 20; territories 30–60 ha. **Movements:** overlapping home ranges 74–123 ha. **Habitat:** grassland, forest savannah, woodland, farmland; always near water. **Food:** grass and shoots. **Breeding:** throughout year; gestation 230–240 days; 1 young; inter-birth interval 9–14 months; sexual maturity 9–24 months. **Longevity:** 10 years captive. **Status:** declined in range and numbers.

■ HISTORY OF INTRODUCTIONS

AFRICA

South Africa

Southern reedbuck were unsuccessfully re-introduced in the 1960s to Addo Elephant National Park, Bontekok National Park, Golden Gate Highlands National Park and Mountain Zebra National Park, all

from Northam, Transvaal (Penzhorn 1971; Novellie and Knight 1994). The main reason for the failures appears to have been unsuitable habitat.

The Transvaal Department of Nature Conservation liberated two reedbuck at McIlwaine National Park and one at Matopos National Park before 1963, where they were surviving in 1963 (Riney 1964). Between 1981 and 1984, 108 were released in Pilanesberg National Park, Bophuthatswana, where they became established and in late 1984 there were about 100 (Anderson 1986).

■ DAMAGE

Southern reedbuck will raid crops (Whitfield 1985).

MOUNTAIN REEDBUCK
Redunca fulvorufula (Afzelius)

■ DESCRIPTION

HB 1100–1250 mm; T 170–260 mm; SH 600–800 mm; WT 19–38 kg.

Coat greyish or grey-brown on upper parts; under parts white; head and neck yellowish; eye stripe, lips, chin and throat yellow to greyish white; front legs without dark band; tail bushy white; males horned, short, stocky, five to eight rings, slightly curved forwards, 130–380 mm. Female larger than male, has two pairs inguinal mammae.

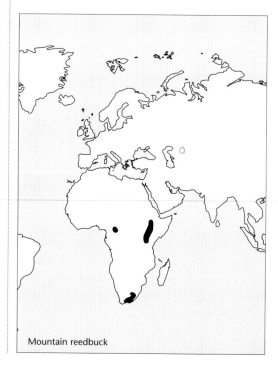

Mountain reedbuck

■ DISTRIBUTION
Africa. South, east and central Africa. Isolated populations in northern Cameroon and Nigeria, north-eastern Africa, southern Mozambique and southern Africa.

■ HABITS AND BEHAVIOUR
Habits: solitary or small groups; males territorial; feeds mainly at night. **Gregariousness:** territorial males, unstable herds females with young 3–8, bachelor herds or solitary males; territories 10–76 ha. **Movements:** no information. **Habitat:** dry slopes of hills; open mountain grassland. **Food:** grass and plants, and foliage bushes. **Breeding:** throughout year, most births November–March; 1 young; young reproductively mature at 12–14 months, males at about 1 year; gestation 225–251 days; females leave group to lamb, and rejoin group in about 3 months. **Longevity:** to 12 years. **Status:** reduced in range and numbers and vulnerable to further loss.

■ HISTORY OF INTRODUCTIONS
AFRICA
South Africa
Mountain reedbuck have been re-introduced to Addo Elephant National Park and Golden Gate Highlands National Park from Cape Province, Orange Free State and the Transkei (Penzhorn 1971). They were successfully established in Golden Gate Highlands National Park, but failed to become established at Addo Elephant National Park, probably because of unsuitable habitat (Novellie and Knight 1994).

■ DAMAGE
No information.

GREY RHEBOK
Rhebok, vaal rhebok
***Pelea capreolus* (Forster)**

■ DESCRIPTION
HB 1050–1250 mm; T 100–300 mm; SH 700–800 mm; WT 15–30 kg.
Slender with long neck and legs; narrow snout with bulbous nose; body hairy, wooly, curly; upper parts brownish grey; face and legs yellowish; muzzle with dark blaze; under parts and tail white; ears erect, long pointed; skin of nose, inside ears and eyes black; lower legs with dark front stripe; male horns straight, upright, 150–270 mm. Female has four mammae.

■ DISTRIBUTION
Africa. From Cape Province, Transvaal, south-east Botswana, and Lesotho, South Africa.

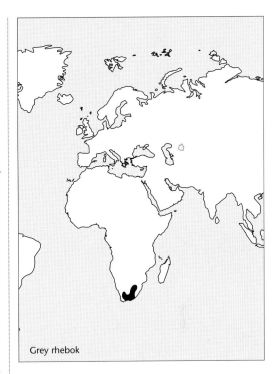

Grey rhebok

■ HABITS AND BEHAVIOUR
Habits: mainly diurnal, also nocturnal; males territorial. **Gregariousness:** family parties to 12 to 30; immature males solitary until win territory. **Movements:** home range 33–75 ha. **Habitat:** hilly grassland; rocks and tangled growth on mountains and plateaus. **Foods:** grass and herbs, leaves of shrubs. **Breeding:** mates January–April, births September–January; gestation 261–285 days; 1–2 young; weaned 6–8 months; sexual maturity 18–30 months. **Longevity:** 8–10 years (wild). **Status:** fairly common.

HISTORY OF INTRODUCTIONS
AFRICA
South Africa
Re-introduced to Addo Elephant National Park and Bontebok National Park (Penzhorn 1971; Haltenorth and Diller 1994) grey rhebok appear to be still established in Bontebok National Park but have failed to become established in Addo Elephant National Park (Novellie and Knight 1994).

It is not reliably known whether grey rhebok ever occurred naturally in Kruger National Park, but they were introduced to the south end of the park in the 1970s and have established a small breeding population there (Macdonald and Frame 1988).

■ DAMAGE
No information.

SABLE
Sable antelope
***Hippotragus niger* (Harris)**

■ DESCRIPTION
HB 1880–2670 mm; T 370–760 mm; SH 1000–1650 mm; WT 150–300 kg.
Compact build, thick neck, and upstanding mane; coat short, glossy, shiny black and reddish; face white with dark markings; eyebrows and long muzzle white, divided by cheek stripe; forequarters black; belly and rump patch white; tail long with tufted tip; ears large, long and pointed; horns massive, heavily ringed, curved backwards 510–1650 mm, both sexes horned. Males tend to be black, while females tend to be sorrel or chestnut.

■ DISTRIBUTION
Africa. South-eastern Kenya, eastern Tanzania, and Mozambique to Angola and southern Zaire.

■ HABITS AND BEHAVIOUR
Habits: active day and night, but mainly crepuscular; territorial; usually near water. **Gregariousness:** one male, females and young 30–75; territorial bulls may be solitary; temporary groups 10–20 and up to 200; territories 3.9–9 km^2; density 0.4–9.2/km^2. **Movements:** sedentary in dry season, disperse in wet season; 1.2 km/day; home range 10–25 km^2. **Habitat:** savannah woodland, grassland, floodplains, wooded country near water. **Foods:** grasses, herbs and foliage. **Breeding:** seasonal, but varies regionally, mainly at end of rains; gestation 240–281 days; 1 calf; calf

concealed at birth for 3 weeks; weaned 8–9 months; sexual maturity 2–3 years. **Longevity:** to 17 years captive. **Status:** declining in numbers due to hunting and agricultural development.

■ HISTORY OF INTRODUCTIONS
AFRICA
South Africa
Sable have been re-introduced into Swaziland, Orange Free State and the Transvaal where they had become extinct (Haltenorth and Diller 1994). In the Transvaal they have been translocated to mainly private lands and reserves (Smithers 1984). From 1981–84, 77 were released in Pilanesberg National Park, Bophuthatswana, where they were surviving and breeding in late 1984 (Anderson 1986).

Zimbabwe
Sable have been translocated in Zimbabwe where 16 were introduced to McIlwaine National Park and 11 to Matapos National Park, by authorities before 1963. They are surviving in both areas (Riney 1964; Smither 1983).

■ DAMAGE
Not recorded as a pest in Africa.

ROAN ANTELOPE
Roan
***Hippotragus equinus* (Desmarest)**

■ DESCRIPTION
HB 1900–2650 mm; T 370–480 mm; SH 1260–1600 mm; WT 223–300 kg.
High shoulders, powerful neck and upstanding mane; coat grey to rich chestnut; under parts grey to yellowish white; mane grey brown with blackish edge; facial pattern brown-black; ears long white inside with dark tufts at tips; muzzle white; tail brown-black; horns backward curved, massive, heavily ridged, 550–1000 mm. Female with shorter horns than male.

■ DISTRIBUTION
Africa. Range fragmented south of the Sahara. Cameroon, Senegal, western Ethiopia south to South Africa.

■ HABITS AND BEHAVIOUR
Habits: nocturnal and diurnal; dominance hierarchy among older males. **Gregariousness:** singly or pairs, occasionally groups 6 or so to 12–60; bachelor herds 2–5; older males solitary; density 0.15–0.8/km^2. **Movements:** sedentary but wander in dry season; territories 25–50 ha; home range 50–100 km^2. **Habitat:** open grassland, floodplains, wooded grassland, savannah, dry bush near water. **Foods:** grass,

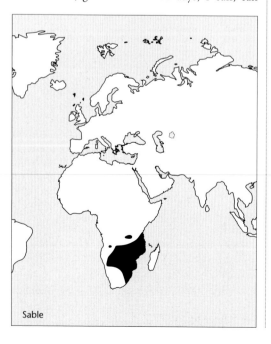
Sable

Roan antelope

herbs and foliage. **Breeding:** all year, peaks in October–November; gestation 255–286 days; 1 calf; female stays with young after birth for few days to weeks; weaned at 6 months; inter-birth interval 10.5 months; sexual maturity females 2–3 years, males 6 years. **Longevity:** to 17 years. **Status:** reduced in numbers and range.

■ HISTORY OF INTRODUCTIONS
AFRICA
Namibia
Seventy-four roan antelope were introduced and 70 released (four died) in 1970 and are now breeding in wild and noted regularly in Etosha National Park (Hofmeyer 1975; Haltenorth and Diller 1994).

South Africa
Introduced in northern Transvaal, Natal (Haltenorth and Diller 1994). Roan antelope have been introduced to nature reserves in other parts of the province, South Africa, Transvaal (Smithers 1983).

Tanzania
Roan antelope were introduced to Rubondo Island in the south of Lake Victoria (Haltenorth and Dillier 1994).

Zimbabwe
Two roan antelope were introduced to McIlwaine National Park by Southern Rhodesian authorities before 1963, but failed to become established (Riney 1964).

■ DAMAGE
No information.

SCIMITAR-HORNED ORYX
Oryx dammah (Cretzschmar)

■ DESCRIPTION
HB 1530–2350 mm; T 450–900 mm; SH 900–1400 mm; WT 100–210 kg.
Head short and blunt; ears short, broad and rounded; upper parts and flanks pale fawn to cream; mid to dark markings on centre of face and beneath the eyes; mane extends from head to shoulders; tail tufted. Both sexes have horns from 600–1500 mm in length, those of females usually longer and more slender; horns curving back in a large arc.

■ DISTRIBUTION
Northern Africa. Originally found in the semi-desert areas of Morocco and Senegal to Egypt and Sudan.

■ HABITS AND BEHAVIOUR
Habits: linear dominance hierarchy in males. **Gregariousness:** usually found in herds of 20–40, but formerly occurred in herds of up to 1000 in areas of fresh pasture, or surface water or during wet season migrations. **Movements:** Unknown. **Habitat:?** Food: grass. **Breeding:** oestrous cycle 21–22 days, oestrus 1 day, gestation 222–253 days; attains sexual maturity at 11 months; 1 calf born weighing 10 kg; **Longevity:** to 20 years. **Status:** Greatly reduced and close to extinction; single wild population in Chad and more than 500 animals in captivity around the world.

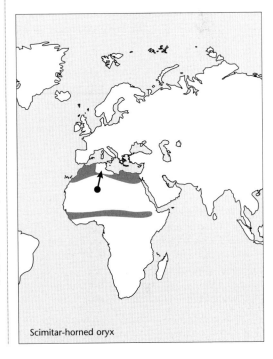
Scimitar-horned oryx

■ HISTORY OF INTRODUCTIONS
AFRICA
Tunisia

Close to extinction in North Africa. Populations formerly roamed the desert edges of the northern Sahara from Morocco and Senegal across Egypt and Sudan. Desertification, competition with domestic animals and hunting contributed to a drastic decline. Scimitar-horned oryx disappeared from Tunisia in 1922 and most other countries soon after. Now perhaps only a few dozen survive in the wild in Chad and Niger, but there have been no reports for 20 years. Captive stocks in zoos number some 3000–4000 animals.

In December 1985, 10 animals from Britain were shipped to Tunisia and re-introduced in the wild in Bou-Hedma National Park (Bertram 1988; Theobald 1999). They were kept in enclosures for four and a half months then released into a 10-ha pre-release enclosure for 14 months. They were released into the 6000-ha park in July 1987.

■ DAMAGE
None known.

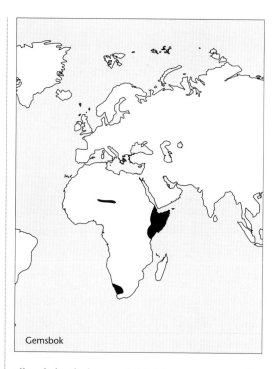

Gemsbok

GEMSBOK

Oryx, beisa
Oryx gazella (Linnaeus)

■ DESCRIPTION

HB 1600–2350 mm; T 450–900 mm; SH 850–1400 mm; WT 85–255 kg.
Head short and blunt; ears small; upper parts and flanks fawn grey; mane dark brown; shoulders to mid back band dark brown and widens to saddle on rump and narrows to base of tail; tail dark brown; under parts white; muzzle, between horns, front of horns through eyes to lower jaw with dark brown patches; flanks have dark brown band; throat with distinctive dewlap; horns straight, 15–30 prominent rings, about 1050 mm.

■ DISTRIBUTION

Africa. South-west and eastern Africa from Ethiopia and Somalia to Namibia and eastern South Africa. Two discreet populations: Namibia to Botswana, and from southern Somalia, parts of Sudan, north-eastern Uganda, Kenya and to north-eastern Tanzania. Formerly most of Africa except high mountains, swamps and closed forest.

■ HABITS AND BEHAVIOUR

Habits: diurnal and nocturnal. **Gregariousness:** female groups 30–40, but aggregations or mixed herds of several hundred may form; males may be solitary;

all male herds; harems 1–12. **Movements:** nomadic; 1–6.5 km/day; home range 10–127 km². **Habitat:** stony plains, alkaline flats, steppe; arid plains and desert, rocky hillsides and thick bush-forest. **Foods:** grasses, forbs and foliage; occasionally browse. **Breeding:** any time of year (breeds seasonally but varies regionally?); females post-partum oestrus; gestation 260–300 days; inter-birth interval 9 months; 1 young, twins rare; calves hide for first 6 weeks; young weaned 3–4 months; sexually maturity females 1–2 years, males 5 years. **Longevity:** 18–22 years captive. **Status:** common in parts, but exterminated in many localities.

■ HISTORY OF INTRODUCTIONS
AFRICA
South Africa

Gemsbock were unsuccessfully re-introduced to Mountain Zebra National Park (Penzhorn 1971; Novellie and Knight 1994) from Kalahari Gemsbok National Park. Between 1981 and 1984, 158 were released in Pilanesberg National Park, Bophuthatswana, where they were surviving in late 1984 (Anderson 1986).

Namibia

Twenty-three gemsbok (*Oryx gazella gazella*) captured with drugs in the Namib Desert were translocated in about 1975 (Ebedes 1975). They are apparently caught throughout south-west Africa and sold to farmers for restocking purposes (Hofmeyer 1975).

NORTH AMERICA
United States
Two gemsbok were released into a pasture paddock at Red Park, Texas, in September 1965; two more were released in 1966, six in 1967 and three in 1968. A total of five males and 10 females are now present and breeding in the area and it is intended to release them in the wild some time later (Wood *et al.* 1979).

Introduction of gemsbok occurred as early as 1962 in New Mexico (Haltenorth and Diller 1994). There is a small introduced population in the White Sands area of southern New Mexico (Reid and Patrick 1983) which are the result of several releases at the White Sands Missile Range Military Reservation beginning in 1969 (Upham 1980).

■ DAMAGE
No information.

ARABIAN ORYX

Oryx

Oryx leucoryx (Pallas)

=*Oryx gazella leucoryx*
In some older taxonomic works is held to be a subspecies of the gemsbok.

■ **DESCRIPTION**
HB to 1400–1650 mm; T about 280 mm; SH 840–1016 mm; WT 35–113 kg
Coat pure white; legs chocolate coloured and chocolate facial markings; male and female horned, straight to 700 mm.

■ **DISTRIBUTION**
Middle East. Formerly throughout Sinai, southern Palestine, Jordan, Iraq and most of Arabian Peninsula. Extinct in the wild since 1972. (May have only survived in the Rub-al-Khali desert of southern central Arabia for some time.)

■ **HABITS AND BEHAVIOUR**
Habits: mainly crepuscular; dominance hierarchy; largely independent of free water. **Gregariousness:** mixed groups of 8–20; 1 male and several females and young; bachelor males; density 0.035/km^2. **Movements:** sedentary and highly nomadic. **Habitat:** desert, rocky desert, dunes and arid areas, rocky hillsides, gravel plains, shallow woods and depressions; dunes edging sand deserts. **Foods:** herbs, grass, wild melons, fallen fruits, seed pods, roots, tubers, leaves and shoots of trees and bushes. **Breeding:** seasonal; gestation 260–265 days; 1 young; weaned 3–4 months; sexual maturity 1–2 years. **Longevity:** about 20 years. **Status:** endangered in wild; number in captivity.

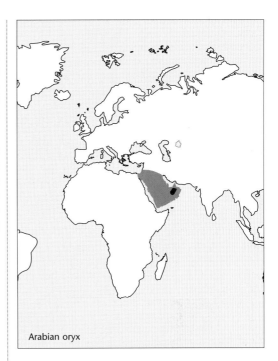
Arabian oryx

■ **HISTORY OF INTRODUCTIONS**
MIDDLE EAST
Populations began to disappear in the wild due to overhunting in the mid-nineteenth century and by 1914 there were few left outside Saudi Arabia. The decline was accelerated after World War 1 by the spread of firearms and four-wheel drive vehicles. They were extinct in Egypt and a large part of the Sudan as well as parts of Arabia, Syria and Gojjam Mountains (Haltenorth and Diller 1994). By the early 1960s the species was confined to two small areas, one near the borders of Saudi Arabia, Yemen and Oman, and the other in the Duru and Wahiba country in north-east Oman. The last wild oryx were probably killed in 1972 in the Jiddat al-Harasis of Oman, although rumours of sightings persist (Daly 1988).

Some were captured for breeding in captivity in about 1963 (Woodford 1963). By 1965 the population in the wild was less than 100 (Dorst 1965). Three wild oryx (one male and two females) were captured in 1962 and taken first to Kenya and then to Phoenix Zoo, United States. These were added to five from the Riyadh Zoo and one from the Zoological Society of London (Crouch 1987; Grimwood 1988; Jones 1988). By the 1980s oryx had been returned to the wild in several countries, with Oman staging the first re-introduction in the wild in 1982. The successful re-introduction of captive-bred animals to the central desert (Jiddat-al-Harasis) of Oman is now an introduction classic.

Oman

The oryx had become so reduced by hunting in the 1960s (the last wild one was killed in 1972) that survival as a wild animal was unlikely. A plan to re-establish the species in the desert began in 1962 by the Fauna Preservation Society. Some eight or more were imported by New Mexican authorities in about 1962–67 and kept in Albaquerque Zoo for breeding and later released if successful (Gordon 1967).

In March 1981, five arrived in Oman from the San Diego Zoo, United States, and after acclimatisation in pens these were to be released at Jiddat-al-Harasis. (Encycl. Brit. 1970–80; Walker 1992).

In January 1982, four males, four females, a yearling and a calf born in Oman, were released at Yalooni in the Jiddat-al-Harasis. Supplementary feeding continued for them up until the next rains. By 1984 the herd had increased to 13 and expanded its range to occupy 1500 km² near Yalooni (Lever 1985) and another 10 were released. Further releases were made in 1984, 1988 and 1989. From the initial re-introduction of 12 oryx in 1982 and the subsequent releases, the herd grew to 109–110 animals by 1990, of which 80 per cent were wild-born. This herd roamed over 10 000 km² of country and a further 670 animals were in captivity in various places in the United States and on the Arabian Peninsula (Spalton 1990; East 1992; Walker 1992).

Jordan, Saudi Arabia, Israel

Oryx were formerly found throughout the desert regions of the Arabian Peninsula north to the Syrian desert but by the 1930s they were extinct in Jordan (Abu-Jafar and Hays-Shahin 1988). In 1978, four males and four females from the world herd and three from Qatar were placed in enclosures and by 1983 they had increased to 31. These animals were then released into the wildlife reserve at Shaumari where they established successfully. By 1987 there were 70 animals present. Oryx have also been re-introduced to fenced reserves in Saudi Arabia (Mahazet As Said) and to Israel (Hai-Bar) in the 1980s (Burton and Pearson 1987; Abu-Zinada et al. 1988; East 1992; Walker 1992). Those re-introduced in Jordan in 1983 came from the United States. Thirty-one oryx were released by the Royal Jordainian Conservation Society into an enclosed 22 km² reserve near Azraq in October 1983 where they have survived. Seventy-one oryx were re-introduced to the 2200 km² Mahazet As Said Reserve in Saudia Arabia in 1990. Since 1995, 125 captive-bred oryx (along with 270 sand gazelle and nearly 100 mountain gazelle) have been released at Saudi Arabia's Rub al Khali (Empty Quarter), a 11 780 km² sanctuary (Wacher 1998).

■ **DAMAGE**

No information.

BONTEBOK
Blesbok, blesbuck
Damaliscus dorcas (Pallas)
=Damaliscus pygargus

■ **DESCRIPTION**

HB 1400–1600 mm; T 300–450 mm; SH 850–1000 mm; WT 55–80 kg.

Back red-brown; crown, sides of face and neck, flanks, thighs and front of rump and upper parts of limbs are dark brown to nearly black; sides of face and neck, flanks and upper limbs glossed purplish; front of face white; under parts white; ears brown; tail white basally and black or brown terminally; both sexes horned, S-shaped 350–500 mm. Female paler than male.

■ **DISTRIBUTION**

Africa. Southwest Cape Province. Formerly more widespread in South Africa and extinct in a wild state. Now found mainly on ranches and in reserves.

■ **HABITS AND BEHAVIOUR**

Habits: diurnal; active early morning and late afternoon; agile; territorial behaviour declines after rut. **Gregariousness:** mixed herds 20–500; territorial males; females in herds of adults and young; young males form bachelor groups; territorial males 4–40 ha. **Movements:** migratory. **Habitat:** coastal plains

Bontebok

and steppe, open grassland. **Foods:** grass. **Breeding:** mates autumn, ruts mid-March–April, calves born spring (September–February); gestation 238–254 days; 1 calf; born in high grass, females in herds; weaned 4 months; sexual maturity females 1–2 years, males 2.5 years, but fully territorial 3–6 years. **Longevity:** 14–16 years wild, 17 years captive. **Status:** extinct in wild, only exists on farms and in reserves.

■ HISTORY OF INTRODUCTIONS
AFRICA
South Africa
Hunting probably brought the bontebok close to extinction in the early 1900s. In 1931 the Bontebok National Park was created and by 1969 there were about 800 head there. Many introductions have since been made to many reserves and private ranches (Burton and Pearson 1987). The species is now only found in reserves and on game farms where they are increasing, and being introduced more widely, so that at present there are good numbers in all states of South Africa (Estes 1993; Haltenorth and Diller 1994). However, being fenced is said to have kept them in small in-bred units (Estes 1993).

The Transvaal Department of Nature Conservation liberated 1250 bontebok between 1952 and 1961 in various areas (Riney 1964). They were successfully re-introduced in Golden Gate Highlands National Park from Orange Free State, Transvaal, Transkei and other parks, and to Mountain Zebra National Park from Orange Free State and Transvaal (Penzhorn 1971; Novellie and Knight 1994).

In about 1957–58, 84 bontebok were translocated a distance of 96.5 km. The transfer was successful and the herd numbered 69 head in 1961 (Barnard and Van Der Walt 1961).

■ DAMAGE
No information.

TOPI
Tsessebe, sassaby, korrigum, tiang
Damaliscus lunatus (**Burchell**)

■ DESCRIPTION
HB 1500–2000 mm; T 400–600 mm; SH 1130–1200 mm; WT 108–140 kg.
Coat dark reddish brown or tan with purplish sheen; black on top of head and muzzle; lower shoulders and upper parts forelimbs darker than body; upper parts of hind legs and thighs darker than body; tail basally yellowish white with black or dark brown tassel

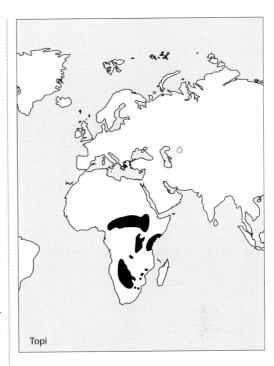

Topi

towards tip; back of ears, hind parts of rump, inside of hind legs and abdomen yellowish white; throat patch yellowish white above darker part inside front legs; lower parts legs brownish yellow; both sexes horned, strongly ridged, 300–600 mm.

■ DISTRIBUTION
Africa. Formerly south of the Sahara from Senegal to Ethiopia and south to South Africa, but range now very discontinuous. Extinct in Mauritania, Mali, Guinea-Bissau, and has disappeared from most of West Africa.

■ HABITS AND BEHAVIOUR
Habits: mainly crepuscular, but active day and night with rest in middle day. **Gregariousness:** harems 5–20 females and offspring; young males in bachelor groups, older males form breeding leks; larger aggregations or herds at feeding areas or in dry season from 100s–1000s; some males solitary; density 47/km^2. **Movements:** some local movements; males territorial 1–400 ha; migratory. **Habitat:** savanna, open plains, floodplains, grassland fringes with open woodland. **Foods:** grass and herbage. **Breeding:** ruts mainly January–April; births mainly July–December; gestation 7–8 months; 1 calf; young join nursery herds; weaned 6 months; males sexually mature at 40–42 months, females 16–36 months. **Longevity:** 12–15 years. **Status:** numbers and range reduced greatly by hunting and habitat destruction, but locally common some areas.

■ HISTORY OF INTRODUCTIONS
AFRICA
South Africa
The Transvaal Department of Nature Conservation released 26 topi between 1952 and 1961 (Riney 1964). The subspecies *D. l. hunteri* was released in 1963, when 30 animals were introduced into Tsavo National Park (Haltenorth and Diller 1994). Between 1981 and 1984 some 70 were released in Pilanesberg National Park, Bophuthatswana, where they were surviving and breeding in late 1984 (Anderson 1986).

■ DAMAGE
No information.

COMMON HARTEBEEST
Red hartebeest
Alcelaphus buselaphus (Pallas)

■ DESCRIPTION
HB 1500–2450 mm; T 300–700 mm; SH 1100–1500 mm; WT 120–200 kg.
Tall with high shoulders, sloping back and elongated head; coat varies reddish brown, fawn through to grey-brown or chestnut, paler below and on rump and white mark on face between eyes; prominent white patches on hips; forehead black; muzzle, shoulders and thighs with black; rump lower than shoulders; legs slender; tail tufted; both sexes horned, bracket-shaped, 300–700 mm. Female with two mammae.

■ DISTRIBUTION
Africa. Formerly almost everywhere in Africa except the Sahara. Now from Senegal and Somalia and northern Tanzania and western Zimbabwe, southern Angola, Namibia, Botswana to South Africa. Extinct in Gambia, Sierra Leone, Egypt, Arabia, Palestine (doubtful records?) and parts of Kenya.

■ HABITS AND BEHAVIOUR
Habits: territorial; diurnal; active mainly early morning and late afternoon. **Gregariousness:** density 1.3–1.4/km²; herds 15–300; aggregations in dry season to 10 000; adult males territorial; bachelor groups 35–100; herds female and young. **Movements:** daily movements 3–5 km; irregular movements related to rainfall; territories 0.35–4/km²; overlapping home range 370–550 ha. **Habitat:** grassland, savannah woodland, dry savannah near water. **Foods:** grass and small amount of browse. **Breeding:** throughout year, mainly mates February–May, calves October–January; gestation 214–242 days; 1 calf, twins rare; calve in cover alone;

Common hartebeest

hides calf 1–2 weeks; lactation 4 months; females calve at 3 years age; inter-birth interval 9–10 months sexual maturity 1.5–2.5 years. **Longevity:** 11–20 years (captive). **Status:** greatly reduced in numbers and range through hunting and cattle grazing.

■ HISTORY OF INTRODUCTIONS
AFRICA
Common hartebeest were exterminated in Egypt about 1850, in northern Arabia and Palestine about 1900, in Atlas and Tripoli about 1920 and in Rio de Oro about 1950. They have been widely exterminated elsewhere and in many places are a threatened species. In South Africa they are extinct in the wild though they have been re-introduced into game reserves in Cape Province, Orange Free State and Transvaal (Haltenorth and Diller 1994).

South Africa
Within recent years common hartebeest have been re-introduced to many farms and reserves in South Africa, especially in Cape Province (Smithers 1983).

Re-introductions have occurred in Addo Elephant National Park, Bontebok National Park, Mountain Zebra National Park and Golden Gate Highlands National Park. In Addo Elephant National Park they were re-introduced from Kalahari Gemsbok National Park and Sommerville Game Reserve in Orange Free State. Common hartebeest from the Kalahari and Cape Province have also been re-introduced to Mountain Zebra National Park (Penzhorn 1971).

They were only successful in establishing in the Addo Elephant and Mountain Zebra National Parks (Novellie and Knight 1994).

The Transvaal Department of Nature Conservation liberated 14 common hartebeest (*A. b. caama*) between 1952 and 1961 (Riney 1964). Cape hartebeest were re-introduced to Pilandesberg National Park, Bophuthatswana, when 902 were released there between 1981 and 1984. There were 600 there in late 1984, but over 500 were removed before this date because the population increased so prolifically (Anderson 1986).

Kenya

In about 1975 Swaynes hartebeest (*A. b. swaynei*) were moved from the central plains, Kenya, to Awash National Park and to Nechisar; 90 animals to Awash and 120 head to Nechisar (Encycl. Brit. 1970–80).

Namibia

Common hartebeest have been caught throughout south-west Africa and sold to farmers for restocking purposes (Hofmeyer 1975).

■ DAMAGE

No information.

BLACK WILDEBEEST

White-tailed gnou, wilderbeest, white-tailed gnu
***Connochaetes gnou* (Zimmerman)**

■ DESCRIPTION

HB 1700–2200 mm; T 800–1000 mm; SH 900–1500 mm; WT males140–250, females 110–122 kg.
Shoulders humped, hind quarters lightly built; mane upstanding; head massive, elongated, broad at nostrils and lips; chin has beard of long black hair; coat buffy brown to blackish brown with dirty white or buff brown, tail long-haired, black and white, almost reaching ground; muzzle with tufts black hair; both sexes horned, 450–780 mm.

■ DISTRIBUTION

Africa. Formerly in Cape Province, Orange Free State, Natal and parts of Transvaal. Now extinct as a wild animal.

■ HABITS AND BEHAVIOUR

Habits: active afternoon and night, but lies up in heat of day; adult males territorial. **Gregariousness:** separate male and female herds; males territorial; female herds (adults and young) 11–32; bachelor groups. **Movements:** migratory, nomadic and sedentary; formerly extensive local movements if not migrations which are now impossible; home range about 89 ha. **Habitat:** open plains (Karoo) and grasslands, near

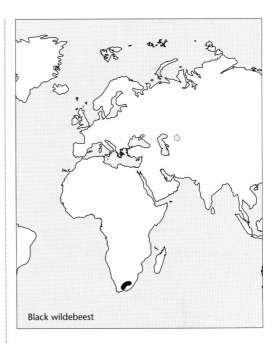

Black wildebeest

water; dwarf shrubland. **Foods:** grazes on grass, occasionally shrubs and bushes. **Breeding:** mates March–June, young born November–January; gestation 240–276 days; single calf; gains feet shortly after birth and follows female; weaned 6–9 months; female sexually mature in second or third year, male 3 years. **Longevity:** about 20 years captive. **Status:** no longer in a wild state (fair numbers in preserves).

■ HISTORY OF INTRODUCTIONS

Aᴀfᴜʀɪᴄᴀ
South Africa

Numbers of black wildebeest were reduced by hunting and habitat destruction from hundreds of thousands to about 300 in 1938. With re-introduction in reserves and the restocking of private farms this total increased to 1800 in 1965, but since then it has not been endangered and surplus stock has been made available for many introductions and re-introductions both within and beyond its former range. By 1970 there were 3000 and shortly after 3500, and now with continued re-introduction and restocking there are in the vicinity of 10 000 head in reserves and on private preserves or ranches (Smithers 1983; Burton and Pearson 1987; East 1989; Haltenorth and Diller 1994). There is now no extensive part of its former natural range where they could live and survive in a natural state.

Eighty-three black wildebeest were successfully re-introduced to Golden Gate Highlands National Park from Orange Free State and Transvaal, and to Mountain Zebra National Park from Orange Free

State, Transvaal and Cape Province (Penzhorn 1971; Novellie and Knight 1994).

An introduction to the Huhluwe-Umfolozi Game Reserve, probably in the 1960s or 1970s, failed (Macdonald and Frame 1988).

■ DAMAGE
No information.

BLUE WILDEBEEST
Common wildebeeste, brindled gnu
***Connochaetes taurinus* (Burchell)**

■ DESCRIPTION
HB 1700–2400 mm; T 600–1000 mm; SH 1150–1450 mm; WT 165–290 kg.
Large bearded antelope; neck short; shoulders high; legs thin; muzzle blunt; coat short and glossy, slate grey to dark brown, narrow stripes darker; mane black; tail and beard black; forehead, nose, chin, back and lower mane black-brown; both sexes horned; horns cow-like, smooth, enlarged boss, 450–800 mm.

■ DISTRIBUTION
Africa. East Africa from southern Kenya to Orange River, South Africa; and from Mozambique to Namibia and southern Angola. Extinct in Malawi and South Africa.

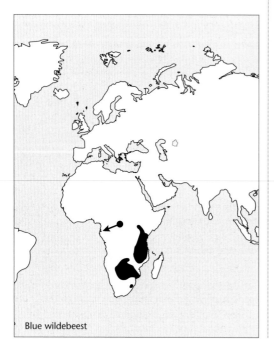

Blue wildebeest

■ HABITS AND BEHAVIOUR
Habits: active day and night, but mainly crepuscular; partly territorial. **Gregariousness:** females and young herds to 25; bachelor male herds; males territorial for part of year; formerly enormous herds to 400 000. **Movements:** sedentary part of year; migratory or nomadic for remainder; wander after rains; male territories 10–20 ha; female and young territories 6–30 ha. **Foods:** grass. **Breeding:** mainly September–January; gestation 8–8.5 months; 1 calf; calve in herds; lactation 4 months or more; sexual maturity females 3 years, males 4 years. **Longevity:** wild 18 years to captive 20 years. **Status:** drastically reduced in numbers and range.

■ HISTORY OF INTRODUCTIONS
AFRICA
Gabon
Blue wildebeest have been introduced to the Wonga-Wongue Presidential Reserve (4800 km²) in north-west Gabon from South Africa. They have had initial success in becoming established (Nicoll and Langrand 1986; Blom *et al.* 1990).

South Africa
In South Africa blue wildebeest have been re-introduced into some national parks and to some game reserves (Haltenorth and Diller 1994). The Transvaal Department of Nature Conservation released 45 blue wildebeest between 1952 and 1961. An unspecified 15 wildebeest were released in McIlwaine National Park before 1963 and in 1963 there were 27 of them present (Riney 1964). Between 1981 and 1984, 822 were re-introduced to Pilandesberg National Park, Bophuthatswana, where they increased so well that 1361 were removed to keep the population at a reasonable level in the park (Anderson 1986).

Zimbabwe
Twelve wildebeest were taken to Matapos National Park in 1963 by the Southern Rhodesian authorities and 15 were there in 1963 (Riney 1964).

EUROPE
Russian Federation
A breeding herd of blue wildebeest was established in southern Russia (Askania Nova) in 1910 (Haltenorth and Diller 1994).

■ DAMAGE
No information.

ORIBI
Oorbietjies
Ourebia ourebi (Zimmerman)

■ DESCRIPTION
HB 920–1400 mm; T 40–150 mm; SH 500–700 mm; WT 12–22 kg.
Hair fine and silky; coat upper parts sandy rufous to reddish brown or tawny; under parts and chin white; tail short, bushy, black, underside white; ears large, bare glandular area beneath; legs with tufts of long hair on knees; horns ringed at base, 75–125 mm. Female has dark crown patch, is hornless, legs slender, four mammae and is larger than male.

■ DISTRIBUTION
Africa. Africa south of the Sahara Desert from Senegal to Ethiopia and south to Kenya and Tanzania; coastal East Africa and southern Africa from Malawi and Mozambique to Angola and Cape Province.

■ HABITS AND BEHAVIOUR
Habits: mainly diurnal, but also nocturnal; territorial; defecate in regular areas; lies up in thickets during heat of day. **Gregariousness:** solitary, pairs or small parties females and young 3–20; bachelor herds 100 or more; density 2.3/km². **Movements:** sedentary. **Habitat:** tall grass with trees; open grassland, savannah woodland, floodplains, grassy plains with low bush near water. **Foods:** grass, herbs and browses foliage of shrubs. **Breeding:** in rainy season, August–September, calve September–December or any time in dry season; gestation 180–210 days; 1 calf; 2 young/year; hidden at birth; weaned 2–4 months; sexual maturity 10–14 months. **Longevity:** 8–12 years in wild, 12–14 years captive. **Status:** range and numbers declining through agriculture and pastoralism.

■ HISTORY OF INTRODUCTIONS
AFRICA
Oribi have been extensively introduced and in both South and East Africa, where they are kept on ranches (Burton and Pearson 1987).

South Africa
Oribi occurred in Kruger National Park, but probably became extinct before an attempt was made to re-introduce them. They were also re-introduced to Golden Gate Highlands National Park some time after 1963 (Penzhorn 1971). They were successful in becoming established in Golden Gate Highlands National Park, but failed in Kruger National Park probably because of unsuitable habitat (Novellie and Knight 1994).

The Transvaal Department of Nature and Conservation released 65 oribi between 1952 and 1961 in various areas (Riney 1964).

■ DAMAGE
No information.

SUNI
Neotragus moschatus (von Dueben)

■ DESCRIPTION
HB 570–620 mm; T 80–130 mm; SH 300–410 mm; WT 4–9 kg.
Small, brownish grey to chestnut; under parts white; flanks and legs reddish brown; males horned to 65–133 mm, heavily ringed last three-quarters of length. Female has four mammae.

■ DISTRIBUTION
Africa. From Kenya to eastern South Africa; also Zanzibar and Mafia islands.

■ HABITS AND BEHAVIOUR
Habits: shy, secretive; crepuscular(?); territorial; uses regular paths through undergrowth. **Gregariousness:** solitary, pairs or family groups. **Movements:** sedentary(?); territories about 3 ha. **Habitat:** gallery forest, dry bush country with underbrush, bush thickets. **Foods:** fallen leaves, fruit, flowers, buds, shoots, grass and weeds. **Breeding:** breed November–December; gestation 180 days; sexual maturity 6 months. **Longevity:** captive 6–10 years 2 months. **Status:** range reduced, and probably endangered.

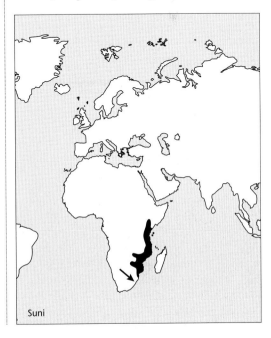
Suni

■ **HISTORY OF INTRODUCTION**

AFRICA

South Africa

Introduced to the Hluhluwe-Umfolozi Game Reserve, South Africa but failed to become established (Macdonald and Frame 1988).

■ **DAMAGE**

Suni will visit plantations and market gardens if they have cover nearby and they will eat field crops and vegetables (Haltenorth and Diller 1994)

BLACKBUCK

Indian antelope, kaljar

***Antilope cervicapra* (Linnaeus)**

■ **DESCRIPTION**

HB 840–1200 mm; T 82–180 mm, SH 672–838 mm; WT male 19.5–56.7 kg, female 19–33 kg.

Upper parts blackish brown; under parts white; albinos fairly common; sharply defined white patch around eyes; muzzle narrow and white; ears long, white or cream; patch on chin white, sometimes extending onto dorsal tip of muzzle between nostrils; forehead sometimes with ill-defined whitish spots; hair on inner ear white or cream; tail short; horns ringed, spirally twisted, 456–813 mm, sharply pointed from bases to form 'V' above head; outside of legs blackish brown. Female yellowish fawn and lacks horns.

■ **DISTRIBUTION**

Asia. Formerly Pakistan and India from Sind, Kathiawar and the Punjab eastwards to Bengal, and southward to Cape Comorin. Range now very discontinuous within this area. There are now probably more blackbuck in the United States than in the wild.

■ **HABITS AND BEHAVIOUR**

Habits: fleet; mainly diurnal, but also nocturnal; territorial in breeding season. **Gregariousness:** single sex groups, mixed groups, pseudo harems, and single; density 1/2–3 ha. Herds of 15–50 to hundreds, occasionally several hundred and formerly a few thousand; during non-breeding loose aggregations of males and females, or breeding 1 male + 3–8 females; herds of bachelor bucks. **Movements:** sedentary, males territorial 1–100 ha; overlapping home ranges for part of time 3.25–13.50 km². **Habitat:** open plains, steppe, dry deciduous forest, river banks, scrub and grassland, salty flatlands, undulating, stony hills with bushes and cultivated areas. **Foods:** grass and cereal crops, leaves, forbs, browse. **Breeding:** throughout year, but mainly ruts January–April and August–October; gestation 150–180 days (5–6 months); 1 young, rarely 2; female bears young at 2.5–3 years. **Longevity:** 18 years in wild, 15–16 in captivity. **Status:** range greatly reduced, numbers reduced, and becoming rare in many areas.

■ **HISTORY OF INTRODUCTIONS**

ASIA

India, Pakistan, Nepal

Numbers of blackbuck in India fell from 4 million to 8000 by 1964 and there are now very few wild animals left; the species is almost extirpated in Nepal, Bangladesh and Pakistan, and only remnant populations exist in India.

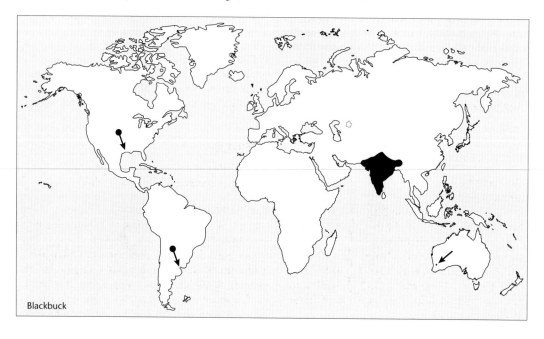

Blackbuck

At Samonagar, Tahangir (1909) and his followers killed blackbuck on Sundays and Thursdays, but on days of no killing the animals were netted live: of 641 live captives, 488 were sent to the plains of Fathpur for release. Evidently the releases were made for hunting as the large herds were reduced and the animal became less abundant in many areas because of hunting and land changes. The present populations are only remnants of the teaming herds familiar to past generations. It is estimated that there was a population of about 4 million at one time. Now there are few outside sanctuaries and the unofficial estimate in 1977 indicated a total population of about 5000–10 000 head. Pakistan had only a few left by 1970 near Fort Abbas (Bokhari 1970). In 1970 the Texas Hill country ranches donated stock (three males and seven females) to start a herd on the Cholistan wildlife reserve, Lal Suhanra.

In India a number of relocation efforts have been attempted to save the species. In 1970 the Fauna Preservation Society sponsored the collection of nine bucks to stock the Bandhavghar National Park (Wright 1972). Others were translocated into this area from Shantineketan. In 1971, 10 were re-introduced. Some were also re-introduced into the Cholistan Desert, Sind, Pakistan. Introductions have also been made to the Bardia Wildlife Sanctuary (East 1993).

SOUTH AMERICA
Argentina
Blackbuck were introduced to Santa Fé, Cordoba, and Buenos Aires provinces in 1912 and again to Santa Fé in the 1960s (Petrides 1975), but little appears to have been documented. There were further releases in Buenos Aires in 1940, and more recently the blackbuck are reported to be established over large areas of the province. Some were released in La Pampa in 1906 on the estancia of Senor Pedro Luro, who brought in oxen, red and fallow deer and wild boar (Barrett 1968) and where his descendents graze wild domestic cattle.

On some ranches in eastern Argentina blackbuck have become so numerous that thinning 'by the hundreds' has been necessary (Barrett 1968).

NORTH AMERICA
United States
Blackbuck were introduced in Texas in the 1930s or 1940s (Schreiner 1968) and by the mid-1950s were said to be not well established, but had spread somehow beyond the original fenced areas (Presnall 1958). The first known release was in 1932 in Kerr County (Jackson 1964) and some 15 animals were introduced by the H. B. Zackry Ranch, Laredo, Texas, in 1959–60 (Sanders 1963).

In 1955 there were probably 1000–1500 head in Texas (Stilwell 1955). In 1970 it was reported that there were 50 or more head in 11 counties and less than 50 head in at least 22 counties of Texas, all fenced in (Ramsey 1970). By 1974, there were 7339 in 57 counties with 80.7 per cent on Edwards Plateau (Harmel 1975), although most were said to be confined behind fences (Ables and Ramsay 1974). However by 1979 there were 9639, of which 2593 were free-ranging in 55 counties.

More recently in 1988 there were at least 20 000 blackbuck on 326 ranches in Texas (East 1993).

AUSTRALASIA
Australia
Blackbuck were introduced into Western Australia by W. McKenzie in about 1900 (Allison 1970). They were liberated at Kojarina, Wiluna, Roelands and Newmarracarra, late in the nineteenth and early twentieth century. Released 300 miles north of Perth they had done well and required thinning out (Colebatch 1929). Blackbuck were reported to have been introduced with success to the Murchison area and a small but steadily increasing herd was seen near Wiluna (Kingsmill 1920). In about 1969 blackbuck were said to number 100–150 (presumably at Newmarracarra) (Allison 1969). In 1965 some were being kept at Newmarracarra and Coolyala in the Geraldton region and protected by the managers of properties and were strictly confined because of concerns about their decreasing numbers and the danger of them being completely wiped out (Tomlinson 1955). They were recorded at Newmarracarra up until the mid-1980s (Bentley 1978; Long 1988) but eventually died out.

■ DAMAGE
Blackbuck nibble mainly the young shoots of various cereal and pulse crops and damage is not great (Chauhan and Singh 1990) in Haryana, India.

IMPALA
Black-faced impala
Aepyceros melampus (Lichtenstein)

■ DESCRIPTION
HB 1100–1600 mm; T 220–450 mm; SH 700–1000 mm; WT males 53–80 kg, females 30–60 kg;
Upper parts fawn or reddish brown; under parts white; distinctive black rump pattern; legs long and slender; flanks pale fawn, tinged reddish; white patches above eyes which extend in front of these as narrow white bands; throat white; forehead with dark

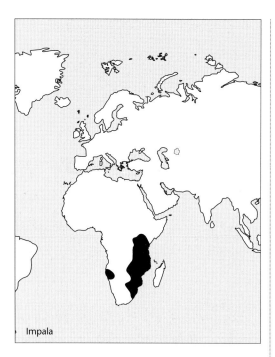

Impala

brown or nearly black patch high up; lower back of hind legs have conspicuous oval tufts of black hair just above the ankle joints; tail bushy; top of tail black, underside white; ears tipped black; horns males lyrate 450–917 mm, ridge on front surface; hooves lack clefts; glandula tufts of black hair on hind feet.

■ DISTRIBUTION
Africa. South-east Africa from Kenya and southern Angola south to Natal and western Namibia.

■ HABITS AND BEHAVIOUR
Habits: active mainly in day, but also at night; territorial; dominance hierarchy among males; complex social behaviour; fleet of foot and jump well. **Gregariousness:** small mixed herds which congregate to 200 in dry season; males in bachelor herds to 60; females and young 6–20, but formerly to 1000; breeding season harems 15–20. **Movements:** sedentary; wanders widely in dry season; territories 0.2–0.9 km^2; home range 2–6 km^2; female home range 8–180 ha. **Habitat:** savanna, open woodland, sandy bush country; acacia savannah. **Foods:** grass and herbs; browse foliage, shoots, flowers, seed pods from shrubs. **Breeding:** breeds all year, but varies regionally; birth peak September–October; gestation 6–7 months (194–200 days); 1 young; newborn calf hidden for 5 days; weaned 4–7 months; sexual maturity females 1–1.5 years, males 4 years. **Longevity:** 13 years wild, 17 years 5 months captive. **Status:** range and numbers reduced.

■ HISTORY OF INTRODUCTIONS
AFRICA
Widely introduced and re-introduced to privately owned lands and game reserves in Zimbabwe, the Transvaal, Natal and parts of KwaaZulu (Smithers 1983), Orange Free State and Namibia (Haltenorth and Diller 1994).

Namibia
Extinct in northern Namibia (Haltenorth and Diller 1994), black-faced impala (*A. m. petersi*) have been released in Etosha National Park; 81 in 1970 and 127 in 1971, and are now well established and breeding there (Hofmeyer 1975).

South Africa
Impala have disappeared in many areas in South Africa, but have been extensively introduced in some areas beyond their original range (Ansell in Meester and Setzer 1977; MacDonald and Frame 1988). They have been re-introduced in Natal (Smithers 1983) and also successfully introduced and re-established in Kruger National Park (MacDonald and Frame 1988). They have been widely introduced and established in Orange Free State, Transvaal and Natal, and in the latter two areas both inside and outside their natural range (Haltenorth and Diller 1994).

The Transvaal Department of Nature Conservation introduced 700 impala between 1952 and 1961.

Impala were successfully established in Hluhluwe-Umfolozi Game Reserve and increased to such an extent that a population reduction operation had to be initiated to prevent permanent habitat alteration (Bourquin *et al.* 1971; Brooks and Macdonald 1983; MacDonald and Frame 1988). They were also introduced to Pilanesberg National Park where 1937 were released between 1981 and 1984. Some 1345 had to be removed up until 1984 because of over-population (Anderson 1986).

Zimbabwe
Twenty-three impala were introduced in Matapos National Park, Zimbabwe, in 1963 by authorities there and were noted in 1964 (Riney 1964). Impala have also been introduced into Kyle and Lake McIllwaine National Parks (Smithers 1983).

■ DAMAGE
Impala increased after introduction to such an extent in Hluhluwe-Umfolozi Game Reserve that they were considered to be altering the habitat and population reduction operations had to be initiated (Bourquin *et al.* 1971; Brooks and Macdonald 1983).

GOITRED GAZELLE
Persian gazelle
Gazella subgutturosa (Güldenstaedt)

■ DESCRIPTION
HB 930–1100 mm; T 120–200 mm; SH to 600–750 mm; WT 14–33 kg.
Inflated throat in breeding season; only male horned to 140–760 mm; horns black and lyrate.

■ DISTRIBUTION
Arabia–central Asia. From Palestine, central Arabia and eastern Caucasus through Iran, Baluchistan, southern Turkestan and Sinkiang to the Gobi Desert, the Ordos Plains and Tsaidam, Mongolia. Formerly widespread from the Levant to the Gobi Desert and northern China, but declined markedly particularly in the west of range.

HABITS AND BEHAVIOUR
Habits: no information. **Gregariousness:** small groups or herds of 5–10; near breeding season herds of 10–30. **Movements:** often migrate over long distances (e.g. hundreds of kilometres) **Habitat:** deserts, arid plains (steppe), rocky valleys and treeless areas. **Foods:** grasses and herbaceous plants, forbs and saltworts. **Breeding:** no information. **Longevity:** 5–6 years up to 8–9 years captive. **Status:** declined in range and numbers, particularly China; extinct or precariously surviving in many parts of range (Syria and Iraq); endangered.

■ HISTORY OF INTRODUCTIONS
ARABIAN PENINSULA
Bahrain and United Arab Emirates
There are now more than 1000 goitred gazelles in introduced populations on various islands in Bahrain and in the United Arab Emirates. These gulf populations are of uncertain identity and purity as the captives released have been of different species and subspecies from the mainland. Some have probably originated from mixed releases of *G. s. marica* and the Persian Gulf race *G. s. subgutturosa* (East 1992).

Saudi Arabia
Goitred gazelles disappeared from Arabia due to hunting and degredation of habitat from overgrazing by livestock.

Experimental re-introductions of captive-bred animals began at Mahazat As Said Reserve (2200 km^2 and fenced) in south-west Arabia in the 1990s. Between February 1990 and May 1992, 97 gazelles were translocated to Mahazat As Said from King

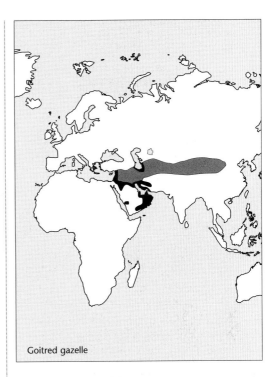

Goitred gazelle

Khalid Wildlife Research Centre and from Al-Sudairy Gazelle Research Centre.

To date, 91 gazelles have been released from quarantine pens into the reserve and breeding in the wild has been confirmed. The most recent estimate of numbers is 110–120 in the reserve.

ASIA
Russian Federation
The goitred gazelle may have been a resident of Barsa-Kel'mes Island, Aral Sea, before 1929, but nine were imported in 1930. By 1948 there were 1000 on the island, but large-scale deaths reduced the population in 1949–50, and in the early 1960s there were only 30–40 of them left (Bannikov 1963). They have also become established on the islands of Bulla and Glinyany in the Caspian Sea (Lever 1985).

NORTH AMERICA
United States
New Mexico authorities imported goitred gazelles in about 1965–67 and kept them in the Albuquerque Zoo. The progeny were to be released some time later (Gordon 1967), but there appear to be no further records.

■ DAMAGE
No information.

Springbok

SPRINGBOK

Spring-buck

Antidorcas marsupialis (Zimmermann)

■ DESCRIPTION

HB 1200–1500 mm; T 150–300 mm; SH 680–900 mm; WT 20–48 kg.

Coat cinnamon fawn above; upper foreleg to edge of hip a reddish brown horizontal band separating from white undersides; from mid back to base of tail a glandular pouch; inside of legs, back of thighs, tail and patch from rump are white; both sexes have black ringed, lyre-shaped horns 160–500 mm. Sexes similar, female horns smaller than male.

■ DISTRIBUTION

Africa. Angola, Namibia, Botswana and South Africa. Exterminated over much of range in Cape Province, Orange Free State and Transvaal.

■ HABITS AND BEHAVIOUR

Habits: crepuscular, but mainly feeds at night in hot weather. **Gregariousness:** formerly herds half million or more, now few number 1500; herds to 100 females plus young; bachelor males to 50; males establish territory and harems in breeding season 10–30. **Movements:** migratory 'treks' across range to where rain has fallen; males territorial in breeding season 10–70 ha. **Habitat:** dry savanna, veldt (grassland), stony plains and hilly country. **Foods:** grass and other herbage, leaves of shrubs. **Breeding:** breeds after rain, but any time of year, peaks in autumn and spring; gestation 167–171 days; 1 young, twins rare; female

sexually mature at 6–7 months, males 1–2 years; bulk of lambs born in summer; lambs graze at 6 weeks. **Longevity:** 10 years in wild, 19 years captive. **Status:** considerably reduced in numbers, range reduced but still fairly common.

■ HISTORY OF INTRODUCTIONS

Extensively re-introduced throughout South Africa (Whitfield 1985; Burton and Pearson 1987) within its natural range and outside its historical range (Estes 1993).

AFRICA
South Africa

Although to a large extent exterminated in South Africa, the springbok has generally regained its former range through re-introductions (De Graff and Penzhorn 1976; Smithers 1983; Haltenorth and Diller 1994). They were caught throughout south-west Africa and sold to farmers for restocking purposes (Hofmeyer 1975). This widespread re-introduction to all provinces of South Africa makes it impossible to be certain now of the original limits of occurrence of the species (Smithers 1983). They now occur in numerous reserves, parks, and on private game farms within their former range and in areas extra-limital to this (De Graff and Penzhorn 1976; Smithers 1983).

Between 1952 and 1961, 250 springbok were released by the Transvaal Department of Nature Conservation in various areas (Riney 1964).

Re-introductions have occurred to Addo Elephant National Park from Mountain Zebra National Park and a reserve at Umtata, Transkei; to Bontebok National Park from Mountain Zebra National Park and to Golden Gate Highlands National Park from Kalahari Gemsbok and Mountain Zebra National Parks and Cookhouse, Cape Province; and to Mountain Zebra National Park from Grahamstown (Penzhorn 1971; Novellie and Knight 1994).

In Mountain Zebra National Park springbok were re-introduced in the 1940s, when a number from Grahamstown were released and these were extremely successful. Since then, this population has supplied many for release to other national parks (Penzhorn 1971; Novellie and Knight 1994).

In Addo Elephant National Park, 11 were translocated in 1956 from Mountain Zebra National Park and 16 from the same park in 1958 and from Umtara. These increased to over 100 head in 1962, but by 1964, 60 per cent had died through heartwater. Since then, numbers have decreased further until only a single female remained in 1975.

Twenty springbok were re-introduced to Bontebok National Park from Mountain Zebra National Park in 1960. Seven died, but by 1974 some 246 head were there. Some of these were transferred to Golden Gate Highlands National Park in 1966.

Ten springbok were re-introduced to Golden Gate Highlands National Park from Mountain Zebra National Park in 1964. Numbers here were augmented by some from Mountain Zebra National Park and Bontebok National Park in 1966–67. In 1968–69 the population had reached 86, and by 1974 had increased to 200 head (Graaff and Penzhorn 1976).

An initial re-introduction of 24 springbok to Pilanesberg National Park, Bophuthutswana, failed when the animals died of disease, but a second attempt of 112 animals was more successful and there were 75 there in late 1984. However, the long-term success for the project was uncertain (Anderson 1986).

■ DAMAGE
Springbok formerly made mass migrations that tended to ruin crops in their pathway (Walker 1992).

CENTRAL ASIAN GAZELLE
Zeren, Mongolian gazelle
Procapra gutturosa (Pallas)

■ DESCRIPTION
HB 950–1480 mm; T 20–120 mm; SH 540–840 mm; WT 20–40 kg.
Coat orange buff above, with pinkish cinnamon sides, but paler in winter; under parts white; horns males 200–250 mm, backwards deflection not conspicuous.

■ DISTRIBUTION
Asia. Mongolia, inner Mongolia and a small area of the Russian Federation (southern Siberia) adjacent to north-western Mongolia and in the southern Nerchinsk Mountains (northern China).

■ HABITS AND BEHAVIOUR
Habits: no information. **Gregariousness:** no information. **Movements:** migrates north in spring in large herds 6000–8000. **Habitat:** dry steppe and sub-desert, grassland. **Food:** vegetation. **Breeding:** mates in autumn or winter; throat swells at breeding; seasonally polyoestrous; oestrous cycle 29 days, oestrus 1 day; gestation 186 days; 1–2 young. **Longevity:** 7 years captive. **Status:** uncommon, range and numbers reduced.

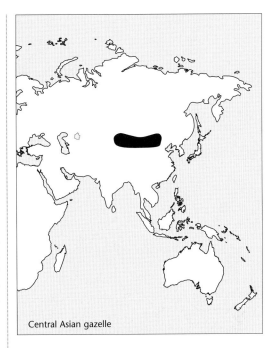
Central Asian gazelle

■ HISTORY OF INTRODUCTION
ASIA
Russian Federation
In 1949 central Asian gazelles were introduced to Bulla Island, Caucasus, where they were established locally (Yanushevich 1966).

■ DAMAGE
None known.

SAIGA
Saiga antelope
Saiga tatarica (Linnaeus)

■ DESCRIPTION
HB 1200–1700 mm; T 60–120 mm; SH 600–800 mm; WT 26–69 kg.
Coat buff in summer and white in winter; nose and sides of face dark; muzzle enlarged and puffy; nostrils overhang mouth, downward pointing openings; horns heavily ringed, lyrate, amber, males only 203–255 mm; hair heavy and wool like; fringe of long hairs on chin and throat; under parts, tail and rump patch are white. Female without horns.

■ DISTRIBUTION
Asia. The lower reaches of the Volga River and Russian steppes from Kalmyckia east across Kazakhstan to Dzungaria in central Asia (western Mongolia). Formerly more widespread in Europe, the south-eastern Russian Federation and central Asia.

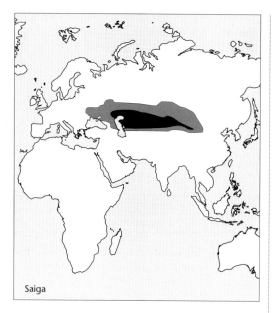

Saiga

Note: In the glacial episodes the saiga lived in central Europe and England; in the middle of the nineteenth century it was present in Poland and the Carpathians, and in 1865 it still remained in the Kalmucks steppe between the Volga and the Don River. Some years ago it was present in the territory extending between the Manic and the Volga, whereas it had been exterminated between the Volga and the Ural (Marcuzzi 1990). By the 1920s and 1930s it was reduced to few isolated areas near the Volga and Lake Balkhash but has now recovered and is found on much of its former range.

■ HABITS AND BEHAVIOUR

Habits: active early morning and late evening; no fixed home range; territorial in mating season. **Gregariousness:** rams solitary or in small groups 2–6; herds often to 1000 at rut, usually 50–100; migrating herds to 100 000; density 0.8/km², but locally 14–40/km². **Movements:** move several dozen km daily; some populations migrate southwards in August–September (autumn). **Habitat:** steppe, grassy plains often in arid areas. **Foods:** grass, herbs, low growing shrubs, leaves of forest trees, lichens, buds, shoots. **Breeding:** ruts December–January, calves April–May; gestation 139–152 days; 1 young, occasionally two; graze at 4–8 days; weaned 4 months; female sexual maturity 1 year, males 19–20 months. **Longevity:** males 5–7 years, females 11–12 years in wild. **Status:** range and numbers considerably reduced; formerly numbered thousands; now fully protected.

■ HISTORY OF INTRODUCTIONS

EURASIA

Russian Federation and adjacent independent republics

Exterminated in the Crimea by the thirteenth century AD, the saiga was scarce in the Ukraine until the eigh-

teenth century. By the early twentieth century there were fewer than 1000 left. In 1920 only a few hundred were left due to over-hunting and cultivation for agriculture (Dorst 1965). They were given total protection in 1919 in Europe and central Asia in 1923 and had by 1958 increased to an estimated two million head.

The saiga has certainly increased in numbers in the Russian Federation due to conservation measures (Bannikov 1961). Numbers were reduced to about 1000 in 1930, but under protection and assisted by reintroductions, there were 900 000 in Kazakhstan in 1951 and today there are probably more than twice this number (Bannikov 1963; Sludskii and Afanas'ev 1964; Burton and Burton 1969).

In the Caspian Sea in 1955, 19 saigas and in 1956, 34 were set free on the island of Glinyanyi (Aliev 1960). By 1958 these had increased to some 90 head, but in 1960 at least 20 died of starvation and it was found necessary to limit the numbers on the island. The saiga has always lived on Barsa Kel'mes Island in the Aral Sea, but the last male was killed in 1922 and by 1929 few remained. In 1929–30 a few were imported and now about 3000 inhabit the preserve on the island (Bannikov 1963).

Saigas have also been introduced in the Caucasus and to Bulla Island (Yanushevich 1966).

■ DAMAGE

In the Russian Federation the saiga is reported to be a pest in drought years (Yanushevich 1966) and in years of deep snow, when they eat the leaves, buds and shoots of forest trees one to three years old (Dinesman 1959). They do not compete strongly with domestic stock and rarely frequent cultivated crops (Bannikov 1961). It is generally concluded that they do not cause substantial damage to sowings in the Russian Federation (Bakeev 1964).

MOUNTAIN GOAT
Rocky Mountain goat
***Oreamnos americanus* (Blainville)**

■ DESCRIPTION

HB 1200–1600 mm, T 100–200 mm, SH 868–1200 mm, WT 28–136 kg.

Coat long, coarse, shaggy and white; chin with beard of long hairs; ears pointed; muzzle black; eyes and hooves black; horns black, spiked, 275–300 mm; large leathery gland behind each horn; legs short and muscular; hooves two-toed. Females, horns more slender and rise two-thirds of length before curving back at tip; four mammae.

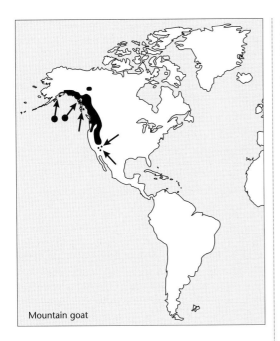

Mountain goat

■ DISTRIBUTION
North America. Western North America from southern Alaska, south to northern Idaho and Montana, United States.

■ HABITS AND BEHAVIOUR
Habits: climbs steep inclines and leaps from ledges; males defend territories; active early morning and late afternoon and at night. **Gregariousness:** solitary or small groups (males), or family groups (female and young); density 0.03–14.0/km^2; **Movements:** moves to areas with less snow in winter (1.7–11.1 km); daily movements of several hundred metres; home range 81 ha –21.5 km^2. **Habitat:** mountain slopes and cliffs at or above tree line, alpine meadows. **Foods:** grasses, sedges, rushes, forbs, moss, lichens, ferns; browse shrubs and trees including conifers, woody plants and grass-like plants. **Breeding:** ruts November–January, kid in spring (May–June); gestation 147–186 days, young 1–2, rarely 3; probably polygamous; male matures 39 months, female 27–30 months; female seasonally polyoestrous; oestrous cycle 20 days; oestrus 48 hours; young follow female in 1 week; weaned 3–4 months. **Longevity:** 14–18 years (in wild). **Status:** considerably reduced numbers; now mainly in reserves where fairly common.

■ HISTORY OF INTRODUCTIONS
North America
Mountain goats now occur in a number of areas of the Rocky Mountains from Montana to Mexico. There have been numerous translocations and introductions including several to Alaskan Islands and Olympic National Park. A number of depleted herds have also been restocked in Alaska (Smith and Nichols 1984).

Alaska
Translocations of mountain goats to Baronof and Kodiak islands have been successful, as have those on the mainland (Franzmann 1988).

In 1923, 18 mountain goats from Tracy Arm were released on Baranof Island, and in 1952–53, 17 from Anchorage and Kenai Peninsula were released at Hidden Basin and on Kodiak Island, and in 1953–57, 22 *O. a. kennedyi* were released at the Basket Bay area and on Chicagof Island (Nelson 1953, 1958; Troyer 1960; Burris 1965; Burris and McNight 1975). They still occur on Kodiak, Baranof and Chicagof islands (Walker 1992), although the Chicagof Island introduction is reported by others to have been unsuccessful (Franzmann 1988).

In 1983 mountain goats were translocated to two areas on the Alaskan mainland. Twelve were released on Kenai Peninsula in the Cecil Rhode Mountains and 17 in the Ketchikan area (Revillagigedo Island). Both introductions were successful (Smith and Nichols 1984) and are persisting (Franzmann 1988).

Canada
In January 1924, four mountain goats were transferred by the Game Commission and National Parks Branch from Banff, Alberta, to Cowichan Lake on Vancouver Island (Lloyd 1925). These animals were last seen in 1936 (Carl and Guiguet 1972).

United States
Introduced populations of mountain goats are successfully established in Colorado, central Montana, the Black Hills of South Dakota, northeastern Oregon, Olympic National Park, Washington, California, Idaho (Pond Oreille Lake) and on the Alaskan islands of Kodiak, Baranof and Chicagof, and unsuccessfully in Oregon (de Vos *et al.* 1956; Sayre 1981; Adams and Bailey 1982; Lever 1985; Naylor 1988; Walker 1992).

In Montana, where mountain goats are native to the major western mountains, they have successfully been translocated and hunting is now allowed (Hoffman *et al.* 1969). At least 228 have been trapped and subsequently moved to 12 new areas (Foss and Rognrud 1971). In 1941, six females and four males were translocated and released in the Sweetgrass Creek, Crazy Mountains area, and in 1943 another seven females and four males in the same area (Lentfer 1955). By 1953, 278 *O. a. missoulae* were present there in an area of some 16 km in width. Other successful introductions in the 1940s and 1950s included: 44

animals to the Beartooth Mountains betwen 1942 and 1953, nine to the Gallatin area in 1947–50, 14 to the Tobacco Root Mountains in 1955, 42 to the Madison Range in 1950–59, four to the Highwood Mountains in 1943, 19 to the gates of the mountains in 1950–51 and 17 or 23 to the Absaroka Mountains, south-central Montana in 1956–58; five in 1956, 10 in 1957, and eight in 1958 from south-western Montana (White 1946; Foss and Rognrud 1971; Swenson 1985). Those in the Absaroka Mountains now occupy about 500 km² and have been hunted since 1964. Numbers increased, then remained stable until 1972 when they declined until 1974, then remained stable to 1978. Thereafter they increased to 1983, when 96 were present (Swenson 1985). All these populations are well established and were regularly hunted about 10 years after the original releases. There were introductions of 20 mountain goats to the Snowy Mountains in 1953–54, 17 to the Elkhorn Mountains in 1956–58, five to the Highland Mountains in 1962, and 13 to the Bridger Mountains in 1969 and these are established but have not yet been hunted.

Twelve mountain goats from Canada were released in the Olympic Mountains, Washington, in the 1920s and there are now about 700 present (Sayre 1981).

In Colorado, mountain goats have been successfully established through introductions (Hibbs 1967). Some were moved from Montana to Colorado in about 1947 (Gulbreath 1947). The Sheep Mountain–Gladstone Ridge herd was established in 1950 when six from Montana were released in Sawatch Range, central Colorado (Adams and Bailey 1982; Kohlmann 1987). The estimated population was 36 head in 1965, reached a peak of 130 in 1979 and has remained at 120 since then. Harvesting of the herd occurred from 1967–69 and 1973–82. This herd occupies 73 km² area, 11 km west of Buena Vista, Colorado, at an elevation of 2775–4031 m in an alpine habitat (Adams and Bailey 1982, 1983).

Introductions have occurred in Montana at Beartooth Plateau, in Wyoming 48 km east of Yellowstone National Park on the Montana–Wyoming State line (Long 1965), in the Colorado–Collegiate Range about 32 km north-west of Salida (Hibbs 1967), and in the South Dakota–Mt. Rushmore–Needles–Harney Peak area of the Black Hills (Turner 1974 in Hall 1981).

Mountain goats have also been successfully translocated to Santa Catalina Island in the Channel Islands off the coast of California (Lever 1985).

■ DAMAGE
About 60 years after their introduction into Washington, United States, it has been claimed that there are now too many mountain goats present in the Olympic Mountains. Studies by the University of Washington indicate that the animals have become a potential menace to the survival of endemic plants in some areas (Sayre 1981). It is now proposed to remove 40–50 of them per year for three years and then to re-evaluate the effects of those remaining.

CHAMOIS
Gems, gemzen
Rupicapra rupicapra (**Linnaeus**)

■ DESCRIPTION
HB 900–1300 mm; T 30–40 mm; SH 600–900 mm; WT 19–50 kg.
Coat long, stiff and coarse with thick underfur, chestnut-brown or fawn in summer and dark brown to black in winter; under parts light and throat white; face has some white patches; from nose around eyes to base of horns a dark brown or black band; horns (in both sexes, not shed) close set, vertical with backward bend in the form of a hook, sharply pointed, black, 152–250 mm; ears pointed, white inside, darker outside; dorsal stripe black; rump white or pale fawn. Male and female similar in appearance, males generally heavier.

■ DISTRIBUTION
Europe and Asia Minor. Southern and central Europe to Asia Minor. Range now discontinuous and fragmented, but present in Cantabrian Mountains, the Pyrenees, Alps, Appenines, Sudeten, Tatra, Carpathian

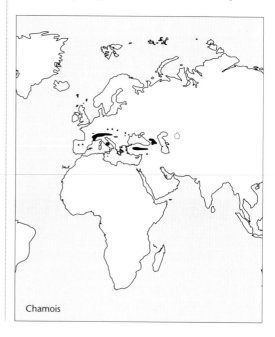

Chamois

Jarva, Abruzzi, Balkan, Taurus and Caucasus mountains and in Turkey and northern Asian Minor.

■ HABITS AND BEHAVIOUR

Habits: mainly diurnal, but some nocturnal activity in spring and summer; agile in steep terrain; shy. **Gregariousness:** matriarchal social system; females and young in loose, unstable groups 15–30; loose social hierarchy among resident females; old males solitary except at rut; herds of several hundred animals at rut; density 3.2–5.0/100 ha (NZ). **Movements:** altitudinally migratory (to lower levels in winter); males wander less predictably than females; home range varies with sex and season (70–207 ha). **Habitat:** alpine bluffs, steep rugged areas, ridges and spurs on mountain tops in alpine forest and upper edges along tree line. **Foods:** grasses, herbs, lichen, flowers, moss; browse including the shoots and flowers of trees, bushes and sub-alpine shrubs. **Breeding:** ruts October–December (early May to June in NZ), kid April–June; gestation 153–180 days (5.5–6 months); young 1–3, usually 1; kids born agile and follow female 1 hour after birth; sex ratio 0.66 males to 1 female; sexually mature at 6–18 months; males fully mature at 8–9 years. **Longevity:** 22 years as captive; 14 for males and 19 years for females in wild. **Status:** much reduced in numbers and range.

■ HISTORY OF INTRODUCTIONS

EUROPE

The chamois has been introduced and re-introduced in various parts of Europe (Masini and Lovari 1988; Walker 1992), including France, Austria, Switzerland and Czechoslovakia.

Czechoslovakia

Chamois were introduced to the Jeseniky area in about 1905 and although surviving, the result is not considered very satisfactory (Hvlas 1965). Attempts were also made to acclimatise them in the Gadersky Valley area in 1955, and by 1962 they were surviving, but were not increasing much in numbers (Sokol 1955). It is reported that there were 350 chamois in the Brdy Forest as introduced animals around 1964 (Ganzak 1964).

There may now be 300 and 100 head respectively in the Jeseniky area and Luzicke mountains of north-western Czechoslovakia (Lever 1985).

Austria

In Austria chamois have been introduced outside their native range (Lever 1985).

France

In France chamois have been introduced outside their native range (Lever 1985). They are established in the Vosges where they were introduced in the 1950s (Miller 1987).

Germany

A colony of chamois are established in the Black Forest in Germany, where they were established in the 1920s (Miller 1987).

Switzerland

Between 1950 and 1962 chamois were re-acclimatised in the Jura Mountains when 84 from the Alps were released in five different areas (Salzmann 1975). By 1963 there were at least 200 chamois in Switzerland (Anon. 1963) and by the mid-1970s the population had grown to some 3000–3500 and they were spread over the whole region between the junctions of the rivers Hare and Rhine and the mountains west of Geneva (Salzmann 1975).

NEW ZEALAND

The original stock of chamois introduced to New Zealand were a gift from Emperor Franz Josef of Austria to the New Zealand government.

Chamois (two males and six females) were introduced from Austria to New Zealand in 1907 and 1913 near Mount Cook. They have spread more rapidly than the introduced thar (*Hemitragus jemlahicus*), about 10 km/year, and now occupy the South Island alps from Wairau River to Lake Wakatipu; by 1970 they occurred over an area of 36 136 km^2 (Wodzicki 1950, 1965; Gibb and Flux 1973; Wodzicki and Wright 1982).

Most of the major mountain ranges along the southern axis of the South Island from about Lake Wakatipu to Lake Rotoiti, Nelson, were colonised by 1960 and they were spreading into Fiordland and other areas.

Next to red deer (*Cervus elaphus*), chamois are probably the most numerous and widespread ungulate in the South Island. They are throughout the high country mostly in alpine areas, but extending into lower elevations in some places. They often wander widely and the limits of their range are difficult to define in some areas (Barnett 1985; King 1990), but they occupy about 33 per cent of the South Island and are slowly expanding (Fraser *et al.* 1996).

■ DAMAGE

From the 1930s to 1982 about 90 000 chamois were removed during culling programs in New Zealand, without making any difference to the population or the distribution. Together with other exotic species, they have contributed to the modified state of mountain vegetation (King 1990), but have some recreational, game meat and trophy value in New Zealand.

MUSK-OX
Muskox, Musk-oxen
Ovibos moschatus (Zimmermann)

■ **DESCRIPTION**
HB 1900–2450 mm; T 60–171 mm; SH 1200–1380 mm; WT 182–660 kg.
Upper parts coat shaggy, dense and long, deep brown to blackish, with a light patch in mid-dorsal region of back; under parts black; body stocky, hump on shoulders; neck, legs and tail short; ears pointed; fore and hind limbs yellowish white; hooves rounded with tuft of hair between; upper side of tail black, underside white; horns massive down curved, joined at bases to form bony boss. Female horns generally shorter than male; have four mammae.

■ **DISTRIBUTION**
North America and Greenland. Northern Canada and north-western Greenland. In prehistoric times occurred across arctic Europe and Asia, and the north coast of Alaska.

■ **HABITS AND BEHAVIOUR**
Habits: withstands low temperatures (–40°F); slow moving; swims well. **Gregariousness:** females and young in herds 3–100 with perhaps a single male; males solitary or in groups 2–3, and in winter small herds 15–20; density 0.30–0.44/km². **Movements:** mainly sedentary, but moves between summer and winter feeding grounds (up to 80 km). **Habitat:** arctic tundra; in summer river valleys, lake shores and wet meadows, and in winter hilltops, slopes and plateaus. **Foods:** grass, moss, lichens, sedges, heath, forbs, leaves, willow and pine shoots. **Breeding:** ruts July–September; calves April–June; gestation 8–9 months; male polygamous; 1 young every second year, occasionally 2 in 2 years, twins rare; calf precocious and follows female within 1 day; eats vegetation in 1 week; weaned about 1 year; inter-birth interval 1–2 years; mature at 3–5 years in wild. **Longevity:** 20–23 (captivity), 24 years (wild). **Status:** considerably reduced in numbers and in danger of extinction.

■ **HISTORY OF INTRODUCTIONS**
As a result of introduction, re-introduction and translocation, small populations of the musk-ox now occur in various parts of Alaska, the Russian Federation, Svarlbard, Norway, Sweden, Canada (northern Quebec), and west Greenland (Gray 1984; Whitfield 1985; Walker 1992).

ASIA
Russian Federation
Negotiations were initiated prior to 1974 for the translocation and re-establishment of musk-oxen to Russia (Uspenski 1984).

In 1974, 12 musk-oxen (six cows and six bulls) from Banks Peninsula, Canada, and in 1975 an additional 40 from Nunivak Island, Alaska, were flown to the Taimyr Peninsula, north-western Siberia, where they were placed in an enclosure. Four years later, 13 were

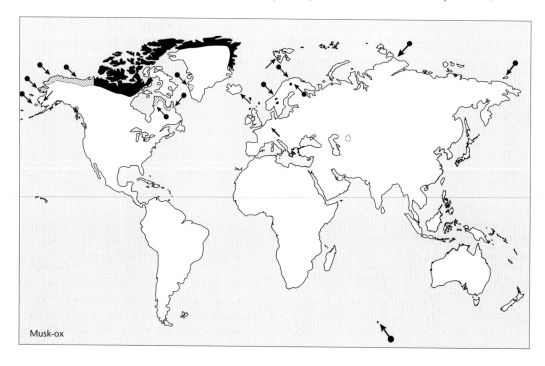

Musk-ox

released and three years after this another 20 were also released in the wild (Uspenski 1984; Klein 1988).

Four to five years after their release in the wild, the population had increased to 83 head (Uspenski 1984). By 1986 there were 175–177 and they had dispersed from the Strenk River in the north tundra of the Taimyr Peninsula, south to the border of the forest tundra (Klein 1988).

Twenty of the musk-oxen from Alaska were taken to Wrangel Island in the Siberian Sea (probably in 1978–79 when the releases occurred on the mainland) and released. In 1982 there were 21 present, and the herd was said to be stable and slowly increasing in numbers (Uspenski 1984; Klein 1988).

EUROPE
Germany
Some musk-ox were introduced into Germany, but failed to become established (Lever 1985).

Greenland
At present there is a natural population of musk-oxen in north-eastern and northern Greenland, which appears to be stable (Thing 1984; Klein 1988). However, at least three releases have occurred: the first two on the west coast of Greenland in 1962 of 13 head and in 1965, 14 head from north-east Greenland were released near Sondrestrom at the head of Sondre Stromfjord. By 1986 they had increased to about 1500 head. The third release occurred in 1986 when 47 yearlings from west Greenland were released in the north-west at Cape Atholl (seven head), Smith Sound (20), McCormick Fjord north of Qanaq (six), and Renselear Bay, Inglefield Land (14), where they are persisting (Vibe 1967; Thing *et al.* 1984; Klein 1988). Animals captured in both Canada and northern Greenland have been translocated to west Greenland (Gray 1984).

Iceland
Seven musk-oxen were taken to Iceland from north-east Greenland in 1930, but all died after a few months and the attempt failed (Barrow 1963; Allendal 1980).

Norway
Several musk-oxen were taken to Norway between 1920 and 1940 and subsequently released when zoos did not want them (Klein 1988). Musk-oxen are now successfully established in Norway (Whitfield 1985).

In 1925, six musk-oxen and in 1926, three from north-east Greenland were released on Gursk Island, but failed to become established there (Alendal 1980). Also from north-east Greenland, eight head were released at Bardu, but these again failed to become established (Alendal 1980).

Musk-oxen from north-east Greenland were released in the Dovrefjell mountains in southern Norway in 1932 and 1938. In 1932, 10 were released and in 1938, two were released (Barrow 1963; Corbet 1966; Alendal 1980). Apparently five died soon after in an avalanche and few of the remainder survived for long (Barrow 1963). Some reports indicate that they were exterminated during the German occupation (Hvass 1961). A population was established in the Dovre Mountains by translocation (27 head) from north-east Greenland in 1947–53 and has remained stable since then; recently it was reported that a herd of about 30–35 still exist in the area (Hvass 1961; Corbet 1966; Burton and Burton 1969; Alendal 1980; Klein 1988).

Svalbard (Norway)
Seventeen musk-oxen (11 calves and six yearlings) from north-east Greenland were released in September 1929 at Moskushamn, Adventfjorden, in Svalbard, Norway, to increase the diversity of wildlife as a source of food (Alendal 1976, 1980). The first calves were born in 1932 and the species continued to increase in numbers until 1939 when they decreased, probably due to hunting and dogs. In 1950 there were 50 there, but by 1974, probably due to adverse weather in 1973–74, only 30 remained. The population declined as the reindeer population built up and contributed to their final disappearance by 1982 (Klein and Staaland 1984; Klein 1988).

Sweden
There is mention of six musk-oxen which were released in Sweden, but all apparently died and the introduction was unsuccessful (Barrow 1963; Banfield 1977). However, five animals emigrated to Sweden from the Norwegian herd in the Dovre Mountains in 1971 and established a Swedish population which increased to 40 head by 1984 (Alendal 1980; Lundh 1984).

INDIAN OCEAN ISLANDS
Kerguelen
Musk-oxen have been introduced on this island (Bonfield 1977).

NORTH AMERICA
Alaska
Musk-oxen once roamed the entire arctic slope of Alaska, but the last were killed near Barrow, Alaska, in 1850–60. Re-introductions were initiated in 1930 when 34 musk-oxen were brought to Alaska from Greenland. As a result of releases over the last 69 years, five populations are now located on Nunivak Island, Nelson Island, Seward Peninsula, Cape Thomson and the Arctic National Wildlife Refuge. At least three populations exist within the Arctic

National Wildlife Refuge, which are stable and slowly increasing in numbers (Hone 1934; Grauvogel 1984; Smith 1984; Franzmann 1988; Klein 1988).

Some musk-oxen from Nunivak Island were released on Nelson Island in 1967 (eight head) and 1968 (15), where they became established and in 1986 numbered about 213 head. Some were released on Barter Island in north-eastern Alaska in 1969 (52 head) and in the Kavik River area, about 130 km west of Barter Island in 1970 (13). Here they became established along the Sadlerochit River drainage area and numbered 450 head in 1985. In 1970 (36) and 1977 (34) animals from Nunivak were released at Cape Thomson in north-western Alaska, and in 1985 there were 96 head present (Jingfors 1982; Jingfors and Klein 1982; Grauvogel 1984; Townsend 1986; Klein 1988).

The musk-ox was re-introduced to Fairbanks, Alaska, using semi-domestic animals (Palmer and Rouse 1963; Jones 1966). They were introduced on the Seward Peninsula from Nunivak Island in 1970 when 36 were released in the Feather River area and in 1981 when another 36 were released (Jingfors and Klein 1982; Townsend 1986). They have become established and were increasing in numbers in 1984–86 (Grauvogel 1984; Klein 1988).

Canada
The decision to bring musk-oxen to northern Quebec was in response to the Inuit Indians who wished to establish musk-ox farms for wool and also to introduce them to the area. From 1973 until 1983 some 54 animals from Ellesmere Island were released at three locations at Kuujjuak (Fort Chimo) at the head of Ungava Bay in northern Quebec as part of the domestication project (Le Henaff 1985; Le Henaff and Crete 1989). Musk-oxen are now well established in the tundra of northern Quebec; the last estimates of numbers indicated that between 290 and 350 head were in the area (Klein 1988; Le Henaff and Crete 1989).

Nunivak Island (Aleutian Islands, Alaska)
Musk-oxen, (*O. m. wardi*) (Hall 1981), were returned to Alaska in 1930 as a result of United States Congressional action that appropriated money for their purchase and transport (Palmer and Rouse 1936; Murie 1940). Thirty-four musk-oxen were transported to Norway following their capture in north-eastern Greenland and then shipped to the United States. They were held in quarantine in New Jersey for one month and then taken to Seward, Alaska, and thence to College where they were released in paddocks of the Biological Experimental Station. Finally, in 1935 and 1936, the remaining 31, including calves born during the trip, were shipped to Nenana, thence to the Bering Sea and Nunivak Island

where they were released (Barrow 1963; Burris 1965; Gottschalk 1967; Spencer and Lensink 1970; Jingfors and Klein 1982; Gray 1984).

The 1935 release consisted of four animals and in the 1936 release some 27 musk-oxen. By 1938 the population numbered 50; in 1939 it was 60; in 1943 it was 100; in 1947, 49; in 1957, 143; in 1962, 340; by 1967–68 there were 651–714 head there and in 1970 there were 750 (Young 1941; Barrow 1963; Burton and Burton 1969; Spencer and Lensink 1970). In 1985 there were about 616 head and many had been translocated to mainland Alaska.

◼ DAMAGE
None known.

HIMALAYAN TAHR
Tahr, thar
Hemitragus jemlahicus (Smith)

◼ DESCRIPTION
HB 910–1580 mm; T 90–120 mm; SH 610–1060 mm; WT males 20–50 kg and up to 100 kg, females 16–45 kg.
Generally dark brown to reddish brown; under parts lighter than upper parts; goat-like appearance; face, muzzle and dorsal stripe brown black; coat long and shaggy with long grey to yellowish mane on neck and chest; lacks glands on face and feet; rump patch red-brown; ears narrow and pointed; horns short and laterally flattened, curved backwards, 300–375 mm, prominently keeled along front edge; underside of tail bare. Female lacks rump patch and mane, has four mammae. Differ from goats in having a naked muzzle, glands on feet and non-twisted horns. Yearlings resemble female, but are smaller.

◼ DISTRIBUTION
Asia. Northern India, Nepal, Sikkim and Bhutan in Himalayas. Range now very discontinuous.

◼ HABITS AND BEHAVIOUR
Habits: mainly diurnal, rests in middle of day; climbs well. **Gregariousness:** females in herds with juveniles (2–40); old males may be solitary for periods; bachelor groups of younger males; mixed sex groups (to 10) before onset of rut; density 4.5–6.8/km². **Movements:** sedentary, but little information. **Habitat:** steep hillsides and rock bluffs with scrub and precipices, mountains to tree line, elevated forest clearings and grasslands. **Foods:** grass and sub-alpine shrubs, herbs, flowering heads, seed heads. **Breeding:** mates October–January (Himalayas) (rut lasts 6 weeks May–mid July in NZ); gestation 6–8 months (or 165–168 days?), 1 kid, rarely 2; lactation 6 months;

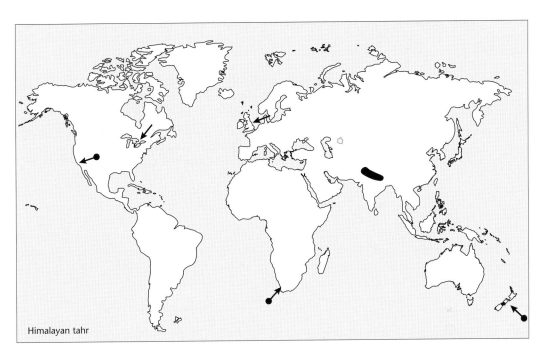

Himalayan tahr

kids nurse within half hour, walk in 2–3 hours; bulls mature by 2.5 years, but rarely mate before 4.5 years. **Longevity:** 16–21.75 years (captive). **Status:** now reduced to isolated populations by settlement, hunting and agriculture; still occur in some parts of Nepal and common in New Zealand.

■ HISTORY OF INTRODUCTIONS
Successfully introduced into New Zealand, Africa and California.

AFRICA
South Africa
The Himalayan tahr were introduced by Cecil Rhodes to the Cape Peninsula where they were kept in captivity. A pair was sent to the zoo at 'Groote Schuur' on the lower slopes of Table Mountain, Cape Town, from the National Zoological Gardens in 1935 (Bigalke 1977). Between this date and 1939 a pair escaped from a zoo on the Groote Schuur estate, by jumping over a 1.5 m fence. By the early 1960s there were about 50 descendants roaming the area, and by the early 1970s their numbers had increased to around 300–500 (Burton and Burton 1969; Hey 1974). In 1971 there were 264, and in 1972, 330. Shooting campaigns in 1973 killed 31, in 1976 some 228 and in 1977 about 191 (Brooke *et al.* 1986), although other estimates suggest between 1975 and 1981 some 600 were removed (Smithers 1983).

Culling was instigated in 1973 because of the damage cause by the tahr to vegetation and also erosion on mountain slopes, and by 1981 there were thought to

be only 88 left (Smithers 1984). However, there are probably now as many as there were before the shooting campaigns (Bigalke and Pepler 1991).

EUROPE
United Kingdom
Prior to 1914 some tahr were said to have been liberated at Cairnsmore of Fleet, Kirkcudbrightshire, Scotland, but they died out (Whitehead 1972).

PACIFIC OCEAN ISLANDS
New Zealand
Tahr were introduced to New Zealand in 1904 (three females and two males) and 1909 (eight females and six males) near Mount Cook for hunting. In 1919, four more were released – two other liberations in 1909 and 1913 failed. By the mid-1960s they were reported to be locally common on the South Island (Wodzicki 1950, 1965) and by the mid-1970s occurred over an area of the alps stretching from the Waimakarri River to Lake Wanaka (Gibb and Flux 1973). However, control efforts between 1936 and 1968 eliminated 30 000 tahr (Wodzicki and Wright 1982).

Himalayan tahr reached their maximum distribution in New Zealand in the 1970s, but have since declined somewhat due to aerial hunting, especially on the north-western and southern limits of their range; they are still established in the Canterbury and Westland areas of the Southern Alps (Barnett 1985; King 1990; Fraser *et al.* 1996).

NORTH AMERICA

Canada

Some Himalayan tahr from Toronto Zoo were released in Peterborough, Ontario, where they survived for a while, but did not spread (Lever 1985).

United States

Himalayan tahr are reported to occur in the vicinity of the Hearst Ranch, San Luis Obispo County, California, from whence they were released and became established (Barrett 1966; Williams 1979).

■ DAMAGE

In South Africa Himalayan tahr are denuding the vegetation and causing soil erosion on the steep slopes of Table Mountain, Cape Province (Hey 1974; Smithers 1983; Bigalke and Pepler 1991).

They may also have caused some damage to alpine vegetation in New Zealand in the past, but their numbers are now so reduced that they probably do little damage (King 1990).

CAUCASIAN TUR

Caucasian goat, West Caucasian tur, kuban
***Capra caucasica* Guldenstaedt and Pallas**

■ DESCRIPTION

HB 1200–1650 mm; SH 780–1090 mm; WT 50–100 kg.
Body stout; neck massive; legs short; tail very short; beard of males short; upper parts rusty grey to rusty

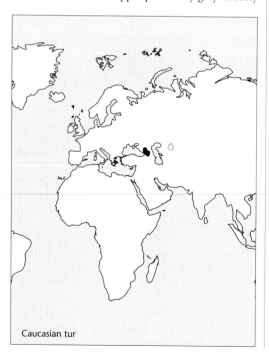

Caucasian tur

chestnut, grey-brown in winter, lighter on flanks; under parts greyish or whitish; horns scimitar-shaped, to 740 mm in males, shorter in females.

■ DISTRIBUTION

Asia. In the Russian Federation occurs in the western Caucasus from 39° 55′ E to the headwaters of the River Psygansu at 43° 30′ E.

■ HABITS AND BEHAVIOUR

Habits: seasonal migrations between 800–2400 m in altitude; herds of up to 500, mixed herds during breeding season, otherwise normally only a few dozen females and their young; males live in separate herds during non-breeding season. **Gregariousness:** herds to 500; groups of a few dozen; males usually separate; mixed herds in breeding season; density 50–60/100 ha. **Movements:** seasonal migrations 1500–2000 m; move up in May, down in October–November; daily movements 15–20 km. **Habitat:** mountains, steep rocky slopes, alpine meadows, barren areas, forest. **Foods:** grass, leaves or trees and shrubs. **Breeding:** mates late November–early January; gestation 150–160 days; 1 young, rarely 2; births mid-May–June; eats grass at 1 month; suckles to end of summer; female sexual maturity second year, male 4–5th year. **Longevity:** most die before 10 years in wild, but maximum may be 22 years. **Status:** declined in nineteenth and twentieth centuries, recovering in parts of its range following protection.

■ HISTORY OF INTRODUCTIONS

EUROPE

Russian Federation

The Caucasian tur has been introduced in the Caucasus where it became established, but later died out (Yanushevich 1966). Attempts to introduce this species outside its natural range have not generally been successful in the Russian Federation.

■ DAMAGE

No information.

MARKHOR

***Capra falconeri* (Wagner)**

■ DESCRIPTION

HB 1400–1800 mm; T 80–150 mm; SH to 650–1040 mm; WT females 32–40 kg, males 80–110 kg.
Beard long, black; flowing ruff of white to grey on chin, shoulder and stifle; flank stripe dark and separates white belly from brown and grey sides; rump patch small and white; lower legs white except for dark haired wedge below knee; shaggy mane of long dark hairs extending down neck and chest in males.

Markhor

Males have long horns to 1650 mm, horns spiral-shaped, but variable in shape and size. Female much smaller than male; lacks beard; horns shorter. Yearlings resemble females, but are two-thirds their size.

■ DISTRIBUTION
Asia. Mountains from eastern Kashmir to the Hindu Kush and south in west Pakistan to the Quetta Region; southern Uzbekistan and Tadzhikistan and possibly in north-west Afghanistan.

■ HABITS AND BEHAVIOUR
Habits: mainly active early morning and late afternoon. **Gregariousness:** mixed herds (3–35) or female and sub-adults (2–10); males in herds or solitary; sometimes aggregations 30–100; density 1–9/km². **Movements:** moves to summer range 10–15 km; home range 80 km². **Habitat:** arid habitats 600–3600 m elevation; steep gorges, rocky areas, scrub forest, grassy meadows. **Foods:** grass, leaves and twigs of shrubs in summer. **Breeding:** mates mid-December–January, births in April–June; gestation 155 days; young 1–2; sexual maturity 30 months. **Longevity:** 11–13 years in wild. **Status:** range now highly fragmented; possibly endangered.

■ HISTORY OF INTRODUCTIONS
ASIA

Russian Federation
Almost certainly the markhor has contributed to some breeds of domestic goat and has itself been introduced and domesticated in the Caucasus

(Couturier 1962; Corbet 1978). They will hybridise with feral goats which could lead to the loss of pure-bred populations (WCMC 1998).

■ DAMAGE
No information.

GOAT
Feral goat, domestic goat, wild goat
Capra hircus Linnaeus
=the principal ancestor of the feral goat, if not conspecific with, was the Persian wild goat or bezoar (C. aegagrus Erxleben) which is sometimes only given subspecific status: C. hircus aegagrus.

■ DESCRIPTION
HB 610–1620 mm; T 64–170 mm; SH 356–914 mm; WT 15–79 kg.
There are numerous breeds and a multitude of colours, horned or hornless; coat colour varies from black, brown to white and variations of all three colours; original wild animal was reddish brown in summer and greyish brown in winter, with black marks on body and limbs; horns scimitar shaped, 300–1300 mm, laterally compressed and inner front edge bears large knobs. Female horns small and slender.

■ DISTRIBUTION
Eurasia. From south-eastern Europe, through Asia Minor to Pakistan.

■ HABITS AND BEHAVIOUR
Habits: diurnal, peak activity crepuscular; play is conspicuous feature of behaviour, particularly at dusk. **Gregariousness:** small flocks (matriarchal groups) 3–20; mature males separate except at breeding; young adult males form distinct mixed age bands continually forming and dissolving; density 11.8–68/km². **Movements:** sedentary; daily movements erratic around home range; home range 1–5 km². **Habitat:** steep rocky areas, woodland, steppe, farmland, scrub, forest; exploit areas (ledges and steep crags) that deer and sheep can not reach. **Foods:** grazes and browses; grass, leaves, twigs, shoots, flowers, berries, bark, ferns; browse shrubs and trees; herbs. **Breeding:** breeds all year; ruts mainly August–November and kid January–March; gestation 150 days; young 1–2, rarely 3 in wild; seasonally poly-oestrous; females may breed twice per year; kids hidden for first few days while female forages; kids reach adult size at 3 years; females about to give birth are solitary; sexual maturity 6 months. **Longevity:** 12–13 years (wild), to 16 years (captive). **Status:** common and abundant.

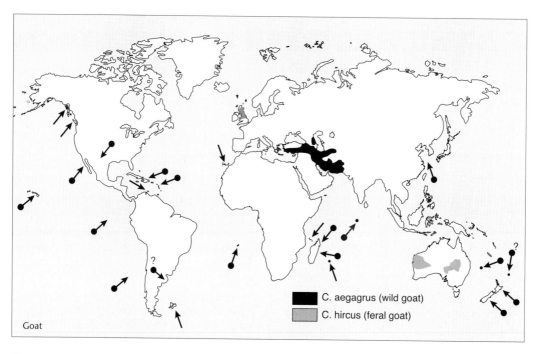

C. aegagrus (wild goat)

C. hircus (feral goat)

Goat

■ HISTORY OF INTRODUCTIONS

DOMESTICATION

Present evidence indicates that domestication of goats took place in south-west Asia between 8000 and 9000 years ago (Walker 1968), probably in the Neolithic 7000–6000 BC in the Middle East (Zeuner 1963). The wild form still exists in the mountains from Asia Minor and Caucasus to Kopet Dag, western Afghanistan, Baluchistan and Sind. Isolated populations occur in Oman and on some Aegean Islands and Crete. Formerly found in northern Syria, Lebanon, and Palestine.

There are now few wild populations that have not been affected by inter-breeding with feral domestic goats. They are found as feral animals in Australia, New Zealand, Hawaiian Islands, St. Helena, Mauritius, Christmas Island, Juan Fernández, Galápagos Islands, Skokholm, Channel Islands, British Isles, Canada, Greek Islands and most small islands in the Mediterranean Sea.

ISLANDS ON WHICH GOATS HAVE BEEN INTRODUCED

Achill (Eire), Adams (NZ), Aegean Is (Greece), Agil (Greece), Aguijan (Marianas), Ailsa Craig (UK), Aiwa (Fiji), Albemarle (see Isabela), Albingdon (see Pinta), Aldabra (Seychelles), Allejandro Selkirk (Juan Fernández), Althorpe (Australia), Amsterdam, Anglesey (UK), Antipodes (NZ), Arapawa (NZ), Arran (UK), Ascension, Assumption, Auckland (NZ), Austral Is (Pacific O.), Azores (Atlantic O.).

Balearic Is (Spain, Mediterranean), Baltra (Galápagos), Barington (see Santa Fé), Bonaire (Netherland Antilles), Bonin (Japan), Bowen (Australia), Broughton (Australia), Bugio (Desertas), Burgess (Mokohinau, NZ), Bute (UK).

Calf of Man (UK), Campbell (NZ), Canary Is, Capricorn Group (Australia), Cara (UK), Cavalli (NZ), Cerro Azus (Isabela, Galápagos), Channel Is (United States), Chão (Desertas), Charles (see Floreana), Charles (see Santa Maria), Chatham (see San Cristobal), Christmas, Cocos (Costa Rica), Coll (UK), Colonsay (UK), Columbrete Grande (Mediterranean), Congo (Virgin Is), Corsica (France, Mediterranean), Cosmoledo (Aldabra), Crete (Greece), Crozet (France), Cull (Recherche Arch., Australia), Curacao (Netherlands Antilles), Curvier (NZ), Cyprus (Greece).

Davaar (UK), D'Urville (NZ), Davaar (UK), Desecheo (Puerto Rico), Deserta Grande (Desertas), Desertas, Devilan (Fiji), Dia (Greece), Dirk Hartog (Australia), Dutch Cap (Virgin Is).

Easedale (UK), East (NZ), East Falkland, Eigg (UK), Enderby (NZ), Eorsa (UK), Erimomilos (Greece), Ernest (NZ), Espanola (Galápagos), Europa (Madagascar), Ewing (NZ).

Falkland Is, Fatuhiva (Marquesas), Faure (Australia), Fijian Is, Floreana (Galápagos), Forsyth (NZ), French (Australia).

Galápagos Is (Ecuador), Galiana (Canada), Gambier (Australia), Garaina (New Guinea), Gigha (UK), Giour

(Greece), Goat (Fiji), Goira (Greece), Gough (Tristan de Cunha), Grand Jason (Falklands), Grand Terre (Aldabra), Great (Three Kings, NZ), Great Barrier (NZ), Great Hans Lollick (Virgin Is), Great Mercury (NZ), Guadalupe (Mexico), Guam, Gunna (UK).

Harris (Hebrides), Hawaiian Is, Hebrides (UK), Hen (Canada), Herekopare (NZ), Heron (Capricorn Group, Australia), Hispaniola (West Indies), Holy (UK), Hualalai (Hawaii, United States).

Ibiza (Spain), Isla de los Estados, Inaccessible (Tristan da Cunha), Indefatigable (see Santa Cruz), Inisteeraght (Ireland), Inisvickillane (Ireland), Isabela (Galápagos), Islay (UK), Isle of Man (UK).

James (see Santiago, Galápagos), Jarvis (United States, Pacific O.), Juan Fernández (Chile), Jura (UK).

Kadavu (Fiji), Kahoolawe (Hawaiian), Kangaroo (Australia), Kapiti (NZ), Kauai (Hawaiian), Kerrera (UK).

Lanai (Hawaiian), Lasquiti (Canada), Lesser Hans Lollick (Virgin), Lismore (UK), Little Colonsay (UK), Lord Howe (Australia), Los Estados (Chile), Luf (PNG), Lundy (UK), Lynton Rocks (UK).

Macauley (Kermadecs, NZ), Macquarie (Australia), Madagascar, Mahurangi (NZ), Majorca (Balearic), Makodroga (Fiji), Makogai (Fiji), Makoia (NZ), Malabar (Aldabra), Malden (Line Is), Man (UK), Mangere (NZ), Marchena (Galápagos), Marianas (Pacific O.), Marion (PNG), Marquesas (Pacific O.), Más Afuera (Juan Fernández), Más á Tierra (Juan Fernández), Maud (NZ), Mauna Kea (Hawaii, United States), Mauritius, Mayne (Canada), Meetia (Mehetia), Menai (Cosmoledos), Mingo (Virgin), Mistaken (Australia), Molokai (Hawaiian), Mona (Puerto Rico), Montecristo (Italy), Montagu (Australia), Monuriki (Fiji), Moreton (Australia), Motuoruhi (Goat, NZ), Moutohora (NZ), Mull (Hebrides).

Namara (Kadavu, Fiji), Netherlands Antilles, New Caledonia (Pacific O.), New Zealand, Niihau (Hawaiian), Norfolk (Australia), North East (Cosmoledos), North-East (Australia), North Goulbourne (Australia), Nouvelle Amsterdam (see Amsterdam), Nukutaunga (Cavalli, NZ).

Oahu (Hawaiian), Ocean (Auckland), Oronsay (UK), Otaheite (Society), Outer Brass (Virgin).

Pantes (Greece), Philip (Norfolk, Australia), Picard (Aldabra), Pinta (Galápagos), Pitcairn (Pacific O.), Plaza Sur (Galápagos), Possession (Crozet), Pourewa (NZ), Prevost (Canada), Prince of Wales (Australia), Providence, Puerto Rico (West Indies).

Rabida (Galápagos), Raivavae (Austral), Ramsay (Canada), Raoul (Kermadecs, NZ), Rapa (Austral),

Rathlin (Eire), Rhum (UK), Rimatara (Austral), Robinson Crusoe (Juan Fernández), Round (Mauritius), Rurutu (Austral).

Salt (Puerto Rico), Saltspring (Canada), Samothrace (Greece), San Clemente (Channel, United States), San Cristobal (Galápagos), Santa Catalina (Channel, United States), Santa Clara (Juan Fernández), Santa Cruz (Galápagos), Santa Cruz (Channel, United States), Santa Fé (Galápagos), Santa Maria (Azores), Santa Maria (Galápagos), Santiago (Galápagos), Saturna (Canada), Scarba (UK), Scolpay (UK), Seil (UK), Shark (Virgin), Shunna (UK), Sidney (Canada), Sierra Negra (Galápagos), Skellig Rocks (UK), Skokholm, Skye (UK), Snares (NZ), Society Is (Pacific O.), South East (NZ), South Seymour (see Baltra), St. Helena, St. Paul (Indian O.), St. Pierre (Providence), Staffa (UK), Steep-to (NZ), Summer Is (UK).

Tahiti (Society), Tenerife (Canary Is), Texa (UK), Texada (Canada), Theodoru (Greece), Three Kings Is. (NZ), Tiree (UK), Tomogasima (Japan), Trinidade (Brazil), Tristan da Cunha, Tubuai (Austral).

Ulva (UK), Vancouver (Canada).

Virgin Is (USVI), Viti Levu (Fiji).

Wakaya (Fiji), Walpole (New Caledonia), West Cay (Virgin), Whale (NZ), Woody (Australia).

Yadua (Fiji), Yaduataba (Fiji), Yaqaga (Fiji), Yasavas (Fiji).

ATLANTIC OCEAN ISLANDS
Ascension Island
Goats were introduced by the French some time before 1701 as they were reported to be common on the island at this date (Duffey 1964).

Azores
In the 1960s some were reported to be on the ledges and coves on steep hillsides on the island of Santa Maria (Bannerman and Bannerman 1966).

Canary Islands
Feral goats were present on these islands in the 1960s (Zeuner 1963). They were noted to be present on Tenerife above the tree line in the late 1940s (Lack and Southern 1949).

Desertas
Domestic goats had become feral on Deserta Grande in the sixteenth century. They were present in fair numbers in the 1920s and also occurred on Chão and Bugio, and are still there on Deserta Grande (Bannerman and Bannerman 1965; Cook and Yalden 1980).

Falkland Islands

First introduced in 1764 by the French explorer de Bougainville, goats were later introduced on some offshore islands by nineteenth-century sealers and whalers. By 1846 goats were established in the mountains of East Falkland, and in 1870 several hundred were exported from Grand Jason Island. There are apparently no feral goats on any of the islands today (Lever 1985).

St. Helena

The Portuguese are thought to have introduced goats to St. Helena in 1513 (Brooke 1824), but in fact they may have been released there as early as 1502 (Cronk 1986). By 1588 the goat population had exploded and numbered thousands (Wallace 1911). Flocks nearly a mile long were noted by Hal Kluyt in 1589.

Further goats were introduced in 1676 in attempts to improve the strain of the domestic goats on the island.

Because of damage to the vegetation, attempts were first made to exterminate goats in 1731. These attempts lasted for 10 years and were repeated in 1745. In 1809 the Governor recorded a goat population of 2887 head and revived the policy of extermination (Cronk 1986). By 1810 most of the islands forests had disappeared, cut down by sailors for timber, and prevented from regeneration by the grazing of the goats (Wallace 1902; George 1962; Holdgate 1967).

Despite the early extermination attempts the population again increased. During the 1960s and 1970s an extermination policy was again followed and now only a few goats exist in inaccessible places (Cronk 1986).

The goats on Île St. Paul were liberated by fishermen in the nineteenth century on the island where they were used to provide meat (Jeannel 1941). However, they had disappeared by 1847, apparently all killed for meat.

Tristan da Cunha

Goats could have been introduced to Tristan as early as 1506 when the islands were discovered (Lever 1985), but certainly some time before 1790 (Holdgate and Wace 1961; Holdgate 1967) probably by the Portuguese as they were numerous there at this time (Munch 1945). Goats were present on the island in 1829 (Morrell 1832), but have now been exterminated (Holdgate and Wace 1961).

Inaccessible Island once also supported goats, but they had dwindled to a small remnant population by 1873 (Moseley 1892). More may have been introduced in 1882–83 (Lever 1985).

Some goats were introduced on Gough Island in 1958, but their destruction was ordered immediately after they were landed by the Tristan da Cunha Administration (Holgate and Wace 1961). They are reported to have died out in the 1960s (Lever 1985).

Islands off the coast of Australia with feral goats

Island	State	Date	Status, notes and source
Althorpe	SA		present (Mahood 1978)
Bowen	NSW		present (Lane 1976)
Broughton	NSW		released for milk but later removed (Lane 1976)
Cull, Arch. de Recherche	WA	1935	still there (Lane 1982)
Dirk Hartog	WA		800 shot 1972–73 (Burbidge & George 1978)
Faure	WA	early 1900s	present (Clarke 1976)
French	Vic	?	present (Flux & Fullagar 1992)
Gambier	SA		present (Mahood 1978)
Heron, Capricorn Group	Qld		removed long ago (Kikkawa & Boles 1976)
Kangaroo	SA	?nineteenth century	established 1967 (Condon 1967)
Lord Howe	NSW	nineteenth century	extermination in progress (Pickard 1976)
Mistaken	WA	before 1841	died out or removed (Abbott 1978)
Montagu	NSW	?	present (Smith & Dodkin 1989)
Moreton	Qld	1860s	liberated by navy, now c.1000
North-East	NT	unknown	transferred from Umba Kumba mission by F. Gray as food for aborigines (Bd Enquiry 1979)
Philip (off Norfolk)	NSW	early settlers ?	died or shot out c. (1850) (Coyne 1982)
Prince of Wales	NT		present (Draffan et al. 1982)
Woody	Qld		present (Mahood 1978)

AUSTRALASIA
Australia

Imported to Australia in 1788 and on many subsequent occasions by European settlers, goats were abandoned, released or escaped from early pastoralists, miners, and construction workers who used them as a source of meat and milk (Holst 1981; Mahood 1983; Strahan 1995). Feral flocks soon formed and goats began to spread as a feral animal.

Feral goats now occur in all states and territories except the Northern Territory, although there are herds on North Goulbourne Island and other islands. The largest populations are in Western Australia, western New South Wales, southern South Australia and south-central Queensland (Wilson *et al*. 1992). Isolated populations occur elsewhere. Most feral goat populations occur where the topography provides refuge from predation by dingoes.

By 1912 on many stations in the Northern Territory breeding groups escaped from time to time but have failed to survive (Letts 1964).

Fifty thousand goats still occur in the wild in Queensland but numbers have decreased in last 20 years due to culling for the pet meat industry and export. Some colonies exist in the far south-west, originating from escapees from pet food abatoirs (Mitchell *et al*. 1982). There are numberous colonies in New South Wales (Mahood 1978). In Tasmania some general escapes occur from time to time and there are about 23 small populations.

Papua New Guinea

Goats are reported to be feral on Marion and Luf islands in the North Western Island group, and in Garaina in the Morobe Province (Herrington 1977).

CENTRAL AMERICA
Costa Rica

Goats were usuccessfully introduced on Cocos Island, Costa Rica.

EUROPE

Feral goats occur on most small islands in the Mediterranean, in several places in Britain, western Ireland, North Wales, the Cheviot Hills, in the Scottish Highlands and some of the Hebrides (Mull) (Corbet 1978). They occur over most of Mediterranean region and several offshore islands: on Crete, Cyprus, Theodoru, Goira, Erimomilos and Samothrace (Greece); on Montecristo (Italy); on Corsica (France); on Ibiza and Majorca in the Balearics (Spain) (Lever 1985).

In 1910 the Bezoar goat was introduced from Asia Minor to the High Tatra Mountains of Czechoslovakia, where this species and the Nubian ibex (*C. ibex nubiana*) bred with the native ibex (*C. ibex ibex*) which had also been re-introduced since 1901 (de Vos *et al*. 1956).

Feral goats occur on some Greek islands (Hvass 1961). They are evidently present in some afforestation areas in Yugoslavia where they are controlled (Ziani 1964), but no recent information is available (Rudge 1984).

Corsica

Goats are occasionally feral in mountainous areas where strays are left on summer pastures (Rudge 1984).

Spain

Goats are present on Ibiza Island and Majorca where they are controlled by shooting (Tegner 1971). They are also present on Columbrete Grande, in the Mediterranean Sea, where they were introduced about 1855.

United Kingdom and Ireland

Goats were introduced to Britain during the Neolithic era and feral animals were established in many districts in the seventeenth century. Evidence of them in Wales dates from the eighteenth century onwards. They were observed in the Bachwy Ravine, Radnorshire, in 1803 and on the north coast of Pembrokeshire as late as 1942 (Fitter 1959; Matheson 1959; Milne *et al*. 1968; Lever 1977).

In Scotland goats were known from the fourteenth century and there were many more records from the eighteenth century. Whitehead (1972) listed 134 populations containing less than 10–400 animals, with extensive distributions on the mainland and western islands.

In the late 1960s remnant herds were confined to North Wales in the counties of Caenarvonshire and Merionethshire in mountainous areas such as Yr Eifl, Carneddan, Glyder, Snowdon, Moelwyn, Rhinog, Arenig, Rhobell, Fawr, Cader Idris and Craig Aderyn (Southern 1964; Milner *et al*. 1968). Groups occurred in the past on Aron Fawddwy and Moel Siabod (Milner *et al*. 1968). In the late 1970s they occurred in Carnarvonshire, Merionethshire, Pembrokeshire, and on the Isle of Anglesey (Lever 1977).

A flock in Mamlon, Perthshire, Scotland, dates from 1661–79 and others in Inverness-shire at least since 1745 and 1835 and others from Ross-shire in the same era (Lever 1977). Some occurred on An Teallach in 1937 (Darling 1937), probably from introductions in about 1927 (Fitter 1959). Feral herds are still to be found in remote parts of the Scottish Highlands and some Hebridean islands (Southern 1964). In the late 1970s they were thinly but widely distributed throughout the highlands and islands as well as in a number of lowland and border counties. These counties included Argyllshire, Ayrshire, Dumfriesshire, Inverness-shire, Kirkudbrightshire, Morayshire, Nairnshire, Perthshire, Ross and Cromarty, Roxburgshire, Selkirkshire, Stirlingshire, Sutherland and the isles of Cara, Colonsay, Islay, Jura, Little Colonsay, Mull, Oronsay, Texa (south-east of Islay) and Holy Island (east of Arran) (Lever 1977). Those in Wester Ross are of ancient stock either introduced or indigenous from early times (Fitter 1959).

In England there were goats on Lundy Island in the eighteenth century, but these disappeared before 1914, were re-introduced in 1926 by M. Harman, and thereafter increased to a herd of 200. However, by 1975 they had been reduced to around 30 (Lever 1977). There have been goats on Lynton Rocks, Devon, while those on the Isle of Man may have originated from introductions by lighthouse keepers in 1818 and 1875 (Lever 1977). A number of flourishing flocks occur on the Cheviot Hills, Northumberland (Southern 1964), and in Staffordshire they have existed for some 600 years (Lever 1977). Today they occur in Devon, on the Isle of Man, Lancashire and Northhumberland (Lever 1977).

In Eire goats occur on Achill Island and in Northern Ireland at Rathlin Island and Fair Head, County Antrim (Southern 1964). More recently they have been reported from counties Clare, Donegal, Dublin, Galway, Kerry, Mayo, Sligo, Tipperary, Waterford, Wicklow, Antrim, Armagh and Fermanagh.

There were probably feral goats in the forests of England in 1615 and doubtless they were common in Scotland from earliest times, certainly 1661–79. Many escaped or were abandoned in the eighteenth century when breeding of goats waned. Around 1900 some were deliberately released for stalking. At least 134 populations containing several thousand goats in total occur in Scotland. Some populations are very old and may date from the fourteenth century and mid-eighteenth century (Rudge 1984). In Wales feral herds have occurred there since the mid-nineteenth century at least. There are at least seven populations with a total of 300 goats. In Ireland some have been known to be feral for 200 years, but many have an unknown origin and some are of twentieth century origin; in all, 17 feral herds and 11 park herds are listed (Rudge 1984).

Goats were present in the 1970s on a number of islands off the west coast of Scotland including: Harris (Hebrides) (end nineteenth century), Rhum (1772 on), Cara (eighteenth century on), Colonsay,

Islay (eighteenth century on), Jura (early twentieth century on), Little Colonsay (1922 on), Mull (before 1914 on), Oronsay (?1947 on), Texa (off Islay) (twentieth century), Davaar (recent?), Kerrera (recent?), Holy (1874 on), Bute (ancient–), and formerly on Scolpay (eighteenth and nineteenth centuy), Skye (seventeenth century–1952), Eigg (before 1914–1946/47), Coll (eighteenth and nineteenth century), Gunna (1934–?), Lismore (?1845–?1952), Seil (?–1939–45), Shuna (?–?1964), Ulva (?1930s–?1940s), Easdale, Eorsa, Gigha, Scarba, Staffa, Tiree, Ailsa Craig (eighteenth century–1925), Arran (nineteenth century) and Summer Islands (1936–37).

Goats are present on islands off England, Wales and Ireland on: Lundy (eighteenth century on), Anglesay, Isle of Man (1905 on), Calf of Man (1818–75 on), Rathlin (1760 on), Achil (eighteenth century on) and Skellig Rocks off St. Finans Bay; formerly on Innisvickillane (c. 1880), and Innisteeraght islands (c. 1895), (County Kerry) (Whitehead 1972). They were introduced on Skokholm by lighthouse keepers for milking and six were seen there in 1968 (Berry 1968).

Crete and Greece

The Creton wild goat (*C. h. aegagrus*) formerly occurred in most countries around the Mediterranean but during the nineteenth century became confined to a small region of the Gorge of Samaria in the White Mountains of Crete. After World War 2 these were threatened with extinction, so in 1954 the Greek government translocated a number of them to islets off the north coast of Crete where they flourished in semi-captivity and at present number almost 100 on various islets (Anon. 1963). They are now present on Agil, Pantes, and Dia islands (Rudge 1984). Goats are still present and declining on Crete. Feral goats (Agrimi wild goat) also occur on Giour, Erimomilos and Samo-Thrace islands (Greece) (Schultze-Westrum 1963; Rudge 1984). *C. aegagrus* is said to survive on Crete and has been introduced to Theodoru Island just to the north. Populations on other Aegian Islands have hybridised with *C. hircus*. There are still 100 on Theodoru and 300 on Crete (Walker 1992).

INDIAN OCEAN ISLANDS
Aldabra Island (Seychelles)
Possibly introduced by Arab seafarers in the seventeenth or eighteenth century or by early European mariners (Mason 1979; Lever 1985), goats are first recorded on the island in 1878. Some 40 to 50 head were present on Île Picard in 1895 (Baty 1895 in Stoddart 1981) and it was claimed that they were

placed there in 1890. Baty released eight more from Assumption Island on Picard.

In 1900, 20–30 goats were noted on Picard (Bergne 1900 in Stoddart 1981) and in 1906 they were thriving on both Île Picard and on the mainland (Dupont 1907). They were reported to be present there in the hundreds in 1916 and several thousand in 1929. In 1959 there were considerable herds in the southern sand dunes (Travis 1959), and in 1957 they were recorded on the eastern end of Île Malabar. In 1967 (Stoddart 1967) they were found on Grand Terre, Malabar, and Picard. More recently (Diamond 1981) it was reported that they were still on Picard, where they have possibly been introduced and exterminated more than once.

Assumption Island

Goats were introduced to the island by Captain Bidenfield of the HMS *Wasp* (Bergne 1900 in Stoddart 1981) before 1893 from Europa Island in the Mozambique Channel. Others (Dupont 1907) suggest that they were introduced in about 1887. However, there was 500–600 there in 1878, reportedly left by passing ships. In 1895 there were 300–400 goats on the island (Baty 1895; Bergne 1900; Stoddart 1981). Seventy were collected in two days in 1905 (Tonnet 1905, in Stoddart 1981). By 1906 several thousand were reported (Dupont 1907), although in 1908 only 20 were noted (Nicoll 1908). Apparently there were only a few left in 1916 (Dupont 1907 in Stoddart 1981).

There is no further mention of goats until the late 1930s when they are reported as present again on the island (Vesey-Fitzgerald 1942). There is no mention of them in 1942 and there were none there in 1964 or 1967 (Stoddart *et al.* 1970). Although none were seen in the 1960s they may still be present (Stoddart 1981).

Cosmoledo Island (Aldabra group)

There appear to have been a few goats on Menai Island in 1878 and larger numbers there in 1900 (Bergne 1900, in Stoddart 1981). There may have been large numbers in 1892, few on Menai in 1895, but none in 1968, although some may have existed on North East Island in 1961 (Diamond 1981; Stoddart 1981).

Amsterdam (Nouvelle Amsterdam)

A few goats were present on the island in 1823 (Goodrich 1843), probably introduced by early visitors, but numbers gradually dwindled until 1957 when only an old male remained (Holdgate and Wace 1961). They have caused much damage to the island's flora (Holdgate 1967).

Crozet Archipelago

Goats were reported on Crozet in 1875 (Kidder 1876), but there appear to be few further records of them.

They are reported to have been introduced to Possession Island where they failed to become established (Watson 1975).

St. Paul

Feral goats were formerly present on this island (Lever 1985).

Madagascar

Goats were introduced to Île Europa (in Mozambique Channel) about 1860 and still occurred there in 1966, when 150–200 of them were reported (Malzy 1966).

Mauritius

Goats were introduced to Mauritius as early as 1512 when Pedro Mascarenhas released 'hogs' and other animals on the island (Lever 1985). They were released by Thomas Corby on Round Island, off Mauritius, in 1844 (Temple 1974) or between 1840 and 1865 (Bullock 1977), but all but two of them were shot in 1976 (Bullock 1977) or 1978 (North and Bullock 1986). In 1950 there were about 100 there, but only old sign was noted in 1982 and no live goats have been seen since then (Bullock and North 1984; North and Bullock 1986).

Providence group

Feral goats are present in this group on the island of St. Pierre (Lever 1985).

NORTH AMERICA
Canada

Feral goats occurred on some islands in the Straits of Georgia, British Columbia including Saturna, Galiana, and Saltspring, and were also reported to be on Lasquiti Island, Texada Islands, Vancouver Island and islands in the Queen Charlotte Sound in the 1960s (all nineteenth century introductions); Mayne Island possessed a small herd introduced in 1942, but they had died out by 1955–56 (Geist 1960).

In 1960 Mt. Bruce, on Saltspring Island, was inhabited by the progeny of goats introduced in 1925 and which had been maintained for a small cheese-making factory until 1940 when they were abandoned (Geist 1960). They are thought to have been extirpated by deer hunters in about 1957.

On Galiano Island a small herd of 25 inhabited Mt. Sutall and Mt. Galiano where they have been in a feral state for about 40 years. Another herd of 15 goats on Saltspring Island originated from those released in the 1930s by J. Whimms. About 100 goats inhabited Saturna Island and probably originated from an introduction in the 1890s by B. Dyne and J. Pain (Geist 1960; Shank 1972). Further animals were added to this herd in 1925 from some grazed on the island by L. C. Harris (Geist 1960). They were well established on this island in 1919, but many were kept there in the 1920s and doubtless many of these joined feral herds or became feral (Shank 1972).

In the 1970s domestic goats were found feral on Prevost Island, Saturna Island, Sidney Island, Galiano Island and Saltspring Island. Those which occurred in the Sooke area and the Highland District of Vancouver Island had by this time disappeared, however, goats no doubt occur on other gulf islands settled by humans (Carl and Guiguet 1972). Several attempts have been made to exterminate them on Saturna, but they still occur there on the southern coast of the island (Shank 1972) and it was confirmed there were about 70 there in 1982 (Rudge 1984). Some occurred on Ramsay Island, introduced in the nineteenth century(?) and there were about 40 there in 1982 (Rudge 1984).

Some goats occurred on Hen Island, Ontario, in the 1960s (McKnight 1964).

United States
Feral goats are reported from 27 states of the western United States (McKnight 1964). Most herds occuring now are small, but there appear to be few recent surveys of numbers or range.

A small population of feral goats exists in the Sacramento Mountains of south-central New Mexico where they have been present for 30 years. At present there are about 35 head which occur in a 44 km^2 area, but this population appears to be slowly declining because of a high juvenile mortality (Watts and Conley 1984).

Goats (*Capra* spp.) are established in the Florida Mountains in south-west New Mexico, near Deeming (Upham 1980). The New Mexico Department of Game and Fish released them in 1972 with subsequent releases to 1977 and there are now *c.* 300 wild there.

Released in Channel Islands, California, goats were established and persisting in 1933 (Storer 1934). By the 1970s there were large feral populations on Santa Clemente, Santa Cruz and Santa Catalina (Coblentz 1980).

Goats were liberated on Santa Catalina Island, California, by Spanish explorers in the seventeenth century or by English pirates in the eighteenth century (Coblentz 1978). The first record of them may have been in 1827 (Dunkle 1950), and their progeny still occurred on the island in 1975 (Johnson 1975). In recent years there has been action to eliminate them and some 1500–1800 have been removed by shooting from the air. They are now confined to one part of the island by a fence (Brooke 1984) and only a few are left.

S. Ramirez introduced goats to San Clemente Island in 1875 and they still occur there (Johnson 1975). In 1980 they were reported to number some 1000 head (Steinhart 1981). A removal program began in 1973 and 16 000 were removed by 1977 leaving 1500 (Johnson 1975; Ferguson 1979). Goats occurred in Pinnacles National Monument when *c.* 30 were eliminated in 1983 (Macdonald *et al.* 1988).

PACIFIC OCEAN ISLANDS
Goats are present on the Marquesas, Austral (present on five islands), Society (three), Malden, Mariana, Fiji (two), and New Caledonia (present on Walpole I.) islands, where they were introduced in the nineteenth(?) century (Douglas 1979).

Other islands in the Pacific Ocean
At one time or another goats were established on: Île de la Possession, Pitcairn Island, Fatuhiva (Marquesas), Rapa, Raivavae, Tubuai, Rurutu and possibly Rimatara (Austral or Tabuai and Rap Islands), Meetia (Mehetia) and Tahiti (Society Islands), Malden Atoll (Line Islands), Aguijan (Marianas), Walpole Island (New Caledonia), possibly on Marion and Luf islands (Papua New Guinea).

On Ascension Island goats died out in 1944 (Lever 1985).

Aguijan Island (Marianas)
Goats were being removed from the island in the early 1990s (Rice 1991).

Antipodes
Goats were introduced to the island in 1887 (Atkinson and Bell 1973) or 1888 and there were further introductions in 1903 (Rudge 1976). These goats all later died out (Atkinson and Bell 1973).

Arapawa Island
Introduced by Cook in 1777, goats later died out. However, a feral poulation is known there from 1939 (Rudge 1984).

Auckland Islands
Goats were liberated in at least 10 places in the late nineteenth century as food for castaways (Rudge and Campbell 1977). Most authors indicate the earliest introduction was about 1865–66 or before (Holdgate and Wace 1961; Rudge 1976).

Early introductions include those on Ewing Island in the 1890s, Ocean Island in the 1880s, Adams Island in the 1880s and 1890s, and Enderby Island in the 1860s (Taylor 1968). However, the goats only survived permanently on Auckland Island and lasted for a short period on the others (Taylor 1968). There may have been another introduction of goats on Auckland Island in 1915 (Rudge 1976).

Some goats frequented the north end of Auckland Island in 1960 (Holdgate and Wace 1961), and a small population was present in the early 1970s (Gibb and Flux 1973; Challies 1975; Rudge 1976), but this was reduced to a single colony of about 100 animals inhabiting the north-west side of Port Ross a few years later (Rudge and Campbell 1977). This population is thought to be slowly declining (Rudge 1984).

Bonin Islands (Japan)
Goats have been established on Ogasawara Gunto, 900 km south-east of Tokyo. In 1968, 1000 were there which were said to be descended from some landed by American Admiral Perry in 1853 (Lever 1985).

Campbell Island
The release of goats occurred on Campbell Island in 1868 (Rudge 1976) or 1883 (Holdgate and Wace 1961) and in 1890 (Rudge 1976), but they failed to become permanently established (Holdgate and Wace 1961; Rudge 1976).

Cuvier Island
On Cuvier Island goats were introduced in the late 1890s (Rudge 1976). They remained there until the 1960s converting the coastal scrubs to grassland (Atkinson and Bell 1973; Rudge 1976). The population was exterminated mainly between 1951 and 1961, but a few remained until about 1966 when the last were eliminated.

Fijian Islands
Feral goats were reported from Makogai and Makodroga islands and elsewhere in Fiji in the 1940s (Turbet 1941). They were certainly present on Viti Levu, and in some parts of Kadavu group in small numbers in the 1970s (Pernetta and Watling 1978). Also at this time they were present on Goat Island in the Yasavas and Namara Islands in the Kadavu group. They were reported on Wakaya Island in the late 1970s (Chapman and Chapman 1980).

Goats are now feral on Yaqaga, Devilan, Yadua, Yaduataba and Monuriki, and perhaps also on Aiwa, Wakaya and Makodroga (Lever 1985).

Galápagos Archipelago (Ecuador)
Goats (four) were brought to the Galápagos in the frigate 'Essex' in 1813 and probably later released there. They were certainly taken to the Galápagos in the seventeenth century by the Corsairs (Dorst 1965), or English buccaneers (Lever 1985), but do not appear to have been there in 1709. However, they were also probably introduced to the archipelago any time between 1700 (or earlier) and 1835 when Darwin observed them on Isla Santa Maria (formerly Charles Island) (Koford 1966). More could have been introduced after 1832 when settlers from Ecuador established settlements on Floreana (Charles); after 1869 on San Cristóbal (Chatham) and after 1893 on Isabela (Albemarle). In 1865 the Viceroy of Peru introduced dogs to Isla Santiago (James) in an attempt to control goats (Lever 1985). Last century goats may have been introduced to Santiago from Baltra (South Seymour) in 1906; to Santa Cruz (Indefatigable) from Santa Fé (Barrington), Baltra and Santiago some time after 1925; and to Pinta (Abingdon) in 1957 (Lever 1985).

One male and two female goats from Isla San Cristobal (formerly Chatam I.) were released on Isla Pinta (Abingdon Island) by fishermen in 1959 to provide fresh meat on future voyages (Weber 1971; Eckhardt 1972; Hamann 1975). Here, they increased rapidly and by 1964 were said to be numerous, certainly by 1968 there was a population of between 300 and 5000, by 1970 between five and 10 thousand (Weber 1971; Eckhardt 1972), and by 1971 numbered some 20 000 individuals (Harman 1975).

In the 1970s feral goats occurred on Pinta, Marchena, Sierra Negra and Cerro Azus (Isabella Islands), Santiago, Baltra, Santa Cruz, Floreana, Española, Santa Fé and San Cristóbal (Eckhardt 1972). However, at this time active extermination campaigns began (Mason 1979) and continued into the 1980s (Calvopina 1985). They were eliminated from Isla Santa Fé in about 1971 by the National Parks Service.

Isla Santiago had the largest population of goats and in 1975, 80 000–100 000 were present. Intensive shooting was carried out in the 1980s on most of the islands in order to exterminate them and prevent further damage to the flora and fauna. They were exterminated on Plaza Sur and Rábida (Rudge 1984), and there no longer appear to be any on Española, Santa Fé, Baltra and Marchena (Lever 1985). Between 1971 and 1982, between 40 000 and 60 000 were destroyed by the National Parks Service on Isla Pinta. Recently hunters hired by the park service aimed to rid Isabela and Santiago islands of nearly 200 000 feral goats with the help of trained dogs (Benchley 1999). As of the mid-1990s more than 100 000 wild goats roamed the area of Volcan Alcedo on Isabela Island consuming the food and destroying the habitats of dozens of endemic species.

Great Barrier Island
Some time in the nineteenth century goats were introduced to this island (Rudge 1984). They were still present in the 1970s (Gibb and Flux 1973), but

were being shot with the intention of complete elimination from the island (Rudge 1976; Ogle 1981).

Guadalupe Island (Mexico)

Whalers introduced goats to this island some time after 1830 to provide fresh meat for their subsequent visits. They became established and increased substantially in numbers, causing extensive damage to the native flora (Huey 1925). During the first part of the twentieth century large herds were present and vast numbers were slaughtered for their pelts and tallow. By 1923 there was still a large population present on the island.

The largest numbers estimated to be present at any one time was 21 000 goats, but in 1971 about nine per cent of these were killed in an effort to allow the island to become re-afforested (Rudge 1984).

Guam

Domestic goats were introduced to Guam by the Spanish during the early colonisation of the island. A feral population became established during the early 1700s but was decimated by overhunting by 1801. At present a small feral population exists in the northern cliff-line areas but the species is not an important game resource (Conry 1988). Any damage caused by the goats is undocumented and probably restricted to private property.

Hawaiian Islands

Captain Cook landed two goats on at least two Hawaiian islands in 1778 (Baker and Reeser 1972). However, it appears that Cook left one male and two females on Niihau in February 1778 and that Captain Vancouver left one male and one female on Kauai in March 1792 (Bryan 1930; Yocom 1967). The first feral goats may have been those released by Vancouver on Niihau, as they were not released in the charge of people as were those introduced by Cook (Fisher 1951).

Those goats multiplied rapidly and their progeny were transported to other islands and soon spread into the steep and inaccessible mountain areas of all the Hawaiian Islands (Yocom 1967). On Niihau, however, the residents made every effort to eliminate them and the last goat had disappeared by 1911–12 (Forbes 1913; Fisher 1951). There have been no goats on Niihau since then (Kramer 1971). They were feral on Kahoolawe before 1863 (Juad 1916) and on Hualalai and Mauna Kea in 1867 (Baker 1916). It is not known when they were introduced to Lanai, but by 1870 the vegetation had been decimated (Kramer 1971). By 1905 there were 9500 head feral on Oahu, about 2000 head on Molokai (Marques 1905), and large numbers on Lanai (Kramer 1971). In 1908 it was

estimated that there were 10 000 on Lanai and 100 000 on Hawaii (Judd 1930). There were an estimated 5000 on Kahoolawe in 1909 and from 1906–16 some 4300 were slaughtered (Judd 1916). On Kahoolawe from 1918 to 1928 the lessees killed another 13 000, at which time it was thought that goats had been eliminated from the island (Henke 1929). However they were still present in the late 1960s (Kramer 1971).

In 1850, some 78 years after the first introduction, 26 519 skins were exported. From 1853 to 1884, 245 862 goat skins were exported and over a 66-year period (1844–1900) 1 581 000 were exported (Henke 1929).

From 1924 to 1930 it was estimated that 40 000 goats were killed on Hawaii, but at the end of this period it was suggested that 75 000 still remained (Bryan 1930). Between 1921 and 1946 some 134 551 were killed during a forestry program (Bryan 1947) and in an eight-year program from 1933 to 1940, foresters on other islands eliminated 21 000 goats (Tinker 1941).

On Lanai the goat population remained at a low level, but in 1967 efforts were made to eradicate them completely from the island. In the late 1960s Oahu had a reduced population of about 100 goats, mainly in the Koolau and Waianae ranges, about 2800 still existed on Kauai, mainly in the Waimea Canyon and on the Na Pali coast, and Molokai had a population of about 800. On Maui there were about 600 in the Haleakala National Park in 1963, but from 1946 to 1964 park rangers killed at least 11 870 goats.

The Hawaiian Volcanoes National Park, Hawaii, had over 70 000 goats removed from 1927 to 1980 (Yocom 1967; Kramer 1971; Stone and Keith 1987). In 1971 at least 14 000 were estimated to be in this park (Berger 1972). Over 16 000 were removed from 1970 to 1980 using all methods available, including hunting, organised drives, helicopter shooting and 'Judas' goats and fencing. Many were also eliminated from Haleakala National Park by the same methods.

Although extensive efforts have been made to eradicate them, goats were still present on Mauna Kea in August 1981 (Anon. 1981).

Herekopare Island

Introduced in 1975–76 by mutton birders (Rudge 1976), feral goats may still be present on the island.

Jarvis Island (Line Is)

Goats were probably introduced to Jarvis Island some time in the period 1858–79 when settlement and mining for phosphate were in progress. By 1935 they had been extirpated (Rauzon 1985).

Juan Fernández (Chile)

A Spanish navigator, Juan Fernández, stocked these islands with goats (Angora types) between 1563 and 1574 (Vicuna Mackenna 1883). He lived on the island of Más á Tierra until about 1580 when he left, abandoning the goats. Goats may have been left on other islands at later dates as a food supply for stranded sailors (Petrides 1975).

Goats were abundant on the islands in 1616 when the navigators Jakob le Maire and William Schouten called there for provisions. They were present and plentiful on Más Afuera in 1866 (Holdgate and Wace 1961).

The goats on Más á Tierra thrived and were an important source of food for the buccaneers of the seventeenth century. To prevent buccaneers from using the goats the Spanish are said to have introduced mastiffs in 1686 and these successfully reduced the goat population. The Spanish also stationed a garrison on the island in 1750, presumably for the same reasons.

After the destruction of the mastiffs in 1830 and while the islands were uninhabited between 1837 and 1877, the goats again increased and 300 were recorded there in 1877 (Holgate and Wace 1961).

Feral goats now occur on all the islands in the group: Robinson Crusoe (estimated numbers in 1980, 250), Allejandro Selkirk (4000) and Santa Clara (40) (Mason 1979; Rudge 1984).

Raoul Island and Macauley Islands (Kermadecs)

Feral goats have been present in the Kermadecs probably since 1836 (Atkinson and Bell 1973), but may have been released in 1842 from Samoa. Goats have been on Macauley Island in this group since 1836 and until their extermination in 1966 had serious effects on the vegetation. In 1887 there were about 100 of them, but by 1908 there were thousands.

In 1955 the New Zealand Wildlife Service eradicated 1422 goats from the Kermadecs, but it was estimated that there were 300–400 still there. Some 3200 were destroyed in 1966 (Williams and Rudge 1969) and in 1966–67 the population was estimated as 3000 head.

Further control work to eliminate feral goats began in 1972 (Rudge 1976) when up until 1978 some 1286 (Rudge and Clark 1978) were destroyed and it was thought that equal numbers still survived. In 1973 another 627 were destroyed and the eradication program was to continue (Smuts-Kennedy 1975), and did so until at least the 1980s (Anon. 1980).

Lord Howe Island

Introduced in nineteenth century, probably before 1851 by whalers, goats have in recent years become locally abundant. They inhabited most areas in 1978–80, except for the settlement, Intermediate Hill, Big Slope, Little Slope, Little Pocket and Mt. Lidgbird summit. Early in the twentieth century they caused extensive damage to the vegetation (Pickard 1982).

At least 300 were shot in 1955 in the Little Slope area (Miller and Mullette 1985). Shooting in 1970 removed most of them from the island except for the most inaccessible parts (Recher and Clark 1974). In 1975 some 50 remained (Pickard 1976), but efforts were to continue. Since 1979 many have been destroyed and the residual population hunted and the population remains at about 40–50 animals.

Macquarie Island

Goats were imported and released on this island in 1947, but were later destroyed because of the harmful effects on the vegetation (Taylor 1955). It was later reported that they did not become established on the island (Watson 1975).

Mangere Island

Introduced before 1900, goats were finally removed by the owners just prior to 1967 (Bell 1975). There are now none on the island (Rudge 1976).

New Zealand

Captain Cook sent goats ashore in Queen Charlotte Sound in New Zealand in 1773 and 1778 (Kippis 1904; Baker and Reeser 1972; Gibb and Flux 1973; King 1990), but was later informed that they had been destroyed. There were also introductions as early as the 1830s on some offshore islands as a source of food.

During the last 200 years the descendants of goats released by the early explorers and settlers have become feral throughout New Zealand and on many offshore islands (Williams and Rudge 1969).

In Canterbury in 1843 there were over 100 domestic goats (Lamb 1964). In 1867 the Canterbury Acclimatisation Society released three goats, but nothing more was heard of them. There were numerous introductions from 1850 on by British and Australian settlers, and goat-shooting parties hunted in the hills about Banks Peninsula in 1850s (Lamb 1964). Goats were liberated in the Rimutaka Range in 1858 where they have at times become exceedingly numerous and still extend from Palliser Bay to Tairarua range some 30 miles to the north (Rudge 1969). Goats still existed there in 1970 (Rudge 1970). In 1876 a flock of 120 angoras was sold and dispersed.

Goats on outlying and offshore islands of New Zealand

Island	Date introduced	Present status
Adams	c. 1885	died out by 1888
Antipodes	1887–1903	died out
Arapawa	1777 & pre-1839	early died out; later still present
Auckland	1860s or c.1900	still present, but declining 1980s; attempted eradication 1992 incomplete
Burgess (Mokohinau group)	?	exterminated in 1973
Campbell	1888 & 1890	died out
Cavalli	?	exterminated by 1972
Cuvier	1890s	exterminated 1959–61
D'Urville	?	still present 1990
East	?	exterminated 1959–60, again in 1968 by Maoris
Enderby	1850 on	died out by 1889
Ernest (Masons Bay)	<1948	present 1976, eradicated about 1980
Ewing	1850 on	died out
Forsyth	19th century	still present
Great (Three Kings)	pre-1830 & 1889	exterminated by 1946
Great Barrier	19th century	still present
Great Mercury	pre-1868	still present 1990
Herekopare	1973–75	eradicated 1976, but still present 1990?
Kapiti	c. 1830	exterminated in 1928
Macauley	pre-1836	exterminated by 1966–70
Mahurangi	late 19th century	removed in 1915
Makoia	1987	exterminated 1989
Mangere	pre-1900	removed about 1967
Maud	?	exterminated in 1970s
Motuoruhi (Goat)	?	none present now
Moutohora	c. 1890	eradicated 1964–77
Nukutaunga (Cavalli)	?	eradicated 1972
Ocean (Auckland)	1865	eradicated 1941–42
Pourewa	<1950	attempted extermination (1992) incomplete
Raoul	pre-1836	exterminated 1972–84
Snares	1890 or 1900	died out
South East	pre-1900	exterminated 1914–16
Steep-to	?	none present now
Whale	?	present 1976, none present now?

In 1867 the Otago Acclimatisation Society imported and released four Angoras and in 1869 the Auckland Acclimatisation Society introduced many. In 1903 the Department of Agriculture bought some at Wereroa, Hawkes Bay. The department also sent some to the Ureweri Maori, who protected them initially (Thomson 1922).

As settlers opened up new areas, goats were often translocated from place to place. Miners may have played a large part in their initial distribution (Rudge 1976). They have now been exterminated on many offshore islands for conservation reasons.

Goats were restricted but abundant in both North and South islands (Wodzicki 1965) but have increased in range (Wodzicki and Wright 1982). They are now found in forested ranges and scrubby hills throughout the North Island and in the Nelson/Marlborough, West Coast and southern lakes regions of the South Island (Rudge 1984). New populations are continually appearing and currently they occupy 39 000 km^2 of New Zealand (Fraser et al. 1996).

The distribution of goats has changed little in the last 30 years but numbers are now lower. Goats are scattered in low density with locally high concentrations in forested ranges and scrubby hill country in both main islands (King 1990). They are now exterminated from 12 of 23 outlying islands where they were liberated (Veitch 1955; Wodzicki and Wright 1982). Inshore island populations now remain only on Auckland, Arapawa, Forsyth, D'Urville and Great Barrier (King 1990).

Norfolk Island
Introduced in the eighteenth century, goats were eradicated in the nineteenth century, although it is possible that some may still be present (Marsh and Pope 1967).

Society Islands
Captain James Cook gave two goats to natives on Otaheite, Society Islands (Kippis 1904).

Three Kings Islands (group)
On Great Island the feral goat stock may have originated in part from those introduced early in the nineteenth century (Turbott 1948), but undoubtedly from four animals released in November 1889 to provide food for castaways.

In 1946, 393 feral goats were removed from the island when the total population was destroyed (Turbott 1948). There have been no goats on the island since this time (Rudge 1976) and the vegetation has recovered substantially (Atkinson 1964).

Tomogasima Island (Japan)
In 1955 six Tokara males and four females were placed on Tomogasima Island where they increased to 20 by 1958 (Asahi 1960). In 1968 it was estimated that there were about 100, but there is little information about them (Numata and Ohsawa 1970; Rudge 1984).

SOUTH AMERICA
Argentina
Goats escaped into the wild prior to 1700 in Argentina and have become established in some areas. At least 300 were reported south of Buenos Aires in the late 1970s (Chapman and Chapman 1980).

Brazil
Goats have been introduced on Trinidade Island, Brazil (Holdgate 1967), but little information could be found.

Chile
In Chile goats have an extensive distribution in the wild (Petrides 1975).

Goats were introduced to Guayapa (La Rioja) in the eighteenth century (Hayward 1967) but their present status not known (Rudge 1984). They were also introduced to Isla de los Estados in the eighteenth(?) century and 60 were observed there in 1971 (Pine *et al.* 1978).

Costa Rica
Goats have been unsuccessfully introduced on Cocos Island, Costa Rica.

Venezuela
Goats have probably been introduced to Venezuela (Lever 1985).

WEST INDIES
Hispaniola
Although feral goats were not plentiful in 1535 (de Oviedo y Valdes 1851–55) they were common by the 1780s (Moreau de Saint-Mèry 1797–98) mainly in the arid, rocky, limestone areas of Anse-à-Pitre. In the early 1800s (Descourtilz 1809) snares were set on paths to watering holes to catch wild goats.

In the 1960s feral goats were to be found on the Peninsula of Barnadères, in the Tapion de Papaye region of the Northwest Peninsula, and on the coast near Grand-Gosier and Anse-à-Pitre; also on Morne Bienac near Gonaives and in the western most part of the Massif de la Hotte in 1952–53 (Street 1962).

Netherlands Antilles
Goats were introduced to Curaçao and Bonaire islands in about the eighteenth century. There are now at least 1000 on each island and they have had some effect on the vegetation (Rudge 1984).

Puerto Rico
It was an early custom in this area to place goats on small islands for later use by mariners. The islands of Mona and Desecheo still have populations derived in this manner (Heatwole *et al.* 1981). Mona is now a reserve to hunt wild pigs and goats. The flora on Salt Island has now been reduced to low open vegetation by the goats.

Virgin Islands (United States)
Goats exist on numerous cays where they are periodically hunted. In 1980 they were present on seven of 27 cays examined. Dutchcap and Congo cays have four to eight goats on their 32 and 25 acres respectively (Dewey and Nellis 1980). Today small goat populations are present on Shark, Congo, Dutchcap, Mingo, West Cay, Outer Brass, Greater Hans Lollik and Lesser Hans Lollik.

■ DAMAGE
The most serious problems associated with feral goats occur on tropical and subtropical islands with a pronounced dry season, variable precipitation and frequent droughts (Brooke 1984). The black goat (*C. h. mambrica*) is said to be the principal culprit for the degeneration of wild plant growth in Israel and consequently for the grave erosion damage in the last

centuries (I.P.S.T. 1964: 279). They strip most of the foliage from shrubs, weakening the plant and damage the taller shrubs and small trees by bending them down to reach the upper foliage. High usage of such areas as cliffs renders them barren and soil erosion begins in the vicinity of traditional bedding grounds (Coblentz 1975).

Since their introduction in the early nineteenth century on Round Island, Mauritius, goats have irrevocably damaged the vegetation – hardwood forests have disappeared, palm savanna has been reduced, large scale soil erosion is evident, and the endemic reptile population has been affected. Since their extermination the vegetation is recovering (Bullock and North 1984).

Through the grazing of feral goats and other animals on San Clemente and Santa Catalina Islands, California, some 48 indigenous plant species and 18 introduced species have now disappeared (Thorpe 1967; Steinhart 1981). In 1975 a study determined that the presence of goats had considerable effect upon the total percentage cover and the species composition of both herbaceous and shrubby components of brushland areas on San Clemente (Coblentz 1975). Endangered endemic species included the larkspur and bush mallow on San Clemente Island (Ferguson 1979).

Irreparable damage was caused to St. Helena by goats and also the destruction of forests by man and introduction of alien plants (Wallace 1902). By 1810 the forests had gone, cut by sailors and kept from regenerating by grazing goats. By 1870 the original 60 species of plants had risen to some 900, nearly all of them exotics fostered by goats (Steinhart 1981). The ravages of goats, together with the stripping of bark from ebony and redwood trees for the tanning industry, robbed the island of much of its vegetation (George 1962). Ebony, *Trochetiopsis melanoxylon*, was formerly one of the main constituents of the island flora. Its decline commenced in the sixteenth century with the introduction of the goat which prevented regeneration by eating the seedlings and killing the older trees. The Gumwood, *Commidendrum robustum*, was also severely affected by grazing goats and also the clearing of timber for fuel by 1659 (Cronk 1986).

In Haleakala National Park, in the Hawaiian Islands, feral goats have increased soil erosion and altered the plant communities (Yocom 1967). The fact that 19 plants were considered rare was partly attributable to overgrazing by feral goats (Baker and Reeser 1972).

In the Galápagos Islands goats are responsible for the extinction of many endemic plants and the alteration of plant communities themselves. On Espanola 10 species of plants have gone. A notable regeneration of woody species on Isla Santa Fé occurred after the removal of goats (Hamann 1975).

On Guadalupe Island three endemic subspecies of birds have become extinct through the grazing of goats (Steinhart 1981). On Juan Fernández they have had a devastating effect on the vegetation and severe soil erosion has ensued (Holdgate and Wace 1961).

Feral goats in New Zealand are reported to effect soil compaction, de-bark trees and shrubs, and to browse tree and shrub crowns, although compacting the soil and causing run-off is not proven and de-barking is probably not dangerous. Browsing the understorey is the greatest threat that prevents regeneration of some canopy species, and the understorey becomes dominated by low preference species (replacement of browse types). Several high-altitude areas have been denuded by goats in New Zealand.

On Cuvier Island the effects on coastal scrub have been catastrophic and after 70 years the scrub has been replaced by grass and sedge lands. On Great Island (Three Kings group), shrublands of low preference Kanuka (*Leptospermum ericoides*) and grassland have developed after 60 years (Atkinson 1964). The history of goats on Great Island provides an example of the effects of a single species on the vegetation. The presence of goats resulted in the establishment of a different plant community replacing the primary forest (Turbott 1948; Holdsworth 1951). By the time the goats were totally removed in 1946 the continued modification of the vegetation was probably responsible for the small number of animal species on the island (Turbott 1948).

It has been suggested that the restricted movements and catholic tastes of goats may explain the propensity for converting forest to grassland despite high rainfall in New Zealand (Riney and Caughley 1959). On Auckland Island at present they do not endanger the plant communities or rare species within their present range and do not seem likely to spread (Rudge and Campbell 1977). Before extermination in 1966 on Macauley Island, in the Kermadecs, goats had reduced the vegetation to short grassland (Atkinson and Bell 1973).

In Volcanoes National Park, goat browsing induces suckering by *Acacia koa* and suckers are then browsed and those between 0.5 and 2 m in height are destroyed (Spatz and Mueller-Dombois 1973).

In Australia feral goats are regarded as pests by pastoralists as they compete with livestock for pasture

(Holst *et al.* 1982), change the composition of plant communities and prevent the regeneration of canopy trees (Harrington 1979), compete with native animals for food, shelter and water, particularly during drought (Dawson and Ellis 1979), and may also cause soil erosion (Mahood 1988). Annual losses have been estimated at A\$2.5 million (Parkes *et al.* 1996), however, occasionally they are valued individually at A\$25 per head (Korn *et al.* 1998) for hair or meat, thus making them a sometimes valuable commodity locally. Commerical exploitation currently supports a A\$29 million industry employing about 500 people (Parkes *et al.* 1996).

On Lord Howe Island goat damage was most evident on exposed slopes and cliffs and forest edges (Recher and Clark 1974).

ALPINE IBEX

Siberian ibex, Himalayan ibex, Nubian ibex, Abyssinian ibex, steinbock

Capra ibex Linnaeus

The alpine ibex (C. i. ibex), Nubian ibex (C. i. nubiana), Siberian ibex (C. i. sibirica) and the Himalayan ibex (C. i. sakeen) are in some classifications held to be separate species. Other species of ibex not mentioned here may also be involved. They are grouped together here because of the interrelationship confusion. The various ibex species interbreed freely and with the wild goat (Capra hircus).

■ DESCRIPTION

HB 1300–1700 mm; T 120–250 mm; SH 650–1100 mm; WT males 55–130 kg, females 24–91 kg.

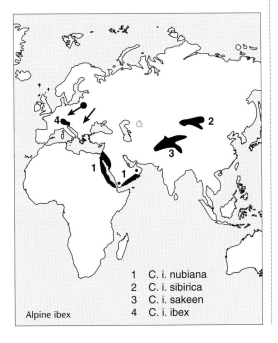

1 C. i. nubiana
2 C. i. sibirica
3 C. i. sakeen
4 C. i. ibex

Alpine ibex

Short fur is greyish or greyish brown in summer, yellowish brown in winter; fore legs shorter than hind legs; the various subspecies differ only in size, thickness and ridging of the horns; horns curved 700–1500 mm; beard of hairs on chin; forehead convex. *C. i. nubiana* is whitish with dark legs; face has black patches; horns 700 mm. Male with short beard (*sibirica*); brown with black dorsal stripe; throat, chest and outer surfaces of legs black; under parts white, white band around each leg above hooves. Female beardless, horns shorter; horns (*sibirica*) long, curved in single plane with broad flat anterior surface and transverse ridge shaped knobs.

■ DISTRIBUTION

Europe and Asia. Formerly from the Alps of France and Switzerland, Bavaria, Italy and Austria, but now inhabit a small part of the Italian Alps and possibly Salzburg, Austria (*C. i. ibex*); central Asia (*C. i. sibirica*), Himalayas in northern India, Pakistan, and Tibet (*C. i. sakeen*), and the Red Sea Coast from Ethiopia to Yemen (*C. i. nubiana*).

■ HABITS AND BEHAVIOUR

Habits: mainly diurnal, but most active in early evening and early morning; agile. **Gregariousness:** mixed herds of 2–40 and up to 200; males and females and young generally in separate flocks; at rut male collects 10–16 females; male herds 2–23 or solitary; density 1–9/km^2. **Movements:** sedentary and local; to higher elevations in spring and summer and lower areas in winter. **Habitat:** mountainous areas; steep ravines, gorges, bluffs, precipitous cliffs, mountain crags, steep hills, rocky outcrops high meadows of rugged mountain country, steppe, desert, semidesert. **Foods:** grass, sedges, lichens, forbs, and browse from shrubs and heather. **Breeding:** most of year; ruts October–January, kid in May–June, gestation 150–180 days (maybe up to 240–280 days for *C. i. ibex*?); young 1–2, rarely 3; young weaned at 6–7 months; sexual maturity in second year. **Longevity:** 12–15 years in wild, possibly to 20; 22 years 3 months captive. **Status:** range and numbers reduced; severely threatened in most of range, but a number of small protected herds survive.

■ HISTORY OF INTRODUCTIONS

EUROPE

Ibex populations declined steadily from the fifteenth century until the eighteenth century due to overhunting and poaching (Ausserer 1946), and only about 100 survived in the Gran Paradiso Mountains in the European Alps in north-west Italy (Grodinsky and Stüwe 1987). Protection of the remaining ibex was initiated in 1821 and has been enforced strictly since 1858 by the Italian government (Stuwe and Scribner

1989). Captive breeding programs, initiated in 1902 and continued until 1942, increased the total population and allowed further colonies to be established (Stüwe and Scribner 1989). Beginning in the 1950s, additional populations were founded. Most Swiss efforts were successful and in 1986 the total Swiss population was estimated at 12 500 individuals.

Ibex have been widely introduced in Austria, Switzerland, France, Italy, Germany, Yugoslavia, Spain and the Russian Federation, as well as the along the Hungary–Poland border (Carpathians) (Lever 1985).

The European population of alpine ibex was almost exterminated, but survivors in the Gran Paradiso National Park, Italy, have been used for translocations throughout the Alps and Yugoslavia (Tosi *et al.* 1986; Burton and Pearson 1987). Alpine ibex still occur in the western Alps at Grand Paradiso, Valle D'Aosta, in northern Italy. In the eastern Alps, mainly in Switzerland, Austria and Yugoslavia, about 4000 head of ibex existed in the wild in the 1960s, most of them having been introduced (Hagen 1967). At Aosta in 1820 there were only 50 animals left and most of the introductions originated from this stock (Hagen 1967). The entire living population of alpine ibex was estimated as 5500–6000 (Couturier 1959).

Austria

The alpine ibex was exterminated in Austria apart from some 50 individuals in the Aosta in the Western Alps. Re-introductions have been carried out over a number of years (F. Spitzenberger *pers. comm.* 1982), and for more than 150 years well over 7000 animals from this herd have been released in the western Alps (Hagen 1967). In 1924 pure-bred ibex were released near Salzburg and in 1936 at Wildalpen (in Styria) in Austria. These and other colonies have flourished so that by the late 1960s there were six separate colonies with well over 250 individuals living in wilderness area of the country. At this time there were three colonies in Tirol, one of 60–70 ibex in the Pitz Valley, and since 1955 a growing population in Styria with more than 60 head (Hagen 1967).

The first releases of ibex in Austria occurred at Salzkammergut (Upper Austria) in 1856, but failed because it was made with goat hybrids which had unfavourable characteristics. Previous failures were thought to be due to using eastern races in western habitat, but now it seems that it happened because of goat crossing (Hagen 1967). Some (34 released) may also have been released in the Tennen Mountains, Salzburg, in 1866, but these failed by 1875 probably due to the fact they were goat–ibex crosses (Niethammer 1963).

In 1953 a re-introduction of ibex occurred in the Pitz Valley near Imst, Tirol, southern Austria, where they had been unknown for 450 years. Six males and 11 females from Switzerland were released in a 6-ha enclosure. After three years only four were left, the rest having escaped or died. Those that were left were released in the wild and with the addition of three males and eight females, between 1956 and 1966 the herd increased and was thriving. Between 1962 and 1966 some 37 kids were born and by 1967 the herd numbered some 60–70 head in the wild (Hagen 1967).

In Achensee (south-west of Achenkirch), also in the Tirol, two small colonies were released in 1953 and 1958. Here in the 1960s there were 40–50 head in the wild. A previous release at Reutte, AuBerfern district (Tirol), died out because of the use of inadequate stock (Hagen 1967). This colony was started in 1951–52 with eight ibex in the Lech Valley, Tirol, but by 1961 there were only five left (Niethammer 1963).

Introduction of alpine ibex occurred in Blühnbach in Salzburg in 1924 with eight ibex (two males and six females). This release was successful and ibex were still established there in the late 1960s. These were enclosed by fences until 1927 and then released and by 1961 there were 64 animals. The Wildalpen introduction of 1936 was successful for many years until 1950 when there was only one animal left, but a few moved from Switzerland were released and there were 46 by 1956 and the herd was still established there in the 1960s. The most recent introductions at Mixnitz (Styria) in 1955 were successsful and herd numbers were 60 head in the late 1960s (Hagen 1967).

Some alpine ibex may still be established in western Austria at Galtür in the south-west Tirol, where in the 1950s and 1960s animals wandered across from Switzerland (Niethammer 1963).

Carpathians

A few introductions appear to have been made in the Carpathians, where the ibex is not known to have occurred in historical times. The first of these appears to have been in 1901 when 23 animals (six males and 17 females) were released, but at least 10 were lost in the same year (Niethammer 1963). They were apparently not pure ibex, but probably goat–ibex crosses.

Additional ibex were continually imported and between 1901 and 1928 some 128 animals including Bezoar goats (*Capra aegagrus*), two Nubian ibex from Sinai, 24 Caucasian tur and some additional ibex from the Italian Alps, Siberia, Turkestan, the Himalayas and Abyssinia were released. The last of these were apparently almost extinct by the beginning of World War 2. In 1951 some pure bred ibex may have been released and in 1953–54, four Caucasian tur were released but the latter did not last long (Niethammer 1963).

Czechoslovakia

In 1901 the alpine ibex (*C. i. ibex*) was re-introduced successfully in the Tatra Mountains (de Vos *et al.* 1956; Dorst 1965), where they became established. The Nubian subspecies (*C. i. numbiana*) was introduced from Asia Minor to the mountains in 1910 where they interbred with the re-introduced native ibex (*C. i. ibex*) and the introduced Bezoar goat (*C. hircus aegagrus*) (de Vos *et al.* 1956). The resulting hybridisation ruined the entire population since the reproductive period was changed and the young were born in mid-winter (Dorst 1965).

France

From migrants from the Italian Schutz district, a small colony of alpine ibex has established itself about the Grande Cass, Savoie. In 1959 they numbered some 20 head (Niethammer 1963). Two males were released in 1959 at Col du Lautaret in the Upper Alps and were later observed some distance away (Couturier 1959).

Germany

The alpine race of the ibex has been re-introduced and established in the Bavarian Alps (Corbet 1966).

An attempt was made to re-introduce them at Röth, south of Obersees, Berchtesgaden, in the south-east of Germany (German Federal Republic) in 1936. Twenty-four were obtained from Switzerland, Italy and the Berlin Zoo and these were given their freedom in 1944. This herd travels back and forth across the Alps and only appears on Bavarian territory in the summer (Niethammer 1963).

Italy

Apart from the Gran Paradiso colony, at least three other established populations of alpine ibex existed in Italy. Since 1920 ibex have been released in a reserve at Valdieri-Entraque in the province of Cuneo, close to the French border (near Grenze). These numbered some 200 in 1959. In north-western Italy, in the valley of Valpelline (north of Aosta), animals which have deserted the Swiss colonies at Mt. Pleureur have become established. Also, near Tirano, deserters from Piz Albris and the national park in Switzerland have become established (Niethammer 1963; Framarin 1976).

From 1920 to 1985 colonies were established in the Italian Alps from animals from Gran Paradiso National Park (Peracino and Bassano 1987). During this time 18 new colonies were successfully established, 16 on Italian territory and two in Triglav National Park and Bovec, Slovenia.

Russian Federation and adjacent independent republics

C. ibex sibirica has been translocated (Naumoff 1950 in de Vos *et al.* 1956) and successfully re-established in the Russian Federation (Bannikov 1963). Some work has also been attempted to extend their range in Kazakhstan (Sludskii and Afanas'ev 1964). *C. i. sibirica* have been released successfully on the Crimea Peninsula (Tschapskij 1957 in Niethammer 1963), but outside European Russia there were no known introductions to 1963.

Switzerland

At the beginning of the fifteenth century alpine ibex still populated parts of the Alps (Fellay 1967). By the sixteenth century they were rare in Switzerland and disappeared some time later (Dorst 1965). The last probably disappeared between 1840 and 1860 (Fellay 1967). Also animals from Gran Paradiso, northern Italy, were re-introduced to a number of places in the Alps (Walker 1992). Attempts to introduce *C. ibex* × *C. hircus* hybrids between 1870 and 1890 failed (Kuster 1966).

Alpine ibex were re-introduced to several parts of the country in 1910–11 (Vaucher 1946; Burckhardt *et al.* 1961; Couturier 1962; Fellay 1967) and in the 1920s (Dorst 1965; Fellay 1967). The population rose from nil in 1910 to about 2000 animals in 1960 (Burckhardt *et al.* 1961). Ibex were successfully introduced by A. Rauch about 1921, when two females appeared on Pis Albris having wandered from Parc National where they had been re-introduced in 1920. Rauch obtained two males from the 'Peter and Paul' Game Park at St. Gallen and by 1972 there were 500 ibex there (Whitehead 1972). In the early 1960s they were well established in Switzerland, and the largest herds were probably established on Pis Albris (Tegner 1963). At the end of 1965 some 55 colonies including 3719 head of ibex were free in the wild (Schenk 1966). The Swiss commenced breeding in 1906 with three ibex kids, and the species is now widespread in its former habitat and is spreading into the eastern Alps (Hagen 1967). In 1972 there were some 3000 ibex in Switzerland (Whitehead 1972).

In spite of a number of successes, two introductions of alpine ibex have failed, one of them after being established for 20 years.

The first reacclimatisation attempt was probably made in 1906 at the park 'Peter and Paul' at St. Gallen (Nievergelt 1966), but it seems that the first kids were born in Switzerland in 1911 (Fellay 1967).

Prior to 1928 alpine ibex were re-introduced at Graue Hörner, St. Gallen, in 1911, Piz Aela, Bergün, in 1914,

Parc National in 1920, Albris-Pontresina and Augsmathorn-Berne in 1921, Schwarzhorn-Berne in 1924, and Wetterhorn-Berne in 1926. These became successfully established and maintained at these sites, except for the colony at Piz Aela which was composed of some 40 animals by 1926, but was decimated and abandoned in 1932 (Fellay 1967).

In June 1928 alpine ibex (two males and three females) from Interlaken and St. Gallen were released at Jeurs-Grasse on Mont Pleureur in Valais and this was followed by further releases in 1929 (two males and two females from Interlaken), in 1933 (six from Italy), and in 1935 (five animals). These releases were so successful that in 1955 there were 380 head and in 1960, 620 head. Translocation of animals to other areas has resulted in the successful establishment of a further 13 colonies, totalling in 1967 some 1130 alpine ibex in the Valais area (Fellay 1967).

In 1952 there were 10 colonies with 1100, by 1957 there were 25 colonies with 1600 head, and in 1960, 35 colonies with about 2400 ibex (Kuster 1958, 1961). At this time there were colonies in the cantons of Bern, Ob-Walden, Mid-Walden, Glarus, Freiburg, St. Gallen, Graubünden, Waadt and Wallis, and the largest colonies were those at Piz Albris, National Park, Mont Pleureur, Augstmatt-horn and Aletsch-Bietschhorn. At Piz Albris there were 604 and at Mont Pleureur 470 alpine ibex (Kuster 1961). These colonies were founded between 1911 and 1960.

The population in 1980 was about 10 000 head, and was expanding into new areas largely by artificial introductions. Ibex are fully protected but about 400 are taken annually in colonies where damage to vegetation occurs (M. Dollinger *pers. comm.* 1982).

United Kingdom
Ibex were introduced to Britain about 1903 when a pair were released on Inverinate by Sir Keith Fraser to improve the heads of feral goats. The introduction was said to have had good results (Whitehead 1972).

Yugoslavia
The alpine race of the ibex has been re-introduced and established in Yugoslavia (Corbet (1966).

Between 1890 and 1896 Baron von Born purchased 20 ibex from Lausanne (said to have come from Gran Paradiso, Italy) and kept them enclosed on his estate in the Karawanken in northern Yugoslavia. A herd of 40 occurred in the area in 1959. An attempt in 1954 was made to establish ibex in Kamniska Bistrica (Steinalpen) when four were released, but shortly after only two females remained (Niethammer 1963).

NORTH AMERICA
United States
Siberian ibex (*C. i. sibirica*) were imported by New Mexican authorities in 1962 (four animals) and at later dates. These have been kept in the Albuquerque Zoo for breeding with the object of releasing their progeny in the wild at some date in the future (Gordon 1967).

Six Siberian ibex were introduced to a pasture along the Gila River, near Red Rock, Grant County, New Mexico, in January 1966. Twelve Iranian ibex were kept in a corral in the south corner of the pasture awaiting release (Wood *et al.* 1970).

Ibex were recently introduced (1980?) in New Mexico at Canadian River Gorge, near Roy (Upham 1980).

SPANISH IBEX
Capra pyrenaica Schinz

■ DESCRIPTION
HB 1000–1500 mm; T 100–150 mm; SH 650–750 mm; WT 35–80 kg.
Coat colour and body size differs between different populations; dark greyish to greyish red or brown to buff; paler below; chest and upper forelegs blackish; forehead dark; dark leg markings and whitish undersides; horns curved outwards and backwards to 750 mm; posterior keel sharp.

■ DISTRIBUTION
Europe. South-east Iberian Peninsula in central Pyrenees; Sierra de Gredos; Sierra Morena, Sierra de Ronda, Sierra Nevada, Sierra de Cazorla (Andalusia); Sierra de Cardo (Valencia).

■ HABITS AND BEHAVIOUR
Habits: similar to *Capra ibex*. **Gregariousness:** mixed herds to 10 in mating season; bachelor males; adult females; young of both sexes; density 0.13–0.29/ha. **Movements:** no information. **Habitat:** mountainous areas; rocky areas. **Foods:** grass, herbs, lichens. **Breeding:** gestation 23–24 weeks; young born May. **Longevity:** no information. **Status:** rare and endangered.

■ HISTORY OF INTRODUCTIONS
EUROPE
Present at least until recently in central Spanish Pyrenees; Sierra de Gredos, Sierra Morena, Sierra de Ronda, Sierra Nevada, Sierra de Cazorla (Andalusia); Sierra de Cardo (northern Valencia).

During the nineteenth century most populations were heavily depleted and several became extinct.

Since then they have recovered and there have been extensive re-introductions and translocations (Burton and Pearson 1987).

Spain

Spanish ibex from Sierra de Cazorla (where 2000 existed) were translocated to the national park Monte Codovonga in Spain in 1957–58 (Niethammer 1963).

Survival of an indigenous race of ibex in the Pyrenees is very dubious (near Mt. Perdido), but animals from Sierra de Gredos and elsewhere have been released nearby (IUCN 1966 in Corbet 1978).

■ **DAMAGE**

None known.

BHARAL

Blue sheep

***Pseudois nayaur* (Hodgson)**

■ **DESCRIPTION**

HB 1150–1650 mm; T 100–200 mm; SH 750–910 mm; WT 25–80 kg.

Body brownish grey to slate grey colour; chest black; ventral surface of neck black; flank stripe black and marks boundary of white under parts; prominent dark markings on front of legs; rump patch, chin and tip of muzzle white; muzzle with grey streaks on top and sides; tail black; horns curve backwards and inwards; 200–780 mm; lacks face glands of true sheep, beard, and smell of goats. Female similar to male but smaller and black on neck and chest less marked; small almost non-functional horns.

■ **DISTRIBUTION**

Asia. Kashmir through the Himalayas and east through most of Tibet to Szechuan in western China, and north to Ordos Plateau, the Ala Shan and extreme south-west of Inner Mongolia.

■ **HABITS AND BEHAVIOUR**

Habits: feeds and rests alternately during day; climbs precipitous cliffs. **Gregariousness:** herds to 200–400; may be mixed or female; males separate after rut to form male groups (to 40) or become solitary; density 0.35–10.0/km². **Movements:** no information. **Habitat:** mountains, avoids woods and forest, alpine meadows; open slopes and plateaus with grass cover at 2500–6500 m. **Foods:** graze grass, forbs, leaves, acorns, herbs and lichens. **Breeding:** ruts October–January; young born spring mid May–July; gestation 160 days; 1 young, sometimes 2; lactation 6 months; sexual maturity 18 months. **Longevity:** 14–20 years wild, 20 years 3 months captive. **Status:** no information.

■ **HISTORY OF INTRODUCTION**

PACIFIC OCEAN ISLANDS

New Zealand

Blue sheep were introduced on Mt. Cook before 1945 (Carter *et al.* 1945), but apparently failed to become established.

■ **DAMAGE**

No information.

AOUDAD

Barbary sheep, arui

***Ammotragus lervia* (Pallas)**

Some authorities now place this species in the genus Capra, *a policy not followed by everyone (see Corbet and Hill 1986)* =Capra lervia (Pallas)

■ **DESCRIPTION**

HB 1300–1900 mm; T 150–250 mm; SH 750–1120 mm; WT males 66–145 kg, females 40–86 kg.

Coat upper parts tawny to whitish on undersides; beard extends to hair on chest and throat; inside of ears, chin, line on under parts, insides of legs white; forelimbs hairy; lack facial glands; tail long and tufted; horns male 330–860 mm; horns heavy, corrugated, sweep out and back again. Female horns not as massive as that of males, 300–500 mm.

■ **DISTRIBUTION**

Africa. In Morocco, Algeria and Tunisia south into the Sahara, rare in Libya, and Upper Egypt east of the Nile

Bharal

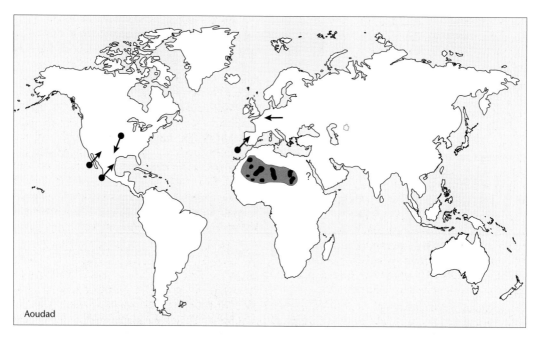

Aoudad

River. Distribution discontinuous. Formerly more widespread range, but now fragmented. Now extinct in Egypt.

■ HABITS AND BEHAVIOUR

Habits: mainly nocturnal; but also diurnal with rest period in middle of day; linnear dominance hierarchy. **Gregariousness:** solitary or small family flocks, bachelor herds common in summer; nursery groups in spring and summer; mixed groups at autumn rut (United States); sexes separate after rut?; density 0.35–2.37 sheep/km² (United States). **Movements:** sedentary(?); home range 719–1567 ha (1–31 km²); daily movements of 0.2–3.4 km. **Habitat:** rough rocky arid hills, mountains and low hills in desert zone; rocky precipices. **Foods:** grass, forbs, and browse from shrubs, bushes and herbaceous plants. **Breeding:** breeds throughout the year, but most young born March–May; in North Africa peak in March; gestation 150–165 days; kids 1–2 (occasionally 3), sometimes twice per year; female stays with kid 1–2 days after birth; female sexual maturity at 18 months, males less maybe 11–15 months? **Longevity:** 10.5 years (wild) and up to 24 years (captive). **Status:** decreasing rapidly in northern edge of range; becoming rare or locally extirpated throughout original range due to human encroachment, overgrazing and hunting.

■ HISTORY OF INTRODUCTIONS

Introduced successfully in Europe, the United States and Mexico. In the United States introduced in New Mexico, California, islands of Lake Erie and many reserves (Haltenorth and Diller 1994). Also introduced on ranches in South Africa (Burton and Pearson 1987).

AFRICA
Tunisia
Preparations were being made to re-introduce Barbary sheep back into the wild (Bertram 1988).

EUROPE
Germany
In 1883 Prince Woldemat placed out 10 Barbary sheep in a 660-ha enclosure. These were said to be well naturalised by 1890, but other records indicate they disappeared after a few years. Some were released at Eenzelfällen, but likewise disappeared about 1902. In about 1900 some were also released in an animal park at Neumarktl, Krain. The reason for their failure has been reported as – semi-domestic animals which died of starvation when run wild (Niethammer 1963).

Spain
An introduced population of Barbary sheep was established in Spain in about 1960 as a game animal (Seegmiller and Simpson 1980; Gray 1985; Lever 1985; Fandos and Reig 1987; Cheylan 1991), where they still increasing in numbers and range (Fandos and Reig 1987, 1992).

NORTH AMERICA
Mexico
Barbary sheep have escaped and become established in three separate localities in Mexico – two east of the Sierra Madre Orientale and one west of this range. The releases were made by private landowners onto

their properties to either add to collections of exotics or as potential meat animals (Rangel-Woodyard and Simpson 1980). They were for some years also established in Coahuila, Mexico (Petrides 1975).

Some were released in the Sierra Pajaros Azules, located along the western boundary of Nuevo Leon with Coahuila. An established population was confirmed present recently on the eastern side of Sierra Madre Orientale. Some also escaped from the Sierra Morena Ranch, Nuevo Leon when the land was subdivided. A small population established in the eastern foothills of Sierra Madre range and has since increased to about 100 head. Twelve Barbary sheep were released in 1975–74 on private land in San Luis Potosi (two adult males, eight adult females and two lambs), where they have remained in a fenced area and are increasing in numbers (Rangel-Woodyard and Simpson 1980). However, several groups released in Mexico have now died out (Walker 1992).

United States
Barbary sheep were introduced in the United States in the early to mid-nineteenth century for sport and hunting. They were originally raised successfully in the New York Zoological Park and National Zoological Park, Washington, and by 1935 they were being raised in most zoos throughout the United States. As numbers increased in these locations they were translocated to private ranches during the 1920s and later released into the wild (Yoakum 1980). By the 1970s there were well established free-roaming populations in California, New Mexico and Texas. However these have now bred into populations of some hundreds or thousands (in 1978, 400 in California; 1700 in New Mexico; 1700 in Texas). Currently free-ranging populations occur only in Texas and New Mexico, those in California having been eliminated (Feldhamer and Armstrong 1993).

Fifty-seven Barbary sheep, four rams and eight ewes from the McNight Ranch, Picacho, and 45 from the Hearst Ranch, California, were released in Canadian River Canyon, New Mexico, in 1950 by the New Mexico Department of Game and Fish (de Vos *et al.* 1956; Presnall 1958; Hibben 1964; Morrison 1980). They became well established and hunting seasons were opened in 1955 (Gordon 1967). In 1955 a further 21 head were purchased by the San Juan County Wildlife Federation from Lois Goebal, California, and released in Canyon Largo, near Farington (Morrison 1980; Upham 1980). In about 1958 there were reported to be 250–350 head in the Canadian River Canyon area (Presnall 1958), but by 1960 they were reported to number more than 1000 (Hibben 1964). However, they could not sustain the hunting pressure applied and numbers dwindled to a

low level. Some were also released in the Largo Canyon area in San Juan County, New Mexico, where they became established, but have not increased in numbers (Hibben 1964; Gordon 1967). However, hunting was again in progress in 1967. Over the years there were many escapes of ranch stock and these established populations in the Hondo Valley and the Guadelupe Mountains. A herd also became established in the Mt. Taylor area near Grant's National Monument. Various other sightings were made and the current population is estimated at about 1750 animals (Morrison 1980). Currently in New Mexico they are found in the Kiowa National Grasslands, Cibola National Forest, Lincoln National Forest and Carson National Forest (Zeedyk 1980).

Over the years numerous escapes occurred from McKnight Ranch in south-east New Mexico and these established wild populations in Hondo Valley and the Guadalupe Mountains. A herd also became established in the Mt. Taylor area near Grant's National Monument. Other sightings of single Barbary sheep and bands have been confirmed in various parts of the National Monument. The current population in the state is 1750 head. They have been hunted since 1955 (1008 head taken up to 1979) (Morrison 1980).

The Hondo Valley animals now occur in two areas within the Roswell District in the Guadalupe Mountains and in the Hondo River area (Upham 1980). In 36 years wild populations extended from Lincoln National Monument, 198 km to the Guadalupe Mountains National Park, Texas (Dickinson and Simpson 1980).

Twenty-one Barbary sheep were released in Largo Canyon in 1956 and Barbary sheep are now spread over 120 000 acres of lands within Farmington Resource Area, Albuquerque District. The total numbers in New Mexico on public lands were 350–450 head in 1978 (Upham 1980).

Barbary sheep have now spread into the Kiowa National Grasslands, Cibola, Lincoln and Carson national forests. A total of 26 males and 37 females were released in Kiowa National Grasslands at Canadian River and by 1956 there were 200–225 head. By 1960 there were over 1275. Some escaped into the Cibola National Forest in the San Mateo Mountains, Valencia County from private ownership in *c.* 1975, and 30–50 were seen there in 1978. Small bands and sighting of individuals are often reported from Lincoln National Forest where small herds exist in Dry Canyon, Otero County, and in Middle Rocky and Indian Creek canyons in 1979. A small herd also exists in Carson National Forest in Rio Arriba County (Zeedyk 1980).

Barbary sheep were introduced to McKnight Ranch in 1940 (three ewes and two rams) from St. Louis Zoological Park, St. Louis, Missouri. One ram died and an additional ram and two lambs were added in 1941 from San Diego Zoological Gardens, San Diego, California. The herd increased rapidly and despite efforts to contain them began escaping in 1943. In 1965 *c.* 100 escaped and in 1977 *c.* 50. It is estimated that *c.* 10–20 escaped every year since 1943 and up to 1979 (i.e. *c.* 510 escaped in the 36 years). They spread south into Texas at rate of 0.3–1.8 km/year (Dickinson and Simpson 1980).

In Texas the Barbary sheep has spread over portions of the Edwards Plateau from introductions by individual ranchers (Presnall 1958). Some were released in Palo Duro Canyon Texas panhandle, just prior to 1958 where they have become permanently established (Presnall 1958; Evans 1967; Yoakum 1980). Forty-four were released in Palo Duro Canyon, in the winter of 1957–58 (Simpson 1980; Gray 1982; Gray and Simpson 1983) and the population increased to 1200–1500 in 1977, but may have been as many as 2500 in 1978. They have been hunted each year since 1963. Other herds occur in the Canadian River Gorge, north-eastern Mexico; Largo Canyon, north-eastern Mexico; Rio Hondo Valley, southern New Mexico; and on the Hearst Ranch, California (Gray and Simpson 1983). By 1988 there were over 20 000 head in Texas of which about half are free-ranging (Feldhamer and Armstrong 1993).

Texas
Barbary sheep were released in Palo Duro Canyon and became established where introduced mule deer had not succeeded significantly. Forty-four were released in the canyon between 1957 and 1958. They increased tremendously and recent estimates suggest between 1400 and 1600 there. Hunting commenced in 1963 and until 1978 some 959 were shot (Dvorak 1980). By 1983 a population of 600–800 head existed in the 500000-acre canyon and since 1963, some 1484 head have been taken by hunters (Dvorak 1983).

California
Barbary sheep were introduced on the Hearst Ranch, 2.5 km east of San Simeon, San Luis Obispo County, California, shortly after 1924. There were 20–30 there in 1937 and 98 there in 1949. Some 81 were sold in 1950–51 to an animal dealer. In 1953 Barbary sheep escaped from a zoo onto nearby property and in 1963 there was an estimated herd of 172 in the area between Hearst Castle and Red Rock. By 1964 the Barbary sheep had been noted as far afield as 30 km north and 65 km south of Hearst Castle. By 1980 there were four herds in the area – west of Red Rock (154 in 1977), Glazier Ridge (50?), Cline Peak (50?) and at Vulture Rock (40–60 head) (Barrett 1980).

Barbary sheep were also been released in California (de Vos *et al.*1956), where they became established in San Luis Obispo County (Presnall 1958). Here, some escaped or were released from the Hearst Ranch about 1953 and although they were known to be breeding they did not increase much in numbers nor spread. At Red Rock, California, they declined from 258 in 1965 to 154 in 1977 (Gray and Simpson 1983), but appear to have been successful for some time (Barrett 1967, 1988).

■ **DAMAGE**
It is suggested that Barbary sheep may represent a strong competitive element to desert bighorn sheep (*Ovis canadensis*) and be a threat to their continued survival (Simpson *et al.* 1978 in Seegmiller and Simpson 1979; Rangel-Woodyard and Simpson 1980). Research indicates that the Barbary sheep may adversely impact or even displace mule deer (*Odocoileus hemionus*) in the Palo Duro Canyon and bighorn sheep in the Trans-Pecos (Valentine 1980).

A study carried out in the Palo Duro Canyon, Texas, indicated that Barbary sheep were competing directly with mule deer and to some extent with livestock (Evans 1967). However, they browse the more difficult rugged terrain and so use vegetation that deer will not or cannot browse.

A herd established in New Mexico has been used to point out that exotic species can be introduced as game without affecting other wildlife (Hibben 1964). However, it has been found that there is considerable dietary overlap between Barbary sheep and mule deer in New Mexico.

The same has been found in Texas, where it is suggested that if food becomes scarce or sheep continue to increase they will displace mule deer from existing sympatric areas (Bird and Upham 1980; Simpson 1980; Krysl *et al.* in Simpson 1980).

In the light of this new evidence policy has changed to one of removal of Barbary sheep and replacement with native bighorn sheep (Feldhamer and Armstrong 1993).

ARGALI

Mouflon, urial, European wild sheep, Corsican wild sheep, parmir argali, Marco Polo's sheep, Asiatic mouflon, nayan

Ovis ammon (Linnaeus)
=*O. musimon* Schreber, *O. orientalis*

It is now generally recognised that O. musimon *is merely a subspecies of* O. ammon. *It has also been suggested recently that it is descended from ancient (c. 5000 years ago) domestic sheep rather than being a relic wild species (see Poplin 1979).*

■ DESCRIPTION

Measurements of the various subspecies of this group vary substantially: HB 1200 mm (orientalis), 1100–1300 mm (musimon); T 70 mm (orientalis), 30–60 mm (musimon), 109–142 (urial); SH 800–1270 mm (ammon), 650–750 mm (musimon); WT 99–159 kg (ammon), 25–57 kg (musimon), 21–41 kg (urial).

Coat reddish-brown (*musimon*) or pale coloured (*ammon*), with conspicuous whitish saddle patch; narrow muzzle; ears pointed; fringe of long hair down front of neck (*musimon*); horns large and wrinkled (*ammon*) or spiral (500–850 mm in *musimon*), black; dark ruff on underside of neck (*musimon*); vertical dark line in front of saddle patch (*musimon*); tail short. Female horned (*ammon*); lacks ruff and not always horned (*musimon*) and has less distinct rump patch. Face often grey or whitish as are lower legs; some females lack horns. Horns variable to 1 m; basal circumference varies 340–530 mm.

■ DISTRIBUTION

Europe and Asia. In Europe on the islands of Corsica and Sardinia and in Asia in all the mountains of central Asia from Great Khingan in Manchuria, the Sayan and Altai, western China and the Himalayas to Iran and Asia Minor, Oman and Cyprus.

■ HABITS AND BEHAVIOUR

Habits: crepuscular and nocturnal; bouts of feeding activity with rest between. **Gregariousness:** female and young in herds 2–23, 40; males in separate herds outside rut; old males solitary; density 0.2–2.2/km² and up to 11–13/km². **Movements:** sedentary (*musimon*); moves higher in summer, lower in winter (*ammon*). **Habitat:** rugged mountainous areas; mountain valleys in thick cover interspersed with grassy glades; steppe. **Foods:** grass, forbs and herbs, flowers, leaves, weeds, browse, shoots and sedges. **Breeding:** mates autumn (*musimon*) and early winter (*ammon*); gestation 146–180 days (*musimon* and *orientalis*); seasonally polyoestrous?; young 1–2 (*musimon*); ewes rejoin flock when newborn a few days old; sexual maturity 1.5–2 years. **Longevity:** 8–15 (wild) to 19 (captivity) years. **Status:** greatly reduced numbers and range in wild; range and numbers increased by introductions.

■ HISTORY OF INTRODUCTIONS
ATLANTIC OCEAN ISLANDS
Desertas
One pair of Corsican mouflon were put on the island of Deserta Grande by S. A. S. Le Prince de Monaco in August 1912 (Bannerman and Bannerman 1965), but their fate does not appear to have been determined.

EUROPE
The argali or mouflon has been introduced in a number of areas in southern and central Europe where it is well established in the wild (Lyneborg 1971). They are now common through introductions in central Europe (Hvass 1961; Niethammer 1963; Cheylan 1991) and feral populations exist in Germany, France, Switzerland, Austria, Hungary, Italy, Czechoslovakia, Romania, Yugoslavia, European Russia (Crimea), Holland, Denmark and Poland, Iberian Peninsula and have been re-introduced in Corsica and Sardinia. *O. a. orientalis* has been introduced in Germany, Hungary, Austria and Czechoslovakia (Whitfield 1985).

Argali have been introduced for hunting in Allemagne (Hesse, Harz, Thuringe), in Austria, in Hungary, in Czechoslovakia and in Romania.

Austria
Several introduced herds are now present in Austria (de Vos *et al.* 1956; F. Spitzenberger *pers. comm.* 1982). In about 1840 the first specimens of mouflon were translocated from Sardinia to a deer park (Lainz) in Austria by Prince Eugen (Zeuner 1963). This colony was used to start another in Hungary.

Corsica and Sardinia
It is possible that relic populations here and on Cyprus are descended from animals introduced by humans, which were domesticated, but still very similar to wild ancestors (Lever 1985).

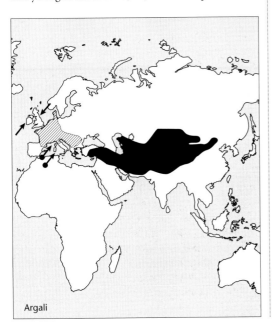

Argali

Czechoslovakia

In 1869, 10 argali were transferred to the Tribek Mountains of Slovakia, where they became well established (de Vos *et al.* 1956). *O. m. sinesella* was introduced to Rostyn in 1931, but bred with sheep and the subspecies is probably now lost (Horacek 1962). Some were also introduced in the Pavlov Hills in South Moravia (Grulih 1979). A number imported from Corsica and Sardinia were successfully introduced in the Brdy Forest (Ganzak 1964). They were first known to have been introduced in Hluboká Reserve in (Bohemia) Czechoslovakia from Vienna, Austria, in 1858 (Lever 1985). In 1977 there were 11 674 argali in reserves and wild in Czechoslovakia (Lever 1985).

Denmark

In 1951–52 argali were released on private lands in Southfyn, Denmark (de Vos *et al.* 1956), but their continued survival is not known.

France

The first attempt at acclimatisation of argali in France occurred in the Mercantous Reserve in 1949 and later to others in the Pyrenees, at Pic du Midi d'Ossau in 1951, at Donon in Vosges in 1953, and 12 were released on Mount Ventoux in 1954. They became established and bred in all these areas except Donon. They also flourish in the Chambord Park where no mountains exist (Dorst and Giban 1954).

Germany

There are now several herds of argali in Germany (de Vos *et al.* 1956) in selected forest areas (Webb 1960).

In Schlewig-Holstein in West Germany three of five introductions (*O. a. musimon*) have been successful and approximately 38 animals live in a wild state while 30 are still kept in enclosures (Heidemann and Witt 1978). Before 1937 argali were introduced to the Harz Mountains (Hesse 1937). Most German stock initially came from an introduction to Hungary in about 1840 and in 1945 it was estimated that there were 10 000–15 000 in Germany (Zeuner 1963). In 1939, eight argali were released in an enclosure near Emkendorf, but later broke out and escaped (Rieck 1954). These animals failed to establish themselves in the wild and disappeared without trace. Shortly before 1950, near Geesthacht on the northern bank of the Elbe River, argali were released in a 130-ha enclosure. There were some 30 living there in 1950, but several years later only one remained and so additional animals were released (Türcke and Schmincke 1965).

Following these attempts three more introductions occurred, of which two were successful in the Schleswig-Holstein area. On an island in the Elbe River four tame argali (two males and two females) were placed out in 1967. These became established and their numbers increased to nine, but the whole herd was annihilated in floods following a storm in January 1976. A successful introduction occurred in the Siehagen district of east Holstein on the Baltic Sea coast. In 1958 a single animal was released in a 15-ha enclosure from which it escaped. However, in 1963, four argali (two males and two females) were added to the enclosure and later allowed their freedom. By 1969 they had increased to 36 head, but continuing severe snowy winters thereafter reduced them to five or six. In the summer of 1976 these had again increased and it was believed that about 19 were present. In an area within a 50 km^2 state forest at Bad Segeberg in 1969, four argali were introduced to a 2.5-ha enclosure. This was opened in 1970 and the animals became established in the vicinity. These were supplied with food especially during the winter period and by 1977 there were 19 in the flock, but only one ram had become completely independent of the supplementary feeding (Heidemann and Witt 1978).

Holland

A herd of argali was started in Holland in 1918–19 and now numbers about 100 head (de Vos *et al.* 1956).

Italy

Argali were introduced to Italy during the mid-1800s, but have long since disappeared (Harper 1945).

Poland

Argali have been introduced to Poland (Suminski 1963), but there appears no recent information on their success.

Romania

Argali were introduced in the Carpathians before 1937 (Hesse 1937). They became established in a hunting park in Transylvania about the middle of the nineteenth century (Almeshan 1959).

Russian Federation and adjacent independent republics

In 1913–14 argali were introduced to the Ukraine where they became established locally (Yanushevich 1966). They were also introduced to the Crimea at about the same time (Turcek 1951) and are still present (Lever 1985). The range of *O. a. polii* has been extended in Kazakhstan (Sludskii and Afanas'ev 1964). Some populations have been established in the Transdanubian Mountains and appear to have been successful on the basis of improved 'trophy' parameters (Nahlik 1989).

Spain
Argali were first introduced to Spain in 1954, and today they are abundant in several national parks and hunting areas (Fandos and Reig 1987, 1992).

Switzerland
Some argali were introduced in the canton of Zürich in 1916 (four animals) and 1918 (two) and these became established and resulted in a population of up to 30 head, but all had disappeared by 1938. An established population, about 30 head in 1980, has resulted from animals immigrating from High Savoie, France, into the western parts of the canton of Valais where they are increasing slowly (M. Dollinger *pers. comm.* 1982).

Tunisia
Argali have been introduced to the island of Zembra in the Mediterranean Sea off Tunisia (Vigne 1988).

United Kingdom and Ireland
A pair of argali escaped from a private zoo at Paignton, South Devon, in 1939 and became established and bred nearby. In 1947, five were observed and at one time the herd numbered 12, but by 1958 only one remained. Two attempted introductions occured in Ireland – at Powerscourt, County Wicklow, in the 1860s and on Lambay Island, County Dublin, in 1906, but both over time were unsuccessful in becoming permanently established (Fitter 1959).

NORTH AMERICA
United States
Argali were introduced into Texas (on King Ranch) in 1946 (Schreiner 1968) and have been established in California since before 1958, but do not appear to be spreading (Presnall 1958). Some 28 were released on the H. B. Zachry Ranch, Laredo, Texas, in 1959–60 (Sanders 1963). They were first introduced to the King Ranch, Texas, in the 1930s and 40s, and some were free-ranging by the 1960s. In 1970 they were present in Texas in 22 counties with over 50 animals each and in 32 counties with less than 50 animals each (Ramsay 1970). In 1979 at least 2538 head were free-ranging, mainly in the Edwards Plateau area (Lever 1985).

PACIFIC OCEAN ISLANDS
Hawaiian Islands
Some argali were imported from a zoo in the United States and released for game on the island of Lanai in 1954 (Walker 1967; Berger 1972). Some were also released on Mauna Kea, Oahu, as they were present there in the early 1980s (Anon. 1981). Argali were introduced to 'fill a vacant niche' on Lanai, however, the main purpose appears to have been to cross-breed with feral sheep and create an animal which is less damaging to the vegetation (Kramer 1971).

The first introduction was in 1954 when two males and three females and five juvenile ewes were released on a ridge between Keone and Naupaka gulches on the western slopes of Lanai. During the next eight years an additional 35 argali were liberated and by 1964 the herd had increased to 250 head and hunting was allowed (Walker 1960, 1961, 1962; Kramer 1971).

In 1958, two males and two females were liberated in the Na Pali region of Kauai, but they disappeared a short time later (Kramer 1971).

From 1962 to 1966 a total of 46 males and 48 females were liberated on Mauna Kea, as were 33 hybrid rams and 66 hybrid ewes. The hybrids were released in the Puu Laau area, 16 pure stock also at Puu Laau and the remaining 78 in the Kahinahina section (Walker 1966).

Argali are now established on Lanai in lowlands mainly at the western end and are thought to number about 100. They have not established well on Hawaii where they remain at the release site.

Kerguelen
Two pairs of argali were set free on Île Blakeney and Île Haute in 1957, but only the pair on Île Haute became established (Lesel 1967, 1969). These came from the Parc Zoologique de Paris, whose animals came from a herd living in the Parc du Château de Chambod (Lesel and Derenne 1977). The pair on Île Haute bred and the herd increased to some 42 by 1968. There were 170 in 1972 and in 1974 the herd was estimated to be about 300 despite the removal of 34 in 1972 and 75 in 1973. Since 1969 argali are to be found in all parts and elevations of the Île Haute.

SOUTH AMERICA
Chile and Argentina
Argali appear to have been a recent introduction into Chile (Miller 1973) where they are ranched in Osorno Province and established in the wild on an island in Lake Rupanco. Some are kept on ranches in Neuquen Province in Argentina (Lever 1985), but are not known there in wild.

■ DAMAGE
In Germany, where they are established, argali do no significant damage to forest regeneration (Webb 1960). In Spain, as introduced animals, they apparently have some negative influences on the environment mainly to do with the alteration of natural vegetation and competition with other ungulates (Fandos and Reig 1987).

Where introduced on Mauna Kea, Oahu, in the Hawaiian Islands, they are said to eat the native mamane and to have food habits similar to sheep (Anon. 1981).

A new race of argali has developed in Slovakia since about 1910 (Allen 1954 in de Vos and Petrides 1967).

FERAL SHEEP
Wild domestic sheep
Ovis aries Linnaeus
=*ancestor O. orientalis*
Domestic sheep arose from the Asiatic mouflon (Ovis orientalis) of Asia Minor (Corbet and Harris 1991).

■ DESCRIPTION
WT 34–57 kg.
Most feral sheep resemble domestic sheep in appearance and size. Both Soay and Boreray sheep are highly variable in colour. Coat is woolly fleece with hair fibres; may be short or long, colour may be white, brown or black or multi-coloured, long tailed; males generally larger, may be horned.

Soay sheep – narrow bodied, long legged, short tail, narrow face; chocolate-brown above with off-white undersides and rump, or pale oatmeal, ewes horned, at shoulder 570–770 mm; weight males 34–50 kg and females 9–41 kg. Boreray is smaller than Soay, the head is larger and the animal often has a dark collar.

■ DISTRIBUTION
Wild sheep formerly occurred from Asia Minor through central Asia to north-eastern Siberia.

DOMESTICATION
Sheep were probably domesticated first in south-west Asia or the Middle East. Bones have been found back to the Neolithic, 5000 BC, in association with human settlement. They may have been derived from one or more *Ovis* spp. (mouflon *O. a. musimon* or *O. a. orientalis*) still surviving, but their ancestry has not been accurately determined (Zeuner 1963, Walker 1968). The wild sheep on Corsica and Sardinia (*O. a. musimon*) appear to be feral relics of sheep that have been under domestication (Corbet and Harris 1991).

■ HABITS AND BEHAVIOUR
Habits: diurnal. **Gregariousness:** in flocks; loose flocks or groups females and young; rams together except at breeding season; density 0.2–2.1/ha (Santa Cruz). **Movements:** sedentary; home range may be up to 45 ha (11–207 ha Santa Cruz), 0.2–0.9ha^{-1} (on Santa Cruz). **Habitat:** rough pasture, bush, scrub, forest, grasslands. **Foods:** pasture and herbaceous plants and shrubs, herbs, grasses, leaves, forbs, ferns.

■ HISTORY OF INTRODUCTIONS
Introduction of sheep on islands

Island name	Date	Status
Adams (Auckland)	1880s	died out after few years
Ailsa Craig (UK)	1930s	?
Amsterdam	1957?	still present?
Antipodes (NZ)	<1887	died out?
Arapawa (NZ)	pre–1880s	
Auckland (NZ)	1840s	died out
Boreray (St. Kilda)	1870,1930	still present
Campbell (NZ)	1895	eradicated 1970–91
Cardigan (UK)	1944	still present
Chatham (Canada)	nineteenth century	?
Chathams (NZ)	1850s	still present?
DeCourcey (Canada)	nineteenth century	?
Enderby (Auckland)	c. 1850, 1890s	died out after few years
Galápagos (Ecuador)	nineteenth century	still present?
Gough (Tristan da Cunha)	1956	confined ?
Grande Terre (Kerguelen)	?	still present?
Hawaii (Hawaiian)	1793–94	?
Hawaiian (United States)	1791–94	still present
Hirta (St. Kilda)	1870,1930	still present
Hispaniola (West Indies)	<1797	none by 1950s
Île de Corbeau (Kerguelen)	?	?
Île Longue (Kerguelen)	?	?
Île Mussel (Kerguelen)	1952	?
Inaccessible (Tristan)	<1938	?
Isabela (Galápagos)	nineteenth century	still present?
Juan Fernández (Chile)	?	?
Kahoolawe (Hawaiian)	eighteenth century	?
Kapiti (NZ)	<1896	eradicated c. 1930–69
Kauai (Hawaiian)	1791–94	?
Kerguelen	1909	?
Lasqueti (Canada)	nineteenth century	?
Lihou (nw Guernsey)	1974	still present
Lilla Karfso (Baltic Sea)	?	?
Linga Holm (UK)	1974	?
Lundy (UK)	1927	still present
Macquarie (Aust)	1947	not feral
Mangere (NZ)	c. 1900	eradicated 1968

Introduction of sheep on islands (*continued*)

Island name	Date	Status
Marion (Sth Africa)	1927	failed
Maui (Hawaiian)	?	?
Middleholm (UK)	?	?
Molokai (Hawaiian)	?	?
New Zealand	1773	still present
North Ronaldsay (Orkneys)	Neolithic	?
Orkney (UK)	prehistoric	?
Otaheite (Society)	1770s	died
Pitt (Chathams)	1850s	still present?
Presque'ile Bouquet de la Grye (Kerguelen)	1911	?
Prince Edward (Sth Africa)	1927	failed
Puerto Rico (West Indies)	?	disappeared
Rose (Auckland)	1890s	died out
Salt (Puerto Rico)	?	disappeared
Saltspring (Canada)	nineteenth century	?
San Clemente (Channel, United States)	<1862	still present
Santa Cruz (Channel, United States)	1850s	still present
Saturna (Canada)	nineteenth century	
Skokholm (UK)	1934	still present?
Skomer (UK)	1985	died out?
Soay (St. Kilda)	Neolithic	still present
Society	1770s	died
South East (NZ)	1915	eradicated 1956–61
St. Kilda (UK)	prehistoric	still present
St. Margaret's (UK)	1932	killed 1959
St. Paul	<1961	still present?
Tristan da Cunha	1824	still present?

Breeding: Soay sheep ruts October–November, lamb April; gestation 150 days; some females breed before 12 months age; lambs born with 10 milk teeth in lower jaw and 6 premolars in upper jaw; full body size reached in 1.5–3 years; females sexually mature at 6 months (Santa Cruz). **Longevity:** 5–11 years (NZ).

ATLANTIC OCEAN ISLANDS
St. Kilda (United Kingdom)
There are two forms of feral sheep in Britain: Soay sheep on St. Kilda and Boreray on Boreray (Corbet and Harris 1991). Soay sheep may have been introduced in prehistoric times as they closely resemble bones (of sheep) from Neolithic deposits. They may also have been brought to the island by Vikings from Scandinavia and Denmark around AD 800, as there is little difference between soay bones and those found in Greenland in the old Viking settlements (Lever 1977). According to Lever most support appears for the latter theory.

However they arrived on the Île of Soay, they have existed there in a semi-feral state for centuries, possibly introduced in Neolithic (Jewell *et al.* 1974). During this time the flock has remained fairly stable at about 200 animals (Lever 1977).

A flock was established on the isle of Hirta from Soay after the residents and stock were evacuated in 1930 (Morton Boyd *et al.* 1964). In 1932, 107 Soay sheep were transferred to Hirta and by 1939 the flock numbered about 500 head (Fitter 1959; Jewell *et al.* 1974; Lever 1977). Between 400 to 450 were there in 1947, 700 in 1955 and about 1013 in 1975 (Fitter 1959; Lever 1977).

In 1930, when the inhabitants left St. Kilda, flocks of black-faced sheep were left on Boreray. These have increased in numbers and there were 150 in 1955, *c.* 300 in 1959, and there are now probably still 300 of them (Fitter 1959; Lever 1977).

Orkney Islands
There are semi-feral sheep on North Ronaldsay Island in the Orkneys and they have possibly been there since the Neolithic (Hall 1975). About 400 were introduced to Linga Holme from North Ronaldsay in 1974 as a conservation measure (Jewell 1978). Some were also taken to Lihou Island (north-west of Guernsey) from North Ronaldsay in 1974 where they are still semi-feral (Wilberley 1979).

On Boreray they were farmed until 1930 when the island was abandoned. In 1971, 466 were there (Jewell *et al.* 1974; Boyd 1981).

The 'Hebridean' was imported to Boreray and Hirta to provide new blood. The sheep on Boreray have doubtless changed by introductions from Hirta, unlike those of Soay which have remained unaltered since the Viking period.

In 1983 (Boyd 1981) or 1870 Scottish 'blackface' sheep from northern England were introduced to Hirta and Boreray. Those present on Boreray were farmed until 1930 when the island was abandonded. They resemble a cross between a Soay and a modern Scottish blackface. These have remained uncontaminated since the inhabitants departed in 1930. In 1980 there were about 700 animals present (Lever 1985).

Soay sheep are now found on Soay and Hirta (St. Kilda) and more recently were introduced to many offshore islands (e.g. Lundy Island in Bristol Channel); Boreray sheep are confined to Boreray, St. Kilda (Corbet and Harris 1991), where there are between 200–500 sheep. On Hirta Island there are between 600 to 1800.

Australia

Feral sheep have become established in Australia from time to time (Lever 1985; Myers 1986). Some small populations have persisted for decades in areas such as the Kimberley, Western Australia, and in the Necoleche Nature Reserve, north-western New South Wales (Wilson *et al.* 1992). Sheep were released from the settlement at Escape Cliffs, north of Darwin, in 1866, as well as from the Coburg Peninsula outpost some 20 years earlier. At present feral sheep occur on vacant crown land and some native reserves in the Northern Territory (Letts 1964). A small flock of about 150 feral sheep also existed near Mount Lynton, Western Australia, until the mid-1980s.

Europe
Norway and Sweden

A semi-feral population of sheep existed in south-west Norway where they were protected from hunting (Mason 1979; Rudge 1984), but their presence has not been confirmed recently. A flock of 100 ewes and six rams of a Swedish breed live on Lilla Karfsö off Gotland in the Baltic Sea (Lever 1985).

United Kingdom

In 1910 some Soay sheep were transferred to Woburn, England, and in 1934 some of these (two males and six females) (Fitter 1959) were sent to Skokholm Island (in the Irish Sea off the Pembrokeshire coast) where in 1944 there were 40. A ram and three ewes were sent to Cardigan Island, Cardigan Bay, in 1944 and by 1959 the flock numbered 70 animals. A ram and three ewes were also sent in 1932 to St. Margaret's Island (near Tenby) and which increased to 20–30 by 1959, but all were killed by lightning late in the same year (Fitter 1959; Lever 1977). Some were introduced to the island of Skomer in 1958 and two years later there were 40 sheep on Skomer, Skokholm and Cardigan islands. However in 1975 Soay sheep occurred only on Cardigan Island where they numbered some 80 animals (Lever 1977). At one time they were also kept on Middleholm between Skomer and the mainland (Fitter 1959).

M. Harman introduced some Soay sheep to Lundy Island in 1927 and these increased to 80 by 1959 and some were still present in 1973 (Fitter 1959; Lever 1977). Soay sheep on Ailsa Craig Island were introduced from St. Kilda in the 1930s and in 1956 there were still 14 there (Lever 1977).

Sheep were introduced to Skokholm after World War 2 and up to 76 existed there until 1959–60. In 1968 only one was seen (Berry 1968).

Pacific Ocean islands
Antipodes

Sheep were present on the islands in 1887, but died out at a later date (Atkinson and Bell 1973).

Arapawa Island

Sheep were introduced to the island pre-1880s by whalers and early settlers and there are about 100 sheep (Merino type) in a reserve for fauna or flora and some were also kept privately (Mason 1979; Rudge 1984).

Auckland Islands

Sheep were introduced a number of times on Auckland Island in the 1840s and 1850s and again in the 1890s and early 1900s, but failed to become permanently established there (Taylor 1968). They were also released on Adams Island in the 1880s and 1890s, but did not survive here for more than a few years. They were introduced to Enderby Island in about 1850 and again in the 1890s, but also disappeared after a few years, and also to Rose Island in the 1890s.

Campbell Island

Domestic sheep were introduced to Campbell Island in 1895, 1901 and 1902 and managed for their wool until 1931 when management was discontinued (Wilson and Orwin 1964; Wilson 1964). During this period the number of sheep present on the island reached 8000 head (Cockayne 1909; Laing 1909; Eden 1955). When abandoned as a farming venture, some 4000 sheep and some cattle were left to run wild (Taylor 1968; Bell and Taylor 1970). Since 1941 shooting for meat reduced their numbers by about 50 sheep per year, but the herd was also declining from 1916 up until 1961 at the rate of five per cent annually (Wilson and Orwin 1964). About 1500 were present in 1950 (Oliver and Sorenson 1951) and there were 950–1000 in 1961 (Holdgate and Wace 1961; Bell and Taylor 1970). However, the population had built up to 3000 by 1969 and the island was fenced across the middle in 1970 and some 1300 sheep on the northern half were killed (Bell and Taylor 1970; Gibb and Flux 1973; Atkinson and Bell 1973). The fence was erected and animals shot in order to allow the measurement of their effects on the vegetation in future years (Bell and Taylor 1970).

The sheep continued to increase in numbers on the southern half of the island and there were 2861 there in 1977 (Dilks and Wilson 1979; Rudge 1986), but there has been some spectacular recovery of the vegetation in the northern half. Those on Campbell Island appear to be resistant to footrot and may be of scientific value (Wodizicki and Wright 1982).

Sheep are now restricted to the fenced half of the island that is a reserve for fauna and flora (Dilks and Wilson 1979). Type of sheep is merino × lincoln, leicester or romney (Van Vuren and Coblentz 1984).

Channel Islands, United States
San Clemente Island
Early reports list sheep on the island and they were probably introduced there before 1862 (Johnson 1967). However they no longer occur there.

Santa Cruz
Sheep ranching began in 1850s and by 1890, there were 50 000 on island. By the 1920s many had become wild and could not be captured, and attempts to regain control of them were abandoned. Many were shot and trapped but despite these efforts many thousands remained in a feral state. The sheep were mainly merino with some Rambouillet and Leicester inbred (Van Vuren and Coblentz 1984) and were still there in 1979–80 (Brooke 1984; Van Vuren 1981; Van Vuren and Coblentz 1987). Between 1979 and 1981 there were about 21 240 sheep on the island (Van Vuren and Coblentz 1989).

Chatham Islands
Feral sheep are present on the Chathams (Atkinson and Bell 1973). Sheep (merino type) were first reported there in 1900 (Whitaker 1976) and there are now two flocks, a small one on the mainland near the south-west corner, and a larger one on Pitt Island of some 2000–3000 animals. They were introduced to Pitt Island in the 1850s and have been feral there for about 70 years. In 1981 there were about 300 in a reserve especially created for them (Whitaker 1976; Rudge 1983). Domestic sheep were introduced to Mangere Island before 1900 and became feral when they were abandoned at a later date. In 1968 these were exterminated (Bell 1975).

Galápagos Islands
Sheep were introduced to Islabela Island in the nineteenth century (Eckhardt 1972), but it is not known if they were feral sheep or whether they are now extinct (Rudge 1984).

Hawaiian Islands
Captain Cook landed sheep on Kauai in 1791 and more introductions followed with the visits of Captain G. Vancouver in 1793 and 1794. Sheep occur on all the islands except Oahu (Brooke 1984).

Sheep introduced in 1793–94 by Vancouver became well established on Mauna Kea by 1822 and have been there ever since (Tinker 1941). There were 300 on Mauna Kea in August 1981, but steps to eradicate them were in progress (Anon. 1981). There were about 2000 there in 1976, probably merino, but some crosses (Mouflon?) (Griffin 1976).

By 1851 it was estimated that there were at least 3000 wild sheep on Hawaii and some were recorded present on Maui, Molokai and Kauai (Bishop 1852).

On the slopes of the Mouna Kea Forest Reserve in 1937 it was estimated that there were 40 000 feral sheep (Brooke 1984). In the period 1921–46 some 46 765 were removed from Mauna Kea alone and another 24 703 from other forest reserves (Bryan 1947). A small number survived and continued to breed and despite heavy hunting efforts there were still nearly 2000 left in 1964 (Kramer 1971).

Sheep were introduced to Kahoolawe Island in the eighteenth century(?) and there are now 300–400 there (McKnight 1964). Following the unsuccessful farming of sheep on Kahoolawe, about 2000 sheep were left there in 1859. Other lessees looked after the flock on Kahoolawe and introduced more sheep, so that in 1909 there were some 3200 there. The private lease was cancelled soon after and efforts were made to remove the sheep, but there were still 300 there in 1913 (Forbes 1913) and 150 in 1916 (Judd 1916). In the years to 1964 the flock has fluctuated between 2000 and 5000 head (Kramer 1971). In 1975 there were 1700–2000 sheep on Hawaii (Giffin 1976; Brooke 1984).

Juan Fernández Island (Chile)
Sheep are mentioned as present (Kunkel 1968), but it is not known if they were feral.

Kerguelen (France)
Sheep were introduced to Kerguelen in 1909 (Holdgate and Wace 1961). Some were imported to Île Longue and in 1911, 1000 were liberated on Presque'Île Bouquet de la Grye (Aubert de la Rue 1930; Jeannel 1941). Those sheep on the islands were largely maintained by shepherds for whaling station staff but this venture was interupted between 1914 and 1921 but recommenced at Port Couvreaux and Île du Corbeau until abandoned in 1932 (Holdgate and Wace 1961). Some sheep were also landed on Île Mussel in 1952.

Controlled populations of sheep now live on two or three islands in the Golfe Morbihan to supply fresh meat to staff at the base at Port-aux-Française (Lesel and Derenne 1977). Some sheep escaped to Grande Terre while they were being moved before slaughter, and in 1973 about 70 were living wild on Grande Terre where they are confined to Péninsule Courbet. The sheep on the island are a mixed race (dominant type 'Bizet').

Macquarie Island
Sheep were introduced in 1947 and a flock of 15 was kept on the island (Taylor 1955), but they did not become established as a feral species (Watson 1975).

New Zealand
Sheep have been feral at times in New Zealand since the early nineteenth century and are now found locally in the North Island and rarely in the South Island (Wodzicki 1965). They have remained feral in some remote areas of the both islands (Gibb and Flux 1973).

At Omahaki and the Mohaka River merino-cross feral sheep have existed at Mohaka from the 1880s. On the Hokonui Range a flock reported to have originally come from Tasmania, Australia, in 1858 existed until recently. In the Clarence and Wairau rivers area sheep introduced as farm stock in the 1880s existed until at least recently. Some were present in the Oxford State Forest, also from farm stock brought in in the 1850s, and may be still present there. A flock also existed on the Raglan Peninsula from the 1930s until they were shot out. Merino crosses were established for 50–60 years after escaping from farm stock in the 1880s in the Waianakarua River area.

In the 1960s feral sheep were in the high country of the South Island and in the drier parts of the North Island. Between 1951 and 1958 some 15 678 were destroyed by official hunters under a bonus scheme.

Captain Cook landed a pair of sheep in Queen Charlotte Sound (Kippis 1904) in 1773, but these did not succeed as they were later found dead. In 1814 S. Marsden brought sheep from Sydney to the Bay of Plenty Islands. In 1834 J. Bell settled on Mana Island with 102 sheep. By the 1840s introductions were commonplace. During the early development of New Zealand, sheep were run on open range and it was inevitable that they would form feral flocks. By the 1880s feral flocks were common in the mountainous districts of the South Island (Thomson 1922), in Hawkes Bay and doubtless elsewhere. By the early 1900s they were only found in inaccessible areas, except during the depression when many farms were abandoned. Many feral flocks were destroyed after World War 2. They now probably exist in about 12 places on the mainland from Hawkes Bay to Southland. In Hawkes Bay they exist in the north-eastern Ruahines and on the Mohaka River, at Marlborough at Wairau and Clarence, and some on Arapawa Island (about 120), some south of Oamaru and in the Hokonui Hills in Southland (Whitaker 1976). In 1922 they were still abundant in the wilder parts of the country especially Marlborough (Thomson 1922).

There are probably at least seven small populations on the main islands at Hawkes Bay, Omahaki and Mohaka rivers, Hokonui Range, Clarence and Wairan rivers, Oxford State Forest, Raglan Peninsula, and the Wainakarua River (Rudge 1984). Some may have originated from as early as 1850 (Hokonui Range 1858) (Oxford State Forest 1850) and 1880s (Hawkes Bay, Omahaki and Mohaka, Clarence and Wairan), and others more recently (Raglan 1930s) (Whitaker 1976; Rudge 1984).

Sheep are still present in isolated local mainland areas (Wodizicki and Wright 1982).

Formerly widespread on main islands but at present only eight discrete flocks on mainland and four on islands (King 1990) (islands include Arapanua, Chatham, Pitt and Campbell).

Society Islands
Captain Cook gave sheep to the natives on Otaheite in the Society Islands, but they died before they could become established (Kippis 1904).

INDIAN OCEAN ISLANDS
Amsterdam (France)
A small flock of sheep existed on the island in 1957 (Holdgate and Wace 1961).

St. Paul
Sheep were imported to this island some time before 1961 but are not numerous there (Holdgate and Wace 1961).

South Georgia
Sheep have at times lived on the island, but are not established there now (Watson 1975).

Tristan da Cunha
Sheep were introduced to the island in 1824 (Holdgate and Wace 1961). Some were present there in 1829 (Morrell 1832) and in 1938 there were seven on Inaccessible Island (Hagen 1952). During the 1940s and 1950s they were grazed on Tristan where they were only partly confined (Munch 1945; Holdgate 1958).

The residents left the island and 740 sheep and other animals in 1961 when a volcano erupted, but before their return in 1963, dogs which had also been left behind ran wild and killed nearly all of them (Anon. 1963).

Marion and Prince Edward Islands
Sheep were imported in 1927 for the South African Weather Bureau Station, but failed to become established (La Grange 1954).

Gough Island
Sheep were introduced to the island in 1956, but are confined in enclosures (Holdgate and Wace 1961).

They appear to have been there in 1975 (Watson 1975).

Canada
Sheep are run on some of the larger islands off British Columbia and sometimes are truly feral (Carl and Guiguet 1972). They were introduced at some time in the nineteenth century and occurred on Lasqueti Island, Saltspring Island, Chatham Island, DeCourcey Island and Saturna Island (British Columbia) in the 1980s (Rudge 1984).

United States (mainland)
Small feral flocks may still exist in Utah, Colorado, Oregon and Alabama. Little is known of their origin (McKnight 1964). Experimental sheep (mouflon × rambouillet) are still present in numbers in Texas (Mason 1980). They do not persist in California probably because of the presence of large predators (Moomey *et al.* 1986).

WEST INDIES
Hispaniola
Moreau de Saint-Mery (1797–98) reported wild sheep at Anse-à-Pitre but there were none present in Haiti in the early 1950s (Street 1962).

Puerto Rico
Feral sheep were present on Salt Island, but have since disappeared because the vegetation has gone (Heatwole *et al.* 1981).

■ DAMAGE
On Mauna Kea, in Hawaii, feral sheep are frequently reported to be destructive to plant forms and to cause soil erosion. The main problem appears to be their effects on the mamane (*Sophora chrysophylla*) forests of Mauna Kea (Atkinson 1977).

An experiment on Campbell Island of confining sheep to the southern half of the island by fencing has shown some spectacular revegetation changes on the northern half where they were removed. Breeding albatrosses also appear to have increased in numbers, more so on the northern half where sheep no longer graze (Dilks and Wilson 1979).

On Santa Cruz feral sheep had a significant impact on vegetation and nesting sea birds. Overgrazing has caused an increase in grasslands and a decrease of coastal sage scrub. The damage is expected to continue until equilibrium is reached (Van Vuren and Coblentz 1984). Besides the effects of defoliation, trampling damage was also high. The two combined has resulted in a moderate to severe impact of about half of the island. The sheep had a severe negative input on native biola of islands and the endemic plants and birds are particularly vulnerable (Van Vuren and Coblentz 1987).

In New Zealand sheep have little effect because they are essentially grazers. They can create a local nuisance by mixing with domestic stock to disrupt breeding and spread ectoparasites (King 1990). On some islands they may have prevented regeneration of plants (e.g. Campbell Island).

BIGHORN SHEEP
Californian bighorn, Rocky Mountain sheep, mountain sheep, desert bighorn, American bighorn
Ovis canadensis Shaw

■ DESCRIPTION
HB 1200–1800 mm; T 70–150 mm; SH 800–1120 mm; WT males 57.7–156 kg, females 33.6–90.9 kg.
Fleece tawny-buff on back; sides brownish; under parts light horn to yellowish white; face, chest and legs chocolate brown; muzzle long and narrow; ears short, pointed and hairy; lower belly, back of legs, muzzle and rump patch ivory white; horns massive, spiralled with transverse rings to 1.15 m; tail short; white tail patch. Female has short horns, erect, and curved backwards.

■ DISTRIBUTION
North America and Asia. Western North America from Alberta, Canada, south to Baja California. Formerly east to western North Dakota and the Blackhills, South Dakota. In Asia (race *O. c. nivicola*) in north-east Siberia east of Lake Baikal and the Lena River; an isolated population may exist in the Putorana Mountains east of the mouth of the Yenesei River.

■ HABITS AND BEHAVIOUR
Habits: diurnal; climbs well. **Gregariousness:** bands up to 10–15, occasionally 100 animals; rams over three years form bands or harems in spring; old rams often solitary. **Movements:** seasonal migration (higher in summer, lower in winter); up to 64 km; home range 3.8–85.7 km². **Habitat:** mountainous regions including desert mountain ranges, rugged rocky cliffs and bluffs, alpine meadows, and grassy mountain slopes. **Foods:** grass, forbs, sedges, moss, lichens, fungi, berries, and browse from shrubs and trees. **Breeding:** ruts July–October; lamb January–April: gestation 171–180 days; young 1–2; female seasonally polyoestrous (?); male mature at 3 years but do not breed until 7 years, females 2–3 years. **Longevity:** 14–15 years in wild. **Status:** range and numbers decreased, but still numerous some areas.

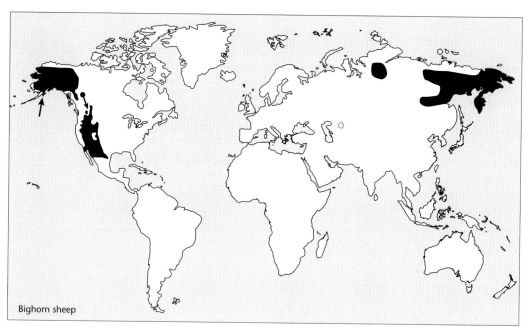

Bighorn sheep

■ HISTORY OF INTRODUCTIONS
NORTH AMERICA
Alaska
In 1964–65 some seven bighorns were released on Kodiak Island, Alaska (Burris 1965). The introduction was evidently unsuccessful.

Canada
Bighorn sheep from eastern British Columbia have been introduced successfully to several points in the Selkirk Mountains (Banfield 1977). From Banff 40 females and 10 males were introduced to Squilax in 1927, although some already existed in the area (Cowan and Guiguet 1960; Carl and Guiguet 1972). Also in 1927, 35 females and 14 males were transferred to Spences Bridge from Banff National Park. Some success was also said to have been achieved with releases at Adams Lake.

In 1955, 18 bighorns from Chilcotin and Fraser River, of which five were taken to Vaseux Lake, two to Taseko Lake and 11 to Westbranch in the Anahim district, were released (Carl and Guiguet 1972). In 1966, 11 bighorn from the same area were transferred to ranges north of Kamloops Lake. Both these introductions were apparently successful. In 1969 introduced herds occurred in the Kamloops Lake and Dog Creek areas of British Columbia (Spalding and Mitchell 1970).

United States
Translocations and re-introductions of bighorn sheep began in Montana in the 1940s and resulted in much improved distribution and numbers (Couey and Schallenberger 1971). In 1939, two bighorns taken to

Wildhorse Island, Lake County, became established and were said to have increased to number 137 animals. In 1942, 14 were translocated to Lewis and Clark counties but were unsuccessful in becoming established. In 1947, 42 bighorns were translocated to Garfield County, but this release failed in 1952, and six translocated to Wildhorse Island became established but did not increase in numbers or spread. Sixteen bighorns released at Sixteen Mile Creek, Gallatin County, were unsuccessful in becoming established. Successful introductions were made to Kootenai Falls, Lincoln County, in 1955 (13 animals), and between 1955 and 1957 to Bull Mountain, Jefferson County (23), but few of these animals survived by 1970. Thirteen bighorn released in the Blue Hills, Custer County, provided limited hunting by 1970, and 19 released at Thompson Falls, Sanders County, in 1959 were also still surviving. Releases at Sheep Creek, Cascade County (21 animals), in 1959, and these at Sheep Creek, Meagher County (18), in 1962 appeared to have had little success.

From 1963 to 1968 a series of translocations in a number of counties either provided limited hunting or the results were not determined: 1963, 14 to Doris Mountain, Flathead County; 1964, 25 to Tobacco Root Mountains, Madison County; 1965, 31 to Fergus County, where they survived in 1970 on a Jane ranch; 1967, 21 to Highland Mountains, Madison County, and 25 to Olson-Foster Gulch, Deer Lodge County; 1968, 34 to Sieben, Lewis and Clark counties, 16 to Pretty Creek, Missoula County, 15 to Teakettle Mountain, Flathead County, 33 to Troy Bull River,

Lincoln County, and 30 to Highland Mountains, Madison County. Bighorn distribution in Montana has improved due to translocations, and 11 herds now exist and 13 other areas have been stocked with variable results (Couey and Schallenberger 1971).

From 1944–52 the Colorado Division of Wildlife engaged in a translocation program with bighorn stock from the Tarryall Mountains. Some 202 were trapped and translocated to other locations in Colorado and 16 to locations in Montana (Jones 1948; Rutherford 1972; Schmidt and Rutherford 1978). Successful translocations of bighorn sheep were made in Colorado in 1978–79 (Bear 1979; Weaver 1986). Releases were made in 1979 at and near the Colorado National Monument and near Grand Junction with stock obtained from Nevada and Arizona (Weaver 1986).

Between 1969 and 1984, about 297 bighorns were translocated to other sites in Nevada from the River Mountains (Leslie and Douglas 1986).

As early as 1935 there were translocations of bighorn sheep in Wyoming (Hume 1935). Some were moved from Colorado to Montana in c. 1947 (Gulbreath 1943). A survey in 1969 indicated that bighorn had been established in British Columbia, Washington, Oregon, Idaho, North Dakota and Nevada, numbering about 784 animals (Spalding and Mitchell 1970). Prior to 1974 in the south-west United States there had been few successful desert bighorn translocations and few restoration programs, largely because of the past re-introduction failures (Wilson 1986). Populations of bighorns declined drastically in the south-west United States since the early 1800s. The decline was put down to such factors as exploitative hunting, disease, habitat destruction and livestock overgrazing (Cooperrider 1986).

Until 1973 in the south-western states the estimated number of re-introduction sites was 91 (four in Arizona, 13 in California, 51 in Nevada, 15 in New Mexico and eight in Utah) (Cooperrider 1986).

Since 1976, 165 successful re-introductions have been made in Nevada by the Department of Wildlife and Bureau of Land Management. Begining in 1968 and to date, 20 re-introductions were made in 15 mountain ranges. Nevada animals were translocated to Utah, Colorado and Texas (Weaver 1986).

Relocation efforts began in Utah in 1973 with the re-introduction of 12 bighorns from Nevada to Zion National Park, with stock released at eleven sites (Weaver 1986).

Introductions have been made in Oregon, Colorado, Montana and New Mexico before 1956 (de Vos et al. 1956).

The bighorn was extirpated in Texas by 1960. The first attempts at re-introduction were made in 1957 with animals from Arizona. Breeding in enclosures is in progress and it is hoped to relocate 20 animals at a time until at least five mountain ranges are fully stocked (Weaver 1986).

By the 1940s New Mexico had only two mountain ranges (San Andreas Mountains) with bighorn sheep. In 1972 a captive-breeding program began which resulted in some being released in 1979 in the Big Hatchet Mountains. In 1980 re-introductions were made to Peloncillo Mountains with animals from Arizona. More re-introductions are planned for New Mexico (Weaver 1986).

Arizona has a successful re-introduction program dating back to 1958. Some 33 translocation sites have been identified and 15 releases into historic ranges have been made (Weaver 1986). Twelve bighorns from the Black Mountains were transferred to the Virgin Mountains in north-west Arizona in early 1980. These were surviving and breeding in an enclosure in 1981 and it was planned to release them when the herd numbered 20–30 (Sayr 1981). Twenty-one were driven out of the enclosure in late 1981 and additionally 41 in two groups from the Black Mountains were successfully released at two other locations in the Virgin Mountains (Morgart et al. 1987). Since 1979, 384 bighorns have been re-introduced to 28 separate sites in Arizona (Dodd 1983).

The California Department of Fish and Game and the National Parks Service attempted to restore bighorn to the Lava Beds National Monument when 10 animals from British Columbia were introduced to Siskiyou County. These animals survived and increased to 43 by 1979. In 1980, 10 were removed and introduced into the Warner Mountains, Modoc County, however, the Lava Beds herd contracted bacterial pneumonia and in two months all were dead (Sayre 1980). To date 273 sheep have been trapped from four mountain ranges for relocation to eight mountain ranges in California (Wehausen et al. 1987).

Some sheep were successfully re-introduced in California in 1983 and since then eight captures have been made and releases made in five mountain ranges. Three releases have been made in the Whipple Mountains at different sites. One release has been made within the San Gabriel Mountains in an attempt to re-establish a population, and one release made in the Sheephole Mountains to augment a declining population of less than five animals. It is anticipated that re-introductions will be made every year until all suitable ranges are restocked (Weaver 1986). Successfully re-introduced in the Sierra Vevada Mountains in California (Chan *et al.* 1993).

Prior to the arrival of Europeans, bighorn were widespread in the Sierra Nevada Mountains but were absent from many areas by the 1880s. In 1979, 1980 and 1982, 61 sheep were moved from the Mount Baxter herd for re-introduction at two sites in the southern Sierra Nevada Mountains (Wheeler Ridge and Mount Langley) and one in the Warner Mountains, in north-east California. In 1986 they were also re-introduced into the Lee Vining Canyon when 27 sheep from the Mount Baxter herd were released (Keay *et al.* 1987).

Some were re-introduced in Capital Reef National Park (Henderson and Rentchler 1986).

■ DAMAGE

Mountain goats (*O. americanus*) and mountain sheep (*O. canadensis*) graze the same alpine sites and eat similar plant species during summer, but not in winter (Dailey *et al.* 1984) and there is the potential for competition for food.

REFERENCES

Abbott, I. (1978) Corella *2(2): 24–5.*

__ (1978) Corella *2(2): 26–7.*

__ (1978) Corella *2(2): 32–3.*

__ (1980) J. R. Soc. West. Aust. *63: 39–45.*

__ (1981) Corella *5(3): 62–63.*

__ and Burbidge, A.A. (1995) CALMScience *1(3): 259–324.*

Abbott, W.L. (1893) Proc. U.S. Nat. Mus. *16: 759–64.*

Ables, E. (1977) Kleberg Studies in Natural Resources: Texas A & M University, Texas Agricultural College Station, Texas, USA.

Ables, E.D. (1975) In Fox, M.W. (ed) The Wild Canids, *15: 216–36. New York.*

__ and Ramsey, C.W. (1974) J. Bombay Nat. Hist. Soc. *71(1): 18–25.*

Abramov, V.K. (1963) In Okhr. prir. na Dal 'nan Vostoke, 1: 113–19. Vladivostok.

Abu Jafar, M.Z. and Hays-Shahin, C. (1988) In Dixon, A. and Jones, D. (eds), Conservation and Biology of Desert Antelopes, *5: 34–40.*

Abu-Zinada, A.H., Habibi, K. and Seitre, R. (1988) In Dixon, A. and Jones, D. (eds), Conservation and Biology of Desert Antelopes, *6: 41–7.*

Ackerman, B.B., Harmon, M.E. and Singer, F.J. (1978) In Studies European Wild Boar in Great Smoky Mountains National Park: *93–137. I: Report to Superintendent Great Smoky Mountains National Park.*

Adams, L.G. and Bailey, J.A. (1982) J. Wildl. Mgmt. *46(4): 1003–9.*

__ and __ (1983) J. Wildl. Mgmt. *47(4): 1237–43.*

__ and Robus, M.H. (1981) Trans. 46th. N. Am. Wildl. Nat. Resour. Conf. *46: 319–28.*

Adamson, G. (1986) My Pride and Joy. *London.*

Adamson, J. (1960) Born Free. *London.*

__ (1969) The Spotted Sphinx. *New York.*

Advani, F. and Rana, B.D. (1981) Acta Theriologica *26(7): 133–4.*

Afanas'ev, Yu.G. (1962) In Voprosy ekologii' *6: 15–16. vyshaya Shkola, Moscow.*

Agoramoorthy, G. and Rudran, R. (1993) Folia Primatol. *61(2): 92–6.*

Aimi, M., Baker, A. and Supriatna, J. (1982) Kyoto Univ. Overseas Res. Rep. Studies on Asian non-human Primates *2: 51–6.*

Airumyan, V.A. (1962) In Voprosy ekologii' *7–8. Vysshaya Shkola, Moscow.*

__ (1962) Izvest. Akad. Nauk. Armyansk SSR. Biol. Nauki *15(11): 69–78.*

Aizin, B.M. (1963) In Acclimatization of Animals in the USSR: *43–4. Akad. Nauk. Kazakhsk. SSR., Alma-Ata.*

Al'shtul, M.P. (1963) Sb. Nauchno. Statei Zapadn. Otd. Vses Nauchno.– Issled. Inst. Zhivotn. Syr'y a Pushniny *2: 3–41.*

__ (1963) Sb. Nauchno. Statei Zapadn. Otd. Vses

Nauchno.–Issled. Inst Zhivotn. Syr 'ya Pushniny *2: 42–98.*

Al-Sanei, K.S., Zaghloul, T.M., Salit, A.M., Omar, M.T. and Balba, M.M. (1984) Proc. 11. Vert. Pest Conf., *March, 1984: 77–81. Sacramento, California.*

Alados, C.L. and Escos, J. (1988) J. Mamm. *69(1): 172–5.*

Albignac, R. (1987) Primate Conservation *8: 44–5.*

Alendal, E (1980) Norsk Polarinstitutt, Meddelelser Nr. *107: 3–22.*

__ (1974) Nor Polarinst Arbok, *1974: 159–74.*

__ (1980) Fauna *33: 49–51.*

Alexander, B. (1898) Ibis *1898: 277–85.*

__ (1898) Ibis *1898: 74–118.*

Aliev, F. (1965) Saugetierkd. Mitt. *17(2): 152–5.*

__ (1966) J. Mamm. *47(2): 353–5.*

Aliev, F.F. (1955) Acad. Sci. Azerbaijan SSR. Rep. *8: 571–8.*

__ (1956) Acad. Sci. Azerbaijan SSR. Rep. *12(8): 21–30.*

__ (1960) Izvest. Akad. Nauk. Azerb. SSR. Ser. Biol. i Med. Nauk. *1: 107–10.*

__ (1962) Izvest. Akad. Nauk. Azerbaidz. SSR.: Ser. Biol. Med. Nauk. *1: 53–7.*

__ (1963) In 1st. All-union Conf. Mammals in USSR, *Alma-Ata.*

__ (1965) J. Mamm. *46(1): 101–2.*

__ (1970) Trans. lX Internat. Congr. Game Biol., *Moscow: 167–9.*

__ and Sanderson, G.C. (1966) J. Wildl. Mgmt. *30(3): 497–502.*

Allen, E. (1971) In Game Management in Montana, *Mussehl, T.W. and Howell, F.W. (eds), 7: 69–79. Game Management Division, Helena, Montana.*

Allen, G.M. (1954) Checklist of African Mammals. *Cambridge, Mass., USA. Boston Mass.*

Allen, P.H. (1965) Bull. (6), Alabama Department Conservation, Division of Game and Fish, State Game Management Section.

Allen, R.B., Lee, W.G. and Rance, B.D. (1994) J. Bot. *32(4): 429–39.*

Allin, A.E. (1950) Canad. Field-Nat. *64(3): 122–4.*

__ (1955) Canad. Field-Nat. *69(1): 25–6.*

Allison, C. (1969) The Australian Hunter. *Sydney.*

__ (1970) Sporting Shooter, *1970: 30–1, 70–1.*

Almeshan, Kh.A. (1959) Soob. Inst. Lesa Akad. Nauk. SSSR. *13: 118–23.*

Alsina, G. and Brandani, A. (1981) Proc. World Lagomorph Conf., *1979, Guelph, Ontario: 486–92. University of Guelph, Ontario.*

Altmann, S.A. (1962) Ann. N.Y. Acad. Sci. *102: 338–435.*

Amaya, J. W. (1981) Proc. World Lagomorph Conf., *1979, Guelph, Ontario: 493–4. University of Guelph, Ontario.*

Amerson, A.B. (1971) Atoll Res. Bull. *(150): 1–383.*

__, Clapp, P.C. and Wirtz, W.O. (1974) Atoll Res. Bull. *(174): 1–306.*

__ and Shelton, P.C. (1976) Atoll Res. Bull. (192): 1–479.

Amon, R. (1931) Die Tierwelt Niederosterreichs. *Verlag Optische Werke Reichert, Wien.*

__ (1955) Der Amblick 10: 302.

__ (1959) Z. Jagdwissenschaft 5(4): 132.

Amor, R.L. and Piggin, C.M. (1977) *In* Exotic Species in Australia: their Establishment and Success . Proc. Ecol. Soc. Aust. 10: 15–26.

Andersen, J. (1961) Terre et Vie 108 (1): 41–53.

Anderson, D. (1977) Proc. Ecol. Soc. Aust. 10: 1–186.

Anderson, J.L. (1981) Biol. Conservation 19: 107–17.

__ (1986) Int. Zoo Yb. 24/25: 192–9.

__ (1992) Reintroduction News 4: 9.

Anderson, M., Meurling, P., Dahlback, M. et al. (1981) Proc. World Lagomorph Conf., 1979, Guelph, Ontario: 175–181. *University of Guelph, Ontario.*

Anderson, P.M. (1923) Canad. Field-Nat. 37(4): 75–6.

Anderson, R.C. (1971) Can. J. Zool. 49: 159–66.

Anderson, R.M. (1934) Proc. 5th. Pacif. Sci. Congr., Vic. and Vanc., B.C., 1933: 769–78. *University Toronto Press, Toronto.*

Anderson, S. and Jones, J.K. (eds) (1967) Recent Mammals of the World: A Synopsis of Families. *New York.*

Andersson, M., Dahlback, M. and Meurling, P. (1979) Viltrevy 11(2): 103–27.

Andreev, I.F. (1963) *In* Acclimatisation of Animals in the USSR.: 53–56. Akad. Nauk. Kuzakhsk. SSR., Alma-Ata.

Andrews, C.W. (1909) Proc. Zool. Soc. London 1909: 101–3.

Andrews, M.A. (1982) The Flight of the Condor. *London.*

Andrews, R.D., Storm, G.L., Phillips, R.L. and Bishop, R.A. (1973) J. Wildl. Mgmt. 37(1): 69–72.

Anon. (1875) Zool. Garten 16: 463.

__ (1885) Planters' Monthly 4: 65.

__ (1885) Deutsche Jagerz. 5: 116.

__ (1886) Deutsche Jagerz. 6: 347.

__ (1886–87) Deutsche Jagerz. 8: 169.

__ (1886–87) Deutsche Jagerz. 8: 296.

__ (1886–87) Deutsche Jagerz. 8: 346.

__ (1887) Planters' Monthly 16: 437–8.

__ (1888) Planters' Monthly 7: 96.

__ (1892) Deutsche Jagerz. 19: 348.

__ (1894–95) Deutsche Jagerz. 24: 812.

__ (1895) Deutsche Jagerz. 25: 143.

__ (1907–08) Deutsche Jagerz. 50: 620.

__ (1908–09) Deutsche Jagerz. 5: 654.

__ (1925) Bernice P. Bishop Mus. Spec. Publ. (10): 1–96.

__ (1932) Deutsche Jagerz. 98: 63.

__ (1935) Maryland Conserv. 12(1): 28.

__ (1935) Deutsche Jagerz. 3: 700.

__ (1935) Deutsche Jagerz. 2: 1078.

__ (1941) Field 177: 32.

__ (1946) Wyo. Wildl. 1(3): 4–16.

__ (1946) Victorian Nat. 62: 205.

__ (1950) Boletin Ganadero 1(7): 21–2.

__ (1951) South Dakota Cons. Digest 18(1): 2–5.

__ (1954) Sci. News Letter 66(20): 310.

__ (1959) Priroda 6: 100–2.

__ (1959) Wild und Hund 62: 776.

__ (1963) Animals 1(19): 30

__ (1963) Animals 1(21): 2.

__ (1963) Animals 2(13): 336.

__ (1963) Animal Life (8): 31–5.

__ (1963) Animals 2(16): 448.

__ (1967) Fauna Bull. 1(1): 32. Dept. Fish. and Fauna: Perth.

__ (1978) Report of the coypu strategy group. *Ministry of Agriculture, Fisheries and Food.*

__ (1980) Counter Pest (4): 9–12.

__ (1981) Elepaio 42(2): 15

__ (1981) J. Agric. West. Aust. 22(3): 107–12.

__ (1987) NT. Rural News, *August, 1987:* 85–7, 88.

__ (1991) Cat News 14: 12.

__ (1991) Cat News 15: 6–7.

__ (1993) Cat News 18: 12.

__ (1994) Cat News 20: 19.

__ (1968) Tenn. Conservationist 34: 16.

Ansell, W.F.H. (1960) Mammals of Northern Rhodesia. *Lusaka.*

Anthony, R.G. and Fisher, A.R. (1977) Wildl. Soc. Bull. 5(3): 107–12.

Anthony, R.M., Flint, P.L. and Sedinger, J.S. (1991) Wilson Soc. Bull. 19: 176–84.

Appleton, P.L. and Norton, J.H. (1976) Qld. Agric. J. 102: 339–43.

Apps, P.J. (1983) S. Afr. J. Zool. 18: 393–9.

__ (1986) J. Mamm. 67(1): 199–200.

Archer, M. (1972) West. Aust. Nat. 12(4): 85–6.

Arentsen, P. (1953) Boletin Ganadero 3(34): 3–4.

__ (1954) Boletin Ganadero 4 (42): 23–5.

Armitage, P., West, B. and Steedman, K. (1984) The London Archaeologist 4: 375–83.

Armstrong, D.M. and Jones, J.K. (1971) J. Mamm. 52(4): 747–57.

Armstrong, P. (1982) J. Biogeogr. 9(4): 353–62.

Armstrong, W.E. and Harmel, D.E. (1981) Texas Parks and Wildlife Magazine, *February, 1981:* 2 pp.

Arraya, M.B. (1983) *In* Proc. Jean Delacour/IFCB Symp. Breed. Birds in Captivity: 125–40. *Los Angeles.*

Asahi, M. (1960) Physiology and Ecology (Japan). 9(1): 44–53.

__ (1962) J. Biol. Osaka City Univ. 13: 119–60.

Ashbrook, F.G. (1948) J. Wildl. Mgmt. 12: 87–95.

Aspinall, A. (1954) The Pocket Guide to the West Indies and British Guiana, British Honduras, Bermuda, the Spanish Main, Surinam, the Panama Canal. *10th edn rev., London.*

Aspisov, D.I. (1959) Soob. Inst. Lesa Akad. Nauk. SSSR. 13: 89–93.

Aston, B.C. (1912) Trans. N.Z. Inst. 44: 19–24.

Atkinson, I.A.E. (1964) Proc. N.Z. Ecol. Soc. 11: 39–44.

__. (1973) J. Roy. Soc. NZ. 3: 457–72.

__ (1977) Pacific Sci. 31(2): 109–33.

__ (1985) *In* Moors, P.J. (ed), Conservation of Island Birds. *ICBP Tech. Publ.* 3: 35–81.

__ and Bell, B.D. (1973) In Williams, G.R. (ed), The Natural History of New Zealand: An Ecological Survey, 15: 372–92. Wellington.

Attebury, J., Kroll, J.C. and Legg, M.H. (1977) Wildl. Soc. Bull. 5(4): 179–84.

Aubert de la Rue, E. (1930) Terres Francaises in connues. Paris.

__ (1955) Les Terres australes. Paris.

Aubry, J.R. (1959) La Vie des Betes 17: 19.

Auffray, J.C. (1988). Universite Sciences et Techniques de Languedoc, Montpellier, These.

Auld, B.A. and Tisdell, C.A. (1986) In Groves, R.H. and Burdon, J.J. (eds), Ecology of Biological Invasions: An Australian Perspective: 79–88. Australian Academy of Science, Canberra.

Ausserer, C. (1946) Der Alpensteinbock. Universum–Vlg., Wien.

Austin, O.L. (1948) Bird Banding 14: 60–5.

Averin, Yu.V. and Uspenskii, G.A.(1962) Izvest. Akad. Nauk. Moldavsk SSR. 3: 57–65.

Avery, D.M. (1985) Mammalia 49: 573–6.

Avila-Pires, F.D. de (1969) Rev. Bras. Biol. 29: 49–64.

Baber, D.W. and Coblentz, B.E. (1986) J. Mamm. 67(3): 512–25.

__ and __ (1987) J. Wildl. Mgmt. 51(2): 306–17.

Backhouse, G.N., Clark, T.W. and Reading, R.P. (1995) In Serena, M. (ed), Reintroduction Biology of Australian and New Zealand Fauna, 33: 209–18. Chipping Norton.

Bagley, L. (1938) Wyo. Wildl. Mag. 3(11): 1–3.

Bailey, H.H. (1924) J. Mamm. 5: 264–5.

Bailey, T.N. and Bangs, E.E. (1985) US Fish Wildl. Serv., Prog. Rep. (1): 1–10.

Baillie, J.L. (1928) Canad. Field-Nat. 42(7): 179.

__ (1929) Canad. Field-Nat. 43(3): 64.

Bajomi, D. (1980) Proc. 9th. Vert. Pest Conf., Fresno, California, 1980: 124–9.

Bakeev, N. (1964) Okhota Okhotn. Khoz. 1: 19.

Bakeev, Yu.N. (1963) Sb. Nauch.-Teckh. Inform. Vses Nauch.-Issled. Inst. Zhivotnogo Syr'ya Pushniny 5(8): 66–77.

Baker, A.L. (1966) The Times; 3rd Sept. 1953: 6.

Baker, A.S. (1916) Thrum's Hawaiian Ann. for 1917: 62–70.

Baker, C.M.A. and Manwell, C. (1981) Z. Tierzüchtg. und Züchtgsbiol. 98: 241–57.

Baker, J.K. and Russell, C. A. (1979) Elepaio 40(4): 51–2.

Baker, K. (1972) N.Z. Wildl. (38): 27–31.

Baker, L.A. (1991) M.S. Thesis, University, Georgia, Athens.

Baker, S.J. (1986) J. Zool. 209: 285–6.

__ (1986) Int. Zoo Yb. 24/25: 200–5.

__ (1990) Mammal Rev. Nos. 2/3: 75–96.

Baldwin, J.D. and Baldwin, J.I. (1976) In R.W. Thorington, and P.G. Heltne (eds), Neotropical Primates: Field Studies and Conservation: 30–31, Nat. Acad. Sci., Washington DC.

Baldwin, L.A. and Teleki, G. (1974) Gibbon and Siamang 4: 17–86.

Baldwin, P.H., Schwartz, C.W. and Schwartz, E.R. (1952) J. Mamm. 33: 335–56.

Ball, W.S. (1960) Bull. Dept. Agric., California, USA. 49(3): 177–85.

Banfield, A.W.F. (1977) The Mammals of Canada. Toronto and Buffalo.

__ (1962) Nat. Mus. Canad. Bull. (177) Biol. Ser. (66): 1–137.

Banks, R.C. (1972) U.S.D.I., Fish and Wildlife Service, Special Sci. Rep. No. 200.

Bannerman, D.A. and Bannerman, W.M. (1965–68) The Birds of the Atlantic Islands. 4 vols. Edinburgh and London.

Bannikov, A. (1963) Okhota Okhotn. Khoz. 3: 22–3.

Bannikov, A.G. (1961) Terre et Vie 108(1): 77–95.

__ (1963) Priroda 52(8): 67–72.

Banwell, D.B. (1966) Wapiti in New Zealand. Sydney and Wellington.

__ (1968) The Highland Stags of Otago. Wellington and Sydney.

__ (1970) The Red Stags of the Rakaia. Wellington and Sydney.

Barabash-Nikiforov, I. (1938) J. Mamm. 19(4): 423–9.

Barbash, L.A. (1961) Novosibirsk: 45–60.

Barbehenn, K.R. (1962) In Storer, I. (ed), Pacific Island rat ecology. B.P. Bishop Mus. Bull. 225.

Barbour, T. (1930) Proc. New Engl. Zool. Club 2: 72–85.

Barker, E.S. (1948) Field and Stream 53(4): 26–7, 121–3.

Barker, H.M. (1964) Camels and the Outback. Pitman and Sons Ltd., 1964. Repr. Melbourne and Perth, 1972.

Barnard, P.J. and Van der Walt, K. (1961) Koedoe 4: 105–9.

Barnes, W.C. (1924) American Forests and Forest Life 30: 643–8.

Barnett, B.D. (1986) Proc. Vert. Pest Control 12: 358–68.

__. and Rudd, R.L. (1983) Int. J. Stud. An. Prob. 4: 44–58.

Barnett, S. (1985) New Zealand in the Wild. Auckland, New Zealand.

Baron, J. (1982) Americ. Nat. 107(1): 202–5.

Barrau, J. and Devambez, L. (1957) Terre et Vie: 324–6.

Barrett, C. (1934) Victorian Nat. 51: 108–10.

Barrett, C. (1955) An Australian Animal Book. 2nd ed. Melbourne.

Barrett, J.W. (1918) The Twin Ideals: An Educated Commonwealth. London. 2 vols.

Barrett, P. (1968) True, March: 52–3, 88–9, 92–5.

Barrett, R.H. (1966) MS Thesis, Univ. Michigan.

__ (1967) Trans, Desert Bighorn Counc. 11: 16–26.

__ (1970) Trans. Calif.– Nevada Sect. Wildl. Soc. 17: 71–8.

__ (1977) In G.W. Wood (eds) Research and Management of Wild Hog Populations, Belle W. Baruch Forest Science Institute, Clemson University, Georgetown, S. Carolina: 111–13.

__ (1978) Hilgardia 46(9): 283–356.

__ (1980) Proc. Symp. Ecol. Mgmt. Barbary Sheep, Nov. 19–21, 1979: 46–50. Department Range and Wildlife Management, Texas Technical University, Lubbock, Texas.

__ (1980) Hilgardia 49: 281–355.

__ and Pine, D.S. (1980) Fish and Game 67(1): 105–17.

Barrette, C. (1986) J. Mamm. 67(1): 177–9.

Barrington, R.M. (1880) Sc. Proc. Roy. Dublin Soc. N.S. 2: 615–31.

Barrow, D. (1963) Animals 1(19): 5–7.

Barry, D.H. and Campbell, P.R. (1977) Occ. Pap. Anthropol. Univ. Qld. (8): 147–78.

Bartlett, I.H. (1944) Mich. Cons. 13(8): 10.

Bashore, T.L. and Bellis, E.D. (1982) Wildl. Soc. Bull. 10(4): 386–8.

Batchelor, R.F. (1955) The Roosevelt Elk in Alaska: Its Ecology and Management. Alaska Department Fish and Game, Juneau.

Bauer, J.J. (1988) Tigerpaper 15(4): 26–32.

Baumann, F. (1949) Die freilebenden Saugetier der Schweiz. Bern, Switzerland.

Baverstock, P.R., Adams, M., Maxson, M.R. and Yosida, T.H. (1983) Genetica 105: 969–83.

Baylis, P. and Yoemans, K.M. (1989) Aust. Wildl. Res. 16: 651–76.

Bayly, C.P. (1976) Sth. Aust. Nat. 51(2): 22–4.

__ (1978) Sth. Aust. Nat. 53(2): 20–8.

Bayne, C.J., Coyan, B.H.; Diamond, A.W. and Frazier, J. et al. (1970) Atoll Res. Bull. (136): 83–99.

Bear, G.D. (1979) Colorado Division of Wildlife, Special Report (45): 1–12.

Beatty, H.A. (1944) J. Agric. 28: 181–5.

Beck, A.M. (1970) The Ecology of Stray Dogs: A Study of Free-ranging Urban Animals. Baltimore, Maryland.

Beck, A.M. (1975) In Fox, M.W. (ed) The Wild Canids, 26: 380–90. New York.

Becker, H.N.R., Belden, R.C., Breault, T., Burridge, M., Frankenberger, W.B. and Nicoletti, T. (1978) J. Am. Vet. Med. Assoc. 173: 1142–82.

Becker, K. (1972) Proc. 5th. Vert. Pest Contr. Conf., Fresno, California: 18–21

Becking, R.W. (1970) Report of the Kipalhula Trip. Haleakala National Park, Maui, Hawaii.

Bednarik, K. (1955) Ohio Conserv. Bull. 19(12): 8–9.

Beebe, W. (1923) Zoologica 5: 3–22.

__ (1924) Voyage of the Arcturus. London.

Behan, R.W. (1978) Trans. 43rd. Nth. Am. Wildl. Nat. Resour. Conf. 43: 424–33.

Beishebaev, K. (1963) Izv. Akad. Nauk. Kirg. SSR., Ser. Biol. Nauk. 5(2): 5–10.

Bejot, G. (1959) Cons. Innter. Chasse, 8 eme. Assem. Vienne, 20–25 Mai, 1959, Paris.

Bekoff, M. (1975) In Fox, M.W. (ed) The Wild Canids, 9: 120–42. New York.

__ (ed.) (1978) Coyotes. New York.

Belden R.C. and Frankenberger, W.B. (1977) In Wood, G.W. (ed) Research and Management of Wild Hog Populations, Belle W. Baruch Forest Science Institute, Clemson University, Georgetown, S. Carolina: 71–85.

__ and __ (1989) Proc. Feral Pig Symp., April 27–29, 1989, Orlando, Florida: 3–10. Livestock Conservation Institute, Madison, Wisconsin.

__ and McCowan, J.W. (1993) Florida panther captive breeding/reintroduction study: annual performance report. U.S. Fish and Wildl. Serv., Gainseville, Florida.

Bell, B.D. (1975) In Wildlife – A Review, N.Z. Department of Internal Affairs July, 1975: 31–4.

__ and Taylor, R.H. (1970) N.Z. For. & Bird 178: 6–10.

Bell, H.L. (1975) Emu 75(2): 77–84.

Bellingshausen, F.F. (1948) In F. Debenham (ed) The Voyage of Captain Bellingshausen to the Antarctic Seas, 1819–21 . London

Belozertsev, I. (1962) Okhota Okhotn. Khoz. 2: 28–9.

Ben'kovskii, L.M. (1963) In Voprosy Geogr. Dal'nego Vostoka 5: 103–17. Khabarovsk.

Benchley, P. (1999) Nat. Geogr. Mag. 195(4): 1–31.

Bencze, L., Walter, V. and Kiss, G. (1977) Z. Jagdwiss. 23(4): 214–18.

Benke, A. (1973) Texas Parks Wildl. 31(1): 6–9.

Bennett, C.F. (1968) Ibero–Americana 51: 1–112.

Benson, D.A. (1959) J. Mamm. 40(3): 451.

Benson, D.W. and Dodds, G.D. (1977) N.S. Dept. Lands and For., Halifax: 1–92.

Bentley, A. (1957) J. Wildl. Mgmt. 21(2): 221–5.

__ (1978) An Introduction to the Deer of Australia. Forest Commission Victoria. Melbourne.

__ and Downes, M.C. (1968) Papua and N.G. Agric. J. 20(1–2): 1–14.

Bentley, E.W. (1964) J. Anim. Ecol. 33: 371–3.

Berdov, A.Z. (1987) Trans. Congr. Int. Game Biol. 18: 20–1.

Berestennikov, D. and Ardamatskaya, T. (1964) Okhota Okhotn. Khoz. 9: 15–17.

Berger, A.J. (1972) Wilson Bull. 84 (2): 212–22.

Berger, J. (1986) Wild Horses of the Great Basin: Social Competition and Population Size. Chicago and London.

Berger, J. and Wehausen, J.D. (1991) Conserv. Biol. 5(2): 244–8.

Berger, N.M. (1944) Zool. Zh. 23: 267–74.

__ (1962) In Voprosy ekologii' 6: 22–3. Vysshaya Shkola, Moscow.

Bergerud, A.T. (1985) Can. J. Zool. 63: 1324–9.

__ and Manuel, F. (1968) J. Wildl. Mgmt. 32(4): 729–46.

__ , __ and Whalen, H. (1968) J. Wildl. Mgmt. 32(4): 722–8.

__ and Mercer, W.E. (1989) Wildl. Soc. Bull. 17: 111–20.

Berkson, G. and Ross, B. A. (1969) In Carpenter, C.R. (ed) Behaviour, Vol. 1: 43–7. S. Karger, Basel.

__ (1987) Proc. Aust. Vert. Pest Control Conf. 8: 256–9.

__ and Jarman, P. (1988) Environmental impact of feral horses in central Australia. Vol. 4. Int. Doc., Cons. Comm. N.T., Alice Springs. 59 pp.

__. and Jarman, P.J. (1987) Ecology of feral horses in central Australia and their interaction with cattle. Vol. 1. Int. Doc., Cons. Comm. N.T., Alice Springs. 188 pp.

Berry, R.J. (1968) J. Anim. Ecol. 37(2): 445–70.

__ (1981) Mammal Rev. 11: 91–136.

__ and Jackson, W.B. (1979) J. Mamm. 60(1): 222–5.

__ , Jakobson, M.E. and Peters, L. (1978) J. Zool., London 185: 73–92.

Berteaux, D. (1993) J. Mamm. 74 (3): 732–7.

Bertram, B.C.R. and Moltu, D.P. (1986) Mammal Rev. 16: 81–8.

Bertram, C.R. (1988) In Dixon, A. and Jones, D. (eds), Conservation and Biology of Desert Antelopes, 14: 136–45.

Beyshebayev, K. (1956) Akklim. pushn. zverey v Kirgizii: 107–12.

Bezzi, A. (1930) The Introduction of the Camel to Libya. Clin. Vet., Milano.

Bigalke, R. (1937) Sth. Afr. J. Sci. *33: 46–63.*

Bigalke, R. (1977) Zoologica Africana *12: 504.*

__ and Pepler, D. (1991) In Groves and Di Castri (eds), Biogeography of Mediterranean Invasions. *Cambridge and Sydney.*

Bigg, M.A. and MacAskie, I.B. (1978) J. Mamm. *59(4): 874–6.*

Birch, L.C. (1965) In Baker, H.G. and Stebbins, G.L. (eds) Genetics of Colonizing Species. *New York & London.*

Bisbal, F.J. (1986) Mammalia *50(3): 329–39.*

Bishop, C.R. (1852) Trans. Roy. Hawaiian Agric. Soc. *1(3): 91.*

Black, H.C. (1965) Ph.D. Thesis, Oregon State University.

__, Dimock, E.T., Dadge, W.E. and Lawrence, W.H. (1969) Trans. 34th. N. Am. Wildl. Nat. Resour. Conf., *Mar. 2–5, 1969: 388–408.*

Blackman, L.G. (1904) Hawaiian Forester and Agriculturist *1: 115–17.*

Blackman, S. (1999) BBC Wildlife *17(2): 28.*

Blake, J.G.; Douglas, C.L. and Thompson, L.F. (1981) J. Mamm. *62(1): 58–63.*

Blanchet, M. (1960) Terre et Vie *107(1): 1–43.*

Blandford, P.R.S. (1987) Mammal Rev. *17(4): 155–98.*

Bleich, V.C. (1974) Murrelet *55: 7–8.*

Blom, A., Alers, M.P.T. and Barnes, R.F.W. (1990) In R. East (compiler), Global Survey and Regional Action Plans. Part 3 West and Central Africa. *22: 113–20. IUCN, Gland, Switzerland.*

Bloomer, J.P. and Bester, M.N. (1990) S. Afr. J. Wildl. Res. *20(1): 1–4.*

__ and __. (1992) Biol. Cons. *60: 211.*

Blouch, P.I. (1954) Mich. Cons. *23(3): 6–9.*

Blume, S. (1911) Wild und Hund *17: 230.*

Boback, A.W. (1957) Deutsche Jagerz., *1957: 185.*

Bobry, G.Ya. (1978) Zool. Zh. *57(2): 253–9.*

Bodenheimer, F.S. (1958) Bull. Res. Counc. Isr., Sect. B. Zool. *7(3/4): 165–90.*

Bonhomme, F., Catalan, J., Britton-Davidian, J. Chapman, V.M., Moriwaki, D., Nevo, E. and Thaler, L. (1984) Biochemical Genetics *22: 275–303.*

Bonner, W.N. and Leader-Williams, N. (1976) Polar Record *18: 512.*

Boorman, L.A. and Fuller, R.M. (1981) J. Anim. Ecol. *50: 241–69.*

Booth, J. (1970) Sunbird *1: 85–92.*

Booth, V.R., Jones, M.A. and Morris, N.E. (1984) Oryx *18: 237–240.*

__ and Coetzee, A.M. (1988) In L. Nielson and R.D. Brown (eds) Translocation of Wild Animals. *Wisconsin Humane Society and Caesar Kleberg Wildlife Research Institute.*

Bopp, P. (1956) Schweizer Naturschutz *22: 29.*

Borisov, A.V. (1963) Izv. Omsk Otd. geogr. Obshchest SSR. *5(12): 161–73.*

Borner, M. (1985) Oryx *19: 151–4.*

__ (1988) In L. Nielson and R.D. Brown (eds) Translocation of Wild Animals. *Wisconsin Humane Society and Caesar Kleberg Wildlife Research Institute.*

Borodin, L.P. (1959) Soob. Inst. Lesa. Akad. Nauk. SSSR. *13: 102–10.*

__ (1965) Byul. Mosc. Ob. shch. Ispytat. Prirody Otd. Biol. *70(1): 20–32.*

Bosch, P.C. and Sverdsen, G.E. (1987) J. Mamm. *68(2): 425–9.*

Bothma, J du P. (1966) Zool. Afr. *2(2): 205–9.*

Bourdelle, E. (1939) J. Mamm. *20(3): 287–91.*

Bourne, W.R.P. (1971) Atoll Res. Bull. *(149): 175–207.*

__ (1975) New Scientist *(963): 422–5.*

Bowers, R.R. (1953) Va. Wildl. *14 (10): 5–7.*

Bowker, G.M. (1980) Corella *4 (4): 104–6.*

Bowling, A.T. and Ouchberry, R.W. (1990) J. Wildl. Manage. *54(3): 424–9.*

Box, T.W. (1968) In Symp. Introduction of Exotic Animals: Ecological and Socioeconomic Considerations, *1967: 17–20. C. Kleberg Research Program on Wildlife Ecology, College of Agriculture, Texas A & M University.*

Boyd Watt, H. (1923) Essex Nat. *20: 189–205.*

Boyd, J.M. (1981) Biol. Conserv. *20(3): 215–77.*

Braithwaite, R.W., Dudzinski, M.L., Ridpath, M.G. and Parker, B.S. (1984) Aust. J. Ecol. *9: 302–22.*

Brander, A. (1905–06) Deutsche Jagerz. *46: 554.*

Brander, J. (1940) Tristan da Cunha, 1506–1902. *London.*

Bratton, S.P. (1974) Bull. Torrey Bot. Club *101(4): 109–206.*

__ (1975) Ecology *56(6): 1356–66.*

__ (1977) In Wood, G.W. (ed) Research and Management of Wild Hog Populations. *Belle W. Baruch Forest Science Institute Clemson University, Georgetown: 47–52.*

Breitenmoser, U. and Haller, H. (1993) J. Wildl. Mgmt. *57(1): 144–54.*

__, Breitenmoser-Würsten, C. and Capt, S. (1994) In Cat News *20: 19.*

Brennan, E.J., Else, J.G. and Altmann, J.A. (1985) Afr. J. Ecol. *23: 35–44.*

Breytenback, G.T. and Skinner, J.D. (1982) S. Afr. Wildl. Res. *12(1): 1–7.*

Bridges, E.L. (1949) Uttermost Part of the Earth. *New York.*

Briederman, L. (1967) Trans. 7th. Congr. Inter. Union Game Biol.: *207–13.*

__. (1968) Zool. Gart. *35(4/5): 224–9.*

Bright, P.W. and Morris, P.A. (1994) J. Appl. Ecol. *31: 699–708.*

Brine, L. (1877) The Geographical Magazine *4: 266–7.*

Brisbin, I.L., Smith, M.W. and Smith, M.H. (1977) In Wood, G.W. (ed) Research and Management of Wild Hog Populations. *Belle W. Baruch Forest Science Institute Clemson University, Georgetown: 71–85.*

Brocke, R.H. and Gustafson, K.A. (1992) Reintroduction News *4.*

__, __ and Major, A.R. (1990) Trans. 55th N.A. Wildl. & Nat. Res. Conf., *Denver, Colorado: 590–8.*

Brockie, R.E. (1958) M.Sc. Thesis, Victoria University, Wellington.

__ (1959) N.Z. J. Sci. *2(1): 121–36.*

__ (1960) Proc. Zool. Soc., London *134: 505–8.*

__ (1974) N.Z. Vet. J. *22(12): 243–7.*

__ (1975) N.Z. J. Zool. *2(4): 445–62.*

Broekhuizen, S. (1977) Intern. Rapport R.I.N., Arnhem: *1–9.*

Bromlei, G.F. (1959) Soob. Dal'nev. Filiala Sirbirskii Otdel Akad. Nauk. SSR. 11: 121–3.

Bronson, E.H. and Tiemeier, O.W. (1958) Trans. Kansas Acad. Sci. 61(2): 226–9.

Brooke, R.K., Lloyd, P.H. and de Villiers, A.L. (1986) In MacDonald, I.A.W., Kruger, F.J. and Ferrar, A.A., The Ecology and Management of Biological Invasions in South Africa: 63–74. Capetown.

Brookes, C. (1984) In Feral Mammals, Workshop Feral Mamm., Capr. Spec. Group, 3rd, Int. Theriol. Conf., Helsinki: 19–30. IUCN: Morges.

Brosh, A. (1979) Corella 3(3): 55–7.

__ (1983) Corella 7(4): 89–90.

__ (1983) Corella 7(4): 85–6.

__ (1983) Corella 7(4): 91–2.

__, Eberhard, I.E., Copson, G.R. and Skira, I.J. (1982) Aust. Wildl. Res. 9: 477–85.

__, Shkolnik, A. and Choshniak, I. (1986) Ecology 67(4): 1086–90.

__ and Skira, I.J. (1987) Corella 11(3): 81–2.

__ and __. (1988) Corella 12(3): 85–6.

__, __. and Copson, G.R. (1985) Aust. Wildl. Res. 12: 425–36.

Brothwell, D. (1981) In Berry, R.J. (ed) Biology of the House Mouse. Symp. Zool. Soc. London, 47(30): 715 pp.

Brouard, N.S. (1963) A History of Woods and Forest in Mauritius. Port Louis, Mauritius.

Browne-Cooper, R. et al. (1990) W. Aust. Nat. 18: 40–51.

Brown, L.M. (1975) J. Mamm. 56(4): 287–91.

Brown, L.N. and McGuire, R.J. (1975) J. Mamm. 56(2): 405–19.

Brown, P.W. (1956) Trans. Roy. Highl. Agric. Soc., Scotland, 1956: 26–35.

Brown, T.L.; Decker, D.J. and Dawson, C.P. (1978) Wildl. Soc. Bull. 6(4): 235–9.

Browne, C.O. (1982) Proc. 10 Vert. Pest Conf., Monterey, California, Feb. 1982: 109–15.

Brownell, R.L. and Rathbun, G.B. (1988) Endangered Species Tech. Bull. 13(4): 1, 6.

Browning, B.M. and Lauppe, E.M. (1964) Calif. Fish Game 50(3): 132–47.

Browning, T.O. (1977) In Exotic Species in Australia: their Establishment and Success. Proc. Ecol. Soc. Aust. 10: 27–38.

Bruce, J. (1941) California Cons. 6(5): 14, 21.

Bryan, E. H. (1938) Paradise of the Pacific 50(4): 38–9.

__ (1930) Paradise of the Pacific 43(12): 31–2.

__ (1937) Paradise of the Pacific 49(12): 71–8.

__ (1947) Hawaiian Planters Rec. 51: 1–80.

Bryan, W.A. (1915) Natural History of Hawaii. Honolulu.

Brygoo, E.R (1972) In G. Richard-Vindard and R. Battistini (eds) Biogeography and Ecology of Madagascar. Monogr. Biolog. 21. The Hague.

Buchanan, G.D. (1955) Anim. Kingdom 58(3): 82–8.

Buckner, C.H. (1966) Ann. Rev. Ent. 11: 449–70.

Buden, D.W. (1986) Flor. Field Nat. 14: 53–84.

Buechner, H.K. (1950) Trans. Nth. Am. Wildl. Conf. 15: 627–44.

Bullock, D. (1977) Oryx 14 (1): 51–8.

__ and North, S. (1984) Oryx 18(1): 36–41.

Bump, G. (1941) Trans. Nth. Am. Wildl. Conf. 5: 409–20.

__ (1968) In Symp. Introduction of Exotic Animals: Ecological and Socioeconomic Considerations. C. Kleberg Research Program Wildlife Ecology, College of Agriculture, Texas A & M University, Texas: 5–8.

Bump, G. (1952) Atlantic Nat. 7(3): 112–17.

Burbidge, A.A. (1971) Dept. Fisheries and Fauna W. Australia Report, No. 9.

__ and Fuller, P.J. (1998) Corella 22(4): 118–21.

__ and George, A.S. (1978) J. R. Soc. W. Aust. 60: 71–90.

__ Johnson, K.A., Fuller, P.J. and Southgate, R.I. (1988) Aust. Wildl. Res. 15: 9–39.

__ and Prince, R.I.T. (1972) Department of Fisheries and Fauna W. Australia Report, No. 11.

Burckhardt, D., Kuster, A. and Schloeth, R. (1961) Terre et Vie 108(1): 101–9.

Burghduff, A. (1935) Calif. Fish Game 21(1): 83–4.

Burley, J.R.W.; Creeper, D.A. and Moulds, G.A. (1983) Vertebrate Pests Control Authority, Sth. Aust., Technical Report (24), May, 1983.

Burris, O.E. (1965) Proc. 45th. a. Conf. W. Assoc. State Game Fish Comm., Anchorage, Alaska, July 7–8, 1965: 93–104.

__ and McKnight, D.E. (1973) Alaska Department Fish and Game, Game Wildl. Tech. Bull. (4): 1–57.

Bursey, J.A. (1933) New Mexico 11: 10–12, 36–7.

Burton Cleland, J. (1911) Report 18th Meeting Australian Association for the Advancement of Science: 314–16.

Burton, J.A. (1991) Field Guide to the Mammals of Britain and Europe. Kingfisher Books, London.

__ and Pearson, B. (1987) Collins Guide to the Rare Mammals of the World. London, Sydney and Toronto.

Burton, M. (1962) A Systematic Dictionary of Mammals of the World. London.

__ (1963) Animals 2(5): 114.

__ (1976) Guide to the Mammals of Britain of Europe. Oxford.

Buskirk, S.W. and Gipson, P.S. (1981) Proc. Worldwide Furbearers Conf., Frostburg, Maryland, 1980, 1: 38–54.

Bussen, O. (1977) Elepaio 38(2): 19.

Butler, W.H. (1975) W. Aust. Nat. 13: 78–80.

Büttner, K. and Worel, G. (1990) Waldhygiene 18: 168–76.

Byrne, S. (1979) Ph D Thesis, Univ. Calif., Berkeley.

Cabrera, A. (1945) Caballos de America. Buenos Aires.

Cahalane, V.H. (1955) Atlantic Nat. 10(4): 176–85.

Calder, G.J. (1981) Tech. Bull. 46. Dept. Prim. Prod., Divis. Agric. and Stock, Northern Territory. 58 pp.

Calhoon, R.E. and Haspel, C. (1989) J. Anim. Ecol. 58: 321–8.

California Department of Game and Fish (1952) Outdoor Calif. 13(25): 4–5.

Californian Department of Food and Agriculture (1975) Laws and Regulations Governing the Importation, Transportation and Possession of Live Wild Animals in California. California Department Food and Agriculture and California Department of Fish and Game.

Calligaris, C., Perco, F. and Perco, F. (1976) Ricerche di Biologia della Selvaggina Suppl. (8): 155–207.

Calvopina, L. (1985) ICBP Tech. Publ. (3): 157–8.

Cameron, A.W. (1959) National Museum of Canada, Bull. (154): 1–165.

Cameroun, A.W. (1962) J. Mamm. 43 (4): 505.

Campbell, A.J. (1888) Vic. Nat. 4: 129–64.

Campbell, D.G., Lowell, K.S. and Lightbourn, M.E. (1991) Conserv. Biol. 4: 536–41.

Campbell, J.L. (1955) Scot. Nat. 67: 122–3.

Canadian Wildlife Service (1997) Swift fox, Vulpes velox. www.ec.gc.ca/cws/swiftfox.

Caras, R. (1978) Brittanica Yearbook, 1978: 724.

Carbyn, L.N. (1989) Recovery 1(1): 8–9.

Cardinell, H.A. (1958) J. Wildl. Mgmt. 22(4): 435–6.

Carl, G.C. and Guiguet, C.T. (1972) British Columbia Provincial Museum, Department of Recreation and Conservation, Canada. 7 Hndbk. (14), 1957, 2nd edn rev., 1972.

Carley, C.J. (1979) US Fish. & Wildl. Serv., Albuquerque, Endang. Spec. Rep. No. 7: 36 pp.

Carne, P.H. (1956) Oryx 3: 200.

Caroline, M. (1968) (ed) Symposium: Introduction of Exotic Animals: Ecological and Socioeconomic Considerations. C. Kleberg Research Program Wildlife Ecology, College of Agriculture, Texas A & M University, Texas.

Carothers, S.W., Stitt, M.E. and Johnson, R.R. (1976) Trans. 41st. N. Am. Wildl. Nat. Resour. Conf., 1976, Washington: 396–406.

Carpenter, C.P. (1972) In Beveridge, W. (ed) Breeding Primates: 76–87. S. Karger, Basel.

Carpenter, C.R. (1942) J. Comp. Psychol. 33: 113–62.

Carpenter, M. (1967) Va. Wildl. 28(5): 8–9.

Carrick, R. and Costin, A.B. (1959) In Biogeography and Ecology in Australia. Monogr. Biol. 8: 605–27.

Carrington, R.P. (1950) J. Ass. School Nat. Hist. Soc. 3: 27–9.

Carson, G. (1980) Nat. Hist. 89(5): 70–5.

Carter, T.D., Hill, J.E. and Tate, G.H.H. (1946) Mammals of the Pacific World. New York.

Caslick, J.W. and Decker, D.J. (1977) Cornell University Department of Natural Resources and Conservation Circ. (15): 5.

__ and __. (1979) Wildl. Soc. Bull. 7(3): 173–5.

Casperson, K. (1968) Proc. Ecol. Soc. Aust. 3: 113–19.

Caughley, G. (1963) Nature 200(4903): 280–1.

__ (1988) Aust. Deer 131: 13–18.

Causey, M.K. and Cude, C.A. (1980) J. Wildl. Mgmt. 44 (2): 481–4.

Cawkell, E.M. and Hamilton, J.E. (1961) Ibis 103a (1).

Centre for Disease Control (1974–78) Trichinosis Surveillance Annual Summaries, 1973–1977. US. Department Health, Education, Welfare, Public Health Service, Atlanta, Georgia. 5 vols.

Chabreck, R.H., Love, J.P. and Linscombe, G. (1981) Proc. Worldwide Furbearers Conf., Frostburg, Maryland, 1980, 1: 531–43.

Chaddock, T.T. (1938) Wis. Cons. Bull. 3(4): 49–52.

Chadwick, D.H. and Sartore, J. (1988) N.G.M. 193(5): 72–99.

Chalbreck, R.H. (1958) J. Wildl. Mgmt. 22(2): 179–83.

Challies, C.N. (1975) N.Z. J. Zool. 2(4): 479–90.

__ (1976) In Whitaker, A.H. and Rudge, R. (eds) Value of Feral Farm Animals in N.Z. N.Z. Department Lands and Surveys, Information Service (l), Wellington.

Chamberlain, P.A. (1980) Proc. 9th Vert. Pest Conf., Fresno, California, 1980: 163–9.

Chamberlin, W.H. (1945) New Mexico Hist. Rev. 20: 14–57, 144–80, 239–68, and 336–57.

Chan, S. (1993) Anim. Keepers' Forum 20(11): 379–80.

Chanin, P.R.F. and Jefferies, D.J. (1978) Biol. J. Linnean Soc. 10: 305–28.

__ and Linn, I. (1980) J. Zool. London 192(2): 205–23.

Chapin, R.E. and White, R.W. (1970) Deer 2: 561–5.

__ and Dangerfield, G. (1973) J. Zool. Proc. Zool. Soc. London 170(2): 150–1.

Chapman, D.I. and Chapman, N. (1969) Biol. Conserv. 2(1): 55–62.

Chapman, F.B. (1954) Ohio Department Natural Resources, Final Report, Federal Aid Project W-80-D-1: 1–10.

Chapman, J.A. and Morgan, R.P. (1973) Wildl. Monogr., Wildl. Soc. 36: 1–54.

__ and Trethewey, D.E.C. (1972) J. Wildl. Mgmt. 36(4): 1221–6.

__ and __. (1972) J. Wildl. Mgmt. 36(1): 155–8.

Chapman, N.G. and Chapman, D.I. (1980) Mammal Rev. 10(2 and 3): 61–138.

Chapple, K. (1999) Wingspan 9(1): 23.

Charbreck, R.E. (1962) J. Mamm. 43: 337–44.

Chasen, F.N. (1940) Bull. Raffles Mus. 15: 1–209.

Chashchin, S.P. (1961) Okhrana Prirody Na Urale 2: 111–19.

Chashchukhin, V.A. (1975) Byull Mosk. O–Va Ispyt. Otd. Biol. 80(6): 21–9.

__ (1987) Trans. Congr. Int. Game Biol. 18: 38.

Chavane, C. (1957) De Tentatives, recentes ou en Cours D'acclimatisation de Gibier dans les Montagnes de France. Conseil International de la Chasse, Comm. de L'Elevage, de L'acclim. et des Madadies du Gibier, Paris.

Cheater, M. (1998) Wolf spirit returns to Idaho. National Wildlife Aug/Sept. 1998. National Wildl. Fed. www.nfw.org/nfw/natlwild/1998/nezperce.html.

Chesemore, D.L. (1970) J. Mamm. 51(1): 162–6.

__ (1975) In Fox, M.W. (ed) The Wild Canids: 143–63. New York.

Chesnokov, N.I. (1976) Ekologiya 6: 63–70.

Cheylan, G. (1991) In Groves, R.H. and Di Castri, F. (eds), Biogeography of Mediterranean Invasions, 17: 227–262. Cambridge and Sydney.

Chichikin, Yu.N. (1965) Tr. Sary – Chelekskogo Zapovednika. 1: 29–39.

Childs, J.E. (1986) J. Mamm. 67(1): 196–9.

Chisholm, E., Gibb, J.A., McIntosh, I.G. and McKelvey, P.J. (1966) Departmental Report on the Opossum Damage in the East Coast Rabbit District 1965–66. Report of Inter-Departmental Committee, East Coast Rabbit Board.

Choquenot, D. (1998) Aust. Vert. Pest Conf. 11: 365.

__, McIlroy, J. and Korn, T. (1996) Managing Vertebrate Pests: Feral Pigs. Canberra.

Christensen, P.E. (1980) Forest Focus 23: 3–12.

__ *and Burrows, N.D. (1986) In Groves, R.H. and Burdon, J.J. (eds),* Ecology of Animal Invasions: An Australian Perspective: 97–105.

Christensen, P. and Burrows, N. (1995) In Serena, M. (ed), Reintroduction Biology of Australian and New Zealand Fauna, 32: 199–207. Chipping Norton.

Christian, J.J., Flyger, V. and Davis, D.E. (1960) Chesapeake Sci. 1(2): 79–95.

Christie, A.B., Chen, T.H. and Elberg, S.S. (1980) J. Inf. Dis. 141(6): 724–6.

Christie, A.H.C. and Andrews, J.R.H. (1965) Tuatara 13(1): 1–8.

__ *and __ (1966)* Tuatara 14 (2): 82–8.

Church, B.M., Jacobs, F.H. and Thompson, H.V. (1953) Plant Pathology 2: 107–12.

__, *Westmacott, M.H. and Jacobs, F.H. (1956)* Plant Pathology 2: 66–9.

Ciani, A.C. (1986) Aggressive Behav. 12(6): 433–9.

Clapp, R.B. (1977) Atoll Res. Bull. (198): 1–8.

__, *Weske, J.S. and Clapp, J.C. (1976)* J. Mamm. 57(1): 180–1.

__ *and Wirtz, W.O. (1975)* Atoll Res. Bull. (186): 1–165.

Clark, A.H. (1949) The Invasion of New Zealand by People, Plants, and Animals – The South Island. New Brunswick.

Clark, B. (1983) Oryx 17: 28–31.

__ *(1983)* Oryx 17: 113.

Clark, C.M.H. (1976) NZ. J. For. Sci. 5(3): 235–49.

Clark, D.B. (1980) Ecology 61(6): 1422–33.

Clark, P., Ryan, G.E. and Czuppon, A.B. (1975) Aust. J. Zool. 23: 411–17.

Clark, R.J. (1970) Mammal Rev. 1(3): 92.

Clark, S.P. (1970) Mammal Rev. 1(2): 1–8.

Clark, T.W., Grensten, J., Georges, M., Crete, R. and Gill, J. (1987) Prairie Nat. 19(1): 43–56.

Clark, W.K. (1958) J. Mamm. 39(4): 574–7.

Clark, W.N. (1841) The Inquirer (56) 25th Aug.

Clarke, C.M.H. (1976) N.Z. Wildl. (52): 5–13.

Clarke, J.R. (1953) J . Mamm. 34 (3): 299–315.

Clarke, W.H.J. (197b) J. Agric. West. Aust. 17(4): 102–6.

Clout, M.N. (1980) N.Z. J. Ecol. 3: 141–5.

Coblentz, B.E. (1976) Natural History 85(6): 71–7.

__ *(1978)* Biol. Conserv. 13(4): 279–86.

__ *(1990)* Conservation Biology 4(3): 261–5.

__ *and Coblentz, B.A. (1985)* Biol. Conserv. 33(3): 281–8.

Cochi, I. (1858) Bull. Soc. Imp. Zool. Acclimatation 5: 479–83.

Cockayne, L. (1909) In Chilton (ed) Article vol. 1 10: 182–235.

Cockrill, W.R. (1976) The Buffaloes of China. Australian Freedom from Hunger Campaign., F.A.O. of United Nations, Rome.

Cockrum, E.L. (1961) The Recent Mammals of Arizona: Their Taxonomy and Distribution. Arizona.

Cohen, R.D.H. (1980) Aust. J. Exp. Agric. Sci. 47: 191–9.

Colebatch, H. (1929) The Story of 100 Years. Western Australia 1829–1929. Chapt. 30 'Zoology and Acclim.' Perth.

Collier, R.V. (1965) J. Northampton Nat. Hist. Soc. 35: 360–4.

Collinson, R.F. and Anderson, J.L. (1984) Acta. Zool. Fennica 172: 169–70.

Colon, E.D. (1930) Datos Sobre la Historia de la Agricultura de Puerto Rico antes 1898. Privately Printed, San Juan, Puerto Rico.

Coman, B. (1973) Aust. J. Zool. 2: 391–401.

__ *(1972)* Aust. Vet. J. 48: 133–6.

__ *(1975)* Victoria's Resources June 1975: 16.

__ *and Brunner, H. (1972)* J. Wildl. Mgmt. 36: 848–53.

Commer, A. (1926) Deutsch Jagerz. 86: 135.

Compton, H., Egan, J. and Trueblood, R. (1971) In Mussehl, T.W. and Howells, F.W. (eds) Game Management in Montana. Chapt. 8: 81–7. Game Mgmt. Divis., Helena, Montana.

Conaway, C.H. and Sade, D.S. (1969) J. Mamm. 50(4): 833–5.

Condon, H.T. (1967) Aust. Nat. Hist., December 1967.

Condry, W.M. (1966) Snowdonia National Park. London.

Conley, R.H. (1977) In Wood, G.W. (ed) Research and Management of Wild Hog Populations: 67–70. Belle W. Baruch Forest Science Institute, Clemson University, Georgetown, South Carolina.

__, *Henry, V.G. and Matschke, G.H. (1972)* Final Report for the European Wild Hog Research Project W-34. Proj. W-34, Tennessee Game and Fish Commission.

__, __ *and __ (1972)* European Wild Hog Research Project W-34. Tennessee Game and Fish Commission.

Conry, P.J. (1988) Trans. West Sect. Wildl. Soc. 24: 26–30.

Considine, M.L. (1985) Ecos 44: 3–11.

Constable, I.D., Mittermeier, R.A., Pollock, J.I., Ratsirarson, J. and Simmons, H. (1985) Primate Conservation 5: 59–62.

Cooke, B.D. (1977) In Exotic Species in Australia: their Establishment and Success, Proc. Ecol. Soc., Aust. 10: 113–20.

Cooper, J. and Brooke, R.K (1982) South Afr. J. Wildl. Res. 12(2): 71–5.

Cooper, M.E. (1993) J. Zoo and Wildl. Med. 24(3): 296–303.

Cooperrider, A.Y. (1986) Trans. 51st. Nth. Am. Wildl. Nat. Resourc. Conf. 51: 45–51.

Cop, J. (1992) In The situation, conservation needs and reintroduction of lynx in Europe. Proc. Symp. 17–19 October: 60–63, Neuchatel, Council of Europe, Strasbourg.

Copley, P.B. (1995) In Serena, M. (ed), Reintroduction Biology of Australian and New Zealand Fauna, 7: 35–42. Chipping Norton.

Copson, G.R. (1986) Aust. Wildl. Res. 13(3): 441–5.

__, *Brothers, N.P. and Skira, I.J. (1981)* Aust. Wildl. Res. 8(3): 597–611.

Corbet, G.B. (1961) Nature 191: 1037–40.

__ *(1978)* The Mammals of the Palaearctic region: a Taxonomic Review. London. 314 pp.

__ *and Hill, W.O. (1980)* A World List of Mammalian Species. London.

Corbett, L.K. (1985) Proc. Ecol. Soc. Aust. 13: 277–91.

Cornelius, G. (1963) Animals 1(8): 11.

Corner, L.A., Barrett, R.H., Lepper, A.W.D., Lewis, V. and Pearson, C.W. (1981) Aust. Vet. J. 57(12): 537–42.

Costin, A.B. and Moore, D.M. (1960) J. Ecol. 48: 729–39.

Cottrell, W. (1978) J. Mamm. 59(4): 886.

Couch, L.K. (1929) J. Mamm. 10(4): 334–6.

__ (1937) USDA Farmers Bull. (1768): 1–18.

Couey, F. and Schallenberger, A. (1971) In Mussehl, T.W. and Howells, F.W. (eds) Game Management in Montana, 10: 97–105. Game Mgmt. Divis., Helena, Montana.

Couturier, M. (1955) Saugetierkundliche Mitt. 3(3): 105.

__ (1955) Z. Jagdwisse. 1: 8.

__ (1955) La Terre et la Vie 3: 447.

__ (1956) La Montagnes et Apinisme, N.F. 82(8): 239.

__ (1957) Le Saint – Hubert (l): 6.

__ (1962) Le bouquetin des Alps. Grenoble.

Cowan, l.T. and Guiguet, C.J. (1960) British Columbia Provincial Museum and Department Recreation and Conservation, Hndbk. (11) Victoria, B.C.

Cowdy, C. (1973) Bird Study 20: 117–20.

Cowen, G.A., Ridley, V. and Tegner, H.S. (1965) Trans. Nat. Hist. Soc. Northumberland 15: 109–20.

Cowling, S.J. (1975) In Papers Presented at Deer Mgmt. Conf., 1974, 1–10. Fisheries and Wildlife Division, Melbourne.

Cox, W.T. (1939) Soil Cons. 5(6): 138–143, 156.

__ (1941) Am. Forests 47(2): 55–7, 93–4.

Coyne, P. (1982) Geo 4 (2): 30–9.

Cramer, W.M. (1940) Pa. Game News 11(2): 6–7, 31 and 11 (7): 6–7, 27.

Crane, C. (1990) Animals 1990: 18–23.

Cresswell, Sir M. (1972) Deer 2: 937–8.

Crichton, M. (1974) Provisional Atlas of Amphibians, Reptiles and Mammals in Ireland. Dublin.

Croft, J.D. and Hone, L.J. (1978) Aust. Wildl. Res. 5: 85–92.

Cronk, Q.C.B. (1986) Biol. Conserv. 35: 159–72.

__ (1986) Biol. Conserv. 35: 173–86.

Crook, J.H. (1961) Ibis 103a(1): 517–48.

Crouch, M.A. (1987) West. Aust. (newspaper) May 9: 35, 1987.

Crowell, K.L. (1973) Am. Nat. 107(956): 535–58.

__ and Pimm, S.L. (1976) Oikos 27(2): 251–8.

Cruz, F. and Cruz, J.B. (1987) Vida Silvestre Neotrop. 12: 3–13.

Csarada, A.V. (1914) Wild und Hund 20: 50.

Cummins, J.N. and Norton, R.L. (1974) New York Agriculture and Experimental Station, N.Y. Food and Life Science Bull. (41).

Cummins, T. and Meek, J. (1851) Roy. Hawaiian Agric. Soc. 1(2): 77–8.

Cumpston, J.S. (1968) Aust. Nat. Antarc. Res . Exp. Sci. Rep., Ser. A(1) Publ. (93) Antarctic Division, Department of External Affairs, Melbourne.

Currie, P.O. and Goodwin, D.L. (1966) J. Wildl. Mgmt. 30(2): 304–11.

Curry, R.L. (1985) Not. Galapagos 43: 13–15.

Curry-Lindahl, K. (1967) Acta Theriol. 12(1/3): 1–15.

Cuthbert, J.H. (1973) Mammal Rev. 3(3): 97–103.

Dabbene, R. (1902) Anales del Museo Nac. de Buenos Aires 8: 341–410.

Dailey, T.V., Hobbs, N. T. and Woodard, T.N. (1984) J. Wildl. Mgmt. 48 (3): 799–806.

Dal', S.K. (1941) Izvestiyah AN SSSR, Armyahnskii Filial, 1941, no. 1(6).

Dalquest, W.W. (1948) Mammals of Washington. University Kansas Publication, Mus. Nat. Hist. 2: 1–444.

Daly, R.H. (1988) In Dixon, A. and Jones, D. (eds), Conservation and Biology of Desert Antelope, 3: 14–17.

Dammerman, K.W. (1929) Mammalia 7: 149–63.

__ (1929) Mammals. In Boeroe Expedite 1921–1922. Resultats Zoologiques de L'expedition Scientifique Neerlandaise. Archipel Drukkerij, Buitzenborg.

Daniel, M. (1984) In M. Archer and G. Clayton (eds) Vertebrate Zoogeography and Evolution in Australia: 1099–102.

Daniel, M.J. (1962) Nature 194 (4829): 527–8.

__ (1967) N.Z. J. Sci. 10: 949–63.

__ (1967) Saugetier. Mitt. 13 (2): 149–55.

Danilkin, A.A. (1986) Doklady Biol. Sci. 283: 350–2.

Danilov, P.I. (1969) Problems Ecol. Biocenol., 9: 148–58.

__ and Ivanter, A.V. (1969) Material Vsesouznoi Nauchno–proizvodstvennei Kon ferentsii, pt. 2: 127–30.

Dards, J.L. (1978) Carniv. Genet. Newsl. 3: 242–55.

Dareste, A. (1857) Acclimatisation of the Dromedary to the Plains of Northern Brazil. Typographia Nacional, Rio de Janeiro.

__ (1857) Bull. Soc. Imp. Zool. Acclimatation 4: 190–215.

Darin, B.A. (1962) In Problems of Ecology 4: 123–5. Kievsk. Univ., Kiev.

Darling, F.F. (1937) J. Anim. Ecol. 6 (l): 21–2.

__ (1947) Natural History in the Highlands and Islands. London.

Darlington, J.M. (1957) Zoogeography: the Geographical Distribution of Animals. New York.

Dartnall, J.A. and Todd, N.B. (1975) Aust. J. Zool. 23: 405–9.

Darwin, C. (1845) The Voyage of the Beagle. Everymans Library Edition, 1906. London.

__ (1905) The Variation of Animals and Plants under Domestication. London.

Dasmann, R.F. (1968). In Symp. Introduction of Exotic Animals: Ecology and Socioeconomic Considerations. 1967: 11–12. C. Kleberg Research Program Wildlife Ecology, College of Agriculture, Texas A & M University, Texas.

Dauphine, T.C. (1975) Canad. Field-Nat. 89(3): 299–310.

Davey, S.P. (1963) Unpublished Report, US Bureau Commerce and Fisheries, Seattle, Washington.

Davids, R.C. (1955) Farm J. 1955: 33, 150.

Davis, D.H.S. (1950) Proc. Zool. Soc. London 120(2): 265–8.

Davis, G.S. (1953) Nat. Mag. 46(7): 370–4.

Davis, M.H. (1983) J. Wildl. Mgmt. 47(1): 59–66.

Davis, R.A. (1956) Agriculture 63: 127–9.

__ (1963) Proc. Assoc. Appl. Biol. Ann. Appl. Biol. 51: 325–50.

__ and Jensen, A.G. (1960) J. Anim. Ecol. 29: 397.

__ and Rowe, F.P. (1963) New Scientist 19: 127–30.

Davis, W.B. (1966) Bull. Texas Parks and Wildlife Department 41: 1–267.

Davydov, M.M. (1962) In Problems of Zoological Investigations in Siberia: *67–69. Knigoizdat, Gorno–Altaisk.*

Dawson, T.J. and Ellis, B.A. (1979) Aust. Wildl. Res. *6: 245–54.*

Day, G.I. (1985) Arizona Game and Fish Dept., Phoenix, Arizona.

Day, J. (1980) Birds *8: 43–45.*

Daycard, L. (1990) Rev. Ecol. (Terre Vie) *45: 35–53.*

de Brichambaut, J.P. (1978) Alauda *46: 272–3.*

De, R.C. and Spillett, J.J. (1966) J. Bombay Nat. Hist. Soc. *66(3): 576–98.*

De Calesta, D.S. and Schwendeman, D.B. (1978) Wildl. Soc. Bull. *6(4): 250–3.*

De Graaf, G. (1981) The Rodents of South Africa. Pretoria.

De Graaffe, G. and Penzhorn, B.L. (1976) Koedoe *19: 75–82.*

De las Casas, B. (1875–76) Historia de las Indias (Coleccion de documentos ineditos). vols. 62–66. Madrid.

De Lime, J.L. (1951) Kentucky Division of Game and Fisheries: 1–35.

De Oviedo y Valdes, G.F. (1851–55) In Amador de los Rios, J. (ed) Historia General y Natural de las Indias. 4 vols. Madrid.

De Roguin, L. (1989) Mammalia *53(3): 480.*

De Vos, A. and Petrides, G.A. (1967) Biological Effects caused by Terrestrial Vertebrates Introduced into Non-native Environments. IUCN Publication, New Series (9): 113–19.

__ , Manville, R.H. and Van Gelder, R.G. (1956) Zoologica *41(4): 163–94.*

Deag, J.M. (1977) In Rainier & Bourne. Primate Conservation, 267–87. New York.

Dean, C.D. and O'Gorman, F. (1969) Ir. Nat. J. *16: 198–202.*

Dean, P.B. and de Vos, A. (1965) Canad. Field-Nat. *79(1): 38–48.*

Debenham, F. (1945) (ed) The Voyage of Captain Bellinghausen to Antarctic Seas, 1819–1821. Vol. 2. Hakluut Soc., London.

Dehnel, A. (1957) Ann. Univ. Mariae Curie – Sklodo wska, *Section C. 10: 269.*

Delany, M.J. and Copland, W.O. (1965) Glasgow Nat. *18: 351–62.*

Delibes, M. (1974) Donana Acta Vertebrata (Spain) *1: 143–99.*

__ (1976) Saeugetierkd Mitt. *24(1): 38–42.*

__ (1978) Zeitschrift fur Saugetierkunde *43: 282–8.*

Dell, J. (1952–53) N.Y. State Conserv. *7(3): 23.*

__ (1957) N.Y. State Conserv. *11(4): 28–9.*

Delroy, L.B., Earl, J., Radbone, I., Robinson, A.C. and Hewett, M. (1986) Aust. Wildl. Res. *13: 387–96.*

Deng Zhi and Wang Cheng-Xim (1984) Proc. 11th Vert. Pest Conf., Sacramento, California, March 1984: *47–53.*

Denhardt, R.M. (1947) The Horse of the Americas. Oklahoma.

Department Conservation and Environment (1989) Atlas of Victorian Wildlife. Wildl. Mgmt. Br., Dept. Cons. and Env., Victoria.

Derenne, P. (1972) Mammalia *36(3): 459–81.*

__ (1974) Eradication du chat a Kerguelen; Campagnes 1973 et 1974. Terres Australes et Antarctiques Francaises, Paris: 29–44.

Derenne, Ph. (1976) Mammalia *40(4): 531–95.*

__ and Mougin, J.L. (1976) Mammalia *40(3): 495–516.*

__. and Mougin, J.L. (1976) Mammalia *40(1): 21–53.*

Descourtilz, M.E. (1809) Voyages d'un Naturaliste, et ses Observations. 3 vols. Paris.

Dewey, R.A. and Nellis, D.W. (1980) Trans. 45th. N. Am. Wildl. Nat. Resour. Conf., Florida *45: 445–52.*

Dezhkin, V.V. and Zharkov, I.V. (1960) Tr. Voronezh. Gos. Zapov. *9: 3–36.*

Dhaliwal, S.S. (1961) J. Mamm. *42(3): 349–58.*

Diamond, A.W., Douthwaite, R.J. and Indge, W.J.E. (1965) Scottish Birds *3: 397–404.*

Diamond, E.P. (1981) Atoll Res. Bull. *(255): 1–10.*

Dickerson, J. (1790) God's Protecting Providence. 7th edn. London.

Dickey, D.R. (1923) J. Mamm. *4: 55–6.*

Dickinson, T.G. and Simpson, C.D. (1980) Proceedings of the Symposium on Ecology and Management., Barbary Sheep, Nov. 19–21, 1979: 33–45. Dept. Range and Wildl. Mgmt., Texas Technical University, Texas

Dicks, B. (1974) Rhodes. Newton Abbot.

Diefenbach, D.R. (1992) Ph.D Thesis, University Georgia, Athens.

Diehl, S.R. (1988) In Nielsen, L. and Brown, R.D. (eds), The Translocation of Wild Animals,: 239–47. Wisconsin Humane Soc. and C. Kleberg Res. Inst..

Dietz, L.A. (1985) Primate Conserv. *(6): 21–7.*

Dilks, P.J. and Wilson, P.R. (1979) N.Z. J. Zool. *6(1): 127–39.*

Dill, H.R. and Bryan, W.A. (1912) U.S. Department Agriculture Biological Survey Bull. *(42): 1–30.*

Dinesman, L.G. (1959) Obshch. Inst. Lesa. Akad. Nauk. SSSR. *13: 5–24.*

Diong, C.H. (1973) Malay. Nat. J. *26: 120–50.*

Dixon, J. (1920) Agric. Exp. Stat. Bull. *(320): 379–97. Berkeley.*

__ (1929) J. Mamm. *10: 358–9.*

do Mar Oom, M. and Santos Reis, M. (1986) Arg. Mus. Bocage Ser. A *3(10): 169–96.*

Dobinskii, O.K. (1968) Okhota i Okhotniche Khozyahistvo *(5): 40–1.*

Dobschova, F. (1972) Osterreichs Weidwerk *(10): 507–13.*

Dodd, N.L. (1983) Texas Desert Bighorn Council, 1983: *12–16.*

Dodge, N.N. (1951) Natl. Parks Mag. *25(104): 10–15.*

Doerges, B. and Heucke, J. (1989) in a consultancy report on the conservation committee of the Northern Territory. Unpublished Rep. 20 pp.

Dollinger, M. (1982) Pers. Comm., Federal Veterinary Officer, Division International Traffic and Animal Welfare, Switzland.

Domm, S.B. (1971) Atoll Res. Bull. *No. 142.*

__ and Messersmith, J. (1990) Atoll Res. Bull. *No. 38.*

Doncaster, C.P. (1981) J. Anim. Ecol. *50(1): 195–218*

Donndorf, J.A. (1792) Zoologische Beitrage zur X111 Ausgabe des linneischen Natursystems Erster Band. Die Saugethiere. Leipzig.

Donne, T.E. (1924) The Game Animals of New Zealand. London.

Dorrance, M.J. (1984) Proc. 11th Vert. Pest Conf., Sacramento, California, 1984: 64–70.

Dorst, J. (1964) In Ecology of Man in the Tropical Environment. Proc. & Papers 9th Tech. Meeting, Nairobi, 1963. IUCN Publ., New Ser. (4): 245–52. Morges, Switzerland.

__ (1965) Avant que Nature Meuse. Delachaux et Niestle, Suisse.

__. and Giban, J. (1954) Terre et Vie 4: 217–29.

Dottrens, E. (1965) Acta Theriol. 10(6/9): 107–9.

Doude van Trooswijk, W.J. (1976) Ph.D. Thesis. Rijksuniversiteit te Leiden.

__ (1978) In Proc. 8th. Vert. Pest Contr. Conf., Sacramento, California: 115–17.

Douglas, C.L. and Hiatt, H.D. (1987) Natl. Park Serv. University Nevada, Las Vegas No. 006/46: 1–25.

Douglas, G. (1969) Micronesia 5(2): 327–463.

Douglas, G.W. (1977) Vermin and Noxious Weeds Destruction Board, Dept. Crown Lands and Survey, Victoria. Pamphlet 41.

__ (1981) Proc. World Lagomorph Conf., 1979, Guelph, Ontario: 822–9. University of Guelph, Ontario.

Douglas, N. (1952) N.Z. Fish. Shooting Gaz. 20(3): 13–20.

Dowding, J.E. (1999) Notornis 46: 167–80.

__ and Murphy, E.C. (1996) Notornis 43: 144–6.

Downes, M.C. (1968) Papua & New Guinea Agric. J. 20(3 & 4): 95–9.

Dozier, H.L., Hardy, T.M.P. and Markley, M. (1948) J. Mamm. 29(4): 383–93.

Draffan, R.D.W., Garnett, S.T. and Malone, G.J. (1982) Emu 83(4): 207–34.

Drees, H. (1953) Nachrbl. Deutsch. Pflanzenschd. 5(3): 35.

Drent, R., van Tets, G.F., Tompa, F. and Vermeer, K. (1964) Canad. Field–Nat.78: 208–63.

Dreux, P. (1970) Terres Australes et Antarctiques Francaises (52–53): 45–6.

du Preez, J.S. (1970) Unpublished report, Ministry of Wildlife Conservation and Tourism, Etosha Ecological Institute, Okaukuejo, Namibia.

du Toit, J. (1998) BBC Wildlife 16(10): 38.

Dudzinski, W., Haber, A. and Matuszewski, G. (1965) Chronmy Przyrode Ojczysta 20(1): 21–30.

Duffy, A.C., Seebeck, J.H., McKay, J. and Watson, A.J. (1995) In Serena, M. (ed), Reintroduction Biology of Australian and New Zealand Fauna, 34: 219–25. Chipping Norton.

Dulic, B. (1987) In Sterling, K.B. (ed) An International History of Mammals. Maryland.

Dulitskii, A.I. and Kormilitsii, A.A. (1970) Vestnik zoologii (4): 25–9.

Duncan, P. (1992) (ed) Zebras, Asses and Horses, An Action Plan for the Conservation of Wild Equids. IUCN, Gland, Switzerland.

Duncan, R.P., Bomford, M., Forsyth, D.M. and Conibear, L. (2001) J. Animal Ecol. 70: 621–32.

Dunlop, J.N. and Storr, G.M. (1981) Corella 5(3): 71–4.

__ et al. (1988) Corella 12: 93–8.

Dunn, F.D. (1965) Trans Northeast Fish and Wildl. Conf. 22: 1–5.

Dunnet, G.M. (1975) In Goodier, R. (ed), The Natural Environment of Orkney, Edinburgh.

Dunsmore, J.D. (1971) Aust. J. Zool. 19: 355–70.

__ (1981) Proc. Worldwide Furbearers Conf., Frostburg, Maryland, 1980, 2: 654–69.

Dupont, R.P. (1907) Report on a Visit of lnvestigation to St. Pierre, Astove, Cosmoledos, Assumption and the Aldabra Group. Mahe.

Duran, J.C. and Cattan, P.E. (1985) Biol. Cons. 34: 141–8.

Durham, N.M. (1972) Symp. Distribution and Abundance of Neotropical Primates, August 1972. Battelle. Seattle Res. Centre and Inst. Lab. Anim. Resour. Nat. Res. Council.

Dvoichenko, G.G., Krivonosov, G.A. and Kulugina, N.M. (1963) Mat. Konferentsii po akklim. jivot. v SSSR., Alma-Ata, Izd–vo Kaz.: 83–5.

Dvorak, D.F. (1980) in Proc. Symp. Ecol. Mgmt., Barbary sheep, Nov. 19–21 1979: 23. Dept. Range Wildl. Mgmt., Texas Technical University, Lubbock, Texas.

__ (1983) Federal Aid Project No W-109-R-6, Texas Parks and Wildlife Department, Austin, Texas.

Dwyer, P.D. (1975) Aust. Wildl. Res. 2(1): 33–45.

__ (1978) Aust. Wildl. Res. 5: 221–48.

__ (1978) Aust. J. Ecol. 3: 213–32.

__ (1983) Science in New Guinea 10(1): 28–38.

Dymond, J. R. (1922) Canad. Field-Nat. 36(8): 142–3.

__. (1928) Canad. Field-Nat. 42(4): 95.

__ (1930) Canad. Field-Nat. 44(8): 199.

Eabry, S. (1969) New York State Cons.: 1–32.

Eadie, W.R. (1961) Cornell University Extension Bulletin (1055).

Earle, A. (1832) A Narrative of Nine Months Residence in New Zealand in 1827, with a Journal of Residence in Tristan da Cunha. London.

East, R. (1989) Antelopes, Global Survey and Regional Action Plans. Part 2. Southern and South–Central Africa. Internatl. Union Conserv. Nat., Gland Switzerland. 96 pp.

__ (1992) Species 19: 23–5.

__ (1993) Species 20: 40–2.

Ebedes, H. (1975) J. S. Afr. Vet. Sci. 46(4): 359–62.

Eberhard, T. (1954) J. Wildl. Mgmt. 18(2): 284–6.

Eberhardt, L.L., Majorowicz, A.K. and Wilcox, J.A. (1982) J. Wildl. Mgmt. 46(2): 367–74.

Ecke, D.H. (1954) J. Mamm. 35(4): 521–5.

Eckhardt, R.C. (1972) Bio. Science 22(10): 585–90.

Eckhert, J. (1970) S. Aust. Ornith. 25: 201–5.

Eden, A.W. (1955) Islands of Despair. London.

Edgar, A.T. (1962) Notornis 10: 1–15.

Edge, W.D. and Olsen-Edge, S.L. (1990) J. Mamm. 71(12): 156–60.

Edmonds, J.W., Backholer, J.R. and Shepherd, R.C.H. (1981) Aust. Wildl. Res. 8(3): 589–96.

Edmonds, J.W. et al. (1978) Vic.. Nat. 93: 110–12.

__, Backholer, J.R. and Shepherd, R.C. (1981) Aust. Wildl. Res. 8: 589–596.

__, Nolan, I.F., Shepherd, R.C.H., Backholer, J.R. and Jackson, R. (1976) Vic. Nat. 93: 112–19.

Egan, J. (1971) Mule Deer. In Mussehl, T.W. and Howell, F.W. (eds) Game Management in Montana, Chapt. 6: 53–67. Game Management Division, Helena, Montana.

Eglitis, V.K. (1963) In Acclimatization of Animals in the USSR: 37–8. Akad. Nauk. Kazakhsk. SSR., Alma-Ata.

Egorov, O.V. (1963) In Problemy Okhr. prir. Yakutsk. *99–106. Yakutsk.*

Egorov, Yu.E. (1963) In Acclimization of Animals in the USSR. *85–6. Akad. Nauk. Kazakhsk. SSR., Alma-Ata.*

___ *(1963) In* Zoogeography of the Land. *91–2. Tashkent.*

Eisenberg, J.F. (1989) Mammals of the Neotropics: Northern Neotropics *vol. 1. Panama, Colombia, Venezuela, Guyana, Suriname, French Guiana. Chicago & London.*

Elder, J.K. and Ward, W.H. (1978) Aust. Vet. J. *54: 297–300.*

Eliot, J.L. (1999) Nat. Geogr. Mag. *195(3): Earth Almanac.*

Elkins, W.A. and Nelson, U.C. (1954) Proc. Alaska Sci. Conf. *5: 1–21.*

Ellerman, J.R. and Morrison-Scott, T.C.S. (1966) Checklist of Palaearctic and Indian Mammals, 1758–1946. B.M. (Nat. Hist.), London, 2nd edn, 810 pp.

Elliott, R.R. (1948) Colo. Cons. Comment *10(5): 3–4, 22.*

Ellis, W.A.H., White, N.A., Kunst, N.D. and Carrick, F.N. (1990) Aust. Wildl. Res. *17: 421–6.*

Ellison, M.F. (1961) Proc. Liverpool Nat. Fld. Clb., 1960: *22.*

Eloff, P.J. and Van Rooyen, E. (1987) S. Geogr. *14(1–2): 88–100.*

Elschner, C. (1915) The Leeward Islands of the Hawaiian Group. *Honolulu.*

Elton, C. (1927) Animal Ecology. *London.*

Elton, C.S. (1958) The Ecology of Invasions by Animals and Plants. *London.*

Ely, C.A. and Clapp, R.B. (1973) Atoll Res. Bull. *(171): 1–361.*

Emmett, C. (1932) Texas Camel Tales. *San Antonio.*

Enderby, C. (1875) In The New Zealand Pilot, 4th. edn: *315. Hydrographic Office, Admiralty, London.*

Enders, R.K. (1952) Proc. Am. Phil. Soc. *96: 691–741.*

England, D. (1955) Zoo Life *10(3): 66.*

Englund, J. (1975) Viltrevy *3: 507–30.*

Epstein, H. (1969) Domestic Animals in China. *Buckinghamshire; Comm. Agric. Bur.*

Erickson, D.W. and McCullough, C.R. (1987) Wildl. Soc. Bull. *15: 511–17.*

Erlinge, S. (1981) Oikos *36(3): 303–15.*

Esler, A.E. (1980) NZ J. Botany *18: 15–36.*

Espeut, W.B. (1882) Zool. Soc. Proc. Lond.: *712–14.*

Estes, J.A. (1980) Enhydra lutris. Mammalian Species No. *133: 8 pp.*

___, *Jameson, P.J. and Johnson, A.M. (1981)* Proc. Worldwide Furbearers Conf., Frostburg, Maryland, 1980 1: *606–41.*

___, *Smith, N.S. and Palmisano, J.F. (1978)* Ecology *59(4): 822–33.*

Estes, R.D. (1993) The Safari Companion. A Guide to Watching African Mammals. White River Junction, Vermont.

Estrada, A. and Estrada R. (1976) Primates *17(3): 337–55.*

Etheredge, D.F. (1949) Texas Game Fish and Oyster Commission, FA Report Series (5): *1–10.*

Etheridge, R. (1916) Mem. Geol. Surv., N.S.W., Ethnol. Ser. *2: 43–54.*

Evans, J. (1970) USDI, Bur. Sport Fish. and Wildl., Denver Wildl. Res. Centre, Resour. Publ. *(86): 65 pp: Washington.*

___, *Hegdal, P.L. and Griffith, R.E. (1970)* Proc. 4th. Vert. Pest Contr. Conf., Calif.: *109–15.*

Evans, P.K. (1967) Proc. lst. Ann. Conf. SE. Assoc. Game and Fish Commrs.: *183–8.*

Evans, P.R. and Flower, W.U. (1967) Scottish Birds *4: 404–45.*

Everett, A. (1986) Novit. Zool. *3: 591–9.*

Evtushevskii, N.N. (1977) Vestn. Zool. *1: 7–11.*

Fa, J.E. (1986) Biol. Conserv. *35: 215–58.*

Fadeyev, E.B. (1969) Vestn. Mosk. Univ. Ser. *6(3): 16–22.*

Fagerstone, K.A., La Voie, G.K. and Griffiths, R.E. (1980) J. Range Mgmt. *33(3): 229–33.*

Fall, M.W., Medina, A.B. and Jackson, W.B. (1971) J. Mamm. *52(1): 69–76.*

Falla, R.A. (1937) B.A.N.Z. Antarctic Res. Exp., Ser. B., *2: 1–304.*

Fandos, P. and Reig, S. (1987) Trans. Congr. Int. Game Biol. *18: 59.*

___ *and* ___ *(1992)* Trans. Congr. Int. Union Game Biol. *18(2): 139–40.*

Farman, A. (1969) Saeugetierk Mitt. *17(2): 152–5.*

Fateev, K. Ya. (1960) Zool. Zhur. *39(8): 1236–8.*

Feare, C.J. (1979) Atoll Res. Bull. *(226): 1–29.*

Fedosov, A.V. (1959) Inst. Lesa Akad. Nauk. SSSR. *132: 80–8.*

Feist, J.D. and McCullough, D.R. (1976) Z. Tierpsychol. *41(4): 337–71.*

Feldhamer, G.A. and Armstrong, W.E. (1993) Trans. 58th N.A. Wildl. & Natur. Resour. Conf .: *468–78.*

___ *and Chapman, J.A. (1978)* Wildl. Soc. Bull. *6(3): 155–7.*

Feldman, B.M. and Carding, T.H. (1973) Health Serv. Rep. *88(10): 956–62.*

Fellay, R. (1967) Bull. de la Murithienne *84: 25–39.*

Fellows, D. and Sugihara, R. (1977) Hawaiian Plant Rec. *59(6): 67–86.*

___, ___ *and Pank, L.F. (1978) In* Proc. 18th a. Meet. Hawaiian Macadamia Prod. Assoc.: *43–54.*

Fenner, F. (1977) In Exotic Species in Australia: their Establishment and Success. Proc. Ecol. Soc. Aust. 10: *39–61.*

Fennessy, B.V. (1958) CSIRO Wildlife Survey Section Technical Paper (1). CSIRO, Melbourne.

Ferguson, H.L. (1979) Fremontia *7(3): 3–8.*

Ferrand, J.P. (1986) Les Oiseaux (Lile de Groix) Penn ar Bed. *16: 136–47.*

Ferriere, G., Cerda, J. and Roach, R. (1983) In Corporacion Nacional Forestal, Ministeria de Agricultura Republika Chile. Gerencia Technica Boletin Technico 8: *1–35.*

Fielder, L.A., Fall, M.W. and Reidinger, R.F. (1982) Proc. 10 Vert. Pest Conf., Monterey, California, February, 1982: *73–9.*

Fiennes, R. and Fiennes, A. (1968) The Natural History of the Dog. *London.*

Filunov, C. (1980) J. Wildl. Mgmt. *44(2): 389–96.*

Findley, J.S., Harris, A.H., Wilson, D.E. and Jones, C. (1975) Mammals of New Mexico. *Albuquerque.*

Finlayson, H.H. (1958) Rec. S. Aust. Mus. *13: 141–91.*

Fisher, H.I. (1951) Condor *53(1): 31–42.*

Fisher, R.A. (1954) New Jersey Outdoors *4(10): 20–3, 26.*

Fitch, H.S., Goodrum, P. and Newman, C. (1952) J. Mamm. *33(1): 21–36.*

Fitter, R.S. (1967) IUCN. Publication New Ser. *(9): 177–80, Switzerland.*

Fitter, R.S.R. (1959) The Ark in our Midst. *London.*

Fitzgerald, B.M. and Karl, B.J. (1979) N.Z. J. Zool. 6(1): 107–26.

__ *and Veitch, C.R. (1985)* N.Z. J. Zool. 12: 319–30.

Fitzpatrick B.M. (1878) Irish Sport and Sportmen. *Dublin.*

Fitzwater, W.D. (1967) Pest Contr. 10(35): 70–8.

Fjeldsa, J. (1975) Dansk orn. Foren. Tidsskr. 69: 89–102.

Flannery, T. (1990) Aust. Nat. Hist. 23: 394–400.

__ *(1995)* Mammals of New Guinea. *Carina, Queensland.*

__, *Kirch, P., Specht, J. and Spriggs, M. (1988)* Archeology in Oceania 23: 89–94.

__ *and White, J.P. (1991)* National Geographic Res. & Explorer 7: 96–113.

__ *and Wickler, S. (1990)* Aust. Mammalogy 13: 127–39.

Fleet, R.R. (1972) Auk 89: 651–9.

Fleming, P.J.S. and Robinson, D. (1986) Proc. Aust. Soc. Anim. Production 16: 84–7.

Flerov, K.K. (1960) Fauna of the USSR, Musk Deer and Deer. *Washington, D.C.*

Flux, J.E.C. (1981) In Proc. World Lagomorph Conf., 1979, Guelph, Ontario: 155–74. University Guelph, Ontario.

__ *and Fullagar, P.J. (1983)* Acta Zool. Fennica 174: 75–7.

Flyger, V. (1960) J. Mamm. 41(1): 140.

__ *and Bowers, R.J. (1958)* Maryland Tidewater News 14(4): 13–14.

__ *and Davis, N.W. (1964)* Chesapeake Sci. 5(4): 212–13.

__ *and Warren, J. (1958)* In Proc. 12th. a. Conf. SE. Assoc. Game Fish Comm. 12: 209–11.

Food and Agriculture Organisation (1986) The Przewalski Horse and Restoration to its Natural Habitat in Mongolia. F.A.O./U.N.E.P. Expert Consultation held Moscow, USSR, 29–31 May, 1985. F.A.O. of U.N.E.P.

Fooden, J. (1975) Fieldiana Zool. 67: 1–169.

Fooks, H.A. (1958) The Roe Deer. Forest Comm. Tech. Notes, London (6).

Forbes, C.N. (1913) Bernice P. Bishop Mus., Occ. Papers 5(3): 17–30.

Forbes, H.O. (1893) Ibis 5: 521–46.

Ford, B.D. and Tulloch, D.G. (1982) The Australian Buffalo. Department Primary Production, Bull. (62), Division of Agriculture and Stock, Darwin.

Foss, A. and Rognrud, M. (1971) In Mussehl, T.W. and Howell, F.W. (eds) Game Management in Montana, Chapt. 11: 107–13. Game Management Division, Helena, Montana.

Fowle, C.D. (1951) 41st. Conv. Internat. Assoc. Game Fish Commrs., Sept. 10–11, 1951, Rochester, New York: 144–52.

Fowler de Neira, L.E. and Johnson, M.K. (1985) J. Wildl. Mgmt. 49(1): 165–9.

Fowler, H.D. (1950) Camels to California. *Stanford, California.*

Fox, B.J. and Pople, A.R. (1984) Aust. J. Ecol. 9: 323–34.

Fox, J.R. and Pelton, M.R. (1977) In Wood, G.(ed) Research and Management of Wild Hog Populations. B.W. Baruch For. Sci. Inst., Georgetown, S.Carolina.

Fox, M.W. (1978) The Dog. Its Domestication and Behaviour. *New York. 296 pp.*

Fradkin, P.I. (1980) Audubon Mag. 82(6): 70–81.

Framarin, F. (1976) Serie Atti Studi WWF No. 2: 17–19.

Francine, J. (1953) New Jersey Outdoors 3(8): 20.

Francis, J. (1964) Aust. Vet. J. 40: 114–18.

Frankenberger, W.B. and Belden, R.C. (1976) Proc. a. Conf. SE. Assoc. Game Fish Commrs.

Franzmann, A.W. (1988) In Nielsen, L. and Brown, R.D. (eds), Translocation of Wild Animals. 210–29. Wisconsin Humane Soc. and C. Kleberg Wildl. Res. Inst.

Fraser, K.W., Cone, J.M. and Whitford, E.J. (1996) The Established Distribution and New Populations of Large Introduced Mammals in New Zealand. Landcare Research Contract Report LC9697/22, Manaaki Whenua, Lincoln, New Zealand.

Freeland, W.J. (1990) J. Biogeogr. 17(4/5): 445–52.

Frei, M.N., Peterson, J.S. and Hall, J.R. (1979) J. Range Mgmt. 32(1): 8–11.

Friend, G.P. and Taylor, J.A. (1984) Aust. Wildl. Res. 11(2): 303–9, 311–23.

Friend, J.A. (1989) Prog. and Abstracts Sci. Meeting, April 24–25, Alice Springs. Aust. Mamm. Soc.: 17.

__ *(1994)* Recovery plan for the Numbat (Myrmecobius fasciata) 1995–2004. Dept. Cons. and Land Management, Perth.

__ *and Thomas, N.D. (1995)* In Serena, M. (ed), Reintroduction Biology of Australian and New Zealand Fauna, 30: 183–8. Chipping Norton.

Frith, H.J. (1970) In Moore, R.M., Australian Grasslands, 74–83. Aust. Nat. Univ., Canberra.

__ *(1973)* Wildlife Conservation. *Sydney. pp 414.*

Fritts, S.H. (1984) J. Wildl. Mgmt. 48(3): 709–21.

Fry, C.H. (1961) Ibis 103a(1): 267–76.

Fujiwawa, E. (1951) Mamm. Soc. Japan (2): 10–12.

Fullagar, P.J. (1973) Aust. Bird Bander 11: 36–9.

__ *(1976)* Aust. Bird Bander 14(4): 94–7.

__ *(1978)* Report on the Rabbits on Philip Island, Norfolk Island. CSIRO Divis. Wildl. Res., Canberra.

__ *(1978)* Corella 2(2): 21–3.

Fuller, P.J. and Burbidge, A.A. (1987) W. Aust. Nat. 16: 177–81.

__ *and Burbidge, A.A. (1998)* Corella 22(4): 113–15.

Funmilayo, O. (1982) In Proc. 10 th. Vert. Pest Conf., Monterey, California, February 1982: 107–8.

Furet, L. (1989) Rev. Ecol. Terre Vie 44 (1): 33–45.

Furness, R.W. and Hislof, J.R.G. (1981) J. Zool. 195: 1–23.

Furon, R. (1958) The Gentle Goat, Archdespoiler of the Earth. UNESCO Courier, 1958, (1): 30–2.

Futuya, Y. (1973) J. Mamm. Soc. Japan 5(6): 199–205.

Gams, H. (1956) Terre et la Vie 1956: 272.

Ganzak, J. (1964) Okhota and Okhotnich'e khoz. (5): 52–3.

Gardiner, J.S. and Cooper, C.F. (1907) Trans. Linn. Soc. Lond., Ser. 2, Zool. 12: 111–75.

Garland, H.P. (1922) The Water Buffalo. *Saco, Maine.*

Garner, M.G. and O'Brien, P.H. (1988) In Rev. Sci. et Office Internat. des Epizootics Techniques 7: 823–41.

Garnett, S.T. and Crowley, G.M. (1987) Corella 11 (3): 77–8.

__ *and Jackes, B.R. (1983)* Qld. Nat. 24: 40–63.

Garrison, R.L. and Lewis, J.C. (1987) Wildl. Soc. Bull. 15: 555–9.

Garrott, R.A. (1991) Wildl. Soc. Bull. 19: 52–58.

__ *and Taylor, L. (1991)* J. Wildl. Manage. *54(4): 603–12.*

Garson, P.J. (1984) Lagomorph Newsletter *4: 25–6.*

__ *and Haig, T. (1986)* Bull. British Ecol. Soc. *17: 135–7.*

Gauthier-Pilters, H. and Dagg, A.I. (1981) The Camel: Its Evolution, Ecology, Behaviour, and Relationship to Man. *Chicago and London.*

Geist, V. (1960) Murrelet *41(3): 34–40.*

Geller, M.Kh. (1959) Tr. NII. selskogo Khozyahistva Krainego Severa, *vol. 9.*

Genov, P. (1987) Trans. Congr. Int. Game Biol. *18: 66.*

Genovesi, P. (2000) In Guidelines for Eradication of Terrestrial Vertebrates: a European contribution to the Invasive Alien Species Issue. *Convention on the Conservation of European Wildlife and Natural Habitats. Strasburg.*

George, G.G. (1987) In Archer, M. (ed), Possums and Opossums. *507–26.New South Wales.*

George, W. (1962) Animal Geography. *London.*

George, W.G. (1974) Wilson Bull. *86(4): 384–96.*

Georges, A., Kennett, R. (1989) Aust. Wildl. Res. *16: 323–35.*

Gerasimova, M.A. (1954) Tret'yah akalo gicheskayah Konferentsiyah Tezis' dokladov, *pt. 3 Kiev, Izd–vo Kievskogo un–ta, 1954: 24–9.*

Gerbil'skii, N.L. (1963) In Acclimatization of Animals in the USSR: *7. Akad. Nauk. Kazakhsk. SSR., Alma-Ata.*

Gerell, R. (1967) Viltrevy *5(1): 1–30.*

__ *(1968)* Viltrevy *5(5): 120–211.*

Gerke, V.A. and Krechmar, A.V. (1963) Tr. Nauch.–Issled. Inst. Sel'sk Kraninego Sev. *11: 79–84.*

Gerlach, R. (1949–50) Wild und Hund *52: 369.*

Gerstell, R. (1936) Pa. Game News *7(7): 6–7, 26.*

__ *(1937)* Pa. Game News *7(12): 6–7, 27, 30; 8(2): 8–11, 32; 8(3): 12–15, 26.*

Ghosh, A. (1988) Telegraph *Sunday Supplement, 23 July. Calcutta, India.*

Gibb, J. (1951) Ibis *93: 109–27.*

Gibb, J.A. and Flux, J.E.C. (1973) In The Natural History of N.Z.: An Ecological Survey, *Chapt. 14: 334–71. Wellington.*

Gibson, D.F., Johnson, K.A., Langford, D.G., Cole, J.R., Clarke, D.E. and Willowra Community (1995) In Serena, M. 9ed), Reintoduction Biology of Australian and New Zealand Fauna, *28: 171–6. Chipping Norton.*

__, *Lundie-Jenkins, G., Langford, D.G., Cole, J.R., Clark, D.E. and Johnson, K.E. (1994)* Aust. Mammalogy *17: 103–7.*

Gibson, J.D. (1976) Aust. Bird Bander *14(4): 100–3.*

Gibson, W.N. and Macarther, K. (1965) Forestry *38(2): 173–82.*

Gier, H.T. (1975) In Fox, M.W. (ed) The Wild Canids, *chapt. 17: 247–61. Van Nostrand Reinhold Co., New York.*

Giffin, J.C. (1976) Pitt. Roberts. proj. W-15-5 Study No. 11., Final Report, State Hawaii Department Lands and Natural Resources, Division of Fish and Game.

Gill, C.L. (1985) Corella *8: 117–18.*

Gillham, M.E. (1957) J. Ecol. *45: 757–78.*

__ *(1963)* J. Ecol. *51: 275–94.*

__ *and Thomson, J.A. (1961)* Proc. R. Soc. Vict. *74: 21–35.*

Gineyev, A.M. (1987) Trans. Congr. Int. Game Biol. *18: 67–8.*

Gizenko, A.I. (1963) In Acclimatization of Animals in the USSR: *76–8. Akad. Nauk. Kazakhsk. SSR., Alma-Ata.*

Glaser, L. (1868) Zool. Garten *9: 146.*

Glauert, L. (1956) In Problems of Conservation (c) Introduction of Exotics. *Fisheries Department, Western Australia, Fauna Bull. (1).*

Glover, I. (1986) Terra Australis *11.*

Goberts, E. and Gaufrey, R. (1932) L'Anthropologiste *42: 449–90.*

Godbey, J. and Biggins, D. (1994) Endangered Species Tech. Bull. *19(1): 10–13.*

Godin, A.J. (1977) Wild Animals of New England. *Baltimore.*

Godin, J. (1978) Final Rep. P-R Proj. W-15-3 (11), Hawaiian Department Lands Natural Resources, Division of Fish and Game.

Godoy, J.C. (1963) Fauna Silvestre. Evaluation de los Recursos Naturals de la Argentina. *CFI, Buenos Aires, Argentina.*

Goethel F. (1955) Abh. Mus. Naturk. Munster/Westf. *17(1/2).*

Gogan, P.J. and Barrett, R.H. (1987) J. Wildl. Mgmt. *51(1): 20–7.*

__ *and* __ *(1988) In Nielsen, L. and Brown, R.D. (eds),* Translocation of Wild Animals: *275–87. Wisconsin Humane Soc. and C. Kleberg Res. Inst.*

__, *Jordan, P.A. and Nelson, J.L. (1990)* Trans. N. Am. Wildl. Nat. Resour. Conf. *55: 599–608.*

__, *Thompson, S.C., Pierce, W. and Barrett, R.H. (1986)* Calif. Fish and Game *72(1): 47–61.*

Goldsmith, A.E. (1988) In Nielsen, L. and Brown, R.D. (eds), Translocation of Wild Animals: *288–97. Wisconsin Humane Soc. and C. Kleberg Wildl. Res. Inst.*

Goldushko, B.Z. (1963) In Acclimatization of Animals in the USSR: *79–81. Akad. Nauk. Kazakhsk. SSR., Alma–Ata.*

Gollan, K. (1984) In Archer, M. and Clayton, G. (eds), Vertebrate Zoogeography and Evolution in Australia: *921–7. Carlisle, Western.Australia.*

Golubeva, A.I. (1961) Novosibersk *57–60.*

Gonzlez, C.B. and Leal, C.G. (1984) Forest Mammals of the Mexican Basin. *Prog. on Man and the Biosphere (UNESCO) and Editorial Limusa, Mexico City. (in Spanish)*

Goodall, R.N.P. (1979) Tierra del Fuego: Argentina, Territorio Nacional de la Tierra del Fuego, *Antartica e islas del Atlantico Sur. Shanamaiim, Buenos Aires.*

Gooding, C.D. and Long, J.L. (1958) Bulletin No. 2525. *J. Agric. West. Aust. 7: 173–8.*

Goodpaster, W.W. and Hoffmeister, D.F. (1952) J. Mamm. *33(3): 362–71.*

Goodrich, C.M. (1843) Narrative of a Voyage to the South Seas and the Shipwreck of the 'Princess of Wales' Cutter, with an account of Two Years Residence on an Uninhabited Island. *W.C. Featherstone, Devon.*

Goodsell, J., Tingay, A. and Tingay, S.R. (1976) Report No. 21 to Dept. Fisheries and Wildlife, Perth.

Gordon, L.S. (1967) In Proc. 47th. a. Conf. W. Assoc. State Game and Fish Comm., *Honolulu, Hawaii, 1967: 77–83.*

Gorman, M.L. (1975) J. Zool. Lond. *175: 273–8.*

__ *(1976)* J. Zool. Lond. *178(2): 237–46.*

__ *(1979)* J. Zool. Lond. *187(1): 65–73.*

Gorshkov, P.K. (1963) In Acclimatization of Animals in the USSR: *81–2. Akad. Nauk. Kazakhsk. SSR., Alma–Ata.*

Gosling, L.M. (1974) Trans. Norfolk & Norwich Nat. Soc. *23: 49–59.*

__ *(1980)* J. Zool. Lond. *192: 546–9.*

__ *(1989)* Trans. Norfolk & Norwich Nat. Soc. *28: 154–7.*

__ *(1989)* New Scientist *121: 44–9.*

__ *and Baker, S.J. (1989)* Biol. J. Linnean Soc. *38(1): 39–51.*

__, *Guyon, G.E. and Wright, K.M.H. (1980)* J. Zool. Lond. *192(2): 143–6.*

__, *Watt, A.D. and Baker, S.J. (1981)* J. Anim. Ecol. *50(3): 885–901.*

Gossow, H. and Honig-Erlenburg, P. (1986) In Cats of the World: Biology, Conservation and Management, *S.D. Miller and D.D. Everett (eds): 77–83.*

Gottschalk, J.S. (1967) In Proc. Paps. 10th. Tech. Meet. IUCN, 1966: *124–40. IUCN Publ. New Ser. (9), Morges, Switzerland.*

Gould, E. and Eisenberg, J.F. (1966) J. Mamm. *47(4): 660–86.*

Government of the United States (1983) Proposal to remove Lynx rufus *(populations of the US and Canada) from Appendix 11. Proc. Conf. of the Parties to CITES 4, CITES Secretariat, Lusanne, Switzerland.*

Graafland, N. (1898) De Minahassa, Haarlem, Batavia.

Graells, M.P. (1854) Bull. Soc. Imp. Zool. Acclim. *2: 109–16.*

Graf, W. (1955) J. Wildl. Mgmt. *19 (2): 184–8.*

__ *(1959)* Am. Philos. Soc., Yrbk.: *236–8.*

__ *and Nichols, L. (1966)* J. Bombay Nat. Hist. Soc. *66(3): 629–734.*

Granadeiro, J.P. (1991) Seabird *13: 30–9.*

Grant, P.J. (1956) N.Z. J. For. *7(3): 111–12.*

Grant, T. and Fanning, D. (1984) The Platypus. *Kensington, N.S.W.*

Grasse, J.E. (1949) Wyo. Wildl. *13(9): 10–17, 34.*

Grauvogel, C.A. (1984) In Klein, D.R., White, R.G. and Keller, S. (eds). Proc. Int. Muskox Symp. 1, Biol. Pap. Univ. Alaska Spec. Rep. 4: 57–62.

Graves, H.B. and Graves, K.L. (1977) In Wood,G.W. (ed) Research and Management of Wild Hog Populations. *B.W. Baruch For. Sci. Inst., Clemson University, Georgetown, S. Carolina: 103–10.*

Gray, D.R. (1984) Muskox. *Canadian Wildlife Service, Ottawa, Canada*

Gray, G.G. (1982) J. Wildl. Mgmt. *46(4): 1096–101.*

__ *(1985) In Hoefs, M. (ed)* Wild Sheep, Distribution, Abundance, Management and Conservation of the Sheep of the World and Closely Related Mountain Ungulates: *95–126. Northern Wild Sheep and Goat Council, Whitehorse, Yukon.*

__ *and Simpson, C.D. (1983)* J. Wildl. Mgmt. *47(4): 954–62.*

Gray, R.L. (1977) Calif. Fish Game *63: 58.*

Green, G. and Catling, P. (1977) In Messel, H. and Butler, S.T., Australian Animals and their Environment: *49–60. Sydney.*

Green, G.A. (1972) Fifth Abrohlos Expedition 1970. Aquinas College, Manning Western Australia.

Green, K.M. (1976) In R.W. Thorington, P.G. Helne (eds) Neotropical Primates: Field Studies and Conservation: *85–98, Washington DC.*

Green, L.G. (1950) At Daybreak to the Isles. Capetown.

Green, R.H. (1965) Tasmanian Naturalist *3: 1–2.*

__ *(1969)* Rec. Queen Victoria Mus., Launceston *No. 34.*

__ *(1979)* Rec. Queen Victoria Mus., Launceston *No. 66.*

__ *and McGarvie, A.M. (1971)* Rec. Queen Victoria Mus., Launceston *No. 40.*

Greenly, A.H. (1952) Paps. Bibliogr. Soc. Am. *46: 359–72.*

Greer, J.K. (1965) Publ. Mus. Michigan State University, Biol. Ser. *3(2): 49–152.*

Greig, J.B. (1886) Letter to the Royal Society of Victoria from the Captain of the SS. Kekeno, 11th. Oct., 1886. N.Z. National Archives.

Greth, A. and Schwede, G. (1993) Captive Breeding Specialist News *4(2): 18–19.*

Griess, J.M. and Anderson, B. (1987) Proc. SE Nongame Wild. Symp. *3: 167–75.*

Griffin, C.R., King, C.M., Savidge, J.A., Cruz, F. and Cruz, J.B. (1988) Proc. Int. Ornithol. Congr. *19: 688–98.*

Griffin, J. (1977) Elepaio *37(12): 140–2.*

Griffith, B., Scott, J.M. and Carpenter, J.W. (1987) Proc. Int. Conf. Zool. Avian Med. *1: 412–13.*

Grigera, D.E. and Rapoport, E.H. (1983) J. Mamm. *64(1): 163–6.*

Grigor'ev, N.D. and Egerov, Yu.E. (1969) Inst. Zhivotn. Syr'ya Pushniny *22: 26–32.*

Grimwood, I.R. (1969) Spec. Publ. Am. Comm. Inst. Wildl. Protection *21: 5–86.*

Grinnell, J. (1914) Univ. Calif. Publs. Zool. *12: 51–294.*

__ *(1915)* Calif. Fish Game *1: 114–16.*

__ *(1933)* Univ. Calif. Publs Zool. *40: 71–234.*

__ *and Dixon, J. (1918)* Calif. State Comm. Hortic., Monthly Bull. *7(11–12): 597–807.*

Grodinsky, C. and Stuwe, M. (1987) Smithsonian Mag. *18: 68–77.*

Grosvenor, M.B. (1965) Nat. Geogr. Mag. *128(3): 398–405.*

Groves, C.P. (1971) Proc. 3rd. Int. Congr. Primat., *Zurich 1: 44–53.*

__ *(1972)* Mammalian Species, *No. 8: 1–6.*

__ *(1976)* Z. Saugetierk. *41: 201–16.*

__ *(1980) Notes on the Systematics of Babyrousa (Artiodactyla, Suidae) Zool. Meded. 55: 29–46.*

__ *(1981) Tech. Bull. No. 3, Dept. Prehistory, Res. School Pacific Studies, Canberra.*

__ *and Grubb, P. (1987) In Wemmer, C.M.* Biology and Management of the Cervidae: *21–29. Washington, DC.*

Groves, R.H. and Di Castri, F. (1991) Biogeography of Mediterranean Invasions. *Melbourne and Sydney.*

Grulih, I. (1979) Zool. Zhur. *58(3): 419–27.*

Gruzdev, V.V. (1967) Bul 1. MOIP, Otd. biol. *77(5): 152–3.*

__ *(1969)* Byull. Mosk. Ob. Isp. Priroda Otd. Biol. *74(6): 19–29.*

Gryaznukhin, A.N. (1958) Tr. Inst. Biol. Yakutsk Fil. Sibirsk Otd. Akad. Nauk. SSR *4: 143–71.*

Grzimek, B. (1966) Afr. Wildl. *20(4): 271–88.*

__ *(1975)* Grzimek's Animal Life Encyclopedia. *Mammals. Vols. 10–13. New York.*

__ *(1990)* Grzimek's Encyclopedia of Mammals. *Vol. 3. New York.*

Guilday, J.E. (1958) J. Mamm. 30(1): 39–43.

Guiler, E.R. (1968) Tasmanian Yearbook 2: 55–60.

Gulbreath, J.C. (1947) Colo. Cons. Comments 10(4): 3–4, 28.

Gunnderson, H.L. (1955) J. Mamm. 36(3): 465.

Haagner, A. (1920) South African Mammals. London.

Haber, A. (1961) Terre et Vie 108 (1): 74–6.

___., Paslawski, T. and Zaborowski, S. (1977) Gospodarstwo Lowieckie. Panstwowe Wydawnictwo Naukowe, Warszawa.

Hachisuka, M. (1953) The Dodo and Kindred Birds. London.

Hagen, K. (1967) Cbl. ges. Forstwesen 84 (2/6): 150–5.

Hagen, Y. (1952) Results Norw. Sci. Exped. to Tristan da Cunha, 1937–38 (20).

Hainard, R. (1969) Le Courrier de la Nature, Paris No. 10: 49–54.

Hakluyt, R. (1598) The Principall Navigations, Voiages and Discoveries of the English Nation. 1st edn facsimile. Cambridge.

___ (1600) The Principall Navigations, Voiages, Traffiques and Discoveries of the English Nation. 2nd edn, republished 10 vols. Dent, London.

Halder, U. (1976) Mammalia Depicta, (10). Oekologie und Verhalten des Banteng (Bos javanicus) in Java: Ein Feldstudie. Hamburg, West Germany.

Hale, H. (1846) United States Exploring Expedition during 1838–1842 under the Command of Charles Willkes, USN Ethnography and Philology. Philadelphia.

Hall, E.R. and Kelson, K.R. (1959) The Mammals of North America. New York. Vol. 1.

Hall, G.P. (1991) Recovery Plan for the Tammar Wallaby, Macropus eugenii in Southern Australia. Dept. Conserv. and Land Mgmt, Perth.

Hall, L.S. (1977) Northwest Science 51(4): 293–97.

Hall, R.E. (1964) Proc. l6th. Intern. Congr. Zool. 1: 267.

Hall, S.J.G. (1975) Mammal Rev. 5(2): 59–64.

___ and Moore, G.F. (1986) Mammal Rev. 16: 89–96.

Hall, W.L. (1904) Planters' Monthly 23: 367.

Halloran, A.F. (1943) J. Wildl. Mgmt. 7(2): 203–16.

___ and Howard, J.A. (1956) J. Wildl. Mgmt. 20(4): 460–1.

Haltenorth, Th. (1953) Saugetierkundliche Mitt. 1(4): 173.

Hamilton, A. (1894) Trans. N.Z. Inst. 27: 559–79.

Hamilton, H. (1948) Svensk Jakt. 98: 66–7.

___ (1960) Svensk Jakt. 98: 66–7.

Hamilton, P.H. (1969) East Afr. Wildl. J. 7: 73–84.

Hamilton, R. (1962) J. Wildl. Mgmt. 26(1): 79–85.

Hamilton, W.J. (1958) Life History and Economic Relations of the Opossum (Didelphis marsupialis virginianus) in New York State. Cornell University, Agriculture Experimental Station, N.Y. State College of Agriculture, Ithaca, Memoir (354).

Hanley, T.A. and Brady, W.W. (1977) J. Range Mgmt. 30(5): 370–3.

___ and ___ (1977) J. Range Mgmt. 30(5): 374–7.

Hansen, E.L. (1965) J. Mamm. 46(4): 669–71

Hansen, R. M. (1976). J. Range Mgmt. 29(4): 347.

___ and Clark, R.C. (1977) J. Wildl. Mgmt. 41(1): 76–80.

Harcourt, C. and Thornback, J. (1990) Lemurs of Madagascar and the Comoros. IUCN – the World Conservation Union, Gland, Switzerland and Cambridge, UK.

Hardberger, F. M. (1950) Turtox News 28(9): 174–7.

Harder, J.D. (1968) Colo. Dep. Game Fish Parks, Spec. Rep. (12): 1–22.

___ (1970) Trans. N. Am. Wildl. Conf. 35: 35–47.

Harmel, D.E. (1975) Habit Preference of Exotics. Job No. (18), Kerr Wildl. Mgmt. Area Res., Fed. Aid Prog. No. W-76-R-18, Job Perform. Rep. Texas Parks and Wildlife Department: 20 pp.

Harper, F. (1945) Extinct and Vanishing Mammals of the Old World. American Committee for International Wild Life Protection, New York.

Harrington, G.N. (1979) Aust. Rangeland Res. 1: 334–45.

Harris, C.J. (1968) Otters: A Study of the Recent Lutrinae. London.

Harris, C.R. (1974) MA Thesis, University of Adelaide.

Harris, D.R. (1962) Proc. Linn. Soc., Lond. 173(2): 79–91.

Harris, M.P. and Deerson, D.M. (1980) Corella 4: 79–80.

___ and ___ (1980) Corella 4(4): 65–6.

Harris, M.T., Palmer, W.L. and George, J.L. (1983) J. Wildl. Mgmt. 47(2): 516–19.

Harrison Matthews, L. (1931) South Georgia. Bristol.

Harrison, J.L. (1950) Malay Nat. J. 5(1): 21–4.

___ (1966) Mammals of Singapore and Malaya. Singapore.

Harrison, M. (1989) Aust. Deer 14(2): 13–19.

Hartman, L. (1964) N.Z. J. Sci. 7(2): 147–56.

Harty, F.M. (1986) Nat. Areas J. 6(4): 20–6.

Harvie-Brown, J.A. (1880–81) Proc. phys. Soc. Edinb. 5: 343–8, 6: 31–63, 115–83.

Hasler, M.J., Hasler, J.F. and Nalbandov, A.V. (1977) J. Mamm. 58(3): 285–90.

Haspel, C. and Calhoon, R.E. (1989) Can. J. Zool. 67(1): 178–81.

Hatcher, R.T. and Wigtil, G.W. (1985) Proc. a. Conf. SE Assoc. Fish and Wildl. Agencies 39: 321–5.

Hattingh, I. (1956) Proc. Zool. Soc. Lond. 127(2): 191–9.

Hausfater, G. (1974) Lab. Primate Newsl. 13(1): 16–18.

Havlas, M. (1965) Jesenikach. Zool. Listy 14(1): 1–8.

Hawkins, R.E. and Montgomery, G.G. (1969) J. Wildl. Mgmt. 33 (1): 196–203.

Hayashi, Y. (1981) Proc. World Lagomorph Conf., 1979, Guelph, Ontario: 926–7. University Guelph, Ontario.

Hayward, K.J. (1967) Acta Zool. Lilloana 22: 211–20.

Heatwole, H., Levins, R. and Byer, M.D. (1981) Atoll Res. Bull. (251): 1–20.

Heck, L. (1943–44) Wild und Hund: 154.

Hefmeyr, J.M. (1975) In Reid, R. (ed), Proceedings 3rd World Conference on Animal Production: 126–31. Sidney.

Heidemann, G. (1973) Mammalia depicta 9.

___ (1973) Zeitschrift fur Saugetierekund 38: 341–7.

___. and Witt, H. (1978) Z. Jagdwiss. 24(1): 24–6.

___ and Vauk, G. (1970) Zeitschrift fur Saugetierkunde 35: 185–90.

Heim de Balsac, H. (1940) Compte rendu hebdomadaire des sciences de l'Academie des sciences 211: 212–14.

___ (1940) Compte rendu hebdomedaire des sciences de l'Academie des sciences 211: 296–8.

Heinrich, D. (1976) Zool. Anz. 196 (3/4): 273–8.

Held, W. (1954–55) Wild und Hund 54: 531.

Helle, E. and Kauhala, K. (1993) J. Mamm. 74(4): 936–42.

Heller, E. (1903) Proc. Wash. Acad. Sci. 5: 39–98.

Helms, R. (1902) J. Dept. Agric. W. Aust. 5: 33–55.

Henderson, N.R. and Rentchler, P. (1986) Conf. Sci. Natl. Parks 4: 69.

Henke, L.A. (1929) Univ. Hawaiian Res. Publ. 5: 1–82.

Hennig, R. (1960) Schweiz. Zeitschr. Forstw. 111(12): 746–56.

Henry, V.G. (1966) in Proc. a. Conf. SE Assoc. Game Fish Commrs. 20: 139–45.

__ (1968) J. Wildl. Mgmt. 32(4): 966–70.

__ (1969) J. Tenn. Acad. Sci. 44(4): 103–4.

__ (1969) Trans. a. Conf. SE Assoc. Game Fish Commrs. 23: 185–8.

__ (1970) J. Tenn. Acad. Sci. 45(1): 20–3.

__ and Conley, R.H. (1972) J. Wildl. Mgmt. 36: 854–60.

Henwood, R.R. and Keep, M.E. (1989) Lammergeyer (40): 30–8.

Hepburn, I.R., Schofield, P. and Schofield, R.A. (1977) Bird Study 24: 25–43.

Hepner, V.H. and Sludskii, A.A. (1972) Mammals of the Soviet Union. Vol. 3, Carnivores (Feloidea), Moscow.

Heptner, V.G. (1967) In Proc. and Pap. 10th. Tech. Meet. IUCN, 1966: 194–6. IUCN Publ., New Ser. (9), Morges, Switzerland.

__ Nasimovich, A.A. and Bannikov, A.G. (1966) Die Saugetiere der Sovjetunion. Paarhufer & Unpaarhufer vol. 1, Jena.

__, __ and __ (1989–1992) Mammals of the Soviet Union. New Delhi.

Herbert, S. (1882) in U.S. Southern Cultivator & Dixie Farmer, 1882: 1–11.

Herington, J.G. (1977) Wildlife Introduced and Imported into Papua New Guinea. Wildlife in Papua N.G., Wildl. Publ. 77/2.

Hernández-Comache, J. and Cooper, R.W. (1976) The Nonhuman Primates of Colombia. In R.W. Thorington, P.G. Helne (eds) Primates: Field Studies and Conservation,: 35–69. Washington DC.

Herold, W. (1921) Beitrage zur Saugetierfauna Usedom-Wollins. Abh. Ber. Pomm. Naturf. Ges. 2: 75.

Herren, H. (1964). Schweizer Naturschutz, Basel 30: 69–70.

Herrero, S., Schroeder, C. and Scott-Brown, M. (1986) Biol. Conserv. 36: 159–67.

Hershkovitz, P. (1969) Quart. Rev. Biol. 4(1): 1–70.

Herter, K. (1936) Z. Saugetierkunde 11(3): 274.

__ (1952) Der Igel. Die neue Brehm-Bucherei, Wittenberg–Leipzig.

Hesse, R. (1937) Ecological Animal Geography. New York.

Hewett, M. (1980) Pearson Island Rock Wallaby Management. S. Aust. Nat. Pks and Wildlife Service, Adelaide.

Hewson, R. (1971) Glasgow Nat. 18: 539–46.

__ (1977) J. appl. E col. 14: 779–86.

Hey, D.D (1964) The Control of Vertebrate Problem Animals in the Province of the Cape of Good Hope, Republic of South Africa. Proc. 2nd. Vert. Pest Contr. Conf., 1964, Anaheim, California: 57–68.

__ (1974) Vertebrate Pest Animals in the Province of the Cape of Good Hope, Republic of South Africa. Proc. 6th. Vert. Pest Contr. Conf., Anaheim, California.

Hibbard, A.H. (1956) J. Mamm. 37(4): 525–31.

Hibbard, E. (1958) J. Wildl. Mgmt. 22(2): 209–11.

Hibben, F.C. (1964) In Proc. 44th. a. Conf. w. Assoc. State Game Fish Commrs., San Francisco, 1964: 125–30.

Hibbs, D.L. (1967) J. Mamm. 48(2): 242–8.

Hickie, P.F. (1939) Mich. Cons. 8(6): 4–5, 8, 10–11.

Highley, K. (1998) BBC Wildlife 16(9): 14–22.

__ (1999) BBC Wildlife 17(1): 24.

Hill, E.P., Sumner, P.W. and Wooding, J.B. (1987) Wildl. Soc. Bull. 15: 521–4.

Hill, G.R. (1961) Nature in Wales 7: 43–5.

__ (1964) Nature in Wales 9: 17–18.

Hill, J.E. (1959) Rats and Mice from the Islands of Tristan da Cunha and Gough, South Atlantic Ocean. Results of the Norwegian Scientific Expedition to Tristan da Cunha 1937–1938 (46): 1–5.

Hindwood, K. A. et al. (1963) CSIRO Wildl. Tech. Pap. No. 3.

__ (1969) Proc. R. Zool. Soc. NSW 1967–8: 46–72.

Hingston, F. (1975) Deer 3: 386–7.

Hinton, H.E. and Dunn, A.M.S. (1967) Mongooses: their Natural History and Behaviour. London.

Hinton, M.A.C. (1933) The Field Dec. 1933: 1421.

Hirai, L.T. (1979) Elepaio 40(6): 91.

Hirotani, A. and Nakatani, J. (1987) Ecol. Res. 2(1): 77–84.

Hobbs, J.A., Miller, M.W., Bailey, J.A., Reed, D.F. and Gill, R.B. (1990) Trans. 55 th N. Am. Wildl. & Nat. Res. Conf.: 620–32.

Hobson, P.D., Proulx, G. and Dew, B.L. (1989) Canad. Field-Nat. 103(3): 398–400.

Hock, R.J. (1952) J. Mamm. 33(4): 464–70.

Hodgkin, S.E. (1984) Biol. Cons. 29: 99–119.

Hoffman, R.S., Wright, P.L. and Newby, F.E. (1969) J. Mamm. 50(3): 579–604.

Hoffmann, M. (1952) Die Bisamratte. Die Neue Brehm–Bucherei, Leipzig.

__ (1956) Wasserwirtschaft – Wassertechnik 6(1): 17.

Hoffmeyer, J.M., Ebedes, H., Fryer, R.E.M. and De Bruine, J.R. (1975) Madoqua 9(2): 35–44.

__ (1975) In Proc. 111 World Conf. Anim. Prod.: 126–31. Sydney.

Hohenberg, M.O. (1902) St. Hubertus 17: 485.

Holdgate, M.W. (1958) Mountains in the Sea. London.

__ (1967) In Proc. and Paps. 10th. Tech. Meet. IUCN, 1966: 151–176. IUCN Publ. New Ser. (9), Morges, Switzerland.

__ and Wace, N.M. (1961) The Influence of Man on the Floras and Faunas of Southern Islands. Polar Record 10: 475–93.

Holdsworth, M. (1951) Rec. Auckland Inst. Mus. 4(2): 113–22.

Hollister,–. (1912) Proc. Biol. Soc., Washington 25: 93.

Holst, P.J. (1981) Aust. Wildl. Res. 8(3): 549–53.

__, Harrington, G.N., Turner, H.N., Clarke, W.T. and Smith, I.D. (eds) (1982) In Harrington, G.N. (ed) The Feral Goat, Standing Committee for Agriculture in Australia, Chap. 1.

Honacki, J.H., Kinman, K.E. and Koeppl, J. W. (1982) Mammal species of the world: a taxonomic and geographic reference. *Lawrence, Kansas.*

Hone, E. (1934) Am. Comm. Int. Wildl. Protection Spec. Publ. 5: 1–87.

Hone, J. (1980) J. Aust. Inst. Agric. Sci. 46: 130–2.

__ and Bryant, H. (1981) In 'Wildlife diseases of the Pacific Basin and other countries', Fowler, M.E. (ed) Proc. 14th Int. Conf. Wildl. Dis. Assoc., *Sydney, Australia, Aug. 25–28, 1981: 79–85.*

__, O'Grady, J. and Perdersen, H. (1980) Dep. Agric. AG. Bull. *(5), Department of Agriculture N.S.W.*

__, Waithman, J., Robards, G.E. and Saunders, G.R. (1981) J. Aust. Inst. Agric. Sci. 47: 191–9.

Hood, G.A., Nass, R.D. and Lindsey, G.D. (1970) Proc. 4th. Vert. Pest Contr. Con.: 34–7.

__, __., __. and Hirata, D.N. (1971) J. Wildl. Mgmt. 35(4): 613–18.

Hope, J.H. (1973) Proc. Roy. Soc., Vict. 85: 163–95.

Hopkins, G.H.E. and Rothschild, M. (1953) An Illustrated Catalogue of the Rothschild collection of fleas (Siphonaptera) in the British Museum (Natural History). *London.*

Horacek, V. (1962) Czechslovakia. Ziva 10(4): 149–50.

Hornocker, M.G. (1992) Nat. Geogr. Mag. 182(1): 52–65.

Horwich, R., Koontz, F., Glander, K., Saqui, E. and Saqui, H. (1994) Responses of black howler monkeys (Alouatta pigra) reintroduced into the Cockscomb Basin, Belize. Am. J. Primatol. 33(3): 215–16.

Householder, V.H. (1930) Arizona Wild Life 3: 4.

Housse, P.R. (1953) Animales Salvajes de Chile. *Ediciones Universitad de Chile, Santiago.*

Howard, B. (1940) Rakiura. A History of Stewart Island, New Zealand. *Wellington.*

Howard, W.E, (1963) N.Z. Farmer 83(27): 2–5.

__ (1953) California. J. Mamm. 34 (4): 512–13.

__ (1957) Proc. N.Z. Ecol. Soc. (5): 13–14.

__ (1958) Dep. Sci. Ind. Res., Wellington, N.Z. Info. Ser. (16).

__ (1959) J. Mamm. 40(4): 613.

__ (1963) N.Z. Farmer 83(38): 10–11.

__ (1964) Proc. 2nd. Vert. Pest Contr. Conf., Agric. Ext. Serv., Univ. Davis, California: 117–26.

__ (1964) J. Wildl. Mgmt. 28(3): 421–9.

__ (1964) Proc. N.Z. Ecol. Soc. 11: 59–62.

__ (1965) N.Z. Dep. Sci. and Ind. Res., Info. Serv. (45): 1–96.

__ (1965) In Baker,H.G. and Stebbins,G.L. (eds) Genetics of Colonizing Species. *New York and London.*

__ (1967) Ecological Changes in New Zealand Due to Introduced Mammals. In Proc. IUCN 10th. Tech. Meet., Lucerne, 1966, Part 3 Changes due to Introduced animal species: IUCN Publ. New Ser. (9): 219–40.

__ and Amaya, J.N. (1975) J. Wildl. Mgmt. 39(4): 757–61.

__ and Marsh, R.E. (1976) The Rat: Its Biology and Control. *Division Agricultural Science, University California.*

__ and __ (1977) The House Mouse: Its Biology and Control. *Division Agricultural Science, University California.*

__ and __ (1984) Ecological Implications and Management of Feral Mammals in California. In 'Feral Mammals – Problems and Potential'. Workshop Feral Mammals, 3rd. Int. Theriol. Conf., 1982: 33–41. IUCN, Morges.

Howe, J.P. (1983) Audubon 85 (1): 34–9.

Howe, T.D. and Bratton, S.P. (1976) Castanea 41(3): 256–64.

Howell, A.B. (1917) Pacific Coast Avifauna 12: 22–4.

Howell, S.N.G. and Webb, S. (1989) Condor 91: 1007–8.

Howery, D.L., Pfister, J.A. and Demarais, S. (1989) J. Wildl. Manage. 53(3): 613–17.

Howes, C.A. (1973) Annual Report, Yorkshire Naturalists' Union 1972: 4–7.

__ (1983) The Naturalist 108: 41–82.

__ (1983) Doncaster Nat. 1: 29–30.

__ (1984) Yorkshire Nats. Un. Bull. 1: 10.

Howitt, H. (1925) Canad. Field–Nat. 39(7): 158–60.

Hubach, P. (1981) The fox in Australia: A review of investigations of predation by the fox (Vulpes vulpes) in Australia. *20 p. Agriculture Protection Board of Western Australia.*

Hubbs, E.L. (1951) Calif. Fish Game 37: 177–89.

Huchison, C.B. (ed) (1946) Californian Agriculture. *Berkeley.*

Hudson, R. (1986) Rare Red Wolves to be released from Captivity. A.A.S. 11/7/86.

Huey, L.M. (1925) Science 61(1581): 405–7.

Hume, B.B. (1935) Outdoor Life 76(4): 20–21, 51.

Hunt, H.M. (1979) Canad. Field–Nat. 93(3): 282–7.

Hunter, C. (1952) Arkansas Game Fish Comm., P.R. Proj. 17–D: 1–17.

Hunter, L. (1998) BBC Wildlife 16(10): 20–5.

Husson, A.M. (1957) Studies on the Fauna of Suriname and other Guyanas 2: 13–40.

Hutchison, B. (1950) The Fraser. *Toronto.*

Hutson, A.M. (1975) Atoll Res. Bull. (175): 1–25.

Hvass, H. (1961) Mammals of the World. *Politikens Forlag 1956. Engl. Trans. London.*

I.U.C.N. (1984) Feral Mammals – Problems and Potential. Proc. Workshop Organised by the Species Survival Comm., Caprinae Specialist Group, Helsinlki, August, 1982. IUCN, Morges.

Il'yushenko, A.F. and Smirnov, K.A. (1979) Lesovedenie (5): 73–9.

Imaizumi, Y. (1949) The Natural History of Japanese Mammals. *Tokyo.*

__ (1973) Mem. Nat. Sci. Mus., Tokyo 6: 129.

Imber, M.J. (1973) Wildlife – A Review: 4–12.

__ (1976) Ibis 118: 51–64.

Imshenetskii, S.B. (1961) Trudy Moscov. Vet. Akad. 33: 173–6.

Inglis, J.M. (1964) Texas Parks and Wildl. Dept. Bull. 45: 1–122.

Inns, R.W. et al. (1979) Mammals. In Tyler, M.J., Twidale, C.R. and Ling, J.K. (eds), Natural History of King Island. *Adelaide.*

Inukai, T. (1949) Trans. Sap. Nat. Hist. Soc. 18(3–4): 56–9.

Ioganzen, B.G. (1963) The Scientific Basis for the Acclimatization of Animals. In Akklimatizatsiya Zhivotna v SSSR: 9 Akad. Nauk. Kazakh. SSR, Alma-Ata.

I.P.S.T., see Israel Program for Scientific Translations.

Irby, L.R. and Andryk, T.A. (1987) J. Environ Manage. 24(4): 337–46.

Israel Program for Scientific Translations (1964) Geography of Israel. I.P.S.T., Israel.

Itoh, M. (1986) Man and Culture in Oceania 2: 1–26.

Ivanov, F.V. (1966) Biol. Nauk. 9(3): 46–9.

Iwano, T. (1989) Primates 30(2): 241–8.

Izawa, M.T. Doi, T., and Ono, Y. (1982) Jap. J. Ecol. 32: 373–82.

Izmailov, I.V. (1969) Uch. Zap. Buryat. Gos. Pedagog. Inst. 31: 13–17.

Jablonska, J. (1965) Prezegl. Zool. 9(3): 299–300.

Jackson, A. (1964) Texas Game and Fish Dept. 22(4): 7–11.

Jackson, H.H.T. (1921) J. Mamm. 2: 234–5.

__ (1961) Mammals of Wisconsin. *Wisconsin. 504 pp.*

Jackson, J.E. and Langguth, A. (1987) pp. 402–9. In Wemmer, C.M. (Ed.) Biology and Management of the Cervidae. *Smithsonian Institution Press. 577 pp.*

Jacobi, J.D. (1976) Proc. 1st. Conf. Nat. Sci. HAVO, CPSU/UH. Botany Dept.: 107–12. *University Hawaii, Honolulu.*

Jacobson, E.R. and Kollias, G.V. (1988) Exotic Animals. *Edinburgh.*

Jahangir (1909) The Tuzuk-i-Jahangiri, or memoirs of Jahangir, I, *Trans. A. Rogers, (ed) H. Beveridge, Oriental Trans. Fund, n.s., xix. London. 478 pp.*

Jaksic, F.M. and Fuentes, E.B. (1980) J. Ecol. 68: 665–9.

__ and Sorihuer, R.C. (1981) J. Anim. Ecol. 50: 269–81.

__ and Yáñez, J.L. (1983) Biol. Conserv. 26(4): 367–74.

__, Fuentes, E.R. & Yáñez, J.L. (1979) J. Mamm. 60(1): 207–9.

__, Schlatter, R.P. and Yáñez, J.L. (1980) J. Mamm. 61(2): 254–60.

Jameson, E.W. and Peeters, H.J. (1988) Californian Mammals. *California Natural History Guides: 52. Berkeley, Los Angeles and London.*

Jameson, R.J., Kenyon, K.W., Johnson, A.M. and Wright, H.M. (1982) Wildl. Soc. Bull. 10(2): 100–7.

Jarman, P. (1986) In Kitching, R.L. Ecology of Exotic Animals and Plants: Australian Case Histories, *Chapt. 5: 62–76. Brisbane and New York.*

Jarman, P.J. and Johnson, K.A. (1977) Proc. Ecol. Soc. Aust. 10: 146–66.

Jarvis, M.J.F. and La Grange, M. (1984) Proc. 10 Vert. Pest Conf., Monterey, California, February 1982: 95.

Javed, S. (1993) Species 20: 56.

Jeannel, R. (1941) Au seuil de l'antarctique. Croisiere du 'Bougainville' aux iles des Manchots et des Elephants de Mer. *Ed. du Museum et PUF, Paris.*

Jefferies, D.J. and Mitchell-Jones, A.J. (1982) Otters, J. Otter Trust, *1981: 13–16.*

__ and Wayre, P. (1983) Otters, J. Otter Trust, *1983: 20–2.*

__, Jessop, R. amd Mitchell-Jones, A.J. (1983) Otters, J. Otter Trust., *1983: 37–40.*

Jellison, W.L. (1945) J. Mamm. 26: 432.

Jenkins, A.C. (1963) Animals 2(7): 170–3.

Jenkins, J.H. and Provost, E.E. (1964) Publ. TID-19562 of U.S. Energy and Development Administration, Washington, D.C.

Jenkins, S.H. (1989) J. Mamm. 70(3): 667–70.

Jepsen, P.V. (1975) Danske Vildtundersogelser 24.

Jerdon, T.C. (1874) The Mammals of India. *London.*

Jewell, P. (1978) Oryx 14(3): 204–205.

Jewell, P.A. (1982) Conservation of cheetah: should cheetah be moved to distant areas? *Unpublished Workshop Report. International Fund for Animal Welfare, Cambridge.*

__, Milner, C. and Morton Boyd, J. (1974) Island Survivors; the Ecology of the Soay Sheep of St. Kilda. *London.*

Jezierski, W. (1968) Acta Theriol. 13(1–7): 1–29.

__ and Myrchia, M. (1975) Polish Ecol. Stud. 1(2): 61–83.

Jingfors, K.T. (1982) J. Wildl. Mgmt. 46(2): 344–50.

__ and Klein, D.R. (1982) J. Wildl. Mgmt. 46(4): 1092–6.

Johnson, A. (1998) Aust. Vert. Pest Conf. 11: 33–4.

Johnson, A.M. (1982) Trans. 47th. N. Am. Wildl. Nat. Resour. Conf., Oregon: 293–9.

Johnson, A.W. (1965) The Birds of Chile. *Buenos Aries.*

Johnson, D.H. (1962) In Storer, T. (ed), Pacific Island rat ecology. *B.P. Bishop Mus. Bull. 225: 1–274.*

Johnson, D.L. (1975) J. Mamm. 56(4): 925–8.

Johnson, G.B. (1942) Va. Wildl. 6(1): 28–34.

Johnson, R.A., Carothers, S.W. and McGill, T.J. (1987) J. Wildl. Mgmt. 51(4): 916–20.

Johnson, W.V. (1964) Proc. 2nd Vert. Pest Contr. Conf.: 90–6. Anaheim, California.

Johnston, D.H. (1962) Proc. Biol. Soc., Washington 75: 317–19.

Joleaud, L. (1920) Bull. de la Societe Zoologique de France 45: 106–12.

__ (1935) Mem. presentes a 1'Institut d'Egypte 27: 1–85.

Jolly, A., Oliver, W.L.R. and O'Connor, S.M. (1982) Folia Primatologica 39: 115–23.

Jones, A.C., Jones, R.E., Meagher, T.D. and Nunn, R.M. (1966) B.Sc. thesis, University of Western Australia.

Jones, C. and Paradiso, J.L. (1971) USDI Fish and Wildl. Serv., Bur. Sport Fisheries and Wildl., Special Sci. Rep. No. 147: 1–33.

Jones, E. (1977) Aust. Wildl. Res. 4: 249–62.

__ and Coman, B.J. (1982) Aust. Wildl. Res. 9: 409–20.

Jones, G.W. (1948) Rocky Mountain Bighorn Sheep Restoration. *Colo. Game Fish Dep., Denver, Colorado.*

Jones, J.K. (1964) University Kansas Publ., Mus. Nat. Hist. 16: 1–356.

Jones, P. (1959) The European Wild Boar in North Carolina. *Game Division, N. Carolina Wildlife Research Commission, Raleigh, North Carolina.*

Jones, P.H. (1980) British Birds 73: 561–8.

Jones, R.D. (1966) J. Wildl. Mgmt. 30(3): 453–60.

Jordan, D.B. (1991) A proposal to establish a captive breeding population of Florida panthers. *Final Supplement environmental assessment, office of Florida recovery coordinator, US Fish and Wildlife Service, Gainesville.*

__ (1994) Florida panther update: Jan–March 1994. *US Fish and Wildlife Service, Gainesville.*

Judd, C.S. (1916) Kahoolawe. Thrum's Hawaiian Annual for 1917: 117–25.

Judd, H.P. (1930) Friend 100: 193–4.

__ (1939) Paradise of the Pacific 51(1): 17–18.

Jurek, R.M. (1977) M.Sc. Thesis, Humboldt State University.

Kaburaki, T. (1934) Proc. 5th. Pac. Sci. Congr., Victoria and Vancouver, B.C., 1933: 801–5.

__ (1940) Proc. 6th. Pac. Congr., Berkeley, 1940: 229–30.

Kadlec, J.A. (1971) J. Wildl. Mgmt. 35(4): 625–36.

Kalinin, M.N. and Korsakov, G.K. (1964) Sb. Nauchno.–Tekhnol. Inf. Vses. Nauchno.–Issled. Inst. Zhivotn. Syr'ya Pushniny 10: 9–13.

Kami, H.T. (1964) Zoonoses Res. 3(3): 165–70.

Kaplin, A.A. (1960) Priroda (6): 93–5.

Karns, P.D. and Jordan, P.A. (1969) J. Wildl. Mgmt. 33(2): 431–3.

Kartchner, K.C. (1940) Ariz. Wildl. Sportsman 2(5): 2, 11.

Kastdalen, A. (1982) Noticias de Galâpagos 35: 7–12.

Katsma, D.E. and Rusch, D.H. (1980) J. Wildl. Mgmt. 44(3): 603–12.

Katzenmeier, Ph. (1959) Z. Jagdwissenschaft 5(2): 64.

Kauffman, S. (1999) Black and white ruffed lemurs return to Madagascar. www.duke.edu/web/primate/reintro.html.

Kawamoto, Y. and Suryobroto, B. (1985) in Kyoto Univ. Overseas Res. Rep., studies on non-human Primates 4: 35–40.

Kazarinov, V.P. (1963) Tr. Vses Nauch Issled Inst. Zhivotn. Sy r'ya Pushn. 20: 85–9.

Kean, R.I. (1951) N.Z. Sci. Rev. 9(8): 146–52.

__ and Pracy, L.T. (1949) Proc. 7th. Pacific Sci. Congr. 4: 696–715.

Keast, J.C., Littlejohns, I.R., Rowan, L.C. and Wannan, J.S. (1963) Aust. Vet. J. 39: 99.

Keay, J.A., Wehausen, J.D., Hargis, C.D., Weaver, R.A. and Blankinship, T.E. (1987) Trans. West. Sect. Wildl. Soc. 23: 60–4.

Keiper, R.R. (1976) Proc. Pennsylvania Acad. Sci. 50: 69–70.

Keiper, R.R. and Keenan, M.A. (1980) J. Mamm. 61(1): 116–18.

Keith, A.R. (1969) Dukes Co. Intellingencer 11(2): 47–98.

Kennamer, J.E. and Braun, C.E.(1990) Trans. 55th N. Am. Wildl. Nat. Resour. Conf.: 545–7.

Kepler, C.B. (1967) Auk 84: 426–30.

__ (1978) Condor 80: 72–87.

Kerr, I.S. (1976) Campbell Island: A History. Wellington.

Kerschagl, W. (1964) In H. St. Furlinger (ed) Jagd in Osterreich. 113–15. Vienna.

Khrustalev, S. (1963) Okhota Okhotn a Khoz. 8: 19–20.

Kidder, J.H. (1876) Mammals. Bull. U.S. Nat. Mus. (2): 1–122.

Kiddie, D.G. (1962) N.Z. Forest Service, Information Service (44).

Kik, P. (1980) Lutra 23: 55–64.

Kikkawa, J. (1959) The Glasgow Naturalist 18: 65–77.

__ (1976) Aust. Bird Bander 14: 3–6.

__ and Boles, W. (1976) Corella 14(1): 3–6.

Kimura, D. and Ihobe, (1985) J. Ethol. 3: 39–47.

King, C. (1984) Immigrant Killers: Introduced Predators and the Conservation of Birds in New Zealand. Auckland, Melbourne and Oxford.

King, C.M. (1979) N.Z. J. Zool. 6: 619–22.

__ (1980) J. Anim. Ecol. 49(1): 127–59.

__ and Moors, P.J. (1979) Oecologia (Berl.) 39: 129–50.

__ and __ (1979) N.Z. J. Zool. 6: 619–22.

King, D.R. and Smith, L.A. (1985) Rec. West. Aust. Mus. 12: 197–205.

King, W.B. (1973) Wilson Bull. 85(1): 89–103.

Kingdon, J. (1974) East African Mammals. Vol. 1–2. London.

Kingsmill, W. (1920) J. Proc. Roy. Soc., W.A. 5: 33–8.

Kinnear, J.E., Onus, M. and Bromilow, R.N. (1988) Aust. Wildl. Res. 15: 435–50.

__, __ and Sumner, N.R. (1998) Wildl. Res. 25: 81–8.

Kippis, A. (1904) The Life and Voyages of Captain Cook. London and New York.

Kirch, P.V. and Yen, D.E. (1982) B.P. Bishop Mus. Bull. 225.

Kirisa, I.D. (1972–74) Akklimatizatsiyah okhotniche – Prom'slo v'kh zverei i ptits v SSSR. Kirov. 2 vols.

Kirk, D.A. (1981) BSc. Hons. Thesis, University Aberdeen, Scotland.

Kirk, G.L. (1923) J. Mamm. 4: 59–60.

Kirk, H.B. (1920) Append. J. House Repres. N.Z. (Sess. 1) 1–28.

Kirkpatrick, J.B. (1973) Vict. Nat. 90: 312–21.

Kirkpatrick, R.D. (1958) Ind. Qu. Prog. Rep. 19(2): 79–107.

__ (1959) Final Report. Indiana Department Conservation, Division of Fish and Game, Indiana: 1–58.

__ (1960) Proc. Indiana Acad. Sci. 69.

__ (1966) J. Mamm. 47(4): 701–4.

__ (1966) J. Mamm. 47(4): 728–9.

__ and Rauzon, M.J. (1986) Biotropica 18(1): 72–5.

Kishida, K. (1950) Mohi Shinbun (17): 1.

Kitchener, A. (1991) The Natural History of the Wild Cats. London.

Kitchener, D.J., Boeadi, Charlton, L. and Maharadatunkamsi (1990) Wild Mammals of Lombok Island. Perth.

__, How, R.A., Iveson, J.B. and Maharadatunkamsi (1996) In Kitchener, D.J. and Sunyanto (eds), Proc. 1st. internat. conf. on E. Indonesia–Australian vert. fauna, Monado, Indonesia, November 22–26, 1994: 147–50.

Kitching, R.L. and Jones, R.E. (eds) (1981) The Ecology of Pests. Melbourne.

Kleberg, C. Research Program (1968) Symposium: Introductions of Exotic Animals: Ecological and Socioeconomic Considerations. Texas A & M University, C. Kleberg Research Program in Wildlife Ecology, College of Agriculture, Texas A & M University, Texas. 24 pp.

Kleiman, D.G. (1989) BioScience 39(3): 152–61.

__, Beck, B.B., Dietz, J.M., Dietz, L.A., Ballou, J.D. and Coimbra-Filho, A.F. (1986) In Benirschke (ed) Road to Self-sustaining Populations.: 959–79. New York.

Klein, D.R. (1959) U.S. Fish and Wildlife Service Special Report, Wildlife (43): 1–48.

__ (1968) J. Wildl. Mgmt. 32(2): 350–67.

__ (1982) J. Wildl. Mgmt. 46(3): 728–33.

__ (1988) In Nielsen, L. and Brown, R.D. (eds), Translocation of Wild Animals,: 298–318. Wisconsin Humane Soc., and C. Kleberg Wildl. Res. Inst.

__ and Staaland, H. (1984) In Klein, D.R., White, R.G. and Keller, S. (eds). Proc. Int. muskox Symp. 1, Biol. Pap. Univ. Alaska Spec. Rep. 4: 26–31.

Klemm, M. (1949) Deut. Pflanz. Nachr. 3(11–12): 201–5.

__ (1957) Nachrbl. Deutsch. Pflanzenschd. N.F. 9(3): 180.

Klingel, H. (1977) Z. Tierpsychol. 44(3): 323–31.

Knap, J.J. (1975) International Wildl. 5(5): 32–5.

Knight, J.E. (1981) Trans. 46th. N. Am. Wildl. Nat. Resour. Conf. 46: 349–57.

Knopff, M.A. (1936) Deutsche Jagd. 4: 92.

Knudsen, G.J. and Hale, J.B. (1965) J. Wildl. Mgmt. 29(4): 685–8.

Koebele, A. (1897) Planters' Monthly 16: 65–85.

Koenders, J.W. (1964) Lutra 6: 68–73.

Kohlmann, S.G. (1987) Trans. Congr. Int. Game Biol. 18: 89.

Kohlmeyer, J. (1959) Natur und Volk 89: 214–23.

Koivisto, I. (1966) Suomen Riista 19: 100–4.

Kokes, 0. (1976) Cas. Nar. Muz. Oddil Prirodoved 45(2): 107–14.

Kokhanovskii, N. (1968) Zemlya Sib. Dal 'Nevost 2: 50–1.

Koller, J., Kabai, P. and Demeter, A. (1988) Z. Jagdwiss. 34(2): 86–97.

Konecny, J.M. (1987) Oikos 50(1): 17–23.

Konecny, M.J. (1987) Oikos 50(1): 24–32.

Kopytov, A. and Kopylov, A. (1962) Okhota Okhotn. Khoz. 2: 28.

Kormilitsin, A.A. and Dulitskii, A.I. (1972) Vestnik Zoologii (1): 38–44.

Kormilitsina V.V. and Ivanovskii, B.N. (1970) Rezultat akkklimatizatsii i reakklimatizatsii mlekopitaushtchikh v Kr'mu (1913–1969). Material Konferentsii molod'kh uchen'kh. Kiev, 'Naukova du mka', 1970.

Korn, T., Saunders, G. and Leys, A. (1998) Aust. Vert. Pest Conf. 11: 69–73.

Korsakov, G.K. (1959) Tr. Vses. Inst. Zhivotn. Syr'ya Pushniny 18: 64–57.

__ (1963) In Resources of Game Animals in USSR and Counting them: 187–90.

Korschgen, L.J. (1957) Pitman-Robertson Prog. Cons. Comm., Missouri (15): 1–63.

Korsh, P.V. (1963) In Data from 1963; Summary Conf. Diseases with Natural Foci: 174–9. Tyumen.

Koryakov, B.F. (1962) Knigoizdat, Gorno-Altaisk: 138–40.

__ (1963) In Acclimatization of Animals in the USSR: 116–17. Akad. Nauk. Kazakhsk. SSR., Alma Ata.

Kotov, V.A. and Ryabov, L.S. (1963) Tr. Kav. Kaz.–skogo Gos. Zapovedn. 7: 1–239.

Kotov, V.P. (1939) J. Caucasian State Preserve (5) Maikop: 192–4.

Kovach, S.D. (1986) Feral Burro Report, 1986. Desert Bighorn Council Trans.: 23.

Kowalski, K. and Rzebik-Kowalska, B. (1991) Mammals of Algeria. Warsaw.

Kozlo, P and Pucek, Z. (1993) Species 20: 57–8.

Kozlo, P.G. (1968) Vestnik Zoologii 6: 53–8.

Kozlovskii, A.A. (1959) Soobshch Inst. Lesa Akad. Nauk. SSSR 13: 97–101.

Kramer, R.J. (1971) Hawaiian Land Mammals. Rutland, Vermont and Tokyo, Japan.

Krapivnyi, A.P. (1963) In Acclimatization of Animals in the USSR: 118. Akad. Nauk. Kazakhsk. SSR., Alma-Ata.

Krasinski, Z. (1963) Poland. Chronmy Przyrode Ojczysta 19(5): 3–8.

Kratochvil, J. and Kratochvil, Z. (1976) Zool. Listy 25(3): 193–208.

Krausman, P.R. and Leopold, B.D. (1986) Trans. 51st. Nth. Am. Wildl. Nat. Resour. Conf. 51: 52–61.

Krecek, R.C., Louw, J.P. and Sneddon, J.C. (1995) J. Helminthol. Soc. Wash. 62(1): 84–6.

Krumbiegel, I. (1955) Saugetierkundl. Mitt. 3(1): 12.

__ (1955) Saugetierkundliche Mitt. 13: 12–18.

Kruuk, H. (1979) Ecology and Control of Feral Dogs in the Galâpagos. Unpublished Report, Institute of Terrestrial Ecology, Banchory, Scotland.

Krzywinski, A., Twardowski, L. and Skotnicki, J. (1987) Trans. Congr. Int. Game Biol. 18: 97–8.

Kubatov, V.P. (1939) Works Uzebskiy Zool. Gdns. 1: 420–5.

Kucheruk, V.V., Kartushin, P.A. and Shilov, I.A. (1959) In Problems of Epidem. and Tularemia Prophylaxis: 159–68. Medgiz.

Kuhbier, H., Alcover, J.A. and D'Arellano Tur, C.G. (1984) Monographiae Biologicae 52: 1–704.

Kuhn, L.W. and Peloquin, E.P. (1974) Proc. 6th. Vert. Pest Conf., Anaheim, California: 101–5.

Kuhn, W. (1935) Z. Saugetierkunde 10(3): 144.

Kumerloeve, H. (1982) Saugetierkundlitche Mitt. 30: 26–30.

Kunkel, G. (1968) Pacific Discovery 21(1): 1–8.

Kurbatov, V.P. (1956) Akklim. pushn. zverey v Kirgizii: 83–106.

Kuroda, N. (1955) J. Mamm. Soc., Japan 1(2): 13–18.

Kurten, B. (1968) Pleistocene Mammals of Europe. London. 317 pp.

Kurz, J.C. and Marchinton, R.L. (1972) J. Wildl. Mgmt. 36: 1240–8.

Kuster, A. (1958) Schweizer Naturschutz 24(3): 63.

__ (1961) Schweizer Naturschutz 27(1): 6.

__ (1966) In Schenk, P., Jagd und Naturschutz in der Schweiz. Basel.

Kuzyakin, A.P. and Panteleev, P.A. (1961) In Materialy Plan.–metod. soveshchaniya po zashchiterastitel'noi zany Urala i Sibiri, 1960: 34–7. Novosibirsk.

Kverno, N.B. (1964) in Proc. 2nd. Vert. Pest Contr. Conf., Anaheim, California: 81.

Kwapena, N. (1975) The Habitat of Deer (Rusa timoriensis) in the Tonda Wildlife Management Area, Western District, Papua New Guinea. In Deer Management Conference, November, 1974: 56–64. Ministry Conservation Fisheries and Wildlife Division, Melbourne, Victoria.

Kydyrbaev, S. (1964) In Acclimatization of Animals in the USSR: 119. Akad. Nauk. Kazakhsk. SSR., Alma-Ata.

La Grange, J.J. (1954) Polar Record 7(48): 155–8.

Laar, V. van (1974) Lutra 16: 34–9.

__ (1981) Report 10. Terrestrial and freshwater fauna of the Wadden Sea area. Final Report 'Terrestrial Fauna', Wadden Sea Working Group. Leiden.

Labat, J.B. (1724) Nouveau voyage aux isles de l'Amerique. 2 vols. The Hague.

Lagaude, V. (1961) Phytoma 132: 30–3.

Lagenbach, J.R. and Beule, J.D. (1942) Pa. Game News 13(8): 14–15, 30.

Laidler, K. (1980) Squirrels in Britain. Newton Abbot.

Laing, R.M. (1909) In Chilton (ed), Subantarctic Islands of N.Z., vol. 2, Article 21: 482–92.

Laird, M. (1963) Proc. 10th. Pacif. Sci. Congr.: 535–42. Honolulu, Hawaii.

Lamb, R.C. (1964) Birds, Beasts and Fishes: The First

Hundred Years of the North Canterbury Acclimatisation Society. *For the N. Canterbury Acclimatisation Society, Christchurch, New Zealand.*

Lamine-Cheniti, T. (1988) Bulletin d'Ecology 19: 403–6.

Lamoureux, C.H. and Stemmerman, L. (1976) *Report of the Kipahula Bicentennial Expedition. CPSU/UH Technical Report (11): 1–50. Botany Department, University Hawaii, Honolulu.*

Lancum, F.H. (1961) Wild Mammals and the Land. *Bulletin (150) 5th. Impress. London.*

Lane, S.G. (1975) Aust. Bird Bander *13: 34–7.*

__ (1975) Aust. Bird Bander *13: 80–2.*

__ (1976) Aust. Bird Bander *14: 10–13.*

__ (1976) Aust. Bird Bander *14: 14–15.*

__ (1976) Aust. Bird Bander *14: 24–6.*

__ (1982) Corella *6(3): 67–8.*

Lanfranco, G.G. (1969) Maltese Mammals *(Central Mediterranean). Malta.*

Lange, K.I. (1960) Am. Midl. Nat. *64: 436–58.*

Langguth, A. (1975) In M.W. Fox (ed), The Wild Canids, *Chapt. 13: 192–206. New York.*

Langkavel, (1894) Deutsche Jagerz. *22: 525*

Lapin, I. (1963) Okhota Okhotn. Khoz. *11: 27–8.*

Larsen, S. (1960) Oikos *11: 276–305.*

Lasocki, W.A. (1963) Animals *1(6): 12–14.*

Lataste, F. (1892) Actes de la Societe scientifique de Chili 11. Notes et Memoirs*: 210–22.*

Latham, R.M. (1951) Pa. Game News *21(11): 4–8 and 21(12): 35–40.*

__ (1954) Pa. Game News *25(9): 20–5.*

__ (1955)Trans. 20th N. Am. Wildl. Conf.*: 406–11.*

Launay, H. (1980) Bull. Mensuel, Office National de la Chasse, Numero Special Scientifique et Technique*: 213–41.*

Laundre, J. (1977) Anim. Behav. *25(4): 990–8.*

Laurent, P. (1963) J. Mamm. *44(4): 574.*

Lauret, M. (1982) Elepaio *43(4): 25–7.*

Laurie, E.M.O. (1946) J. Anim. Ecol. *15: 22–34.*

__ and Hill, J.E. (1954) List of Land Mammals of New Guinea, Celebes and adjacent islands (1758–1952). *London.*

Lavery, H. (1974) Fauna of Queensland. *Qu. Yearbook.: 4–11. Brisbane.*

Lavrov, L. (1965) Okhota Okhotn. Khoz. *(9): 13–16.*

Lavrov, M.A. (1962) In Prob. zool. issledov. v Sibiri*: 149–50. Knigoizdat, Gorno-Altaisk.*

Lavrov, N.P. (1941) Trans. Cent. Lab. Game Indus. *(5): 155–71.*

__ (1958) Sb. Nauchno.–Teknol. Info. Vses Nauchno.–Issled. Inst. Zhivotn. Syr'ya Pushniny *3: 47–61.*

__ (1969) Priroda *8: 64–8.*

__ and Pokrovsky, V.S. (1967) In Proc. and Papers 10th. Tech. Meet. IUCN, *1966: 181–193.IUCN Publ. New Ser.(9) ,Morges, Switzerland.*

Lavsund, S. (1975) Fauna och Flora (Stockh.) *70(5): 216–31.*

Lawson, T. (1998) BBC Wildlife *16(9): 30.*

Lay, D.W. (1944) Texas Game Fish *2(10): 4–5, 13.*

Laycock, G. (1966) The Alien Animals: The Story of Imported Wildlife. *New York.*

__ (1982) Audubon Mag. *84(5): 16–23.*

Lazell, J.D. (1980) Tigerpaper *(UN–FAO Regional Office for Asia and Far East) 7: 31–2.*

__ (1984) J. Mamm. *65: 26–33.*

__, Sutterfield, T.W. and Giezentanner, W.D. (1984) Biol. Conserv. *30(2): 99–108.*

Lazo, A. (1992) PhD Thesis, Universidad de Sevilla.

__ (1994) Anim. Behav. *48: 1133–41.*

Le Hénoff, D. (1985) Canad. Field-Nat. *99: 103–4.*

__ and Crete, M. (1989) Can. J. Zool. *67(5): 1102–5.*

Le Souef, D. (1891) Vict. Nat. *8: 121–31.*

__ (1912) Handbook of Western Australia, *1912: 249–52.*

Leader-Williams, N. (1979) J. Zool., Lond. *188: 501–15.*

__ (1980) J. Wildl. Mgmt. *44 (3): 640–57.*

__ (1988) Reindeer on South Georgia: The Ecology of an Introduced Population. *Cambridge.*

__, Smith, R.I.L. and Rothery, P. (1987) J. Appl. Ecol. *24: 801–22.*

Leberg, P.A. (1990) Trans. 55th N. Am. Wildl. & Nat. Res. Conf.*: 609–19.*

Lee, A. .K. and Martin, R. W. (1987) The Koala. *Kensington, New South Wales.*

__ and __ (1988) In L. Nielson and R.D. Brown (eds), Translocation of Wild Animals,*: 152–90. Wisc. Humane Soc. and C. Kleberg Wildl. Res. Inst.*

Leege, T.A. (1968) J. Wildl. Mgmt. *32(4): 973–6.*

Lefebvre, L.W., Ingram, C.R. and Yang, M.C. (1978) Proc. Am. Soc. Sugar Cane Technol. *7: 75–80.*

Legge, C.M. (1936) J. Manchester Geogr. Soc. *46: 21–48.*

Leiderman, M. (1955) Papers and Proceedings Roy. Soc. Tas. *89: 125–30.*

Lekagul, B. and McNeely, J. A. Mammals of Thailand. *Bangkok.*

Lemke, C.W. and Oshesky, L. (1955) Wis. Cons. Bull. *20(11): 9–11, 40.*

Lensink, G.J. (1960) J. Mamm. *41(2): 172–82.*

Lentfer, J.W. (1955) J. Wildl. Mgmt. *19(4): 417–29.*

Lesel, R. (1967) Terres Australes et Antarctiques Francaises *(38): 3–40.*

__ (1968) Mammalia *32(4): 612–20.*

__ (1969) Mammalia *33(2): 343–5.*

__ (1971) Terres Australes et Antarctiques Francaises *(55–56): 55–63.*

__ and Derenne, P. (1975) Polar Record *17(110): 485–93.*

Lesley, L.B. (1929) Southwestern Hist. Quart. *23: 18–33.*

__ (1929) Uncle Sam's Camels. *Cambridge, Massachusetts.*

Leslie, D.M. and Douglas, C.L. (1979) Wildl. Monogr. *(66): 1–56.*

__ and __ (1986) Trans. 51st N.A. Wildl. & Nat. Res. Conf.*: 62–73.*

Leslie, W.W. (1966) The Covered Wagon, *Fresno Historical Society, Fresno, California, 1966: 49–51.*

Letts, G.A. (1962) Aust. Vet. J. *38: 282–7.*

__ (1964) Aust. Vet. J. *40(3): 84–8.*

Leuthold, B.M. (1979) Afr. J. Ecol. *17(1): 19–34.*

Lever, C. (1977) The Naturalized Animals of the British Isles. *London.*

__ *(1985)* Naturalized Mammals of the World. *London.*

Levi, H.W. *(1952)* Sci. Monthly *74(6): 315–22.*

Lewis, J.C. *(1966)* J. Wildl. Mgmt. *30(4): 832–5.*

Lewis, T. *(1907)* Spanish Explorers in the Southern United States 1528–1543. *New York.*

Liberg, 0. *(1984)* J. Mamm. *65(3): 424–32.*

__ *(1980)* Oikos *35: 336–49.*

Lidicker, W.Z. *(1973)* Ecol. Monogr. *43: 271–302.*

__ *(1991) In Groves, R.H. and Di Castri, F. (eds),* Biogeography of Mediterranean Invasions *18: 263– 271. Cambridge and Sydney.*

__ *and Ziegler, A.C. (1968)* Univ. California Publ. in Zool. *87: 1–60.*

Likyavichene, N. *(1962)* Tartu: *277–80.*

Lincke, M. *(1943)* Das Wildkaninchen. *Neudamm.*

Lindemann, W. *(1956)* J. Wildl. Mgmt. *20(1): 68–70.*

Lindgren, E. *(1972)* Research Report on the Populations of the Java Rusa on the Bansbach Plains, Western District, Papua. *Department of Agriculture Stock and Fisheries, Port Morseby, New Guinea.*

__ *(1975) In Papers Presented* Deer Mgmt. Conf., 1974: *37–51. Fisheries and Wildlife Division, Melbourne.*

Ling, H. *(1961)* Eesti Loodus *5: 281–6.*

Linn, I. and Chanin, P. *(1978)* New Scientist *77: 560–2.*

Linn, I.J. and Birks, J.D.S. *(1981)* Proc. Worldw. Furbearers Conf., Frostburg, Maryland *2: 1088–102.*

Linscombe, G., Kinler, N. and Wright, V. *(1981)* Proc. Worldw. Furbearers Conf., 1980, Frostburg, Maryland *1: 129–41.*

Lister-Kaye, J. *(1972)* The White Island. *London.*

Litjens, B.E.J. *(1980)* Lutra *23: 43–53.*

__ *(1980)* De Pelsdieren fokker *12: 359–61.*

__ *(1981) E.P.P.O. Working Party on the Musk-Rat: Report from the Netherlands by the Wildlife Division of the Ministry of Agriculture. Wildlife Division, Ministry of Agriculture, Netherlands.*

Little, A., McKenZie, A.J., Morris, R.J.H., Roberts, J. and Evans, J.V. *(1970)* Aust. J. Exp. Biol. Med. Sci. *48: 17–24.*

Litvaitis, J.A. and Shaw, J.H. *(1980)* J. Wildl. Mgmt. *44 (1): 62–8.*

Lloyd, B. *(1947)* Trans. Herts. Nat. Hist. Soc. *22: 34.*

Lloyd, D.S., Smith, R.B. and Sundberg, K.A. *(1987)* Murrelet *68(2): 57–8.*

Lloyd, H. *(1925)* Canad. Field-Nat. *39(7): 151–2.*

Lloyd, H.G. *(1962)* J. Anim. Ecol. *31: 157–66.*

__ *(1963)* J. Anim. Ecol. *32: 549–63.*

__ *(1970)* Symposium Zool. Soc., London *26: 165–88.*

__ *(1975) In M.W. Fox (ed),* The Wild Canids, *Chapt. 14: 207–15. New York.*

__ *(1983)* Mammal Review *13: 69–80.*

Lockhart, J.G. *(1930)* Blendon Hall. *London.*

Lockie, J.D. *(1956)* Scottish Agriculture *36: 44–5.*

__ *and Stephen, D. (1959)* J. Anim. Ecol. *28: 43–50.*

Lockley, R.M. *(1952)* Ibis *94: 144–57.*

__ *(1953)* Puffins. *London.*

__ *(1964)* The Private Life of the Rabbit. *London.*

Long, C.A. *(1965)* Kansas Publ., Mus. Nat. Hist. *14: 493–758.*

Long, J.L. *(1981)* Introduced Birds of the World. *Sydney.*

__ *(1988)* Introduced Birds and Mammals in Western Australia. *Technical Series No. 1. 2nd edn. 1988. South Perth, Western Australia.*

__ *and Mawson, P. (1991) In Groves, R.H. and Di Castri, F. (eds),* Biogeography of Mediterranean Invasions.*: 365–75. Melbourne and Sydney.*

__, __, *Hubach, P. and Kok, N. (1988)* J. Agric. West. Aust. *29(3): 104–6.*

Lonnberg, E. *(1936)* Svenska JagForb. Tidskr. *74: 225–31.*

__ *(1937)* Svenska JagForb. Tidskr. *75: 42–6.*

Lons, H. *(1907)* Jahresber Nat. Hist. Ges. Hannover: *28.*

Loudon, A. and Fletcher, J. *(1983)* New Scientist *99 (1366): 88–92.*

Loudon, A.S.I. *(1978)* Forestry (Oxf.) *51(1): 73–84.*

Loughlin, T.R. *(1980)* J. Wildl. Mgmt. *44(3): 576–82.*

Lowe, V.P.W. *(1969)* J. Anim. Ecol. *38: 425–57.*

__ *and Gardiner, A.S. (1989)* J. Zool. (Lond.) *218(1): 57–8.*

Lowery, G.H. *(1974)* The Mammals of Louisiana and its Adjacent Waters. *Baton Rouge, Louisiana.*

Lowrie, W.J. *(1904)* Planters' Monthly *23: 354–5.*

Lowry, D.A. and McArthur, K.L. *(1978)* Wildl. Soc. Bull. *6(1): 38–9.*

Lozan, M.N. *(1965) In* Problems of Ecologic and Economic Importation of Moldavian Birds and Mammals*: 10–24. Kartya Moldovenyaske, Kishinev.*

Lucas, A. *(1968)* Cruising the Coral Coast. *Sydney.*

Lucas, E.G. *(1977) In* Research and Management of Wild Hog Populations. *Belle W. Baruch For. Sci. Inst., Clemson University, Georgetown, South Carolina: 17–21.*

Lueth, F.X. *(1949)* Ala. Cons. *20(12): 4, 22.*

Lund, M. *(1981)* An. Rep. Danish Pest Infestation Laboratory, *Lingby, Denmark 1980: 97.*

__ *(1982) Control of rabbits on the island of Lolland.* An. Rep. Danish Pest Infestation Laboratory, *Lyngby, Denmark.*

Lundh, N.G. *(1984) In Klein, D.R., White, R.G. and Keller, S. (eds).* Proc. Int. Muskox Symp. 1, *Biol. Pap. Univ. Alaska Spec. Rep. 4: 7–8.*

Lundie-Jenkins, G.W. *(1989) The Ecology and Management of the Rufous Hare-wallaby,* Lagorchestes hirsutus *in the Tanami Desert. Report to the Conservation Commission of the Northern Territory.*

Luttringer, L.A. *(1945)* Pa. Game News *15(12): 12–13, 26.*

Lyarski, P.A. *(1961)* Izvest. Akad. Nauk. Belorussk. SSR Ser. Biol. Nauki *4: 123–9.*

Lyneborg, L. *(1971)* Mammals in Colour. *London.*

Ma, Y., Wang, F., Jin, S., Li, S., Lin, Y. and Yie, Z. *(1981)* Acta Theriol. Sinica *1(2): 177–88.*

MacDonald, D. *(1984)* Encyclopaedia of Mammals. *London.*

MacDonald, D.A. *(1975)* Sand Lapper: *58–62.*

MacDonald, I.C. *(1984)* Lagomorph Newsletter *4: 37–8.*

Macdonald, I.A.W. and Frame, G.W. *(1988)* Biol. Cons. *44: 67–93.*

__, *Graber, D.M., DeBenedetti, S., Groves, R.H. and Fuentes, E.R. (1988)* Biol. Conserv. *44: 37–66.*

MacFarlane, J.R.H. *(1887)* Ibis *201.*

MacIntosh, N.W.G. *(1975) In M.W.Fox (ed),* The Wild Canids, *Chapt. 7: 87–106. New York.*

MacKenna, B.V. (1883) Juan Fernandez. Historia verdadera de la isla de Robinson Crusoe.

MacKenzie, A.F. (1953) Proc. Zool. Soc. Lond. *122: 541.*

Mackin, R. (1970) Acta Theriologica 15: 447–58.

MacLeod, C.F. and Lethieco, J.L. (1963) J. Mamm. *44(2): 277–8.*

Macnaghten, R. (1963) Animals 2(18): 502.

MacNamara, L.G. (1955) New Jersey Outdoors 5(12): 4–9.

MacPherson, A.H. (1964) J. Mamm. *45(1): 138.*

MacRoberts, B.R. and MacRoberts, M.H. (1971) Nat. Hist. *80(7): 38–47.*

Mahoney, J.A. and Richardson, B.J. (1988) Muridae. Zoological Catalogue of Australia. *Vol. 5. Mammalia. In Walton, D.W, (ed). Canberra.*

Mahony, D.J. (1937) Proc. R. Soc. Vict. 49: 331–2.

Mahood, I. (1978) In Parks and Wildlife 2(2): 50–1.

Malzy, P. (1966) Mus. Nat. Hist. Natur. Ser. A. Zool. *41: 23–7.*

Manville, R.H. (1957) J. Mamm. *38(3): 422.*

Marcstrom, V. (1964) Viltrevy 2(6): 329–84.

Marcuzzi, G. (1990) In F. di Castri, A.J. Hansen and M. Debussche (eds) Biological Invasions in Europe and the Mediterranean Basin.*: 217–27. Dordrecht, Netherlands.*

Mares, M.A., Streilein, K.E. and Willig, M.R. (1981) J. Mamm. *62(2): 315–28.*

Marlow, B.J. (1962) Aust. Nat. Hist. *14(6): 61–3.*

Marples, R.R. (1955) Pacif. Sci. *9: 69–76.*

Marques, A. (1905) Thrum's Hawaiian Annual for 1906*: 4 8–55.*

Marsh, L. and Pope, E. (1967) Aust. Nat. Hist. *December 1967.*

Marsh, R.E. (1965) In Congr. de la Protect. des Cultures Tropical, Marseilles*: 633–7.*

__ and Salmon, T.P. (1981) Proc. World lagomorph Conf., *1979: 842–57. University Guelph, Ontario.*

Marshall, F. (1970) Deer 2: 514–16.

Marshall, J.T. (1955) J. Mamm. *36(2): 259–63.*

__ (1962) In Storer, I (ed), Pacific island rat ecology. *B.P. Bishop Mus. Bull. 225: 241–6.*

Marshall, L.G. (1977) Mammalian Species 81: 1–3.

Marshall, M. (1975) Atoll Res. Bull. 189: 1–155.

Marshall, P. (1963) Wild Mammals of Hong Kong. *London.*

Marshall, W.H. (1961) N.Z. J. Sci. *4(4): 822–4.*

__ (1963) N.Z. Department of Science and Industrial Research, Information Service (38): 1–32.

Martensz, P.N. (1971) CSIRO. Wildl. Res. 16: 73–5.

Martin, G.R., Twigg, L.E. and Robinson, D.J. (1996) Wildl. Res. 23: 475–84.

Martin, J.T. (1973) NZ. J. Agric. *125(6): 18–22.*

__ (1975) J. Mamm. *56(4): 914–15.*

Martin, M. (1716) A Description of the Western Islands of Scotland. *Edinburgh (Facsimile reprint James Thin, 1970)*

Martin, W. and Sobey, W. (1983) Corella 7: 40.

Martinez, N. (1915) Impresiones de un Viaje. *Talleres de Policia Nacional, Quito, Ecuador.*

Martinez, R., Pablo, J.V. and Sanuy, D. (1976) Inst. Munic. Cienc. Nat. Misc. Zool. *3(5): 243–50.*

Mas-Coma, S. and Feliu, C. (1977) Vie et Milieu *27: 231–41.*

Masini, F. and Lovari, S. (1988) Quarterny Res. *30: 339–49.*

Mason, I.L. (1979) Inventory of Special Herds. *FAO/UNEP project no. FP/1108–76–02 (833). Conservation of animal genetic resources. FAO, Rome: 1–114.*

__. (1980) Prolific Tropical Sheep. *FAO, Rome.*

Massa, B. (1973) Atti Dell'Accademia Gioenia di Scienze Naturali in Catania, Series 7, 5: 63–95.

Masters, K.B. (1979) Feral Pigs in the South-West of Western Australia. *Final Report to Feral Pig Committee. W.A. Department of Agriculture.*

Mate, B.R. (1972) Proc. Ann. Conf. Biol. Sonar Diving Mammals 8: 47–53.

Matheson, C. (1954) West Wales Field Soc., 16th. An. Rep.

__ (1963) Animals 1(24): 21–5.

Matschke, G.H. (1963) Proc. a. Conf. SE. Assoc. Fish Game Comm. *16: 21–34.*

__ (1965) Proc. a. Conf. SE. Assoc. Fish Game Comm. *18: 35–9.*

__ (1966) Proc. a. Conf. SE. Assoc. Game and Fish Comm. *19: 154–6.*

__ and Hardister, J.P. (1967) Proc. a. Conf. SE. Assoc. Game Fish. Comm. *20: 74–84.*

Maude, H. and Maude, H.E. (1952) J. Polynesian Soc. *61: 62–89.*

Mauget, R. (1972) Ann. Biol. Anim. Biochim. Biophys. *12: 195–202.*

__ (1979) Biol. Behav. *4 (1): 25–42.*

Mawson, D. (1934) Geogr. J. *83(1): 1–12.*

__ (1943) Macquarie Island, its Geography and Geology. *Sci. Rep. of Aust. Antarc. Exp., 1911–14, Ser. A. vol. 5.*

Maximov, A.A. and Netsky, G.I. (1966) The Muskrat of Western Siberia. *Acad. Sci. USSR. Izdat. 'Nauka', Novosibirsk.*

Maynes, G.M. (1977) Aust. J. Ecol. *2(2): 207–14.*

__ (1977) Aust. Wildl. Res. *4: 109–25.*

__ (1989) In G. Grigg, P. Jarman and I. Hume (eds) Kangaroos, Wallabies and Rat-Kangaroos.*: 47–66. New South Wales, Australia.*

Mayol, J. (1974) Vida silvestre 8: 207–13.

__ (1978) Naturalia Hispanica 20: 1–34.

McAnich, J.B. (1976) Ms. Thesis, Ohio State University, Ohio.

McCabe, T.R. and Wolfe, M.L. (1981) In Proc. Worldw. Furbearers Conf., 1980, Frostburg, Maryland 2: 1377–91.

McCarley, H. and Carley, C.J. (1979) In US Fish. & Wildl. Serv., Albuquerque, Endang. Spec. Rep. No. 4: 1–38.

McCarthy, P.H. (1980) Aust. Vet. J. 56: 547–51.

McCarthy, T.J. (1992) Mammalia 56: 302–6.

McCool, C.J., Pollitt, C.C., Fallon, G.R. and Turner, A.F. (1981) Aust. Vet. J. 57(10): 444–9.

McCormick, R. (1875) In The N.Z. Pilot. *4th. edn.: 314. Hydrographic Office, Admiralty, London.*

__ (1884) Voyages of Discovery in the Arctic and Antarctic Seas, and Around the World. *Vol. 1. London.*

McDowell, R.D. (1955) J. Wildl. Mgmt. *19(1): 61–5.*

__ and Pillsbury, H.W. (1959) J. Wildl. Mgmt. 23(2): 240–1.

McGill, L. (1972) Afr. Wildl. 26: 66–8.

McGill, T.J. (1984) In Proc. 11th Vert. Pest Conf., Sacramento, California, 1984: 168–73.

McGuire, R.J. and Brown, L.N. (1973) Am. Midl. Nat. *89(2): 498.*

McInnis, M.L. (1985) Diss. Abstr. Int., B. Sci. Eng. *45(12): 3685–8.*

McKee, J. (1988) Scottish Wildlife *4: 9.*

McKenzie, N.L., Burbidge, A.A. and Baynes, A. (1999) *Australian mammal map updates. unpublished. CALM, Wanneroo, W. Australia.*

__, __ and Chapman, A. (1978) Wildl. Res. Bull. W. Aust. *No. 7.*

McKinley, D. (1959) J. Mamm. *40(2): 248–9.*

McKinsey, K. Explorations: Going wild. *Scientific American. http://www.sciam.com. /explorations/1998/042098tamarin/index.html.*

McKnight, T. (1977) Ecos *13: 10–18.*

McKnight, T.L. (1958) J. Wildl. Mgmt. *22(2): 163–79.*

__ (1958) Geogr. Rev. *49: 506–25.*

__ (1969) The Camel in Australia. *Carlton, Victoria.*

__ (1971) Annals Assoc. Am. Geogr. *64(4): 759–73.*

__ (1976) Univ. Calif. Publ. Geogr. *21: 1–104.*

McLaren, F.B. (1948) The Auckland Islands. *Sydney.*

McLaren, I.A. (1972) Canadian Geographical Journal *85: 108–14.*

McLennon, J.A., Potter, M.A., Robertson, H.A., Wake, G.C., Colbourne, R., Dew, L., Joyce, L. McMann, A.J., Miles, J., Miller, P.J. and Reid, J. (1996) New Zealand Journal of Ecology. *23: 279–86.*

McManus, T.J. (1979) Corella *3(3): 52–4.*

McMurray, F.B. and Sperry, C.C. (1941) J. Mamm. *22: 185–90.*

McNab, R. (1913) The Old Whaling Days. A History of Southern New Zealand from 1830 to 1840. *Christchurch.*

McNally, J. (1960) Aust. Mus. Mag. *13(6): 178–81.*

McNamee, T. (1998) The Return of the Wolf to Yellowstone. *Henry Holt & Co.*

McNaught, D.A. (1987) Wildl. Soc. Bull. *15: 518–21.*

McNeely, J.A. (1981) Proc. World lagomorph Conf., 1979: *929–32. University Guelph, Ontario.*

McNeil, R.J. (1964) Proc. N.Z. Ecol. Soc. *11: 44–8.*

Mead, C. (1999) BBC Wildlife *17(2): 33.*

Mead-Briggs, A.R. (1967) Entemologists Monthly Magazine *103: 115–19.*

Meagher, M. (1989) J. Mamm. *70(3): 670–5.*

Meaney, C.A., Bissell, S.J. and Slater, J.S. (1987) Southwestern Nat. *32: 507–8.*

Mech, D. (1992) Species *19: 63.*

Mech, L.D., Fritts, S.H. and Paul, W.J. (1988) Wildl. Soc. Bull. *16: 269–72.*

Medway, L. (1970) In J.R. Napier and P.H. Napier (eds) Old World Monkeys: *513–53. London.*

__ (1978) The Wild Mammals of Malaya and Singapore. *London.*

__ and Yong, H.S. (1976) Malaysian J. Sci. *4(A): 43–53.*

Meester, J. and Setzer, H.W. (1977) The Mammals of Africa: An Identification Manual. *Washington, D.C.*

Meklenburtsev, R. (1963) Okhot. Okhotn. Khoz.*11: 25–6.*

Mel'chinov, M.S. (1958) Tr. Inst. Biol. Yakutsk Fil. Sib. Otd. Akad. Nauk. SSSR. *4: 172–86.*

__ (1962) Nauchn. Soobshch. Yakutsk Fil. Sibirsk Otd. Akad. Nauk. SSSR. *8: 93–6.*

Mel'nikov, V.K. (1962) *In* Problems of Zoological Investigation in Siberia: *162–4. Knigoizat, Gorno-Altaisk.*

Melville, H. (1856) The Encantadas, In 'Four Short Novels'. *New York.*

Menkhorst, K. and Mansergh, I. (1977) Report on the Mammalian Fauna of South Gippsland Study Area. National Museum of Victoria, Melbourne.*

Menkhorst, P.W., Kerry, K.R. and Hall, E.F. (1988) Corella *12(3): 72–7.*

Menzies, J.I. and Dennis, E. (1979) Handbook of New Guinea Rodents. Wau Ecol. Inst. Handbook No. 6.*

Mercer, J. (1974) Hebridean Islands, Colonsay, Gigha, Jura. *Glasgow.*

Mercer, W.E., Hearn, B.J. and Finlay, C. (1981) *In* Proc. World Lagomorph Conf., 1979: 450–68. *University Guelph, Ontario.*

Merriam, H.R. (1965) Proc. 45th. a. Conf. W. Assoc. State Game Fish. Comm. *125–8. Anchorage, Alaska.*

Merton, D. (1998) Eclectus *4: 4–7.*

Meylan, A. (1966) Bull. Soc. Vaud. Sc. nat. Lausanne *69: 233–45.*

Michaux, J., Cheylan, G. and Croset, H. (1990) *In* F. di Castri, A.J. Hansen and M. Debussche (eds),Biological Invasions in Europe and the Mediterranean Basin: *263–84. Dordrecht, Nederlands.*

Middleton, A.D. (1930) Proc. Zool. Soc. London: *809–43.*

__ (1932) J. Anim. Ecol. *1: 166–7.*

__ (1937) Proc. zool. Soc. Lond. A. *107: 471.*

Miers, K.H. (1961) Proc. N.Z. Ecol. Soc. *9: 31–3.*

Migot, A. (1956) The Lonely South. *London.*

Milham, P. and Thompson, P. (1976) Mankind *10: 175–80.*

Millais, J.G. (1904–06) The Mammals of Great Britain and Ireland. *London.*

Miller, B. and Mullette, K.J. (1985) Biol. Conserv. *34: 55–95.*

Miller, C. (1987) Trans. Congr. Int. Game Biol. *18: 123.*

Miller, G.S. (1912) Catalogue of the Mammals of Western Europe. *London.*

__ (1918) Proc. U.S. Nat. Mus. *54: 507–11.*

Miller, J.E. (1974) Proc. 6th. Vert. Pest Contr. Conf., *Anaheim, California: 85–9.*

Miller, R. (1983) J. Rangeland Mgmt. *36(2): 195–9.*

__ (1983) J. Rangeland Mgmt. *36(2): 199–201.*

Miller, S. (1973) Trans. 38th. N. Am. Wildl. Nat. Resour. Conf., *Washington, 1973: 55–68.*

Miller, S.D. and Ballard, W.B. (1982) J. Wildl. Mgmt. *46(4): 869–76.*

Miller, T. (1952) Texas Game Fish. *10(10): 8–11.*

Mills, J.A. and Mark, A.F. (1977) J. Anim. Ecol. *46(3): 939–58.*

Mills, S. (1982) Oryx *16: 411–14.*

Milne, L.J. and Milne, M. (1962) The Balance of Nature. *London.*

Milner, C., Goodier, R. and Crook, I.G. (1968) Nat. in Wales *11(1): 2–11.*

Minamino, M. (1950) Nippon Ryoyu (8): 4, (10): 3.*

Ministry of Agriculture, Fisheries and Food (1972) Wild Mink at Large. *Edinburgh.*

__ *(1973)* Pest Infestation Control, 1968–70. *London.*

__ *(1975)* Pest Infestation Control Laboratory Report 1971–73. *London.*

Mishru, H. R. and Dinerstien, E. *(1987)* Smithsonian *18(6):* 66–73.

Mitchell, F.J. and Behrndt, A.C. *(1947)* Rec. S. Aust. Mus. *9:* 169–79.

Mitchell, J. L. *(1961)* J. Wildl. Mgmt. *25:* 48–53.

Mitchell, J., Greer, K. and Weckwerth, R. *(1971) In Mussehl, T.W. and Howell,F.W. (eds) Game Management in Montana. Chapt. 25: 197–205. Game Management Division, Helena, Montana.*

__, Merrell, P. and Allen, L. *(1982)* Vertebrate Pests of Queensland: Results of a Survey on Feral Animals of Queensland, 1981/1982. *Stock Routes and Rural Lands Protection Board, Department Lands, Queensland.*

__, Staines, B.W. and Welch, D. *(1977)* Ecology of Red Deer: A Research Review Relevant to their Management in Scotland. *National Env. Res. Counc., Inst. Terrestrial Ecol., Cambridge.*

Mitchell, P.J., Bilney, R. and Martin, R.W. *(1988)* Aust. Wildl. Res. *15: 511–14.*

Mitchell, T.D. *(1986) In P.J. Joss, P.W. Lynch and O.B. Williams (eds) Rangelands: A Resource under Siege.:* 200–201.

Mittermeier, R.A., Bailey, R.C., Sponsal, L.E. and Wolf, K.E. *(1977)* Oryx *13(5):* 449–453.

__, Konstant, W.R., Nicholl, M.E. and Langrand, O. (eds) *(1992)* Lemurs of Madagascar: An Action Plan for their Conservation 1993–1999. *IUCN, Gland*

Miura, S. *(1976)* J. Mammal Soc. Japan *6(5/6: 231–7.*

Mobes, W.K.G. *(1946)* Bibliographie des Kaninchens. *Adadmischer Verlag Halle, Halle (Saale), Frankfurt (Main).*

Modinger, B.A. and Duffy, D.C. *(1987)* Cormorant *15: 3–6.*

Moffat, C.B. *(1938)* The Mammals of Ireland. *Proc. Roy. Irish Academy 46B: 61–128.*

Mohr, E. *(1929)* Schriften de Naturwissenschaftlichen Vereins fur Schleswig-Holstein *19: 59–72.*

__ *(1931)* Die Saugetiere Schleswig-Holstein. *Altona*

__ *(1933)* J. Mamm. *14: 58–6 3.*

__ *(1949)* J. Soc. Preserv. *Fauna Empire New Ser. (59):* 29–33.

Monagu, I. *(1965)* Proc. Zool. Soc., London *144: 425–8.*

Monnat, J.Y. *(1982)* Penn ar Bed. *13: 134–43.*

Monson, G. and Sumner, L. *(1980)* The Desert Bighorn. Its Life History, Ecology and Management. *Tucson, Arizona.*

Montfort, N. and Montfort, A. *(1979)* Terre et Vie *33(1):* 27–48.

Mooney, H.A., Hamburg, S.P. and Drake, J.A. *(1986) In Mooney, H.A. and Drake, J.A. (eds), Ecology of Biological Invasions of North America and Hawaii: 250–72. New York.*

Moore, G.C. and Millar, J.S. *(1984)* J. Wildl. Mgmt. *48(3):* 691–9.

Moors, P.J. *(1983)* Ibis *125: 137–54.*

Morgan, R.P. and Chapman, J.A. *(1981)* Proc. World Lagomorph Conf., *1979: 64–72. University Guelph, Ontario.*

__, Willner, G.R. and Chapman, J.A. *(1981)* World. Furbearers Conf. Proc., 1980, Frostburg, Maryland *1: 30–7.*

Morgart, J.R., Smith, D.R. and Krausman, P.R. *(1987)* Trans. Congr. Int. Game Biol. *18: 126–7.*

Morrell, B. *(1832)* A Narrative of Four Voyages. *New York.*

Morris, A.K. *(1974)* Aust. Bird Bander *12: 62–4.*

Morris, D. *(1882)* The Mungoose on Sugar Estates in the West Indies. *Kingston, Jamaica.*

Morris, D. *(1965)* The Mammals: A Guide to the Living Species. *London.*

Morris, K., Alford, J. and Shepherd, R. *(1991)* Landscope *7:* 28–33.

Morris, P.A. *(1986)* Mammal Rev. *16: 49–52.*

Morrison, B., Muller-Using, B. and Cotera, M. *(1987)* Trans. Congr. Int. Game Biol. *18: 127.*

__, __, __ *(1992)* Trans. Congr. Int. Union Game Biol. *18(2): 153–4.*

__ *(1980)* Proc. Symp. Ecol. Mgmt., Barbary Sheep, *Nov. 19–21, 1979: 15–16. Texas Tech University, Lubbock, Texas.*

Morton Boyd, J., Doney, J.M., Gunn, R.G. and Jewell, P.A. *(1964)* Proc. Zool. Soc. Lond. *14 2(1): 129–64.*

Mosby, J.M. and Wodzicki, K. *(1973)* N.Z. J. Sci. *16:* 799–810.

Moseley, H.N. *(1892)* Notes by a Naturalist on H.M.S. 'Challenger'. *London.*

Mountfort, G. *(1962)* Portrait of a River. *London.*

Moureau de Saint-Mery, M.L.E. *(1797–98)* Description topographique, physique, civile, politique et historique de la partie francaise de l'isle Saint-Dominique. *2 vols. Philadelphia.*

__ *(1944) In* Description de la par te espanola de Santo Domingo, trans. C. Amando Rodriguez. Ciudad Trujillo.

Mourik, S. van., Stemasiak, T. and Attwood, V.R.L. *(1987)* Trans. Congr. Int. Game Biol. *18: 129.*

__, __, __ *(1987)* Trans. Congr. Int. Game Biol. *18: 127–8.*

Msangi, A.S. *(1975)* Univ. Sci. J. (Dar-es-Salaam) *1: 8–20.*

Mukhamedkulov, M.M. *(1963) In* Okhotn. – promysl. zhivotn. Uzbekistana*: 36–42. Akad. Nauk. Uzbek. SSR., Tashkent.*

Mukhtarov, P. *(1961)* Game and Game Keeping *7: 19.*

Mukhtarov, R.D. *(1963) In* Commercial Game Animals of Uzbek: 25–35. Akad. Nauk Uzbek SSR., Tashkent.

Mullan, J.M., Feldhammer, G.A. and Morton, D. *(1988)* J. Mammal. *69(2): 388–9.*

Muller, S. and Schlegel, H. *(1839)* Verh. Nat. Gesch. Ned. Overz. Bez. Zool. *1.*

Muller-Using, D. *(1938)* Z. Saugetierkunde *15: 335.*

__ *(1959)* Z. Jagdwissenschaft *5(3): 108.*

__ *(1954)* Z. Saugetierkunde *19: 166.*

Mulloy, F. *(1970)*J. Brit. Deer Soc. *2(2): 502–4.*

Munch, P.A. *(1945)* The Sociology of Tristan da Cunha. *Results Norw. Sci. Exp. to Tristan da Cunha, 1937–38 (13).*

Mungall, E.C. *(1978) in* Wildlife Ecology and Department Wildl. and F. Sci. , *Texas Agricultural Experimental Station, Texas A & M University. 184 pp.*

Munter, W.H. *(1915)* An. Rep. Coast Guard for 1915*: 130–40.*

Munton, P.N. *(1984)* Feral mammals – Problems and Potential. *Papers on Feral Mammals organised by Caprinae Specialist Group, Species Survival Commission, 3rd Internat. Therio. Conf., Helsinki, 1982. IUCN, Helsinki.*

Murie, O.J. *(1935)* U.S. Department of Agriculture (Washington), *Circular (362): 1–24.*

__ *(1941)* Proc. 5th. N. Am. Wildl. Conf *.: 432–6.*

__ and Scheffer, W.B. *(1959)* US Fish Wildl. Serv., North Am. Fauna *61: 1–406.*

Murphy, E.C. and Pickard, C.R. (1990) In King, C.M. (ed), Handbook of New Zealand Mammals.: 225–45. Auckland.

Murray, K.G., Winnett-Murray, K., Eppley, Z.A., Hunt, G.L. and Schwartz, D.B. (1983) Condor 85 (1): 12–21.

Murray, M.D. and Snowdon, W.A. (1976) Aust. Vet. J. 52: 547–54.

Murray, W.R. (1904) Extracts from journals of explorations by R.T. Maurice: Fowler's Bay to Rawlinson Ranges and Fowler's Bay to Cambridge Gulf. South Australian Parliamentary Paper 43. Adelaide.

Murua, R., Neumann, 0. and Dropelmann, I. (1981) Proc. Worldw. Furbearers Conf., 1980, Frostburg, Maryland 1: 544–58.

Musgrave, A. (1926) Aust. Zool. 4: 199–209.

Musgrave, T. (1866) Castaways on the Auckland Islands. London.

Mussehl, T.W. and Howell, F.W. (1971) Game Management in Montana. Montana Fish and Game Department, Helena, Montana.

Musser, G. and Carleton, M.D. (1993) Muridae. In Wilson, D.E. and Reeder, D.M. (eds), Mammal Species of the World: A Taxonomic and Geographic Reference. Washington DC.

Musser, G.G. (1970) J. Mamm. 51: 606–9.

__ (1972) J. Mamm. 53: 861–5.

__ (1973) A. Mus. Novitates 2525: 1–65.

__ (1977) Am. Mus. Novitates 2624: 1–15.

__ (1981) Bull. Am. Mus. Nat. Hist. 168: 225–334.

__ and Califia, D. (1982) Am. Mus. Novitates 2726: 1–30.

__ and Holden, M.E. (1991) Bull. Am. Mus. Nat. Hist. 206: 1–434.

__ and Newcomb, C. (1983) Bull. Am. Mus. Nat. Hist. 174: 327–598.

Myers, J.G. (1931) A Preliminary Report on an Investigation in to the Biological Control of West Indian Insect Pests. Empire Marketing Board, London (42): 1–178.

Myers, K. (1986) In Groves, R. and Burdon, J. (eds), Ecology of Biological Invasions: An Australian Perspective: 120–36. Canberra.

__ and Poole, W.E. (1963) J. Ecol. 51: 435–51.

Myrberget, S. (1967) Acta Theriol. 12(2): 17–26.

__ (1984) Fauna 37: 84.

__ (1987) Fauna 40: 160–2.

Nachtsheim, H. (1949) Vom wildtier zum Hanstier. Berlin and Hamburg.

Nahlik, A. (1989) Vadbiologia (3): 102–15.

Napier, J.R. (1979) Corella 3(3): 50–1.

__ and Singline, T.A. (1979) Aust. Bird Bander 8: 35–40.

Nasimovich, A.A. (1963) In Papers on Theoretical and Special Questions of Scientific and Technical Information. Geogr. Coll., Inst. Sci. Acad. Sci., Moscow, USSR: 105–11.

__ (1966) Zool. Zhur. 45(11): 1593–9.

Nass, R.D. (1963) Pacific Sci. 31(2): 135–42.

Nassimovitsch, A.A. (1961) Zool. J. 40: 957–70.

National Geographic Society (1960) Wild Animals of North America. Washington.

Natoli, E. and De Vito, E. (1991) Anim. Behav. 42: 227–41.

Naude, T.J. (1962) Farming in South Africa 35(6): 38–9.

Naumov, N.P. (1972) In Levine, N.D. (ed), The Ecology of Animals. Urbana, Illinois.

Navarrete, F.S. (1978) Proc. 8th. Vert. Pest Contr. Conf., Sacramento, Calif.: 118–19.

Naveda, B.H. (1950) Galâpagos a la Vista. El Commercio, Quito. Ecuador.

Naylor, K.S. (1988) MSc. Thesis, Univ. Idaho.

Nazarova, I.V. (1959) Soobshch. Inst. Lesa Akad. Nauk. SSSR. 13: 94–6.

Nebe, J. (1928) Qld. Nat. 6: 102–8.

Neill, W.T. (1952) Ecology 33(2): 282–4.

Nellis, D.W., Eichholz, N.F., Regan, T.W. and Feinstein, C. (1978) Wildl. Soc. Bull. 6(4): 249–50.

Nelson, F.P. (1955) South Carolina Wildl. 2(1): 2–3.

Nelson, L. and Hooper, J.K. (1953) California Big Game. Co-op. Extension Service, U.S. Department Agriculture, University California.

Nelson, L.S., Storr, R.F. and Robinson, A.C. (1990) Management Plan for Brush-tailed Bettong in South Australia. ANPWS, Canberra and S. Aust. NP and Wildl. Serv., Adelaide.

__, __ and __ (1992) Plan of Management for the Brush-tailed Bettong (Marsupiala: Potoroidae) in South Australia.Dept. Envir. and Planning, Adelaide.

Nernyshev, V.I. (1959) Trudy Akad. Nauk. Tadzh. SSR. 106: 1–65.

Nesbitt, W.T. (1975) In M.W. Fox (ed), The Wild Canids, Chapt. 27: 391–5. New York.

Newby, E. (1987) Round Ireland in Low Gear. London.

Newhook, F.J., Dickson, E.M. and Bennett, K.J. (1971) Tane 17: 97–117.

Newman, C. (1949) Fla. Wildl. 3(5): 3–5.

Newman, D.M.R. (1975) In Reid, P.L. (ed), Proc. 3rd Wld. Conf. Anim. Prod.: 95–101. Sydney University Press, Sydney.

Newsom, W.M. (1937) J. Mamm. 18(4): 435–60.

Newsome, A.E. and Noble, I.R. (1986) In Groves, R. and Burdon, J. (eds), Ecology of Biological Invasions: An Australian Perspective,: 1–20. Canberra.

__ Corbett, L.K. and Best, L. (1973) A.M.R.C. Rev. 14: 1–11.

Newson, R. (1965) J. Reprod. Fert. 9: 380–1.

Newson, R.M. (1969) In Energy Flow Through Small Mammal Populations. Proc. IBF. Meet. 2nd. Prod. in Small Mammal Populations, Oxford, England,1968: 203–204.

__ and Holmes, R.G. (1968) J. Anim. Ecol. 37(2): 471–81.

Newton, R. (1999) BBC Wildlife 17(2): 53.

Nichols, L. (1960) Proc. 42nd. a. Conf. W. Assoc. State Game Fish Comm.: 90–7.

__ (1960) The History of the Antelope Introduction on Lanai Island, Hawaii. Mimeo. State of Hawaii, Division of Fish and Game, Honolulu.

Nicholson, A.J. and Warner, D.W. (1953) J. Mamm. 34(2): 168–79.

Nicoll, M. and Langrand, O. (1986) Conservation et utilisa-tion rationelle des ecosystemes forestiers du Gabon. Gland, WWF/IUCN.

Nicoll, M.E. (1985) J. Anim. Ecol. 54: 71–88.

Nicoll, M.J. (1908) Three Voyages of a Naturalist, being on Account of many little known Islands in Three Oceans Visited by the 'Valhalla'. London.

Nielson, L. and Brown, R.D. (1988) Translocations of Wild Animals. New York.

Nieminen, M. and Helminen, M. (1987) Trans. Congr. Int. Game Biol. *18: 90.*

Niethammer, G. (1937) Z. Morph. Okol. Tiere *33: 297.*

__ *(1963)* Die Einburgerung von Saugetiere und Vogeln in Europa. *Hamburg & Berlin.*

Niethammer, J. and Krapp, F. (1978–86) Handbuch der Saugetiere Europas. *Vol. 1–3. Wiesbaden.*

Nievergelt, S. (1966) Der Alpensteinbock in seinem Lebensraum. *Hamburg & Berlin.*

Nogales, M., Martin, A., Delgado, G. and Emmerson, K. (1987) Afr. Small-Mamm. Newsl. *(9): 12.*

__, *Abdola, M., Alonso, C. and Quilis, V. (1990)* Mammalia *54(2): 189–96.*

Nonakhov, V. (1987) Trans. Congr. Int. Game Biol. *18: 124.*

Nordberg, S. (1951) Suomen Riista *5: 7.*

Norman, F.I. (1970) Proc. R. Soc. Vict. *83: 193–200.*

__ *(1970)* J. Zool. *162: 493–503.*

__ *(1970)* Aust. J. Zool. *18: 215–29.*

__ *(1971)* J. Appl. Ecol. *8: 21–32.*

__ *(1971)* Proc. Roy. Soc. Vict. *84: 7–18.*

__ *(1975)* Atoll Res. Bull. *(182): 1–13*

__ *(1977)* Corella *3(3): 56–7.*

__ *(1977)* Corella *1(3): 58–9.*

__ *and Brown, R.S. (1980)* Corella *4(4): 89–90.*

__ *et al. (1980)* Proc. R. Soc. Vict. *91: 135–54.*

__, *Harris, M.P., Brown, R.S. and Dearson, D.M. (1980)* Corella *4(4): 77–8.*

Norman, W.H. and Musgrave, T. (1866) Journals of the Voyage and Proceedings of the H.M.C.S. Victoria in Search of Ship-wrecked People at Auckland and other Islands. *Melbourne.*

Norris, J.D. (1967) J. appl. Ecol. *4: 191–9.*

__ *(1967)* J. Appl. Ecol. *4: 167–89.*

North, S.G. and Bullock, D.J. (1986) Biol. Conserv. *37(2): 99–117.*

Northcott, T.H., Payne, N.F. and Mercer, E. (1974) J. Mamm. *55(1): 243–8.*

Norton, J.H. and Thomas, A.D. (1976) Aust. Vet. J. *52: 293–4.*

Notini, G. (1948) Svensk Jakt *86: 68–70.*

Novellie, P.A. and Knight, M. (1994) Koedoe *37(1): 115–19.*

Novick, H.J. and Stewart, G.R. (1982) Calif. Fish Game *67(4): 21–35.*

Novikov, G.A. (1956) Carnivorous Mammals of the Fauna of the USSR. *Transl. by I.P.S.T., Jerusalem, 1962. Moscow*

Novoa, C. (1970) J. Reprod. Fertil. *22: 3–20.*

Nowak, E. (1968) Acta Theriol. *13(5): 75–97.*

Nowak, R.M. (1979) Monogr. Mus. Nat. Hist., *University Kansas, No. 6: 1–154.*

Nowell, K. and Jackson, P. (1996) Wild Cats. Status Survey and Conservation Action Plan. *IUCN, Gland, Switzerland.*

Numata, M. and Ohsawa, M. (1970) Vegetation Succession in Chichijima, Bonin Islands. *In* Nature of the Bonin and Volcano Islands. *Ministry of Education, Tokyo.*

Numerov, K.D. (1958) Tr. Vses. Nauch. Issled. Inst. Zhivotnogo Sy'ra i Pushniny *17: 80–95.*

Nunley, G. (1986) Proc. Great Plains Wildl. Damage Control Workshop *7: 9–27.*

O'Brien, P.H. (1987) Aust. Rangeland J. *9: 96–101.*

O'Brien, S.J., Belden, R.C. and Martenson, J.S. (1990) Nat. Geog. Res. *6(4): 485–94.*

O'Bryan, M.K. and McCullough, D.R. (1985) J. Wildl. Mgmt. *49(1): 115–19.*

O'Farrell, T.P. (1965) J. Mamm. *46(3): 525–7.*

O'Neill, G. (1999) Wingspan *9(1): 1–15.*

O'Rourke, F. (1970) The Fauna of Ireland. An Introduction to the Land Vertebrates. *Cork.*

O'Sullivan, M. (1953) Twenty Years a-Growing. *London.*

Oates, J.F. (1996) African Primates. Status Survey and Conservation Action Plan. *rev. edn., IUCN, Gland, Switzerland.*

Obee, B. (1983) Wildl. Rev.: *16–20.*

Oberline, A. (1940) Paradise of the Pacific *52(11): 32–4.*

Ognev, S.I. (1940) Mammals of the U.S.S.R. and Adjacent Countries. *Vol. 4. Izdat. Akad. Nauk. SSSR, Moskow. Translation I.P.S.T., Jerusalem 1966.*

Ohio Department of Natural Resources (1954) Ohio Cons. Bull. *18(8): 30.*

Ohtaishi, N. and Gao, Y. (1990) Mammal Rev. *20(2/3): 125–44.*

Oka, H. and Takashima, H. (1947) Naturalized Animals in Japan. *Tokyo.*

Oko, Z. and Wlodek, K. (1978) Roczniki Akad. Rolniczej w Poznaniu, C. Zootechnika *24: 113–25.*

Okubo, A., Maini, P.K., Williamson, M.H. and Murray, J.D. (1989) Proc. R. Soc. Lond., B. Biol. Sci. *238(1291): 113–25.*

Ol'kova, N.V. (1962) Dokl. Irkutsk Protivochumnogo Inst. *4: 120–7.*

Oleinikov, N.S. and Vasil'eva, I.N (1963) Izv. Akad. Nauk. Turkm. SSR. Ser. Biol. Nauk *5: 74–6.*

Oliver, A.J., Marsack, P.R., Mawson, P.R., Coyle, P.H., Spencer, R.D. and Dean, K.R. (1992) Feral Pig Species Management Plan. *Agriculture Protection Board of Western Australia.*

Oliver, R.L. and Sorensen, J.H. (1951) Cape Expedition Ser. Bull. *(7).*

Oliver, W.L.R. (1984) In Feral Mammals – Problems and Potential. *Workshop on Feral Mammals, Caprinae Specialist Group, 3rd. Int. Theriol. Conf., Helsinki, 1982: 87–126. IUCN: Morges.*

__ *(1985) In* Proc. Symp. 10, Assoc. of Wild Animal Keepers, *35–52.*

__, *Wilkins, R.H., Kerr, R.H. and Kelly, L. (1986)* Dodo J. Jersey Wildl. Preserv. Trust *23: 32–58.*

Olrog, C.C. (1950) Acta Zool. Lilloana *9: 505–32.*

Olson, S.L. and Pregill, G.K. (1982) In Olsen, S.L. (ed) Fossil Vertebrates from the Bahamas. *Smithson, Contrib. Paleobiol. (48): 1–7.*

Onoyama, K., Kagawa, T. and Karita, Y. (1990) Res. Bull. Obihiro Univ. Ser. *17(1): 57–67.*

Ord, W.M. (1964) Elepaio *25: 3.*

Osgood, W.H. (1904) N. Am. Fauna *24: 1–86.*

__ *(1943)* Field Mus. Nat. Hist., Zool. Ser. *30: 1–268.*

Osman, D.I., Moniem, K.A. and Tingeri, D. (1979) Acta Anat. *104: 164–71.*

Osman Hill, W.C. (1966) Primates: Comparative Anatomy and Taxonomy. *Edinburgh. 6 vols.*

Ostapenko, M.M. (1963) In Okhotn.–promys Zhivotn. Uzbekistana: *43–56. Akad. Nauk. Uzbek. SSR., Tashkent.*

Overbo, C.M. (1951) In Norges pelsdyralslag jubileumsskrift 1926–1951*: 144–58. Oslow.*

Owadally, A.W. (1980) Rev. Agric. Sucr. (Ile Maurice) *59: 76–94.*

Owen-Smith, N. (1981) The White Rhinoceros Overpopulation Problem and a Proposed Solution. *In* Problems in Management of Locally Abundant Wild Mammals, *New York, 361 pp.*

Pack, J.C. and Cromer, J.I. (1981) Proc. Worldw. Furbearers Conf., 1980, Frostburg, Maryland *2: 1431–42.*

Packard, R.L. (1955) J. Mamm. *36(3): 471–3.*

Paker, R.L. (1968) In Symposium on Introductions of Exotic Animals*: 21–2. C. Kleberg Research Program in Wildlife Ecology, College of Agriculture, Texas A & M University, Texas.*

Palii, V.F. (1963) In Akklim. zhivotn. v SSR*: 27–29. Akad. Nauk. Kazakhsk SSR., Alma-Ata.*

__ (1963) In Akklim. zhivotn. v SSR.*: 310–311. Akad. Nauk. Kazakhsk SSR., Alma-Ata.*

Palionene, A. (1965) Acta Theriol. *10(6/9): 111–16.*

Palmer, L.J. and Rouse, C.H. (1963) Progress of muskoxen investigations in Alaska 1930–35. *USDI, Bureau of Sport Fish and Wildlife, Juneau, Alaska, 1936, 35 pp.*

Palmer, T.S. (1898) US Dept. Agriculture Yearbook 1898*: 87–110.*

Palmer, W.L., Kelly, G.M. and George, J.L. (1982) Wildl. Soc. Bull. *10(3): 259–61.*

Panaman, P. (1981) Z. Tierpsychol. *56(1): 59–73.*

Pankrat'ev, A.G. (1959) Akad. Nauk. SSSR. *11: 115–20.*

Pankratov, I.G. (1961) In Knowledge of the Fauna and Flora of Ivanov Oblast *1: 80–4. Ivanovo.*

Panwar, H.S. and Rodgers, W.A. (1986) Indian Forester *112(10): 939–44.*

Papenfus, D. (1990) Is the common brushtail possum still common in South Australia? *Cons. and Parks Management Course, South Australia. College Adv. Educ., Salisbury.*

Paradiso, J.L. and Handley, C.O. (1965) Chesapeake Sci. *6(3): 167–71.*

Parker, R.L. (1968) In Symposium: Introductions of exotic animals; Ecological and socioeconomic considerations: *21–2. C. Kleberg Res. Proj. Wildl. Ecol., College of Agriculture, Texas A & M Univ., Texas.*

Parker, W.T. (1991) Wildl. Soc. Bull. *19: 73–9.*

Parkes, J., Henzell, R. and Pickles, G. (1996) Managing Vertebrate Pests: Feral Goats. *Canberra.*

Parkes, J.P. (1972) Unpublished M.Sc. Thesis, Massey University, Manawatu.

Parmalee, P.W. (1953) J. Wildl. Mgmt. *17(3): 375–6.*

Parmentier, J. and Parmentier, R. (1883) Le discours de la navigation de Jean and Raoul Parmentier de Dieppe (Recueil de voyages et de documents). 4 vols. Paris.*

Paszkowski, L. (1969) The Aust. Zoologist *15: 109–20.*

Paton, D. (1973) S. Aust. Ornith. *26: 77–84.*

Paton, J.B. and Paton, D.C. (1977) Corella *1(3): 65–7.*

__ and __ (1977) Corella *1(3): 68–9.*

Patten, D.C. (1974) Proc. 6th. Vert. Pest Contr. Conf., Anaheim, Calif. *6: 210–16.*

Paulian, R. (1955) Le Nat. Malagache *7(1): 1–18.*

Pavlinin, V.N. and Shvarts, S.S. (1961) Fil. Akad. Nauk. SSSR. *24: 1–43.*

Pavlov, M.P. (1958) Tr. Vses Nauchno.–Issled. Inst. Zhiv. Syr'ya Pushn. *17: 96–116.*

__ (1963) In Acclimatization of Animals in the USSR*: 133–4. Akad. Nauk. Kazakhsk SSR., Alma-Ata.*

__ (1964) In Fish Stock of Aral Sea and Methods for Rational Utilization*: 172–6. Nauka Tashkent.*

__, Korsakova, I.B., Timofeyev, V.V. and Safonov, V.G. (1973) Acclimatization, of Game Animals and Birds in the USSR. *Kirov.*

Pavlov, P., Hone, J., Kilgour, R.F. and Pedersen, H. (1981) Aust. J. Exp. Agric. Anim. Husb. *21: 570–4.*

Pavlov, P.M. (1981) N.Z. J. Ecol. *4: 132–3.*

__ (1983) Feral pig Sus scrofa. *In Strahan, R. (ed),* Australian Museum Complete Book of Australian Mammals*: 495p. Sydney.*

__ (1987) Trans. Congr. Int. Game Biol. *18: 144.*

__ and Hone, J. (1982) Aust. Wildl. Res. *9: 101–9.*

Payne, J., Francis, C.M. and Phillipps, K. (1985) A Field Guide to the Mammals of Borneo. *Sabah Society and W.W.F. Malaysia.*

Payne, N.F. (1976) Canad. Field-Nat. *90(1): 60–4.*

Peacock, E.H. (1933) Field, Lond. *162(4210): 592–3.*

Pearson, O.P. (1983) J. Mamm. *64(3): 476–92.*

__ (1985) In Tamarin, R.H., Biology of New World Microtus,*: 535–566. Am. Soc. Mammalogy Spec. Rep. No. 8.*

Pedersen, J.A. (1964) Norske Landbruk *71(3): 41–51.*

Pedler, L. and Copley, P.B. (1992) Reintroduction of stick-nest rats to Reevesby Island. *Final Rep., WWF Aust. Proj. 175. Biol. Conserv. Br., Dept. Env. and Land Management, Adelaide.*

Pefaur, J.E., Hermosilla, W. di Castri, F., Gonzales, F. and Salinas, F. (1968) Rev. Soc. Med. Vet. (Chile) *18: 3–15.*

Pei, K. and Liu, H. (1994) J. Zool. (Lond.) *233(2): 293–306.*

Peiper, H. (1976) Zeitschrift fur Saugetierkunde *41: 274–7.*

Pemberton, C.E. (1925) Hawaiian Sugar Planters Assoc., Exp. Station, Entomol. Ser. Bull. *(17): 1–46.*

__. (1933) Some Food Habits of the Mongoose. *Hawaiian Planters Rec. 37 (1): 12–13.*

Penicaud, P. (1979) Revue d'Ecologie (La Terre et la Vie) *33: 591–610.*

Pennetier, G. (1905) Actes du Museum d'Histoir naturelle, Rouen *9: 1–108.*

Penny, M. (1974) The Birds of Seychelles. *London.*

Penzhorn, B.L. (1971) Koedoe *14: 145–59.*

Peracino, V. and Bassano, B. (1987) Trans. Congr. Int. Game Biol. *18: 145–6.*

Pererva, V.I. (1987) Trans. Congr. Int. Game Biol. *18: 145–7.*

Perez de Paz, P.L. (1983) Vida Silvestre *46: 88–99.*

Perkins, R.C.L. (1904) Hawaiian Forester & Agriculturist *1: 138–9.*

__ (1925) Hawaiian Planters Rec. *29: 359–64.*

Pernetta, J.C. and Watling, D. (1978) Pacific Science *32(3): 223–44.*

Pernety, D. (1773) The History of a Voyage to the Malouine (or Falkland) Islands, made in 1763 and 1764 under the command of M. de Bougainville, in order to form a settlement there. *Transl. from D. Pernety's Historical Journal in French. London.*

Perry, M.C. (1971) Va. Wildl. *32(5): 17–19.*

Perry, R. (1946) A Naturalist on Lindisfarne. *London.*

__ (1963) Animals *2(11): 282–5.*

Perryman, P. and Muchlinski, A. (1987) J. Mamm. *68(2): 435–8.*

Pescott, T.W. (1976) Aust. Bird Bander *14(1): 29–31.*

Peterle, T.J. (1958) J. Wildl. Mgmt. *22(3): 221–31.*

Peterson, G.D. (1956) J. Mamm. *37(2): 278–9.*

Peterson, R.L. and Downing, S.D. (1956) J. Mamm. *37(3): 431–5.*

__ and Reynolds, T.K. (1954) Contrib. Poy. Ont. Mus. Zool. Paleo ntol. *(38): 1–7.*

Peterson, R.S. (1967) J. Mamm. *48(1): 119–29.*

Petocz, R.G. and Raspado, G.P. (1984) Conservation and develpopment in Irian Jaya: a study for rational resource utilisation. *WWF/IUCN Conservation for Development Prog. in Indonesia, Bogor, Indonesia.*

Petrashov, V.V. (1977) Izv. Timiryazev S-kh. Akad. *(5): 162–9.*

Petrides, G.A. (1950) Texas Game Fish. *8(6): 4–5, 27.*

__ (1959) J. Mamm. *40(2): 248–9.*

__ (1975) Env. Conserv. *2(2): 133–6.*

__ (1975) Environmental Conserv. *2(1): 47–51.*

__ and Leedy, D.L. (1948) J. Mamm. *29: 182–3.*

Petrov, V.I. (1962) *In* Problemy Zool. Issled. v Sibiri: *194–7. Knigoizat, Gorno-Altaisk.*

Petruska, M. (1949) New York State Cons. *4(1): 2.*

Petter, J-J. and Peyriéras, A. (1970) Mammalia *34: 167–93.*

__, Albignac, R. and Rumpler, Y. (1977) Mammafères lèmuriens (Primates prosimiens). *Faune de Madagascar No. 44. Orstom–CNRS, Paris.*

Pettifer, H.L. (1981) *In* Proc. Worldw. Furbearers Conf., *1980, Frostburg, Maryland 2: 1121–42.*

__ (1981) *In* Proc. Worldw. Furbearers Conf., 1980, *Frostburg, Maryland 2: 1001–4.*

Philibosian, R. and Yntema, J.A. (1977) Annotated Checklist of the Birds, Mammals, Reptiles, and Amphibians of the Virgin Islands and Puerto Rico. *Information Service, Frederiksted, St. Croix, US. Virgin Islands.*

Phillips, M.K. (1988) AAZPA Annu Conf. proc.: *426–33.*

Phillpot, C.M. (1965) *Unpublished Hons. Thesis, Univ. Adelaide.*

Pickard, J. (1976) Aust. J. Ecol. *1: 103–14.*

Pickvance, T.J. and Chard, J.S.R. (1960) Proc. Birm. Nat. Hist. Phil. Soc. *19(1): 1–8.*

Pielowski, Z. (1969) Z. Jagdwiss. *15(1): 16–17.*

Pierle, C.B. (1941) W. Va. Cons. *5(9): 4–5, 19.*

Pimlott, D.H. (1975) *In* M.W. Fox (ed), The Wild Canids, *Chapt. 19: 280–5. New York.*

__ and Carberry, W.J. (1958) J. Wildl. Mgmt. *22: 51–62.*

Pimm, S.L. (1987) Trends in Ecology and Evolution (TREE) *2(4): 106–8.*

Pinder, L. (1986) WWF Monthly Report. Project US-441: *295–6.*

Pine, D.S. and Gerdes, G.L. (1973) Calif. Fish Game *59: 126–37.*

Pine, R.H., Angle, J.P. and Bridge, D. (1978) Mammalia *42(1): 105–14.*

__, Miller, S.D. and Schamberger, D. (1979) Mammalia *43: 339–76.*

Pippin, W.F. (1961) J. Mamm. *42(3): 344–8.*

Pitmental, D. (1955) J. Mamm. *36(1): 62–8.*

Plant, J.W., Marchant, R., Mitchell, T.D. and Giles, J.R. (1978) Aust. Vet. J. *54: 426–9.*

Polikhronov, D. St. (1974) The Buffaloes of Bulgaria. In Cockrill, W.P. (ed), The Husbandry and Health of the Domestic Buffalo. *Food and Agriculture Organisation, Rome.*

Pollock, J.J. (1984) Preliminary report on a mission to Madagascar by Dr J.F. Pollock in Aug.–Sept. 1984. *Unpublished report to WWF.*

Polushina, N.A. (1963) *In* Zoogeography of the Land: *228–9. Tashkent.*

Poole, W.E., Wood, J.T. and Simms, N.G. (1991) Wildl. Res. *18: 625–39.*

Poplin, F. (1979) Ann. Genet. Sel. Anim. *11(2): 133–43.*

Popov, M.V. (1963) *In* Problemy okhrany priroda Yakutii: *107–12. Yakutsk.*

Popov, V.A. (1964) *In* Prir. res. Volzhskogo-Kamskogo Kraya: *5–15. Zhivotn. mir. Nauka, Moscow.*

Potts, G. (1942) Northwest Nat. *17: 246.*

Powell, R.A. (1981) *In* Proc. Worldw. Furbearers Conf., *1980, Frostburg, Maryland 2: 883–917.*

Pracy, L.T. (1962) Introduction and Liberation of the Opossum *(Trichosurus vulpecula)* into New Zealand. *N.Z. Forest Service., Information Series (45). Wellington, New Zealand.*

__ (1964) N.Z. Forest Service: *1–14.*

__ and Kean, R.I. (1963) N.Z. Forest Service Publication *(40): 1–19. Wellington, New Zealand.*

Presnall, C.C. (1958) J. Wildl. Mgmt. *22(1): 45–50.*

Pritchard, A.L. (1934) Canad. Field-Nat. *48(6): 103.*

Prosperi, F. (1957) Vanished Continent: An Italian Expedition to the Comoro Islands. *Translated by D. Moore. London.*

Provorov, N.V. (1963) Sb. Nauch. Statei Zapadn. Otd. Vses Nauchno.–Issled. Inst. Zhiv. Syr'ya Pushn. *2: 99–123.*

Prusskii, V.F. (1965) Zool. Zhur. *44(9):1382–95.*

Pullar, E.M. (1953) Mem. Nat. Mus. Melb. *(18): 7–23.*

Pulliainen, E. (1975) *In* M.W. Fox (ed), The Wild Canids, *Chapt. 21: 292–9. New York.*

__ (1981) *In* Proc. Worldw. Furbearers Conf., 1980, *Frostburg, Maryland 1: 580–98.*

Pyabov, L.S. (1979) Zool. Zhur. *58(4):564–9.*

Racey, P.A. and Nicoll, M.E. (1984) Mammals of the Seychelles. In D. Stoddart (ed), Biogeography and Ecology of the Seychelles Islands: *607–26. The Hague, Holland.*

Radetsky, P., Long, W. and Reed, D. (1993) Can he survive out there? *The West Australian, Earth 2000, Monday, September 8: 4.*

Radwan, M.A. and Campbell, D.L. (1968) J. Wildl. Mgmt. *32(1): 104–8.*

Raedeke, K.J. and Simonetti, J.A. (1988) J. Mamm. *69(1): 198–201.*

Ragni, B., Possenti, M., Guidali, F., Mingozzi, T. and Tosi, G. (1992) *In* The situation, conservation needs and reintro-duction of Lynx in Europe. Proc. Symp. *17–19 October: 74–76. Council of Europe, Strasbourg.*

Rahm, U. and Stocker, G. (1978) Feld Wald Wasser *6(4): 36–9.*

Rahn, U. (1976) Die Saugetiere der Schweiz. *Basel.*

Raines, J.A. (1985) Rotamah I. Bird observatory report. *RAOU, Melbourne*

Ramsey, C. (1968) Texas Parks and Wildl. *25(4): 3–7.*

__ (1970) Bulletin No. (49), Texas Parks and Wildlife Department, PR Project W-76-R, Texas.

__ (1972) Texas Agric. Progress (T.A.P.) *18(3): 9–12.*

Ramsey, C.W. (1968) *In Symposium*: Introduction of Exotic Animals: Ecological and Socioeconomic Considerations: *9–10. C. Kleberg Research Project in Wildlife Ecology, College of Agriculture, Texas A & M Univ., Texas.*

__ (1975) *In* Land Use: Food and Living, Proc. 30th An Meet. Soil Conserv. Soc. America, *Aug. 10–13, 1975: 123–7. San Antonio, Texas.*

__ and Anderegg, M.J. (1972) Southwest Natur. *16(3–4): 267–80.*

Rangel-Woodyard, E. and Simpson C. D. (1980) In Proc. Symp. Ecol. Mgmt., Barbary Sheep, *Nov. 19–21, 1979: 30–2. Dept. Range. Wildl. Mgmt., Texas Technical University, Lubbock, Texas.*

Rary, R.M., Henry, V.G., Matschke, G.H. and Murphree, R.L. (1968) J. Heredity *59(3): 201–4.*

Rashek, V.A. (1959) Priroda *7: 97–9.*

__ (1976) Zool. Zhur. *55: 784–7.*

Ratcliffe, F.N. (1959) *In* Biogeography and Ecology in Australia, *Ser. Monogr. Biol. 8: 545–64.*

__ and Calaby, J.H. (1958) Rabbit. *In* Australian Encyclopaedia, *Sydney.*

Ratcliffe, P.R. (1987) Mammal Rev. *17(1): 39–58.*

Rausch, R.A. (1958) J. Wildl. Mgmt. *22(3): 246–60.*

__ (1969) Trans. 34th. N. Am. Wildl. Nat. Resour. Conf., *1969: 117–30. Washington.*

Rausch, R.L. (1953) Arctic *6: 91–148.*

Rauzon, M.J. (1985) Atoll Res. Bull. *(282): 32 pp.*

Rawley, E.V. (1954) Utah State Department of Fish and Game, *Department Information Bulletin (12): 1–15.*

Raynal, F.E. (1874) Harpers Mag. *38: 535–40.*

Razumovskii, B.I. (1962) *In* Problems of Zoological Investigations in Siberia: 208. Knigoizdat, Gorno-Altaisk.

Reading, R.P., Myronuik, P., Backhouse, G. and Clark, T.W. (1993) Species *19: 29–31.*

Recher, H.F. and Clark, S.S. (1974) Biol. Conserv. *6(4): 263–73.*

Redford, K.H. and Eisenberg, J.F. (1992) Mammals of the Neotropics: Southern Cone: *Vol. 2. Chile, Argentine, Uruguay, Paraguay. Chicago & London.*

Redford, P. (1962) J. Mamm. *43(4): 541–2.*

Redhead, T.D., Singleton, G.R., Myers, K. and Coman, B.J. (1991) *In* Groves and Di Castri (eds), Biogeography of Mediterranean Invasions. *293–308. Cambridge and Sydney.*

Rees, M.D. (1989) Endang Spec. Rep. Tech. Bull. *14(1–2): 3*

Reese, S.W. (1944) Wis. Cons. Bull. *9(4): 6–10.*

Reest, P.J. van der and Pelgers, E. (1983) Lutra *26:105–14.*

Rehder, H.A. and Randall, J.E. (1975) Atoll Res. Bull. *(183): 1–40.*

Reid, W.H. and Patrick, G.R. (1983) Southwestern Nat. *28: 97–9.*

Reilly, P.N. (1977) Corella *1(3): 51–3.*

Reimov, R. (1960) Fil. Aad. Nauk. Uzbek. SSR. *1: 50–7.*

__ (1964) *In* Stocks of Fish of Aral Sea and Utilization: *176–83. Nauka, Tashkent.*

Rein, R. (1934–35) Der Naturforscher *11: 350.*

Remfrey, J. (1981) Strategies for Control in the Ecology and Control of Feral Cats. *UFAW, Potters Bar, Hertsfordshire.*

Reppe, X.N. (1957) L'aurore sur 1 'antarctique. *Paris.*

Reumer, J.W.F. (1986) Mammalia *50(1): 118–19.*

__ and Sanders, E.A.C. (1984) Zeitschrift fur Saugetierkunde *49: 321–5.*

Reynolds, J.C. (1985) J. Anim. Ecol. *54: 149 –62.*

Reynolds, J.K. (1955) Canad. Field-Nat. *69(1): 14–20.*

__ and Stinson, R.H. (1959) Canad. J. Zool. *37(5): 627–31.*

Rice, C.G. (1991) Trans. West. Sect. Wildl. Soc. *(27): 42–6.*

Richard, P.B. (1967) Penn. Ar. Bed. *6(49): 45–52.*

Richard-Foy, F. (1887) Rapport du Lieutenant de Vaisseau commandant l'aviso-transport la 'Meurthe' a M. Le Capitaine de vais seau commandant la Division Navale de l'Ocean Indien. *In manuscript: 19 pp.*

Richards, E.C. (1950) Diary of E.R. Chudleigh 1862–1921 *Chatham Islands. Christchurch.*

Richardsion, F. (1963) Elepaio *23: 43–5.*

Ride, W.D.L. (1970) A Guide to the Native Mammals of Australia. *London.*

Ridpath, M.G. and Waithman, J. (1988) Wildl. Soc. Bull. *16: 385–90.*

Rieck, W. (1954) Saugetierkdl. Mitt. *2: 54–60.*

Rijk, J.H. de (1988) Lutra *31: 101–31.*

Rinaldi, G. and Milone, M. (1981) Annuario R. Museo zoologico della Universta di Napoli *23: 25–32.*

Riney, T. (1964) The Impact of Introductions of Large Herbivores in the Tropical Environment. *IUCN. Publication, New Series (4): 261–73.*

__ (1967) Ungulate Introductions as a Special Source of Research Opportunities. *IUCN. Publication, New Series (9): 241–54.*

Rinke, D. (1987) Emu *87: 26–34.*

Ripinsky, M.M. (1975) Antiquity *49(196): 295–8.*

Ritchie, J. (1920) The Infuence of Man on Animal Life in Scotland. A Study in Faunal Evolution. *Cambridge.*

Roberts, B. (1958) Polar Record *9(59): 97–134 and 9(60): 191–239.*

__ (1985) Murrelet *66: 24.*

Roberts, P.J. (1980) Bird Study *27: 116.*

Robinson, A.C. (1980) Trans. Roy. Soc. S. Aust., *104: 93–100.*

__ (1989) *In* Burbidge, A. (ed), Australia and New Zealand Islands: Nature Conservation Values and Management, *Occasional Paper 2/89:168–181. Department of Conservation and Land Management, Western Australia.*

__ and Smyth, M.E.B. (1976) Trans. Roy. Soc. S. Aust. *100: 171–6.*

__, Spark, R. and Halstead, C. (1989) South Aust. Nat. *64(1): 4–24.*

Robinson, D.J. and Cowan, I.Mct. (1954) Canad. J. Zool. *32(3): 261–82.*

Robinson, M. (1986) British Birds *79: 256.*

Robinson, R.W. (1941) Pacific Islands Monthly *12: 45.*

Roche, D. (1989) Geo *11: 68–77.*

Roff, C. (1960) Qld. J. agric. Sci. *17(1): 43–58.*

Rogers, G. (1947) Colo. Cons. Comments *10(3): 9–10.*

Rogers, L.A. (1964) Appalachia *35(2): 292–300.*

Rognrud, M. and Janson, R. (1971) In Mussehl, T.W. and Howell, F.W. (eds) Game Management in Montana, *Chapt. 5: 39–51. Game Mgmt. Divis., Helena, Montana.*

Rolls, E.C. (1969) They All Ran Wild. *Sydney.*

Romer, A.S. (1928) Logan Museum Bulletin, *No. 26.*

Romic, S. (1975) Conspectus Agriculturae Scientificus *34: 13–24.*

Roonwal, M.L. and Mohnot, S.M. (1977) Primates of South Asia. *Cambridge.*

Ross, J.C. (1847) A Voyage of Discovery and Research in the Southern and Antarctic Regions, during the Years 1839–43. *vols. 2. London.*

Rothfuss, E.L. (1986) Conf. Sci. Natl. Parks *4: 192.*

Rotterman, L.M., Simon-Jackson, T. (1988) Mar. Mamm. Comm. *Washington, DC, pp 239–75.*

Rounsevell, D. (1989) In Burbidge, A. (ed), Australian and New Zealand Islands: Nature Conservation Values and Management,. *Occasional Paper 2/89: 157–61. Department of Conservation and Land Management, Western Australia.*

Rounsevell, D.E., Taylor, R.J. and Hocking, G.J. (1991) Wildl. Res. *18: 699–717.*

Rowe, F.P. (1960) Proc. Zool. Soc. Lond *134(3): 499–503.*

___ (1967) J. Mamm. *48(4): 649–650.*

___ (1968) J. Zool. *156: 529–30.*

Rubenstein, D.I. and Hohmann, M.E. (1989) Oikos *55(3): 312–20.*

Rudasill, L.S. (1956) Md. Cons. *33(5): 8–9.*

Rudge, M.R. (1969) N.Z. J. Sci. *12(4): 817–27.*

___ (1970) Z. Tierpsychol. *27(6): 687–92.*

___ (1976) N.Z. Ecol. Soc. Proc. *23: 83–4.*

___ (1976) Feral Goats in New Zealand. In Whitaker, A.H. and Rudge, M.R. (eds), Value of Feral Mammals in New Zealand. *N.Z. Department of Lands and Survey, Information Series (1), Wellington, N.Z.*

___ (1983) NZ. J. Zool.*10: 349–64.*

___ (1984) In Feral Mammals – Problems and Potential. *3rd. Int. Theriol. Conf., Helsinki, 1982: 56–84. IUCN, Morges.*

___ (1986) NZ. J. Ecol. *9: 89–100.*

___ and Campbell, D.J. (1977) N.Z. J. Bot. *15(2): 221–54.*

___ and Clark, J.M. (1978) N.Z. J. Zool. *5(3): 581–9.*

Rudyshin, M.P. (1960) Lesn. Khoz. *8: 38.*

Rue, L.L. (1978) The Deer of North America. *New York.*

Ruhl, H.D. (1941) Trans. N. Am. Wildl. Conf. *5: 424–7.*

Rukovskii, N.N. (1948) Priroda *(6): 65–6.*

___ (1958). Byull. Nauchno.–Tekhnol. Inf. Vses. Nauchno–Issled. Inst. Zhivotn. Syr'y a Pushniny *3: 43–6.*

___ (1963) The Raccoon in the USSR. In Acclimatization of Animals in the USSR: *146–8. Akad. Nauk. Kazakhsk SSR., Alma-Ata.*

Russel, P. (1937) New Mex. *15 (6): 32–3.*

Russell, H. (1910) Zoologist, London *14: 113–15.*

Rutherford, W.H. (1972) Game Information Leaflet *(92). Colo. Divis. Game Fish. Parks, Denver, Colorado: 1–3.*

Ryabov, L.S. (1975) Byull Mosk. O-VA Ispyt. Prir. Otd. Biol. *80(5): 11–22.*

Ryan, G.E. and Croft, J.D. (1974) Aust. Wildl. Res. *1: 89–94.*

Ryan, P. (1972) Encyclopaedia of Papua and New Guinea. *Melbourne University Press and University Papua New Guinea. 3 vols.*

Sade, D.S. and Hildrech, R.W. (1965) Caribbean J. Sci. *5(1–2): 67–81.*

Safanov, V. G. (1963) In Acclimatization of Animals in the USSR: *159–161. Akad. Nauk. Kazakhsk. SSR., Alma-Ata.*

Saint-Hilaire, I.G. (1861) Acclimatation et domestication des animaux utiles. *Paris, France.*

Sale, J.B. (1986) Indian For. *112(10): 945–8.*

___ and Singh, S. (1987) Oryx *21(2): 81–4.*

Salganskii, A.A. (1963) In Acclimatization of Animals in the USSR: *164–6. Akad. Nauk. Kazakhsk. SSR., Alma-Ata.*

Salley, A.S. (1911) Narative of Early Carolina. *New York.*

Salmi, A.M. (1949) Suomen Riista *4: 124.*

Salo, L.J. (1976) Wildl. Soc. Bull. *4(49): 167–74.*

Salter, R.E. and Hudson, R.J. (1978) Nat. Can. (Que.) *105(5): 309–32.*

___ and ___ (1979) J. Range Mgmt. *32(3): 221–5.*

___ and ___ (1980) J. Range. Mgmt. *33(4): 266–71.*

Salvin, O. (1876) Trans. Zool. Soc. London *9: 447–510.*

Salzmann, H.C. (1975) Mitt. Naturf. Ges. Bern *32: 15–35.*

Samosh, V. and Razumovskii, V. (1962) Okhota i Okhotnoe Khoz. *3: 10–11.*

Samosh, V.M. (1962) Problems of Ecology *6: 126–7.*

Samusenko, E.G. (1962) In Voprosy ekologii: *127–8. Vysshaya Shkola, Moscow.*

___ (1963) In Acclimatization of Animals in the USSR: *155–7. Akad. Nauk. Kazakhsk. SSR., Alma-Ata.*

San Jose, C., Braza, F. and Blom, A. (1987) Trans. Congr. Int. Game Biol. *18: 173–4.*

Sand, G.X. (1952) Pa. Game News *22(11): 4–8.*

Sanders, C.L. (1963) Ecology *44(4): 803–6.*

Sanderson, G.C. and Hubert, G.F. (1981) In Proc. Worldw. Furbearers Conf., 1980, Frostburg, Maryland *1: 487–513.*

Sanderson, I.T. (1955) Living Mammals of the World. *London.*

Sanges, M. and Alcover, J.A. (1980) Endins *7: 57–62.*

Sanguilly, W.M. (1869) Harpers Mag. *38: 535–40.*

Santini, L. (1974) Redia *55: 393–408.*

___ (1978) In Proc. 8th. Vert. Pest Contr. Conf., Sacramento, California: *78–84.*

___ (1980) In Proc. 9th. Vert Pest Conf., 1980, Fresno, California: *149–53.*

Sapaev, V.M. (1965) In Voprosy Geogr. Da l'nego Vostoka *7: 236–46. Vladivostok.*

Sartorius, 0. (1970) Starkste Damschaufler der Welt. *Hannover.*

Savage, A. (1988) AAZPA Annu. Conf. Proc.: *78–84.*

Savenkov, V.V. (1987) Trans. Congr. Int. Game Biol. *18: 174.*

Sayre, R. (1980) Audubon Mag. *82(6): 18, 20.*

___ (1981) Tale of Two Bighorn Sheep Transplants. Audubon Mag. *83 (2): 22–3.*

___ (1983) Audubon Mag *85(2): 44.*

Schaff, E. (1892) Deutsche Jagerz. *19: 270.*

Schauenberg, B.C. and Schauenberg, P. (1969) Rev. Suisse de Zool *76(7): 183–210.*

Schauenberg, P. (1970) Rev. Suisse de Zool. *77(8): 127–60.*

Scheinder, K.B. and Scheffer, W.B. (1972) Sea otter report. *Alaska Dep. Fish Game, Fed. Aid Wildl. Restor. Prog. Rep. W-17-4: 1–30. Juneau.*

Schenk, P. (1966) Jagd und Naturschutz in der Schweiz. Stuttgart.

Schiller, E.L. (1956) J. Mamm. 37(2): 181–8.

Schitoskey, F., Evan, J. and La Voie, G.K. (1972) In Proc. 5th. Vert. Pest Contr. Conf., 1972, Fresno, California: 15–17.

Schmidt, F. (1954) Kosmos 50: 73.

Schmidt, K.P. (1928). In Sci. Survey Porto Rico and Virgin Islands 10(1): 3–160. N.Y. Acad. Sci., New York.

Schmidt, R.H. (1985) Wildl. Soc. Bull. 13: 592–94.

Schmidt, R.L. and Rutherford, W.H. (1978) Wildl. Soc. Bull. 6(3): 159–63.

Schmitt, S.M. and Aho, R.W. (1988) In Nielsen, L. and Brown, R.D. (eds), Translocation of Wild Animals: 258–274. Wisconsin Humane Soc. and C. Kleberg Wildl. Res. Inst.

Schneider, C.O. (1936) Comm. del Museo de Concepcion 1: 159–60.

Schreiber, A., Wirth, R., Riffel, M. and Van Rompaey, H. (1989) Weasels, Civets, Mongooses, and their Relatives: An Action Plan for the Conservation of Mustelids and Viverrids. International Union for the Conservation of Nature and Natural Resources, Gland, Switzerland.

Schreiner, C. (1968) In Introduced Exotic Animals: Ecological and Socioeconomic Considerations.: 13–16. C. Kleberg Research Program in Wildlife Ecology, College of Agriculture, Texas A & M University, Texas.

Schultze-Westrum, T. (1963) Saugetierkundliche Mitteilungen 11(4): 145–82.

Schuster, W. (1904) Deutsche Jagerz. 43: 545, 672.

Scott, C.D. and Pelton, M.R. (1976) In Proc. a. Conf. SE. Assoc. Game Fish Comm. 29: 585–93.

Scott, J.P. (1968) Evol. Biol. 2: 243–75.

Scott, M.D. and Causey, K. (1973) J. Wildl. Mgmt. 37(3): 253–65.

Scott, T.G. (1943) Ecol. Monogr. 13: 427–79.

__ (1955) Nat. Hist., Surv. Divis., Urbana, Illinois, Biol. Notes (35):1–16

Seal, U.S. and Plotka, E.D. (1983) J. Wildl. Mgmt. 47(2): 422–9.

Seaman, G.A. (1952) Trans. 17th. N. Am. Wildl. Conf.: 188–97.

__ (1966) Carib. J. Sci. 6(1–2): 33–41.

Seaman, G.R. and Randall, J.E. (1962) J. Mamm. 43(4): 544–5.

Searle, A.K. (1980) Qld. J. Agric. 106(2): 119–24.

__ (1981) Qld. J. Agric. 107(1): 17–20.

__ and Parker, M.S. (1982) The Federal Deerbreeder (J. Aust. Deer Breeders Feder.) 1(3): 9–14.

Sedgwick, L.E. (1968) West. Aust. Nat. 11(1): 1–4.

Seebeck, J.H. (1981) Aust. Wildl. Res. 8: 285–306.

__ (1984) Vic. Nat. 101: 60–5.

__, Bennett, A.F. and Scotts, D.J. (1989) In Kangaroos, Wallabies and Rat-Kangaroos, pp 67–88, G. Grigg, P. Jarman and I. Hume (eds). New South Wales, Australia.

Seegmiller, R.F. and Ohmart, R.D. (1981) Wildl. Monogr. (78).

__ and Simpson, C.D. (1979) Trans. Desert Bighorn Council 23: 47–9.

Seidensticker, J., Lahiri, J.E., Das, K.C. and Wright, A. (1976) Oryx 11: 267–73.

Senzota, R.B.M. (1982) Afr. J. Ecol. 20(3): 211–12.

Sequeira, D.M. (1980) Comparison of the Diet of the Red Fox (Vulpes vulpes L. 1758) in Gelderland (Holland), Denmark and Finnish Lapland. In Zimen, E. (ed), The Red Fox, W. Junk BV Publ., Hague, Boston and London.

Serdyu, V.N. (1978) Vestn. Zool. (2): 78–80.

Serena, M. (1995) Reintroduction Biology of Australian and New Zealand Fauna. Chipping Norton, New South Wales.

Serfass, T.L. and Rymon, L.M. (1985) Trans. NE Sect. Wildl. Soc. 42: 138–49.

Serventy, D.L. (1977) Corella 1(3): 60–2.

Serventy, V. (1987) Aust. Nat. Hist. 22(4): 158–9.

Serventy, V.N. (1953) Australian Geographic Society Report (1).

Seymour, G. (1960) Furbearers of California. Calif. Dept. Fish Game, Sacramento.

Seymour, W. (1961) Qu. J. Forestry 55(4): 293–8.

Shackelford, R.M. (1949) Am. Fur Breeder 22: 12–14.

Shank, C.C. (1972) Zeitschrift fur Tierpsychologie 30(5): 488–28.

Shaposhnikov, L.V. (1939) Bull. Moscow Soc. Nat. 48(1): 65–73.

__ (1941) Trans. Cent. Lab. Game Indus. (5): 207–11.

__ (1941) Trans. Cent. Lab. Game Indus. (5): 129–38.

__ (1960) Okhrana Prirody i Zapov. Delo v SSR. 4: 37–51.

Shaposhnikov, V.D. (1960) Kraevedcheskogo Muz. 2: 125–32.

Shaw, A.C. (1941) Proc. 5th. N. Am. Wildl. Conf.: 436–41.

Shaw, E.B. (1940) Econ. Geogr. 16(3): 223–49.

Shay, R. (1976) Oregon Wildl. 31: 3–5.

Sheail, J. (1971) Rabbits and their History. Newton Abbott.

Sheffield, W.J., Ables, E.D. and Fall, B.A. (1971) J. Wildl. Mgmt. 35(2): 250–257.

__, Fall, B.A. and Brown, B.A. (1983) The Nilgai Antelope in Texas. Kleburg Studies Nat. Res., Texas A & M University.

Sherman, H.B. (1937) Proc. Fla. Acad. Sci., 1936. 1: 102–28.

__ (1937) Fla. Entomol. 26: 54–9.

Sherman, H.B. (1954) J. Mamm. 35(1):126.

Sherrard, J.M. (1966) Kaikoura. A History of the District. Kaikoura.

Shigesada, N. and Kawasaki, K. (1997) Biological Invasions: Theory and Practice. Oxford and Tokyo.

Shigirevskaya, E.M. (1964) Zool. Zhur. 43 (11): 1727–9.

Shimaoka, Y. (1953) Trans. Mamm. Soc., Japan (5): 3.

Shirley, M.G., Chabreck, R.H. and Linscombe, G. (1981) In Proc. Worldw Furbearers Conf., 1980, Frostburg, Maryland 1: 517–30.

Shmit, E. (1959) Okhota i Okhotn. Khoz. 1: 62.

Short, J., Bradshaw, S.D., Giles, J., Prince, R.I.T. and Wilson, G.R. (1992) Biological Conservation 62: 189–204.

__ and Turner, B. (1993) Wildl. Res. 20: 525–34.

__, and __ (2000) Biological Conservation 96: 185–96.

__., __, Parker, S. and Twiss, J. (1995) In Serena, M. (ed), Reintroduction Biology of Australian and New Zealand Fauna, 30: 183–188. Chipping Norton, New South Wales.

Shorten, M. (1957) Forestry 30(2): 151–72.

__ (1957) J. Anim. Ecol. 26: 287–94.

__ (1963) Animals 1(20): 4–7.

Shortridge, G.C. (1934) The Mammals of South West Africa. Vol 1. London.

Shtil'mark, F.R. (1963) Zool. Zhur. 42(1): 92–102.

Shult, M.J. (1973) In Exotic big game short course,: 1–15. Texas Agric. Ext. Serv. and Texas Agric. Exp. Stn., Texas A & M Univ., Texas.

Shulyatyev, A.A. (1987) Trans. Congr. Int. Game Biol. 18: 184.

Shurygin, V.V. and Nikiforov, N.M. (1964) Sb. Nauch-Tekh. Inform. Vses Nauch-Issled. Zhivotn. Syr'ya Pushniny 10: 13–24.

Shvarts, S.S (1963) In Akklim. Zhivotn. v SSR 33–34. Akad. Nauk. Kazakh. SSR., Alma-Ata.

Siebert, B.D. and Newman, D.M.R. (1989) Camelidae: In Walton, D.W. and Richardson, B.J., Fauna of Australia: Mammalia. Vol. 1B, Canberra.

Siegfried, W.R. (1962) Cape Provincial Administration, Dept. Nature Conserv. An. Rep. No. 19: 80–7.

Siivonen, L. (1943) Luonnon Ystava 47: 135–7.

__ (1953). Suomen Riista 8: 177–9.

__ (1958) Suomen Riista 12: 165–6.

__ (1976) Nordeuropas Daggdjur. Stockholm.

Silver, J. (1924) J. Agric. Res. 28(11): 1133–7.

__ (1927) J. Mamm. 8: 58–60.

__ (1928) Bur. Biol. Surv., Divis. Econ. Investigations, leaflet (21): 1–6.

__ (1937) USDA. Circular (423). Washington, DC.

__ (1941) USDF, Fish and Wildlife Service, Wildlife Circular (6): 1–18.

Silverstein, A. and Silverstein, V. (1974) Animal Invaders: The Story of Imported Wildlife. New York.

Simmons, H. (1985) Primate Conservation 5: 59–62.

Simonetti, J.A. (1989) Mammalia 53(3): 363–8.

Simpson, C.D. (ed) (1980) Proceedings of the Symposium on the Ecology and Management of Barbary Sheep, November 19–21, 1979. Department Range and Wildlife Management, Texas Technical University, Lubbock, Texas.108 pp.

__, Krysl, L.J., Hampy, D.B. and Gray, G.G. (1978) Trans. Desert Bighorn Council 22: 23–31.

Sinel'nikov, A.M. (1963) In Acclimatization of Animals in the USSR: 148–9.

Singer, F.J. (1981) Environmental Mgmt. 5(3): 263–70.

__, Swank, W.T. and Clebsch, E.C. (1984) J. Wildl. Mgmt. 48(2): 464–73.

Singh, A. (1981) Tara: a Tigress. London.

__ (1984) Tiger! Tiger!. London.

Sinitsyn, A.A. (1987) Trans. Congr. Int. Game Biol. 18: 185.

Sitwell, N. (1986) Smithonian 17(3): 114–19.

Skead, C.J. (1980) Historical Mammal Incidence in the Cape Province. Vol. 1. The Western and Northern Cape. Department Nature and Environmental Conservation, Capetown.

Skegg, P.D.G. (1963) Notornis 10: 153–68.

Skinner, J.D., Breytenbach, G.T. and Maberly, C.T.A. (1976) S. Afr. J. Wildl. Res. 6(2): 123–8.

Skira, I.J. and Brothers, N.P. (1988) Corella 12(3): 80–1.

__ and __ (1988) Corella 12(3): 82–4.

Skoptsov, V. (1964) Okhota Okhotn. Khoz. 8: 26.

__ (1967) Oryx 19(1): 54–6.

Skudskii, A. (1963) Okhot. Okhotn. Khoz. 10: 24–8.

Skuncke, F. (1968) Zool. Rev. 30(3–4): 105–12.

Slevin, J.R. (1931) Calif. Acad. Sci. 17: 1–162.

__ (1959) Calif. Acad. Sci. 25: 1–150.

Slim, P.A. (1985) Lutra 28: 4–20.

Slough, B.G. (1989) J. Wildl. Manage. 53(4): 991–7.

Sludskii, A.A. and Afanas'ev, Yu. G. (1964) Trudy Inst. Zool. Akad. Nauk. Kazakh. SSR. 23: 5–74.

__ (1963) In Acclimatization of Animals in the USSR: 167–172. Akad. Nauk. Kazakhsk. SSR., Alma-Ata.

Smal, C.M. (1988) Mammal Rev. 18(4): 201–8.

__ and Fairley, J.S. (1984) Mammal Rev. 14: 71–8.

Smale, S. and Owen, K. (1990) In Towns, D.R., Daugherty, C.H. and Atkinson, I.A.E. (eds), Ecological Restoration of New Zealand Islands. Conservation Sciences Publication No. 2;109–112. Department of Conservation, Wellington.

Smallshire, D. and Davey, J.W. (1989) Nature in Devon 10: 62–9.

Smielowski, J.M. and Ravel, P.P. (1988) Oryx 22(2): 85–8.

Smit, C.J. and Van Wijngaarden, A. (1981) Threatened Mammals in Europe. Wiesbaden.

Smith, A.C. and Nichols, L. (1984) In Proc. 4th Symp. Nthn. Wild Sheep and Goat Council, M. Hoefs (ed): 467–80.

Smith, C.A. (1968) In Symposium Introduction Exotic animals; Ecological and socioeconomic considerations: 23–5. C. Kleberg Res. Proj. Wildl. Ecol., College of Agriculture, Texas A & M Univ., Texas.

__ and Nichols, W.B. (1984) Proc. North Wild Sheep and Goat Counc. 4: 467–80.

Smith, C.W. and Diong, C.A. (1977) Coop. Nat. Pk. Resour. Study Unit, Tech. Rep. (19): 1–52. Department of Botany, University Hawaii, Manoa, Honolulu.

Smith, J. (1840) The Winter of 1840 in St. Croix with an Excursion to Tortola and St. Thomas. Privately Printed, New York.

Smith, M.W., Smith, M.H. and Brisbin, I.L. (1980) J. Mamm. 61(1): 39–45.

Smith, P.J. and Dodkin, M.J. (1989) In Burbidge, A. (ed), Australian and New Zealand Islands: Nature Conservation Values and Management. Occasional Paper 2/89:141–156. Department of Conservation and Land Management, Western Australia.

Smith, R. and Phillips, M. (1987) AAZPA Annu. Proc.: 670–7.

Smith, T.E. (1984) Biol. Pap. Univ. Alaska Spec. Rep. 4: 15–18.

Smith, W. (1745) A Natural History of Nevis and the rest of the Charibee Islands in America. Cambridge.

Smithers, R. H. N. (1983) The Mammals of the Southern African Subregion. University of Pretoria, 736 pp.

__ (1971) Memoirs Nat. Mus., Rhodesia 4: 1–340.

Smuts-Kennedy, J.C. (1975) In Wildlife – A Review. N.Z. Department of Internal Affairs, July, 1975: 34–7.

Sneddon, J.L. (1953) Scot. Agric. 33(2): 1.

Snethlage, K. (1967) Das Schwarzwild. Hamburg.

Soderquist, T.R. (1995) In Serena, M. (ed), Reintroduction Biology of Australian and New Zealand Fauna, 26: 159–164. Chipping Norton, New South Wales.

Sody, H.J.V. (1940) Treubia 17: 391–401.

__ *(1941)* Treubia *18: 225–325.*

Sofonov, V.G. *(1981) In* Proc. Worldw. Furbearers Conf., *1980, Frostburg, Maryland 1: 95–110.*

Sokol, J. *(1965)* Fatre *20(6): 440–6.*

Solomatin, A.0. *(1963) In* Acclimatization of Animals in the USSR*: 172–4. Akad. Nauk. Kazakhsk. SSR., Alma-Ata.*

Sorenson, J.H. *(1951)* Wild life in the Subantarctic. *London.*

Southern, H.N. *(1964)* The Handbook of British Mammals. *Oxford. 1965, 465 pp.*

__ *and Watson, J.S. (1941)* J. Anim. Ecol. *10(1): 1–11.*

Southern, W.E., Patton, S.R., Southern, L.K. and Hanners, L.A. *(1985)* Auk *102: 827–33.*

Southgate, R.I. *(1995)* Why reintroduce the bilby? In Serena, M. (ed), *Reintroduction Biology of Australian and New Zealand Fauna, 27: 165–70. Chipping Norton, New South Wales.*

Southwell, B. *(1999)* BBC Wildl. *17(1): 66–72.*

Sowman, W.C.R. *(1981)* Meadow, mountain, forest and stream. The provincial history of the Nelson Acclimatisation Society. *Wellington.*

Spagnesi, M., Cagnolaro, L., Perco, F. and Scala, C. *(1986)* Ricerche di Biologia della Selvaggina *76: 1–147.*

Spalding, D.J. and Mitchell, H.B. *(1970)* J. Wildl. Mgmt. *34(2): 473–4.*

Spalton, A. *(1990)* Species *15: 27–9.*

Spatz, G. and Mueller-Dombois, D. *(1972)* Phytocoenologia *3: 346–73.*

__ *and __ (1973)* Ecology *54(4): 870–6.*

Spaulding, T.M. *(1930)* Paradise of the Pacific *43(11): 8.*

Spencer, D.L. and Lensink, C.J. *(1970)* J. Wildl. Mgmt. *34(1): 1–15.*

Spencer, R.M. *(1929)* Field *154: 871.*

Sperry, C.C. *(1941)* U.S. Department Interior, Fish and Wildlife Service (Washington), Wildl. Res. Bull. *(4): 1–70.*

Spillett, J.J. *(1966)* J. Bombay Nat. Hist. Soc. *63(3): 494–528.*

__ *(1966)* J. Bombay Nat. Hist. Soc. *63(3): 535–56.*

__ *(1966)* J. Bombay Nat. Hist. Soc. *66(3): 599–601.*

__ *and Tamang, K.M. (1966)* J. Bombay Nat Hist. Soc. *66(3): 557–72.*

Spinney, L. *(1995)* New Sci. *145: 35–8.*

Spittler, H. *(1978)* Z. Jagdwiss. *24(1): 33–4.*

Spriggs, J.W. *(1943)* Wyo. Wildl. *8(2): 1–5, 15–17.*

Springer, M.D. *(1972) In* Wood, G.W. (ed) *Research and Management of Wild Hog Populations. Belle W. Baruch Forest Science Institute, Clemson University, Georgetown, S. Carolina.*

Springer, P.F., Byrd, G.V. and Woolington, D.W. *(1978)* Reestablishing Aleutian Canada Geese. In S.A. Temple (ed), *Endangered Birds. University Wisconsin, Madison, Wisconsin.*

St Aldwyn, T.D. *(1955)* Myxomatosis. Second Report. Her Majesty's Stationary Office, *London.*

St John, B.J. and Saunders, G.M. *(1981)* Plan of management for the hairy-nosed wombat in South Australia. *Dept. Env. and Planning, Adelaide.*

St. Catherine's Workshop *(1986) Unpublished reports to participants of the conference on lemur conservation held on St. Catherine's Island, Georgia on 26–27 April, 1986.*

Stahl, P. *(1993) In Proc. Seminar on the Biology and Conservation of the wildcat* (Felis silvestris)*, Nancy, France, 23–25 September, 1992: 16–25 (in Fr.).*

__ *and Artois, M. (1991)* Status and Conservation of the Wildcat in Europe and around the Mediterranean Rim. *Council of Europe, Strasbourg.*

Stains, H.J. *(1975) In* M.W. Fox (ed), *The Wild Canids, Chapt. 1: 3–26. New York.*

Staub, F. *(1970)* Atoll Res. Bull. *(136): 197–209.*

Stead, D.G. *(1935)* The Rabbit in Australia. *Sydney.*

Steen, M.0. *(1951) In* 41st. Conv. Internat. Assoc. Game Fish Cons. Comm., *1951, Rochester, New York: 24–6.*

__ *(1952)* 42nd. Conv. Internat. Assoc. Game Fish Comm., *1952, Dallas, Texas: 89–91.*

Stegeman, L.C. *(1938)* J. Mamm. *19: 279–90.*

Stehlik, J. *(1979)* Folia Venatoria *9: 255–65.*

Steinhart, P. *(1981)* Audubon Mag. *83(2): 8, 10.*

Stelfox, A.W. *(1965)* Ir. Nat. J. *15: 57–60.*

Stephen, D. *(1974)* Highland Animlas. *Inverness.*

Sterna, D.J. and Barrett, R.H. *(1991)* Trans. West. Sect. Wildl. Soc. *(27): 47–53.*

Stevens, D.R. *(1966)* J. Wildl.Mgmt. *30(2): 349–63.*

Stevens, E.F. *(1988)* Anim. Behav. *36(6): 1851–3.*

Stevens, P. *(1981) Vermin and Noxious Weeds Destruction Bo ard, pamphlet (83). Department Crown Lands and Surveys, Victoria.*

Stevens, P.L. *(1981) In Natural Resourses Conservation League of Victoria, Public Forum on Feral Animals, Bairnsdale, 4–5 April, 1981: paper no. 4: 1–7.*

Stevens, W.F. and Weisbrod, A.R. *(1981) In* Proc. World Lagomorph Conf., *1979, Guelph, Ontario: 870–879. University Guelph, Ontario.*

Stilwell, H. *(1955)* Field and Stream *60(1): 68–9, 143–6.*

Stocker, G.C. *(1971)* Wildlife in Australia *8: 10–12.*

__ *(1972)* Aust. For. Res. *5(1): 29–34.*

Stoddart, D.P. *(1981)* Atoll Res. Bull. *(255): 23–26.*

Stoddart, D.R. and Poore, M.E.D. *(1970)* Atoll Res. Bull. *(136): 155–65.*

__ *and __ (1970)* Atoll Res. Bull. *(136): 171–81.*

__, *Benson, C.W. and Peake, J.F. (1970)* Atoll Res. Bull. *(136): 121–45.*

Stokes, J.L. *(1846)* Discoveries in Australia. *vol. 2. London.*

Stone, C.P. and Keith, J.O. *(1987)* pp. 277–287. *In* Control of Mammal Pests *Eds Richards, C.G.J. and Ku, T.Y. London, New York and Philidelphia.*

Stoneham, G. and Johnson, J. *(1987)* Bur. Agric. Econ. Occ. Pap., *Canberra*

Storer, T.E. *(1962)* Bernice P. Bishop Mus. Bull. *(225).*

Storer, T.I. *(1934) In* Proc. 5th. Pacif. Sci. Congr., *Victoria and Vancouver, British Columbia, 1933: 779–784. University Toronto Press, Toronto, Canada.*

__ *(1937)* J. Mamm. *18: 443–60.*

__ *(1947)* California Agriculture Extension Service, Circular *(138): 1–51.*

__ *(1958)* California Agriculture Experimental Station Extension Service, Circular *(434).*

Storr, G.M. *(1960)* J. R. Soc. W. Aust. *43: 59–62.*

__ *(1976)* Aust. Bird Bander *14: 35–8.*

__, *Johnstone, R.E. and Griffin, P. (1986)* Rec. West. Aust. Mus., *Supplement 24.*

Stott, P. (1998) Aust. Vert. Pest Conf. *11: 263.*

Stout, G.G., Lowry, F.C. and Carlile, F. (1973) *In* Proc. 26th. a. Conf. SE. Assoc. Game and Fish Comm.: *202–3.*

Strahan, R. (1995) Mammals of Australia. *Reed Books, Sydney.*

Strautman, E.I. (1963) *In* Resources of Game Animals in the USSR: *172–174. Akad. Nauk. Kazakhsk. SSR., Alma-Ata.*

___ (1963) *In Akad. Nauk. Kazakhsk. SSR., Alma-Ata.*

Street, J.M. (1962) Geogr. Rev. *52(3): 400–6.*

Streever, F. (1947) N.Y. State Cons. *1(4): 11, 29.*

Stribling, H.L. and Brisbin I.L. (1984) J. Wildl. Mgmt. *48(2): 635–9.*

Strickland, M.A. and Douglas, C.W. (1987) *In* Novak, Baker et al. *(1987): 530–547, Wild Furbearer Management and Conservation in North America. Ontario Ministry Nat. Res. 1150 pp.*

Stromberg, M.R. and Boyce, M.S. (1986) Biol. Conserv. *35: 97–110.*

Strong, B.W. (1983) *Conservation Commission N.T., Alice Springs, N.T. Technical Report (3).*

Struhsaker, T. (1998) National Geographic Magazine *194(5): 73–8.*

Stubbe, M. (1980) *In* Zimen, E. (ed), The Red Fox. *Hague, Boston and London.*

Stuttard, R.M. (1981) *In* Proc. World Lagomorph Conf., *1979, Guelph, Ontario: 907–916. University Guelph, Ontario.*

Stüwe, M. and Scribner, K.T. (1989) J. Mamm. *70(2): 370–3.*

Suckling, G.C. and Mcfarlane, M.A. (1983) Aust. Wildl. Res. *10: 249–58.*

Sullivan, T.P. and Sullivan, D.S. (1986) J. Appl. Ecol. *23: 795–806.*

Sumar, J. (1988) Outlook on Agric. *17(1): 23–9.*

Suminski, P. (1963) Chronmy Przyrode Cjczysta *19(3): 13–32.*

Summers, S. (1991) *In* Proc. 5th Aust. Sem. on Nat. Parks and Wildl. Manage., *Hobart.*

Sumption, K.J. and Flowerdew, J.R. (1986) Mammal Review *15: 151–86.*

Suomalainen, E. (1950) Arch. Soc. 'Vanamo' *5: 20–2.*

Sussman, R.W. and Tattersall, I. (1986) Folia Primatol. *46: 28–43.*

Swahn, S. (1947) Svensk Jakt. *85: 264–5.*

Swan, C.A. (1981) Feral Mink. *MAFF (Publications) Northumberland, leaflet (794).*

Swank, W.G. and Petrides, G.A. (1954) Ecology *35(2): 172–6.*

Swann, R.L. and Ramsay, A.D.K. (1978) British Birds *71: 46.*

Swanson, H. and Hudson, R. (1980) Alaska Geogr. *7(3): 123–5.*

Swanson, N.M. (1976) Aust. Bird Bander *14: 88–91.*

Swedberg, G. (1965) Introduction of Exotic Game Mammals. *W-5-R-16, Job 54 (17) mimeo. State of Hawaii, Division Fish Game, Honolulu.*

___ (1966) Introduction of Exotic Game Mammals. *W-5-R-16, Job 54 (18) mimeo. State of Hawaii, Division Fish Game, Honolulu.*

___ (1967) Introduction of Exotic Game Mammals. *W-5-R-16, Job 54 (19) mimeo. State of Hawaii, Division Fish Game, Honolulu.*

___ (1967) The Black-tailed Deer Introduced to Kauai. *State of Hawaii, Division Fish Game, Honolulu.*

Sweeney, J.M., Sweeney, J.R. and Provost, E.E. (1979) J. Wildl. Mgmt. *4 3(2): 555–9.*

Swenson, J.E. (1985) J. Wildl. Mgmt. *49(4): 837–43.*

Sykes, L. (1998) BBC Wildlife *16(9): 86–7*

Sysoev, N.D. (1967) Zool. Zhurv. *46(5): 785–7.*

Szederjei, A. (1959) Z. Jagdwissenschaft *5(3): 81.*

Szunyoghy, J. (1958) Annales Historico-Naturales Musei Nationalis Hungarici *50: 349–58.*

Tappe, D.T. (1942) Division Fish & Game Bull. *(3): 1–59.*

Tate, G.H.H. (1935) Bull. Am. Mus. Nat. Hist. *68: 145–78.*

Tattersall, I. (1976) Anthropol. Pap. Am. Mus. Nat. Hist. *53(4): 369–80.*

___ (1977) Annals N.Y. Acad. Sci. *293: 160–9.*

___ (1977) Anthropol. Pap. Am. Mus. Nat. Hist. *54: 425–82.*

___ (1977) Oryx *13(5): 445–8.*

___ (1982) The Primates of Madagascar. *New York.*

___ and Sussman, P. (1975) Anthropol. Pap. Am. Mus. Nat. Hist. *52(4): 193–216.*

Taub, D.M. (1977) Folia Primatol. *27: 108–33.*

___ (1984) *In* Fa (1984) Barbary Macaque, *71–8. New York.*

Taylor, B.W. (1955) A.N.A.R.E. Reps., Series B., *vol. 2., Botany.*

Taylor, D. (1957) Western Livestock J. *35(45): 40–4 and (52): 80–2.*

___ and Stone, C.P. (1986) Conf. Sci. Natl. Parks *4: 193.*

Taylor, J.M. and Horner, B.E. (1973) Bull. Am. Mus. Nat. Hist. *150: 1–130.*

Taylor, K. (1999) BBC Wildlife *17(1):23.*

Taylor, K.D. (1963) *In* Proc. Assoc. Appl. Biol., Ann. appl. Biol. *51: 325–50.*

Taylor, R.H. (1950) Report No. 15, Animal Ecology Section, *Department of Scienticic and Industrial Research.*

___ (1967) *In* Motunau Island, Cantebury, New Zealand. An Ecological Survey. *N.Z. Department Scientific Industrial Research Bulletin (178): 42–67. Wellington, N.Z.*

___ (1968) N.Z. Ecol. Soc. *15: 61–7.*

___ (1971) N.Z. J. Botany *9: 225–68.*

___ (1976) Feral Cattle in New Zealand. *In* Whitaker, A.H. and Rudge, M.R. (eds) Value of Feral Farm Animals in N.Z. *N.Z. Department Lands and Surveys, Information Series (1). Wellington, New Zealand.*

___ (1978) *In* The Ecology and Control of Rodents in New Zealand Nature Reserves: *135–43. NZ. Dept. Survey. Info. Series No. 4.*

___ (1979) N.Z. J. Ecol. *2: 42–5.*

___ (1981) A Vertebrate Survey of Some Outer Islands of the Marlborough Sounds 26–30 January 1981.File Report *4/15/13, Ecology Division, DSIR, New Zealand.*

___, Bell, B.D. and Wilson, P.R. (1970) N.Z. J. Sci. *13(1): 78–88.*

Taylor, W.L. (1939) J. Anim. Ecol. *8(1): 6–9.*

___ (1948) J. Anim. Ecol. *17(2): 151–4.*

___ (1949) J. Anim. Ecol. *18(2): 187–92.*

Tchernov, E. (1984) *In* Martin, P.S. and Klein, R.G. (eds) Quarternary Extinctions: *528–52. Tucson.*

Teer, J.G., Renecker, L.A. and Hudson, R.J. (1993) Trans. 58th Nth. Am. and Nat. Resour. Conf.: *448–59.*

Tegner, H. (1971) J. Zool., London *164 (2): 263–5.*

Tegner, H. (1976) Wildlife 18(2): 78–9.

Tellkamp, W. (1979) Saugetierkundliche Mitteilungen 27: 206–16.

Temple, R.C. (1914) The Travels of Peter Mundy 1608–1667, 2, Asia 1628–1634. *London.*

Temple, S.A. (1974) Wildlife 16(8): 370–4.

Tenovuo, R. (1963) Suomen Riista 16: 92–9.

Terry, M. (1963) People 14(11): 12–15.

Tetley, H. (1941) Proc. Zool. Soc. Lond. *111b: 23–35.*

Thamdrup, H.M. (1965) Dansk Vildrforskning, 1964–65: 36–9.

Theobald, F.V. (1926) Bull. S.E. Agric. Coll., *March, 1926.*

Theobald, J. (1999) BBC Wildlife 17(2): 28.

Thibault, J.C. (1989) Ois. Rev. Franc. Ornithol. 59: 305–24.

__ and Rives, C. (1975)* Birds of Tahiti. *Les Editions du Pacifique. Papeete, Tahiti.*

Thing, H. (1984) Dan. Rev. Game Biol. *12: 1–53.*

__ , Henrichsen, P. and Lassen, P. (1984)* In Klein, D.A., White, R.G. and Keller, S. (eds). *Proc. Int. Muskox Symp. 1, Biol. Pap. Univ. Alaska Spec. Rep. 4: 1–6.*

Thoday, M. (1999) Best Practice (Australian Minerals and Energy Environment Foundation) Yearbook 1999: 28, 31.

Thomas, B.W. (1976) Mammals. In Collyer, E. (ed), *History and Natural History of the Boulder Bank, Nelson Haven, Nelson, New Zealand. Nelson.*

Thomas, I.M. and Delroy, L.B. (1971) Trans. Roy. Soc. S. Aust. 95: 143–5.

Thomas, L.K. (1986) Conf. Sci. Natl. Parks 4: 194.

Thomas, O. (1927) Ann. Mag. Nat. Hist., Ser. 9(19): 650–8.

Thompson, F. (1974) The Uist and Barra. *London.*

Thompson, H.V. (1951) Ann. appl. Biol. 38(3):725–7.

__ (1953)* Proc. Zool. Soc. Lond. 122: 1017–24.

__ (1955)* J. Wildl. Mgmt. 19(1): 8–13.

__ (1962)* New Scientist 13 (270): 130–2.

__ (1964)* Agriculture: 564–7.

__ (1967)* Agriculture: 114–16.

__ (1968)* The Field, 5 December 1968, 2 pp.

__ (1971)* Agriculture, 1971: 421–5.

__ (1981)* In Proc. World Lagomorph Conf., 1979, Guelph, Ontario: 816–21. University Guelph, Ontario.

__ and Peace, T.R. (1962)* Qu. J. For. 56 (1): 33–42.

__ and Worden, A.N. (1956)* The Rabbit. *London.*

Thompson, R.L. (1977) In Wood, G.W. (ed) Research and Management of Wild Hog Populations: 11–15. B.W. Baruch Forest Science Institute, Clemson University, Georgetown, South Carolina.

Thompson, S.L. (1931) Canad. Field-Nat. 45(8): 192–3.

Thomson, A.P.D. (1951) J. Hist. Med. Allied Sci. 6(4): 471–80.

Thomson, G.M. (1921) Wildlife in New Zealand. *Wellington.*

__ (1922)* The Naturalisation of Animals and Plants in New Zealand. *Cambridge.*

Thomson, P.C. (1984) J. Agric. W.A. 25: 27–31.

Thoreson, A.C. (1967) Notornis 14: 182–200.

Thornton, I. (1971) Darwin's Islands: A Natural History of the Galâpagos. *New York.*

Thorpe, I. (1989) In Waters, D. (ed), University East Anglia Comoro Islands Expedition 1988, Unpublished final report: 61–2.

Thouless, C.R. ansd Sakwa, J. (1995) Biol. Conserv. 72(1): 99–107.

__ , Liang Chengqui and Loudon, R.S.I. (1988)* Int. Zoo Yrbk (27): 223–30.

Tidemann, C.R., Yorkston, H.D. and Russack, A.J. (1994) Wildl. Res. 21: 279–86.

Tilman, H.W. (1960) Geogr. Mag. 33(7): 391–5.

Timm, R.M. (1975) Bell Mus. Nat. Hist., University Minnesota, Occasional Paper (14): 1–56.

Timofeeva, E.K. and Fedotova, V.G. (1973) Zool. Zhur. 52(7): 1046–54.

Timofeeva, V.V. (1961) In Ratsionalizatsiya Okhotna Promysla 9: 34–40.

Tinker, S.W, (1938) Animals of Hawaii. *Honolulu.*

Tipton, A.R. (1977) In Wood (ed) G.W Research and Management of Wild Hog Populations: 91–100. B.W. Bar. For. Sci. Inst., Clemson University, Georgetown, S. Carolina.

Tittensor, A.M. and Lloyd, H.G. (1983) Rabbits. *Forestry Commission, Forest Records 125, HMSO.*

Tjernberg, M. (1981) Holarctic Ecology 4: 12–19.

Tkachenko, A.A. (1961) Zool. Zhur. 40(11):1715–24.

Tokuda, M. (1951) Kagaku – Asahi 11(5): 38–9.

Tomich, P.Q. (1969) J. Wildl. Mgmt. 33(3): 576–84.

Torres, H. (1992) South American Camelids: An action plan for their conservation. *IUCN, Gland, Switzerland.*

Toschi, A. (1953) Ricerche di Zoologia Applicata allacaccia 23: 1–52.

__ (1965)* Fauna d'Italia. Mammalia. Lagomorpha, Rodentia, Carnivora, Ungulata, Cetacea. *Bologna.*

Tosi, G., Scherini, G., Apollonio, M., Ferrario, G., Pacchetti, G., Torso, S. and Guidali, F. (1986) Italy. Ric. Biol. Selvaggina 77: 1–77.

Towne, C.W. and Wentworth, E.N. (1950) Pigs from Cave to Cornbelt. *Norman, Oklahoma.*

Towney, G. and Skira, I.J. (1985) Corella 8: 111–12.

Townsend, B (ed) (1986) Annual report of survey and inventory activities. Part X. Muskoxen. *Alaska Dep. Fish and Game, Fed. Aid Wildl. Restor. Proj. Rep. 16, Proj. W-22-4 and W-22-5, pp 25.*

Trainer, D.O. (1973) J. Wildl. Dis. 9: 376–8.

Trethewey, D.E.C. and Verts, B.J. (1971) Oregon. Am. Midl. Nat. 86: 463–76.

Treus, V.D. and Lobanov, N.V. (1976) Vestn. Zool. 1: 3–9.

Trouessart, E.L. (1917) Bulletin du Museum d'histoire naturelle 1917: 366–73.

Troughton, E. Le G. (1971) Proc. Linn. Soc. NSW 96: 93–8.

Trout, R.C., Tapper, S.C. and Harradine, J. (1986) Mammal Rev. 16: 117–23.

Troyer, W.A. (1960) J. Wildl. Mgmt. 24(1): 15–21.

Tsetsevinskii, L.M. (1963) Sb. Nauchn-Tekh. Inform. Vses Nauchn-Issled. Inst. Zhivot. Syr'ya Pushniny 5(8): 77–81.

Tufts, R.N. (1939) Can. Field-Nat. 53: 123.

Tulloch, D.G. (1969) Turnoff 2: 117–21.

__ (1969)* Aust. J. Zool. 17(1): 143–52.

__ (1970)* Aust. J. Zool. 18: 399–414.

__ (1975) Buffalo in Northern Swamp Lands. In Proc.111 World Conf. Anim. Prod.: 60–6. Sydney University Press, Sydney.

__ (1986) In Kitching, R. (ed), Ecology Exotic Animals and Plants: Australian Case Histories, Chapt. 6: 78–90. Brisbane and New York.

__ and Cellier, K.M. (1986) Aust. Wildl. Res. 13: 433–9.

__ and Grassia, A. A. (1981) Aust. Wildl. Res. 8: 335–48.

Turbet, C.R. (1941) Trans. and Proc. Fiji Soc. Sci. and Ind., 1938–40: 7–12.

Turbott, E.G. (1948) Rec. Auckland Inst. Mus. 3: 253–72.

Turner, J.S., Smithers, C.N. and Hagland, R.D. (1968) The Conservation of Norfolk Island. Australian Conservation Foundation, Spec. Publ. 1.

Turner, M.G. (1984) J. Wildl. Mgmt. 48(4): 1461–4.

__ (1988) J.Range Manage. 41(15): 441–7.

Tustin, K.G. (1974) J. Mamm. 55(1): 199–200.

Tweedie, M. (1963) Animals 1(5): 7–9.

Twyford, G. (1991) Australia's Introduced Animals and Plants. Balgowlah, New South Wales.

Tyndale-Biscoe, C.H. (1955) Aust. J. Zool. 3(2): 162–184.

__ (1960) Aust. Mus. Mag. 13(7): 234–8.

Tyurin, P.S. (1958) Priroda (3): 101–3.

__ (1964) In Lyubite, okhranyaite prirodu Kirgizii, Frunze: 65–70. (from Biol. Abstr. 47(22): 1966).

__ and Busalaeva, M.S. (1963) Acclimatized in Kirghizia. Izv. Akad. Nauk Kirghiziya SSR. – Ser. Biol. Nauk 5(2): 11–15.

U Tin Yin (1967) Wild Animals of Burma. Rangoon

Udagawa, T. (1952) Trans. Mamm. Soc., Japan (3): 4.

__ (1970) In Proc. 4th Vert. Pest Contr. Conf., California: 148–9.

Ueckermann, E. and Hansen, P. (1968) Das Damwild. In Naturgeschichte Hege und Jagd, Hamburg und Berlin.

__, Zander, J., Scholz, H. and Luelfing, D. (1977) Z. Jagd. 23(3): 153–62.

Uloth, W. (1956) Saugetierkundliche Mitt. 4: 1.

United States Fish and Wildlife Service (1997) Black-footed ferret (Mustela nigripes) http://www.fws.gov/~r9extaff/bio-logues/bio_ferr.html.

University of California (1977) Divis. Agric. Sci., University of California. Leaflet No. 2945.

Upham, L.L. (1980) Proc. Symp. Ecol. and Mgmt. Barbary Sheep, Nov. 19–21, 1979: 17–18. Dept. Range and Wildl. Mgmt., Texas Tech. Univ., Lubbock, Texas.

Urich, F.W. (1931) Trop. Agric. (Trinidad) 8(4): 95–7.

Ursin, E. (1948) Flora og Fauna 54: 99–109.

Uspenski, S.M. (1984) In Klein, D.R., White, R.G. and Keller, S. (eds) Proc. Int. Muskox Symp. 1, Biol. Pap. Univ. Alaska Spec. Rep. 4: 12–14.

Uspenskii, G.A. (1963) In Acclim. Anims. in USSR: 194–196. Akad. Nauk. Kazakhsk. SSR., Alma-Ata.

Valdez, R. and Bunch, T.D. (1980) In Proc. Symp. Ecol. Mgmt. Barbary Sheep, Nov. 19–21, 1979: 27–9. Dept. Range and Wildl. Mgmt., Texas Tech. Univ., Lubbock, Texas.

Valencia, G.D. (1980) In Proc. 9th Vert. Pest Contr. Conf., 1980, Fresno, California: 110–13.

Valentijn (1726) Oud and Nieu-Oost Indien 3: 267.

Valentine, G.L. (1980) In Proc. Symp. Ecol. and Mgmt. Barbary Sheep, Nov. 19–21, 1979: 22–3. Dept. Range Management, Texas Tech. Univ., Lubbock, Texas.

Vallaux, C. (1928) Bull. Instit. Oceano 512: 1–20.

Van Aarde, R.J. (1979) S. Afr. J. Antarc. Res. 9: 14.

__ and Skinner, R.D. (1986) Afr. J. Ecol. 24: 97–101.

Van Bemmel, A.C.V. (1949) Treubia 20(2): 195–263.

Van Bloeker, J.C. (1937) J. Mamm. 18: 360–1.

Van de Veen, H.E. (1973) Deer 3: 15–20.

Van Dyk, A. (1991) The Cheetahs of DeWildt. Cape Town.

Van Ee, C.A. (1962) Annals of the Cape Provincial Museum 2: 53–5.

Van Gelder, R.G. (1979) Wildl. Soc. Bull. 7(3): 197–8.

Van Gessel, F.W.C. and Dorward, D.F. (1975) Aust. Bird Bander 13: 80–2.

Van Helvoort, B.E., de Iongh, H.H. and Van Bree, P.J.H. (1985) Z. Saugetierk. 50: 182–4.

Van Peenen, P.F.D., Cunningham, M.L. and Duncan, J.F. (1970) J. Mamm. 51(2): 419–24.

Van Rensburg, P.J.J., Skinner, J.D. and Van Aarde, R.J. (1987) J. Appl. Ecol. 24: 63–73.

Van Vuren, D. (1981) The Feral Sheep on Santa Cruz Island: Status and Impact. The Nature Conservancy, Arlington, Virginia. 131 pp.

__ (1984) Calif. Fish and Game 70(3): 140–4.

__ and Bray, M.P. (1986) J. Mamm. 67(3): 503–11.

__ and Coblentz, B.E. (1987) Biol. Cons. 41(4): 253–68.

__ and __ (1989) J. Wildl. Manage. 53(2): 306–13.

__ and __ (1984) In Feral Mammals – Problems and Potential. 3rd Int. Theriol. Conf., Helsinki, 1982: 45–52. IUCN, Morges.

Van Wijngaarden, A. (1961) Natuurhistorisch Maanblad 50: 54.

__ (1964) Zeitschrift fur Saugetierkunde 29: 359–68.

__ and Morzer Bruijns, M.F. (1961) Lutra 3: 35–42.

__, Van Laar, V. and Trommel, M.D. (1971) Lutra 13: 1–41.

Van't Woudt, B.D. (1951) N.Z. Sci. Rev. 9(8): 146–52.

Vandal, D. (1984) MS Thesis, Univ. Laval, Quebec City, Quebec.

__ and Barrette, C. (1985) In Meredith, T.C. and Martellm M. (eds), Proc. 2nd N.A. Caribou Workshop, McGill University: 199–212. McGill subarc. Res. Pap. 40.

Vander Meulen, J.H. (1977) Sth. Afr. J. Wildl. Res. 7(1): 15–18.

Varona, L.S. (1974) Academie Ciencias, Cuba: 1–139.

Vaseneva, A.Ya. (1964) Sb. Nauchno.–Tekhnol. Info. Vses Nauchno.–Issled. Inst. Zhivotn. Syr'ya Pushniny 69: 60–2.

Vasil'kov, V. (1966) Okhot. Okhotn. Khoz. 4: 22–3.

Vaucher, C. (1946) La Vie suavage en Montagne. Geneve.

Vaughan, H.V. and Carne, P.H. (1971) Deer 2: 767.

Veale, E.M. (1957) Agric. Hist. Rev. 5: 85–90.

Veitch, C.R. (1985) ICBP Technical Publication 3: 125–41.

__ (1995) In Serena, M. (ed), Reintroduction Biology of Australian and New Zealand Fauna, 17: 97–104. Chipping Norton, New South Wales.

__ and Bell, B.D. (1990) In Towns, D.R., Daugherty, C.H. and Atkinson, I.A.E., Ecological Restoration of New Zealand Islands: 137–46. Cons. Sci. Publ. (2), NZ. Dept. Conserv., Wellington.

Velain, Ch. (1877) Arch. Zool. Exp. et Gen. 16: 1–144.

Venables, L.S.V. (1956) Scottish Naturalist 68: 71–3.

Vereshchagin, N.K. (1941) Trav. Inst. Zool. Acad. Sci. RSSG. 4: 3–42.

__ *(1947)* Acad. Sci. Azerbaijan SSR News 5: 68–74.

__ *(1950)* The Coypu: its Raising and Trapping on the Reservoirs of the Transcaucasus. *Acad. Nauk. Azerbaidzhanskoi SSR., Baku.*

Vershinin, A.A. (1962) Tr. Vses Nauchno.-Issled. Inst. Zhivotn, Syr'ya Pushniny 19: 206–20.

__, *Lavrov, N.P. and Khakhin, G.V. (1973)* Vsesouznogo soveshtchaniyah go akklimatizatatsii i reaklimatizatsii okhotnichikh jivotn'kh: 55–81.

Vesey-Fitzgerald, B. (1936) Field 168: 1075.

__ *(1938)* Field 171: 927.

Vesey-Fitzgerald, L.D. (1942) J. Ecol. 30: 1–16.

Vevers, G.M. (1948) Beds. Nat. 2: 42.

Vibe, C. (1967) Medd. om Gronland. 170: 1–227.

Vigiani, A.R. (1960) Agro (Buenos Aires) 2(4): 15–18.

Vigne, J.D. (1992) Mammal Rev 22: 87–96.

__ *and Alcover, J.A. (1985)* Actes 110 th Congres National Societes Savantes (Sciences) 2: 79–91.

Vincent, J. (1962) Annals of the Cape Provincial Museum 2: 110–17.

Vinson, C. (1946) Nature Mag. 39(8): 405–6.

__ *(1947)* Nature Mag. 40(3): 153–4.

Vladimirskii, M.G. (1961) In Materials of the Planning and Methods Conference on Protection of Plant Zones in the Urals and Siberia, 1960: 61–5. *Novosibirsk.*

Volchenko, O. (1964) Sel'skok. Poiz. Sibir. Dal'nego Vost. 2: 83–4.

Von Bloeker, J.C. (1967) In Proc. Symp. Biol. of the Calif. Is.,: 245–63. *Santa Barbara Botanic Garden, Santa Barbara.*

Von Bülow, B. (1981) Zeitschrift fur Angewandte Zoologie 68(1): 67–94.

Von Drygalski, E. (1935) Hund auf Kerguelen. Zeitschrift der Gesellschaft fur Erdkunde *(1/2).*

Von Goldschmidt-Rothschild, B. and Lueps, P. (1976) Rev. Suisse Zool. 83(3): 723–35.

Vorhies, C.T. and Taylor, W.P. (1933) Univ. Arizona Coll. Agric. Tech. Bull. (49): 471–587.

Voris, J.C., Yoakum, J.D. and Yocum, C.F. (1955) J. Mamm. 36(2): 302.

Voronov, V.G. (1963) Tr. Sakhalinsk Komple Nauch-Issled. Inst. 14: 30–8.

Wacher, T. (1998) BBC Wildlife 16(7): 62–9.

Waggoner, D.W. (1946) Wisconsin Cons. Bull. 11(6): 3–5.

Waite, E.R. (1909) In C. Chitton (ed) The Subantarctic Island of New Zealand, 2: 542–600. *Wellington.*

__ *and Wood Jones, F. (1927)* Trans. Roy. Soc., South Australia 51: 322–5.

Walker, D. (1991) Seabird 13: 45–50.

Walker, F.D. (1909) Log of the Kaalokai. *Honolulu.*

Walker, L.W. (1945) Nat. Hist. 54 (11): 396–400.

__ *(1948)* Audubon Mag.: 80–5.

Walker, R.L. (1959) Experimental Introduction and Hybridization of the European Mouflon (Ovis musimon). *W-5-R-10, Job 51 (10) mimeo. State of Hawaii, Division of Fish and Game, Honolulu.*

__ *(1960)* Experimental Introduction and Hybridization of the European Mouflon (Ovis musimon). *W-5-R-11, Job 51 (11) mimeo. State of Hawaii, Division of Fish and Game, Honolulu.*

__ *(1960)* Proc. 40th a. Conf. W. Assoc. Fish Game Comm.: 148–55.

__ *(1961)* Experimental Introduction and Hybridization of the European Mouflon (Ovis musimon). *W-5-R-12, Job 51 (12) mimeo. State of Hawaii, Division of Fish and Game, Honolulu.*

__ *(1962)* Experimental Introduction and Hybridization of the European Mouflon (Ovis musimon). *W-5-R-13, Job 51 (13) mimeo. State of Hawaii, Division of Fish and Game, Honolulu.*

__ *(1966)* Hybridization of the Mouflon (Ovis musimon) with the Feral Sheep. *W-5-R-17, Job (17) mimeo. State of Hawaii, Division of Fish and Game, Honolulu.*

__ *(1967) In* Proc. 47th a. Conf. W. Assoc. State Game Fish Comm., Hawaii, 1967: 94–112.

Wallace, A.R. (1902) Island Life. *London.*

Waller, C. (1982) Wildlife 1982: 48–51.

Wallmo, O.C. and Gallizioli, S. (1954) J. Mamm. 35(1): 48–54.

Walters, J. and Soos, J. (1961) Forest Chron. 37(1): 22–8.

Wapstra, J.E. (1975) In Papers Presented Deer Management Conference, 1974: 23–35. *Fish and Wildlife Division. Melbourne.*

Warkentin, M.J. (1970) Tulane Studies in Zool. 15: 10–17.

Warner, R.E. (1985) J. Wildl. Mgmt. 49(2): 340–6.

Warren, G.L. (1970) In Proc. Tall Timbers Conf. Ecol. Anims. Control Habitat Mgmt. (2, 1970: 185–202.

Warren, R.J., Conroy, M.J., James, W.E., Baker, L.A. and Diefenbach, D.R. (1990) In Trans. 55th N.A. Wildl. & Nat. Res. Conf., Denver, Colorado: 580–9.

Warwick, T. (1934) J. Anim. Ecol. 3: 250–67.

__ *(1940)* Proc. Zool. Soc. London A. 110: 165–201.

Washburn, M. (1949) Louisiana Cons. 2(4): 4–5, 21–22.

Watcher, T. (1999) Nat. Geogr. Mag. 195: Earth Almanac.

Watkins-Pitchford, D. (1963) Country Life November 21, 1963.

Watson, G.E. (1975) Birds of the Antarctic and Sub Antarctic. *Washington, DC.*

Watson, J.S. (1956) N.Z. J. Sci. & Technol. 37(5): 560–70.

__ *(1957)* N.Z. J. Sci. 38(5): 451–82.

__ *(1959)* N.Z. J. Agric.: 365–8.

__ *(1961)* Feral Rabbit Populations on Pacific Islands. *Pacific Sci. 15 (4): 591–3.*

__ *(1961).* N.Z. Dept. Sci. Indust. Res. Bull. (137): 132–5.

__ *(1961) In* Proc. 9th Pacific Sci. Congr., 1957: 15–17.

Watts, C.H.S. and Aslin, H.J. (1981) The Rodents of Australia. *Sydney and Melbourne.*

Watts, T. and Conley, W. (1984) J. Wildl. Mgmt. 48(3): 814–22.

Wayre, P. (1985) Oryx 19(3): 137–9.

Weaver, R.A. (1974) In Proc. 6th Vert. Pest Contr. Conf., Anaheim, California.: 204–9.

__ *(1986)* Trans. Nth. Am. Wildl. Nat. Resour. Conf. 51: 41–4.

Webb, J.W. and Nellis, D.W. (1981) J. Wildl. Mgmt. 45(1): 253–8.

Webb, W.L. (1960) J. Wildl. Mgmt. 24(2): 147–61.

Weber, D. (1971) Biol. Conserv. *4(1): 8–12.*

Weckwerth, R.P. and Wright, P.L. (1968) J. Wildl. Mgmt. *32(4): 977–80.*

Wehausen, J.D. and Elliott, H.W. (1982) Calif. Fish Game *68(3): 132–45.*

__ Bleich, V.C. and Weaver, R.A. (1987) Trans. West. Sect. Wildl. Soc. *23: 65–74.*

Weidenhofer, M. (1977) *Maria Island, a Tasmanian Eden. Hobart.*

Weir, J.S. (1977) Proc Ecol. Soc. Aust. *10: 4–14.*

Weir, T. (1981) *Weir's Way. Edinbourgh.*

Westbury, H.A. (1989) *Viral Haemorrhagic Disease of Rabbits. Report of study tour, Committee of Nature, Conservation Ministers of Australia, Geelong.*

Westerskov, K. (1952) N.Z. Sci. Rev. *10(9): 136–44.*

Westman, K. (1966) Suomen Riista *18: 101–16.*

Wetmore, A. (1925) National Geogr. Mag. *48: 77–108.*

Wettstein, O. (1941) Ann. Naturhist. Mus. Wien *52: 245.*

Weyndling, R. (1998) BBC Wildlife *16(11): 28–9.*

Wharton, J.C.F. and Dempster, J.K. (1981) *In* Nat. Resour. Cons. League of Victoria, Public Forum on Feral Animals, *Bairnsdale, 4–5 April, 1981: paper no. 5: 1–20.*

Wheeler, J. (1988) *In* Proc. Wester Vet. Conf., *14–18 February: 301–10. Las Vegas, Nevada.*

Wheelwright, H.W. (1862) *Bush Wanderings of a Naturalist. London.*

Whinray, J.S. (1971) Vict. Nat. *88: 279–86.*

__ (1976) Feral Sheep in New Zealand. *In A.H. Whitaker and M.R. Rudge (eds) Value of Feral Farm Mammals in New Zealand. N.Z. Dept. Lands Surv., Info Series (1), Wellington.*

Whitaker, A.H. and Rudge, M.R. (1976) The Value of Feral Farm Mammals in New Zealand. *N.Z. Dept. Lands Surv., Info. Series (1), Wellington.*

White, D. (1946) Nat. Hist. *55(1): 20–3.*

White, G. (1980) Islands of South-west Tasmania. *Author, Sydney.*

Whitehead, G.K. (1951) Country Life *110(2859): 1450–2.*

__ (1953) The Field *201(5223): 239.*

__ (1954) The Field *203(5284): 663.*

__ (1962) Gamekeeper and Countryside *(781): 12–13.*

__ (1964) The Deer of Great Britain and Ireland: An Account of their History, Status and Distribution. *London.*

__ (1972) The Wild Goats of Great Britain and Ireland. *Newton Abbott.*

Wiens, H.J. (1962) Atoll Environment and Ecology. *Neew Haven and London.*

Wilberly, E.J. (1979) Ark *6(9): 282–7.*

Wilcove, D.S. (1987) Trends Ecol. Evol. *2(6): 146–7.*

Wildhagen, A. (1949) Fauna *2: 107–28.*

__ (1956) J. Mamm. *37(1): 116–18.*

Wildlife Management Institute (1954) Outdoor News Bull. *8(22): 3.*

Wilkes, C. (1845) *Narratives of the U.S. Exploring Expedition During the Years 1838–1842. Vols. 4 and 5. Philadelphia.*

Wilkins, B.T. (1957) J. Wildl. Mgmt. *21(2): 159–69.*

Willett, J.A. (1970) J. Brit. Deer Soc. *2(2): 498–504.*

__ and Mulloy, F. (1970) J. Brit. Deer Soc. *2(2): 498–504.*

Williams, C.B. (1918) Bull. Dept. Agric. Trinadad and Tobago *17: 167–8.*

Williams, C.K. and Moore, R.J. (1989) J. Anim. Ecol. *58: 249–59.*

__ and Ridpath, M.G. (1982) Aust. Wildl. Res. *9: 397–408.*

Williams, D.F. (1979) Annals Carnegie Museum *48(23): 425–33.*

Williams, G.R. (1960) Ibis *102: 58–70.*

__ and Rudge, M.R. (1969) Proc. Ecol. Soc. N.Z. *16: 13–28.*

Williams, J.R. (1953) Revue Agric. Sucr. Ile Maurice *2(32): 55–56.*

Williams, R.T. (1971) Aust. J. Zool. *19: 41–51.*

Williamson, K. and Boyd, J.M. (1963) A Mosaic of Islands. *Ediburgh and London.*

Willner, G.R., Chapman, J.A. and Pursley, D. (1979) Wildl. Monogr. *(65): 1–43.*

Wilson, D.E. and Reeder, D.M. (1993) Mammal Species of the World. *Smithsonian Institution Press. 1206 pp.*

Wilson, G., Dexter, N., O'Brien, P. and Bomford, M. (1992) Pest Animals in Australia. *Bur. Rural Resour., Canberra, Australia.*

Wilson, G.R. and O'Brien, P.H. (1989) Disaster Management *1: 30–5.*

Wilson, J.W. (1981) *In* Proc. World Lagomorph Conf., *1979, Guelph, Ontario: 99–101. University of Guelph, Ontario.*

Wilson, L.O. (1986) Trans. 51st Nth. Am. Wildl. Nat. Resour. Conf. *51: 39–40.*

Wilson, M.L. and Elicker, J.G. (1976) Primates *17(4): 451–73.*

Wilson, P.R. and Orwin, D.F.G. (1964) N.Z. J. Sci. *7(3): 460–90.*

Wilson, R.T. (1984) The Camel. *London and New York.*

Wingate, D.B. (1985) ICBP Technical Publication *3:225–38.*

Wirth, R. (1993) Newsletter Species Survival Commission *(20): 66.*

Wirtz, W.O. (1973) J. Mamm. *54(1): 189–202.*

Wodzicki, K.A. (1961) La Terre et la Vie *1: 130–57.*

__ (1963) Acta Biol. Cracoviensia, Ser. Zool. *6: 111–34.*

__ (1969) *A Preliminary Survey of Rats and other Land Vertebrates of Niue Island, South Pacific. Wellington, New Zealand.*

__ (1969) Proc. N.Z. Ecol. Soc. *16: 7–12.*

__ (1972) Oleagineux *27(6): 309–314.*

__ and Flux, J.E.C. (1967) Tuatara *15(2): 47–59.*

__ and __ (1967) Aust. J. Sci. *29 (11): 429–30.*

__ and Wright, S. (1984) Tuatara *27(2): 77–104.*

__ (1950) Introduced Mammals of New Zealand: An Ecological Survey. *Bull. (98), Dept. Sci. Ind. Res., Wellington, New Zealand.*

__ (1965) The Status of some Exotic Vertebrates in the Ecology of New Zealand. *In Baker and Stebbins (eds)* Genetics of Colonising Species.

Wolcott, G.N. (1953) J. Agric. Puerto Rico *37(3): 241–7.*

Wolf, T.F. (1971) Calif. Fish Game *57(3): 219–20.*

Wolfe, J.L. (1968) Q. J. Fla. Acad. Sci. *31(3): 209–12.*

Wolfe, L.D. and Peters, E.H. (1987) Florida Sci. *50(4): 234–45.*

Wolfe, M.L. (1980) J. Range Mgmt. *33: 354–60.*

__, Ellis, L.C. and Macmullen, R. (1989) J. Wildl. Manage. *53(4): 916–24.*

Wolfheim, J. (1983) Primates of the World. Distribution, Abundance and Conservation. *Seattle and London.*

__ (1912) Coral and Atolls. *London.*

__ (1922) Trans. Proc. R. Soc. S. Aust. *46: 181–93.*

__ (1923) Trans. Proc. R. Soc. S. Aust. *47: 82–94.*

Wood, G.W. (1977) In Research and Management of Wild Hog Populations. *B.W. Baruch For. Sci. Inst., Clemson University, Georgetown, South Carolina.*

__ and Barrett, R.H. (1979) Wildl. Soc. Bull. *7(4): 237–46.*

__ and Brenneman, R.E. (1977) Research and Management of Feral Hogs on Hobcaw Barony. *In Wood, G.W. (ed) Research and Management of Wild Hog Populations, B. W. Baruch For. Sci. Inst., Clemson University, Georgetown, South Carolina.*

__ and __ (1980) Feral Hog Movements and Habitat Use in Coastal Carolina. *In Wood, G.W. (ed) Research and Management of Wild Hog Populations, B.W. Baruch For. Sci. Inst., Clemson University, Georgetown, South Carolina.*

__ and Lynn, T.E. (1977) J. Appl. For. *1(2): 12–17.*

__ and Roark, D.N. (1980) J. Wildl. Mgmt. *44 (2): 506–11.*

__, Hendricks, J.B. and Coagman, D.E. (1976) J. Wildl. Dis. *12: 579–82.*

Wood, J.E., White, R.J. and Durham, J.L. (1970) New Mexico Dept. Game and Fish, Bull *(15): 1–57.*

Woodall, P.F. (1988) Qld. Nat. *29: 10–13.*

Woodford, M.H. (1963) Animals *1(2): 19–22.*

Woodroffe, G.L., Lawton, J.H. and Davidson, W.L. (1990) Biol. Conserv. *51: 49–62.*

Woodward, S.L. (1979) Z. Tierpsychol. *49(3): 304–16.*

__ and Ohmart, R.D. (1976) J. Range Mgmt. *29(6): 482–5.*

Worcester, D.E. (1945) Pacif. Hist. Rev. *14: 409–47.*

__ (1945) New Mexico Hist. Rev. *20: 1–13.*

World Conservation Monitoring Centre (1998) Species under threat: African wild ass Equus africanus Linnaeus 1758. *WCMC/WWF Species sheet.*

__ (1998) Species under threat: Arabian oryx Oryx leucoryx (Pallas 1777). *WCMC/WWF Species sheet.*

__ (1998) Species under threat: Aye-aye Daubentonia madagascariensis (Gmelin 1788). *WCMC/WWF Species sheet.*

__ (1998) Species under threat: Black rhinoceros Diceros bicornis (Linnaeus 1758). *WCMC/WWF Species sheet.*

__ (1998) Species under threat: Black-footed ferret Mustela nigripes (Audubon and Bachman 1851). *WCMC/WWF Species sheet.*

__ (1998) Species under threat: Maned sloth Bradypus torquatus Illiger 1811. *WCMC/WWF Species sheet.*

__ (1998) Species under threat: Markhor Capra falconeri (Wagner 1839). *WCMC/WWF species sheet.*

__ (1998) Species under threat: Pere David's deer Elaphurus davidianus Milne-Edwards 1866. *WCMC/WWF Species sheet.*

__ (1998) Species under threat: Red wolf Canis rufus Audubon and Bachman 1851. *WCMC/WWF Species sheet.*

__ (1998) Species under threat: Ruffed lemur Varecia variegata (Kerr 1792). *WCMC/WWF Species sheet.*

__ (1998) Species under threat: Siberian musk deer Moschus moschiferus Linnaeus 1758. *WCMC/WWF Species sheet.*

__ (1998) Species under threat: Wild dog Lycaon pictus (Temminck 1820). *WCMC/WWF Species sheet.*

World Health Organisation (1968) Intern. Pest Contr. *5: 11.*

Wridgley, R.E., Drescher, H. and Drescher, S. (1973) J. Mamm. *54(3): 782–3.*

Wright, B.S. (1959) The Ghost of North America – The Story of the Eastern Panther. *New York.*

WWF-Canada (1997) Black-footed ferret factsheet. *http://www.wwfcanada.org/facts/ferret.html.*

Wykes, B.J., Pearson, D. and Maher, J. (1999) Fauna Survey of Garden Island, WA, 1996–1997. *HMAS Stirling Env. Working Pap. No. 12.*

Wyman, W.D. (1945) The Wild Horse of the West. *Idaho.*

Yahner, R.H. (1979) J. Mamm. *60(3): 560–7.*

Yakushkin, G.D. (1958) Sb. Nauchno. Tekhnol. Inf. vses. Inst. Zhivotn. Syr'ya Pushniny *3: 28–33.*

Yalden, D.W. (1988) J. Zool. *215: 369–74.*

Yanushevich, A.I. (1966) In Proc. Conf. Acclim. Anims. USSR, *Frunze, 1966. I.P.S.T., Jerusalem.*

Yaoting, Gao (1981) Acta Theriol. Sinica *1(1): 19–26.*

Yazan, Yu. P. (1959) Trudy Inst. Eliol. Ural. Fil. Akad. Nauk. SSSR. *18: 153–68.*

__ (1963) Sb. Nauchn.–Tekh. Inform. Vses Nauchn-Issled Inst. Zhivot. Syr'ya Pushniny *5(8): 81–5.*

Yoakum, J.D. (1980) In Proc. Symp. Ecol. Mgmt. Barbary Sheep, *Nov. 19–21, l979: 9–14. Dept. Range Wildl. Mgmt., Texas Tech. Univ., Lubbock, Texas.*

Yocom, C.F. (l967) Am. Midl. Nat. *77(2): 418–51.*

__, Stumpf, W.A. and Perkins, K.M. (1956) Murrelet *37(2): 18–19.*

Yosida, T.H. and Harada, M. (1985) Proc. Jap. Acad., *ser. B, 61: 208–11.*

__, Udagawa, T., Ishibashi, M., Moriwaki, K., Yabe, T. and Hamada, T. (1985) Proc. Japan Acad. *61, ser. B: 71–4.*

Young, C. (1981) SWANS *11: 13–16.*

Young, S.P. (1941) Am. Forests *47(8): 368–72.*

__ (1945) Mountain Lion Trapping. *USDI Fish Wildlife Service (Washington).*

Youngman, P.M. (1962) Canad. Field-Nat. *76(4): 223.*

__ (1975) Nat. Mus. Sci. Canad., Publ. Zool. *(10): 1–22.*

Yurkin, M.V. (1961) In Okhrana prirody Komi ASSR: *87–93. Syktyvkar.*

Zabinski, J. (1949) J. Soc. Preserv. Fauna Empire New Ser. *59: 11–28.*

Zablotskaya, L.V. (1961) Trudy Prioksk-Terransk Gosudar Zapovednik *3: 77–84.*

Zablotskii, M.A. (1960) Akad. Nauk. SSSR., Kom. Okhr. Prir. Byull. *4: 52–70.*

Zamakhaev, V.A. (1963) Trudy Vses Nauchno-Issled. Inst. Zhivotn. Syr'ya Pushniny *20: 90–7.*

__ (1963) Sbornik Nauchno-Tecknol. Inform. Vses Nauchno-Issled. Inst. Zhivotnogo Syr'ya Pushniny *6 (9): 49–59.*

Zeedyk, W.D. (1980) In Proc. Symp. Ecol. Mgmt. Barbary Sheep, *Nov. 19–21, 1979: 20–21. Dept. Range Wildl. Mgmt., Texas Technical University, Lubbock, Texas.*

Zeuner, F.E. (1963) A History of Domesticated Animals. *London.*

Zexun, Luo (1981) Acta Theriol. Sinica *1(2): 149–57.*

Zharkov, I.V. (1961) Tr. Voronezh. Gos. Zapovednika *12: 5–23.*

Zharkov, V. and Sokolov, V.E. (1967) Acta Theriol. *12(1/3):* 27–45.

Zharov, V.K. (1963) Sel 'sk. Proizod-stvo Sib. Dal'nego Vost *1: 86.*

__ *and Vinichenko, V.V. (1962)* Zool. Zhur. *41(6): 957–9.*

Zhdanov, A. (1963) Sel'skok. Proi. Sibiri Dal'nego Vost *11: 76.*

Zhdanov, A.P. (1962) In Voprosy ekologii *6: 59–60. Vysshaya Shkola, Moscow.*

__ *(1963) In* Acclimatization of Animals in the USSR*: 94–96. Akad. Nauk. Kazakhsk. SSR., Alma-Ata.*

Ziani, P. (1964) Sumarski List *88 (7/8): 277–306.*

Ziegler, A.C. (1982) An ecological checklist of New Guinea recent mammals. In Flannery, T.F. (1995) Mammals of New Guinea. *Reed Books.*

Zimen, E. (1980) The Red Fox. In Symposium on Behaviour and Ecology. *Hague, Boston and London.*

Zimmermann, K. (1953) Zeitschrift fur Saugetierkunde *17: 21–51.*

Zurowski, W. (1979) Acta Theriol. *24(1–11): 85–91.*

__ *(1987)* Trans. Congr. Int. Game Biol. *18: 226.*

__ *(1992)* Trans. Congr. Int. Union Game Biol. *18(2): 163–6.*

INDEX TO SCIENTIFIC NAMES